The South Atlantic

Present and Past Circulation

With 346 Figures and 42 Tables

Springer

Prof. Dr. Gerold Wefer
Universität Bremen
Fachbereich Geowissenschaften
Klagenfurter Straße
28359 Bremen
Germany

Prof. Dr. Wolfgang H. Berger
Scripps Institution of Oceanography
Geological Research Division
University of California, San Diego
La Jolla, CA 92093-0215
USA

Prof. Dr. Gerold Siedler
Institut für Meereskunde an der
Christian-Albrechts-Universität
Abteilung Meeresphysik
Düsterbrooker Weg 20
24105 Kiel
Germany

Dr. David J. Webb
Southampton Oceanography Centre
Empress Dock
Southampton SO14 3ZH
United Kingdom

ISBN-13: 978-3-642-80355-0

CIP-Data applied for

Die Deutsche Bibliothek - CIP-Einheitsaufnahme

The South Atlantic: present and past circulation /
Gerold Wefer ... – Berlin; Heidelberg; New York; Barcelona;
Budapest; Hong Kong; London; Mailand; Paris; Santa Clara;
Singapur; Tokio: Springer 1996
ISBN-13: 978-3-642-80355-0   e-ISBN-13: 978-3-642-80353-6
DOI: 10.1007/978-3-642-80353-6
NE: Wefer, Gerold

Typesetting: cameraready by authors
Cover Design: E. Kirchner, Heidelberg

SPIN: 10050853   32/3136 – 5 4 3 2 1 0

Printed on acid free paper

The South Atlantic

Gerold Wefer · Wolfgang H. Berger
Gerold Siedler · David J. Webb

Springer

Berlin
Heidelberg
New York
Barcelona
Budapest
Hong Kong
London
Milan
Paris
Santa Clara
Singapore
Tokyo

Preface

In the summer of 1994 (Aug. 15 - 18) about 220 oceanographers and paleoceanographers from Western Europe, the USA, Africa, South America, Russia and Lithuania met at Bremen University to present and discuss the results of recent research on present and past circulation in the South Atlantic. The symposium was organised by Gerold Siedler (University of Kiel) and Gerold Wefer (University of Bremen), assisted by a „Scientific Programme Committee", which included W.H. Berger, K. Herterich, N.G. Hogg, Y. Ikeda, J.H.F. Jansen, J.R.E. Luetjeharms, J. Merle, R.G. Peterson, and D.J. Webb. The local organisation was in the hands of Barbara Donner.

A symposium on the South Atlantic seemed timely and appropriate because the results from a number of recent projects in oceanography and marine geosciences needed to be integrated into existing knowledge. Thus, monitoring programs have been in operation since 1991 as part of the World Ocean Circulation Experiment (WOCE). These programs included studies of the meridional water-mass and heat transports, the western boundary currents, and deep- and bottom-water transports in the Brazil Basin. Also comprehensive studies are being carried out in the context of IGBP Core Projects PAGES (Past Global Changes) and JGOFS (Joint Global Ocean Flux Study), generating new data for the various current and production systems of the South Atlantic.

Since 1989 the Colloborative Research Centre (Sonderforschungsbereich) No. 261 at the University of Bremen („The South Atlantic in the Late Quaternary: Reconstruction of material budget and current system") has carried out extensive geological, geophysical and geochemical investigations on the various current and production systems of the South Atlantic . The central theme of this research is the reconstruction of current and productivity systems in the South Atlantic for the last 300,000 years.

The International Ocean Drilling Program is currently planning operations in the South Atlantic, for example off the coasts of Angola and Namibia, using the drilling vessel „Joides Resolution" (Leg 175). We hope the results presented here will help in the planning of the expedition and later in evaluating the results.

A key aim of the organisers was to provide a forum in which people could integrate the results coming from both oceanographic and geological studies. The South Atlantic is an ideal region for this purpose. As the connecting link between the Antarctic and the North Atlantic, the South Atlantic plays a crucial role with regard to the heat budget of the North Atlantic and to the biogeochemical budget of the global ocean. Water from the South Atlantic moves north across the equator at shallow depths replacing North Atlantic Deep Water (NADW), which forms in the northern North Atlantic due to cooling of salty surface waters, and moves south at depth.

Investigations of sediments have revealed that the production of NADW decreased during glacials when, in its place, there was probably more intermediate water production reaching to greater depths. The reduction or possibly the complete interruption of NADW production during glacial periods may have been directly caused by a decrease in energy and salt inputs from the South Atlantic, resulting from changes in surface circulation as well as changes in the Agulhas Current system and the Antarctic Circumpolar Current. Thus changes occurring further south and in other ocean can influence the climate of the North Atlantic via the South Atlantic heat transfer system.

The South Atlantic also plays a decisive role in the coupling of oceanic processes between the Antarctic and lower latitudes. The Antarctic water belt is of major significance for global climate as it supplies large regions of ocean with intermediate and bottom water. While the North Atlantic is influenced by the South Atlantic through the input of warm near-surface water, the input of NADW into the Southern Ocean and the Antarctic Circumpolar Current is one of the determining factors of the oceanography of the Antarctic. It is probable that the formation of Antarctic Intermediate and Bottom Water is greatly enhanced by the upward flow of warm, salt-enriched deep water, which hinders sea-ice formation and delivers salt for increased density after cooling.

Approximately 40 lectures were given during the four-day symposium on various aspects of oceanography and paleoceanography of the South Atlantic, and over 140 poster presentations helped to illustrate recent research findings and provided a basis for discussions. Oceanographic and paleoceanographic topics were deliberately mixed in the morning sessions so as to present as broad a spectrum of information as possible to the neighbouring disciplines. Lively interdisciplinary discussions took place after the lectures, during the breaks as well as during the poster sessions, whereby geologists gained insight into processes, and oceanographers expanded their appreciation for the range of possible states of the system.

Encouragement of this type of exchange guided selection of the papers for publication in this book. Thus, there is an emphasis on the more general contributions, and on those with interdisciplinary implications. The papers have also benefited from detailed reviews by P.M. Saunders, R.G. Peterson, A. Gordon, L. Talley, K. Herterich, B. Barnier, W. Roether, D. Smythe-Wright, M. Rhein, J.R.E. Lutjeharms, W. Curry, A. Mix, R. Schneider, F. Jansen, L. Labeyrie, T. Bickert, C. Robert and V. Ittekkot.

The symposium on the South Atlantic was supported by several international organisations, including the Oceanography Society (TOS), the international WOCE Scientific Steering Group and the Scientific Committee on Oceanic Research (SCOR). Financial backing was provided by the German Research Foundation (DFG) as well as by the Senate of the Free and Hanseatic City of Bremen. Technical support was given by Dr. B. Donner, S. Middendorf, A. Grimm-Geils, G. Meinecke and V.Diekamp. To each and all of those involved, our sincere thanks.

Gerold Wefer, Bremen
Wolfgang H. Berger, La Jolla
Gerold Siedler, Kiel
David J. Webb, Southampton

Contents

Contents

Central Themes of South Atlantic Circulation

W. H. Berger[1] and G. Wefer[2]

[1]Scripps Institution of Oceanography, La Jolla, Ca. 92093, USA
[2]Universität Bremen, Fachbereich Geowissenschaften, 28334 Bremen, Germany

Abstract: The central problem of South Atlantic oceanography and paleoceanography is the exchange of heat between South and North Atlantic. More specifically, it is the nature of the North Atlantic heat piracy. Without this heat transfer, northern Europe would have an entirely less clement climate. The necessity to understand the operation of the Atlantic Heat Conveyor provides the rationale for studying both physical and historical oceanography of the South Atlantic. The study of present conditions allows a synoptic view of dynamic interaction of surface- and deep circulation, and associated productivity patterns. The study of past patterns provides clues to the range of possible states and rates of change, and to the long-term stability of the system.

Physical and Geological Oceanography: An Emerging Symbiosis

A central problem in oceanography concerns the efficiency and stability of the heat transfer from South to North Atlantic, within the general framework of global thermohaline circulation (e.g., Gordon 1986). The question is vital for climatic conditions in northern Europe. In recent years, a sense that ocean circulation is subject to considerable change on rather short time scales (decades to centuries) has emerged from high-resolution studies of marine sediments deposited at high rates of accumulation. Studies of corals also have proved useful in this regard. A mounting concern for the sensitivity of the ocean-climate system to the continuing large-scale introduction of greenhouse gases has focussed attention of both physical and historical oceanographers on this type of evidence. There is no question that this is a vital interface, where the physical and historical approach will thrive in symbiosis.

Here we argue for a broadening of this interface, to include a greater range of time scales in the overlap. But why should physical oceanographers take interest in time scales beyond their limits of measurement? After a few thousand years the ocean (except for its sediment) essentially loses all memory of its past. None of their measurements from the water column call for interpretation on time scales beyond the mixing time.

The reason why history is important is that the past allows us to see a much greater range of possible states of the ocean-climate system than we could reasonably imagine when extrapolating from the present ocean. The present ocean emerges as a special case of many possible oceans: It is up to the modelers to invent ways to move from one state to another without violating basic physics and physically-based common sense.

The discovery of the (poorly known) boundaries of reality, then, is the pay-off to the physical oceanographers who enter the realm of paleoceanography. At the same time, their skills at providing explanations are being severely tested: The conceptual models of paleoceanographers may be good and necessary, but they need to be transformed to reflect our quantitative understanding of the system.

The contents of the symposium here presented reflect the tentative nature of the incipient symbiosis between physical and geological oceanographers: Each group has started to take interest in the

From WEFER G, BERGER WH, SIEDLER G, WEBB DJ (eds), 1996, *The South Atlantic: Present and Past Circulation*. Springer-Verlag Berlin Heidelberg, pp 1-11

other's work, but collaboration is rare, as yet. The South Atlantic, because of its central position in the global thermohaline system, is a great place to begin the process. Paleoceanographers already know that they need the physicists to help solve their puzzles. One aim of the conference, and of this symposium, is to convince the physical oceanographers that the historical approach has much to offer them, in exchange.

The Central Problem: North Atlantic Heat Piracy

Of the many interesting problems arising when studying the oceanography of the South Atlantic, the most important and intriguing is the question of how much heat is delivered to the North Atlantic, and from where, and how the transport takes

place. On average, heat transport in the ocean is from the tropics and subtropics to the poles (Fig. 1). This is required by the fact that it is warm in the tropics and cold at the poles. The South Atlantic provides a striking exception to the pattern (Fig. 1, dashed line, southern latitudes). Remarkably, in this case the heat is driven equatorward, from cold to warm regions, across the equator.

The fundamental reason for the anomalous heat transfer is the presence of a self-stabilizing feedback system. The two most important ingredients, presumably, are the northern position of the Intertropical Convergence Zone (ITCZ) and the ready access to a silled high northern basin (Nordic Sea between Greenland, Iceland and Norway), where deepwater production can occur on a large scale. The North Atlantic trades cold deep water for warm upper water. About one half of the heat imported

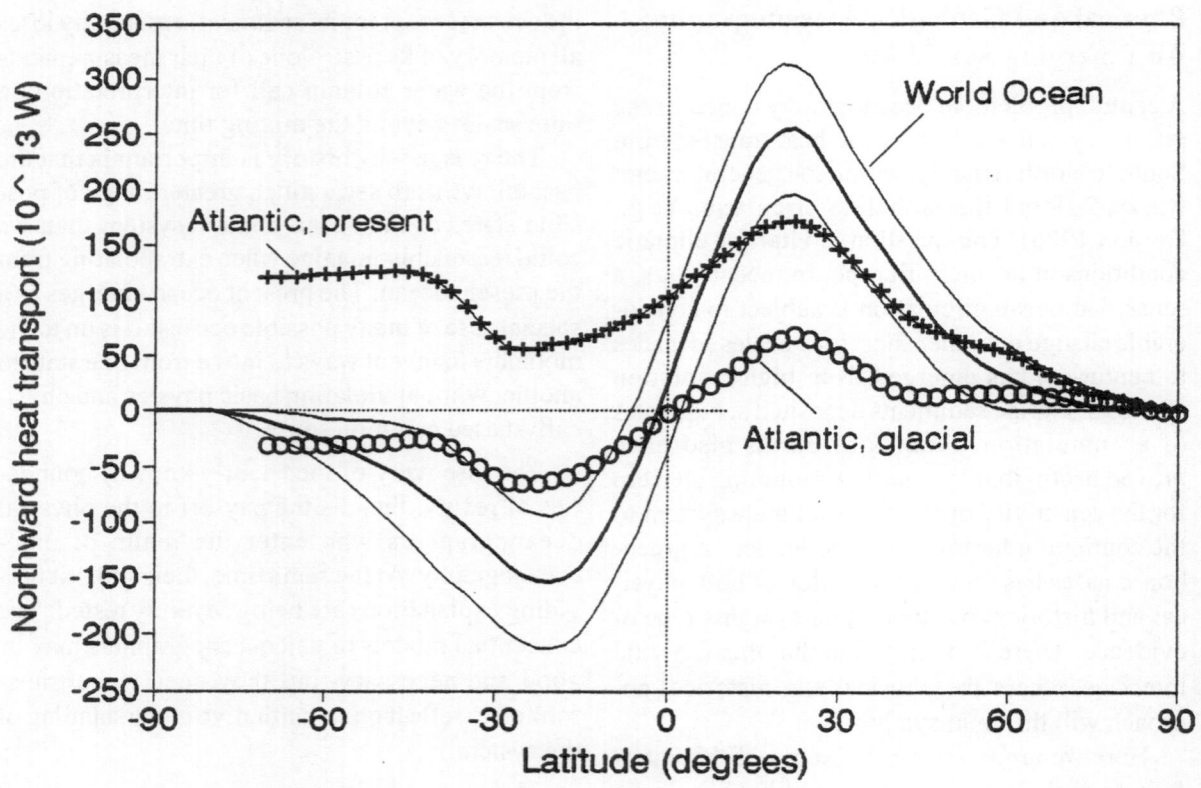

Fig. 1. Estimates for annual heat transports for the present global ocean (solid line with approximate error bounds) and for the Atlantic, for the present and for the last glacial maximum (as labeled). Source: Miller and Russell (1989). Note the anomalous pattern for the present South Atlantic, and the more symmetric pattern for glacial conditions.

is used for evaporation of surface water in the sub-tropics, with much of the vapor lost to the Pacific, bringing salinity to the highest open-ocean values anywhere in the ocean. The other half is available for warming western Europe and the boreal regions. Upon cooling, the high-salinity water reaches the necessary density to fall to a depth of several kilometers, in the boreal ocean, mixing with deep water there. The outflow of this cold water into the South Atlantic and from there to the Circumpolar Deep Water sets in motion the Atlantic Heat Conveyor, that is, the self-stabilizing system responsible for northern heat piracy (Bryan 1986; Reid, this volume).

The heat loss by the South Atlantic across the equator has a number of corollaries concerning the surface circulation and the productivity in the South Atlantic. One is the absence of a strong poleward jet in the west of the basin (Fig. 2). There is no analog to Kuroshio or Gulf Stream in the South Atlantic: the Brazil Current starts out weak, well south of the equator, and picks up flux mainly through recycling within the basin. In contrast, the Gulf Stream has its roots over a much larger warm area, including the Caribbean, and closer to the equator, benefitting from the fact that the ITCZ is well north of the equator much of the year. In contrast to the weak Brazil Current, the northwestward

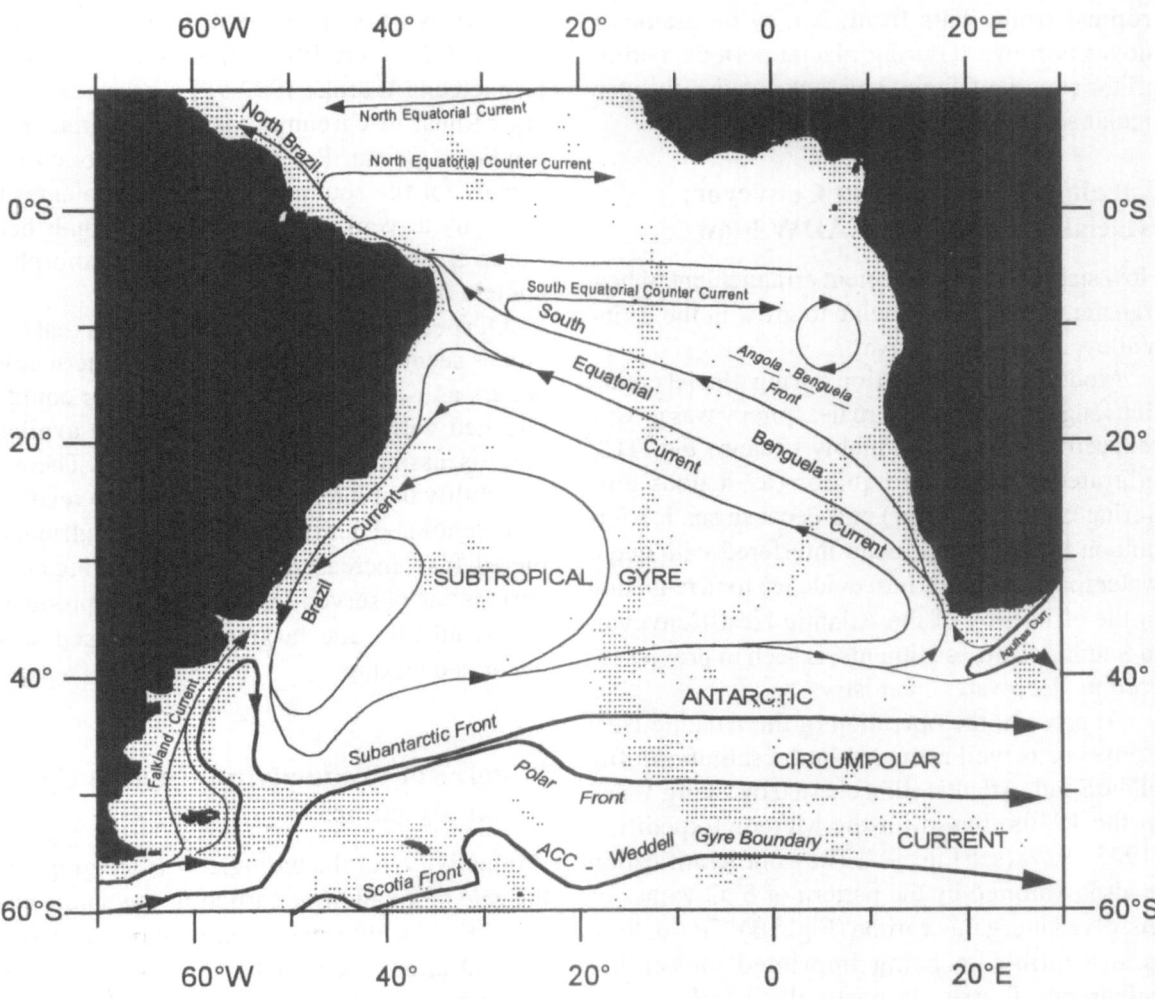

Fig. 2. Schematic representation of the large-scale, upper-level geostrophic currents and fronts in the South Atlantic Ocean. Source: Mainly Peterson and Stramma (1991), with minor additions from several other compilations.

flowing Benguela Current is quite vigorous. Indeed, according to geostrophy, this flow would seem to dominate much of the tropical and subtropical region in the basin. Thus, heat derived from sunlight is moved northward quite efficiently while the return flow is modest.

To what degree subsurface currents are involved in transporting heat across latitudes is a matter of contention. Of special interest is the observation that warm eddies enter the South Atlantic around the tip of Africa at the Agulhas Current retroflection (Gordon 1985; Lutjeharms, this volume). It would appear that the effectiveness of this retroflection in generating warm eddies for the South Atlantic depends on the position of the subtropical front. This front, it may be assumed, moves northward during glacial periods, cutting off the supply of eddies and thus denying this particular source of heat to the South Atlantic.

Turning Down the Heat Conveyor: Glacial Reduction of NADW Flow

How stable is this convenient arrangement of heat transfer, which allows wine to grow in the Rhine Valley, and rye in Norway?

Modelling of circulation during glacial conditions suggests that northern heat piracy was greatly reduced (Fig. 1), presumably because the ITCZ migrated toward the equator (as it does now during northern winter) and because sea ice formation in the boreal realm interfered with deepwater formation. The best evidence for a reduction in the efficiency of the Atlantic Heat Conveyor, in South Atlantic sediments, is seen in proxies recording deepwater chemistry.

At present, the operation of the Atlantic Heat Conveyor is well reflected in the salinity profile of the South Atlantic (first charted by Georg Wüst, in the 1930s, based on the Meteor Expedition (1925-1927) (see Fig. 3A). The same stratification is also exhibited by the pattern of $\delta^{13}C$-values of dissolved inorganic carbon (Fig. 3B). This pattern is susceptible to being imprinted on benthic calcareous fossils, in particular benthic foraminiferans that live on the surface of the sediment (such as *Planulina wüllerstorfi*). By recovering such fossils from different depths and different

latitudes, in glacial-age sediments, it is possible to obtain a general idea about the patterns of $\delta^{13}C$, and hence the distribution of water masses (Fig. 3C).

During the last glacial period, the bottom-near layer of cold Antarctic Water (AABW) tends to thicken, as the overlying NADW flow is reduced. Thus, the NADW-AABW boundary moves upward, as seen in an associated shift in the preservation of calcareous fossils, and especially in the $^{13}C/^{12}C$ ratio recorded by benthic foraminiferans (Fig. 3C). The $\delta^{13}C$ proxy for relative deepwater age has become perhaps the single most important clue for the reconstruction of deepwater circulation (Zahn et al. 1994; Bickert and Wefer, this volume). One important corollary of a reduction of NADW flux, and much discussed (see in Hsü and Weissert 1985; Bleil and Thiede 1990; Kennett and Warnke 1992/93) is a reduced delivery of heat to Circumpolar Deep Water, in the Southern Ocean. Presumably this is one of the signals for the southern hemisphere to enter the (boreally driven) ice age; the other signals being a drop of global sea level and a drop of atmospheric content of CO_2.

Of special interest is the observation that $\delta^{13}C$-values seem to be relatively high in intermediate waters as well (Fig. 3C), as far as this could be checked with the limited information available (see discussions in Curry, this volume). There is a possibility that intermediate waters are relatively well ventilated, and correspondingly nutrient depleted. If so, increased productivity during glacial periods (as observed in areas of high productivity) would be due largely to increased wind-generated mixing.

Patterns of Productivity: Clues to Upper Water Dynamics

Productivity, for the biological oceanographer, is the rate of fixation of carbon in the sunlit water layer. For the physical oceanographer, patterns of concentration of chlorophyll and its derivatives at the surface provide a means to recognize regions of mixing and upwelling from reflectance, as measured by satellites. For the paleoceanographer, productivity is measured by what accumulates in

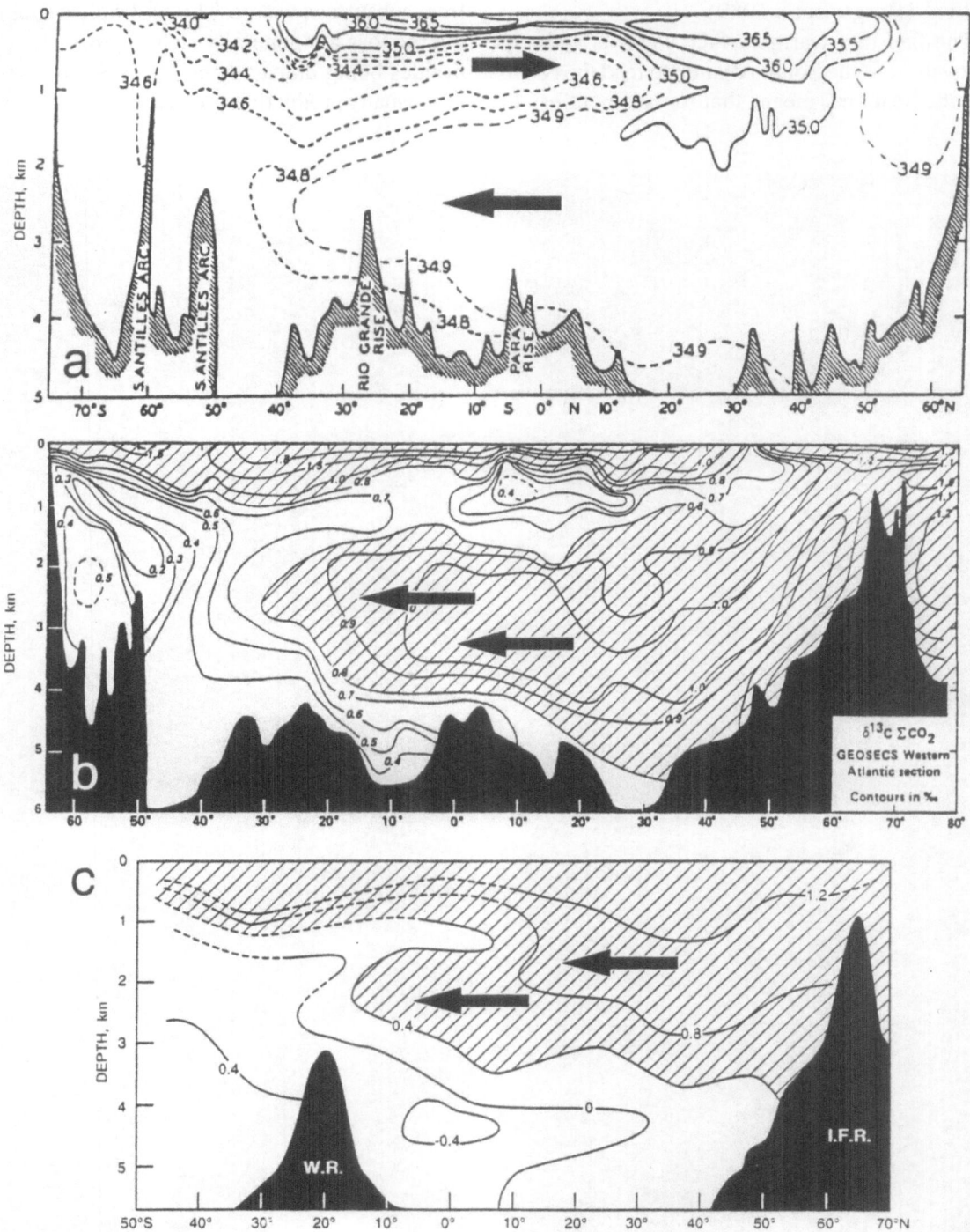

Fig. 3. Deepwater patterns and flow in the South Atlantic, present and last glacial maximum. A, present salinity distributions; source: Sverdrup et al. (1942), in large part according to G. Wüst. B, distribution of δ¹³C-values in dissolved inorganic carbon; source: Kroopnick (1980). C, distribution of δ¹³C-values in dissolved inorganic carbon, 20 kyr ago, inferred from δ¹³C-values in benthic foraminiferans of that age; source of data: Duplessy et al. (1988) and Sarnthein et al. (1994).

traps and on the sea floor, that is, the „export production" (Berger et al. 1989).

The fact that warm surface water is actively removed from the South Atlantic to feed the North Atlantic heat sink means that replacement has to be found, from across the southern boundary and from subsurface waters. The cool waters supplanting the loss are generally rich in nutrients. Thus, besides being unusually cool, the South Atlantic is unusually productive as ocean basins go (as seen,

Fig. 4. Pigment distribution in surface waters, inferred from color scanning data aboard CZCS satellite (November 1978 - July 1986). Source: NASA Goddard Space Flight Centre and University of Miami, compiled by B. Davenport, Bremen. Orange: high produktivity (> 150 g/m²/yr); deep blue: low produktivity (< 50 g/m²/yr)

for example, in color reflectance; Fig. 4). There are only two other areas that can compare with the eastern South Atlantic in this respect: the eastern tropical Pacific (which is the largest center of high ocean productivity) and the northern Indian Ocean (where monsoonal stirring is important).

The present patterns of productivity in the South Atlantic represent an integration of a large number of dynamic properties of the system. Factors include seasonal winds, geostrophic currents, and nutrient content of subsurface waters. Given the right proxies, some of these factors can be reconstructed, at least semi-quantitatively. It can be argued that, in essence, export production is a function of the product of mixing intensity (upward transport of subsurface water) and nutrient content of subsurface water. If both export production and nutrient concentration can be constrained by proxies, the ratio yields wind intensity (Herguera and Berger 1994).

To a first approximation, it is seen that wind intensity is perhaps the single most important factor in producing the overall productivity pattern (Fig. 4). Both stirring and the thickness of the mixed layer are related to wind and current strength. Also, there are feedback processes coupling the concentrations of nutrients in subsurface waters to the productivity of overlying waters. However, divergences along the equator and along the eastern coastal strip have additional dynamic ingredients, and nutrient contents in subsurface waters are influenced by the patterns of intermediate water sourcing and flow, as well. Near the southern boundary of the South Atlantic, beyond the Subantarctic Convergence Zone, the availability of sunlight becomes an important limiting factor.

It is well to remember that most of what we know about surface circulation and upwelling in the past has to be extracted from proxies related to ocean productivity, that is, shelled plankton. An important task in this respect is the proper calibration of proxies, by comparing surface sediments with present-day oceanographic conditions, and by trapping material on the way to the sea floor (Fig. 5).

The various planktonic remains are not equally useful; some are not preserved at all, others are subject to selective removal by dissolution in the water column and on the sea floor. The most widely used fossils are remains of foraminiferans, coccolithophorids, polycystine radiolarians, and diatoms. Supply of the various remains is highly seasonal (Fig. 5), so that it is difficult to extract average or typical conditions of any kind. Without exception, organisms are „biased reporters", providing information in a selective manner (Wefer and Berger, 1991). The task of calibration is to detect the bias and to correct for it, using multiple proxies.

Productivity Cycles and Milankovitch Dynamics: Ice versus Wind

The importance of productivity cycles in the context of ice age fluctuations and changing intensity of trade winds was first recognized by Arrhenius (1952) when studying pelagic sediments in the eastern equatorial Pacific. The analysis of cyclic phenomena in the deep-sea record has since made great strides, especially with a view to the type of orbital climate forcing proposed by Milankovitch, 1930 (Hays et al. 1976; Berger et al. 1984; McIntyre et al. 1989; Crowley and North 1991; Imbrie et al. 1993). Much can be learned about the dynamics of the system by studying fluctuations in the state of the ocean and the possible forcing over the last, say, 200,000 to 300,000 years. We next illustrate this point, highlighting the productivity record from a site in the eastern equatorial Atlantic (see Schneider et al., this volume; for sites in the Benguela system, see Wefer et al., this volume, and Summerhayes et al. 1995).

It appears well-established that the eastern equatorial Atlantic is more productive during glacial periods than during interglacial ones (e.g., McIntyre et al. 1989; Mix 1989). Two questions are of special interest in this context: (1) To what degree is the course of productivity defined by the usually available proxies of past physical conditions, that is, $\delta^{18}O$ and paleotemperature of surface waters, and (2) which part of the signal is of global significance („ice-related"), and which is regional („wind-related"). Inasmuch as the ice-related global signal also contains a wind factor, the regional signal refers to *additional* wind-related information, beyond the global effect. In the area studied, we expect this additional information to be a matter of monsoon-tradewind interaction contingent upon insolation of the North African land mass (cf. McIntyre et al. 1989; Molfino and McIntyre 1990; Schneider et al. 1995, and this volume).

To address these two questions, we consider three proxies determined in Core GeoB 1105-4, as well as a „forcing function" (Fig. 6). The proxies are (1) the δ^{18}O-values of the planktonic foraminifer *G. ruber* (labelled rbr18), (2) the estimated cold-season temperature (labelled SST, and based on foraminiferal assemblages, calculated by the MAT algorithm; Meinecke 1992), and (3) the concentration of organic matter (labelled TOC, percent C_{org}). As „forcing function" we use the reconstructed series of insolation in July at 15°N (Berger and Loutre 1991; labelled Jul15N). As concerns the „forcing functions" in general, we believe that the δ^{18}O record in Site 806 is the best record available for representing ice mass in the northern hemisphere: It was taken from below the

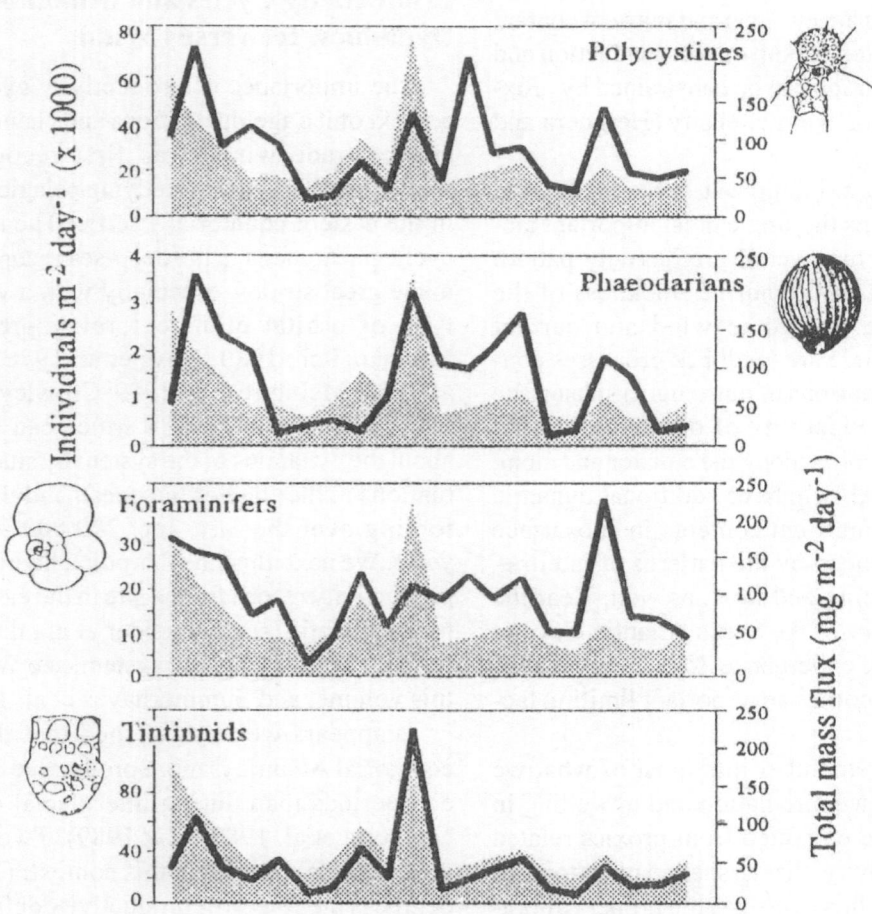

Fig. 5. Seasonal variations in the flux of potential proxies (polycystine radiolarians, foraminiferans) and other shelled plankton, according to trapping results in the equatorial Atlantic. Trap station is in eastern equatorial Atlantic at about 2°N, 11°W, water depth 853 m, set by the Collaborative Research Program of the University of Bremen (SFB 261). Source: Boltovskoy et al. (1993).

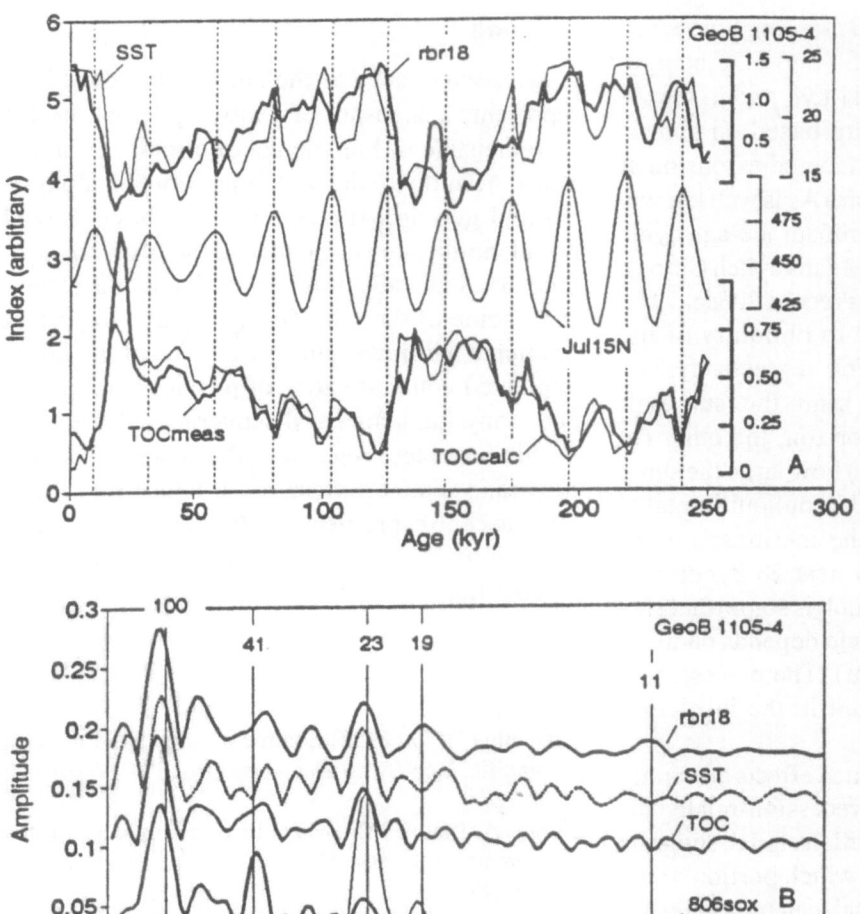

Fig. 6. Productivity record of the eastern equatorial Atlantic, as seen in Core GeoB 1105-4. Proxies are: $\delta^{18}O$ of *G. ruber* (rbr18, negative permil), most likely cold-season analog temperature based on foraminiferal assemblages (SST, degrees), total organic carbon (TOC, percent). „Forcing function" is insolation received in July at 15°N (Jul15N,watts/m²). A, time series of proxies and „forcing function". B, spectral representations of the same proxies and „forcing function". Vertical grid: maxima of the insolation series Jul15N. For comparison, $\delta^{18}O$ of *G. sacculifer* in the western equatorial Pacific, representing ice mass (806sox),is shown also. Sources: Proxies from Wefer et al. (this volume); 806sox from Berger et al. (1995), modified; Jul15N from Berger and Loutre (1991).

warm water pool in the western equatorial Pacific, where tropical temperatures are thought to be quite stable. This „ice-mass" or „sealevel" signal, then, should represent the global change. The July insolation at 15°N represents the regional forcing, which has to do with tradewind and monsoon intensity, above and beyond the global effect.

Inspection of the proxies rbr18 and TOC (Fig. 6A) immediately reveals a strong correlation between the two, such that interglacial conditions (high rbr18 values) coincide with low productivity (low TOC values) and vice versa (r^2=0.62). Also, a general agreement is seen between rbr18 and SST, as expected (interglacials being warmer than glacial periods). By combining rbr18 and SST, the measured TOC values (TOCmeas) can be repre-

sented with a high degree of accuracy, using multiple linear regression (TOCcalc= f[rbr18, SSt]; r^2=0.73). The one outlier is the spike of TOC centered on the last glacial maximum. Presumably, this is in large part a matter of preservation, with relatively fresh and reactive organic matter still being present.The series rbr18, in turn, can be represented as consisting approximately of 75% 806sox (global sealevel) and 25% Jul15N (regional SST).

The answer to the first question, then, regarding whether productivity fluctuation is closely tied to the standard proxies of ocean state, is that yes, it is: productivity is well integrated into the changing dynamics of the system, and glacials are indeed characterized by increased productivity.

The spectra of proxies and „forcing functions" (Fig. 6B) reveal that rbr18, SST and TOC contain strong power near 100 kyr, 41 kyr, and in the vicinity of 23 kyr. (Spectra are based on simple Fourier expansion, with 5% tapered extension at each end, using autocorrelation.) As is well known, the 100-kyr period is the dominant ice-age cycle in the last 600,000 years (the Milankovitch Chron). It is prominent in the ice-mass record 806sox. The other two bands are related to obliquity of the Earth's axis and to its precession, respectively. The first effect controls how high the sun will maximally rise above the horizon, the other (in conjunction with eccentricity) how large the sun's disk will appear, in summer. The obliquity-related 41-kyr cycle is strong in the ice-mass record 806sox. However, the power near 23 kyr and 19 kyr is weak in this record (which is somewhat surprising, since the 100-kyr cycle depends on input from precession, it is thought). The precessional effect is completely dominant in the insolation series (Jul15N).

The fact that obliquity-related effects are strong in the ice-mass record, and precession-related effects in the low-latitude insolation signal, suggests that the question concerning which portion of the record is global, which regional, can be readily addressed. One simply notes the relative importance of obliquity-related and precession-related periodicity. Based on the spectral analysis, the $\delta^{18}O$ record of *G. ruber* can be represented as a mixture between global forcing (806sox) and regional forcing (Jul15N) in the proportion 3 to 1, that is, rbr18 contains 25% regional monsoon-tradewind signal. The cold-season faunal temperature estimate, SST, contains 30% regional effect, by the same criterion. Thus, we conclude that the productivity fluctuations are governed by driving forces which are 70 to 75 percent global, related to the ramifications of ice dynamics in the boreal realm, and 25 to 30 percent regional, related to changing insolation of the North African land mass.

These results emphasize the interplay between polar and tropical processes in modifying the state of the ocean, and hence in governing the central problem of South Atlantic circulation: the heat transfer north across the equator.

Outlook

A symposium such as the one we introduce here represents a snapshot of a moving target; in this case the state of knowledge of circulation in the South Atlantic. In the past, practitioners of physical and geological oceanography have moved ahead rapidly, and quite independently from each other. Undoubtedly, this trend will continue. However, some of the most intriguing and important problems will increasingly be found at the thinly populated boundary zone of physics and history, and many fundamental discoveries will be made at this interface, regarding the workings of the ocean and climate system. In particular, we believe that the central problem, northern heat piracy and its stability, cannot be solved by either physics or history alone.

References

Arrhenius GOS (1952) Sediment cores from the east Pacific. Rep Swed Deep-sea Exped 1947-1948 5: 1-227

Berger A, Loutre MF (1991) Insolation values for the climate of the last 10 million years. Quat Sci Rev 10:297-317

Berger A, Imbrie J, Hays J, Kukla G, Saltzman B (eds) (1984) Milankovitch and Climate. 2 vols Reidel, Dordrecht, Holland

Berger WH, Smetacek VS, Wefer G(eds) (1989) Productivity of the Ocean: Present and Past. Wiley and Sons, Chichester, 471 pp

Berger WH, Yasuda M, Bickert T, Wefer G (1995) Brunhes-Matuyama boundary: 790 k.y. date consistent with ODP Leg 130 oxygen isotope records based on fit to Milankovitch template. Geophys Res Letters 22(12):1525-1528

Bleil U, Thiede J (eds) (1990) Geological History of the Polar Oceans: Arctic Versus Antarctic. Kluwer Academic, Dordrecht, Holland, 823 pp

Boltovskoy D, Alder VA, Abelmann A (1993) Annual flux of Radiolaria and other shelled plankters in the eastern equatorial Atlantic at 853 m: seasonal variations and polycystine species-specific responses. Deep-Sea Res, 40(9):1863-1895

Bryan F (1986) High-latitude salinity effects and interhemispheric thermohaline circulations. Nature 323: 301-304

Crowley TJ, North GR (1991) Paleoclimatology. Oxford University Press, New York, 339 pp

Duplessy J-C, Shackleton NJ, Fairbanks RG, Labeyrie L, Oppo D, Kallel N (1988) Deep-water source variations during the last climatic cycle and their impact on the global deep-water circulation. Paleoceanogr 3:343-360

Gordon AL (1985) Indian-Atlantic transfer of thermocline water at the Agulhas retroflection. Science 227:1030-1033

Gordon AL (1986) Inter-ocean exchange of thermocline water. J Geophys Res 91:5037-5046

Hays JD, Imbrie J, Shackleton NJ (1976) Variations in the Earth's orbit: Pace-maker of the ice ages. Science 194:1121-1132

Herguera JC, Berger WH (1994) Glacial to postglacial drop of productivity in the western equatorial Pacific: Mixing rate versus nutrient concentrations. Geology 22(7):629-632

Hsü K J, Weissert HJ (eds) (1985) South Atlantic Paleoceanography. Cambridge University Press, Cambridge, 350 pp

Imbrie J, Berger A, Boyle EA, Clemens SC, Duffy A, Howard WR, Kukla G, Kutzbach J, Martinson DG, McIntyre A, Mix AC, Molfino B, Morley JJ, Peterson LC, Pisias NG, Prell WL, Raymo ME, Shackleton NJ, Toggweiler JR (1993) On the structure and origin of major glaciation cycles 2. The 100,000-year cycle. Paleoceanogr 8: 699-735

Kennett JP, Warnke DA (eds) (1992/93) The Antarctic Palaeoenvironment: A Perspective on Global Change. American Geophysical Union, Washington, D.C., Part One, 385 pp; Part Two, 273 pp

Kroopnick P (1980) The distribution of ^{13}C in the Atlantic Ocean, Earth Planet. Sci Lett 49:469-484

McIntyre A, Ruddiman WF, Karlin K, Mix AC (1989) Surface water response of the Equatorial Atlantic to orbital forcing. Paleoceanogr 4:19-55

Meinecke G (1992) Spätquartäre Oberflächenwassertemperaturen im östlichen äquatorialen Atlantik. Thesis Fachber Geowiss, Univ Bremen 29, 181 pp

Milankovitch M (1930) Mathematische Klimalehre und astronomische Theorie der Klima-schwankungen. Handbuch der Klimatologie, Bd 1, Teil A. Bornträger, Berlin, 176 pp

Miller JR, Russell GL (1989) Ocean heat transport during the last glacial maximum. Paleoceanogr 4:141-155

Mix AC (1989) Influence of productivity variations on long-term atmospheric CO_2. Nature 337:541-544

Molfino B, McIntyre A (1990) Precessional forcing of nutricline dynamics in the equatorial Atlantic. Science 249:766-769

Peterson RG, Stramma L (1991) Upper-level circulation in the South Atlantic Ocean. Progess in Oceanography 26:1-73

Sarnthein M, Winn K, Jung SJA, Duplessy J-C, Labeyrie L, Erlenkeuser H, Ganssen G (1994) Changes in east Atlantic deepwater circulation over the last 30,000 years: Eight time slice reconstructions. Paleoceanogr 9:209-267

Schneider RR, Müller PJ, Ruhland G (1995) Late Quaternary surface circulation in the east equatorial South Atlantic: Evidence from alkenone sea surface temperatures. Paleoceanogr 10:197-219

Summerhayes CP, Emeis K-C, Angel MV, Smith RL, Zeitzschel B (eds) (1995) Upwelling in the Ocean. Modern Processes and Ancient Records. John Wiley and Sons, Chichester, 422 pp

Sverdrup HU, Johnson MW, Fleming RH (1942) The Oceans. Prentice-Hall, Englewood Cliffs, 1087 pp

Wefer G, Berger WH (1991) Isotope paleontology: growth and composition of extant calcareous species. Mar Geol 100:207-248

Zahn R, Kaminski M, Labeyrie LD, Pederson TF (eds) (1994) Carbon Cycling in the Glacial Ocean: Constraints on the Ocean's Role in Global Change. Springer Verlag, Berlin Heidelberg, 580 pp

On the Circulation of the South Atlantic Ocean

J.R. Reid

Marine Life Research Group, Scripps Institution of Oceanography,
La Jolla, CA 92093-0230, U.S.A.

Abstract: In a recent study (Reid, 1994) maps of the general circulation of the Atlantic Ocean were presented and discussed. The emphasis was placed upon the exchange of waters between the North and South Atlantic as indicated by the patterns of characteristics.

The exchange with the waters entering through the Drake Passage involves further discussion of the effect of the Atlantic waters in providing a major defining characteristic of the Circumpolar Water — the layer of warm and saline water. The associated features - the high oxygen and low nutrient concentrations - are also defined well for a long distance down stream in the Antarctic Circumpolar Current. However it is only the vertical maximum in salinity that persists all along the flow around Antarctica and through the Drake Passage into the South Atlantic.

This returning salinity maximum is much lower than the salinity maximum from the north that it meets and mixes with in the South Atlantic. Part of their mixture turns back northward into the South Atlantic where it appears as the vertical maximum in salinity seen in the east, but lower in salinity than the maximum seen along the western boundary. The rest of the mixture continues eastward, with its salinity increased well above that of the Drake Passage water and begins another circuit of Antarctica.

Introduction

I have discussed the South Atlantic Ocean and its relation to the rest of the world ocean in various papers (Reid 1981; 1984; 1989; Reid and Lynn 1971). I will review those results briefly, and the exchange with the North Atlantic and the other oceans, and the consequences of these exchanges.

The most important points that I want to make are:

Water from the South Atlantic crosses northward across the equator at all depths, not just as the bottom water and the Intermediate water that enter the North Atlantic from the south. The northward flow of circumpolar water does not take place only beneath the southward flow along the western boundary, but also offshore of it.

The high temperature and salinity observed at mid-depth in the North Atlantic and extending southward through the South Atlantic have two sources. One is the outflow from the Mediterranean Sea, which is directly effective to more than 2500m.

The other, denser source of warm and saline water is not convective overturn in the northern North Atlantic, as was supposed for a long while. Instead it is formed by the mixture that takes place at the Iceland-Scotland sills between the upper layers of the North Atlantic, with the very dense waters overflowing from the Norwegian Sea.

And last, the effect of this warm and saline layer is to contribute salinity to the other oceans, and to set up the formation of the Weddell Sea Bottom Water.

The South Atlantic receives mid-depth waters from the North Atlantic. From the south it receives its densest water from the Weddell Sea, from the circumpolar flow through the Drake Passage at all depths, and the shallower layer of Intermediate water from near the surface along about 50°-55°. The circumpolar waters and the waters from the North Atlantic have widely different characteristics and overlap in density. They are caught up in the

From WEFER G, BERGER WH, SIEDLER G, WEBB DJ (eds), 1996, *The South Atlantic: Present and Past Circulation.* Springer-Verlag Berlin Heidelberg, pp 13-44

circulation imposed by the winds and thermohaline processes. Interleaving takes place and leads to various vertical extrema as they spread and mix laterally.

While no new or special layers are formed within the South Atlantic, it has its own system of gyral and boundary flows, and waters passing through this system are changed, little by little, by contact with their neighbors.

The patterns along vertical sections

First, consider a vertical section of salinity along the western Atlantic, with the Norwegian-Greenland Sea on the right and the Weddell Sea on the left (Fig. 1).

The major feature is the thick layer of high salinity extending southward from the central North Atlantic into the Antarctic Circumpolar Current. The upper part of this saline layer, down to about 2500 m, drives from the Mediterranean outflow of warm and saline water. The North Atlantic is the warmest and most saline of the open world ocean's waters.

Waters of low salinity from the south enter as Intermediate Water near 1000m and near the bottom, from the Drake Passage inflow and the Weddell Sea.

In the north the Labrador Sea overturns to about 1500m and provides relatively low salinity, and the Denmark Strait provides a thin stratum of low salinity along the slope.

It appears that the warm and saline layer from the North Atlantic penetrates into the lower-salinity circumpolar waters from the south, which extend from top to bottom near 60°S and split them into two layers, above and below the saltier water from the North Atlantic.

This western section and similar sections in the central and eastern Atlantic had led to the concept of a northward flow at the bottom of colder, lower salinity water from the south, called Antarctic Bottom Water (though most of it is not from the bottom), a southward flow of warm and saline water from the North Atlantic at mid-depths, and a northward flow of Intermediate Water — the salinity minimum. That is, three layers, two from south one

from north. It was assumed that each layer filled its depth range from west to east.

Wüst made the point, based upon the oxygen pattern, that the southward flow was taking place along the western boundary. This was assumed to be a coherent flow from the Labrador Sea to about 45°N to 50°S, where it turned eastward to join the Antarctic Circumpolar Current. But he did not recognize any influence from the south in the depth range of the great salinity maximum. In the South Atlantic the saline layer is partly interrupted by lower salinity that must have entered from the south.

This sort of interruption is seen more clearly in the fields of oxygen and nutrients (Figs. 2 and 3). The circumpolar waters, that is those that have come with the circumpolar current through the Drake Passage from the Pacific Ocean, are much lower in oxygen and much higher in nutrients than those from the North Atlantic.

The southward extension of this high-oxygen layer is much like the layer of high salinity. But there is a different feature in this maximum that is not quite so evident in the salinity. This great maximum has a minimum in the middle. The characteristics at the minimum — the lower oxygen and higher nutrients, and here and there a slightly lower salinity — indicate that it has some component of circumpolar water from farther south. That is, as well as splitting the circumpolar water into an upper and lower minimum, as it certainly does, the layer of high salinity and oxygen and low nutrients is itself split to some extent by the circumpolar water of the same density.

Remember that the section has two dimensions only, and the flow is not just along this section. The circumpolar water extends northward also east of this section. As we shall see, it takes a sinuous path, diving into and out of this section at various latitudes.

The penetrations are seen also in phosphate, which is almost the mirror-image of oxygen. The high values of the circumpolar water — which extend top-to-bottom in the circumpolar water in the far south — survive the split by moving northward east of this section and penetrating it from the east here and there. And lest any modern chemist not accept the traditional characteristics, we see the same features in carbon 14 (Fig. 4).

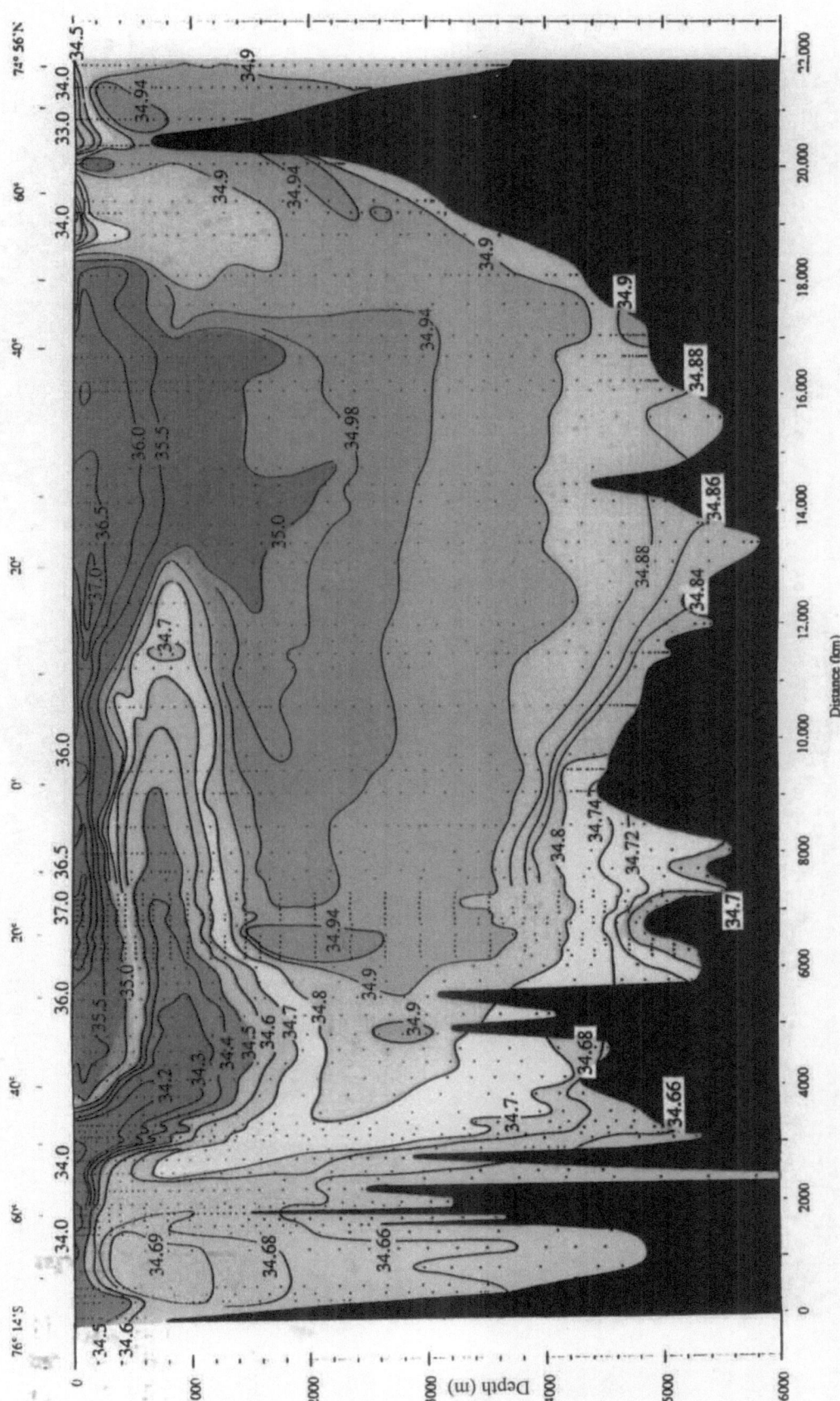

Fig. 1. Salinity on the north-south section. (Reid 1994).

J. L. Reid

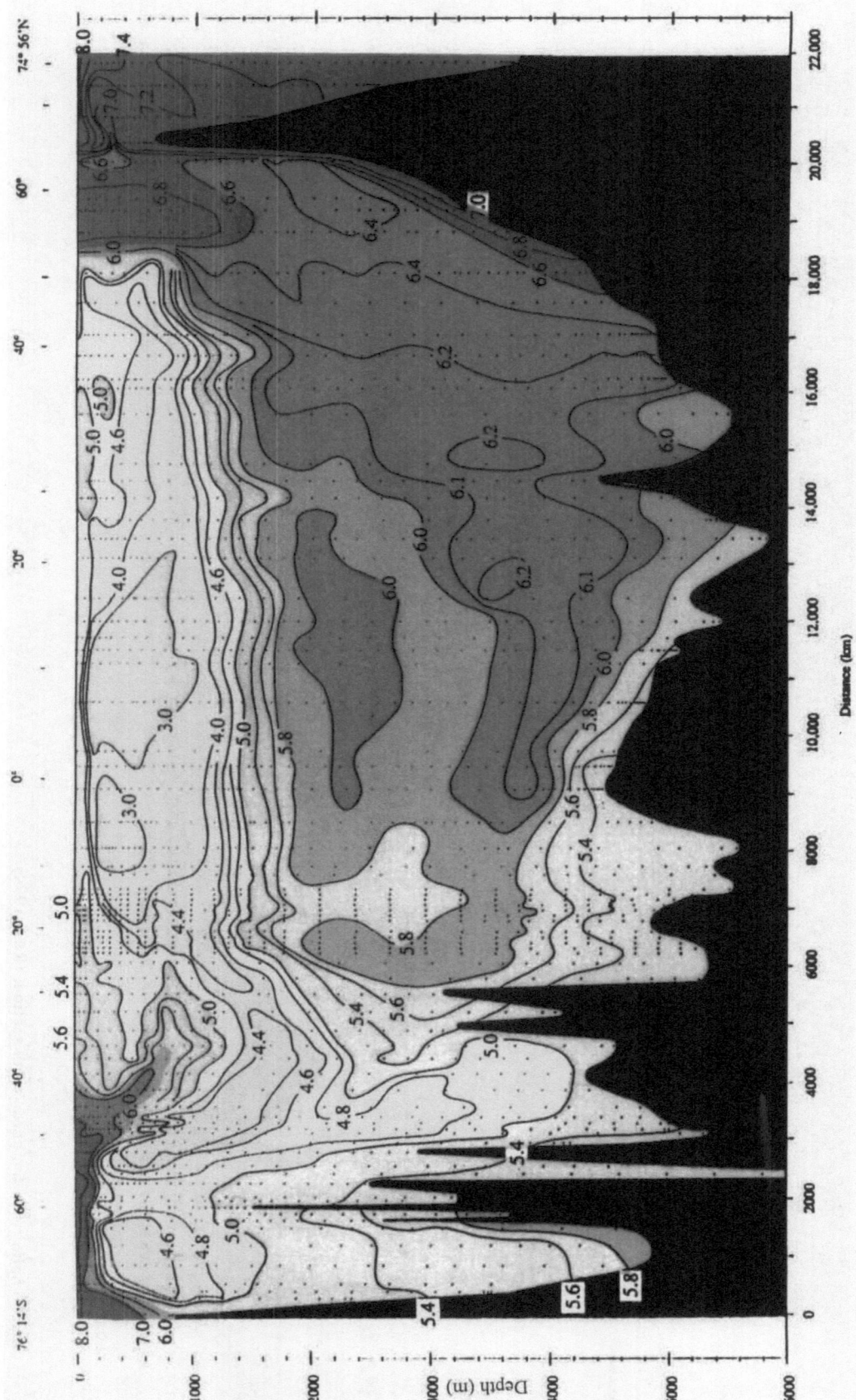

Fig. 2. Oxygen (ml/l) on the north-south section. (Reid 1994).

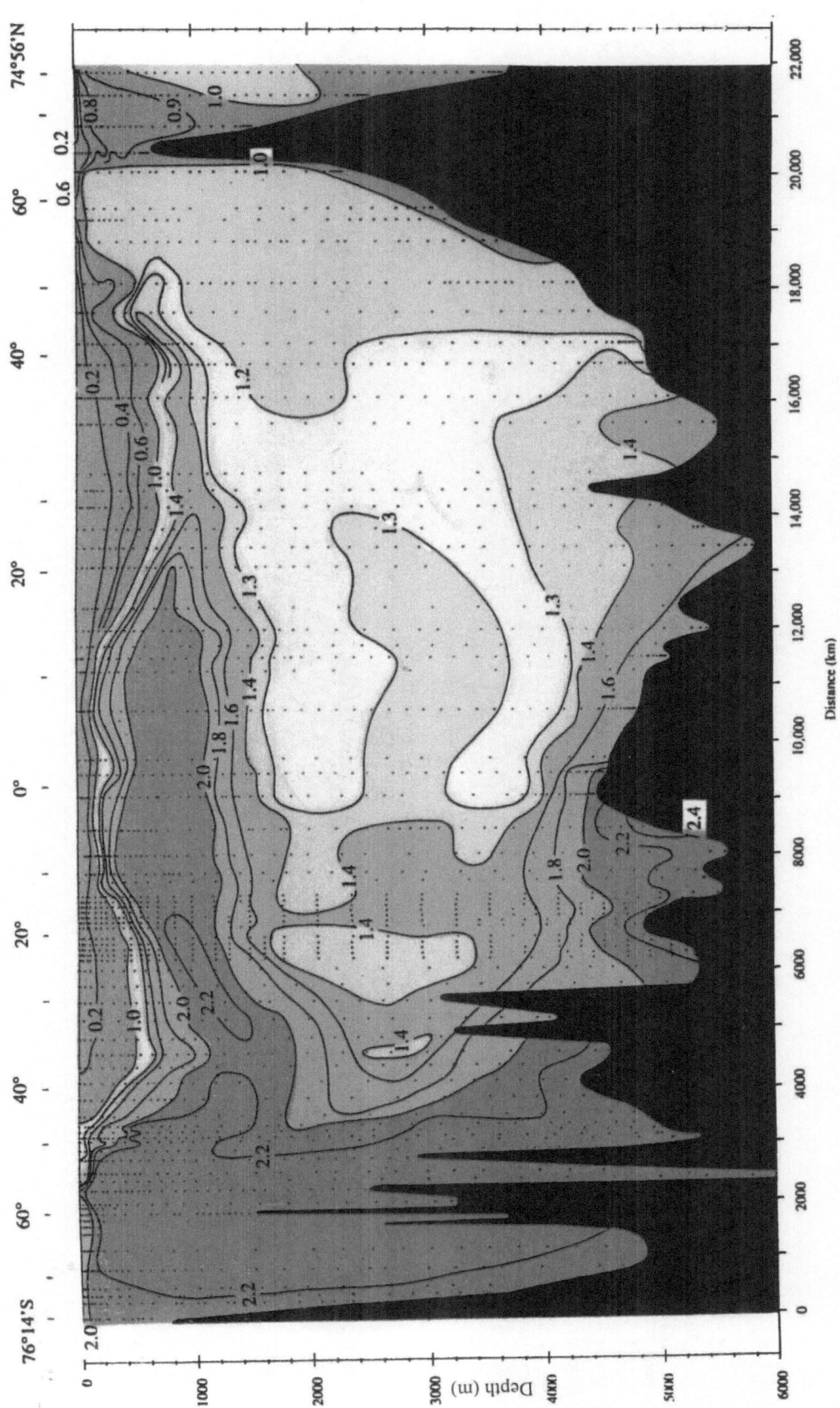

Fig. 3. Phosphate (μm kg^{l}) on the north-south section. (Reid 1994).

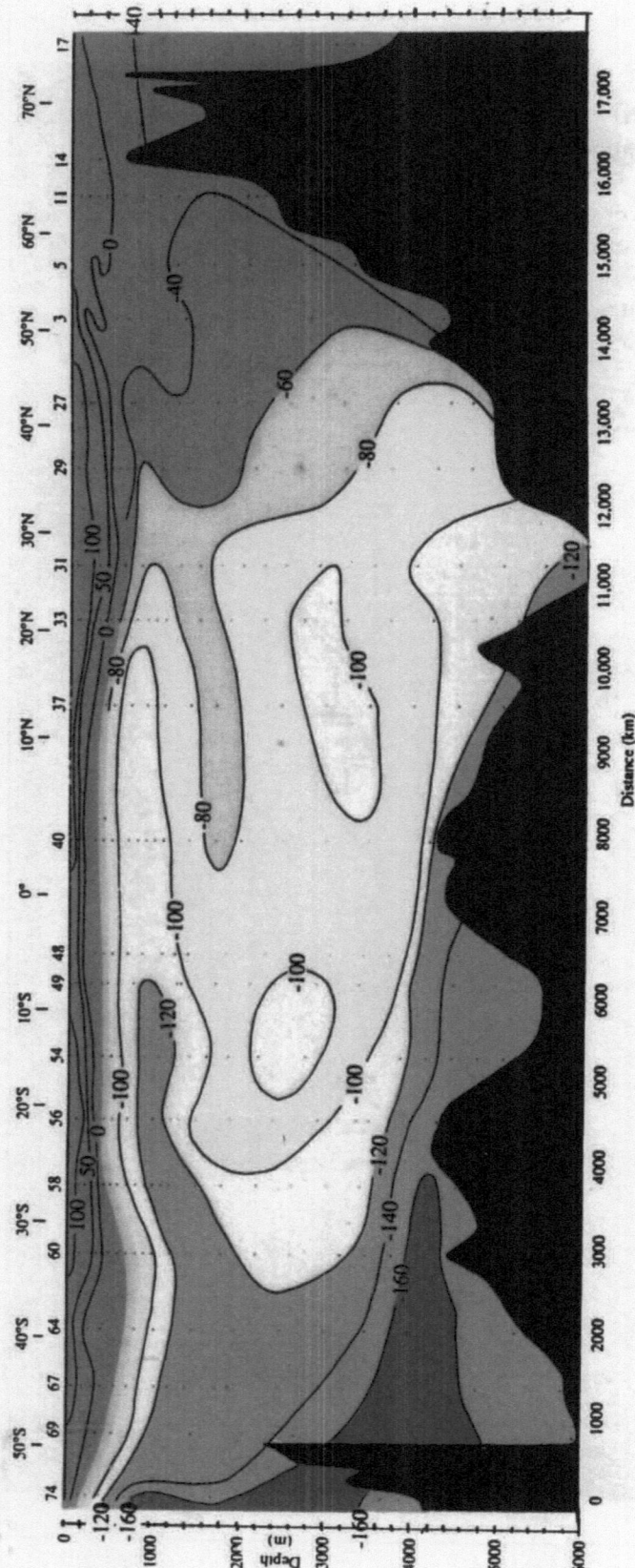

Fig. 4. The meridional distribution of $\Delta^{14}C$ (‰) along the western Atlantic Geosecs section. Adapted from STUIVER and OSTLUND (1980) (Reid 1989).

These figures have shown that circumpolar waters and the North Atlantic waters are quite different from each other. The North Atlantic is warm, saline, oxygen-rich and nutrient-poor. The South Atlantic is cooler, lower in salinity and oxygen and higher in nutrients. Their depth and density ranges overlap, and where they meet and circulate in the South Atlantic they interleave at various levels, and several vertical extrema are created in the fields I have shown. We can look at these characteristics on maps also.

The patterns along isopycnals

The Intermediate Water ($\sigma_1 = 31.938$)

The isopycnal along which the Subantarctic Intermediate Water salinity minimum lies as it crosses the equator northward lies a little deeper than 900m in low latitudes but outcrops south of 50°S in winter and south of Greenland in the north. It lies deepest within the great anticyclonic gyre centered near 30°S (Fig. 5).

The low salinity of the Intermediate Water from the far south extends across the equator along the western boundary (Fig. 6). It appears as a vertical salinity minimum to about 24°N, where it meets the upper part of the highly-saline Mediterranean Outflow.

Along 30°S, the range of salinity in the Intermediate Water is not very strong. This is about the axis of the anticyclonic gyre of the South Atlantic, and there is some recirculation of the Intermediate Water. Part of it turns back south across 35°S with the Brazil Current.

The oxygen field appears to have a rough north-south symmetry (Fig. 7). Where the isopycnal lies shallow in high latitudes the oxygen is replenished in both hemispheres. There are roughly symmetric minima just north and south of the equator in the east, separated by an eastward flow along the equator. They are the result of the circulation — the wind-driven divergence in those areas.

As with the salinity, there is a suggestion of northward flow across the equator in the west and of southward flow south of 30°S along the western boundary.

The Labrador Sea water ($\sigma_{1.5} = 34.64$)

This isopycnal represents the Labrador Sea water. It outcrops just north of Iceland and south of 60°S and lies near 1600 to 1800 meters in lower latitudes. The low salinity of the Labrador Sea upper waters extends eastward through the subarctic cyclonic flow and southward along the western boundary to beyond 40°N (Fig. 8). There is also some warm and saline water coming over the Iceland-Scotland Ridge.

South of there the dominant feature is the warm and saline water reflecting of the Mediterranean outflow. It extends both northward into the subarctic gyre and westward across the Atlantic and turns southward along the western boundary.

Lower-salinity water from the south crosses the equator at this depth but now it crosses in the east, and turns northward and westward into the great salinity maximum.

The southward flow along the western boundary appears to be continuous from the Labrador Sea across the equator to 40°S, but south of 25°N much of the water flowing southward along the western boundary has not come directly from the Labrador Basin. Instead it has come from the southern, westward-flowing limb of the anticyclonic gyre, and contains a large component of the heat and salt from the Mediterranean outflow. Because of this the maximum salinity along the southward flow is at about 30°N.

This is the beginning of the great oxygen feature that led Wüst to propose a deep southward flow along the western boundary from the subarctic gyre across the equator into the southern hemisphere and the eastward flow along the equator (Fig. 9).

In the North Atlantic the eastern tropical minimum plays an important role in the shallow oxygen minimum seen on the vertical section in the west. In the South Atlantic the eastern tropical minimum does not extend so deep, and it is the circumpolar water that contributes the low oxygen.

To represent the flow I have first calculated the geostrophic flow relative to the bottom along selected lines of stations. I have then used the patterns of characteristics along vertical sections and isopycnals to infer a bottom velocity between each

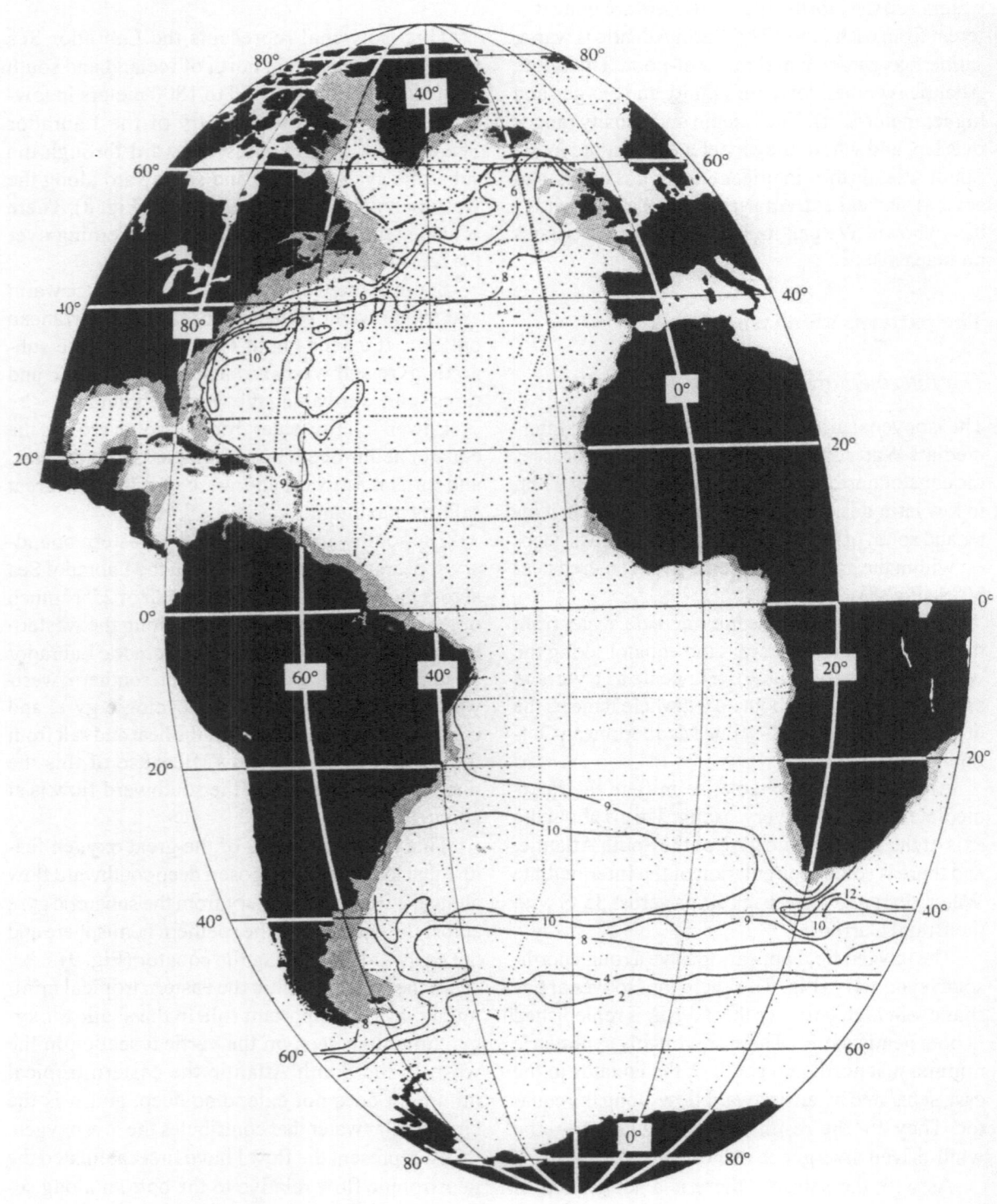

Fig. 5. Depth (hm) of the isopycnal defined by 31.938 in σ_1 below 500m. (Reid 1994).

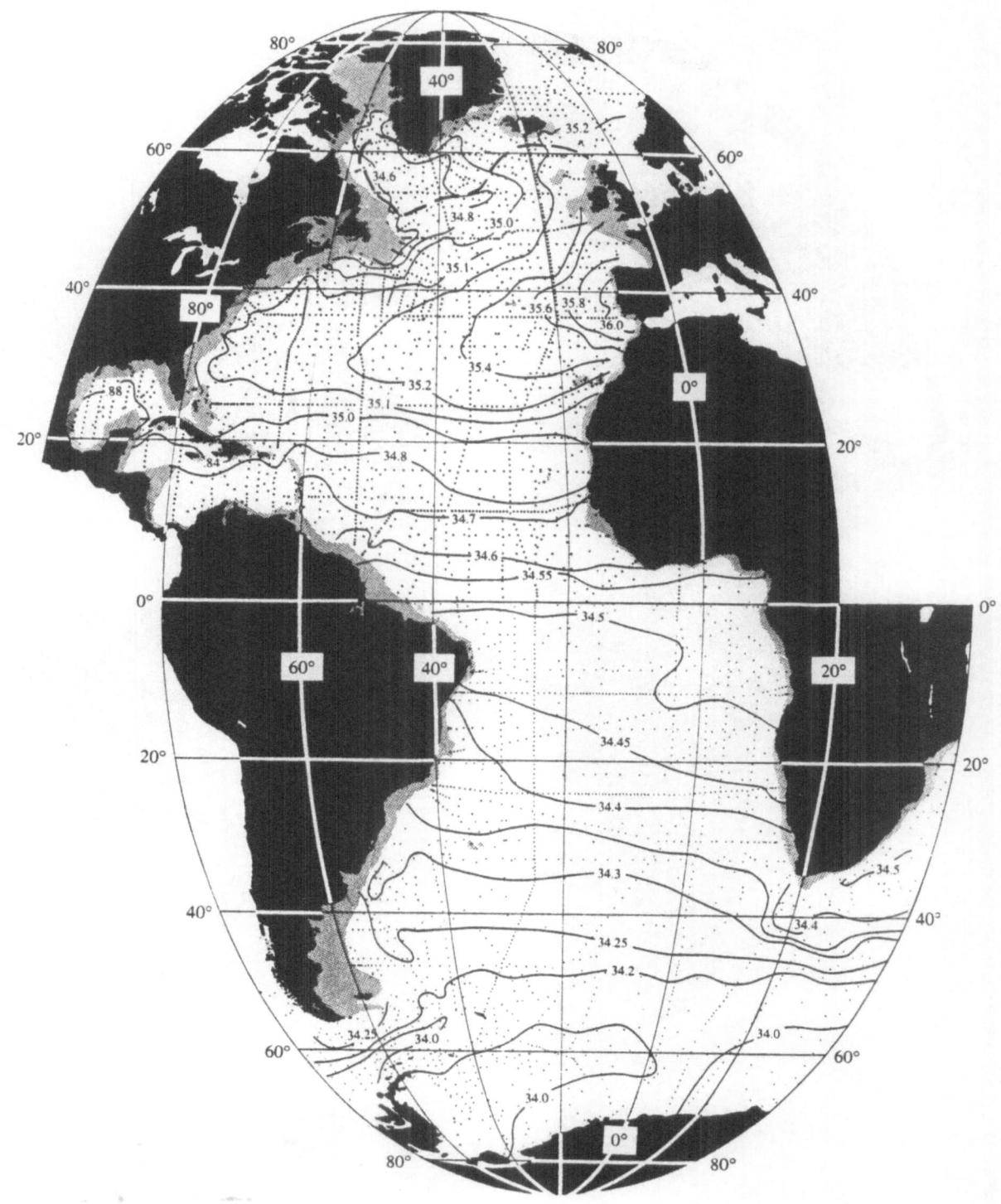

Fig. 6. Salinity on the isopycnal defined by 31.938 in σ_1 below 500m. (Reid 1994).

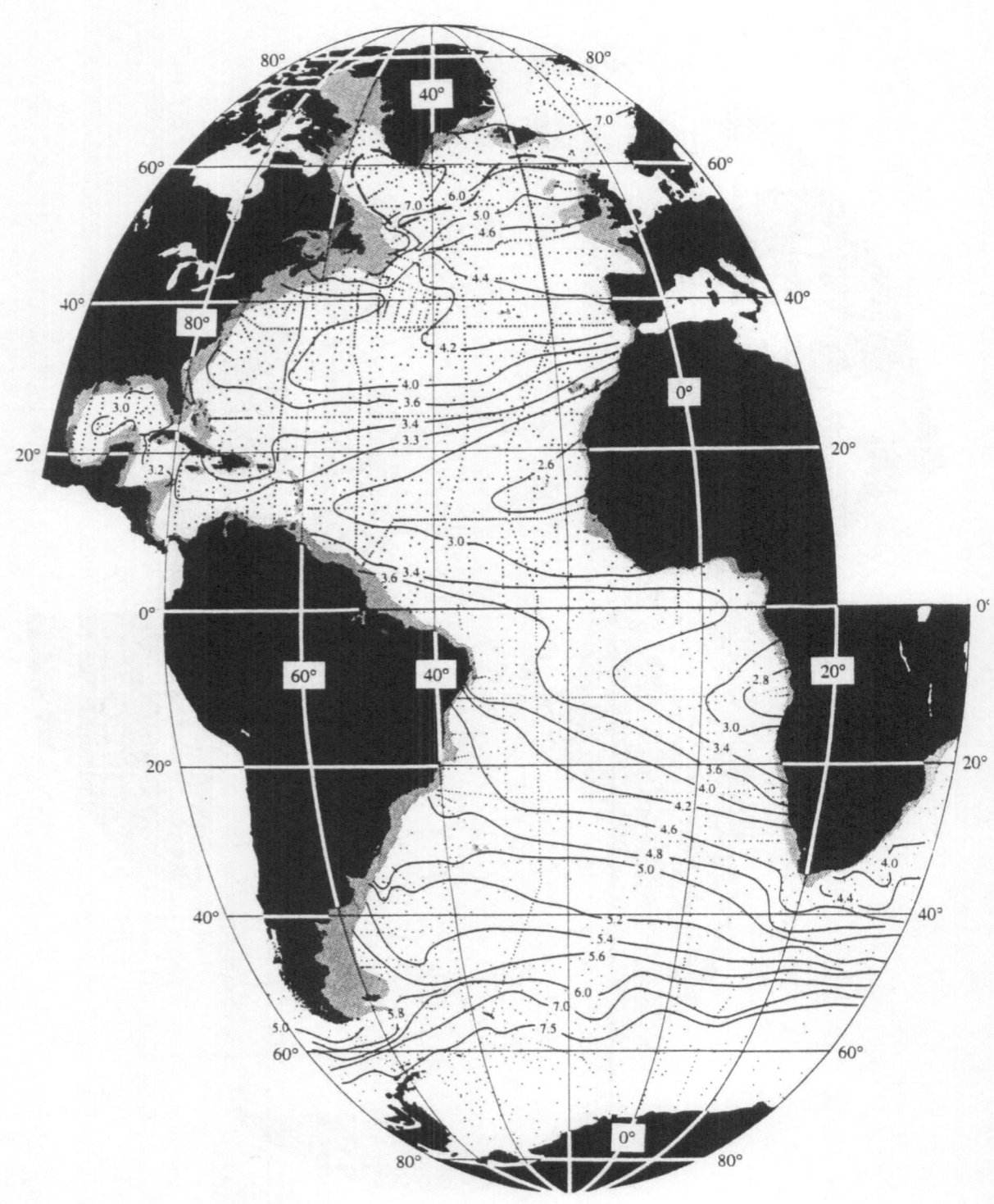

Fig. 7. Oxygen (ml/l) on the isopycnal defined by 31.938 in σ_1 below 500m. (Reid 1994).

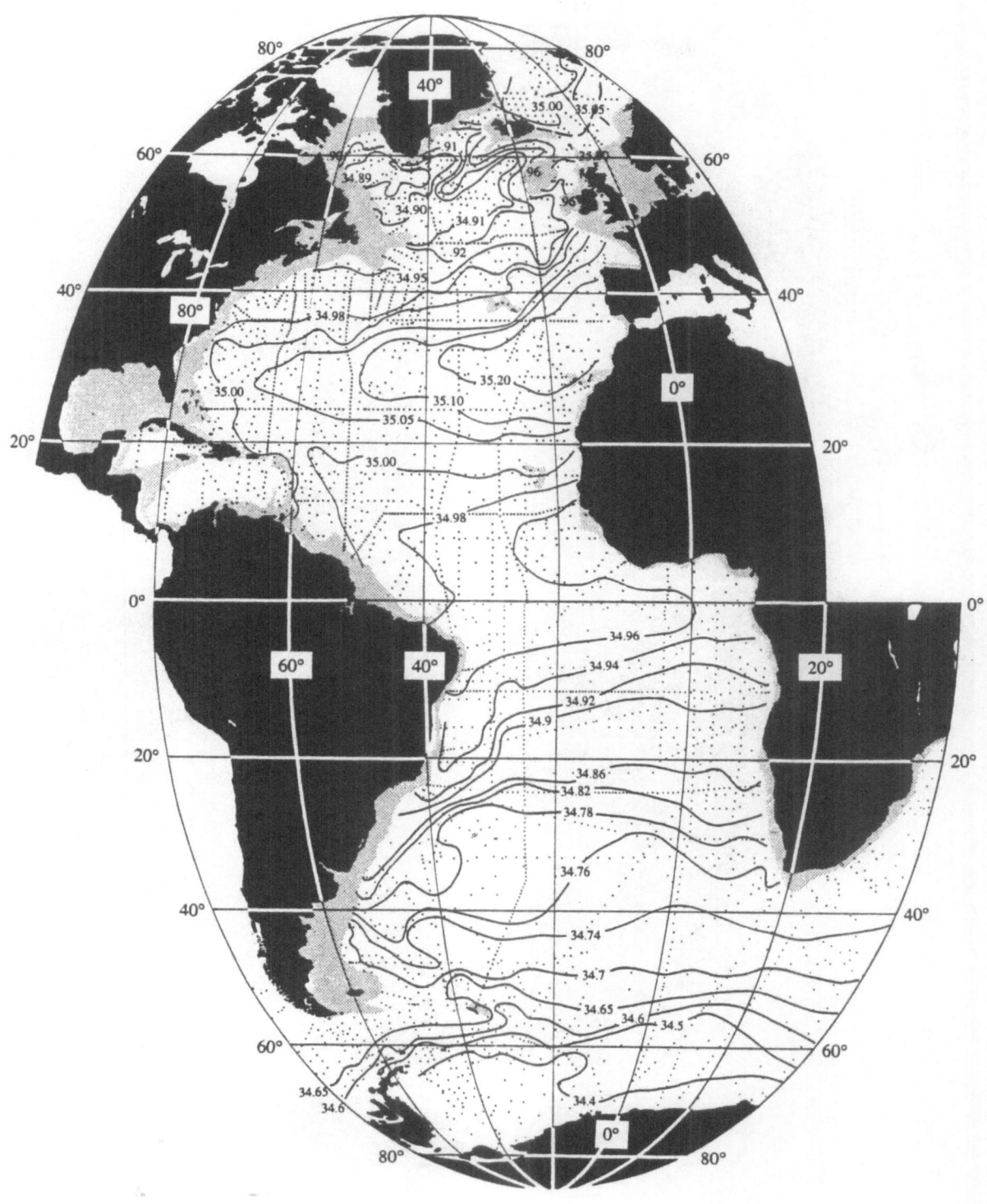

Fig. 8. Salinity on the isopycnal defined by 34.64 in $\sigma_{1.5}$. (Reid 1994).

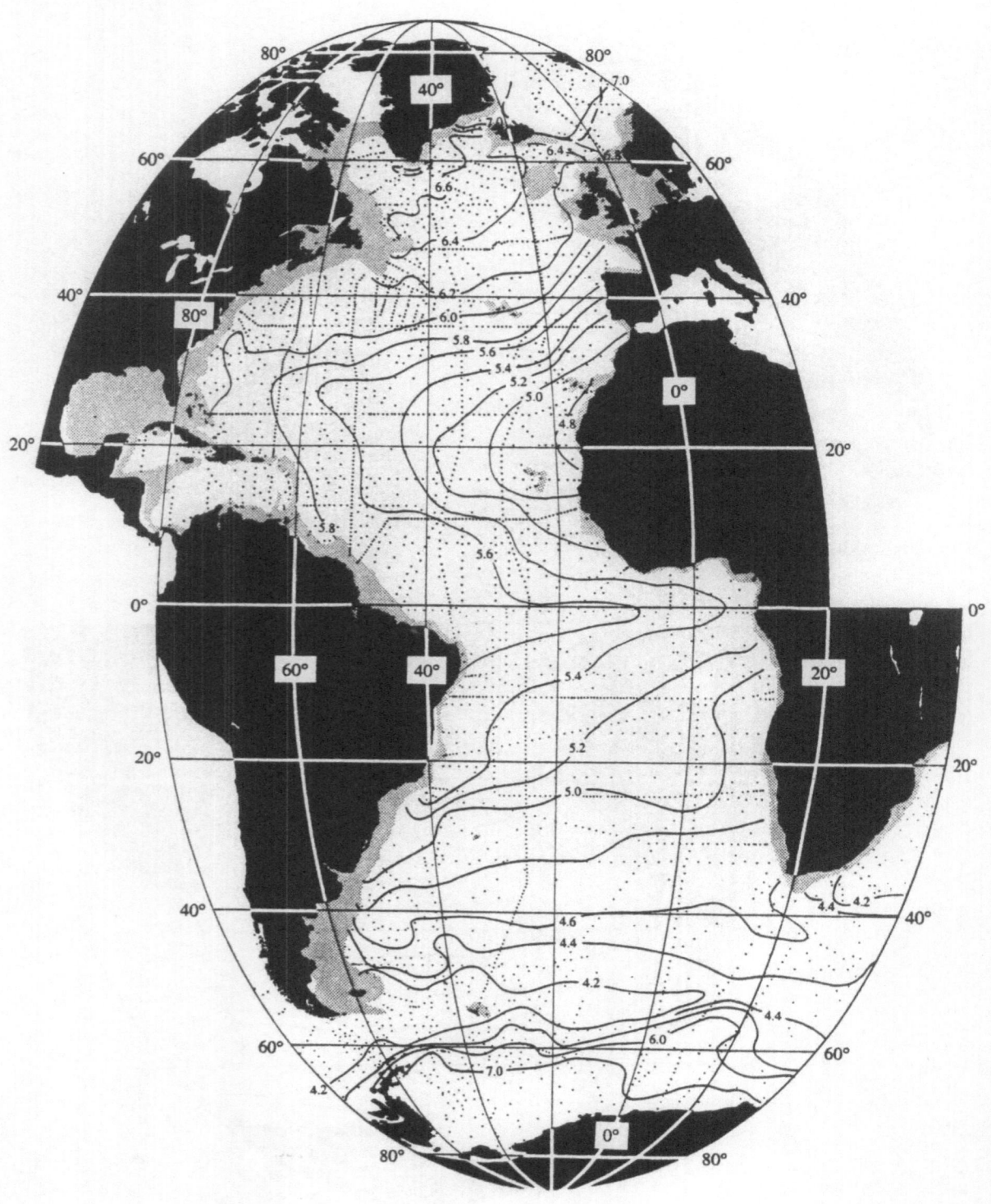

Fig. 9. Oxygen (ml/l) on the isopycnal defined by 34.64 in $\sigma_{1.5}$. (Reid 1994).

station pair along each line. This gives the total geostrophic glow at every depth. I call this the adjusted geostrophic flow.

This is the geostrophic flow along the isopycnal (Fig.10). As it represents the flow at the various depths of the isopycnal it is more useful than the flow at single mean depth. It shows the recognized features and some that are not so familiar. From north to south we see the subarctic cyclonic gyre, part of which extends down to about 35°N along the western boundary before turning back to the northeast.

In the South Atlantic we see the same features — the Weddell Sea cyclonic flow and the anticyclonic gyre, with the Antarctic Circumpolar Current flowing eastward between them — and the southward flow of the Brazil Current meets the Circumpolar Current and makes two great loops before turning eastward and around the cyclonic gyre.

The less-familiar features are the cyclonic flows north and south of the equator near 30°N and 30°S.

We see the southward flow everywhere along the western boundary to about 40°S, where it meets the northward flow from the circumpolar current and turns eastward with it. But offshore in the South Atlantic we see a northward flow from 50°S to the equator and then turning eastward crossing the equator in the east and then turning westward, northwestward, and eastward again near 20°N as part of the cyclonic gyre. Much of this general pattern extends down to the crest of the Mid-Atlantic Ridge, which separates the eastern and western basins at greater depth.

The salinity maximum in the South Atlantic

This isopycnal where $s_2 = 36.98$ (Figs. 11 and 12) lies at or near the vertical salinity maximum in the South Atlantic from the equator to about30°S. It lies near 2100m in low latitudes and outcrops north of 60°N and south of 60°S.

There is still a diversion from the western-boundary flow along the equator. The less-saline circumpolar water crosses the equator in the east and moves northwestward toward the salinity maximum.

There is a stronger overflow of saline water from the Iceland-Scotland Ridge but the maximum temperature and salinity along the western boundary current are still found near 20°N, deriving from the Mediterranean source.

The oxygen pattern that revealed the western boundary current is even stronger on this isopycnal (Fig. 12). The westward intrusion of low oxygen near 20° to 24°N is partly from the circum-polar water, as the salinity pattern indicates, and partly from the low oxygen of the eastern tropical minimum. The southern intrusion, near 15°S, is almost entirely from circumpolar water.

As the depth of this isopycnal in mid-latitudes is near 2000 meters it can be compared with the 2000 decibar flow(Fig. 13). This figure shows the intrusions near 15°S and 15°N that we saw on the vertical section. It has much the same pattern as the waters just above. The anticyclonic gyres are narrower, and the southern one is at least partly interrupted by the Rio-Grande Rise. The northern cyclonic feature off northern Africa appears to be nearly continuous with the subarctic gyre at this depth. And the southern cyclonic gyre off South Africa has been joined to the great northward loop from the Circumpolar Current.

And the carbon 14 pattern (Fig. 14) is much the same as the oxygen pattern.

The characteristics near the depth of the Mid-Atlantic Ridge

The Mid-Atlantic Ridge separates the North Atlantic into an eastern and western basin. The densest water that can pass westward through the Charlie Gibbs Fracture Zone, near 53°N, has a density of about 41.50 in s_3. Water along this isopycnal lies at about 3200m to 3400m in mid-ocean and outcrops in the Norwegian-Greenland and Weddell seas (Fig. 15).

At the sill between Iceland and Scotland the upper waters of the North Atlantic meet the very dense waters of the Norwegian-Greenland Sea. They mix and pour down into the Iceland Basin as a warm and saline layer. The cyclonic gyre carries these mixtures through the Fracture Zone, around

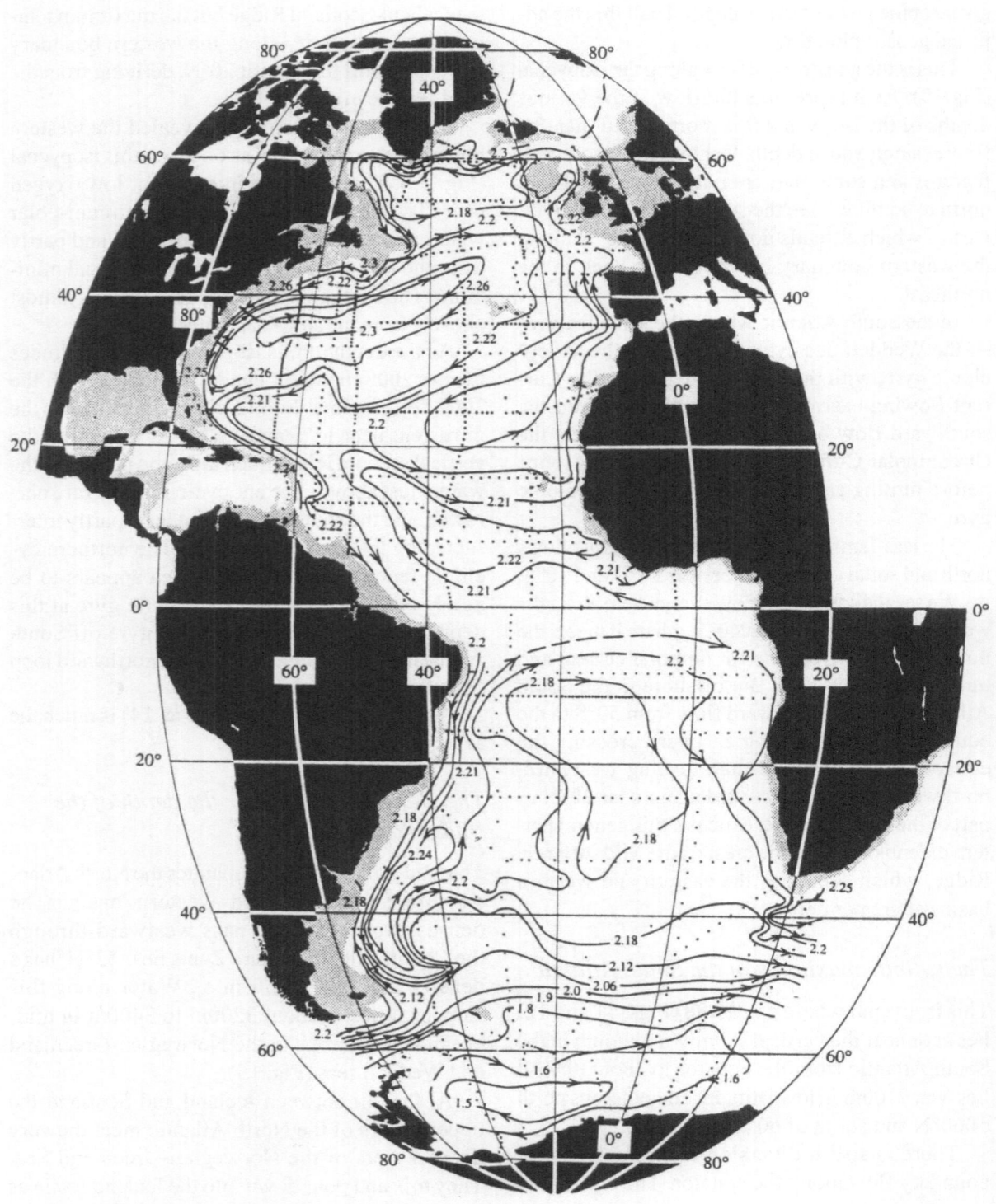

Fig. 10. Adjusted steric height along the isopycnal defined by 34.64 in $\sigma_{1.5}$. (Reid 1994).

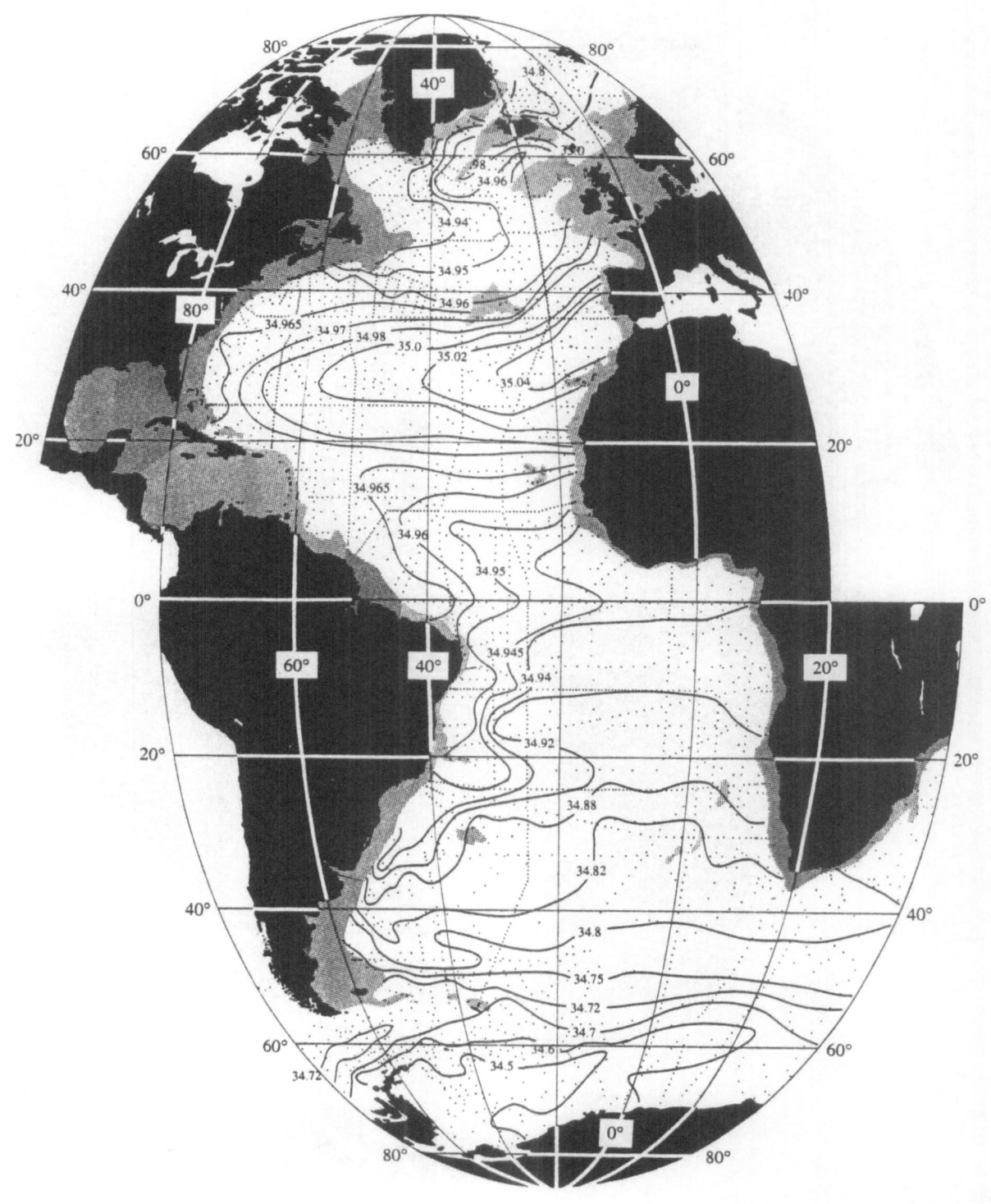

Fig. 11. Salinity on the isopycnal defined by 36.98 in σ_2. (Reid 1994).

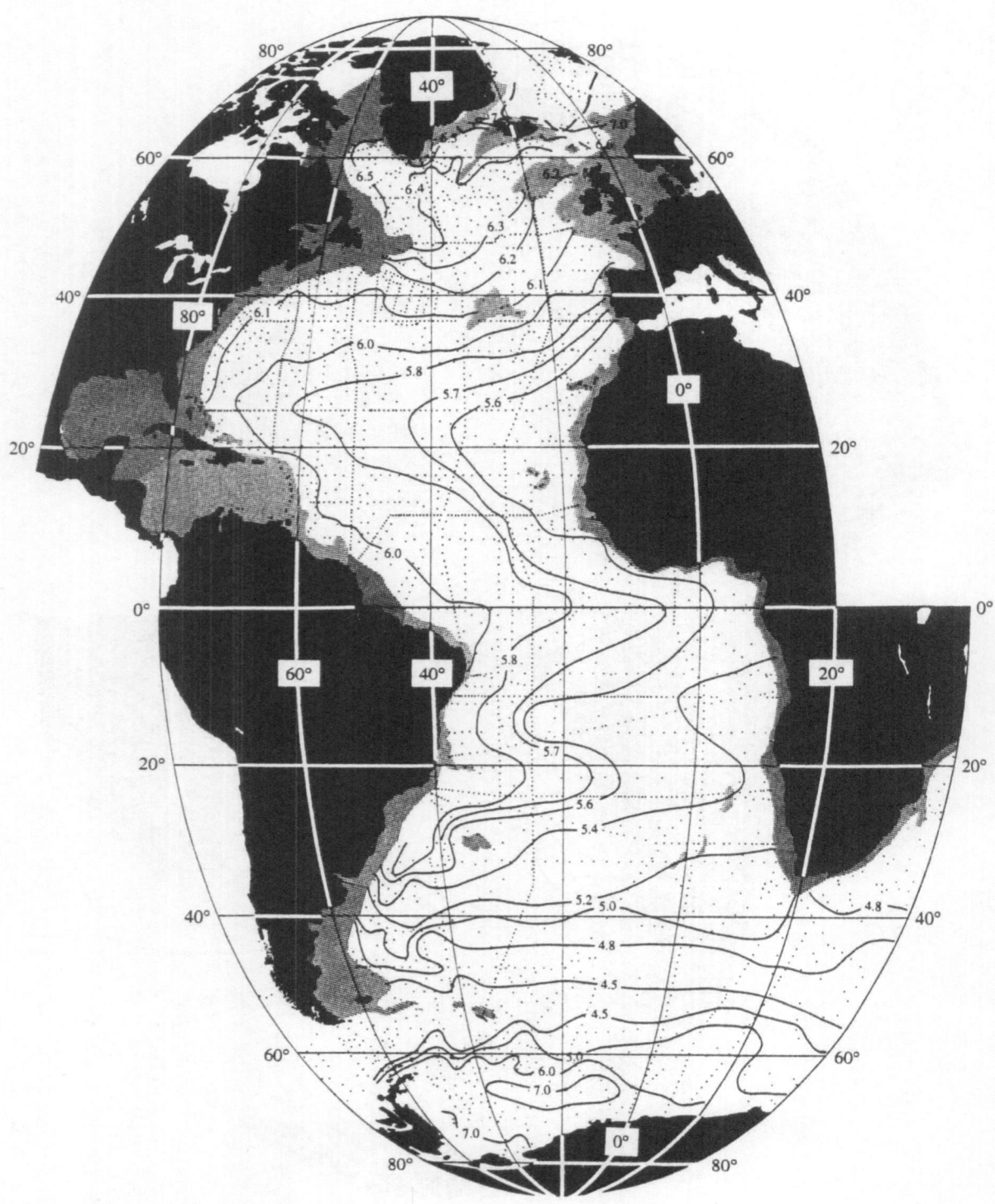

Fig. 12. Oxygen (ml/l) on the isopycnal defined by 36.98 in σ_2. (Reid 1994).

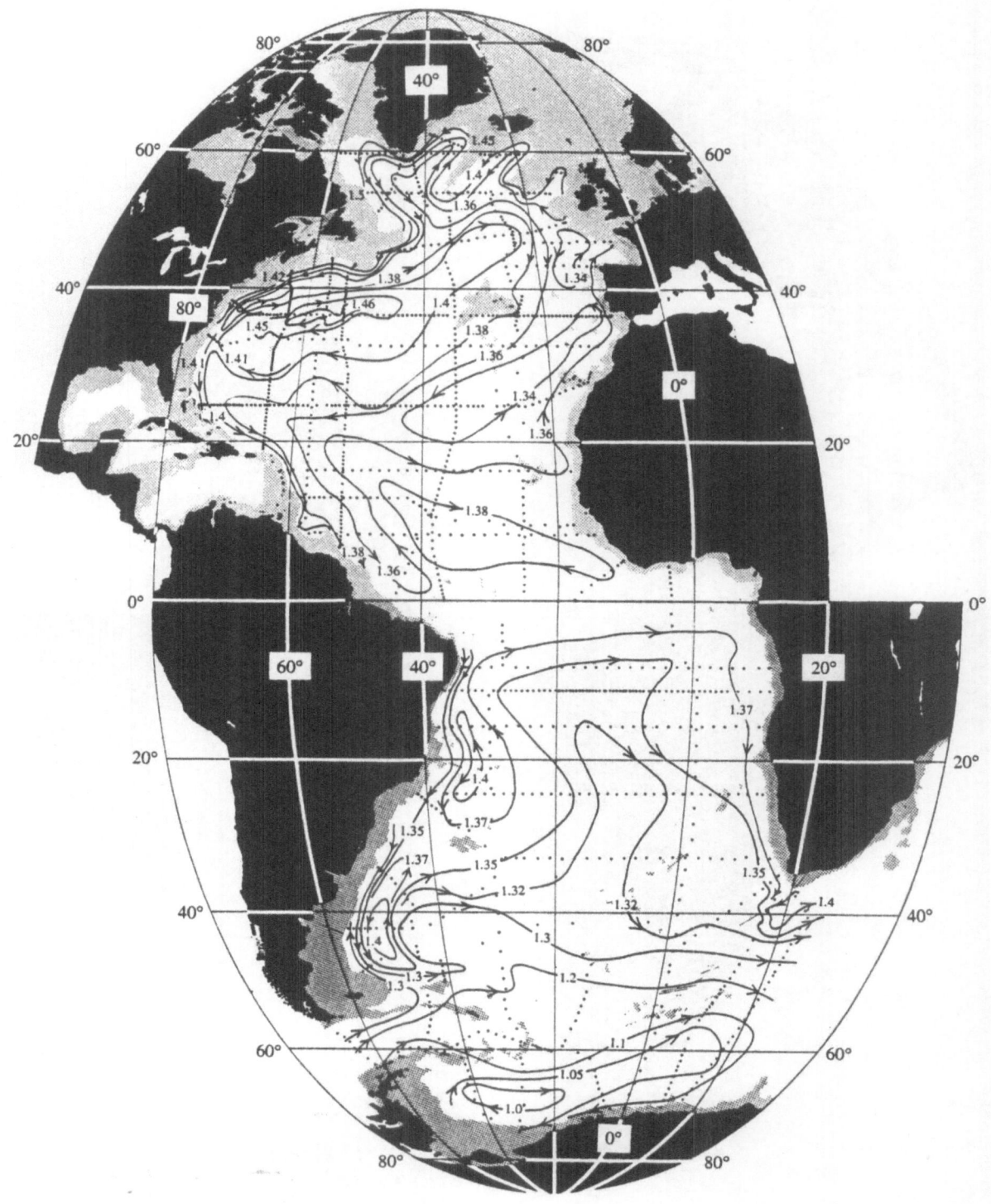

Fig. 13. Adjusted steric height at 2000 db (10m²s⁻² or 10Jkg⁻¹). (Reid 1994).

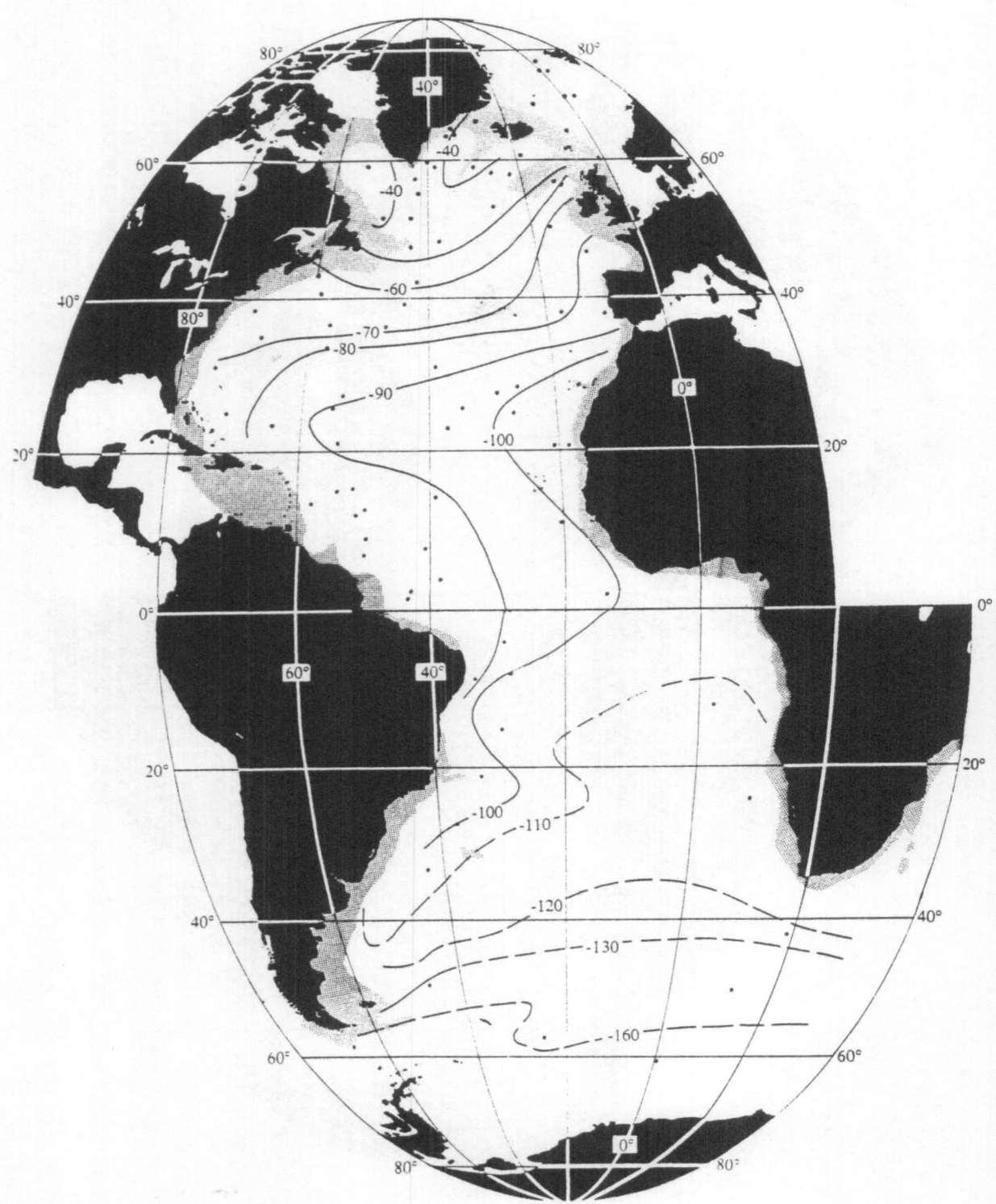

Fig. 14. $\Delta^{14}C$ (‰) on the isopycnal defined 36.98 in σ_2.

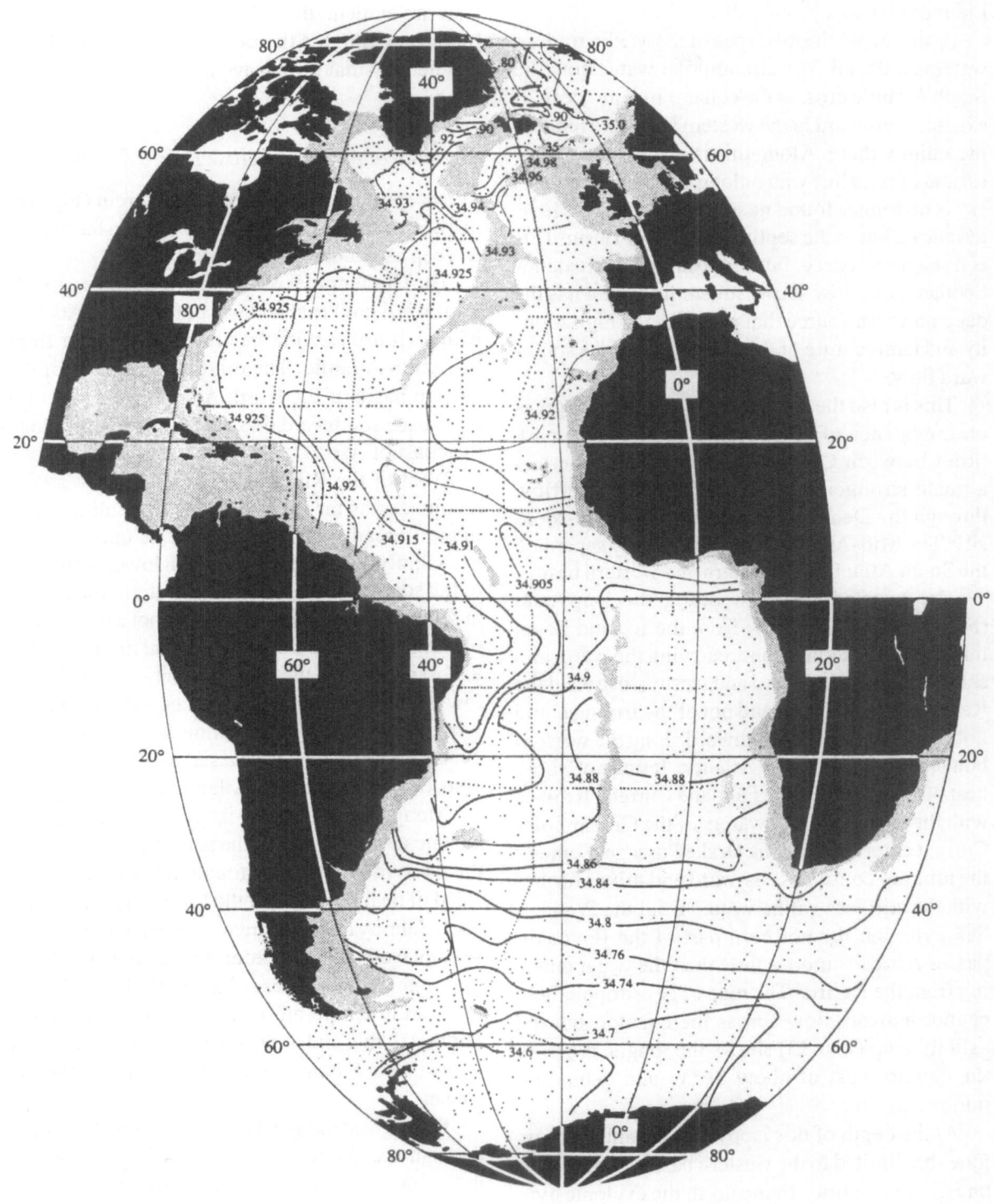

Fig. 15. Salinity on the isopycnal defined by 41.50 in σ_3. (Reid 1994).

the Reykjanes Ridge and into the Irminger and Labrador basins.

At this depth the influence of the Mediterranean water is reduced. The circumpolar water from the South Atlantic crosses the equator in mid-ocean. It extends northward in the western basin and reduces the salinity there. Along this isopycnal the warmest and most saline water along the western boundary is no longer found near 20°N, from the Mediterranean, but in the southern Labrador Basin. This is not a new source, but the result of the Iceland-Scotland overflow. Downstream from here it is the deep northern source that provides the high salinity and temperature of the deeper part of the southward flow.

This is also the shallowest isopycnal that gives clear evidence of the overflow through Denmark Strait between Greenland and Iceland. There is a much stronger oxygen signal in the overflow through the Denmark Strait (Fig. 16) at this density. The Mid-Atlantic Ridge has nearly separated the South Atlantic into eastern and western basins.

The geostrophic flow along the isopycnal (Fig. 17) shows the flow from the Iceland Basin through the Fracture Zone, carrying the warm and saline mixture southward and westward around the Reykjanes Ridge, into and out of the Irminger and Labrador basins, and southward along the western boundary into the South Atlantic. It turns offshore near 40°S as it meets the Falkland Current. It mixes with the lower-salinity waters of the Circumpolar Current entering from the Drake Passage. Some of the mixture continues eastward and mixes further with the still less-saline waters from the Weddell Sea gyre, but the northern part of the flow as it passes Africa is more saline than the water entering from the Pacific. The map of geostrophic flow cannot indicate flow across the equator, but the salinity map (Fig. 15) shows the tongue crossing the equator east of about 30°W and extending northwestward.

At the depth of this isopycnal the anticyclonic gyres are limited to the western basins of the North and South Atlantic. In the north the cyclonic gyre extends northward into the Iceland Basin. In the South Atlantic part of the anticyclonic gyre is confined north of the Rio Grande Rise. Even at this depth the Agulhas eddies and the Agulhas return flow are evident, though the characteristics of the Indian and South Atlantic oceans are so similar at this density that they leave no signal in their patterns.

The characteristics below Ridge depth

The deepest isopycnal illustrated herein (Fig. 18) lies near 3400m to 3600m in the western basins and as deep as 4000m in the eastern basins.

The Circumpolar Water at this density extends northward both east and west of the Mid-Atlantic Ridge. It mixes with the more saline water from the North Atlantic. The mixture fills most of the eastern basin of the North Atlantic and part of it turns back southward into the South Atlantic along the coast of Africa.

West of the Ridge the Circumpolar Water is recognized as far as 50°N by its low salinity. But in the Labrador Basin, which in its upper waters has salinity and temperature much lower than those to the south, there is an isolated lateral maximum in salinity and temperature. This is not a new source of warm and saline water but vertical mixing within the basin between the bottom water — the cold and low-salinity Denmark Strait water — and the overlying warmer and more saline Iceland-Scotland overflow-water that has passed westward at lower densities above the Mid-Atlantic Ridge.

Along this and deeper isopycnals the warmest and most saline water of the southward flow along the western boundary is found in the far north instead of near 20°N, from the Mediterranean influence. Salinity and temperature both decrease downstream from the Labrador Basin, but this layer is recognizable by these characteristics — lateral maxima in temperature and salinity — from the South Atlantic and along the Circumpolar Current to the Drake Passage and to the equatorial Pacific Ocean.

The highest oxygen (Fig. 19) is in the overflow through the Denmark Strait, which at this depth is the only lateral source of oxygen north of the Weddell Sea. Oxygen gives the clearest picture of the Denmark Strait source of the southward flow along the western boundary, and at this depth the pattern is more smoothly continuous. It is not im-

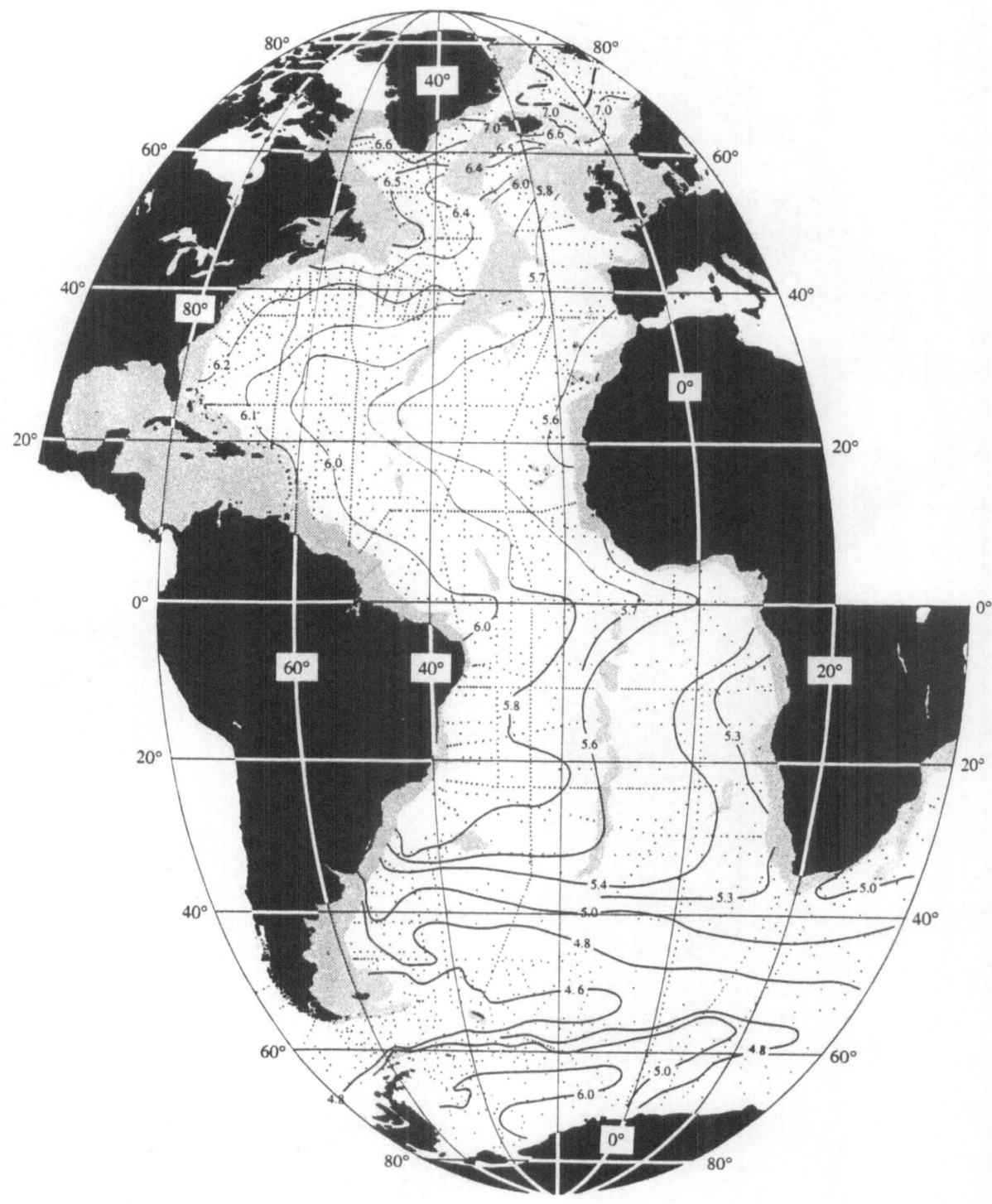

Fig. 16. Oxygen (ml/l) on the isopycnal defined by 41.50 in σ_3. (Reid 1994).

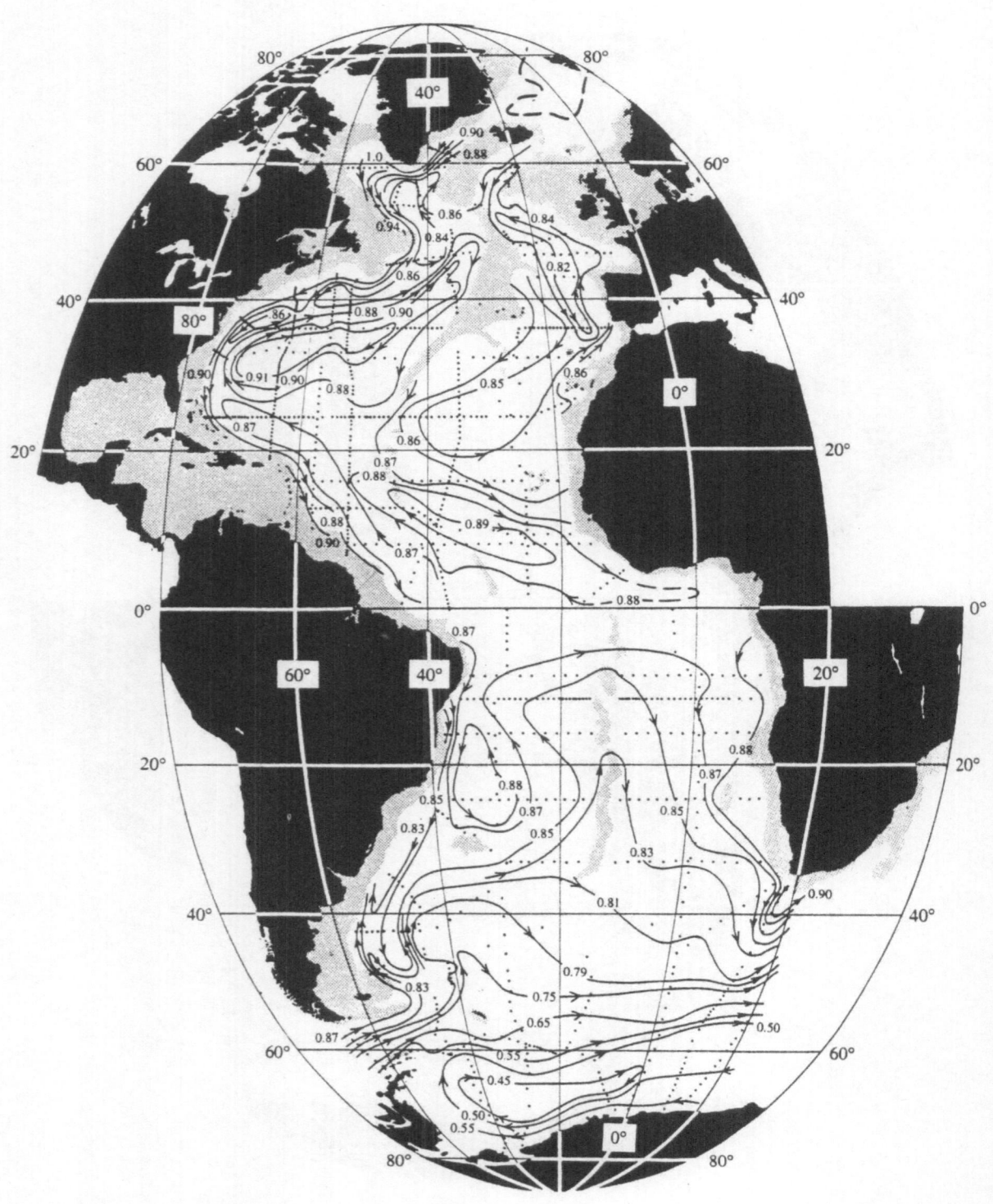

Fig. 17. Adjusted steric height along the isopycnal defined by 41.50 in σ_3. (Reid 1994).

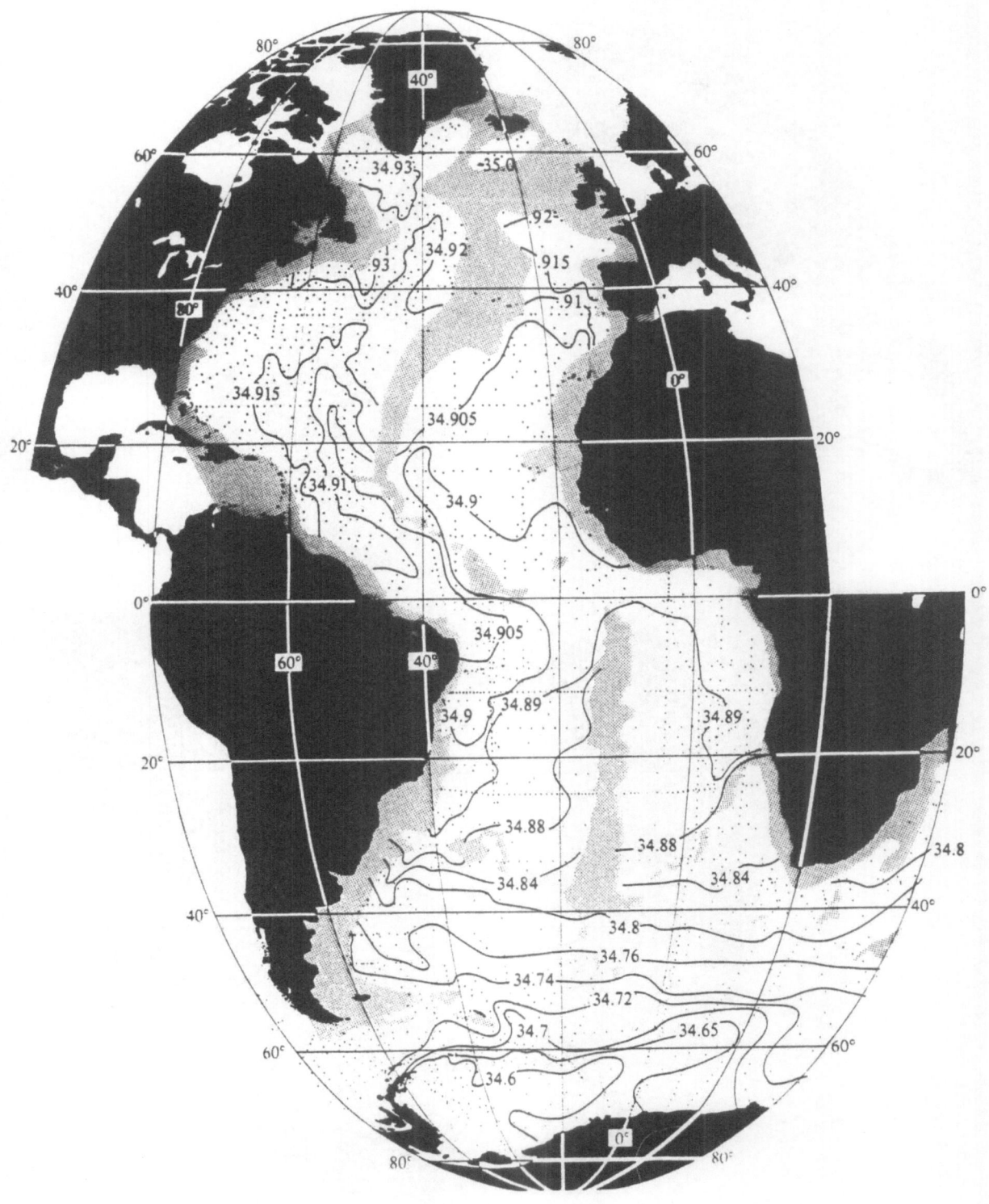

Fig. 18. Salinity on the isopycnal defined by 45.86 in σ_4 below 3500m.

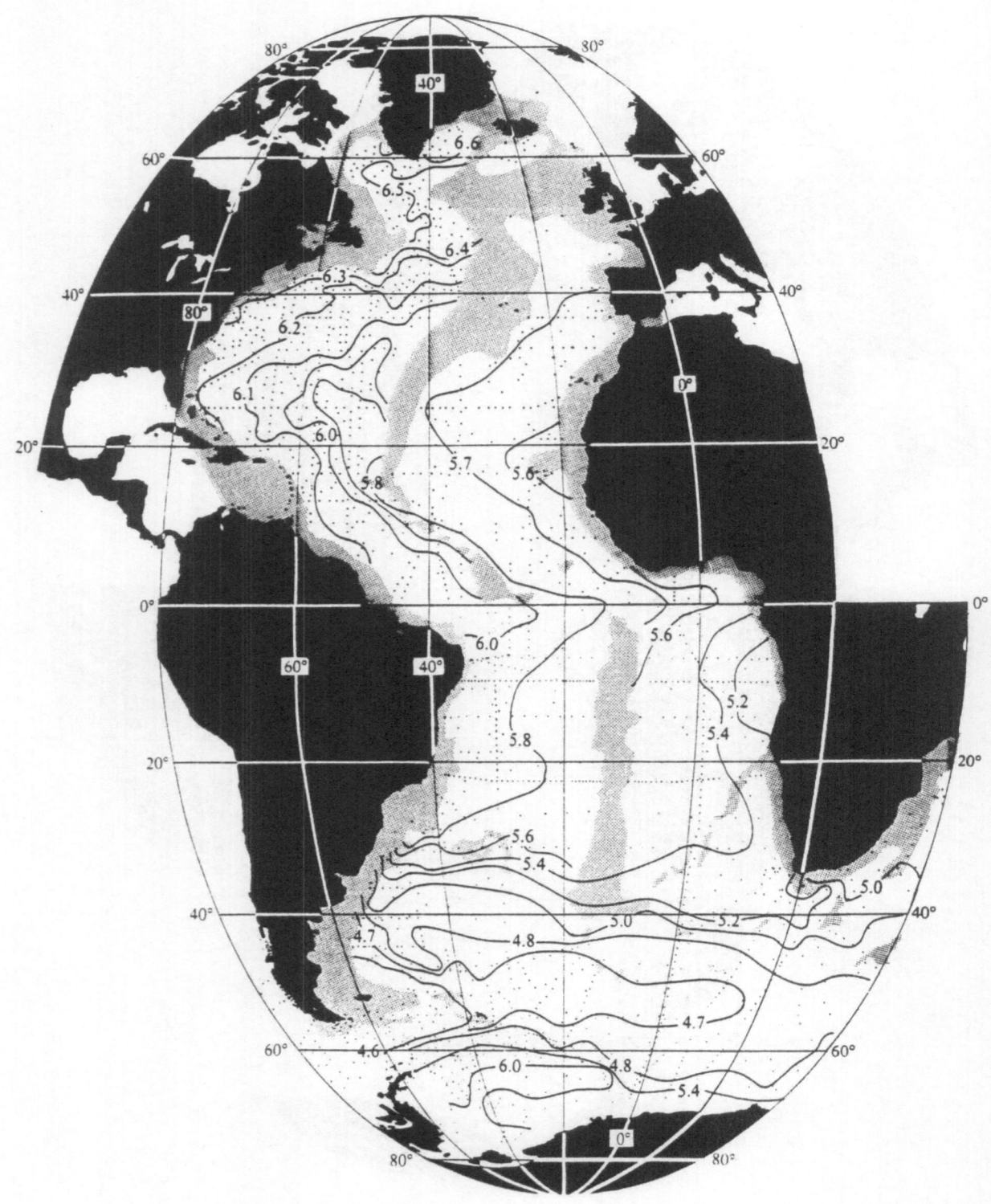

Fig. 19. Oxygen (ml/l) on the isopycnal defined by 45.86 in σ_4 below 3500m.

pinged upon so much by the westward flow of lower oxygen from the eastern basin of the South Atlantic.

The flow at 3500 decibars (Fig. 20) shows the anticyclonic patterns in both the North and South Atlantic, the continuous southward flow along the western boundary, the northward path of the lower-salinity water in both hemispheres, and the southward flow along the eastern boundary of the South Atlantic.

South of 30°S some of the water from the south turns back southward again in the Argentine and Cape basins, and cyclonic flows develop.

The deeper circulation and the total transport

Near 4000 decibars (Fig. 21) the South Atlantic shows little or no southward flow along the western boundary, but northward flow instead. In the north some of the southward flow may turn back northward, but the anticyclonic gyre in the north and the cyclonic gyre east of the Ridge are still evident. There is still cyclonic flow in the Argentine and Cape basins.

At 4500 decibars (Fig. 22) the Atlantic is cut up into several separate basins but we can still recognize some patterns of northward flow from the Weddell Sea to the Argentine Basin.

The total geostrophic transport (Fig. 23), is the top-to-bottom sum of the sorts of circulation maps shown in the maps of adjusted geostrophic flow. Its large-scale pattern is very much like the flow patterns seen from about 1000 decibars down to 2500 decibars, though between 25°N and 35°N the southward flow along the western boundary is masked by the shallow Gulf Stream. As most of the transport is above the crest of the Mid-Atlantic Ridge the deeper and abyssal flows are not obvious here.

Exchange with other oceans

The most obvious contribution of the Atlantic to the rest of the world ocean is the layer of warm and saline water that passes southward through the South Atlantic and joins the Antarctic Circumpolar Current.

An isopycnal that lies in the layer of high salinity in the Atlantic would follow the Atlantic water from the maximum value in the North Atlantic extending southward along the western boundary not only to the Antarctic Circumpolar Current but eastward around the world. It would still show a lateral salinity maximum as it passes through the Indian and Pacific oceans and returns to the Atlantic.

It would be much lower in salinity as it flows through the Drake Passage, and is replenished from the Atlantic waters, but is still a vertical salinity maximum as it reenters the Atlantic.

The great salinity maximum of the Atlantic extends southward, rises across the denser water in the Circumpolar Current, and appears to end as it turns eastward.

But an isolated patch of relatively high salinity is found south of 60°S from about 500m to 1500m (Fig. 1). This does not have a separate origin. It is merely the return flow of some of the more saline water that has joined the Weddell Sea gyre and turned southward and westward along the coast of Antarctica. This can be seen on some of the isopycnal maps (Figs. 11, 15, and 18). That is, the interleaving of circumpolar and water from the North Atlantic that was seen in the South Atlantic also takes place in the Antarctic domain.

The effect of this turn-back is to bring relatively saline water at shallow depths into the high latitudes where salinity is nearly everywhere low. By low, I mean so low that even at the freezing point it would not be dense enough to penetrate and renew the abyssal waters of the Weddell Sea.

Because of the vertical exchanges and the flow field the dissolved oxygen is very unequally distributed throughout the world ocean. Along such an isopycnal oxygen is very high in the Atlantic and very low in the Pacific as the Drake Passage inflows suggest (Figs. 9, 12, 16, and 19).

Circulation in different periods

What is this thing that has been commonly called North Atlantic Deep Water? We have seen that the waters in the North Atlantic have several local sources, and although some of the sources are cold

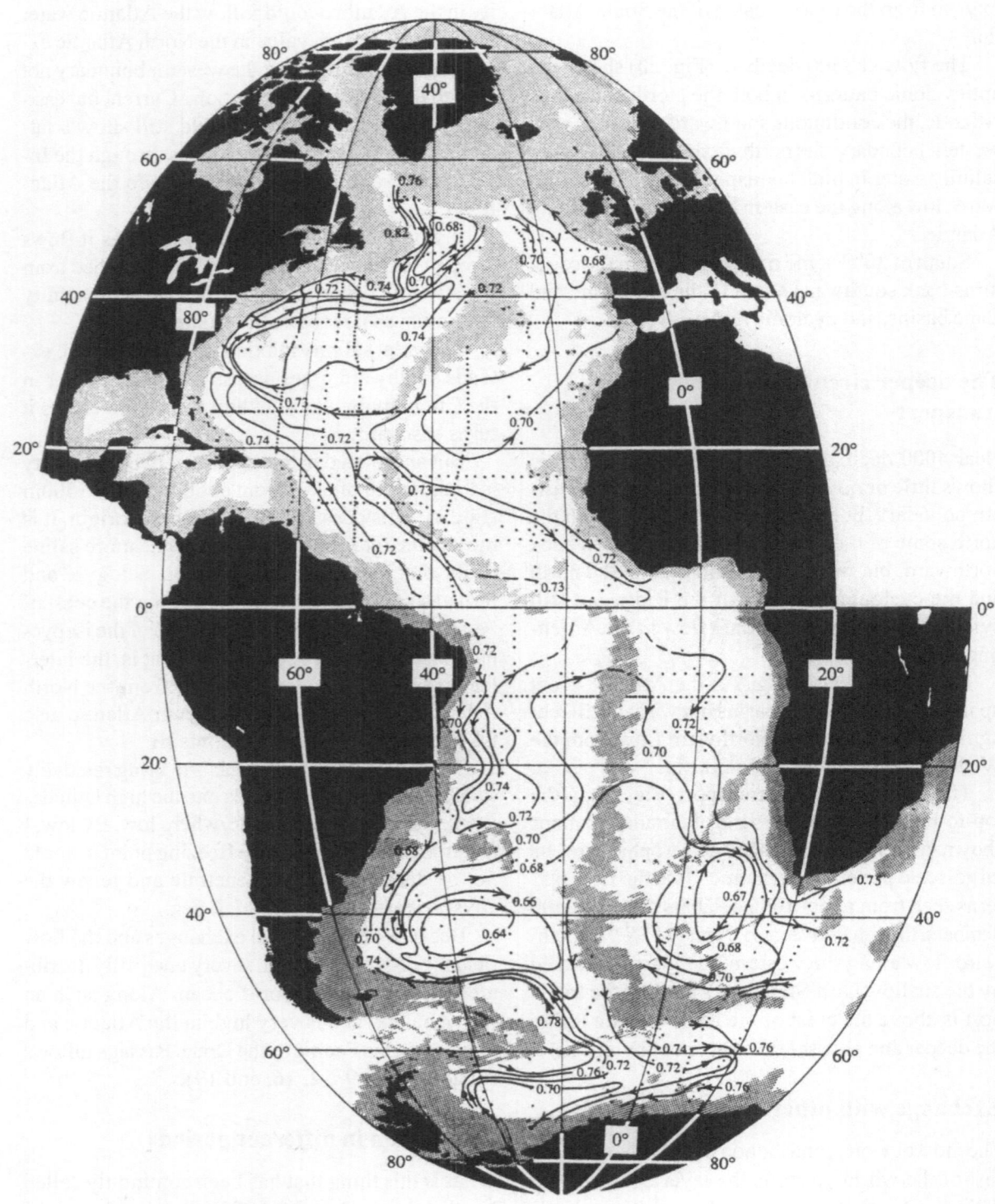

Fig. 20 Adjusted steric height at 3500 db (10m²s⁻² or 10Jkg⁻¹). (Reid 1994).

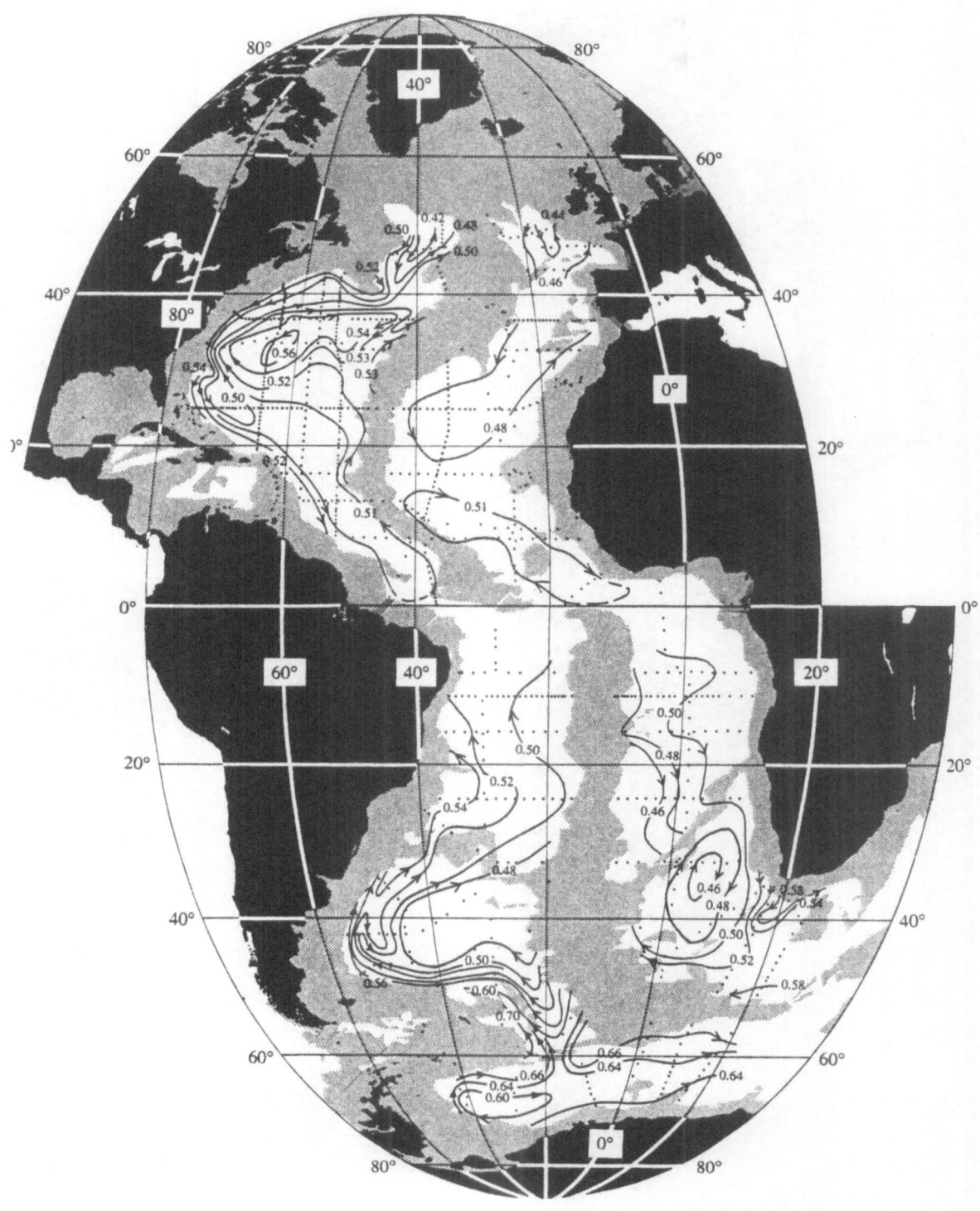

Fig. 21 Adjusted steric height at 4000 db (10m²s⁻² or 10Jkg⁻¹). (Reid 1994).

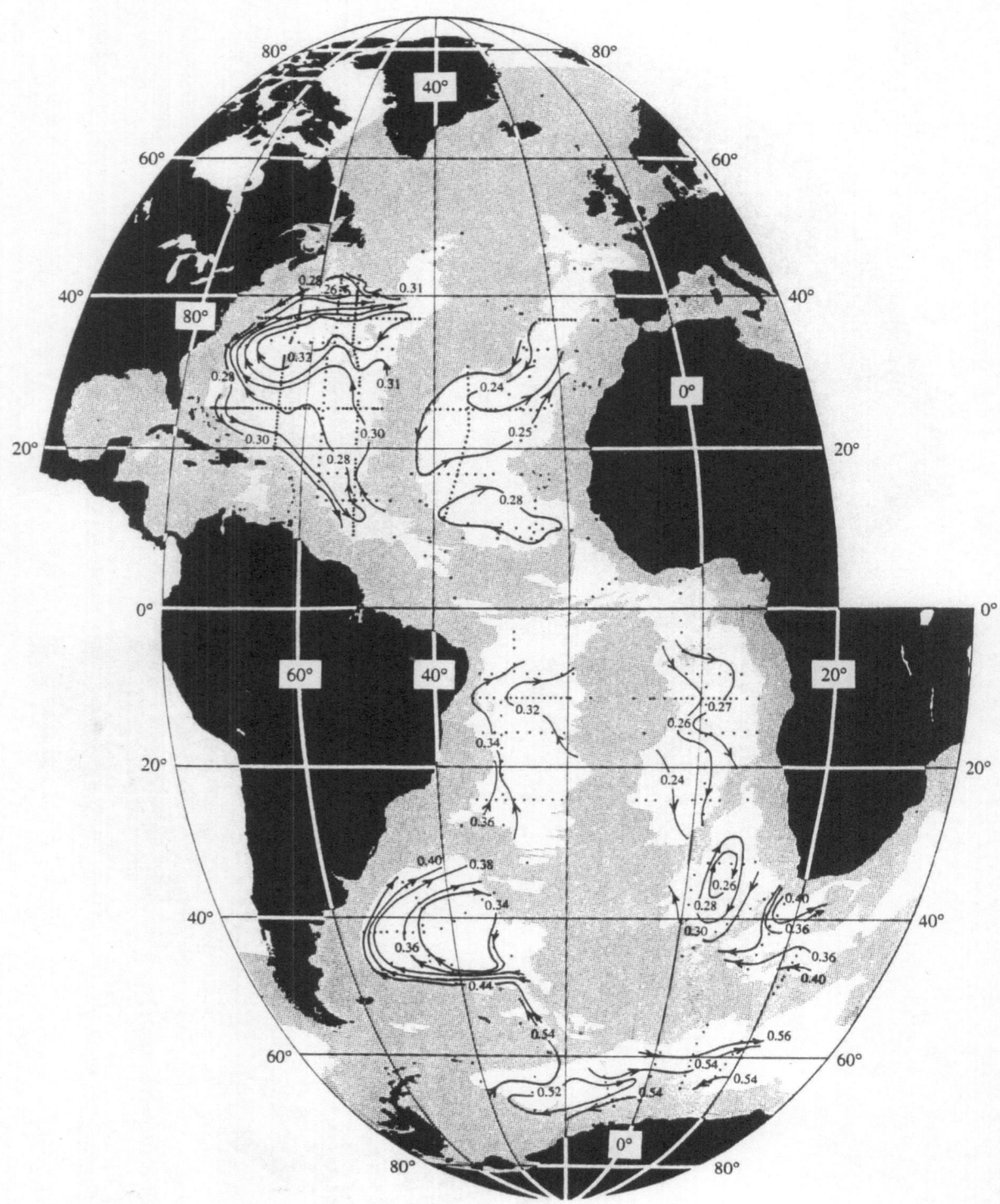

Fig. 22. Adjusted steric height at 4500 db ($10m^2s^{-2}$ or $10Jkg^{-1}$). (Reid 1994).

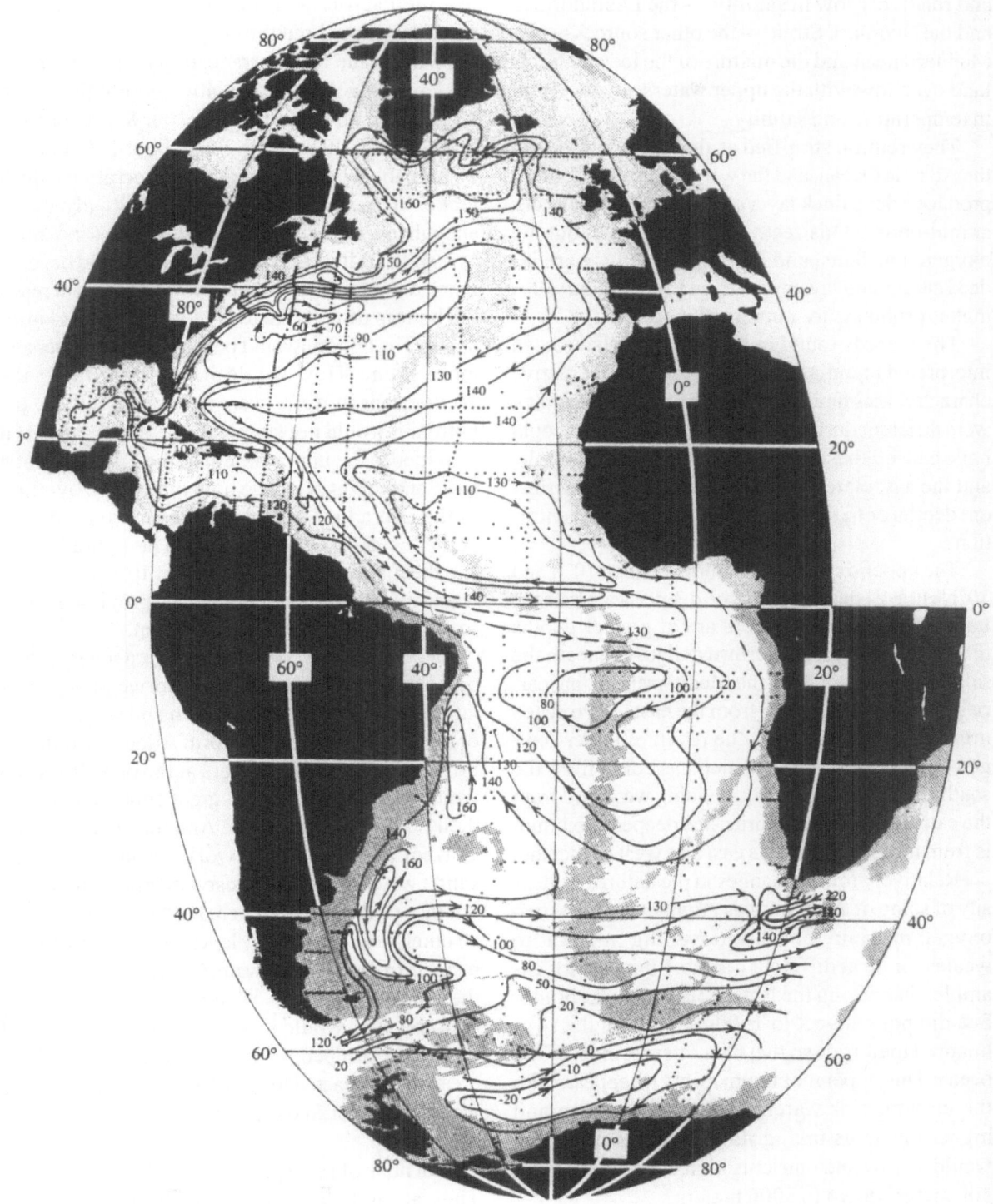

Fig. 23. Transport ($10^6 m^3 s^{-1}$). Integration is from a zero value at the coast of Antarctica and reaches 130×10^6 $m^3 s^{-1}$ everywhere along the coast of the American continents and Greenland. Along the coasts of Europe and Africa it reaches $132 \times 10^6 m^3 s^{-1}$, as $2 \times 10^6 m^3 s^{-1}$ are assumed entering the Atlantic across the Greenland-Scotland sills. The shaded area represents depths less than 3500m. (Reid 1994).

and relatively low in salinity — the Labrador Sea and the Denmark Strait — the other sources — the Mediterranean and the mixture of the Iceland-Scotland overflow with the upper waters are very high in temperature and salinity.

They remain stratified as they circulate within the Atlantic Ocean, and the warm and saline sources produce a deep thick layer of warm and saline water at mid-depth. This seems simple enough, but the oxygen, nutrients, and ^{14}C show that there are indeed alternating layers of maxima and minima. This makes problems for numerical modelers.

The records found in the sediments have been interpreted to indicate that these non-conservative characteristics have varied over time, and that they were different during the last glaciation. What could have caused these layers of oxygen, for example, and the associated nutrient layers, to lie at different depths or to increase or decrease in concentration?

The upper oxygen maximum between 10°S and 30°N (Fig. 2) derives from the deeper waters of the Labrador Basin, but others are at greater depths than the 1500-meter overturn and just beneath the salinity maximum from the Mediterranean that carries lower-oxygen water from the east. The oxygen minimum near 2600m is the result of lower-oxygen circumpolar water, which has come from the south, and the low oxygen extending westward from the eastern tropical Atlantic. The deeper maximum is from the two overflows east and west of Iceland.

Relatively minor changes in the nature or intensity of some of these sources might cause these two oxygen maxima, or the intervening minima, to weaken or lie at different depths. Suppose, for example, that during the last glaciation the Labrador Sea did not convect to 1500m, but only the two-hundred meters or so that characterize most of the ocean. This upper maximum might disappear, with the circumpolar water, with lower oxygen and higher nutrients taking its place so that oxygen would be low and nutrients high everywhere from 500 meters down to 3000 meters.

And suppose also that the exchange with the Norwegian-Greenland Sea were reduced, perhaps by lowered sea level and ice sheets covering much of the passages. Then we would lose the deeper oxygen maximum and the Atlantic would be much like the Pacific, with lower oxygens and higher nutrients throughout the deep water.

What if the Mediterranean Sea became much cooler during the last glaciation — not the 12° to 13° we find now but down to 7° or 8°? At present it pours directly down to only about 1100m (Fig. 1), probably because of its high temperature, which makes it less compressible. It is already dense enough to sink to the bottom, but it mixes too much as it spills. If it were a little colder it would become much denser as it sinks downward and might reach the bottom directly instead of leveling off at mid-depth. The bottom would become much denser and more saline. How would it get back up? If the Mediterranean water were to flow directly to the bottom it would not supply the heat and salt that it provides to the upper waters of the North Atlantic now, and might preclude both the 1500m overturn of the Labrador Sea and the deep mixture formed at the Iceland-Scotland Ridge. The Labrador Sea and the Intermediate water might fill the mid-depth waters with lower salinity, the one providing lower oxygen and the other higher oxygen.

Worse still, one consequence of reducing or shutting off the exchange with the Norwegian-Greenland Sea would be that the warm and saline waters of the upper levels of the North Atlantic would not be mixed with denser waters at the overflows and would not extend down to great depth. At present their product extends to the Antarctic Circumpolar Current and around the world. And part of this saline water turns back westward along the southern part of the Weddell Sea. It adds enough salt to an otherwise low-salinity layer so that when cooled it can form the densest waters of the open ocean — the Antarctic Bottom Water. The sources of the denser waters would be weakened or destroyed. If they were stopped the abyssal density would decrease. If this lasted long enough the deeper waters would be lower in oxygen and higher in nutrients everywhere.

But none of these processes could happen alone. They are in balance now, giving us a relatively steady circulation over at least the period in which our measurements were made.

But we should not imagine that one part can change independently of the others. And for this reason we should not think of the system only in terms of North Atlantic Deep Water — the infamous NADW. It has several components, each from different sources, and our understanding is at best only qualitative.

This makes it very hard to advance our understanding. I believe we should use every method we can think of. Perhaps the most difficult are numerical modeling and the meticulous examination and interpretation of sediment cores. Those investigators have my support, my encouragement, and my sympathy.

Acknowledgements

The work reported here represents one of the results of research supported by the National Science Foundation, the Office of Naval Research, and the Marine Life Research Program of the Scripps Institution of Oceanography.

TABLE 1. SPECIFICATIONS OF THE ISOPYCNAL SURFACES.

The potential density is expressed as σ_0 from 0-500 db, as σ_1 from 500-1500 db, as σ_2 from 1500-2500 db, as σ_3 from 2500-3500 db, and as σ_4 from 3500 db to the bottom. Along the isopycnal defined by 34.64 in $\sigma_{1.5}$, $\sigma_{0.5}$ was used from 250-750 db and $\sigma_{1.5}$ below 1250 db. The potential density is given in units of σ, which is $\rho - 1000$, where ρ is in kg m^{-3}. This table lists the different numbers used for each isopycnal as it extends to the different pressure ranges. The numbers in bold-face type are those used in the text and figures to identify each isopycnal.

North Atlantic

σ_0	$\sigma_{.5}$	σ_1	$\sigma_{1.5}$	σ_2	σ_3	σ_4
27.440		**31.938**				
27.777	30.082	32.376	**34.640**			
27.824		32.456		**36.980**	41.395	
27.847*		32.466*				
27.874		32.523		37.067	**41.500**	45.838
27.915*		32.547*		37.073*		
27.884		32.536		37.083	41.520	**45.860**

*used east of the Reykjanes Ridge

South Atlantic

σ_0	$\sigma_{.5}$	σ_1	$\sigma_{1.5}$	σ_2	σ_3	σ_4
27.300		**31.938**				
		31.948 60°-90°S				
27.675	30.035	32.355	**34.640**			
27.775		32.425		**36.980**	41.400	
27.787		32.476		37.041	**41.500**	45.840
27.794		32.482		37.051	41.515	**45.860**

References

This work is mostly a review of the results of my own published studies. Each of those studies drew upon the work of other investigators and included many pages of citations of their work. Those citations are not repeated here but are available in the publications listed below.

Reid JL (1981) On the mid-depth circulation of the world ocean. In: Warren BA and Wunsch C (ed) Evolution of Physical Oceanography. The MIT Press Cambridge MA and London England. pp 70-111

Reid JL (1989) On the total geostrophic circulation of the South Atlantic Ocean: Flow patterns, tracers, and transports. Prog Oceanogr 23(3):149-244

Reid JL (1994) On the total geostrophic circulation of the North Atlantic Ocean: Flow patterns, tracers, and transports. Prog Oceanogr 33(1):1-92

Reid JL, Lynn RJ (1971) On the influence of the Norwegian-Greenland and Weddell seas upon the bottom waters of the Indian and Pacific Oceans. Deep-Sea Res 18(11):1063-10

Transient-Tracer Information on Ventilation and Transport of South Atlantic Waters

W. Roether and A. Putzka

Institut für Umweltphysik, Universität Bremen, Fachbereich 1,
Postfach 33 04 40, 28334 Bremen, GERMANY

Abstract: We report distributions of the chlorofluorocarbon CFC 11, of CCl_4, and of terrigenic ^{3}He, along three zonal sections across the South Atlantic (WOCE WHP sections A8 - A10, 11.7°S, 19°S, 30°S). The distributions fully reflect the water mass structure of the sections. They reveal a region of comparably much slower water renewal in the range of the Central and Antarctic Intermediate Waters, northeast of the Angola-Benguela Front. For all water masses further down, the distributions demonstrate that renewal is very much slower still, and that it occurs via advective cores adjacent to the respective western boundaries of the basins. The oldest waters are found to be present in a wedge centered in about 3000 m depth and extending from the African slope westward across the Midatlantic Ridge. The tracers are characterized by different input time scales, and these are well apparent in their distributions. The observed correlation between tracers indicates rather steady formation of Upper North Atlantic Deep Water over the past several decades. CCl_4 is powerful in tracing relatively older waters, but in the upper South Atlantic waters our data confirm the decomposition of CCl_4 reported previously.

Introduction

The attraction of oceanic transient tracers rests on their ability to elucidate and quantify the current thermohaline circulation of the ocean on decadal time scales on the one hand, and on their role as analogues with respect to oceanic transport for any dissolved substance on the other. Recent accounts of tracer methodology have been given by Broecker and Peng (1982), Roether and Rhein (1989), and Roether (1994), among others. Transient-tracer-based information complements classical means of studying ocean circulation, such as determination of currents by either directly or by the geostrophic method, and hydrographic analysis.

Classical hydrographic properties have quasi-steady-state distributions, and thus give information on the turnover time scale of a water mass. Some substances, like steady-state radioactive tracers and oxygen, additionally feel time scales of internal processes - radioactive decay and consumption. Contrary to both these groups, the distributions of the transient tracers are governed by time scales set by their time-dependent input into the ocean (see below), except for parts of the upper ocean. They can be looked upon as dyes that have been introduced in a known time pattern at the ocean surface, from where they gradually penetrate into the ocean interior. To invert tracer-based information into ocean circulation and mixing fields in a quantitative fashion, it is generally necessary to invoke ocean circulation models. A variety of models can be used, but a model must be capable of reproducing the Lagrangean transfer of the tracers from the ocean surface layer into the interior, over the tracer input period. Information is extracted by comparing tracer distributions simulated in this way with observed ones (Roether et al. 1994). However, many features of the subsurface circulation are apparent directly in tracer distributions, without following up the tracer transport in detail. The present communication illustrates this capability for tracer data in the S Atlantic.

We start by outlining tracer geochemistry and the status of tracer studies in the S Atlantic (section 2). Subsequently, we present and interpret new

From WEFER G, BERGER WH, SIEDLER G, WEBB DJ (eds), 1996, *The South Atlantic: Present and Past Circulation.* Springer-Verlag Berlin Heidelberg, pp 45-62

tracer sections from S Atlantic WOCE (World Ocean Circulation Experiment) cruises (section 3), followed by a discussion (section 4) that addresses special features of the S Atlantic circulation in the light of the new data. The final section gives some conclusions and a brief outlook on future S Atlantic tracer work.

Geochemical and other features of transient tracers, available observations in the S Atlantic, and previous results

Oceanographic transient tracers include bomb ^{14}C, tritium (^3H, in the form of tritiated water), tritiugenic ^3He (decay product of tritium, $T_{1/2}$ = 12.43 years), and certain halocarbons (HCs), i. e. the chlorofluoro-carbons (CFCs, or freons) CFC 11, CFC 12, and CFC 113, and CCl_4 (for literature on the following, see references above). Bomb ^{14}C and

tritium were produced in atmospheric nuclear weapon testing largely during the early 1960s. Input to the ocean mixed layer for ^{14}C occurs by slow equilibration with the atmosphere (about 10 year time scale). Tritium is transferred into the ocean quickly (time scale of removal from the troposphere a few weeks), and is added to the ocean essentially as an imposed flux. ^{14}C ($T_{1/2}$ = 5730 years) has a large natural background, which makes separation of the bomb component non-trivial. Tritium has but a small natural background (presently contributing on the order of 20 % in subtropical southern-hemisphere upper waters). Surface-ocean concentrations of both nuclides went through maximum levels in the 1970s and have since been decreasing. The HCs named are anthropogenic compounds that have found large-scale commercial use, such that atmospheric mixing ratios have been rising continuously, up to current levels of the order of 10^{-10}.

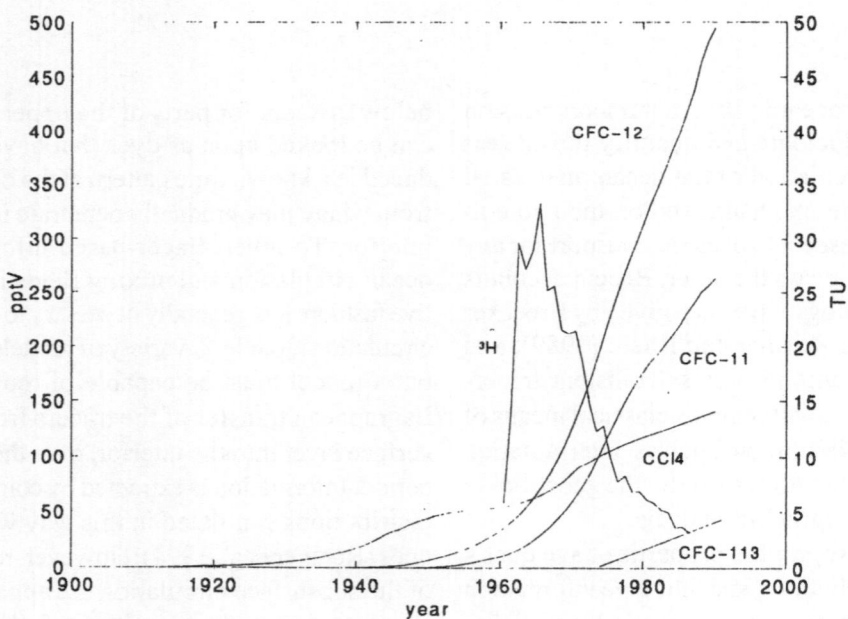

Fig. 1. Time histories of transient tracers for the period 1900 to 2000. For the halocarbons CFC 11, CFC 12, CFC 113, and CCl_4, atmospheric mixing ratios (pptv) are shown (Weiss and Salameh 1992); mixed-layer concentrations are near to a solubility equilibrium with these values. For tritium, yearly-mean values characteristic of marine precipitation in about 35°S are given (Doney et al. 1992). Tritium input to the mixed layer is proportional to these values. Input strongly increases with latitude and is larger in the vicinity of continents (Weiss and Roether 1980; Doney and Jenkins 1993); 1 TU corresponds to a $[^3H]/[H]$ ratio of 10^{-18}. Compared to the tritium curve shown, the northern-hemisphere tritium input has been much higher and even more peaked in time.

Their mixed-layer concentrations correspond to a solubility near-equilibrium with atmospheric mixing ratios. Surface-ocean source functions for CFCs and tritium are given in Fig. 1.

Mixed-layer ^{3}He concentrations are held in a near-equilibrium with atmospheric He by ocean-atmosphere gas exchange, but deeper down, interior-ocean sources produce moderate excesses over such values. Owing to low tritium levels in the southern hemisphere, the tritiugenic component is small. ^{3}He from terrigenic sources is generally more prominent. Its principal sources are release at mid-ocean ridges, largely in the Pacific, so that Circumpolar Deep Water carries a distinct ^{3}He signal. Information gained from terrigenic ^{3}He, which is a steady-state tracer with an inhomogeneous ocean-bottom source, and a sink in the mixed layer, suitably complements that derived from the transient tracers. The different input time histories of the tracers mentioned, and/or their different modes of input, make studies involving more than one tracer particularly powerful (Roether et al., 1994).

Measurement of ^{14}C was relieved recently from the requirement of large-volume water sampling by the accelerator mass spectrometry technique (AMS), but is still the most elaborate one among the tracers. Tritium and ^{3}He are determined by He isotope mass spectrometry, yielding detection limits for tritium (measured by ^{3}He grown in from tritium decay) as low as about 10^{-3} Bq/kg. HC measurement, by ECD gas chromatography, is comparably the least elaborate, can be done on shipboard, and is very sensitive and precise. The compounds CFC 11 and CFC 12 have been measured routinely since about 1980, whereas measurement of CFC 113 and CCl_4 has become possible only quite recently. Water age information in cores of subsurface water flow on the basis of concentration ratios of CFC 11 and CFC 12 has found some applications (Weiss et al. 1985), but to convert such ages into flow velocities of deep-water cores, mixing with more stagnant waters adjoining the core has to be accounted for (Pickart et al. 1989; Rhein 1994).

The first coherent tracer survey in the S Atlantic was that of the GEOSECS (GEochemical Ocean SECtionS) program (1972-73), which included ^{14}C, tritium, and He isotopes (GEOSECS, 1987). Essen-

tially two meridional sections, with considerable station spacing, but unusually detailed depth spacing, were made. The follow-up program TTO (Transient Tracers in the Ocean), basically a N Atlantic program, included several tropical stations in 1982-83 (Brewer et al. 1985), and measured CFC 11 and CFC 12 for the first time. TTO had a narrower station spacing, as had its S Atlantic counterpart, the SAVE (South Atlantic Ventilation Experiment) program, 1987-1990. The latter provided coverage of the entire S Atlantic. Since 1990s, the WOCE program (WCRP 1988) has provided a third survey, which will be completed in 1996. It has put little emphasis on ^{14}C, while some of its hydrographic lines have data also of CFC 113 and CCl_4. Fig. 2 is a station map of the latter two programs. Upon completion of WOCE, up to three time slices of S Atlantic tracer distributions will be available. There have been furthermore regional programs, such as in the Agulhas Retroflexion region (Fine et al. 1988).

Results from these programs are numerous. GEOSECS tritium data demonstrated swift transfer from the mixed layer into the upper ocean waters (Broecker et al. 1986). From the ^{14}C data, inventories of bomb ^{14}C were derived, and from them, large-scale horizontal transfers of upper waters (Broecker et al. 1985). Jenkins et al. (1983) studied renewal of Antarctic Bottom Water (AABW) using tritium data. Ventilation of Antarctic Intermediate Water (AAIW) was studied by Warner and Weiss (1992) on the basis of SAVE CFC data, and Gordon et al. (1992) used these and further data to study the inflow of waters from the Indian Ocean. In addition there have been several studies of the Weddell Sea and with the Antarctic Circumpolar Current (ACC) region (Bayer and Schlosser 1991; Foster and Weiss 1988; Michel 1978; Roether et al. 1993; Roether et al. 1995; Schlosser et al. 1987; Schlosser et al. 1991; Schlosser et al. 1994; Weiss et al. 1979).

South Atlantic tracer sections from WOCE cruises

The Bremen tracer group has been in charge of transient tracer measurements on various S Atlantic WOCE cruises (Fig. 2). For ^{3}He and tritium the

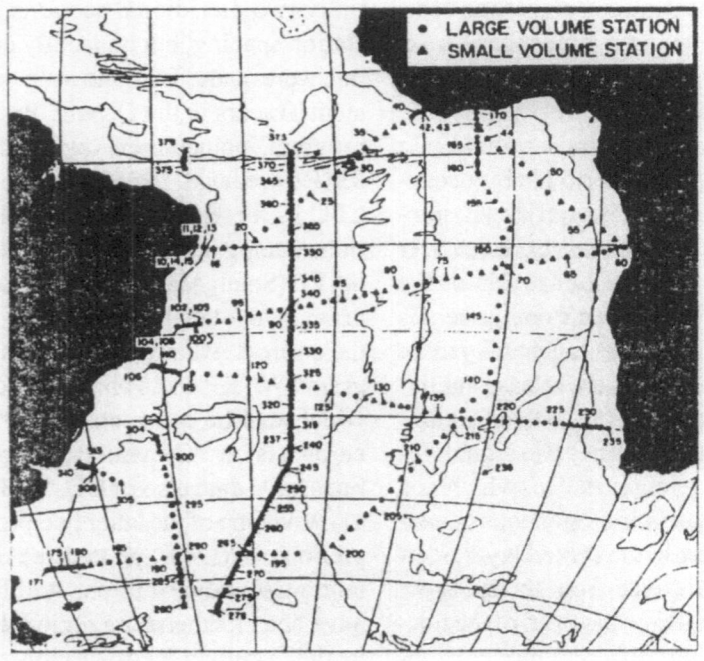

Fig. 2a. Station map of SAVE (South Atlantic Ventilation Experiment, 1987-1989).

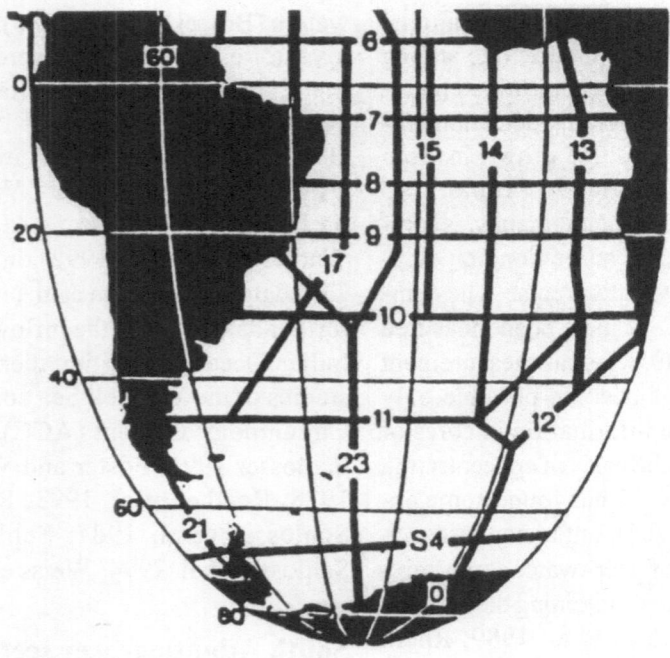

Fig. 2b. Planned hydrographic sections for WOCE WHP (World Ocean Circulation Experiment Hydrographic Programme), 1990-1996.

method is helium isotope mass spectrometry, and for the HCs seagoing ECD gas chromatography (for an outline of procedures, see Roether 1993). One achievement has been successful CCl_4 measurement since 1993. HC data are available in preliminary form at the end of a cruise. We have presently only part of the He isotope and tritium data available, as measurements for these, and especially for tri-tium, take some time to be completed. For the Southern Ocean part of the S Atlantic, tracer results have been published previously (Roether et al. 1993; 1995). We present in the following HC and some ^{3}He data from the zonal WOCE lines A8, A9, and A10 in the subtropical S Atlantic (11.7°S, 19°S, 30°S; Fig. 2).

Fig. 3 gives the ^{3}He distribution along the A9 line (19°S). ^{3}He is lowest in the ocean surface layer, where exchange with the atmosphere keeps excess He at a minimum. There is a layer of somewhat enhanced levels in a few hundred m depth, which is ascribed to tritiugenic ^{3}He. The effect (no more than about 2 % in δ^3He) is much lower than found in the N Atlantic (up to 10 %; Jenkins 1987). Further down, the ^{3}He levels in Fig. 3 reflect specific sources of terrigenic He, as well as different degrees of ventilation (for a description of the various water masses mentioned below see Reid (1989)): A first maximum layer is found near 1000 m depth, essentially all along the section. This reflects the influence of Upper Circumpolar Deep Water (UCDW) which as mentioned carries elevated ^{3}He. Upper North Atlantic Deep Water (UNADW) shows up as a core of ^{3}He being depleted relative to the surrounding waters, between 1500 and 2000 m depth. The signal is strongest toward the western boundary, but at least in traces is coherent across the entire section. A similar, although less pronounced depletion is found in the depth range of lower NADW (LNADW; 3000 - 3500 m). Enhanced values, in turn, in fact the highest of the section, characterize Antarctic Bottom Water (AABW), which fills the depth range below about 3700 m west of the Midatlantic Ridge (MAR), with some emphasis toward the western slope. In this depth range, the eastern basin is lower in 3He throughout. As there is no sink of ^{3}He in the ocean interior, the low concentrations suggest that the prominence of AABW in this basin is restricted.

This conclusion is enabled by ^{3}He signals in northern and southern deep water sources being of opposite sign. Near the ridge crest of the MAR, concentrations are again high, which is undoubtedly caused by a ^{3}He source on the MAR in the vicinity of the section. Rather high concentrations at similar depths reappear toward the east of the basin, and are assumed to arise from recirculation of the ^{3}He-enriched waters. Such recirculation can thus be studied using this type of data. We note in passing that ^{3}He sections of such a data density high as in Fig. 3, that allows detailed contouring across an entire ocean basin, have only recently become available .

Fig. 4 presents the corresponding section of CFC 11. High upper-ocean concentrations decrease to quite low values in about the AAIW depth range (around 800 m). But in contrast to level isolines in Fig. 3, the isolines between 500 and 1000 m now display a distinct upward slope toward the east. The explanation offered is that, while advection and lateral mixing have rather homogenized the quasi-steady-state tracer terrigenic ^{3}He, homogenization is not achieved for the CFC 11 because of its time-dependent input. Much of the deep (> 1000 m) interior in Fig. 4 away from the western boundary has undetectable CFC 11, characterizing these waters as the oldest ones along the section. At the western boundary, deep core signatures are apparent that correspond to those in Fig. 3. However, all cores now carry tracer *maxima*, independent of their hemisphere of origin, and signatures are more concentrated toward the boundary, due to the effect of tracer time scales .

Tracer signatures in the deep-water cores at the other two WHP lines correspond to conditions further upstream or downstream compared to Figs. 3 and 4. Figs. 5 and 6 give the CFC 11 and CCl_4 sections along the southerly line A10 (30°S). The center of vertical decline of CFC 11 (Fig. 5) from the upper waters both lies deeper than in Fig. 4 (around 1200 m) and is more level. UNADW (1500-2000 m depth) no longer shows a CFC 11 signal that is clearly separated from that in the waters further up. LNADW (2500-3500 m) is also somewhat less prominent, and it appears to have left the western boundary to adhere to the western slope of the Rio Grande Rise (near 38°W). AABW

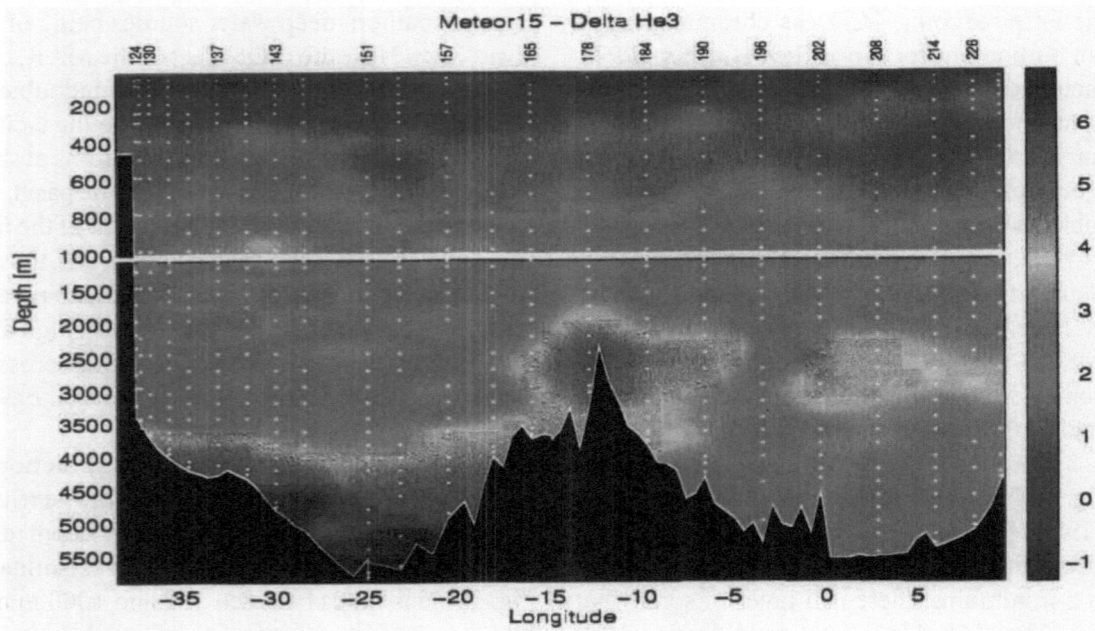

Fig. 3. Distribution of ³He on WOCE WHP line A9, 19°S (Fig. 2b), METEOR cruise M15/3, 1991. Data points are indicated by dots, values are in δ³He notation (percent difference in isotopic ratio from that of atmospheric He), scale is given besides figure. Data precision is approximately 0.2%.

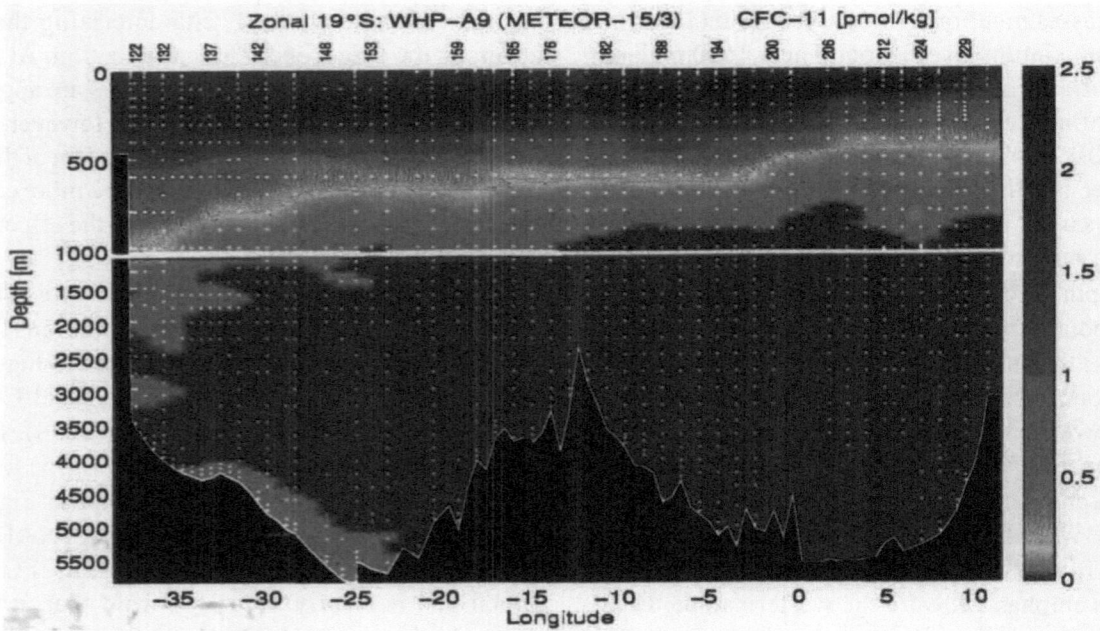

Fig. 4. Distribution of CFC 11 (pmol/kg) on A9, see Fig. 3; data precision is about 1% or 0.01 pmol/kg, whichever is greater (Beining 1993).

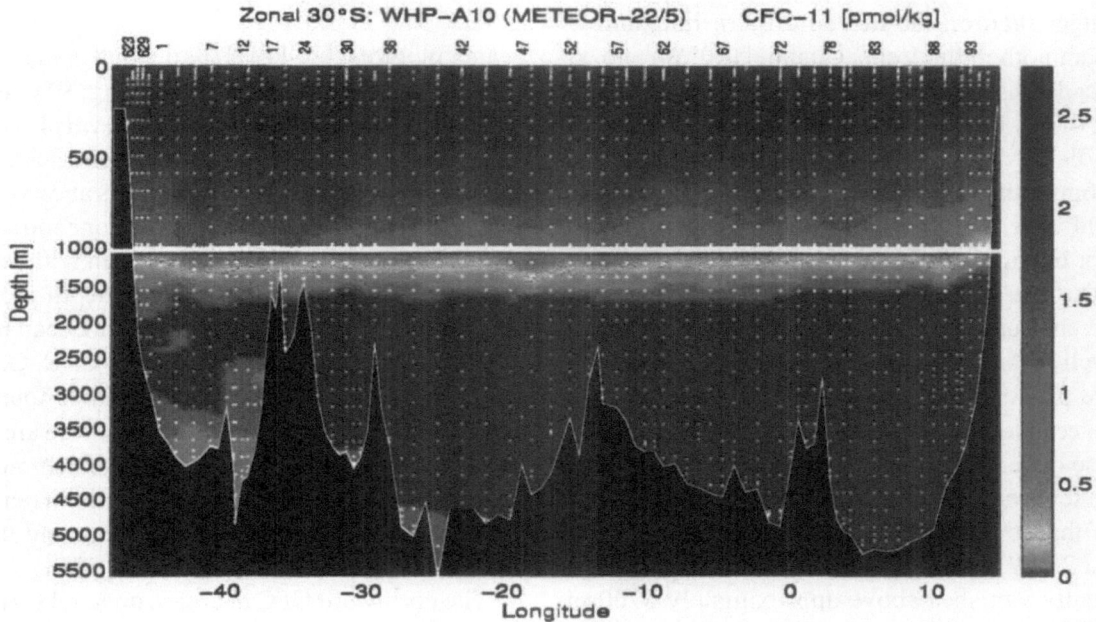

Fig. 5. Distribution of CFC 11 (pmol/kg) on WOCE WHP line A10, 30°S, METEOR M22/5, 1993, see Fig. 4. Data are preliminary, but precision is rather better than in Fig. 4. The regions of maximum depth correspond to the Vema Channel (near 40°W), the Hunter Channel (near 25°W), the Angola Basin (about 0° to 10°W) and the Cape Basin (about 5° to 10°E).

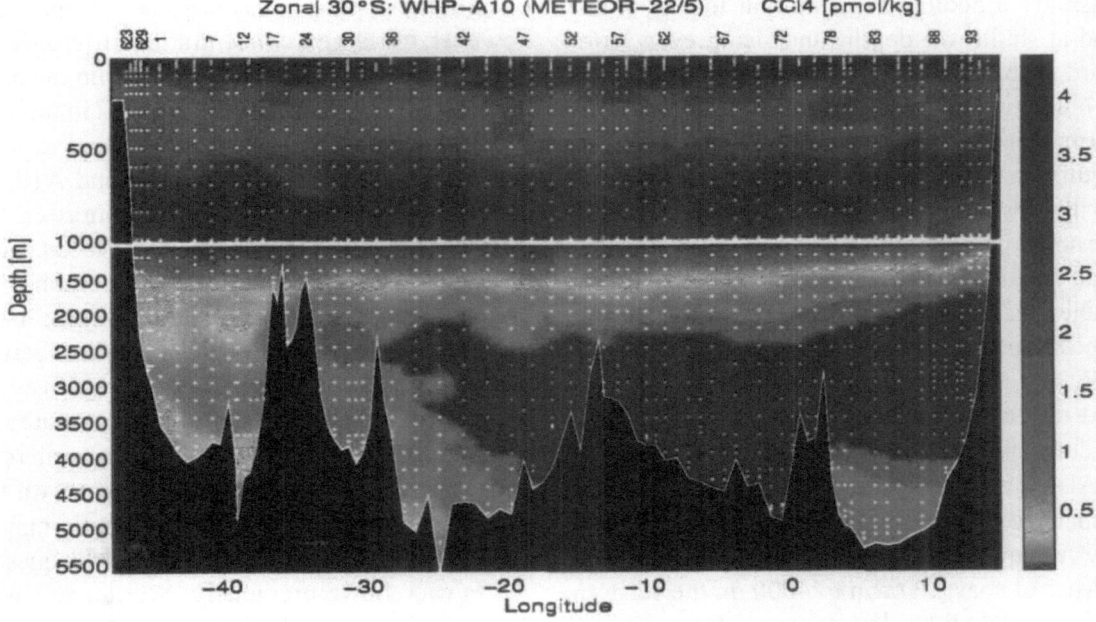

Fig. 6. Distribution of CCl_4 (pmol/kg) on A10, see Fig. 5. Data are preliminary; there is a calibration uncertainty, but precision should be about 2% or 0.02, whichever is greater.

(below 3700 m) on the other hand as expected is stronger than on the more northern line, and is present both in the Vema Channel (40°W) and, at larger depths, north of the Hunter Channel (25°W). East of the MAR, CFC 11 is undetectable in the Angola Basin, while there are traces of it near the bottom in the Cape Basin (near 5°E). The wedge of old waters is simular to that in Fig. 4, but its upper boundary is found at greater depths.

The corresponding section of CCl$_4$ (Fig. 6) is different from Fig. 5, in that the deep western cores, as well as AABW in the Cape Basin, are distinctly more pronounced, and the range of waters with undetectable concentrations is considerably smaller. These differences are caused by the longer input time scale of CCl$_4$ relative to CFC 11 (Fig. 1). We note that to the northeast of the Hunter Channel (near 28°W), i.e. west of the MAR, CCl$_4$ between virtually vanishes above approximately 3700 m depth and up to the ridge crest. That this region does not have an appreciable signal is surprising, as it indicates that horizontal recirculation within the Brazil Basin is insufficient to replenish the waters in the southeastern corner of the Brazil Basin on the CCl$_4$ time scale.

On the northerly line A8 (11.7°S), CFC 11 (Fig. 7) displays a decline zone simular to Fig. 4, but found at shallower depths and rising even more toward the east. The rise is interrupted near 11°W by a bulge of enhanced CFC 11 reaching down to 1000 m. Not unexpectedly, UNADW and LNADW are quite pronounced, the former being detectable from the western boundary to 17°W, but despite the more northern location AABW still exhibits a clear signal. The deep Angola Basin mostly has non-detectable CFC 11 concentrations, except for marginally positive values close to the bottom towards the MAR.

At the latter location, in contrast, the corresponding CCl$_4$ section (Fig. 8) shows a distinct signal. Also the other features noted in Fig. 7 appear in enhanced form. UNADW extends well beyond the MAR up to 3°W (and in traces perhaps up to the African coast). Down to 4000 m, the western slope of the MAR has low concentrations, similar to Fig. 6. Below 4000 m, some recirculation of AAWB is indicated in the Brasil Basin, in that concentrations rise again toward the MAR.

Discussion

Scatter plots of CFC 11 versus depths for the northern and southern lines (A8, A10; Fig. 9) demonstrate extreme differences in CFC levels between upper and deep waters. Evidently, identification of the deep cores in the CFC 11 sections above (Figs. 4, 5, and 7) has to rely on very low concentrations (at most 1 % of a solubility equibrium with the atmosphere). Throughout Fig. 9, the depth profiles show a clear break between an upper-ocean high-CFC regime, and a deep one low in CFC. On the northern line, the apparent boundary is found at considerably shallower depths, and these are also more variable. Fig. 9 provides ample demonstration of an order-of magnitude difference in renewal time scales between the upper-ocean and deep-ocean regimes.

The northward CFC decrease on level surfaces the apparent from Fig. 9, reflects the renewal of the Central Waters and the AAIW from outcrops in the south. As lateral transport in the ocean is believed to follow isopycnal, rather than level, surfaces, a plot of CFC 11 versus density for all three lines combined is presented in Fig. 10. The expected trend of concentrations on isopycnals to decrease towards the north is in fact observed. Surprisingly, however, the largest concentration differences are found not between the lines, but within the northernmost line (A8). While the stations in the western part of A8 follow the regular trend expected from the more southerly lines A9 and A10, distinctly lower CFC 11 prevails at the more easterly stations of line A8. These stations are located north of the Angola-Benguela Front in a region of cyclonic circulation (Peterson and Stramma 1991). The conclusion from low CFC 11 at these stations is that lateral exchange between this region and the subtropical gyre to the south and west is quite slow. The region of delayed access for young waters that is delineated by our data, is consistent with flow trajectories in Reid's (1989) S Atlantic maps of adjusted geostrophic flow for the layers in question.

As was shown previously (Warner and Weiss 1992), a similar reduced lateral exchange is also found in the AAIW. Fig. 11 presents a horizontal map of CFC 11 on $\sigma_0 = 27.25$, which isopycnal represents the center of the AAIW layer. The con-

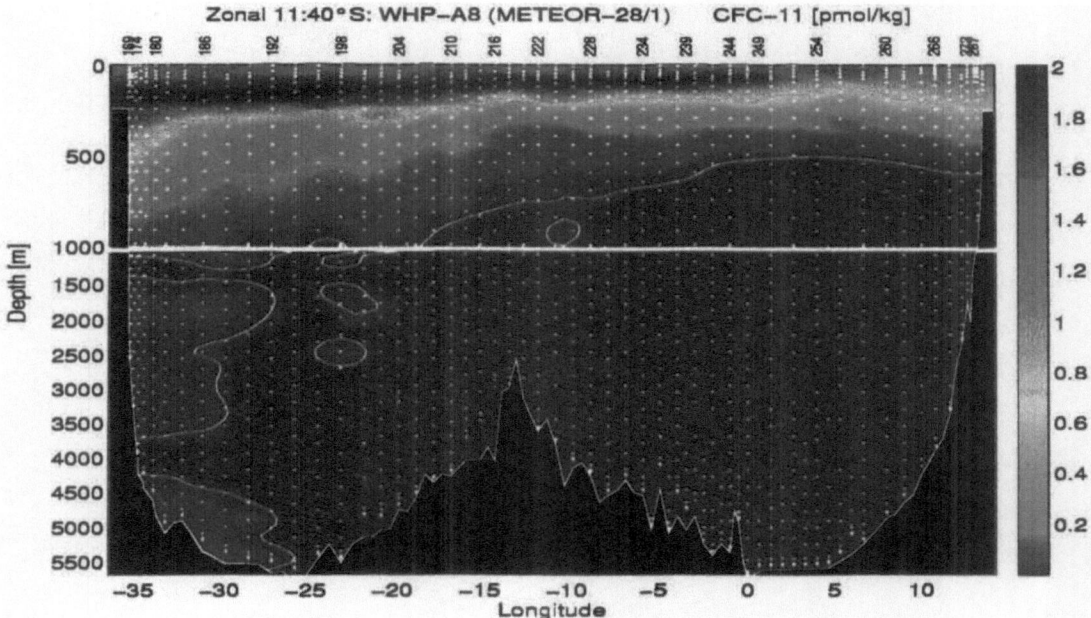

Fig. 7. Distribution of CFC 11 (pmol/kg) on A8, 11.7°S, METEOR M28/1, 1994, see Fig. 5.

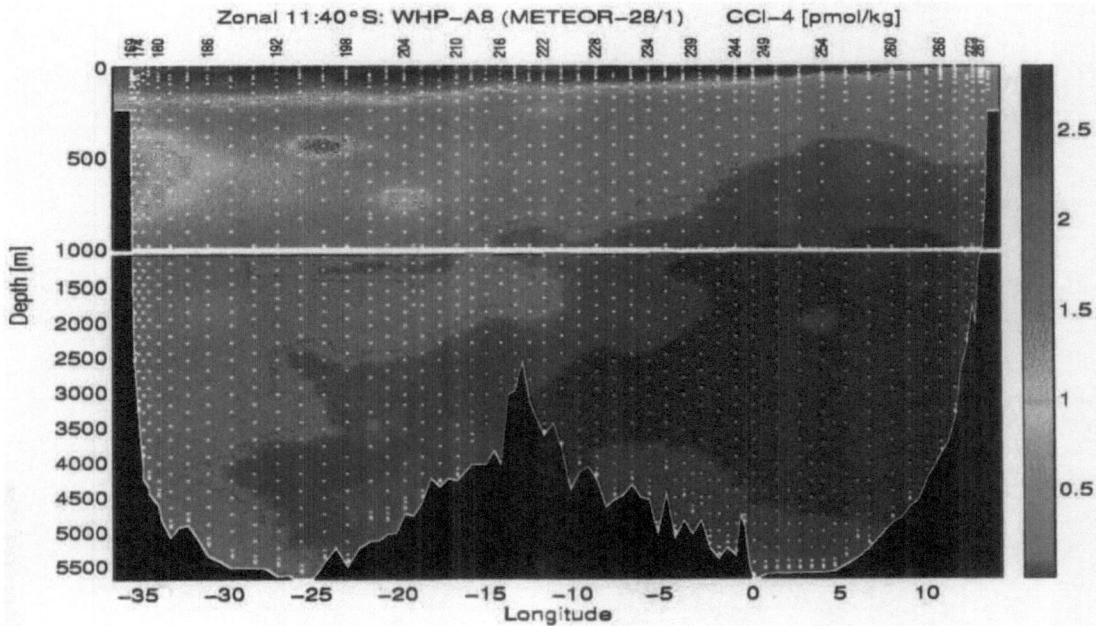

Fig. 8. Distribution of CCl$_4$ (pmol/kg) on A8, see Fig. 6.

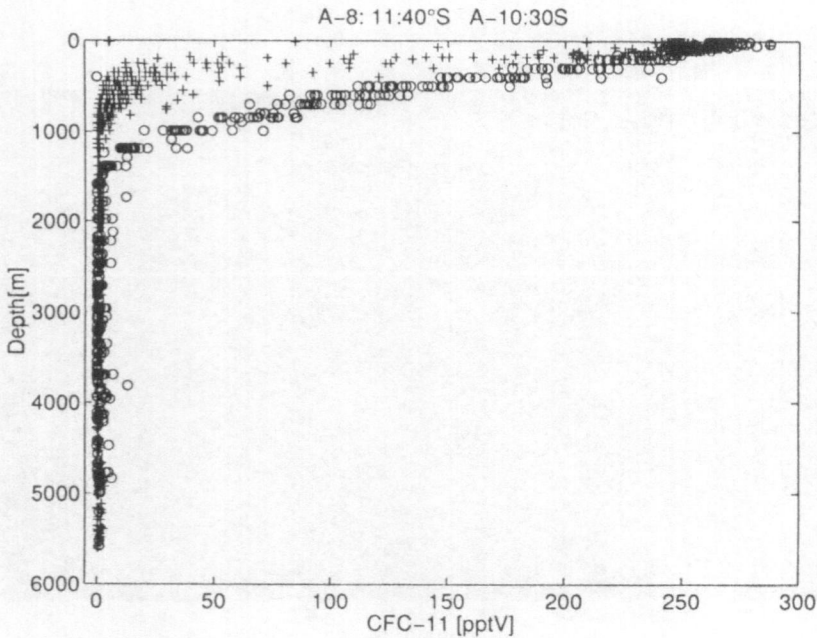

Fig. 9. Scatter plots of CFC 11 (pptv) versus depth (m) for the stations of WOCE WHP lines A10, 30°S (open circles) and A8, 11.7°S (crosses). The abscissa is in pptv, i.e. the partial pressure equivalent to the measured concentration at the potential temperature and salinity of the sample, thus avoiding effects from temperature-dependent solubility (Warner and Weiss 1985).

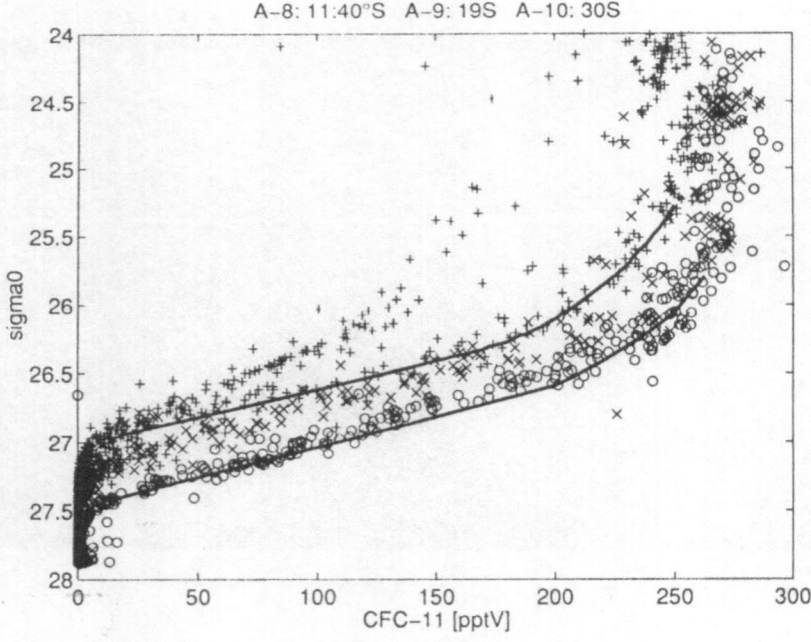

Fig. 10. Scatter plots of CFC 11 (pptv) versus potential density for WOCE WHP lines A8 and A10 (see Fig. 9) and for A9, 19°S (inclined crosses). Full lines drawn by eye to illustrate general relationship for lines A10 and A9.

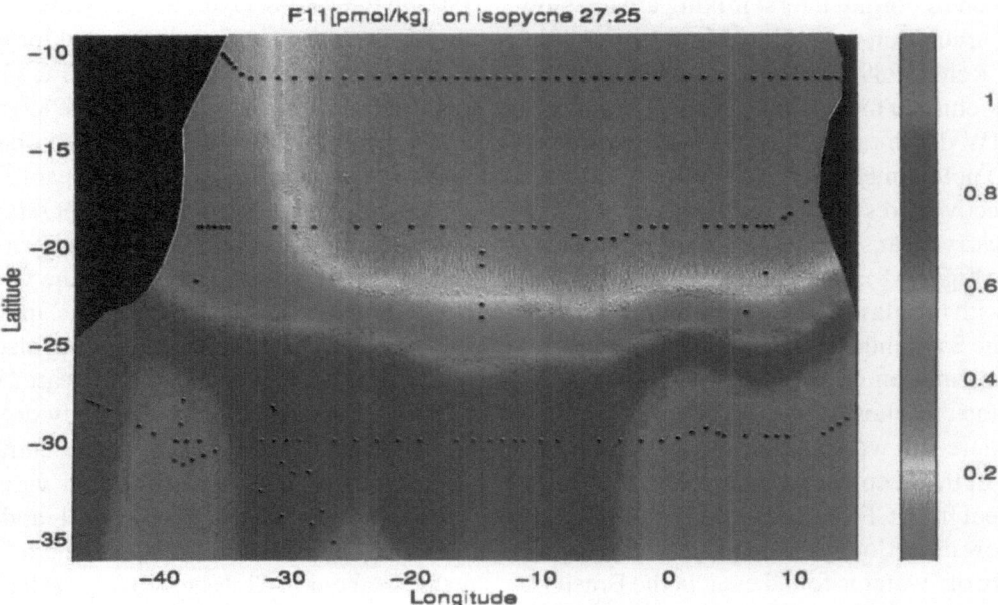

Fig. 11. Isolines of CFC 11 (pmol/kg, see scale besides figure) on the isopycnal $\sigma_0 = 27.25$ that characterizes Antarctic Intermediate Water. Dots give positions of stations used in contouring, see text.

touring is based on data from the three zonal lines with gaps in between (apart from just a few stations, see figure). Therefore the CFC distribution in Fig. 11 has to be taken with a grain of salt, but the following features are clearly robust: CFC 11 concentrations south of about 25°S are rather high and fairly uniform, while north of this latitude, one finds moderate concentrations at the western boundary, which decrease toward the east to almost vanishing values. As noted for the upper waters, the low values in Fig. 11 point to slow lateral exchange, and the presence of the region of slow exchange, as well as its indicated size, are consistent with Reid's (1989) mentioned flow maps. A difference to the Central Water range is that the region of slow exchange in the AAIW range appears to be more extended, as is manifest both in the flow maps and in the present data (Figs. 4, 5, 7).

The rather homogeneous high-CFC pool south of 25°S in Fig. 11, on the other hand, is in keeping with Reid's (1989) anticyclonic recirculation, as well as with trajectories of subsurface floats obtained recently (Boebel et al. 1994) which indicate

southward return flow in the west, at the appropriate latitudes in the AAIW layer. Nevertheless a flow of AAIW must continue north toward the equator, and it is clear that north of about 25°S this flow follows a route at or near to the western boundary (Reid 1989; Fig. 11). We maintain that such flow cannot be supplied from the high-CFC pool, along trajectories either branching off from the anticyclonic recirculation or following the western boundary all along from the south. The reason is that such trajectories would be incompatible with the sudden decline in CFC 11 concentrations observed near the western boundary between 25°S and 20°S; an effect of transit time in view of CFC 11 increasing in time (Fig. 1) must be much smaller than the decrease actually observed (Beining and Roether 1995). Therefore the low CFC 11 concentrations downstream are to be explained by substantial admixture of waters from the low-CFC 11 pool that is present toward the northeast. We believe that the rather sudden downstream decline of CFC 11 results from mixing between streamlines recirculating through the northeastern pool and subsequently rejoining those from the south, since other mechanisms such as normal lateral mixing are too slow. Such recirculation is exhibited in the circulation maps of Reid (1989; his Fig. 17), and is consistent with the return of higher CFC 11 concentrations in the AAIW depth range at mid-section in line A8 (11.7°S) noted in Fig. 7.

Advective cores of deep water masses below the AAIW carry clear signatures in all the tracers considered (Figs. 3-8). The signatures are fully consistent with the classical view of water mass structure in the S Atlantic (Reid 1989), and provide further confirmation of the notion that transport is concentrated in, mostly western, boundary currents, from where the waters advected spread into the interior of the deep-ocean basins. Such spreading is apparent from CFC concentrations falling gradually below detection limits in the interior (Figs. 4-8). Fairly old water is found even in the Brasil Basin west of the MAR, which as mentioned indicates that recirculation in this basin is rather slow. We observe only one case of an eastern boundary current (LNADW on line A10; Fig.5, 39°W), and nowhere do we find evidence of younger waters at the African continental slope (Figs. 4-8). We note that the distribution of CCl_4 in the deep waters is fully compatible with that of CFC 11, but that the former tracer is detectable distinctly further into the old waters (Figs. 5-8); the enhanced detectability is due to the fact that CCl_4 has had comparably more time to enter the interior (Fig. 1).

Effects of tracer input time scales are adressed further in Fig. 12. Fig. 12a presents a correlation of CFC 11 and salinity in the AAIW layer (see Fig. 11). Data for the three sections fall into fairly distinct groups (with strongest signatures for line A10, 30°S), reflecting the different distances from the formation regions in the south. The well-known northward salinity increase (Molinelli 1981), due to a reduction of the salinity extremum by mixing, is apparent. It is accompanied by a CFC 11 decrease, but for mid-salinities in Fig. 12a, CFC 11 concentrations are less than corresponding to a linear mixture of data points for the more extreme salinities. Such curvature reflects the transient intrusion of CFC 11 into the AAIW, i.e. the fact that the CFC 11 signal farther into the AAIW is delayed. As the curvature arises from an interplay of the time

history of tracer input (Fig. 1) with the spreading rates in the AAIW layer, it represents information on this spreading in time-integrated form.

A similar correlation, between CFC 11 and CCl_4 in the UNADW ($s_2 = 36.84$; Reid, 1989) along line A8, is presented in Fig. 12b. The highest values shown correspond to data points near to the western boundary (see Figs. 7 and 8). For reference, the inset in the figure gives the mixed-layer time history of the relationship between the two tracers, according to which not only the mixed-layer CFC 11 and CCl_4 concentrations but also the concentration ratios have markedly increased with time. UNADW concentrations evidently are small in comparison. In fact, the largest concentrations observed correspond to mixed-layer values in approximately the year 1930 for CCl_4 and 1950 for CFC 11, while their ratio equals the mixed-layer value in about 1962. Respectively earlier years are obtained for the lower concentrations found away from the boundary.

One may assume that the tracer signature of the UNADW is derived entirely from its northern source waters, as other contributions presumably represent entrained background waters which should be sufficiently old to be tracer-free (tracer-free for CCl_4 implies that the water was formed prior to about 1920; Fig. 1). Under such conditions, as was shown by Weiss et al. (1985) for CFC 11 and CFC 12, the observed HC ratio indicates the year of formation of the northern component. The formation year at the western boundary is thus 1962, corresponding to an age of of about 32 years (i.e. 1962-1994). This age represents an average as there will be contributions spread out over several years of formation. Furthermore, because younger such contributions have higher HC concentrations, and hence carry a stronger weight in the mixture, the age is biased (too young). In total, a northern-component mean travel time of approximately 35 years up to line A8 may be a realistic value. The formalism (Weiss et al., 1985) furthermore allows one to determine the fraction of northern component remaining, by comparing the observed concentration (7.5 pptv for CCl_4; Fig. 12c) to that in the mixed layer for the formation year (60 pptv; Fig. 1). The fraction obtained amounts to about 1/8 only. We note that the deduced age represents a mean

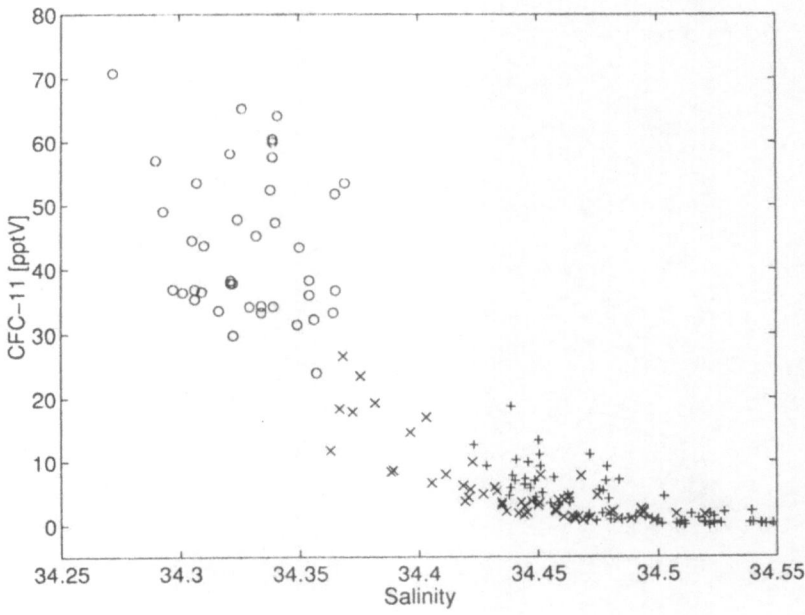

Fig. 12a. CFC 11 (pptv) versus salinity (psu) from data points nearest to isopycnal $\sigma_0 = 27.25$ (range 27.2 - 27.3), on a subset of the stations used in Fig. 11. Data points denoted as in Fig. 9 and 10 (A10: open circles, A9: inclined crosses, A8: crosses).

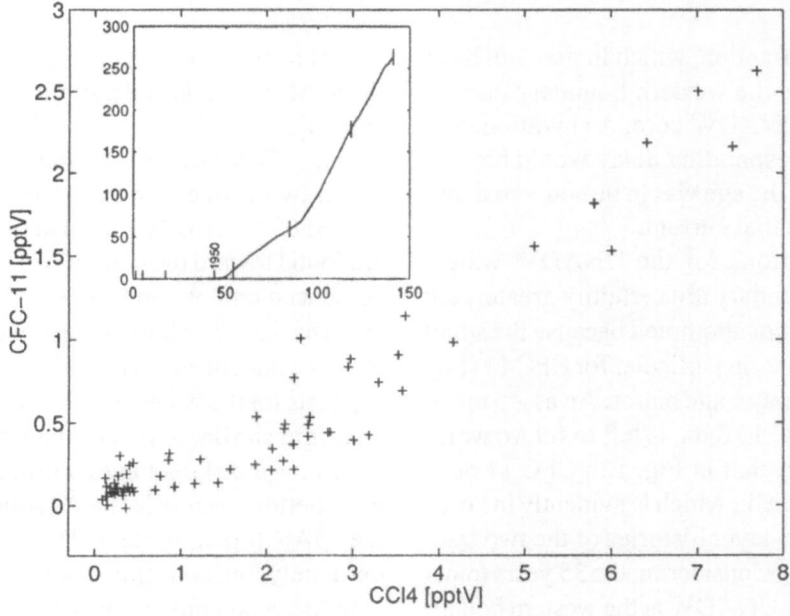

Fig. 12b. CCl_4 versus CFC 11 (pptv) on line A8 from data points nearest to the core of the Upper North Atlantic Deep Water ($\sigma_2 = 36.84$). The mixed-layer relationship (Fig. 1) up to 1990 is given in the inset; short vertical bars at the inset curve give time in 10-year steps (1950 is indicated). Note difference in scales between main diagram and inset.

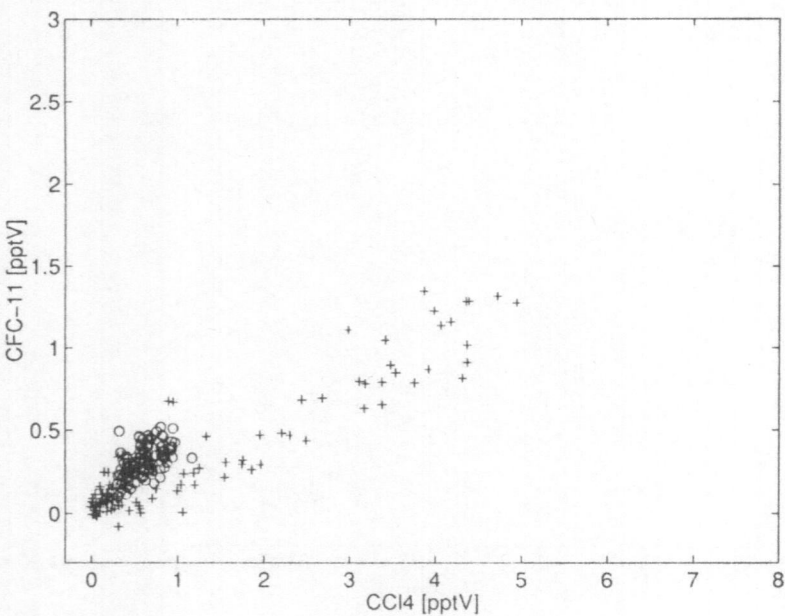

Fig. 12c. Same as Fig. 12b, but for data points near to the center of the tracer minimum layer in about 3000 m depth (crosses), and near to the base of the eastern slope of the Midatlantic Ridge in the Angola Basin (open circles).

travel time of this fraction, which in part will have been spent outside the western boundary current representing the UNADW core. As mentioned in section 2, the corresponding delay would have to be corrected for, if the age was to be converted into a mean velocity in that current.

Ages and dilutions for the UNADW waters away from the boundary are certainly greater, but a quantification is not attempted because the small tracer concentrations, in particular for CFC 11 (Fig. 12b), may induce major age biases. An assessment, adressing also CFC 12 data, is left to future work. We note, however, that in Fig. 12b CFC 11 does fall off faster than CCl_4, which is evidently in keeping with the mixed-layer histories of the two tracers (Fig. 12b, inset). Considering the 35 years transfer age deduced for UNADW at the western boundary, we interpret this consistency as suggesting a rather steady supply of UNADW from the northern source, through at least most of the CCl_4 transient period (i. e. since several decades, Fig. 1). Support for such a supply is provided by the find-ings of penetration of the CCl_4 signature well across the MAR in Fig. 8, and a similar one of ^{3}He at 19°S (Fig. 4).

Fig. 12c addresses the same correlation along A8 in two further ranges. One covers the depths near 3500 m (crosses), i.e. where the oldest waters are found toward the east; like in Fig. 12b, the highest concentrations correspond to the western boundary. The data distribution is about the same as found at low concentrations in Fig. 12b. This finding suggests for the waters in the tracer minimum layer an origin similar to that of the UNADW, but with a shift towards larger ages. Furthermore shown are near-bottom waters in the Angola Basin adjoining the MAR (open circles). These data points have extremely low concentrations, in fact near enough to the detection limits to be somewhat uncertain, but a comparably steeper slope is indicated. By the formalism above, these values correspond to a young component that is small indeed, with a mean age of less than about 30 years. Presence of very small amounts of comparably rather young water

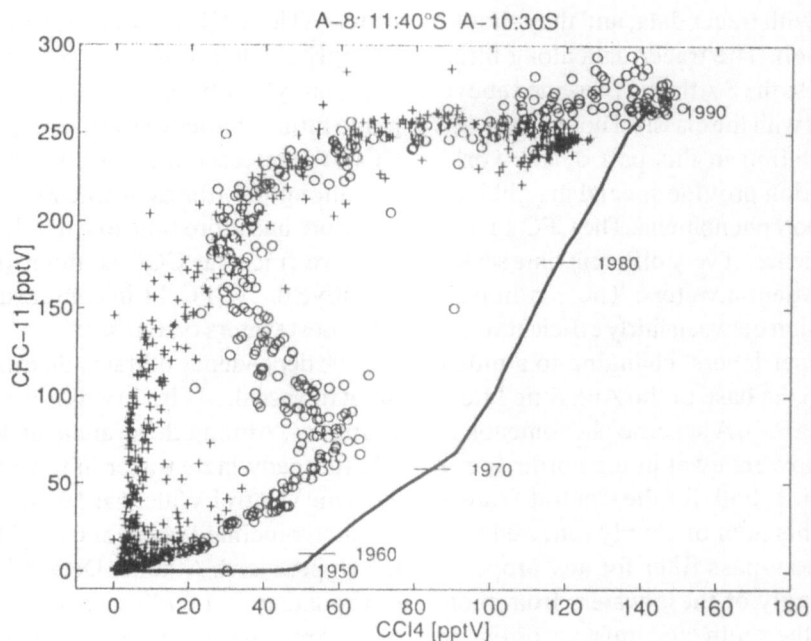

Fig. 13. CCl$_4$ versus CFC 11 (pptv) for the entire water column of WOCE WHP lines A10, 30°S (open circles) and A8, 11.7°S (crosses). The full curve gives the mixed layer history (like in inset of Fig. 12b, but on the same scale as the data).

is thus suggested (as NADW appears to be older even in the west, i.e. upstream, the younger age might indicate AABW addition).

The final topic is instability of CCl$_4$ in upper waters that was previously reported using data from the A9 line (Wallace et al. 1994). Such instability is underlined by Fig. 13 which presents correlations of CCl$_4$ and CFC 11 for lines A10 and A8, together with the mixed-layer relationship versus time for the two tracers (full curve). Relative to the latter curve, one would expect that, for stable behaviour, mixing between waters descending from the mixed layer in different years will draw inner-ocean data points toward the CFC 11 axis in Fig. 13, i.e. will tend to smooth out the curvature of the mixed-layer relationship. For the line A10 data points (open circles), such behaviour is in fact observed for CFC 11 concentrations up to about 60 pptv and again for CCl$_4$ concentrations above about 100 pptv. In between, however, the data display a pronounced dip toward the CFC 11 axis. These data points cannot be composed by any conceivable mixture of mixed-

layer values, so non-conservative behaviour is obvious. The A8-line data (crosses) display an even more dramatic effect. They behave similarly to the A10 data above about 180 pptv CFC 11, but below this value, CCl$_4$ is very much lower except near the origin and for a small fraction of the data points up to at most 50 pptv. Correlation with oxygen (not shown) provides evidence that the loss of CCl$_4$ is definitely related to the oxygen minimum, which is particularly pronounced in the east of the A8 line. The underlying mechanism remains to be determined; it has been argued that temperature does not exert a primary direct influence (Beining 1993). The apparent upper-water loss of CCl$_4$ is in strong contrast to apparently very stable behaviour in the deeper ocean, so that the usefulness of CCl$_4$ as a deep-water tracer does not seem to be affected.

Conclusions and outlook

The present communication was written with a view to introduce a non-specialist to the tracer oceanography of the S Atlantic. Other contributions in this volume also deal with tracer data, and they should be consulted as well. The tracer data along three zonal sections across the S Atlantic presented above are fully consistent with the classical notion of water masses and circulation in this part of the world ocean, but in addition provide special insight into a number of transport phenomena. The CFC 11 and CCl$_4$ data give evidence of very different time scales of renewal of S Atlantic waters. There is in particular a clear division between fairly efficient water renewal in the upper layers, changing to a much slower one near to the base of the Antarctic Intermediate Water (AAIW). A specific phenomenon is the comparably slow renewal in the northeastern part of the S Atlantic, both for the Central Waters and the AAIW. This pool of slowly renewed waters will act as a low-pass filter for any property changes in the supply of these waters from their source regions in the south, and must accordingly be taken into account in evaluating property distributions in this region. The boundary of the region appears to be related to a front between the S Atlantic Subtropical Gyre and cyclonic circulation in the northeast (the Angola-Benguela Front; Petersen and Stramma 1991). In our view the feature so far has not attracted the attention it deserves.

Renewal of the waters below the AAIW occurs via western boundary currents, which carry distinct tracer signatures. But even at these boundaries, the CFC 11 concentrations do not exceed 1% of current mixed layer concentrations. CCl$_4$ has had more time to enter the deep ocean (Fig. 1), and accordingly, its deep-water concentrations are moderately higher. We observed only one case of a tracer signature at an eastern boundary (Lower North Atlantic Deep Water against the Rio Grande Rise), and no such signature at the African slope. In the Angola Basin, bottom water intrusion across the Walvis Ridge appears to be small, but intrusion from further north is indicated in the section in 11.7°S. Recirculation within the basin at depths near to the top of the Midatlantic Ridge is traced by ^{3}He released from the ridge. A wedge of old water is found centered around 3000 m depth and toward the African slope.

The new data confirm previous findings (Wallace et al. 1994) that CCl$_4$ is a highly useful tracer in the deep ocean, but is unstable in upper waters. The CCl$_4$ decomposition appears to be strongly related to the oxygen minimum, i.e. to marginally aerobic conditions. This finding could guide future studies of the decomposition process. CCl$_4$ was released into the environment the earliest among the transient tracers (Fig. 1), and has therefore had more time to enter the deep ocean. In fact, we find that CCl$_4$ is about five times more sensitive than CFC 11 in contouring the extent of the oldest waters on our sections.

The dependence of tracer distributions on tracer input time scales (Fig. 1) was illustrated by several examples. Among these are near-level isolines of ^{3}He (a steady-state tracer in this case, see section 2) in the Central Waters at 19°S, accompanied by distinctly inclined ones for CFC 11 (Figs. 3, 4). In the Upper North Atlantic Deep Water at the western boundary in 11.7°S, an average advection time of approximately 35 years from the northern source region was deduced. Other features in the tracer distributions were interpreted as suggesting that formation of this water mass was reasonably steady over an extended period in the past (Figs. 3, 12b). These examples illustrate the usefullness of combining data for several tracers.

The present work clearly provides no more than an overview of oceanographic information that can be gained from the tracer distributions shown. Many features in the distributions deserve more detailed assessment so that much work remains to be done. This includes extracting information from the tracer distributions by employing ocean circulation models. There is a consensus that such evaluation has great potential, both to quantify tracer-based information, and to verify and adjust thermohaline circulation in general circulation models (Roether et al. 1994).

We shall have additional information available on the above sections shortly when our tracer measurements are completed. There will be tritium and some CFC 113 data, as well as ^{3}He for all three sections. Because CFC 113 has the shortest input time scale among the tracers (Fig. 1), CFC 113 data are believed to be particularly valuable for waters of rather short turnover, as is the case in the Central Waters. A further step will be to combine the tracer data for all S Atlantic WOCE sections

(Fig. 2). This should be possible quite soon, as the WOCE sections in the S Atlantic sector are to be completed by 1996. In addition, tracer data from the previous surveys named in section 2 can be employed, allowing one to assess also the temporal development of the transient-tracer distributions. Such a multi-tracer, repeated-time-slice approach will enable one to study the circulation in considerable detail. Despite the large quality of tracer data available, however, there is still a need for additional tracer surveys in future. The point is that in the *deep* waters of the S Atlantic, transient tracer concentrations are still small to undetectable. Valuable information on the renewal and turnover of these waters will be missed unless observations will be taken that outline a substantial further part of the tracer transients in these waters.

Acknowledgements

We are grateful to the masters and crew of METEOR cruises M15/2, M22/5, and M28/1 for generous assistance in the field work. For tracer measurements and sample collections at sea, and tracer measurement and raw-data editing at Bremen, thanks are due to Peter Beining, Klaus Bulsiewicz, Gerd Fraas, Wilfried Plep, Henning Rose, Jürgen Sültenfuß, and Roland Well. Goran Martinic prepared the figures. Development of the techniques used, sample collection, measurements, and data evaluation were supported by grants from the Bundesminister für Forschung und Technologie, the Deutsche Forschungsgemeinschaft (both Bonn-Bad Godesberg), and the Kommission für Forschungsplanung und wissenschaftlichen Nachwuchs of the Unversity of Bremen.

References

Bayer R, Schlosser P (1991) Tritium profiles in the Weddell Sea. Mar Chemistry 35:123-136

Beining P (1993) Darstellung und Interpretation ozeanischer FCKW-Verteilungen. PhD dissertation, University of Bremen, p 135

Beining P, Roether W (1995) Temporal evolution of CFC 11 and CFC 12 concentrations in the ocean interior. J. Geophys Res Oceans, in press

Boebel O, Schmid C, Zenk W (1994) Direct observation of Antarctic Intermediate Water recirculation in the deep subtropical South Atlantic. Annales Geophys submitted

Brewer PG, Sarmiento JL, Smethie WM Jr (1985) The Transient Tracers in the Ocean (TTO) Program: The North Atlantic Study, 1981; the Tropical Atlantic Study, 1983. J Geophys Res 90:6903-6905

Broecker WS, Peng T-H (1982) Tracers in the Sea. Eldigio Press, Palisades, p 690

Broecker WS, Peng T-H, Östlund HG (1986) The distribution of bomb tritium in the ocean. J Geophys Res 91:14331-14344

Broecker WS, Peng T-H, Östlund HG, Stuiver M (1985) The distribution of bomb radiocarbon in the ocean. J Geophys Res 90:6953-6970

Doney CD, Jenkins WJ (1993) A tritium budget for the North Atlantic. J Geophys Res 98:18,069-18,081

Doney CD, Glover DM, Jenkins WJ (1992) A model function of the global tritium distribution in precipitation. J Geophys Res 97:5481-5492

Fine RA, Warner MJ, Weiss RF (1988) Water mass modification at the Agulhas Retroflexion: chlorofluormethane studies. Deep-Sea Res 35:311-332

Foster TD, Weiss RF (1988) Antarctic Bottom Water formation in the northwestern Weddell Sea. Antarc J US 74-76

GEOSECS (1987) GEOSECS Atlantic, Pacific and Indian Ocean Expeditions, Shorebased Data and Graphics. GEOSECS Atlas Series, Vol. 7, National Science Foundation, Washington, D.C., p 200

Gordon AL, Weiss RF, Smethie WM, Warner MJ (1992) Thermocline and intermediate water communication between the South Atlantic and Indian Oceans. J Geophys Res 97:7223-7240

Jenkins WJ (1987) ^{3}H and ^{3}He in the beta triangle: observations of gyre ventilation and oxygen utilization rates. J Phys Oceanogr 17:763-783

Jenkins WJ, Lott DE, Pratt MW, Boudrau RD (1983) Anthropogenic tritium in South Atlantic bottom water. Nature 305:45-46

Michel RL (1978) Tritium distributions in Weddell Sea water masses. J Geophys Res 83:6192-6198

Molinelli EJ (1981) The antarctic influence on Antarctic Intermediate Water. J Mar Res 39:267-293

Peterson RG, Stramma L (1991) Upper-level circulation in the South Atlantic. Progr Oceanogr 26:1-73

Pickart RS, Hogg NG, Smethie WM Jr (1989) Determining the strength of the deep western boundary current using the chlorofluoromethane ratio. J Phys Oceanogr 19:940-951

Reid JL (1989) On the total geostrophic circulation of the South Atlantic ocean: flow patterns, tracers and transports. Progr Oceanogr 23:149-244

Rhein M (1994) The deep western boundary current: tracers and velocities. Deep-Sea Res 41:263-281

Roether W (1993) Workshop Report Development of Ocean Tracer Measurements, Univ. of Bremen, Germany, 16 - 18 June, 1983 (mimeographed manuscript)

Roether W (1994) Studying thermohaline circulation in the ocean by means of transient tracer data. In: Malanotte-Rizzoli P, Robinson AR (eds) Ocean Processes in Climate Dynamics: Global and Mediterranean Examples. Kluwer Academic Publ., pp 157-171

Roether W, Rhein M (1989) Chemical Tracers in the Ocean. In: Sündermann J (ed) Landolt-Börnstein, New Series, Group V, Vol. 3b, Chapter 4.3 pp 59-122

Roether W, Schlitzer R, Putzka A, Beining P, Bulsiewicz K, Rohardt G, Delahoyde F (1993) A chlorofluoromethane and hydrographic section across Drake Passage: Deep water ventilation and meridional property transport. J Geophys Res 98:14,423-14,435

Roether W, Roussenov V, Well R (1994) A tracer study of the thermohaline circulation of the Eastern Mediterranean. In: Malanotte-Rizzoli P, Robinson AR (eds) Ocean Processes in Climate Dynamics: Global and Mediterranean Examples. Kluwer Academic Publ. pp 371-394

Roether W, Sültenfuß J, Putzka A (1995) Spreading of newly formed waters along the northern rim of the Weddell Basin. J Geophys Res Oceans, in revision

Schlosser P, Roether W, Rohardt G (1987) ^{3}He balance of the upper layers of the northwestern Weddell Sea. Deep-Sea Res 34:365-377

Schlosser P, Bullister JL, Bayer R (1991) Studies of deep water formation and circulation in the Weddell Sea using natural and anthropogenic tracers. Mar Chem 35:97-122

Schlosser P, Kromer B, Weppernig R, Loosli HH, Bayer R, Bonani G, Suter M (1994) The distribution of ^{14}C and ^{39}Ar in the Weddell Sea. J Geophys Res 99:10,275-10,287

Wallace DWR, Putzka A, Beining P (1994) Carbon tetrachloride and chlorofluorocarbons in the South Atlantic Ocean, 19°S. J Geophys Res 99:7803-7819

Warner MJ, Weiss RF (1985) Solubilities of chlorofluorocarbons 11 and 12 in water and seawater. Deep-Sea Res 32:1485-1497

Warner MJ, Weiss RF (1992) Chlorofluoromethanes in South Atlantic Antarctic Intermediate Water. Deep-Sea Res 39:2053-2075

WCRP (1988a, b) World Ocean Circulation Experiment Implementation Plan, a: Vol. I, Detailed Requirements. WCRP-11; b: Vol. II, Scientific Background. WCRP-12, World Climate Research Programme, July 1988

Weiss RF, Östlund HG, Craig H (1979) Geochemical studies of the Weddell Sea. Deep-Sea Res 26:1093-1120

Weiss RF, Bullister JL, Gammon RH, Warner MJ (1985) Atmospheric chlorofluoromethanes in the deep equatorial Atlantic. Nature 314:608-610

Weiss W, Roether W (1980) The rates of tritium input to the world oceans. Earth Planet Sci Lett 49:435-446

The Circulation and its Variability of the South Atlantic Ocean: First Results From the TOPEX/POSEIDON Mission

L.-L. Fu

Jet Propulsion Laboratory,
California Institute of Technology,
Pasadena, CA 91109, U.S.A.

Abstract: The sea surface height observations made by the radar altimetry system aboard the TOPEX/POSEIDON satellite, the first satellite dedicated to the study of the global ocean circulation, were used to study the circulation and its variability of the South Atlantic Ocean. Preliminary results from the first 18 month's worth of data are presented on the mean ocean dynamic topography (1 year average) and its seasonal and mesoscale variabilities. Due to the large uncertainty in the geoid model at small scales, the utility of the dynamic topography is limited to scales larger than 2000 km. Comparisons were made with historical hydrographic observation and contemporaneous simulation by a computer model. Both annual and semiannual harmonics (amplitude and phase) were estimated for the sea level variations at scales larger than 500 km. The strongest seasonal variabilities (both annual and semiannual) were found in the Brazil/Malvinas Confluence and the Agulhas Retroflection, with secondary maxima in the Gulf of Guinea (semiannual) and the western tropical Atlantic (annual). The mesoscale variability revealed by the standard deviation of sea level is similar to previous observations. The effects of eddies on the mean flow, estimated from the eddy Reynolds stress, are important in the Brazil/ Malvinas Confluence and the Agulhas Retroflection.

Introduction

The ocean is a turbulent fluid system of global scales. Space observations provide a unique approach to understanding the ocean circulation with a global perspective. Described in this paper are the first results of the circulation and its variability of the South Atlantic Ocean from a new satellite called TOPEX/POSEIDON. Launched on August 10, 1992, this United States/France joint satellite mission has been making continuous observations of the height of the sea surface using a state-of-the-art radar altimeter system. The height of the sea surface is a manifestation of the surface pressure field of the ocean and is the only observable from space that is directly linked to the ocean circulation.

If the ocean were at rest, the sea surface would coincide with a gravitational equipotential surface (the specific surface is designated the "geoid"). With the removal of such high frequency phenomena as tidal variations, the elevation of the sea surface relative to the geoid is the oceanic "dynamic topography". Simple theory based upon the equations of fluid dynamics (e.g. Pedlosky 1979) shows that the dynamic topography is a manifestation of the movement of the entire oceanic water column, extending in some circumstances to the sea floor. Sufficiently accurate measurements of the sea surface elevation thus provide very powerful constraints upon the large scale circulation and its variability (e.g. Wunsch 1992).

Space measurements of the sea surface topography are based upon radar altimeters in a simple measurement geometry illustrated in Fig. 1. The height of the sea surface relative to the Earth's center of mass is obtained by subtracting the altimeter range measurement from the geocentric altitude of the spacecraft (the radial orbit height). But the ac-

From WEFER G, BERGER WH, SIEDLER G, WEBB DJ (eds), 1996, *The South Atlantic: Present and Past Circulation*. Springer-Verlag Berlin Heidelberg, pp 63-82

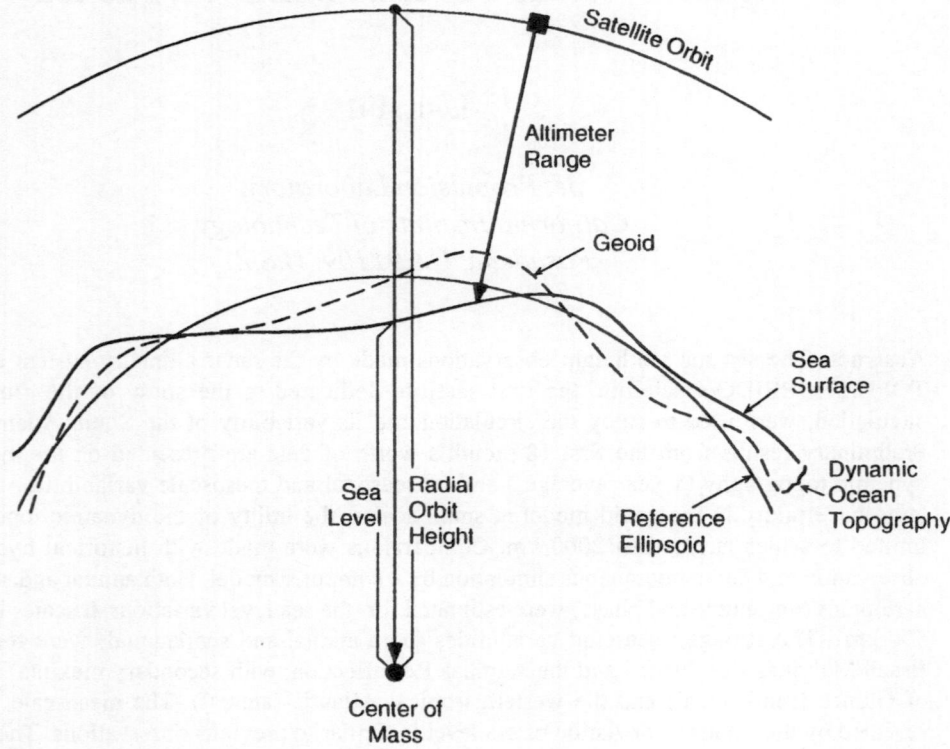

Fig. 1. The geometry of the technique of satellite altimetry.

curacies required to be useful must be significantly better than 10 cm, and achieving them requires solutions to a host of technical problems in both satellite altimetry and orbit determination (Wunsch and Gaposchkin 1980; Fu et al. 1994).

The first satellite altimeter designed for demonstrating the utility of such measurements for ocean circulation applications was flown on Seasat in 1978 (Journal of Geophysical Research Vol. 87 No. C5 1982 and Vol. 88 No. C3 1983). With the error in the altimetry measurement being 10 cm and the orbit error about 1 m, patterns of global ocean dynamic topography and its variability began emerging from measurement errors (Tai and Wunsch 1983 and 1984). However, the results are not useful quantitatively, especially at basin scales where the orbit errors dominate the measurement. Moreover, Seasat demised prematurely after only 3.5 months' operation.

The Geosat Mission launched by the U.S. Navy for geodetic mapping of the sea surface for military

applications has nontheless provided the first multiyear global altimeter data for oceanographic studies. Although the measurement accuracies are not satisfactory for ocean circulation studies, a great deal has been learned about the ocean from the Geosat data (Journal of Geophysical Research Vol. 95 Nos. C3 and C10 1990).

After Seasat's premature demise, serious planning began in both the US and France for an altimetric mission which would be specifically designed for meeting the objective of substantially improving the knowledge of the global ocean circulation. TOPEX/POSEIDON is the result of this joint endeavor (Fu et al. 1991 and 1994). TOPEX is an acronym standing for Ocean Topography Experiment, the name originally used by the U.S.; POSEIDON is the original French name for the mission (a French acronym). This joint acronym is abbreviated as T/P hereafter.

The lifetime of T/P was designed for a minimum of 3 years, with a possible extension of another 3

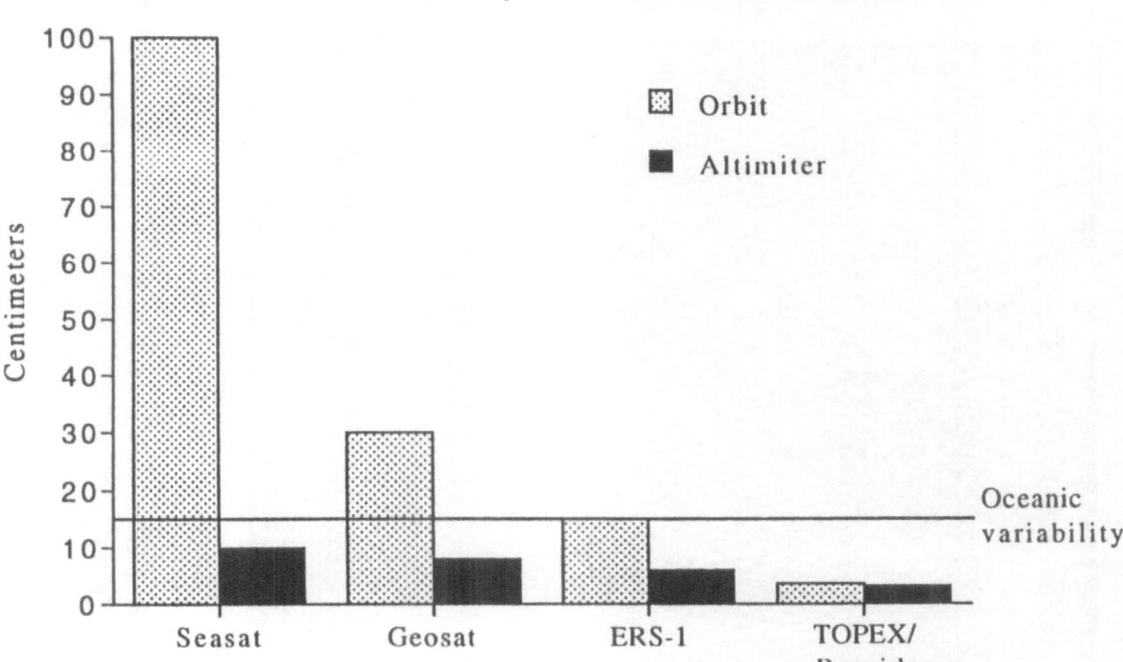

Fig. 2. The evolution of the measurement accuracy of various satellite altimeters. The horizontal line denotes a representative sea level change (15 cm) caused by the variations of large-scale ocean currents.

or more. The top priority during the first 6 months of the mission was to test and calibrate the overall system. Numerous field campaigns were conducted by the mission's project and science teams for that purpose. The results of this "verification phase" were published in the Journal of Geophysical Research Vol. 99 No. C12 in December 1994. The primary conclusions indicated that both the altimeter measurement and the orbit determination had exceeded pre-launch requirements. The root-mean-square errors in both measurements were in the range 3-4 cm, resulting in an uncertainty in the range of 4-5 cm (primarily random error) for the sea level measurement. Shown in Fig. 2 is a comparison of the performance of T/P with three other recent satellite altimeters. Only T/P has sufficient accuracy for detecting the large-scale oceanic signals. The calibrated mission data are now available without restriction to all users (Callahan 1994) on magnetic tapes and CD-ROMs within two months from data reception. "Quick-look" data are available via electronic media to operational users within seven days from reception. The CD-ROMs have been distributed by both the French space agency CNES (AVISO 1992) and the U.S. NASA (Benada 1993).

The satellite's orbit has an altitude of 1336 km, an inclination of 66 degrees, and a repeat period of 9.9 days. Shown in Fig. 3 are the ground tracks of the observations. These tracks have been maintained within +/- 1 km from their nominal locations, minimizing the effects of the uncertainties of small-scale geoid on detecting the minute temporal changes of the sea surface. This particular orbit configuration was chosen in part for determining the ocean tides, which are a major source of error for circulation studies. Preliminary tide models derived from the T/P data have accuracies on the order of 3-4 cm (Shrama and Ray 1994; Ma et al. 1994; Egbert et al. 1994). It is expected that the accuracies will approach 2 cm with more data included in the analysis, perhaps reaching the practical limit of tidal observation (Le Provost et al. 1995).

L.-L. Fu

TOPEX/POSEIDON Ground Tracks

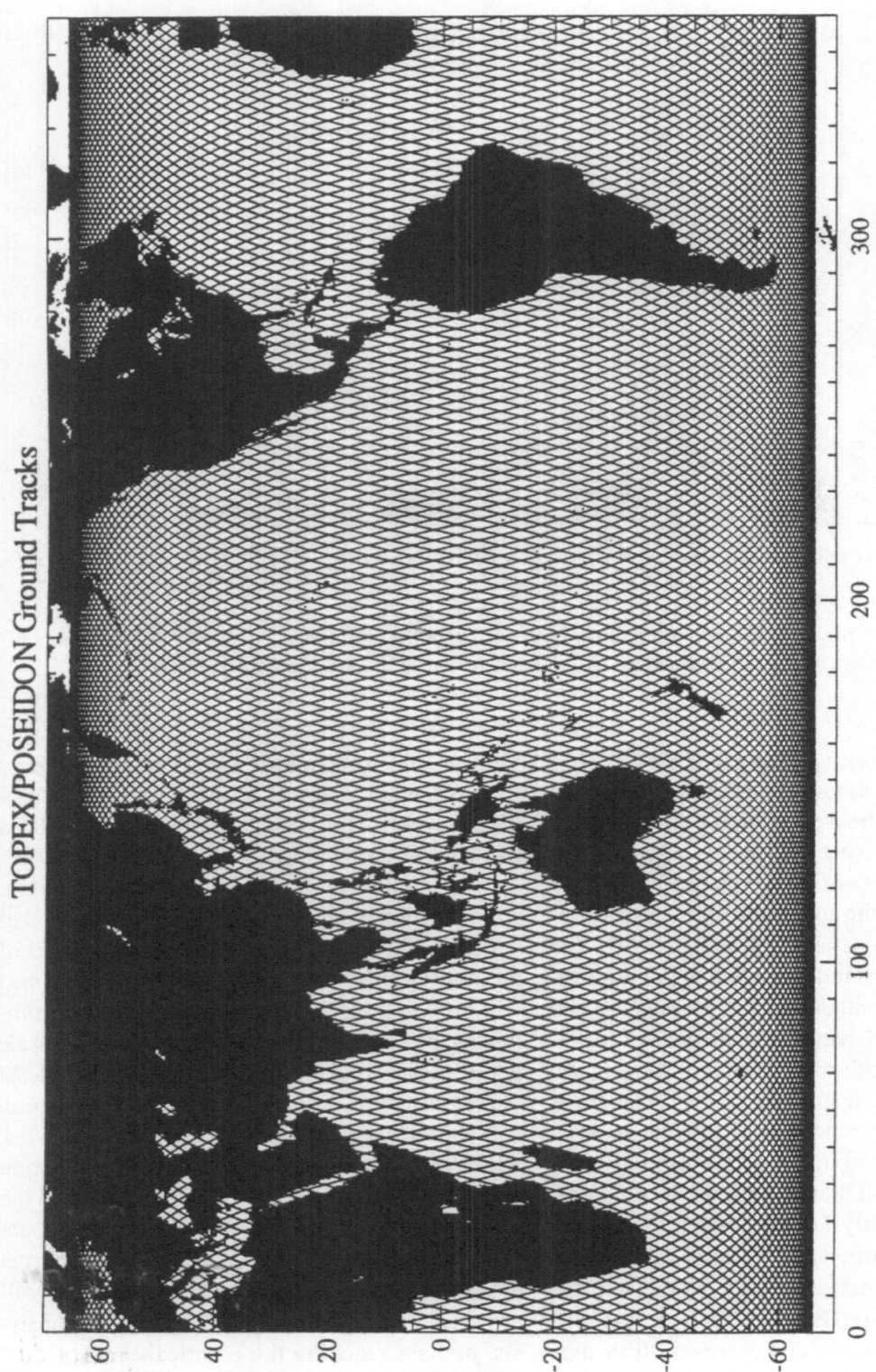

Fig. 3. The ground tracks along which the TOPEX/POSEIDON satellite is making sea level observations at a rate of one record every second of flight time (or every 6.2 km along the track). The satellite repeats these tracks every 9.96 days and produces a "snap shot" of the sea level of the world's oceans.

Described in the following sections is a survey of the T/P observations of the South Atlantic Ocean. Topics include the mean circulation, the seasonal variabilities, and the mesoscale eddy field. The results presented serve as a sample of the scientific content of the data. They are not exhaustive by any means. The discussions and interpretations offered are preliminary. There is a great deal of more science to be learned of the South Atlantic circulation from this continuously growing data set.

Dynamic Topography

Because of the high accuracy of the T/P measurement, the estimate of the dynamic topography is limited by the accuracy of the geoid (e.g. Stammer and Wunsch 1994). A long-lead effort to improve the Earth's gravity field model was initiated by the T/P Project before the launch of the satellite for the purpose of precision orbit determination. After launch the model was improved further by incorporating the T/P satellite tracking data, resulting in the Joint Gravity Model-2 (JGM-2 Nerem et al. 1994). Geoid models are usually expressed in terms of spherical harmonics; so are the uncertainties of the models. The rms cumulative error of the JGM-2 geoid model is about 10 cm for spherical harmonics up to degree and order 20, corresponding to a wavelength of 2,000 km. This error does not include those at higher degrees and orders. At higher degrees and orders, the geoid errors overwhelm the oceanic signals; therefore, we examine only the dynamic topography up to degree and order 20.

Displayed in Fig. 4 is the time-averaged dynamic topography using the first year's observations from T/P (see Appendix A for a description of the data processing procedures). Only the large-scale features of wavelengths larger than 2000 km are visible in this low-pass filtered view. For instance the details of the Brazil Current and the Malvinas Current (also called the Falkland Current) are lost and they appear as rather broad flows off the coast of South America, with the Confluence region of the two currents appearing as a saddle point at 40°S and 60°W. The subtropical gyre emerges as a well defined triangular plateau. The Antarctic Circumpolar Current (ACC) manifests itself as a broad current south of the subtropical gyre with its

multiple fronts not distinguishable. A cyclonic gyre of somewhat unexpected strength is situated at the eastern tropical Atlantic in the Angola Basin.

Although such a "blurred view" of the topography does not delineate the details of many of the familiar currents, it is the first direct observation of the low-wavenumber portion of the topography with an uncertainty of only 10 cm. This large-scale view of the topography has never been adequately sampled by in-situ observations. Nevertheless, when compared to the topography of Reid (1994, Fig. 8a), a fair degree of similarity is found at the large scales. A point-to-point comparison cannot be made to Reid's map because it is based on widely separate hydrographic sections.

Computer models of the global ocean circulation have become increasingly realistic as a result of the speedy advancement of computer technology over the past several years (Semtner and Chervin 1992). A recent model developed at the Los Alamos National Laboratory based on the Bryan-Cox codes (Bryan 1969) has a global average resolution of 1/6 degrees, capable of resolving the energetic mesoscale eddies over most of the global oceans (Smith et al. 1992). The time-averaged dynamic topography simulated by the model (courtesy of Richard Smith of the Los Alamos National Laboratory) was also low-pass filtered to retain spherical harmonics up to degree and order 20 for comparison to the T/P result. The reader is referred to Appendix B for a brief description of the model. The resulting simulated topography of the South Atlantic Ocean is shown in Fig. 5. There is a high degree of resemblance between Figs. 4 and 5, with a correlation of 0.98. The rms difference between the two maps is 16 cm, slightly greater than the estimated error of the altimetry result. Shown in Fig. 6 is the difference between the two. The observed subtropical gyre has a steeper northern flank, indicating a stronger South Equatorial Current, whereas the model has a stronger Brazil Current and a sharper Subtropical Front. The model also shows a stronger Benguela Current but lacks a cyclonic eddy in the Angola Basin.

A comparison was also made between the T/P dynamic topography and that of Levitus (1982) shown in Fig. 7. The choice of the Levitus result for comparison is not because it is the best, simply

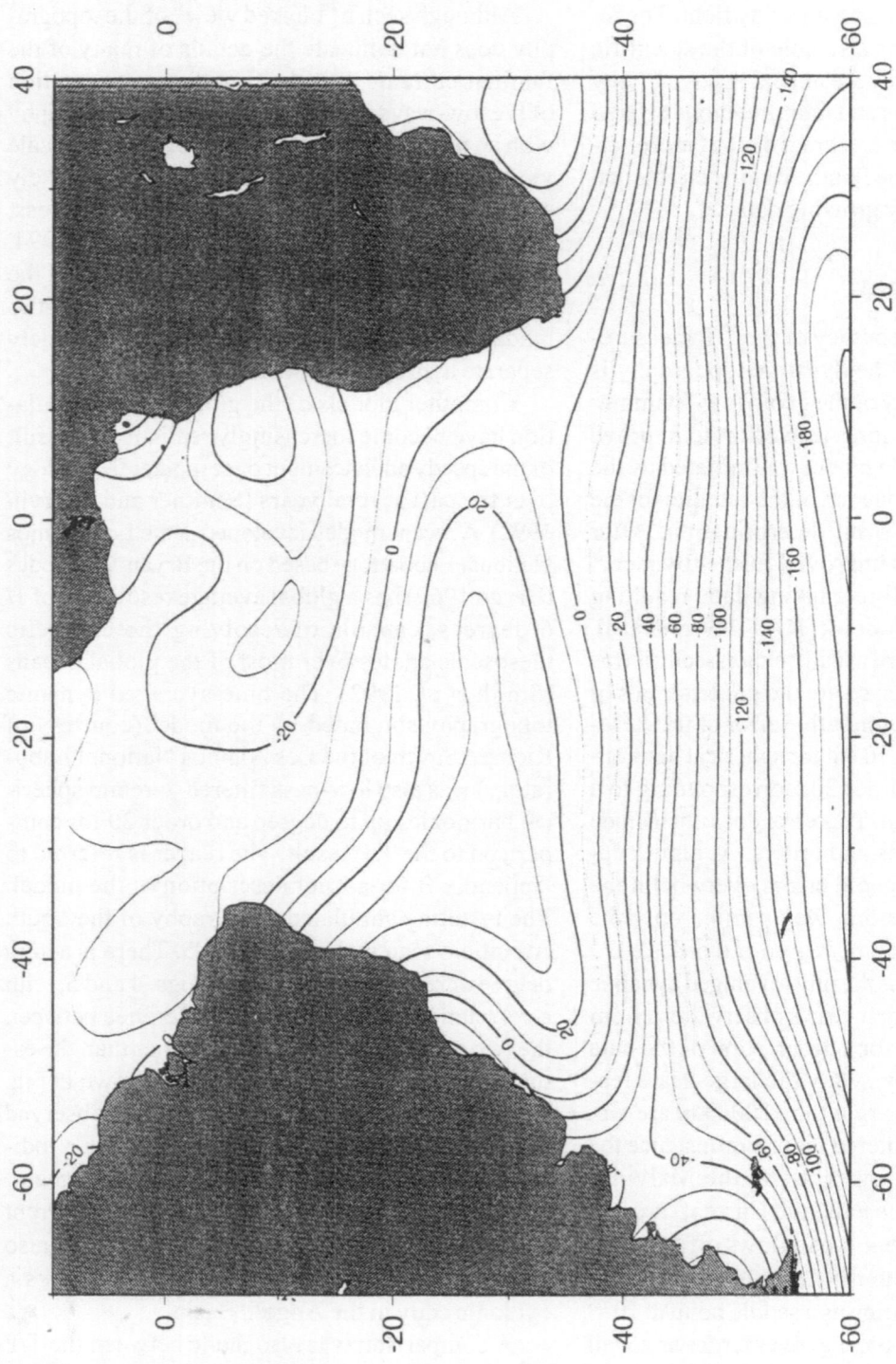

Fig. 4. Mean dynamic topography based on the first year of the T/P data and the JGM-2 geoid. Only the spherical harmonics of degree and order up to 20 are shown.

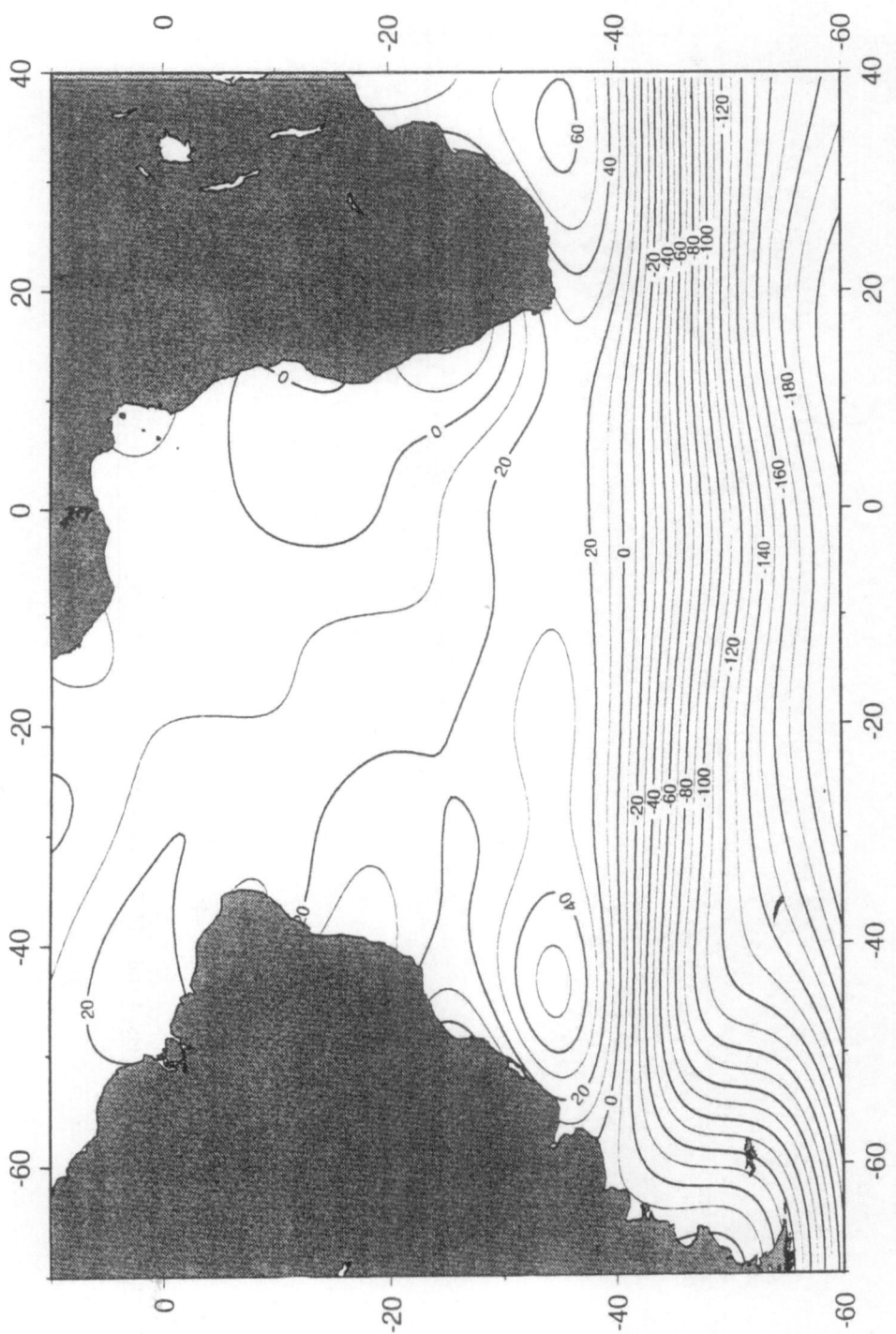

Fig. 5. Same as Fig. 4 except based on the ocean model for the same period of time.

L.-L. Fu

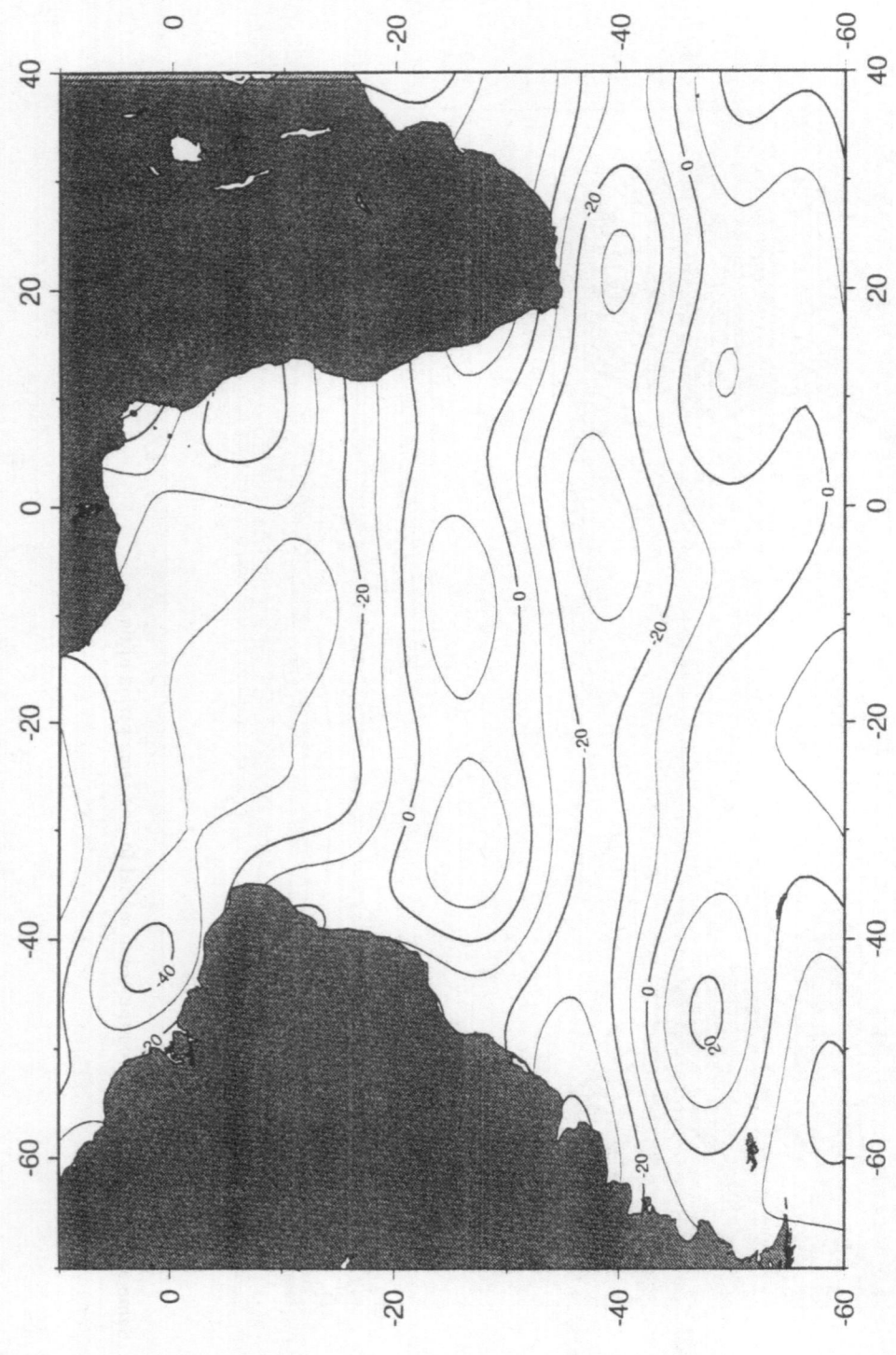

Fig. 6. The difference between Figs. 4 and 5 (T/P minus model).

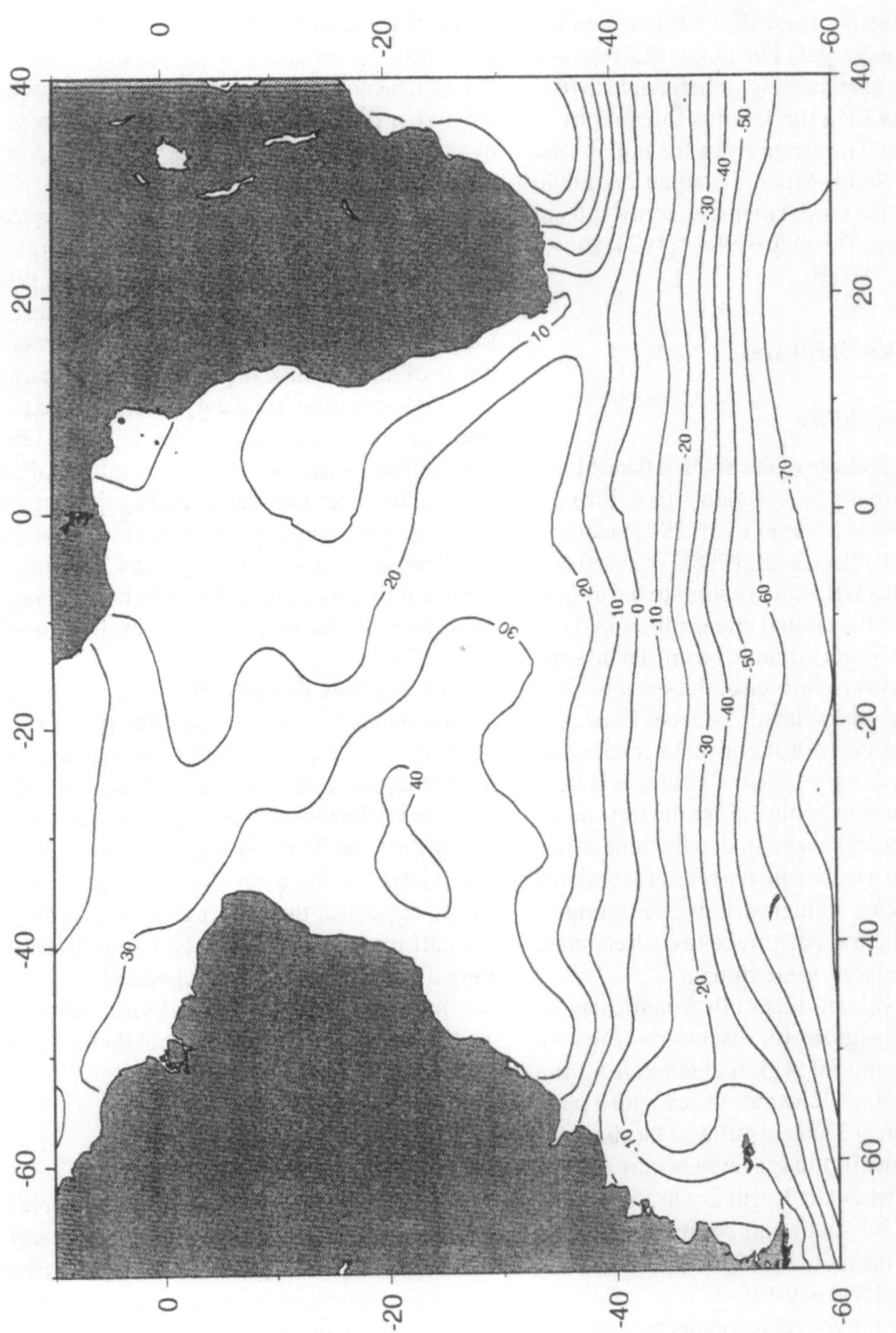

Fig. 7. Same as Fig. 4 except based on the Levitus climatology.

because it is still the only available globally-gridded hydrographic analysis widely in use. The rms difference is 31 cm, much larger than the error of the T/P estimate as well as the difference between the T/P result and the model. The major discrepancy occurs in the sea surface height difference across the ACC. The ACC in the Levitus topography is simply too weak. The range from the high to the low in the entire South Atlantic is about 220 cm in both the T/P and the model maps, but only 110 cm in the Levitus map. The map of Reid (1994) shows a range of about 180 cm.

The Seasonal Variabilities

The Annual Variability

The seasonal variabilities of the South Atlantic have been studied by many investigators (e.g. Provost and Le Traon 1993; Matano et al. 1993; Garzoli and Garaffo 1989; Olson et al. 1988). We used 640 days' worth of the T/P data to estimate the amplitude and phase of the annual and semi-annual cycles of the entire South Atlantic basin. To investigate the temporal variabilities of the sea level, the altimeter data are interpolated to a set of fixed normal points along each of the ground tracks displayed in Fig. 3. A time series of sea level is constructed at each normal point. After the time mean is removed, a sinusoid was fitted to the time series of the residuals in a least-squares sense for both the annual and the semi-annual periods. A running 5° x 5° box averaging was performed to each estimate to retain only the large-scale signals.

The annual cycle of the South Atlantic can be divided into four regions for discussion (Figs. 8a and 8b). The equatorial region extends from the mouth of the Amazon River to Africa with a peak amplitude of 8 cm at 3°N and 40°W. The phase indicates that the maximum sea level occurs in September/October when the North Equatorial Counter Current is in its maximum strength along the northern slope of the region of high amplitude (Katz 1993; Arnault and Cheney 1994).

The subtropical gyre corresponds to the triangular region roughly bounded by the 90-degree phase line extending from 10°S and 37°W to 40°S and 15°E and by the 40°S latitude to the south.

Within this region the amplitude is generally greater than 3 cm and the maximum sea level takes place in March. The peak amplitude is about 13 cm just north of the Brazil/Malvinas Confluence region, reflecting the intensity of the annual cycle of the Brazil Current. The annual cycle of the subtropical gyre is primarily forced by the wind stress curl over the region (Matano et al. 1993; Garzoli and Giulivi 1994).

The region northeast of the subtropical gyre has two distinct subregions. A region of moderate amplitude of 3-4 cm at 13°S and 5°W with a 120 degree phase, reflecting the annual cycle of the South Equatorial Current. North and east of the region is a strip of rapid change in phase, with the maximum sea level occurring in January/February next to the west coast of Africa and the Gulf of Guinea (Arnault and Cheney 1994). The pattern of phase change indicates that the annual cycle propagates westward in the region. The phase in the mid ocean lags behind the near-shore region by 3 months. The details of the annual variability in both subregions were not revealed in the Geosat result of Jacobs et al. (1992).

The Agulhas Retroflection region has a peak amplitude greater than 10 cm. The phase is about the same as the subtropical gyre, but the Retroflection region is clearly separate from the subtropical gyre by the 90 and 120 degree phase lines bordering the gyre. There is slight westward propagation within the the Retroflection region, with the annual cycle near the peak region leading the western part by about 60 days. The amplitude of the annual cycle south of 45°S is generally less than 3 cm, indicating that the annual variability of the ACC is fairly weak in the South Atlantic (Chelton et al. 1990).

The Semiannual Variability

There are basically three regions of appreciable semiannual variability in the South Atlantic (Figs. 9a and 9b). The most prominent is the Confluence region characterized by three clustered areas with amplitudes greater than 5 cm and a phase in the range from 180 to 240 degrees (maximum occurring in April and October). The southwestern one of the three areas is located in the region of the

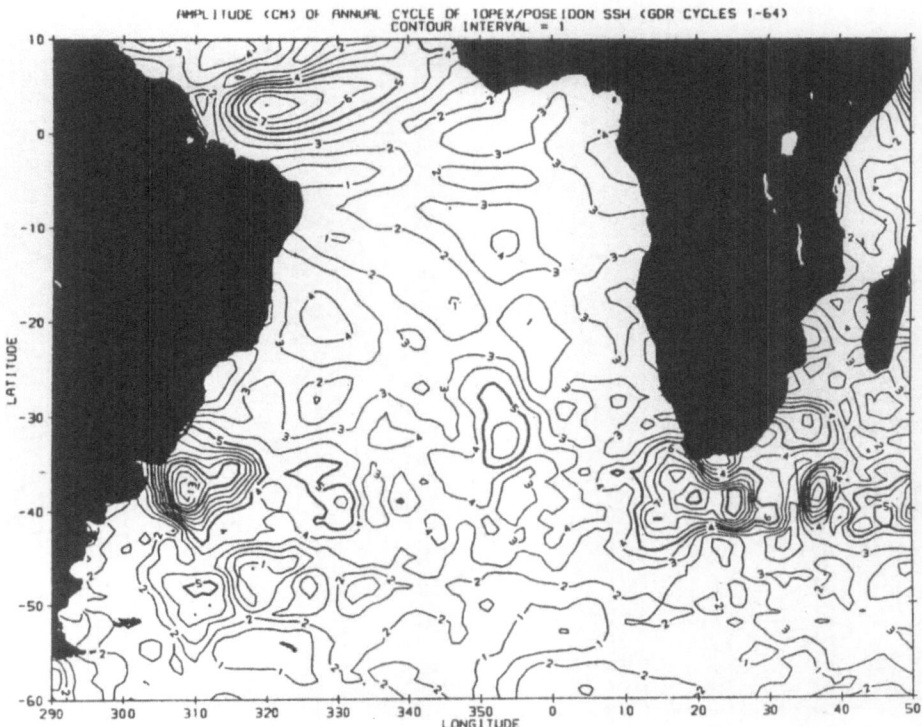

Fig. 8a. The amplitude of the annual cycle based on 640 days' worth of T/P data.

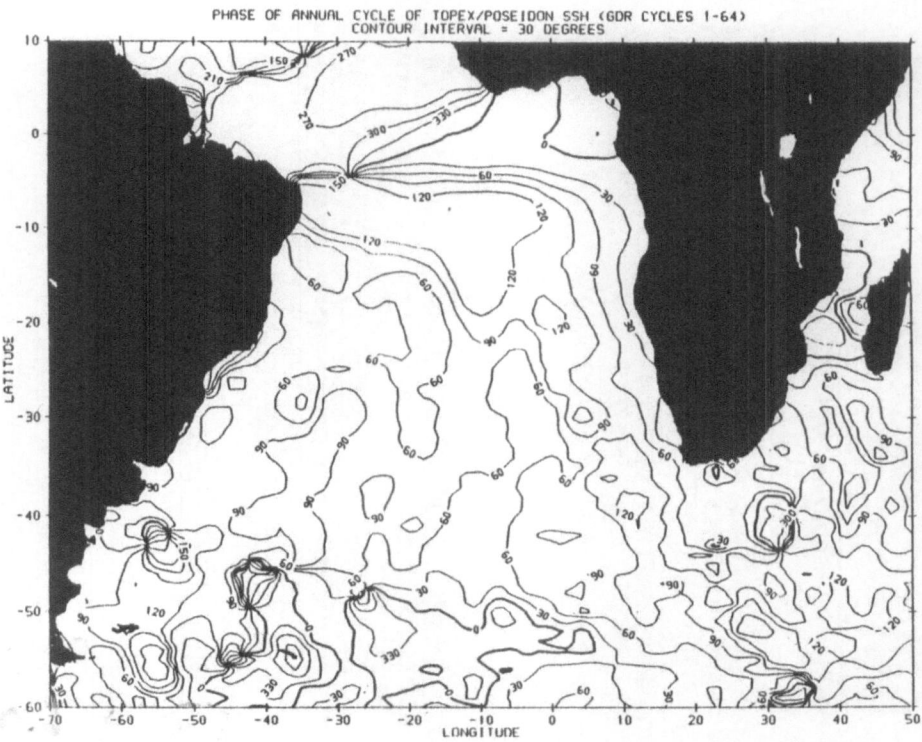

Fig. 8b. The phase (in terms of the year day when the sea level is maximum) of the annual cycle based on 640 days' worth of T/P data.

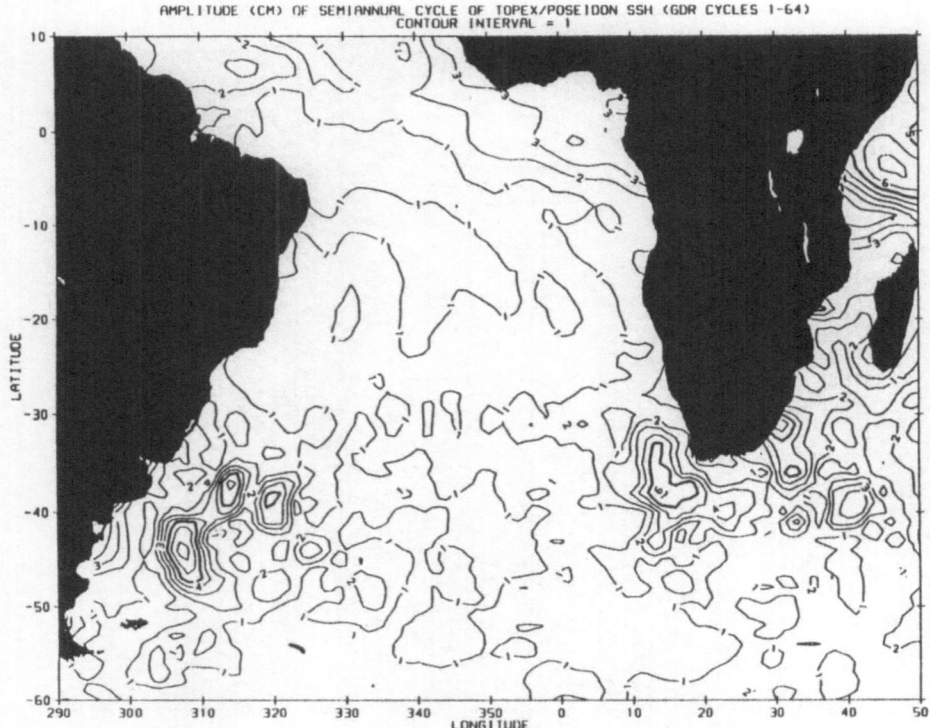

Fig. 9a. Same as Fig. 8a except for the semiannual cycle.

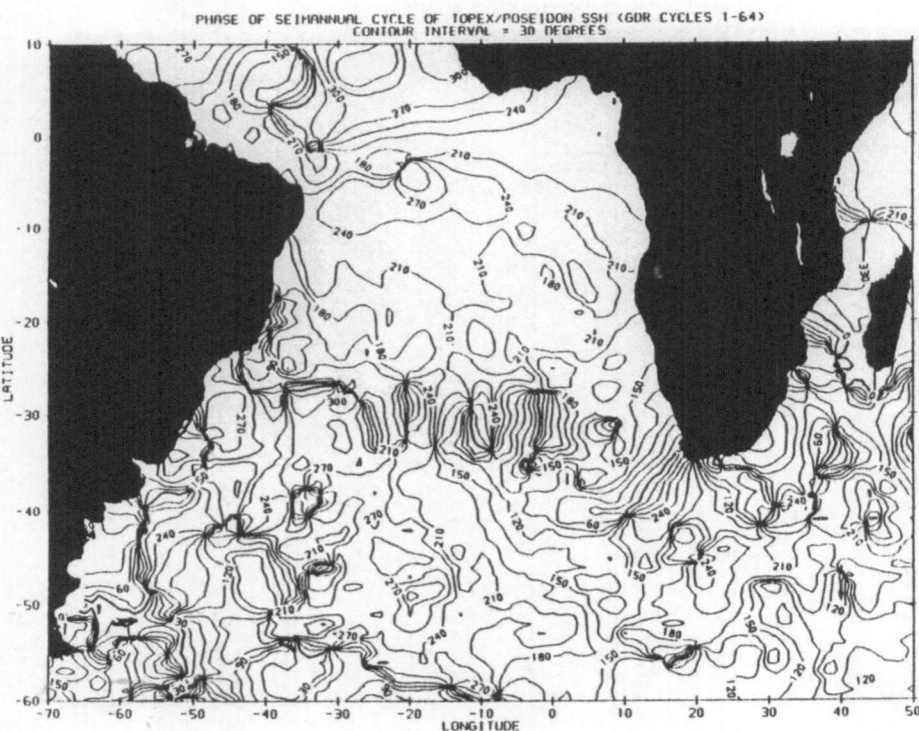

Fig. 9b. Same as Fig. 8b except for the semiannual cycle.

southward extension of the Confluence consisting of the overshooting of the Brazil Current and the retroflection of the Malvinas Current, indicating that the southward extension of the Confluence has a distinct semiannual periodicity. Note that the annual variability has a local minimum here. Although there is a semiannual component in the wind variability over the Southern Ocean in general (Large and van Loon 1989; van Loon and Rogers 1984), it is not clear why the response of the ocean at the semiannual period is different from the annual period at this location.

The Agulhas Retroflection region also has significant semiannual variability with the maximum amplitude located near the western tip of the Retroflection region. This finding is different from the Geosat result of Jacobs et al. (1992), who reported no significant semiannual variability in the region. Judging from the phase pattern, there appears to be westward propagating waves radiating from the region across the basin toward South America along 30°S. This finding is similar to Le Traon and Minster (1993), who reported westward propagating Rossby waves of the semiannual period with a wavelength of 450 km and an amplitude of 2-3 cm. Such waves have been smeared by the spatial smoothing process employed in the study to reveal the large-scale variabilities. See the next section for a clearer evidence (Fig. 11) for these waves.

The third region of semiannual variability is the Gulf of Guinea, where the amplitude is 2-3 cm with a phase of 210 degrees (with the maximum occurring in April and October). This finding is consistent with the Geosat result of Arnault and Cheney (1994) and Arnault et al. (1992). The variability is associated with the twice-per-year wind-driven upwelling in the region.

The Mesoscale Variability

The most energetic variabilities of the ocean occur at the mesoscales (circa 100 km and 100 days) and are often called the mesoscale eddies (Robinson 1983). Shown in Fig. 10 are two types of information about the mesoscale variability of the South Atlantic. The grey-scale shading shows the rms sea level variability calculated from the first year's worth of the T/P data, reflecting primarily the in-

tensity of the mesoscale variability. One might also view it as a representation of the potential energy of the eddies, as the sea level variability at this scale is highly correlated to the thermocline variability. The ellipses display the principal components (e.g. Freeland et al. 1975) of the geostrophic velocities calculated from the sea surface slopes at the cross-over points of the satellite ground tracks. The size of the ellipses is proportional to the kinetic energy of the mesoscale motions, while the orientation of the ellipses exhibits the directional preference of the motions. A circle corresponds to isotropic motions and a collapsed line segment corresponds to linear motions.

The pattern and magnitude of the current ellipses are similar to the Geosat results (Morrow et al. 1994; Provost and Le Traon 1993), indicating that the mesoscale variability is a fairly robust component of the ocean circulation. There is more meridional energy than zonal in both the Brazil/Malvinas Confluence and the Agulhas Retroflection regions.

The rms variability is also very similar to the Geosat results (e.g. Gordon and Haxby 1990; Zlotnicki et al. 1989; Forbes et al. 1993; Le Traon and Minster 1993; Chelton et al. 1990). There is a large area between the equator and 20°S where there is a minimal amount of eddy energy, sometimes called the "eddy desert." The characteristic "C-shaped" region of high eddy energy in the Confluence region is clearly shown. The northern arm of the "C" is associated with the intense meandering and eddy shedding of the Brazil/Malvinas Confluence. The southern arm with much less energy is associated with the convergence of the subantarctic and the subpolar fronts of the ACC (see Fig. 1 of Peterson and Stramma 1991). The middle part of the "C" is associated with the southward extension of the Confluence, where a strong semiannual variability is present as discussed in the preceding section.

There is considerable mesoscale variability associated with the ACC when it is hugging against the continental slope before flowing over the Scotia Ridge at about 50°S and 50°W. There is very little variability in the ACC when it is over the Scotia Ridge, whereas the variability recurs after it passes the ridge and reenters the deep ocean basin, as re-

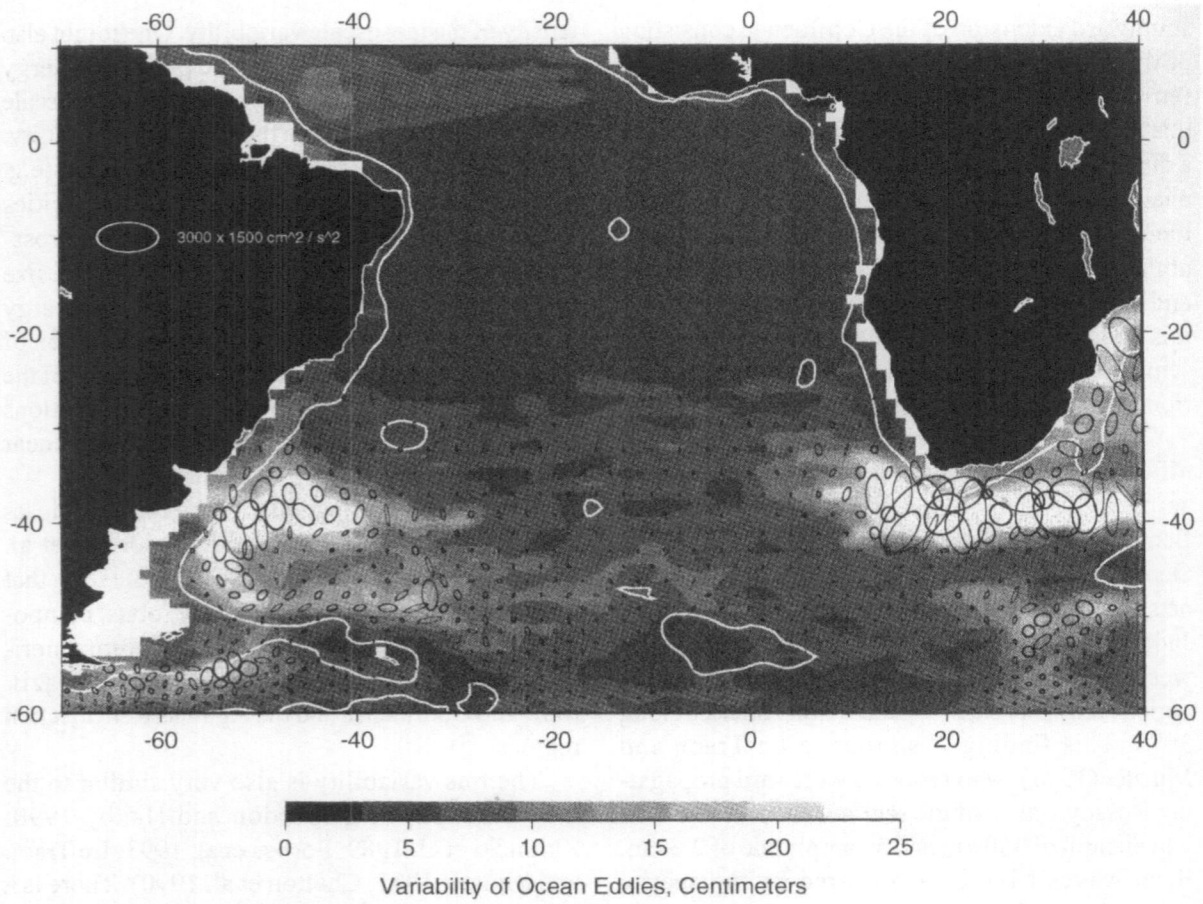

Variability of Ocean Eddies, Centimeters

Fig. 10. The rms sea level variability (in grey shades) and the geostrophic current ellipses. (with the scales for the semi-major and semi-minor axes shown in the upper left corner). Also shown in white are the 3000 m bathymetry contours.

vealed by the southern arm of the "C".

The Agulhas Retroflection is another region of strong mesoscale variability. The variability is confined to a region surrounded by the retroflection of the Agulhas Current to the west and the Agulhas Return Current to the south. This band of mesoscale variability continues into the Indian Ocean and eventually merges with the path of the ACC. Eddies shed from the Retroflection region drift to the west into the South Atlantic and have been traced all the way to the western basin (Gordon and Haxby 1990; Byrne et al. 1995). This train of eddies are partly responsible for the moderate eddy energy level in the northern half of the zone between 25°S

and 50°S across the entire basin. As noted above, there are also Rossby waves of the semiannual period in the region. Shown in Fig. 11 is a time-longitude display of the sea level variabilities along 30°S. A train of westward propagating waves with periods close to the semiannual and zonal wavelengths about 500-800 km is clearly revealed, consistent with the findings of Le Traon and Minster (1993) using the Geosat data.

The southern half of the 25°S- 50°S zone is the core of the ACC, whose meandering and eddies are the source for the mesoscale variability. There is indication of eastward propagation of mesoscale features east of 20°W along 50°S (Fig. 12). The speed inferred is about 1-2 cm/sec. These might be Rossby waves whose westward propagation is overwhelmed by the eastward mean flow of the ACC (Hughes 1995). The cause for the fairly high energy level at about 50°S and 30°E is not clear. The presence of the Prince Edward Fracture Zone might cause some fluctuations of the ACC, but there are similar topographic features upstream from the region such as the Mid-Atlantic Ridge, where no significant eddy energy is found.

In a turbulent fluid the effects of the temporal fluctuations (or the eddies) on the time-averaged flow can be expressed in terms of the divergence of the Reynolds stress tensor (e.g. Pedlosky 1979), whose elements consist of <u'v'>, <u'^{2}> and <v'^{2}>, where u' and v' are the time-dependent zonal and meridional velocity components, and the angled bracket represents time averaging. The resultant acceleration of the zonal mean flow caused by the eddies can be expressed by

$$-\frac{\partial}{\partial x}\left\langle u'^2\right\rangle - \frac{\partial}{\partial y}\left\langle u'v'\right\rangle$$

and that of the meridional mean flow by

$$-\frac{\partial}{\partial x}\left\langle u'v'\right\rangle - \frac{\partial}{\partial y}\left\langle v'^2\right\rangle$$

The vectors representing the acceleration are shown in Fig. 13 superimposed on the dynamic topography of Fig. 4. The order of magnitude of the accelerations in the regions of maximum eddy activities is comparable to that due to the Coriolis force acting on a flow of 1 cm/sec at mid latitudes. If integrated over the upper 1000 m of the ocean assuming that the surface values are representative, the depth-integrated eddy force amounts to 1-5 (cm/ sec)^2, comparable to the wind stress forcing and consistent with the results of Morrow et al. (1994). The spatial patterns of the acceleration vectors are mostly dominated by the distribution of the eddy kinetic energy (i.e., 1/2 (<u'^{2}>+<v'^{2}>)). The patterns suggest a generally eastward acceleration of the mean flow of the Agulhas Return Current as well as the Brazil/Malvinas Confluence.

Conclusions

A survey of the first results of the TOPEX/ POSEIDON observations in the South Atlantic Ocean is described in the paper. The large-scale (greater than 2000 km) dynamic topography is directly observed by the satellite altimeter with an estimated accuracy of 10 cm. Comparisons to both the climatological atlas of Levitus (1982) and the result of an ocean general circulation model (Smith et al. 1992) have shown significant differences. The rms difference is 31 cm with the former, mainly caused by the differences in the ACC, which is simply too weak in the Levitus atlas. The rms difference with the latter is 16 cm, resulting from differences in the subtropical gyre, the Benguela Current, and the Angola Basin.

The seasonal variabilities are studied using 640 days' worth of data. The phase of the annual cycle reveals four distinctively different regions: the equatorial region, the subtropical gyre, the South Equatorial Current and the Agulhas Retroflection. Amplitudes greater than 10 cm are found in the Brazil/ Malvinas confluence and the Agulhas Retroflection. There is also large semiannual variability in these two regions. However, the geographic patterns are somewhat different from those of the annual cycle.

The mesoscale variability is examined in terms of the rms variability of the sea surface height and the geostrophic current ellipses calculated from the sea surface height slopes. The results are very similar to those from the Geosat data. The acceleration of the mean flow resulting from the mesoscale eddies is estimated from the divergence of the eddy Reynolds stress tensor. The effects of eddies on the mean flow are important in the regions of the Brazil/Malvinas confluence and the Agulhas Retroflection.

The preliminary results discussed in the paper have demonstrated the potential of the TOPEX/ POSEIDON data for ocean circulation studies. The data is of global extent and its volume is continuously growing. Significant advancement is expected in the understanding of the dynamics of the circulation of the world ocean. Many of the questions about the circulation of the South Atlantic will benefit from the global perspective provided by the data set.

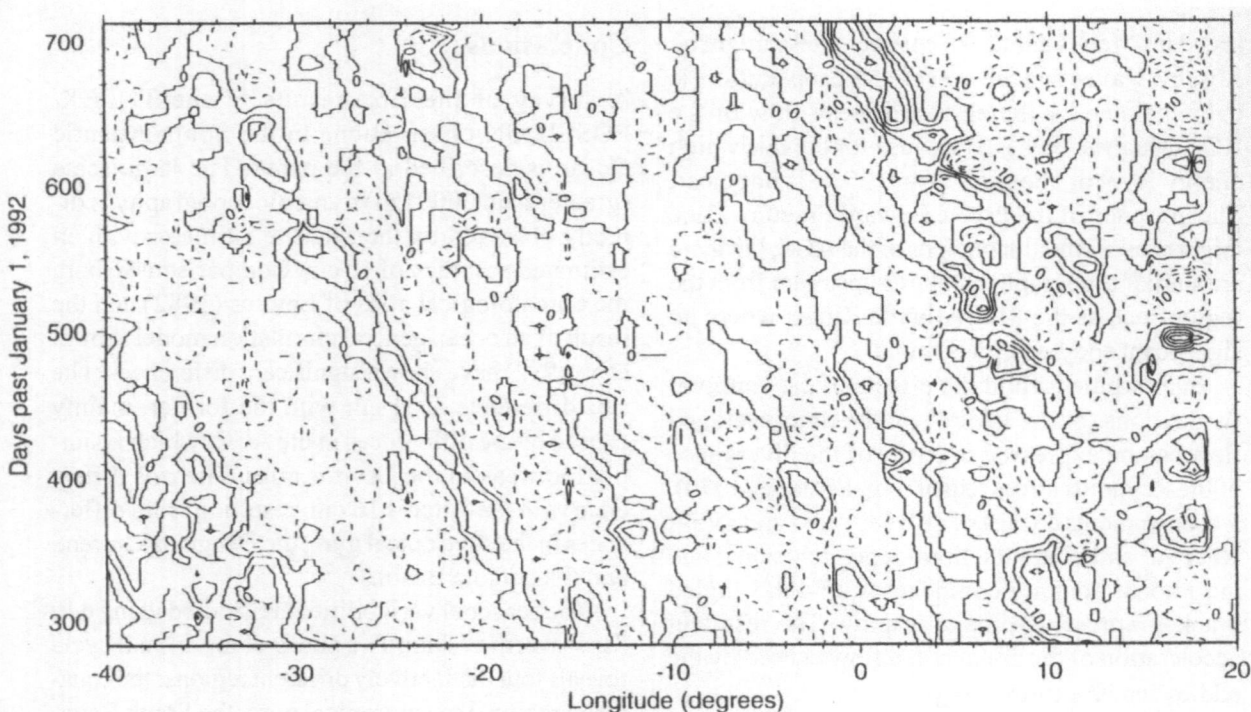

Fig. 11. The sea level anomalies as a function of time and longitude along 30°S.

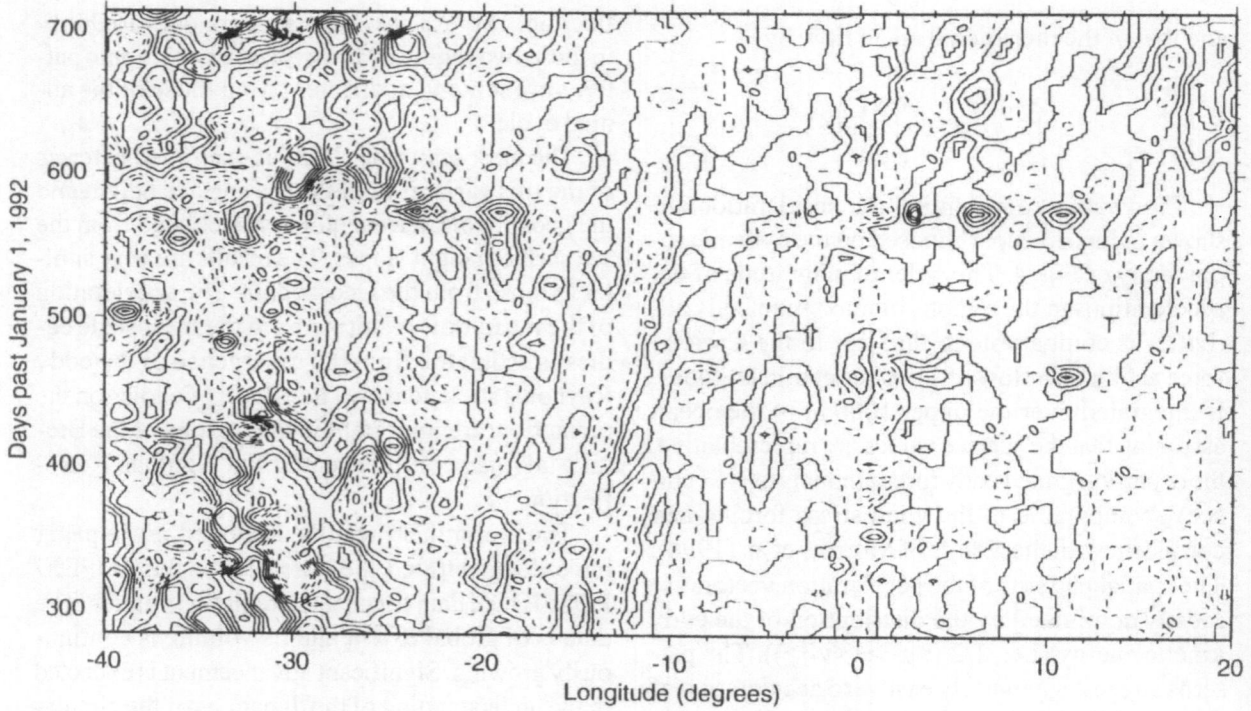

Fig. 12. Same as Fig. 11 except for 50°S.

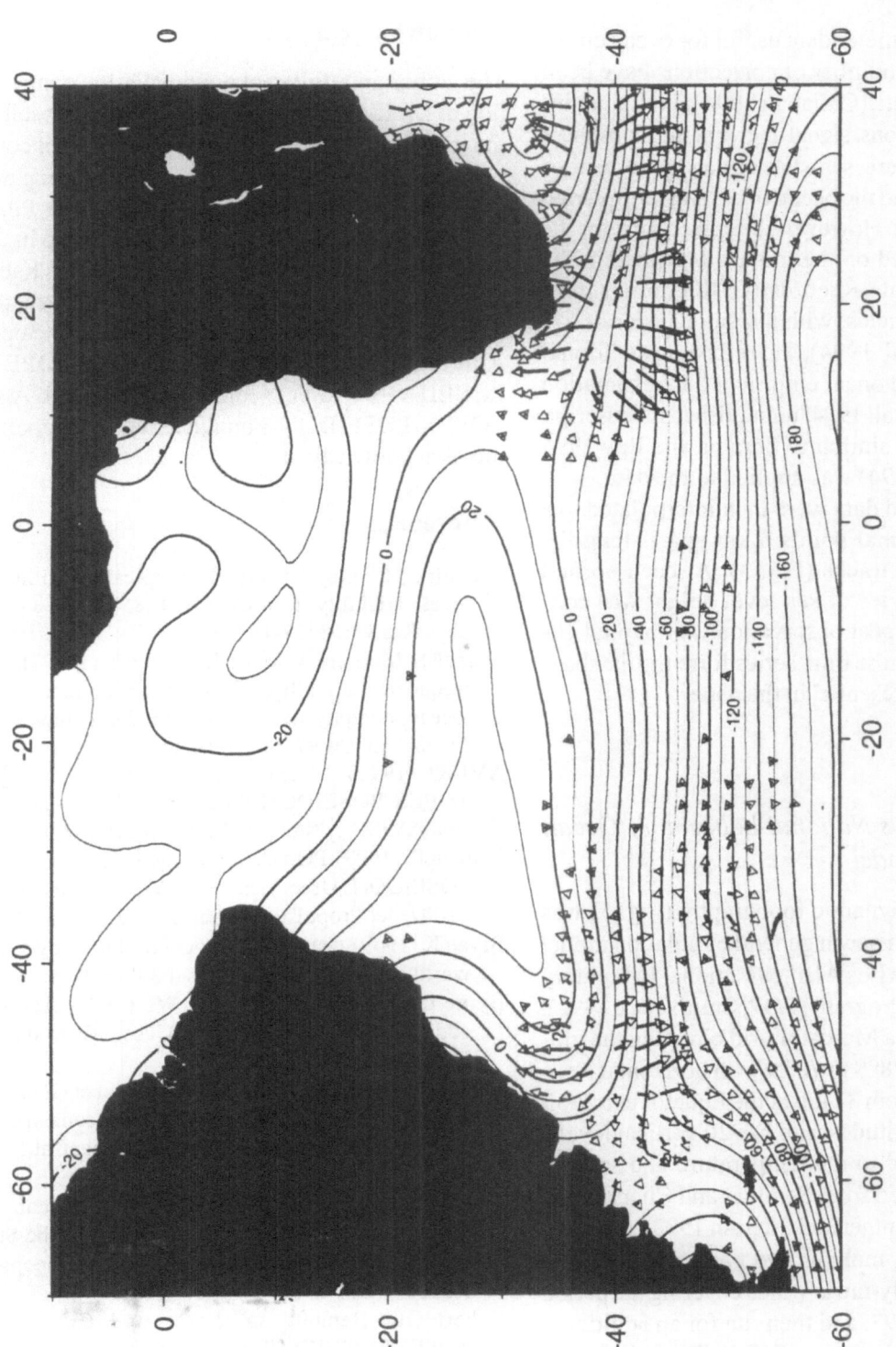

Fig. 13. The force per unit mass exerting on the mean flow by the eddy field (shown by the arrows). Only the arrows with force magnitudes greater than 5×10^{-7} cm/sec^2 are shown. The dynamic topography is the same as in Fig. 4.

Appendix A

Altimeter Data Processing

To make the altimeter data useful for ocean circulation studies, a number of corrections have been applied to the data (Callahan 1994), including instrument corrections, signal delays in the ionosphere and the troposphere, sea-state effects, the effects of the solid earth and the ocean tides, and the inverted barometer effect. However, the corrections for the ocean tides based on the models provided in the Geophysical Data Records (GDR) do not have sufficient accuracies, with a residual error of 5-6 cm (Molines et al. 1994). Therefore, an additional correction based on an empirical tidal estimation procedure (Fu et al. 1994, unpublished manuscript; the procedure is similar to Shrama and Ray 1994 and Ma et al. 1994) was applied to the data.

The corrected data were then interpolated to a set of fixed normal points 6 km apart along the satellite ground tracks (Fig. 3). At each normal point, a time series of sea level height was constructed from repeat observations at nominal 10-day intervals. These time series form the database for the results presented in the paper.

Appendix B

The Los Alamos National Laboratory Ocean Circulation Model

The simulated dynamic topography (Fig. 5) was obtained from an ocean general circulation model developed by the Los Alamos National Laboratory Parallel Ocean Program (POP, Smith et al. 1992). The model uses a Mercator grid covering the global ocean from 78°S to 78°N with horizontal resolution ranging from 31 km at the Equator to 6.5 km at the highest latitudes, and has 20 vertical levels. It was initialized to the temperature and salinity fields interpolated from Semtner and Chervin's 1/4 degree run (Semtner and Chervin 1992), and spun up for 18 years, making two passes through the ECMWF monthly-mean winds covering the period from 1985 to 1993, and then run for an additional 9 years using 3-day mean ECMWF winds for the same period. Surface heat and freshwater fluxes were approximated by restoring to the seasonal Levitus climatology (Levitus 1982).

Acknowledgements

The author gratefully acknowledges the contribution of Dr. Richard Smith in providing the result of the model simulated dynamic topography for comparison in the paper. Computer programming and graphics assistance was provided by Greg Pihos and Denis Leconte. The research described in the paper was carried out by the Jet Propulsion Laboratory, California Institute of Technology, under contract with National Aeronautics and Space Administration. Support from the TOPEX/POSEIDON Project funded under the NASA TOPEX/POSEIDON Announcement of Opportunity is acknowledged.

References

Arnault S, Cheney RE (1994) Tropical Atlantic sea level variability from Geosat (1985-1989). J Geophys Res 99:18207-18223

Arnault S, Morliere A, Merle J, Menard Y (1992) Low-frequency variability of the Tropical Atlantic surface topography: altimetry and models comparison. J Geophys Res 97:14259-14288

AVISO (1992) AVISO User Handbook: Merged TOPEX/POSEIDON Products, AVI-NT-02-101-CN, AVISO Altimetrie, CNES, Toulouse, France

Benada R (1993) PO.DAAC Merged GDR (TOPEX/POSEIDON) Users Handbook, Version 1.0, JPL D-11007, Jet Propulsion Laboratory, Pasadena, CA

Bryan K (1969) A numerical model for the study of the world ocean. J Comput Phys 4:347-376

Byrne DA, Gordon AL, Haxby WF (1995) Agulhas eddies: a synoptic view using Geosat ERM data. J Phys Oceanogr 25:902-917

Callahan PS (1994) Topex/Poseidon Project GDR Users Handbook, JPL D-8944 (internal document), rev A, Jet Propulsion Laboratory, Pasadena, Calif., 84 pp

Chelton DB, Schlax MG, Witter DL, Richman JG (1990) Geosat Altimeter Observations of the Surface Circulation of the Southern Ocean. J Geophys Res 95:17877-17903

Egbert GD, Bennett AF, Foreman MGG (1994) TOPEX/POSEIDON tides estimated using a global inverse model. J Geophys Res 99:24821-24852

Forbes C, Leaman K, Olson D, Brown O (1993) Eddy and wave dynamics in the South Atlantic as diagnosed from Geosat altimeter data. J Geophys Res 98:12297-12314

Freeland HJ, Rhines PB, Rossby T (1975) Statistical observations of the trajectories of neutrally buoyant floats in the North Atlantic. J Mar Res 33:383-404

Fu LL, Lefebvre M, Christensen EJ (1991) TOPEX/POSEIDON: The Ocean Topography Experiment. EOS, Transactions, American Geophysical Union. 72(35):369-373

Fu LL, Christensen EJ, Yamarone C, Lefebvre M, Menard Y, Dorrer M, Escudier P (1994) TOPEX/POSEIDON Mission Overview. J Geophys Res 99:24369-24381

Garzoli SL, Garraffo Z (1989) Transports, frontal motions and eddies at the Brazil-Malvinas confluence as revealed by inverted echo sounders. Deep-Sea Res 36:681-703

Garzoli SL, Giulivi C (1994) What forces the variability of the south western Atlantic boundary currents? Deep-Sea Res 41:1527-1550

Gordon A, Haxby W (1990) Agulhas eddies invade the S. Atlantic: evidence from Geosat Altimeter and Shipboard Conductivity-Temperature -Depth Survey. J Geophys Res 95:3117-3126

Hughes CW (1995) Rossby waves in the Southern Ocean: a comparison of TOPEX/POSEIDON altimetry with model predictions. J Geophys Res 100:15933-15950

Jacobs GA., Born GH, Parke ME, Allen PC (1992) The Global Structure of the Annual and Semiannual Sea Surface Height Variability from Geosat Altimeter Data. J Geophys Res 97:17813-17828

Katz EJ (1993) An interannual study of the Atlantic North Equatorial Counter Current. J Phys Oceanogr 23:116-123

Large WG, van Loon H (1989) Large scale, low frequency variability of the 1979 FGGE surface buoy drifts and winds over the Southern Hemisphere. J Phys Oceanogr 19:216-232

Le Provost C, Bennett AF, Cartwright DE (1995) Ocean tides for and from TOPEX/POSEIDON. Science 267: 639-642

Le Traon PY, Minster JF (1993) Sea Level Variability and Semiannual Rossby Waves in the South Atlantic Subtropical Gyre. J Geophys Res 98:12315-12326

Levitus S (1982) Climatological Atlas of the World Ocean. NOAA Professional Paper 13, NOAA, Rockville, Md

Ma XC, Shum CK, Eanes RJ, Tapley BD (1994) Determination of ocean tides from the first year of TOPEX/POSEIDON altimeter measurements. J Geophys Res 99:24809-24820

Matano RP, Schlax MG ,Chelton DB (1993) Seasonal Variability in the Southwestern Atlantic. J Geophys Res 98:18027-18035

Molines, Le Provost, Lyard, Ray, Shum, Eanes (1994) Tidal corrections in the TOPEX/ POSEIDON GDRs. J Geophys Res 99:24749-24760

Morrow R, Coleman R, Church J, Chelton D (1994) Surface eddy momentum flux and velocity variances in the Southern Ocean from Geosat altimetry. J Phys Oceanogr 24:2050-2071

Nerem S, et al. (1994) Gravity model development for TOPEX/POSEIDON: joint gravity models 1 and 2. J Geophys Res 99:24421-24447

Olson DB, Podesta GP, Evans RH, Brown OB (1988) Temporal variations in the separation of Brazil and Malvinas Currents. Deep-Sea Res 35:1971-1990

Pedlosky J (1979) Geophysical Fluid Dynamics. Springer New York

Peterson RG, Stramma L (1991) Upper-level circulation in the South Atlantic Ocean. Progr Oceanogr 26:1-73

Provost C, Le Traon PY (1993) Spatial and Temporal Scales in Altimetric Variability in the Brazil-Malvinas Current Confluence Region: Dominance of the Semiannual Period and Large Spatial Scales. J Geophys Res 98:18037-18051

Reid JL (1994) On the total geostrophic circulation of the North Atlantic Ocean: flow patterns, tracers, and transports. Progr Oceanogr 33:1-92

Robinson AR (ed) (1983) Eddies in Marine Science. Springer New York

Schrama EJO, Ray RD (1994) A preliminary tidal analysis of TOPEX/POSEIDON altimetry. J Geophys Res 99:24799-24808

Semtner AJ, Chervin RM (1992) Ocean general circulation from a global eddy-resolving model. J Geophys Res 97:5493-5550

Smith RD, Dukowicz JK, Malone R (1992) Parallel ocean general circulation modeling. Physica D 60:38-61

Stammer D, Wunsch C (1994) Preliminary assessment of the accuracy and precision of TOPEX/POSEIDON altimeter data with respect to the large-scale ocean circulation. J Geophys Res 99:24584-24604

Tai CK, Wunsch C (1983) Absolute measurement by satellite altimetry of dynamic topography of the Pacific Ocean. Nature 301(5899):408-410

Tai CK, Wunsch C (1984) An estimate of global absolute dynamic topography. J Phys Oceanogr 14:457-463

van Loon H, Rogers JC (1984) Interannual variations in the half-yearly cycle of pressure gradients and zonal wind at sea level in the Southern Hemisphere. Tellus 36A:76-86

Wunsch C (1992) Observing ocean circulation from space. Oceanus 35(2):9-17

Wunsch C, Gaposchkin EM (1980) On Using Satellite Altimetry to Determine the General Circulation of the Oceans with Application to Geoid Improvement. Rev Geophys Space Phys 18:725-745

Zlotnicki V, Fu LL, Patzert W (1989) Seasonal variability in global sea level observed with Geosat altimetry. J Geophys Res 94:17959-17969

The Zonal WOCE Sections in the South Atlantic

G. Siedler[1], T. J. Müller[1], R.Onken[1] M. Arhan[2], H. Mercier[2]
B.A. King[3], P. M. Saunders[3]

[1]*Institut für Meereskunde an der Universität Kiel Düsternbrooker Weg 20,
24105 Kiel, GERMANY*
[2]*Laboratoire de Physique des Océans (CNRS - IFREMER - UBO)
IFREMER / Centre de Brest B.P. 70, 29280 Plouzané, FRANCE*
[3]*Southampton Oceanography Centre Empress Dock,
Southampton SO14 3ZH, ENGLAND*

Abstract: The data from six zonal sections in the World Ocean Circulation Experiment (WOCE) in the tropical and southern Atlantic are used to describe the distribution of water masses. Due to the high spatial resolution, the structure of temperature, salinity, oxygen, silicate and nitrate displays details related to transport and mixing in this region. Temperature-salinity diagrams are also presented which indicate the effects of branching and recirculation loops in the water mass flow.

Introduction

The South Atlantic has a specific function in the global thermohaline circulation. It provides the passage for the North Atlantic Deep Water on its course from the source regions in the polar and subpolar North Atlantic to the Southern, Indian and Pacific Oceans, and the passage for the return flow of Intermediate and Central Waters on their way to the North Atlantic, after having been formed in these oceans by upwelling and mixing. The basic properties of this global vertical circulation cell were already deduced from the early models by Stommel (1957) and Stommel and Arons (1960). The thermohaline circulation loop is now usually called the „global conveyor belt" (Gordon 1986; Broecker 1991; Schmitz 1996). It is well established that the return transport of cool and warm water to the South Atlantic has two sources: the flow from east to west south of Africa and the flow from west to east through the Drake Passage south of South America. The relative contributions of these two flows, however, is still in question (Rintoul 1991; Saunders and King 1995b).

Ocean-wide hydrographic section data are most appropriate for describing large-scale water mass distributions and for determining property transports. Zonal sections are particularly useful in the South Atlantic because they allow the determination of the total meridional fluxes and their changes in this well-defined passage of the conveyor belt water. The early investigations of the METEOR expedition from 1925 to 1927 (Wüst 1936; Defant 1936) supplied such zonal section data, as did several cruises performed during the International Geophysical Year in 1957 and 1958 (Fuglister 1960). The station separations were usually much larger than the baroclinic Rossby radius in this ocean. Such large distances between observations can introduce considerable errors in the determination of cross-sectional fluxes.

More recent measurements therefore aimed at reducing these deviations by choosing sufficiently small station separations. Two high-resolution zonal sections were obtained on OCEANUS in 1983 (Warren and Speer 1991), and the South At-

From WEFER G, BERGER WH, SIEDLER G, WEBB DJ (eds) 1996, *The South Atlantic: Present and Past Circulation*. Springer-Verlag Berlin Heidelberg, pp 83-104

lantic Ventilation Experiment (SAVE) in 1987 - 89 provided several ocean-wide sections (see Durrieu de Madron and Weatherly 1994). Most of these data were already used by Reid (1989) for his comprehensive description of the South Atlantic circulation.

The World Ocean Circulation Experiment (WOCE) has the goal to collect the data necessary for developing and testing models which are useful for predicting climate changes, with an emphasis on determining the global oceanic circulation and its relation to climate. As part of the global WOCE Hydrographic Programme a set of ocean-wide zonal and meridional sections was occupied in the tropical and South Atlantic by groups from France, Germany, the UK and the USA. In Figure 1 we present the locations of all zonal WOCE one-time sections in this region which were occupied between 1991 and 1995. Sections A6 and A7 were done with the research vessel L'ATALANTE in January - March 1993 (Le Groupe CITHER-1 1994), sections A8, A9 and A10 with METEOR in March - May 1994, February - March 1991 and December 1992 - January 1993, respectively, and section A11 with DISCOVERY in January - February 1993. The particu-

lar goals of this part of the WOCE programme were to observe the pathways and water mass transports of the large-scale circulation and to determine the meridional transports of heat and other properties. Subsets from the Brazil Basin are also used for the WOCE Deep Basin Experiment which has the goal of improving the understanding of processes controlling the deep flow in that basin and particularly the relative importance of boundary and interior mixing and the role of passages.

While the results of the WOCE data analysis to address these goals will be published in various journals, the present book contribution is aimed at describing the water mass distributions from these new high-resolution zonal sections, also for the benefit of marine researchers from other disciplines. Great efforts were made in the WOCE programme to ensure high quality data. The target accuracy is 0.003°C in temperature, 0.002 in salinity, 1% for oxygen and nitrate, and better than 3% for silicate. At the time when this article was written, not all the data used here had gone through the final stage of international quality control. But all these data can be expected to be either at or close to WOCE standards, and are certainly of sufficiently high quality to allow the description which is given in the following.

Water mass distributions

We present all the zonal temperature and salinity sections from WOCE covering the equatorial belt and the subtropical and temperate South Atlantic between 7°30'N and 45°S (Figure 1) on the colour plates in Figures 2 to 7. Oxygen and nutrient distributions for the sections between 11°20'S and 45°S are displayed in Figures 8 - 13. In addition temperature-salinity diagrams are presented for these four sections in Figures 14 -16. It should be noted that the latitudes given for the individual sections are nominal only. Tracks partly deviate from the indicated latitudes, particularly in the western and eastern boundary regions where attempts were made to orient the sections perpendicular to the mean currents. The largest deviation from the zonal orientation occurs at the southernmost section of the set which runs along 45°S from the South American shelf to the Mid-Atlantic

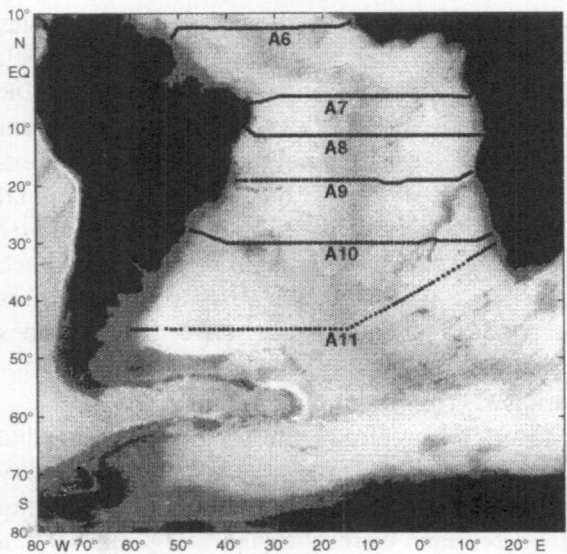

Figure 1: Station positions of the six zonal WOCE sections

Ridge and from there in a northeasterly direction to southern Africa at 30°S.

The description will follow the main water masses from their entry at the northern or southern boundary on their course through the region. The following water masses will be considered:

Antarctic Bottom Water (AABW)
Weddell Sea Deep Water (WSDW)
Lower Circumpolar Deep Water (LCDW)
North Atlantic Deep Water (NADW)
Upper Circumpolar Deep Water (UCDW)
Antarctic Intermediate Water (AAIW)
South Atlantic Central Water (SACW)
North Atlantic Central Water (NACW)

We start with the description of the lowest layer. The Weddell Sea Deep Water (WSDW) which is characterized approximately by $\theta < 0.20$°C and $S < 34.70$ (Reid 1989) enters the region of study in the southwest through the Argentine Basin. It is found between the bottom and 3800 - 4300 m (Figure 7). After complex recirculation within the Argentine Basin (Saunders and King 1995a), WSDW propagates northward into the Brazil Basin after passing through the Vema Channel at 30°S 39°W (see Figures 3 and 6 to appreciate its northward penetration). There is also a small amount of WSDW flowing northward into the Brazil Basin through the Hunter Channel at 35°S 27°W (see Speer et al. 1992, Speer and Zenk 1993), but no trace of it is found in the 30°S section where near-bottom temperatures northeast of the Rio Grande Rise are ~ 0.6°C (Figure 6). This suggests strong mixing north of the Hunter Channel sill resulting in an increase in the near-bottom temperatures. The WSDW properties gradually change through the Brazil Basin from south to north (Mantyla and Reid 1983), and the water mass has essentially disappeared at 4.5°S where the lowest near-bottom temperature found in the western basin is ~ 0.2°C.

On top of the WSDW we find the Lower Circumpolar Deep Water (LCDW) which also has its origin south of the region of study. It is characterized by approximately $\theta < 2.00$°C and $S < 34.86$ (Reid 1989). The WSDW and LCDW are often lumped together into the Antarctic Bottom Water (AABW). This combined water mass can be rec-

ognized in all the sections in Figures 2 to 7 and by its silicate maximum (Figures 8 to 11) spreading northward in the western South Atlantic, crossing the Rio Grande Rise through the Vema Channel (Hogg et al. 1982) and partly through the Hunter Channel (Speer et al. 1992). In the Brazil Basin, the westward shoaling of the near-bottom isotherms and isohalines at the approach to the South American continental slope characterizes the geostrophic western boundary current that transports AABW northward. The circulation pattern in the deep Brazil Basin has been described by Speer and Zenk (1993) and Durrieu de Madron and Weatherly (1994). It includes the western boundary current but also recirculation cells that render the interpretation of the tracer fields more difficult. A check on the somewhat different circulation patterns in these two analyses will be possible once all the zonal and meridional WOCE Brazil Basin sections are available.

North of the Brazil Basin, near-bottom tracer fields (Mantyla and Reid 1983) indicate a bifurcation of the AABW. Part of the AABW does not occur as a western boundary current in the northern hemisphere, but this water mass crosses the equator to the west in a zonally oriented channel opening to the Ceara Abyssal Plain (McCartney and Curry 1994), and another part flows eastward in the northern Brazil Basin, crosses the Mid-Atlantic Ridge through the Romanche and Chain Fracture Zones (Mercier et al. 1994) and feeds the Sierra Leone and Guinea Basins. As pointed out by Warren (1981), the northward flow of AABW does not occur in a western boundary current, rather this water mass is found at the western flank of the Mid-Atlantic Ridge (Figure 2). During the transit from 4.5°S to 7.5°N in the western equatorial Atlantic, the AABW experiences considerable property changes, with the lowest near-bottom temperatures increasing from ~ 0.2°C at 4.5°S to ~ 1.1°C at 7.5°N. This is due to the blocking by bottom topography and mixing. AABW property changes are even larger during the transit through the equatorial fracture zones. This is mainly caused by strong vertical mixing occurring after the water crosses the fracture zone sills (Mercier et al. 1994; Polzin et al. 1996). The coldest near-bottom water found in the eastern equatorial basin at

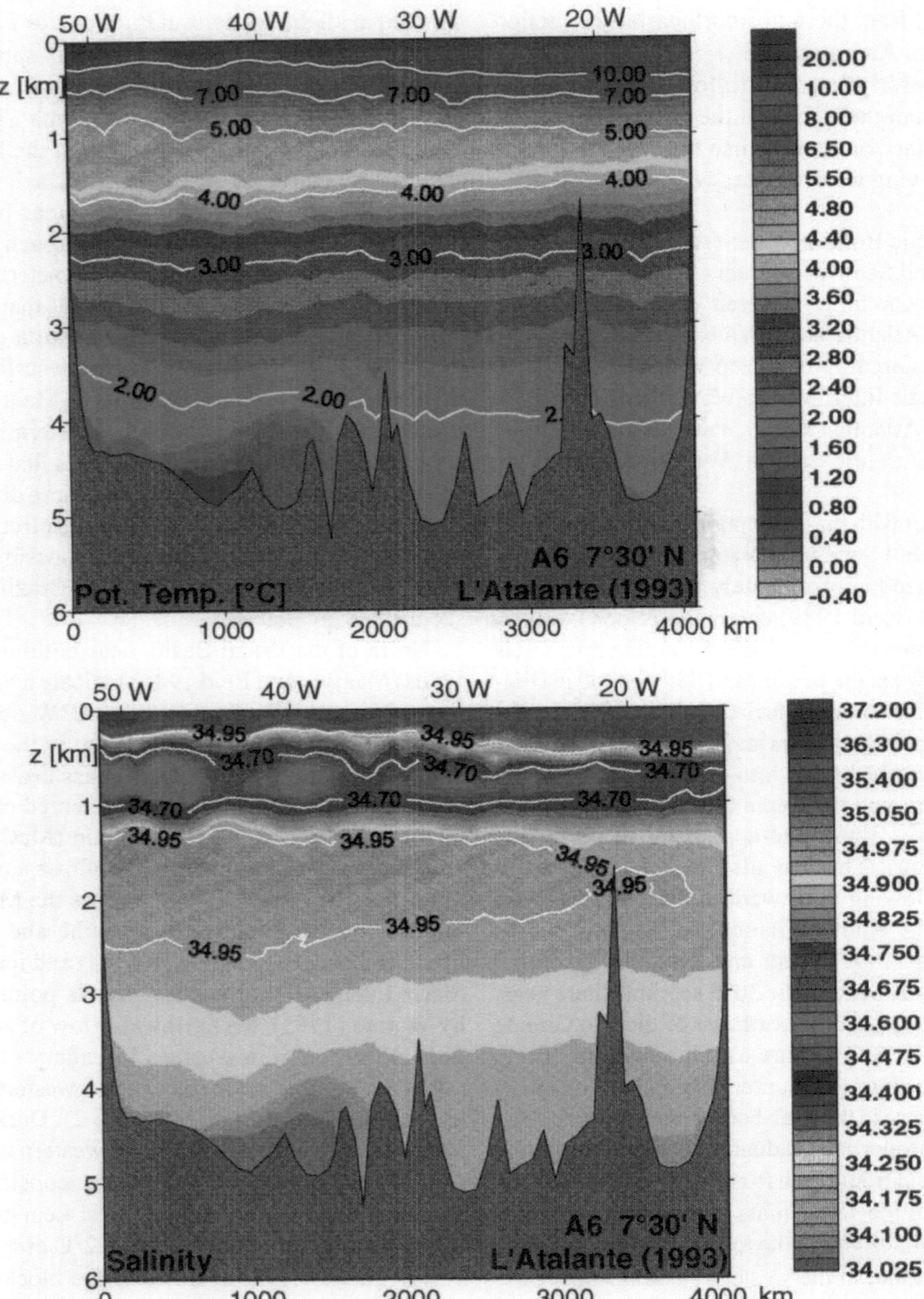

Figure 2 - 7: Zonal sections of potential temperature (referenced to the surface) and salinity from CTD- measurements at the indicated nominal latitudes. Positions are given in Figure 1.

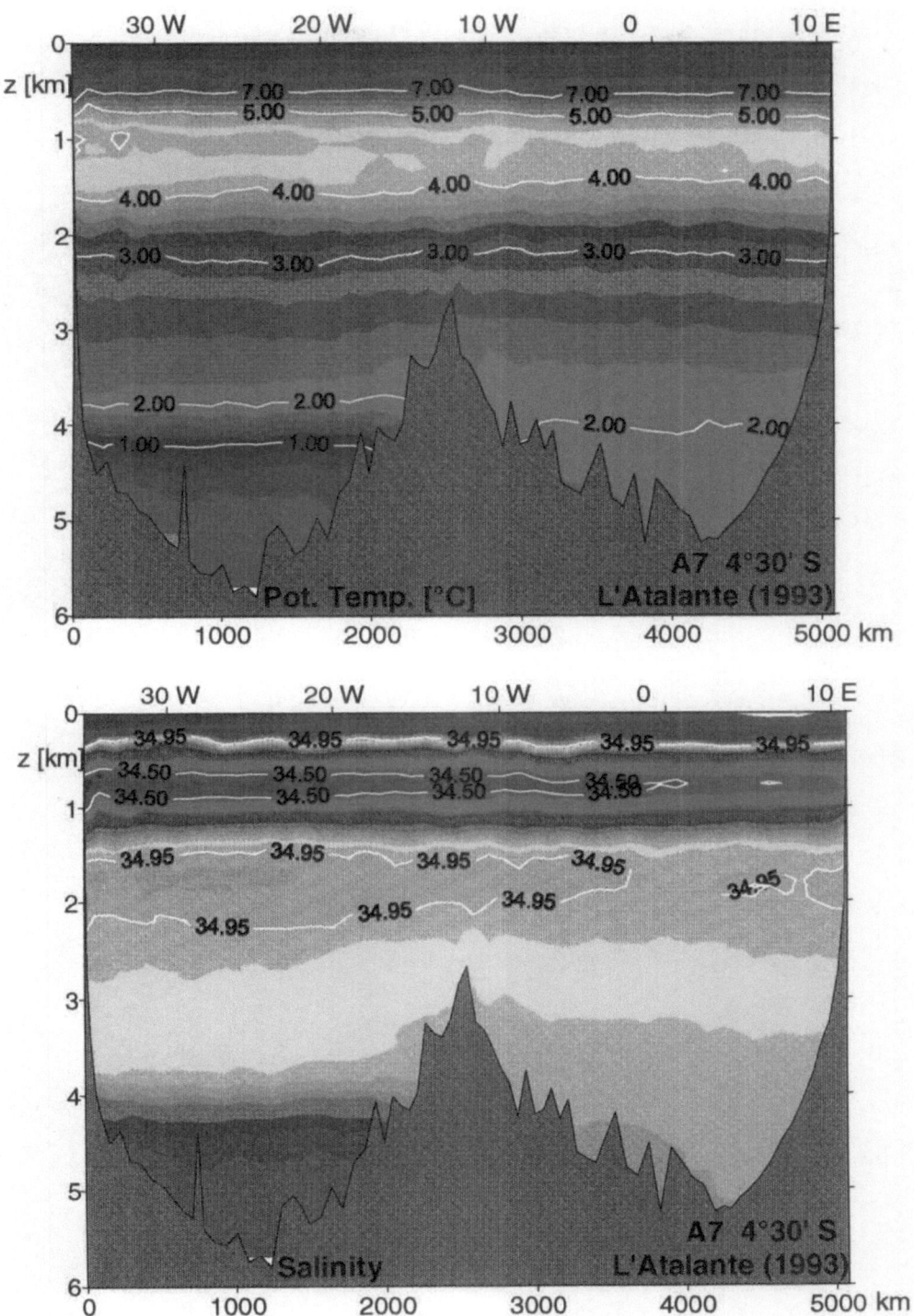

Figure 3

G. Siedler et al.

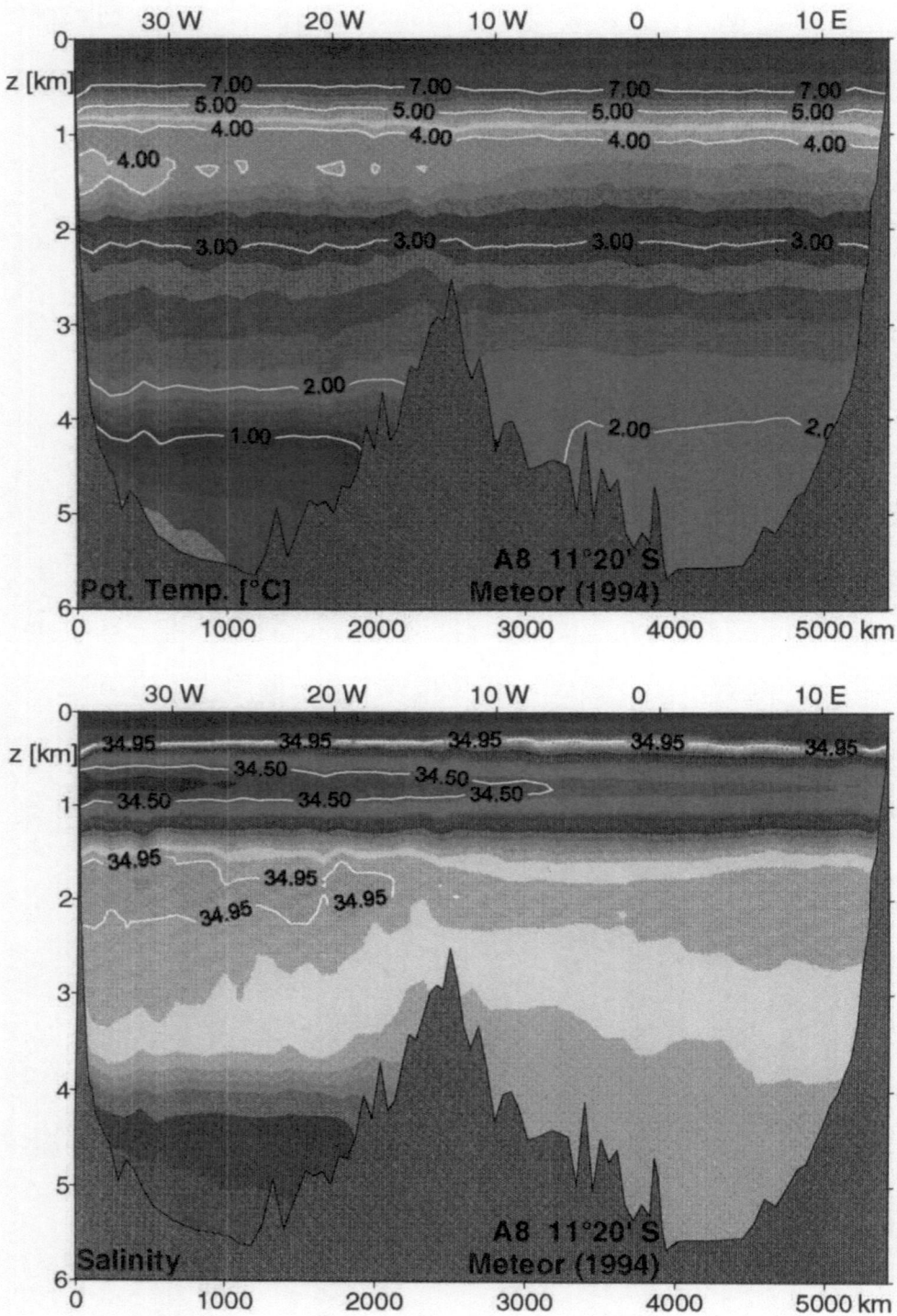

Figure 4

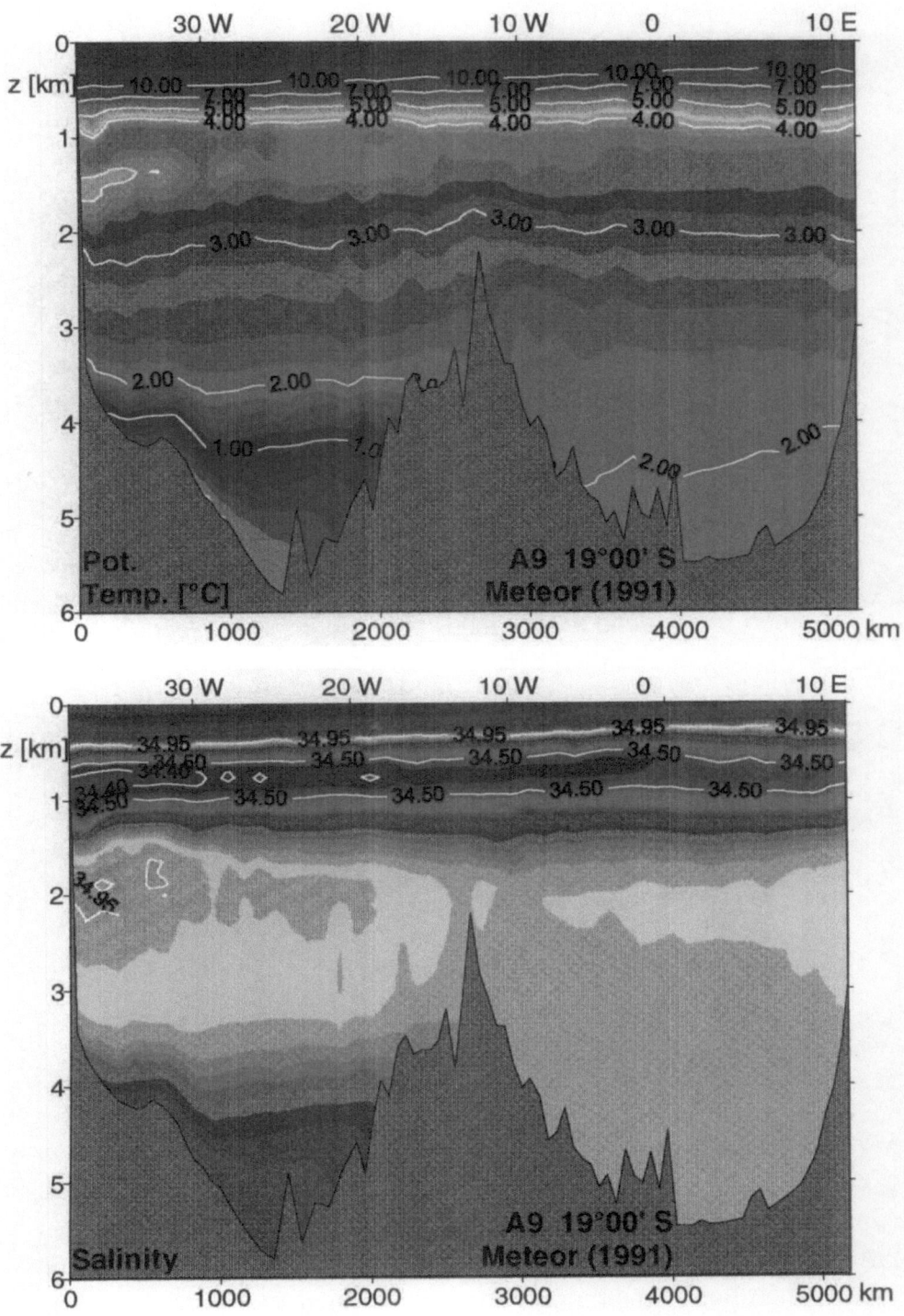

Figure 5

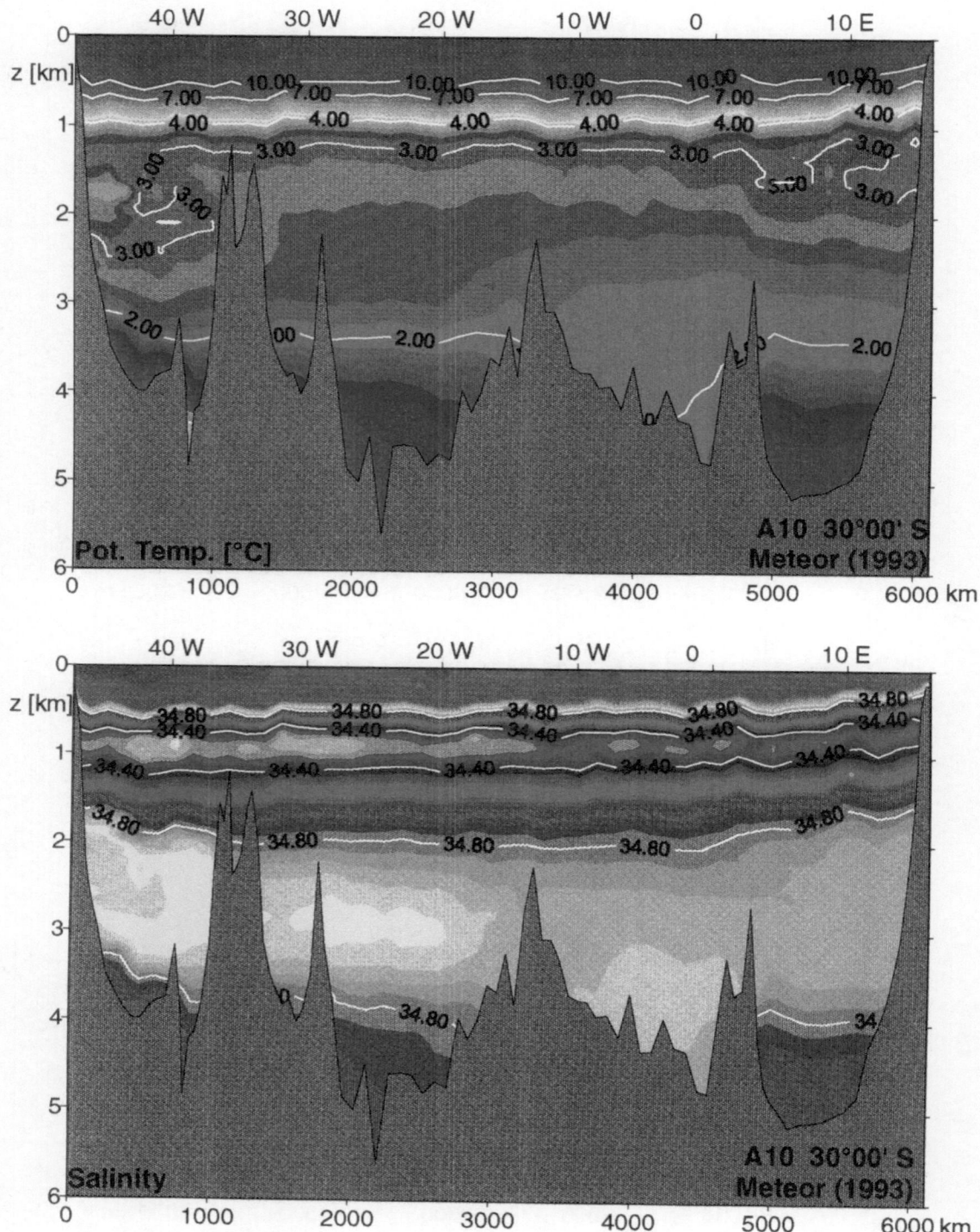

Figure 6

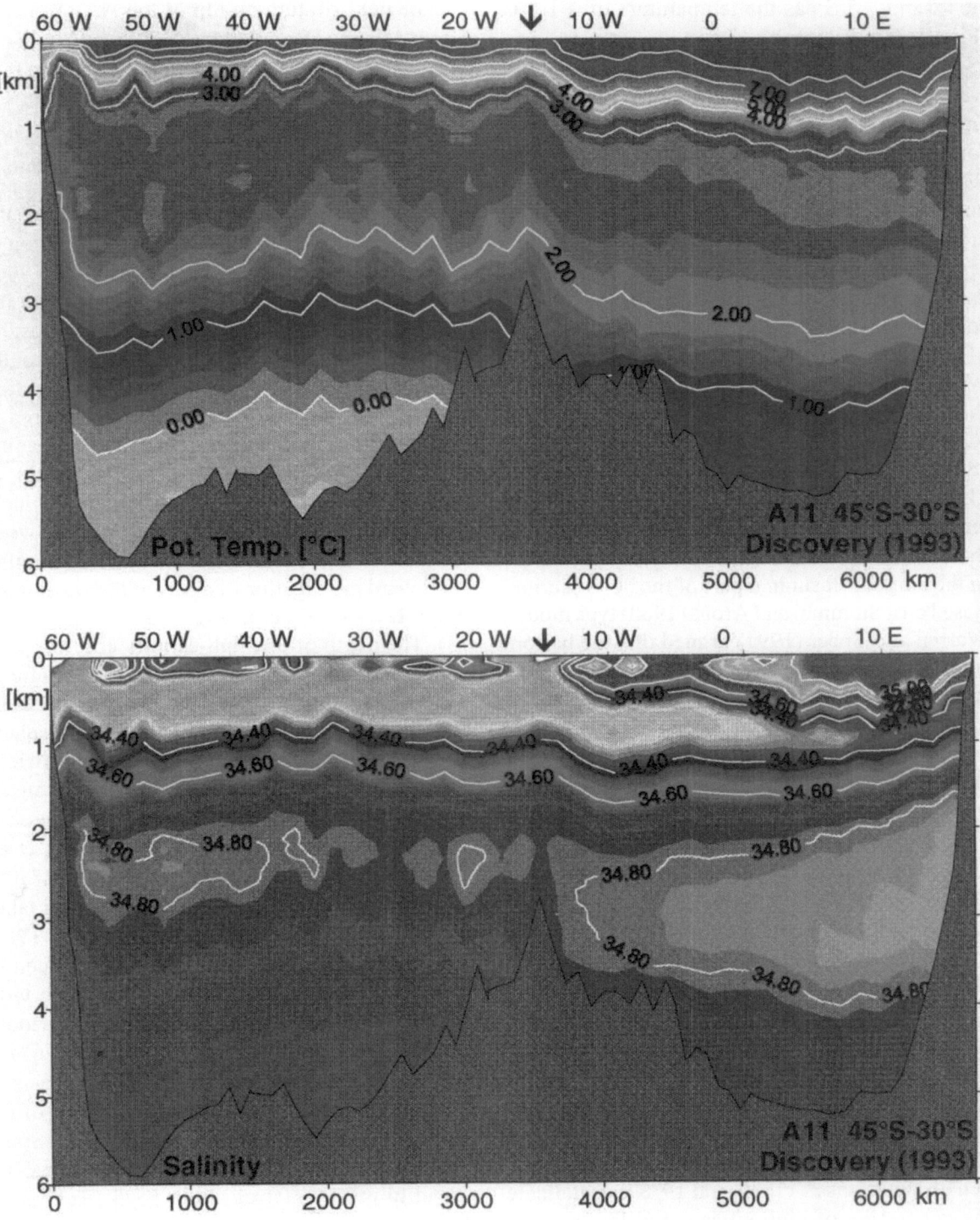

Figure 7

7.5°N and 4.5 S has the temperature of ~ 1.7°C (Figures 2 and 3).

Warren and Speer (1991) described the bottom water circulation in the Angola Basin in the east and argued that the main source of bottom water for this basin is at the equator through the Romanche Fracture Zone. This spreading from the north fits well with the observed decrease in the cross-sectional area of water less than 2.0°C observed from 11°20'S to 19°S (Figures 4 and 5). The AABW is also entering the Cape Basin from the south (right side of Figure 7), and a small amount of its warmest component passes through the Walvis Ridge into the southern Angola Basin through the Walvis Passage near 37°S 7°W (Connary and Ewing 1972) and the Walvis Kom (Shannon and Chapman 1991) near 32°S 0°E. Shannon and Chapman's data show a near-bottom temperature of 1.22°C at 32°S, 3°W, much colder than the ~ 1.8°C near-bottom temperature found in the Angola Basin in a part of the 30°S section. Based on a Stommel and Arons (1960) type model, Warren and Speer (1991) argued that the bottom water flow in the southern Angola Basin is to the southwest.

Silicate concentrations are high in the AABW and this property is therefore particularly well suited to identify the water mass. Near 3500 km in Figure 10 we note a maximum of silicate well below 4000 m, resulting from leakage from the Cape Basin into the southern Angola Basin, with similar temperature and salinity patterns in Figure 6. Farther north at 19°S, however, the silicate maximum is centered on the African continental slope at about 4000 m (Figure 9), and similarly at 11°20'S (Figure 8). The configuration corresponds to an oxygen minimum and nutrient maximum pattern at the slope which was related by Van Bennekom and Berger (1984) to the decomposition of suspended material from the Congo River plume which begins at 6°S. The silicate and oxygen distributions at 11°20' and 19°S thus indicate the generation of high silicate and low oxygen values in the AABW below the Congo River outflow area and a southward spreading of the water mass at these latitudes.

The next stratum on top of the AABW is the North Atlantic Deep Water (NADW). We will not discuss the sublayers of this water mass here (see Wüst 1936; Whitworth and Nowlin 1987; Peterson and Whitworth 1989), but rather describe the spatial distribution of the total water mass which is identified by its salinity and oxygen maxima and nitrate minimum. The water mass is seen in the 7° 30'N section (Figure 2) entering the region in the depth range of approximately 1200 m - 4000 m. Although enhanced property anomalies at the Guyana continental slope mark the Deep Western Boundary Current (DWBC) that carries it southward, the water mass signatures extend eastward well beyond the Mid-Atlantic Ridge centered at about 35 W at that latitude. This zonal spreading implies a partial northward recirculation of the NADW (McCartney 1993). Farther south the maintenance of the property maxima at the western side of all sections indicates the continuing southward propagation of a part of the water mass in the DWBC (Figures 3 to 7, 8 to 11 and 12 to 13). The depth of the high-salinity core which is seen at about 1500 m depth at 7°30'N (Figure 2) increases southward to depths in excess of 2000 m at 45°S (Figure 7). In the southern hemisphere a weak second core is apparent along the African continental slope, indicative of a partial transfer of the southward flow from the western to the eastern boundary of the basin (see Friedrichs et al. 1994). The existence of the eastern core at 4°30'S (Figure 3) confirms that part of the transfer takes place in the equatorial region (Weiss et al. 1985; Tsuchiya et al. 1994). The western core becomes patchy at 19°S (Figure 5) in accordance with Zemba's (1990) suggestion that recirculation loops exist in the Brazil Basin. At 30°S (Figure 6) the structure with two cores in the western basin is more likely associated with the separation of the NADW into a southward branch along the South American continental slope which was earlier found to exist as far as about 55°S (Reid et al. 1977), and a deeper branch which is related to a more zonal flow across the Brazil Basin (Zangenberg and Siedler, pers. communication). That branch causes an additional transfer from the western to the eastern core, and the bulk of NADW is finally found at approximately 2000 - 3500 m in the eastern part of the 45°S section (Figures 7 and 11) where the track cuts across the Cape Basin to 30°S (Saunders and King 1995b).

Potential temperature is a less characteristic tracer of NADW than salinity or oxygen, yet both coastally trapped cores stand out on the 30°S potential temperature section (Figure 6). At this latitude the deep waters of the ocean interior are much influenced by the signature of the cold Circumpolar Deep Water, providing a sharp contrast with the warmer NADW on either side.

This cold water signature can be seen attenuated in the northward extension at 19°S, 11°S and 4°30'S (Figures 3 to 5) in the form of an upward bulging of the mid-depth isotherms at longitudes near the Mid-Atlantic Ridge crest. A similar pattern exists in the salinity sections. In their study of the Angola Basin deep and bottom waters which was based on two sections at 11°S and 23°S, Warren and Speer (1991) suggested a Stommel-Arons type circulation with a northward boundary current at the upper slope of the Mid-Atlantic Ridge.

The meridional and zonal changes in the properties of NADW can be well recognized in the temperature-salinity (T/S) diagrams. In Figure 14 we summarize the distributions of potential temperature, referenced to the surface, versus salinity for the whole water column in the western and eastern basins, separated by the crest of the Mid-Atlantic Ridge. There are characteristic differences between the western and eastern basins. The meridional changes in the NADW in the western basins are much larger than in the eastern basins while the opposite occurs in the near-surface waters. The diagram for the lower layers in Figures 15 and 16 with potential temperature referenced to 3000 dbar indicate a multimodal character of NADW properties, particularly at 19°S and 30°S in the western basin which is associated with branching or recirculation patterns. The diagram for the eastern part of the 30°S section actually covers two basins, the southern Angola and the northern Cape Basins. The higher-salinity AABW part of the Angola Basin values corresponds to the

distribution in the western basin as can be expected with a bottom water supply from the Brazil Basin through the Romanche Fracture Zone.

Returning to the section presentation, we note an oxygen minimum and a silicate maximum near 1500 m at the 45°S and 30°S sections (Figures 10 and 11), particularly at the eastern side, which require a southern source. The silicate maximum is associated with an oxygen minimum at 45°S. These extrema indicate the inflow of Upper Circumpolar Deep Water (UCDW) from the south which separates the NADW from the southern-origin Antarctic Intermediate Water (AAIW). A minimum of potential temperature above the western core of the NADW is sometimes also attributed to this water mass (Reid 1989). The oxygen minimum disappears farther north. This is due to the presence of the equatorial oxygen minimum which, although shallower and more pronounced in the east, is of such a high intensity that it dominates all other watermass-associated signals at the upper and intermediate levels. For a discussion of the UCDW we refer to Whitworth and Nowlin (1987), Peterson and Whitworth (1989), and Tsuchiya et al. (1994).

The AAIW above the UCDW is identified by its low salinity core and can best be distinguished from the underlying UCDW by higher oxygen values which are indicative of a more recent contact with the atmosphere. The salinity minimum and oxygen maximum are most intense and closest to the surface in the western part of the 45°S section. This is an indication for the subduction of this water mass in the neighbouring circumpolar region (McCartney 1977). At the other latitudes the low-salinity layer occupies the full width of the ocean in the depth range of approximately 700 - 1100 m. The oxygen signal is reduced to a narrow core at the western boundary of the 19°S and 11°S sections (Figures 8 and 9), being apparently annihilated at the other longitudes by the equatorial oxygen minimum. A northward AAIW transport all along the western South Atlantic was suggested by Wüst (1936), but later investigators (Taft 1963; Buscaglia 1971) doubted this conclusion, and Reid (1989) determined an ocean-wide anticyclonic

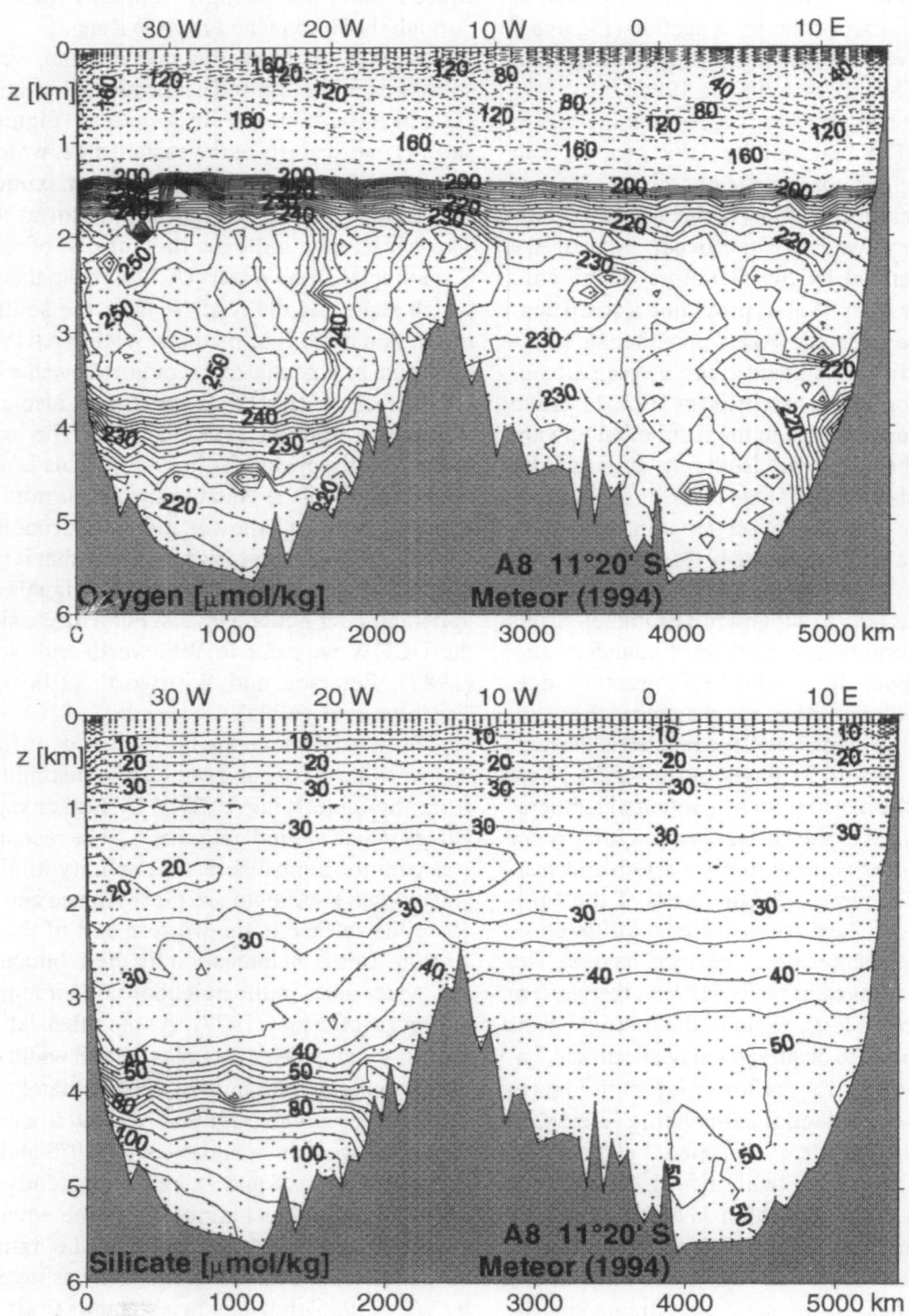

Figures 8 - 11: Zonal sections of oxygen and silicate concentrations at the indicated nominal latitudes. Positions are given in Figure 1. The dots indicate the sampling depths.

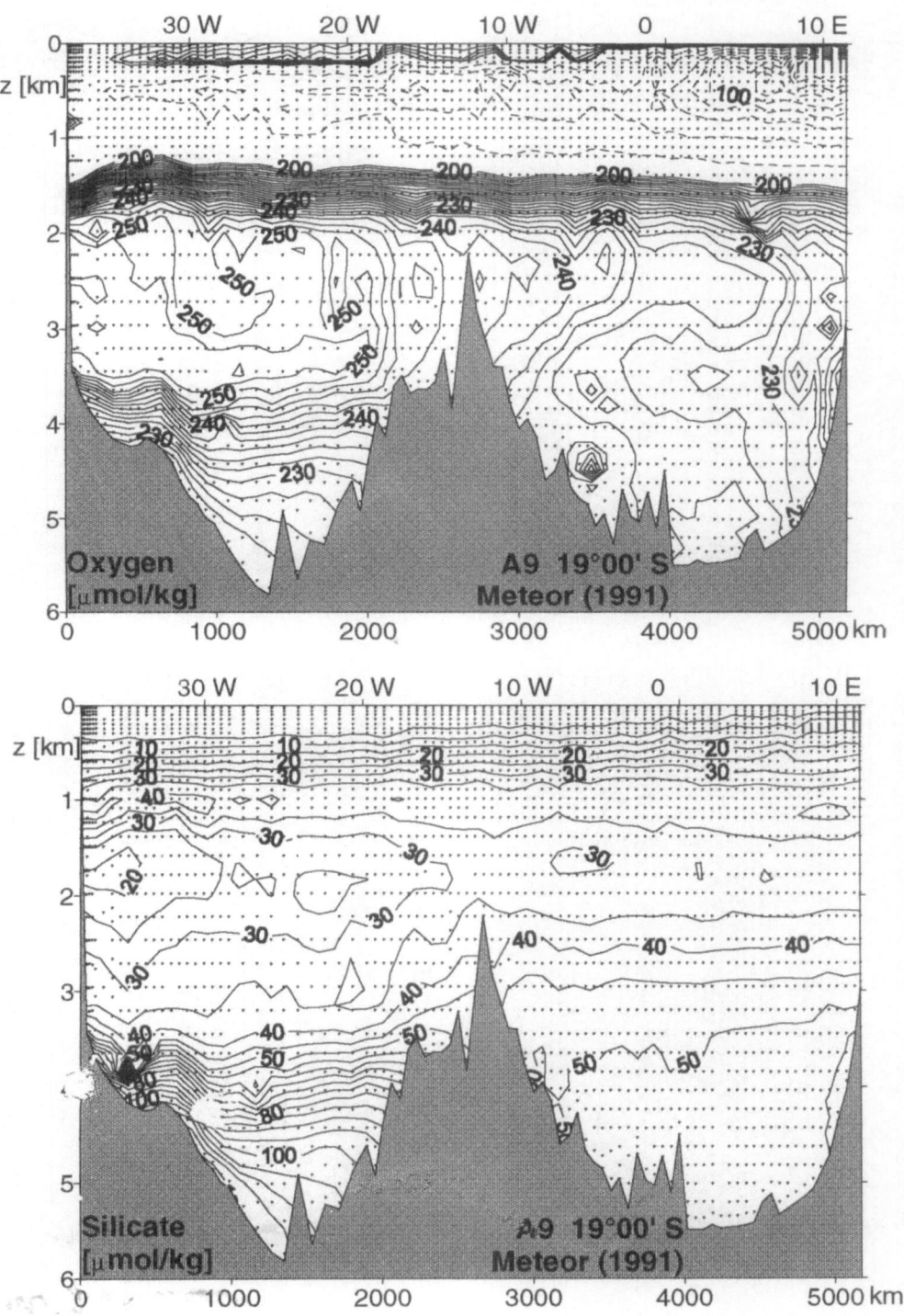

Figure 9

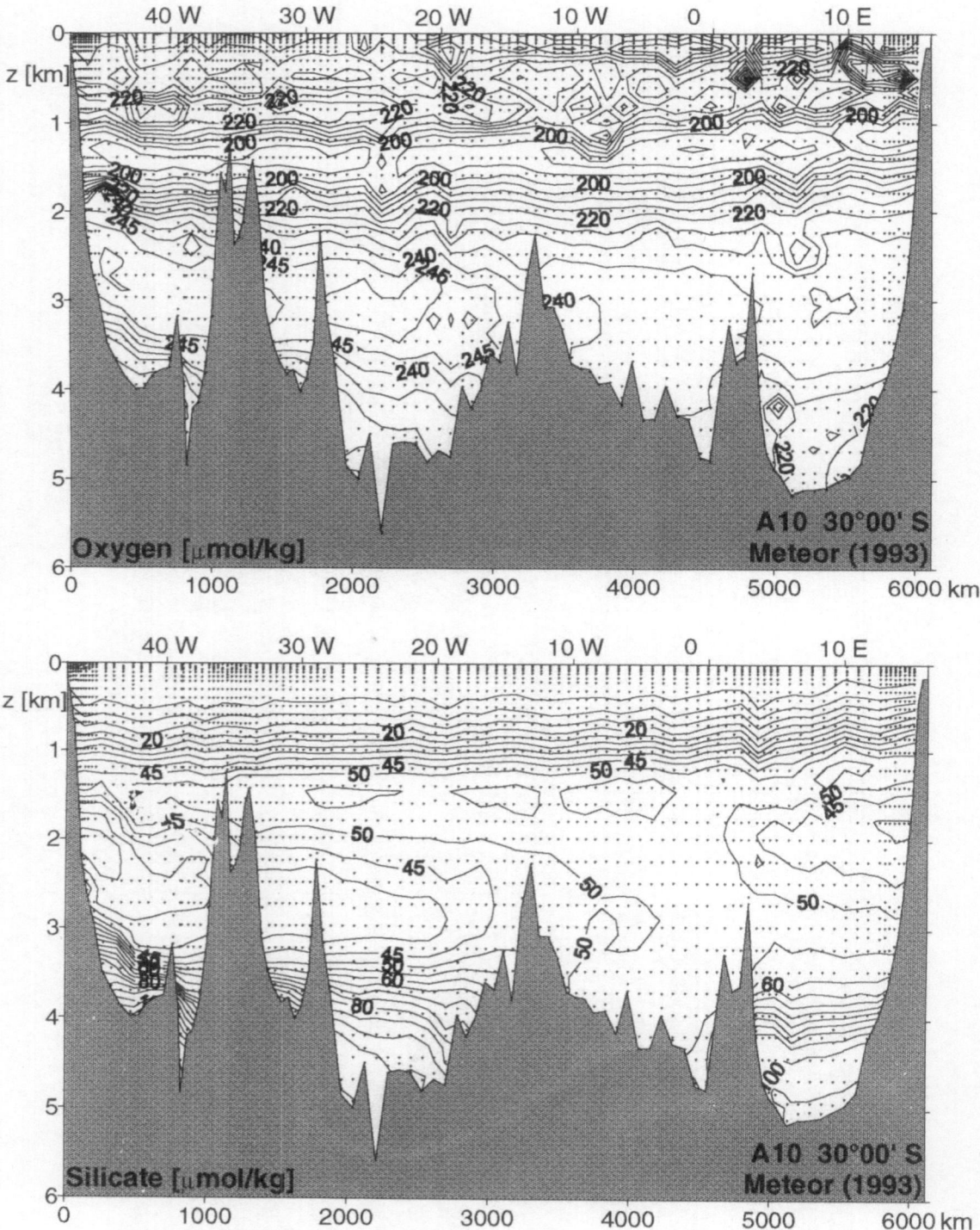

Figure 10

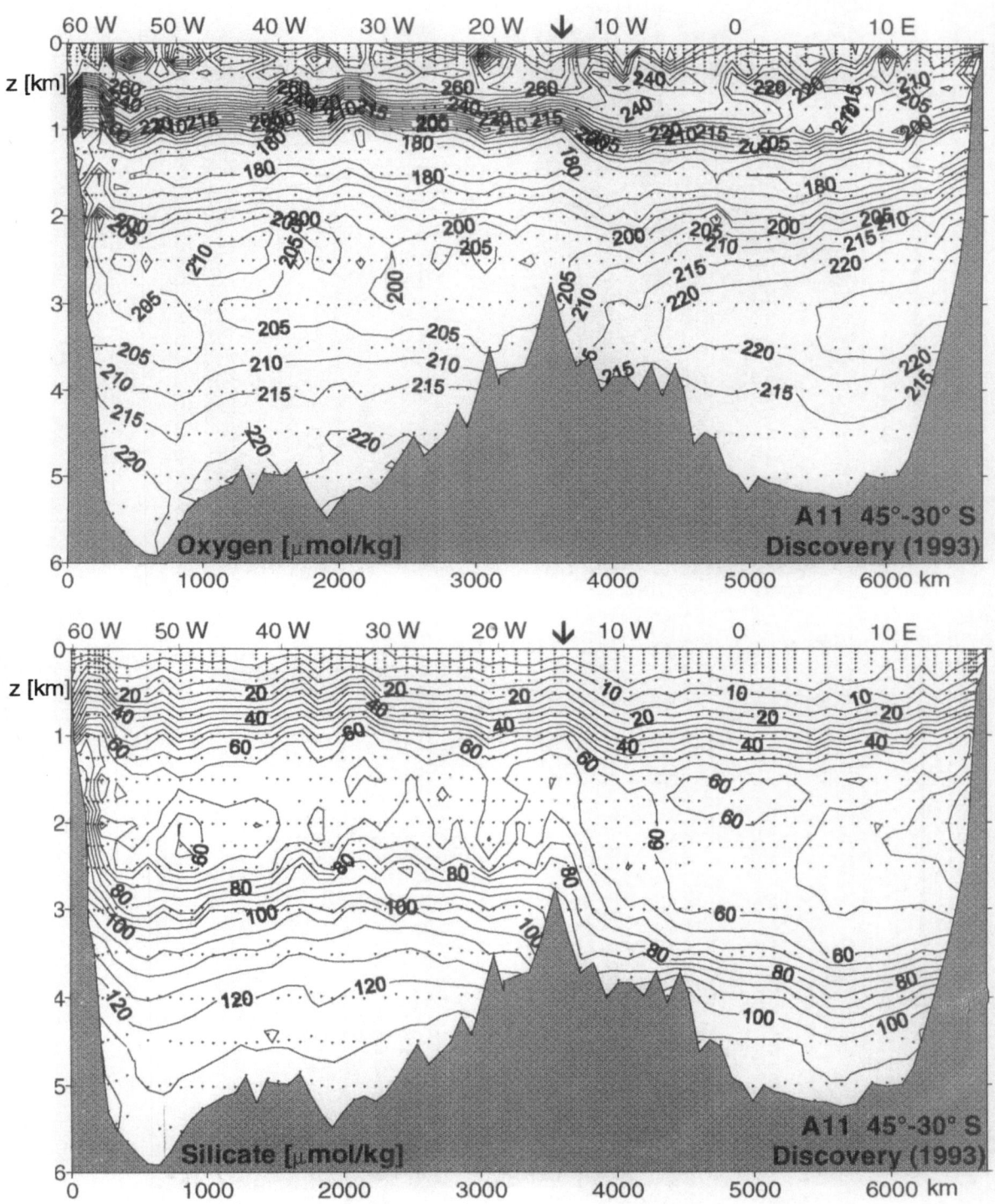

Figure 11

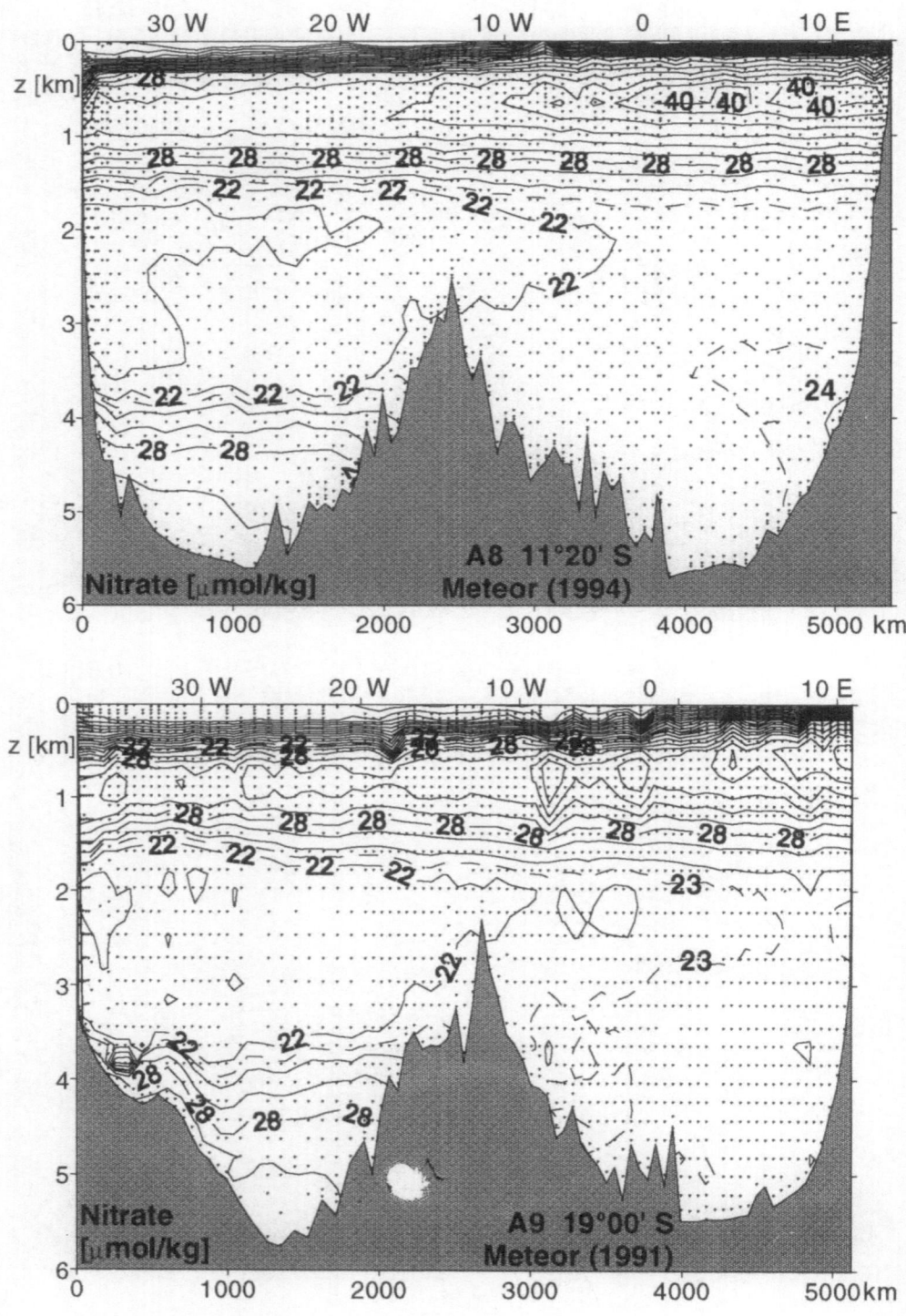

Figures 12 - 13 : Zonal sections of nitrate concentrations at the indicated nominal latitudes. Positions are given in Figure 1. The dots indicate the sampling depths.

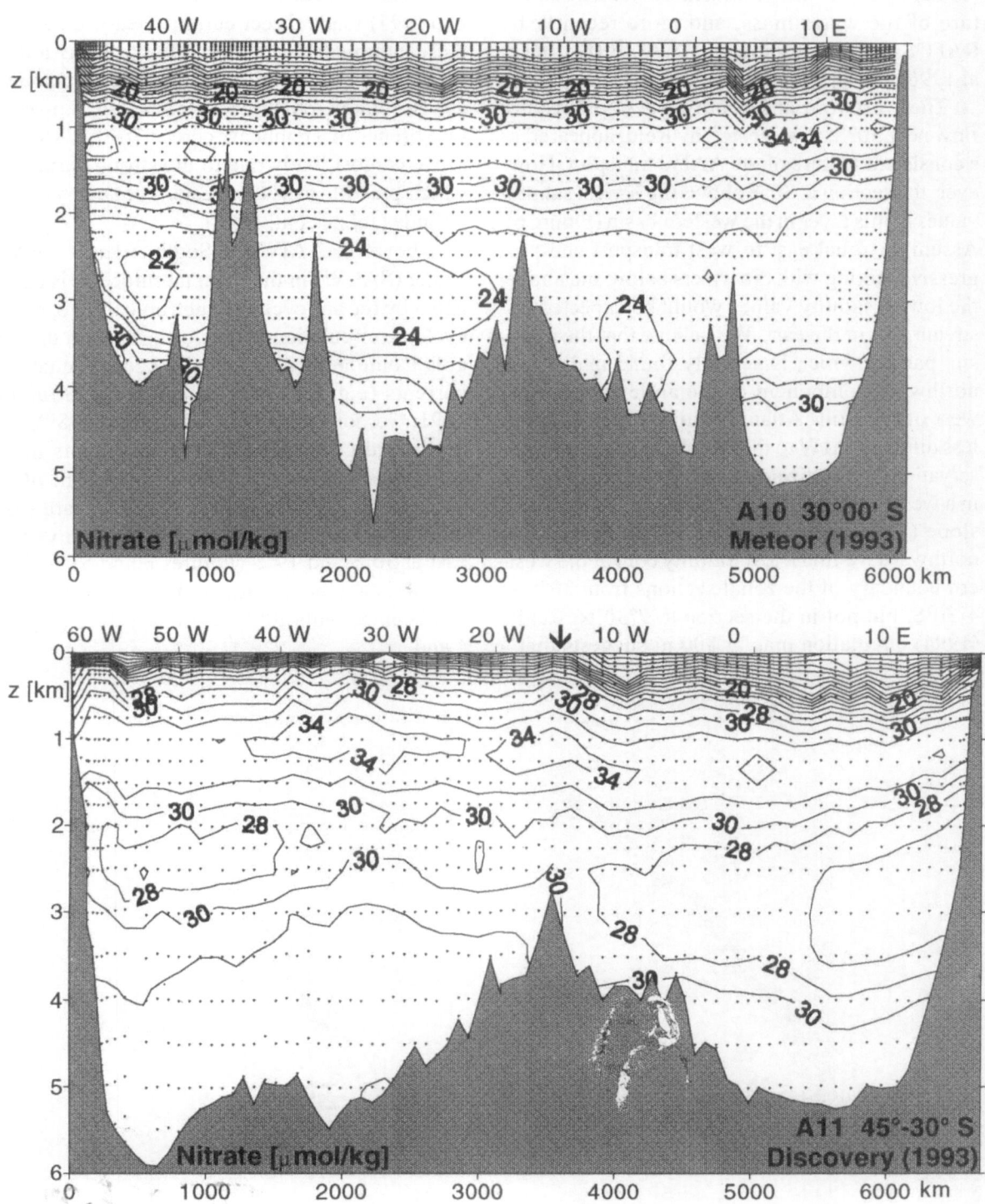

Figure 13

loop between about 45°S and 25°S. Reid's circulation pattern was confirmed by Warner and Weiss (1992) who used the chlorofluoromethane signature of the water mass, and more recently by RAFOS float observations in WOCE (Boebel et al 1996).

The westward and southward AAIW return flow near 30°S, as expected by Reid, appeared as a consistent feature of the AAIW transport. However, the section at 30°S shows the lowest salinity values in this layer in the western basin (Figure 6). Assuming zonal east-to-west transport and progressive mixing with the waters below and above, the lowest salinity values would be expected upstream, i.e. in the east. We believe that the opposite pattern which is actually found is due to a northward component of the anticyclonic flow west of the Mid- Atlantic Ridge which supplies less diluted AAIW to the western basin. Some observations of northward flow have also been made in a western boundary current along the Brazilian slope (Evans and Signorini 1985). Proceeding northward we find a low-salinity core at the western boundary of the zonal sections from 30°S to 4°30'S, but not in the section at 7°30'N. Reid's (1994) circulation map at 800 m suggests that a

fraction of the AAIW entering the equatorial zone at 4°30'S should turn eastward near the equator, and Richardson and Schmitz (1993) and Schott et al. (1995) made direct current measurements of these intermediate equatorial flows. Eastward bifurcation of part of the northward western boundary current of AAIW does not explain, however, the absence of a boundary core at 7°30'S. This absence is most likely due to the strong variability which prevails near the western boundary at this latitude (Johns et al. 1990).

Above the AAIW the South Atlantic Central Water (SACW) in the main thermocline is dominated by the anticyclonic subtropical gyre, by the cyclonic circulation in the northeast with the Angola Dome in its centre, and by zonal equatorial currents (e.g. Fu 1981; Peterson and Stramma 1991; Gordon and Bosley 1991; Reid 1989). The northward geostrophic flow components in the subtropical gyre and in the transition zone of the two gyres can be recognized in the upward sloping of the 10°C and 7°C isotherms from west to east at 30°S and 19°S (Figures 6 and 5), and the reverse behaviour of these isotherms due to southward components at 11°20'S and 4°30'S (Figures 4 and 3).

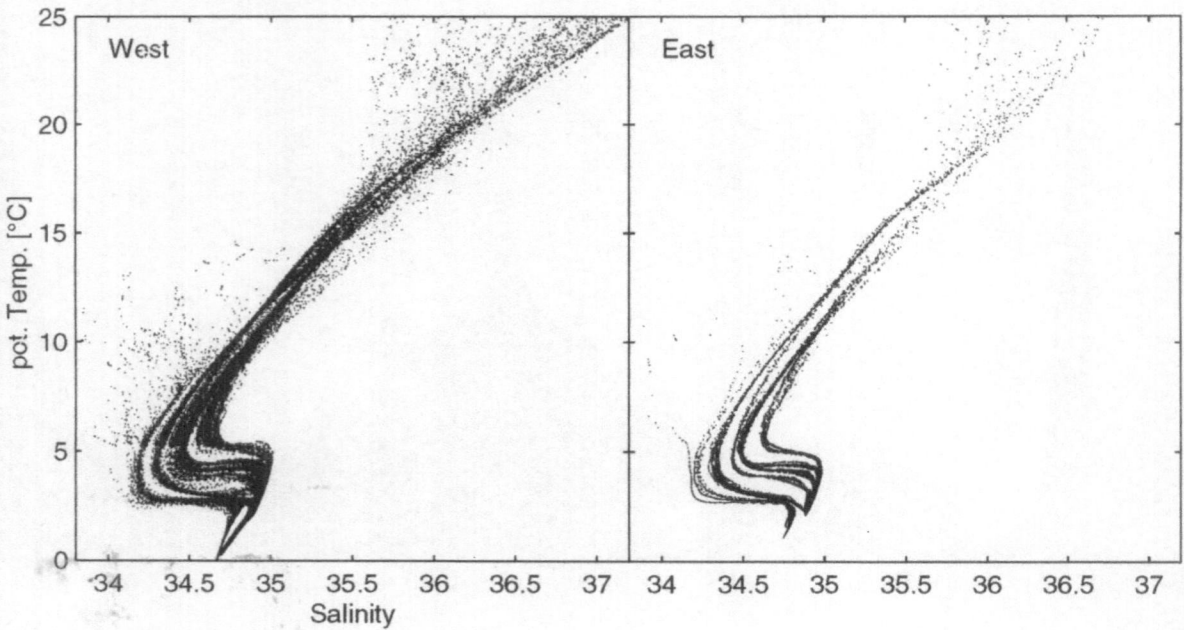

Figure 14 : Potential temperature (referenced to the surface) versus salinity of all six zonal sections.

The θ-S diagrams from all six sections in the western and eastern basins (Figure 14) illustrate the increase in salinity of the SACW from south to north, with salinity values at 10°C at 7°30'S being higher by 0.2 and 0.3 psu in the western and eastern basins, respectively, than at 45°S. This change is associated with the same tendency of the underlying AAIW. The section at 7°30'N is strongly influenced by NACW (Reverdin et al 1993) and has a distribution which is more scattered than the other ones. An analysis of the 100 m values shows a striking west-east asymmetry of the highest temperature and salinity values attained in each basin, with the extreme eastern values of about 19°C and 36 psu being significantly lower than their western counterpart of 24°C and 37 psu.

The latter values match the highest 100 m values of the tropical high salinity water observed at 25 W at latitudes between 13 S and 25 S by Tsuchiya et al. (1994).

While the basic patterns of the water masses in the South Atlantic are known today, further analysis is still required to elucidate their detailed pathways. In this connection tracers such as CFCs, tritium-helium and radiocarbon which have not been utilised here can be expected to play an important part. In addition the strength of the meridional and zonal transports will need to be estimated before the lifetime of the water masses we have described can be determined. Post WOCE we anticipate a wealth of new information about the circulation and physical and chemical properties of the South Atlantic.

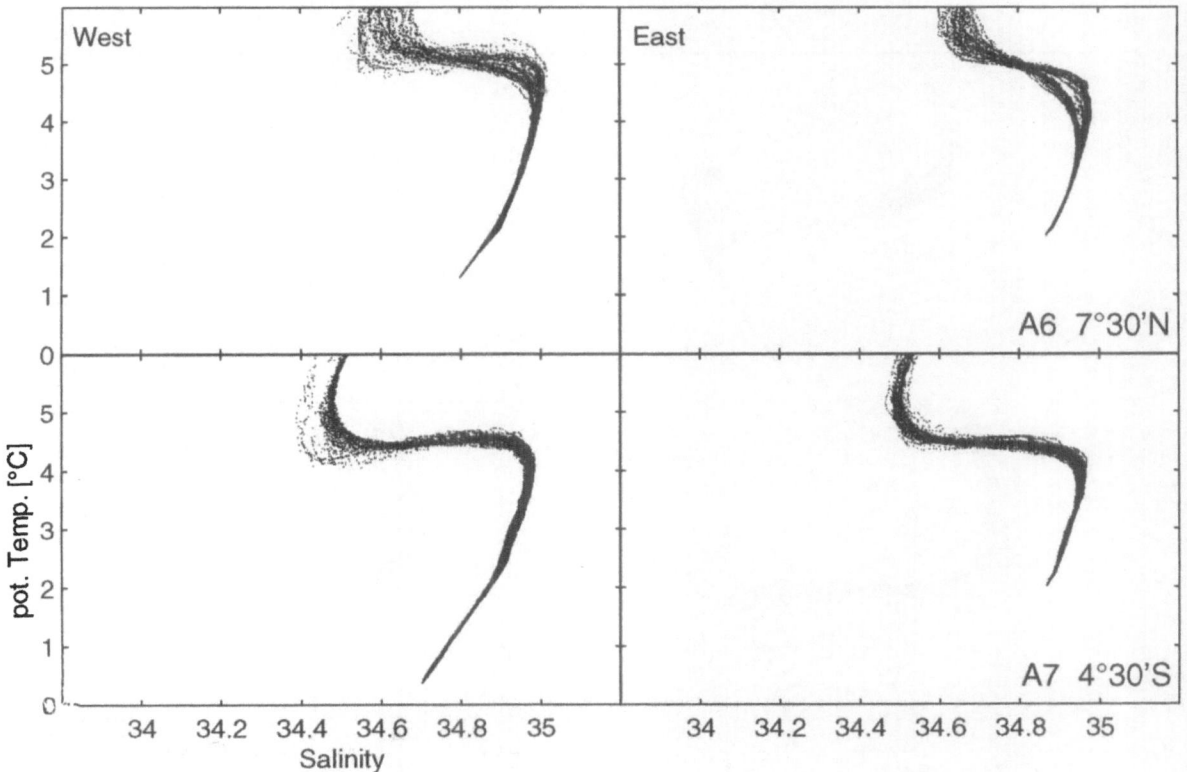

Figures 15 - 16: Potential temperature (referenced to 3000 dbar) versus salinity in the deep layers at the indicated nominal latitudes.

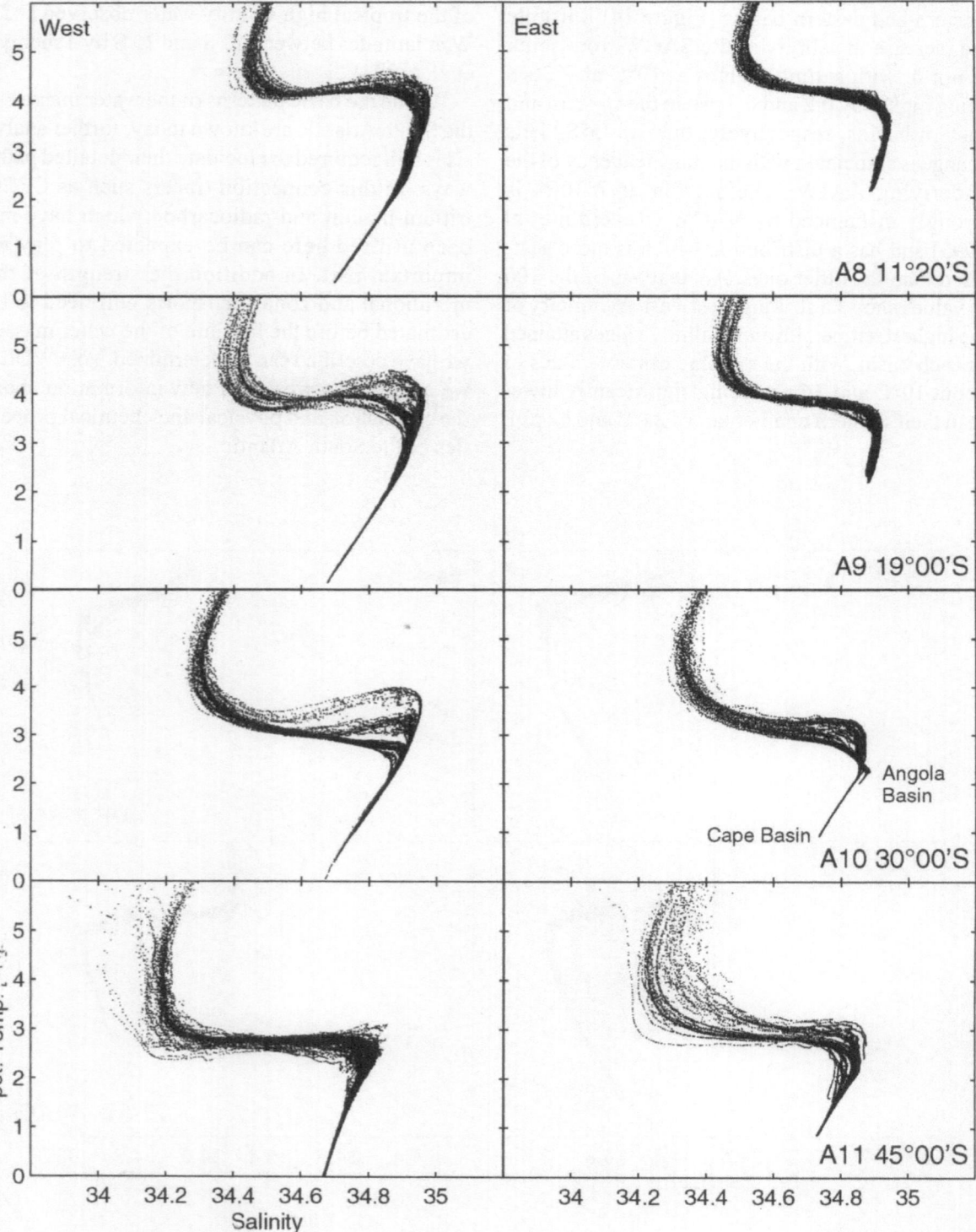

Figure 16

Acknowledgements

Data for the present analysis were assembled during a stay of the first author at the Laboratoire de Physique des Océans (L.P.O.) at IFREMER, Brest, France from May to July 1994. The assistance of the staff of that laboratory was much appreciated. Most of the figures for the analysis were prepared by Catherine Lagadec at L.P.O. Brest, and the final set of figures was plotted by Peter Beining and Wolfgang Erasmi at the Institut für Meereskunde in Kiel. Their most valuable help is thankfully acknowledged. The observational programmes were funded by the Ministry for Science and Technology (BMBF) and by the German Research Foundation (DFG) in Bonn, by the Institut Français de Recherche pour l'Exploitation de la Mer (IFREMER), the Centre National de la Recherche Scientifique (CNRS), and the Institut Français de Recherche Scientifique pour le Developpement en Cooperation (ORSTOM), in the framework of the Programme National d'Etude de la Dynamique du Climat (PNEDC) and by the Natural Environment Research Council (UK). This is a WOCE contribution.

References

Bennekom van AJ, Berger GW (1984) Hydrography and silica budget of the Angola Basin. Neth J Sea Res 17(2/4):149-200

Boebel O, Schmid C, Zenk W (1996) Flow and recirculation of Antarctic Intermediate Water across the Rio Grande Rise. J Geophys Res (in press)

Broecker WS (1991) The great ocean conveyor. Oceanogr 4(2):79-89

Buscaglia JL (1971) On the circulation of the Intermediate Water in the southwestern Atlantic Ocean. J Mar Res 29:245-255

Connary SD, Ewing M (1972) The nepheloid layer and bottom circulation in the Guinea and Angola Basins. In: Gordon AL (ed) Studies in Physical oceanography: a tribute to George Wüst on his 80th birthday. Gordon and Breach, New York, 2, pp169-184

Defant A (1936) Die Troposphäre. Deutsche Atlantische Expedition „Meteor" 1925-1927. Wiss Erg 6(1):289-411

Durrieu de Madron X, Weatherly G (1994) Circulation, transport and bottom boundary layers of deep currents in the Brazil Basin. J Mar Res 52:583-638

Evans DL, Signorini SR (1985) Vertical structure of the Brazil Current. Nature 315:48-50

Friedrichs MAM, McCartney MS and Hall MM (1994) Hemispheric asymmetry of deep water transport modes in the western Atlantic. J Geophys Res 99 (C12):25,165-25,179

Fu L (1981) The general circulation and meridional heat transport of the subtropical South Atlantic determined by inverse methods. J Phys Oceanogr 11: 1171-1193

Fuglister FC (1960) Atlantic Ocean atlas of temperature and salinity profiles and data from the International Geophysical Year of 1957-58. Woods Hole Oceanographic Institution Atlas Series, 1, 209 pp

Gordon A (1986) Interocean exchange of thermocline water. J Geophys Res 91:5037-5046

Gordon AL, Bosley KT (1991) Cyclonic gyre in the tropical South Atlantic. Deep Sea Res 38 (Suppl. 1): pp323-343

Groupe CITHER-1 (Le) (1994) Campagne CITHER-1, Recueil de donnes, Volume 2: CTD-02. Rapport Interne 94-04, Laboratoire de Physique des Océans, Brest

Hogg NG, Biscaye P, Gardner W, Schmitz Jr WJ (1982) On the transport and modification of Antarctic Bottom Water in the Vema Channel. J. Mar. Res. 40 (Suppl.): 231-263

Johns WE, Lee TN, Schott FA, Zantopp RJ, Evans RH (1990): The North Brazil Current retroflection seasonal structure and eddy variability. J Geophys Res 95(C12):22103-22120

Mantyla AW, Reid JL (1983) Abyssal characteristics of the world ocean waters. Deep Sea Res 30:805-833

McCartney MS (1977) Subantarctic Mode Water. pp 103-119 in Angel M (ed) A voyage of DISCOVERY, Geoge Deacon Anniversary Volume, Pergamon Press, 696 pp

McCartney MS (1993) Crossing of the equator by the Deep Western Boundary Current in the Western Atlantic Ocean. J Phys Oceanogr 23(9):1953-1974

McCartney MS, Curry RA (1993) Transequatorial flow of Antarctic Bottom Water in the western Atlantic Ocean: Abyssal geostrophy at the equator. J Phys Oceanogr 23(6):1264-1276

Mercier H, Speer KG, Honnorez J (1994) Tracing the Antarctic Bottom Water through the Romanche and Chain Fracture Zones. Deep Sea Res 41:1457-1477

Peterson RG and Stramma L (1991) Upper level circulation in the South Atlantic Ocean. Prog Oceanogr 26:1-73

Peterson RG, Whitworth III T (1989) The subantarctic and polar fronts in relation to deep water masses through the southwestern Atlantic. J Geophys Res 94(C8):10817-10838

Polzin KL, Speer KG, Toole JM, RW Schmitt (1996) Intense mixing of Antarctic Bottom Water in the equatorial Atlantic. Nature 380:54-57

Reid JL (1989) On the total geostrophic circulation of the South Atlantic Ocean: flow patterns, tracers, and transports. Prog Oceanogr 23:149-244

Reid, J.L. (1994) On the total geostrophic circulation of the North Atlantic Ocean: Flow patterns, tracers and transports. Prog Oceanogr 33:1-92

Reid JL, Nowlin WD, Patzert WC (1977) On the characteristics and circulation in the southwestern Atlantic Ocean. J Phys Oceanogr 7:62-91

Reverdin G, Weiss RF, Jenkins WJ (1993) Ventilation of the Atlantic Ocean Equatorial Thermocline, J. Geophys. Res. 98 (C9): 16289-16310

Richardson PL, Schmitz WJ (1993) Deep cross-equatorial flows in the Atlantic measured with SOFAR floats. J Geophys Res 98(C5):8373-8387

Rintoul SR (1991) South Atlantic interbaasin exchange. J Geophys Res 96(C2):2675-2692

Saunders P, King BA (1995) Oceanic fluxes on the WOCE A11 section. J Phys Oceanogr 25(9):1942-1958

Schmitz WJ (1996) On the World Ocean circulation: Volume 1: Some global features / North Atlantic circulation. Technical Report WHOI-96-03, Woods Hole Oceanographic Institution, Woods Hole, Mass.

Schott FA, Stramma L, Fischer J (1995) The warm water inflow into the western tropical Atlantic boundary regime, spring 1994. J Geophys Res 100 (C12):24745-24760

Shannon LV, Chapman P (1991) Evidence of Antarctic Bottom Water in the Angola Basin at 32° S. Deep Sea Res 38(Suppl.):1299-1304

Speer KG, Zenk W (1993) The flow of Antarctic Bottom Water into the Brazil Basin. J Phys Oceanogr 23(12):2667-2682

Speer KG, Zenk W, Siedler G, Pätzold J, Heidland C (1992) First resolution of bottom water flow through the Hunter Channel in the South Atlantic. Earth and Planet. Sci Lett 113:287-292

Stommel H (1957) A survey of ocean currents theory. Deep Sea Res 4:149-184

Stommel H, Arons AB (1960) On the abyssal circulation of the world ocean. II. An idealized model of the circulation pattern and amplitude in ocean basins. Deep Sea Res 6:217-233

Taft BA (1963) Distribution of salinity and dissolved oxygen on surfaces of uniform potential specific volume in the South Atlantic, South Pacific, and Indian Ocean. J Mar Res 21(2):129-146

Tsuchiya M, Talley LD, McCartney MS (1994) Water mass distribution in the western South Atlantic; a section from South Georgia Island (54°S) northward across the equator. J Mar Res 52:55-81

Warner MJ, Weiss RF (1992) Chlorofluoromethanes in the South Atlantic Antarctic Intermediate Water. Deep Sea Res 39:2053-2075

Warren BA (1981) Deep circulation of the World Ocean. In: Warren BA and Wunsch C (eds) Evolution of Phys Oceanogr: 6-41

Warren BA, Speer K (1991) Deep circulation in the eastern South Atlantic Ocean. Deep Sea Res 38 (Suppl.): pp 281-322

Weiss RF, Bullister JL, Gammon RH, Warner MJ (1985) Atmospheric chlorofluorocarbons in deep equatorial Atlantic. Nature 314:604-610

Whitworth III T, Nowlin Jr WD (1987) Water masses and currents of the Southern Ocean at the Greenwich meridian. J Geophys Res 92 (C6):6462-6476

Wüst G (1936) Schichtung und Zirkulation des Atlantischen Ozeans. Die Stratosphäre des Atlantischen Ozeans. Deutsche Atlantische Expedition auf dem Forschungs- und Vermessungsschiff METEOR 1925-1927. Band VI, 1. Teil. Verlag von Walter de Gruyter & Co., Berlin und Leipzig, pp 110-288 (Also available in English edition, edited by W.J. Emery (1978) National Science Foundation, Amerinc Publishing Co. Pvt. Ltd., New Delhi 1981)

Zemba JC (1990) The Brazil Current at 31°S: Another look at its velocity structure and transport. EOS 71(17):544

South Atlantic Heat Transport at 11°S

K.G. Speer[1], J. Holfort[2], T. Reynaud[1] and G. Siedler[2]

[1]*Laboratoire de Physique des Océans*
IFREMER, B.P. 70, 29280 Plouzané, FRANCE
[2]*Institut für Meereskunde*
Düsternbrooker Weg 20, 24105 Kiel, GERMANY

Abstract: Hydrographic data along 11°S occupied in 1983 by the R.V. OCEANUS are used together with various wind climatologies to estimate the annual average transport of heat at this latitude. Some motivation for expecting fairly well-defined estimates at this latitude compared to others comes from the absence of a strong western boundary current. Results include flow in four layers representing the thermocline, Antarctic Intermediate Water, North Atlantic Deep Water, and Antarctic Bottom Water, using zero velocity reference level choices based on property distributions. The annual average heat transport is estimated to be $0.6 \pm 0.17 \times 10^{15}$ W. Previous estimates of the transport at 8-16°S range from 0.2 PW to greater than 1 PW. Interannual variability from the wind field alone leads to interannual heat transport variability of about 0.05 PW. Comparisons with other recent studies at 45-30°S and 11°N are made.

Introduction

Bryan (1962) demonstrated that meridional heat transport in the Atlantic Ocean is dominated by the thermohaline circulation, with the wind-driven geostrophic flow contributing less because it is weakly correlated with temperature. On the other hand, the ageostrophic Ekman layer flow is also an important element of heat transport (Roemmich 1983). The basic variability is seasonal (Hsiung et al. 1989), and the particularly strong seasonal signal in the South Atlantic was used by Peterson and Stramma (1991) to explain some of the scatter in the numerous estimates of northward heat flux.

A naive average of the South Atlantic heat transport estimates, reported by Peterson and Stramma (1991) (Fig.1), may be a better representation of the mean transport for it brings them into agreement with recent values based on air-sea heat exchange (Bunker 1988; Isemer and Hasse 1987) within about 0.2 PW (PW = 10^{15} W). Straightforward mass conserving (in deep layers) circulation solutions derived from combining hydrographic sections at 11°S, 19°S, and 23°S were calculated; results fall close

to the average historical heat transport values. However, the simple mass conservation constraint for flow in deep layers is arbitrary and undoubtably incorrect because of vertical motion between layers. This, and dissatisfaction with certain deep circulation elements in our solutions (though it should be noted that heat transport is not always sensitive to the quality of the flow constraints imposed upon inverse model solutions) led us to concentrate on the 11°S section, with the goal of estimating various sources of error and obtaining a plausible circulation with relatively few, simple reference level choices. Such results could be the basis for initial reference level choices in further statistical studies involving more sections.

A potential problem with estimates of heat transport, or the transport of any other quantity, from hydrographic data is the unresolved barotropic flow, especially at the western boundary, where dynamics places the strongest currents. A wide continental shelf in the South Atlantic south of 15°S accentuates the problem (Peterson and Stramma

From WEFER G, BERGER WH, SIEDLER G, WEBB DJ (eds), 1996, *The South Atlantic: Present and Past Circulation*. Springer-Verlag Berlin Heidelberg, pp 105-120

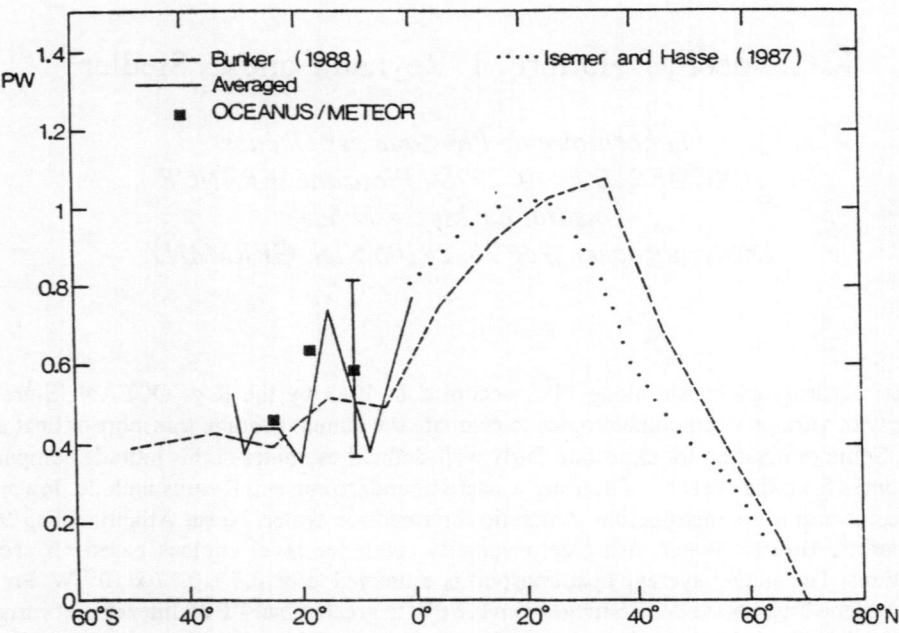

Fig. 1. Estimates of heat flux in the Atlantic Ocean. Includes average of historical values reported by Peterson and Stramma (1991), and annual mean results of an inverse-type calculation using recent <u>Meteor</u> (19°S) and <u>Oceanus</u> (11°S and 23°S) data [error bar applies to results at all three latitudes].

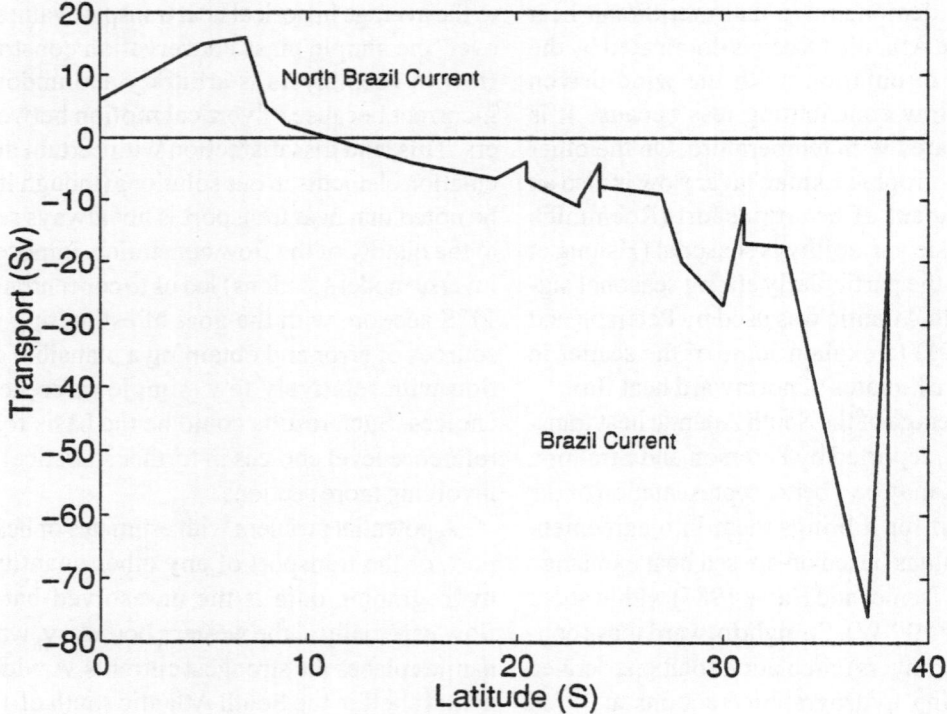

Fig. 2. Transport estimates of the North Brazil Current and the Brazil Current (reported in Peterson and Stramma 1991).

1991). At 11°S the shelf width is minimal; moreover, simple wind-driven circulation calculations of Sverdrup transport suggest that no shallow western boundary current is present near this latitude (Hellerman and Rosenstein 1983). A plot of reported North Brazil Current transport and Brazil Current transport (Fig. 2) supports the idea of a null point and also partially motivates the focus on this latitude.

In this study the CTD data taken by OCEANUS along 11°S in March-April 1983 are analysed; salinity, potential temperature, and oxygen are shown in Fig. 3. Here, the emphasis will be on the volume transport and heat flux across the section as determined from hydrography and climatological wind fields. The contribution of various components of the circulation is considered by dividing the ocean into layers, and by examining longitudinally cumulative transports, but no attempt is made to distinguish mesoscale eddy-driven components of heat or volume transport.

Following, for example, Hall and Bryden (1982) or Friedrichs and Hall (1993), the heat transport across a section of width L and thickness h might be expressed as :

$$H = H_g + H_e. \qquad (1)$$

H_g is the geostrophic heat transport and H_e the Ekman heat transport. These are defined as:

$$H_g = \int_0^L \int_{-h}^0 \rho C_p(\theta - \bar{\theta})v\,dxdz, \qquad (2)$$

and,

$$H_e = \int_0^L \int_{-h}^0 \rho C_p(\theta_e - \bar{\theta})v_e\,dxdz. \qquad (3)$$

ρ, C_p, θ, θ_e, $\bar{\theta}$, v and v_e are respectively the in-situ density, the heat capacity, the potential temperature, the potential temperature in the Ekman layer, the average potential temperature of the section, the geostrophic north-south current component and the north-south component of the Ekman velocity. Rewriting (3):

$$H_e \approx C_p(\bar{\theta}_e - \bar{\theta})\Psi_e \qquad (4)$$

where Ψ_e the Ekman transport across the section, is given by :

$$\Psi_e = -\frac{1}{\rho_o f}\int_0^L \tau_{x0}\,dx \qquad (5)$$

τ_{x0} being the eastward component of the windstress and f the Coriolis parameter. We will discuss later how the choice of the windstress dataset changes the Ekman transport and the associated Ekman heat flux.

Circulation Elements

Western Boundary Current

The location of the zero Sverdrup transport streamfunction at the western boundary near 11°S (Hellerman and Rosenstein 1983) does not necessarily imply zero transport over the Brazilian shelf at this latitude. Direct evidence from ship drift observations does, however, suggest the absence of mean flow near the western boundary at this latitude (Richardson and Walsh 1986; Arnault 1987). Geostrophic flow calculations (Arnault 1987; Stramma et al. 1990; Stramma 1991) all suggest zero shallow transport at the boundary above the shelf near 10°S. Stramma et al. (1990) show schematically 1-2 Sv (1 Sv = 1 x 10^6 m^3 s^{-1}) flowing south near 11°S, using data taken during a season (February-March) of expected weak southward flow based on ship drift. They point out that the shelf is narrow north of 15°S, and express doubt that significant flow is missing from their estimates in this area. Silviera et al.(1994) describe a northward flowing current near the western boundary at 10.5°S, but this flow is farther offshore and is well resolved by our data.

Our best estimate of unresolved annual mean western boundary transport at 11°S is thus zero. An average temperature at depths less than 200 m in the western-most station pair of 22.9°C leads to a maximum heat transport error of 0.08 PW per 10^6 m^{3}s^{-1} of shelf flow.

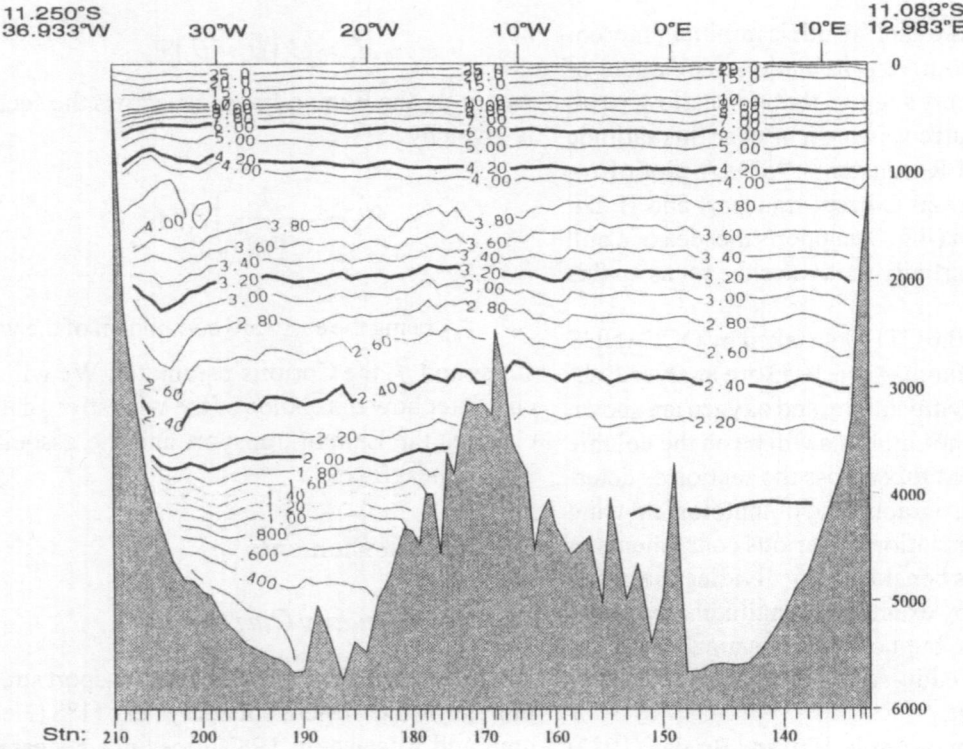

Fig. 3a. Potential temperature at 11°S (°C). Heavy lines mark layer boundaries (see Table 4).

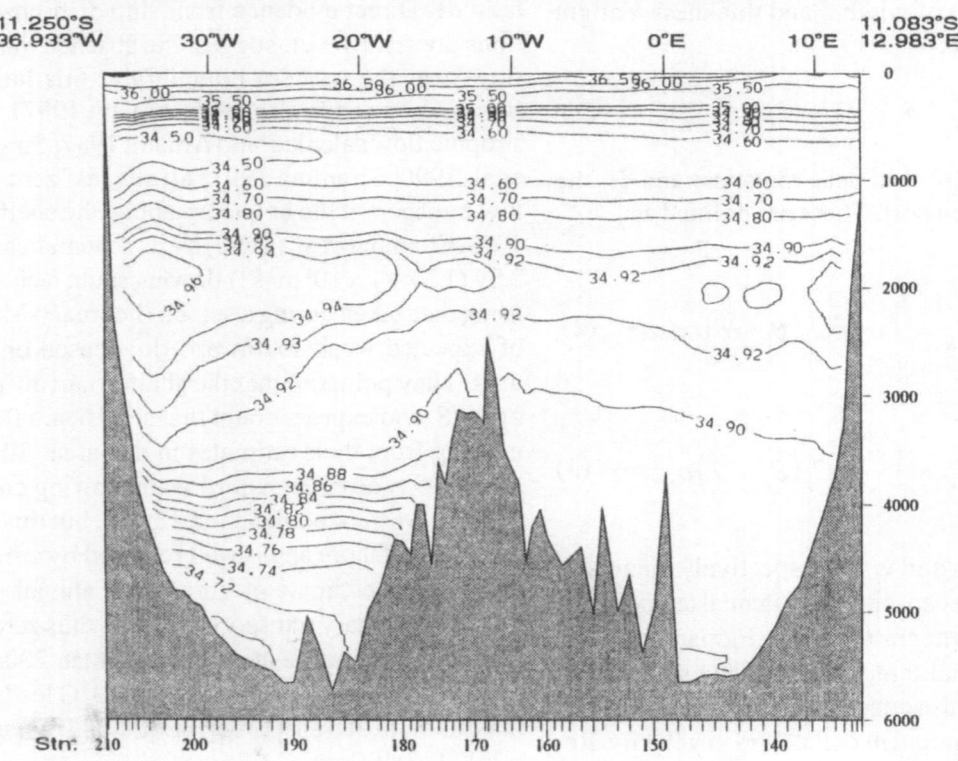

Fig. 3b. Salinity at 11°S (‰).

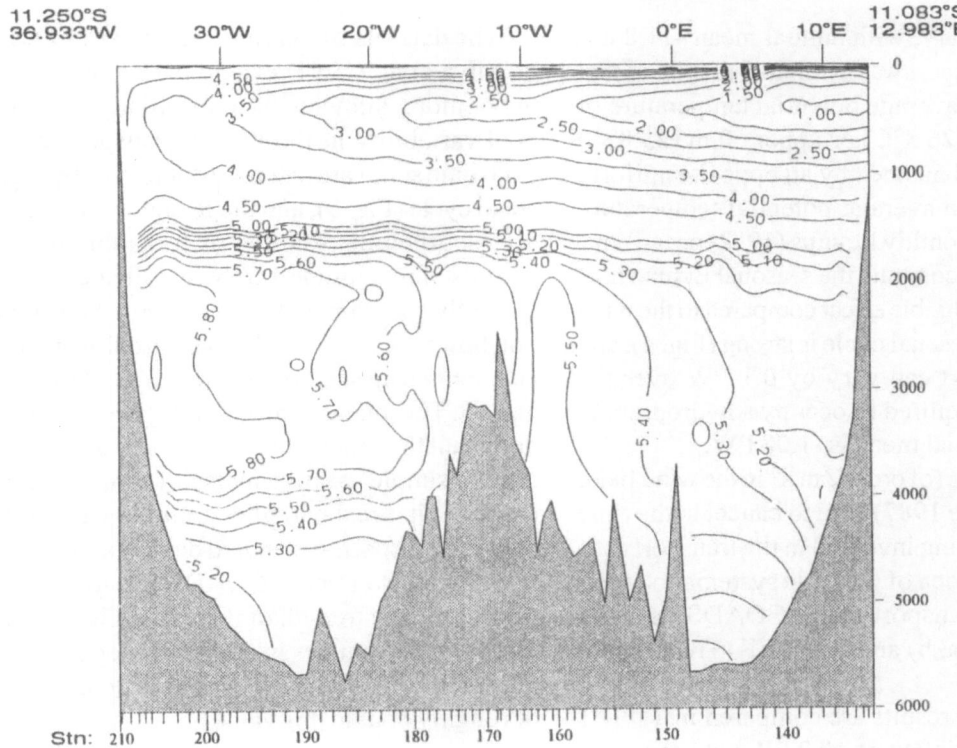

Fig. 3c. Oxygen concentration at 11°S (ml l⁻¹).

Eastern Boundary Current

We are missing information east of station 131, at roughly the 1000 m isobath, to the coast, a distance of 87 km. Wacongne and Piton (1992) describe a Gabon-Congo Undercurrent above the shelf break near 6°S, with an austral winter mean southward speed of 8 cm/s, and thickness of 250 m. Akademik Kurchatov (April-June 1968) stations described by Moroshkin et al. (1970) extended onto the shelf to the south of 6°S and showed a southward surface intensified flow referred to as the Angola Current, from 9°-16°S over the slope and shelf. Near 11°S this current was roughly one half its maximum intensity, as observed near 16°S, with a speed greater than 50 cm/s. Wacongne and Piton (1992) emphasize the high variability of the surface flow at 6°S, and although variability may be weaker at 11°S, it is not clear what the annual mean flow of the Angola Current is at 11°S from these data. Furthermore, the flow is not uniformly southward; we cal-

culated geostrophic transport using Akademik Kurchatov stations near 11°S to fill in the area east of our section and found negligible net transport of mass and heat (bottom reference level).

These considerations lead us to represent an eastern boundary current solely as a southward extension to 11°S of the Gabon-Congo Undercurrent. At 8 cm/s, width 87 km, and mean thickness of 250 m, the seasonal transport is 1.7 Sv, giving an annual mean of roughly 1 Sv to the south. The average temperature of this flow at 6°S is 15°C (Wacongne and Piton 1992) producing a heat transport contribution of -0.05 PW to the total.

Ekman transports

Ekman transport was calculated using climatological wind fields at 11°S (Hellerman and Rosenstein 1983). Values vary from a maximum of magnitude of -15.7 Sv in August to a minimum magnitude of

-8.5 Sv in February, with annual mean -11.8 Sv. Ekman heat transport was estimated assuming that the flow had the average potential temperature of the upper 40 m (26.5°C) or upper 60 m (25.8°C) for comparison, balanced by an opposite uniform flow with section average potential temperature (4.0°C). Using monthly Levitus (1982) ocean temperature fields to compute the seasonal Ekman heat transport had negligible effect compared to the wind variation. The seasonal cycle is strong (Fig. 4), and the heat transport can vary by 0.1 PW over the several weeks required to occupy a hydrographic section. The annual mean is -1.08 PW.

Random errors (of order 2 m s^{-1} in the wind field; Isemer and Hasse 1987) tend to cancel in the time and space averaging involved in the transport estimate. To get an idea of possible systematic errors we examined transport using COADS data (da Silva et al. 1994a, b) and ECMWF (Trenberth et al. 1990).

Annual mean results are compared in Table 1 and suggest a variation of ±0.2 PW between datasets of diverse origin. We believe improvements and corrections such as applied by Isemer and Hasse (1987) to the historical North Atlantic data (n.b. COADS data here have not been corrected according to their method) would reduce the variation between datasets, and also estimates of systematic error. We expect a plausible total error to be ±0.1 PW in the annual mean.

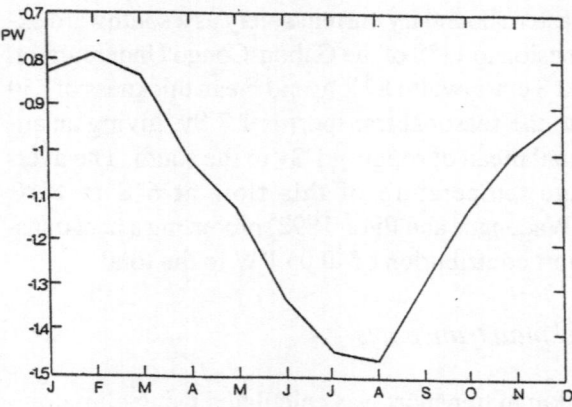

Fig. 4. Annual cycle of Ekman heat transport at 11°S (Hellerman and Rosenstein winds).

The datasets of Servain and Lukas (1990) and da Silva et al. (1994) extend over a period of 30 yrs or more, allowing us to examine the inter-annual variability in the Ekman transport at 11°S. Both time series are naturally dominated by the seasonal cycle (Fig. 5), and the overall trend is that of a strengthening southward transport over the records of magnitude 1.5 Sv (da Silva et al. 1994) to 1.7 Sv (Servain and Lukas 1990). This trend is not, however, supported by geostrophic wind calculations using sealevel pressure from the COADS dataset. The interannual variability appears to have dominant time scales of 6-8 yrs and 2-3 yrs according to a simple spectral analysis (with low significance). If it is real it implies a similar variability in the heat transport. A standard deviation of 0.4 x 10^6 m^{3}s^{-1} for the de-trended COADS data gives an idea of the overall strength of the variability. Average year-to-year changes in the associated heat transport are about 0.05 PW, with occasional changes of roughly double this value.

Geostrophic transport

To compute geostrophic velocity, zero velocity reference levels were chosen following Warren and Speer (1991) for stations in the Angola Basin east of the crest of the Mid-Atlantic Ridge (hereafter MAR), located near 12°W. Water property distributions were also used when choosing reference level in the west. These choices are summarized in Table 2. Above the western flank of the ridge a mid-depth reference level is used to insure some northward transport of deep water in the same depth range, and for similar reasons, as that above the eastern flank (Warren and Speer 1991). Farther west, in the Brazil Basin, a reference level between deep and bottom water produces reasonable northward flow of bottom water. This level (Table 2) was raised several hundred meters from deeper initial choices to improve the overall mass balance of the section. Next to the western boundary the reference level switches to a level between deep water and intermediate water, a natural choice supported by direct measurement at 5°S (Rhein et al. 1994). At shallower depths the velocity was set to zero at the bottom.

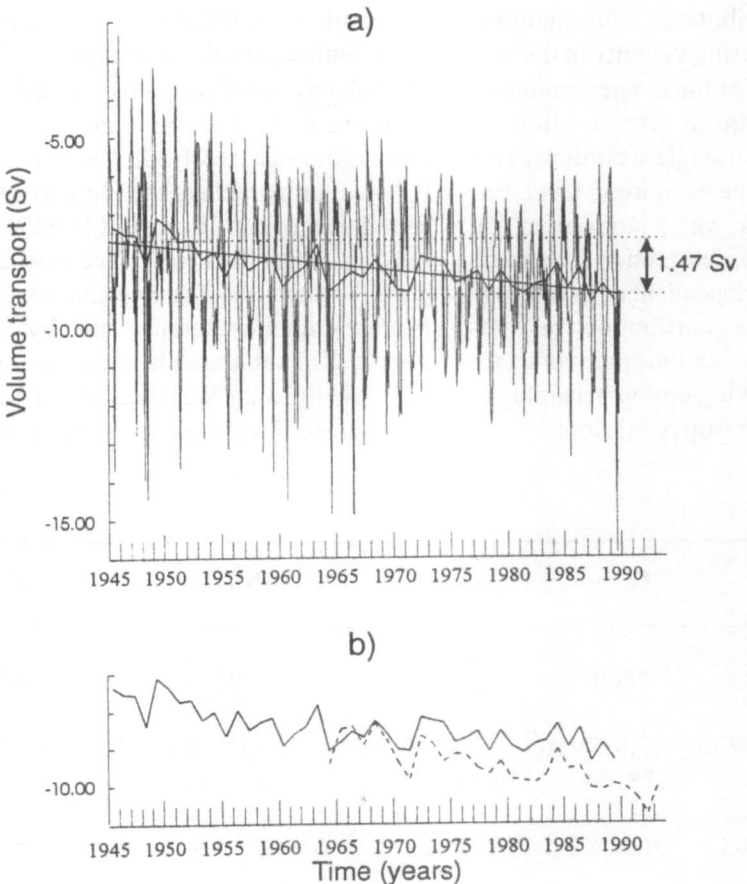

Fig. 5.
a) Time series of the Ekman volume transport across 11°S using COADS wind stress data, monthly and annual mean.
b) Annual mean Ekman volume transport: COADS (solid), Servain (dotted).

Data sets	Annual Mean	March Mean	March 83	Heat
HR	-11.745	-8.960		-1.08
TR	-11.935	-9.870		-1.10
DS	- 8.448	-6.866	- 9.334	-0.78
SL	- 9.474	-8.050	-10.150	-0.87
ECMWF	- 8.307	-5.669		-0.76

Table 1. Ekman Transport (10^6 m³ s⁻¹) and annual mean heat transport (PW) across 11°S using various data sets for the wind stress (HR: Hellerman and Rosenstein 1983; TR: Trenberth et al. 1990; DS: da Silva et al. 1994a, b; SL: Servain and Lukas 1990; and the 1986-88 analysis of the ECMWF).

Bottom triangle contributions were included using several methods: setting velocity in the bottom triangle equal to that at the deepest common level, setting velocity there to zero, or allowing velocity to vary within the triangle according to the isopycnal slope at the deepest common level. Excluding the first station pair, which jumps from 250 m depth to 1493 m depth, the section integrated results were essentially independent of the method used. The first station pair contribution was simply set to zero; subsequent stations pair contributions were estimated allowing bottom triangle velocity to vary according to isopycnal slope.

Results are tabulated for a number of reference level choices, similar to those described above, to investigate their effect on the volume and heat transports (Table 3). The combinations shown are ones that produced a geostrophic volume transport imbalance comparable to the Ekman transport. It was found that volume transport is rather sensitive to the deep Brazil Basin reference level and that transports 3-4 times the Ekman value could be obtained by changing the reference level by a few hundred meters. The tabulated heat transports range from 1.74 PW to 1.82 PW. It was found that changes in the reference levels have little effect on the total heat

Regions	WB	BB	WMAR	EMAR	AB
Level (m)	1100	3800	2400	2400	4000
Longitude	35.5°W to 37.0°W	17.67°W to 35.5°W	12.17°W to 17.67°W	10.17°W to 12.17°W	10.17°W to 12.98°E
Stations	210-205	205-178	178-167	167-163	163-131

Table 2. The reference levels for each region. Regions are western boundary (WB), Brazil Basin (BB), western Mid-Atlantic Ridge (WMAR), eastern Mid-Atlantic Ridge (EMAR) and Angola Basin (AB).

WB	BB	WMAR	EMAR	AB	ψ_g (Sv)	H_g (PW)
1100	3800	2200	2200	4000	12.64	1.77
1100	3800	2400	2400	4000	9.01	1.75
1100	3800	2600	2600	4000	4.03	1.74
1100	4400	2400	2400	4000	13.29	1.82
1100	4400	2600	2600	4000	8.31	1.81
1200	3800	2600	2600	4000	16.97	1.82

Table 3. Other reference level choices for each region and their associated net geostrophic volume transport (ψ_g) and heat flux (H_g).

transport. This is because they primarily affect only the horizontal structure of the circulation, not the zonally averaged vertical structure. We infer an rms error associated with these reference level choices (0.05 PW) and bottom triangle contributions (±0.05) to be ±0.1 PW.

To display the horizontal structure of the flow, water masses were defined by potential temperature classes (Table 4) and volume transport was summed cumulatively in each layer from the western boundary (Fig. 6). The thermocline circulation has been reviewed by Peterson and Stramma (1991), and our surface layer transport is consistent with previous estimates. Deep water flow in the Angola Basin is described by Warren and Speer (1991), and in the interior part of the Brazil Basin Zhang and Hogg (1992) discussed flow based on calculations with climatological data in comparison to other studies. Our reference level choices in the Angola Basin east of the Mid-Atlantic Ridge are the same as Warren and Speer (1991). In the Brazil Basin, Zhang and Hogg's (1992) interior flow calculation from climatological data over Longs. 31°-19°W shows essentially no significant meridional deep water transport, consistent with our results (Fig. 6). Their transport at shallower levels is northwards in all layers at 11°S, similar to our surface or thermocline layer, but our intermediate

water transport has northward and southward components over this longitude range, with no significant net (interior) transport to the north.

Further deep flow characteristics west of the Mid-Atlantic Ridge are discussed by Durrieu de Madron and Weatherly (1994), especially bottom boundary layer effects, and Friedrichs et al. (1994), who emphasized deep recirculation in the Brazil Basin. Our 11°S transport results in deep layers in the Brazil Basin, including boundary currents, recirculation, and interior flow, are similar to the latter study; only net western boundary transports can be compared to Durrieu de Madron and Weatherly (1994), whose deep water transport (20 x 10^6 Sv) is about one-third smaller than ours.

Flow direction is also quite similar to Reid (1989), except for deep water flow above the eastern flank of the Mid-Atlantic Ridge, at which location Warren and Speer (1991) describe a northward net transport of deep water between 2400-4000 m depth (LNADW) while Reid's (1989) flow at 3000 db is southward. Upper deep water also shows a net northward transport from 9°W to 0°W, of 2-5 Sv, bringing the total deep water northward flow above the eastern flank of the ridge to 4-7 Sv. Southward net transport does appear in the interior of the Angola Basin east of the ridge (Fig. 6), stronger at middle and upper deep water levels, with

Layers	Potential temperature (°C)	SK
Surface water	$\theta > 6.0$	$\sigma_\theta < 26.8$
AAIW	$4.2 < \theta \leq 6.0$	$\sigma_\theta < 27.61$
Upper NADW	$3.2 < \theta \leq 4.2$	$\sigma_4 < 45.95$
Middle NADW	$2.4 < \theta \leq 3.2$	
Lower NADW	$2.0 < \theta \leq 2.4$	
AABW	$\theta \leq 2.0$	$\sigma_4 > 45.95$

Table 4. Definition of water masses in terms of potential temperature. SK denotes layer interface definitions of Saunders and King (1995).

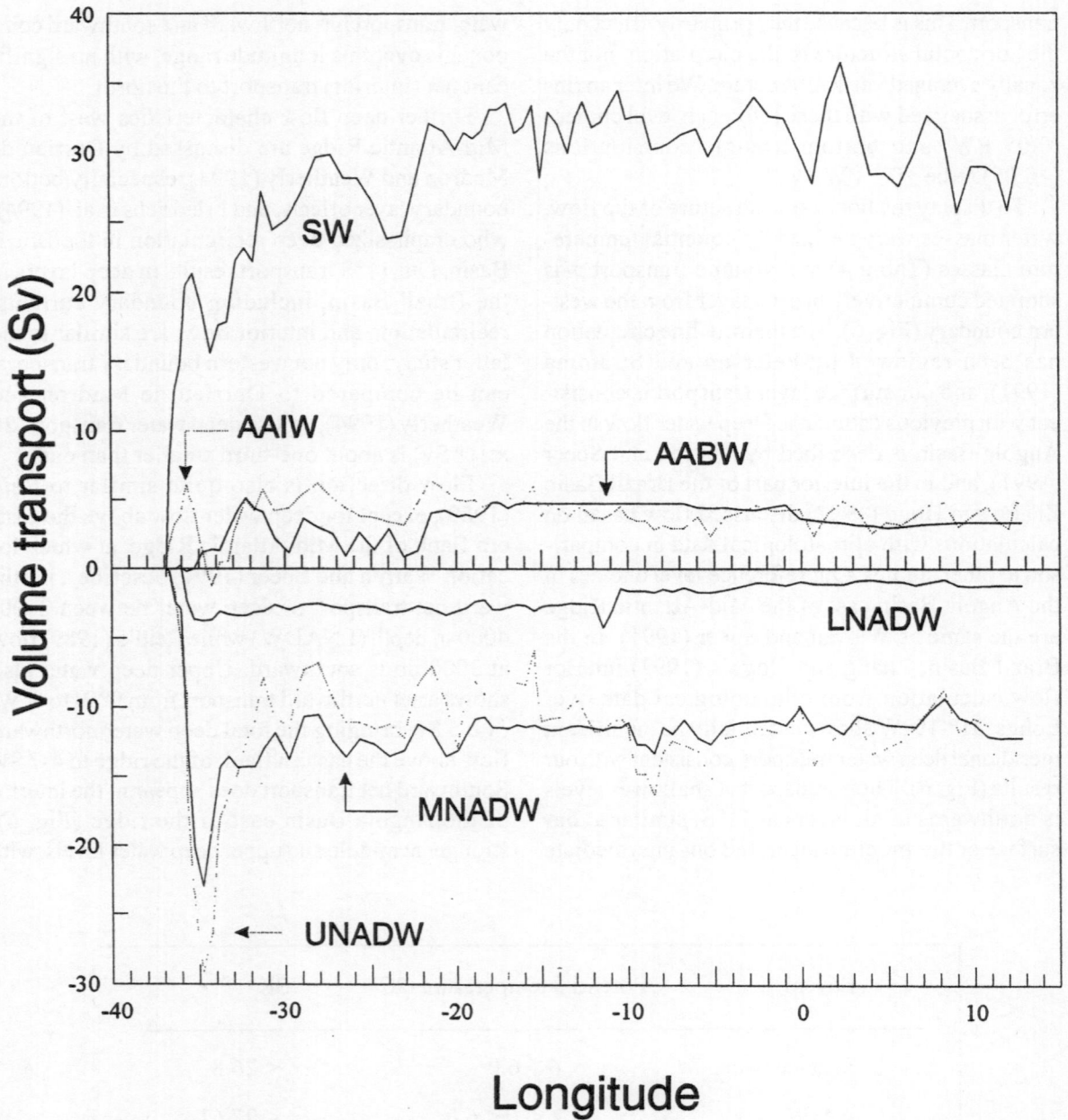

Fig. 6. Volume transport across 11°S by layers, plotted cumulatively from west to east (see Table 2).

an interesting concentration next to the eastern boundary, correlated with high salinity (Fig. 3). Net transports east of 3°E are roughly -2 Sv (UNADW), -2 Sv (MNADW), and -0.5 Sv (LNADW), consistent with the interior flow scheme and property distributions at 11°S described by Warren and Speer (1991).

The net northward transport between 25°W and 15°W in middle deep water (Fig. 6) is consistent with Reid's (1989) northeastward flow. It exists across a region in which salinity decreases to the east (Fig. 3) owing to the presence of deep water that has made a large southward excursion in the western boundary current before returning north, losing salt along the way by vertical and lateral mixing. Between 25°W and 12°W, upper deep water shows a northward net transport, followed by a compensating southward flow above the western flank of the ridge. Salinity and dissolved oxygen concentration both decrease markedly eastward between 18°W and 15°W (Fig. 3), corresponding roughly to the reversal from northward to southward transport in this layer (Fig. 6).

Flow above the crest of the ridge, roughly within the longitude range 10°-17°W, is dominated by a southward transport of upper deep water (Fig. 6). Over this longitude range the net transport is about 5 Sv to the south, making this flow second only to the 10 Sv net western boundary current transport in this layer. Here, as elsewhere, the rough division of the layers into flow regions at the section boundaries and above the flanks and crest of the ridge is guided by property distributions; numerous reversals of flow are present, though, often obscuring clear boundaries between flow regions. The overall structure of high salinity above the crest (Fig. 3), decrease to the east, and the reappearance of high salinity farther east are consistent with the above circulation pattern.

The strength of the lower deep water current at the western boundary (Fig. 6) is about 3 times stronger than that found by Friedrichs et al. (1994), (their schematic shows a value of -1 Sv). However, the middle deep water flow is in agreement. The net bottom water transports in the Brazil Basin, reported here, are similar to those of Durrieu de Madron and Weatherly (1994), McCartney and Curry (1993), and Speer and Zenk (1993).

The horizontally integrated vertical structure of transport (Fig. 7) bears some resemblance to Roemmich's (1983) and Fu's (1981) results at 8°S based on IGY data. Roemmich (1983) shows an overturning cell strength (sum of deep water components up to the base of intermediate water) of 21 Sv which compares with our result of 25 Sv. A layer breakdown shows that our bottom water, upper deep water, and surface layer components are larger, while our lower deep water flow is much weaker. Fu's (1981) vertical structure is somewhat closer to ours, with its weaker lower deep water component, and similar cell strength, but his lower deep water transport is still too strong and there is

no net bottom water transport in his solution. Both of these studies constrained their results to conserve mass (within error) in several deep layers across a wide latitude range.

Total heat and salt transport

Adding together the heat transport components (Table 5), the total annual mean heat transport at 11°S is estimated to be 0.60 PW +0.17 PW. The horizontal structure of temperature transport (Fig. 8) shows the dominant role of the deep western boundary currents at this latitude, compared to the relatively greater role eastern boundary flow plays near 30°S (Rintoul 1991; Saunders and King 1995).

The salt flux across the section is 6 Svpsu (0.13 x 10⁶ kg s⁻¹) with zero net volume transport. To keep the total amount of salt between 11°S and the Bering Strait constant, a total salt flux of -26 Svpsu is needed (-0.8 Sv x 32.5 psu). This can be accomplished at 11°S by a net volume transport of - (26 + 6) /34.89, or -0.9 x 10⁶ Sv, using the section mean salinity. This is in rough agreement with expectations based on climatological surface evaporation and precipitation (Wijffels et al. 1992). Similar estimates at 25°N (Saunders and King 1995) and 11°N (Friedrichs and Hall 1993) are also in rough agreement. On the other hand, Saunders and King (1995) find a southward net transport of 0.75 Sv at 35°-40°S, in strong disagreement with integrated surface fluxes. Cumulative errors makes the integrated surface fresh water flux results suspect in the South Atlantic.

Discussion

Our results show that the net heat transport at 11°S is relatively insensitive to reference level variation when these lie within a range of reasonable choices based on property distributions. The horizontally integrated vertical structure is somewhat more sensitive, while the horizontal structure of the flow is naturally quite dependent on reference level choice. The fluxes have been estimated using a flow field with plausible bottom, deep, and thermocline circulation fields. This enables us to show how various circulation elements contribute to the net flux. The dominant errors in the heat transport calcula-

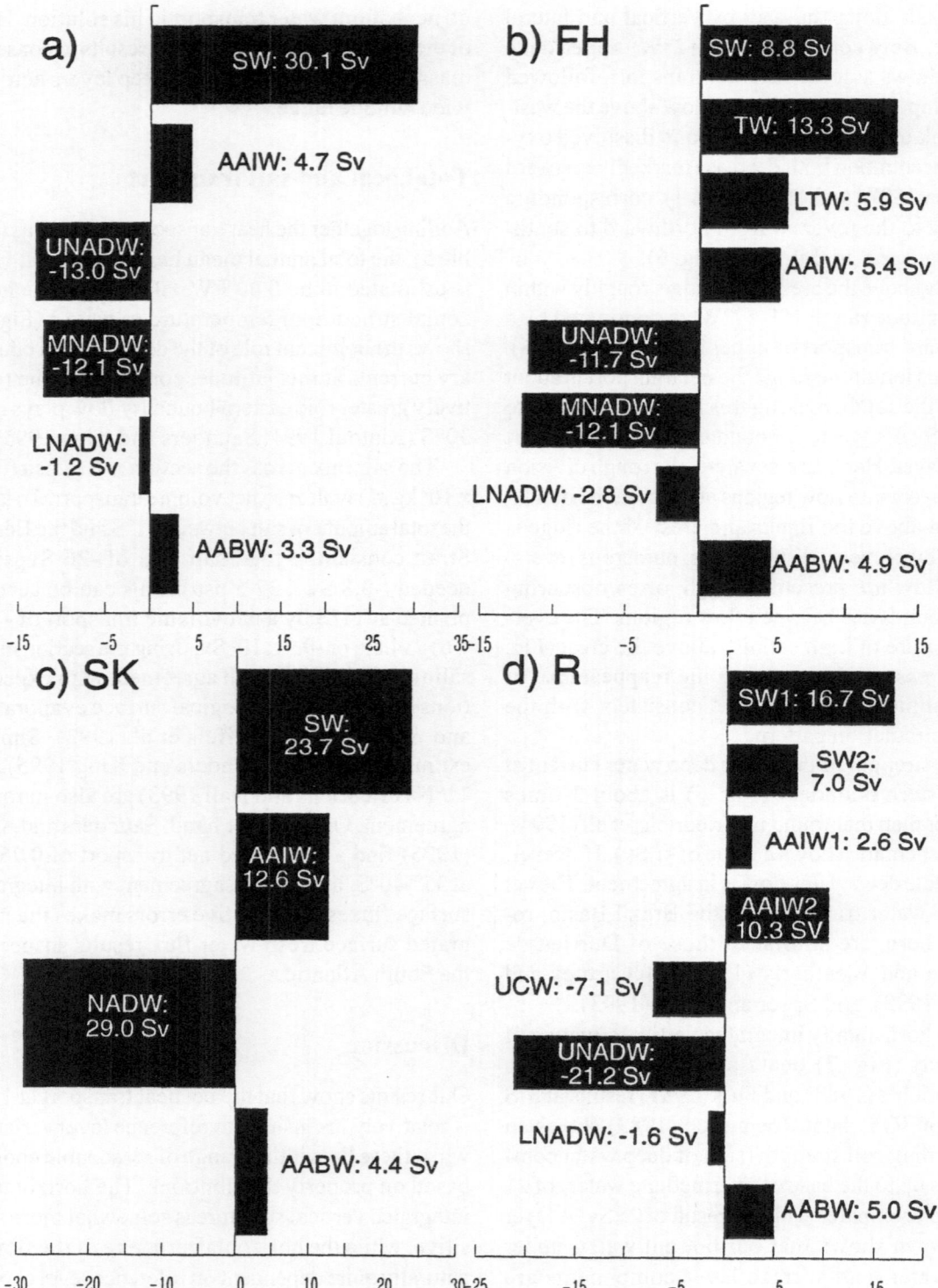

Fig. 7. a) Vertical structure of geostrophic volume transport. Decomposition of volume transport at 11°S (Fig. 6 and Table 4, potential temperature classes) into different layer definitions following: (a) Table 4, (b) Friedrichs and Hall 1993, (c) Saunders and King 1995, (see Table 4) (d) Roemmich 1983. Apparent discrepancies in watermass transport (all at 11°S) arise simply from different definitions.

	H$_g$	H$_e$	WBC	EBC	Total
Annual	1.75 ± 0.1	-1.1 ± 0.1	0.00 ± 0.08	-0.05 ± 0.05	0.60 ± 0.17
March	1.75 ± 0.1	-0.9 ± 0.1	-0.1 ± 0.08	-0.05 ± 0.05	0.70 ± 0.17

Table 5. Heat transport components (PW) at 11°S.

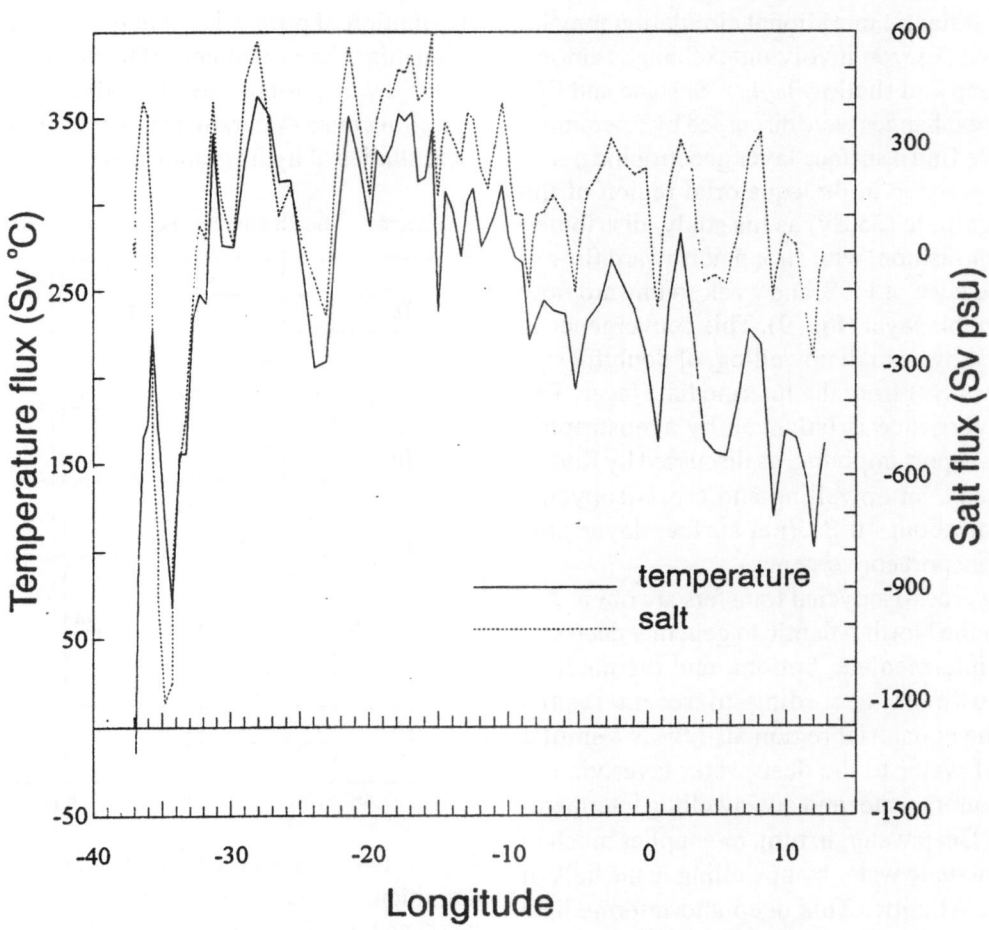

Fig. 8. Horizontal structure of temperature flux and salt flux at 11°S.

tion are due to reference level choices and wind field variability, but a further, unknown, source of error is associated with seasonal and interannual variability in the baroclinic structure. Outside a narrow equatorial band and eastern boundary layer, the baroclinic structure is conventionally thought to adjust too slowly (Gill and Niiler 1973; Anderson and Kill-worth 1979) to modify significantly the local baro-tropic response to seasonal winds (Fig. 4). In this dynamical limit a single hydrographic section is sufficient, and our annual mean result is well determined, except for presumably small nonlocal effects or interannual variability.

Taking recent similar calculations at other latitudes at face value, without the benefit of an analysis of the combined dataset, a simplified picture of the South Atlantic meridional circulation may be constructed (Fig. 9) involving exchanges among several deep and shallow layers. Surface and Ekman layer exchanges were discussed by Roemmich (1983). We find a surface layer geostrophic transport convergence in the equatorial region of the same magnitude (25 Sv) as his study, distributed in a similar fashion, with strong northward flow in the surface layer at 11°S, and weak southward flow at 14°N in this layer (Fig. 9). This convergence is augmented by a small upwelling, of doubtful significance (1 Sv) from the intermediate layer. The total convergence is balanced by ageostrophic Ekman transport implying, as discussed by Roemmich (1983), an upwelling and cross-isopycnal transfer of about 10 Sv (net surface layer plus Ekman transport convergence).

Strong cross-isopycnal transfers are obviously present in the North Atlantic to generate deep water from intermediate, bottom, and thermocline water; however, according to recent results (Fig. 9) the equatorial region also feeds a similar amount of water to the deep water layer via upwelling bottom water and downwelling intermediate water. Deep water, in turn, re-supplies much of the intermediate water by upwelling in the bulk of the South Atlantic. This deep and intermediate recirculation cell seems excessively intense, with an upwelling of on average 6×10^{-5} cm s^{-1} at about 1000 m depth (though certainly this estimate of its strength suffers from large errors) and a similar equatorial downwelling (7×10^{-5} cm s^{-1}). We only

wish to demonstrate the tendency of the results to produce a deep upwelling in the South Atlantic, that is, upward cross-isopycnal transport. Finally, the southward transport of deep water supplies bottom and intermediate layers in the Antarctic region through, as elsewhere, some combination of interior and boundary layer mixing and formation processes.

This combination of layer transport results at several latitudes highlights likely problems with reference level choices, leading especially to a deep water transport south of the equator which is probably too high, and one north of the equator which could be too low in the mean. The datasets themselves may be unrepresentative of the mean fields. Nevertheless these choices reflect the large-scale distribution of properties, and the resulting circulation might have elements of truth lacking in data analyses with pre-conceived or arbitrarily assigned upwelling rates or parameterizations. Reconciling local studies of hydrography and flow at high ver-

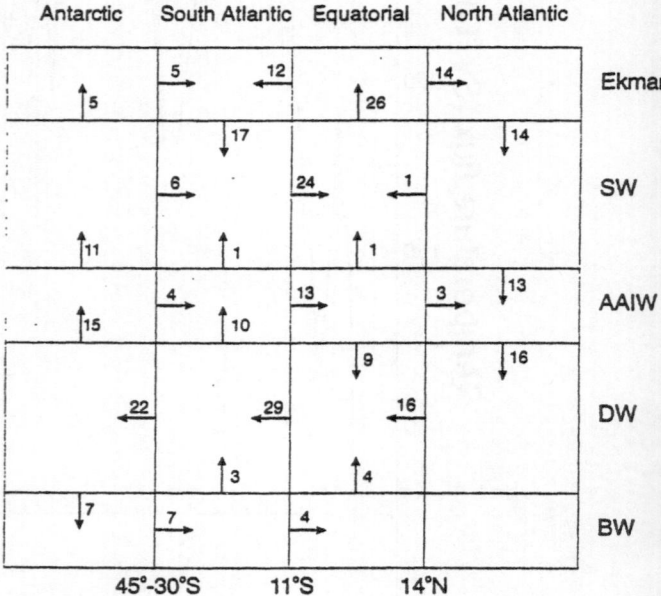

Fig. 9. Layer transports (using Saunders and King's definitions of four layers [Table 4] plus the Ekman layer) and implied vertical transfer between layers to conserve mass. Results are from Saunders and King (1995) at 45-30°S and Klein et al. (1995) at 14°N. The Antarctic box includes the rest of the World Ocean ; the North Atlantic box includes the Nordic Seas, the Arctic connection is neglected.

tical resolution with global property budgets may help to constrain rates of interior and boundary mixing and fluxes.

Acknowledgements

CTD data at 11°S and 23°S collected during OCEANUS cruise 133, Chief Scientists M. McCartney (23°S) and B. Warren (11°S). Data at 19°S were obtained as part of the German WOCE program. P. Saunders and B. Klein kindly shared results from their studies at other latitudes in the Atlantic Ocean. Wind data from J. Servain and A. da Silva were appreciated, as was helpful comment by S. Wacongne and Y. Gouriou. Support for this study came from the CNRS, BMFT and LPO.

References

Anderson DLT, Killworth PD (1979) Non-linear propagation of long Rossby waves. Deep-Sea Res 26:1033-1050

Arnault S (1987) Tropical Atlantic geostrophic currents and ship drifts. J Geophys Res 92:5076-5088

Bryan K (1962) Measurements of meridional heat transport by ocean currents. J Geophys Res 67:3403-3414

Bunker AF (1988) Surface energy fluxes of the South Atlantic Ocean. Month Weather Rev 116:809-823

Durrieu de Madron X, Weatherly G (1994) Circulation, transport and bottom boundary layers of the deep currents in the Brazil Basin. J Marine Res 52:583-638

Friederichs MAM, Hall MM (1993) Deep circulation in the tropical North Atlantic. J Marine Res 51:697-736

Friederichs MAM, McCartney MS, Hall MM (1994) Hemispheric asymmetry of deep water transport modes in the western Atlantic. J Geophys Res 99:25,165-25,179

Fu L-L (1981) The general circulation and meridional heat transport of the subtropical South Atlantic determined by inverse methods. J Phys Oceanogr 11:1171-1193

Gill AE, Niiler PP (1973) The theory of the seasonal variability in the ocean. Deep-Sea Res 20:141-178

Hall MM, Bryden HL (1982) Direct estimates and mechanisms of ocean heat transport. Deep-Sea Res 29:339-359

Hellerman S, Rosenstein M (1983) Normal monthly wind stress over the world ocean with error estimates. J Phys Oceanogr 13:1093-1104

Hsiung J, Newell RE, Houghtby T (1989) The annual cycle of oceanic heat storage and oceanic meridional heat transport. Quarterly J Royal Met Soc 115:1-28

Isemer H-J, Hasse L (1987) The Bunker climate atlas of the North Atlantic Ocean. Vol.2, Air-Sea Interactions, Springer-Verlag, 256 pp

Klein B, Molinari RL, Siedler G, Müller TJ, Jones (1995) Water mass distribution and the meridional transport of heat and mass at 14.5°N in the Atlantic. J Marine Res 53:929-957

Levitus S (1982) Climatological Atlas of the World Ocean. NOAA Prof Pap, 13, Natl Oceanic and Atmos Admin, Washington DC

McCartney MS, Curry RA (1993) Transequatorial flow of Antarctic bottom water in the western Atlantic Ocean : abyssal geostrophy at the equator. J Phys Oceanogr 23:1264-1276

Moroshkin KV, Bubnov VA, Bulatov RP (1970) Water circulation in the eastern South Atlantic Ocean. Oceanology 10(1):27-34

Peterson RG, Stramma L (1991) Upper level circulation in the South Alantic Ocean. Prog in Oceanogr 26:1-73

Reid JL (1989) On the total geostrophic circulation of the South Atlantic Ocean: flow patterns, tracers, and transports. Prog in Oceanogr 23:149-244

Rhein M, Stramma L, Send U (1995) The Atlantic deep western boundary current: water masses and transport near the equator. J Geophys Res 100:2441-2457

Richardson PL, Walsh D (1986) Mapping climatological seasonal variations of surface currents in the tropical Atlantic using ship drifts. J Geophys Res 91:10537-10550

Rintoul SR (1991) South Atlantic interbasin exchange. J Geophys Res 96:2675-2692

Roemmich D (1983) The balance of geostrophic and Ekman transports in the tropical Atlantic Ocean. J Phys Oceanogr 13:1534-1539

Servain J, Lukas S (1990) Climatic atlas of the tropical Atlantic wind stress and sea surface temperature 1985-89, IFREMER, Brest

Saunders PM, King BA (1995) Oceanic fluxes on the WOCE A11 section. J Phys Oceanogr (in press)

da Silva AM, Young CC, Levitus S (1994a) Revised wind stress over the global oceans with bias corrections : climatology and anomalies for 1945-1989. J Phys Oceanogr (submitted)

da Silva AM, Young CC, Levitus S (1994) Atlas of

surface marine data 1994, Volume 1: algorithms
and procedures. NOAA Atlas NESDIS 6, Dept. of
commerce, Washington, 83 pp

Speer KG, Zenk W (1993) The flow of Antarctic bottom water into the Brazil Basin. J Phys Oceanogr
12:2667-2682

Stramma L, Ikeda Y, Peterson RG (1990) Geostrophic
transport in the Brazil Current region north of 20°S.
Deep-Sea Res 37:1875-1886

Stramma L (1991) Geostrophic transport of the south
equatorial current in the Atlantic. J Marine Res
49:281-294

Trenberth KE, Large WG, Olson JG (1990) The mean

annual cycle in global ocean wind stress. J Phys
Oceanogr 20:1742-1760

Wacongne S, Piton B (1992) The near surface circulation in the north eastern corner of the South
AtlanticOcean. Deep-Sea Res 39:1273-1298

Warren BA, Speer KG (1991) Deep circulation in
the eastern South AtlanticOcean. Deep Sea Res
38(suppl):5281-5322

Zhang H, Hogg NG (1992) Circulation and water mass
balance in the Brazil Basin. J Mar Res 50:385-420

Wijffels SE, Schmidt RW, Bryden HL, Stigebrandt A
(1992) Transport of freshwater by the oceans. J Phys
Oceanogr 22:155-162

Comment on the South Atlantic's Role in the Global Circulation

A.L. Gordon

*Lamont-Doherty Earth Observatory Palisades,
New York, 10964-8000 USA*

Abstract: The role of the South Atlantic in the global climate system, specifically as it concerns the North Atlantic Deep Water driven thermohaline meridional overturning cell, has drawn much attention in recent years. I propose that the configuration of the continents bounding the South Atlantic, specifically the contrast in the southern extreme of South America and Africa relative to the maximum westerlies wind, allow the development of a salty Atlantic Ocean, preconditioning the region for production of North Atlantic Deep Water, whose positive feedbacks further energizes the process. Much of the low salinity upper layer water carried by the South Pacific Current towards the coast of Chile, turns northward into the subtropical South Pacific as the Peru Current. Pacific surface water that does flow through the Drake Passage tends to transverse the South Atlantic, picking up more freshwater enroute drawn from the evaporative regime of the South Atlantic subtropics, into the Indian Ocean. Only a small component of the South Atlantic Current turns northward into the Benguela Current. Warm, saline Indian Ocean thermocline water can slip along the southern rim of Africa into the South Atlantic to further increase Atlantic salinity.

Introduction

This note is meant as a thought exercise, which presents some of the issues raised in my key note address at the South Atlantic Symposium in Bremen, August 1994.

Thermohaline circulation may be defined as that part of the ocean circulation driven by the buoyancy flux between ocean and atmosphere. While it is more sluggish than the wind-driven circulation (resulting from wind stress on the sea surface) the thermohaline circulation may be equally or perhaps even more important to the global climate system, as it couples the full volume of the ocean to the atmosphere and forms a global circulation network of heat and freshwater (Gordon 1986; Schmitz 1995). Of course, it is artificial to divide the circulation into wind-driven and thermohaline-driven components, as nature satisfies all of the forcing mechanisms with a single integrated circulation system. Thermohaline and wind-driven circulation are linked by the wind action. The wind not only imparts momentum flux but also regulates ocean-atmosphere buoyancy fluxes, and the wind

speed and direction is dependent to some extent on the sea-surface temperature, whose pattern is shaped by overall horizontal and vertical circulation.

Present-day, bottom-reaching thermohaline circulation is driven by water-mass formation in the North Atlantic and Southern Ocean. In the North Atlantic, thermocline water with a long history of contact with the atmosphere, is cooled and sinks to the deep ocean as the warm salty North Atlantic Deep Water (NADW). In the southern hemisphere south of the Antarctic Circumpolar Current (ACC) upwelling of deep water, long removed from direct contact with the atmosphere, is quickly converted to cold dense Antarctic Bottom Water (AABW), often under the cover of sea ice, chilling the global ocean. About 57% of the abyssal ocean is cooler than NADW temperature (roughly 2°C), attesting to the influence of the Southern Ocean thermohaline processes (Gordon 1991). These water masses, each with unique properties, spread throughout the ocean forcing the resident

From WEFER G, BERGER WH, SIEDLER G, WEBB DJ (eds), 1996, *The South Atlantic: Present and Past Circulation.* Springer-Verlag Berlin Heidelberg, pp 121-124

deep ocean water, which has been made more buoyant by vertical turbulent mixing, to slowly upwell. Eventually the upwelled water migrates back to the sinking regions to close the thermohaline circulation loop.

Estimates of NADW formation are approximately 14 Sv (Schmitz 1995), while AABW formation rates are less well known but global scale budget studies suggest AABW formation rates in the range of 18 Sv to 40 Sv (Saunders and Thompson 1993; Gordon and Taylor 1975). Thus up to 50 Sv of near-surface water ventilates the deep ocean, though the AABW source may not be fully in equilibrium with the atmospheric gases owing to the involvement of sea ice which inhibits gas exchange. In equilibrium conditions the global convection rate must balance the global downward turbulent buoyancy flux. Except for short periods of non-equilibrium (of the order of the residence time of the deep ocean, about 1000 years) the rate of deep-water ventilation of the non-isolated sectors of the global ocean may not have changed very much in the geological past, even under vastly different climate situations (though the temperature and salinity of the convection has changed). This rather drastic and controversial statement is based on a suspected climate-invariant nature of the oceanic vertical mixing processes as the mixing process to a large measure may be a consequence of the dissipation of tidal energy and associated internal wave-induced mixing. Thus the primary changes in vertical buoyancy flux and ocean ventilation may be related to changing ocean basin geometry and the ratio of ocean volume to continental margin contact and perhaps very low frequency changes in the tide-raising forces.

Interocean Exchange and the South Atlantic

In the last decade there has been much attention given to the NADW component of the thermohaline circulation, specifically to the nature of transfer of upper layer water from the Pacific and Indian Oceans into the South Atlantic Ocean as required to compensate for the Atlantic export of NADW. Schmitz (1995) provides an excellent review of the literature and his opinion as to the

most likely global pathways of the NADW-driven thermohaline circulation „conveyor belt".

A fundamental question is: why is the North Atlantic so salty? or stated another way, what selects the North Atlantic as the site for relatively warm saline deep water formation? And a second question, already a focus of numerous published observational and modeling studies: Along what path does the returning, upper layer, limb of the NADW-driven conveyor belt enter the South Atlantic?

[A] Why is there deep convection in the North Atlantic?

Here I consider the answer lies in the distribution of the continents. Models that relax the T/S properties to the observational base are physically forced into the present-day climate condition and confuse cause and effect. Hopefully, the development of coupled climate models will provide the proper simulations of the climate system. An example of cause-and-effect confusion is the statement that the net export of atmospheric moisture from the North Atlantic basin to the Pacific basin is responsible for NADW formation. Well yes it is, but I see it as a positive feedback response to NADW formation as warm upper-limb water is drawn into the North Atlantic feeding large evaporative fluxes. I do not believe that the export of water vapor initiates NADW production, it maintains it. Then, what does initiate it? Here I suggest we look to the South Atlantic, specifically to its western and eastern continental borders. South America extends to approximately 56°S while Africa extends to 37°S (the southern limits of the continental shelf are considered, as that boundary controls the flow of the open ocean boundary currents), while the maximum westerlies wind generally follows the 40 to 45°S belt. Thus subtropical gyre features (western boundary currents and the polar edge of the gyre) of the South Atlantic and Indian Oceans are linked far more effectively than are those of the South Pacific and South Atlantic Oceans. Linking of the South Atlantic and Indian Ocean permits the export of freshwater from the South Atlantic by the ACC and the introduction of saline Indian Ocean water around the southern rim of Africa (Gordon and Piola 1983; Gordon et al. 1992). In contrast, South America blocks much of the Pacific's excess freshwater from entering the Atlantic and does not allow any of the South Atlantic's subtropical water from exiting to the west.

South America turns the low salinity subpolar water advected eastward by the South Pacific Current into the South Pacific subtropical gyre via the Peru Current, while Africa's low latitude permits the similar water within the South Atlantic Current to continue on to the Indian Ocean. Gordon and Piola (1983) show that the excess precipitation of the subpolar South Atlantic belt freshens the surface layer of ACC which then advects it out of the Atlantic basin. As the amount of the freshening is roughly equivalent to the net evaporation of the South Atlantic subtropical region (Gordon and Piola 1983) this process would tend to make the Atlantic saltier. In addition, the low latitude of the southern limit of Africa allows subtropical water from the Indian Ocean to slip into the South Atlantic within the complex Agulhas Retroflection (Gordon et al. 1992). Both of these processes make the South Atlantic saltier than it would be if Africa forced the Agulhas Current to separate before reaching the southern rim of Africa and allowing the Benguela Current to be totally fed by the South Atlantic Current.

[B] What is the primary return pathway of NADW?

In the South Atlantic NADW is drawn into the Antarctic Circumpolar Current (ACC) and is swept into the Indian and Pacific Oceans. Somewhere NADW upwells and is altered to near-surface water (see Schmitz 1995). It is likely that the primary upwelling site occurs in the massive upwelling south of the ACC. Some upwelled water feeds sinking AABW along the margin of Antarctica and some may be advected to the north where it contributes to Antarctic Intermediate Water (AAIW) which then spreads along the thermocline base, upwelling into the thermocline where it is altered into warmer, more saline water thus closing the conveyor-belt loop (Gordon 1986). The return of upper layer water may be derived from the Drake

Passage (the cold route) and from the southern rim of Africa (the warm route). It is important to stress that these routes are not mutually exclusive. The 'argument' of warm (Gordon 1986) vs. cold (Rintoul 1991) routes is somewhat artificial as both pathways are active (Gordon et al. 1992; Schmitz 1995; Garzoli and Gordon 1995). What is important is the dependence of the Atlantic salinity and susceptibility to NADW formation on the cold-to-warm pathway transport ratio.

Studies suggest that the bulk of the Atlantic export of NADW is balanced by lower thermocline water and AAIW. However, Gordon et al. (1992) show that the cold water path follows a rather unexpected convoluted course: rather than follow a route along the western boundary of the South Atlantic, the bulk of the cold path (which provides for AAIW) flows across the South Atlantic and is drawn into a tight, intense gyre in the southwest Indian Ocean, returning to the Atlantic via the Agulhas Retro-flection. Within the Indian Ocean the AAIW is increased in salinity by 0.25 ppt over the initial characteristics of AAIW in the Argentine Basin. In addition, the injection of salty Indian Ocean thermocline water into the South Atlantic produces a saltier Benguela Current than would occur if it were fully supplied by the South Atlantic Current, with no Indian Ocean thermocline link.

Discussion

Most of the net meridional flow within the South Atlantic heading towards the North Atlantic occurs in the AAIW (cooler than 9°C but shallower than 1500 db) layer rather than the thermocline (above 9°C) layer, with a ratio of 7 to 2 Sv (Gordon et al. 1992). Boddem and Schlitzer (1995) in their model study find similar values. Yet in the North Atlantic the northward transport within the upper limb of the meridional overturning cell resides in the thermo-cline (Schmitz and Richardson 1991). Clearly there must be transfer of cold AAIW water to warmer thermocline water in the Atlantic tropics. Equatorial upwelling is not a new idea, but perhaps the large cyclonic gyre of the eastern tropical Atlantic (Gordon and Bosley 1991) may play an important role in this upwelling.

Recent studies find that the northward heat flux in the South Atlantic may be supported by Indian Ocean heat. Saunders and King (1995) suggest 0.5 PW (Petawatt) passes across the WOCE A11 section, with 10 Sv of upper thermocline water. Döös (1995) using the FRAM results states: „A heat transport study is carried out for the Atlantic, showing that the northward heat flux into the Atlantic comes 85% from the Indian Ocean and the rest from the Drake Passage". However, Cai and Greatbatch (1995) find that while 35% of export of NADW from the Atlantic is compensated by Indian Ocean import, suppressing Indian Ocean input to the Atlantic, while it leads to a cooler and fresher Atlantic, does not alter the density field and the NADW formation rate.

The Benguela Current inverted echo-sounder array of 1992-1993 (Garzoli and Gordon 1995) suggests that 25% to 40% of the 13 Sv Benguela Current is drawn from the Indian Ocean water. However, it must be pointed out that the IES data only provide transport information, not water-mass properties, so the source-water inference is drawn from transport patterns. Water-mass studies in progress indicate that Indian Ocean water masses contribute to the transport that Garzoli and Gordon attribute solely to the South Atlantic.

The salt transfer from the Indian to the Atlantic around the rim of southern Africa most likely varies with time. Gordon et al. (1992) estimate that the Indian salt boosts the upper layer salinity of the Atlantic by 0.5 ppt. If this were reduced to 0.3 ppt or to 0.0 ppt, would the formation rate of NADW change? Might there be a sudden shift to another stable mode for NADW (Stocker and Wright 1991)? These are questions that may be addressed by coupled climate models.

References

Boddem J, Schlitzer R (1995) Interocean exchange and meridional mass and heat fluxes in the South Atlantic. J Geophys Res 100 (C8):15821-15834

Cai W J, Greatbatch RJ (1995) Compensation for the NADW outflow in a global ocean circulation model. J Phys Oceanogr 25(2):226-241

Döös K (1995) Interocean exchange of water masses. J Geophys Res 100 (C7):13499-13514

Garzoli SL, Gordon AL (1995) Origins and Variability of the Benguela Current. J Geophys Res (in press)

Gordon AL (1986) Interocean exchange of thermocline water. J Geophys Res 91:5037-5046

Gordon AL (1991) The Southern Ocean - its involvement in global change. In: Weller G et al. (eds) Proceedings of the Conference: „Role of the Polar Regions in Global Change", June 1990, University of Alaska, Fairbanks, Alaska, Vol. 1, pp 249-255

Gordon AL, Taylor HW (1975) Heat and salt balance within the cold waters of the world ocean; supplement to General Ocean Circulation, Numerical Models of Ocean Circulation: Symposium, Durham, New Hampshire, Oct. 17-20, 1972. National Academy of Sciences. Publ., pp 54-56

Gordon AL, Piola A (1983) Atlantic Ocean upper layer salinity budget. J Phys Oceanogr 13(7):1293-1300

Gordon AL, Bosley K (1991) Cyclonic gyre in the tropical South Atlantic. Deep Sea Res 38 (Suppl. 1):S323-S343

Gordon AL, Weiss RF, Smethie WM Jr, Warner MJ (1992) Thermocline and intermediate water communication between the South Atlantic and Indian Oceans. J Geophys Res 97 (C5):7223-7240

Rintoul S (1991) South Atlantic interbasin exchange. J Geophys Res 96 (C2):2675-2692

Saunders PM, Thompson SR (1993) Transport, heat and freshwater fluxes within a diagnostic numerical model (FRAM). J Phys Oceanogr 23(3):452-464

Saunders PM, King BA (1995) Oceanic Fluxes on the WOCE A11 Section. J Phys Oceanogr 25(9):1942-1958

Schmitz WJ, Richardson PL (1991) On the sources of the Florida Current. Deep Sea Res 38 (supplement):379-409

Schmitz WJ (1995) On the interbasin-scale thermohaline circulation. Reviews in Geophysics 33(2):151-173

Stocker TF, Wright DG (1991) Rapid transition of the ocean's deep circulation induced by changes in surface water fluxes. Nature 351:729-732

The Exchange of Water Between the South Indian and South Atlantic Oceans

J.R.E. Lutjeharms

*Institut für Meereskunde an der Universität Kiel,
Düsternbrooker Weg 20, D-24105 Kiel, Germany
and Department of Oceanography, University of Cape Town,
7700 Rondebosch, South Africa*

Abstract: The circulation at mid-latitudes of the South Atlantic Ocean accommodates a substantial, though intermittent, leakage of tropical and subtropical water from the South Indian Ocean past the southern tip of Africa. This transferral of water masses is for the most part, but not exclusively, due to large Agulhas rings that are shed at the Agulhas retroflection. Direct transfer in the form of filaments also occurs. These inter-basin exchanges of waters influence global climate, local ocean dynamics, biogeographic patterns and possibly even local fish recruitment.

Agulhas rings are major mesoscale features that average 250 km in diameter and extend to more than 1000 m depth. They drift into the South Atlantic at rates of about 7 km/day (8 cm/s). An estimated 6 to 9 are formed each year with a range of dimensions. Recent studies have estimated the total inter-basin heat flux due to Agulhas rings as being 0.05 PW per annum, and the salt flux as 78 x 10^{12} kg per annum.

A secondary, and minor, exchange between the ocean basins is due to Agulhas filaments. These are detached from the landward edge of the southern Agulhas Current and are then advected into the South Atlantic by the Benguela drift. These features are only about 50 m deep on average, lose most of their heat to the atmosphere, but contribute a salt flux that is estimated to be 29 x 10^{12} kg per annum, or about 12% of that due to Agulhas rings.

Agulhas rings are formed through complex processes. Perturbations in the flow paths of the Agulhas Return Current, and that of the Agulhas Current proper, cause the opposing flows to coalesce and thus to pinch off a ring. In this process quantities of cold, Subantarctic Surface Water are forced into the South Atlantic, further complicating the array of water types already present. The driving forces that bring about perturbations in the Agulhas system are poorly understood. Meanders in the southern Agulhas Current are similar to those observed in other western boundary currents, but are, by comparison, small. A large soliton meander on the current, the Natal Pulse, occurs at irregular intervals, about 8 times per year. Recent analyses suggest that the passage of each Natal Pulse is followed by the spawning of an Agulhas ring. The triggering of a Natal Pulse occurs far upstream in the Natal Bight and may be caused by the impingement of deep-sea eddies on the Agulhas Current, or by fluctuations in the velocity of the current leading to barotropic instabilities.

Introduction

The circulation of the South Atlantic Ocean is remarkable in at least two major ways. First, there is a net northward heat flux in the low- and mid-latitude (Hastenrath 1982); second, it receives an intermittent supply of heat and salt from the subtropical gyre in the South Indian Ocean.

The leakage of water from the South Indian Ocean, presumably for the greater part by way of the Agulhas Current, has attracted considerable interest, particularly in the past decade. This has led to a large number of research cruises (e.g. Camp et al. 1986; Lutjeharms 1987b; Read et al. 1987; Bennett 1988; Rigg et al. 1992; Lutjeharms et al.

From WEFER G, BERGER WH, SIEDLER G, WEBB DJ (eds), 1996, *The South Atlantic: Present and Past Circulation.* Springer-Verlag Berlin Heidelberg, pp 125-162

1994), modelling efforts (e.g. Boudra and Chassignet 1988; Chassignet and Boudra 1987; de Ruijter and Boudra 1985; Lutjeharms et al. 1991b; Holland et al. 1991; van Ballegooyen et al. 1991) and theoretical studies (e.g. de Ruijter et al. 1994; de Ruijter 1982; Darbyshire 1972). The reason for this focused research is that this transfer of water between the two oceans is critical in the global ocean heat and salt balance and hence the coupled ocean-atmosphere circulation that controls the global climate and climate change (IPP 1990).

Although most of the recent attention has been concentrated on the westernmost termination of the Agulhas Current and on the dynamical processes responsible for the exchange of water above the thermocline, it is most likely that an improved understanding of the nature of the general circulation in the Southwest Indian Ocean is a prerequisit to a proper interpretation of what occurs south of Africa. Mesoscale perturbations to the flow regime in the Southwest Indian Ocean are likely to have a controlling influence on the dynamical behaviour of the Agulhas retroflection (e.g. Lutjeharms 1989).

The aims of this paper are to present a comprehensive review of what is presently known about the nature and characteristics of the full Agulhas Current circulation regime (Fig. 1) and to discuss what influence this may have on the exchange of water masses between the South Indian and South Atlantic Oceans. Attention will be given to suggestions made by recent theoretical results and by results of numerical modelling in areas where our present knowledge is poorest.

The Agulhas Current lies on what was a major commercial route for sailing vessels of the seventeenth to nineteenth centuries and has therefore attracted research attention from a very early stage in the development of marine science (Rennell 1778; KNMI 1857). The first modern study of the southern Agulhas Current, using all available hydrographic data, was carried out by Dietrich only in 1935, for his doctoral dissertation, and he showed, in a very coarse form, the main dynamic features of the current through the entire water column. The data at that time could not be used to estimate the amount of Agulhas water penetrating into the South Atlantic.

This perennially recurring question has been addressed by a number of ocean scientists over the years (viz. Lutjeharms et al. 1992a) and it continues to be asked up to the present. For instance, in an investigation on the atmospheric pressure differences along the southwest coast of Africa and the corresponding sea surface temperatures, Schell (1968) has contended that stronger southeasterlies are related to a weaker penetration of Agulhas Current water into the South Atlantic. It was only with the acquisition of a first full, and detailed, hydrographic data set for the whole Agulhas Current (Bang 1970) that the richness of mesoscale detail of the circulation in this region became apparent and that the complexity of interchange processes between the ocean basins was clearly demonstrated. The leakage of water from the Agulhas Current into the South Atlantic and the resulting presence of anomalously warm water to the southwest of Africa has a number of wide-ranging effects, both on the dynamical processes in the South Atlantic, and on the biogeography of the area. The overlying atmosphere is also affected.

One of the dynamical effects is on the upwelling regime of the Southeast Atlantic. The major eastern boundary feature of the South Atlantic is namely the Benguela upwelling system (Shannon 1985) that is the major source of surface nutrients and hence of primary productivity for this ocean. It has been recognized that the front between this upwelling regime and the open ocean is a convoluted and complex system with many spatial and temporal scales. A major part of the primary productivity probably takes place at this front. The observed effect of the northward penetration of Agulhas filaments on the stability of this front (Lutjeharms 1981b) has suggested an important role for products of the Agulhas Current on the upwelling system. With a much more intensive and far-ranging study, Duncombe Rae et al. (1992a, b) have described the interaction of an Agulhas ring with the South Atlantic upwelling system. Very long upwelling filaments, probably driven by intense, land to sea, berg winds (Lutjeharms et al. 1991a), have been observed to be incorporated into the borders of an Agulhas ring. It has even been speculated that this process may contribute to the overall level of productivity of the coastal upwelling system (Duncombe Rae et al. 1992a). The quantification of the influence on the upwelling system of a greater or lesser influx of Agulhas water into the South Atlantic (e.g. Taunton-Clark and Shannon 1988; Shannon et al. 1989; Shannon et al. 1990b; Shannon et al. 1990a) is fraught with difficulties. Nevertheless, it seems apparent that these products of the Agulhas Current have an important local role in the ocean dynamics of the southeastern Atlantic Ocean.

It would also seem immediately evident that the penetration of a large body of essentially foreign water, laden with tropical and subtropical organisms, would also have a notable effect on the local biogeography of the South Atlantic Ocean. In the waters over the Agulhas Bank (viz. Fig. 1) and along the west coast of southern Africa typical Indo-Pacific chaetognath species, for example, such as *Sagitta enflata* that are normally found off Durban (Schleyer 1985), have been considered unambiguous indicator species for the presence of Agulhas water (de Decker 1973). Surface sampling of copepoda (de Decker 1984) over a distance of 26000 nautical miles has given similar results. The biogeograhpical effect is therefore another important result of the inter-basin leakage of water.

The hydrographic characteristics of the Agulhas Current as well as of the products of its penetration into the South Atlantic may furthermore play a role in the distribution of organisms that find these characteristics congenial or not. Such characteristics include temperature, nutrient availability and vertical stability of the water column. So, for instance, the distribution of small cetaceans south of Africa has been considered to be due to the wide range of zoogeographical regimes created here by the Agulhas Current system (Findlay et al. 1992). Whereas the ecological nature of Gulf Stream rings has received considerable attention (e.g. Ring Group 1981), the fate of organisms in Agulhas rings (Cockroft et al. 1990) in the South Atlantic has, to date, engendered minimal interest. By contrast, a great deal of interest has been shown in the effect of the Agulhas Current leakage into the South Atlantic on the overlying atmosphere.

Brundrit and Shannon (1989) have speculated that an increase of inter-basin flux of Agulhas water past the Cape of Good Hope may have an immediate and measurable effect on the intensity of storms in this region. [This research problem of the influence of the Agulhas on storms is not new, having been broached long before by van Gogh (1858).] The area southwest of Africa is the major region of heat flux from the ocean to the atmosphere in the South Atlantic Ocean (Bunker 1988; Walker and Mey 1988, viz. Fig. 9). It therefore comes as no surprise (e.g. Jury and Walker 1988; Mey et al. 1990) that his heat flux leads to major modifications of the air-masses over elements of the Agulhas circulation system. These in turn have been shown to affect directly the rainfall over the adjacent coastline (Jury et al. 1993) and are linked to changes in the summer rainfall over the southern African subcontinent (Walker 1990). Mason (1990) has even put forward the view that such sea surface temperature variations may be responsible for the hypothesized 18-year rainfall oscillation over southern Africa.

It is clear that the southern termination of the Agulhas Current and inter-basin leakage of its waters play an important role in a number of processes that affect the southern Atlantic and adjacent land masses. From an oceanic viewpoint, however, the exchange of heat and salt between ocean basins as part of a global circulatory system is, arguably, the process that demands most attention at present.

Global ocean thermohaline circulation cell

One of the components of the circulation system of the global ocean that has received much attention recently is the thermohaline circulation cell, the so-called oceanic conveyor belt (Broecker et al. 1985). This circulation appears to be driven by the excess salt left behind as the result of a net atmospheric water vapour transport from the Atlantic to the Pacific (Broecker 1991). A result of the operation of the cell, drawing warm water northwards in the Atlantic, is the heat that maintains the anomalously warm climate of northern Europe. The cessation of its operation was suggested (Broecker 1987) as the possible cause of the cold conditions that persisted over western Europe during the Younger Dryas event 10000 years ago, clearly therefore of substantial climatological interest.

The global oceanic thermohaline circulation involves numerous recirculations at all levels and is highly complex; the conveyor belt is a grossly simplified idealization that consists of just a near-surface and a deep component. The deep part starts where warm, salty water spreads into the northern reaches of the North Atlantic and is cooled by evaporation (Gordon 1986). These waters' increased density cause them to sink to the deep ocean, forming North Atlantic Deep Water (NADW) that spreads southward mainly within a deep western boundary current. This layer stands out as a body of high salinity water with low nutrient content in any vertical section of water properties in the Atlantic. On approaching the equator the NADW overrides a wedge of Antarctic Bottom Water moving northwards below about 4 000 m. South of 40°S this NADW becomes part of the Antarctic Circumpolar Current where it is rapidly, and progressively, mixed downstream (Broecker 1991). This diluted NADW then advects into both the Indian and Pacific Ocean while part of it continues on in the Antarctic Circumpolar Current to re-enter the South Atlantic via the Drake Passage. In the Southern Ocean, the Indian as well as in the Pacific, the NADW and its mixed products gradually mix vertically to form the surface part of the conveyor belt. This simplification of global circulation excludes the important role played by the Mediterranean outflow.

Part of the surface return route carries water from the Pacific equatorial region to the Indian Ocean through the Indonesian archipelago (Gordon 1986). Most of this water then flows down the southeastern coast of Africa, probably as part of the Agulhas Current, and leaks into the South Atlantic. This warm shallow water eventually finds its way across the equator and ends up in the Norwegian, Greenland and the Labrador Seas where the subduction process again takes over and the cycle continues.

Of considerable importance in this very simplified depiction of this "conveyor belt" system are the contributions by the various components. So, for instance, there has been some debate about the relative contributions of the surface component into the South Atlantic via the route south of Africa and via the route south of America. From radio-carbon

measurements Broecker et al. (1990) have come to the conclusion that the Agulhas Current pathway accounts for only about one quarter of the return flow while that through the Southern Ocean via the Drake Passage for the rest. Gordon (1986), on the other hand, has considered the flow through the Drake Passage as insignificant, and the Indian-Atlantic transfer of thermocline water at the Agulhas termination as the major pathway (Gordon 1985). An analysis by inverse methods of hydrographic data (Rintoul 1991) suggests, by contrast, that this global thermohaline circulation cell is closed primarily by the intermediate water entering the South Atlantic past Cape Horn. A large transfer of warm thermocline water of Indian Ocean origin is not, according to this box scheme, required to balance the export of heat by deep and bottom waters out of the Atlantic. Using recent tracer chemistry Gordon et al. (1992) have nevertheless concluded that there is strong evidence for a significant interbasin exchange of thermocline and intermediate water between the South Indian and South Atlantic Oceans.

Since the circulation system of the Southwest Indian Ocean does seem to play a significant role in the global thermohaline cell, any shortcomings to the argument or gaps in the data in this region may well be important. Such deficiencies do exist. In his analysis of water mass exchanges in the South Indian Ocean Gordon (1985) has for instance come to the conclusion, from temperature-salinity relationships, that a sufficient flow is maintained through the Mozambique Channel to maintain this component of the global cell. Since the continuity of a Mozambique Current is now seriously being called into question (e.g. Saetre and Jorge da Silva 1984; Zahn 1984), the mechanism by which this transport could take place is unclear.

Similarly, all the flow of water from east of Madagascar (Gordon 1985) is usually assumed to be part of the East Madagascar Current. Many recent results have indicated that this is not necessarily the case (e.g. Lutjeharms et al. 1981; Lutjeharms 1988a; Gründlingh et al. 1991). In fact, there is increasing evidence that the Agulhas Current may be driven for the greater part by recirculation in a sub-gyre west of 15°E (Wyrtki 1971; Harris 1972; Lutjeharms 1976) and in a larger component of the subtropical gyre that closes at about 80°E (Stramma and Lutjeharms 1995).

These results do not necessarily mitigate against either the Gordon (1985) or indeed the Rintoul (1991) portrayal. They do suggest, however, that a further understanding of the processes and circulating mechanisms that are part of the Agulhas Current circulation system is required in order to interpret correctly the leakage into the South Atlantic and hence the influence of the Agulhas on many aspects of the behaviour of this ocean.

Agulhas circulation system

Southwest Indian Ocean circulation

The circulation of the South Indian Ocean is unique in the sense that its subtropical gyre is separated from the seasonally changing monsoonal gyre to the north by a very pronounced front at about 10°S (Wyrtki 1973); this front partitions the waters of low oxygen content and high nutrients in the North Indian Ocean from those with high oxygen and low nutrients in the South. The South Indian subtropical gyre is limited in the south by the Subtropical Convergence at about 45°S (e.g. Lutjeharms 1985; Stramma 1992).

The anti-cyclonic subtropical gyre consists of the South Equatorial Current in the north, the Agulhas Current system to the west and an eastward flow, including the South Indian Ocean Current (Stramma 1992), along the Subtropical Convergence, to the south. The eastward closure of this gyre is complicated by the presence of the seasonally varying Leeuwin Current along the west coast of Australia (Cresswell and Golding 1980). Northward movement across the gyre is present at all longitudes (Wyrtki 1971), but most pronounced at about the longitude of the mid-Indian Ocean ridge south of India (Stramma and Lutjeharms 1995).

The warm water in the gyre is deep, the 12°C isothermal surface being depressed to more than 600 m depth in the Southwest Indian Ocean (Wyrtki 1971). The anticyclonic flow generally penetrates to the intermediate water at 1000 m, but much deeper in the Southwest Indian subgyre (Wyrtki 1973), which forms a pivotal point for the whole circulation of the subtropical gyre. The South Equa-

torial Current transports about $55 \times 10^6 \, m^3/s$ with highest transports occurring during the Southwest Monsoon. At the northern tip of Madagascar about one third of this water is diverted southward along the east coast of Madagascar.

The Agulhas Current is the western boundary current of the South Indian Ocean being most intense in the region between Durban and Port Elizabeth along the eastern coastline of southern Africa (Fig. 1). It is highly baroclinic, extends deeper than 2000 m, has surface speeds that often exceed 2 m/s (Wyrtki 1973) and a mean transport of about 100 $\times 10^6 \, m^3/s$. South of Africa it turns abruptly eastwards (Fig. 1) in a tight retroflection loop. The southern part of this loop is known as the Agulhas Return Current which becomes the South Indian Ocean Current farther downstream. While the Subtropical Convergence is very pronounced where the Agulhas Current water is forced up against it, to the south of Africa, it becomes gradually weaker and more indistinct towards Australia (Stramma 1992).

This above description gives a portrayal of the general circulation of the Southwest Indian Ocean, but of particular relevance to a proper understanding of the dynamic processes involved, is also knowledge of the mesoscale processes (Lutjeharms 1989). The distribution and intensity of mesoscale variability in the Indian Ocean is shown very effectively from the sea height variability, as measured by SEASAT, in Fig. 2. From this figure it becomes immediately apparent that the Agulhas Current system has by far the highest mesoscale variability in the Indian Ocean. High values are found south of Mozambique, but not in the Mozambique Channel. Values are even higher in the Agulhas Current proper and in the South Indian Ocean Current, but by far the highest are in the Agulhas retroflection region. These results agree roughly with those found previously from an analysis of ships' drift (Wyrtki et al. 1976).

This mesoscale variability is due to a number of processes, many of which have become more easy to study with the advent of remote sensing of thermal infrared by satellite (Harris et al. 1978; Lutjeharms 1981a). Harris et al. (1978) have for instance observed plumes of warm water south of Madagascar, meanders on the southern Agulhas

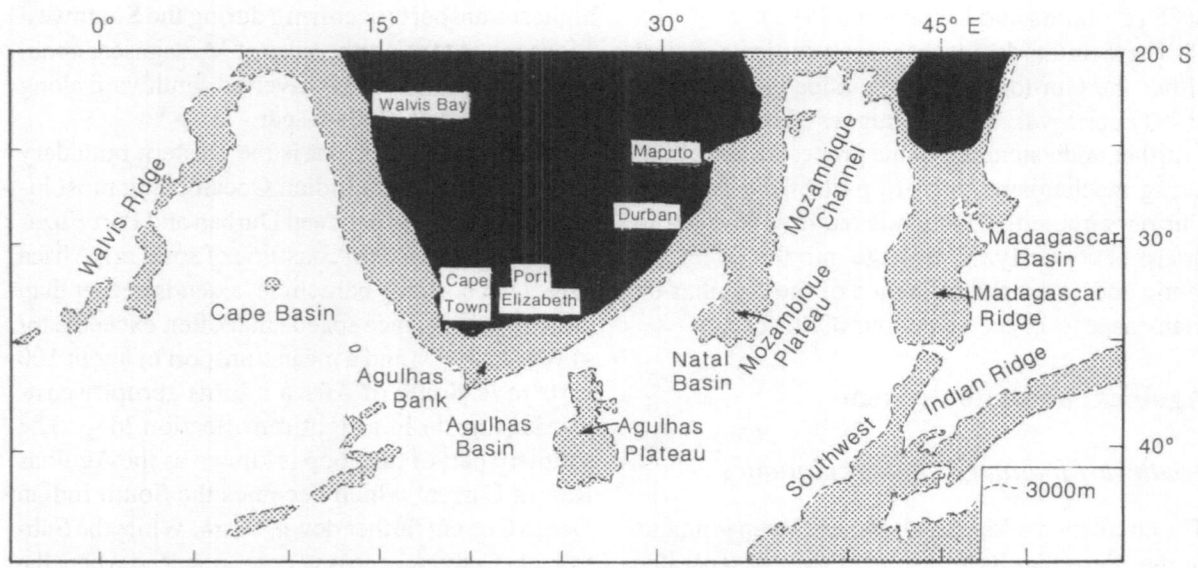

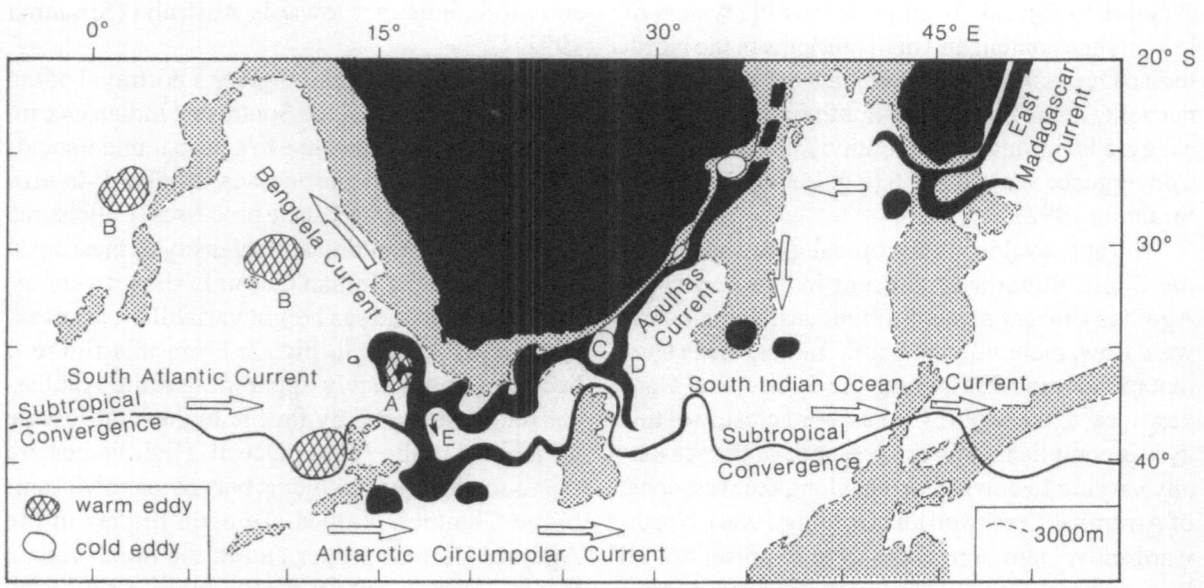

Fig. 1. The upper panel displays the main features of the bottom topography of the South Indian and South Atlantic oceans. In the lower panel the main mesoscale and larger features of circulation in the upper ocean layers of this region are portrayed. **A** is an Agulhas ring, recently shed from the Agulhas retroflection (**E**), being encircled by an Agulhas filament. **B**s are Agulhas rings drifting off into the South Atlantic. **C** is a cyclonic eddy forming part of a well-developed Natal Pulse on the Agulhas Current causing an upstream retroflection (**D**) at the Agulhas Plateau. **F** is the retroflection of the East Madagascar Current. Note the trapped lee eddies near Durban and Maputo. The large-scale, general background circulations of the subtropical gyres are shown by open arrows. (Modified from van Ballegooyen et al. 1991)

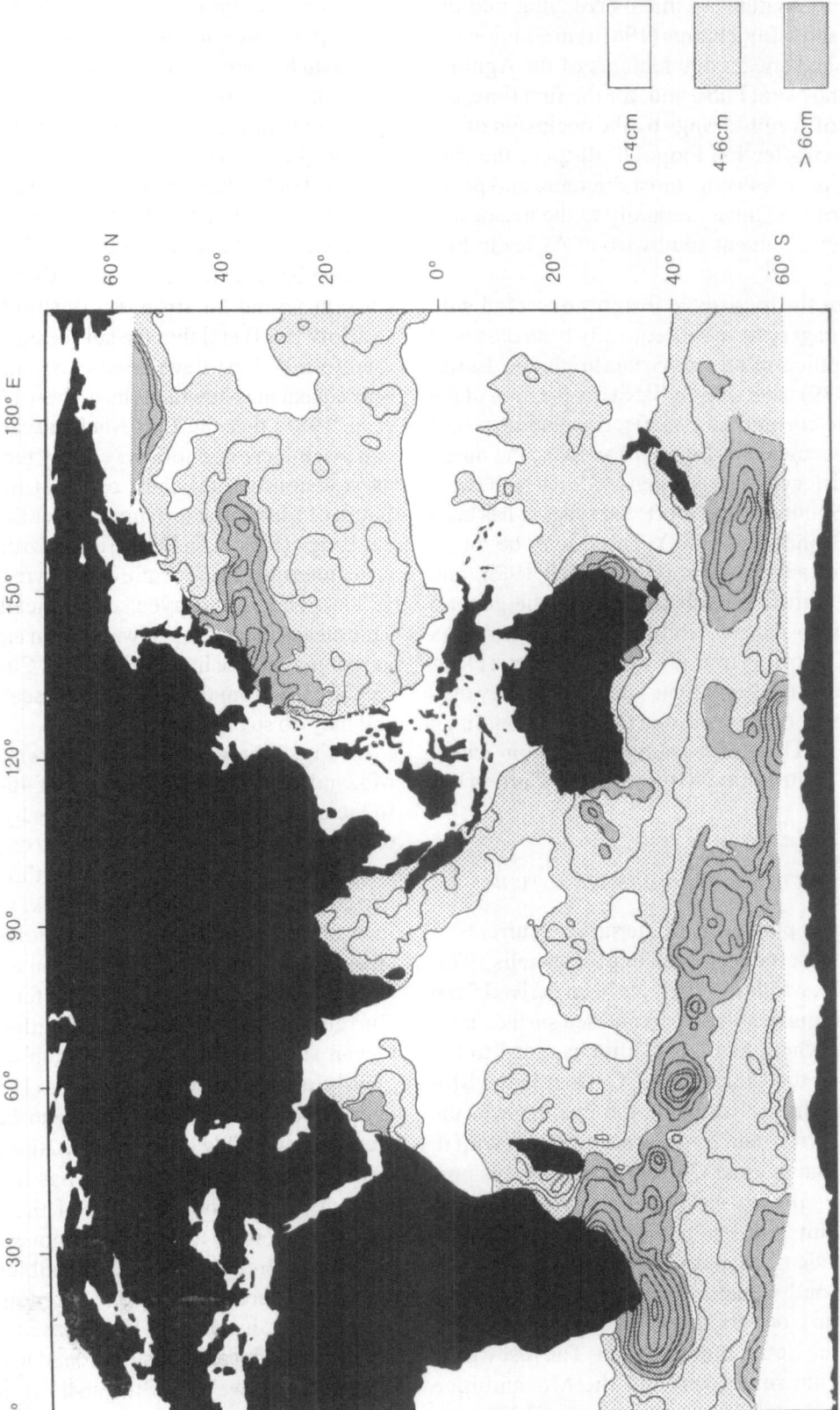

Fig. 2. Sea height variability on the mesoscale as measured by the SEASAT altimeter from 15 September to 10 October 1978. (Modified from Cheney et al. 1983)

Current and Agulhas filaments reaching into the South Atlantic. Lutjeharms (1981a) in addition has distinguished shear edge features of the Agulhas Current, the Natal Pulse and, for the first time, the shedding of Agulhas rings by the occlusion of the Agulhas retroflection loop. Of all these the ring spawning process is the most dramatic and probably contributes most markedly to the mesoscale variability so evident southwest of Africa in Fig. 2.

Most of the mesoscale features observed with satellite imagery have subsequently been observed hydrographically at sea. Gründlingh and Lutjeharms (1979) have substantiated the location of the core of the current, its retroflection and its return flow, as portrayed in Fig. 1. Moreover, the unexpected existence of large intense eddies in the centre of the Southwest Indian Ocean subgyre has been shown (Gründlingh 1988). These may be either cyclonic or anticyclonic (Gründlingh 1989) and probably contribute substantially to the general level of mesoscale variability found in this region (Fig. 2). The origin of these eddies is not yet clear, but altimetric measurements indicate that they may come from both the north and the east (Gründlingh et al. 1991). These areas may also be from where the flows contributing to the Agulhas Current derive.

Source currents of the Agulhas Current

The classical portrayals of the surface currents of the Southwest Indian Ocean (e.g. Michaelis 1923; Paech 1926; Möller 1929) have been derived from compendiums of ships' drift and sea surface temperatures, where available. This has lead to the portrayal of the Mozambique Current in the Mozambique Channel as being the precursor to the Agulhas Current and forming a continuum with it. The East Madagascar Current would, in this portrayal, flow directly westward south of Madagascar to feed into the Agulhas-Mozambique Current at the latitude of Maputo (viz. Fig. 1). Of the water of the South Equatorial Current only that of the East African Coastal Current would flow north and be lost to the South Indian Ocean. The rest would take separate routes through the Mozambique Channel and southwards to the east of Madagas-

car, but would rejoin to form the Agulhas Current. Subsequent measurements have shown that this portrayal has serious shortcomings.

An analysis of all the available hydrographic measurements taken during the Southeast Monsoon Season (Figure 3) has shown that at depth water from east of Madagascar does indeed join the Agulhas Current, but that in the upper layers there is no evidence of such a connection. This season was selected because it was believed that the Agulhas Current would be strongest during this season (Barlow 1933) and that the source currents would therefore be best developed (Lutjeharms 1976). Subsequent measurements have shown (Lutjeharms et al. 1981) that the East Madagascar Current is narrow, intense and follows closely the edge of the narrow continental shelf off East Madagascar. South of Madagascar it moves over the Madagascar Ridge (Fig. 1) and retroflects, with most of its water returning to the centre of the gyre (Lutjeharms 1988a). This is portrayed schematically in Fig. 4. This means that although water from east of Madagascar may move into the Agulhas Current, water specifically from the East Madagascar Current itself may do so only intermittently.

In Fig. 3 there is little flow shown through the Mozambique Channel at depths of 400 to 500 m. In fact, analyses of all the available hydrographic data in the channel (Saetre and Jorge da Silva 1984) has shown no clear and continuous throughflow as part of a Mozambique Current. These authors have instead suggested that a range of eddies of various sizes circumscribe the circulation in the channel.

A third possible source of water for the Agulhas Current is recirculation in a Southwest Indian Ocean subgyre (viz. Fig. 1). Such a subgyre is clearly evident in Fig. 3. Harris (1972) has attempted to estimate the relative contributions from these three possible sources to the Agulhas Current during the spring of 1964. He has found that 49 percent of the total transport of the Agulhas at Durban of 67×10^6 m^3/s, comes from east of Madagascar, 14 percent from the Mozambique Channel and that 38 percent is recycled. Gordon (1986) has surmized (as has Harris 1972) based on the temperature-salinity relationships found in the Agulhas Current, that water on the inshore side of the Agulhas comes from the Mozambique Channel

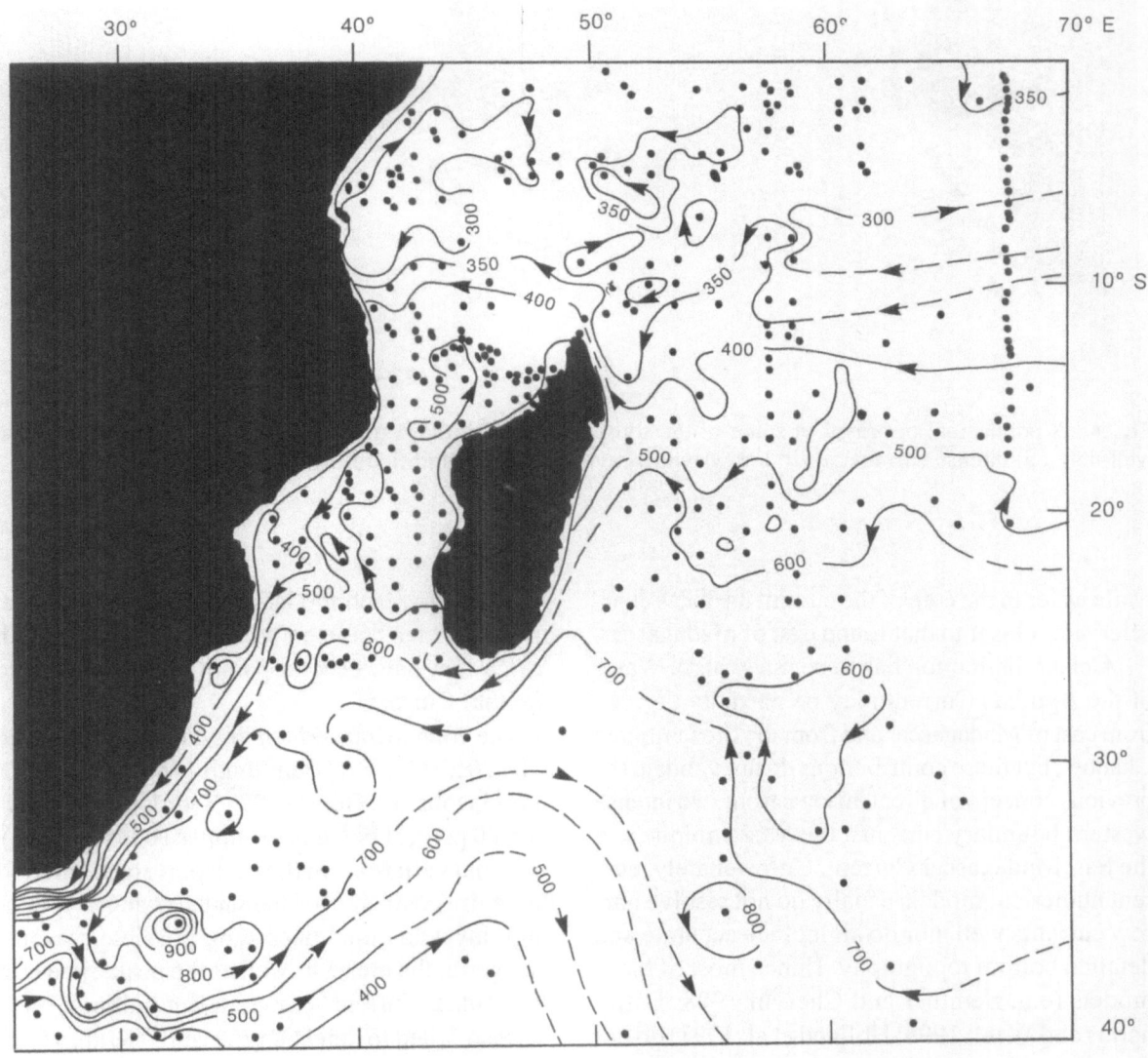

Fig. 3. Circulation of the Southwest Indian Ocean during the Northeast Monsoon Season as expressed by the depth of the 26.80 sigma-t surface. Contouring intervals are at every 100 m except for one 350 m contour inserted to clarify some features. Hydrographic stations on which the analysis is based are shown as dots. Note the intense subgyre off southeastern Africa. (Modified after Lutjeharms 1976)

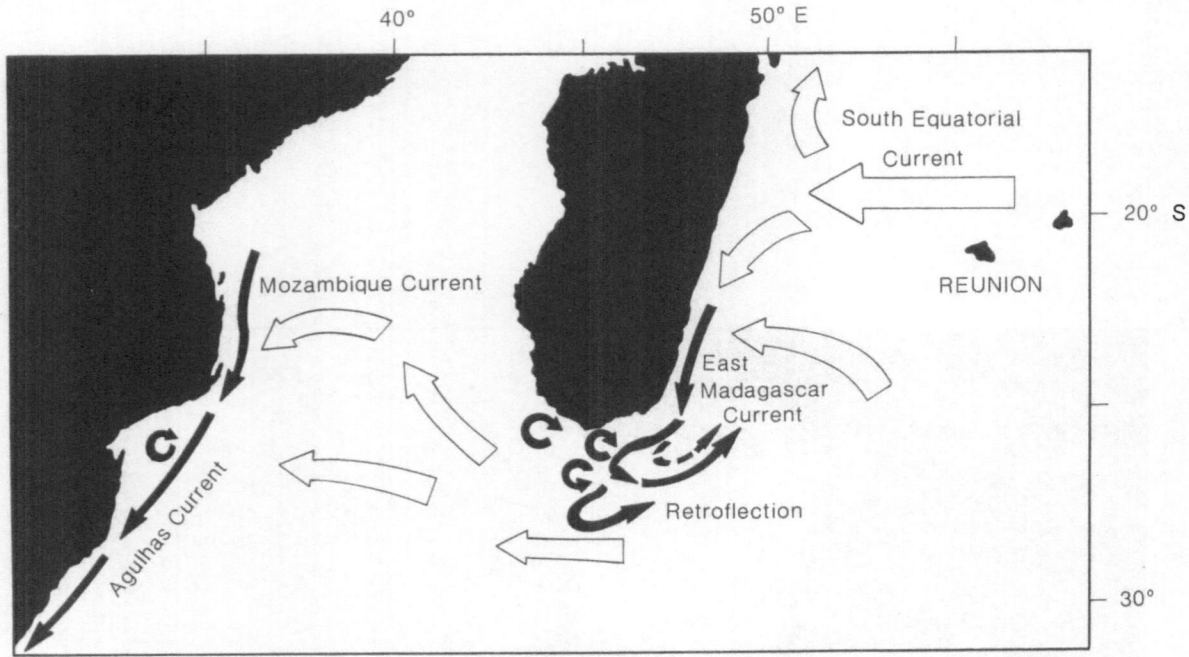

Fig. 4. A conceptual portrayal of some of the source currents of the Agulhas Current. Solid arrows indicate well-defined, intense currents: open arrows wider, average drift. (Modified from Lutjeharms et al. 1981)

while water in the core of the current itself has characteristics closer to that found east of Madagascar.

A clear distinction has to be made here. Water of the Agulhas Current may be partially derived from east of Madagascar and from the Mozambique Channel, but these contributions do not validate the previous concept of direct inflows from two intense western boundary currents, the Mozambique and the East Madagascar Current. Unfortunately, current numerical models usually do not resolve narrow currents well, nor do all include accurate and detailed bottom topography. Hence most of these models (e.g. Semtner and Chervin 1988; Lutjeharms and Webb 1995; Holland et al. 1991) do not show a retroflection of the East Madagascar Current, but show this current in a direct tributary role to the Agulhas Current.

Northern Agulhas Current

Notwithstanding our incomplete knowledge about the sources of the Agulhas Current, by the time these waters reach the latitude of Ponto do Oura, on the border between Mozambique and South Africa, they have become part of a fully constituted Agulhas Current.

The volume transport of this current is estimated to be 70×10^6 m³/s (Gründlingh 1980) to 140×10^6 m³/s (Jacobs and Georgi 1977; Gordon et al. 1987) with 80 percent being in the upper 1000 m (Fig. 5). Seasonal variations in the transport seem small, if they exist, compared to interannual changes. A thorough investigation concerning possible seasonal changes in the mean, as well as the peak, speeds of the Agulhas Current (Pearce and Gründlingh 1982) has also failed to detect any seasonal pattern (Fig. 6).

The current follows the edge of the narrow continental shelf closely (Harris and van Foreest 1978), all the way to the latitude of Port Elizabeth (Fig. 1). In fact, it has been shown that, with some infrequent exceptions, the current core meanders less than 15 km to either side (Gründlingh 1983) along this part of its trajectory. This invarient flow path

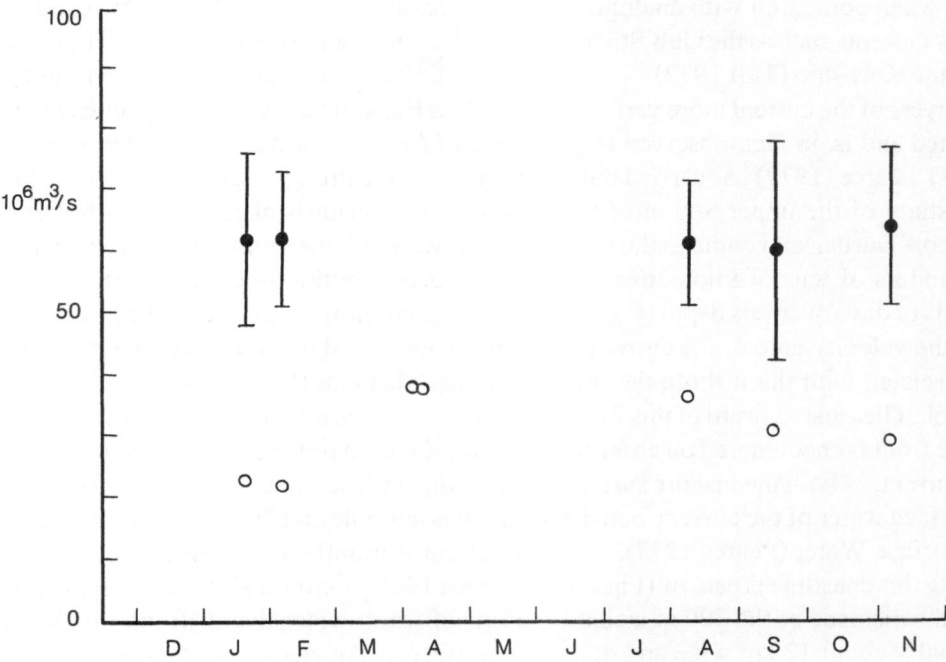

Fig. 5. Measurements of volume transport of the upper 1000 m of the Agulhas Current (dots with error bars) and geostrophically calculated transport (circles) over the calender year. (Modified from Gründlingh 1980)

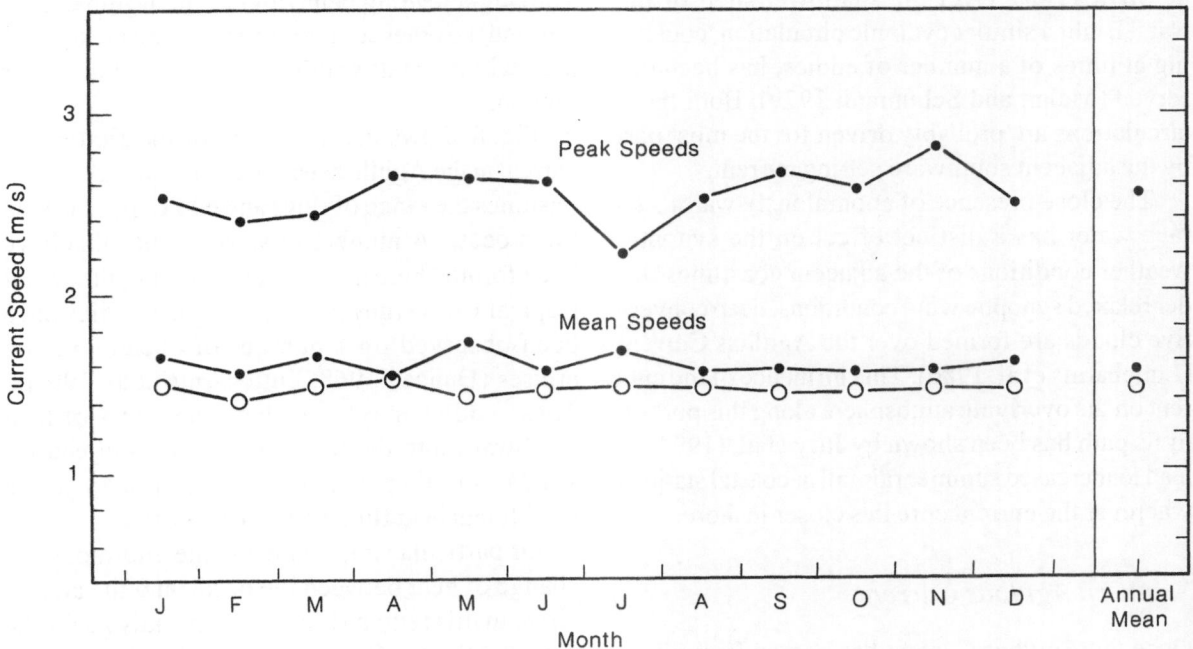

Fig. 6. Monthly mean and peak speeds in the core of the Agulhas Current based on ship-drift data. Annual averages are shown on the right. Curves with dots and with circles represent the results from two independent data sets. (Modified from Pearce and Gründlingh 1982)

is most unusual, when compared with analogous western boundary currents such as the Gulf Stream (Watts 1983) or the Kuroshio (Taft 1972).

In the upper layers of the current more variability can be expected and is, in fact, observed (e.g. Gründlingh 1974). Pearce (1977) has carried out a comprehensive study of the upper 500 m of the Agulhas Current off Durban and come to the conclusion that meanders of tens of kilometres can occur over periods of days over this depth (Fig. 7). The location of the velocity core of the current is usually well-correlated with the inshore thermal front of the current. It lies just seaward of this front. No distinct saline front is encountered on crossing the edge of the current, as both the inshore surface water and the surface water of the current consist of Subtropical Surface Water (Pearce 1977).

The shelf along this coastline is narrow (Fig. 1); as measured by the distance of the 200 m isobath offshore it is usually about 12 km wide and does not exceed 25 km except in two regions. North of Durban, in the Natal Bight, the shelf is 50 km wide, and an even greater width is found farther north, in the Delagoa Bight. In the latter region a trapped lee eddy can usually be found (Lutjeharms and Jorge da Silva 1988). Over the shallower shelf of the Natal Bight a similar cyclonic circulation, consisting at times of a number of eddies, has been observed (Malan and Schumann 1979). Both these circulations are probably driven for the most part by the adjacent southward setting current.

The close presence of anomalously warm surface water has a distinct effect on the synoptic weather conditions of the adjacent coastline. Under relaxed synoptic wind conditions, deep convective clouds are formed over the Agulhas Current (Lutjeharms et al. 1986). This influence of the current on the overlying atmosphere along this portion of its path has been shown by Jury et al. (1993) to lead to increased summer rainfall at coastal stations wherever the current core lies closer inshore.

Southern Agulhas Current

Once the Agulhas Current has passed Port Elizabeth and runs along the larger Agulhas Bank (Fig. 1) it starts meandering with progressively larger amplitudes (Lutjeharms et al. 1989). This leads to the formation of inshore shear edge features such as eddies and warm surface plumes. These contribute to the heat content of the surface waters of the Agulhas Bank. It has even been speculated that this input of Agulhas surface water makes a major contribution to the strongly developed seasonal thermoclines found on the bank (Swart and Largier 1987). Warm water plumes formed farther downstream, close to the southernmost point of the Agulhas Bank, have on the other hand been observed to advect northward into the South Atlantic Ocean as Agulhas filaments (Lutjeharms and Cooper 1995).

When the current leaves the shelf at the southern termination of the Agulhas Bank it retroflects in a tight, anticyclonic loop (Fig. 1). This configuration is unstable and the loop occludes at intervals of about 2 months (Lutjeharms and van Ballegooyen 1988a; Feron et al. 1992), thus pinching off rings of warm water that drift into the South Atlantic Ocean. As seen above, this region is characterized by extreme mesoscale variability (e.g. Cheney et al. 1983, Fig. 2). The nature of this variability is also evident in a portrayal of sea surface temperature fronts (Fig. 8) and in the mesoscale trajectories of surface drifters (e.g. Patterson 1985). The ring-shedding behaviour of the Agulhas Current and the other complex processes that take place here will be dealt with in greater detail in a later section.

Fig. 8 shows the instability of location of the current at the Agulhas retroflection, but it also demonstrates the range of rings and eddies produced in the process. A number of warm eddies that have been formed here are seen to lie south of the Subtropical Convergence (Fig. 8). Such eddies have been observed on a number of oceanographic cruises (Duncan 1968, Lutjeharms et al. 1994a). These eddies may carry large amounts of heat southward into the Southern Ocean (Lutjeharms 1987a) and thus contribute substantially to the meridional heat flux in the world ocean.

Of particular importance is the enormous exchange of heat between the ocean and the atmosphere in this region (Fig. 9). In an analysis for the whole Atlantic Ocean (Bunker 1988) this area stands out for its exceptionally high values of heat loss from the ocean. Walker and Mey (1988) have demonstrated that, in contrast to western boundary

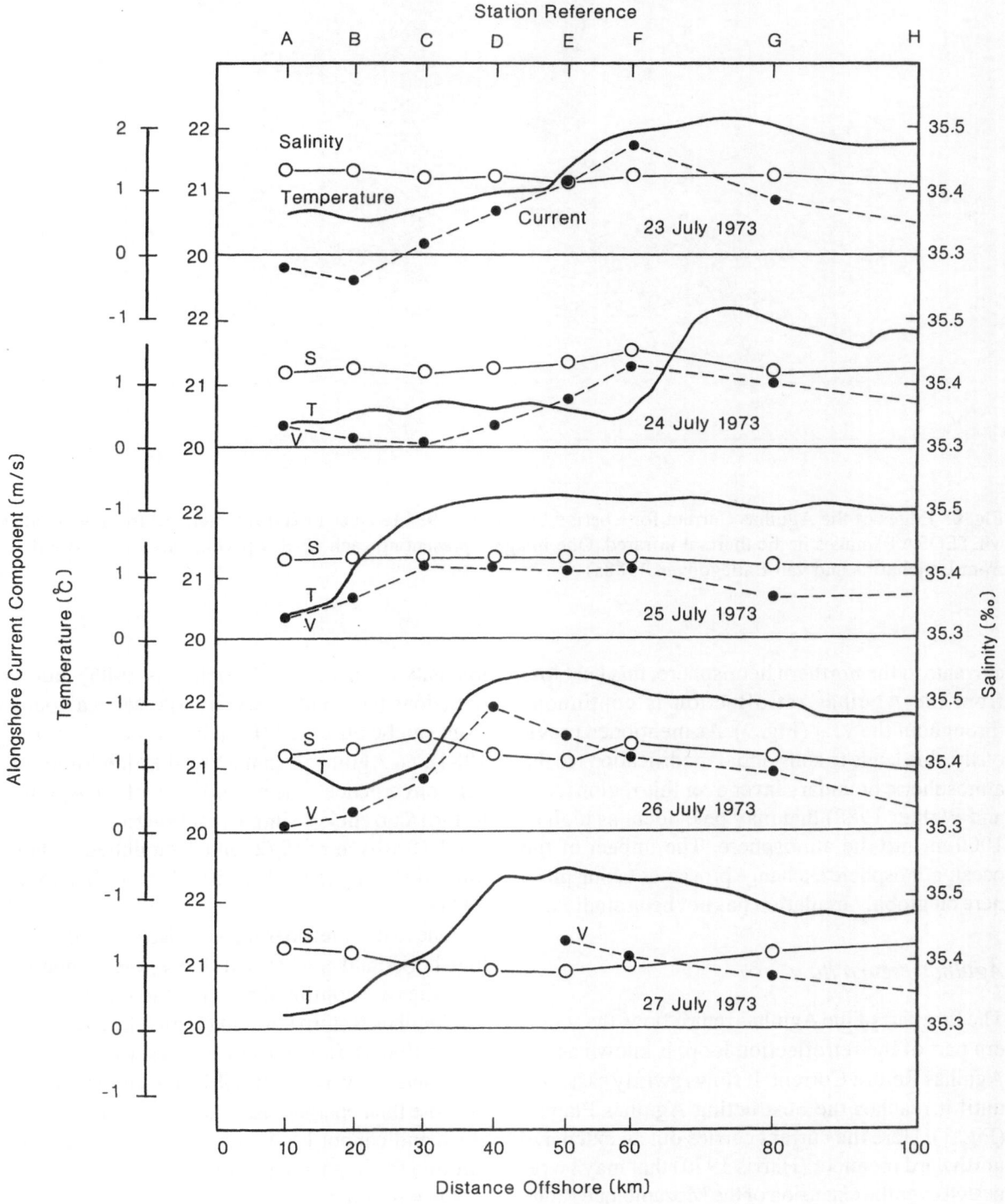

Fig. 7. Surface characteristics of the Agulhas Current as measured off Durban on the dates shown. (Modified from Pearce 1977)

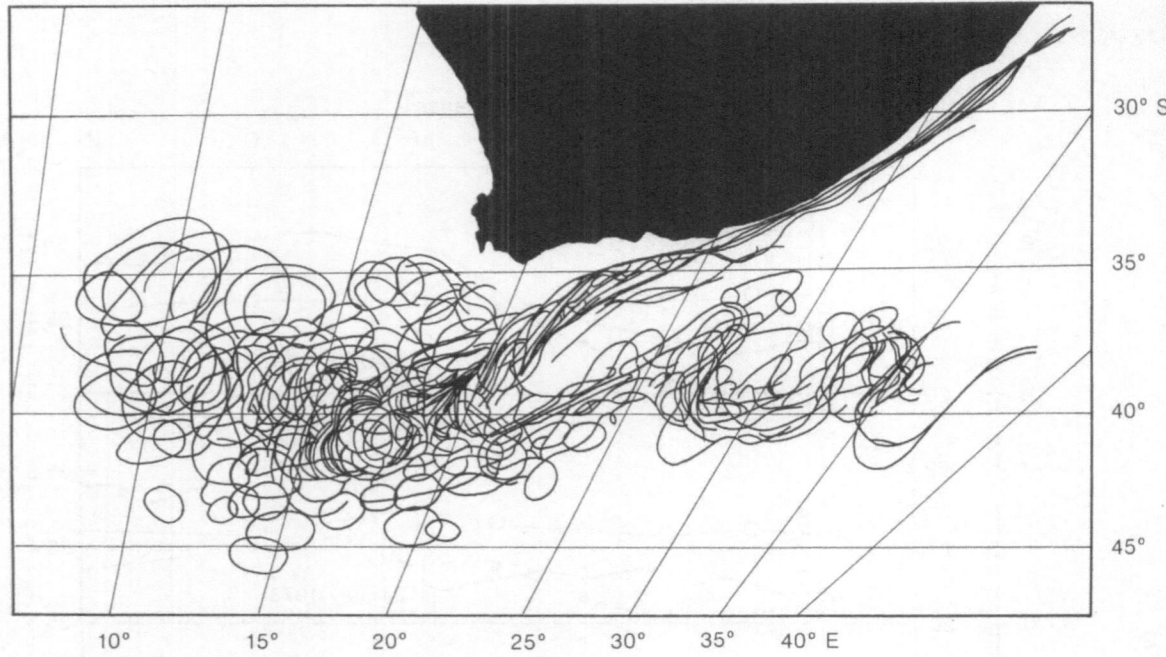

Fig. 8. Edges of the Agulhas Current for a period December 1984 to December 1985, derived from declouded METEOSAT images in the thermal infrared. One image representing each 12-day period was used. (Modified from Lutjeharms and van Ballegooyen 1988a)

currents in the northern hemisphere, this heat loss from the Agulhas retroflection is continuous throughout the year (Fig. 9). As mentioned previously, this leads to substantial modifications to the atmospheric boundary layer over this region (Jury and Walker 1988) that may be evident as high as 1000 m into the atmosphere. The impact of the ocean-atmosphere exchange processes taking place here on global circulation has not been studied.

Agulhas return flow

The first part of the Agulhas return flow, the southern part of the retroflection loop, is known as the Agulhas Return Current. It flows swiftly eastward until it reaches the obstructing Agulhas Plateau (Fig. 1). Here the current carries out an extensive northward meander (Harris 1970) that may be repeated over the extension of the Mozambique Ridge (Fig. 1) and the Mid-Indian Ridge south of Madagascar. The Agulhas Return Current as a rule follows the Subtropical Convergence, enhancing the

intensity of this front (Lutjeharms 1985), but on occasions their paths may diverge when a double front can be observed (Lutjeharms and Valentine 1984); an Agulhas Front as well as the Subtropical Convergence. The meandering of the Agulhas Return Current is a simple reaction to the variable depth (Darbyshire 1972) and is modelled without much difficulty (Lutjeharms and van Ballegooyen 1984).

Due to this meandering, and also due to the very high horizontal and vertical shears, a large number of eddies are continually formed to either side of the Agulhas Return Current (Lutjeharms and Valentine 1988a). Both the cold eddies drifting northward and the warm eddies drifting southward rapidly lose their characteristic surface thermal expressions and cannot be tracked very far by thermal infrared sensing from satellite.

On entering the eastward flow of the southern South Indian subtropical gyre this flow of Agulhas water becomes known as the South Indian Ocean Current and can be traced across the greater part

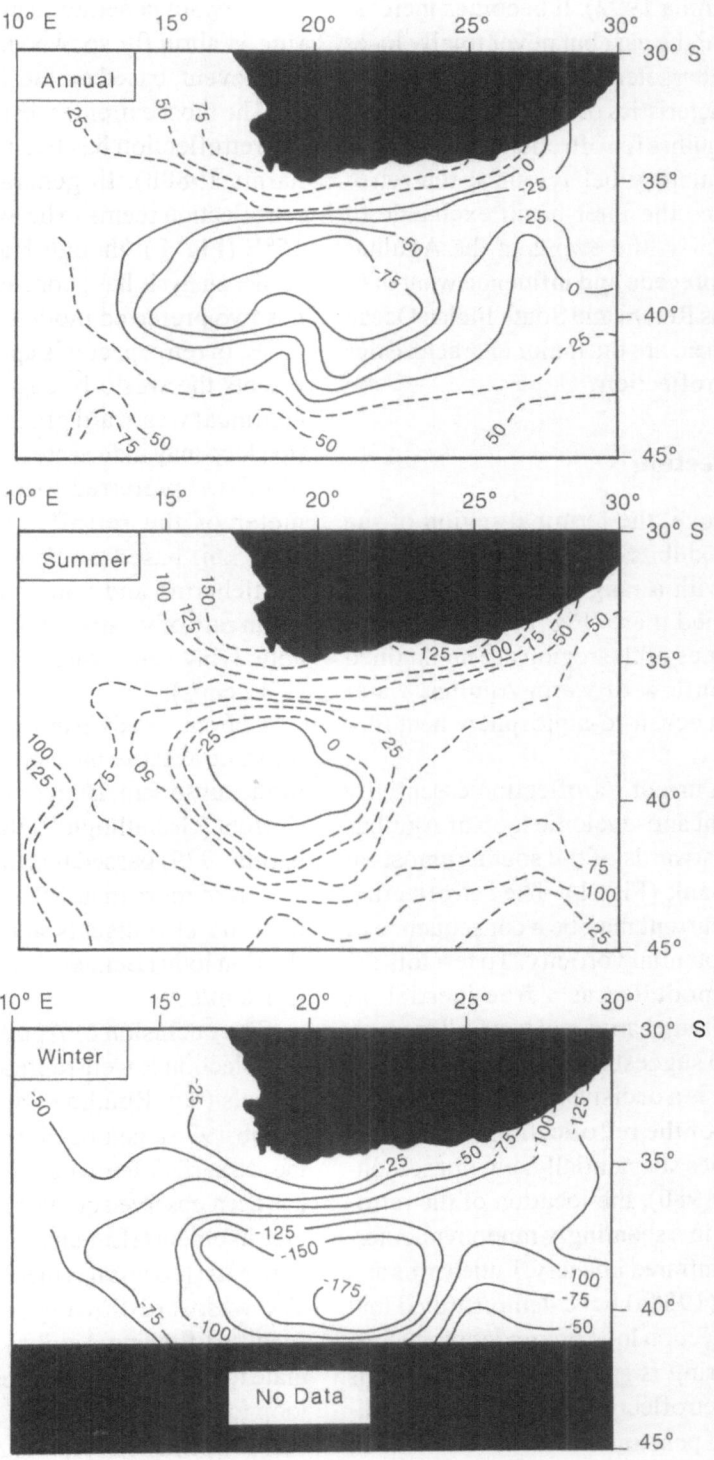

Fig. 9. Averages of net heat fluxes into the ocean from the atmosphere (in W/m²) over the western termination of the Agulhas Current. Negative values indicate heat lost from the ocean to the atmosphere and are shown by unbroken lines. Summer months used here are December to February; winter months, June to August. (Modified from Walker and Mey 1988)

of the basin (Stramma 1992). It becomes increasingly dilute toward the east but never totally loses its hydrographic characteristics.

The flow characteristics of the Agulhas Current upstream of the Agulhas retroflection will be shown to affect the circulatory behaviour at the retroflection, and hence the inter-basin exchange of water here. Similarly, the events at the Agulhas retroflection will precede and influence what happens to the Agulhas Return and South Indian Ocean Currents. What, then, are the major characteristics of the Agulhas retroflection?

Agulhas retroflection

As mentioned above, the terminal region of the Agulhas Current exhibits extreme mesoscale variability (Fig. 2), with a range of warm rings and eddies being formed there (Fig. 8). The high sea surface temperatures of the region are maintained by a continuous inflow of warm Agulhas water resulting in a high ocean-to-atmosphere heat flux (Fig. 9).

The Agulhas Current retroflection essentially consists of the tight anti-cyclonic loop of Agulhas water that lies westwards of the southernmost tip of the Agulhas Bank (Fig. 1). The retroflecting behaviour of the current may be a consequence of conservation of potential vorticity. To test this the current may be modelled as a free inertial jet (Darbyshire 1972; Lutjeharms and van Ballegooyen 1984). This model suggests that the volume transport of the current is a decisive factor which regulates the position of the retroflection. As the volume transport varies substantially, but not seasonally (Gründlingh 1980), the location of the retroflection may vary in a seemingly random manner.

Using satellite infrared imagery, Lutjeharms and van Ballegooyen (1988a) have demonstrated that the Agulhas retroflection loop progrades westwards until an Agulhas ring is spawned (Fig. 10). This event returns the retroflection to an eastward position. An analysis of penetrations of the retroflection loop into the South Atlantic over a period of three years has shown a number of consistent characteristics in this behaviour (Fig. 10). The progradations occur as identifiable events of variable duration and at irregular intervals. Within each event, the rate

of westward penetration increases until an Agulhas ring is abruptly spawned. The mean duration of each event, based on these data, is 39 days.

The most extreme westward longitude at which the retroflection has been observed is 5°E (Lutjeharms 1988b). In general the Agulhas Current retroflection seems to be located between 20°E and 15°E (Fig. 8), though Harris et al. (1978) in an earlier analysis have concluded that the retroflection has two preferred modes, one at 19°E and one at 13°E. In retrospect it is apparent that the latter represents the westerly edge of Agulhas rings. The continual westward progradation of the retroflection loop may indeed create the impression (e.g. Fig. 8) of two preferred locations. The average diameter of the retroflection loop is about 340 ± 60 km, based on data for a four-year period (Lutjeharms and van Ballegooyen 1988a). The mean rate of westward progradation is quite variable, lying somewhere between 7 and 15 km/day (8-17 cm/s).

Current speeds in the retroflection loop are much the same as those measured in the Agulhas Current farther upstream. Buoys drifting down the Agulhas Current (Gründlingh 1978, Gründlingh and Lutjeharms 1979) passed through the retroflection with speeds of more than 1.0 m/s. Geostrophic velocity maxima at transects across the Agulhas retroflection loop (Bennett 1988) lie in the range of 0.7 to 1.8 m/s.

The occlusion of Agulhas rings at the Agulhas retroflection is well-represented in most numerical models (e.g. Boudra et al. 1989, Lutjeharms and Webb 1995), can be clearly seen in portrayals of the sea surface temperature (Lutjeharms 1981a) and has been observed at sea with detailed hydrographic measurements (Lutjeharms and Gordon 1987). The kinematic process has therefore been well described. The westward-flowing Agulhas Current and the eastward-flowing Agulhas Return Current amalgamate to form a new retroflection loop, the occluded loop to the west thus being pinched off to form an independently circulating Agulhas ring.

This process sets up a number of secondary reactions in the regional flow. First, a wedge of cold, Subantarctic Surface Water is usually observed to flow northward through the gap left between the newly formed ring and the new retroflection loop

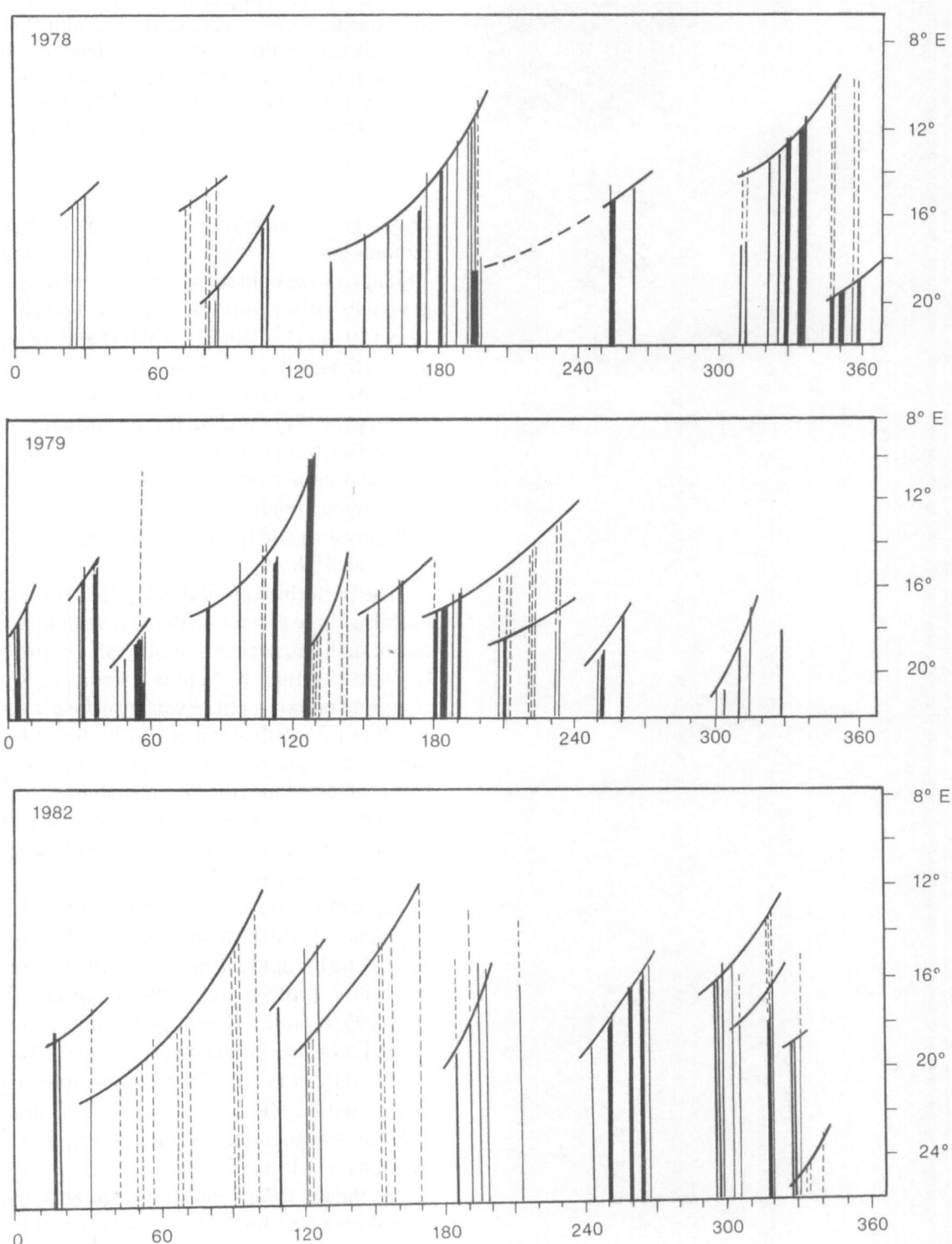

Fig. 10. The westernmost limit of the Agulhas Current retroflection during three years from METEOSAT thermal infrared imagery. Abscissas are in year days. Solid lines denote an intact retroflection loop. Dotted lines denote the detachment or possible detachment of an Agulhas ring. (After Lutjeharms and van Ballegooyen 1988a)

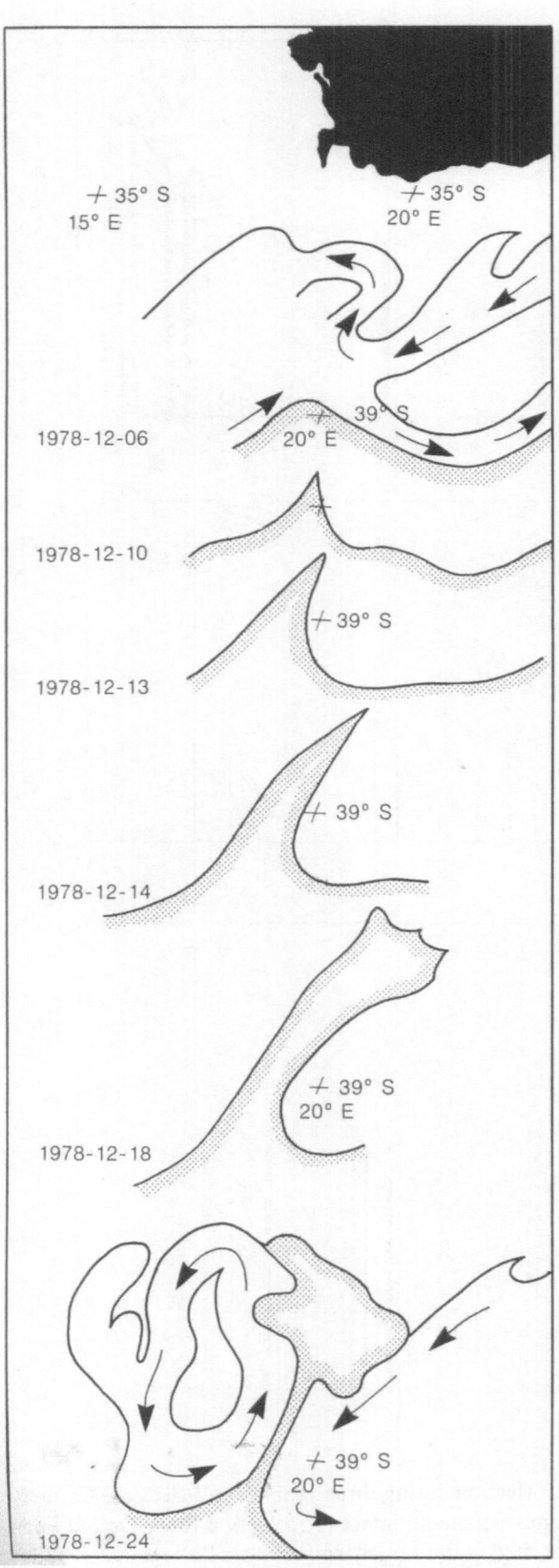

Fig. 11. Northward penetration of a cold wedge of Subantarctic Surface Water during the formation of an Agulhas ring at the Agulhas retroflection. This occurred in December 1978 and is characteristic of this process. Thermal data were from METEOSAT I. (After Lutjeharms and van Ballegooyen 1988a)

(Fig. 11). Usually the width of this throughflow remains narrow (Lutjeharms and van Ballegooyen 1988a) with the cold surface water covering a larger area only to the north. On occasion it can be wider than 150 km (Lutjeharms 1988b) and the surface area to the north can extend as far as 33°S into the South Atlantic and last for up to two months (Shannon et al. 1989). Such major perturbations to the sea surface temperature field have been suggested as significant contributors to intra-annual and interannual variability in the marine environment, both physical and biological, of the region (Shannon et al. 1990a).

The perturbations at the Agulhas retroflection can also come about due to other secondary processes. In the austral spring of 1985 a combination of increased flux by Agulhas filaments, Agulhas rings and entrainment of warm surface water into Agulhas rings (Shannon et al. 1990b) led to a prodigious amount of warm Agulhas surface water being advected into the South Atlantic. This documented intrusion was the cause of 1986 being the warmest year on record this century in the extreme southeastern Atlantic Ocean.

As can be expected for a region such as the Agulhas retroflection, in which a variety of water masses are brought together, the temperature-salinity relationships are unusually complex (Valentine et al. 1993, Table 1) exhibiting tropical, subtropical and subantarctic characteristics. Based on high-quality data collected during a mutliple-ship survey of the Agulhas retroflection (Gordon et al. 1987), it was found that the flow of Agulhas water into the retroflection was about 95×10^6 m³/s relative to the sea floor consisting of Subtropical Surface Water, Southwest Indian Ocean Central Water and Antarctic Intermediate Water. About 10 percent of this was lost to the South Atlantic directly. About 40×10^6 m³/s (relative to 1500 db) became included in a newly formed ring. Indian

Table 1. *Summary of the water masses of the southern Agulhas region (after Valentine et al. 1993)*

	Temperature range (°C)	Salinity range
Surface Water	16.0—26.0	>35.50
Central Water		
(i) Southeast Atlantic Ocean	6.0—16.0	34.50—35.50
(ii) Southwest Indian Ocean	8.0—15.0	34.60—35.50
Antarctic Intermediate Water		
Characteristic *T/S*	2.2	33.87
(i) Southeast Atlantic	2.0—6.0	33.80—34.80
(ii) Southwest Indian	2.0—10.0	33.80—34.80
Deep Water		
North Atlantic Deep Water (Southeast Atlantic)	1.5—4.0	34.80—35.00
Circumpolar Deep Water (Southwest Indian)	0.1—2.0	34.63—34.73
Antarctic Bottom Water	-0.9—1.7	34.64—34.72

Ocean water introduced into the retroflection region by the Agulhas Current was confined to levels shallower than 1500 to 2000 m. At greater depths the waters within the retroflection regime are of Atlantic or Circumpolar Deep origin (Gordon et al. 1987).

In terms of interbasin exchange, the most important features of the Agulhas retroflection system, based on their frequency of production and their volume flux discussed above, are without doubt the Agulhas rings.

Agulhas rings and eddies

Agulhas rings have, at the sea surface, very distinct annular shapes as observed with satellite remote infra-red sensing (Lutjeharms and Gordon 1987). Other eddies produced at the terminal region of the Agulhas Current (e.g. Lutjeharms 1987a) and farther downstream (e.g. Lutjeharms and Valentine 1988a) do not exhibit this structure. One may therefore distinguish between Agulhas **rings**, shed by

loop occlusion, and other Agulhas **eddies** formed in a number of different ways. Due to rapid heat loss to the atmosphere, the characteristic surface features of rings are rapidly attenuated and do not serve as good distinguishing characteristics for a long time.

Agulhas rings can be distinguished in other hydrographic ways. The portrayal in Fig. 12 is instructive in this regard. First, the dynamic height characteristics of the Agulhas ring closest to the current on this occasion were very similar to those of the Agulhas retroflection loop. Since the characteristics of this loop cannot be expected to be constant, the characteristics of the rings being shed could also vary considerably. Second, the dynamic height anomalies of the other ring (Fig. 12) were considerably reduced from those found in the retroflection loop. This may have come about through a large heat loss to the atmosphere (Lutjeharms and Gordon 1987; Walker and Mey 1988) and due to frictional effects by which rings will start to spin down (Byrne et al. 1995, Fig. 14). On the other hand

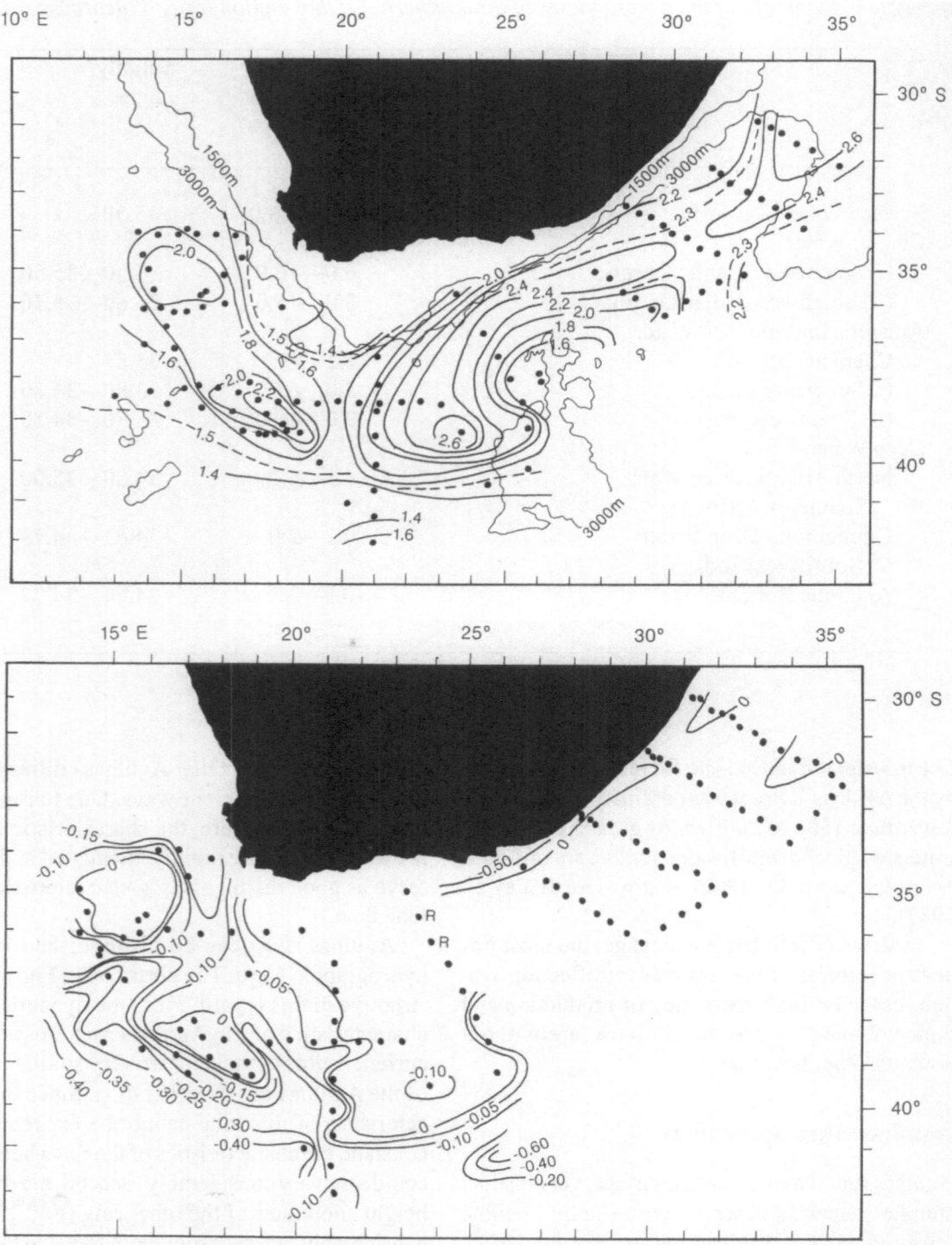

Fig. 12. The dynamic height anomaly of the sea surface relative to the 1500 decibar surface (upper panel) over the southern Agulhas region during a two-ship cruise undertaken during the period October to December 1983. The salinity anomaly at the sigma-theta 26.5 surface, using as reference two stations marked **R** in the core of the Agulhas Current (lower panel); from the same cruise. (Modified from Gordon et al. 1987)

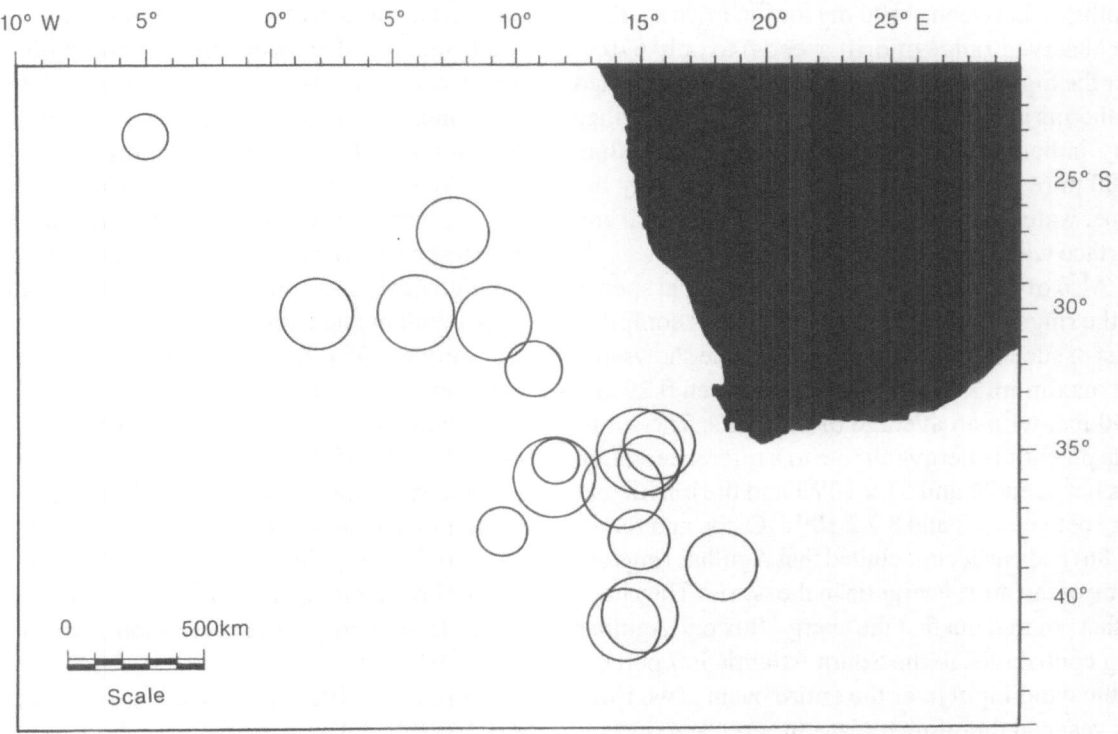

Fig. 13. The geographic location and dimensions of seventeen Agulhas rings that have been detected hydrographically in the southeastern Atlantic Ocean from 1969 to 1990. (Modified from Duncombe Rae 1991).

the hydrographic contrasts between the rings and their environment are considerably enhanced when these warm, saline bodies of Indian Ocean origin drift into the less saline environment of the South Atlantic Ocean (Fig. 12, lower panel).

The two rings shown in this figure had different shapes and sizes. The older, more northern ring, was more circular and had a diameter of about 200 km. The southern one had a more elliptical form with minor and major diameters of 100 to 250 km (Gordon et al. 1987). Duncombe Rae (1991) has carried out a review of 17 rings observed in the general vicinity of Cape Town (Fig. 13). The average diameter of maximum radial velocity for Agulhas rings, according to his data, is 240 ± 40 km; the depth of the 10°C isotherm at the ring centre 650 ± 130 m. These data include independent observations of Agulhas rings by Olson and Evans (1986), by McCartney and Woodgate-Jones (1991), Bennett (1988), Lutjeharms (1987b), Gordon and

Haxby (1990), Duncombe Rae et al. (1992b) as well as by Gordon et al. (1992), thus probably giving as reliable an estimate of the dimensions of these features as can currently be obtained. The depth of the 10°C isotherm has been shown to be a good proxy for the dynamic topography of these rings (e.g. van Ballegooyen et al. 1994). But to what depth can these features really be observed?

According to Flierl (1981), fluid particles may be trapped in, and thus advect with, a flow disturbance such as an eddy, or they may oscillate in position as an eddy passes. The depth to which particles are actually trapped in the eddy and move with it depends on the ratio between the maximum azimuthal speed and the translational speed of the eddy. Duncombe Rae (1991) has demonstrated that for Agulhas rings with the sizes and translational speeds of those observed by Olson and Evans (1986), Gordon and Haxby (1990) and McCartney and Woodgate-Jones (1991), the calculated trapped depths lie between 1 100 m (for the lowest end of the observed range of drift speeds) to only 670 m (for the highest drift speeds). Although depressed isotherms in vertical sections across Agulhas rings may indicate depths of penetration greater than 4000 m (e.g. Gordon and Haxby 1990), only the upper water masses (i.e. Intermediate, Central and Surface waters) might move with the rings.

Also of particular interest are the radial speeds of the rings. Duncombe Rae (1991) has compiled a list of such measurements in which he shows that the maximum radial speeds lie between 0.29 and 0.90 m/s with an average of 0.56 m/s. The available potential energy relative to a reference station lies between 26 and 51 x 10^{15} J and the kinetic energy between 2.3 and 8.7 x 10^{15} J. Olson and Evans (1986) had earlier concluded that Agulhas rings are some of the most energetic in the world. They have in fact pointed out that the energy flux one Agulhas ring contributes to the South Atlantic is 7 percent of the wind input over the entire basin. Two rings per year can therefore replace the eddy energy calculated for the basin outside the direct influence of the Agulhas retroflection.

It is consequently of considerable importance to estimate how many rings are shed per year. Using imagery from satellite infrared sensors, Lutjeharms and van Ballegooyen (1988a) have calculated an annual production of 6 to 9 rings. Chassignet and Boudra (1987) have suggested fewer, using similar data for a different period. Using altimetric data, Feron et al. (1992) have shown that the rate of ring shedding may be quite variable, ranging from 4 to 8 per year. It is not immediately clear whether all the features in the altimetric data are rings or whether some are eddies that go south of the Subtropical Convergence instead. This would limit the number even further. With altimetric observations, Gordon and Haxby (1990) have in fact estimated that only five Agulhas rings are shed annually. Even at the lower end of this suggested range the energy input into the South Atlantic, by the estimates of energy per ring by Olson and Evans (1986), is huge. How this energy is distributed in the South Atlantic depends on the duration of these Agulhas rings and their translational paths.

Based on the energy losses measured in other rings, outside the influence of strongly sheared currents, Olson and Evans (1986) have come to the conclusion that these rings should have a life-time of between 5 and 10 years. McCartney and Woodgate-Jones (1991) have observed a ring 3000 km from its point of inception and estimated that it was 2 years old. The high temperature contrast between ocean and atmosphere at the Agulhas retroflection, as well as other local atmospheric conditions, leads to a high heat loss from the sea, mentioned previously (Walker and Mey 1988, Fig. 9). Lutjeharms and Gordon (1987) have estimated a mean net heat loss from an Agulhas ring in early spring of 157 W/m^2, while Duncombe Rae et al. (1989) have estimated a loss of 80 W/m^2 in autumn. Such heat losses lead to convective overturning and considerable modifications to the upper layers of these rings, including enhanced salinities (Olson et al., 1992). Nevertheless, these effects are limited to the surface layers and the gradual erosion at depth takes a much longer time.

Byrne et al. (1995) have calculated the dissipation rates of Agulhas rings based on both altimetric data and on the potential energy estimated from fortuitous hydrographic measurements (Fig. 14). They have estimated that over a distance of 5000 km the surface amplitudes of rings decrease by 85 percent. The residence time of Agulhas rings in the South Atlantic has thus been calculated as being 3-4 years (Byrne et al. 1995). Since they last so long and have been observed to drift away from the retroflection area at a rate of about 5 to 8 km/day (6-9 cm/s) (Olson and Evans 1986) their general drift patterns and region of occurrence is of interest.

During the SCARC (Subtropical Convergence and Agulhas Retroflection Cruise) (Lutjeharms 1987b) six rings and eddies were observed clustered around the Agulhas retroflection, more or less as implied by the portrayal in Fig. 8. A study in one particular sector of this region, directly southwest of Cape Town (Lutjeharms and Valentine 1988b), has shown the moderately high frequency with which Agulhas rings are found there. That particular location is important because rings enhance the shelf edge current as well as the advection of Agulhas filaments into the South Atlantic (Lutjeharms and Cooper 1995). In his review of all rings that had been hydrographically observed up to this time

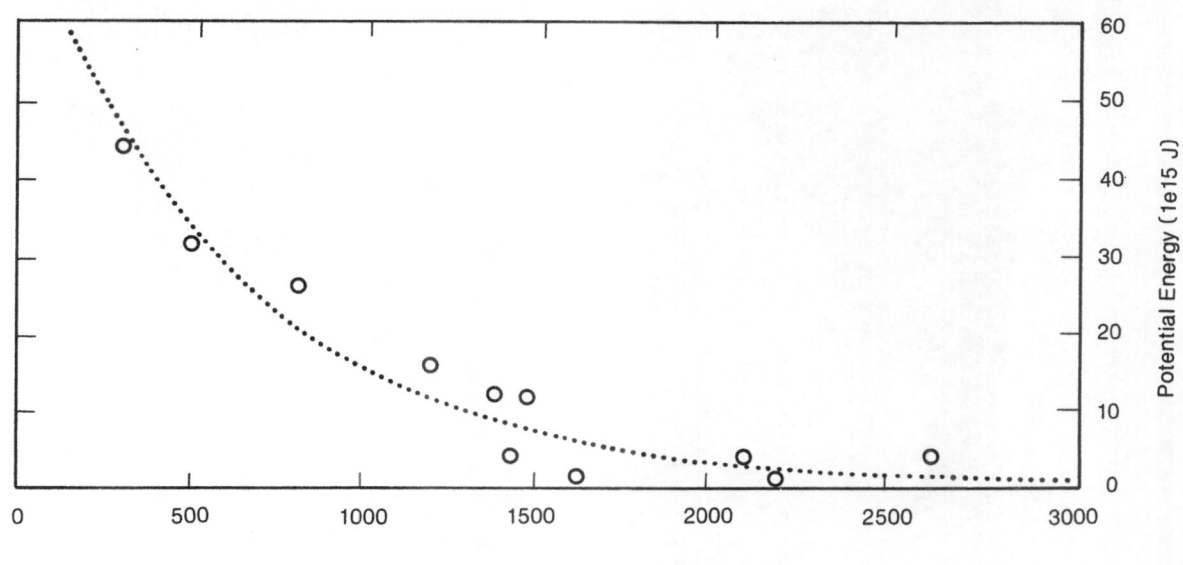

Fig. 14. The available potential energy of twelve Agulhas rings that have been surveyed hydrographically, plotted as a function of distance from the Agulhas retroflection. (Modified from Byrne et al. 1995)

Duncombe Rae (1991) has sketched the location and dimensions of these features (Fig. 13). The rings lie in a broad band stretching in a northwesterly direction, very similar to that produced by some modelling results (e.g. Lutjeharms and Webb 1995).

Satellite altimetry has been a very useful tool in tracking Agulhas rings, particularly since they have strong signals in sea level measurements and drift into a comparatively quiescent region. Wakker et al. (1990) have tracked a number of these features. They have found that near the retroflection, rings are closely packed but that most move off at speeds of 4-8 cm/s in a general northwesterly direction that corresponds roughly with the local flow direction of the South Atlantic subtropical gyre. These rates of progression agree well with, for example, those that have been found by Olson and Evans (1986) from drifter tracks. This has also been found by Gordon and Haxby (1990) who have tracked such features over the greater part of the width of the South Atlantic and, perhaps even more important, have been able to identify with hydrographic measurements one of these sea surface height anomalies as indeed representing an Agulhas ring. Van Ballegooyen et al. (1994) have established such correspondences with even greater success, using eight rings and eddies observed hydrographically at sea and following their subsequent progression. During a 2-year period half of the rings observed migrated no farther than 1500 km from the Agulhas retroflection. Eddies shed to the south of the Agulhas retroflection were all observed ultimately to move westward rather than southward across the Subtropical Convergence.

The most extensive study of this kind to date has been undertaken by Byrne et al. (1995). They have shown (Fig. 15) that all the sea height anomalies, assumed to be Agulhas rings, moved across the South Atlantic Ocean slightly to the left of the mean flow pattern. As a result, none crossed to north of a latitude of 25°S (Fig. 15). It is thought that these trajectories are due both to the general existing flow pattern (e.g. Reid 1989, Peterson and Stramma 1991) as well as to the internal dynamics of the rings

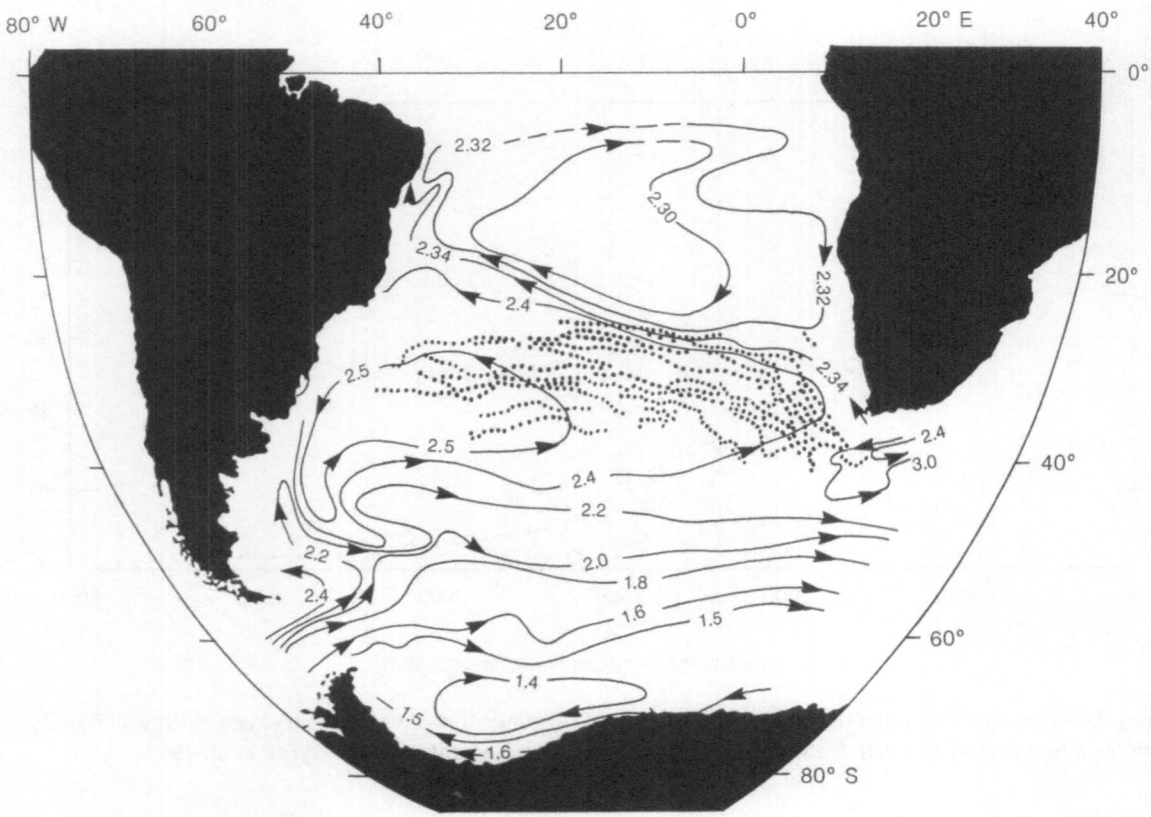

Fig. 15. The movement of anomalies in GEOSAT ERM altimetric data, inferred to be Agulhas rings, across the South Atlantic Ocean from 1986 to 1989 (lines of triangles). The general direction of movement is from the steric height anomaly at 500 dbar according to Reid (1989). (Modified from Byrne et al. 1995)

themselves. Some anomalies shown in Fig. 15 lay far to the south and moved in a direction opposite to that of the South Atlantic Current (Stramma and Peterson 1990) and the average circulation.

Over shallower bottom topography the rings have been observed to advect more slowly (Byrne et al. 1995) such, as for instance, over the Walvis Ridge (Fig. 1), although others (Wakker et al. 1990) have observed some rings to advect faster on crossing this ridge. Of particular interest are results suggesting that Agulhas rings select preferred routes across the Walvis Ridge (Byrne et al. 1995) on the whole crossing at deeper parts of the bathymetry (Fig. 16). Since the Walvis Ridge is deeper to the south and since all rings in this study took a more southerly route even before they reached the Walvis

Ridge (Fig. 15) it seems that the exact influence of the Walvis Ridge on Agulhas rings, if any, still has to be rigorously determined.

Perhaps the major findings about Agulhas rings during the recent past have been the following: first, that a variable number of rings are shed per year with a probable average of about one every two months; second, that these drift across the full width of the South Atlantic retaining enough of their dynamic and hydrographic characteristics to allow them to be identifiable as Agulhas rings in the western part of the basin; third, that they do not as a rule cross 20°S latitude. The most important aspect remaining is then, perhaps, to estimate what interbasin exchange of heat and salt they, and other flow features, bring about.

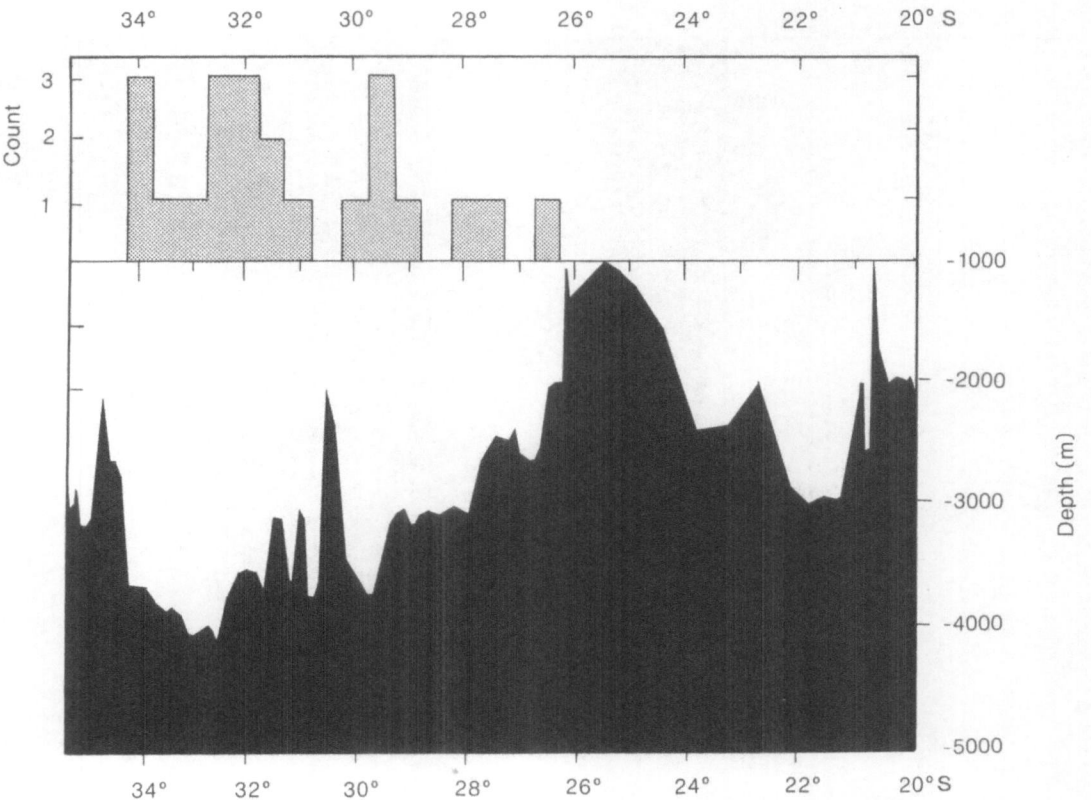

Fig. 16. The latitudinal distribution of Agulhas rings as they cross the Walvis Ridge. The bathymetry of the ridge follows its crest (viz. Figure 1). The rings and their positions are inferred from anomalies in satellite altimetric data. (Modified from Byrne et al. 1995)

Exchange of water at the Agulhas retroflection

One of the minor, but in no way negligible (e.g. Shannon et al. 1990b), contributors to inter-basin exchange of Indian and Atlantic Ocean water are Agulhas filaments. These offshoots of the landward border of the Agulhas Current are rapidly carried into the South Atlantic by the shelf edge current (Bang and Andrews 1974) and by the presence of Agulhas rings (Lutjeharms and Valentine 1988b). They are only about 50 m deep, 50 km wide and annually carry excess salt of about $3 - 9 \times 10^{12}$ kg into the South Atlantic (Lutjeharms and Cooper 1995), their excess heat rapidly being lost to the overlying atmosphere.

A more substantial exchange may also take place in the deeper, Antarctic Intermediate Water (Shannon and Hunter 1988). Using averages from historical hydrographic data it has been shown (Fig. 17) that Antarctic Intermediate Water in the Agulhas Current is modified by Red Sea Water, but becomes fresher along its track due to mixing. The water not entering the South Atlantic is further freshened outside the retroflection region (Fine et al. 1988; Fine 1993). This pattern bears a strong similarity to that observed in the movement of the upper layers and in the proposed role and trajectories of Agulhas rings.

What then is the presently understood role of Agulhas rings in the interbasin exchange south of Africa of specific water types? Valentine et al.

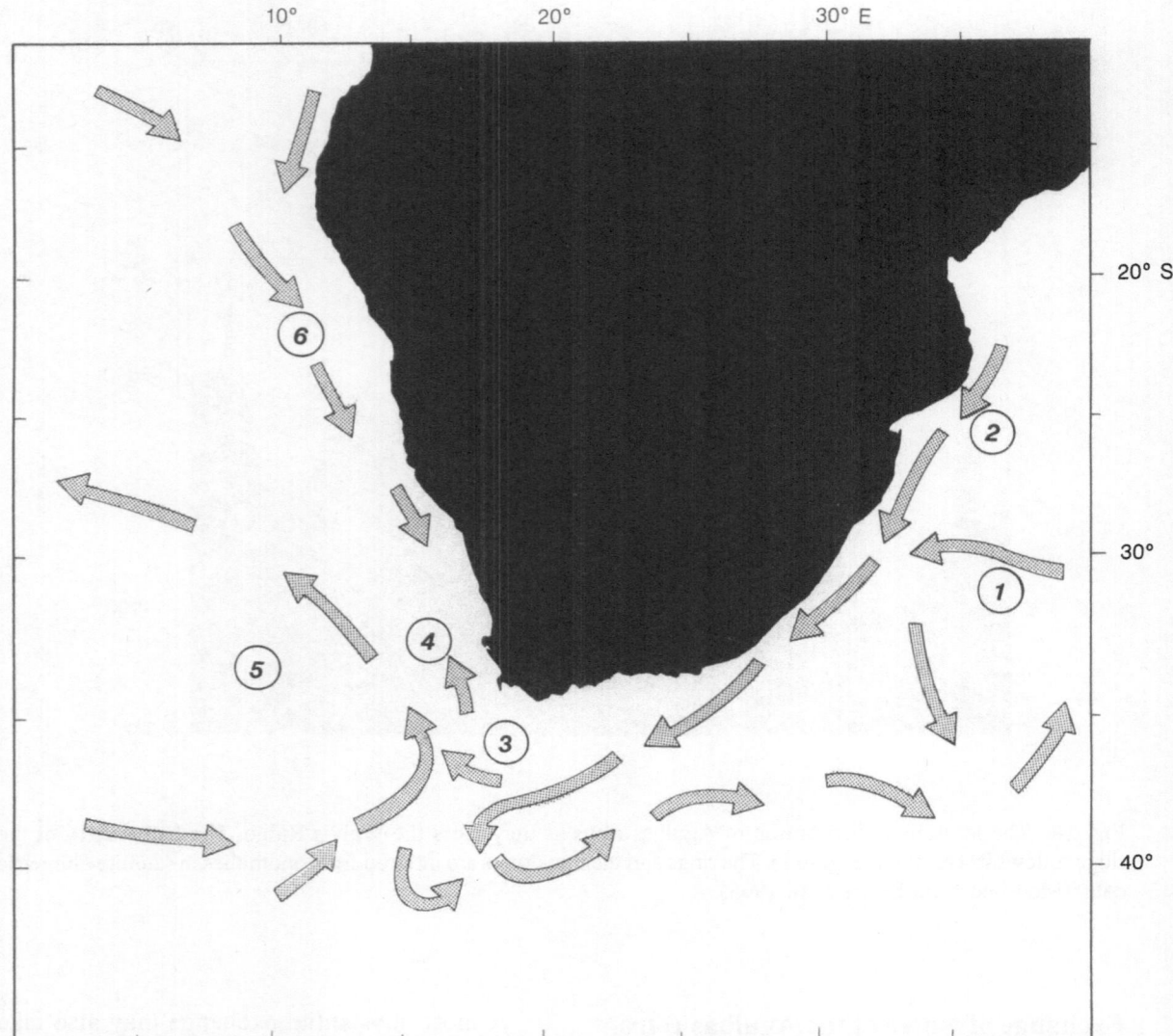

Fig. 17. Inferred flow directions of Antarctic Intermediate Water around southern Africa from the available hydrographic data set. **1** is the primary source of Antarctic Intermediate Water to the Agulhas Current; **2** is the secondary source with Red Sea Water; **3** inter-basin flux from the Agulhas Current; **4** relatively fresh Antarctic Intermediate Water from the southwest; **5** anticyclonic gyre of Antarctic Intermediate Water; **6** poleward flow from the tropical eastern Atlantic. (Modified from Shannon and Hunter 1988)

(1993) have drawn up a list of the water masses and their volumes in the Agulhas retroflection region (Table 2). Since, as was discussed above, Agulhas rings will only carry water in the layers above the Deep Water, their effect will be restricted to Surface Water, Central Water and Antarctic Intermediate Water (Table 1). Of the total volume of the water masses, these three represent only 39 %, the major water mass by volume being the Deep Water (Table 2). Rintoul (1991) has in fact shown that it is the flow of Deep and Bottom Water that is mainly responsible for the water mass balance in the South Atlantic. Attempts at quantifying the transport function of Agulhas rings without atten-

Table 2. Volumes of the water masses of the southern Agulhas region (after Valentine et al. 1993)

Water mass	Volume x 10^3 km^3	Percentage of total
Surface water	82.1	3
Central water	310.9	13
Antarctic Intermediate Water	548.8	23
North Atlantic Deep Water	954.1	40
Circumpolar Deep Water		
Antarctic Bottom Water	457.9	19
Others	39.7	2
Total	2393.5	100

tion to detail of the water type components has nonetheless been made.

Gordon and Haxby (1990) have calculated a mean, effective volume flux of 2 - 3 x 10^6 m^3/s per ring, above the Antarctic Intermediate Layer. Olson and Evans (1986) have even estimated that the replacement time for water above 10°C in the South Atlantic by Agulhas rings alone would take only 70 years. A number of other estimates of the effective inter-basin volume flux by Agulhas rings have been made. Gordon (1985) has estimated 15 x 10^6 m^3/s, but adjusted it to 15 - 14 x 10^6 m^3/s (Gordon et al. 1987). Byrne et al. (1995), using altimetric data and independent, but not concurrent, hydrographic data have estimated it to be 5 x 10^6 m^3/s. They have also estimated an average potential energy flux per year of 18 x 10^{16} J and an average kinetic energy flux per year of 18 x 10^{16} J. Stramma and Peterson (1990), from a compilation of historical data, have found a transfer of 8 x 10^6 m/s between the Indian and Atlantic. A recent model by Boddem and Schlitzer (1994), constrained with the historical hydrographic data, shows that an inflow into the South Atlantic of 4 - 8 x 10^6 m^3/s can be accommodated, but not larger values. A strong correlation is found between the meridional heat transport, the strength of the global thermohaline cell and inflow from the Indian Ocean in this model.

An analysis of inter-basin exchange by Agulhas rings, based on the largest number of hydrographically measured rings (van Ballegooyen et al. 1994), gives a volume flux of 6.2 x 10^6 m^3/s for water warmer than 10°C and 7.3 x 10^6 m^3/s for water warmer than 8°C. Van Ballegooyen et al. (1994) have also estimated a heat flux by Agulhas rings of 0.045 PW and a salt flux of 78 x 10^{12}kg/year. For the present these last mentioned estimates must be considered the best available.

Nonetheless, the precise role played in the total exchange between the South Indian and South Atlantic oceans by the various water types of the region (Tables 1, 2) is at present still unclear. Bennett (1988), reporting on a cruise that accurately determined the flow path of Agulhas water, has come to the conclusion that heat flux westward due to the difference in heat content between warm Agulhas upper level water and Agulhas Return Current water was fully accounted for by the loss of heat to the atmosphere. She suggested that it was the eastward flow of North Atlantic Deep Water that increased the westward heat transport. The use of a layer of depleted oxygen has also been put forward as a tracer for Indo-Atlantic exchange (Chapman 1988), about 3.8 x 10^6 m^3/s of lower oxygen water entering the South Atlantic each year.

Gordon et al. (1992) have used two lines of high-quality hydrographic sections at 30°S (zonal) and 10°W (roughly meridional) to estimate the Atlantic-Indian exchange. They have noted that strong evidence exists for there being a significant inter-ocean exchange of thermocline and intermediate water between the South Indian and South Atlantic basins. In fact, based on chlorofluoromethane tracers, they have found that about two thirds of the northward transport in the upper Benguela Current originates from the Indian Ocean. South Atlantic Central water passes into the Indian Ocean by a route south of the Agulhas Return Current, and according to Gordon et al. (1992) this particular exchange between the two ocean basins decreases rapidly with depth.

These analyses will no doubt continue as more high-quality data become available as part of the WOCE hydrographic programme. What is additionally of importance in more fully understanding the inter-basin exchange are the mechanisms that bring about the spawning of Agulhas rings and that may therefore control the rate of the global thermohaline circulation.

Inter-basin exchange mechanisms

Based on all present information, the leakage of thermocline water from the South Indian into the South Atlantic would seem to be largely a function of the number and dimensions of Agulhas rings formed at the Agulhas retroflection. Although modelling results (e.g. Lutjeharms and van Ballegooyen 1984) have suggested that a decreased volume flux of the Agulhas Current may induce the retroflection loop to penetrate farther into the South Atlantic (Fig. 18) and that shifts in the latitude of minimum wind stress curl (de Ruijter and Boudra 1985) would allow a greater or lesser degree of Agulhas water to leak into the South Atlantic, the actual mechanisms that precipitate the shedding of an Agulhas ring are not clear. Recent results achieved by van Leeuwen (personal communication) have suggested that perturbations to the flow path of the Agulhas Current that progress downstream will cause the amalgamation of this current and the Agulhas Return Current at the retroflection, thus triggering a ring spawning event.

What might these perturbations be? As discussed previously, the trajectory of the Agulhas Current is extreme stable between the point of its inception and Port Elizabeth (Gründlingh, 1983). Both Pearce (1977) and Gründlingh (1983) have, however, reported that in their investigations at sea they were on numerous occasions not able to find the Agulhas Current within 100 km of the continental shelf edge. These unusual events were not included in their statistical results that indicated high path stability. Gründlingh (1979) has in fact reported the observation of at least one unusually large meander progressing downstream with the current in this domain of otherwise high lateral stability.

Results from investigations using satellite remote sensing have shown that these large solitary meanders, Natal Pulses, occur about 20 percent of the time, always move downstream with the current (Fig. 19) at very constant speeds of about 20 km/day and always grow rapidly during their downstream progression (Lutjeharms and Roberts 1988). They all originate just north of Durban in that small offset to the otherwise nearly rectilinear coastline (viz. Fig. 1), the Natal Bight. Further studies of such meanders (Gründlingh 1992) have shown that they may on very rare occasions also originate in the Delagoa Bight, a similar, but much larger, coastal offset at Maputo (Fig. 1), but that these more northern meanders dissipate rapidly offshore and have no subsequent effect on the course of the Agulhas Current.

The similarity between the behaviour of the current at these two offsets in the coast suggest a possible mechanism for the Natal Pulse. In the Delagoa Bight the recurring presence of a trapped lee eddy has been observed (Lutjeharms and Jorge da Silva 1988). Such an eddy might escape from that bight much as a vortex might be shed from behind an obstacle in a flow. Do such lee eddies also occur in the Natal Bight? Circulation patterns inferred from sediment traces (Malan and Schumann 1979) as well as current measurements on the much shallower shelf (Pearce et al. 1978) north of Durban indicate one, or a series of, cyclonic eddies. If such an eddy were to escape from the Natal Bight and move downstream, inshore of the Agulhas Current, substantial coastal countercurrents might

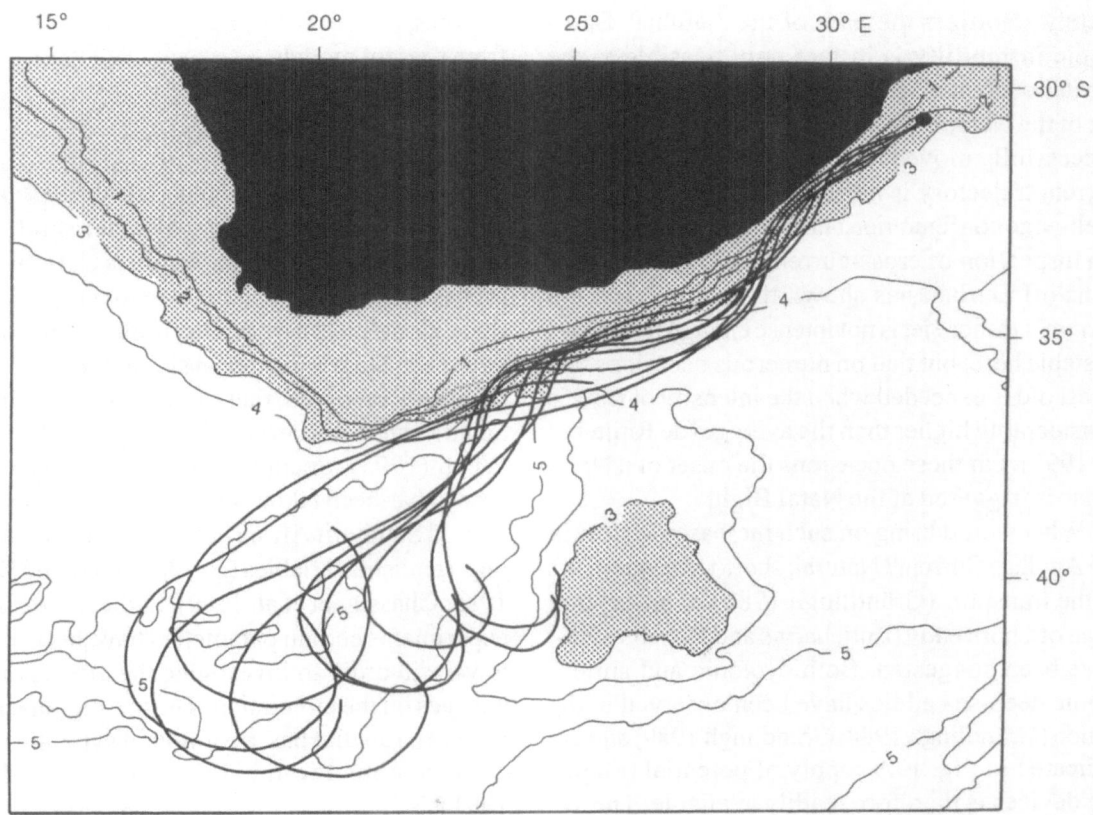

Fig. 18. The trajectories of the Agulhas Current modelled as a free inertial jet initiated off Durban with a range of velocity profiles. Areas shallower than 3000m have been hatched. The tendency to retroflect at about 40°S is evident. (Modified from Lutjeharms and van Ballegooyen 1984)

occasionally be observed. This suggestion is consistent with current meter records from this shelf (e.g. Schumann 1982) which do show counter-current events. On one occasion at least, very strong inshore countercurrents downstream of Durban could be unambiguously related to the passage of a Natal Pulse as seen in satellite thermal imagery (Lutjeharms and Connell 1989).

Apart from the likelyhood that these large, solitary meanders can trigger the spawning of an Agulhas ring far downstream at the retroflection, they have also been observed to create secondary circulation patterns that may slow down the leakage of warm Agulhas water into the South Atlantic. On moving downstream, Natal Pulses may grow so rapidly (Fig. 19) that the Agulhas Current core is

shifted sufficiently far offshore to intersect the Agulhas Plateau (Fig. 1) causing an early retroflection (Lutjeharms and van Ballegooyen 1988b), drawn in Fig. 1. Through this process warm Agulhas water is diverted back into the South Indian gyre at an early stage (Lutjeharms et al. 1992b) and therefore does not participate in the normal, downstream retroflection, thus reducing the amount of water available for ring formation south of Africa and conceivably slowing down the inter-basin exchange of water.

The question now arises why these Natal Pulses, that seem to play such an important role, arise only at the Natal Bight. De Ruijter et al. (1995) have shown that the shelf width and slope gradient along the east coast of South Africa is such that it completely stabilizes the path of the Agulhas. Barotropic instabililty is in fact only possible at the greater width and reduced continental slope gradient of the Natal Bight. Once such an instability has successfully moved the current core offshore, the current trajectory is no longer constrained by the shelf-edge configuration and the meander can grow. An inspection of cross-current hydrographic sections off Durban has shown that on average the current's inshore jet is not intense enough to become unstable here, but that on numerous occasions this threshold is exceeded when the intensity of flow is considerably higher than the average (de Ruijter et al. 1995). On these occasions the onset of a Natal Pulse is triggered at the Natal Bight.

What would bring on such increases in flow in the Agulhas Current? Natural, short-term variations in the transport (Gründlingh 1980) or adsorption of an offshore eddy (Lutjeharms and Roberts 1988) have been suggested. Both cyclonic and anticyclonic deep-sea eddies have been observed in the region (Gründlingh 1988; Gründlingh 1989) and are indicated in Fig. 1. A supply of potential triggering devices is therefore readily available. The absorption mechanism between a current and a deep-sea eddy has been described by Nof (1986) for the Gulf Stream and is therefore entirely conceivable for the Agulhas.

Whence do these observed eddies come? Gründlingh (1988) has hypothesized that they originate from a Mozambique Ridge Current along the eastern flank of that ridge (viz. Fig. 1); Lutjeharms (1989) has proposed cold eddies shed from the Subtropical Convergence at the Agulhas Plateau as being a source (Lutjeharms and Valentine 1988a). Recent results (Gründlingh et al. 1991; de Ruijter et al. 1995) indicate that both these propositions are probably incorrect and that these eddies have their origins to the north and to the east.

The importance of these results concerning the Natal Pulse and other interactions within the Agulhas system (Lutjeharms et al. 1995) is that they suggest that mesoscale processes smaller than those that are presently resolved by numerical models may play a triggering role in the inter-basin water exchange and may possibly even play a significant role in the global thermohaline circulation.

This has, for the reason mentioned, not yet emerged from present models.

Exchange models and theory

Numerical modelling and theoretical investigations of the inter-basin exchange of water south of Africa, and of the role of the Agulhas Current in this process, have consisted of three main endeavours. First, global, eddy-resolving, multilayered models have been fairly effective in simulating some of the processes involved; these have been described by Semtner and Chervin (1988) and by Lutjeharms and Webb (1995). Second, a more experimental approach has been taken where a model of the region around Southern Africa has been run with increasingly realistic variables (e.g. de Ruijter and Boudra 1985; Chassignet et al. 1989). In these latter model experiments certain parameters have been allowed to vary in order to investigate the effects of these changes on the circulation. Third, a regional model has been run that has assimilated synoptic data to drive continual re-initializations (Holland et al. 1991).

The global, primitive-equation model, of which the results have been reported by Semtner and Chervin (1988), was one with a 1/2° latitude-longitude spatial resolution, realistic basin and coastal geometry, 20 vertical levels and annual-mean wind forcing. It was run for 10 model years with a weak constraint towards the Levitus (1982) data set. For the last years of the model-run only the lower and surface layers were so constrained while the thermocline was allowed to develop freely. One of the major results of this model is its distinct simulation of the global thermohaline circulation cell. The retroflection of the Agulhas Current and ring shedding are portrayed fairly realistically as part of this cell. Both warm and cold vortices move into the South Atlantic, the warm ones directly westward, contrary to observation (e.g. Naeije et al. 1992; van Ballegooyen et al. 1994).

The physics of the FRAM (Fine Resolution Antarctic Model) model was based on the above-mentioned one, had increased spatial resolution, a larger number of horizontal layers, but was restricted to the southern hemisphere. Here, the retroflection and the ring spawning process has been

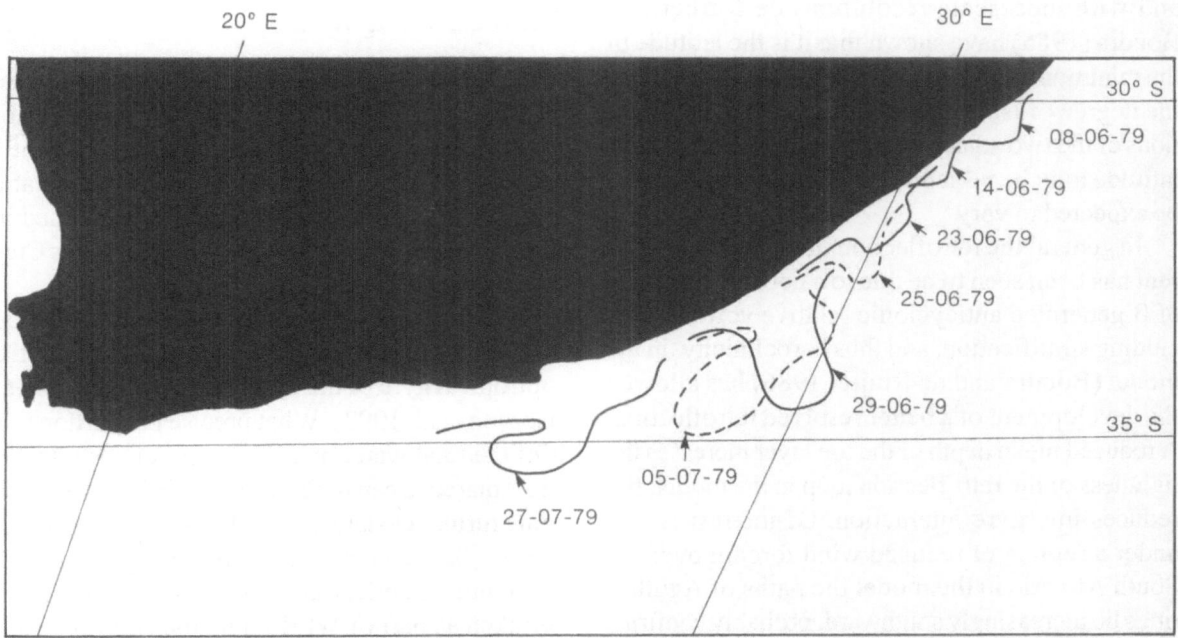

Fig. 19. The shape and geographic location of a perturbation on the in-shore front of the Agulhas Current over a two month period in 1979. This shows the growth and downstream progression of a characteristic Natal Pulse. (Modified from Lutjeharms and Roberts 1988)

even more realistically simulated (Lutjeharms and Webb 1995). Shortcomings in the FRAM representation of the inter-basin exchange have been most evident in that the retroflection has occured too far upstream. This may be due to the smoothed bathymetry employed in the model and the known sensitivity of the Agulhas Current to bottom topography. Also, too few Agulhas rings were produced in the FRAM, their sizes were too uniform and they drifted off in a too undeviating direction. The latter two deficiencies may have been due to the interannually invariant wind stress used.

The only quasi-geostrophic regional model for the Agulhas system by Holland et al. (1991) had 1/4° latitude-longitude spatial resolution and 5 layers. This model was able to assimilate altimetric data at regular intervals in order more closely to approximate reality.

Modelling the Agulhas Current as a free inertial jet has also been attempted (Lutjeharms and van Ballegooyen 1984) in order to investigate in par-

ticular the topographic control on the current system (Fig. 18). This model has predicted that high vertical shears, bottom currents exceeding 5 cm/s and low volume transport will lead to greater leakage and further penetration of the retroflection loop into the Atlantic Ocean. Ou and de Ruijter (1986) also have shown that for an inertial boundary current the separation from a curved coastline occurs sooner if the flux is greater.

De Ruijter, Boudra and Chassignet have carried out a long series of experiments in which they have modelled the inter-basin flow with increasing detail and with increasing configural realism. Using an asymptotic analysis of wind-driven, large-scale circulation in the subtropics of the South Atlantic and South Indian Oceans de Ruijter (1982) has demonstrated that a meridional decrease in the wind stress curl is required to intensify the Agulhas retroflection. With a weak north-south decrease the Agulhas moves as a jet into the Atlantic. Using a similar wind-driven model: (barotropic, non-linear and with mesoscale resolution) de Ruijter and Boudra (1985) have shown that it is the latitude of the minimum in the wind stress curl that determines the degree of isolation between the gyral circulations of the two adjacent basins. Since in nature this latitude may be variable, the inter-basin flux may be expected to vary.

In general the retroflection of the Agulhas Current has been seen to be due to a net accumulation of β-generated anticyclonic relative vorticity. Including stratification, and thus baroclinicity, in the model (Boudra and de Ruijter 1986) has allowed the development of a better resolved retroflection. A reduced mean depth of the top layer increases the tightness of the retroflection loop in the model, but reduces intergyre interaction. Of interest is that under a regime of reduced wind forcing over the South Atlantic in the model the paths of Agulhas rings lie increasingly southward, probably confirming the role of the gyral circulation in advecting these rings.

In subsequent refinements of this model (Boudra and Chassignet 1988) the role of friction has become more evident. The Agulhas Current appears to experience a substantial viscous stress curl along the coast that ends abruptly at separation. The accompanying change of the vorticity balance contributes to retroflection of the current in the model. This coastal friction, as well as the southward inertia and baroclinicity in the overshooting Agulhas, play decisive roles in the frequency of ring formation. At high Rossby numbers ring formation in this model is associated with mixed barotropic instability (Chassignet and Boudra 1988). By allowing outcropping of density layers in a further refinement of the model (Boudra et al. 1989) ring formation has simulated even more realistically (Fig. 20). Rings spawn more frequently and are found closer to the proper location. Chassignet et al. (1989) have furthermore shown that this version of the model simulates the subsequent translation of rings well. As rings round Africa their paths are little influenced by large-scale flows. Once in the South Atlantic gyre, however, advection by the large-scale flow dominates the ring trajectories.

Future investigations

Although enormous progress has been made over the past decade in understanding the exchange of water masses between the Indian and the Atlantic Oceans, it is clear from the above review that a number of key questions need to be addressed in order to learn the precise role of the Agulhas Current system in the exchange.

The fact that substantial quantities of Indian Ocean water are eventually incorporated into the subtropical gyre of the South Atlantic seems clear (Gordon et al. 1992). What precise proportion this is of the total water mass transfer between Africa and Antarctica can probably be quantified best with some further closely-spaced lines of hydrographic stations between these two continents and across the South Atlantic. Such lines have now been undertaken as part of WOCE and the careful analysis of these data may help in answering this question.

The actual mass, heat and salt flux of the Agulhas Current is known only in estimate. Temporal variations in these fluxes are recognized as probably playing a crucial role in inter-basin water exchange, but have not been measured. Plans to place strings of current meters across the current and to use vertical current profilers during the next five years, will, if successful, give the first good data on these important variables.

Many of the mesoscale and even large-scale components of the Agulhas Current system are even now poorly understood, or not known at all. Most of these may play a role in the facility and frequency with which Agulhas rings are shed. Investigations to study these problems would most probably have to start with well-designed cruises supported by satellite remote sensing. Questions that need to be addressed include: What role does the Mozambique Current play in feeding the Agulhas Current? What are the dynamics of the southern termination of the East Madagascar Current? What part of the volume flux of the Agulhas Current is involved in its upstream retroflections? How rapidly do Agulhas rings loose their heat and salt to the ambient water masses in the South Atlantic Ocean? What, if any, is the role of deep-sea eddies in triggering the Natal Pulse?

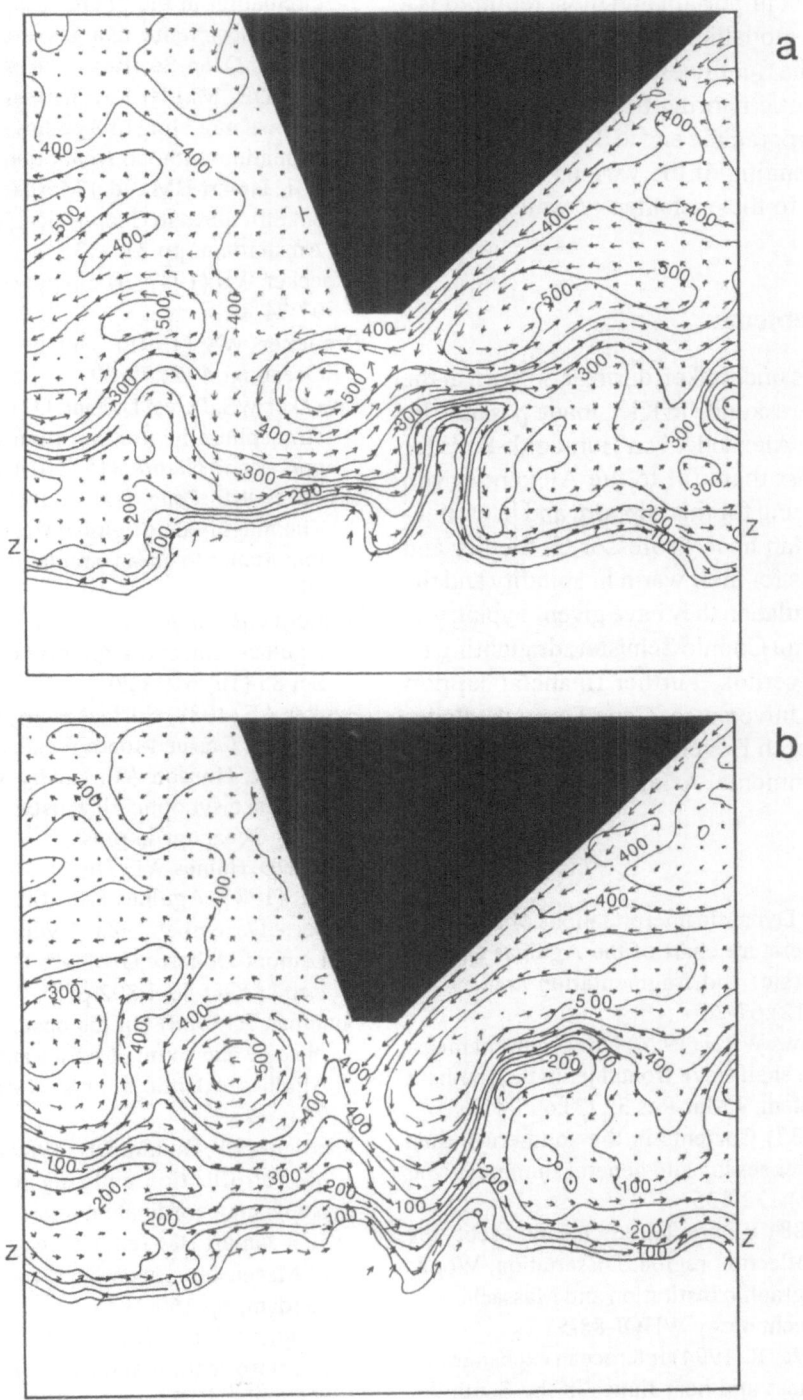

Fig. 20. The thickness and velocity of the southern Agulhas region in a quasi-isopycnic model simulation. The upper panel is 25 model days before the lower one. Depth intervals are at 50 m; a full length arrow and each additional barb represents 0.25 m/s; the latitude of no wind stress curl is indicated by **Z**. The unmarked region is where the layer below has cut the sea surface. The formation and subsequent advection of an Agulhas ring is clearly shown as is the northward penetration of Subtropical Surface Water. (Modified from Boudra et al. 1989).

Perhaps most important and most required is a proper regional modelling effort on the detail of the circulation in the Southwest Indian Ocean and the Southeast Atlantic in order to fascilitate a proper observational programme and to derive a solid theoretical understanding of the various mechanisms that contribute to the exchange of water south of Africa.

Acknowledgements

This review was undertaken during a sojourn at the Institut für Meereskunde in Kiel, made possible by the award of the Alexander von Humboldt-Preis for 1994. I am most thankful to the Alexander von Humboldt-Stiftung for this support, and in particular to my German host, Professor G. Siedler and other colleagues for their warm hospitality and the intellectual stimulation they have given. Typing was ably done by Frau Connie Schuster; draughting by Herr Tiberiu Csernok. Further financial support came from the University of Cape Town, the Foundation of Research Development and the Department of Environmental Affairs in South Africa.

References

Bang ND (1970) Dynamic interpretations of a detailed surface temperature chart of the Agulhas Current retroflexion (sic) and fragmentation area. S Afr Geogr J 52 (12):67-76

Bang ND, Andrews WRH (1974) Direct current measurements of a shelf-edge frontal jet in the southern Benguela system. J Mar Res 32 (3):405-417

Barlow EW (1933) Currents in the southern Indian Ocean, summer season and general summary. Mar Obsv 10 (112):132-135

Bennett SL (1988) Where Three Oceans Meet: the Agulhas retroflection region. Dissertation, Woods Hole Oceanographic Institution and Massachusetts Institute of Technology WHOI-88-51

Boddem J, Schlitzer R (1994) Interocean exchange and meridional mass and heat fluxes in the South Atlantic. J Geophys Res submitted

Boudra DB, Chassignet EP (1988) Dynamics of Agulhas retroflection and ring formation in a numerical model. I. The vorticity balance. J Phys Oceanogr 18 (2):280-303

Boudra DB, de Ruijter WPM (1986) The wind-driven circulation in the South Atlantic - Indian Ocean - II. Experiments using a multi-layer numerical model. Deep-Sea Res 33 (4):447-482

Boudra DB, Maillet KA, Chassignet EP (1989) Numerical modeling of Agulhas retroflection and ring formation with isopycnal outcropping. In: Nihoul JCJ, Jamart BM (eds) Mesoscale/Synoptic Coherent Structures in Geophysical Turbulence. Elsevier, Amsterdam, pp 315-335

Broecker WS (1987) The biggest chill. Nat Hist Mag 97:74-82

Broecker WS (1991) The great ocean conveyor. Oceanogr 4 (2):79-89

Broecker WS, Peteet D, Rind D (1985) Does the ocean-atmosphere have more than one stable mode of operation? Nature 315 (6014):21-25

Broecker WS, Peng T-H, Jouzel J, Russell G (1990) The magnitude of global fresh water transports of importance to ocean circulation. Clim Dynam 4:73-79

Brundrit GB, Shannon LV (1989) Cape storms and the Agulhas Current: a glimpse of the future? S Afr J Sci 85 (10):619-620

Bunker AF (1988) Surface energy fluxes of the South Atlantic Ocean. Month Weath Rev 116 (4):809-823

Byrne DA, Gordon AL, Haxby WF (1995) Agulhas Eddies: a synoptic view using Geosat ERM data. J Phys Oceanogr in press

Camp DB, Haines WE, Huber BA, Rennie SE, Gordon AL (1986) Agulhas Retroflection Cruise, November - December 1983. Hydrographic (CTD) data. Lamont-Doherty Geol Obs Columbia Univ Techn Rep LDGO-86-1 392 pp

Chapman P (1988) On the occurrence of oxygen-depleted water south of Africa and its implications for Agulhas-Atlantic mixing. S Afr J Mar Sci 7:267-294

Chassignet EP, Boudra DB (1987) Dynamics of Agulhas retroflection and ring formation in a quasi-isopycnic coordinate numerical model. In: Nihoul JCJ, Jamart BM (eds) Three Dimensional Models of Marine and Estuarine Dynamics. Elsevier, Amsterdam, pp 169-194

Chassignet EP, Boudra DB (1988) Dynamics of Agulhas retroflection and ring formation in a numerical model. II. Energetics and ring formation. J Phys Oceanogr 18 (2):304-319

Chassignet EP, Olson DB, Boudra DB (1989) Evolution of rings in numerical models and observations. In: Nihoul JCJ, Jarmart BM (eds) Mesoscale/Synoptic Coherent Structures in Geophysical Turbulence. Elsevier, Amsterdam, pp 337-356

Cheney RA, Marsh JG, Beckley BD (1983) Global mesoscale variability from colinear tracks of SEA-SAT altimeter data. J Geophys Res 88 (C7):4343-4354

Cockroft VG, Peddemors VM, Ryan PG, Lutjeharms JRE (1990) Cetacean sightings in the Agulhas Retroflection, Agulhas Rings and Subtropical Convergence. S Afr J Antarc Res 20 (2):64-67

Cresswell GR, Golding TJ (1980) Observations of a south-flowing current in the southeasterly Indian Ocean. Deep-Sea Res 27 (6):449-466

Darbyshire J (1972) The effect of bottom topography on the Agulhas Current. Rev Pure Appl Geophys 101 (9):208-220

De Decker AHB (1973) Agulhas Bank plankton. In: Zeitzschel B (ed) The Biology of the Indian Ocean. Springer, Berlin, pp 189-219

De Decker AHB (1984) Near-surface copepod distribution in the south-western Indian and south-eastern Atlantic Ocean. Ann S Afr Mus 93 (5):303-370

De Ruijter W (1982) Asymptotic analysis of the Agulhas and Brazil Current systems. J Phys Oceanogr 12 (4):361-373

De Ruijter WPM, Boudra DB (1985) The wind-driven circulation in the South Atlantic - Indian Ocean - I. Numerical experiments in a one-layer model. Deep-Sea Res 32 (5):557-574

De Ruijter WPM, van Leeuwen PJ, Lutjeharms JRE (1995) Triggering mechanisms for Natal Pulses in the Agulhas Current. J Geophys Res in press

Dietrich G (1935) Aufbau und Dynamik des südlichen Agulhasstromgebietes. Veröffentl Instit Meeresk Univ Berlin nf A (27):1-79

Duncan CP (1968) An eddy in the Subtropical Convergence southwest of South Africa. J Geophys Res 73 (2):531-534

Duncombe Rae CM (1991) Agulhas retroflection rings in the South Atlantic Ocean; an overview. S Afr J Mar Sci 11:327-344

Duncombe Rae CM, Shannon LV, Shillington FA (1989) An Agulhas ring in the South Atlantic Ocean. S Afr J Sci 85 (11):747-748

Duncombe Rae CM, Boyd AJ, Crawford RJM (1992a) "Predation" of anchovy by an Agulhas ring: a possible contributory cause of the very poor yearclass of 1989. S Afr J Mar Sci 12:167-173

Duncombe Rae CM, Shillington FA, Agenbag JJ, Taunton-Clark J, Gründlingh ML (1992b) An Agulhas Ring in the South Atlantic Ocean and its interaction with the Benguela upwelling frontal system. Deep-Sea Res 39 (11/12):2009-2027

Feron RCV, de Ruijter WPM, Oskam D (1992) Ring-shedding process in the Agulhas Current system. J Geophys Res 97 (C6):9467-9477

Findlay KP, Best PB, Ross GJB, Cockroft VG (1992) The distribution of small odontocete cetaceans off the coasts of South Africa and Namibia. S Afr J Mar Sci 12:237-270

Fine RA (1993) Circulation of Antarctic Intermediate Water in the South Indian Ocean. Deep-Sea Res 40 (10):2021-2042

Fine RA, Warner MJ, Weiss RF (1988) Water mass modification at the Agulhas retroflection: chlorofluoromethane studies. Deep-Sea Res 35 (3):311-332

Flierl GR (1981) Particle motions in large-amplitude wave fields. Geophys Astrophys Fluid Dynam 18:39-74

Gordon AL (1985) Indian-Atlantic transfer of thermocline water at the Agulhas retroflection. Sci 227 (4690):1030-1033

Gordon AL (1986) Inter-ocean exchange of thermocline water. J Geophys Res 91 (C4):5037-5046

Gordon AL, Haxby WF (1990) Agulhas eddies invade the South Atlantic - evidence from GEOSAT altimeter and shipboard conductivity-temperature-depth survey. J Geophys Res 95 (C3):3117-3125

Gordon AL, Lutjeharms JRE, Gründlingh ML (1987) Stratification and circulation at the Agulhas Retroflection. Deep-Sea Res 34 (4):565-599

Gordon AL, Weiss RF, Smethie WM, Warner MJ (1992) Thermocline and intermediate communication between the South Atlantic and Indian Oceans. J Geophys Res 97 (C5):7223-7240

Gründlingh ML (1974) A description of inshore current reversals off Richards Bay based on airborne radiation thermometry. Deep-Sea Res 21 (1):47-55

Gründlingh ML (1978) Drift of a satellite-tracked buoy in the southern Agulhas Current and Agulhas Return Current. Deep-Sea Res 25 (12):1209-1224

Gründlingh ML (1979) Observation of a large meander in the Agulhas Current. J Geophys Res 84 (C7):3776-3778

Gründlingh ML (1980) On the volume transport of the Agulhas Current. Deep-Sea Res 27 (7):557-563

Gründlingh ML (1983) On the course of the Agulhas Current. S Afr Geogr J 65 (1):49-57

Gründlingh ML (1988) Review of cyclonic eddies in the Mozambique Ridge Current. S Afr J Mar Sci 6:193-206

Gründlingh ML (1989) Two contra-rotating eddies of the Mozambique Ridge Current. Deep-Sea Res 36 (1):149-153

Gründlingh ML (1992) Agulhas Current meanders: review and a case study. S Afr Geogr J 74 (1):19-28

Gründlingh ML, Lutjeharms JRE (1979) Large-scale flow patterns of the Agulhas Current system. S Afr J Sci 75 (6):269-270

Gründlingh ML, Carter RA, Stanton RC (1991) Circulation and water properties of the Southwest Indian Ocean, Spring 1987. Progr Oceanogr 28 (4):305-342

Harris TFW (1970) Planetary-type waves in the South-West Indian Ocean. Nature 227 (5262):1043-1044

Harris TFW (1972) Sources of the Agulhas Current in the spring of 1964. Deep-Sea Res 19 (9):633-650

Harris TFW, van Foreest D (1978) The Agulhas Current in March 1969. Deep-Sea Res 25 (6):549-561

Harris TFW, Legeckis R, van Forest (sic) D (1978) Satellite infra-red images in the Agulhas Current System. Deep-Sea Res 25 (6):543-548

Hastenrath S (1982) On meridional heat transports in the world ocean. J Phys Oceanogr 12 (8):922-927

Holland WR, Zlotnicki V, Fu L-L (1991) Modelled time-dependent flow in the Agulhas retroflection region as deduced from altimeter data assimilation. S Afr J Mar Sci 10:407-427

IPP (1990) Climate Change; the IPCC Scientific Assessment. Houghton JT, Jenkins GJ, Ephraums JJ (eds), Cambridge University Press, Cambridge

Jacobs SS, Georgi DT (1977) Observations on the Southwest Indian/Antarctic Ocean. Deep-Sea Res 24 (Suppl):43-84

Jury MR, Walker N (1988) Marine boundary layer modification across the edge of the Agulhas Current. J. Geophys Res 93 (C1):647-654

Jury MR, Valentine HR, Lutjeharms JRE (1993) Influence of the Agulhas Current on summer rainfall on the southeast coast of South Africa. J Appl Met 32 (7):1282-1287

KNMI (1857) De Agulhas stroom, afgeleid uit de temperatuur van het zeewater aan de oppervlakte; en de invloed dien deze op de atmosfeer uitoefent. In: Uitkomsten van Wetenschap en Ervaring, aangaande winden en zeestromingen in sommige gedeelten van de oseaan. Koninklijk Nederlandsch Meteorologisch Instituut, Bosch en Zoon, Utrecht, pp 40-50

Levitus S (1982) Climatological atlas of the world oceans. NOAA Prof Pap 13, US Govern Print Off, Wash, DC

Lutjeharms JRE (1976) The Agulhas Current system during the Northwest Monsoon Season. J Phys Oceanogr 6 (5):665-670

Lutjeharms JRE (1981a) Features of the southern Agulhas Current circulation from satellite remote sensing. S Afr J Sci 77 (5):231-236

Lutjeharms JRE (1981b) Satellite studies of the South Atlantic upwelling system. In: Gower JFR (ed) Oceanography from Space. Plenum, New York, pp 195-199

Lutjeharms JRE (1985) Location of frontal systems between Africa and Antarctica: some preliminary results. Deep-Sea Res 32 (12):1499-1509

Lutjeharms JRE (1987a) Meridional heat transport across the Subtropical Convergence by a warm eddy. Nature 331 (6153):251-253

Lutjeharms JRE (1987b) Die Subtropiese Konvergensie en Agulhasretrofleksievaart (SCARC). S Afr J Sci 83 (8):454-456

Lutjeharms JRE (1988a) Remote sensing corroboration of retroflection of the East Madagascar Current. Deep-Sea Res 35 (12):2045-2050

Lutjeharms JRE (1988b) Examples of extreme circulation events of the Agulhas Retroflection. S Afr J Sci 84 (7):584-586

Lutjeharms JRE (1989) The role of mesoscale turbulence in the Agulhas Current system. In: Nihoul JCJ, Jamart BM (eds) Mesoscale/Synoptic Coherent Structures in Geophysical Turbulence. Elsevier, Amsterdam, pp 357-372

Lutjeharms JRE, Connell AD (1989) The Natal Pulse and inshore counter currents off the South African east coast. S Afr J Sci 85 (8):533-535

Lutjeharms JRE, Cooper J (1995) Inter-basin leakage through Agulhas Current filaments. Deep-Sea Res in press

Lutjeharms JRE, Gordon AL (1987) Shedding of an Agulhas Ring observed at sea. Nature 325 (7000):138-140

Lutjeharms JRE, Jorge da Silva A (1988) The Delagoa Bight eddy. Deep-Sea Res 35 (4):619-634

Lutjeharms JRE, Roberts HR (1988) The Natal Pulse; an extreme transient on the Agulhas Current. J Geophys Res 93 (C1):631-645

Lutjeharms JRE, Valentine HR (1984) Southern Ocean fronts south of Africa. Deep-Sea Res 31 (12):1461-1476

Lutjeharms JRE, Valentine HR (1988a) Eddies at the Sub-Tropical Convergence south of Africa. J Phys Oceanogr 18 (5):761-774

Lutjeharms JRE, Valentine HR (1988b) Evidence for persistent Agulhas rings southwest of Cape Town. S Afr J Sci 84 (9):781-783

Lutjeharms JRE, van Ballegooyen RC (1984) Topographic control in the Agulhas Current system.

Deep-Sea Res 31 (11):1321-1337

Lutjeharms JRE, van Ballegooyen RC (1988a) The retroflection of the Agulhas Current. J Phys Oceanogr 18 (11):1570-1583

Lutjeharms JRE, van Ballegooyen RC (1988b) Anomalous upstream retroflection in the Agulhas Current. Science 240 (4860):1770-1772

Lutjeharms JRE, Webb DJ (1995) Modelling the Agulhas Current system with FRAM (*Fine Resolution Antarctic Model*). Deep-Sea Res 42 (4):523-551

Lutjeharms JRE, Bang ND, Duncan CP (1981) Characteristics of the currents east and south of Madagascar. Deep-Sea Res 28 (9):879-899

Lutjeharms JRE, Mey RD, Hunter IT (1986) Cloud lines over the Agulhas Current. S Afr J Sci 82 (11/12):635-640

Lutjeharms JRE, Catzel R, Valentine HR (1989) Eddies and other border phenomena of the Agulhas Current. Continent Shelf Res 9 (7):597-616

Lutjeharms JRE, Shillington FA, Duncombe Rae CM (1991a) Observations of extreme upwelling filaments in the South East Atlantic Ocean. Science 253 (5021):774-776

Lutjeharms JRE, Webb DJ, de Cuevas BA (1991b) Applying the Fine Resolution Antarctic Model (FRAM) to the ocean circulation around southern Africa. S Afr J Sci 87 (8):346-349

Lutjeharms JRE, de Ruijter WPM, Peterson RG (1992a) Inter-basin exchange and the Agulhas retroflection; the development of some oceanographic concepts. Deep-Sea Res 39 (10):1791-1807

Lutjeharms JRE, Weeks SJ, van Ballegooyen RD, Shillington FA (1992b) Shedding of an eddy from the seaward front of the Agulhas Current. S Afr J Sci 88 (8):430-433

Lutjeharms JRE, Lucas MI, Perissinotto R, van Ballegooyen RC, Rouault M (1994) Oceanic processes at the Subtropical Convergence; report of research cruise SAAMES III. S Afr J Sci 90 (7):367-370

Lutjeharms JRE, de Ruijter WPM (1995) The influence of the Agulhas Current on the adjacent coastal zone: the possible impacts of climate change. J Mar Syst in press

Malan OG, Schumann EH (1979) Natal shelf circulation revealed by Landsat imagery. S Afr J Sci 75 (3):136-137

Mason SJ (1990) Temporal variablity of sea surface temperatures around Southern Africa: a possible forcing mechanism for the 18-year rainfall oscillation. S Afr J Sci 86 (5/6):243-252

McCartney MS, Woodgate-Jones ME (1991) A deep-reaching anticyclonic eddy in the subtropical gyre

of the eastern South Atlantic. Deep-Sea Res 38 (S1):S411-S443

Mey RD, Walker ND, Jury MR (1990) Surface heat fluxes and marine boundary layer modification in the Agulhas retroflection region. J Geophys Res 95 (C9):15997-16015

Michaelis G (1923) Die Wasserbewegung an der Oberfläche des Indischen Ozeans im Januar und Juli. Veröff Inst Meeresk Univ Berlin nf A (16):1-32

Möller L (1929) Die Zirkulation des Indischen Ozeans; auf Grund von Temperatur- und Salzgehaltstiefenmessungen und Oberflächenstrombeobachtungen. Veröff Inst Meeresk Univ Berlin nf A (21):1-48

Naeije MC, Wakker KF, Scharroo R, Ambrosius BAC (1992) Observations of mesoscale ocean currents from GEOSAT altimeter data. ISPRS J Photogram Rem Sens 47 (   ):347-368

Nof D (1986) The collision between the Gulf Stream and warm-core rings. Deep-Sea Res 33 (1):359-378

Olson DB, Evans RH (1986) Rings of the Agulhas Current. Deep-Sea Res 33 (1):27-32

Olson DB, Fine RA, Gordon AL (1992) Convective modifications of water masses in the Agulhas. Deep-Sea Res 39 (S1):S163-S181

Ou HW, de Ruijter WPM (1986) Separation of an inertial boundary current from a curved coastline. J Phys Oceanogr 16 (2):280-289

Paech H (1926) Die Oberflächenströmungen um Madagascar in ihrem Jährlichen Gang. Veröff Inst Meeresk Univ Berlin nf A (16):1-39

Patterson SL (1985) Surface circulation and kinetic energy distribution in the southern hemisphere oceans from FGGE drifting buoys. J Phys Oceanogr 15 (7):865-884

Pearce AF (1977) Some features of the upper 500 m of the Agulhas Current . J Mar Res 35 (4):731-753

Pearce AF, Gründlingh ML (1982) Is there a seasonal variation in the Agulhas Current? J Mar Res 40 (1):177-184

Pearce AF, Schumann EH, Lundie GSH (1978) Features of the shelf circulation off the Natal Coast. S Afr J Sci 74 (9):328-331

Peterson RG, Stramma L (1991) Upper-level circulation in the South Atlantic Ocean. Progr Oceanogr 26:1-73

Read JF, Pollard RT, Smithers J (1987) CTD and SeaSoar data from the Agulhas retroflection zone. Inst Oceanogr Scs Deacon Lab Rep 245, 91 pp

Reid JL (1989) On the total geostrophic circulation of the South Atlantic Ocean: flow patterns, tracers and transports. Prog Oceanogr 23 (3):149-244

Rennell J (1778) Chart of the Bank of Lagullas: and

southern Coast of Africa. London

Rigg GM, van Ballegooyen RC, Attwood C, Newton S, Lucas M, Lutjeharms JRE (1992) Data report on the first cruise of the South African Antarctic Marine Ecosystem Study (SAAMES-I), April - May 1992. Univ Cape Town Dept Oceanogr Rep DO-92-01, 30 pp

Ring Group (1981) Gulf Stream cold-core rings: their physics, chemistry, and biology. Science 212 (4499):1091-1100

Rintoul SR (1991) South Atlantic interbasin exchange. J Geophys Res 96 (C2):2675-2692

Saetre R, Jorge da Silva A (1984) The circulation of the Mozambique Channel. Deep-Sea Res 31 (5):485-508

Schell II (1968) On the relation between the winds off southwest Africa and the Benguela Current and the Agulhas Current penetration in the South Atlantic. Deut Hydrogr Z 21 (3):109-117

Schleyer MH (1985) Chaetognaths as indicators of water masses in the Agulhas Current system. Investl Rep Oceanogr Res Inst 61, 1-20

Schumann EH (1982) Inshore circulation of the Agulhas Current off Natal. J Mar Res 40 (1):43-55

Semtner AJ, Chervin RM (1988) A simulation of the global ocean circulation with resolved eddies. J Geophys Res 93 (C12):15502-15522

Shannon LV (1985) The Benguela Ecosystem. 1. Evolution of the Benguela, physical features and processes. Oceanogr Mar Biol Ann Rev 23:105-182

Shannon LV, Hunter D (1988) Notes on Antarctic Intermediate Water around Southern Africa. S Afr J Mar Sci 6:107-117

Shannon LV, Lutjeharms JRE, Agenbag JJ (1989) Episodic input of Subantarctic water into the Benguela region. S Afr J Sci 85 (5):317-322

Shannon LV, Lutjeharms JRE, Nelson G (1990a) Causative mechanisms for intra-annual and interannual variability in the marine environment around Southern Africa. S Afr J Sci 86 (7/8/9/10):356-373

Shannon LV, Agenbag JJ, Walker ND, Lutjeharms JRE (1990b) A major perturbation in the Agulhas retroflection area in 1986. Deep-Sea Res 37 (3):493-512

Stramma L (1992) The South Indian Ocean Current. J Phys Oceanogr 22 (4):421-430

Stramma L, Lutjeharms JRE (1995) The western circulation cell of the subtropical gyre in the South Indian Ocean. Deep-Sea Res submitted

Stramma L, Peterson RG (1990) The South Atlantic Current. J Phys Oceanogr 20 (6):846-859

Swart VP, Largier JL (1987) Thermal structure of Agulhas Bank water. S Afr J Mar Sci 5:243-253

Taft BA (1972) Characteristics of the flow of the Kuroshio south of Japan. In: Stommel H, Yoshida K (eds) Kuroshio; physical aspects of the Japan Current. University of Washington Press, London, pp 165-214

Taunton-Clark J, Shannon LV (1988) Annual and interannual variability in the South-East Atlantic during the 20th century. S Afr J Mar Sci 6:97-106

Valentine HR, Lutjeharms JRE, Brundrit GB (1993) The water masses and volumetry of the southern Agul-has Current region. Deep-Sea Res 40 (6): 1285-1305

Van Ballegooyen RC, Valentine HR, Lutjeharms JRE (1991) Modelling of the Agulhas Current system. S Afr J Sci 87 (11/12):569-571

Van Ballegooyen RC, Gründlingh ML, Lutjeharms JRE (1994) Eddy fluxes of heat and salt from the southwest Indian Ocean into the southeast Atlantic Ocean: a case study. J Geophys Res 99 (C7):14053-14070

Van Gogh J (1858) De stormen nabij de Kaap de Goede Hoop in verband beschouwd met de temperatuur der zee. Verh Mededel Koninkl Akad afd Natuurk 8:1-23

Wakker KF, Zandbergen RCA, Naeije MC, Ambrosius BAC (1990) Geosat altimeter data analysis for the oceans around South Africa. J Geophys Res 95 (C3):2991-3006

Walker ND (1990) Links between South African summer rainfall and temperature variability of the Agulhas and Benguela Current systems. J Geophys Res 95 (C3):3297-3319

Walker ND, Mey RD (1988) Ocean/atmosphere heat fluxes within the Agulhas Retroflection region. J Geophys Res 93 (C12):15473-15483

Watts DR (1983) Gulf Stream variability. In: Robinson AR (ed) Eddies in Marine Science. Springer, Berlin, pp 114-144

Wyrtki K (1971) Oceanographic Atlas of the International Indian Ocean Expedition. National Science Foundation, Washington, DC, 531 pp

Wyrtki K (1973) Physical oceanography of the Indian Ocean. In: Zeitzschel B (ed) The Biology of the Indian Ocean. Springer, Berlin, pp 18-36

Wyrtki K, Magaard L, Hagar J (1976) Eddy energy in the oceans. J Geophys Res 81 (15):2641-2646

Zahn W (1984) Influence of bottom topography on currents in the Mozambique Channel. Trop Ocean-Atm Newsl 26:22-23

The Benguela: Large Scale Features and Processes and System Variability

L.V. Shannon and G. Nelson

Sea Fisheries Research Institute, Private Bag X2,
Rogge Bay 8012, Cape Town, SOUTH AFRICA
and
Oceanography Department, University of Cape Town,
Rondebosch 7700, SOUTH AFRICA

Abstract: The Benguela is one of four major eastern boundary current regions of the World Ocean. The oceanography off the west coast of southern Africa is dominated, like the regions off California, Peru and North West Africa, by coastal upwelling; but the Benguela is unique in that it is bounded on both the equatorward and poleward ends by warm water regimes. In this paper we build on review articles which were published during the 1980s by highlighting the main advances in the conceptual understanding of the system since 1985. A large amount of research on this eastern boundary domain has been conducted in recent years by the institutes to which the authors are affiliated. This has given clear definition to four aspects of shelf dynamics, these being poleward flow at depth across the shelf and out into the slope region, the existence of baroclinic shelf-edge jets in the vicinity of the shelf break, barotropic shelf waves and the importance of variation in wind rather than constant strength of wind as a factor controlling upwelling. Short discussions of some, as yet, unpublished findings supplement published works by ourselves and other authors. Of particular importance are the events such as Benguela Niños in the north and intrusions of Agulhas water in the south. The latter, which can take the form of filaments or rings, influences the oceanography of the region of the "Greater Agulhas Current", with interactions between Agulhas rings and Benguela shelf waters.

Introduction

The Benguela is one of four major eastern boundary current regions of the World Ocean. The oceanography off the western coast of southern Africa is dominated, like the regions off California, Peru and North West Africa by a coastal upwelling system; however, the Benguela is unique in that it is bounded at both the equatorward and poleward ends by warm water regimes.

Much of the oceanographic research which has been conducted in the Benguela has been in support of biological studies associated with southern African west coast fisheries. Prior to 1970 most of the oceanographic literature was of a descriptive nature. Studies of mesoscale upwelling processes commenced during the late 1960s and were by the early 1980s a major focus of attention. This work has been reviewed by Nelson and Hutchings (1983),

Shannon (1985) and Chapman and Shannon (1985). Since then the research has become more quantitative, with the emphasis being placed on shelf dynamics and ocean variability. These studies, together with larger scale investigations in the Agulhas retroflection area and in the South-east Atlantic, have provided some important new insights into this small region of the World Ocean.

In this paper we build on the reviews cited above by providing an overview of the principal characteristics of the Benguela, including its overlying wind fields, oceanic boundaries and fronts, deep and shelf circulation and system variability. The main advances in the conceptual understanding of the system since 1985 are highlighted. This article does not purport to be an encyclopaedic review of the oceanographic literature on the Benguela region of

From WEFER G, BERGER WH, SIEDLER G, WEBB DJ (eds), 1996, *The South Atlantic: Present and Past Circulation.* Springer-Verlag Berlin Heidelberg, pp 163-210

the South Atlantic, but rather the authors' perspective of the most important elements of the system and its dynamics emerging from a synthesis of the now several hundred papers in existence on the subject.

Geomorphology

The sediment composition and structure and bathymetry of the South-east Atlantic have been described by Birch and Rogers (1973), Birch et al. (1987) and Dingle et al. (1987) in the form of a detailed bathymetric map, and Rogers and Bremner (1991). The salient features of the bathymetry, following Shannon (1985), are as follows.

The Cape and Angola Basins, which comprise the abyssal plain in the South-east Atlantic Ocean, are separated by the Walvis Ridge, which runs from its abutment with the continental shelf at about latitude 20°S in a southwesterly direction for more than 2 500 km towards the Mid-Atlantic Ridge (Fig. 1). The Walvis Ridge forms a barrier to the northward and southward flow of water below a depth of about 3 000 m, although some leakage of Antarctic Bottom Water across the Ridge does occur (Shannon and Chapman 1991). Prominent geological features of the Cape Basin, which is bounded in the south by the Agulhas Ridge, are the numerous seamounts of volcanic origin, of which Discovery and Vema are two examples.

The bathymetry of the western continental margin off southern Africa is variable, with the shelf being narrow off southern Angola (20 km), south of Lüderitz (75 km) and off the Cape Peninsula (40 km). The widest zones occur off the Orange River (180 km) and in the extreme south (Agulhas Bank, 230 km). Double shelf breaks are common off the west coast (Siesser et al. 1974), and near Walvis Bay (23°S) there are inner and outer breaks beginning at depths of about 140 and 400 m (see insets in Fig. 1). At about 31°S, Childs Bank, another shallow feature, is situated about 150 km offshore. Farther south, in particular between 32°S and 35°S, the shelf is variable in width. Between 31°S and 33°S (Cape Columbine) there is an inner and an outer shelf break (200-380 m and 500 m) which merge south of 33°S to form a single, deep shelf break. Between 31°S and 35°S several submarine canyons cut into the shelf, the most prominent being the Cape Canyon, which runs from a head 60 km west of Cape Columbine in a southerly direction. The Agulhas Bank, a relatively wide and shallow feature, forms the southernmost margin of the continental shelf.

Of additional importance to the oceanic conditions of the region is adjacent distribution of land. The west coast of southern Africa consists of a relatively narrow coastal plain rising to the main continental escarpment, situated between 50 and 200 km inland. Much of the coastal region is arid. North of 32°S the coastline is regular and, except for Walvis Bay and Lüderitz, is devoid of significant embayments. South of this latitude, several prominent headlands occur; two granitic outcrops in the vicinity of Cape Columbine and Cape Town form the southern boundaries of bays, so leaving the bays exposed at their northern ends (St Helena Bay and Table Bay), while to the south of the outcrops there are enclosed bays (Saldanha Bay and False Bay). A large peninsula lies south of Cape Town (Cape Peninsula), along which mountains attain heights up to 1 000 m. The topography from Cape Town to Cape Point at the southern extremity of this peninsula play an important role in wind forcing over the adjacent shelf sea.

The Wind Field

Winds in the Benguela are controlled by anticyclonic motion round the South Atlantic high pressure system, the seasonal low pressure field over the subcontinent, and by east-moving cyclones which cross the southern part of the continent. The South Atlantic anticyclone is maintained throughout the year, with seasonal differences in pressure being on the order of 3-4 mb. It shifts seasonally over 6° of latitude, reaching northern and southern extremities in May and February respectively, and 13° of longitude, reaching an extreme westward position in August (Tyson 1986). The scale of the movement is shown in Fig. 2c. It forms part of the discontinuous belt of high pressure systems which encircle the subtropical southern hemisphere. Pressures over the continent change radically from well developed lows

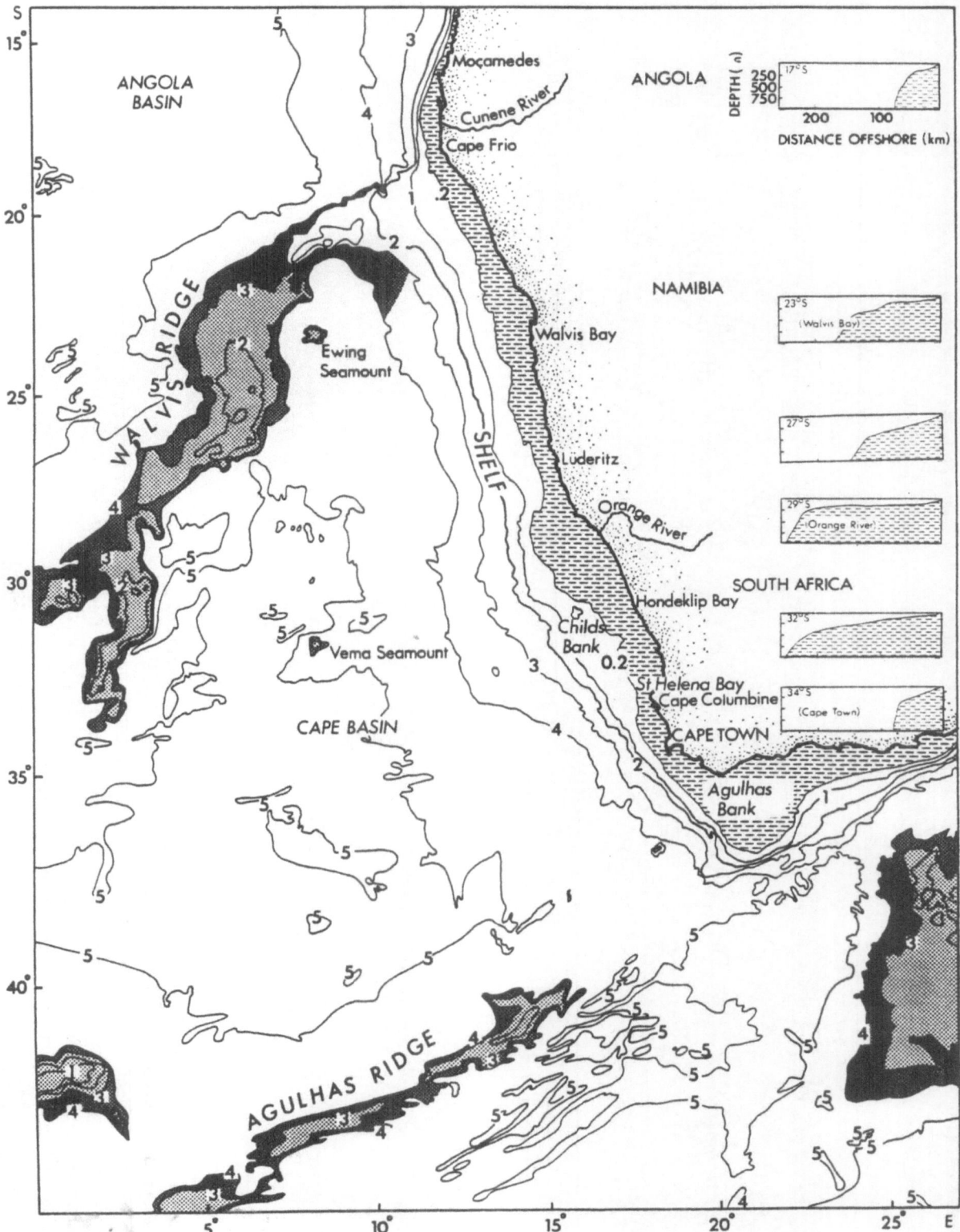

Fig. 1. Bathymetry of the South-east Atlantic Ocean with profiles of the shelf at selected latitudes.

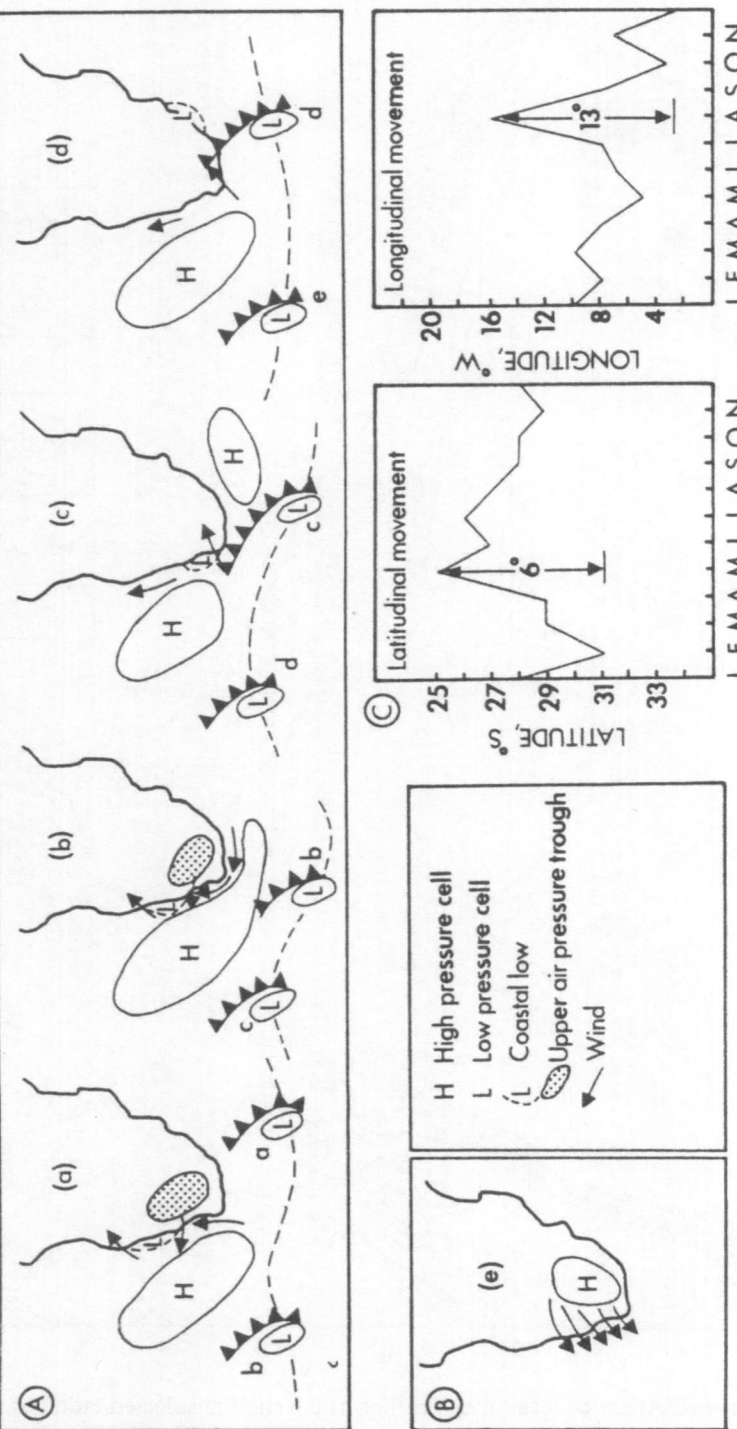

Fig. 2.

A: Cyclic weather pattern over then Benguela system typical of summer conditions.

a - South Atlantic high established, coastal low forms near Lüderitz.

b - High pressure cell ridges round continent with strong south-easterly wind near Cape Town; low starts southward migration.

c - High pressure cell splits; coastal low migrates to Cape Town area as synoptic frontal system passes.

d - High pressure cell strengthens.

B: Berg wind conditions resulting from high pressure cell over land.

C: Annual variation in the positions of the South Atlantic anticyclone (after Tyson 1986)

during summer to weak highs in winter as the continental heat low and Intertropical Convergence Zone (ITCZ) move northwards, with resultant seasonality in the pressure gradients. The curved anticyclonic flow associated with the South Atlantic high is steered along the coast by the interior thermal barrier set up by desert-like conditions in the coastal plain (Nelson and Hutchings 1983).

The comparative climatology of the Benguela and other boundary current regions has been described by Parrish et al. (1983), Pearce (1991) and others. The work of Parrish et al. (1983) was recently extended to a consideration of seasonal changes in wind-stress and wind-stress curl in subtropical eastern boundary current regions by Bakun and Nelson (1991), the results of which are highlighted in Fig. 3. The area of strongest longshore winds in all seasons lies just south 25°S, which is the approximate latitude of the coastal wedge-shaped area of strongest cyclonic wind-stress curl. The wedge is approximately 200 km wide and reaches its most limited poleward extent in June and July, while from late spring to early autumn it extends past Cape Columbine and around the Cape Peninsula. Other coastal areas of cyclonic wind-stress curl are located north of Cape Frio near 15°S during April-May and from October to January and on the south coast near 25°E in all seasons. These latter areas are approximate boundaries of the Benguela upwelling system. The offshore areas, like other eastern boundary current regions, are influenced by anticyclonic wind-stress curl. South of Africa, the strong westerly zonal winds blow during all seasons, reaching a maximum in the austral winter (Fig. 3g). In summer, winds in this area are weaker and are more consistent along the west coast (Fig. 3h).

The essential differences in the seasonal wind regimes between the northern and southern parts of the Benguela were investigated by Hart and Currie (1960) with data from four coastal sites. The longshore and seasonal differences are illustrated by Boyd (1987), whose diagram is reproduced in Fig. 4. The principal perennial centre of upwelling-favourable winds lies near Lüderitz (27°S) with a secondary centre near Cape Frio (18°S), both of which are just south of coastal cyclonic wind-stress curl maxima (Fig. 3). In winter the northward shift

in the pressure system has a stronger influence in the south where there is a greater frequency of westerly winds. These do not produce upwelling. Therefore upwelling- favourable winds reach a maximum during spring and summer in the southern Benguela (cf Shannon 1966; Andrews and Hutchings 1980) with the upwelling season between September and March. Fig. 4 shows this clearly at 33°S during summer. North of 31°S there is less evident seasonality in winds, and upwelling is perennial there, but with a spring-summer maximum and an autumn minimum as far north as 25°S (Stander 1964; Schell 1968). In central Namibia there are lower speeds and little seasonality in wind. Near Cape Frio in northern Namibia the longshore wind is strongest during autumn and spring (Stander 1958, 1963; Boyd 1987). This can be seen in Fig. 4.

In the southern region modulation of upwelling with periods of 3-10 days occurs under wind relaxation or reversal associated with the passage of cyclones south of the continent during the upwelling season. The fundamental mechanism is illustrated in Fig. 2a. In the belt of westerly winds between 35°S and 45°S, low pressure cells form ahead of planetary waves in the westerlies. The associated cyclonic rotation of air produced as these cells advect eastwards causes the wind field as far north as 30°S to be modulated with an intensity which increases southwards to Cape Point. In summer the effect is usually, but not necessarily, weak, manifesting itself as a periodic weakening of the South Atlantic High and slackening or abatement of south easterly winds along the coast. In winter the effect may be, but is not necessarily, strong, bringing in its extreme form gale force north westerly to south westerly winds of some hours duration, in cycles lasting 3-10 days. Associated with the approach of cyclonic systems is the appearance of cells of low pressure which form near Lüderitz and travel round the subcontinent as trapped waves. They are best observed in summer under conditions of weak modulation of the South Atlantic high pressure cell. The cyclonic rotation of air about these cells suppresses upwelling locally as the wave travels along the coast. Ahead of the cell, hot air blows off the escarpment. This is followed the next day by a weak onshore flow and in late summer and autumn.

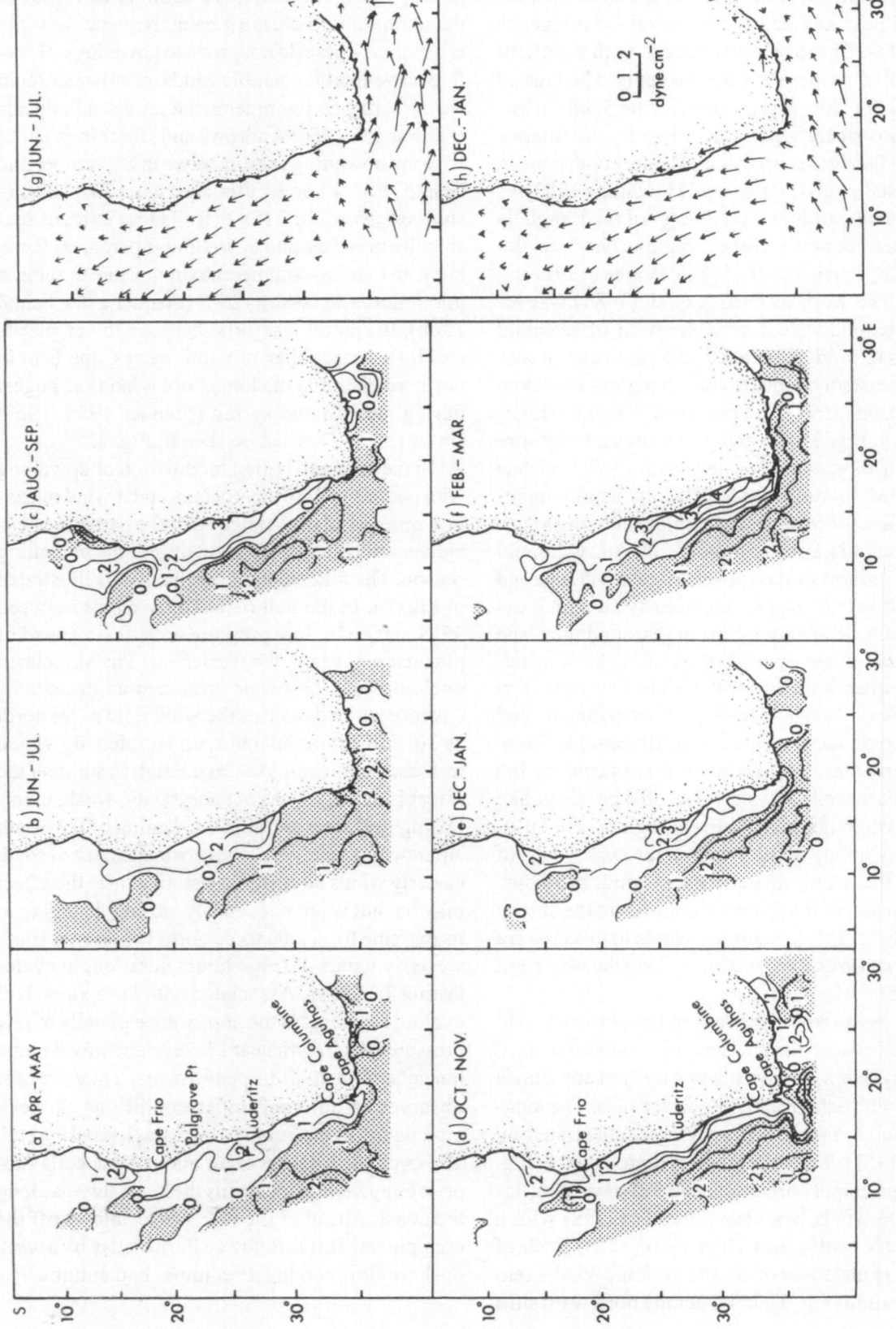

Fig. 3. Wind-stress curl (10^{-8} dyn.cm^{-3}) in the Benguela region (a-f) and surface winds (g-h) (modified from Bakun and Nelson 1991). Regions of anticyclonic wind-stress curl in a-f are shaded.

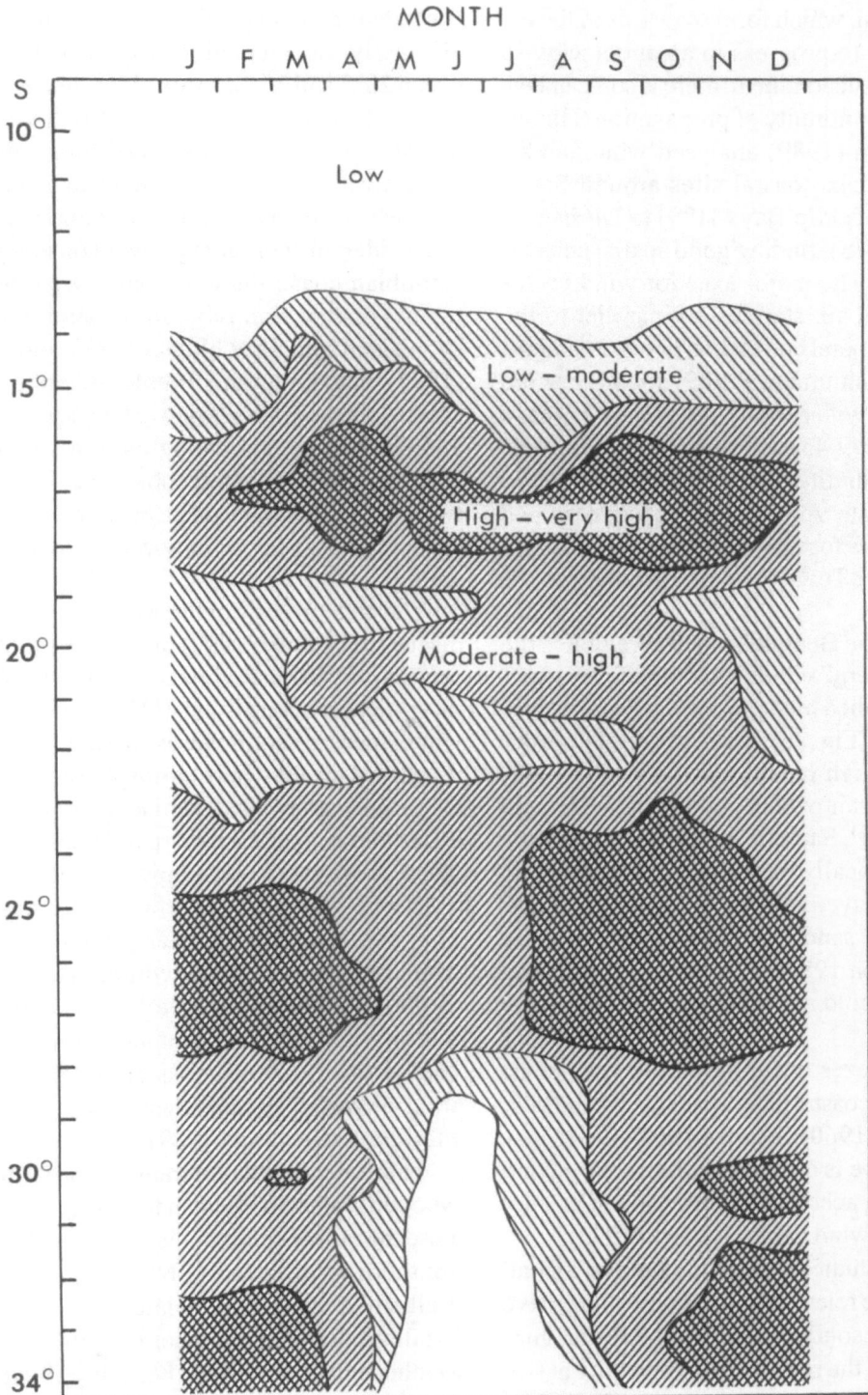

Fig. 4. Monthly wind stress along the west coast of Africa, centred 40 nautical miles offshore at whole degrees of latitude after Boyd (1987). Shading (dynes.cm⁻²): low <0.2; low-moderate 0.2-0.4; moderate-high 0.4-0.8; high-very high >0.8.

While these cells, or coastal lows as they are commonly known, which form seawards of the escarpment appear to progress in an anticlockwise sense around the subcontinent, there is some uncertainty about the continuity of propagation (Hunter 1987). Schumann (1989) analysed wind and air pressure data at six coastal sites around South Africa from Hondeklip Bay (31°S) to Durban on the east coast (30°S), finding good spatial correlation in pressure. The major axes for wind vector distribution at all six stations lay parallel to the coastline, but temporal correlation between stations was not good. Schumann (1989) found that the pressure systems with periods of typically 3-9 days propagated faster (9-45 m.s^{-1}) than the wind speeds (2-20 m.s^{-1}), attributing this unexpected result to differences in scale. A 4.8 day peak was obtained in pressure spectra for the region between Hondeklip Bay and Cape Town, with propagation speeds of 30-45 m.s^{-1}.

A feature of the Benguela coastal region is the occurrence of "berg" winds blowing off the western escarpment when high pressure cells form over the subcontinent (Fig. 2b). The anticyclonic circulation around a high results in a flow of dry adiabatically heated air off the plateau with a typical velocity of 15 m.s^{-1}. Satellite imagery indicates that these winds are locally intensified by topographic features such as river valleys, and possibly transport quantities of sand and dust out to sea (Shannon and Anderson 1982). The warm air will tend to flow over the cold marine boundary layer after passing the coast.

Land-sea breezes occur along the west coast where an interior coastal plain exists (Jackson 1947; Hart and Currie 1960; Stander 1958, 1964; Jury et al. 1985). There is a strong diurnal rotary wind component often accounting for as much as one-third of the total wind vector (Nelson 1992).

A number of studies of the wind field on a small scale, as would be relevant to local upwelling, have been undertaken south of the Orange River. Jury (1988) examined the response of upwelling at two semi-permanent centres (29-31°S and 33-34°S). Variation in the depth of the marine atmospheric boundary layer affects the structure of equatorward wind and, hence, upwelling. Kamstra (1985), on a

statistical basis, showed the difference between consistent equatorward wind in summer and the disorderly winter condition for the shelf region between 28°S and 35°S. Jury (1985a, b) discussed the effects of coastal topography on longshore wind and examined cases of deep and shallow south easterly winds over the Cape Peninsula and Cape Columbine, where under conditions of shallow flow, there is considerable curvature of wind onshore. For the Namibian coast, there are only a few incidental reports on the wind field, for instance those cited by Nelson and Hutchings (1983) and Shannon (1985). Hutchings and Taunton-Clark (1990) examined time-series of wind from Cape Columbine and Cape Point, stretching over nearly 30 years. They found little spatial coherence between these two sites and concluded that mesoscale effects such as topography play an important role in longshore wind.

The meteorology of the Agulhas Bank has been discussed by Jury (1992) and Hunter (1987). The former provides a review of the subject. Composite vertical sections in Jury (1992) show that zonal winds increase southwards, particularly over the Agulhas Current where atmospheric turbulence is extreme. Easterly wind dominance, typical of spring and summer, exhibits a sharp offshore velocity gradient. These winds cause upwelling along the south coast and tend to move surface water southwards. This contrasts with westerly wind dominance, which is characteristic of winter in the area south of the Agulhas Bank (see Fig. 3). There is, of course, not a seasonal dichotomy in this simple way. The weather systems crossing the Agulhas Bank bring alternately easterly then westerly wind within a few days (Schuman 1989).

There are distinct interannual variations in the synoptic scale wind field in the larger Benguela. Of these the most significant is related to the *El Niño*-Southern Oscillation (ENSO). The changes in Walker Circulation associated with the Southern Oscillation and the impact on southern African weather and climate have been described by Preston-Whyte and Tyson (1988), and they are illustrated diagrammatically in Fig. 5. During the low (*El Niño*) phase of the Southern Oscillation the longshore (upwelling) winds in the extreme south-

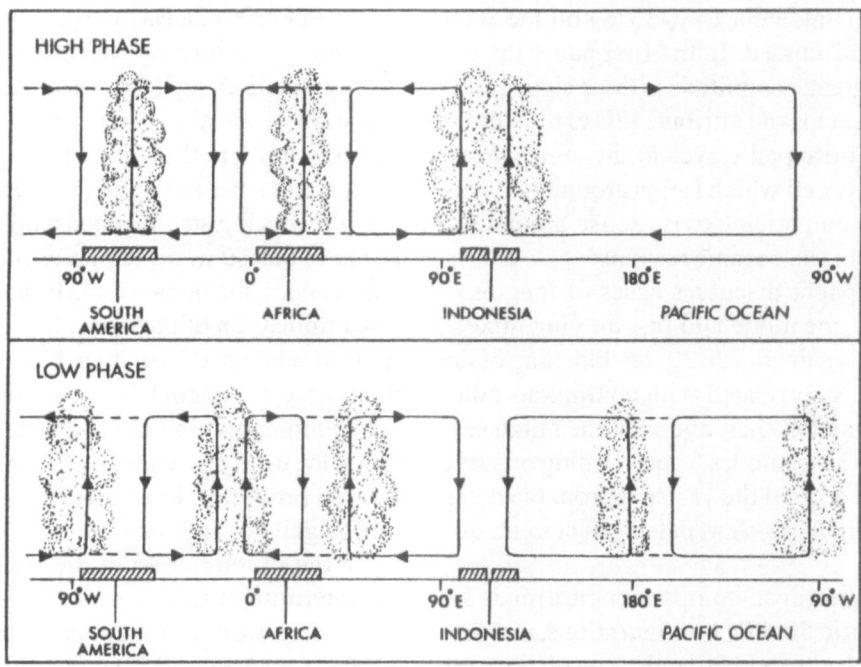

Fig. 5. The Walker Circulation during high and low phases of the Southern Oscillation (after Lindesay 1987).

ern Benguela tend to weaken. Conversely, during the high phase, longshore winds tend to strengthen, particularly along the south coast during *La Niña*.

The mechanism which causes the modulation of longshore wind, and therefore the strength and frequency of upwelling and generation of coastal-trapped waves, is found in the Rossby wave structure in the westerly jet stream. Peaks occur in pressure spectra obtained from various sites along the west coast. This was first demonstrated by Preston-Whyte and Tyson (1973). Following the classic work of Rossby (1939), a barotropic modal analysis gives clues to the origin of periods observed in west coast spectra. If one mode is held stationary by way of an appropriate jet velocity, travelling modes four and higher generate periods typical of those observed in pressure spectra (Nelson 1992).

Although the hemispheric behaviour of synoptic events influencing southern Africa have yet to be given a clear theoretical formulation, current thinking on the subject seems to be crystalizing as follows:

Flow in the upper air jet stream in the southern hemisphere has stable characteristics. The mode 1 wave results from the eccentricity of Antarctica, the mode 3 wave from average thermal distribution in the ocean basins, and the mode 4 wave is the 20-35 day cycle which is dominant in the southern hemisphere geopotentials. No strong mode 2 wave exists. The mode 3 wave is influenced by circulation from tropical to subtropical regions. In *El Niño* years, a negative pressure anomaly forms at the 200 - 500hPa level between Gough Island and Marion Island, which results in a much flatter wave 3 formation. The reverse is true of the *La Niña* years. Mobile mode 5, 6 and 7 waves form in the vicinity of South Africa as a result of smaller scale baroclinic features, and these are the cause of the pressure variation at periods less than 10 days, resulting in concomitant variation in longshore wind on the west coast.

In the papers by Jury et al. (1990) and Jury and Brundrit (1992) the connection between Rossby waves and longshelf variation in current (Cape Columbine, 33°S), temperature (Melkbos, 34°S) and sea level (Saldanha Bay, 33°S) on the west coast shelf is discussed. In the first paper the authors point to greater amplitude in these signals than would be expected, and attribute this to the formation of coastal-trapped waves in the atmosphere. The coastal low cell which forms around Lüderitz and moves in an anticlockwise sense around the continent forces the oceanic response.

The 1990 paper discusses cases of long-term dominance of one mode and fast moving mixed-mode patterns. In the first, a 20-day blocking of the mode 4 wave is correlated with continuous poleward flow along the shelf and sustained high sea-level. In the second, modes 5 and 6, riding on wave 1, contributed 45% of the variance from the mean geopotential, inducing short pulsing in currents and sea level.

In the second paper, comparisons are made for two hemispheric 500hPa configurations, and single pulse, slow pulse and fast pulse correlations are shown against current, sea temperature, sea level and wind. There is a complexity in these variables which is concomitant with the complexity of Rossby waves. Two upper ridge formations corresponding to the two cases are discussed together with a seven-day sequence for the 500hPa and surface pressure fields south of Africa.

Lindesay and Jury (1991) describe a persistent wave 4 formation between January and March 1988, with below normal geopotential heights over the South Atlantic and central Indian Oceans.

Heat Budget

The sea surface heat exchange in the upwelling area along the southern African west coast contrasts strongly with that in the area immediately south of the continent, which is influenced by the warm Agulhas Current system. The differences are illustrated by the climatological sea surface temperature anomalies (Fig. 6). The largest negative anomaly computed from zonal means; it is 5-6°C in the vicinity of Lüderitz corresponds to the region of strong longshore winds and cyclonic wind-stress curl (Fig. 3), whereas south of the continent positive anomalies are typically between 2° and 4°C.

For the Benguela upwelling area, Guastella (1992) showed that the latent and sensible heat fluxes in St Helena Bay (Fig. 1) during the spring of 1986 approximately cancelled out. Extrapolating her results along the coast and assuming a back-radiation of 70 W.m^{-2} and a midsummer incident radiation level in the south of 325 W.m^{-2}, she concluded that a net radiation (255 W.m^{-2}) was capable of producing an integrated temperature increase in the upper 10 m mixed layer of 0.5°C per day. Farther north the figure would be marginally higher. The implication of these results is that, during periods of relaxion of winds such as occur in the central northern Benguela near Walvis Bay in mid to late summer and in the southern Benguela in autumn, the upwelled water rapidly warms and stratifies, so providing ideal conditions for blooms of dinoflagellates and red tides.

In the southern part of the Benguela, a strong but intermittent flux of heat occurs in the form of rings shed from the Agulhas retroflection area. Walker and Mey (1988) were the first to provide estimates of surface heat flux, showing that, although net flux was seasonal, there was no seasonality in maximum or minimum radiation levels, which ranged from 225 to 25W.m^{-2}. A summary of this work together with that of other authors has been given in a paper by van Ballegooyen et al. (1994). In that work, they present the results of a cruise undertaken specifically to study the retroflection area estimate the flux of Indian Ocean water warmer than 10°C into the Atlantic Ocean to be 1.05Sv per eddy with an average of six eddies per year. Satellite altimetry, used to identify and track eddies, showed that the tendency was to move westwards rather than into the Benguela area, although this does occur (see Figs. 7 and 9 of that paper). The combined altimetric and hydrographic analyses gave a volume flux of water into the South-East Atlantic of 6.3Sv, with a surface heat flux of 0.045PW and salt flux of 78 1012kg.year^{-1}.

System Boundaries and Fronts

For the purpose of this review we take the Benguela region to be that part of the South-east Atlantic lying between about 14°S and 37°S, with a western boundary at the 0° meridian. As such it encompasses the coastal upwelling system, frontal jet and the eastern part of the South Atlantic gyre. The

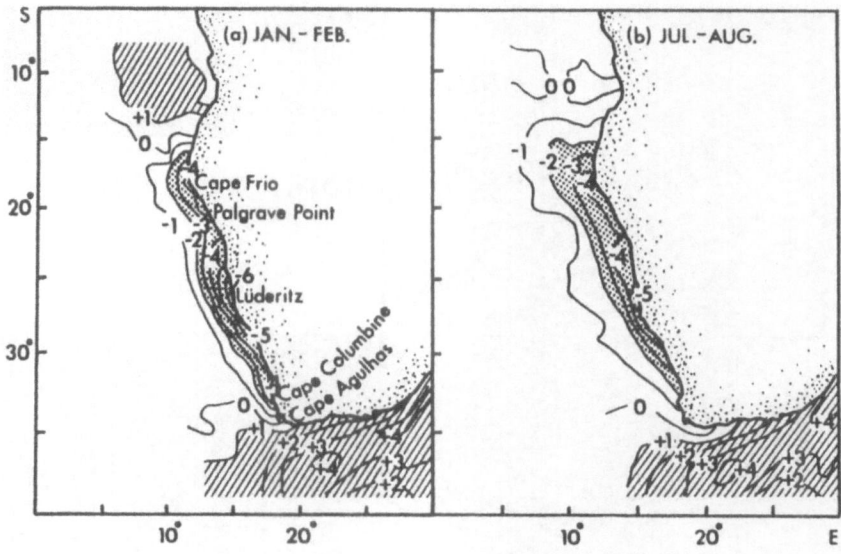

Fig. 6. Climatological sea surface temperature anomaly in the Benguela during (a) austral summer and (b) winter (after Parrish et al. 1983).

Benguela Current will be regarded as the eastern boundary current of the South Atlantic gyre in keeping with the generally accepted international terminology of Stramma and Peterson (1989), rather than the more limited definition adopted by Hart and Currie (1960), Stander (1964), Shannon (1966), Bang (1971) and Nelson and Hutchings (1983). Many of the essential boundary and frontal features of the Benguela region are illustrated in Fig. 7.

The Angola-Benguela Front

The northern boundary of the Benguela is marked by the Angola-Benguela surface frontal zone. This front is most intense within 250 km of the coast and can be traced as far west as the Greenwich maridian. The first comprehensive accounts of the Angola-Benguela front were provided by Shannon et al. (1987) and Shannon and Agenbag (1987). Those authors analysed a combination of large oceanographic and meteorological data sets, as well as measurements made from dedicated cruises, airborne radiation data, thermometry measurements and satellite imagery. This work was subsequently extended by Meeuwis and Lutjeharms' (1990)

analysis of satellite-derived weekly maps of sea surface temperature for the period 1982-1985. In the discussion which follows we summarise the essential results of the above mentioned authors.

The Angola-Benguela front is most marked in the upper 50 m, where it is well defined in terms of both temperature and salinity, but it is identifiable to a depth of at least 200 m. A horizontal sea surface temperature gradient near the coast of 4°C per 1° of latitude is typical in synoptic data. The frontal zone is a permanent feature and is maintained throughout the year within a narrow band of latitudes, characteristically between 14°S and 17°S, and is approximately coincident with the 30 m².s⁻² pseudo wind-stress isopleth (Fig. 8). Its average position migrates seasonally over 2° of latitude, the front being farthest north in winter (August) and south in late summer (March). Seasonally averaged sea surface temperatures and satellite-derived frontal positions (in Shannon et al. 1987) show the general extent of the variability. On shorter time-scales of days to weeks, significant frontal movement and variability is evident, with frontal eddies and tongue-like features being prevalent. Multiple fronts are evident on occasions, most commonly during

Fig. 7. Sea surface temperature distribution from satellite infrared imagery on 15 June 1994 (corrected and calibrated).

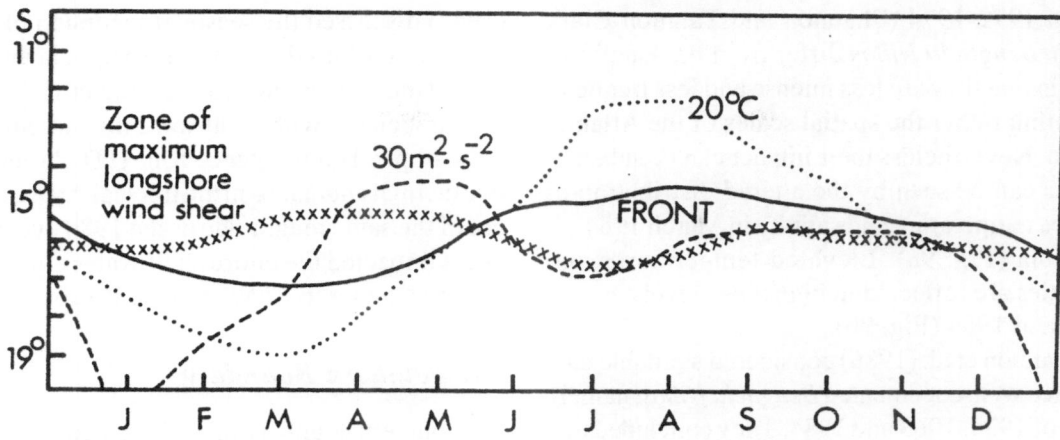

Fig. 8. Monthly mean positions of the 20° C surface isotherm, the Angola-Benguela thermal front and the 30 m².s⁻² isopleth of wind-speed squared near the African coast (after Shannon et al. 1987).

summer when the poleward flow in the Angola Current is strongest (Shannon et al. 1987). A recent study by Kostianoy (1994) suggests that two or three fronts are the norm rather than the exception and that it would be more appropriate to consider the region as a frontal zone rather than a single front. Of these multiple fronts, one is associated with the Angola Current and another with a tongue from the Northern Namibian upwelling cell. The frontal orientation is usually west to east, particularly so in winter, although other orientations such as WNW-ENE are quite common. The Angola-Benguela front is apparently maintained through a combination of factors. These include coastal orientation, bathymetry, stratification, wind stress and opposing flows of the Benguela and Angola Currents. Although Meeuwis and Lutjeharms (1990) considered the front to be "almost exclusively" controlled by the last mentioned, it is difficult to understand how this could constrain the front within such a narrow range of latitudes. It is our view that the curl in wind stress is most important (see Fig. 3).

Benguela Niños

The annual southward migration of the Angola-Benguela front and the intrusion of saline water

from the Angola Basin (Boyd et al. 1987) coincide with the partial relaxation in the equatorward wind stress, in the northern Benguela, and a reduction in upwelling of nutrient-rich water to the surface (Boyd 1987). The situation off northern Namibia and southern Angola is in some respects analogous to the seasonal cycle off Peru in the Pacific during some years, when anomalously warm water appears. These are the El Niño Southern Oscillation events (ENSO).

There is a South Atlantic counterpart of the Pacific ENSO which manifests itself as an episodic extreme warming in the tropical eastern Atlantic and the advection of tropical water southwards along the coast of Namibia. These events, which are superimposed on the normal annual cycle, have been documented by various authors (Walter 1937; Stander and De Decker 1969; Boyd and Thomas 1984), while Shannon et al. (1986) provided a comprehensive review of the abnormal intrusions, which they termed *Benguela Niños*, suggesting a possible causative mechanism. *Benguela Niños* are not necessarily in phase with ENSO, although authors such as Hisard (1990) have discussed these extreme events in the tropical Atlantic in relation to the ENSO.

Benguela Niños occurred in 1934, 1949, 1963, 1984 and probably around 1910, in the mid-1920s and in 1972-1974 (Shannon and Taunton-Clark 1989). *Benguela Niños* differ from the Pacific El Niño in that they are less intense and less frequent, reflecting rather the spatial scales of the Atlantic Ocean. Nevertheless their impact can be substantial, as can be seen by the altered distribution of surface temperature and salinity in March 1984 off Namibia (Fig. 9a). Elevated temperatures and salinities are reflected in higher sea levels, e.g. in 1963 and 1984 (Fig. 9b).

Shannon et al. (1986) considered available data for three well-documented *Benguela Niños*, namely those of 1934, 1963 and 1984. They concluded that these major perturbations were associated with increased flow in the Atlantic Equatorial Counter-current-Undercurrent system, resulting in abnormal advection of equatorial and tropical water eastwards and southwards along the coast of Namibia during periods of low or reduced zonal wind stress in the equatorial western Atlantic off Brazil. The correspondence between sea level at Walvis Bay between 1982 and 1984 and zonal wind anomaly in the equatorial Atlantic is clear from Fig. 9b, and in this respect Horel et al. (1986) suggested that the sudden relaxation in wind stress is more important than the absolute value, as did Hisard (1990). In a recent paper on warm events in the tropical Atlantic, Carton and Huang (1994) also concluded from their simulation of ocean circulation during the 1980s that the relaxation of the stronger than normal trade winds during late 1983 resulted in a surge of warm water eastwards along the equatorial wave guide.

Local winds change during *Benguela Niños*, and the warmings during 1949, 1972 and 1963 were indeed accompanied by significantly lower equatorward wind stress (Taunton-Clark 1990). While this will obviously impact coastal upwelling, the *Benguela Niño* events are characterized by large-scale advection of tropical and equatorial water from the **north west** rather than just reduced upwelling. All the evidence suggests that this movement is triggered by processes in the distant equatorial western Atlantic. There is evidence to suggest that *Benguela Niños* are not related to southward incursions of the Angola Current but rather result from an influx of warm saline water from the west and north-west. In this respect, Boyd et al.

(1987) discussed the seasonal incursion of warm saline water into the northern Benguela and concluded that events such as occurred in 1986, while also associated with southward movement of the inter-tropical convergence zone (ITCZ) and positive thermal anomalies in the tropical Atlantic, were not of the same magnitude as the 1984 event. That event impacted the entire shelf water column in a similar manner to *El Niño* in the Pacific.

The Southern Boundary

The southern boundary of the Benguela system can be considered as being the Agulhas retroflection area. Fronts associated with the Agulhas Current and its retroflection are illustrated in Fig. 10, while a synoptic view of the frontal characteristics is shown in Fig. 7. This definition of the southern boundary is in keeping with Shannon et al. (1981) and Nelson and Hutchings (1983) and is preferred to a definition based on the extent of coastal upwelling. Hart and Currie (1960), Andrews and Hutchings (1980) and others have regarded the Cape Peninsula - 34°S - as the southern limit of significant coastal upwelling, although Shannon et al. (1983), Schumann et al. (1982) and others have shown that upwelling extends along much of the south coast, certainly as far as 25°E. In the area east of 20°E, the edge of the Agulhas Current and the shelf break of the Agulhas Bank are approximately coincident, with the current being topographically steered. In addition, south of latitude 37°S there are marked changes in the wind field with westerlies predominating on average during all seasons (Fig. 3). Therefore, the southern boundary of the Benguela is produced by a combination of oceanographic, topographic, meteorological and biological factors.

Bang (1973) noted that there are nearly always vestiges of Agulhas water west of the Cape Peninsula. This water is, however, generally modified rather than pure Agulhas Current water. The advent of satellite remote sensing has provided a perspective on the retroflection of the Agulhas Current ring-shedding mechanism, hitherto not possible, and considerable advances have been made during the past decade in documenting and understanding the kinematics of the retroflection and its

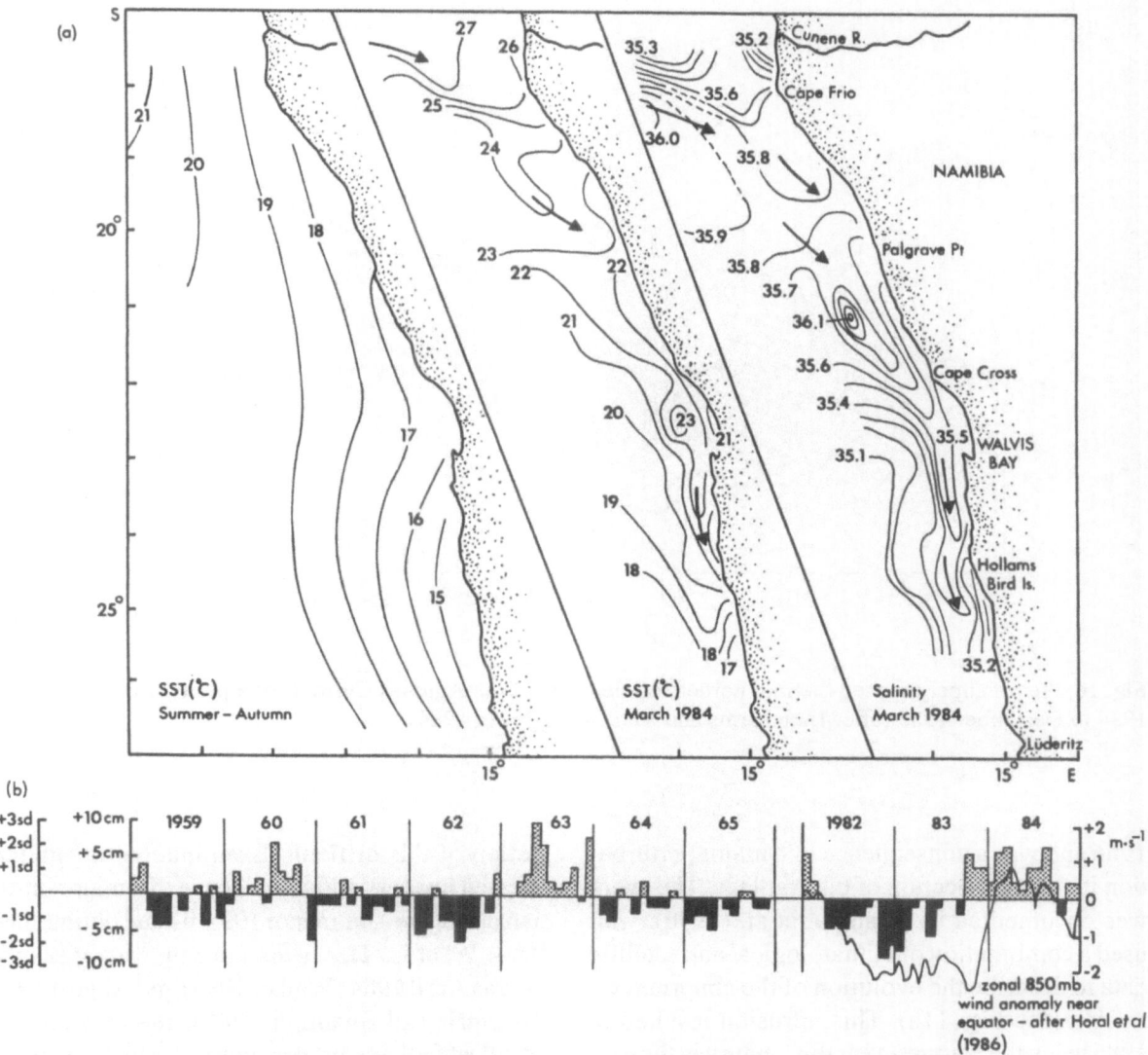

Fig. 9. (a) Mean positions of surface isotherms off Namibia during autumn and summer contrasted with the distribution of SST and surface salinity during March 1984, (b) Changes in sea level at Walvis Bay, 1959-1965 and 1982-1984 (after Shannon et al. 1986).

implications, e.g. Lutjeharms and van Ballegooyen (1988). The retroflection area shows variability (Fig. 10) which is high by global standards, a veritable 'hot spot', comparable with the Gulf Stream, Kuroshio and Brazil/Falkland Current confluence. Walker (1986) and Lutjeharms (1988) have documented some extreme circulation events in the retroflection in 1984 and 1985, while Lutjeharms and van Ballegooyen (1988) have provided a comprehensive overview of its kinematics and proc-

esses. The shedding, advection and structure of Agulhas rings have been investigated using combined *in situ* measurements and remote sensing by a number of authors (e.g. Lutjeharms and Gordon 1987), while Gordon and Haxby (1990) have provided evidence from satellite altimeter data for the passage of rings across the South Atlantic to 30°W.

On occasions there have been substantial incursions of Agulhas Current water into the Southern Benguela. One such *Agulhas intrusion* occurred in

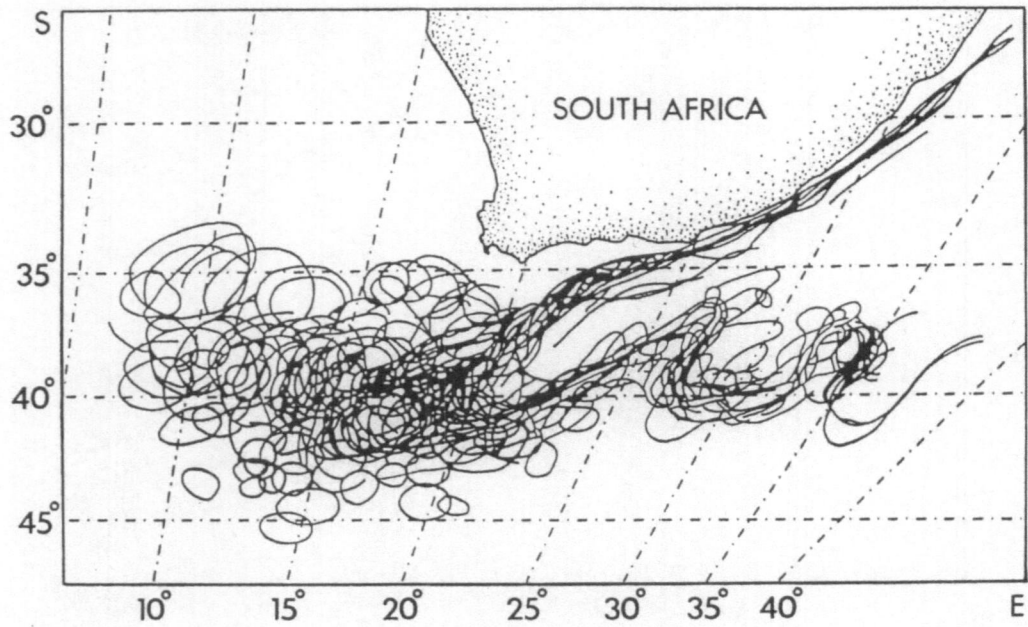

Fig. 10. Set of superimposed thermal borders of elements of the Agulhas Current for a period from December 1984 to December 1985 (after Lutjeharms and Van Ballegooyen 1984).

1986 and was a consequence of a major perturbation in the retroflection of the Agulhas. The event was documented by Shannon et al. (1990), who used a combination of climatological and satellite data to describe the evolution of the abnormal intrusion (see Fig. 11a). This intrusion resulted in 1986 being the warmest year this century in the near South-east Atlantic. The intrusion began during the latter part of 1985 and reached a maximum during late 1986. Shannon et al. (1990) have examined possible causes of the 1986 intrusion, and have provided evidence which suggests that it was linked to changes in winds in the South-west Indian Ocean and south of Africa, which would have resulted in lower volume transport of the Agulhas and a southward displacement of the zero of wind-stress curl, i.e. consistent with model predictions of increased leakage of Agulhas water into the Atlantic (Olsen and Evans 1986). A study of hydrographic data by the authors suggest that Agulhas intrusions may have occurred between 1957 and 1964, although the lack of large-scale synoptic data makes confir-

mation of this difficult. Examination of satellite infrared imagery shows evidence of a major intrusion during the summer of 1992-93 and during June 1994. What is clear is that, during some periods, such as April 1984 (Walker 1986) and August 1989 (Brundrit and Shannon 1989), the current can retroflect farther west than normal, which again has important implications for inter-ocean transfer of heat and salt.

In the process of ring spawning at the Agulhas retroflection, a pulse of cold Sub-antarctic surface water is injected into the South Atlantic (Shannon et al. 1990a, b). Such an event, which turned into a major anomalous perturbation in the Subtropical Convergence south of Africa, occurred during December 1986 and January 1987, following the abnormal flux of Agulhas Current water around the Cape Peninsula in 1986 referred to above. This perturbation resulted in a substantial equatorward flow of Subantarctic Surface Water in the form of cold filaments (Shannon et al. 1989). One of these filaments extended into the Benguela as far north

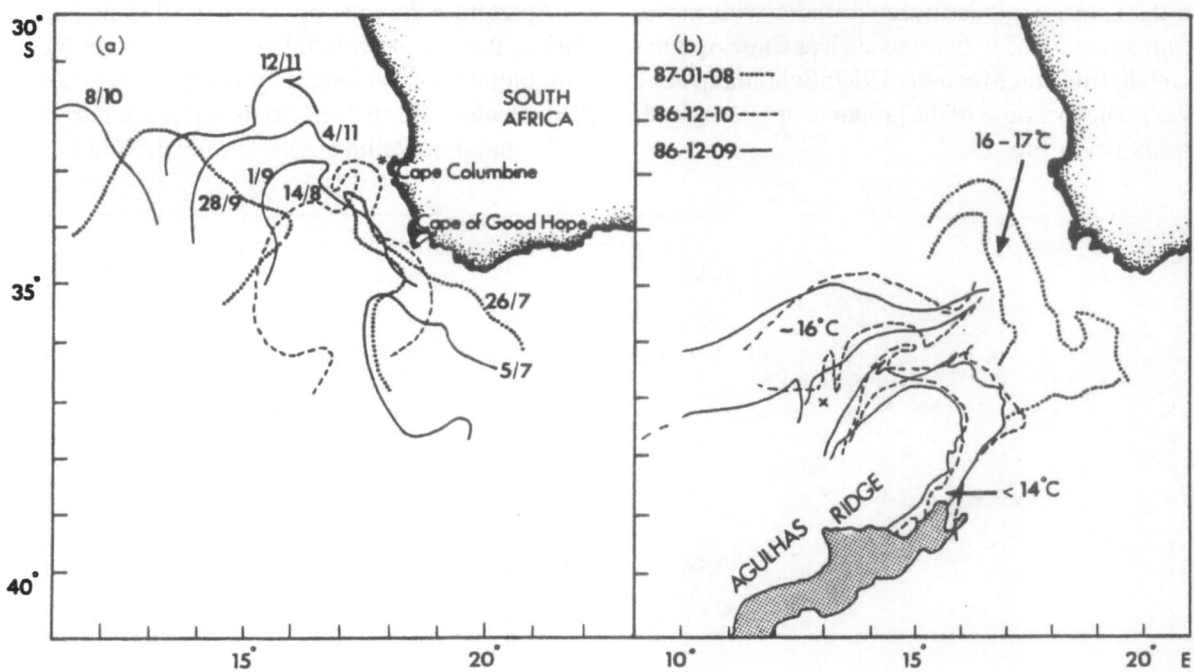

Fig. 11. (a) Superimposed snapshots of the northern boundary of Agulhas Current water, July - November 1986 (after Shannon et al. 1990). (b) Changes in the location of fronts associated with the intrusion of Subantarctic Water during December 1986 and January 1987 (after Shannon et al. 1989). The Agulhas Ridge is shown (shaded).

as 33°S during January 1987 (Fig. 11b). The feature persisted for at least two months during the summer of 1986-87 and appeared to restrict temporarily the flow of surface Agulhas water into the South-east Atlantic. Using a combination of ship and satellite data, the structure and kinematics of the cold intrusion were documented by Shannon et al. (1989). Those authors discussed the apparent association between the filament and the Agulhas Ridge.

Upwelling and Filaments

The climatological sea surface temperature anomalies (Fig. 6) give some idea of the spatial extent of upwelling along the west coast of southern Africa. Imbedded within this continuum are a number of upwelling centres or cells. The principle semi-permanent cell is in the vicinity of Lüderitz (27°S), effectively dividing the Benguela into northern and southern sections. The location and characteristics of the main upwelling cells have been documen-

ted and described by authors such as Nelson and Hutchings (1983), Shannon (1985) and Lutjeharms and Meeuwis (1987). Shannon (1985) identified six such cells along the west coast, two cells in the northern Benguela near and south of Cape Frio (18°S), and one each at Lüderitz (27°S), Namaqua (30°S), Columbine (33°S) and the Cape Peninsula (34°S). The last two are more seasonal than those within the central Benguela. The upwelling cells are normally located near regions of cyclonic wind-stress curl, and are in most cases in regions where there is a change in orientation of coastline. Lutjeharms and Meeuwis' (1987) analysis of satellite-derived sea surface temperature maps for the period 1982-1985 was in agreement with Shannon (1985), although they did identify an additional cell near Walvis Bay (23°S). The typical westward extent of the influence of the upwelling, excluding major filaments, is between 150 and 250 km, the latter figure being associated with the Lüderitz cell, where the frequency of upwelling is also significantly higher than in the other centres. Upwelling on the south coast is more sporadic with minor centres associated with capes such as Cape Agulhas (Lutjeharms and Meeuwis 1987; Schumann et al. 1982). The locations of the principal upwelling cells are shown in Fig. 12.

Apart from the equatorward and poleward boundaries, there exist a number of semi-permanent "discontinuities" within the upwelling system. Zonally oriented "fronts" tend to develop equatorward of the major upwelling cells. One such front exists

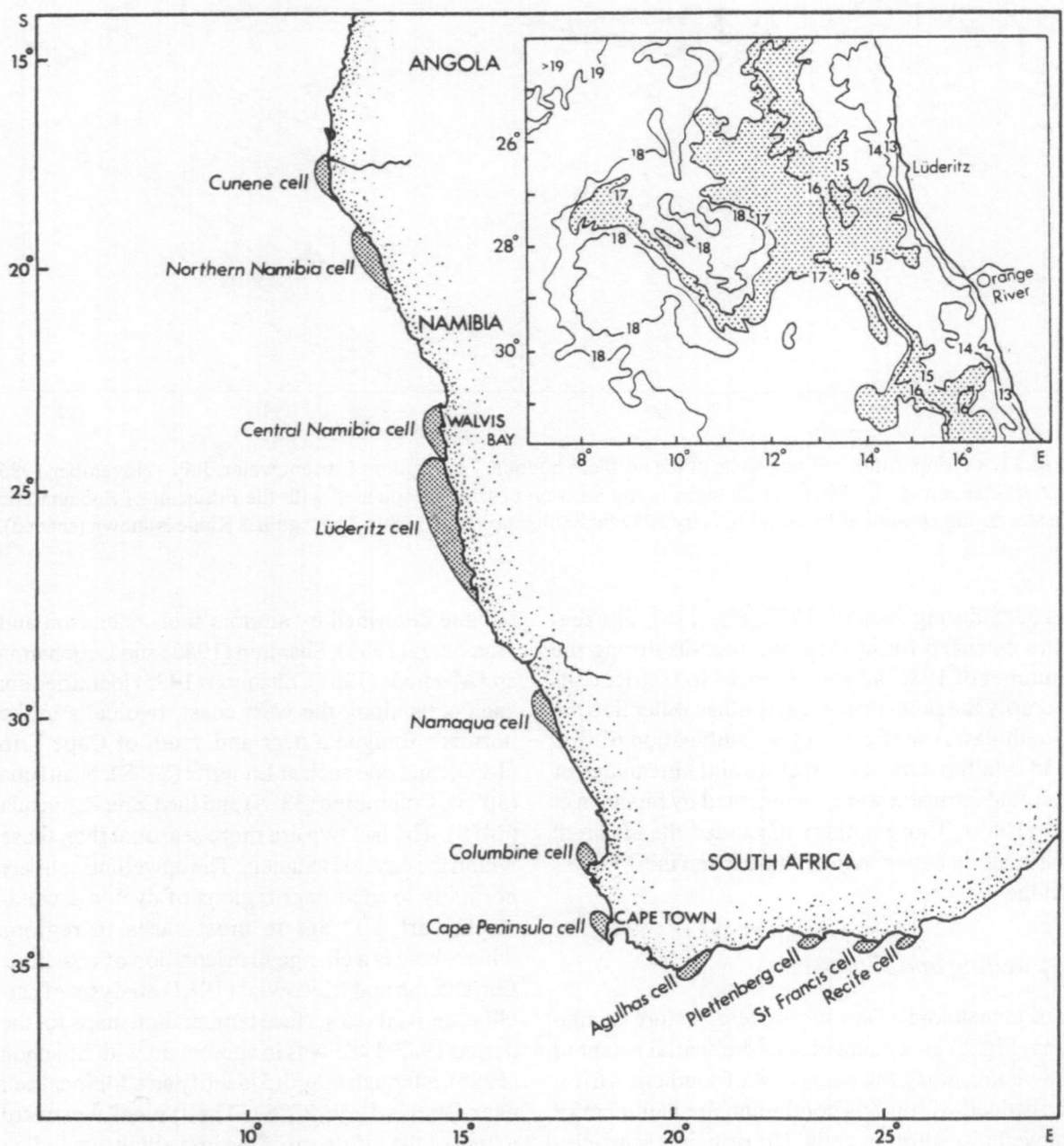

Fig. 12. Principal upwelling cells in the Benguela and the Agulhas Bank. The cells near the extremity of the region tend to be more seasonal and ephemeral. Inset is an interpretation of the surface isotherms for 15 June 1989, showing cool fontal water (shaded) entrained around the Agulhas ring (after Duncombe Rae et al. 1992b).

near 25°S, just north of the perennial Lüderitz upwelling cell. It is approximately coincident with the zone of maximum cyclonic wind-stress curl (Fig. 3), and lies immediately to the south of an area where wind-induced turbulence is significantly lower and stratification of the shelf waters is stronger (Agenbag and Shannon 1988). While this discontinuity is perhaps not significant in terms of the physics of the system, it does have important biological consequences (Agenbag and Shannon 1988).

There exists over much of the area between Cape Frio (18°S) and Cape Point (34°S) a well developed longshore thermal system of fronts demarcating the seaward extent of the upwelled water. The southern extremity of the front extends polewards in spring and eastwards around the Cape of Good Hope in summer. South of Lüderitz a single front is usually well defined, and although it is spatially and temporally variable it coincides approximately with the run of the shelf break. Farther north the surface manifestation of the front is more diffuse and multiple fronts are evident on occasions. The meandering nature of the front was first commented on by Currie (1953), who suggested that this might be related to upwelling cells and resulting mesoscale eddy systems. The highly convoluted form of the oceanic front is apparent in satellite thermal and ocean colour images of the Benguela, as seen for example in Fig. 7. A number of articles have examined the mesoscale dynamics of the front and its apparent response to atmospheric forcing (Shannon et al. 1985; Jury et al. 1985; Armstrong et al. 1987; Jury 1988). The larger scale kinematics of the front were documented by Lutjeharms and Stockton (1987), who based their study on a series of Meteosat II thermal images collected during the period 1983-1985.

Upwelling filaments are characteristic of the Benguela oceanic front (Van Foreest et al. 1984; Lutjeharms and Stockton 1987). These have a typical life-span of a few days to several weeks and have axes which are generally orientated perpendicular to the coast. Although there is no published evidence filaments at preferred geographic locations, recent work by Kostianoy (1994) shows that there appear to be preferred sites for filament formation. Dipole eddy pairs, associated with the filament structure, have been observed in satellite imagery by Stockton and Lutjeharms (1988), most commonly west of the Lüderitz and Namaqua upwelling cells, with the former showing the greatest development and extent. These dipole eddies are not a dominant part of the mesoscale eddy spectrum, but together with the melange of filaments, plumes and frontal eddies they contribute to the extreme complexity of the frontal region. Filament axes have a typical zonal dimension of 200 km, but in extreme cases they may extend 1 000 km or more offshore (Lutjeharms et al. 1991). An *in situ* investigation of a filament in the Benguela was carried out by Shillington et al. (1990), who found it to be a relatively shallow feature confined to the upper 50 m and may not have been typical of a Benguela filament. The net mass flux associated with this particular filament appears to have been small. Duncombe Rae et al. (1989, 1992a) documented the interaction between an Agulhas ring and the Benguela frontal system at 26°-30°S based on shipboard and satellite observations during 1989. A cool filament, which extended some 450 km offshore, had become entrained around the warm core ring (Fig. 12-inset). This process could have drawn off 5×10^{12} m^3 of upwelled surface water from the Benguela shelf over a period of 2-3 months, representing on average volume flux of 1.5×10^6 m^3.s^{-1} (Lutjeharms et al 1991; Duncombe Rae et al. 1992 a, b).

In most of the published literature on the Benguela, authors tend to refer to the oceanic front, shelf-break front or upwelling front without distinction. While it is true that this may well be appropriate in areas where the three are coincident, as off the Cape Peninsula (34°S), it is equally true that two or three longshore fronts can be resolved at times in the northern Benguela. There have been very few synoptic studies on the three-dimensional nature of fronts. This calls for extensive fieldwork. A conceptual three-dimensional model, which was inferred from biological studies, was proposed by Barange and Pillar (1992) for the northern Benguela and is reproduced here in Fig. 13. As a general schematic, it may be applicable also to upwelling systems other than the Benguela.

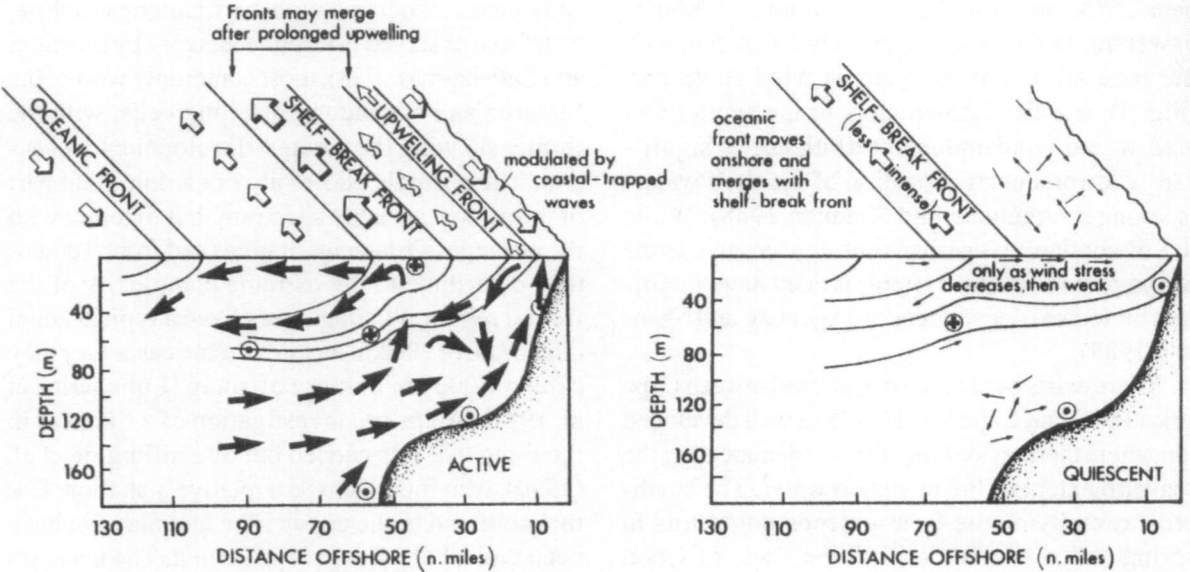

Fig. 13. A conceptual three-dimensional model of cross-shelf circulation during periods of active and quiescent upwelling in the northern Benguela region. Encircled dots indicate poleward current, encircled crosses equatorward current (after Barange and Pillar 1992).

Water masses and large scale circulation

The principal characteristics of the water masses present in the Benguela region and the general circulation have been reviewed by Shannon (1985) and Chapman and Shannon (1985) in their articles, and readers are referred to the numerous publications cited by those authors. Comment here will be restricted to highlighting the principal features and discussing them in the light of more recent findings.

There are a number of water masses present off the west and south coasts of southern Africa, including tropical and subtropical surface waters, thermocline waters (comprising South Atlantic central water, South Indian central water, tropical Atlantic central water), Antarctic intermediate water (AAIW), North Atlantic deep water (NADW) and Antarctic bottom water (AABW). The principal water masses and the typical "core" characteristics are annotated in Fig. 14. The linear part of the θS curve (approximately 6°C, 34.5 psu - 16°C, 35.5 psu) spans the thermocline water stratum, and this is the water mass which upwells along the coast and

which constitutes, often in highly modified form, the shelf waters of the Benguela. In Fig. 14 the principal differences between the more coastal waters, marked Benguela in the figure, and the true South Atlantic thermocline and surface waters are clearly evident, with the water warmer than 10°C in these strata associated with the upwelling system being generally fresher by 0.2 psu or more for the same temperature over much of the range. In the lower portion of the thermocline stratum (cooler than 10°C), the presence of tropical Atlantic central water which is saltier that its South Atlantic counterpart is apparent. At the core of the stratum (around 10-12°C) it is extremely difficult to distinguish between these central waters and that originating from the Indian Ocean, which is advected into the South-east Atlantic in modified form by Agulhas rings and filaments. The advent of high-density CTD data together with better quality nutrient analyses and chlorofluorocarbon (CFC) data has simplified the task, and various authors have been successful during the past decade in differentiating between the thermocline waters of differing

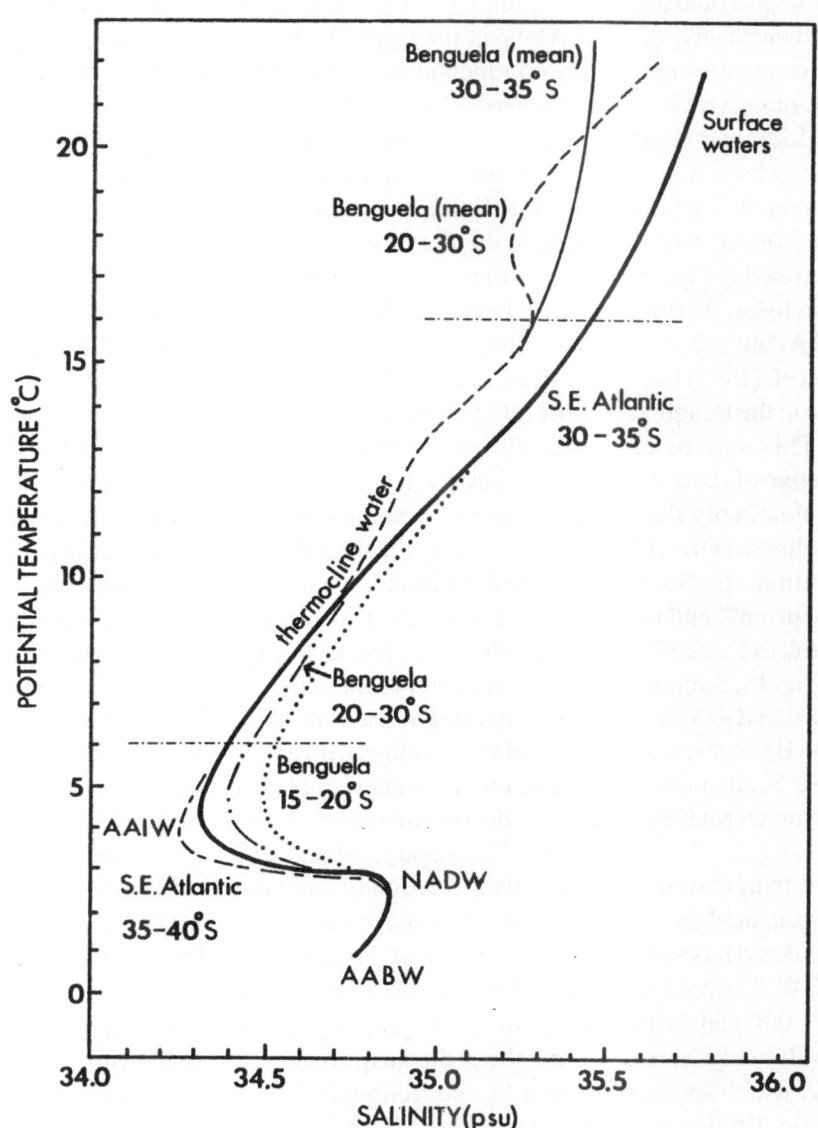

Fig. 14. The principal water masses and potential temperature - salinity characteristics of the South-east Atlantic and Benguela system.

origins (e.g. Gordon et al. 1987; Chapman et al. 1987; Chapman and Largier 1989; Valentine 1990; Gordon and Bosley 1991; Gordon et al. 1992; Valentine et al. 1993). The last-mentioned provides a particularly useful description of water masses in the Agulhas retroflection area. The utility of CFCs in differentiating between the thermocline waters of the Benguela, the Agulhas and the South Atlantic was clearly demonstrated by Gordon et al. (1992).

Thermocline water overlies AAIW, the core of which in the Benguela is in the range 34.2-34.5 psu,

with potential temperatures of 4-5°C. Its core is present at an average depth of 700-800 m in the South-east Atlantic, and somewhat deeper at about 1 100 m in the South Indian Ocean and in the Agulhas retroflection area. Valentine et al. (1993) have estimated that, on a volumetric basis, AAIW accounts for 52% of the water masses present above 1 500 m off southern Africa. The general features of the AAIW and its circulation around southern Africa have been described by Shannon and Hunter (1988). AAIW of Indian Ocean (Agulhas Current) origin is more saline (typically >34.45 psu) and less well oxygenated (typically < 4.5 ml/l than that originating from the South Atlantic Current and which is dominant in the South-east Atlantic. AAIW of the tropical Atlantic likewise is saltier than South Atlantic AAIW, and is characterized by low concentrations of dissolved oxygen (typically < 3-4 ml/l). The differences in the core properties of AAIW and the general circulation are illustrated in Fig. 15. The interchange of AAIW and also lower thermocline water between the Indian and Atlantic Oceans is clearly significant, and Gordon et al. (1992) have estimated that about 50% of AAIW in the Benguela area is of Indian Ocean origin. This may be an overestimate, in our view. Irrespective of the absolute values, the flow of AAIW is remarkably similar to that of the overlying thermocline water, with the Agulhas Current, its retroflection, the South Atlantic Current, the "Benguela Current" and the poleward flow west of the shelf break as far as 30°S clearly discernible in the insert on Fig. 15. Stramma and Peterson (1989) have estimated that 4-5 Sv of AAIW is carried northwards by the Benguela Current, broadly defined, at 28°S and 32°S, all of which turns westwards to cross the Greenwich Meridian south of 24°S (see Fig. 15).

NADW, which has a potential temperature < 3°C and a salinity typically > 34.8 psu, lies beneath the AAIW stratum, and in the Cape Basin it is sandwiched between the latter and AABW. The NADW comprises a thick layer between 1 000 and 3 500 m at the equator (Reid 1989) of relatively warm, saline and well-oxygenated water, which spreads southwards from the North Atlantic. While input into the Cape Basin via the South Atlantic Current is possible, its flow along the continental margin appears to be generally polewards. Readers are referred to the excellent monograph of Reid (1989) for a comprehensive treatment of NADW.

The Walvis Ridge acts as a barrier to the equatorward flow of AABW, which in the Cape Basin has typical θS characteristics of < 1.4°C and < 34.82 psu and a silicate concentration of > 80 mmol.m^{-3}. and lies below the NADW at a depth in excess of about 3 800 m. Some leakage of AABW across the Walvis Ridge was documented by among others Shannon and Van Rijswijk (1969), Connary and Ewing (1972, 1974) and Shannon and

Chapman (1991). The general circulation of AABW in the Cape Basin is cyclonic, and along the continental margin of the Benguela the flow is polewards (Reid 1994; Nelson 1989).

Shannon (1985) and Chapman and Shannon (1985) addressed various aspects of surface and thermocline water circulation in the Benguela region in their review articles. Since 1985 a number of significant papers have appeared in the literature which have contributed to an improved understanding of the larger scale processes. Moreover, several cruises, such as the Benguela Sources and Transport (BEST) series, have been undertaken in the area during the past five years, the full results of which have yet to be published. Therefore, it is anticipated that some important new concepts will emerge during the next few years. During the past decade there have been four important advances in the understanding of circulation in the greater Benguela. These are: an improved quantification of the sources and volume fluxes of water in the system; appreciation and quantification of the influence of the Agulhas Current, in particular Agulhas rings, on the oceanography of the South-east Atlantic; documentation of the poleward undercurrent on the Benguela shelf and, at the continental margin, quantification of shelf circulation as a consequence of extensive current measurement.

The paper of Stramma and Peterson (1989) provides a useful overview of the greater Benguela Current. Their geostrophic calculations suggested a northward transport of surface (thermocline) water by the Benguela Current at 32°S of 21 Sv and at 28°S of about 18 Sv. Beneath this is a flux of AAIW of 4-5 Sv, i.e. a total surface-thermocline-AAIW equatorward transport of about 25 Sv. This figure is somewhat higher than previous estimates which are of the order of 15Sv (e.g. Defant and Wust 1938; Reid 1989), and higher than the most recent estimates of Garzoli et al. (1984). However, the value of 25 Sv for the Benguela Current is supported by estimates made by Gordon et al. (1992). Stramma and Petersons' (1989) geostrophic transport field of surface water relative to σ_0=27.75 kg.m^{-3} is shown as an insert in Fig. 16. Surface velocities estimated by those authors were of the order of 30 cm.s^{-1}. Of the 15-25 Sv equatorward transport in the Benguela Current, some 7 Sv is of In-

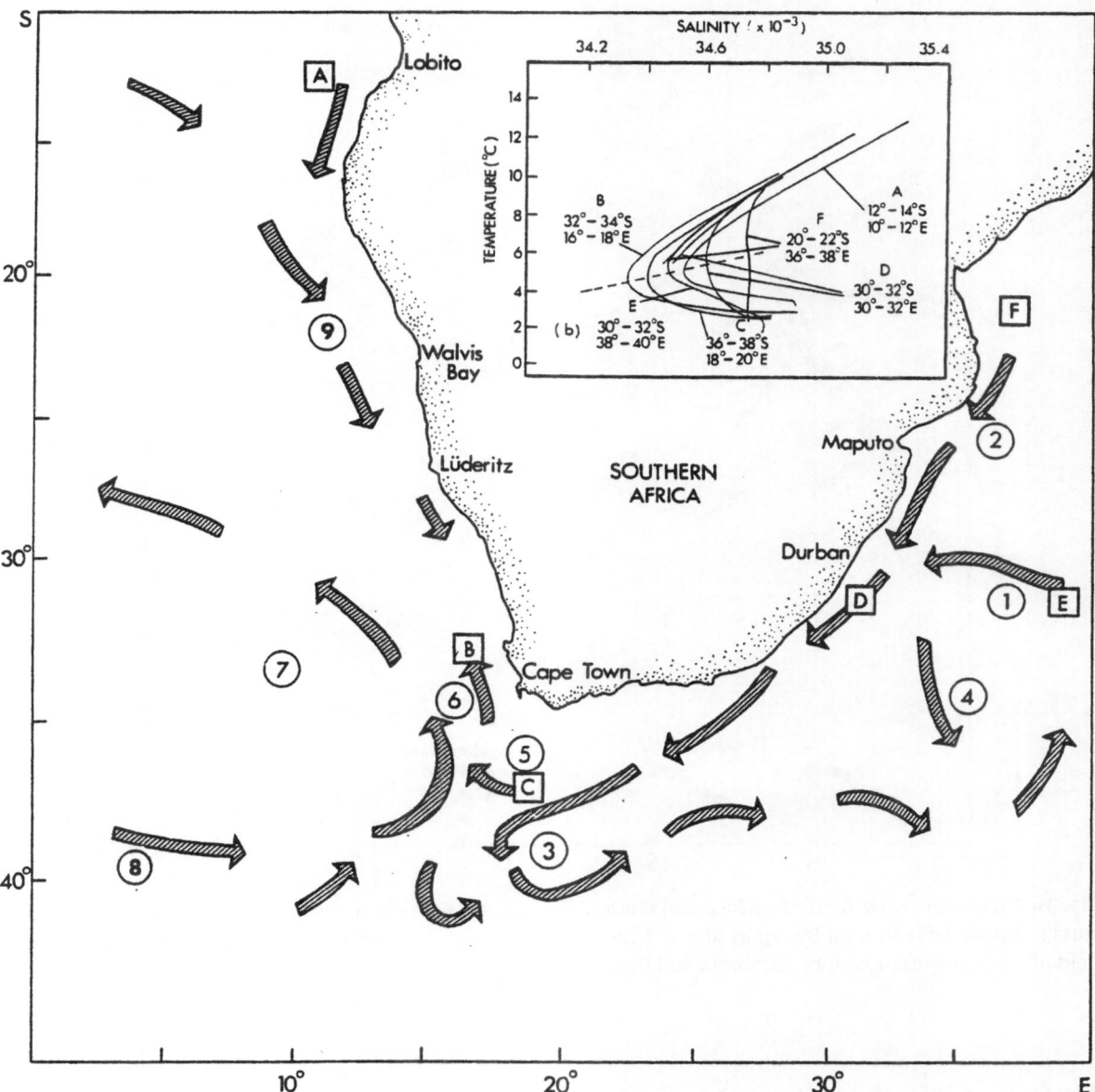

Fig. 15. Temperature-salinity characteristics in selected blocks around southern Africa (after Shannon and Hunter 1988), showing the essential characteristics of the Antarctic Intermediate water core. (1) Primary Indian Ocean source of AAIW; (2) Secondary Agulhas Current source of AAIW plus Red Sea Water; (3) Retroflection; (4) Short-circuit; (5) Flux from the Agulhas into the South-East Atlantic; (6) Relatively fresh AAIW from southwest; (7) Anti-cyclonic AAIW gyre; (8) Inflow from South Atlantic current; (9) Poleward flow from the tropical eastern Atlantic (after Shannon and Hunter 1988).

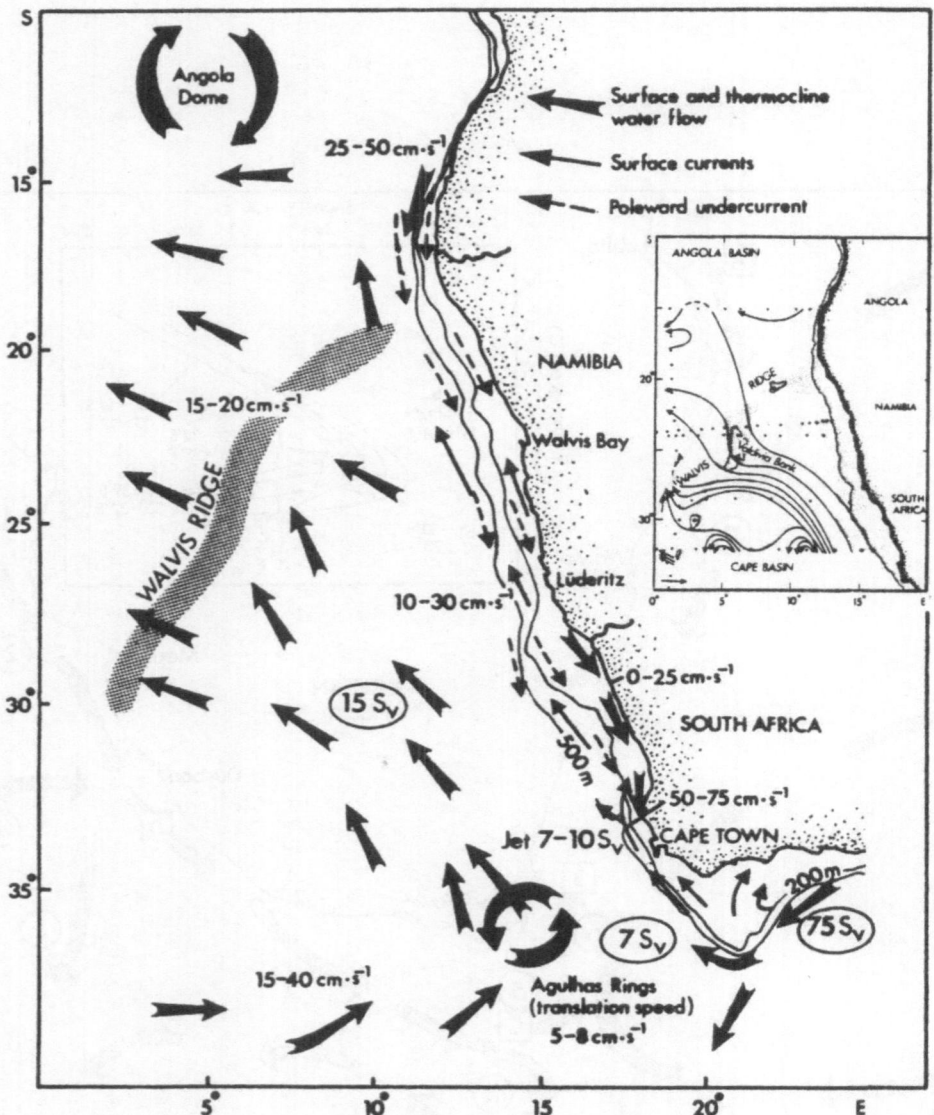

Fig. 16. Schematic flow filed of surface and thermocline waters. Current speeds refer to surface values. Transports (circled) refer to total transport above 1 500 db (i.e. includes AAIW). Inset is the geostrophic transport field of surface water given by Stramma and Peterson (1989).

dian Ocean origin (Van Ballegooyen et al. 1994), the transport being principally via Agulhas rings. These rings also entrain about 5 Sv of Atlantic water (Gordon et al. 1987). Agulhas rings have a major impact on the South-east Atlantic and provide a mechanism for injecting heat and salt into the region. Their translation rate is typically 5-8 cm.s⁻¹ (Olson and Evans 1986; Gordon and Haxby

1990). Generally these rings do not appear to impact on the Benguela shelf, exceptions being extreme events such as documented by Shannon et al. (1990), Shillington et al. (1990) and Duncombe Rae et al. (1992a), describing the interaction of rings with shelf waters.

Whereas the Agulhas Current influences the oceanography of the southern Benguela, the Angola

Current and South Equatorial Current - Undercurrent system influence the northern Benguela. Indeed these pulsating warm boundaries are in some respects rather like mirror images. As in the case of the Agulhas Bank the far northern Benguela region is strongly stratified during summer, and less so during winter when upwelling in the northern Benguela is at a maximum. The poleward flow in the Angola Current is, however, small in comparison with the equatorward flow from the Agulhas Current. Nevertheless the tropical Atlantic is a major source of thermocline water in the Benguela. The circulation proposed by Moroshkin et al. (1970) is shown in Fig. 17. Shannon (1985) has discussed this is some detail and we will not repeat the discussion here. However, we draw attention to the work of Gordon and Bosley (1991) on the cyclonic gyre shown in Fig. 17, viz. the Angola Dome. Thermocline water and AAIW from the tropical Atlantic are advected polewards into the Benguela,

certainly as far south as 27°S. The tropical origin is evident from the work of Gordon et al. (1995) and Stander (1964), De Decker (1970), Chapman and Shannon (1985) and Boyd et al. (1987), who show a clear seasonal signal. Gordon et al. (1985) have estimated the poleward transport on the shelf at 27°S as 1.6 Sv, with an equatorward flow at the surface of upwelled water of 0.7 Sv. The balance (0.9 Sv) would feed the shelf upwelling farther south with a resultant net offshore flux of upwelled water there. In this respect it should be noted that surface shelf waters are very responsive to the wind in the classical Ekman sense, and Shillington et al. (1990) show westward surface flow of 30 cm.s^{-1} associated with the Lüderitz upwelling cell at 27°S.

The general net poleward subsurface flow on the Benguela shelf and over the continental slope is now well established (Nelson 1989), and it is further identified by its low oxygen content. The poleward flow of this low oxygen water from the tropical

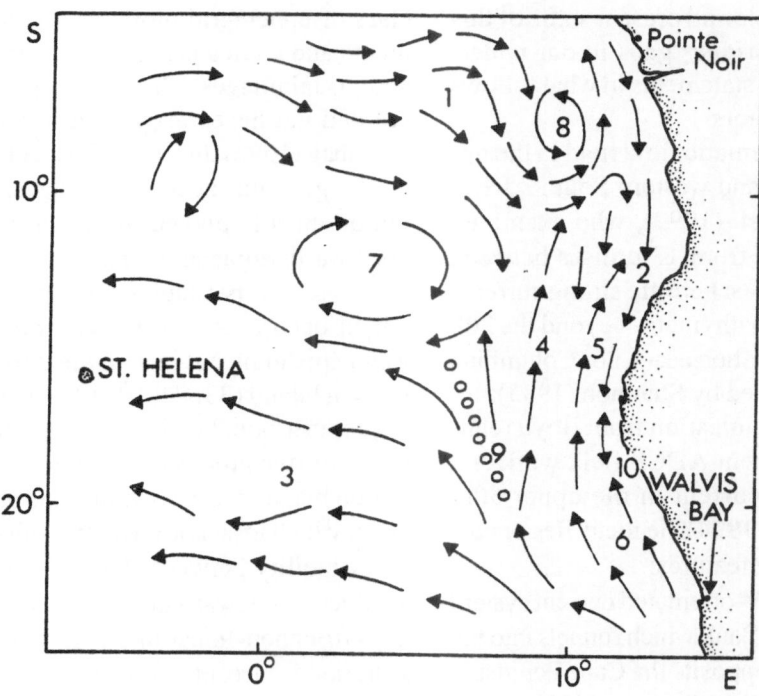

Fig. 17. Geostrophic water circulation in the 0-100 m layer off Angola and Namibia (after Moroshkin et al. 1970): 1, South Equatorial Countercurrent; 2, Angola Current; 3, West (main) branch of Benguela Current; 4, 5, 6, North branches of Benguela Current; 7, eddies in inner region of cyclonic gyre; 8, anticyclonic curl; 9, Benguela Divergence; 10, merging zone of Angola Current and north littoral branch of Benguela Current.

Atlantic is shown conceptually in Fig. 18. If these findings are taken together with the evidence for poleward flow along the continental margin of AAIW as far as 30°S, and the poleward flow of NADW and AABW, then it must be concluded that, at its eastern boundary, except for the water flowing equatorwards at the surface under the influence of the wind and in the shelf-edge jet, the net flow over the greater part of the water column is polewards. It is only much farther west where the flow in the top 1 500 m of the South-east Atlantic is to the north. The contribution of upwelled water to this flow is relatively small and, like the influence of the Agulhas Current, it decreases with depth.

Shelf circulation

Surface Currents

Because the authors are in posession of a substantial body of information on the shelf dynamics of the Benguela system which is as yet unpublished, it seems appropriate here to depart from the strict format of a review article and present some of this information where it bridges gaps in our understanding. Unreferenced statements in what follows are the work of the authors.

Fig. 19 shows a schematic flow field in the upper 50 m off southern and western South Africa. This is from Boyd et al. (1992), who examined ADCP current vectors from 13 cruises between 1989 and 1991. Of interest here are strong currents running parallel to the bathymetry beyond the 200 m contour and the bifurcation near Cape Columbine (33°S), as conceptualized by Shannon (1985).

In a personal communication from Boyd relating to this and subsequent ADCP field work and analysis of averaged current in the upper 50m (Boyd and Oberholster 1994), he identifies or corroborates the following features:

* A convergent NW-orientated current system on the western Agulhas Bank which funnels into the West Coast jet current opposite the Cape Peninsula (Bang and Andrews 1974; Shelton and Hutchings 1982; Boyd et al. (1992); Largier et al. 1992; Boyd and Shillington 1994).

* A sluggish flow inshore of the 500 m isobath as the inner shelf widens between Cape Town and Cape Columbine, before speeding up again to approximately 35cm.s^{-1}.

* The bifurcation of this current into an offshore flow and a longshore northward flow partially into St Helena Bay (Shannon 1985; Nelson 1991).

* Upwelling tongues in both the Cape Peninsula and Cape Columbine regions in summer, with a narrow band of onshore flow evident north of each plume (Agenbag 1992). North of St Helena Bay in the midshelf region there is a mean onshore curvature towards Hondeklip Bay bounding a broad area of weak mean current to the east. Flow patterns on the outer shelf and around the Orange bank have a northward trend, but better resolution is needed.

* A southward current often occurring near the surface close inshore over the entire region, particularly during winter and when barotropic reversals in the longshore direction take place on a time-scale of several days throughout the year (Nelson and Hutchings 1983; Holden 1987; Nelson 1989; Nelson and Polito 1987).

In addition to dynamically structured flow which can be identified by ADCP work of this nature, a class of ephemeral motions which includes filaments and eddies is apparent on satellite thermal and visual images. This is a property of any shelf sea and not necessarily eastern boundary zones. Agenbag (1992) discusses the technique of feature tracking, using successive images, and shows the important role played by warm water spreading northwards adjacent to the west coast shelf.

There is a particular semi-permanent feature which occurs as a large westward excursion of water spreading out in a filament from a point near Dassen Island (33.4°S, 18.1°E), sometimes forming a mushroom head beyond the shelf. This is the result of dynamic processes on the narrow shelf region between Cape Columbine and Cape Point, about which little is known. It results possibly from an instability between the shelf-edge jet and the subducted poleward undercurrent (Fig. 20) or perhaps from non-linear interactions between locally generated currents and pre-existing free-wave modes. A similar but smaller excursion often occurs near the tip of the Cape Peninsula. Both seem to differ in character from the filaments seen north of Cape Columbine, which are less persistent and narrower.

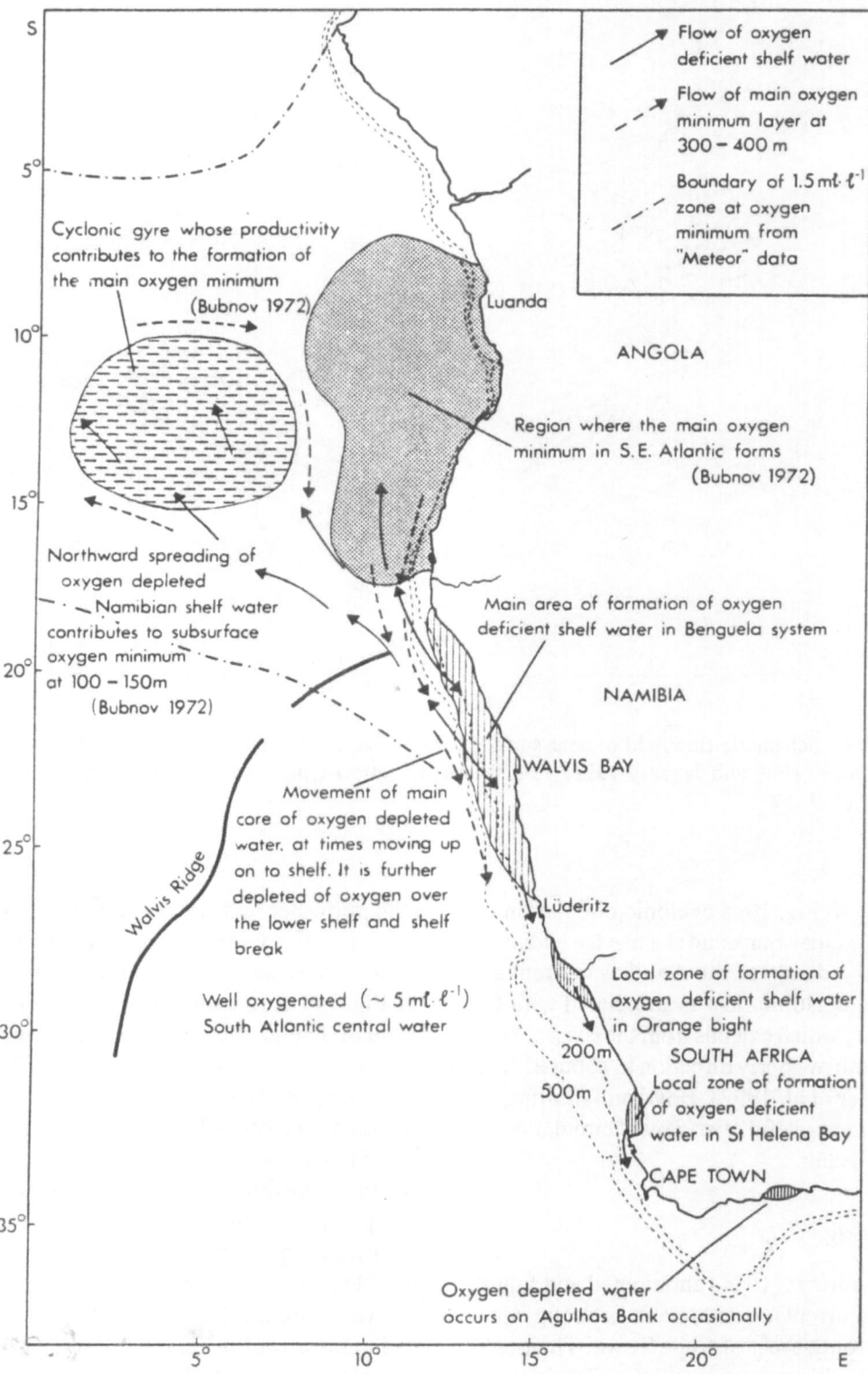

Fig. 18. Conceptual model showing areas where the low oxygen water is formed in the South-east Atlantic and its inferred movement (after Chapman and Shannon 1987).

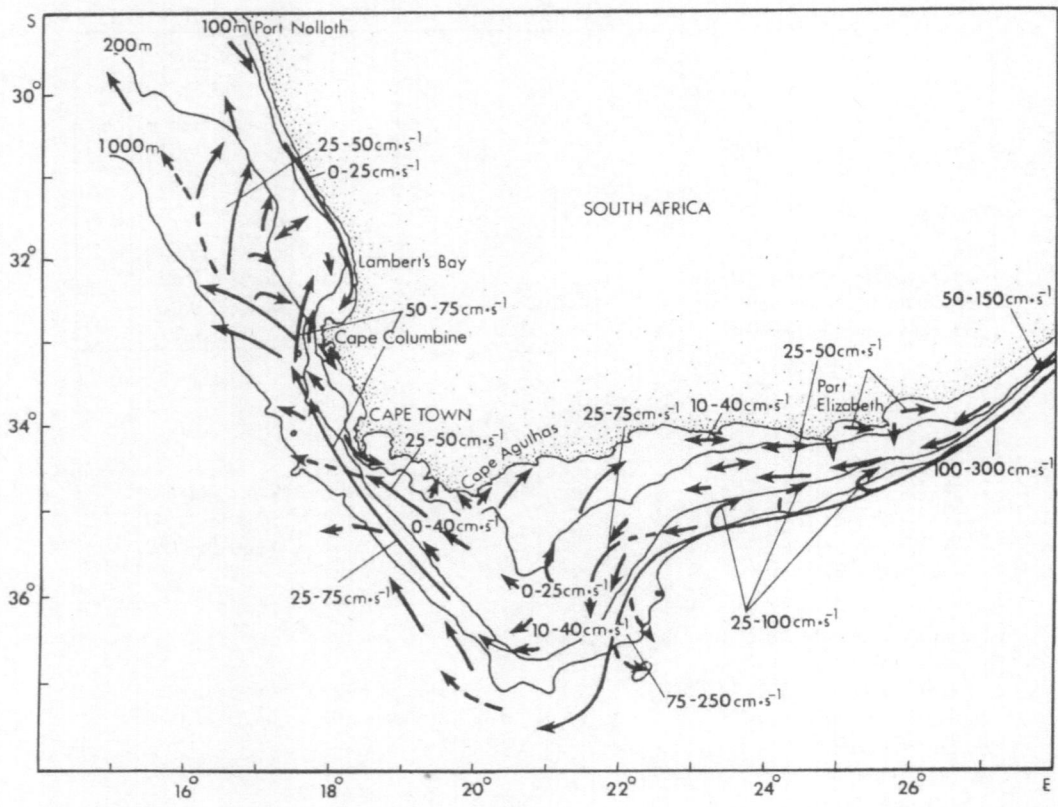

Fig. 19. Schematic flowfield of near-surface currents based on ADCP data collected between November 1989 and January 1992. Velocity ranges reflect typical values, not extremes (after Boyd et al. 1992).

Referring to Fig. 19, a cyclonic circulation on the eastern Agulhas Bank and shear-edge eddies on the equatorward side of the Agulhas Current are evident. The cyclonic flow is associated with the "cool tongue", which extends from the coast at 24°E in a west-south-westerly direction to about 21°E at times (Largier et al. 1992). Boyd and Shillington (1994) provide a useful overview of circulation on the Agulhas Bank.

The Shelf-Edge Jets

Bang and Andrews (1974) anticipated and found, using direct current measurements, a strong equatorward jet south-west of Cape Town. Their estimated flow in this current was 7 Sv. This estimate was based on the belief that the jet reached to the bottom, whereas it is now known that it is confined

to the midwater and surface layers, and to a strip some 5 n. miles wide. Taking an average speed of 58 cm. s⁻¹ over such a strip and to a depth of 200 m (see Fig. 7 of their paper) a more conservative estimate of 1Sv is obtained. The incentive for this came from the classical idea of frontal formation west of an upwelling zone. It was thought that where strong coastal Ekman divergence occurred, as in the case of intense wind episodes off the Cape Peninsula, the westward movement would be halted at some point, with the formation of a front and sinking of water, to be entrained into a longshore jet. This idea was then furthered by the conceptual model of Hart and Currie (1960), which was applied to the upwelling cell west of Cape Town by Bang (1971, 1973) and Andrews and Hutchings (1980). Oceanic frontogenesis is a complicated and as yet not well understood process. Some authors,

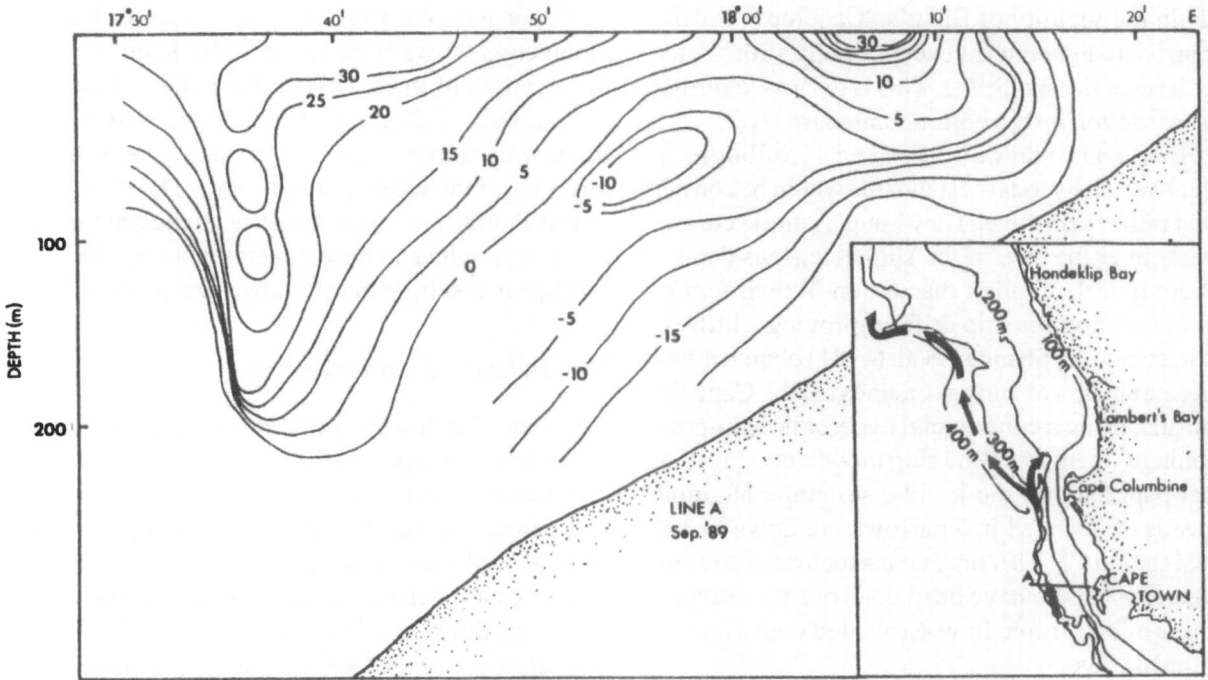

Fig. 20. Vertical velocity section (cm.s⁻¹) on line A (inset) obtained from ADCP measurements. Positive values are equatorwards. Note the shelf-edge jet and poleward undercurrent (after Nelson 1991).

for example Mooers et al. (1976), have modelled the formation of longshore jets at upwelling fronts in this way. It is not entirely clear in the Bang and Andrews (1974) work whether the Ekman transport formed a progressively westward migrating front, interacting in some way with the topography through relative vorticity, or whether a pre-existing jet simply acted as a barrier, limiting westward growth. The latter now seems to be the case, at least for the narrow shelf region west of the Cape Peninsula.

Could there be other reasons for the existence of shelf-edge jets? It seems a worthwhile exercise to explore possibilities. There is always an upward displacement of isopycnals at the shelf break, not just in the Benguela system, but along the whole west coast shelf and, it seems, along many a continental shelf. In the case under discussion, the feature is ubiquitous and independent of upwelling. Thermohaline data with a fine spatial resolution is required to resolve the details of the density structure at the shelf edge, and for this reason perhaps it

has not enjoyed much attention. In extreme cases, there may be a drop of one or two degrees in surface temperature, and this has often been referred to as "dynamic upwelling", and attributed to vortex stretching.

That may occur, but we ask here whether other processes are not also involved and tentatively propose two which might better account for the spatial regularity of the feature. There will be a baro-clinic jet where the density field is distorted in this way, and the question then moves from why the jet is there to what causes the bulge in isopycnals over the shelf break. Vertical advection at the shelf edge resulting from bottom friction and increased speed is a possibility. Two processes present themselves as possible causes for this. One is the semi-diurnal internal tide. With the right geometry and stratification, transient bottom boundary layer currents may be sufficiently strong to thicken the boundary layer. The second possibility is that transient long-shore wind can generate strong currents at the shelf edge. Both the shelf geometry and cross-shelf wind-stress profile in the vicinity of Bang's "Cape upwell cell" contrive to generate an exceptionally strong jet at the base of the shelf edge. This is discussed further in the section on the bottom boundary layer.

Bang and Andrews (1974) used a profiling technique which by today's standards would be considered rather primitive. They hung a simple current meter over the side of the ship at various depths. There is little explicit discussion in their text of compensation for ship drift. Improving a little on that technique, Shannon et al. (1981) obtained spot measurements of current on lines off the Cape Peninsula, using a commercial hyperbolic ship-positioning system to define ship movement. Fig. 9 of that paper shows the jet-like structure, attaining speeds of 80cm.s[-1] in a narrow core down to 100 m at station C3. That line, which took a day to complete, appears to have been done during a reverse phase of barotropic flow associated with a coastal trapped wave.

A further improvement was effected by Nelson (1985), who repeated Bang and Andrews' work, on this occasion using acoustic current meters suspended at various depths for 20-minute periods and a coastal position-fixing system to obtain ship drift. The results did not show any striking dissimilarities with Bang and Andrew's earlier work, either in dimension or strength (see Fig. 18 of Nelson 1985). The latter mission was undertaken during a slackening phase of the wind, whereas Bang and Andrews' work was done during a strong acceleration phase, a pointer to the fact that upwelling is not the sole cause, if a cause at all, of the "frontal jet", as had been assumed. At the same time, Nelson (1985) identified jets at other points along the shelf-edge in the vicinity of the Cape Peninsula.

The suspected permanence of the jet was finally verified by the authors using a ship-mounted ADCP in 1989. The jet was shown over the 300-400 m isobaths and to stretch from Cape Point to Childs Bank (Fig. 1) in varying strength (Fig. 20). The region north of this still awaits investigation, but the ADCP section of Gordon et al. (1995) (Fig. 21a) shows a weak jet feature of 20 cm.s[-1] over the 200-300 m isobaths. The isopycnal distortion at the shelf edge is much weaker than that observed south of Childs Bank.

Some of the early profiling work done with current meters shows evidence of a double jet, one on the shoulder of the shelf and one in the midwater region. The ADCP cannot see the bottom structure because of its inherent acoustic shadow zone - usually 15% of the water column depth. The presence of this double structure strengthens the argument that the shoulder feature generates the isopycnal distortion, resulting in the midwater baroclinic jet.

The poleward undercurrent

The idea of poleward flow in upwelling regions seems to have originated from a conceptual diagram of Hart and Currie (1960; their Fig. 96). Some compensation was considered necessary for the vertical displacement of water on the inner shelf and its movement equatorwards, and so a narrow strip of poleward flow was shown at the base of the shelf. Such flow does exist (Nelson 1989), but only as part of more extensive poleward motion stretching from the coast across the shelf and out into the Cape Basin.

Prior to the first deployment of current meters in 1978 as part of a continuous project on direct current measurements on the shelf, the Benguela Current was thought to be a broad sweep of water moving equatorwards and cooled by episodes of upwelling in summer, as in the Hart and Currie model. That is now known to be true only of the upper layer, which varies in depth from a few metres close to the coast to 120m at points on the outer shelf and several hundred metres out in the Cape basin. Below this layer, poleward motion occurs with varying strength and varying degrees of seasonal dependence. Measurements from more than 150 instruments moored at depths between 40 m and 600 m over the shelf between Cape Point (34°30S) and Cape Cross (21°55) have shown such a trend. Some results of this work have been presented by Nelson (1985, 1989), Holden (1987) and Nelson and Polito (1987).

South of 33°S, where the shelf is narrow, the flow is not seasonally dependent. Averaged over months, a displacement of approximately 5 km. day[-1] (5.78cm.s[-1]) is observed. In the far north, the flow is variable over periods of weeks, flowing al-

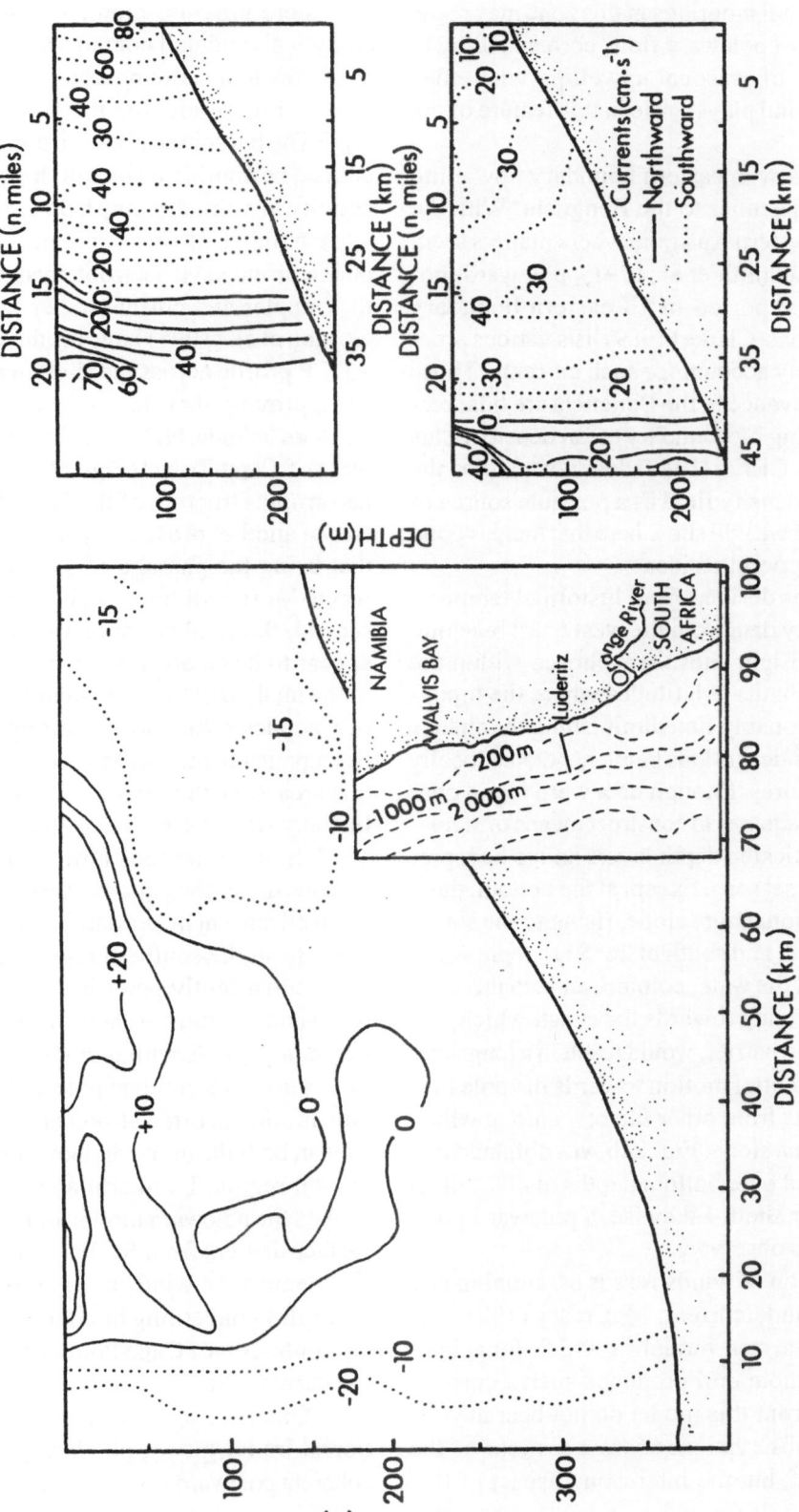

Fig. 21. Isotach sections (cm.s[-1]) across the poleward undercurrent.
(a) (Left panel) From the DISCOVERY 165b cruise (Adapted from Gordon et al - 1995). Negative values are polewards.
(b, c) (Right panels) Transects near Cape Town, November 1978 (Nelson 1985).

ternately equatorwards and polewards along the same axis. Coastal moorings in this zone may show a sudden onset of poleward flow, corresponding to a sudden onset of seasonal upwelling wind, thus showing that wind plays a role in this feature of the system.

Poleward flow in eastern boundary upwelling regions is not peculiar to the Benguela. With the exception of western Australia where matters seem to be reversed (Smith et al. 1991), poleward motion has been reported in all eastern boundary upwelling regions. Clarke (1989) lists various processes which may account for such currents. Those of possible relevance to the Benguela are tidal rectification, forcing by boundary mean density fields and windstress. Clarke (1989) does not mention the internal mean density field as a possible source of the motion, and we will show here that there is good evidence to support this idea.

Fig. 22a was derived from historical temperature and salinity data from the west coast reaching back to 1920. Eight sites were chosen within the 150-180 m isobaths at latitudes where the topography was reasonably flat, diminishing the chance of large baroclinic motions being generated locally by bottom features. Enough data were found over quarter-degree squares to construct means of sigmat within the indicated depth layers and to be representative of all seasons. Except at the bottom, there is seen to be a longshore slope, rising to the south. The slopes north and south of 26°S are representative of much of the water column, and would result in a current moving towards the coast, which, because of the land barrier, would result in a longshore current. Preferential motion towards the pole presumably results from other factors, among which are wind and sea slope. Fig. 22b was obtained in a similar way, but over bottom depths of 300-500 m along the outer shelf. Likewise, a poleward pressure gradient is observed.

The inclusion of windstress is essential in any model of the undercurrent. McCreary (1981) developed a linearized model in which longshore windstress without curl produced such a current. The isotachs from this model do not bear any resemblance to observations of cross sections on the Benguela shelf, but the interesting aspect of this model is that no poleward flow develops without a longshore pressure gradient. Although a simple analytical model, it contains the essential dynamics. Is the longshore pressure gradient then an essential ingredient for the poleward undercurrent? The baroclinic pressure component could, of course, be dominated by sea slope, and the current activity in altimetry may help to resolve this.

A number of cross sections of the poleward undercurrent have been obtained using both acoustic Doppler and current meter profiling devices. Gordon et al. (1995) have produced an interesting ADCP profile across the shelf near Lüderitz (Fig. 21a), proving that the current exists even at this northern latitude. Nelson (1985) shows two sections west of Cape Town (Figs. 21b, c). The strongly barotropic structure of the first of these is indicative of another process superimposed on the flow, that being the passage of a coastal trapped shelf wave. More will be said about this below. In the second, the weaker core current of 30cm.s^{-1} may appear to be an error, but is real within the limits of the method of measurement and the assumption of synopticity. Another process involving wind may be responsible for a shelf-bottom reverse current in this area. (See the section on baroclinic motions). In many ADCP sections on the narrow shelf south of 33°S, it can be seen how the poleward current extends under the equatorward flow (e.g. Fig. 20).

In all current meter time-series, but particularly those from the southern region where the flow is more consistently polewards, filtered series with tidal signals removed show the presence of oscillatory components with periods of 3-10 days superimposed on the ambient poleward flow. This periodic motion is often strong enough to reverse the flow in both the upper and lower layers of the near coastal region. Lamberth and Nelson (1987) described the poleward motion of five DAVIS drogue surface drifters from St Helena Bay into Saldanha Bay against the wind and against all expectations of the day concerning the Benguela Current.

South-east of Cape Point, little is known about the current. An array of three moorings on a line from Quoin Point out to the 400 m contour, positioned by Largier et al. (1992), did not reveal a coherent poleward flow over a period of two months

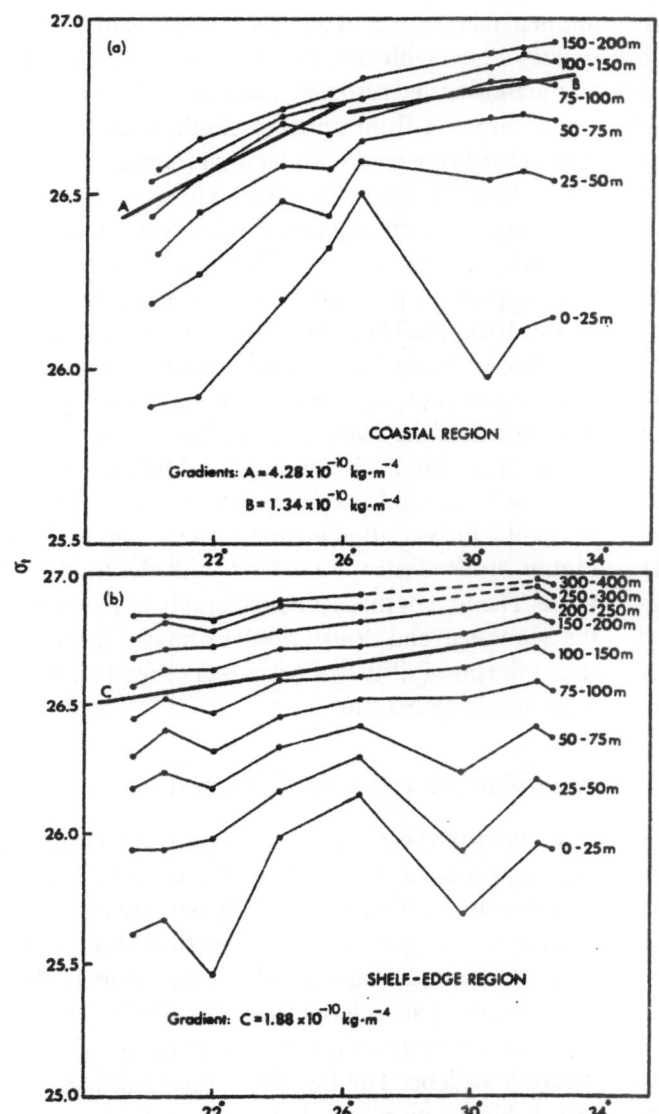

Fig. 22. Average densities within the indicated depth intervals at eight sites from historical data dating back to 1920 along the shelf over bottom depths of (a) 150-180m and (b) 300-500m. Lines marked A, B and C depict alongshore density gradients.

in summer. Boyd et al. (1985) and Chapman and Largier (1989) reported poleward movement at the base of the shelf on the western Agulhas Bank which Largier et al. (1992) ascribe to local density currents.

It may be that the ten-times broadening of the shelf in passing from the Cape Peninsula to the Agulhas Bank diffuses the poleward undercurrent and that it is then not identifiable from the relatively sparse arrays such as were used by Largier et al. (1992). There is also evidence that the current moves off the shelf into the Cape Point Valley (Shannon et al. 1981).

Numerical simulations of coastal-trapped waves

A suite of algorithms developed by Brink and Chapman (1985) provides a means of searching for dispersion relationships for barotropic and baroclinic waves with different shelf geometries and bottom roughness and stratification, and a means of simulating deep shelf-edge jets which develop under the action of transient or periodic longshore windstress profiles. Both rigid lid and free surface cases can be examined, and where convergence in the numerical scheme is difficult for three-dimensional waves a long-wave approximation can be used.

Realistic dispersion relationships are found for barotropic waves in terms of observed current meter and water-level recorder data. Nelson (1989) obtained dispersion relationships for the narrow shelf south of 33°S and a wide shelf section normal to 30°S. Over a wide band of wavelengths, corresponding periods for the narrow and wide shelf of about 3.8 days and 5 days were obtained.

In a paper focused mainly on the south and east coasts, Schumann and Brink (1990) examined five segments with characteristic shelf profiles (Fig. 23). They obtained southward phase speeds ranging from 5.0 to 13.29 m.s^{-1} (432 to 1148km.day^{-1}) and e-folding frictional decay times between 2.7 and 7.9 days for first mode barotropic waves with no ambient flow. Using 260-day water-level data from six tide gauges, two being on the west coast at Port Nolloth and Cape Town separated by 530 km, they constructed space-time contours which clearly show the anti-clockwise progression of waves around the coast. The range of water level on the west coast was 49 cm, on the south coast 93 cm and on the east coast about 76 cm. The greater range on the south coast was attributed to more energetic weather systems crossing the Agulhas bank.

Mixed modes should be anticipated in current-meter records. Observed motions are generally quite complex and higher modes with slower phase speeds will be mixed in with the first mode. Coastal trapped waves will also be altered as the shelf profile, bottom friction and Coriolis parameter change, and therefore will not propagate coherently over great distances, even though the length scale for such waves is usually hundreds of kilometres. Moreover, if one examines the pressure structure for a theoretical wave, a few centimetres seem to be adequate for the observed currents. The large amplitude in the tide-gauge data is difficult to accommodate in the mathematical model.

The tide-gauge data have received attention from Brundrit (1984) and Brundrit et al. (1987). Six west coast stations and one south coast station were examined over periods of 57-204 months. A high frequency signal with an approximate two-day period was identified and associated with synoptic weather events moving from north to south. Enfield and Allen (1983) observed similar high frequency variation along the west coast of Mexico.

Using two bottom pressure recorders separated by 330 km on the west coast, Nelson (1989) showed that the peak-to-peak time difference varied from -4 to +10 hours. These figures partition fairly well into three sets, one for times greater than 7 hours, one for 0-4 hours and one for the negative value. The means of the two non-negative sets are 2.8h and 8.7h, giving 2 830 km.day^{-1} and 905 km.day^{-1} (118 and 88 km.h^{-1}) as corresponding phase speeds. No doubt, the variation in time is an indication that higher order modes are mixed with the mode 1 wave. The first of these is considerably higher than the Schumann and Brink (1990) estimates, and this raises the possibility that a weather-forced Kelvin wave might also occur.

Stratification and baroclinic motions

Schumann and Brink (1990), using the same bathymetric profiles as for the barotropic work, along with density profiles presumably obtained from a local data base giving average surface to abyssal values, obtained phase speeds as shown in Table 1. The sections are referenced in Fig. 23. The speed slackens where the shelf narrows and speeds up where it widens. The kinetic to potential energy ratio is high everywhere, but particularly so on the wider section from Lüderitz to St Helena Bay. The e-folding period ranges from 4.0 to 8.4 days. Temperature sensors on upper-layer current-meter moorings placed by these authors show complicated time-series, again indicating the presence of mixed modes. Although spectral peaks are not observed at any preferred frequency, the periods predicted by Schumann and Brink (1990) can be identified in discreet time-series segments.

Stratification and the slope of the west coast shelf and its extension onto the Agulhas Bank promotes the generation of large amplitude internal tides. Sigma-t profiles are similar along the whole outer shelf region from 20°S to 33°S. Transient

SECTION	c(m s⁻¹)	Ts(days)	R
A	6.32(3.00)	6.6	8.1
B	8.32(4.52)	4.0	11.1
C	12.10(4.74)	8.3	16.6
D	6.50(2.83)	8.4	8.3
E	6.14(3.42)	4.6	8.8

Table 1. Phase speeds for first (and second) mode baroclinic waves, e-folding periods (Ts) and kinetic to potential energy ratios (R) (Schumann and Brink 1990) [10 m.s⁻¹ = 864 km.day⁻¹].

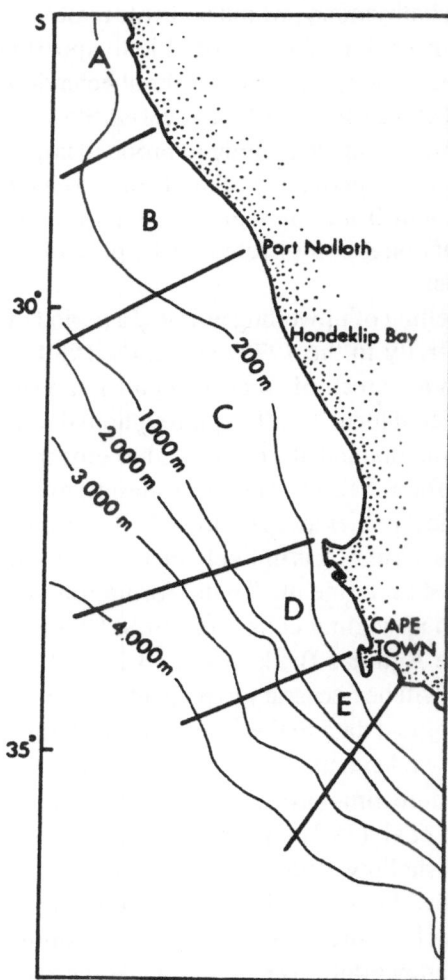

Fig. 23. The division according to shelf profile and stratification as used by Schumann and Brink (1990) for the analysis of coastal-trapped waves.

distortions of profiles occur at sub-inertial periods, indicating the presence of internal tides at the shelf edge. The authors have observed changes in a time-series of CTD casts at 28°S over a tidal period together with a series of wave packets from a thermistor string nearby. The occurrence of internal tides is manifest by irregular striations on calm seas, separated by 200-600 m and progressing with typical periods of 20 minutes, and is a common feature of most of the west coast.

Largier (1994), using thermistor chain data on the western Agulhas Bank, has shown the existence there of a whitened Garrett-Munk spectrum with peaks in the inertial, tidal and high frequency bands. The internal tide, which dominates the temporal signal, transports energy of the order of 0.04 kw.m⁻¹ onto the shelf. A barotropic-baroclinic coupling arising from resonance at multiple generation sites is proposed as a mechanism to account for the unexpectedly large amplitude of the tide. Stratification deepens westward across the Agulhas Bank (Largier and Swart 1987), reflecting a switch from an advectively controlled structure in the east to stronger atmospheric influence in the west.

The bottom boundary layer

In a statistical analysis of near-bottom temperatures and salinities from the historical data base, Dingle and Nelson (1993) mapped average values for the west coast shelf from 17°S to 36°S. Progressively colder water moves onto the shelf southwards to the middle of the Cape Peninsula, after which the trend reverses rapidly. Salinity shows the same pattern. This is a separate analysis from that used to demonstrate the existence of a longshore pressure gradient in the section on the poleward undercurrent and confirms that finding.

There are two features of interest in the temperature field. One is a large "flooded" area of 9°C water between 30°S and 33°S, reaching nearly to the coast at one point. This water seems to move onshore in the vicinity of a canyon a little north-west of the Olifants River. Its T-S characteristics indicate that it is unmixed South Atlantic central water originating from about 500 m. It appears to be the source for all upwelling water farther south. Commonly, upwelling water is thought to cross the shelf. In the Benguela system, it appears that up-welled water moves longshore in the poleward undercurrent in most cases.

The second feature of note is the occurrence of coastal "hot spots" where a surface convergence appears to subduct warm water. The largest of these is near Walvis Bay. It is unlikely that they arise from downwelling, because their permanence seems to appear in a statistical analysis covering all phases of the wind cycle. It is more likely that they occur because of semi-permanent convergence in surface flow, possibly in relation to the cross-shelf flow associated with coastal trapped waves. Offshore and return motions of drogued buoys have been documented by Nelson (1985) and Brown and Hutchings (1987).

Verticle profiles of CTD data at various points on the shelf in summer, when storms do not occur and waves do not affect the bottom, usually show a bottom boundary layer on the mid to outer shelf some 10 to 20 m thick, identifiable by constant temperature and salinity. The thickness depends on bottom roughness and current speed, and in quiescent areas the boundary layer may be absent.

Under the action of a periodic longshore wind, strong jets near the shelf break are predicted. Again, using the programs of Brink and Chapman (1985), deep shelf jets can invariably be obtained for the Benguela topography and stratification. On the wider part of the shelf, its position is sensitive to the cross-shelf profile of wind. Where the shelf is narrow, south of 33°S, the jet is less sensitive to the wind profile, lying consistently just off the shelf-edge. With a cross-shelf wind profile reaching a maximum of 5 dynes.cm^{-2} 50 km from the coast, and having a period of 6 days and wave length of 500 km, speeds of 34 cm.s^{-1} are attained in a narrow jet at 200 m depth south-west of Cape Town. The phase lag with the wind is 15°.

Upwelling cells and frontal formation

We have mentioned the distinction between the front at the shelf-edge and those created by local upwelling episodes. The formation of fronts as a consequence of upwelling and the cross-shelf transport of upwelled water can be modelled only in relation to specific initial and boundary conditions. Apart from the bathymetry, which controls relative vorticity, it is seldom if ever possible to specify the conditions. The spatial and temporal behaviour of the wind alone shows sufficient variance from one upwelling episode to another to produce large differences in the geometry of the front and its thermal gradient from one wind episode to the next. This situation is complicated by the periodicity of wind events.

Modelling of upwelling enjoyed a period of intense activity in the 1970s and early 1980s, and although a plateau of some 150 papers a year on the subject of local upwelling throughout the world is being maintained, the emphasis has moved from striving for a theoretical understanding to applying physical processes to biology. In the case of the Benguela system, it is unlikely that models will be developed to define the frontal formation associated with upwelling cells north of St Helena Bay or on the Agulhas Bank in a rigid format. More likely, it will be necessary to present both physical and biological data in the form of case studies for specific wind events.

An interesting case study by Jury is found in Shannon et al. (1981). He distinguishes between deep and shallow south-easterly wind regimes and shows how the orography of the mountain ranges can affect both the wind field and the ensuing sea surface temperature. Temperature was determined by aerial radiation thermometry and winds by aircraft drift, so that both fields are derived from densely packed and near-synoptic data. These two cases are shown in Fig. 24. Similar work by Jury

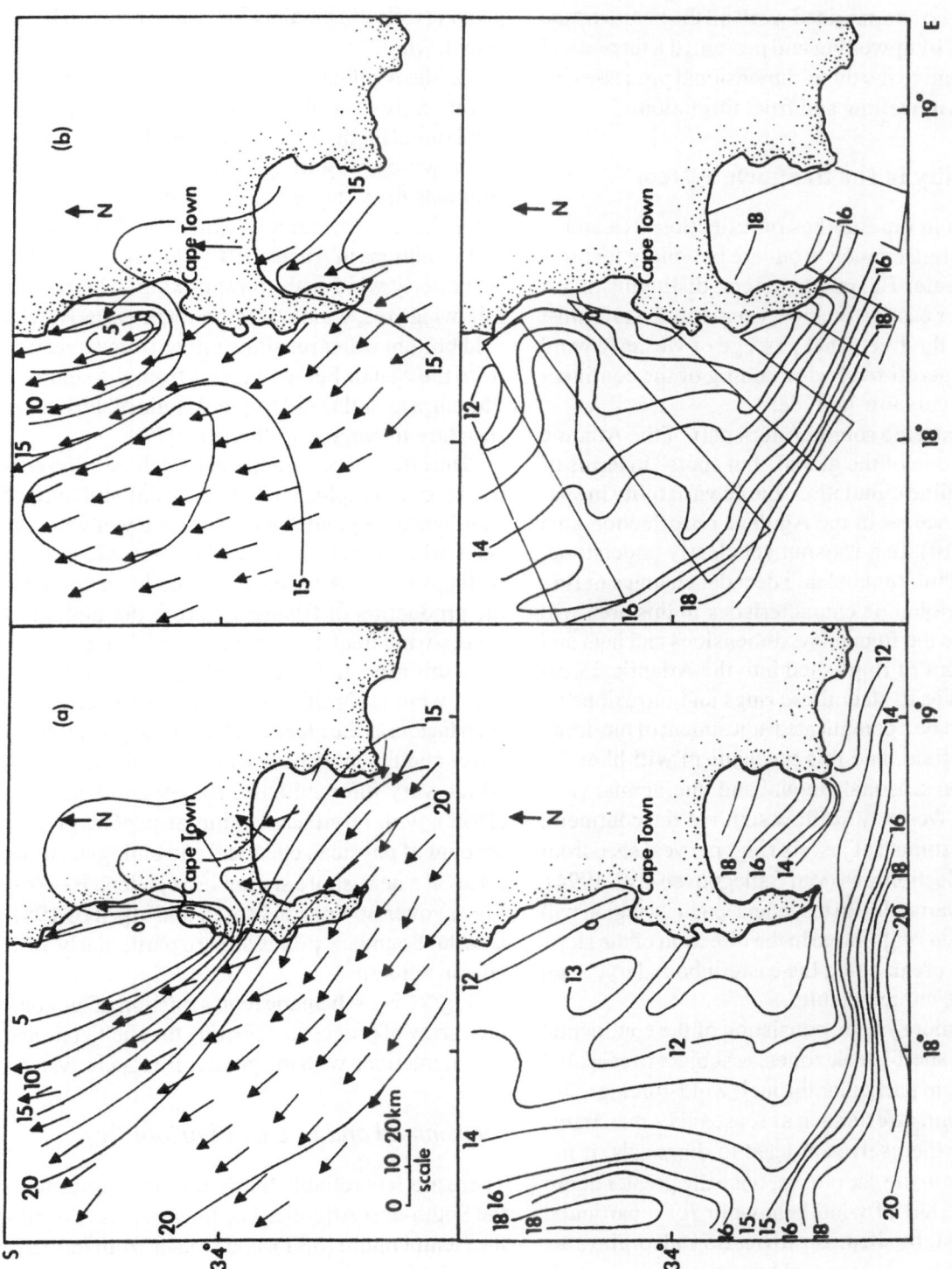

Fig. 24. Wind (upper, m.s⁻¹) and sea-surface temperature (lower, °C) during upwelling off the Cape Peninsula (Shannon et al. 1981) for (a) deep southeasterly wind, and (b) shallow southeasterly wind.

(1985a, b) has been reported for the west coast as far north as 30°S.

Some of the complexities in mesoscale upwelling and frontal formation were presented by Brink (1983), who summarized work on the near-surface dynamics of upwelling and presented a theoretical background to the three-dimensional processes involved in upwelling and front formation.

Variability in the Benguela system

Variation in the Agulhas retroflection area and at the Angola-Benguela front are boundary features of the greater Benguela system, distinguishing it from other eastern boundary regions. To this must be added the free zonal passage of westerly wind which is permitted by the ending of the continent at a relatively low latitude.

The extreme south-eastern part of the Atlantic Ocean is one of the global "hot spots" in terms of thermohaline circulation. Large variability in spatial flow occurs in the Agulhas retroflection area (see Fig.10). In a way not yet clearly understood, seasonal, interannual and decadal changes in flux and thermohaline characteristics in this area will determine the frequency, dimensions and heat and salt content of rings shed into the Atlantic. Local wind serves to steer these rings and intrusions by what must be a complicated adjustment of the ocean pressure field, and this mechanism will likewise depend on seasonal, annual and interannual variability in westerly air flow south of the continent. With an estimate of just six rings per year shed from the retroflection area (van Ballegooyen et al. 1994), episodic variability in the inner Benguela region will occur when rings move in the direction of the slope and shelf break along the eastern boundary. Such events are unpredictable.

The inner system, consisting of the continental shelf and shelf-break zones, is subject to seasonal variation, in particular through wind-forcing. To a large extent, effects such as seasonal Ekman transport lend themselves at least to diagnosis, if not prognosis in the short term, once the greater hemispheric detail of wind behaviour for a particular year is established. Stratification also plays an important part in seasonal behaviour, for example in the generation of internal tides and in geostrophic transport in the midwater zone on the continental shelf. Advection from sources such as the *Benguela Ninos* plays a role in the detail of stratification. Seasonal differences at the northern boundary are marked, reflecting mainly insolation and, to a lesser extent, wind.

On the Agulhas Bank, seasonal changes in stratification are controlled primarily by advection and insolation (Pugh 1982; Swart and Largier 1987), as shown by way of example in Fig. 25, which contrasts the winters of 1977 and 1978 and compares these to the summer situation. The isothermal conditions in August 1977 were a consequence of westerly wind with storm mixing, whereas the following year, wind was more south-easterly, with cold bottom water resulting either from advection from the east or being remnant from the summer. The corresponding chlorophyll isolines show more structure in August 1978 than 1977.

The effect of periodicity in longshore wind cannot be over emphasized. The modal structure of shelf waves depends on the frequency of wind acceleration and relaxation, and shelf waves interact with upwelling. Also. they are possibly involved in the production of filaments. From the biological aspect, wind frequency is of great importance. Also from this aspect, the seasonal behaviour of cross-coast wind is significant, and this will depend on such factors as the albedo of the adjacent land, cloud cover and the longshore wind pattern, all three of which vary interannually. Bailey and Chapman (1991), with reference to similar papers, give an account of physical, chemical and biological time-series at a near-shore station in St Helena Bay (see Fig. 1) over 30 days in March and April of 1987. Coastal boundary processes are particularly variable in autumn.

Fig. 26 is a schematic representation of the complex array of processes affecting the shelf zone and the interactions with the greater Benguela system.

Interannual and Decadal Variability

There are few reliable long-term data series from the South-east Atlantic. The longest is for sea surface temperature (SST), and it dates from the early part of this century. The marine data base is fragmented both in time and in space, however, and as

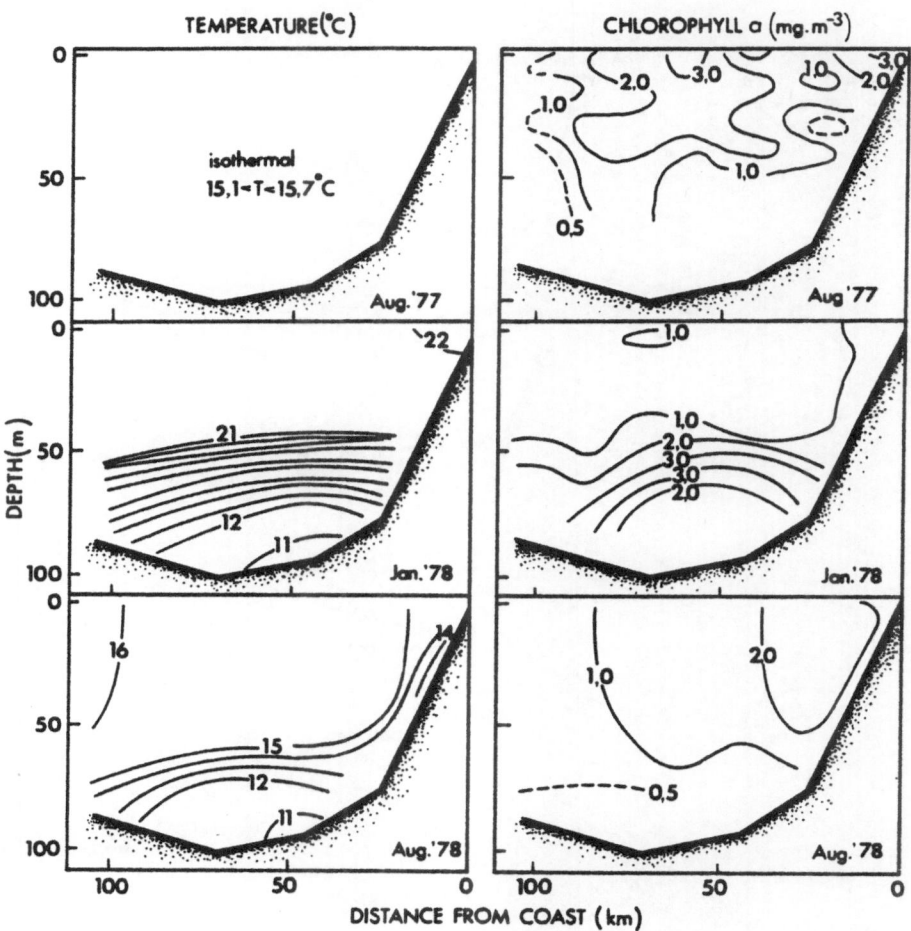

Fig. 25. Temperature and chlorophyll-*a* along a line of stations off the south coast at approximately 20 30'E during two winters and a summer (1977-1978).

a consequence much of our understanding of regional ocean variability on longer time-scales has been deduced from case studies of extreme events (Shannon et al. 1990). The Climate Diagnostics Bulletin series, published monthly by NOAA since August 1983, has provided a useful perspective on some of the more recent changes in the region. The record has highlighted the spatial scales of the variability and the flip-flop between warm and cool conditions. Because of the relatively coarse resolution of the data and the limited length of the time-series, Walker (1987), Taunton-Clark and Shannon (1988) and Shannon and Taunton-Clark (1989) supplemented that information with data from vol-

untary observing ships (VOS) to investigate variability in SST and winds.

Fig. 27 shows examples of interannual changes in SST between 1906 and 1993, except for the periods of World Wars I and II. The changes have been discussed by Shannon and Taunton-Clark (1989), Shannon, Agenbag et al. (1990) and others, and the principal conclusions are as follows: Both SST and wind-stress indices show major change this century. Periods prior to 1911 and after 1974 are characterized by significantly higher equatorward wind stress, with lower values during the 1920s, 1930s and immediately after World War II. An increasing trend between the early 1950s and the early

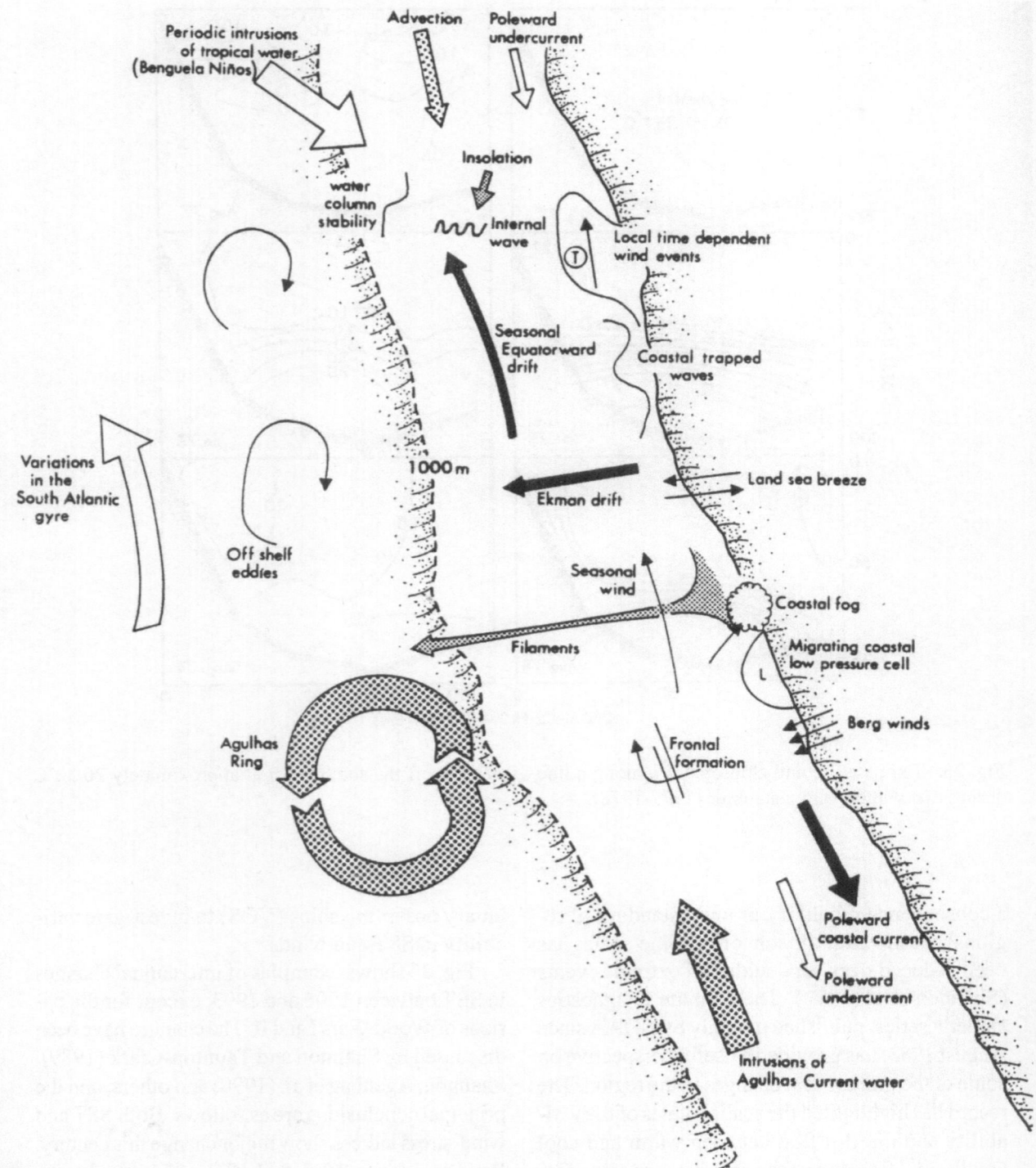

Fig. 26. Schematic showing the great complexity of processes to be considered in the Benguela eastern boundary system.

1980s is clear. Comparable changes are apparent in SST. An increasing trend in SST this century is clear, the post World War II era being typically 0.8°C warmer than earlier periods. SST anomalies in area 3 (Fig. 27) correlated significantly with those in the five other areas over the period of the record, but at times, such as 1982-83, the coastal and oceanic anomalies were not coherent.

Cool and warm periods appear in the record. Benguela *Niños* (Shannon et al. 1986) - already discussed - are the main interannual warm events in the northern Benguela, being superimposed on the normal intra-annual changes in the structure and location of the Angola-Benguela front. Although there is some suggestion that warming in area 3 (Fig. 27) between 1957 and 1964 may have been associated with changes in retroflection of the Agulhas Current, the only well-documented Agulhas intrusion into the Benguela was the recent one which peaked around 1986 (Shannon et al. 1990b). The cool events, while probably as important as the warm ones, have received little attention. Taunton-

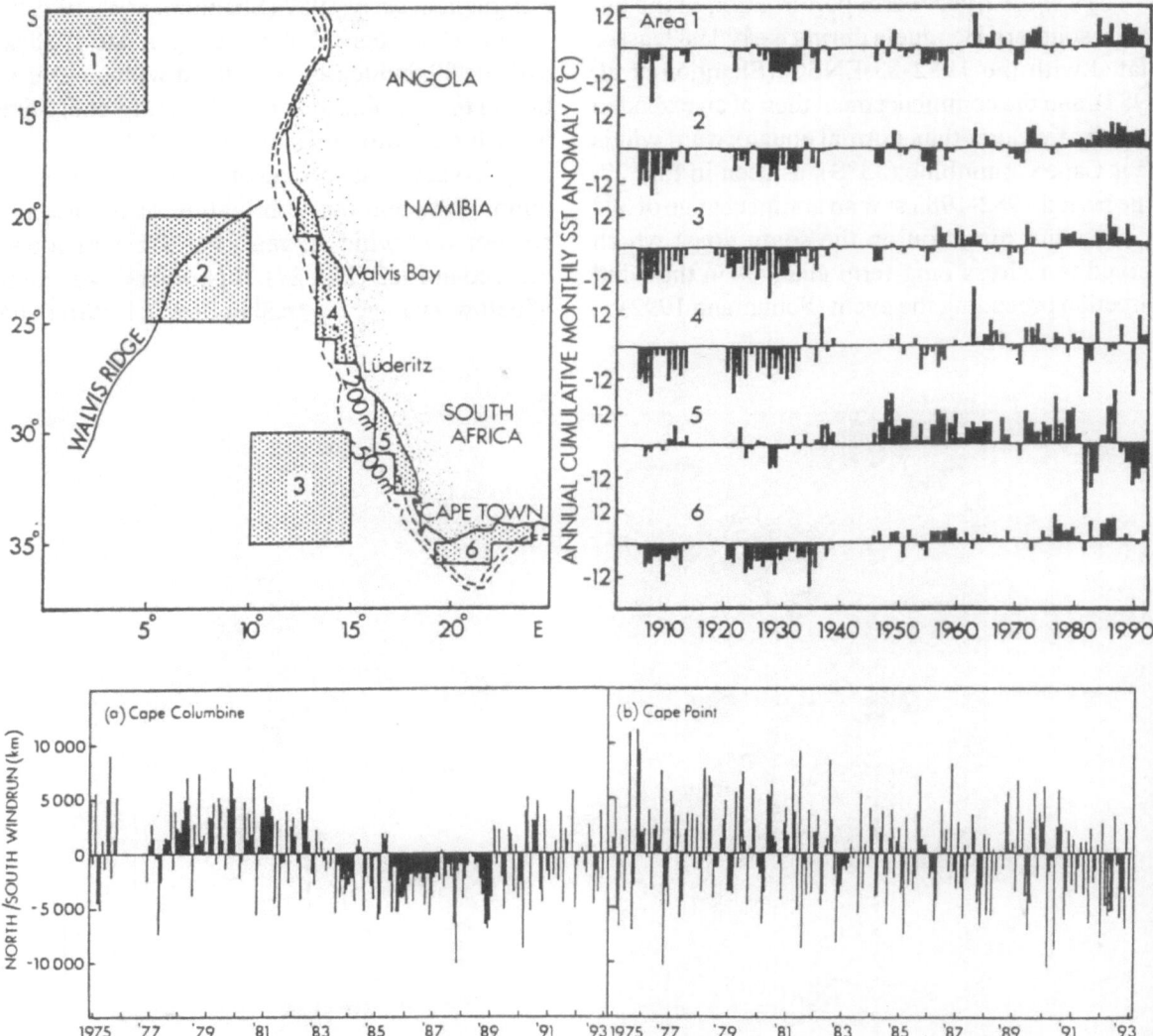

Fig. 27. Annual cumulative monthly anomalies of SST in six areas around southern Africa (after Shannon and Taunton-Clark 1989) North-south cumulative displacement (windrun) anomalies, expressed as departures from the 1960 - 1991 mean, at Cape Columbine (33°S) and Cape Point (34,3°S) (after Shannon and Taunton-Clark 1989 extended to 1993 - Pers. Comm. J. Taunton-Clark).

Clark (1990) has commented that the coolings in 1928 and 1955 may have been associated with changes in the subtropical convergence and in the easterly winds. The 1970-71 cool event was oceanic in origin, whereas the 1982 perturbation was confined to the shelf area.

Shannon et al. (1992) identified marked variability in the Benguela after 1980. The principal features may be summarized as follows:

(i) below average sea surface temperature in the shelf area of the northern and southern Benguela during 1982 and 1983 (Walker 1987);

(ii) a short-lived warm perturbation in the extreme southern Benguela during early 1983, associated with the 1982-83 ENSO (Shannon et al. 1984), and the commencement then of an extended period of weaker than normal equatorward winds near Cape Columbine (33°S), as seen in Fig. 27. The period 1982-1983 saw an abrupt change of 30° in the wind direction on the south coast which served to redress long-term changes in the wind direction preceding the event (Schumann 1992);

(iii) the 1984 *Benguela Niño* that followed the cool period in the northern Benguela (Shannon et al. 1986), but which had little impact on the southern region;

(iv) the warm anomaly in the south in 1986, which coincided with the intrusion of Agulhas Current water into the South-east Atlantic, especially during winter (Shannon et al. 1990b);

(v) an onset of cooling in 1987 (Anon. 1991), i.e. a reversal of the warming trend (Agenbag and Shannon 1987; Shannon and Agenbag 1990) with the influx of Subantarctic water into the southern Benguela in early 1987 (Shannon, et al. 1989);

(vi) a brief period of warming in late 1988 and early 1989, evidently associated with a perturbation in the Agulhas retroflection and a short-lived intrusion of warm water (Anon. 1991);

(vii) sustained cooling of shelf waters from autumn 1989, and the termination of the negative equatorward wind anomaly at 33°S which commenced in 1983 (Fig. 27). A period of two months of below-average reversal in flow at a current me-

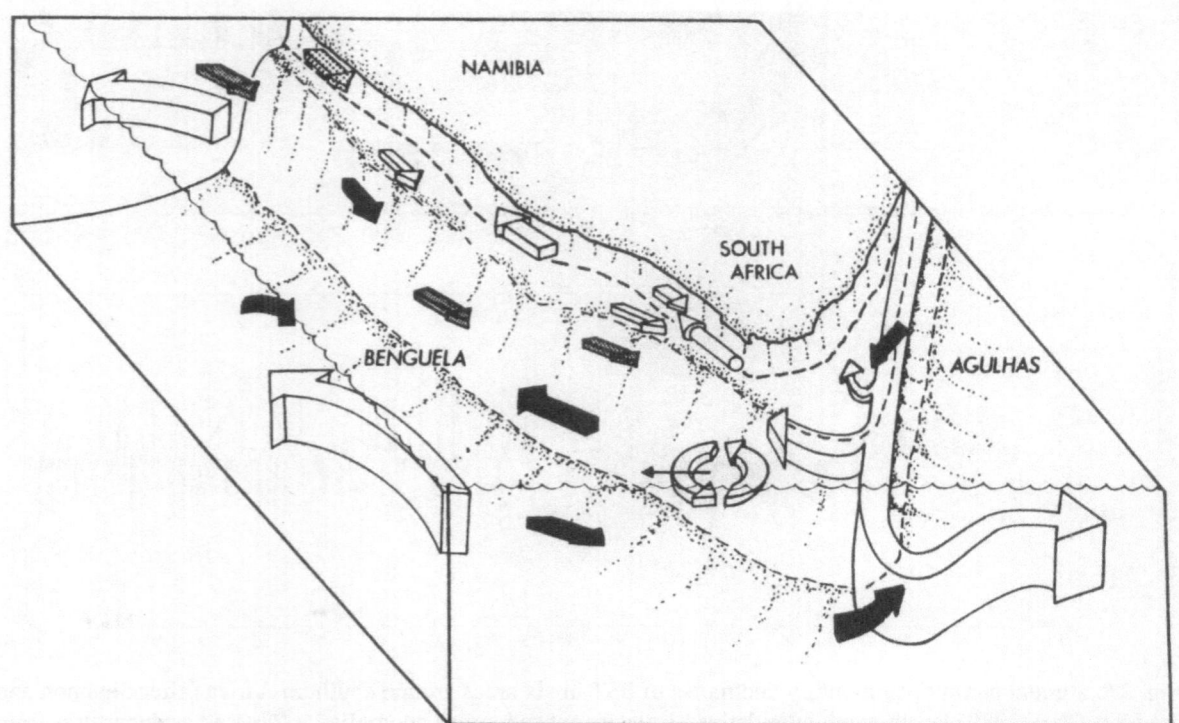

Fig. 28. Conceptualized three-dimensional flow field in the Benguela system.

ter site at 33°S, completely uncharacteristic of the record, occurred during winter 1989 (Shannon et al. 1992);

(viii) an extended period of below-average trade winds in the extreme south of Cape Point, (34.3°S) that commenced in the latter part of 1990 and was associated with the most recent ENSO;

(ix) intrusion of an Agulhas ring during November and December 1992 adjacent to the coast near Cape Point;

(x) anomalously strong south-easterly winds in the southern Benguela during the summer of 1993/94, which resulted in strong upwelling along the west and south coasts, and coinciding with a widespread negative SST anomaly in the South Atlantic.

Concluding remarks

In this paper we have attempted to summarize the main features of the Benguela and to highlight the more important recent advances in knowledge of its structure and functioning. Finally, our present understanding of the primary circulation in the Benguela area and the remote influences is shown conceptually in Fig. 28.

In terms of the larger scale dynamics, it is likely that the series of multinational cruises undertaken in the South-east Atlantic during recent years will result in substantial advances at the process and conceptual levels before the end of the decade. The high quality altimeter data from Topex-Poseidon now available and the relative ease of access to models, modelling expertise and super-computers, lead us to anticipate with confidence that by the year 2000 a regional predictive capability will have been developed which will be both locally useful for resource management and globally relevant.

The authors wish to thank Mr A.P. van Dalsen and his staff for preparing the diagrams and Mrs Anne van Heerden for her assistance in preparing the manuscript.

References

Agenbag JJ, Shannon LV (1987) A preliminary note on a recent perturbation in the Agulhas Current retroflection area. Trop Ocean-Atmos Newsl 37:10-11

Agenbaag JJ, Shannon LV (1988) A suggested physical explanation for the existence of a biological boundary at 24°30'S in the Benguela system. S Afr J Mar Sci 6:119-132

Agenbag JJ (1992) A procedure for the computation of sea surface advection velocities from satellite thermal band imagery, with applications to the south east Atlantic ocean. PhD thesis, University of Cape Town, 394 p

Andrews WRH, Hutchings L (1980) Upwelling in the southern Benguela Current. Prog Oceanogr 9(1):81 pp+ 2 Figures

Anon (1991) An investigation into the causes of the recent fluctuations in pelagic fish stocks off South Africa. Unpublished Report, Sea Fisheries Research Institute, South Africa, 8 pp + 18 Figures (mimeo)

Armstrong DA, Mitchell-Innes BA, Verheye-Dua F, Waldron H, Hutchings L (1987) Physical and biological features across an upwelling front in the southern Benguela. In: Payne AIL, Gulland JA, Brink KH (Eds). The Benguela and Comparable Ecosystems. S Afr J Mar Sci 5:171-190

Bailey GW, Chapman P (1991) Chemical and physical oceanography. In: Short-Term Variability during an Anchor Station Study in the Southern Benguela Upwelling System. Prog Oceanog 28:9-37

Bakun A, Nelson CS (1991) The seasonal cycle of wind-stress curl in subtropical eastern boundary current regions. J Phys Oceanogr 21:1815-1834

Bang ND (1971) The southern Benguela Current region in February, 1966. 2. Bathythermography and air-sea interactions. Deep-Sea Res 18(2):209-224

Bang ND (1973) Characteristics of an intense ocean frontal system in the upwell regime west of Cape Town. Tellus 25(3):256-265

Bang ND, Andrews WRH (1974) Direct current measurements of a shelf-edge frontal jet in the southern Benguela system. J Mar Res 32(3):405-417

Barange M, Pillar SC (1992) Cross-shelf circulation, zonation and maintenance mechanisms of the euphausiids *Nyctiphanes capensis* and *Euphausia hanseni* (Euphausiacea) in the northern Benguela upwelling system. Continent Shelf Res 12(9):1027-1042

Birch GF, Rogers J (1973) Nature of the sea floor between Lüderitz and Port Elizabeth. S Afr Shipping News and Fish Ind Rev 28:56-65

Birch GF, Rogers J, Bremner JM, Moir GJ (1976) Proc First Interdisciplinary Conf Mar Freshw Res S Afr, Fiche 20A (C10D12)

Boyd AJ, Thomas RM (1984) A southward intrusion of equatorial water off northern and central Namibia in March 1984. Trop Ocean-Atmos Newsl 27:16-17

Boyd AJ (1987) The oceanography of the Namibian shelf. PhD. thesis, University of Cape Town, [xv] + 190 pp + [i]

Boyd AJ, Salat J, Masó M (1987) The seasonal intrusion of relatively saline water on the shelf off northern and central Namibia.In: Payne AIL, Gulland JA, Brink KH (eds) The Benguela and Comparable Ecosystems S Afr J Mar Sci 5:107-120

Boyd AJ, Taunton-Clark J, Oberholster GPJ (1992) Spatial features of the near-surface and midwater circulation patterns off western and southern South Africa and their role in the life histories of various commercially fished species. In: Payne AIL, Brink KH, Mann KH, Hilborn R (eds) Benguela Trophic Functioning. S Afr J Mar Sci 12:189-206

Boyd AJ, Oberholster GPJ (1994) Currents off the west and south coasts of South Africa. S Afr Shipping News and Fish Ind Rev 49: 26-28

Boyd AJ, Shillington FA (1994) Physical forcing and circulation patterns on the Agulhas Bank. S Afr J Sci 90:114-122

Brink KH (1983) The near-surface dynamics of coastal upwelling. Prog Oceanogr 12(3):223-257

Brink KH, Chapman DC (1985) Programs for computing properties of coastal-trapped waves and wind-driven motions over the continental shelf and slope. Tech Rep Woods Hole Oceanogr Instn 85-17: unnumbered

Brown PC, Hutchings L (1987) The development and decline of phytoplankton blooms in the southern Benguela upwelling system. 1. Drogue movements, hydrography and bloom development. In: Payne AIL, Gulland JA, Brink KH (Eds) The Benguela and Comparable Ecosystems. S Afr J Mar Sci 5:357-391

Brundrit GB (1984) Monthly mean sea level variability along the west coast of southern Africa. S Afr J Mar Sci 2:195-203

Brundrit GB, De Cuevas BA, Shipley AM (1987) Long-term sea-level variability in the eastern South Atlantic and a comparison with that in the eastern Pacific. In: Payne AIL, Gulland JA, Brink KH (eds) The Benguela and Comparable Ecosystems. S Afr J Mar Sci 5:73-78

Brundrit GB, Shannon LV (1989) Cape storms and the Agulhas Current: a glimpse of the future. S Afr J Sci 84:584-586

Carton JA, Huang B (1994) Warm events in the tropical Atlantic. J Phys Oceanogr 24:888-903

Chapman P, Shannon LV (1985) The Benguela ecosystem 2. Chemistry and related processes. In: Barnes M (ed) Oceanography and Marine Biology. An Annual Review. Aberdeen, University Press 23:183-251

Chapman P, Shannon LV (1987) Seasonality in the oxygen minimum layers at the extremities of the Benguela system. In: Payne AIL, Gulland JA, Brink KH (eds) The Benguela and Comparable Ecosystems. S Afr J Mar Sci 5:85-94

Chapman P, Duncombe RAE CM, Allanson BR (1987) Nutrients, chlorophyll and oxygen relationships in the surface layers at the Agulhas Retroflection. Deep-Sea Res 34(8A):1399-1416

Chapman P, Largier JL (1989) On the origin of Agulhas Bank bottom water. S Afr J Sci 85(8):515-519

Clarke AJ (1989) Theoretical understanding of eastern ocean boundary poleward undercurrents. In: Neshyba SJ, Mooers Ch NK, Smith RL, Barber RT (eds) Poleward Flows along Eastern Ocean Boundries. Springer-Verlag, Berlin, pp 110-130

Connary SD, Ewing M (1972) The nepheloid layer and bottom circulation in the Guinea and Agnola Basins. In: Gordon AL (ed) Studies in Physical Oceanography: Tribute to George Wüst on his 80th Birthday. Gordon, Breach New York 2:169-184

Currie R[I] (1953) Upwelling in the Benguela Current. Nature 171(4351):497-500

De Decker AHB (1970) Notes on an oxygen-depleted subsurface current off the west coast of South Africa. Investl Rep Div Sea Fish S Afr 84:24 pp

Defant A, Wust G (1938) Wiss Ergebn Dt Atlant Exp „Meteor" 6(3):105-181

Dingle RV, Birch GF, Bremner JM, De Decker RH, Du Plessis A, Engelbrecht JC, Fincham MJ, Fitton T, Flemming BW, Gentle RI, Goddlad SH, Martin AK, Mills EG, Moir GJ, Parker RJ, Robson SH, Rogers J, Salmon DA, Siesser WG, Simpson ESW, Summerhayes CP, Westall F, Winter A, Woodborne MW (1987) Deep-sea sedimentary environments around southern Africa (South-East Atlantic and South-West Indian Oceans). Ann S Afr Mus 98(1):1-27

Dingle RV, Nelson G (1993) Sea bottom temperature salinity and dissolved oxygen on the continental margin off South-Western Africa. S Afr J Mar Sci 13:33-49

Duncombe RAE CM, Shannon LV, Shillington FA (1989) An Agulhas ring in the South Atlantic ocean. S Afr J Sci 85:747-748

Duncombe RAE CM, Boyd AJ, Crawford RJM (1992b)

„Predation" of anchovy by an Agulhas ring: a possible contributory cause for the very poor year-class of 1989. In: Payne AIL, Brink KH, Mann KH, Hilborn R (eds) Benguela Trophic Functioning. S Afr J mar Sci 12:167-173

Duncombe RAE CM, Shillington FA, Agenbag JJ, Taunton-Clark J, Gründlingh ML (1992a) An Agulhas ring in the South Atlantic Ocean and its interaction with the Benguela upwelling frontal system. Deep-Sea Res 39(11/12):2009-2027

Enfield DB, Allen JS (1983) The generation and propagation of sea level variability along the Pacific coast of Mexico. J phys Oceanogr 13:1012-1033

Garzoli SL, Fordon AL, Pillsbury D (1994) BEST: Benguela Source and Transport. The South Atlantic: Present and past circulation, Bremen, Germany 15-19 August 1994. Berichte, Fachbereich Geowissenschaften, Universität Bremen 52:167 pp

Gordon AL, Lutjeharms JRE, Gründlingh ML (1987) Stratification and circulation at the Agulhas retroflection. Deep-Sea Res 34(4A):565-599

Gordon AL, Haxby WF (1990) Agulhas eddies invade the South Atlantic: evidence from Geosat altimeter and shipboard. Conductivity-Temperature-Depth survey. J Geophys Res 95(C3):3117-3125

Gordon AL, Bosley KT (1991) Cyclonic gyre in the tropical South Atlantic. Deep-Sea Res 38 (Suppl 1A):S323-S343

Gordon AL, Weiss RF, Smethie WM, Warner MJ (1992) Thermocline and intermediate water communication between the South Atlantic and Indian oceans. J Geophys Res 97(C5):7223-7240

Gordon AL, Bosley KT, Aikman F (1995) Tropical Atlantic water within the Benguela up-welling system at 27°S. Deep-Sea Res 42 pp 1-12

Guastella LA-M (1992) Sea surface heat exchange at St Helena Bay and implications for the southern Benguela upwelling system. In: Payne AIL, Brink KH, Mann KH, Hilborn R (eds) Benguela Trophic Functioning. S Afr J Mar Sci 12:61-70

Hart TJ, Currie RI (1960) The Benguela Current. „Discovery" Rep 31:123-297

Hisard P (1990) Variabilité des précipitations dans l'Atlantique tropical sud-est pendant un El Niño. Hydrol-continent 5(2):87-104

Holden CJ (1987) Observations of low-frequency currents and continental shelf waves along the west coast of South Africa. In: Payne AIL, Gulland JA, Brink KH (eds) The Benguela and Comparable Ecosystems. S Afr J Mar Sci 5:197-208

Horel JD, Kousky VE, Kango MT (1986) Atmospheric conditions in the Atlantic sector during 1983-1984. Nature 322:248-251

Hunter IT (1987) The weather of the Agulha and Cape south coast. CSIR research report 634:148 pp

Hutchings L, Taunton-Clark J (1990) The monitoring of gradual change in areas of high mesoscale variability. S Afr J Sci 86:9-37

Jackson SP (1947) Air masses and the circulation over the plateau and coast of South Africa. S Afr Geogr J 29:1-15

Jury MR (1992) A review of the meteorology of the eastern Agulhas Bank. S Afri J Sci 90:109-113

Jury MR (1985a) Case studies of alongshore variations in wind-driven upwelling in the southern Benguela region. In: Shannon LV (ed) Cape Town; South African Ocean Colour and Upwelling Experiment. Sea Fisheries Research Institute: 29-46

Jury MR (1985b) Mesoscale variations in summer winds over the Cape Columbine - St Helena Bay region, South Africa. S Afr J Mar Sci 3:77-88

Jury MR, Kamstra F, Taunton-Clark J (1985) Diurnal wind cycles and upwelling off the northern portion of the Cape Peninsula in summer. S Afr J Mar Sci 3:1-10

Jury MR, Kamstra F, Taunton-Clark J (1985) Synoptic summer wind cycles and upwelling off the southern portion of the Cape Peninsula. S Afr J Mar Sci 3:33-42

Jury MR (1988) Case studies of the response and spatial distribution of wind-driven upwelling off the coast of Africa: 29-34° south. Continent Shelf Res 8(11):1257-1271

Jury MR, MacArthur CI, Brundrit GB (1990) Pulsing of the Benguela upwelling region: large-scale atmospheric controls. S Afr J Mar Sci 9:27-41

Jury MR, Brundrit GB (1992) Temporal organization of upwelling in the southern Benguela ecosystem by resonant coastal trapped waves in the ocean and atmosphere. In: Payne AIL, Brink KH, Mann KH, Hilborn R (eds) Benguela Trophic Functioning. S Afr J Mar Sci 12:219-224

Kamstra F (1985) Environmental features of the southern Benguela with special reference to the wind stress. In: Shannon LV (ed) Cape Town; South African Ocean Colour and Upwelling Experiment. Sea Fisheries Research Institute: 13-27

Kostianoy AG (1994) Remote sensing of the Angola Benguela front. The South Atlantic: present and past circulation, Bremen, Germany 15-19 August 1994. Berichte, Fachbereich Geowissenschaften, Universitat Bremen 52:167 pp

Lamberth R, Nelson G (1987) Field and analytical drogue studies applicable to the St Helena Bay area off South Africa's west coast. In: Payne AIL, Gulland JA, Brink KH (eds) The Benguela and

Comparable Ecosystems. S Afr J Mar Sci 5:163-169

Largier JL,Swart VP (1987) East-west variation in thermocline breakdown on the Agulhas Bank. In: Payne AIL, Gulland JA, Brink KH (eds) The Benguela and Comparable Ecosystems. S Afr J Mar Sci 5:263-272

Largier JL, Chapman P, Peterson WT, Swart VP (1992) The western Agulhas Bank: circulation, stratification and ecology. In: Payne AIL, Brink KH, Mann KH, Hilborn R (eds) Benguela Trophic Functioning. S Afr J Mar Sci 12:319-339

Largier J (1994) The internal tide over the shelf inshore of Cape Point Valley, South Africa. J Geophys Res 99(C5):10023-10034

Lindesay JA (1987) Relationships between the Southern Oscillation and atmospheric changes over southern Africa, 1957 to 1982. PhD Thesis, University of Witwatersrand, 341 pp (as cited by Tyson, 1988)

Lindesay JA, Jury MR (1991) Atmospheric circulation controls and characteristics of a flood event in central South Africa. Int J Climatology 11:609-627

Lutjeharms JRE, van Ballegooyen RC (1984) Topographic control in the Agulhas Current system. Deep-Sea Res 31(11):1321-1337

Lutjeharms JRE, Gordon AL (1987) Shedding of an Agulhas ring observed at sea. Nature 325:138-139

Lutjeharms JRE, Meeuwis JM (1987) The extent and variability of South-East Atlantic upwelling. In: Payne AIL, Gulland JA, Brink KH (Eds) The Benguela and Comparable EcosystemsS Afr J Mar Sci 5:51-62

Lutjeharms JRE, Stockton PL (1987) Kinematics of the upwelling front off southern Africa. In: Payne AIL, Gulland JA, Brink KH (eds) The Benguela and Comparable Ecosystems. S Afr J Mar Sci 5:35-49

Lutjeharms JRE, Valentine HR (1987) Water types and volumetric considerations of the South-East Atlantic upwelling regime. In: Payne AIL, Gulland JA, Brink KH (eds). The Benguela and Comparable Ecosystems. S Afr J Mar Sci 5:63-71

Lutjeharms JRE (1988) Examples of extreme circulation events at the Agulhas retroflection. S Afr J Sci 84:584-586

Lutjeharms JRE, van Ballegooyen RC (1988) The retroflection of the Agulhas Current. J Phys Oceanogr 18(11):1570-1583

Lutjeharms JRE, Shillington FA, Duncombe RAE CM (1991) Observations of extreme upwelling filaments in the Southeast Atlantic Ocean. Science 253(5021):774-776

McCreary JP (1981) A linear stratified ocean model of the coastal undercurrent. Phil Trans Roy Soc of London A302:385-413

Meeuwis JM, Lutjeharms JRE (1990) Surface thermal characteristics of the Angola-Benguela front. S Afr J Mar Sci 9:261-279

Mooers CNK, Collins CA, Smith RL (1976) The dynamic structure of the frontal zone in the coastal upwelling region off Oregon. J Phys Oceanogr 6:3-21

Moroshkin KV, Bubnov VA, Bulatov RP (1970) Water circulation in the eastern South Atlantic Ocean. Oceanology 10(1):27-34

Nelson G, Hutchings L (1983) The Benguela upwelling area. Prog Oceanogr 12(3):333-356

Nelson G (1985) Notes on the Physical Oceanography of the Cape Peninsula upwelling system. In: Shannon LV (ed) Cape Town; South African Ocean Colour and Upwelling Experiment. Sea Fisheries Research Institute: 63-95

Nelson G, Polito A (1987) Information on currents in the Cape Peninsula area, South Africa. In: Payne AIL, Gulland JA, Brink KH (eds) The Benguela and Comparable Ecosystems. S Afr J Mar Sci 5:287-304

Nelson G (1989) Poleward motion in the Benguela area. In: Neshyba SJ, Mooers CNK, Smith RL, Barber RT (eds) Poleward Flows along Eastern Ocean Boundaries. Springer, New York, 110-130 (Coastal and Estuarine Studies 34)

Nelson G (1991) An equatorward jet west of Cape Town. Abstract IAPSO Proceedings XX Assembly Vienna 1991, p189

Nelson G (1992) Equatorward wind and atmospheric pressure spectra as metrics for primary productivity in the Benguela system. In: Payne AIL, Brink KH, Mann KH, Hilborn R (eds) Benguela Trophic Functioning. S Afr J Mar Sci 12:19-28

Olson DB, Evans RH (1986) Rings of the Agulhas Current. Deep-Sea Res 33(1A):27-42

Parrish RH, Bakun A, Husby DM, Nelson CS (1983) Comparative climatology of selected environmental processes in relation to eastern boundary current pelagic fish reproduction. In: Sharp GD, J Csirke (eds) Proceedings of the Expert Consultation to Examine Changes in Abundance and Species Composition of Neritic Fish Resources, San José, Costa Rica, April 1983. FAO Fish Rep 291(3):731-777

Pearce AF (1991) Eastern boundary currents of the southern hemisphere. J R Soc Western Aust 74:35-45

Preston-Whyte RA, Tyson PD (1973) Note on pressure

oscillations over South Africa. Mon Weath Rev 101(8):650-659

Preston-Whyte RA, Tyson PD (1988) The Atmospheric and Weather of Southern Africa. Oxford University Press, Cape Town, 375 pp

Pugh J (1982) Erosion of the seasonal thermocline on the Agulhas Bank. Internal Report, Dept of Physical Oceanography, University of Cape Town, 16 pp

Reid JL (1989) On the total geostrophic circulation of the South Atlantic Ocean: flow patterns, tracers and transports. Prog Oceanogr 23(3):149-244

Reid JL (1994) On the total geostrophic circulation of the North Atlantic Ocean: Flow patterns, tracers and transports. Prog Oceanogr 33:1-92

Rogers J, Bremner JM (1991) The Benguela ecosystem 4. In: 29 Barnes M (ed) Aberdeen; Oceanography and Marine Biology. University Press: 1-85

Rosby CG and collaberators (1939) Relation between variations in the intesity of the zonal circulation of the atmosphere and the displacements of the semipermanent centers of action. J Mar Res 2:38-54

Schell II (1968) On the relation between the winds off Southwest Africa and the Benguela Current and Agulhas Current penetration in the South Atlantic.Dt Hydrogr Z 21(3):109-117

Schumann EH, Perrins L-A, Hunter IT (1982) Upwelling along the south coast of the Cape Province, South Africa. S Afr J Sci 78(6):238-242

Schumann EH (1989) The propagation of air pressure and wind systems along the South African coast. S Afr J Sci 85:382-385

Schumann EH, Brink KH (1990) Coastal-trapped waves off the coast of South Africa: generation, propagation and current structures. J Phys Oceanogr 20:1206-1218

Schumann EH (1992) Interannual wind variability on the south and east coasts of South Africa. J Geophys Res 97(D18):20397-20403

Shannon LV (1966) Hydrology of the south and west coasts of South Africa.Investl Rep Div Sea Fish S Afr 58:22 pp + 30 pp of Figures

Shannon LV, van Rijswijck M (1969) Physical oceanography of the Walvis Ridge region. Investl Rep Div Sea Fish S Afr 70:19 pp

Shannon LV, Nelson G, Jury MR (1981) Hydrological and meteorological aspects of upwelling in the southern Benguela Current. In: Richards FA (ed) Washington, DC; Coastal and Estuarine Sciences. 1. Coastal Upwelling. American Geophysical Union: 146-159

Shannon LV, Anderson FP (1982) Applications of satellite ocean colour imagery in the study of the Benguela Current system. S Afr J Photogramm Remote Sens Cartogr 13(3):153-169

Shannon LV, Mostert SA, Walters NM, Anderson FP (1983) Chlorophyll concentrations in the southern Benguela Current region as determined by satellite. (*Nimbus-7* coastal zone colour scanner). J Plankt Res 5(4):565-583

Shannon LV (1985) The Benguela ecosystem. 1. Evolution of the Benguela, physical features and processes. In: Barnes M (ed) Aberdeen; Oceanography and Marine Biology An Annual Review 23. University Press: 105-182

Shannon LV, Walters NM, Mostert SA (1985) Satellite observations of surface temperature and near-surface chlorophyll in the southern Benguela region. In: Shannon LV (ed) Cape Town; South African Ocean Colour and Upwelling Experiment. Sea Fisheries Research Institute: 183-210

Shannon LV, Boyd AJ, Brundrit GB, Taunton-Clark J (1986) On the existence of an *El Niño*-type phenomenon in the Benguela system. J Mar Res 44(3):495-520

Shannon LV, Agenbag JJ (1987) Notes on the recent warming in the South-East Atlantic, and possible implications for the fisheries of the region. Colln scient Pap int Commn SE Atl Fish 14(2):243-248

Shannon LV, Agenbag JJ, Buys MEL (1987) Large- and mesoscale features of the Angola-Benguela front. In: Payne AIL, Gulland JA, Brink KH (eds) The Benguela and Comparable Ecosystems. S Afr J Mar Sci 5:11-34

Shannon LV, Hunter D (1988) Notes on Antarctic intermediate water around southern Africa. S Afr J Mar Sci 6:107-117

Shannon LV, Taunton-Clark J (1989) Long-term environmental indices for the ICSEAF area. Sel Pap int Commn SE Atl Fish 1:5-15

Shannon LV, Lutjehamrs JRE, Agenbag JJ (1989) Episodic input of Subantarctic water into the Benguela region. S Afr J Sci 85(5):317-322

Shannon LV, Agenbag JJ (1990) A large-scale perspective on interannual variability in the environment in the South-East Atlantic. S Afr J Mar Sci 9:161-168

Shannon LV, Lutjeharms JRE, Nelson G (1990) Causative mechanisms for intra-annual and interannual variability in the marine environment around southern Africa. S Afr J Sci 86(7-10):356-373

Shannon LV, Agenbag JJ, Walker ND, Lutjeharms JRE (1990) A major perturbation in the Agulhas retroflection area in 1986. Deep-Sea Res 37(3):493-512

Shannon LV, Chapman P(1991) Evidence of Antarctic bottom water in the Angola Basin at 32°S. Deep-Sea Res 38(10):1299-1204

Shannon LV, Crawford RJM, Pollock DE, Hutchings L, Boyd AJ, Taunton-Clark J, Badenhorst A, Melville-Smith R, Augustyn CJ, Cochrane KL, Hampton I, Nelson G, Japp DW, Tarr RJQ (1992) The 1980s - a decade of change in the Benguela ecosystem. In: Payne AIL, Brink KH, Mann KH, Hilborn R (eds) Benguela Trophic Functioning. S Afr J Mar Sci 12:271-296

Shelton PA, Hutchings L (1982) Transport of anchovy, *Engraulis capensis* Gilchrist, eggs and early larvae by a frontal jet current. J Cons perm int Explor Mer 40:185-198

Shillington FA, Peterson WT, Hutchings L, Probyn TA, Waldron HN, Agenbag JJ (1990) A cool upwelling filament off Namibia, southwest Africa: preliminary measurements of physical and biological properties. Deep-Sea Res 37(11A):1753-1772

Siesser WG, Scrutton RA, Simpson ESW (1974) In: Burke CA, Dranke CL (eds) New York; The Geology of Continental Margins. Spinger-Verlag: 641-654

Smith RL, Huyer A, Godfrey JS, Church JA (1991) The Leeuwin current off western Australia, 1986-1987. J Phys Oceanogr 21(2):325-345

Stander GH (1958) The variations of temperature in the surface layer of the sea near Walvis Bay during 1954-57 with an analysis of some wind data from Pelican Point. Investl Rep Div Fish S Afr 35: 40 pp

Stander GH (1963) The pilchard of South West Africa (*Sardinops ocellata*). Temperature: its annual cycles and relation to wind and spawning. Investl Rep Mar Res Lab SW Afr 9: 57 pp

Stander GH (1964) The Benguela Current off South West Africa. Investl Rep Mar Res Lab SW Afr 12: 43 pp + Plates 5-81

Stander GH, de Decker AHB (1969) Some physical and biological aspects of an oceanographic anomaly off South West Africa in 1963. Invest Rep Div Sea Fish S Afr 81: 46 pp

Stockton PL, Lutjeharms JRE(1988) Observations of vortex dipoles on the Benguela upwelling front. S Afr Geographer 15(1/2):27-35

Stramma L, Peterson RG (1989) Geostrophic transport in the Benguela Current region. J Phys Oceanogr 19:1440-1448

Swart VP, Largier JL(1987) Thermal structure of Agulhas Bank water. In: Payne AIL, Gulland JA, Brink KH (eds) The Benguela and Comparable Ecosys-

tems. S Afr J Mar Sci 5:243-253

Taunton-Clark J, Shannon LV (1988) Annual and interannual variability in the South-East Atlantic during the 20th century. S Afr J Mar Sci 6:97-106

Taunton-Clark J (1990) Environmental events within the South-East Atlantic (1906-1985) identified by analysis of sea surface temperature and wind data. S Afr J Sci 86(7-10):470-472

Tyson PD (1986) Climate Change and Variability in Southern Africa. Cape Town; Oxford University Press: 220 pp

Valentine HR (1990) A fine-scale volumetric census of the water masses of the Agulhas Retroflection area. Rep S Afr Coun scient ind Res EMA-R 691: 105 pp

Valentine HR, Lutjeharms JRE, Brundrit GB (1993) The water masses and volumetry of the southern Agulhas Current region. Deep-Sea Res 40(6):1285-1305

van Ballegooyen C, Gründlingh ML, Lutjeharms JRE (1994) Eddy fluxes of heat and salt from the south west Indian Ocean into the southeast Atlantic Ocean: a case study. J Geophys Res 99(C7):14053-14070

van Foreest D, Shillington FA, Legeckis R (1984) Large scale, stationary, frontal features in the Benguela Current system. Continent Shelf Res 3(4):465-474

Visser GA (1969) Analysis of Atlantic waters off the west coast of southern Africa. Investl Rep Div Sea Fish S Afr 75:26 pp

Walker ND, Taunton-Clark J, Pugh J (1984) Sea temperatures off the South African west coast as indicators of Benguela warm events. S Afr J Sci 80(2):72-77

Walker ND (1986) Satellite observations of the Agulhas Current and episodic upwelling south of Africa. Deep-Sea Res 33(8A):1083-1106

Walker ND (1987) Interannual sea surface temperature variability and associated atmospheric forcing within the Benguela system. In: Payne AIL, Gulland JA, Brink KH (eds) The Benguela and Comparable Ecosystems. S Afr J Mar Sci 5:121-132

Walker ND, Mey RD (1988) Ocean/atmosphere heat fluxes within the Agulhas Retroflection region. J Geophys Res 93(C12):15473-15483

Walter H (1937) Die ökologischen Verhältnisse in der Namibnebelwüste (Südwestafrika) unter Auswertung de Aufzeichnungen des Dr G Boss (Swakopmund). In: Pringsheim, N (ed) Leipzig; Jahrbücher für Wissenschaffliche Botanik. Gebrüder Borntraeger: 58-222

The Southern Boundary of the South Atlantic

D.J. Webb

Southampton Oceanography Centre, Empress Dock,
Southampton SO14 3ZH, U.K.

Abstract: The boundary region between the South Atlantic and Southern Ocean is discussed making use of results from the FRAM model and hydrographic sections. Although, at first, the system appears to be a simple one, with a sub-tropical gyre to the north abutting a zonal jet to the south, the detailed structure is more complex. Part of the complexity is well known and arises from the large scale thermohaline circulation, water being transported north in the surface Ekman layer and at depth, and returning south at intermediate levels. More recent work has shown that, in addition, the sub-tropical gyre and the Antarctic Circumpolar Current combine to form the Deacon Cell which transforms surface stress due to the wind down to mid-depths. Further complications arise from the Agulhas eddies which pass through the regions and the constraints on the circulation due to the bottom topography.

Introduction

It is usual to define the boundary between the South Atlantic and the Southern Ocean to lie along the Subtropical Convergence, near 40°S. The convergence also represents the boundary between the Antarctic Circumpolar Current and the South Atlantic Gyre. At 30°S the boundary is defined by surface temperatures near 16°C and salinities near 35.0 (Olbers et al. 1992). The warmer, more saline water to the north appears to have its source in the Brazil Current, the colder fresher water to the south coming from the Falkland Current branch of the Antarctic Circumpolar Current. The results from FRAM (The FRAM Group 1991; Webb et al. 1991), a fine-resolution numerical model of the Southern Ocean, indicate that the front is strongest in the west, due to the large contrast between the water masses of the Brazil and Falkland Currents, and that the front becomes weaker as the water moves eastward.

The traditional picture of the Antarctic Circumpolar Current has always been of a broad current, essentially baroclinic with possible local maxima associated with the Antarctic and Subtropical Convergences (Sverdrup et al. 1942; Wüst 1950; Defant 1961; Tchernia 1980). Sverdrup showed transports of about 100 Sv through Drake Passage but more recent estimates based on current meters and hydrography lie nearer 130 Sv (Reid and Nowlin 1971; Bryden and Pillsbury 1977; Fandry and Pillsbury 1979; Whitworth and Peterson 1985).

Recent numerical models have shown a more complex picture in which the reduced stratification of the Southern Ocean results in a current field with a significant barotropic component. They also show much larger transports. Cox (1975) for example, using a low resolution model finds a transport of 200 Sv. Semtner and Chervin (1988; 1992) and FRAM (The FRAM Group 1991), both using high resolution models, find 202 Sv and 195 Sv respectively. These large figures have tended to throw suspicion on the reliability of the model results.

A large fraction of the total transport in FRAM was found to be due to the barotropic flow. It is thus possible that the differences between the model estimates and observations arise because of difficulties in measuring the ocean's barotropic transport. If the model transport is calculated from the density field in the same way as many experimental estimates, assuming geostrophy and a zero current at the ocean bottom, then the transport calculated from FRAM drops to 127 Sv (Thompson, private communication).

From WEFER G, BERGER WH, SIEDLER G, WEBB DJ (eds), 1996, *The South Atlantic: Present and Past Circulation.* Springer-Verlag Berlin Heidelberg, pp 211-217

One area where the models have indicated that the barotropic current is particularly large is in the Falkland Current region. This prediction has been qualitatively confirmed by float data (Davis 1990; Davis et al. 1994) and by ADCP measurements made during the A11 hydrographic section (Saunders and King 1994). This has not explained the much smaller transports found in the Drake Passage current meter array but it has helped restore some confidence in the model results.

FRAM shows that the Antarctic Circumpolar Current (ACC) has a large barotropic component across much of the South Atlantic sector. The deep currents are generally strongest between 45°S and 50°S, in the region of the Antarctic Convergence. One result of this is that the ACC in this region is greatly influenced by the position of topographic features such as the gaps in the Mid-Atlantic Ridge. Thus in FRAM (Fig. 1) there is a concentration of stream lines in the region where the ridge narrows slightly at 45°S. In the model this also results in a sharpened Antarctic Convergence, as seen in the temperature and salinity fields. A similar sharpening is seen in the field observations (Olbers et al. 1992) but the latter also show a marked sharpening of the Subtropical Convergence near the Tristan de Cunha fracture zone. The latter effect is not seen in the FRAM results.

The meridional fluxes

Our information on the large scale meridional flows through the region has come primarily from hydrographic measurements and their analysis using the ideas of core layers (Wüst 1935; Wüst 1936; Defant 1941; Wüst 1950; Defant 1961). More recent analysis of the FRAM results give support to this picture (Saunders and Thompson 1993; Döös and Webb 1994).

Saunders and Thompson investigated the flux of mass, heat and fresh water in the South Atlantic across 32.5°S (just north of the latitude of Cape Town). They found that circulation in the top 1500m was dominated by the gyre scale circulation with its southward flowing western boundary region and northward flow elsewhere. The latter had a maximum between 0°E and 10°E. The net flow was northward with a magnitude of 19 Sv. Of

this, 2 Sv was due to the surface Ekman layer and the rest appeared to correspond to the inflow of Antarctic Intermediate Water.

Between 1500m and 3900m, where the water is primarily North Atlantic Deep Water, the flow was southward. The total transport was 20.5 Sv and was concentrated in the western basin with little flow further east.

Below 3900m the model showed a weak northward flow (1.5 Sv) concentrated against the South African coast. As this value was unexpectedly small they investigated further and found that in the western basin, along the normal route of Antarctic Bottom Water, the model showed none of the expected deep baroclinic structure. This may be partly due to the lack of such structure in the Levitus data set used to initalise the model (Levitus 1982) and partly to deficiencies in the model physics and model topography.

They found that the heat flux through the section was dominated by the meridional overturning (0.72 PW) in which the northward flowing surface water was replaced by cooler North Atlantic Deep Water. This flux was partly balanced by the southward heat flux due to the sub-tropical gyre (-0.14 PW). In this, the warm southward flowing water in the Brazil Current is balanced by cooler northward flowing water in the rest of the gyre. The total heat transport, including the effect of eddies, was 0.56 PW northwards.

The transport of salt (or fresh water) was found to be dominated by the high salinity of the surface waters. The gyre scale circulation gave a southward transport of $7*10^9$ kg s^{-1}, the net southward flow resulting from the excess salinity of the Brazil Current. The thermohaline circulation gave a smaller northward flux of $2.7*10^9$ kg s^{-1}.

Recently Saunders and King (1994; 1994) have investigated the same fluxes making use of hydrographic and acoustic doppler current meter data collected on WOCE section A11. This crosses the Argentine Basin at 45°S and, from the mid-Atlantic ridge, heads north-eastwards reaching South Africa at 30°S.

If they made the standard assumption of a level of no motion (they chose the bottom of the North Atlantic Deep Water, $\sigma_3 = 41.58$ kg m^{-3}) they de-

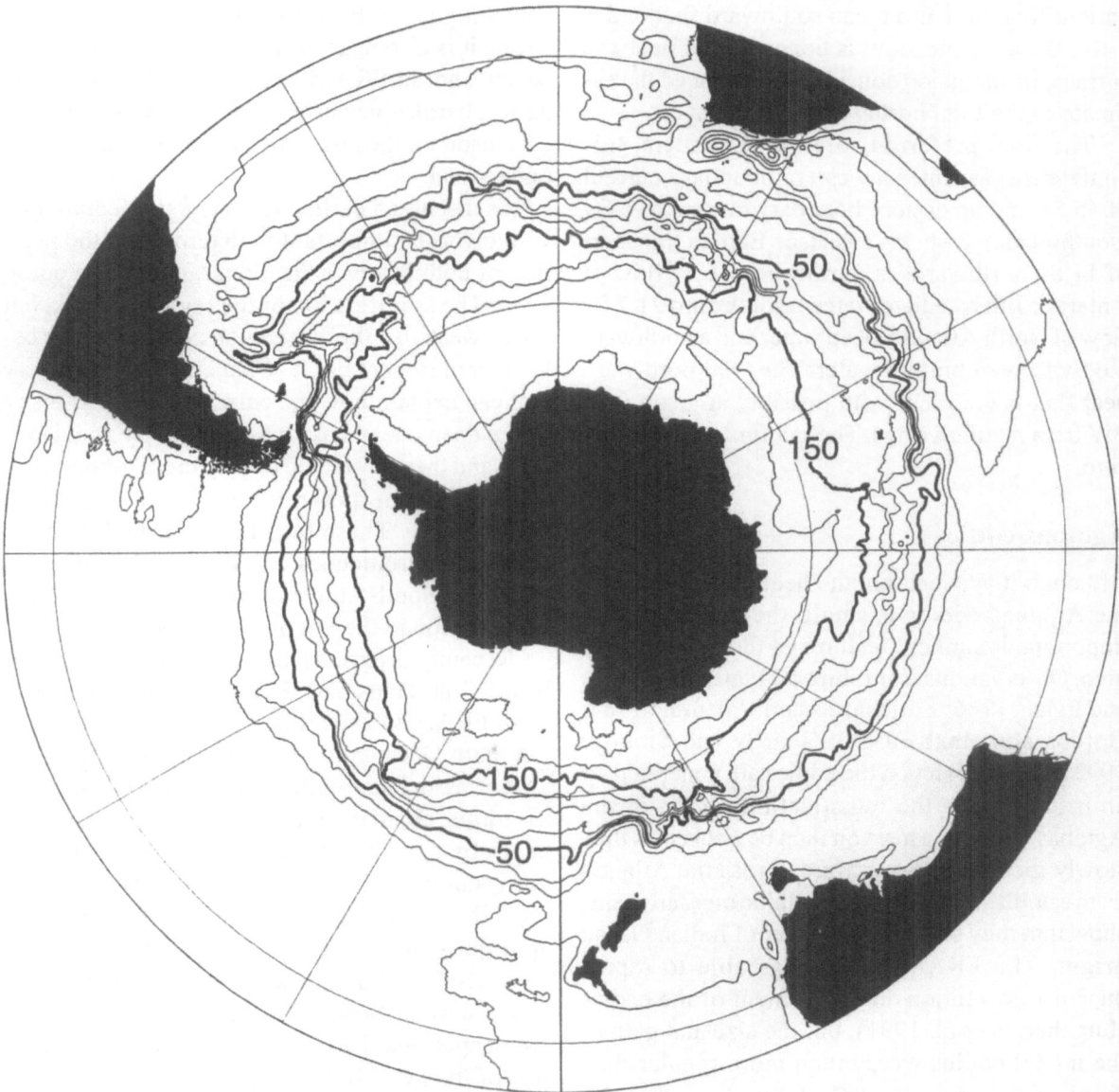

Fig. 1. The transport stream function in the Southern Ocean as calculated by the FRAM model at the end of six years. Contours are intervals of 25 Sv.

duced a northwards surface Ekman transport of 14.6 Sv, a southward intermediate water transport of 7.3 Sv, a southward North Atlantic Deep Water transport of 8.7 Sv and a 1.4 northward transport of bottom water.

The two figures giving most concern are the flux of intermediate water, which is in the wrong direction, and the resulting total northward heat transport through the section which, at 0.94 PW, is much larger than expected.

The solution does not take into account any barotropic current. In fact, for the Falkland Current, whereas the hydrographic calculation gave a transport of 13 Sv, their ADCP data showed that the total transport was nearer 45 Sv, the level of no motion approximation being a very poor one. The other main factor that had to be taken into account was the transport in the Eastern Boundary current. Although it did not show up during the WOCE A11 section, it is known from other observations that there is a mean southward flux of 5 ± 4 Sv. Because the flow is concentrated near the surface, its inclusion considerably reduced the estimate of the total northward heat transport.

The final preferred solution (Saunders, this conference) assumes a western boundary current of 45 Sv and an eastern boundary current of 5 Sv (southwards). It shows a surface Ekman transport of 11 Sv northwards, a northward 4.1 Sv flow of Antarctic Intermediate water, a southward 21.7 Sv flow of North Atlantic Deep water and a northward 6.6 Sv flow of bottom water. The total northward heat flux is 0.53 PW with possibly an extra 0.05 PW from Agulhas rings. The salt flux is essentially zero.

Agulhas eddies

Although it appears that the heat transport due to the Agulhas eddies is small, they are still very important dynamical features of the boundary region. Observations from ships and satellites (Olsen and Evans 1986; Lutjeharms and Valentine 1988; Gordon and Haxby 1990; Quartly and Srokosz 1993) have shown that the eddies are generated by an instability in the retroflection region of the Agulhas Current. They can then be tracked as they slowly move north-westward across the Atlantic from satellite images. Hydrographic measurements show that they still contain water of Indian Ocean origin. The FRAM model was able to reproduce the generation and movement of the eddies (Lutjeharms et al. 1991), but the size and path of the model eddies were much more regular than those of the real ocean (Lutjeharms and Webb 1995).

Dynamics

As these results show, we are still at an early stage in describing the features of the Southern Ocean and its boundary with the South Atlantic. Two of the features long associated with the boundary, the Subtropical Convergence and the Antarctic Convergence further south, are still not properly understood. The sub-tropical convergence is roughly at the latitude of the zero zonal wind stress, but this is only approximately true. The Antarctic Conver-

gence is just north of the maximum in the wind stress, it is also near the northward extent of winter ice and, in FRAM, the region of some of the largest barotropic currents, but again the dynamical reason for the position of the current is not well understood.

As discussed earlier, the one extra feature that is likely to be important in determining the position of both fronts is the topography of the ocean floor. The stratification of the region is weak, but not so weak that the currents show the extreme behaviour of Boyer (1993). What is needed is a theory that can explain how the currents can cross many of the deep ocean basins as essentially zonal currents and then be steered as they cross topography.

Changes in potential vorticity appear to be necessary. These may be associated with the mesoscale turbulence observed in the confluence region off South America and in similar transition regions around the Southern Ocean.

Recently progress has been made in two areas. In the first, Döös and Webb, (1994) used results from FRAM to show that the Deacon Cell in the Southern Ocean and the South Atlantic (see Fig. 2) was associated with the downward transfer of the torque about the Earth's axis transferred to the ocean by the wind stress. In the Southern Ocean, the water masses in the top 1500m of the northward flowing Falkland Current return southwards at slightly greater depths around the rest of the Southern Ocean. In the South Atlantic, water in the Brazil current flows southwards at depths which are deeper than its return flow northwards in the rest of the gyre. In both cases the result is an organised downward transfer of angular momentum.

The reason why the ocean is organised in this way is not so clear. At the latitude of Drake Passage the downward transfer of angular momentum is required because there are no meridional barriers to support a near-surface pressure gradient. This problem was discussed by Munk and Palmen (1951) and one of the mechanisms they proposed was similar to the mechanism found in FRAM. However the Drake Passage stops near 56°S whereas the Deacon Cell itself extends to 36°S.

The other area of progress concerns the factors determining the strength of the Antarctic Cir-

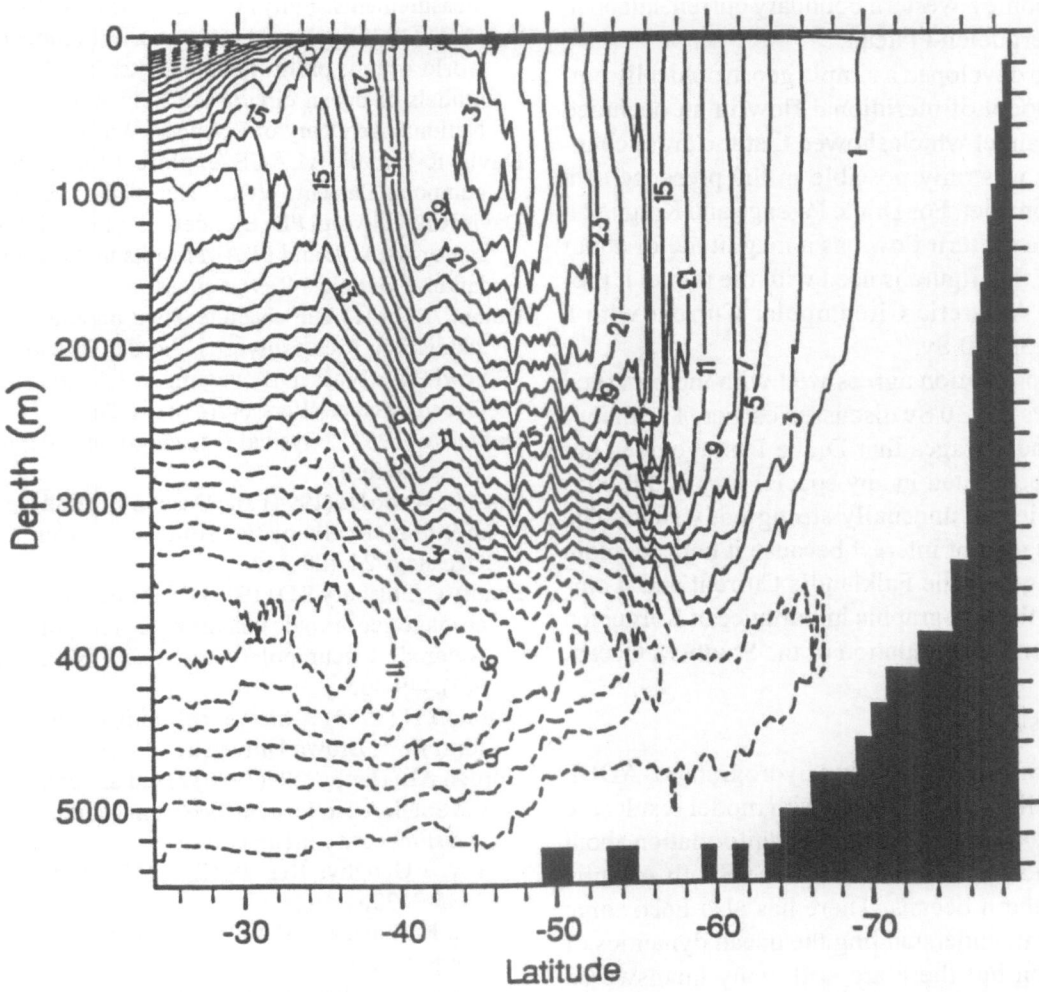

Fig. 2. The meridional stream function (from Doos and Webb 1984). The Deacon Cell extends between the surface and 3000m between 36°S and 70°S.

cumpolar Current. Many ideas have been put forward (Nowlin and Klinck 1986), including treating the system as an open zonal channel (Hidaka and Tsuchiya 1953; Fofonoff 1955), as a distorted Sverdrup Gyre (Stommel 1957; Veronis 1973; Baker 1982), and one in which mesoscale eddies produce the downward transport of surface stress (McWilliams et al. 1978; Johnson and Bryden 1989). However the FRAM not only showed that the Deacon Cell transferred the surface torque down to depths of below 1600m but it also emphasised the way in which the Kerguelen Plateau blocked the latitudes of Drake Passage at deeper

depths. In fact the northern limit of Kerguelen is such that it appeared to be one of the factors determining the position of the sub-tropical convergence in the South Atlantic sector.

These observations led Webb (1993) to propose a model of the Antarctic Circumpolar Current in which the current arose as the ocean's response to the return Ekman transport crossing an occluded zonal channel. The return Ekman transport is seen in figure 2 as the southward flowing branch of the Deacon Cell. In the north it can flow south as a western boundary current attached to South America but at the latitudes of Drake Passage it must become a western boundary current attached to the Kerguelen Plateau.

Webb developed a simple geostrophically balanced model of meridional flow in an occluded zonal channel which showed that the cross channel flow was only possible in the presence of a strong zonal jet. For Drake Passage and Kerguelen the Ekman return flow has a magnitude of about 25 Sv. If this figure is used with the model it predicts an Antarctic Circumpolar Current with a strength of 150 Sv.

This prediction agrees well with the observed transports of 130 Sv discussed earlier. The theory has the advantages that Drake Passage does not have to be treated in any special way and it does not require an unusually strong eddy field. The theory is also of interest because it helps explain the strength of the Falkland's Current and it emphasises the topographic importance of Kerguelen Plateau on the circulation of the Southern Ocean.

Conclusions

Field measurements, using hydrographic, ADCP and satellite data, together with model results are starting to give us a much better information about the boundary region between the South Atlantic and Southern oceans. There has also been some progress in understanding the ocean dynamics of the region but there are still many unanswered questions and the region remains a fertile area for new research.

Acknowledgements

Thanks to Dr Peter Saunders and Dr Brian King for allowing use of results from their WOCE A11 section.

References

Baker DJ (1982) A note on the Sverdrup balance in the Southern Ocean. J Mar Res 40(suppl.):21-26

Boyer DL, Chen R, Tao L, Davies PA (1993) Physical model of bathymetric effects on the Antarctic Circumpolar Current. J Geophys Res 98(C2):2587-2608

Bryden HL, Pillsbury RD (1977) Variability of deep flow in the Drake Passage from year-long current measurements. J Phys Oceanogr 7:803-810

Cox MD (1975) A baroclinic numerical model of the world ocean: preliminary results. In: Numerical models of ocean circulation Reid RO et al. (eds) National Academy of Sciences, Washington, D.C.

Davis R (1990) ALACE explores Antarctic Circumpolar Current. WOCE Newsletter, 10, 5

Davis RE, Killworth PD, Blundell, JR (1994) Comparison of ALACE and FRAM results in the South Atlantic. J Geophys Res (in press)

Defant A (1941) Die absolute Topographie des physikalischen Meeresniveaus und der Druckflächen, sowie die Wasser-Bewegungen im Atlantischen Ozean. Meteor-Werk 6(2(5)):191-260

Defant A (1961) Physical Oceanography. Pergamon Press, Oxford

Döös K, Webb DJ (1994) The Deacon Cell and the other meridional cells of the Southern Ocean. J Phys Oceanogr 24:429-442

Fandry C, Pillsbury RD (1979) On the estimation of the absolute geostrophic volume transport applied to the Antarctic Circumpolar Current. J Phys Oceanogr 9(3):449-455

Fofonoff N (1955) A theoretical study of zonally uniform flow. Brown University, Providence, R.I.

Gordon AL, Haxby WF (1990) Agulhas eddies invade the South Atlantic: evidence from Geosat altimeter and shipboard conductivity-temperature-depth survey. J Geophys Res 95(C3):3117-3125 & plates 3435-34336

Hidaka K, Tsuchiya M (1953) On the Antarctic Circumpolar Current. J Mar Res 12:214-222

Johnson GC, Bryden HL (1989) On the size of the Antarctic Circumpolar Current. Deep-Sea Res 36(1):39-53

Levitus S (1982) Climatological Atlas of the World Oceans. NOAA professional Paper. U.S. Government Printing Office, Washington, D.C. 173 pp

Lutjeharms JRE, Valentine HR (1988) Eddies at the subtropical convergence south of Africa. J Phys Oceanogr 18(5):761-774

Lutjeharms JRE, Webb DJ (1995) Modelling the Agulhas Current system with FRAM. (Unpublished manuscript)

Lutjeharms JRE, Webb DJ, de Cuevas BA (1991) Applying the Fine Resolution Antarctic Model (FRAM) to the ocean circulation around South Africa. South African Journal of Science 87(8):346-349

McWilliams JC, Holland WR, Chow JHS (1978) A description of numerical Antarctic Circumpolar Currents. Dynamics of Atmospheres and Oceans

2:213-291

Munk WH, Palmén E (1951) Note on the dynamics of the Antarctic Circumpolar Current. Tellus 3:53-55

Nowlin WD, Klinck JM (1986) The physics of the Antarctic Circumpolar Current. Reviews of Geophysics 24(3):469-491

Olbers D, Gouretski V, Seiss G, Schroter J (1992) Hydrographic Atlas of the Southern Ocean. Alfred Wegener Institute, Bremerhaven

Olsen DB, Evans RH (1986) Rings of the Agulhas Current. Deep-Sea Res 33(1):27-42

Quartly GD, Srokosz MA (1993) Seasonal variation in the region of the Agulhas retroflection: studies with Geosat and FRAM. J Phys Oceanogr 23(9):2107-2124

Reid JL, Nowlin WD (1971) Transport of water through the Drake Passage. Deep-Sea Res 18:51-64

Saunders PM, King BA (1994) Bottom currents derived from a shipborne ADCP on WOCE cruise A11 in the South Atlantic. (unpublished manuscript)

Saunders PM, King BA (1994) Oceanic fluxes on the WOCE A11 section. J Phys Oceanogr 25(9):1942-1958

Saunders PM, Thompson SR (1993) Transport, heat and freshwater fluxes within a diagnostic numerical model (FRAM). J Phys Oceanogr 22:452-464.

Semtner AJ, Chervin RM (1988) A simulation of the global ocean circulation with resolved eddies. J Geophys Res 93(C12):15502-15522

Semtner AJ, Chervin RM (1992) Ocean general circulation from a global eddy-resolving model. J Geophys Res 97(C4):5493-5550

Stommel H (1957) A survey of ocean current theory. Deep-Sea Res 4:148-184

Sverdrup HU, Johnson MW, Fleming RH (1942) The Oceans, their Physics, Chemistry and General Biology. Prentice-Hall, Englewood Cliffs.

Tchernia P (1980) Descriptive regional oceanography. Pergamon Press, Oxford.

The FRAM Group (Webb DJ et al.) (1991) An eddy-resolving model of the Southern Ocean. EOS 72(15):169, 174-175

Veronis G (1973) Model of the world ocean circulation. J Mar Res 31:228-288

Webb DJ (1993) A simple model of the effect of the Kerguelen Plateau on the strength of the Antarctic Circumpolar Current. Geophys Astrophys Fluid Dynamics 70:57-84

Webb DJ, Killworth PD, Coward AC, Thompson SR (1991) The FRAM Atlas of the Southern Ocean. Natural Environment Research Council, Swindon. 67 pp

Whitworth T; Peterson RG (1985) The volume transport of the Antarctic Circumpolar Current from three-year bottom pressure measurements. J Phys Oceanogr 15(6):810-816

Wüst G (1935) Die Stratosphäre. Deutsch Atl Exped „Meteor" 1925-27, 6(1).

Wüst G (1936) Schichtung und Zirkulation des Atlantischen Ozeans. Meteor-Werk, 6(1)

Wüst G (1950) Blockdiagramme der Atlantischen Zirculation auf Grund der „Meteor"-Ergebnisse. Kieler Meeresforschungen 7(1):24-34

Antarctic Intermediate Water in the South Atlantic

L.D. Talley

Scripps Institution of Oceanography
La Jolla, CA 92093-0230 USA

Abstract: Maps of the Antarctic Intermediate Water (AAIW) in the Atlantic, and on a global isopycnal which intersects the AAIW in the south, show the location and properties of the salinity and oxygen extrema associated with the AAIW, and the likely sources of AAIW. These are primarily the surface waters in the southeastern Pacific, which produce the South Pacific AAIW, and surface waters in northern Drake Passage and the Falkland Current loop, which produce the South Atlantic AAIW. This latter source is the primary one for AAIW of the Indian Ocean as well. Winter surface properties and annual-averaged Ekman pumping and Sverdrup transport for the southern hemisphere suggest that the formation density of the AAIW is the highest density which can be subducted in the South Pacific. The higher density of AAIW in the South Atlantic may result from more complex processes. The connection between the subtropical gyres of the Atlantic and Indian and between the Indian and Pacific Oceans contributes to modification of AAIW as it spreads tortuously northward around the subtropical gyres. Potential vorticity and AAIW salinity and oxygen illustrate the near barrier between the subtropical and tropical regimes, at about 20° to 25° north and south of the equator. Communication between the regimes is primarily through the western boundary currents.

Introduction

The intermediate layer originating from near the sea surface near the Antarctic Circumpolar Current (ACC) and extending northward through the Atlantic, Pacific and Indian Oceans has long been identified by the salinity minimum lying at about 800 to 1000 meters, and in more local regions by oxygen, silica and potential vorticity extrema. The earliest salinity minimum observations are from the Challenger expedition in the last century (Buchanan 1877). Merz and Wüst (1922) presented a more complete meridional section of salinity and summarized previous work and concluded that the low salinity layer contradicted hypotheses of upwelling from the deep ocean to the surface near the equator.

Wüst's (1935) meridional section of salinity and his core layer maps based on the *Meteor* data also clearly show the subtropical circulation of AAIW, intensification of flows along the western boundary, northward penetration of AAIW in the western boundary current in the tropics, and eastward penetration near the equator. Taft (1963) deduced from maps of salinity and oxygen for the globe on an isopycnal intersecting the AAIW that the southeast Pacific and South Atlantic are the sources of AAIW, with no source in the Indian Ocean. He showed that in the South Atlantic most of the AAIW from the southwestern corner spreads northward around the subtropical gyre rather than directly northward along the western boundary. He pointed out the important boundaries in AAIW properties near 20°S in all oceans.

In the Atlantic and Pacific, the intermediate and bottom layers, which are of Antarctic origin, flow northward on average while the deep water between them flows southward, as first suggested by Merz and Wüst (1922). AAIW is part of the return of upper layer water to the northern North Atlantic and

From WEFER G, BERGER WH, SIEDLER G, WEBB DJ (eds), 1996, *The South Atlantic: Present and Past Circulation.* Springer-Verlag Berlin Heidelberg, pp 219-238

is one of the sources of North Atlantic Deep Water (NADW). At 4°to 5°C, AAIW is too cold to furnish directly the warm portion of the observed northward heat transport in the central South Atlantic (Gordon 1986). However, Rintoul (1991) suggests that part of the AAIW originating in the southwest Atlantic upwells due to net surface heat gain in the southern Atlantic, and that thus the formation and circulation of this intermediate layer might be important for the NADW cell. AAIW is also the dense end-member of a shallow overturning circulation in which surface layers north of the ACC lose buoyancy and sink. A test of whether numerical models include the proper physics to reproduce the circulation is whether they successfully produce these large-scale layers. England et al. (1993) show the importance of surface salinity near Antarctica in enabling a numerical model to produce Antarctic Bottom Water and a reasonable AAIW layer.

The following is a summary of some results concerning AAIW with emphasis on the Atlantic, drawn from a larger work in preparation which treats the global distribution. AAIW should not be defined narrowly in terms of property extrema; rather it should be defined as a coherent layer which originates near the ACC. Based on all properties the complete AAIW layer, as distinguished from the overlying surface layer and the underlying NADW/Upper Circumpolar Water, could be taken as 27.0-27.4σ_θ (31.6-32.0σ_1) or possibly to about 0.1σ_θ denser. Nevertheless the primary emphasis here is on identification of the extrema which characterize AAIW because of what they reveal about its sources, and about properties on an isopycnal which intersects the AAIW at its sources near Drake Passage. A core layer approach is used, with full understanding that these are not surfaces of flow. (Neither are isopycnal/neutral surfaces, although locally they are more likely to be such.) A partial discussion of the formation of AAIW follows in the last section, including the relationship between winter surface properties in the southern hemisphere and the winds.

The principal South Atlantic fronts are major boundaries in AAIW properties. I follow Peterson and Stramma's (1991) and Tsuchiya et al.'s (1994) nomenclature. The most important are the Polar

and Subantarctic Fronts, defining the ACC. The Polar Front (about 2°C at the surface) is the northern boundary of the near-surface temperature minimum, and the Subantarctic Front (about 4°C at the surface) is the southern boundary of the AAIW at most longitudes. A maximum horizontal density gradient is the best indicator of both. North of the Subantarctic Front lies the Subantarctic Zone, which is bounded to the north by the Subtropical Front. In the Subantarctic Zone is found Subantarctic Mode Water (SAMW) (McCartney 1977; 1982), a thick near-surface layer with temperature decreasing towards the east around the ACC. McCartney (1977) concluded that SAMW is the primary precursor of AAIW. North of the Subtropical Front is a lighter type of SAMW (Tsuchiya et al. 1994, for the southwestern Atlantic). This is bounded to the north by the Brazil Current Front in the west (Tsuchiya et al. 1994) and the Benguela Current Front in the east (Gordon et al. 1992). These might be considered to be a single front (BCF). North of the BCF is a front which marks the northern boundary of the subtropical circulation. Tsuchiya et al. (1994) call this the Subtropical-Subequatorial Front (STSEF) for want of a better term. It is the deep, poleward-shifted expression of the surface Angola-Benguela Front. Frontal structure which mirrors this pattern is found in the North Atlantic.

AAIW core layer and isopycnal properties

Data. The high quality discrete bottle dataset used in various studies by Reid (1986; 1989; 1994) plus other bottle and CTD data sets collected in the 1980's and 1990's in the Atlantic and Pacific were used for this study. Vertical extrema of salinity, oxygen, and silica were determined from the discrete bottle data. To map salinity and isopycnic potential vorticity, $Q = (f / \rho)\delta \rho / \delta z$ (where f is the Coriolis parameter and relative vorticity is ignored), the station data were interpolated to isopycnals using an Akima cubic spline which avoids introducing spurious vertical extrema. Densities were calculated relative to the sea surface (σ_θ) or 1000 dbar (σ_1). Vertical stability, using the pressure between isopycnals as a reference, was used to compute Q.

The isopycnal chosen for display is $31.7\sigma_1$, which corresponds closely with $27.1\sigma_\theta$. It characterizes the top of the AAIW layer in the Atlantic/Indian and the middle of it in the Pacific. It was chosen for global mapping as it better illustrates AAIW formation than a slightly denser isopycnal which would better characterize the AAIW core layer at midlatitudes and in the tropics.

AAIW extrema (Figs. 1 and 2). The vertical section of salinity along about 25°W from South Georgia Island to Iceland (Fig. 1a) is the modern equivalent of one of Wüst's long meridional sections. Tsuchiya et al. (1992; 1994) describe the water masses on this section. The principal salinity layers identified by Merz and Wüst (1922) and Wüst (1935) are clearly recognized. The main AAIW salinity minimum at 25°W extends northward from the Subantarctic Front (SAF) at 45°S. AAIW of salinity less than 34.3 psu is found as far north as the Brazil Current Front, at 32°S where the low salinity core reaches its maximum depth The minimum can be traced across the equator, ending at about 21°N (here the section lies east of the Mid-Atlantic Ridge). Properties on a Greenwich meridian section (Fig. 13b in Reid 1989) are very similar (Whitworth and Nowlin 1987).

The AAIW is also a vertical oxygen maximum in the South Atlantic's subtropical gyre and in a small region of the tropics (see below) indicating relatively recent surface origin. Wüst (1935) showed an AAIW oxygen maximum reaching the equator along the western boundary, but only to about 20°S in the central and eastern Atlantic. Along 25°W (Fig. 1b) the AAIW oxygen maximum is found between the Subtropical Front and the STSEF at about 20°S (Tsuchiya et al. 1994). An AAIW oxygen maximum is absent south of the Subtropical Front due to the high oxygen of the overlying SAMW. Oxygen at the maximum between the Brazil Current Front and the STSEF is greatly reduced compared with south of the Brazil Current Front.

Scatter plots (Fig. 2) of all vertical salinity minima, oxygen maxima and silica maxima in the Atlantic are based on the broadly distributed bottle data set used in Figs. 3-7. The AAIW salinity minimum clusters around 31.8 to $31.9\sigma_1$ between 40°S and 20°N. South of about 40°S the salinity

minima are more scattered - these are in the surface layer above the temperature minimum. North of about 15°N, the density increases from about 31.85 to $32.0\sigma_1$, and north of about 25°N the salinity minimum layer becomes indistinct. (The cluster in the north at about $32.4\sigma_1$ is the Labrador Sea Water.) The densest salinity minima are the Lower Circumpolar Water relative to the overlying saline NADW. The characteristic density for the Atlantic AAIW salinity minimum thus is approximately $31.8\sigma_1$ south of about 25°S. North of this it is denser, about 31.85 to $31.9\sigma_1$; this corresponds to about $27.3\sigma_\theta$, used in Suga and Talley (1994) to study the tropical circulation of AAIW. Reid (1989; 1994) selected $31.938\sigma_1$ which also is more typical of the tropics and North Atlantic. The main locus of the salinity minimum in the Indian Ocean is $31.8\sigma_1$; in the Pacific Ocean's subtropical gyre it is $31.7\sigma_1$ (not shown).

An AAIW oxygen maximum is apparent south of about 25°S, where the salinity minimum is less dense. The oxygen maximum is slightly less dense than the salinity minimum; this is much more apparent when actually mapping the two extrema. The arc of high oxygens north of 40°N is the Subpolar Mode Water there, and the layer at $32.4\sigma_1$ is the Labrador Sea Water. The densest high oxygens mark the North Atlantic Deep Water.

A silica maximum is evident at densities much higher than AAIW in the South Atlantic's subtropical gyre; this is actually the Upper Circumpolar Water and not the AAIW. This layer is truncated from below in the tropics by the North Atlantic Deep Water, and the silica maximum in the tropics is closely associated with AAIW. The clear silica maximum in this zonally-undifferentiated view ends at 25-30°N, although it can be traced on individual sections far to the north (Tsuchiya, 1989), unlike the salinity minimum. The densest silica maxima mark the Lower Circumpolar Water.

Salinity minimum (Fig. 3a and 4) and isopycnic salinity. The AAIW salinity minimum is bounded to the south by the Subantarctic Front, where isopycnals plunge steeply from 100-200m to greater than 1000 m in the western subtropical South Atlantic. The minimum terminates in the North Atlantic between 20-25°N, with a rather ill-defined

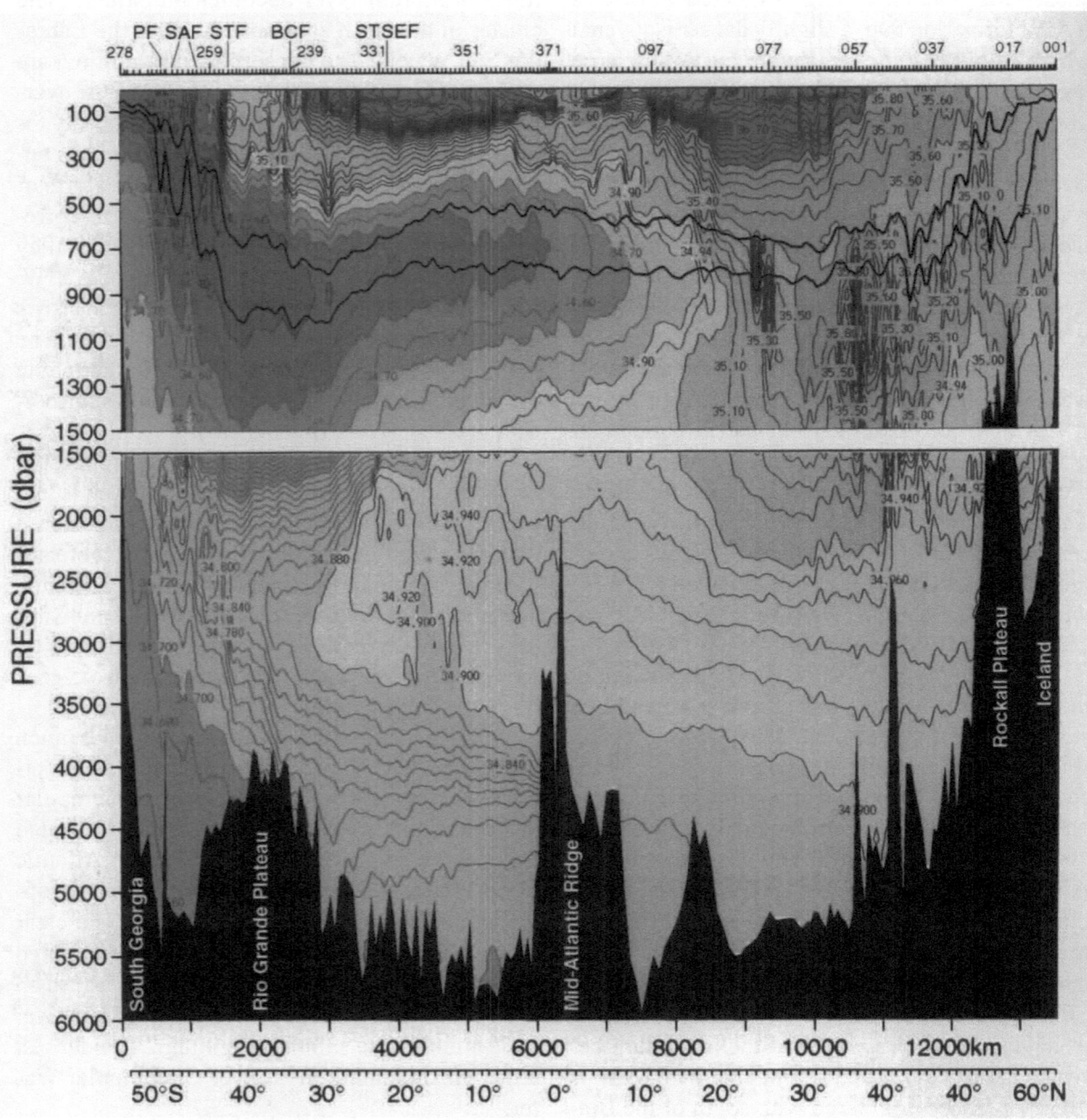

Figure 1. (a) Salinity and along approximately 25°W, from South Georgia Island to Iceland, from 1988-1989. The two curves which pass through the AAIW are the 31.7 and 31.9 σ_1 isopycnal contours. The North and South Atlantic portions were described in Tsuchiya et al. (1992, 1994) respectively. Annotations above the section indicate the Polar Front (PF), Subantarctic Front (SAF), Subtropical Front (STF), Brazil Current Front (BCF), Subtropical-Subequatorial Front (STSEF). CTD data were smoothed with a Gaussian with an 11 dbar half-width and then optimally mapped to a 20km x 40dbar grid for the deep sections and 20km x 10dbar for the shallow sections.

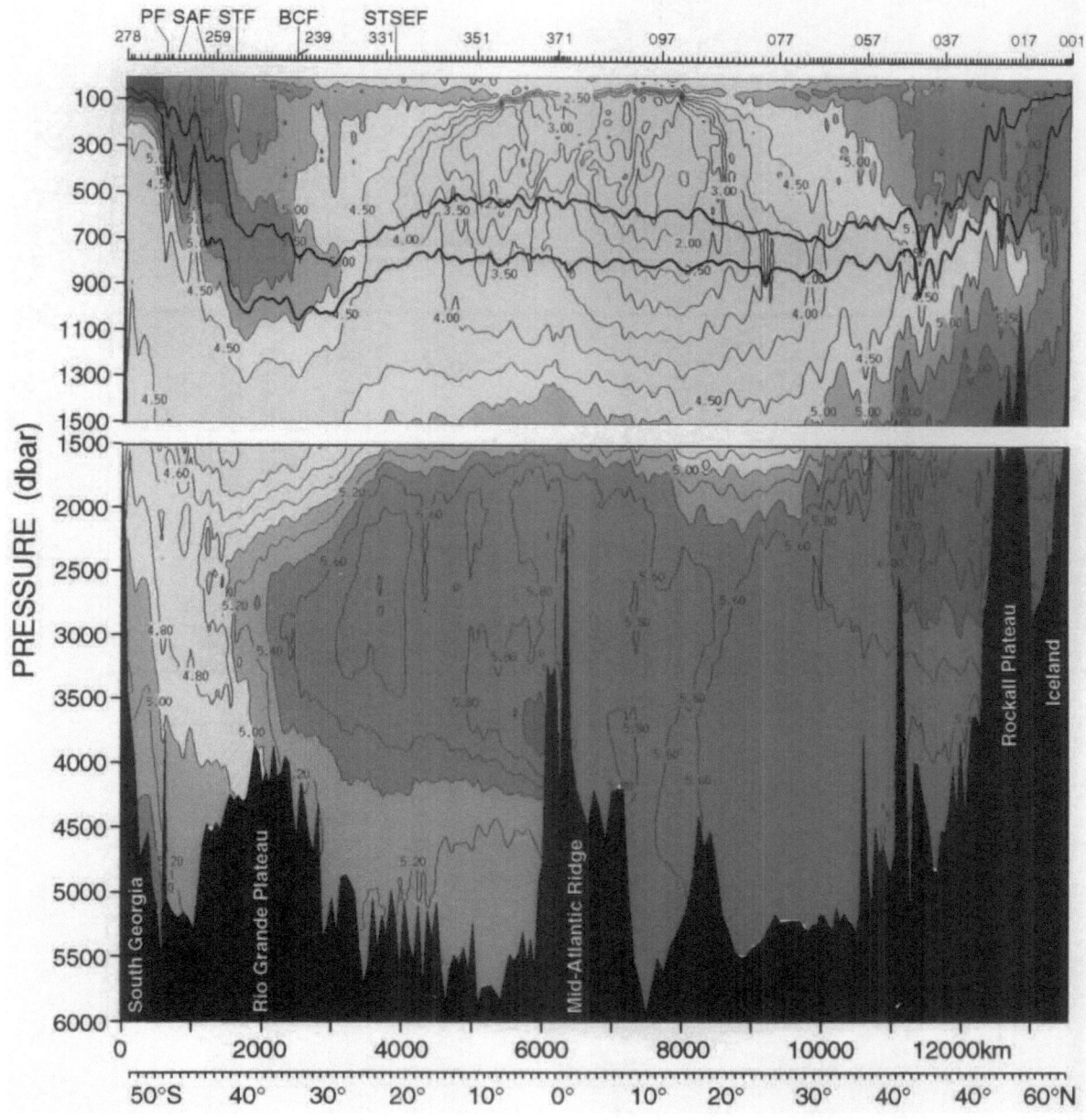

Figure 1. (b) oxygen (ml/l) along approximately 25°W, from South Georgia Island to Iceland, from 1988-1989. See fig. 1. (a) for details.

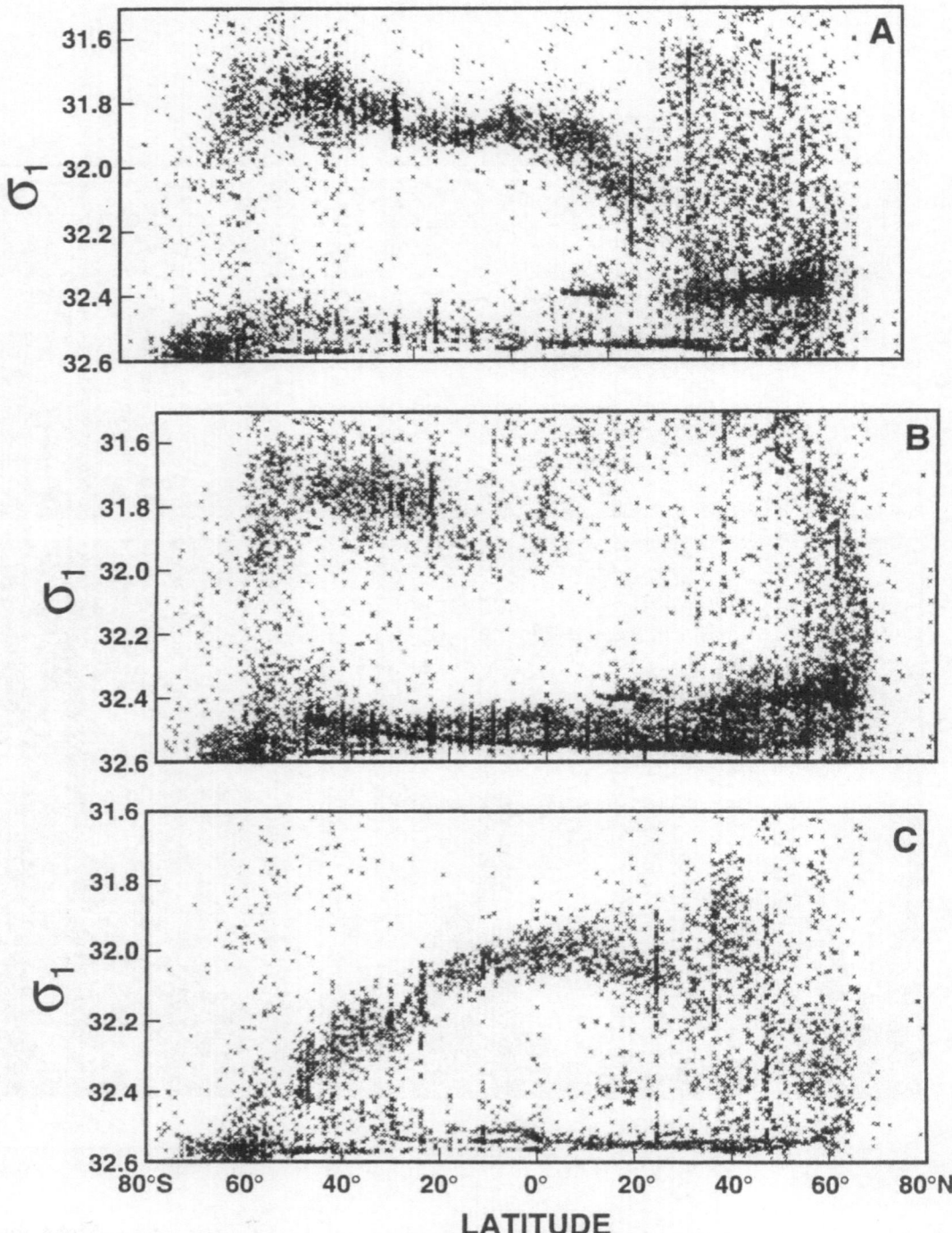

Figure 2. (a) All salinity minima in the potential density anomaly range 31.6 to 32.6 σ_1, referenced to 1000 dbar. **(b)** All oxygen minima in the same potential density anomaly range. **(c)** All silica maxima in the same potential density range. Station distribution is evident from the maps of Figs. 2-6 . The data were not interpolated; all vertical extrema were selected.

boundary where it meets the Mediterranean Water in the eastern Atlantic. The depth distribution is similar to Wüst's (1935), with the modern data adding a shallow ridge in the Benguela Current and extending northwestward near 20°S. The minimum is remarkably flat in the tropics (750± 50m). North of 20°N the core deepens to 1000m as the Mediterranean Water chews into the top of the salinity minimum. AAIW density is less than $31.8\sigma_1$ in Drake Passage and around the subtropical gyre, with an isolated region of denser minima in the central subtropical gyre centered at 40°S. Density increases northward and is nearly uniform between 20°S-20°N, at $31.85\text{-}31.95\sigma_1$. At the northern boundary it increases to greater than $32.0\ \sigma_1$.

Salinity at the minimum increases monotonically northward, being less than 34.2 psu in the south and Drake Passage. Agulhas salinities are greater than 34.4 psu, with no obvious connection to the 34.4 psu minimum found north of 35°S. In the tropics, salinity is lower in the western boundary current. Along the equator, salinity increases eastward, connecting with saline water in the tropical cyclonic gyre (Gordon and Bosley 1991; Suga and Talley 1994). Salinity increases to more than 35.1 psu at the northern boundary. Oxygen at the salinity minimum core mirrors salinity in many important aspects, indicating ventilation in the south, an increase in age northward, a tongue of high oxygen at the western boundary in the tropical South Atlantic, and an eastward tongue of high oxygen along the equator. Low oxygen is found in the usual tropical locations at the eastern boundary. Oxygen increases towards the north in the North Atlantic, because the saline water which mixes with the southern hemisphere water outcrops in the northern North Atlantic (see Tsuchiya 1989; Reid 1994).

The major features of the salinity and oxygen distribution at the AAIW salinity minimum were already quite clear in Wüst's (1935) presentation. However, it is useful to see that our modern and much larger data set does not contradict these features, and provides sharpened focus for important features. These include: (1) a southern boundary which is essentially the Subantarctic Front,

(2) continuity of the southernmost AAIW salinity minimum with that in Drake Passage or with the thick surface layer north of the Subantarctic Front in Drake Passage, (3) possible continuity of mid-latitude AAIW with that of the Agulhas retroflection, (4) the narrow, northward western boundary current in the tropics, eastward flow in the equatorial zone (although possibly not right on the equator - Suga and Talley 1994), (5) northward spreading in the North Atlantic in a less well-defined western boundary current, (6) the presence of AAIW in the Caribbean and Gulf of Mexico, the northern boundary along about 25°N, and (7) the elimination of the southern source signatures by age, mixing with Indian Ocean AAIW, mixing with overlying saline waters, and mixture with northern North Atlantic waters which increase both the salinity and oxygen.

Salinity on isopycnals which characterize AAIW (Fig. 4) has a similar pattern to that in the various core layers (salinity, oxygen extrema) and also on the denser isopycnals mapped in Reid (1989) and Suga and Talley (1994), so only global features and Atlantic features which differ from the core layer description (Fig. 3a) are mentioned here. In the Atlantic equatorial region, relatively low salinity spreads eastward at about 5°S, as also seen at $27.3\sigma_\theta$ ($31.9\sigma_1$) in Suga and Talley (1994). The high salinity near the eastern boundary at 10°S, in the cyclonic gyre (Taft 1963; Gordon and Bosley 1991; Suga and Talley 1994) is isolated, unlike at the salinity minimum (Fig. 3a) or on the $31.938\sigma_1$ surface shown by Reid (1994). The isolated high salinity indicates the presence of vertical mixing, as noted in Gordon and Bosley (1991) and Suga and Talley (1994). In the North Atlantic, the saline Mediterranean and fresh Labrador and GIN Sea influences are notable. A large meridional gradient occurs between 10° and 20°N, separating the AAIW and Mediterranean Water.

In the Indian Ocean, the overall salinity is higher than in either the Atlantic or Pacific Oceans. The salinity of the Agulhas is higher (> 34.5 psu) than in the western subtropical gyres of either the South Atlantic or South Pacific. The high salinity apparently comes from the tropics. The dominant salinity in the subtropical region is between 34.4

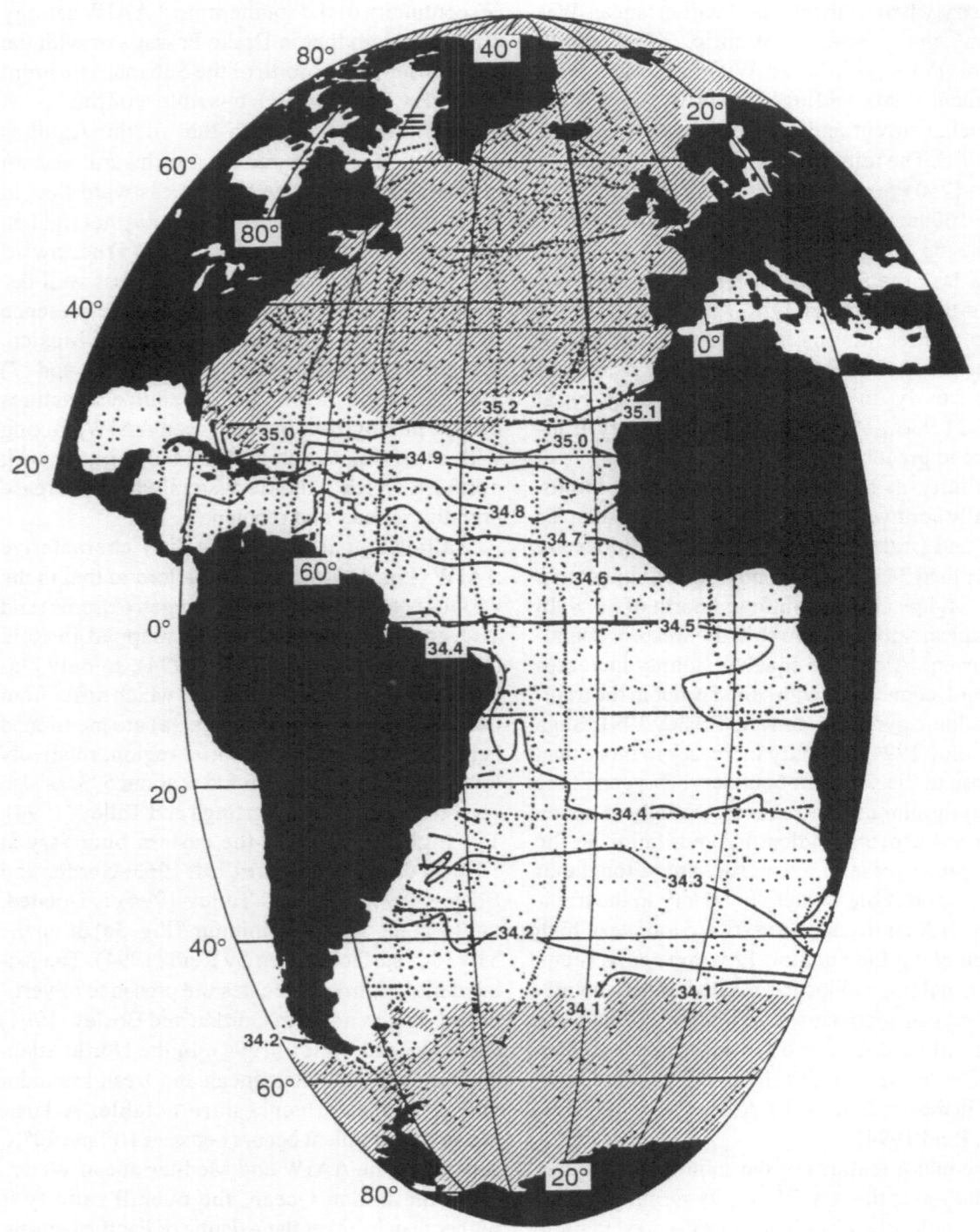

Figure 3. (a) Salinity at the AAIW salinity minimum. The southern boundary is taken to be the 200 m depth contour for the salinity minimum, and coincides well with the Subantarctic Front. The northern boundary is the edge of the uniform occurrence of the salinity minimum; to the north, minima in the desired density range occur at some stations. The northern boundary coincides roughly with the change from tropical to subtropical regimes evident in the potential vorticity distribution (Fig. 5).

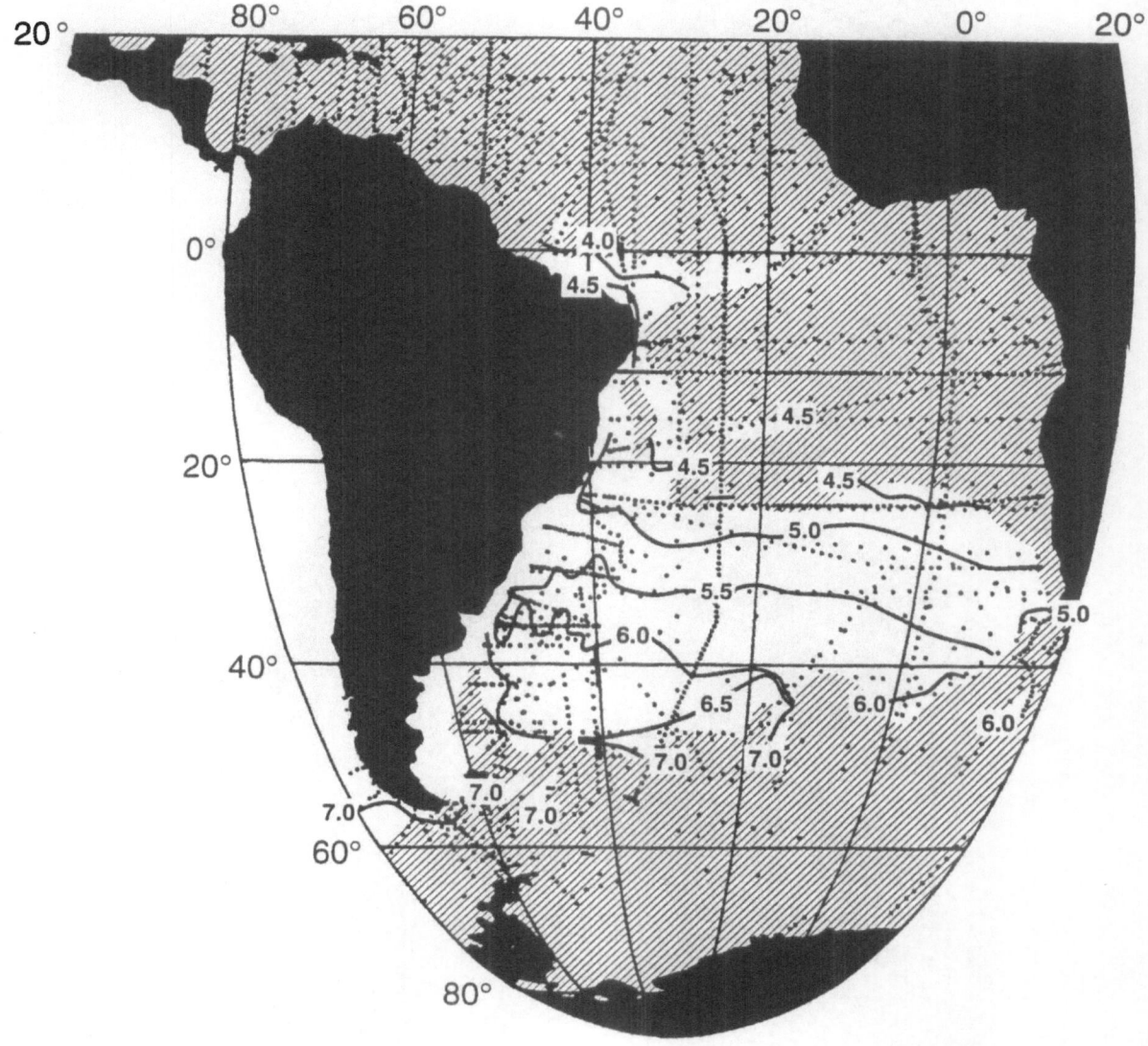

Figure 3. (b) Oxygen (ml/l) at the AAIW oxygen maximum. The southern boundary is somewhat aligned with the Subtropical Front. The northern boundary across most of the South Atlantic is the Subtropical-Subequatorial Front, which corresponds with the change from tropical to subtropical regimes in the potential vorticity distribution (Fig. 5).

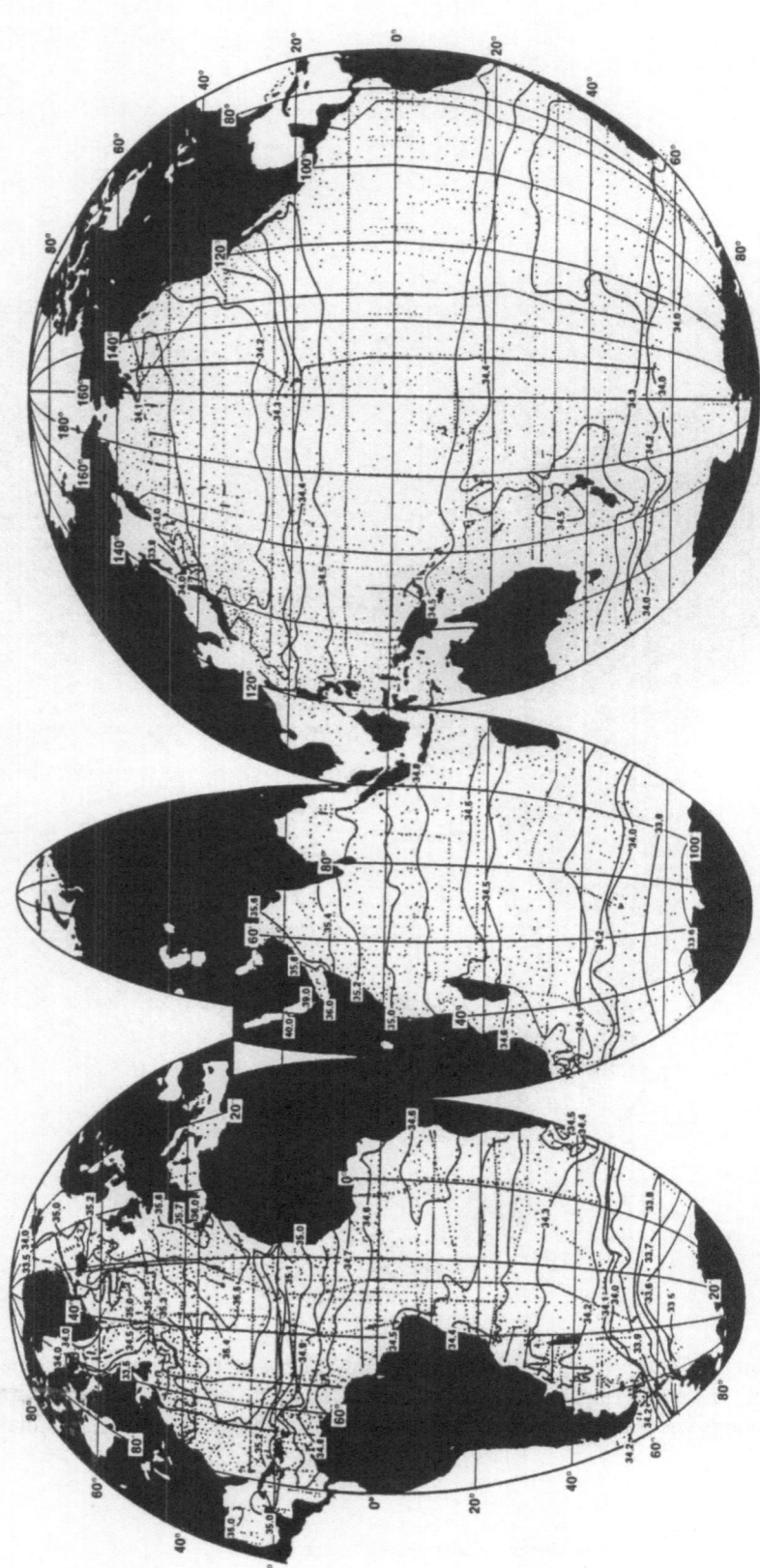

Figure 4. Salinity at $31.7\sigma_1$ for the globe, based on discrete bottle data inter-polated to the isopycnal using an Akima cubic spline.

and 34.5 psu, and is continuous with the region south of Australia and the Tasman Sea. The Red Sea and Arabian Sea are the most saline regions. The AAIW salinity minimum lies below $31.7\sigma_1$ throughout the Indian Ocean.

In the South Pacific, salinity at $31.7\sigma_1$ shows the clear signs of subduction in the eastern subtropical gyre: the low salinity tongue penetrates northward and counterclockwise around the gyre. In the follow-on study of the global AAIW, it will be shown that the $31.7\sigma_1$ isopycnal outcrops just south of the Subantarctic Front in the southeastern Pacific, and that this is located north of the climatological zero of wind stress curl, and hence in the subtropical gyre. Salinity on this isopycnal in the Pacific increases towards the equator, where it is highest, and then decreases to the north, due to the influence of the fresh water of the northern North Pacific and especially the Okhotsk Sea.

Oxygen maximum and isopycnic oxygen (Fig. 3b and Fig. 5). In the South Atlantic the oxygen maximum associated with AAIW is found in the subtropical gyre and extending northward along the western boundary and eastward along the equator. As shown above for 25°W (Fig. 1a), its area is much more restricted than the salinity minimum. Its northern boundary is the STSEF at about 20°S and it penetrates into the tropics only along the western boundary. It is found penetrating weakly into the Indian Ocean between 35°S and the SAF, with oxygen values decreasing monotonically eastward from the southwestern South Atlantic. The Agulhas does not contain an AAIW oxygen maximum. This suggests that the Indian Ocean maximum is advected from the South Atlantic rather than being local (Taft 1963).

The oxygen maximum is about 100 meters shallower than the salinity minimum in the subtropical gyre. Its density is thus lower than the salinity minimum density, being less than $31.7\sigma_1$ in the southeastern Pacific, Drake Passage, and along the western boundary to 38°S. In most of the subtropical gyre it is between 31.7 and $31.8\sigma_1$, and increases to greater than $31.8\sigma_1$ in the tropical western boundary region.

Oxygen at $31.7\sigma_1$ is displayed for the Atlantic Ocean only (Fig. 5). Very high oxygen (> 7.5 ml/l) occurs where the isopycnal lies close to the surface south of the Polar Front, with a reduction to about 6 ml/l through the Falkland loop, and high meridional gradients across the Polar Frontal Zone east of 40°W. Oxygen decreases towards the north. A split of the westward flow between 20° and 30°S occurs at the western boundary, and a tongue of high oxygen extends northward along the western boundary and then eastward near the equator. This eastward tongue appears to be more centered on the equator than at slightly higher density, such as the $27.3\sigma_\theta$ in Suga and Talley (1994). The eastern tropical minima are obvious both north and south of the equator.

Warner and Weiss (1992) showed no measurable chlorofluoromethanes in the eastern tropical regions, coincident with the very low oxygen regions. Thus low oxygen in the tropics is due to long residence time rather than especially high consumption. They also showed high oxygen along the western boundary and eastward along the equator, with the latter slightly displaced to the south.

In the northern hemisphere, the highest oxygen values are at the western boundary and decrease northward, reaching a minimum at 10°N and through the Caribbean. Oxygen is high in the subpolar North Atlantic because this isopycnal outcrops; the high values enter the subtropical gyre in the east, and increase the western boundary oxygen somewhat. The Gulf Stream is an axis of low oxygen, although at 3.5 ml/l, it is higher than where the water from the South Atlantic crosses the equator. Relatively high silica (Tsuchiya, 1989) in the Gulf Stream is due to South Atlantic influence, suggesting that at part of the reason for low Gulf Stream oxygen is South Atlantic water, although tropical North Atlantic water also has low oxygen.

Silica maximum. In the North Atlantic, far from the Antarctic source and beyond the boundary of the recognizable salinity minimum, AAIW can be tagged by a silica maximum (Tsuchiya 1989). In the South Atlantic subtropical gyre the silica maximum is identified with Upper Circumpolar Water which lies below the AAIW; its silica is enriched over that of the North Atlantic Deep Water by admixture of silica-rich Pacific

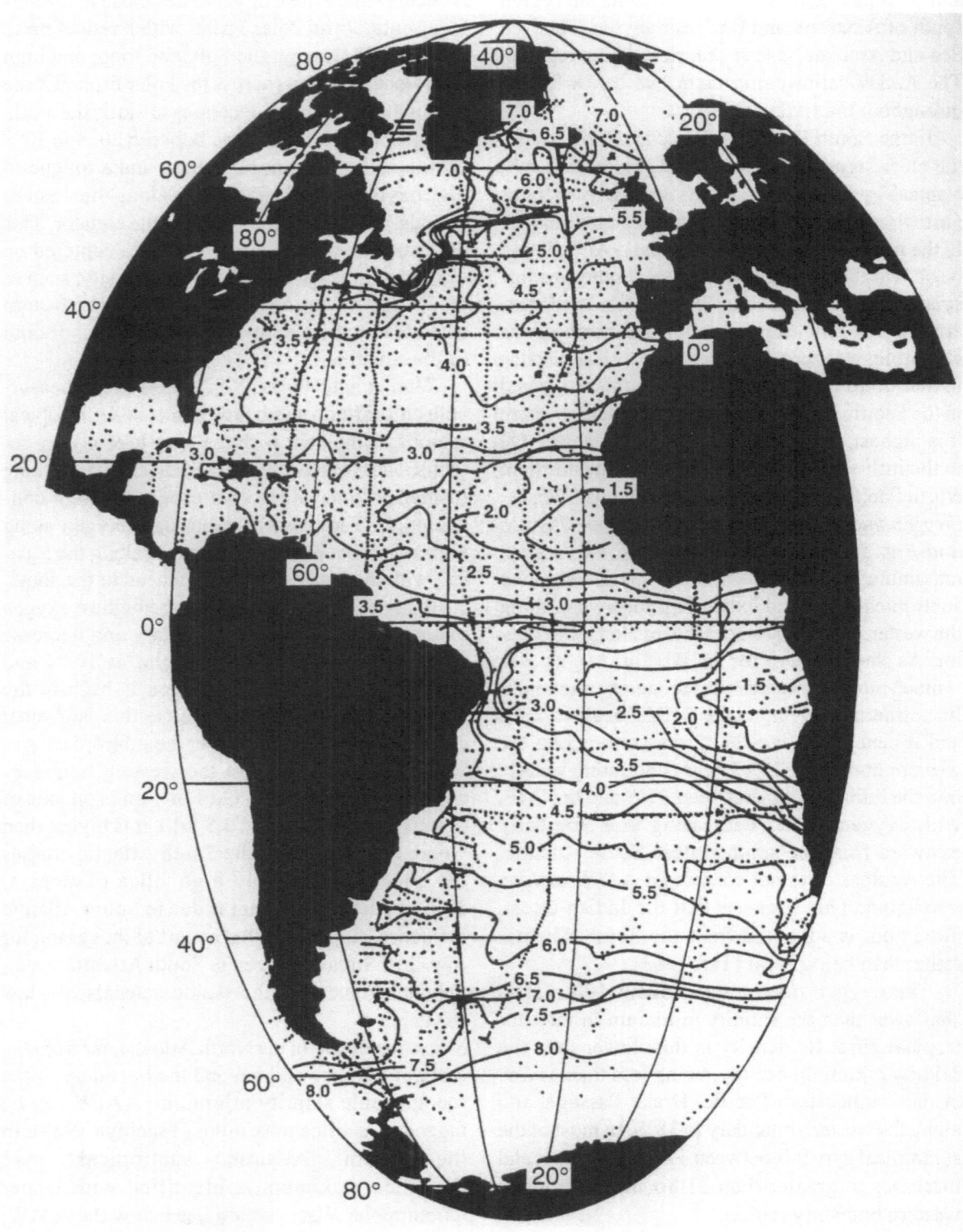

Figure 5. Oxygen (ml/l) at $31.7\sigma_1$ for the Atlantic Ocean, as in Fig. 3.

and Antarctic waters. Its silica is higher than that of AAIW since the latter originates from silica-poor surface waters. In Tsuchiya et al. (1994), it was shown that the Upper Circumpolar Water is truncated from below by NADW north of the STSEF at about 20°S; to the north the density of the silica maximum is only slightly higher than that of the salinity minimum (Fig. 2). It appears on virtually every station in the tropics and sporadically on stations at higher northern latitudes, being apparent even on the zonal section at 48°N. Tsuchiya (1989) mapped the silica as a lateral maximum (on an isopycnal) far northward, showing the southern influence on the intermediate waters in the North Atlantic's subpolar gyre.

Potential vorticity (Q) minimum and isopycnic Q (Fig. 6). The AAIW core can be identified as a weak stability/potential vorticity minimum in the southwestern South Atlantic (Tsuchiya et al. 1994), where isopycnic potential vorticity is $f/\rho \, \delta\rho/\delta z$, and ρ is the locally-referenced potential density. A more persistent AAIW-related potential vorticity signature in the South Atlantic is its position at the top of the low potential vorticity of the deep ocean. Thus it marks the base of the main pycnocline. The strongest AAIW Q minimum is just on the northern side of the SAF in the western region; it underlies a stronger near-surface Q minimum associated with SAMW (McCartney 1977; 1982). In the Polar Frontal Zone, between the Subantarctic and Polar Fronts, the AAIW density is that of the surface outcrop in winter, marked by the potential vorticity minimum which is closest to the sea surface. Extremely low potential vorticity at AAIW densities is found in northern Drake Passage and along the western boundary, suggesting convective overturn (McCartney 1982) enhanced by the baroclinicity of these strong currents. The AAIW Q minimum even in these regions is however much weaker than in the southeastern Pacific, where AAIW is associated with a strong and spatially-persistent potential vorticity minimum (McCartney and Baringer 1993; Talley, in preparation).

East of the mid-Atlantic Ridge, the northern boundary of the Q minimum is the Benguela/Brazil Current Front (BCF), which supports the suggestion that Indian Ocean AAIW is entering the northern subtropical gyre and moving westward (Taft 1963; Gordon et al. 1992). At the western boundary, the BCF is about 5° south of the northern limit of the potential vorticity minimum. North of 20°S, weak potential vorticity minima reappear, but identification as AAIW is not clear.

The global map of potential vorticity at $31.7\sigma_1$ (Fig. 6) is based on individual station data rather than a smoothed average, and thus reveals more features and more detail than Keffer's (1985) map for the layer 27.0-27.3σ_θ (roughly 31.6-31.9σ_1), based on Levitus (1982) data. Keffer showed the dominance of the β-effect in the tropics, where $\beta = \delta f / \delta y = 2\Omega\cos\theta / R$ where θ is latitude and R is the earth's radius. Keffer showed homogenization of potential vorticity in the North Pacific, and also the South Pacific and Indian Oceans. He drew attention to the signature of ventilation in the North and South Atlantic subtropical gyres. Fig. 6 shows the same strong contrast between the β-dominated tropics and the subtropics. On the other hand, *only* the North Pacific has homogenized Q. Water of this density subducts from the sea surface in the subtropical regions of the other oceans; potential vorticity variations along the isopycnal's surface outcrop lead to subsurface variations and hence no overall homogenization.

The lowest Q outside the tropics is on either side of South America, around Drake Passage. The lowest is over Burdwood Bank ($< 2\times10^{14}$ cm^{-1}sec^{-1}), with the South Atlantic tongue extending northward in the Falkland loop and to the east between 40° and 45°S just north of the SAF. High potential vorticity appears to stream westward in the northern part of the gyre, originating from the Agulhas retroflection at the tip of Africa, and suggesting that there is indeed at least some Indian Ocean influence in the South Atlantic. The high Q tongue is separated from the β-dominated tropics by high meridional gradients.

West of Drake Passage is the other region of relatively low Q. The 31.6σ_1 outcrops here and has very low potential vorticity, creating the South Pacific SAMW. The 31.7σ_1 might not outcrop as vigorously. Nevertheless it has low potential vorticity ($< 6\times10^{14}$cm^{-1}sec^{-1}) in the southeast which

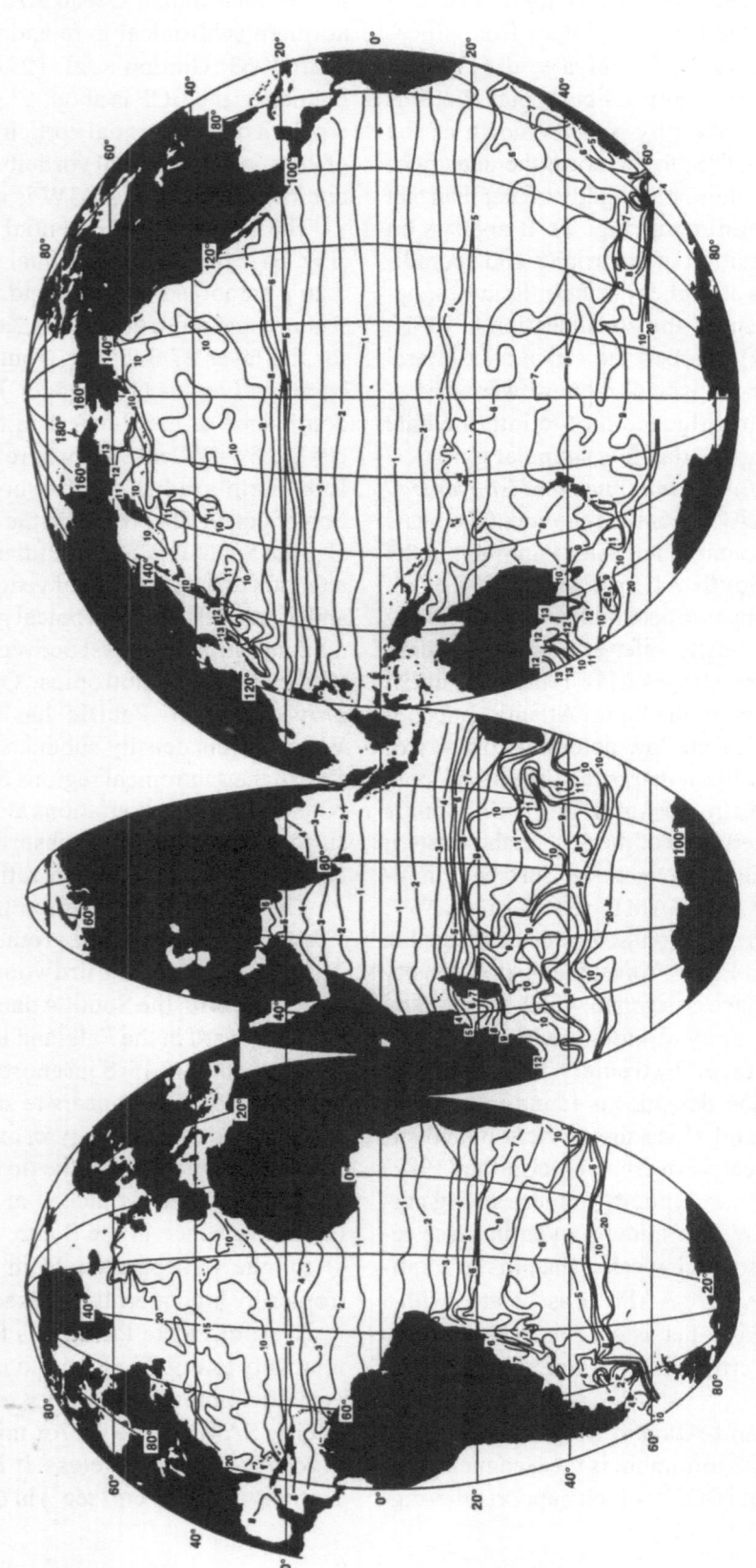

Figure 6. Isopycnic potential vorticity (10^{14}cm^{-1}sec^{-1}), ignoring relative vorticity, at $31.7\sigma_1$ for the globe. (It is negative in the northern hemisphere.)

sweeps around the outer side of the subtropical gyre, and is separated from the β-dominated tropics by a high meridional gradient at about 20°S. The low potential vorticity is thus a signature of subduction of maximum density surface waters near the SAF. High potential vorticity enters from south of Tasmania and affects the southern and western parts of the subtropical gyre. A patch of high Q occurs at the eastern boundary at 30°S; this could be the eastern shadow zone of the subtropical gyre at this density.

Potential vorticity in the Indian Ocean subtropical gyre is overall higher but of a nearly identical pattern to the South Atlantic. Relatively low potential vorticity enters from the South Atlantic just north of the ACC. It branches into a core which enters the tight western part of the subtropical gyre near 70°E, where it corresponds to a high freon patch in the AAIW (Fine 1993), and an eastward core north of the SAF. Another tongue of relatively low potential vorticity extends northward into the subtropical gyre at 90° to 100°E. Along the northern side of the Indian Ocean's subtropical gyre, high potential vorticity streams westward from the southern side of Australia. This creates a higher meridional Q gradient bordering the tropics than in the other oceans. The highest potential vorticity in the southern hemisphere north of the ACC is in the South Australia Basin along the southern side of Australia. A constriction in the pattern south of Tasmania and New Zealand suggests that only the southern eastward limb of the anticyclonic flow can pass through.

Circulation

The geostrophic circulation in the South Atlantic at $31.7\sigma_1$ relative to 3000 dbar is represented by the pressure anomaly streamfunction (PAS) (Fig. 7) (Zhang and Hogg 1993), which is slightly more accurate than the acceleration potential. This reference pressure is deep enough to permit fairly accurate depiction of the flows, which have much in common with Suga and Talley's (1994) for the $27.3\sigma_\theta (\sim 31.9\sigma_1)$, and Reid's (1989; 1994) for the $31.938\sigma_1$. Figure 6 lies in complexity between those two, being entirely a matter of taste in hand

contouring. The most vigorous flows are the ACC in Drake Passage, the Brazil and Falkland Currents at 35° to 50°S, the Agulhas retroflection, and the Gulf Stream. Between 25°S and 25°N, there is virtually no relief to the surface: all values lie between 1.2 and 1.3 dyn m. (This is not to say that the circulation is much weaker, since the vertical shear is 1/f times the lateral PAS gradient where f is the Coriolis parameter.)

The subtropical gyres of both hemispheres are clearly delineated. In the South Atlantic the southward overshoot of the Brazil Current at the western boundary is clearly apparent with the looped Falkland Current lying offshore. The PAS contrast across the Brazil Current is strongest south of 35°S, as shown by Zemba (1991); this major strengthening of the Brazil Current towards the south is likely related to the Sverdrup transport (Godfrey 1989) which is strongly affected by the southernmost location of Africa. The Agulhas retroflection is almost as strong as the Brazil Current, with a contrast of 0.3 dyn m compared with 0.5 dyn m. The double cyclonic gyre in the tropics (Suga and Talley 1994) is a splitting of the cyclonic gyre described by Gordon and Bosley (1991). Westward flow right along the equator and eastward flow to the south are suggested. However, floats released at the equator in the AAIW layer appear to have gone eastward (Richardson and Schmitz 1993), matching the tracer patterns better than would a westward flow.

The North Atlantic subtropical gyre is longitudinally less extensive than the South Atlantic's. A narrow cyclonic recirculation appears on its southeastern flank which is not apparent in Reid's (1994) adjusted circulation at 1000 dbar or Lozier et al.'s (1994) streamlines at $31.85\sigma_1$ which were based on averaged data. The circulation around the outer perimeter of the saline Mediterranean tongue appears weakly cyclonic, as seen more robustly in Lozier et al. (1994), and also as an offshore feature in Reid (1994). The tropical circulation in the North Atlantic is confusing since the total variation in PAS is so small between about 20°N and 20°S, but appears to contain a cyclonic gyre centered at 15°N; this is opposite to Reid's (1994) adjusted circulation.

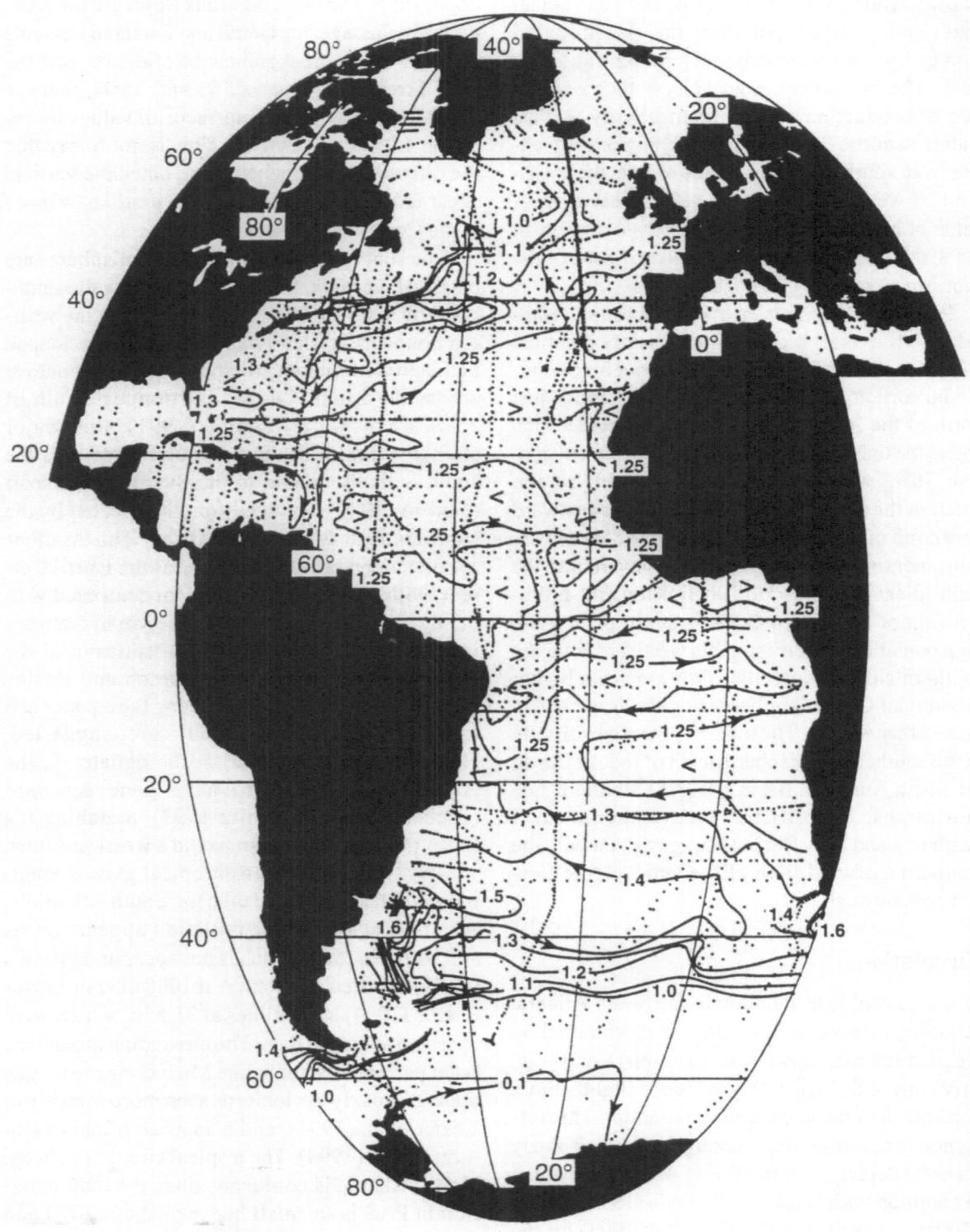

Figure 7. Pressure anomaly streamfunction (dyn m) at $31.7\sigma_1$ relative to 3000 dbar. A mean pressure of 750 dbar was subtracted prior to calculation. (Following Zhang and Hogg 1992).

Formation of AAIW

The following questions arise:
1) Where and how is AAIW principally injected from the surface? (2) What sets its density? (3) What are the significant modification processes both near its formation region and downstream?

1) *AAIW injection.* All recent work, dating from Taft (1963), asserts the importance of the region around Drake Passage for injection of AAIW (McCartney 1977; 1982; Gordon et al. 1977; Georgi 1979; Molinelli 1981; Sievers and Nowlin 1984; Peterson and Stramma 1991; Piola and Gordon 1989). McCartney (1977; 1982) showed the linkage of new AAIW with the surface pycnostad (SAMW) in the southeastern Pacific; a portion becomes the Pacific AAIW and the part which flows through Drake Passage is further modified and becomes the Atlantic and Indian Ocean AAIW. These concepts replace the earlier one of continuous injection all around Antarctica (Merz and Wüst 1922; Wüst 1935). The current debate is centered on the relative importance of cross-SAF exchange through and east of Drake Passage compared with air-sea interaction in modifying the properties of waters which become Atlantic AAIW.

There are two distinct types of AAIW: that produced west of Drake Passage and entering the South Pacific subtropical gyre through subduction (Luyten et al. 1983) and that produced east of Drake Passage in the confluence of the Falkland and Brazil Currents. All other variations in properties are due to mixing and differential advection (and biological processes for oxygen). The southeast Pacific process creates a subsurface pycnostad which can be identified as SAMW separately from AAIW but which I think should be very closely identified with AAIW. This process is identical to Subpolar Mode Water subduction, injecting thick pycnostads into the eastern North Atlantic subtropical gyre and SAMW subduction in the eastern South Indian subtropical gyre (both due to McCartney 1982), and a much less pronounced pycnostad subduction in the eastern North Pacific (Suga et al. 1996; Nakamura 1996). AAIW salinity minimum formation in the southeast Pacific is analogous to shallow salinity minimum formation through subduction in the eastern North Pacific (Yuan and Talley 1992) and to formation of the short-lived salinity minimum at the subarctic front in the North Atlantic (Subarctic Intermediate Water) (Arhan 1990).

The southwest Atlantic process creates the Atlantic and Indian Ocean AAIW; their subtropical gyres are connected in their eastward flowing branches (but more broken in their westward flow due to Africa). The thick winter surface layer south of the SAF (Falkland Current) and inshore of the Falkland Current is injected into the subtropical gyre through ring formation and mixing at the Brazil/Falkland confluence. This process is like that of North Pacific Intermediate Water (NPIW) formation where the Kuroshio and Oyashio waters meet (Talley 1993; Talley et al. 1995). It is marked by enhanced fine structure, as documented by Georgi (1981). An hypothesis is that because mixing is more central to this process than to the gentle subduction of the eastern South Pacific, the potential vorticity minimum associated with the original surface pycnostad is much less persistent in the South Atlantic than in the South Pacific.

(2) *AAIW density.* The debate here centers on why the Atlantic AAIW salinity minimum is denser than the southeastern Pacific AAIW minimum: 31.8 vs. $31.7\sigma_1$ (27.1 vs. $27.0\sigma_\theta$). This increase could be due to air-sea exchange and mixing across the Polar Front in Drake Passage as the surface water makes its way through Drake Passage and around the Falkland Current loop (Georgi 1979). The contrast between Falkland Current and Brazil Current waters could also account for some increase in density of the AAIW through cabbeling (Martineau 1953) as the cold, fresh southern waters mix with the warmer, saltier subtropical waters; this process is important in NPIW formation in the North Pacific (Yun and Talley 1996).

The larger question is what sets the overall density of AAIW. We have two processes to consider: subduction (eastern Pacific) and mixing of the dominant near-surface mode from the Falkland Current. Maps of winter surface density and wind parameters will be presented in the more complete work. Based on the Hellerman and Rosenstein (1983) annual averaged winds and the winter surface density based on all individual winter stations, the only places where the surface isopycnals 27.0-27.1σ_θ (31.6-31.7σ_t) lie north of the zero wind-stress curl, so could be subducted, are in the southeastern Pacific and southwestern Atlantic. The densest, mass-important subduction in a subtropical gyre is of water in its poleward-eastern corner if isopycnal outcrops are relatively zonal. There are only two such southeastern corners in the Southern Ocean: the southeastern Pacific and the southeastern Indian Ocean, since the eastward flow from the South Atlantic continues unobstructed into the Indian Ocean. The subducted water in the southeastern Pacific is the AAIW which is equivalent to the SAMW there. The densest subduction for the Indian Ocean would then be the 26.8-26.9σ_θ SAMW outcropping near the zero wind-stress curl south of Australia.

If the Falkland/Brazil Current AAIW formation is analogous to NPIW formation, then subduction is not the central issue, but rather the density of the winter surface water in and inshore of the Falkland loop. This water is then mixed into the subtropical gyre at the confluence. The thick winter surface layer west of and in the Falkland loop is of density 27.1σ_θ (31.7σ_t).

(3) *Modification processes.* Is there exchange across the SAF which modifies AAIW around Antarctica? Molinelli (1981) looks at changes in AAIW salinity and transport around the ACC, showing that there might be an additional injection of some low salinity at Kerguelen Plateau, but he retains the central importance of Drake Passage for AAIW formation. Laterally minimum potential vorticity for the Atlantic/Indian AAIW is found north of the SAF; its downstream increase could be due to mixing with higher potential vorticity waters to either the north or south.

On the largest scale, AAIW is obviously modified and diluted considerably moving away from its two sources; this is in fact how the two sources are identified. It is not possible at this point in the study to assess the relative importance of isopycnal and diapycnal mixing. There is strong evidence for locally vigorous mixing in the AAIW within a few degrees of the equator in the Atlantic (Suga and Talley 1994), but this is a very small part of the overall modification process. Diapycnal mixing is the only possible process which can create the lateral salinity maximum in the AAIW layer in the tropical Pacific since there are no surface sources of high salinity water at this density.

Discussion

This extended abstract concerning AAIW in the Atlantic is part of work in progress, to describe the global distribution of AAIW, with the ultimate aim of further clarifying its formation, modification, and overall transports as related to deep water formation in the North Atlantic (by overturn) and North Pacific (by diffusion). A clear differentiation was made between the Pacific type of AAIW, formed through subduction in the southeastern corner, and the Atlantic/Indian type of AAIW, formed through injection of Falkland Current surface waters into the subtropical gyre at the Falkland/Brazil Current confluence. The Atlantic and Indian subtropical gyres function as one nearly continuous gyre in the poleward/eastward flow, with important mixing and injection of lower latitude waters occurring in the westward branch at the Agulhas. Westward flow past Africa and southern Australia creates large plumes of high potential vorticity which dye the northern side of the Atlantic and Indian subtropical gyres.

The strong barrier in potential vorticity between the subtropical and tropical regions in each of the oceans is accompanied by a jump in AAIW properties across the barrier. Circulation in the tropics does not cease, but is largely zonal rather than gyral (Richardson and Schmitz 1993; Suga and Talley 1994). Communication between the subtropics and tropics occurs in the western boundary current in both the Atlantic and Pacific; the property variation across the potential vorticity barrier is enhanced in the eastern parts of the oceans. It cannot be a coincidence that this barrier coincides with the cessation of the AAIW oxygen maximum in the South Atlantic and the cessation of the AAIW salinity minimum in the North Atlantic, (South) Indian, and North Pacific. The important dynamical barrier to meridional flow of AAIW density water in all oceans is not the equator but the tropical/subtropical boundary, where potential vorticity patterns change from fairly well mixed (subtropical) to β-dominated (tropics).

Acknowledgments

 This work was supported by the National Science Foundation's Ocean Sciences Division, Grants OCE86-14486 and OCE92-01315, and by NOAA/OGP through the JIMO Consortium (NA47GP0188). Unpublished data from 1980 were made available by M. McCartney, who also made available an unpublished account of formation of AAIW in and east of Drake Passage. Discussions with J. Reid and R. Peterson were valuable. The invitation and support to present this work at the South Atlantic symposium in Bremen is gratefully acknowledged.

References

Arhan M (1990) The North Atlantic current and subarctic intermediate water. J Mar Res 48: 109-144

Buchanan JY (1877) On the distribution of salt in the ocean, as indicated by the specific gravity of its waters. J Roy Geogr Soc 47: 72-86

England M, Godfrey J.S., Hirst A.C., Tomczak M. (1993) The mechanism for Antarctic Intermediate Water renewal in a world ocean model. J Phys Oceanogr 23:1553-1560

Fine RA (1993) Circulation of Antarctic Intermediate Water in the South Indian Ocean. Deep-Sea Res 40: 2021-2042

Georgi DT (1979) Modal properties of Antarctic Intermediate Water in the southeast Pacific and the South Atlantic. J Phys Oceanogr 9:456-468

Georgi DT (1981) On the relationship between the large-scale property variations and fine structure in the Circumpolar Deep Water. J Geophys Res 86: 6556-6566

Godfrey JS (1989) A Sverdrup model of the depth-integrated flow for the world ocean allowing for island recirculations. Geophys. Astrophys. Fluid Dynamics 45:89-112

Gordon AL, Georgi DT, Taylor HW (1977) Antarctic polar front zone in the western Scotia Sea - summer 1975. J Phys Oceanogr 7:309-328

Gordon, A L (1986) Interocean exchange of thermocline water. J Geophys Res 91:5037-5046

Gordon AL, Bosley KT (1991) Cyclonic gyre in the tropical South Atlantic. Deep-Sea Res 38 (Suppl): 323-343

Gordon AL, Weiss RF, Smethie WM, Warner MJ (1992) Thermocline and intermediate water communication between the South Atlantic and Indian Oceans. J Geophys Res 95:7223-7240

Hellerman S, Rosenstein M (1983) Normal monthly wind stress over the world ocean with error estimates. J Phys Oceanogr 90:7087-7097

Keffer T (1985) The ventilation of the world's oceans: maps of the potential vorticity fields. J Phys Oceanogr 15:509-523

Levitus S (1982) Climatological Atlas of the World Ocean. NOAA Professional paper 13, U.S. Government Printing Office, Washington, DC 173 pp

Lozier MS, Owens WB, Curry RG (1994) The climatology of the North Atlantic. Progr in Oceanogr submitted

Luyten JR, Pedlosky J, Stommel H (1983) The ventilated thermocline. J Phys Oceanogr 13:292-309

McCartney MS (1977) Subantarctic mode water. In: Angel M (ed) A voyage of discovery: George Deacon 70th anniv. Vol., Deep-Sea Res (Suppl), pp 103-119

McCartney MS (1982) The subtropical recirculation of mode waters. J Mar Res 40 (Suppl.):427-464

McCartney MS, Baringer MO (1993) Notes on the S. Pacific hydrographic section near 32°S - WHP P6. WOCE Notes, 5

Merz A, Wüst G (1922) Die Atlantische Vertikalzirkulation. Zeit. Gesell. d. Erdkunde zu Berlin, 1. Vorträge und Abhandlungen, pp 1-34

Molinelli EJ (1981) The Antarctic influence on Antarctic Intermediate Water. J Mar Res 39:267-293

Nakamura H (1996) A pycnostad on the bottom of the ventilated portion in the central subtropical North Pacific: its distribution and formation. J Oceanogr in press

Peterson RG, Stramma L (1991) Upper-level circulation in the South Atlantic Ocean. Prog Oceanogr 26:1-72

Piola AR, Gordon AL (1989) Intermediate water in the southwest South Atlantic. Deep-Sea Res 36:1-16

Reid JL (1986) On the total geostrophic circulation of the South Pacific Ocean: flow patterns, tracers, and transports. Prog Oceanogr 16:1-61

Reid J.L. (1989) On the total geostrophic circulation of the South Atlantic Ocean: flow patterns, tracers, and transports. Prog Oceanogr 23:149-244

Reid J.L. (1994) On the total geostrophic circulation of the North Atlantic Ocean: flow patterns, tracers, and transports. Prog Oceanogr 33:1-92

Richardson PL, Schmitz WJ (1993) Deep cross-equatorial flow in the Atlantic measured with SOFAR floats. J Geophys Res 98:8371-8387

Rintoul SR (1991) South Atlantic interbasin exchange. J Geophys Res 96:2675-2692

Sievers HA, Nowlin WD Jr. (1984) The stratification and water masses at Drake Passage. J Geophys Res 89:10489-10514

Suga T, Talley LD (1994) Antarctic Intermediate Water circulation in the tropical and subtropical South Atlantic. J Geophys Res 100:13411-13453

Suga T ,Takei Y, Hanawa K (1996) Thermostad distribution in the North Pacific subtropical gyre: the central mode water and the subtropical mode water. J Phys Oceanogr, submitted

Taft BA (1963) Distribution of salinity and dissolved oxygen on surfaces of uniform potential specific volume in the South Atlantic, South Pacific, and Indian Ocean. J Mar Res 21:129-146

Talley LD (1985) Ventilation of the subtropcal North Pacific: the shallow salinity minimum. J Phys Oceanogr 15:633-649

Talley LD (1993) Distribution and formation of North Pacific Intermediate Water. J Phys Oceanogr 23: 517-537

Talley LD, Nagata Y, Fujimura M, Iwao T, Kono T, Inagake D, Hirai M, Okuda K(1994) North Pacific intermediate water in the Kuroshio/Oyashio mixed water region in spring, 1989. J Phys Oceanogr (in press)

Tsuchiya M (1989) Circulation of the Antarctic Intermediate Water in the North Atlantic Ocean. J Mar Res 47:747-755

Tsuchiya M, Talley LD, McCartney MS (1992) An eastern Atlantic section from Iceland southward across the equator. Deep-Sea Res 39:1885-1917

Tsuchiya M, Talley LD, McCartney MS (1994) A western Atlantic section from South Georgia Island (54°S) northward across the equator. J Mar Res 52:55-81

Warner MJ, Weiss RF (1992) Chlorofluoromethanes in South Atlantic Antarctic Intermediate Water. Deep-Sea Res 39:2053-2075

Whitworth T, Nowlin WD (1987) Water masses and currents of the southern ocean at the Greenwich meridian. J Geophys Res 92:6462-6476

Wüst G (1935) Die Stratosphäre. Wissenschaftliche Ergebnisse der Deutschen Atlantischen Expedition auf dem Vermessungs- und Forschungsschiff „Meteor" 1925-1927, 6:109-288

Yun J-Y, Talley LD (1996) Cabbeling and the density of North Pacific Intermediate Water. In preparation

Zemba JC (1991) The structure and transport of the Brazil Current between 27° and 36° south. Ph.D. thesis, Woods Hole Oceanographic Institution/Massachusetts Institute of Technology, 160 pp

Zhang H-M, Hogg NG (1992) Circulation and water mass balance in the Brazil Basin. J Mar Res 50: 385-420

Lagrangian Measurements in the Malvinas Current

R.G. Peterson[1], C.S. Johnson[1], W. Krauss[2], R.E. Davis[1]

[1]Scripps Institution of Oceanography, University of California - San Diego,
La Jolla, CA 92093-0230, U.S.A.
[2]Institut für Meereskunde an der Universität Kiel, Düsternbrooker Weg 20,
24105 Kiel, GERMANY

Abstract: Direct measurements of magnitude of the northward flow of the Malvinas (Falkland) Current have recently been made with two types of Lagrangian platforms: ALACE floats which cycled between 750-m depth and the sea surface, and 100-m drogued surface drifters. Each data set clearly delineates the path of the Malvinas Current, and the vertical shears inferred from them are commensurate with historical geostrophic shears. Velocities from the surface drifters are used here to adjust geostrophic shears from historical measurements, and the results confirm a large transport of the current, as previously implied by numerical models and a regional inverse calculation. At 42°S, the northward transport of the Malvinas Current in the upper 3000 m appears to be about 70 Sv, several times larger than estimates obtained by adjusting geostrophic shears to assumed levels of no motion. This large barotropic component may have significance in the cross-frontal transfer of intermediate and deep waters from the circumpolar current to the adjacent flow regimes in the South Atlantic, and thus on the inter-basin exchange of water masses.

Introduction

The Antarctic Circumpolar Current in Drake Passage is known to consist of three permanent deep-reaching fronts, each associated with strong eastward flow (Whitworth 1980; Nowlin and Clifford 1982). The northernmost of these fronts, the Subantarctic Front, turns sharply northward just east of Drake Passage to enter the Malvinas (Falkland) Current (Peterson and Whitworth 1989). This current flows north along the eastern coast of Argentina to latitudes of about 35°S (Garzoli 1993), carrying waters of similar temperature-salinity characteristics as those found in northern Drake Passage (Piola and Gordon 1989). The Subantarctic Front does not extend quite so far north, as it retroflects cyclonically back toward the south at about 40°S (Peterson and Whitworth 1989) where its associated flow is joined by southward moving waters of the Brazil Current. The Malvinas Return Current and the Brazil Current then flow side-by-side from the confluence toward the south, mixing actively

(Gordon and Greengrove 1986), before the warm and saline waters of the Brazil Current bend back toward the north and east at latitudes of 46°-47°S. The Subantarctic Front turns east at the southern margin of the Argentine Basin near 49°S (Peterson and Whitworth 1989) to resume its circumpolar course.

Analyses of hydrographic data show the internal density and baroclinic (depth-dependent) velocity fields of the Malvinas Current as being highly stable (Piola and Bianchi 1990), while drifter and altimeter data show it to have minimal values of surface variability (Piola et al. 1987; Gordon and Haxby 1990). The volume of water transported north by the Malvinas Current has traditionally been estimated as being in the range of 10 to 15 Sv (1 sverdrup = 10^6 m^3 s^{-1}) (Gordon and Greengrove 1986; Piola and Bianchi 1990; Garzoli et al. 1990), values that have been obtained by adjusting vertical profiles of geostrophic shear to assumed zero

From WEFER G, BERGER WH, SIEDLER G, WEBB DJ (eds), 1996, *The South Atlantic: Present and Past Circulation*. Springer-Verlag Berlin Heidelberg, pp 239-247

velocities at depths of 1000-1500 m. Characteristics of water masses at depth, however, indicate that the northward flow over the Patagonian slope extends from the sea surface to the bottom (Reid 1989), and geological evidence suggests that the bottom flow is significant (Flood and Shor 1988). Selecting a zero-reference at the bottom or anywhere in the water column can therefore result in serious underestimates of velocities and transport, unlike much of the World Ocean where strong, depth-independent flow does not exist.

Mass conservation in the region also points toward the presence of significant bottom velocities; intense horizontal pressure gradients exist along the continental slope in the Brazil-Malvinas Confluence, and by using these in an inverse calculation the top-to-bottom northward transport of the Malvinas Current appears to be on the order of 70-75 Sv at 42°S and more than 80 Sv at 46°S (Peterson 1990; 1992). Such transports are not inconsistent with what appears in published figures from several numerical models, which have long indicated a Malvinas transport several times larger than the geostrophic shears alone would suggest; some model calculations have yielded transports of 100 Sv or more at 50°S (Cox 1975; Semtner and Chervin 1992). The Fine Resolution Antarctic Model gives more conservative Malvinas transports of 67 Sv at 48°S and 42 Sv at 44°S (Webb et al. 1991; The FRAM Group 1991).

Direct observations of velocity in the southwestern South Atlantic have been few. A set of short-term current meter measurements was made along the southern boundary of the Argentine Basin from the INDOMED expedition in 1978 (JL Reid, personal communication; see Peterson 1992). One instrument was located on the continental slope northeast of the Falkland Islands in about 1500-m depth, and over a ten-day period the average speed recorded by it was 17 cm s⁻¹ toward the west. The flow was remarkably steady in both direction and speed, with variations being mainly due to tides. Instruments have also been moored in the Brazil-Malvinas Confluence (Garzoli et al. 1990; Garzoli 1993) and have provided evidence for there being important seasonal variations in the flow (Garzoli and Giulivi 1994). However, the only direct measurements of northward velocity in the main part of the Malvinas Current have come from the temporal displacements of drifting platforms.

Lagrangian observations

Ship-drift data have existed from the region for over two centuries (Krümmel 1882), but these are of marginal use for studying details of a narrow jet in a region characterized by strong winds. The same limitation holds true for a set of undrogued meteorological drifters that passed through the area in 1979-80 (Peterson 1992). The first reliable observations were made with 4 ALACE (Autonomous Lagrangian Current Explorer) floats deployed from the German research vessel Meteor as part of a larger array in the western Drake Passage in early 1990 (Davis et al. 1992). These floats followed the currents at about 750-m depth for periods of 14 days, after which they rose temporarily to the surface for a day to report via Argos satellite. Their speeds in the core of the Malvinas Current were typically between 30 and 40 cm s⁻¹, three times or more than those that result from geostrophic calculations at 750-m referred to a deeper level of no motion.

The next direct measurements of the absolute strength of the Malvinas Current came from seven surface drifters released in the current from the German research vessel / ice breaker Polarstern in August and September 1992. These were cylindrical spar buoys 11 cm in diameter and 260 cm long. Each was drogued at 100-m depth with a cylindrical holey sock measuring 1 m in diameter and 10 m in length, weighted at the free end. With this configuration, the upper 40 cm of each buoy was exposed to air in calm conditions. In rough conditions the buoys were often completely submerged, exposed to the air only within wave troughs. Non-linear and non-periodic forces imparted on the buoy hulls by surface gravity waves are thought to be the largest slip forces transmitted to the drogues; other sources for slip force are direct windage on the exposed parts of the buoys and vertical shears of velocity acting on the tethers. Tests performed on low-profile buoys drogued with holey socks at 15-m depth show that for a wind speed of 10 m s⁻¹ a drogue-to-buoy drag-area ratio of 40 is required to keep the drogue slippage to less than 1 cm s⁻¹

(Sybrandy and Niiler 1991; Niiler et al. 1994). The drag-area ratio for the configuration employed here was 35 (using the total buoy length), resulting in a somewhat greater slippage. Since the buoys were exposed to wind primarily in wave troughs where turbulent reversals of wind direction often occur, the associated average slip force was likely quite small as compared with that from the surface waves. Probably of lesser importance as well were forces on the tether due to current shear; most of each line was beneath the base of the Ekman layer (typically 30 m or less) and thus acted as additional drogue area in the weakly-sheared geostrophic regime. Weather systems pass through the region frequently, causing large variations in wind direction and speed to occur, so the effects of Ekman layer motion can presumably be further reduced by temporal filtering of the data. As yet, there have not been any engineering tests performed on this particular buoy-drogue configuration, but these buoys appear to have followed the near-surface geostrophic flow in this region quite well.

The trajectories of the surface buoys have been smoothed with a 60-hour running-mean filter, which also removes the effects of inertial motions and diurnal and semi-diurnal tides. These trajectories, and those from the ALACE floats (Fig. 1), demonstrate how closely the Malvinas Current is guided by bottom bathymetry toward the north until its confluence with the Brazil Current. This in turn implies the existence of significant barotropic (depth-independent) components in the flow. Note that the ALACE trajectories are straight-line displacements between the points at which they sank and returned to the surface, so details of the actual motions at depth are lost. This can be important in regions of rapidly changing bathymetric orientation, such as in the far southwestern corner of the Argentine Basin near 48°S. Speeds inferred from the ALACE trajectory segments thus represent minimal values.

The speeds of the 100-m drogued drifters were in excess of 60 cm s^{-1} in the core of the Malvinas Current from 46°S to 39°S (Fig. 1), sometimes approaching 80 cm s^{-1}, again three times or more the magnitudes observed by referencing geostrophic shears to the bottom. The largest (smoothed) speeds experienced in the region by any of the drifters were in excess of 100 cm s^{-1} in the southward frontal jet

set up by the combined Brazil-Malvinas Return flow; the Subantarctic Front follows this course (Peterson and Whitworth 1989), and unsmoothed drifter speeds reached 140 cm s^{-1} in a narrow central core. The ALACE floats, in contrast, tended not to follow the southward combined flow along the near-surface front, but instead tended toward the east. Once that happened, the ALACEs reflected a great deal of mesoscale eddy variability, and while at the surface they were displaced from the motions they had been following at depth. Although the trajectories of the ALACEs away from the western boundary can not be uniquely interpreted, they show that there is most likely a substantial amount of cross-frontal flow of Antarctic Intermediate Water and Upper Circumpolar Deep Water, thus representing a loss of these waters from the Antarctic Circumpolar Current. The details of how this happens, whether by subsurface eddies or continuous flow, are unknown.

No hydrographic data exist that are synoptic with the drifter data, so archive data are used to get an estimate of the absolute transport of the Malvinas Current. The most suitable hydrographic section, in terms of depth of stations and in crossing the greatest number of buoy trajectories, was made at 42°S in late 1979 by A. Gordon on the Atlantis II (Guerrero et al. 1982) (Fig. 1). Reference speeds at 100-m depth are obtained by fitting a Gaussian curve along 42°S to the northward components of buoy velocity and extracting the average values between the Atlantis II stations. The curve is forced to go to zero at the edge of the continental shelf and in the center of the cyclonic trough described by the Malvinas Current and its return; the maximum v-component, in the center of the Malvinas, is 62 cm s^{-1} (corresponding to an absolute velocity of 75 cm s^{-1} rotated 35° from north), while the maximum station-to-station average northward component is 52 cm s^{-1} (63 cm s^{-1} rotated). East of the cyclonic trough a potential density surface referenced to 4000 decibars, sigma-4 = 45.85 kg m^{-3} (2800-3300 m deep) is used as zero reference, since it lies near the transition between the North Atlantic Deep Water, moving south beneath the Brazil Current, and the underlying Lower Circumpolar Deep Water moving north (Reid 1989; Peterson 1992). Unlike the highly stable Malvinas Current, where non-synop-

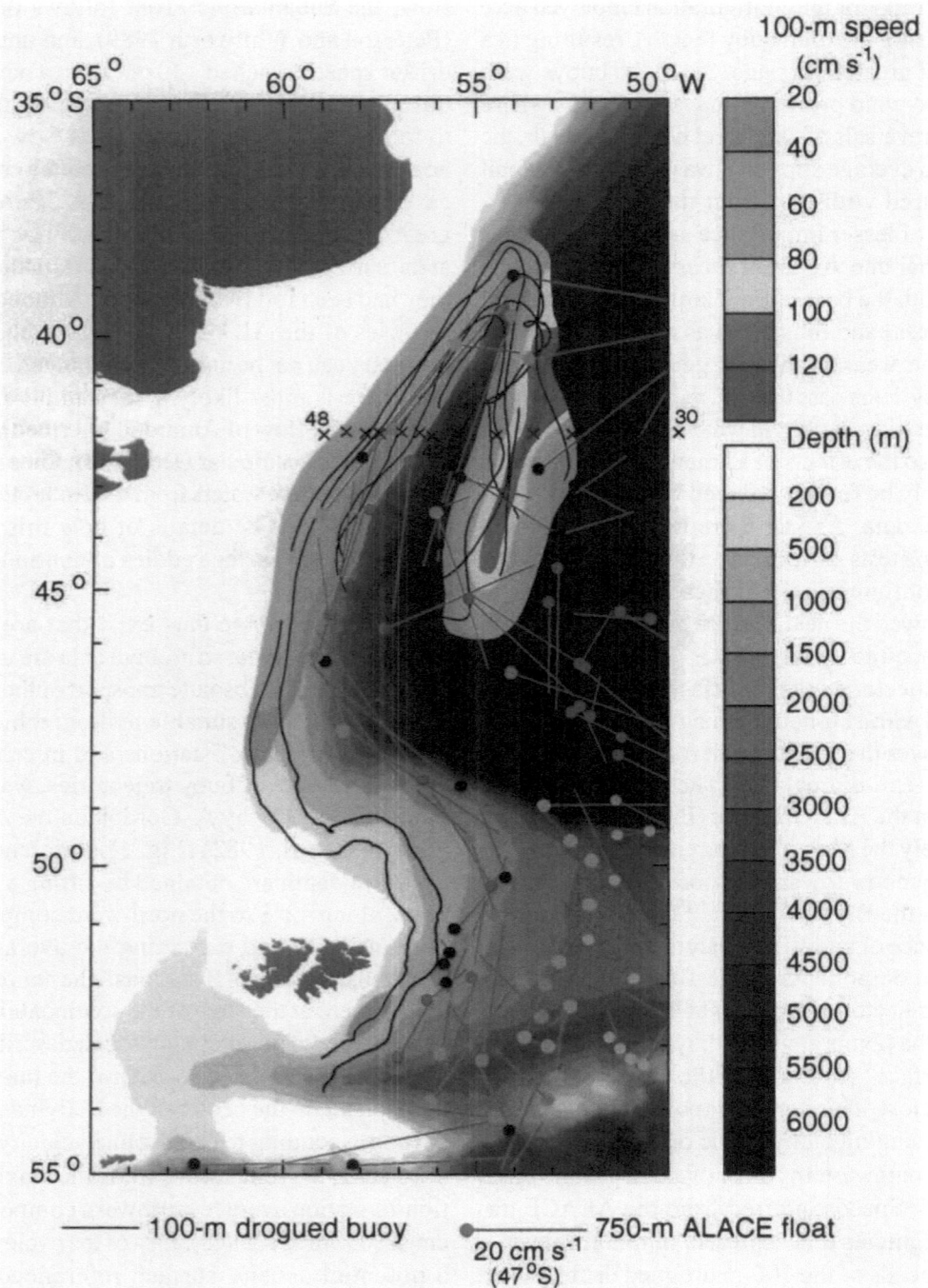

Fig. 1. Trajectories of surface buoys deployed in August and September 1992 and drogued at 100-m depth (solid black curves; smoothed with a 60-hr running mean filter; truncated in the east where the trajectories became eddy-dominated) and of ALACE floats deployed in January 1990 at 750-m depth (magenta lines; segments are displacements over 14-day intervals; dot colors identify individual floats). Colored contouring portrays isotachs of surface drifter scalar speeds only where drifter trajectories are shown. Crosses along 42°S denote hydrographic stations occupied by R/V Atlantis II in December 1979.

tic reference information is useful, the Brazil-Malvinas Return is among the most variable regions in the southern hemisphere (Piola et al. 1987; Gordon and Haxby 1990), making non-synoptic velocities inappropriate as reference.

The adjusted northward geostrophic speeds, gridded and mapped onto a vertical section (Fig. 2), show the Malvinas Current as being compressed over the upper portions of the continental slope, its core centered in about 1400-m of water. The resulting bottom velocities are 20-30 cm s^{-1} through the depth range of 400-1600 m, and are in excess of 10 cm s^{-1} to depths of 2400 m. In the center of the current, the vertical shears in the upper 1000-m are about 50% less than those in the southward flow of the Brazil-Malvinas Return flow (note that the position of the southward jet as indicated by the buoys is nearly the same as that by hydrography). At greater depth, a weak reversal appears near the continental slope at 3000-4000 m depth, perhaps an artifact of using non-synoptic reference information; however, immediately above this was conspicuous interleaving between the North Atlantic Deep Water and the Circumpolar Deep Water (Peterson and Whitworth 1989), which may have altered the density field and vertical stucture of velocity enough to create such a feature.

Because the Malvinas Current is narrow and strongly steered by bathymetry, and because the vertical shears appear to diminish rapidly away from the center of the current, it is difficult to use the ALACE float trajectories in concert with the drifter data to accurately estimate the geostrophic shear. The three ALACEs appear to have intersected 42°S at nearly the same place, but exactly where and at what speeds can not be determined. The bathymetric contours have a slight seaward bulge in the region, so it is likely that they crossed the line somewhat east of where the straight segments would indicate. Following the bathymetry between the appropriate ALACE endpoints, there is an implied average difference in northward components between 100-m and 750-m depth of 11 cm s^{-1}, comparable to the 9 cm s^{-1} measured between Atlantis II stations 44 and 45.

Conclusions

These two sets of Lagrangian trajectories provide the first convincing direct measurements of the magnitude of northward flow in the Malvinas Current proper, and they confirm, in principal, what numerical models had been indicating (Cox 1975: Webb et al. 1991; The FRAM Group 1991; Semtner and Chervin 1992) and the results of a regional inverse calculation using hydrographic data (Peterson 1990; 1992). Integrating the adjusted field of velocities (Fig. 2) from the continental shelf to the center of the cyclonic trough, the total top-to-bottom northward component of transport is 60 Sv. Rotating by 35° for total transport along the bathymetry, the figure becomes 73 Sv. This transport is within the upper 3000-m of the water column, and it is larger than the 50-60 Sv estimated to be supplied directly by the upper 2000-2500 m of the northern Antarctic Circumpolar Current in Drake Passage (Peterson 1992). A recirculation cell probably exists between the northward Malvinas Current and its return, and this is also implied by the orange-coded ALACE trajectory (Fig. 1).

There are sources for error in this analysis that can not be estimated. These arise mainly from the non-synopticity of the data, the marginal drifter resolution, and incomplete knowledge of the water-following characteristics of the buoys. The magnitudes obtained, however, provide persuasive indications that the Malvinas Current has a large barotropic component, as predicted by earlier work. More recently, acoustic Doppler current profiler measurements were made along 45°S in January 1993 from the British research vessel Discovery, and when merged with contemporaneous hydrographic data they indicate a northward Malvinas transport of about 50 Sv (Saunders and King 1994). Although baroclinic variations in the Malvinas Current appear small, with the flow being tightly steered by bathymetry and thus rendering small values of surface variability, it may be that the current undergoes significant barotropic variations in its total transport in response to atmospheric forcing (Peterson and Stramma 1991; Matano et al. 1993). Not until time series are obtained with moored instruments can this question be suitably addressed.

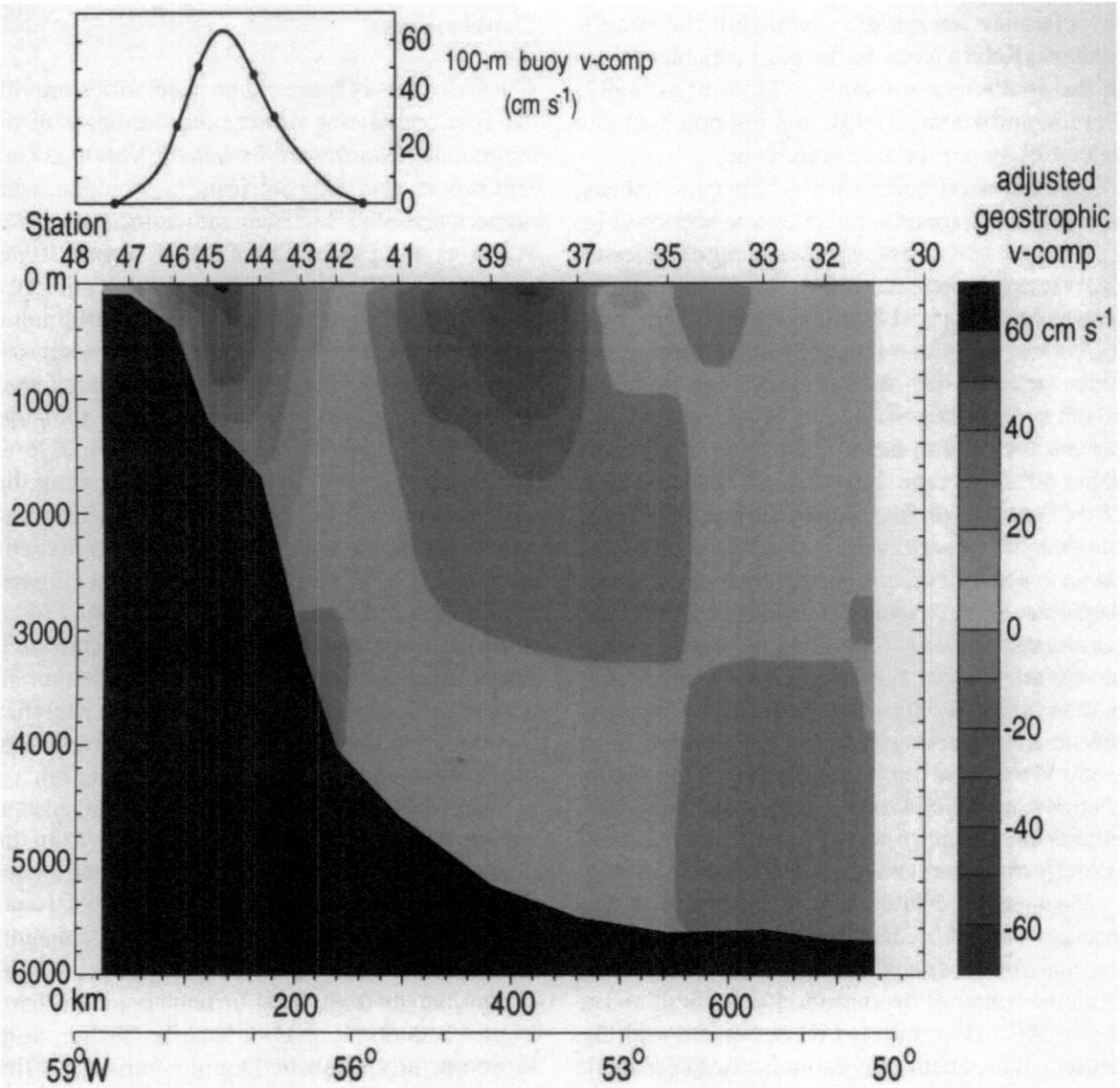

Fig. 2. Vertical section of northward geostrophic velocity along the line of stations depicted in Fig. 1. North-ward components of buoy speeds at 100-m depth in Fig. 1 are used as reference west of station 41, the isopycnal surface sigma-4 = 45.85 kg m^{-3} (2800-3300 m depth) is used as zero reference east of station 41. Vertical shears above the deepest common depths between stations are used to extrapolate the adjusted profiles of velocity to the bottom, and the velocities are then gridded with a fifth-order polynomial fit.

The potential importance of the Malvinas Current having a large transport is that this flow comprises a thick layer having the characteristics of Antarctic Intermediate Water and Upper Circumpolar Deep Water, and it seems evident that at the confluence of the Malvinas and Brazil currents that substantial quantities of these waters are lost from the Antarctic Circumpolar Current to the adjacent circulation regimes to the north. This is suggested by distributions of salinity at the salinity-minimum of the intermediate water and of oxygen at the oxygen-minimum of circumpolar water (Gordon and

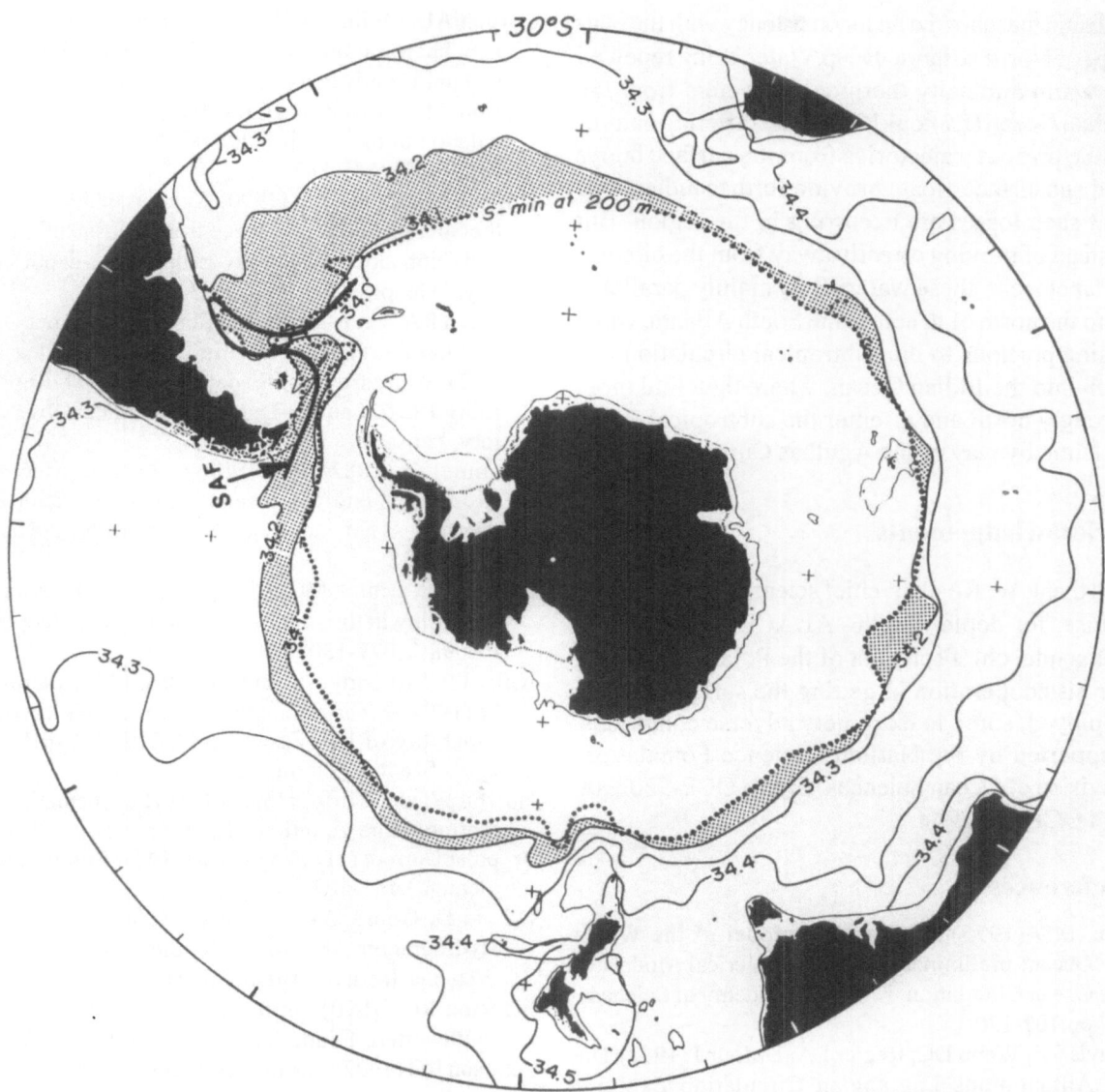

Fig. 3. Hemispheric distribution of salinity at the salinity-minimum of Antarctic Intermediate Water for latitudes south of 30°S (adapted from Gordon and Molinelli 1982). Shaded area is between the 34.1 and 34.2 isohalines. Shown schematically as the heavy line in the southwestern Atlantic sector is the Subantarctic Front (SAF) (after Peterson and Whitworth 1989).

Molinelli 1982; Olbers et al. 1992) (Fig. 3 is an example). The envelopes of isopleths that pass through the northern Drake Passage in the vicinity of the Subantarctic Front widen dramatically northward in the western Argentine Basin to latitudes well north of the front and to the southern subtropi-

cal gyre; the core extrema then attenuate downstream through mixing as these waters are carried eastward with the general circulation.

That there are important losses of water from the circumpolar flow in this region has also been suggested by an inverse calculation for the South

Atlantic that showed an inconsistency with the concept of North Atlantic Deep Water being renewed by warm and salty thermocline waters from the Indian Ocean (Rintoul 1991). The differing natures of the present trajectories from the surface buoys and subsurface floats provide further indications that such losses are occurring in the region. But instead of moving directly away from the circumpolar current, these waters flow mainly parallel to it, to the north of it, across the South Atlantic while losing portions to the subtropical circulation and then into the Indian Ocean, where they fold more strongly north and re-enter the subtropical South Atlantic by way of the Agulhas Current.

Acknowledgements

We thank W. Roether, chief scientist of the Meteor cruise, for deploying the ALACE floats, and R. Gersonde, chief scientist of the Polarstern cruise, for his cooperation in getting the surface drifters deployed, some in extremely adverse conditions. Supported by the National Science Foundation, Division of Ocean Sciences, Grants OCE-9202909 and OCE-9017744.

References

Cox MD (1975) A numerical model of the World Ocean: preliminary results. Numerical Models of Ocean Circulation, National Academy of Sciences, pp 107-120

Davis RE, Webb DC, Regier LA, Dufour J (1992) The Autonomous Lagrangian Circulation Explorer (ALACE). Journal of Atmospheric and Oceanic Technology 9:264-285

Flood RD, Shor AN (1988) Mud waves in the Argentine Basin and their relationship to regional bottom circulation patterns. Deep-Sea Res 35:943-971

Garzoli SL (1993) Geostrophic velocity and transport variability in the Brazil-Malvinas Confluence. Deep-Sea Res 40:1379-1403

Garzoli SL, Giulivi C (1994) What forces the variability of the South Western Atlantic boundary currents? Deep-Sea Res (in press)

Garzoli SL, Brown O, Evans R, Olson D, Podesta G, Provost C, Garcon V, Maillard M, Memery L, Takahashi T, Piola A, Bianchi A (1990) Confluence 1988-1990. An intensive study of the southwestern Atlantic. Eos 71:1131-1137

Gordon AL, Molinelli EJ (1982) Southern Ocean Atlas: Thermohaline and Chemical Distributions. Columbia University Press, New York

Gordon AL, Greengrove CL (1986) Geostrophic circulation of the Brazil-Falkland Confluence. Deep-Sea Res 33:573-585

Gordon AL, Haxby WF (1990) Agulhas eddies invade the South Atlantic: evidence from Geosat altimeter and shipboard conductivity-temperature-depth survey. J Geophys Res 95:3117-3125

Guerrero RA, Greengrove CL, Rennie SE, Huber BA, Gordon AL (1992) Atlantis II cruise 107 leg III CTD & hydrographic data. Technical Report LDGO 82-2, Lamont-Doherty Geological Observatory, Palisades

Krümmel O (1882) Bemerkung über die Meeresströmungen und Temperaturen in der Falklandsee. Aus dem Archiv der Deutschen Seewarte, V, Hamburg, Nr. 2

Matano RP, Schlax MG, Chelton DB(1993) Seasonal variability in the southwestern Atlantic. J Geophys Res 98:18027-18036

Niiler PP, Sybrandy AS, Bi K, Poulain PM, Bitterman D (1994) Measurements of the water-following capability of holey sock and TRISTAR drifters. Deep-Sea Res (submitted)

Nowlin WD Jr., Clifford MA (1982) The kinematic and thermohaline zonation of the Antarctic Circumpolar Current at Drake Passage. J Mar Res 40 (Supplement):481-507

Olbers D, Gouretski V, Seiß G, Schröter J (1992) Hydrographic Atlas of the Southern Ocean. Alfred Wegener Institute, Bremerhaven

Peterson RG (1990) On the volume transport in the southwestern South Atlantic Ocean. Eos 71:542

Peterson RG (1992) The boundary currents in the western Argentine Basin. Deep-Sea Res 39:623-644

Peterson RG, Whitworth T III (1989) The Subantarctic and Polar fronts in relation to deep water masses through the southwestern Atlantic. J Geophys Res 94:10817-10838

Peterson RG, Stramma L (1991) Upper-level circulation in the South Atlantic Ocean. Progress in Oceanography 26:1-73

Piola AR, Gordon AL (1989) Intermediate waters in the southwest South Atlantic. Deep-Sea Res 36:1-16

Piola AR, Bianchi AA (1990) Geostrophic mass transports at the Brazil/Malvinas Confluence. Eos 71:542

Piola AR, Figueroa HA, Bianchi AA (1987) Some aspects of the surface circulation south of 20°S revealed by First GARP Global Experiment drifters. J Geophys Res 92:5101-5114

Reid JL (1989) On the total geostrophic circulation of the South Atlantic Ocean: flow patterns, tracers, and transports. Progress in Oceanography 23:149-244

Rintoul SR (1991) South Atlantic Interbasin Exchange. J Geophys Res 96:2675-2692

Saunders PM, King BA (1994) Oceanic fluxes across the WOCE section A11 (Punta Arenas to Cape Town). Berichte aus dem Fachbereich Geowissenschaften der Universität Bremen, Nr. 52, p 130

Semtner AJ Jr, Chervin RM (1992) Ocean general circulation from a global eddy-resolving model. J Geophys Res 97:5493-5550

Sybrandy AL, Niiler PP (1991) The WOCE/TOGA SVP Lagrangian Drifter Construction Manual. University of California, Scripps Institution of Oceanography, SIO Reference 91/6, WOCE Report Number 63

The FRAM Group (1991) An eddy-resolving model of the Southern Ocean. Eos 72:169-184

Webb DJ, Killworth PD, Coward AC, Thompson SR (1991) The FRAM Atlas of the Southern Ocean. National Environmental Research Council, Swindon

Whitworth T III (1980) Zonation and geostrophic flow of the Antarctic Circumpolar Current at Drake Passage. Deep-Sea Res 27:497-507

Circulation in the Deep Brazil Basin

N.G. Hogg[1], W.B. Owens[1],
G. Siedler[2] and W. Zenk[2]

[1]*Woods Hole Oceanographic Institution,
Woods Hole, MA 02543-1541, U.S.A.*
[2]*Institut für Meereskunde an der Universität Kiel,
Düsternbrooker Weg 20, 24105 Kiel,
GERMANY*

Abstract: The Deep Basin Experiment (DBE), a part of the World Ocean Circulation Experiment (WOCE), is presently underway in the Brazil Basin of the South Atlantic. The program objectives and design philosophy are reviewed and early results are presented. Observations from a moored array along the southern boundary and neutrally buoyant float trajectories in the North Atlantic Deep Water and Antarctic Bottom Water are described with emphasis on their relationship to the recent flow schemes offered by Reid (1989). Also discussed are the process of cross isotherm mixing within the intense flow regime of the Vema Channel and observations of long period warming of the bottom water.

Introduction

Under the auspices of the WOCE Core Project 3 a large international program has been focussed on the deep circulation of the Brazil Basin. Scientists from Brazil, France, Germany and the U.S. are involved in a cooperative exploration of the three-dimensional flow of the various water masses with a special emphasis on the Antarctic Intermediate Water (AAIW), the North Atlantic Deep Water (NADW) and the Antarctic Bottom Water (AABW). The program has four main objectives:

1) to observe and quantify the deep circulation within an abyssal basin;
2) to distinguish between boundary and internal mixing processes;
3) to understand how passages affect the water flowing through them; and
4) to study the means by which deep water crosses the equator.

The Brazil Basin was chosen as the site of this experiment because it has relatively simple geometry with a smooth bottom and a small number of passages to adjoining basins, and it exists in a region of expected low eddy kinetic energy, an important attribute in designing a program aimed at the "mean" circulation.

At this point, in late 1994, the program is still underway but some data have been returned and it is an appropriate time to consider what we have learned so far and what the likelihood is of achieving the goals outlined above. In what follows we will first describe the field program and then discuss each of the objectives given above in the context of what is known, what we have learned and what might be expected at the completion of the project . A summary will complete the paper .

The Field Program

It is almost 40 years since Stommel (1957) first outlined a scheme for the deep circulation which consisted of a series of deep western boundary currents carrying water from two polar regions of the North Atlantic and Antarctic where deep water was believed to be formed through winter cooling, evaporation and freezing. These boundary currents then fed slower interior flows of basin scale that returned

From WEFER G, BERGER WH, SIEDLER G, WEBB DJ (eds), 1996, *The South Atlantic: Present and Past Circulation*. Springer-Verlag Berlin Heidelberg, pp 249-260

the water poleward to conserve potential vorticity, and upward through the thermocline to complete a meridional convective circulation. Although the boundary current part of this scheme has been verified (see the review by Warren 1981) the interior part is a much more difficult observational task involving, as it does, very weak mean flows contaminated by mesoscale eddies, and much weaker vertical or diapycnal processes. Swallow (1971) first searched for this component of the general circulation in the late 1950s using neutrally buoyant floats tracked concurrently from a ship but soon found himself (and most of the rest of the physical oceanography community) overwhelmed by the eddy field. Some consistency checks have been made (e.g. Warren and Owens 1988; Warren and Speer 1991; Johnson et al. 1991) but it is fair to say that the slow (order mm sec^{-1}), poleward return flow has never been observed with any certainty by direct velocity measurements.

The advent of new technology in the form of long-lived current meter moorings and neutrally buoyant floats now makes it possible to attempt such a measurement. Within the Brazil Basin it was anticipated that eddy energy levels would be of order 1 cm^{2}sec^{-2} implying that one could achieve accuracies of about 1 mm sec^{-1} in the estimation of the mean flow by tracking of order 100 floats for 5 years and then averaging the results over 500 km scales. Hence the DBE plans include the deployment of large numbers of floats at the levels of the three major water masses each of which is to be tracked by a network of acoustic sources (Fig. 1).

The floats will provide a direct estimate of interior mean *horizontal* velocities. The estimation of vertical or diapycnal velocities is a harder problem that we intend to attack through a network of hydrographic sections which divide the Brazil Basin into a grid of interconnected boxes. By balancing various property fluxes through the sides of these boxes and using the additional constraints provided by the direct velocity measurements it is hoped that meaningful estimates of the vertical processes can be made. Virtually all the hydrographic measurements have been completed (see Siedler et al. this volume, for a review).

Below the level of the bounding ridge systems the flow of water into and out of the basin is strongly restricted to be through a small number of channels: the Vema and Hunter Channels along the southern boundary, the Romanche-Chain Fracture Zones on the equator and an additional narrow equatorial passage near 35°W which allows bottom water to enter the North Atlantic. Across each of these constrictions an array of current meters has been set with emphasis on measuring the transport of bottom water. In addition, an array has been set across the deep western boundary current system near 19°S, at the mid-basin latitude. At this writing the Vema Channel, Hunter Channel and the equatorial passage moorings at 35°W have all been recovered.

Finally, a deliberate tracer release has been proposed for two sites at the 4000 m level. This would be similar to the successful experiment in the North Atlantic known as NATRE (North Atlantic Tracer Release Experiment, Ledwell et al. 1993), although at a much deeper depth, and would allow direct estimation of cross-isopycnal diffusion and a distinction between boundary and interior processes.

Results

Deep Circulation

The only direct velocity measurements presently available from the DBE are those deriving from the moored array along the southern boundary and a small number of neutrally buoyant float deployments. The array of current meters is superimposed on a full depth potential temperature section made during the deployment cruise in Fig. 2. This cruise, *Meteor* 15 leg 1, occurred in January, 1991 and some 57 conventional current meters and 2 Acoustic Doppler Current profilers were deployed on 13 moorings in a joint effort by the mooring groups at the Institut für Meereskunde, Kiel (IfM) and Woods Hole Oceanographic Institution (WHOI) (see Tarbell et al. 1994 for details). Two of the moorings failed to return when so commanded during the recovery in late 1992 but part of one of these was rescued on a cruise of opportunity later in the year; Fig. 2 shows those instruments that were deployed. In this report we will concentrate on the mean flow

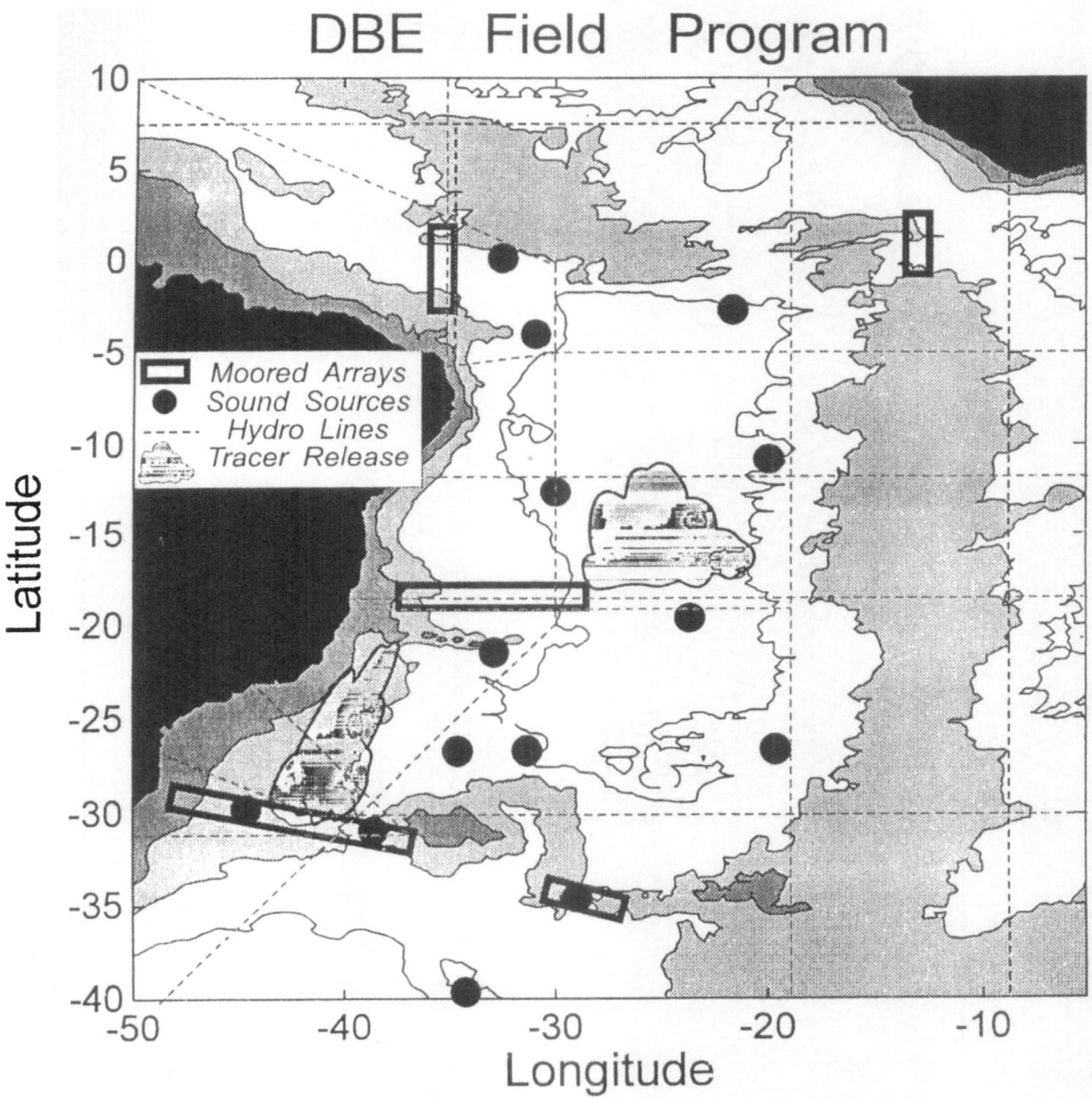

Fig. 1. The Deep Basin Experiment Field program showing the major elements of hydrography, current meter arrays, neutrally buoyant floats and a deliberate tracer release.

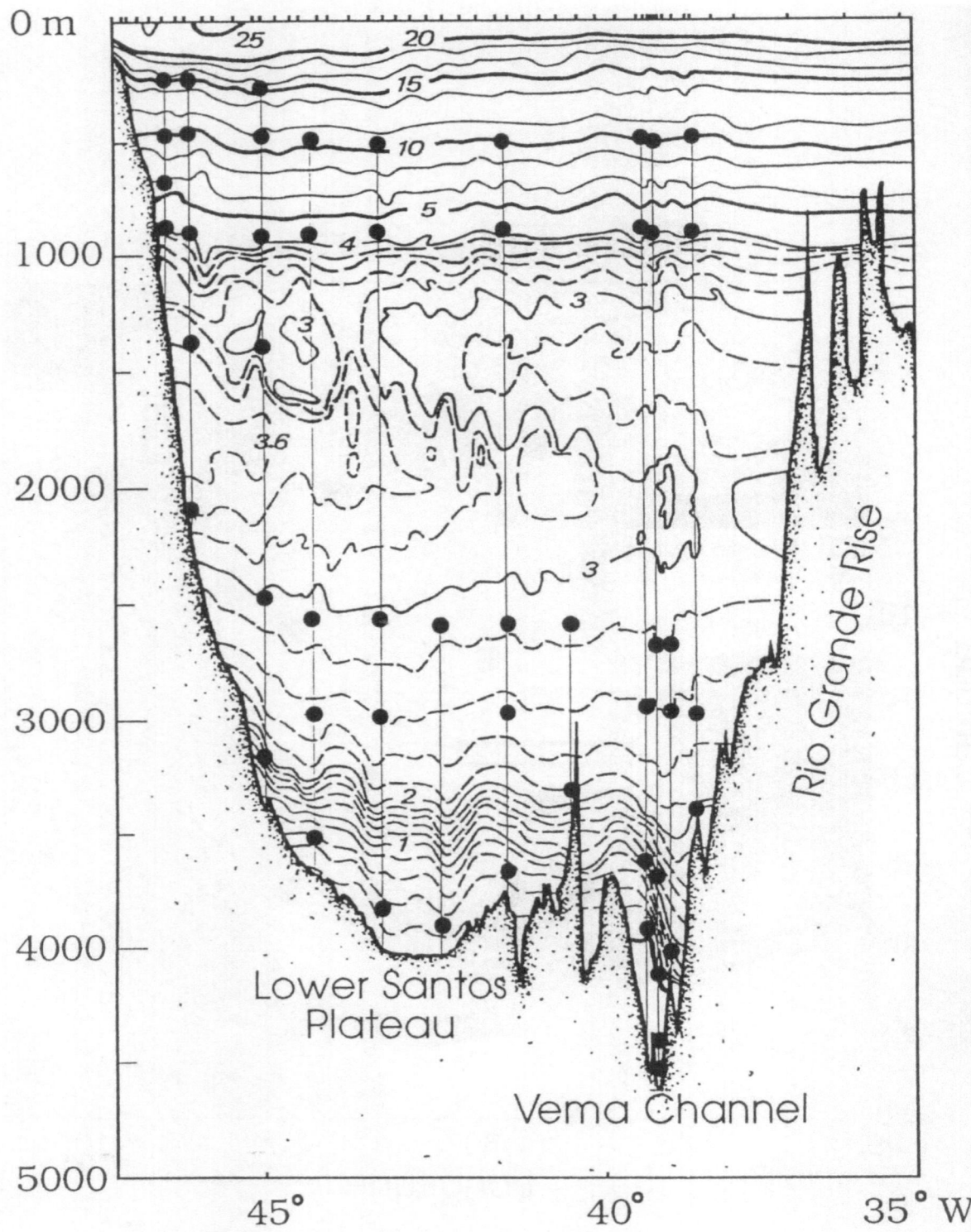

Fig. 2. The moored array along the southern boundary of the Brazil Basin superimposed on a section of potential temperature constructed from CTD data which were collected during the first deployment cruise.

vectors from the measurements made in the three subthermocline water masses and compare these with the recent synthesis of Reid (1989).

Antarctic Intermediate Water (AAIW): North of the Antarctic Circumpolar Current conventional wisdom has the AAIW circulating in two basin scale gyres, a subtropical anticyclone centered near 35°S and a tropical cyclone centered near 15°S. These are not uncoupled and part of the anticyclone feeds water northward near the western boundary into the cyclonic circulation and this eventually makes its way into the North Atlantic (e.g. Schmitz and McCartney 1993) as part of the thermohaline circulation. In Reid's (1989) scheme the separation point between the southward and northward flowing deep western boundary currents for the AAIW is about 30°S, the latitude of the Santos Plateau - Rio Grande Rise ridge system that forms the southern boundary of the Brazil Basin and the location of the Kiel–WHOI array (Fig. 3a). The mean current vectors at this depth are broadly consistent with this concept. They all show flow toward the coast and a veering toward the southwest, as though the separation point is farther to the north. Additional information from neutrally buoyant float tracks at this level suggests that the separation point may be as far north as 25°S.

North Atlantic Deep Water (NADW): The direct current measurements pertinent to this water mass are in the 2000 m to 3000 m range and their means are shown in Fig. 3b with the Reid scheme at this level superposed. According to Defant (1941), Reid (1989) and others (e.g. Zhang and Hogg 1992) NADW flows southward to about 50°S as a deep western boundary current before turning back north in a very elongated loop eventually making its way across the Mid-Atlantic Ridge north of 20°S. Within the Brazil Basin this implies mainly meridional motion with southward flow near the boundary and northward flow in the interior. Recently DeMadron and Weatherly (1994) have proposed a more structured scheme, using the additional hydrographic coverage offered by the South Atlantic Ventilation Experiment. Their scheme has a tight recirculation just offshore of the deep western boundary current and more recirculations in the interior of the Brazil Basin.

The mean vectors (Fig. 3b) provided by the moored array show generally southward flow from the Brazil Basin to the Argentine Basin across the area, the most surprising exception being the one closest to the coast where the flow is northward. (It is also the shallowest of the instruments in this layer but still within the NADW.) The NADW flow divides over the Lower Santos Plateau with that to the west of a shallow depression heading southwestward toward the coast (similar to the AAIW) and that to the east flowing southeastward. There is also evidence for some guiding by the Vema Channel consistent with Hogg et al.'s (1982) finding of a negative silicate anomaly along the channel axis. If anything, the flow at this level is orthogonal to the large scale pattern of Reid.

Also shown in Fig. 3b are three trajectories from RAFOS neutrally buoyant floats deployed at 2500 m and tracked for periods up to 400 days. These are over the floor of the Brazil Basin and suggest weak northward flow, consistent with Reid and the DeMadron and Weatherly (1994) ideas. The three floats are part of a larger scale seeding of the basin and Fig. 4 displays what trajectories are currently available at this time. These show mainly zonal motion with a dominance of westward flow east of 30°W again orthogonal to the Reid scheme. A very recent revision of the DeMadron and Weatherly scheme (G Weatherly, personal communication 1994) which incorporates hydrographic information obtained on the new WOCE section along 19°W is also dominantly zonal east of 30°W but banded: eastward north of 8°S, westward south to 12°S, eastward to 18°S and then westward once again. Our preliminary information in Fig. 4 is not inconsistent with this view but is badly aliased in terms of its meridional sampling. Future float deployments are intended to fill out this picture.

Also displayed on Fig. 4 are the values of the eddy kinetic energy estimated from the float motions. These range from 0.4 to 12.7 $cm^2 sec^{-2}$ with the larger values being near the western boundary. Our design for the float component was based on expectations of an eddy kinetic energy level of around 1 $cm^2 sec^{-2}$ in the interior and this seems to have been realized although the present data set is quite fragmentary.

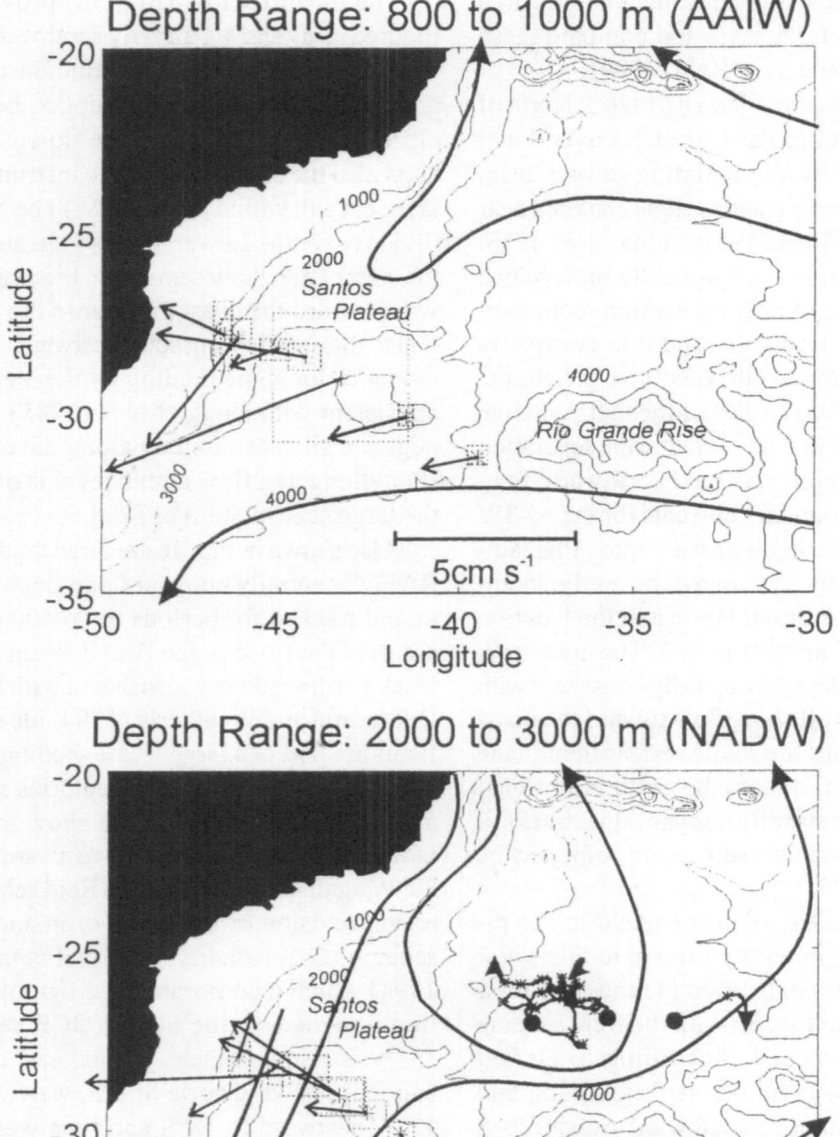

Fig. 3. Mean current vectors (straight arrows), trajectories from deep floats (convoluted curves with arrow heads) and the corresponding circulation scheme from Reid (1989) (larger scale, smooth curves) for the (a,upper) Antarctic Intermediate Water, (b,lower) North Atlantic Deep Water and (c,next page) Antarctic Bottom Water. The symbols along the float trajectories are spaced at 1-month intervals. The velocity scale applies only to the current meter vectors. Boxes on these vectors give the standard error of the mean assuming a 10-day integral time scale.

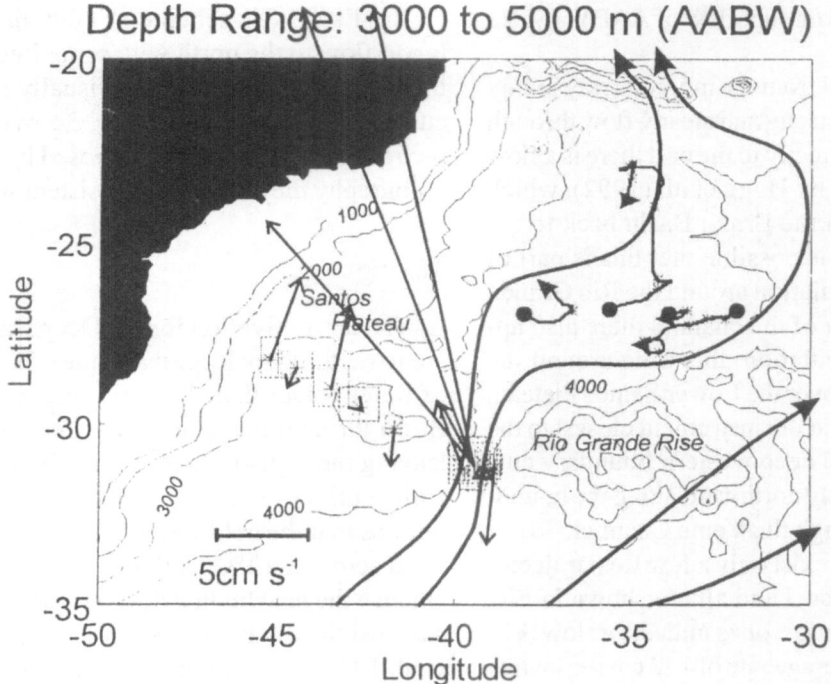

Depth Range: 3000 to 5000 m (AABW)

Antarctic Bottom Water (AABW): Fig. 3c shows the mean current meter vectors from moorings and RAFOS float trajectories that are available below 3000 m. Beneath this depth one finds the 2°C potential temperature isoline (cf. Fig. 2) which is often chosen as the upper limit of AABW (e.g. Speer and Zenk 1993).

At the southern boundary of the Brazil Basin the flow in the AABW layer is controlled by the bottom topography of the Rio Grande Ridge: the Vema Channel cuts through this ridge west of the Rio Grande Plateau while the Hunter Channel connects the Argentine Basin with the Brazil Basin farther east near the Mid-Atlantic Ridge. To the west of the Vema Channel the mooring line shown in Fig. 2 runs over the Lower Santos Plateau which includes an unnamed shallow depression outlined by the 4000 m isobath on Fig. 3c. Most of the bottom water throughflow occurs in the Vema Channel and Lower Santos Plateau region.

There is, however, an additional contribution by the Hunter Channel as identified by Speer et al. (1992) which amounts to approximately 20–25%

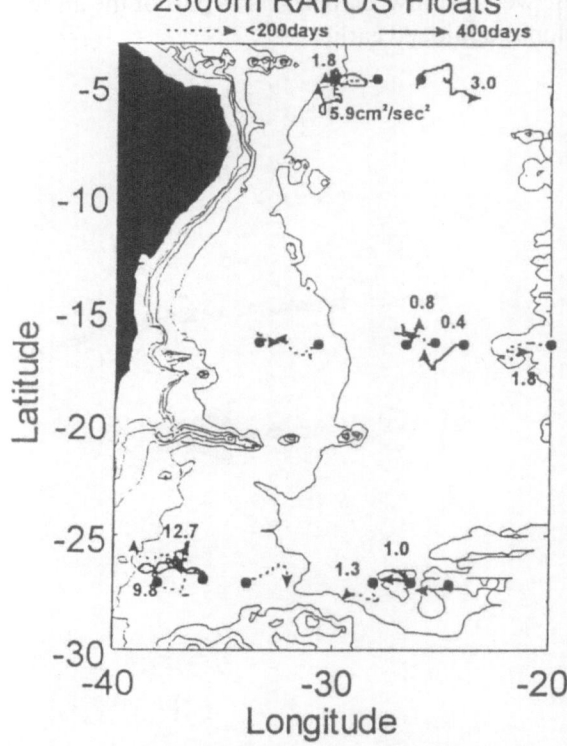

Fig. 4. Float trajectories at 2500 m depth from an initial deployment of RAFOS floats. These range from about 60 days to 400 days in duration. The numbers associated with the trajectories give estimates of eddy kinetic energy for the longest lived floats.

of the Vema Channel throughflow of AABW (Speer and Zenk 1993).

The flow deduced from the moored array in Fig. 2 is dominated by a strong and steady flow through the channel itself. Directly to the east there is a flow reversal, also noted by Hogg et al. (1982), which supplies water from the Brazil Basin back to the Argentine Basin. It is possible that this is part of an anticyclonic circulation around the Rio Grande Plateau. To the west of the channel there also appears to exist recirculation, this time around the shallow depression over the Lower Santos Plateau. The northward flow at the instrument closest to the coast is a part of the deep western boundary current that flows directly northward along the boundary rather than through the Vema Channel.

At present, there exist only a few float trajectories at the 4000 m level and all are shown in Fig. 3c. The two westernmost ones indicate a slow drift to the north at the average rate of 1–2 cm sec^{-1} while the two easternmost ones indicate a flow more to the west-southwest, perhaps as part of the anticyclone mentioned earlier.

Reid's (1989) scheme at 4000 m suggests a broad flow to the north with some intensification in the Vema Channel. If one visually smooths the current meter vectors of Fig. 3c over the local recirculations that may be imposed by small scale topography they are not inconsistent with Reid.

Mixing

A primary motivation for the Deep Basin Experiment was the knowledge that some 4 Sverdrups (Sv) of water colder than 1°C was entering the Brazil Basin through the Vema Channel and the water leaving through all the known exits was warmer. Unless this bottom layer is expanding in time this implies that the cold water must be migrating upward across the 1°C isotherm (Fig. 5). In order to balance the heat budget downward diffusion with a diffusivity of about 5 cm^2 sec^{-1} is implied (Hogg et al. 1982), approximately 50 times the rate that has been measured in the main thermocline (Ledwell et al. 1993). Because this estimate is derived from basin-wide budgets it represents an average rate for

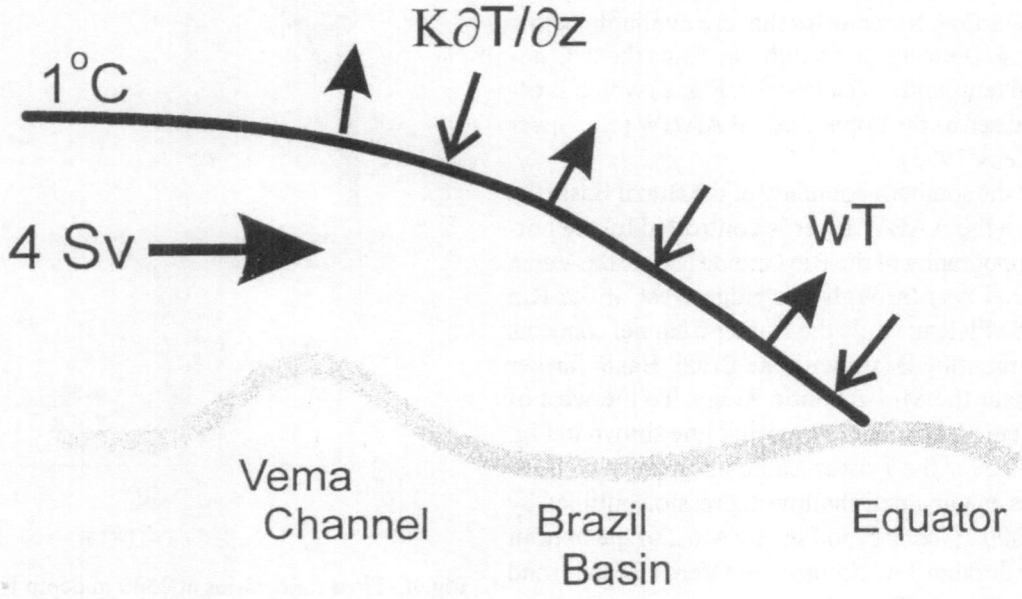

Fig. 5. A schematic view of the basin averaged heat and mass budgets for the Brazil Basin illustrating how the two balances permit estimation of the cross isotherm diffusivity.

the basin: it is unknown whether the mixing processes are more or less uniform across the basin or concentrated in areas such as the western boundary regions or within passages where the mean flow is most intense.

It is possible to make a crude estimate of the mixing rate occurring within the Vema Channel. In Fig. 6a we show the horizontal eddy temperature flux vectors at those sites within the channel where appropriate data are available. These data come from two different experiments of about 2 years duration. The southern two moorings are placed very close to the sill (Zenk et al. 1993) and both have eddy temperature flux directed toward the south, more along the mean temperature contours than across. On the other hand, directly downstream from the sill the heat flux in the swiftly flowing part of the channel is westward or down the mean temperature gradient (Fig. 6b). Using an estimate of the mean cross-channel temperature gradient of 0.2°C in 10 km yields a horizontal eddy diffusion coefficient, $K_H \approx 5 \times 10^6$ cm^2 sec^{-1}. Because of the steep slope of the isotherms, especially at 4000 m, this horizontal mixing can be projected into the cross-isotherm coordinate to obtain one component of the cross-isotherm or diapycnal diffusivity (the other comes from the vertical flux which is unmeasureable from our current meters). We do this by multiplying K_H by the square of the isotherm slope to rotate both the mean temperature gradient and the horizontal velocity into this frame. The isotherms slope at a rate of about 100 m in 10 km and this gives:

$$K_d \approx 50 \text{ cm}^2 \text{ sec}^{-1}$$

a value which is an order of magnitude larger than the basin average quoted earlier. It is interesting to note that the area of the basin at this level is about 2500 km × 1500 km (N-S × E-W) while the area of the Vema Channel along which mixing might be occurring at this rate is 400 km × 20 km. The ratio of these areas is about 500 while the ratio of the above diffusivity to the basin average is about 10 implying that much of the mixing must occur elsewhere although the enhanced mixing within the passage suggests that boundary mixing could be the

important process. Of course the above is an incomplete estimate of K_d: it is possible that the vertical eddy temperature flux could contribute substantially.

Long Period Changes

The above arguments are based on the assumption of steady state so that conservation statements for heat and mass can be employed. The water flowing through the Vema Channel has been observed a number of times over the past 20 years (Fig. 7) and the coldest water, observed to hug the eastern channel wall, has been found to be about -0.18°C ± 0.005 on six occasions from as far back as the first GEOSECS (Geochemical Ocean Sections Study) observation in 1972 to the time of the first DBE mooring deployment in early 1991 (*Meteor* 15 leg 1). However, on the recovery cruise (*Meteor*

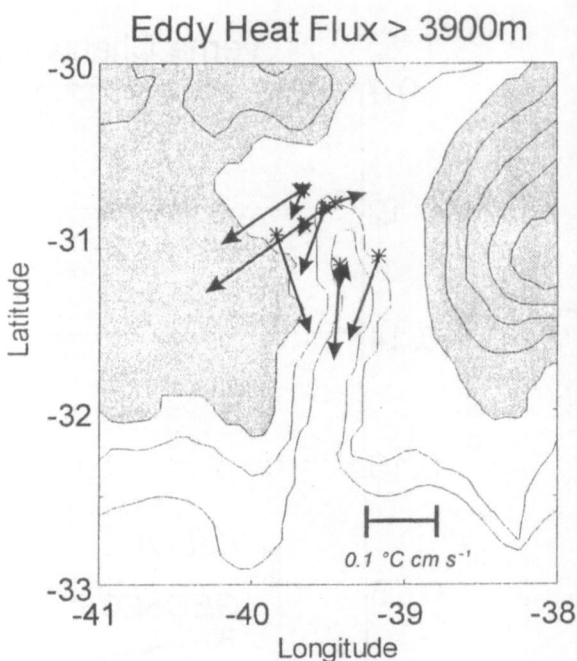

Fig. 6a. Estimates of eddy temperature flux from moored velocity and temperature measurements within the Vema Channel. Those farthest south are near the sill of the Channel and were made in 1991-1992 while the others are farther downstream and were obtained in 1979-1981.

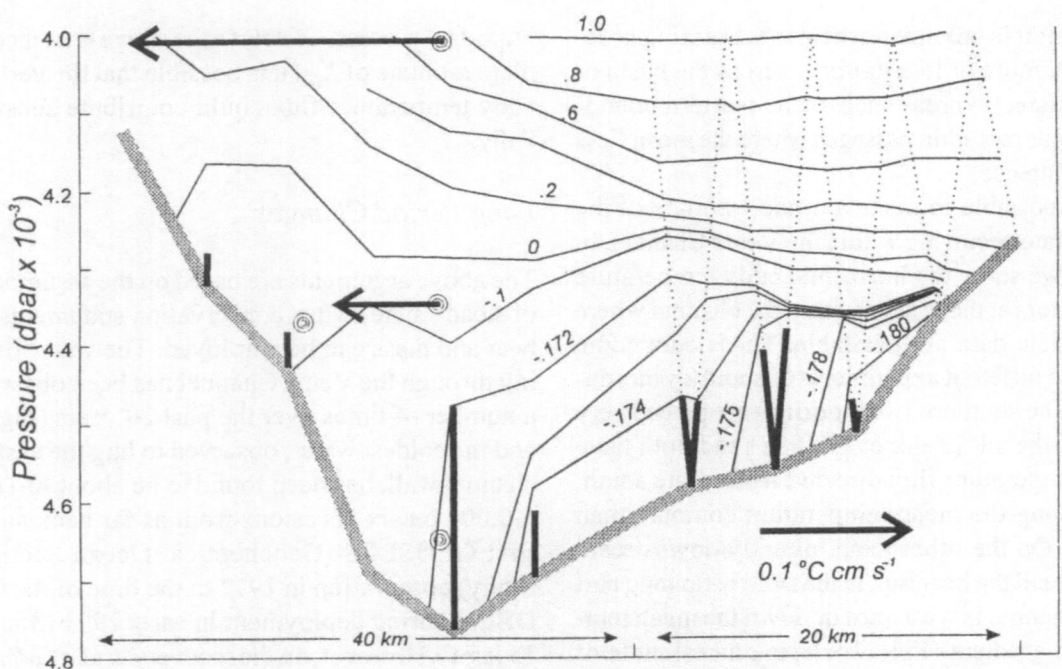

Fig. 6b. A cross channel section of potential temperature from Hogg et al. (1982) to which the cross channel eddy temperature flux has been added showing that it is down gradient.

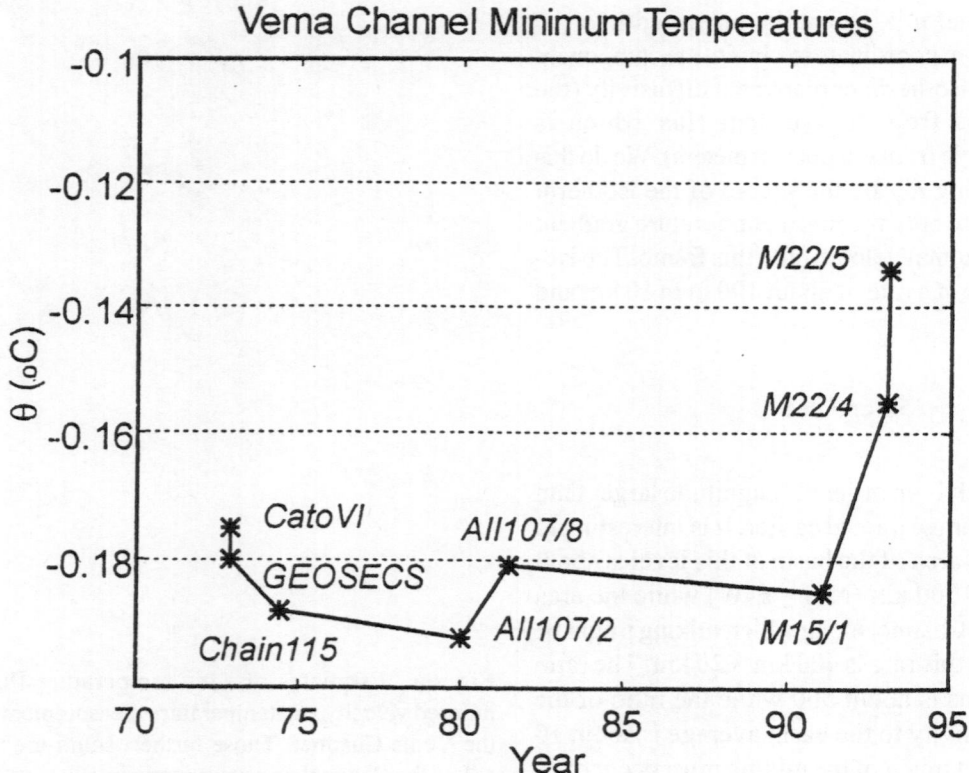

Fig. 7. Warming of the bottom water flowing through the Vema Channel into the Brazil Basin (see text, from Zenk and Hogg 1996).

22 leg 4) in late 1992 the coldest water to be found was -0.155°C and the next leg (*Meteor* 22 leg 5) measured the coldest water to be -0.135°C. This is consistent with temperature trends inferred from the bottom-most instruments on the moored array and indicates a warming of the AABW during this period over the whole of the southern boundary of the Brazil Basin. A similar warming of deep and bottom water in the Argentine Basin has been found in a comparison of late 1980s data with that taken a decade earlier (Coles et al. 1996). More details are given in that work and by Zenk and Hogg (1996).

Summary

Results from the DBE field program are just now returning and it is somewhat premature to make any definitive conclusions. The presently available information is concentrated along the southern boundary of the Brazil Basin. However, with those caveats, it does appear that the mean circulation in the AAIW layer looks much like Reid's (1989) scheme with an anticyclonic gyre centered south of the basin and flow incident on the Brazilian coast, part of which continues north to cross eventually into the North Atlantic. This part of the circulation will be better observed by float data yet to come. The observations in the NADW layer suggest a circulation which is more zonal than was anticipated but is not inconsistent with a new (and unpublished) scheme based on very recent hydrographic observations in the eastern part of the Brazil Basin (G Weatherly, personal communication 1994). The flow of AABW across the southern boundary is generally toward the north, as expected, but there are smaller scale reversals induced by the Rio Grande Rise and a small depression within the Lower Santos Plateau.

Estimates of the horizontal temperature flux from moorings within the Vema Channel have been used to estimate a cross-isothermal diffusivity of $50 \text{ cm}^2 \text{ sec}^{-1}$, a number an order of magnitude larger than the basin averaged value of $5 \text{ cm}^2 \text{ sec}^{-1}$. The ratio of the area of the basin to the area of the channel is about 100. It is possible that most of the mixing could be occurring in the intense flow regions

of the channel and along the western boundary rather than uniformly throughout the basin.

The notion of a steady deep circulation is also challenged by our observation of a warming of the bottom water flowing over the southern boundary. Although little change had been observed through the 1970s and 1980s it appears that a fairly rapid warming is occurring in the early 1990s but this is restricted to the Antarctic Bottom Water layer.

Acknowledgments

We are grateful for the support provided by the National Science Foundation through grants OCE 90-04396 (NGH) and OCE 90-04864 (WBO & NGH), the Deutsche Forschungsgemeinschaft (Si 111/38-1, Si 111/39-1, Ze 145/6-1, GS & WZ) and the Bundesministerium für Forschung und Technologie (03F0535A, 03F0050D, GS & WZ). Reviewers' comments helped clarify the text.

This is contribution number 8880 from the Woods Hole Oceanographic Institution.

References

Coles VJ, McCartney MS, Olson DB, Smethie Jr WM (1996) Changes in Antarctic Bottom Water properties in the western South Atlantic in the late 1980's. J Geophys Res (in press)

Defant A (1941) Quantitative Untersuchungen zur Statik und Dynamik des Atlantischen Ozeans. Die relative Topographie einzelner Druckflächen im Atlantischen Ozean. In: Wissenschaftliche Ergebnisse der Deutschen Atlantischen Expedition auf dem Forschungs- und Vermessungsschiff "Meteor" 1925–1927, 6, 2nd Part, 4, pp 183-190

DeMadron XD, Weatherly G (1994) Circulation, transport and bottom boundary layers of the deep currents in the Brazil Basin. J Mar Res 52:583-638

Hogg N, Biscaye P, Gardner W, Schmitz Jr WJ (1982) On the transport and modification of Antarctic Bottom Water in the Vema Channel. J Mar Res Suppl to 40:231-263

Johnson GC, Warren BA, Olson DB (1991) Flow of bottom water in the Somali Basin. Deep-Sea Res 38:637-652

Ledwell JR, Watson AJ, Law CS (1993) Evidence for slow mixing across the pycnocline from an open-ocean tracer release experiment.

Nature 364:701-703

Reid JL (1989) On the total geostrophic circulation of the South Atlantic Ocean: flow patterns, tracers, and transports. Prog Oceanog 23:149-244

Schmitz WJ Jr, McCartney MS (1993) On the North Atlantic Circulation. Rev Geophys 31(1):29-49

Siedler G et al. this volume

Speer KG, Zenk W (1993) The flow of Antarctic Bottom Water into the Brazil Basin. J Phys Oceanogr 23:2667-2682

Speer KG, Zenk W, Siedler G, Pätzold J, Heigland C (1992) First resolution of flow through the Hunter Channel in the South Atlantic. Earth Planet Sci Lett 113:287-292

Stommel H (1957) The abyssal circulation of the ocean. Nature 180(4589):733-734

Swallow JC (1971) The *Aries* current measurements in the western North Atlantic. Phil Trans Roy Soc London A 270:451-460

Tarbell S, Meyer R, Hogg N, Zenk W (1994) A moored array along the southern boundary of the Brazil Basin for the Deep Basin Experiment - Report on a joint experiment 1991-1992. Woods Hole Oceanographic Inst Tech Rept WHOI-94-07 and Berichte aus dem Institut für Meereskunde an der Christian-Albrechts-Universität-Kiel, IfM-Kiel 243, 107 pp

Warren BA (1981) Deep water circulation in the world ocean. In: Warren BA, Wunsch C (eds) Evolution of Physical Oceanography, Scientific Surveys in Honor of Henry Stommel. The MIT Press, Cambridge, Massachusetts, pp 6-41

Warren BA, Owens WB (1988) Deep currents in the central subarctic Pacific Ocean. J Phys Oceanogr 18:529-551

Warren BA, Speer KG (1991) Deep circulation of the eastern South Atlantic Ocean. Deep-Sea Res 38(suppl):281-322

Zenk W, Hogg NG (1996) Warming trend in Antarctic Bottom Water flowing into the Brazil Basin. Deep-Sea Res (submitted)

Zenk W, Speer KG, Hogg NG (1993) Bathymetry at the Vema Sill. Deep-Sea Res 40:1925-1933

Zhang H, Hogg NG (1992) Circulation and water mass balance in the Brazil Basin. J Mar Res 50:385-420

The Deep Water Regime in the Equatorial Atlantic

M. Rhein, F. Schott, J. Fischer,
U. Send and L. Stramma

Institut für Meereskunde an der Universität Kiel
Düsternbrooker Weg 20, 24105 Kiel,
GERMANY

Abstract: The importance of the Deep Western Boundary Current (DWBC) in the Atlantic for the interhemispheric exchange of water masses and of heat is well known, but data to estimate transports and to follow its pathways are sparse, especially in the equatorial Atlantic. New insight into the distribution of water masses in the DWBC, their transports and their variability off Brazil were gained in a contribution to the WOCE (World Ocean Circulation Experiment) program. In these studies, moored current meter measurements were combined with shipboard data from three cruises in the years 1990, 1991 and 1992. Besides tracer (Chlorofluoromethanes CFMs, components F11 and F12) and hydrographic data, direct velocity measurements were carried out using a lowered ADCP attached to the CTD, and the Pegasus profiling system.

The estimated transports of deep water in the equatorial Atlantic, net eastward transport of 19-22 Sv at 44°W, and 26.8 ± 7.0 Sv at 35°W, net southward transport of 19.5 ± 5.3 Sv at 5°S, are in the range of previously published estimates farther west and south. The data show significant spatial and temporal variability of the flow field and of the estimated transports as well as variability in the tracer distributions. This can lead to large uncertainties in the interpretation of single cruise observations.

The transient tracer distributions (CFMs and tritium) along the DWBC from the northern North Atlantic to 10°S have been used within a box model to estimate the mean spreading velocities of the tracer bearing water masses of the DWBC. Together with assumptions about the mean spatial extent of the DWBC and the vertical velocity structure a mean transport of the DWBC of 9-12 Sv is obtained. These numbers represent the spatially and temporally averaged net DWBC transport, i.e. the deep part of the thermohaline circulation. Thus, the tracer derived estimates are comparable to the estimates from inverse calculations. The different results between the transports obtained from direct observations on one side and from inverse calculations and tracer distributions on the other side are thought to be caused by various recirculation cells along the path of the DWBC. Indications for various recirculation paths of part of the NADW have also been found off Brazil.

Introduction

The Deep Western Boundary Current (DWBC) in the North Atlantic Ocean is a major component of the global oceanic thermohaline circulation, but its transports and pathways are not yet fully resolved. Flowing along the continental margin of the American continent, the DWBC transports water from northern origin to the south across the equator. In proceeding south from its formation region, the DWBC encounters water of southern origin, form-ing a northward flow of Intermediate Water above 1200m. The near-bottom flow is also northward and carries Antarctic Bottom Water (AABW) into the northern hemisphere. The DWBC is important not only for interhemispheric water mass exchange but also for the climate relevant meridional heat flux.

The contributions to NADW come from various sources and the NADW has been separated in the literature into one to three water masses. Recently,

From WEFER G, BERGER WH, SIEDLER G, WEBB DJ (eds), 1996, *The South Atlantic: Present and Past Circulation*. Springer-Verlag Berlin Heidelberg, pp 261-271

measurements of the anthropogenic tracers tritium ³H and CFMs lead to a distinction of two water types within the upper part of the NADW, resulting in a total of four separate NADW components (Pickart, 1992).

Estimates of the DWBC transport crossing the equator are sparse. There are only a few direct velocity data and the closeness of the equator makes geostrophic computations difficult. Reported estimates in the general area of the tropical Atlantic, evaluated from geostrophic computations, but also from current meter moorings, are in the order of 17-25 Sv of NADW flowing towards and across the equator (Speer and McCartney 1991; Molinari et al. 1992; Schott et al. 1993; Schmitz and McCartney 1993). Direct observations tend to obtain higher values than the estimates derived from inverse calculations (about 10-15 Sv, Bryden and Hall 1980; Roemmich 1980; Hall and Bryden 1982; Wunsch and Grant 1982). The difference between these estimates are explained by the presence of various recirculation cells along the spreading path of the DWBC (Schmitz and McCartney 1993; McCartney 1993; Friedrichs et al.1994).

The distributions and transports of deep water masses at the western boundary in the tropical Atlantic off Brazil have been studied during three surveys along 44°W, 35°W, and 5°S and one at 10°S

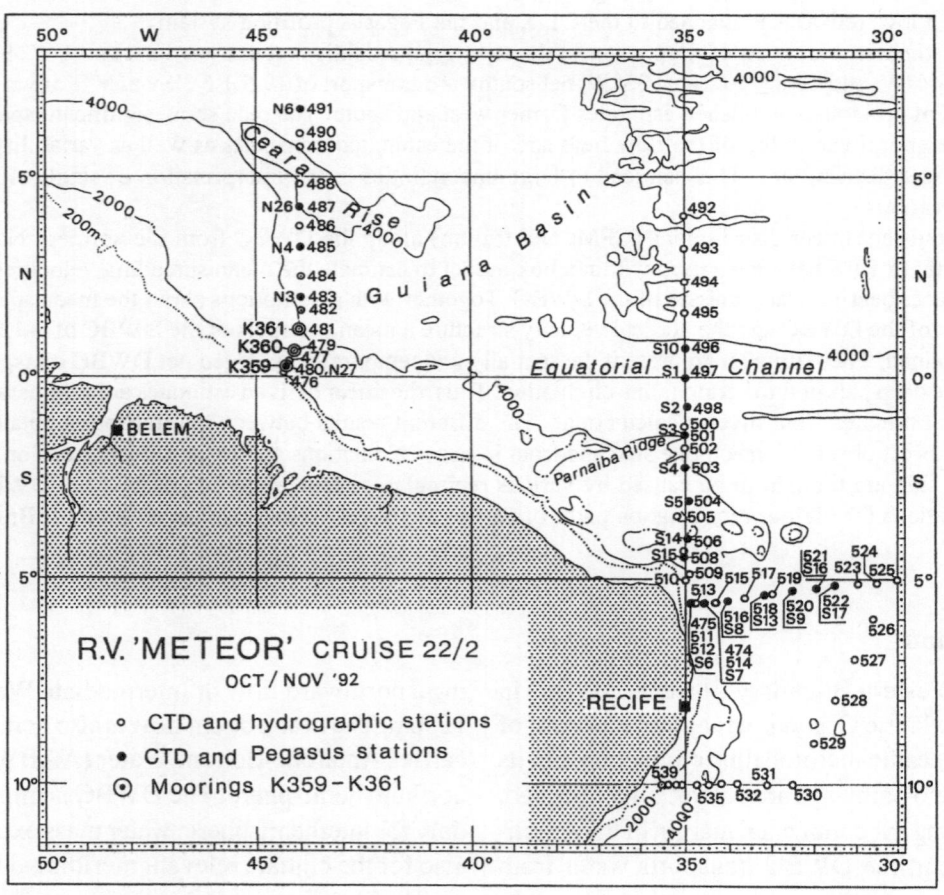

Fig. 1. CTD-stations (o) and CTD and Pegasus stations (●) of RV Meteor cruise M22/2 in Oct-Nov. 1992. Also shown are the positions of the moorings K359 - K361 at 44°W. The sections along 44°W from the Brazilian shelf to south of Ceara Rise, along 35°W and along 5°S were also carried out in October 1990 and May 1991.

(Fig.1). The cruises were conducted in 1990, 1991 and 1992.

Transports were obtained from direct measurements of the velocity fields with the Pegasus profiling system (Spain et al. 1981) and with the lowered Acoustic Doppler Current Profiler (LADCP, Fischer and Visbeck 1993) and from geostrophic computations. At 44°W, moorings were deployed for several years in the DWBC (Schott et al. 1993).

Water Masses

Using CFMs (components F11 and F12) and hydrographic distributions, four water masses could be identified forming the North Atlantic Deep Water (NADW) system. The water mass boundaries and the characteristics of these four components in the tropical Atlantic are shown in Table 1. Two of these water masses have a high CFM content - the 'shallow upper NADW' (SUNADW) and the 'overflow lower NADW', (OLNADW). In the tropical Atlantic, the upper tracer maximum of the SUNADW is found in depths around 1600m (Figs. 2a, b). This water mass is presumably formed by convection to about 600m depth in the southern Labrador Sea (Pickart 1992) or the Irminger Sea (Rhein 1995) rather than being a modified form of the classical Labrador Sea Water (LSW).

Below the SUNADW, water with the T-S characteristic of classical LSW is found in depths between 1900 and 2400m. The comparatively low CFM and tritium load of LSW might be due to incomplete convection in the formation region during the 1960s and 1970s (e.g. Read and Gould 1992). Below 2400m, there is a low oxygen, low CFM layer, consisting mainly of water which spilled through the Gibbs Fracture Zone from the eastern into the western North Atlantic (LNADW-old). This water mass leaves the formation region with relatively low tracer concentrations and in general proceeds more slowly towards the equator than the tracer bearing water masses (Fine and Molinari 1988; Molinari et al. 1992; Watts 1991; Schott et al. 1993). The deepest part of the NADW, the OLNADW is centered around 3800m and exhibits an oxygen and CFM maximum (Figs. 2a, b). Both maxima are caused by the convective renewal of this

water mass north of Iceland before overflowing through the Denmark Strait. The increase in the CFM concentrations in SUNADW and OLNADW between 1990 (Fig. 2a) and 1992 (Fig. 2b) reflects the arrival of water, which had been renewed more recently.

The bottom layer is occupied by the Antarctic Bottom Water (AABW). The denser layers of AABW originate in the Weddell Sea, the lighter parts come from the Circumpolar Current. At 35°W, AABW enters the Guiana Basin through the Equatorial Channel between 1°30'S and 0°50'N (Fig. 1). This is the only path, where AABW can enter the western North Atlantic. In 1991, the first advent of the F11 load of AABW at the equator was observed, i.e. 'younger' water from the South Atlantic had arrived: farther upstream at 10°S, the F11 signal is about 4 to 5 times higher than at 35°W.

Transports

At 44°W, the current meter measurements from two consecutive yearlong deployments showed a maximum annual mean southeastward speed of 30 cms⁻¹ at about 1500m depth in the DWBC (Fig. 3). The current core (velocities > 10 cms⁻¹) between 1300 m and 3100 m was located along the continental slope. The estimated transport of upper NADW (SUNADW and LSW) was 14.2-17.3 Sv, depending on the extrapolation used between the mooring in the core and the continental slope. This transport is higher than cross-equatorial estimates and suggests near equatorial recirculation in the upper NADW level, in agreement with northwestward mean flow found about 140 km offshore. The recir-culation was also observed in the float data of Richardson and Schmitz (1993).

Below 3100m and the 1.8°C isotherm, the core of the lower NADW had speeds of 10-15 cms⁻¹ (Fig. 3). It is found over the flat part of the basin near 1.5°N, clearly separated from the continental slope by a zone of near zero mean speeds. The estimated transport of 4.5 Sv is significantly below other estimates farther to the north (Molinari et al. 1992) and suggests that a major fraction of lower NADW may cross the 44°W meridian north of the Ceara Rise. The intraseasonal variability is large, and occurs at a period of about 2 months. At the upper

Table 1: Limits and characteristics of the water masses in the deep equatorial western Atlantic (from Rhein et al. 1995). Abbreviations are: σ, density; θ, potential temperature; max, maximum; S, salinity; min, minimum

water mass or density boundary	property	range	approx.depth mean 35°W, 5°S
σ_1 = 32.15 $\sigma_{1.5}$ = 34.42			1200 m
SUNADW	θ-max F11-max S-max	3.4 - 4.5°C 34.80-35.00	
σ_2 = 36.94 $\sigma_{1.5}$ = 34.70			1900 m
LSW	θ O_2-max S	2.8 - 3.4°C 34.94-34.98	
σ_2 = 37.00 $\sigma_{1.5}$ = 34.755			2400 m
LNADW-old	θ O_2-min F11 min S	2.2 - 2.8°C 34.91-34.95	
σ_4 = 45.83			3450 m
OLNADW	θ F11 max O_2-max S	1.75 - 2.2°C 34.85-34.91	
σ_4 = 45.90			3900 m
AABW	θ decrease in S,O_2,Θ S F11 increase towards the bottom	< 1.75°C <34.85	

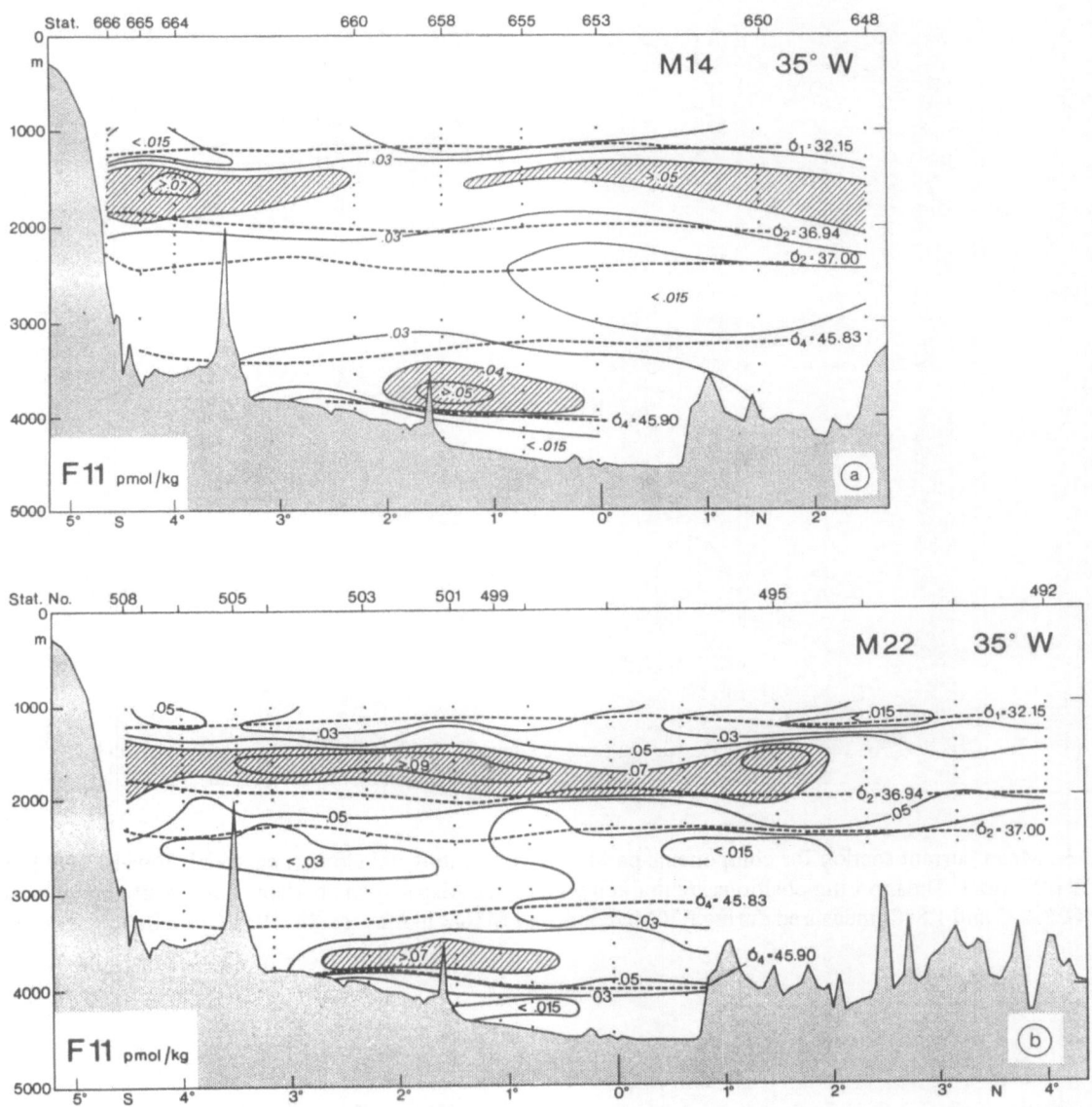

Fig. 2. F11 section along 35°W off Brazil. The data have been taken during the RV Meteor cruises M14 in October 1990 (a) and during M22 in November 1992 (b). The F11 maxima are shaded. Also shown are the isopycnals, which define the water mass boundaries according to Table 1 (from Rhein et al. 1995).

NADW level, a decrease and even a reversal of the DWBC occurred in April 1991 and again in late May to June of the same year.

We combined transport estimates from three cruises (1990, 1991 and 1992) and from both measurement methods (direct, i.e. Pegasus and lowered ADCP) to look for persistent flow fea-

tures and to obtain mean transports and their variability (Rhein et al. 1995). At 35°W, the two water masses with a tracer tongue (SUNADW and OLNADW) showed the highest velocity signals (Fig. 4a). The greater part of the southeastward transport of the SUNADW (centered around 1600 m) flows about 300 km offshore between 3°09'S

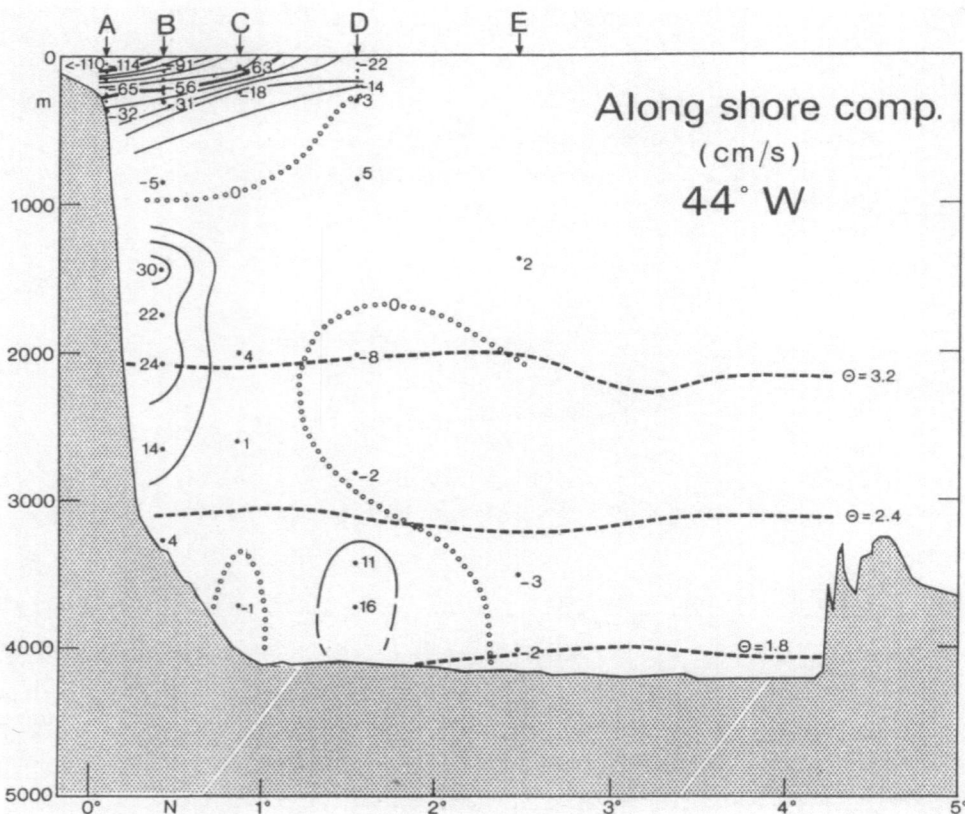

Fig. 3. Mean current section for components parallel to the continental slope (negative is toward 308°) along 44°W off Brazil. The mooring positions are marked with A - E. Also shown are depths of potential temperatures 3.2°C, 2.4°C and 1.8°C, measured during RV Meteor cruise M16/3 in May 24-28, 1991 (from Schott et al. 1993).

and 1°50'S (9.7±3.3 Sv); farther north in that section, a highly variable reversing flow is observed in a second velocity maximum.

In contrast to the SUNADW, the OLNADW (centered around 3800 m) seems to be topographically guided by the Parnaiba Ridge at 1°45'S, 35°W (4.6±2.6 Sv). The water masses located between the two CFM maxima, the Labrador Sea Water (LSW) and the LNADW-old water mass , did not show any persistent flow feature however, a rather constant transport of 11.1±2.6Sv was observed for these two layers. The total southeastward flow of the NADW at 35°W showed a transport of 26.8±7.0 Sv, if one neglects the reversing SUNADW north of 1°50'S.

At 5°S, the flow of all deep water masses shows vertically aligned cores (Fig. 4b); the main southward transport occurred near the coast (19.5±5.3 Sv). The boundary current is limited offshore by a flow reversal, present in all three surveys, but located at different longitudes. At 10°S, a southward transport of 4.7 Sv was observed in November 1992. However, the section extended only to 32°30' W, so that probably a significant part of the flow has been missed.

An important result is the large transport variability between single cruises as well as variability of the spatial distribution of the flow at 35°W, which could lead to large uncertainties in the interpretation of single cruise observations. Despite these uncertainties, a circulation pattern of the various deep water masses near the equator is suggested by combining our mean transport estimates with other observations (Fig. 5a-d). In our cartoon, the observations are presented in solid arrows, and they have been complemented by the open arrows, calculated from mass balance considerations. Although the LSW and LNADW-old water masses differ in their formation and in their formation region, they have a similar flow field in the equatorial Atlantic and are thus combined.

The eastward transport of SUNADW in the DWBC (Fig. 5a) seems to increase from 4 Sv west of 46°W to 11 Sv at 35°W (neglecting the highly variable flow at 35°W north of 1°50'S). This flow pattern could be caused by various recirculation paths. Recirculation of this water mass is also suggested in the CFM and salinity distributions as well as in the float data of Richardson and Schmitz (1993). South of the equator, the SUNADW transports decrease from 7 Sv at 5°S to 3 Sv at 10°S, but the presence of strong recirculation cells farther offshore makes it impossible to decide whether the missing 4 Sv at 10°S flow south or whether they are carried into the interior of the Brazil Basin.

Similar to the flow field of SUNADW the transports of the combined LSW and LNADW-old (Fig. 5b) increase from 5-6 Sv at 44°W to about 11 Sv at 35°W. At 10°S, no net southward flow of these water masses was observed in 1992. The section at 10°S was, however, limited to the region west of 32°30'W and only occupied once, so that probably the major part of the flow had been missed. In contrast to the DWBC at 44°W, the main transport of the upper part of the NADW (SUNADW and LSW) is not trapped to the coast, but occurs about 320 km offshore.

The influence of the topography on the flow field of the OLNADW is evident (Fig. 5c): At 44°W, the Ceara Rise seems to split the OLNADW, with part of the flow on the southern side of the Rise, and part of it leaves the DWBC and flows east on the northern flank of the Ceara Rise. At 35°W, the OLNADW flow is guided by the Parnaiba Ridge and about 5 Sv flow farther to the east and south. This indicates, that part of the water recirculates west of 35°W and north of Ceara Rise. The difference of the tracer signals and the θ - S characteristic between the 35°W and the 5°S section suggest that a major modification of this water mass occurs between these two sections.

About 2.6 Sv of AABW enter the Guiana Basin (Fig. 5d) and the rising topography towards the west prevents the AABW from proceeding west south of Ceara Rise. Instead it has to turn to the northern side of Ceara Rise.

The total eastward transport for the four water masses (SUNADW, LSW, LNADW-old, OLNADW) forming the NADW are 19-22 Sv at 44°W (Schott et al. 1993) and 26.8 ± 7.0 Sv at 35°W. At 5°S, the southward transport of the DWBC was estimated to 19.5 ± 5.3 Sv. These numbers are in the range of previously published estimates from observations farther west and south, but they show the important influence of spatial and temporal variability.

Estimating mean velocities and transports of the DWBC using CFM and Tritium distributions

The time information of the transient tracer distributions (CFMs, ³H) along the DWBC in the North Atlantic can be used to estimate the mean net DWBC transport. Estimates derived from this method are closer to the calculation by inverse methods, because the tracer field integrates over temporal and spatial scales.

Therefore, a simple box model for the tracer bearing components of the DWBC (OLNADW and SUNADW) is developed, which uses the ideas of Pickart et al. (1989) to parameterize turbulent diffusion of the current with its surrounding. Turbulent diffusion has to be included in the calculation, because it modifies the tracer signal in the DWBC significantly. In contrast to previous models, the boundary conditions include all water masses forming the lower part of the DWBC (Denmark Strait Overflow Water, Iceland Scotland Overflow Water, Northeast Atlantic Water). The model-derived mean velocity of the DWBC lead to tracer concentrations, which have to fit the observed F11 and F12 distributions, the F11/F12 ratios, and the tritium distributions. The model area extends from south of the Faeroe Bank along the continental margin of the American continent to 10°S.

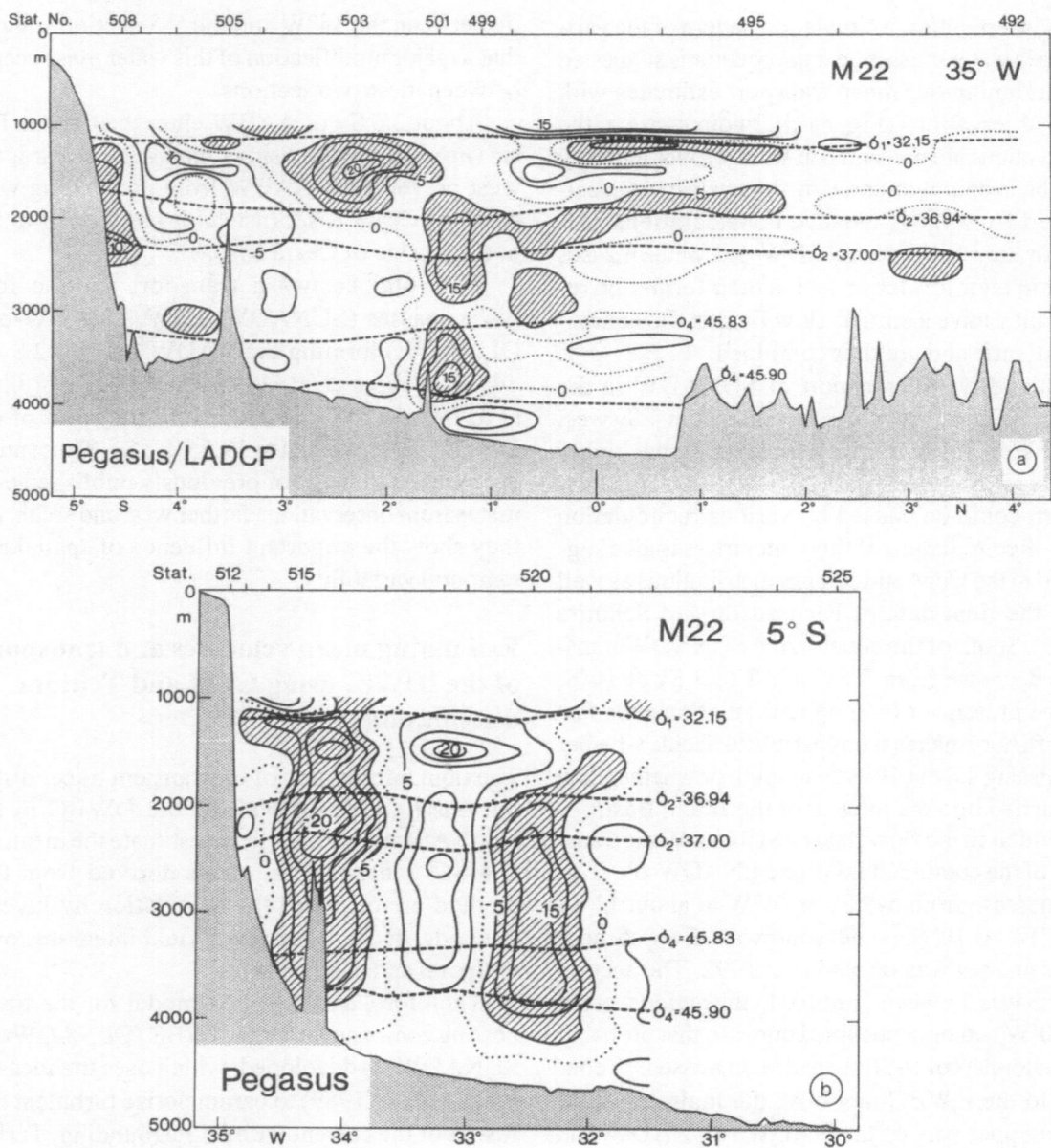

Fig. 4. Combined Pegasus and LADCP velocity fields below 1000m at a): 35°W (zonal component), November 1992. Also shown are the station locations and the isopycnals defining the water mass boundaries. Eastward flow exceeding 5 cms^{-1} is shaded. b): at 5°S (meridional component) for November 1992 . Southward flow exceeding 5 cms^{-1} is shaded. Also shown are the isopycnals defining the water mass boundaries according to Table 1 (from Rhein et al. 1995).

The model assumes uniform velocity and uniform turbulent mixing along the flow path of the DWBC. Enhanced turbulent mixing in the vicinity of the current compared to the ocean's interior allows the water adjacent to the current, which stays at rest, to accumulate tracers. The highest mean velocity of OLNADW, which results in model F12, F11, and ^3H distributions as well as F11/F12 ra-

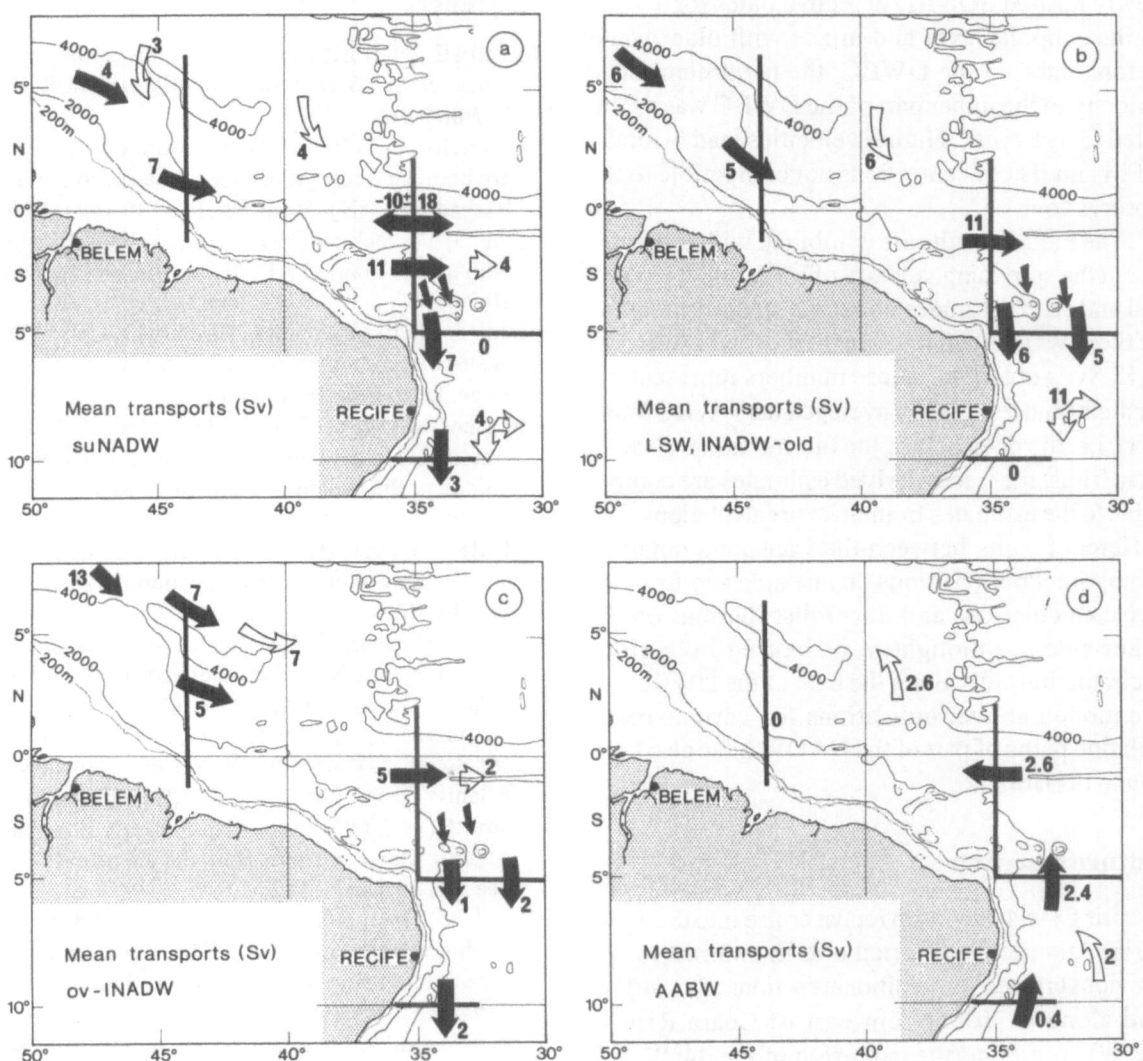

Fig. 5. Circulation patterns and transports (Sv) of the DWBC water masses a) SUNADW, b) combined LSW and LNADW-old, c) OLNADW, d) AABW. The observed estimates are presented by dark arrows, light arrows are deduced from mass balance considerations, from Rhein et al. (1995). Data are from Molinari et al. (1992): west of 44°W and at 44°W, Schott et al. (1993): 44°W north of Ceara Rise, lADCP measurements in October, 1992 (M22), 35°W, 5°S, 10°S (from Rhein et al. 1995).

tios, compatible to measurements of these tracers along the western boundary, is 4.8 cms⁻¹. Variations in the composition of the DWBC as well as changes in the time history of the source water masses do not increase the range of the model velocities (Rhein 1994).

The box model, fed with 'realistic' tracer boundary conditions for the formation of SUNADW and with uniform self mixing was also applied to this water mass (Rhein 1995). The model area extends from the deep water formation region in the northern Atlantic to 10°S . With the assumption, that newly formed SUNADW recir-culates for 4 years in the subpolar gyre and mixes with older water before entering the DWBC, the maximum mean velocity in the upper part of the DWBC was calculated to be 5 cms^{-1}. Higher velocities lead to model CFM and ^{3}H concentrations not compatible to the observations.

The model results are combined with estimates about the spreading velocity of LSW and lNADW-old and with estimates about the spatial extension of the DWBC. A mean transport of the DWBC of 9-12 Sv is obtained. These numbers represent the spatially and temporally averaged net DWBC transport, i.e. the deep part of the thermohaline circulation. Thus, the tracer derived estimates are comparable to the estimates from inverse calculations. The different results between the transports obtained from direct observations on one side and from inverse calculations and tracer distributions on the other side are thought to be caused by various recirculation cells along the path of the DWBC. As mentioned above, indications for various recirculation paths of part of the NADW have also been found off Brazil.

Future research

In spring 1994 the fourth repeat cruise into the equatorial Atlantic was carried out. In contrast to our previous cruises, the additional sections along 40°W and along 4°30'N (from east of Ceara Rise to 35°W), as well as the extension of the 44°W section to the Midatlantic Ridge, provides a better spatial coverage of the Guiana Basin and allows a more detailed study of the recirculation in the equatorial Atlantic. Moreover, CTD stations with LADCP measurements have been carried out in the Vema Fracture Zone at 11°N to estimate the Bottom Water transport into the Northeast Atlantic (Fischer et al. 1995). Regarding the deep equatorial Atlantic, the research will focus on the water mass modification and spreading paths of the Antarctic Bottom Water and on the possible recirculation of OLNADW north of Ceara Rise. From three moorings recovered at 44°W (K359 - K361 in Fig. 1) the mooring time series in the DWBC was extended for another 1.5 years, and changes between different deployment years will be investigated.

References

Bryden HL, Hall MM (1980) Heat transport by currents across 25°N in the Atlantic Ocean. Science 207:884-886

Fine RA, Molinari RL (1988) A continuous deep western boundary current between Abaco (26.5°N) and Barbados (13°N). Deep-Sea Res. 35:1441-1450

Fischer J, Visbeck M (1993) Deep velocity profiling with self-contained ADCPs. J Atm Ocean Techn. 10:764-773

Fischer J, Rhein M, Schott F, Stramma L (1995) Deep Water masses and transports in the Vema Fracture Zone. Deep-Sea Res (submitted)

Friedrichs MAM, McCartney MS, Hall MM (1994) Hemispheric asymmetry of deep water transport modes in the Atlantic. J Geophys Res 99:25196-25179

Hall MM, Bryden HL (1982) Direct estimates and mechanisms of ocean heat transport. Deep-Sea Res 29:339-359

McCartney MS (1993) Crossing of the equator by the Deep Western Boundary Current in the Western Atlantic Ocean. J Phys Oceanogr 23:1953-1974

Molinari RL, Fine RA, Johns E (1992) The Deep Western Boundary Current in the western tropical North Atlantic Ocean. Deep-Sea Res 39:1967-1984

Pickart RS (1992) Water mass components of the North Atlantic Deep Western Boundary Current. Deep-Sea Res 39:1553-1572

Pickart RS, Hogg NG, Smethie WM Jr (1989) Determining the strength of the Deep Western Boundary Current using the chlorofluoromethane ratio. J Phys Oceanogr 19:940-951

Read JF, Gould WJ (1992) Cooling and freshening of the subpolar North Atlantic since the 1960s. Nature 360:55-57

Rhein M (1994) The Deep Western Boundary Current: Tracers and Velocities. Deep-Sea Res 41:263-281

Rhein M, Stramma L, Send U (1995) The Atlantic Deep Western Boundary Current: Water masses and transports near the equator. J Geophys Res 100:2441-2457

Rhein M (1995) The Shallow Component of the Atlantic Deep Western Boundary Current: tracers, velocities and transports. Deep-Sea Res (submitted)

Richardson PL, Schmitz WJ (1993) Deep cross-equatorial flow in the Atlantic measured with SOFAR floats. J Geophys Res 98:8371-8387

Roemmich D (1980) Estimation of meridional heat flux in the North Atlantic by inverse methods. J Phys Oceanogr 10:1972-1983

Schmitz WJ Jr, McCartney MS (1993) On the North

Atlantic Circulation. Rev of Geophys 31:29-49

Schott F, Fischer J, Reppin J, Send U (1993) On mean and seasonal currents and transports at the western boundary of the equatorial Atlantic. J Geophys Res 98:14353-14368

Spain PF, Dorson DL, Rossby HT (1981) PEGASUS: A simple acoustically tracked velocity profiler. Deep-Sea Res 28:1553-1567

Speer KG, McCartney MS (1991) Tracing lower North Atlantic Deep Water across the equator. J Geophys Res 96:20,443-20,448

Watts DR (1991) Equatorwards currents in the temperature range 2°-6°C on the continental slope, Mid Atlantic Bight. In: Chu H, Gascard JG (eds) Deep water convection and deep water formation. Elsevier, New York, pp 183-196

Wunsch C, Grant B (1982) Towards the general circulation of the North Atlantic Ocean. Prog Oceanogr 11:1-59.

Chlorofluoromethanes in the Deep Equatorial Atlantic Revisited

C. Andrié

LODYC Laboratoire d'Océanographie DYnamique et de Climatologie,
CNRS/ORSTOM/Université Paris 6, Tour 14 - 2° ét., 4 place Jussieu, Case 100,
75252 Paris Cedex 05, FRANCE

Abstract: CFM (CCl₃F or *freon* F-11 and CCl₂F₂ or *freon* F-12) measurements were made on board the R.V. Atalante as part of a quasi-synoptic survey of the tropical Atlantic Ocean. The work was carried out during the CITHER 1 cruise, part of the French CITHER program (CIrculation THERmohaline) between January and March 1993. Two zonal sections at 4°30 S and 7°30N (the A7 and A6 WOCE sections) and two meridional sections at 35°W and 3°50 W were sampled for CFMs between the African and South American continents. The results reported here deal primarily with the North Atlantic Deep Water. The CITHER 1 sections were made just 10 years after the first CFMs snapshot of the tropical Atlantic ocean obtained during the Transient Tracers in the Ocean Program.

The detection limit was approximately 0.0025 pmol.kg⁻¹ for both F-11 and F-12. This is sufficient to allow the determination of "apparent" ages and dilution factors for the *freon*-enriched tongues of the Upper and Lower North Atlantic Deep Water (UNADW centered around 1600 m and LNADW centered around 3800 m).

Both zonal sections clearly show the propagation of UNADW into the eastern basin. The eastern meridional section at 3°50W shows the CFM cores extending from 4°S to 3°N with a maximum around 2°S. From the equatorial CFM ratios in the eastern basin, we estimate an eastward velocity close to 2 cm/s.

The CFM distributions at the levels of the UNADW and LNADW show a large variability, probably linked to northern and southern deep recirculation gyres. In both sections, CFM enriched cells are clearly the result of reversed currents.

The net decrease of CFM in the LNADW between 7°30N and 4°30S is partly the result of the bifurcation of the deep flow north and at the equator. This is confirmed by data taken in November 1992 in the region of the Equatorial Romanche Fracture Zone by the ROMANCHE 2 cruise. The influence of Antarctic Bottom Water is noticeable along the South American continent in the southern section. There is no evidence, through CFM data alone, for a northward flow of this "young" bottom water mass, which is probably blocked by the topography near the sampled areas.

Introduction

The 1983 TTO data set, (Weiss et al. 1985), is often used as a benchmark for F-11 studies of the Atlantic. The data shows the flow of the Deep Western Boundary Current (DWBC) from 50°N to 10°S and indicates a southward and eastward splitting of the flow near the equator at the level of the "young" Upper North Atlantic Deep Water (UNADW). Ten years after TTO, one objective of the CITHER 1 cruise was to follow the evolution of the tracer concentration in the tropical region, in particular concentrating on the trans-equatorial and equatorial transport of the NADW.

After 1985, several studies concerning NADW transports and CFMs distributions in the tropical Atlantic ocean have been reported. They principally concern tracers description into the DWBC (Fine and Molinari 1988; Pickart et al. 1989; Pickart 1992; Molinari et al. 1992; Pickart and Smethie 1993; Smethie 1993; Rhein 1994; Rhein et al. 1995). Centered around 1600-1700 m, the enriched

From WEFER G, BERGER WH, SIEDLER G, WEBB DJ (eds), 1996, *The South Atlantic: Present and Past Circulation*. Springer-Verlag Berlin Heidelberg, pp 273-288

CFM core, classically corresponding to the Upper NADW (UNADW), is more precisely identified as the recently ventilated water mass originating from convection in the southern Labrador sea. At a denser level, the Lower NADW (LNADW), centered around 3800 m, originates from the Denmark Strait overflow. Mc Cartney (1993) discusses the previous considerations concerning the question of the DWBC bifurcation at the equator and the zonal advection velocities inferred from tracer distributions. Recirculation processes are assumed to be responsible, in part, for the tongue shape of the CFM distributions and for the underestimation of the DWBC flow velocity along the western boundary of the tropical Atlantic.

During the CITHER 1 cruise, in January-March 1993, four sections were sampled at 7°30N, at 4°30S (A6 and A7 sections of the World Ocean Circulation Experiment WOCE), and at 35°W and at 3°50 W (Fig. 1).

This paper reports informations inferred from CFM data only. Hydrological and other tracers (oxygen, nutrients) data are reported in the CITHER 1 Reports (Le Groupe CITHER-1 1994 a, b, c).

The analytical procedure is presented in section 2. In section 3, we describe and discuss CITHER 1 CFM distributions in order to infer a schematic circulation pattern of the NADW, principally concerning recirculation and eastward bifurcation near the equator. In section 4, informations relative to the use of the "apparent age" evaluated from the CFM ratio method are presented and discussed, and the CITHER1 F-11 sections are compared to the TTO section (Weiss et al. 1985).

Methods

Seawater samples were obtained using the IFREMER rosette equipped with 32 8l-bottles. 188 stations were sampled for F-12 and F-11, at 32 levels regularly spaced from the surface to the bottom, with a separation of less than 60 n.m.

The technique used for CFM measurements on board was the extraction-trapping method coupled to gas chromatography, with electron capture detection, described by Bullister and Weiss (1988). The data reported here are based on the Scripps Institution of Oceanography (SIO) 1986 calibration

scale using a primary standard provided by SIO. The accuracy of the secondary standard used on board is 0.9% for F-12 and 0.8% for F-11.

Atmospheric measurements were performed daily all along the cruise. A weak latitudinal gradient was observed between 7°30N and 4°30S (around 0.45 ppt/°lat for F-12 and 0.3 ppt/°lat for F-11). No noticeable zonal gradient was observed. The mean atmospheric mixing ratios are 513.8 ± 4.2 ppt for F-12 and 276.0 ± 3.2 ppt for F-11 at 7°30N and 508.3 ± 4.9 ppt for F-12 and 272.4 ± 3.1 ppt for F-11 at 4°30S.

The mean deviation of the measured surface concentrations to the theoretical solubility values (Warner and Weiss 1985) was found to be close to 1% for both F-12 and F-11. This value, near the solubility equilibrium, is expected in the tropical area in the northern winter season. The values relative to the stations close to the African continent have been excluded from this global mean. This is because under-saturation, sometimes exceeding - 15%, were found near the coast due to coastal upwelling.

The results of five test-stations for bottle "blanks" determinations (32 bottles sampled at one same deep level, assumed to be *freon*-free) were used to determine the detection limit of the method. The precision of the method for deep samples is given by the standard deviation of the mean bottle contamination level. It is around 0.0025 pmol.kg^{-1} for F-12 and for F-11. The mean bottle contamination levels measured by this method vary in the ranges 0.001-0.008 pmol.kg^{-1} for F-12 and 0.007-0.013 pmol.kg^{-1} for F-11. A net decrease of the contamination level occurred during the cruise as the bottles became cleaner. The final CFMs concentrations are corrected for this contamination using a "blank" level interpolated between the test-stations. The reproducibility has been checked at least once at each station by sampling two bottles at the same depth. In addition, some stations were reoccupied during the cruise; the deviation between the profiles of stations 119-156 (15 days apart) at 35°W, 7°30S being less than 0.5%.

For the deep levels, direct velocity measurements (Pegasus profiles have been performed on the 35°W section) or indirect geostrophic calculations relative to the zonal sections are not presently available.

STATIONS CITHER

cruise code: 35A3 CIT1

number of profiles: 223

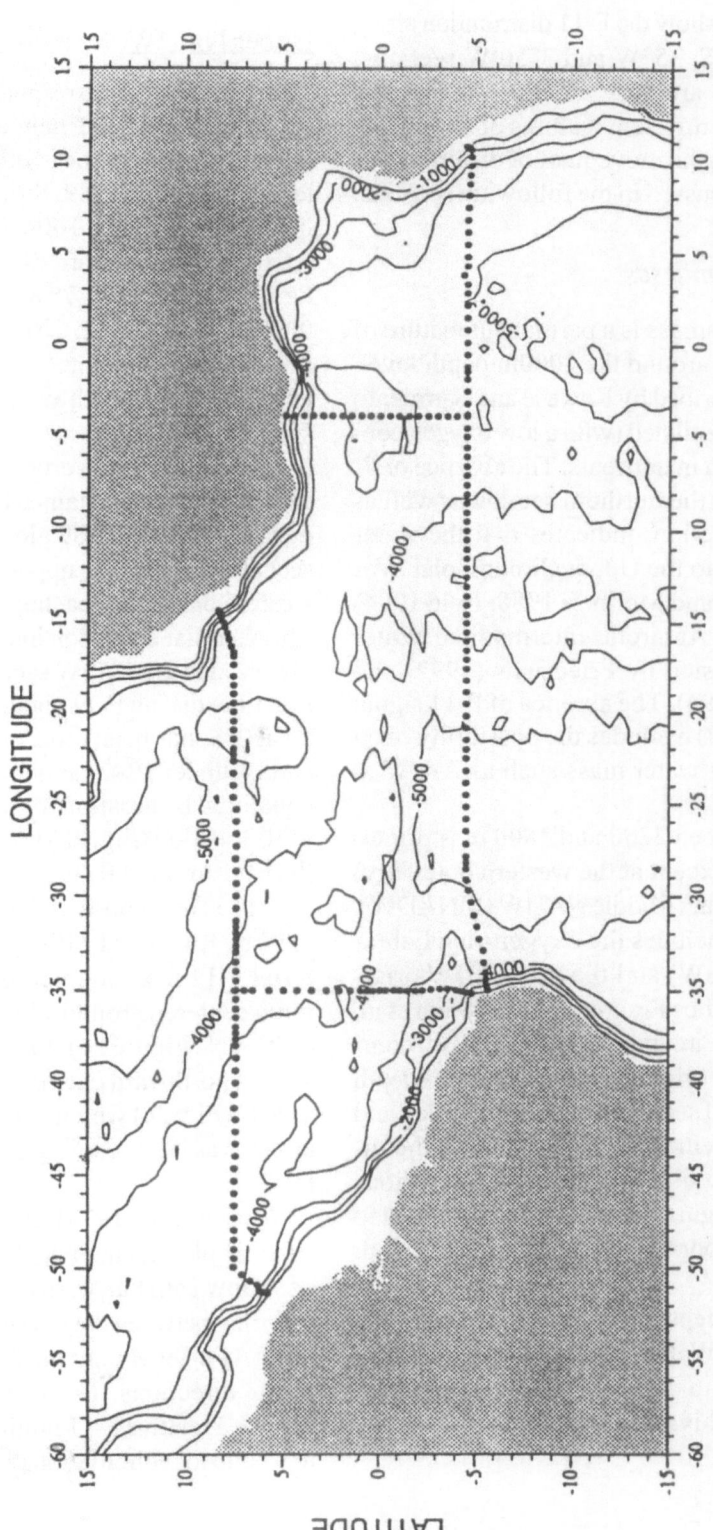

Fig. 1. Stations location of the CITHER 1 cruise in January-March 1993 surimposed on a bathymetric chart showing isobaths at each 1000 m interval.

F-11 distributions

Figs. 2 a, b, c and d show the F-11 distribution along the 7°30 N, 4° 30 S, 35°W and 3°50 W sections. F-12 distributions are very similar (Andrié and Ternon 1994). The different features observed, often associated with the movement of different water masses, are discussed in the following sections.

freon-free water masses

A *freon*-free water mass is a permanent feature of the Atlantic ocean around the 1000m depth level. This level was described by Kawase and Sarmiento (1986) as poorly ventilated (with a low oxygen concentration) and rich in nutrients. The absence of F-11 at this depth, in the northern section as well as in the southern section, indicates that the water mass corresponds to the Upper Circumpolar Water (UCPW) (Fine and Molinari 1988; Reid 1994) rather than to the Antarctic Intermediate Water (AAIW) as suggested by Friedrichs (1992) and Tsuchiya et al. (1994). The absence of F-11 signal (and oxygen signal) excludes the possibility of an input of a younger water mass such as AAIW at this level.

The layer between 2200 and 3800 m is principally *freon*-free (except at the western boundary) and corresponds to the Middle NADW (MNADW). This water mass includes the oxygen-rich Labrador Sea Water (LSW) and the old NADW originating from the Gibbs Fracture Zone (Rhein et al. 1995; Reid 1994). Circumpolar Water (CPW) coming from a southern origin is also found in this depth range (Reid 1994; Tsuchiya et al. 1994; Talley and Johnson 1994; Rhein et al. 1995). Due to the diversity of these possible origins, it is not obvious to determine the origin of some small CFM enriched structures which sporadically appear in this depth range.

Below 4000m depth and above the level of the Antarctic Bottom Water (AABW), the CFM-free waters correspond to the Lower Circumpolar Water (LCPW) (Tsuchiya et al. 1994).

CFM-enriched tongues

Upper NADW

The upper tongue corresponds to shallow UNADW (or suNADW after Rhein et al. 1995). This water mass originates in the southern Labrador Sea Water (Pickart et al. 1989; Smethie 1993) and is found in the depth range 1600-1700 m in the tropical Atlantic. The eastward extension of the 0.01 pmol/kg isoline reaches 27°W at 7°30 N (Fig. 2a) and 0°W at 4°30 S (Fig. 2b). On the western boundary, the maximum intensity of the core exceeds 0.2 pmol/kg in the north while it is half this value at 4°30 S.

The 7°30N F-11 vertical section (Fig. 2a) is very similar to the one obtained by Molinari et al. (1992) during February 1989 along a 200 km zonal transect at 14°30 N. The upper core extends across the western basin and reaches 27°W in the eastern basin. A similar extension into the eastern basin is also observed in the 3°50W section (Fig. 2d). This is the first time that an F-11 signal, exceeding 0.04 pmol/kg at 2°S, has been reported east of 18°W (Doney and Bullister 1992) and provides evidence for the equatorial bifurcation at the flow. On the southern 4°30S section (Fig. 2b) the upper core also crosses the Mid Atlantic Ridge.

At 35°W, as observed in October 1990 and June 1991 by Rhein et al. (1995) in the DWBC, the equatorial F-11 maximum core is split into one maximum centered around 1°S, and a second one north of 0°30N (Figure 2c). In addition our data clearly show a northward extension of the F-11 core as far as 6°N (Fig. 2c) which could be due to the input of an eastward flow coming at the latitude of the Ceara rise.

McCartney (1993) does not attribute the offshore displacement of the core to a shift of the dominant flow path but, partly, to the erosion processes occuring between the overlying UCPW (just below the AAIW level) and the UNADW flowing in opposite directions. The isolation of the northward core by a relative F-11 minimum around 3°N seems to be due to recirculation process discussed below.

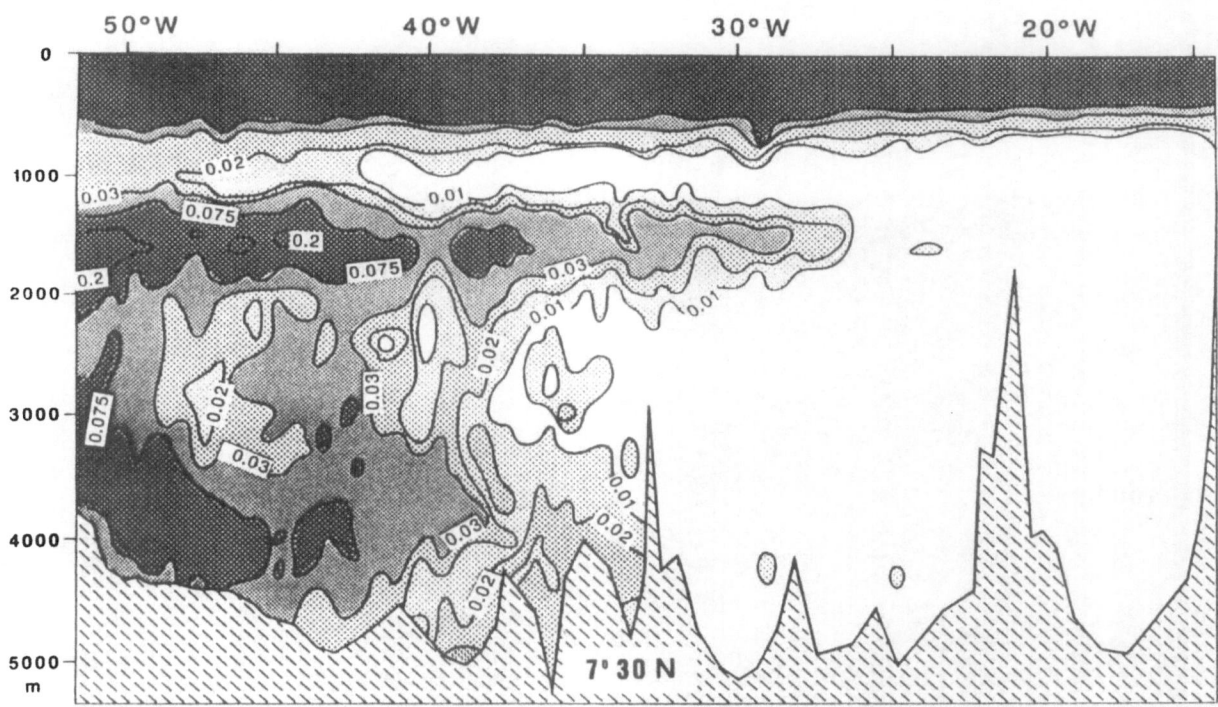

Fig. 2. F11 distributions during CITHER1 **a)** northern F11 section (WOCE A6) at 7°30N (contours in pmol/kg)

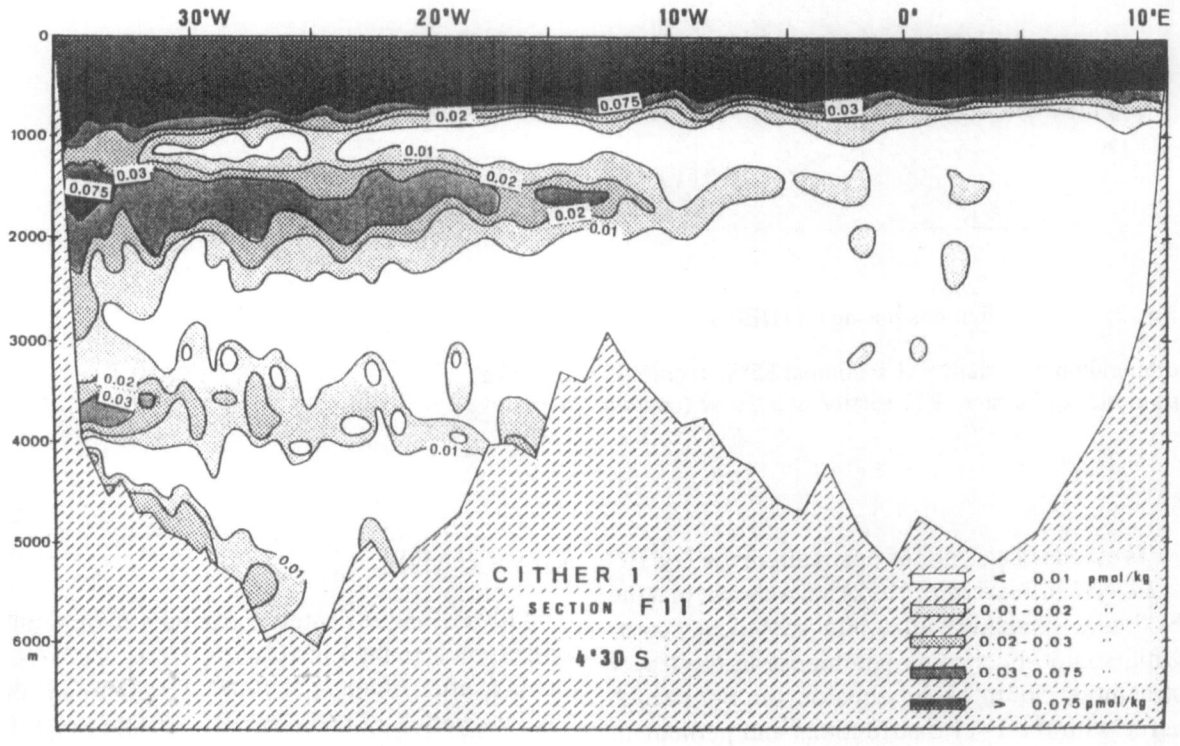

b) southern F11 section (WOCE A7) at 4°30S (contours in pmol/kg)

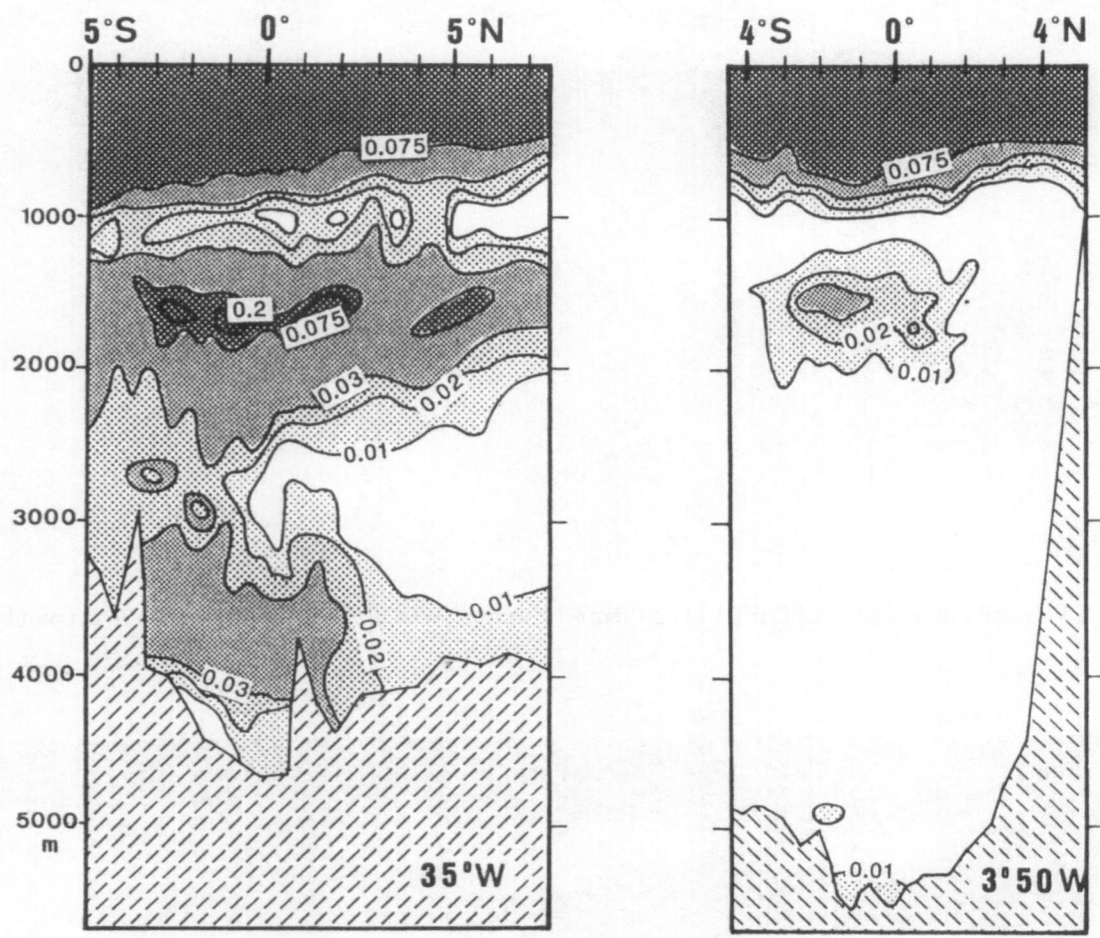

Fig. 2. F11 distributions during CITHER 1

c) meridional western F11 section at 35°W (contours in pmol/kg)
d) meridional eastern F11 section at 3°50 W (contours in pmol/kg)

Previous observations upstream in the DWBC at 14°N (Molinari et al. 1992) have shown that the F-11 cores coincide with southward velocity cores with, sometimes, some patchiness in the tracer structures. South of the equator, the F-11 cores agree with the F-11 distributions and permanent eastward transports described by Rhein et al. (1995) for the 1990-1992 period.

Lower NADW

The lower tongue centered around 4000m depth corresponds to the LNADW overflow water (ov-NADW after Rhein et al.1995), originating from the Denmark Strait (Pickart et al. 1989; Smethie 1993). At 7°30N (Fig. 2a), the extension of the 0.02 pmol/kg isoline observed as far as 30°W for the upper tongue is less pronounced (40°W) for the lower tongue.

At 4°30S (Fig. 2b), the F-11 concentrations in the lower core are considerably reduced compared to the northern section (Fig. 2a). This is probably the result of two successive blockings of the LNADW. The first, just north of the Ceara rise at 5° N, diverts the core offshore (see the 4000 m isobath on Fig. 1). The second, near the equator, is due to the equatorial channel. This is the valley (deeper than 4000 m) located between the southern wall of the Mid Atlantic Ridge, near the equator, and the eastward extension of the continental rise (Parnaiba Ridge), near 3°S, (Speer and McCartney 1991; McCartney and Curry 1993).

On the western boundary, the F-11 content maximum in the lower core reaches 0.1 pmol/kg at 7°30N while it is only 0.04 pmol/kg at 4°30S. At 35°W (Fig. 2c), the F-11 core corresponds to the permanent eastward flow centered around 1°40 S as already mentioned by Rhein et al. (1995). The F-11 maximum is principally concentrated in the equatorial channel between the Parnaiba Ridge at 2°S and the equator.

The implications concerning the bifurcation of the flow along the equator and the deep recirculation gyres are discussed later in the paper.

The results also shed some light on a question proposed by Bennekom (this issue) concerning the origin of the Angola basin deep waters. He argues, from silica data, that the bottom water observed in the Angola basin north of 30°S is predominantly NADW coming from Chain and Romanche fractures. Wallace et al.(1994) report in the deep Angola basin, at 19°S, significant CCl_4 concentrations but no-trace of F-11 on the eastern side of the Mid Atlantic Ridge : the CCl_4 signal is assumed to be derived from waters coming from the North (through the Romanche fracture zone) or from the South (through or over the Walvis ridge). The low F-11 concentrations observed during CITHER 1 on the southern section (Fig. 2b) and on the equator along the 3°50 W section (Fig. 2d) and at the exit of the Romanche Fracture Zone during the ROMANCHE 2 cruise (Messias 1994) do not indicate a southward flow of NADW along the eastern side of the Mid Atlantic Ridge. The weakness of the F-11 signal is not incompatible with a CCl_4

signal at 19°S due to the shorter F-11 story in the atmosphere (injected since 1945) compared to the CCl_4 one (injected since 1920). The question of the origin of the Angola basin deep waters is still open.

Antarctic Bottom Water

The variability in F-11 found at the AABW level is as high as that observed at upper levels. The southern, 4°30S, section (Fig. 2b) exhibits a F-11 enriched core, lying on the bottom below 4500 m, corresponding to the AABW. This core is most noticeable on the western side of the Brasil basin, near the South American continent. As mentioned by Rhein et al. (1995), this water mass is CFM enriched by mixing with the underlying more recently ventilated Weddell Sea Water component. Tushiya et al. (1994) describe more precisely the distinction between the AABW (from LCPW origin) and the WSDW.

On the 35°W section (Fig. 2c), the results confirm the Rhein et al. data of 1990-1992, the AABW signal does not seem to cross the equator, being blocked by the topography. McCartney and Curry (1993) propose two features that probably restrict the northward flow of AABW. The first is the sill (4500 m) at the entrance of the equatorial channel. The second is the Ceara rise, which acts as a barrier for waters deeper than 4500 m between 1°N and 4°N. At 35°W, there is a small F-11 signal in the northern side of the equatorial channel near the equator suggesting an AABW westward flux near the bottom (around 4600m depth). The weakness of the F-11 signal near the equator can be associated with the absence of a F-11 signal in the depth range of the AABW at the entrances of the Romanche and Chain fractures as mentioned by Messias (1994). The bottom water near and north of the equator has a predominant LCPW influence.

Bifurcation of the NADW flow at the equator

Fig. 3 shows the horizontal distribution of the 0.015 pmol/kg and 0.05 pmol/kg isolines inside the tropical area at the UNADW level (1600 m), obtained by interpolating between the four sections reported above. Superimposed are the F-11 isolines corresponding to the 1983 TTO data set. Compared to the TTO distributions, the new data emphasizes the reality of the UNADW core bifurcation along the equator.

The F-11 data at the 3°50W section in the eastern basin give evidence for a bifurcation of the UNADW flow near the equator (Fig. 2 d). However, the core maximum is somewhat shifted south of the equator (around 2°-3° S). The bifurcation could be induced by the rise (around 2000 m depth) of the Parnaiba ridge near 3°S (Fig. 2 c). In addition, we observe on the 35°W section (Fig. 2c) a secondary F-11 enriched core centered around 5°N which can be due to an input of an eastward flow coming along the Ceara rise.

The eastward bifurcation just south of the equator is in agreement with Tsuchiya et al. (1992, 1994) who report through two different salinity sections at 20°W and 25°W an eastward core centered around 2°S.

The circulation pattern inferred from our data at the UNADW level (Fig. 4) shows the eastward bifurcation of the DWBC flow near the equator and reversed flows both sides of the equator. It is similar to the general circulation scheme proposed by Richardson and Schmitz (1993) from drifting floats, on the 1800 m level (Fig. 5).

There is also evidence for an eastward bifurcation of the LNADW core, particularly through the strong north-south latitudinal gradient observed in F-11 concentrations (Fig. 2 c). As discussed above, this bifurcation is principally induced by bathymetric effects of the Ceara rise and then by the equatorial channel. The location of the F-11 core corresponds to the mean flow described by Rhein et al. (1995) i.e. guided by the topography and concentrated on the southern side of the equatorial channel. There is no F-11 enrichment north of 2°N on the 35°W meridional section (Fig. 2c). This means that the LNADW flow does not continue

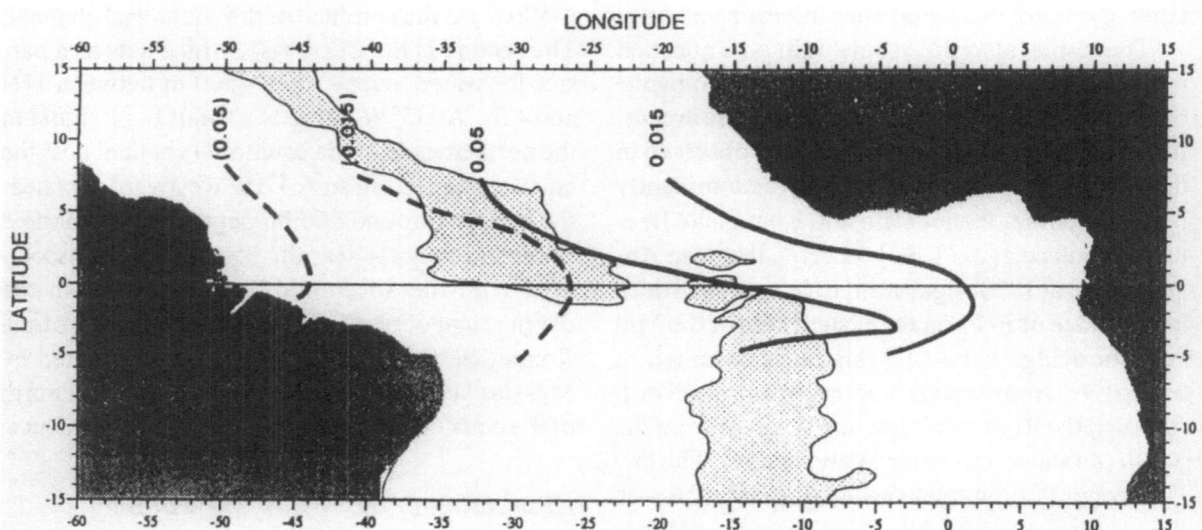

Fig. 3. TTO-CITHER 1 F11 distributions comparison (0.015 and 0.05 pmol/kg isolines). The 0.015 and 0.05 isolines relative to the 1983 data set (broken lines) are indicated in parenthesis. Surimposed is the 4000m depth isobath at the location of the Mid Atlantic Ridge.

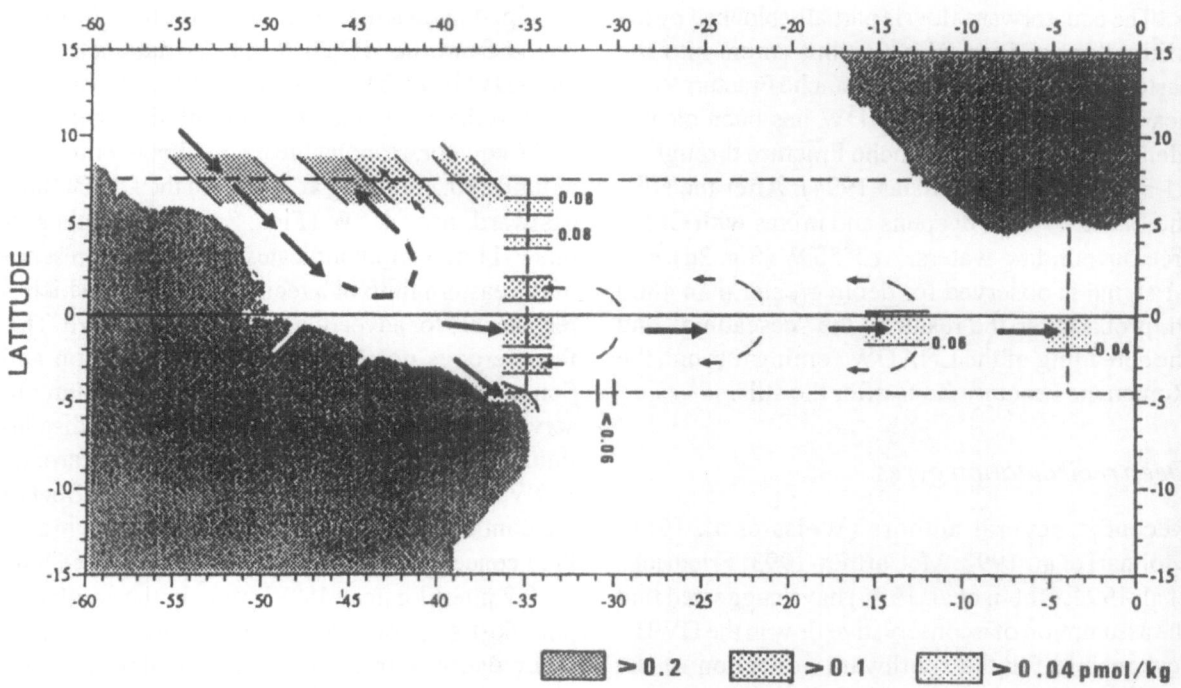

Fig. 4. Schematic circulation pattern inferred from the F-11 distributions at 1600 m depth.

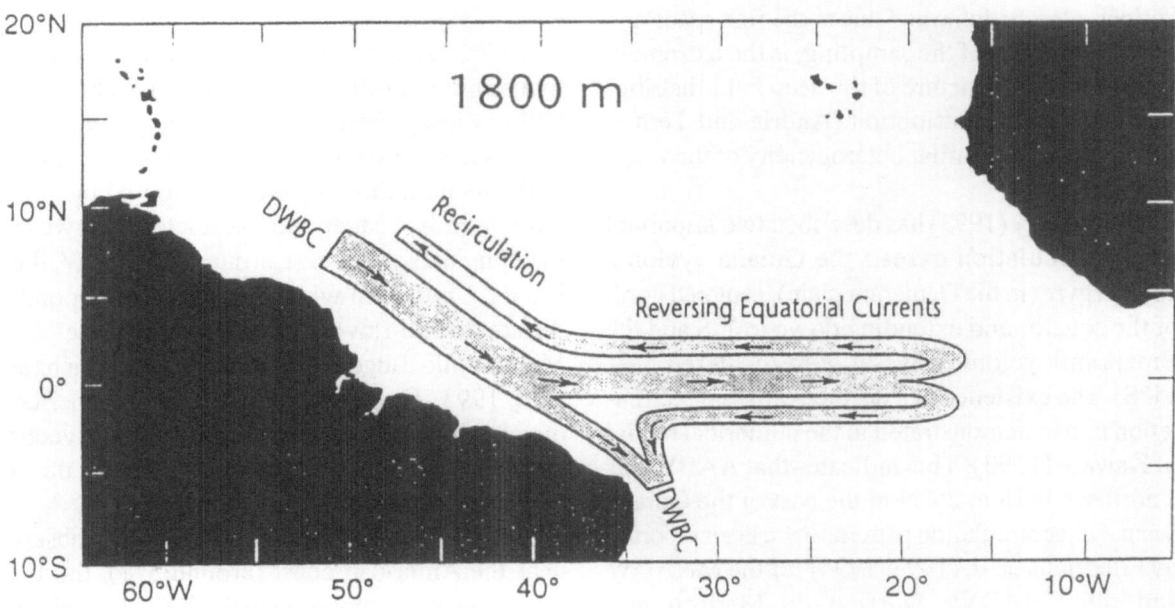

Fig. 5. Recirculation scheme proposed by Richardson and Schmitz (1993) at 1800 m.

eastward (east of 40°W) but turns southward just after the Ceara rise blocking near 5°N.

The equatorward flow is partially blocked by the Mid AtlanticRidge (Fig. 2b) but enters into the eastern basin through the Romanche Fracture Zone near the equator. The LNADW has been clearly identified into the Romanche Fracture through F-11 and F-12 data (Messias 1994). After the sills, the LNADW core deepens and mixes with CFM-free surrounding waters. At 3°50 W (Fig. 2d), a F-11 signal is observed for depth greater than 4800 m, probably as the result of the "cascading" and the spreading of the LNADW coming through the Romanche Fracture Zone after the sills.

Deep recirculation gyres

Recently, several authors (Weiss et al. 1989; Molinari et al. 1992; McCartney 1993; Friedrichs et al. 1994; Rhein et al. 1995) have suggested that the assumption of a conservative flow in the DWBC responsible for the southward advection of the nordic ventilated waters was too simplistic to explain the shape of the tracer distributions : effects of recirculation gyres have been discussed. The most striking feature in the CITHER 1 F-11 distributions, clearly delimited due to the fine space and depth resolution of the sampling, is the extremely heterogeneous structure of the deep F-11 distributions. The F-12 distribution (Andrié and Ternon 1994) indicate a similar heterogeneity of the water masses.

McCartney (1993) has described two important deep recirculation gyres : the Guiana cyclonic abyssal gyre (in the Demerara plain), centered north of the equator and extending down to 1°S and the Brazil anticyclonic abyssal gyre (centered near 11°S). The existence of a northern abyssal recirculation is also demonstrated in the numerical model of Kawase (1993). This indicates that AABW has a northern limit at 25°N in the east of the Guiana basin. Other circulation patterns have been reported by Friedrichs et al. (1993, 1994) for the MNADW and the LNADW. Durrieu de Madron and Weatherly (1994) give circulations patterns for NADW and AABW south of the equator and an eventual connection between northern and southern recirculation cells; they describe the interior

upwelling occurring between the AABW and the NADW.

Fig. 4 shows the main circulation features inferred from the CITHER 1 F-11 data set at the UNADW level. The locations of the F-11 maxima suggest the existence of a recirculation cell north of the equator, responsible for F-11 concentrations as high as 0.2 pmol/kg at 7°30 N in the DWBC and, eastward, near 45°W (Figs. 2a and 4). This second F-11 maximum indicates the possible presence of the eastern limb of a recirculation gyre which is responsible for advecting the tracer northward. This feature does not appear in the Richardson and Schmitz pattern (Fig. 5). The F-11 maximum observed at 1°N on the 35°W section and the discontinuity (> 0,06 pmol/kg minimum at 4°30S around 31°W, Fig. 4) could be intrepreted as the effect of the continuity of the eastern limb : but the higher F-11 concentration in the northern recirculated limb (> 0.2 pmol/kg near 45°W) than in the south (0.1 pmol/kg), suggests that the northward turning of the recirculation gyre must occur north of the equator. The wide extension of the F-11 core (as far as 49°W, Figs. 2b and 4) is partly due to the superimposed inputs of the flow of the DWBC itself and of the western limb of the cyclonic Guiana recirculation gyre.

At 7°30 N, the location of the western F-11 relative minimum (49°- 52 °W, Fig. 4) is about 400-500 km away from the DWBC core and corresponds to the area located between both branches of the recirculation pattern: the flow of the water mass here is weaker and associated to a weaker tracer input. Farther east, around 39°- 42°W, there is a F-11 mininum which seems to correspond to the southward flow along the western side of the Mid Atlantic Ridge, as mentioned by Friedrichs and Hall, 1993. The discontinuities in the upper core may be due to disruptions in the zonal advection flow or to the transient tracer input during the last 20 years (Weisse et al. 1994).

On the 35°W section (Fig. 2c and 4) we observe, near the American coast (around 3°S), the F-11 maximum corresponding to the DWBC core. Farther north (around 2°S), another F-11 enriched core, corresponds to the eastward bifurcation of the mean flow. In between, a third F-11 core could be explained by a reverse westward recirculated flow south of the equator. Another reverse flow near the equator seems evident around 1°-2°N (Fig. 2c and 4). Features like these have already been identified in floats trajectories at 1800 m (Richardson and Schmitz 1993) but not in exactly the same area (Fig. 5). The intensity and the location of the direct and reverse equatorward flows could be variable.

The major circulation features which affect the deep F-11 distributions in the LNADW core are, as mentioned above, the result of topographic effects. The LNADW turns eastward near 5°N due to the blocking effect of the Ceara rise, then flows southward after the Ceara rise and is deflected eastward into the equatorial channel to cross the Mid Atlantic Ridge through the Romanche Fracture Zone.

Recirculation processes due to the Guiana cyclonic gyre are also noticeable at the LNADW level (Figs. 2a and 2b). This shows two separated cores on both northern and southern sections. At 7°30 N the eastern limb is located between 42°W and 44°W, somewhat further offshore than for the UNADW level. In the south at 4°30 S (Fig. 2b), the western core, located around 3800 m, lies adjacent to the coast. A second core is located between 26°W and 28°W, around 750 km offshore from the western limb. The recirculation effect is also noticeable down to the level of the AABW, where two F-11 cores appear at the same longitudes as the LNADW cores.

In conclusion, the Guiana abyssal recirculation gyre is well identified through the whole water column between the bottom and the UNADW level. In addition at the depth of the MNADW, on both northern and southern sections, the *freon*-free water mass (F-11 concentration less than 0.04 pmol/ kg), indicate two F-11-enriched "chimney-like" structures, which may identify the center of the recirculation gyre.

Apparent ages and dilution factors

CFMs are often used as tools to determine the "age" of water masses (Bullister 1989; Fine et al. 1988). The knowledge of the time evolution of F-11 and F-12 since the time of their injection into the atmosphere allows one to evaluate the "age" of previously ventilated water masses i.e the elapsed time since the water parcel left the atmosphere-ocean interface. This method is only valid for the period between 1945 and 1975, since the beginning of the CFMs injection into the atmosphere to the limitation of their production.

Time evolutions of the F-11 content and the corresponding F-11/F-12 ratio in the UNADW source waters are shown in Fig. 6. These theoretical values correspond to the solubility equilibrium values at the temperature and salinity characteristic of the UNADW. They are calculated from the solubility functions of F-11 and F-12 reported by Warner and Weiss (1985) and the atmospheric mixing ratios time evolution (Weiss, personal communication) for the northern hemisphere.

Assuming a CFM-enriched water mass flowing into a *freon*-free environment, the F-11/F-12 ratio of the water parcel is conservative along the flow. So, its measurement allows to determine the "apparent age" of the water mass. The knowledge of the apparent "age" (F11/F12 curve from Fig. 6) allows determination of the theoretical F-11 concentration (F-11°) that the water mass should have at the time of the water mass formation (from Fig. 6): the ratio F-11°/F-11 (F-11 represents the F-11 concentration at the time of the measurement) allows to determine the "apparent" dilution factor.

Recent work on transient tracer distributions, particularly in the Deep Western Boundary Current, have shown difficulties inherent to turbulent mixing processes, occuring between the mean flow and surrounding water, and responsible for a bias in apparent ages or velocities determination (Pickart 1992; Smethie 1993; McCartney 1993; Rhein 1994). In this paper we deal primarily with the CFM eastward penetration into a completely *freon*-free environment. Under these conditions, apparent ages and dilution factors can be considered valid without much correction for mixing. An apparent zonal velocity along the equator can thus be deduced. This evaluation has to be considered as a mean "integrated" value as it takes into account the whole water mass history from the CFM source area.

Due to the uncertainties in the following effects, no corrections have been made for the CFM undersaturations in the formation areas (Wallace and Lazier 1988; Rhein 1991), for the residence times in the sources reservoirs (Pickart et al. 1989 ; Rhein

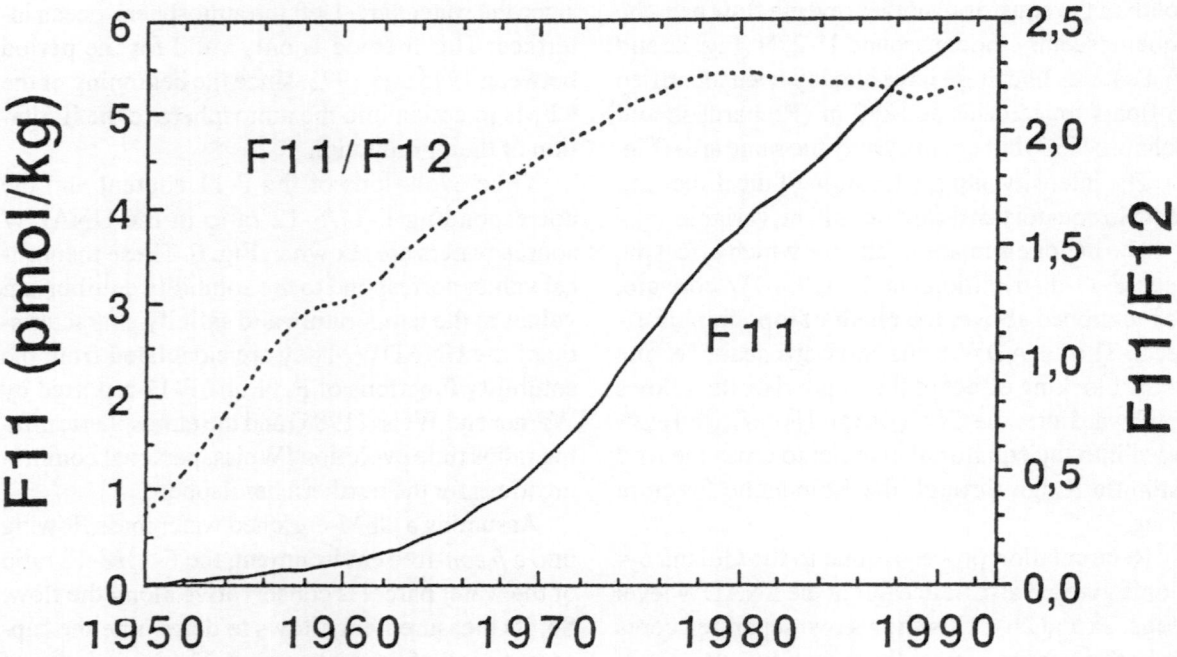

Fig. 6. Evolution with time of the theoretical oceanic F11 concentration evaluated from the time history of the atmospheric F11 concentrations (Weiss personal communication). The temperature and salinity of the UNADW are used in the solubility function of Warner and Weiss (1985).

1991) nor for self-mixing (Pickart et al. 1989; Rhein 1994). Consequently, the reported evaluations have to be considered as minimal values.

Table 1 gives the upper core mean characteristics between the DWBC and the 3°50W section, the F-11 content at the core maximum, the associated F-11/F-12 ratio, the apparent velocity and the apparent dilution factor.

As mentioned above, the use of F-11/F-12 ratios in apparent velocity and dilution factor calculations gives considerably biased estimations when self-mixing of an advected water mass occurs (apparent velocity is underestimated): we have shown in section three that it is typically the case for the NADW inside the DWBC. Consequently, the results relative to apparent age, apparent velocity and dilution factor reported in Table 1 for 50°W and 35°W are presented only for comparison with 3°50 W estimations but must not be considered as representative values. The speed flow deduced by the F-11/F-12 age method in the DWBC is well below the values obtained by direct observations : Johns

et al. (1993) give a mean value inferred from a current meter mooring at 8°N, 52°W, in the DWBC, of about 28 cm/s (annual mean) below 4000 m. Schött et al. (1993) describe a highly variable deep currrent from moored stations at 44°W, near the equator, with an annual mean speed around 30 cm/s. Colin and Bourlès (1994) report Pegasus velocity measurements greater than 50 cm/s in the area 5°N-10°N within the upper core of the DWBC. The same order of magnitude is inferred from SOFAR floats trajectories (velocities higher than 50 cm/s at 1800 m) in the DWBC. On the opposite, we consider that the results concerning the 3°50 W section are significant because the calculations are made near the eastern edge of the CFM tongue, for a water mass assumed to have been, from its northern source to the eastern tropical Atlantic, in a completely *freon*-free environment. They agree well with the few direct estimates inferred from current meters measurements (Ponte et al. 1990) or floats (Richardson and Schmitz 1993) which, despite a great seasonal and interannual variability, lead to

Table 1: Apparent ages and dilution factors inferred from the F11/F12 method for the UNADW.

eastward from the DWBC	7°30N-50°W (DWBC)	35°W	3°50 W
F11 max	0,26 pmol/kg	0,11 pmol/kg	0,05 pmol/kg
F11/F12	2,1 ± 0,05	1,72 ± 0,14	1,7 ± 0,4
apparent age	1974 ±1 19 years	1968 ± 3 25 years	1967 ± 6 26 years
apparent velocity	1,6 cm/s	1,5 cm/s	1,9 cm/s
dilution factor	8 ± 0,5	12 ± 5	60 ±30

a mean eastward value in relative agreement with our estimated value. The indirect approaches reported by Kawase et al.(1992) or Böning and Schott (1993) agree with the existence of a mean, weak, eastward current along the equator of the same order of magnitude as our estimate. However, taking into account the previously mentioned limitations of the F11/F12 aging method, our age and velocity estimations must be considered as tentative.

Comparison between 35°W and 3°50 W (Table 1) cannot indicate wether the increase of the dilution factor eastward is true, or is due to the underestimation linked to the F-11/F-12 method in areas where mixing and/or recirculation occur.

TTO data and CITHER 1 results comparison :

Previously reported comparisons with the first F-11 1983 TTO data set (Weiss et al. 1985) have been made by Weiss et al. (1989) and Molinari et al. (1992). We derive, by a similar approach, completely independent from the F-11/F-12 ratio method, a second estimate of the zonal advection velocity.

For comparison, the CITHER 1 1993 F-11 distribution obtained by interpolation between our 4 sections are superposed to the TTO 1983's one (Fig. 3). Despite the very coarse resolution of this scheme, it allows us to observe the eastward progression of the transient tracers between 1983 and 1993. The southward shift of the UNADW core is well noticeable on the 0.05 pmol/kg contour.

We have choosen to follow the eastward displacement of the 0.05 pmol/kg instead of the 0.015 pmol/kg contour. The latter is very close to the detection limits of the analytical systems and the comparison between 1983 and 1993 data could be biased by a possible difference between both experiments. The choice of the 0.05 pmol/kg contour, still a low-concentration level not really affected by dilution with non-*freon*-free waters, seems more reasonable.

In these conditions, we deduce a mean zonal velocity of 1.4 cm/s inferred from the 4440 kms eastward displacement (from 43°50 W to 3°50 W) of the 0.05 pmol/kg isoline during 10 years. This evaluation yields a value close to the values obtained by the F-11/F-12 ratio method in the east (1,9 cm/s at 3°50 W, Table 1). The small discrepancy can be due to the fact that the second method, considering the F-11 concentrations alone, does not take into account the dilution effect and, so, can be responsible for an underestimation of the advection velocity. Anyway, the deviation between both evaluations can, in part, be explained by the detection limits of the two methods.

Conclusion

The CITHER 1 CFM data set contains significant new information on the deep circulation of the tropical Atlantic. This is particularly true for the North Atlantic Deep Water.

The data set provides evidence for an eastward bifurcation of the DWBC and a zonal advection of the UNADW along the equator. This eastward flow can be followed as far as 4°W. It is shifted south-ward around 2°30 S and opposing currents are identified north and south the main core. For the LNADW, the CFM content decrease between the 7°30N and the 4°30S sections. This is principally a result of the topography. An initial eastward bi-furcation occurs north of the Ceara rise near 5°N, after which the DWBC seems to flow south-ward again. Later, an eastward bifurcation of the LNADW flow is identified in the south of the equa-torial channel near 2°S.

A new observation coming from the CITHER 1 data is the great spatial variability in the F-11 distribution due to recirculation processes and interior upwelling. The existence of the abyssal Guiana cyclonic gyre (McCartney 1993) is confirmed. It shows up particularly well on the UNADW and LNADW levels through the presence of two distinct F-11 cores at the locations of the western and eastern limbs of the gyre. The south-ern limit of this gyre is found to occur north of the equator.

The CFMs allow determination of the "appar-ent age" of the NADW. From the F11/F12 ratio method we determine an eastward "apparent veloc-ity" close to 2 cm/s. An evaluation of similar mag-nitude is obtained when comparing the 1983 TTO F-11 data with CITHER 1 ones taken ten years later. The discrepancy observed between velocities inferred from the F11/F12 method and direct meas-urements is due to the particular approach concern-ing the use of transient tracers such as CFMs: the F-11/F-12 method leads to a mean velocity of the flow, from the source region to the studied area, which includes the effects of turbulent mixing and recirculation. From a general-circulation point of view, the tracer approach is more realistic than a direct measurement of the advection velocity (Doney and Jenkins 1994).

A study including examination of hydrological, nutrients data and geostrophic transports is in progress (Andrié et al. in prep.). It should greatly improve our first conclusions on the deep circula-tion in the tropical Atlantic derived from F-11 and F-12 measurements alone.

Acknowledgements

The author specially thanks C. Oudot, responsible for the CITHER 1 program and A. Morlière and C. Colin, chief scientists of the two legs. The cap-tains and all the crew of the R.V. Atalante have made possible the *freon* measurements in very good conditions on board. The help of M.J. Messias and L. Memery for CFCs measurements has been greatly appreciated on board and particularly the long work of J.F.Ternon who also participated to the final data validation.

This work was supported by ORSTOM (Institut Français de Recherche Scientifique pour le Développement en Coopération) and IFREMER (Institut Français de Recherche pour l'Exploitation de la MER). The program was part of the PNEDC (Programme National d'Etude de la Dynamique du Climat).

References

Andrié C, Ternon JF (1994) Mesures des chorofluoro-méthanes in: Campagne CITHER-1. Recueil de données, Volume 3/4: Traceurs géochimiques. Documents scientifiques N°OP.15 du Centre ORSTOM de Cayenne, 67-77

Bennekom AJ (1995) Silica signals in the South At-lantic, this issue.

Böning C, Schott F (1993) Deep currents and the east-ward salinity tongue in the equatorial Atlantic: re-sults from an eddy-resolving, primitive equation model. J Geophys Res 98:6991-6999

Bullister JL (1989) Chlorofluorocarbons as time-dependent tracers in the ocean. Oceanography nov.89:12-17

Bullister JL, Weiss RF (1988) Determination of CCl_3F and CCl_2F_2 in seawater and air. Deep-Sea Res 35:839-853

Le Groupe CITHER 1 (1994a) Campagne CITHER-1. Recueil de données. Volume 1/4: Mesures "en route", Courantométrie ADCP et PEGASUS. Docu-ments scientifiques N°OP 14, Centre ORSTOM de Cayenne, 161 pp

Le Groupe CITHER 1 (1994b) Campagne CITHER-1. Recueil de données. Volume 2/4: CTD-O2. Rap-port interne LPO 94-04, Laboratoire de Physique des Océans, Brest

Le Groupe CITHER 1 (1994c) Campagne CITHER-1. Recueil de données. Volume 3/4: Traceurs géochimiques (1). Documents scientifiques N°OP 15, Centre ORSTOM de Cayenne, 525 pp

Colin C, Bourlès B (1994) Western boundary currents and transports off French Guiana as inferred from Pegasus observations. Oceanologica-Acta 17: 143-157

Doney SC, Bullister JL (1992) A chlorofluorocarbon section in the eastern North Atlantic. Deep-Sea Res 39:1857-1883

Doney SC, Jenkins WJ (1994) Ventilation of the Deep Western Boundary Current and abyssal western North Atlantic: estimates from tritium and ³He distributions. J Physical Oceanogr 24:638-659

Durrieu de Madron X, Weatherly G (1994) Circulation, transport and bottom boundary layers of the deep currents in the Brazil basin. J Mar Res 52:583-638

Fine RA, Molinari RL (1988) A continuous deep western boundary current between Abaco (26,5°N) and Barbados (13°N). Deep-Sea Res 35:1441-1450

Fine RA, Warner MJ, Weiss RF (1988) Water mass modification at the Agulhas Reteroflection: chlorofluoromethane studies. Deep-Sea Res 35:311-332

Friedrichs MAM (1992) Meridional circulation in the tropical North Atlantic. Master's Thesis, MIT/WHOI, 88 pp

Friedrichs MAM, Hall MM (1993) Deep circulation in the tropical North Atlantic. J Mar Res 51:697-736

Friedrichs MAM, McCartney MS, Hall MM (1994) Hemispheric asymmetry in the vertical structure of Atlantic deep waters. J Geophys Res 99:25165-25179

Johns WE, Fratantoni DM, Zantopp RJ (1993) Deep western boundary current variability off northeastern Brazil. Deep-Sea Res 40:293-310

Kawase M (1993) Topographic effects on the bottom water circulation of the western tropical north Atlantic ocean. Deep-Sea Res 40:1259-1283

Kawase M, Sarmiento JL (1986) Circulation and nutrients in middepth Atlantic waters. J Geophys Res 91:9749-9770

Kawase M, Rothstein LM, Spriinger SR (1992) Encounter of a deep western boundary current with the equator: a numerical spin-up experiment. J Geophys Res 97:5447-5463

McCartney MS (1993) Crossing of the equator by the Deep Western Boundary Current in the Western Atlantic Ocean. J Phys Oceanogr 23:1953-1974

McCartney MS, Bennett SL, Woodgate-Jones ME (1991) Eastward flow through the Mid-Atlantic Ridge at 11°N and its influence on the abyss of the eastern basin. J Phys Oceanogr 21:1089-1121

McCartney MS, Curry RA (1993) Transequatorial flow

of Antarctic bottom water in the western Atlantic ocean: abyssal geostrophy at the equator. J Phys Oceanogr 23:1264-1276

Messias MJ (1994) Les chlorofluoromethanes, traceurs des eaux profondes dans les zones de fracture Romanche et Chain en Atlantique équatorial. Thesis, Université Paris 6

Molinari RL, Fine RA, Johns E (1992) The Deep Western Boundary Current in the tropical North Atlantic Ocean. Deep-Sea Res 39:1967-1984

Pickart RS (1992) Water mass components of the North Atlantic deep western boundary current. Deep-Sea Res 39:1553-1572

Pickart RS, WM Smethie Jr. (1993) How does the deep western boundary current cross the Gulf Stream? J Phys Oceanogr 23: 2602-2616

Pickart RS, Hogg NG, Smethie WM (1989) Determining the strength of the deep western boundary current using the chlorofluoromethane ratio. J Phys Oceanogr 19:940-951

Ponte RM, Luyten J, Richardson PL (1990) Equatorial deep jets in the Atlantic ocean. Deep-Sea Res 37:711-713

Reid JL(1994) On the total geostrophic circulation of the North Atlantic ocean: flow patterns, tracers, and transports. Progr Oceanogr 33:1-92

Rhein M (1991) Ventilation rates of the Greenland and Norwegian seas derived from distributions of the chlorofluoromethanes F11 and F12. Deep-Sea Res 38:485-503

Rhein M (1994) The Deep Western Boundary Current: tracers and velocities. Deep-Sea Res 41: 263-281

Rhein M, Stramma L, Send U (1995) The Atlantic Deep Western Boundary Current: water masses and transports near the equator. J Geophys Res 100:2441-2457

Richardson PL, Schmitz Jr WJ (1993) Deep cross-equatorial flow in the Atlantic measured with SOFAR floats. J Geophys Res 98:8371-8387

Schott F, Fisher J, Reppin J, Send U (1993) On mean and seasonal currents and transports at the western boundary of the equatorial Atlantic. J Geophys Res 98:14353-14368

Smethie Jr W (1993) Tracing the thermohaline circulation in the western North Atlantic using chlorofluorocarbons. Progr Oceanogr 31:51-99

Speer KG, McCartney MS (1991) Tracing lower north Atlantic deep water across the equator. J Geophys Res 96:443-448

Talley LD, Johnson GC (1994) Deep, zonal subequatorial currents. Science 263:1125-1128

Tsuchiya M, Talley LD, McCartney MS (1992) An

eastern Atlantic section from Iceland southward across the equator. Deep-Sea Res 39:1885-1917

Tsuchiya M, Talley LD, McCartney MS (1994) Water-mass distributions in the western South Atlantic; a section from South Georgia Island (54S) northward across the equator. J Mar Res 52:55-81

Wallace DWR, Beining P, Putzka A (1994) Carbon tetrachloride and chlorofluorocarbons in the South Atlantic Ocean, 19°S. J Geophys Res 99:7803-7819

Wallace DWR, Lazier JRN (1988) Anthropogenic chlorofluoromethanes in newly formed Labrador sea water. Nature 332:61-63

Warner MJ, Weiss RF (1985) Solubilities of chloro-fluorocarbons 11 and 12 in water and seawater. Deep-Sea Res 32:1485-1497

Weiss RF, Bullister JL, Gammon RH, Warner MJ (1985) Atmospheric chlorofluoromethanes in the deep equatorial Atlantic. Nature 314:608-610

Weiss RF, Warner MJ, Harrison KG, Smethie WM (1989) Deep equatorial Atlantic chlorofluorocarbon distributions. EOS, Trans American Geophysical Union 70:1132

Weisse R, Mikolajewicz U, Maier-Reimer E (1994) Decadal variability of the North Atlantic in an ocean general circulation model. J Geophys Res 99:12411-12421

Modelling the Ocean Circulation in the South Atlantic: A Strategy for Dealing with Open Boundaries

B. Barnier, P. Marchesiello & A.P. de Miranda

Laboratoire des Ecoulements Géophysiques et Industriels,
Institut de Mécanique de Grenoble, BP 53X, 38041, Grenoble, FRANCE

Abstract: The South Atlantic ocean is widely open to the Indian Ocean and the North Atlantic Ocean, and has a large inflow from the Pacific Ocean through the Drake passage. Strategies of modelling the ocean circulation in this area require to consider inter-ocean exchanges. The paper discusses various numerical approaches of the problem. One consists in modelling the world ocean and to study the South Atlantic as a sub-domain. Exchanges between the various oceans are thus determined by the model, and the circulation obtained for the South Atlantic depends upon the overall model behaviour. It is a consistent way to diagnose the circulation of the water masses in the ocean with coarse resolution models. This approach is also possible at eddy resolving resolution since the global ocean modelling effort undertaken by Semtner and Chervin (1988, 1992), and the simulation of the southern ocean circulation realised by the Fine Resolution Antarctic Model experiment (FRAM, Webb et al. 1991). However, there still are large differences between the various hydrographic or model estimates of the fluxes at the limits of the South Atlantic.
Another approach is to limit the modelled domain to the South Atlantic basin, and then have to deal with the complicated problem of open boundaries. In that case, the simulated circulation of the South Atlantic will largely depend upon the fluxes prescribed at the boundaries. This approach requires an a-priori knowledge of incoming mass, heat and salt fluxes, which is generally derived from climatologies of hydrographic data. It appears suited to process studies which characteristic time scales are short compared to the time evolution of the deep circulation, and to sensitivity studies investigating the impact of the fluxes at the boundaries on the estimates of the meridional mass and heat transports.

Introduction

The hydrographic program of the World Ocean Circulation Experiment (WOCE) is sampling the mass field of the South Atlantic at a hitherto unprecedent level. Intensive experiments (using hydrography, moorings, surface drifters and sub-surface floats), in special study areas such as Confluence, the Brazil Basin and the Cape Basin, will provide additional information on many physical processes relevant to the general circulation. The density and quality requirements of WOCE measurements will mean a tremendous increase in our knowledge of the water mass properties and the ocean circulation in this basin.

However, as in the case of the North Atlantic, which is considered to be the best sampled ocean in the world, field experiments by themselves will not be able to provide all the information necessary for a complete understanding of the ocean circulation and its variability in the South Atlantic. The main reason is that ocean observations, (this is particularly true for hydrography), do not have the synoptic character required for the calculation of turbulent fluxes in the ocean. This was acknowledged in the definition of WOCE, whose field program is tied to the following modelling objectives: to develop and test ocean models that will be useful for predicting climate change and to analyse the WOCE field data.

A numerical ocean model is expected to correctly represent the vertical and horizontal transport and

From WEFER G, BERGER WH, SIEDLER G, WEBB D J(eds), 1996, *The South Atlantic: Present and Past Circulation*. Springer-Verlag Berlin Heidelberg, pp 289-304

mixing of properties, by parameterizing or resolving explicitly the relevant physical processes (including the turbulent fluxes). Such a model will constitute a powerful tool for analysing and interpreting ocean observations, and will contribute greatly to the most realistic picture of the general ocean circulation. It will also identify and help to understand the physical mechanisms which control the evolution and variability of ocean flows. It will also make an indispensable contribution to the definition of original field experiments designed to implement monitoring of the general ocean circulation, and to investigate the physical processes which are responsible for long term changes in the general circulation. While numerical ocean models have not yet reached this level of accuracy in their simulation of the ocean circulation, they are currently capable of qualitatively representing many features of the large scale circulation, including the eddy field.

Modelling the South Atlantic remains a difficult challenge. This ocean is very much open to the Indian Ocean to the east, and receives a large inflow from the Pacific Ocean to the west through the Drake passage. It is also the only ocean basin wherein the meridional heat flux is towards the equator in subtropical regions. Rintoul (1991) has recently estimated this flux to be 0.25×10^{15} W across 32°S, in relation with a "cold water path" described as a northward flow of Intermediate Water (IW) coming from the Drake passage to compensate for a southward flow of North Atlantic Deep Water (NADW). Gordon (1986) and Gordon et al. (1992) suggested a different hypothesis for the closing of the thermohaline circulation associated with the formation of NADW. He proposed a route such that the return flow is made of warm surface waters entering the Atlantic via a leakage from the Agulhas Current. Thus, a strategy for modelling the ocean circulation in this area will need to consider carefully inter-gyre and inter-ocean exchanges.

Consequently, the physical processes which trigger inter-gyre exchanges have to be correctly parameterized or explicitly resolved. This militates for high-resolution numerical simulations in order to generate a realistic mesoscale eddy field capable of mixing tracers efficiently across fronts whilst maintaining the sharpness of the fronts. It also calls for an original treatment of the bottom topography and

overflows, since current ocean numerical models are known to misrepresent the major dynamical effects of the bottom topography as well as the exchanges between deep basins.

Two main numerical approaches to the problem of inter-ocean exchanges in the South Atlantic are worthy of consideration. The first one, which is developed in section 2, consists of modelling the world ocean and to study the South Atlantic as a sub-domain of this. The second approach, (presented in section 3), is to limit the model domain to the South Atlantic basin, and then tackle the complicated problem of open boundaries.

The global ocean modelling approach

This approach consists of modelling the world ocean and studying the South Atlantic as a sub-domain. Exchanges between the various oceans are thus determined by model dynamics, and the circulation obtained for the South Atlantic depends upon the overall model behaviour. It is a consistent way of diagnosing the circulation of the water masses in the ocean using coarse resolution models.

Coarse resolution simulations

The global modelling approach has recently been used by England and Garçon (E&G) (1994) to diagnose the circulation of the South Atlantic ocean. The ocean model employed is from the Geophysical Fluid Dynamics Laboratory (GFDL), and is based upon the primitive equations (PE) for the ocean circulation. The numerical code is described by Bryan (1969) and Cox (1984). The resolution is coarse (1.8 degree in longitude and 1.6 degree in latitude) and the eddy field is not resolved. There are 33 unequally spaced vertical levels and a realistic bottom topography is allowed for. The model is driven by Hellerman and Rosentein (1983) winds, and the model temperature, T, and salinity, S, are relaxed to the climatological fields of Levitus (1982) at the surface, and below 700 m. The global model qualitatively reproduces the inflow of the various water masses into the South Atlantic basin; the NADW from the North Atlantic, the Antarctic Circumpolar Water through the Drake passage, and a significant inflow (8.7 sv) of warm and saline

Agulhas water. The major features of the South Atlantic circulation at the upper and intermediate levels are also qualitatively reproduced, including the Benguela current and the confluence of the Brazil and Malvinas Currents. E&G used model simulations to diagnose the return path of the NADW. Their model results are such that 65% of the NADW production returns through the Drake passage as cold water converted into thermocline water as it circulates in the eastern South Atlantic, in agreement with Rintoul's (1991) proposition of a "cold water path". The contribution of the warm Agulhas water to the return flow is almost 25%, but may be largely underestimated owing to the absence of the Agulhas rings in the coarse resolution model.

However, coarse resolution simulations still produce weakly energetic mean circulations, because the eddy fluxes are parameterized as a sink of energy which does not account for the "negative viscosity effect" of the eddies which act to strengthen the mean flow. This appears to limit the ability of such model simulations to provide quantitative estimates of the meridional fluxes in the ocean.

Eddy-resolving simulations

It is now also possible to study the South Atlantic as a sub-domain of the world ocean at eddy-resolving resolution, since the global ocean modelling effort undertaken by Semtner and Chervin (S&C) (1988, 1992), and the simulation of the Southern Ocean circulation carried out by the FRAM Group (Webb et al. 1991a, 1991b). S&C initialized a series of simulations of the world ocean with the GFDL model (the MOM code), at marginally eddy-resolving resolution (1/2°×1/2°). The FRAM Group did a numerical simulation of the Southern Ocean with the same model, at a sligthly higher resolution (1/2° in longitude and 1/4° in latitude). Comprehensive studies of model results demonstrated that the major features of the South Atlantic circulation were qualitatively reproduced. Model data have since been made available to the scientific community.

It appears, from calculations performed with Semtner's 1/2° model results and the FRAM data, that there are still large differences between the various hydrographic and model estimates of

the fluxes at the boundaries of the South Atlantic. Models exhibit a very large volume transport (near 200 sv) in the Antarctic Circumpolar Current (ACC), and the zonal fluxes of heat and mass are consequently large (Table 1). The average temperature, defined as the ratio of the total heat flux to the total volume transport (Table 1), is an index which has been corrected somewhat for the bias induced by the large transport in the ACC. It shows that models are too warm, probably because a lack of cold bottom waters, as indicated in the balance of the water masses shown in Fig. 1. Note that models indicate that south of 32°S, the Atlantic is cooled by the atmosphere, whereas Rintoul's analysis requires a small warming.

A summary of the major water masses performed with Semtner's 1/2° and the FRAM data is shown in Fig. 1a and Fig. 1b, and is compared to Rintoul's analysis (1991) (Fig. 1c and Fig. 1d). All analyses show similar imports of NADW across the northern boundary (near 20 Sv). Both models form a significant amount of IW (4.7 Sv in Semtner and 8 Sv in FRAM), and have a significant amount of surface water entering the South Atlantic from the Indian ocean, two characteristics of the circulation which are more consistent with Gordon's proposition of an Indian Ocean inflow to compensate for the export of NADW, a scenario named "warm water path" by Rintoul (1991). Indeed, Rintoul's "warm water path" scenario (Fig. 1d) requires both the formation of IW and the inflow from the Indian Ocean. In contrast, Rintoul's "cold water path" (Fig. 1c) shows a conversion of 8 Sv of IW into surface and deep waters, and no inflow of surface waters. Note that almost no bottom water is formed in the model simulations. This is probably related to the absence of sea-ice and seasonal forcing, necessary for the formation of Antarctic Bottom Waters (AABW).

S&C have since increased the horizontal resolution of the model to 2/5°×1/4° (Semtner, personnal communication), and recently a research group at Los Alamos National Research Laboratory started a simulation of the world ocean circulation with a resolution of 1/4° in longitude and an average of 1/6° in latitude (Smith et al., personal communication). Their model results show a significant improvement in their representation of western bound-

Table 1: Transport and fluxes at the boundaries of the South Atlantic from the simulations of Semtner and Chervin (1/2°) and FRAM. The hydrography reference is Rintoul (1991).

Section		68°W	20°E	32°S
Total Volume Transport	Rintoul	129 Sv	130 Sv	0 Sv
	Semter 1/2°	209 Sv	209 Sv	0 Sv
	FRAM	185 Sv	185 Sv	0 Sv
Total Heat Flux	Rintoul	1.30 PW	1.12 PW	0.24 PW
	Semter 1/2°	2.45 PW	1.72PW	0.60 PW
	FRAM	1.95 PW	1.15 PW	0.56 PW
Average Temperature[*]	Rintoul	2.40 °C	2.06 °C	
	Semter 1/2°	2.80 °C	1.97 °C	
	FRAM	2.53 °C	1.49 °C	
Surface Heating		Rintoul	Semtner	FRAM
A positive value is a heat gain for the ocean		0.06 PW	-0.13 PW	-0.24 PW

[*]The average temperature is defined as following ratio: $\dfrac{\text{Total Heat Flux}}{\rho_0 C_P \times \text{Total Volume Transport}}$

ary currents, in the location and strength of the important fronts, and in the intensity and behavior of the eddy field (Semtner, personal communication). The picture of the general circulation of the South Atlantic drawn by these simulations appears to be quite realistic. As an example, the large eddies shed at the retroflection of the Agulhas Current drift westward across the subtropical gyre and sometimes into the Brazil current, as recently observed from altimetric data (Gordon and Haxby 1990). Future analyses of model results may be expected to provide various flux and transport estimates of quantitative interest. However, global models are not tools which are easy to use. They require substantial computing and storage resources that many research teams cannot afford. Models take a long time to spin-up and the potential for sensitivity studies is limited.

To study the South Atlantic circulation, we expect global models to provide the best available boundary conditions on the edge of the South Atlantic basin. However, global models are still characterised by major weaknesses in their ability to form the important water masses in the region (AAIW, AABW, and NADW). There are two main reasons for this. One is the limited ability of numerical models to resolve the physical processes which condition the mixing of properties in the ocean. Consequently, model results still depend on parameterisations of the mixed layer, turbulent vertical mixing, and deep convection which are not fully satisfactory. Another reason stems from the complexity of the air, ocean and sea-ice interactions, and the difficulty of building the effect of sea-ice into numerical simulations.

The basin-scale modelling approach

The basin-scale modelling approach limits the model domain to the South Atlantic basin, and means that the complicated problem of open boundaries has to be dealt with. In this case, the

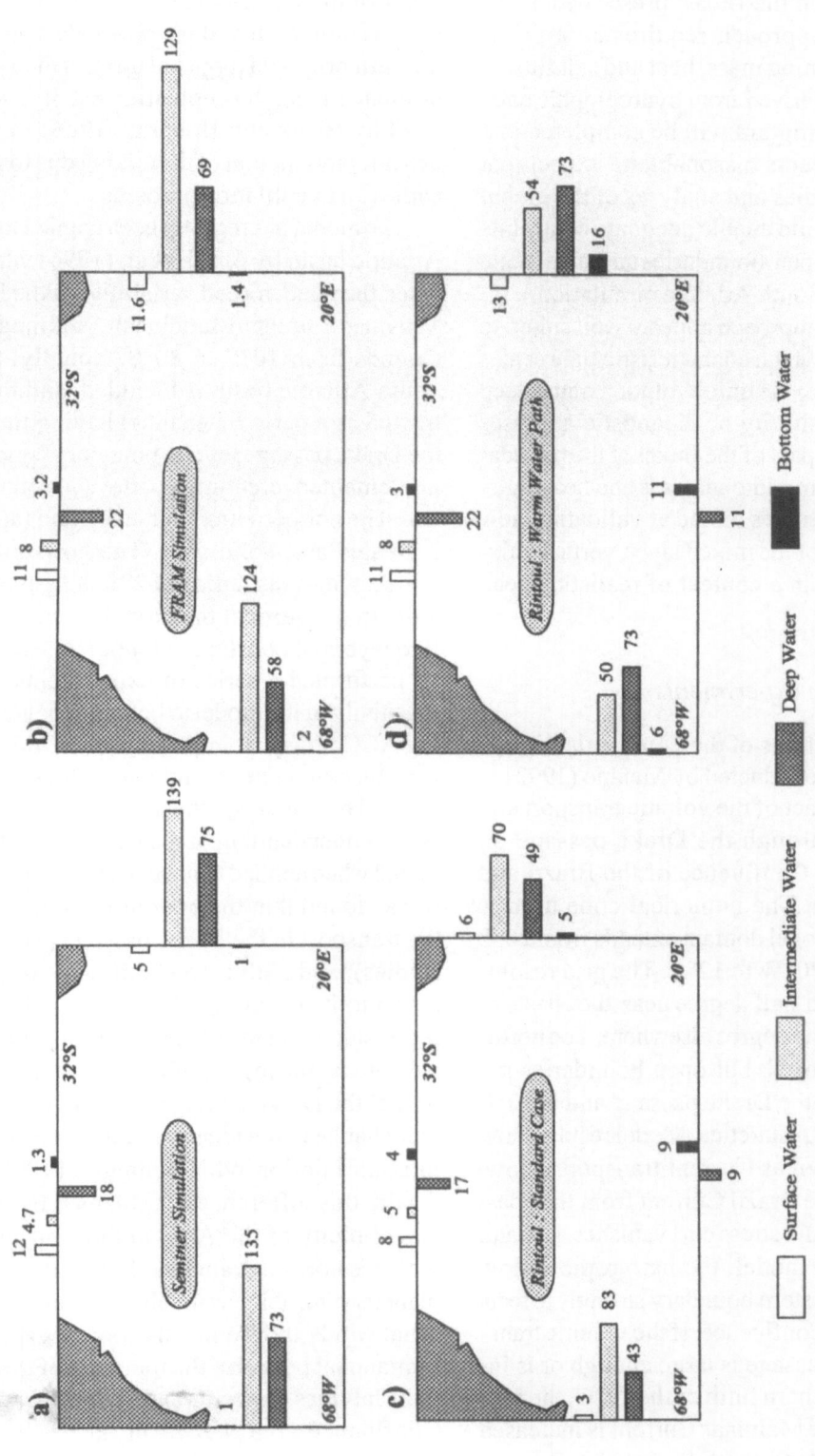

Fig. 1. Transport (in Sv, 1 Sv=10^6 m^{3}s^{-1}) of the major water masses in the South Atlantic from the mean fields produced by the simulations of Semtner and Chervin (1/2°) and FRAM. The water masses are defined from their potential density σ (with the usual definition of σ0, σ1, σ2, σ3, σ4), according to Rintoul (1991); Surface Water σ <26.80, Intermediate Water 26.80< σ <32.36, Deep Water 32.36< σ <41.66, Bottom Water 41.66< σ.

simulated circulation of the South Atlantic will largely depend upon the fluxes prescribed at the boundaries. This approach requires an a-priori knowledge of incoming mass, heat and salt fluxes, which is generally derived from hydrographic data. The WOCE field program will be completed in a few years, and it seems reasonable to expect that comprehensive studies and analyses of the global WOCE data set should enable adequate contraints to be provided at open boundaries in basin-scale simulations of the South Atlantic circulation.

This modelling approach appears well suited to process studies for which characteristic time scales are short compared to the time evolution of the deep circulation, to sensitivity or diagnostic analyses investigating the impact of the fluxes at the boundaries on estimates of meridional mass and heat transports, and to case studies aimed at validating new parameterizations (of the mixed layer, vertical mixing or convection) in a context of realistic ocean simulations.

Coarse resolution experiments

Basin-scale simulations of the South Atlantic circulation have been conducted by Matano (1993) to investigate the impact of the volume transport and velocity profile through the Drake passage on the location of the Confluence of the Brazil and Malvinas Currents. The numerical code used is from GFDL. The model domain extends from 20°S to 70°S, and from 70°W to 12°E. The grid resolution varies from one-half degree near the coasts of South America to one degree elsewhere. The northern boundary is closed, but open boundaries are applied at 70°W in the Drake passage and at 12°E between Africa and Antarctica. Model results show that when the Malvinas Current transport is low, the separation of the Brazil Current from the coast occurs where the wind stress curl vanishes. Matano showed that in the model, the barotropic inflow prescribed at the western boundary strongly affects the location of the Confluence. If the volume transport at the Drake passage is large enough or is intensified in the northern limb of the ACC, the volume transport of the Malvinas Current is increased and the model is able to correctly simulate the latitude at which the Brazil current leaves the coast.

Recent years have witnessed the development of new primitive equation models and their application to simulation of the large-scale ocean circulation. An original isopycnal model (the vertical coordinate of which is potential density) was developed by Bleck and Boudras (1986) to provide a better representation of the mixing due to mesoscale eddies and ventilation processes.

The model has recently been applied to the South Atlantic basin by Smith et al. (1994) who investigated the wind-forced variability of the Brazil and Malvinas Currents. Meridionally, the model domain extends from 10°S to 80°S. Zonally, the entire South Atlantic basin is included, and the ACC is treated as a periodic channel having the width of the Drake passage; cyclic boundary conditions are implemented, creating a pathway around Antarctica. The northern limit of the domain (along 10°S) is treated as a solid wall. The horizontal grid is coarse, with a resolution of 2° in longitude and latitude. In the vertical the model is configured with five layers of constant potential density. Smith et al. performed a series of experiments to test the sensitivity of the model to bottom topography, varying ACC transport imposed at the Drake passage, variable wind stress, and lateral boundary conditions. The paper synthesizes model results for a better understanding of the behavior of this type of model when applied to general circulation problems. It was found that the bottom topography reduces the transport in the Drake passage (as in previous studies), and shifts the Confluence region to the North to its usual location (40°S). This latter effect results from the baroclinisation of the subtropical gyre by the topography. Forcing greater transport at the Drake passage does not cause any further change in the location of the Confluence area, in contradiction with Matano's (1993) results. Again, this difference is explained by the greater baroclinicity of the ACC in the isopycnal model which is not constrained to follow the contours of planetary vorticity as much. Simulations with seasonal winds show a locally forced response at the semiannual period in the transport of the ACC and the Malvinas Current, and in the location of the Confluence area. However, such variability is not observed in the Brazil Current because the Malvinas waters (represented by the middle layer of the model) are not able to penetrate the subtropical gyre which is confined to the upper layers.

A topography-following coordinate model of the South Atlantic Basin

It appears from most numerical studies that bottom topography plays a key role in the South Atlantic ocean, since it obviously places a limit to the ACC transport. Process studies have also shown that the topographic roughness increases the baroclinicity of the flows, which reduces the tendency of strong currents to follow the bathymetry (Barnier and Le Provost 1993). Numerical models generally use staircase bottom topography following the three-dimensional grid of the model, and the ocean "feels" the bathymetry as a mask where a constraint of no normal flow is applied. The bottom slope does not appear directly as a component of the vorticity balance, as it does in the bottom layer of quasi-geostrophic models, or in the equations of PE models using a topography following ("sigma") vertical coordinate.

"Sigma" coordinate models have been successfully applied to lakes, estuaries and coastal oceanography (Blumberg and Mellor 1987; Ezer and Mellor 1992). Application to an entire ocean basin is more difficult because the height of the major topographic features is of the order of the ocean depth. Consequently, a number of numerical problems, which have arisen in the course of meteorological applications of this type of coordinate (particulary the pressure gradient calculation and the hydrostatic inconsistency), are more severe, and it is harder to fit numerical solutions.

In the context of the French participation in WOCE, our modelling group in Grenoble has begun a long-term modelling program (MOCA, Modélisation de la Circulation Atlantique) to simulate the circulation in the South Atlantic with a sigma-coordinate model. The original code is the Semi-spectral Primitive Equation Model (SPEM, Haidvogel et al. 1991), which has been modified for ocean basin applications. In the long term, the aim of our research is to investigate the role of mesoscale dynamics in the transformation and circulation of the upper and intermediate water masses, and to quantify the importance of eddies

in the process of cross-front mixing of tracers, as well as in the winter convective processes related to the formation of water masses.

As a first step, we set up a coarse resolution model of the South Atlantic basin. This relatively low cost model was used to define proper open boundary conditions, and to investigate the conditions of application of the sigma coordinate (Marchesiello et al. 1993). This latter study was essential since it meant that we could estimate and minimise the errors introduced by the pressure gradient calculations and the hydrostatic inconsistency. The coarse resolution model was also used to test the sensitivity of the circulation patterns produced by the model to various flux conditions at the open boundaries and to seasonal wind, heat and sea-ice forcings. The paper presents results obtained using this coarse resolution model. However, plans for the high resolution model are also discussed.

The coarse resolution model domain (Fig. 2a) extends meridionally from 16°S to 76°S, and zonally from 68°W to 20°E. The elementary horizontal grid is square, with a resolution of 1.375° in longitude (λ) and 1.375° cos φ in latitude(φ). Twenty sigma levels are used in the vertical, with higher resolution near to the surface. The bottom topography (Fig. 2b) has been smoothed according to a criterion involving the resolution of the horizontal and vertical grid, in order to reduce errors in the pressure gradient calculation and to minimize the hydrostatic inconsistency. We noticed that if our smoothing criterion is not verified, those errors can drive significant currents (a few centi-meters). The open boundary conditions (Marchesiello 1995) include a radiation condition (the value of the model variables at boundary points are calculated from the inner points using a wave equation), applied only in case of outflow, and a relaxation to climatological values. The relaxation constant is 15 days in the Drake passage, and is 5 years at other boundaries.

With the coarse resolution model we carried out sensitivity studies, and investigated the model response to changes in the ACC in the Drake passage, and to changes in the incoming flow from the Agulhas current. We also investigated the model's response to seasonal forcing using wind stress and heat fluxes from a 3-year climatology obtained

a) MODEL GRID

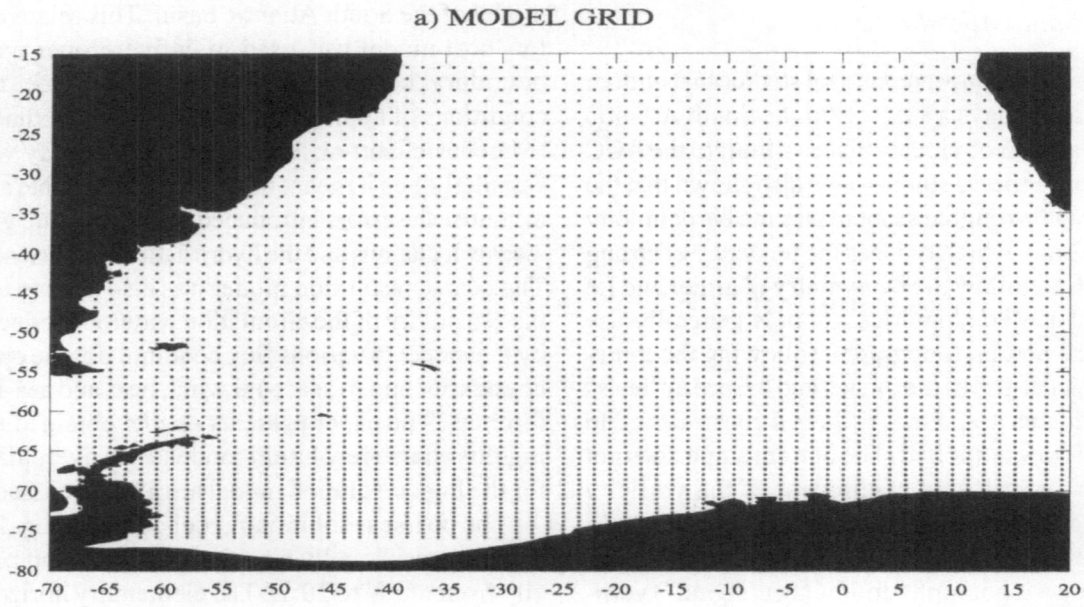

b) MODEL BATHYMETRY

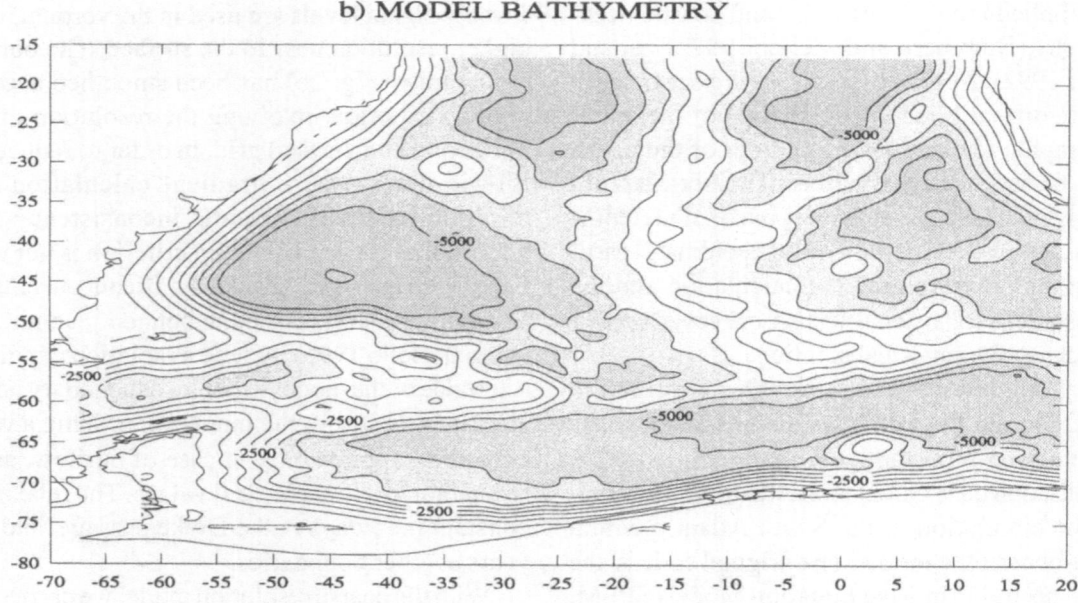

Fig. 2. South Atlantic Basin configuration of the SPEM model for the coarse resolution experiments. a) The horizontal grid, and b) the smoothed bottom topography derived from the ETOPO5 data set (NOAA, 1988). Minimum depth is 750 m. Contour interval is 500 m. Open boundary conditions are applied at the Drake Passage, between Brazil and Angola, and between South Africa and Antarctica.

from analyses performed at the European Center for Medium Range Weather Forecast (ECMWF) (Barnier et al. 1995), and a salt flux in the Weddell Sea estimated using a semi-diagnostic sea-ice model (Darr 1994).

The results presented here concern the circulation produced by a 10-year long diagnostic simulation, during which the model was forced by Hellereman and Rosenstein's (1983) annual wind stress, and where model temperature, T, and salinity, S, are relaxed to the climatological fields of Levitus (1982), at the surface with a relaxation constant of 30 days, and at depth with a relaxation constant of 5 years. The volume transport at Drake is 150 Sv, a value for which the model gives a correct position for the Confluence. A zero inflow of Agulhas waters is prescribed at the southern tip of Africa.

After 10 years the barotropic streamfunction (Fig. 3) is consistent with the schematic presented by Reid (1989) and Peterson and Stramma (1991). The Subtropical Gyre occupies the northern part of the domain with a transport of 34 Sv. The Brazil Current separates from the coast at 35°S, then flows southward down to 45°S before it turns to the northeast and joins the subtropical front, identified as the -150 Sv line and running zonally along 40°S. The ACC enters the domain at 68°W with a transport of 150 Sv, as expected from the open boundary conditions. Between 60°W and 50°W, the northern part of the current (100 Sv to the north of 60°S) turns northward, and splits into two branches just after passing the Falkland Islands. One branch becomes the Malvinas Current (45 Sv) flowing northward along the coast of Argentina until it meets the Brazil Current in the Confluence region. The second branch (55 Sv) turns eastward along

MODEL STREAMFUNCTION

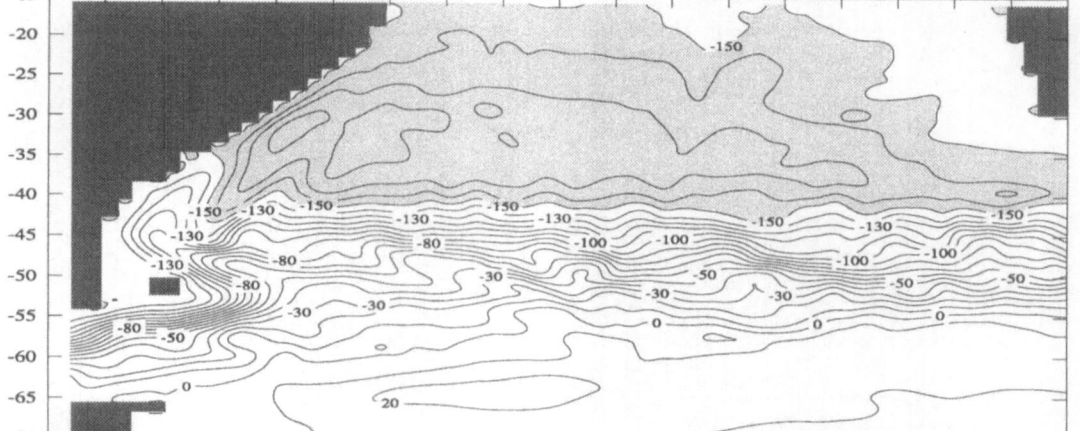

Fig. 3. SPEM South Atlantic. Barotropic streamfunction after 10 years of integration in the diagnostic run. Units are Sverdrups (1 Sv=10⁶ m³s⁻¹). Contour interval is 10 Sv.

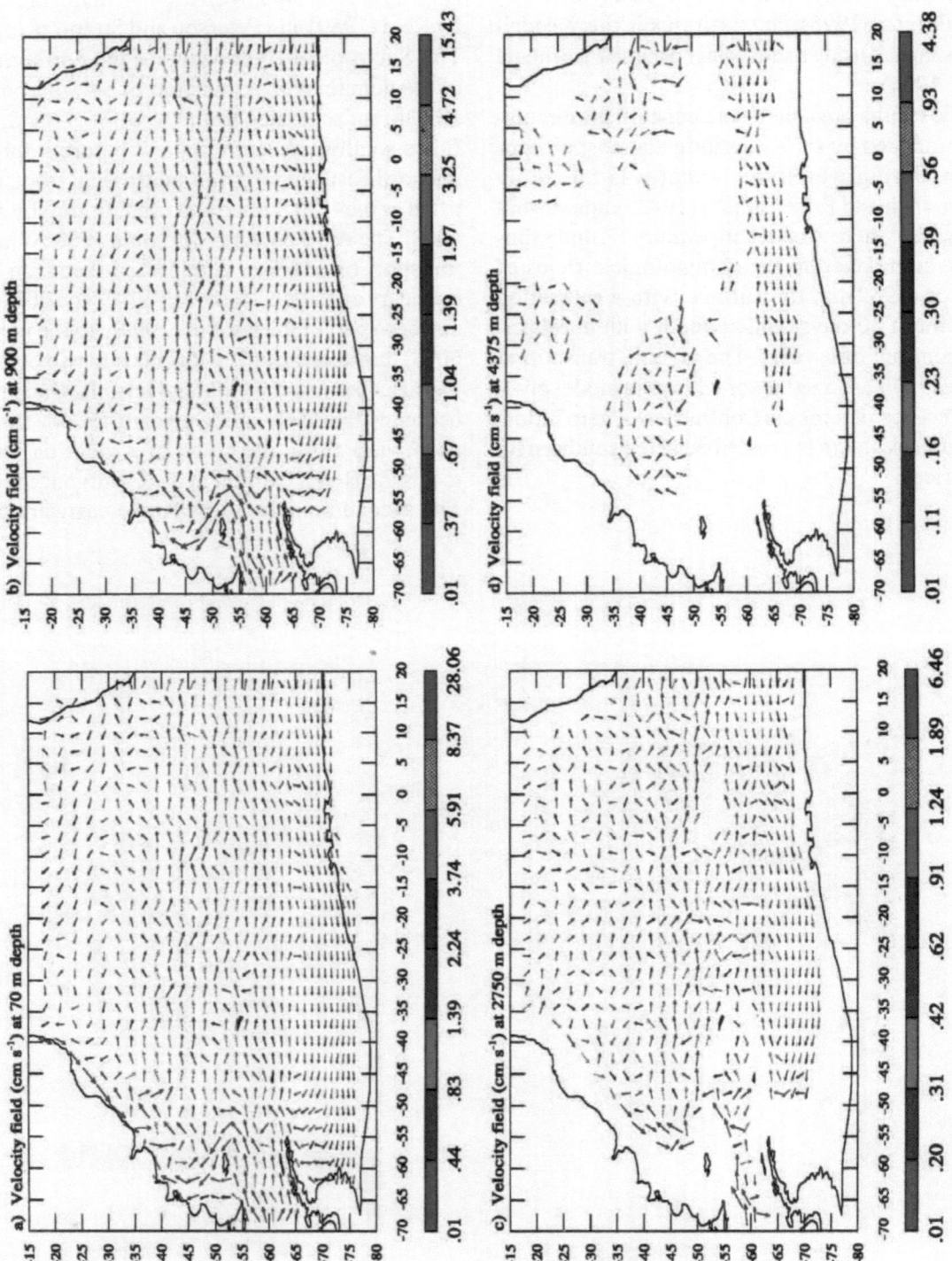

Fig. 4. SPEM South Atlantic. Horizontal velocities after 10 years of integration in the diagnostic run, at depth a) 70 m, b) 900m, c) 2750 m, and d) 4375 m. Only one vector every two is plotted. Vectors are colored according to their amplitude. Units are cm s⁻¹.

the Falkland escarpment and continues to run zonally between 46 and 48°S as the Sub-Antarctic Front (SAF). The part of the ACC which enters the South Atlantic to the South of 60°S (50 Sv) veers sharply northwards near 55°W, but does not cross the Falkland escarpment and moves eastwards as a broad and weak flow. It concentrates again at 50°S as it passes between the Orcadas Rise and the South Sandwich Islands. It then flows eastward at latitudes between 48°S to 52°S as the Polar Front (PF). The Weddell gyre pumps 10 Sv into the South Atlantic at the southeastern corner of the domain to give a total volume transport of 20 Sv.

Fig. 4 shows horizotal velocity fields at four different depths which are representative of the Thermocline Waters (70 m), Intermediate Waters (900 m), Deep Waters (2750 m) and Bottom Waters (4375 m). The most barotropic pattern appears to be the Weddell gyre, where the currents are in phase at all depths with little vertical shear as shown in Fig. 5. The vertical coherence of the ACC is pronounced over the first 1000 m in a broad latitude band (between 35°S to 55°S), but there is considerable vertical shear: between 70 m (Fig. 4a) and 900 m (Fig. 4b) velocities are reduced by a factor of 2. At mid-depth (Fig. 4c), this barotropic character is maintained for the jets which compose the mainstream of the ACC. The bottom circulation (Fig. 4d) is more constrained by the bathymetry than by the flow of the ACC above. An important flow (over 1 cm s^{-1}) can be seen from the Weddell sea into the Argentine basin. The Brazil Current is confined to the surface with velocities of 20 to 30 cm s^{-1} (Fig 4a, b). It flows above 500 m between 15°S and 25°S, but deepens further south (it reaches 1000 m at 30°S, as shown in Fig. 6). The path of IW toward the equator (flow at 900 m, Fig. 4b) is westward within the ACC along 40°S; it returns eastward within the subtropical gyre, and finally turns northward at 30°S near the coast of South America. The Deep Western Boundary Current is a well marked feature at 2750 m (Fig. 4c) and flows

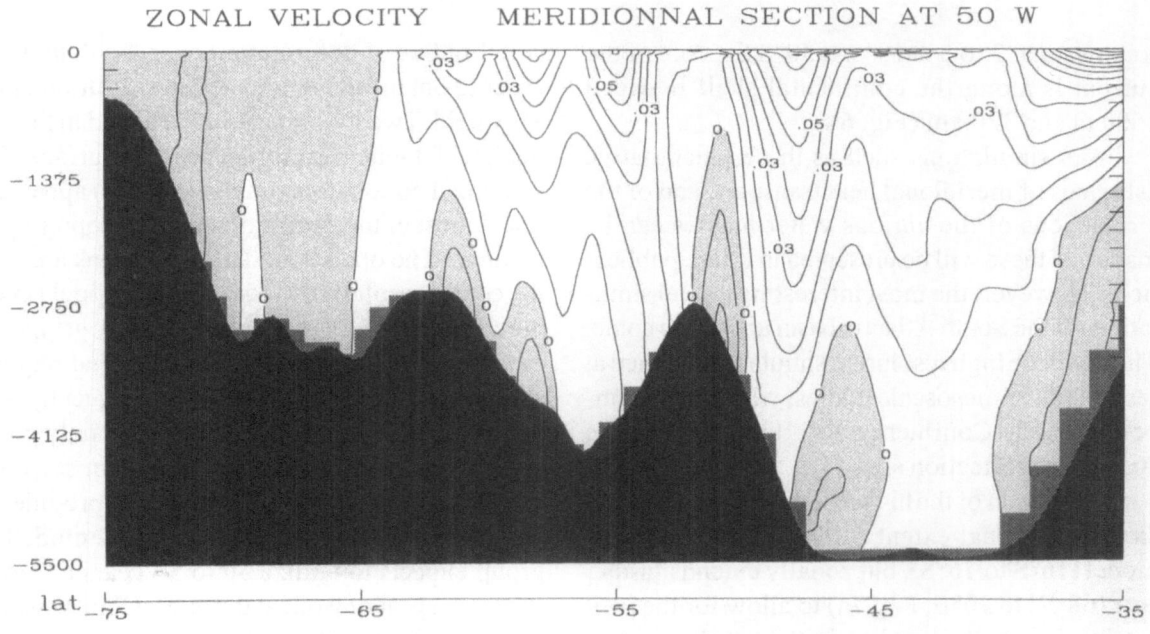

Fig. 5. SPEM South Atlantic. Zonal velocity as a function of depth along 50°W. Positive values indicate eastward velocities. Units are ms^{-1}.

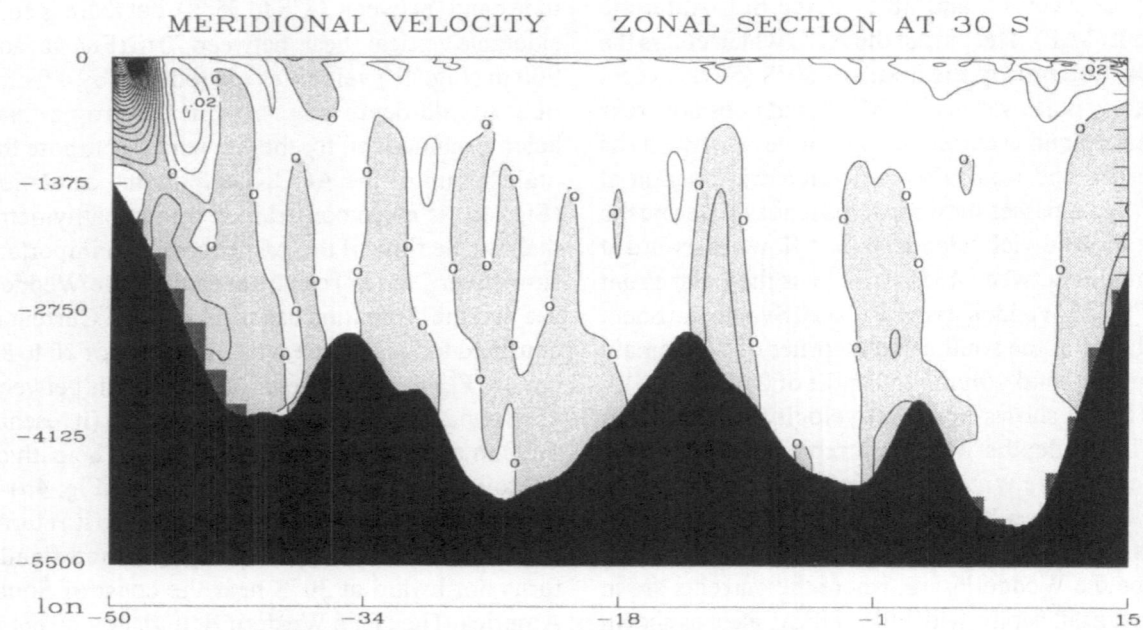

Fig. 6. SPEM South Atlantic. Meridional velocity as a function of depth along 30°S. Positive values indicate northward velocities. Units are ms⁻¹.

southward at 3 to 5 cm s⁻¹. At 30°S the core of the current is along the continental shelf between 1500 m and 3200 m (Fig. 6).

From simulations such as these, quantitative estimates of meridional heat transport, and of the circulations of the various water masses can be made and these will be presented in future publications. However, the most interesting model simulations of the South Atlantic basin are still to come. These will be high resolution simulations, aimed at resolving the mesoscale eddies, which are so important in the Confluence area, the ACC and the Agulhas retroflection area.

The domain of the high resolution model has the same meridional extent as the coarse resolution model (16°S to 76°S), but zonally extends further east (68°W to 30°E, Fig. 7a) to allow for the generation of Agulhas eddies in the Agulhas retroflection area. The horizontal grid has an average resolution of 0.3°, but varies in such a way that the elementary grid size is of the order of the first in-

ternal radius of deformation over the domain, in order to obtain an homogeneous resolution of the eddy field. Twenty sigma levels are used in the vertical, with higher resolution near the surface. The smoothed bottom topography (Fig. 7b) appears to retain most of the details of the major topographic features. The open boundary conditions are as in the coarse resolution model. Such a model is very much a heavy duty one (it has as many gridpoints as a global weather forecast model), and requires substantial computer resources. Consequently, productive research with such a model is only possible within the framework of long-term scientific programs such as WOCE which can provide the indispensable continuous support. Our modelling group expects to realize 30 to 50 years of model run over a year of work at the most. Objectives will be to quantify the effects of eddy dynamics on cross-front mixing, as well as on the winter convective processes related to the formation of mode waters. At present the model has been run for several years

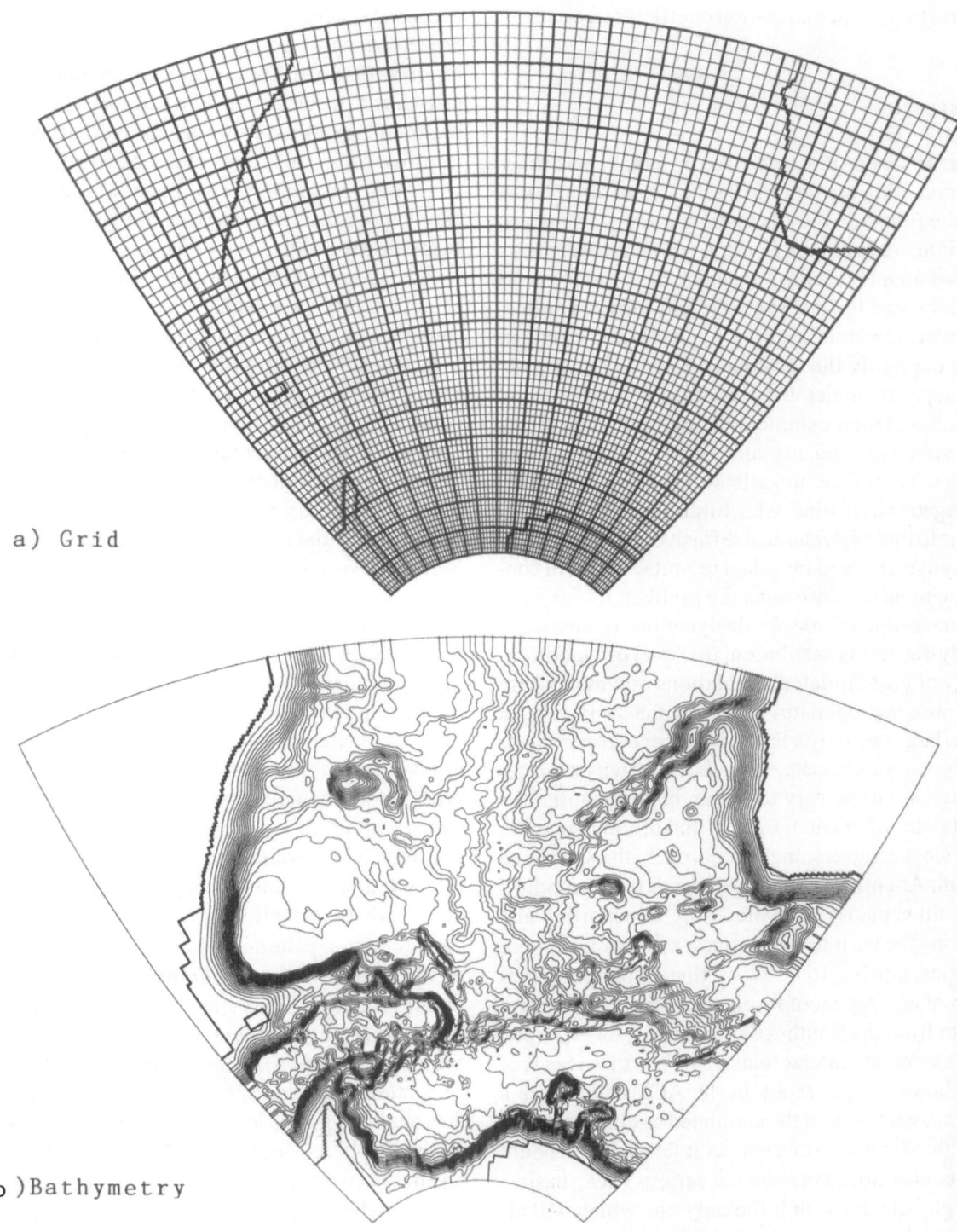

a) Grid

b) Bathymetry

Fig. 7. South Atlantic Basin configuration of the SPEM model for the high resolution experiments. a) The horizontal grid (only one point every four is plotted). b) The smoothed bottom topography derived from ETOPO5. Minimum depth is 250 m. Contour interval is 250 m.

to determine the optimal set of numerical parameters (time step, dissipation coefficients, etc.).

Discussion

The article has presented two different approaches to modelling the South Atlantic Ocean circulation. In the first one, the circulation of the world ocean is simulated, and the South Atlantic is studied as a sub-domain. It is the one which has been the most widely used to date, at coarse and eddy resolution. However, coarse resolution models do not calculate explicitly the turbulent fluxes, which are the most poorly understood aspect of the ocean circulation, and their parameterization, including recent isopycnal mixing, are such that they act as energy sinks. Therefore, models still produce a weakly energetic circulation, whose transport will be analogous to that of dynamical diffusive regimes instead of convective and turbulent regimes. This will continue to be the case until the problem of the parameterization of mesoscale dynamics is solved. At eddy-resolving resolution, the level of kinetic energy of the simulated circulations moves closer to the observed estimates. However, the computer cost is so large that only a limited number of experiments are possible. Consequently, the sensitivity analyses which are necessary to obtain better quantitative estimates of ocean transport and mixing are making slow progress. In this approach, the simulated South Atlantic circulation depends on the model's overall capacity for generating, transporting and mixing the various water masses. Under these conditions, one has to be aware that a misrepresentation of any aspect of forcing functions in areas remote from the South Atlantic basin, (for example, the air-sea-ice interactions in the Northern Seas, or the bottom topography in the ACC), will have a negative impact on the simulated circulation in the South Atlantic. However, as it takes into account the connections between the various ocean basins, the global approach is the only one which will allow the long term variability of the ocean circulation in the South Atlantic to be investigated satisfactorily.

The second approach seeks to simulate the circulation of the South Atlantic basin, and requires numerical solutions to be applied to the problem of the open boundaries. The recent simulations, briefly presented in this paper, performed at coarse resolution with the SPEM model indicate that this approach can be successful. Furthermore, the WOCE program is expected to provide better data and thus improve flow constraints at the open boundaries, while high-resolution eddy resolving simulations should provide quantitative estimates of turbulent fluxes in the South Atlantic basin.

The modelling strategy decision has to match scientific objectives and available computer resources. The circulation of the South Atlantic is rich in basin-scale features (the Weddell sea, the ACC, the subtropical gyre) separated by thermohaline fronts. Consequently, the meridional transport of properties depends upon eddy-driven exchanges between gyres, and could best be studied with numerical studies using a limited basin model allowing higher spatial resolution. On the other hand, the long term variability of the circulation is largely dependent upon the evolution of the deep water masses entering the basin, and numerical studies investigating this aspect will require world ocean model simulations. In fact, both approaches appear to be complementary.

We conclude with some considerations on the atmospheric forcing data from which both approaches will benefit. It will be only when we know enough about the ways in which the circulation is driven, (i. e. which atmospheric variables are relevant and how the forcing penetrates into the ocean), that we shall be in a position to say why the circulation patterns and the distribution of water masses are as they are. Only few VOS observations (Voluntary Observing Ships) are available in the southern hemisphere, and NWP (Numerical Weather Prediction) Centers probably provide the best estimates of atmospheric surface data, although they are known to be less accurate than in the northern hemisphere. However, no long term climatology has been derived from NWP data, because frequent model changes have introduced a non-geophysical variability into the time series of the analyses which is almost impossible to correct: and the re-analysis planned or being performed at the National Meteorological Center (NMC) and ECMWF will be very useful in this regard.

Great expectations are aroused by Earth observations from space. Wind field measurements over the world ocean are already available from the ERS-1 spaceborne scatterometer, and this type of observation will be carried on continuously over the next ten years with the coming space programs ERS-2 and NSCAT/ADEOS. While satellites are expected to provide data on most of the variables necessary for estimating the radiative fluxes, there are still serious difficulties ahead for the estimation of the turbulent fluxes, and the fresh water flux (which require data concerning the precipitation field over the ocean).

Acknowledgments

The authors are grateful to Albert Semtner and David Webb who kindly provided results and data from their recent simulations. The authors are supported by the Centre National de la Recherche Scientifique and the Ministère de l'Enseignement Superieur et de la Recherche. This research was funded by the Institut National des Sciences de l'Univers and IFREMER through the Programme National d'Etude de la Dynamique du Climat. Support for computations was provided by the Institut du Développement et des Resources en Informatique Scientifique.

References

Barnier B, Le Provost C (1993) Influence of a rough bottom topography on the jet and inertial recirculation of a mid-latitude gyre. Dyn Atm Oceans 18:29-65

Barnier B, Siefridt L, Marchesiello P (1995) Thermal Forcing for a Global Ocean Circulation Model From a Three-Year Climatology of ECMWF Analyses. J Mar System 6:363-380

Bleck R., Boudra DB(1986) Wind-driven spin-up in eddy-resolving ocean models formulated in isopycnic and isobaric coordinates. J Geophys Res 91:7611-7621

Blumberg A F, Mellor G L (1987) A description of a three dimensional coastal ocean circulation model. In: Heaps N (ed) Three dimensional Coastal Ocean Models, Vol 4, N Amer Geophys Union, p 208

Bryan K (1969) A numerical method for the study of the circulation of the world ocean. J Comp Phys 4:347-376

Cox MD (1984) A primitive equation 3-dimensional model of the ocean. GFDL Ocean Group Tech Rep No 1, GFDL/Princeton University p 147

Darr D (1993) A semi-diagnostic sea-ice model. Internal report, Equipe MEOM, LEGI-IMG, BP53X, 38041 Grenoble, France

England MH, Garçon VC (1994) South Atlantic circulation in a world ocean model. Annales Geophysicae 12:812-825

Ezer T, Mellor GL (1992) A numerical study of the variability and the separation of the Gulf Stream, induced by surface atmospheric forcing and lateral boundary flow. J Phys Oceanogr 22:660-682

Gordon AL (1986) Interocean exchange of thermocline waters. J Geophys Res 91:5037-5046

Gordon AL, Haxby WF (1990) Agulhas eddies invade the South Atlantic: Evidence from Geosat altimeter and shipboard conductivity-temperature-depth survey. J Geophys Res 95:3117-3125

Gordon AL, Ray FW, Smethie WM, Warner MJ (1992) Thermocline and intermediate water communication between the South Atlantic and the Indian Oceans. J Geophys Res 97:7223-7240

Haidvogel DB, Wilkin JL, Young R (1991) A semi-spectral primitive equation ocean circulation model using vertical sigma and orthogonal curvilinear horizontal coordinates. J Comp Phys 94:151-185

Hellerman S, Rosenstein M (1983) Normal monthly wind stress over the world ocean with error estimate. J Phys Oceanogr 13:1093-1104

Levitus S (1982) Climatological atlas of the world ocean. NOAA Prof Paper No 13, US Govt Printing Office, Washington DC

Matano RP (1993) On the separation of the Brazil current from the coast. J Phys Oceanogr 23:79-90

Marchesiello P (1995) Simulation de la circulation océanique dans l'Atlantique sud, avec un modèle numérique à coordonnée sigma. Thèse de Doctorat, Université Joseph Fourier, Grenoble

Marchesiello P, Nguyen T, Barnier B (1993) Modeling the South Atlantic Ocean: On the control of the circulation by the fluxes at the open boundaries. Proceedings of the Fourth International Conference on Southern Hemisphere Meteorology and Oceanography, 29 March-2 April 1993, Hobart, Tasmania

NOAA (1988) NGDC data announcement 88-MG-02: Digital relief of the surface of the Earth, US department of Commerce, NOAA, NGDC, Boulder, Colorado

Peterson RG, Stramma L (1991) Upper-level circulation in the South Atlantic Ocean. Prog Oceanogr 26:1-73

Reid JL (1989) On the total geostrophic circulation of the South Atlantic Ocean: Flow patterns, tracers, and transports. Prog Oceanogr 23:149-244

Rintoul SR (1991) South Atlantic interbasin exchange. J Geophys Res 96:2675-2692

Semtner AJ, Chervin RM (1988) A simulation of the global ocean circulation with resolved eddies. J Geophys Res 93:15502-15522

Semtner AJ, Chervin RM (1992) Ocean general circulation from a global eddy-resolving model. J Geophys Res 97:5493-5550

Smith LT, Chassignet EP, Olson DB (1994) Wind-forced variations in the Brazil-Malvinas confluence region as simulated in a coarse resolution numerical model of the South Atlantic. J Geophys Res 99:5095-5117

Webb DJ, et al (The FRAM Group) (1991a) Using an eddy resolving model to study the Southern Ocean. EOS, 72, p 15

Webb DJ, et al (1991b) The FRAM atlas of the Southern Ocean. Natural Environnement Research Council, p 67

Mass and Heat Transports in the South Atlantic Derived from Historical Hydrographic Data

R. Schlitzer

Alfred-Wegener-Institute for Polar and Marine Research,
Bremerhaven, GERMANY

Abstract: Mass and heat transports in the South Atlantic as well as exchange flows with the South Pacific and the Indian Ocean are determined by driving a conservative, steady box-model towards the historical temperature (θ) and salinity (s) observations. The optimal model circulation searched for is required (a) to approximately preserve the vertical velocity shear as given by geostrophic calculations and (b) to correctly reproduce the measured distributions of θ and s. Information contained in the θ/s data on baroclinic flows is exploited through constraint (a) and the unknown reference velocities are determined by the model in a way such that the resulting absolute flow velocities produce realistic θ and s fields (constraint (b)). The model is mass, heat and salt conserving and has realistic topography. The adjoint method is applied as an efficient means for calculating cost function gradients needed during the optimization process.

Model experiments show that indeed realistic θ and s model distributions can be obtained with flows that are consistent with geostrophy. Moreover, close agreement between measurements and model is obtained for a variety of model velocity fields that differ considerably with respect to strength of the meridional overturning cell and magnitude of meridional heat transports. The maximal acceptable meridional heat transport across 30°S (based on an evaluation of θ/s misfits and deviations from geostrophic shear) amounts to 0.4 PW. Forcing the model to produce larger heat fluxes results in systematic property misfits in the upper layers of the South Atlantic. Contrary to most published heat transport estimates the model also accommodates poleward (southward) heat fluxes of up to -0.5 PW. The best model property fields are obtained for a heat transport across 30°S close to zero. All acceptable model solutions show a dominance of northward flow of Antarctic Intermediate Water (AAIW) over warmer, upper layer waters, and all model solutions show net heat gain of the ocean from the atmosphere in the South Atlantic. The model results suggest that the upper-layer, northward flowing waters compensating the export of North Atlantic Deep Water (NADW) mainly consist of intermediate waters which enter through Drake Passage and which are modified and gradually warmed within the Atlantic. Direct, net inflow of large amounts of warm water from the Indian Ocean is not found in the model solutions and obviously not required. Forcing the model to produce such inflows actually compromises the model temperature and salinity fields.

Introduction

In recent years the South Atlantic has received increased attention and various research programs have been carried out to study the physical, geochemical and biological processes in this important part of the world ocean. The South Atlantic is unique because of its geographical location linking the Pacific, Indian and Atlantic Oceans. Various major water masses enter the South Atlantic. These water masses interact within the South Atlantic and are ultimately exported again with modified characteristics. Deepwater formed in the North Atlantic (NADW) passes through the South Atlantic and is exported to the Indian and Pacific oceans. This export of NADW must be compensated by northward flows of upper layer (warm surface and/or colder intermediate) waters and (Antarctic) bottom

From WEFER G, BERGER WH, SIEDLER G, WEBB DJ (eds), 1996, *The South Atlantic: Present and Past Circulation.* Springer-Verlag Berlin Heidelberg, pp 305-323

water in the South Atlantic. Because of the required northward upper layer flow it is generally believed that the South Atlantic transports heat northward, contrary to the general poleward heat transport in the other oceans (Bennett, 1978; Hastenrath, 1982).

Different and controversial pictures have emerged from numerous investigations on the origin, depth distribution and associated heat transport of the upper layer northward flow in the South Atlantic. Based on an upper layer heat budget calculation and using the relatively large heat flux estimate of Hastenrath (1982) at 30°S in the South Atlantic (0.69 PW to the north) Gordon (1986) concludes that the northward return flow in the South Atlantic occurs mainly in the warm thermocline layer and because no warm water is entering through Drake Passage is derived primarily from Indian Ocean Central Water (IOCW) entering the South Atlantic by a branch of the Agulhas Current that does not complete the retroflection and/or by warm-core eddies shed by the Agulhas in the retroflection region (warm water route). Because of the relatively low temperatures of the circumpolar waters flowing into the South Atlantic from the South Pacific Gordon (1986) estimates their contribution to the northward flow to be less than 25% of the Indian Ocean inflow.

The dominance of the warm surface water to the northward flow in the South Atlantic as proposed by Gordon (1986) has been questioned by Rintoul (1991) who uses hydrographic data from five non-synoptic sections in the South Atlantic and derives mass and heat transports by inverse calculations. He finds a substantial net inflow of upper IOCW ($\sigma_0=26.80$) to be incompatible with the hydrographic data. Contrary to Gordon (1986), Rintoul suggests that the northward flow in the South Atlantic is mainly derived from waters entering through Drake Passage (cold water route) and is equally split between the surface layers, and the intermediate and bottom waters. Rintoul (1991) estimates a meridional heat transport at 32°S of 0.25±0.12 PW, a value that is considerably smaller than Hastenrath's (1982) result. In a subsequent paper Gordon et al. (1992) use hydrographic and chlorofluoromethane (CFM) data from the southeast Atlantic and arrive at a circulation pattern that is different from the Gordon (1986) and Rintoul

(1991) pictures. The northward flow in the South Atlantic now is composed of a large component of intermediate water entering through Drake Passage and partly flows eastward into the Indian Ocean before returning to the Atlantic. In addition to the intermediate water inflow the Gordon et al. (1992) scheme shows a contribution of about 10 Sv of water warmer than 9°C from the Indian Ocean.

In the present study the historical database of temperature and salinity measurements in the Atlantic is exploited, and the mean circulation in the South Atlantic obtained by driving a steady model towards the observations (using techniques from non-linear optimization theory) is investigated and discussed. Emphasis is on meridional mass and heat transports and the controversy on the origin and composition of the upper layer transport in the South Atlantic is addressed. Results from experiments with widely different values of meridional heat fluxes are presented and consistency of the various solutions with the data is investigated.

The Model

Fig. 1a shows the map of hydrographic stations used in the present study. The data collection contains more than 40,000 stations and has been obtained by merging individual datasets from J. Reid, W. Nowlin (personal communication), Fukumori, et al. (1991), the Southern Ocean Atlas of Gordon et al. (1986), and data from the U.S. NODC archive. All stations contain temperature and salinity data from the surface to the bottom. In general, data coverage is satisfactory, but in the South Atlantic data-gaps are evident some of which will be filled by the World Ocean Circulation Experiment (WOCE 1988) in the near future. The temporal distribution of the data in the South Atlantic is displayed in Figs. 1b to 1e and shows that the hydrographic data are mainly from the period 1950 to 1990 but that older data like, for instance, the Meteor and Discovery expeditions during the 1920's and 1930's (Wüst 1935; Deacon 1933) are included in the dataset as well (Figs. 1b and 1d). For the region between 55°S and the Equator all seasons are well represented, with stations from southern hemisphere spring and summer dominating only slightly over the fall and winter seasons (Fig. 1c). Southward of 55°S the

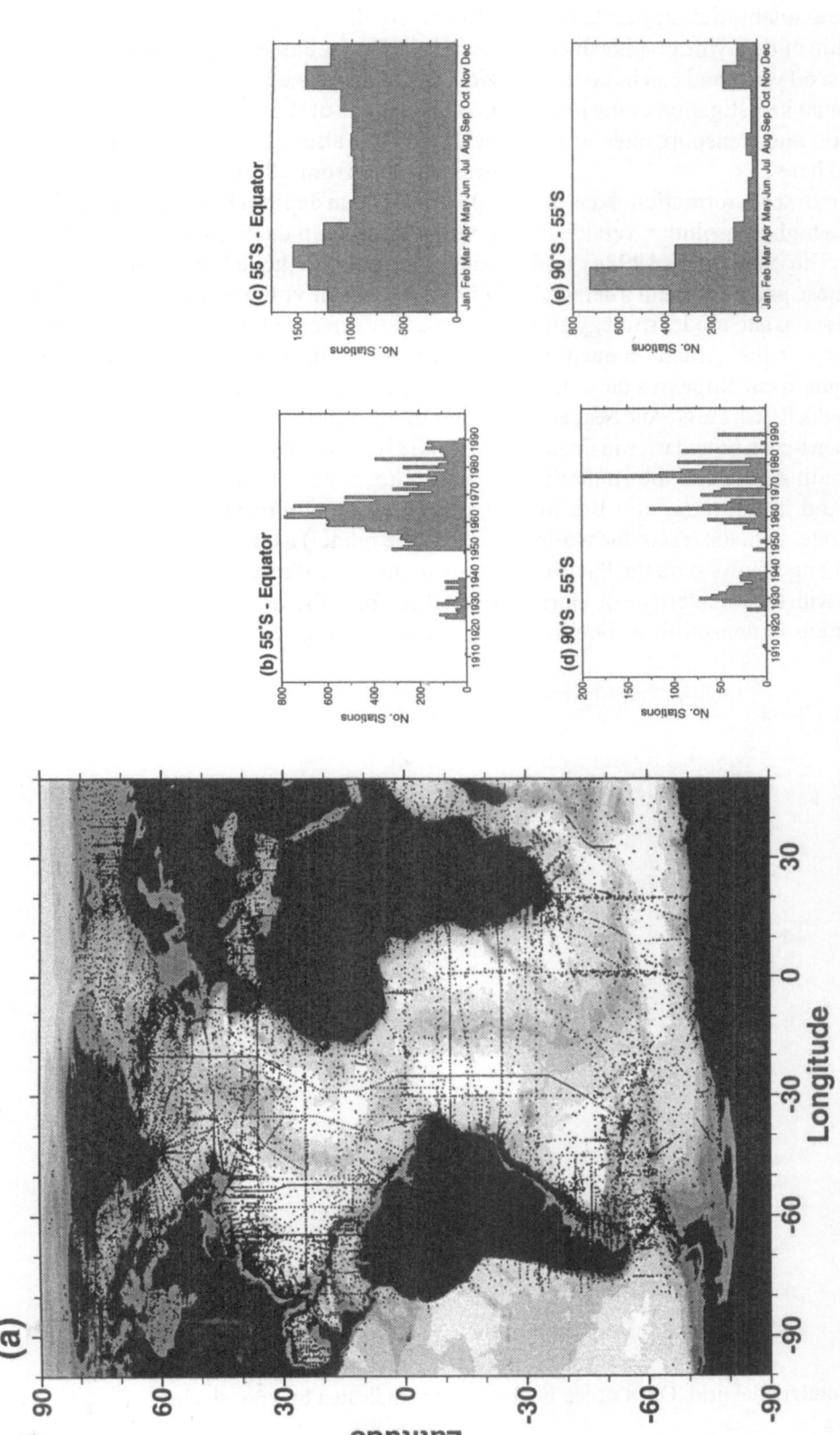

Fig. 1. Map of hydrographic stations used in this study (a) and temporal distribution of stations in the South Atlantic between 55°S and the Equator (b, c) and south of 55°S (d, e). In (a) bottom topography is indicated by gray-shading.

data are predominantly from the summer season (Fig. 1e). Overall, the available data appear to provide a good description of the hydrographic distributions during the last 60 years and can be considered a sound basis for an investigation of the long-term mean circulation and transport rates in the Atlantic as attempted here.

The tool used to extract information from the hydrographic data is a higher resolution version of the model described in Schlitzer (1993a) and Schlitzer (1993b). These papers contain a detailed description of model setup and model strategy and here only an overview of the general concept is given. The model domain encompasses the entire Atlantic including Weddell Sea, the Nordic Seas and the Arctic Ocean. Open ocean boundaries in Drake Passage, between South Africa and the Antarctic continent at 40°W and in Gibraltar and Bering Straits connect the model with the rest of the world ocean and allow exchange-flows with the Pacific, the Indian Ocean and with the Mediterranean. Horizontal model resolution is non-uniform ranging

from 1x0.75° in areas with narrow currents (Gulf Stream, Agulhas, Antarctic Circumpolar Current (ACC) in Drake Passage, overflow region in the North Atlantic, western boundary currents) to 4x3° in most regions of the ocean interior (Fig. 2). Vertical resolution also is non-uniform with layer thickness varying from 60 m at the ocean surface to 500 m at 5000 m depth. The model has realistic topography based on the 5'x5' resolution US Navy bathy-metric data. Altogether the model consists of n_b=19776 control volumes (boxes). Neighboring boxes are connected via horizontal and vertical advective and diffusive fluxes, the respective velocities and mixing coefficients being defined in the center of the interfaces between the boxes.

For a given set of horizontal velocities u, v, air-sea heat fluxes Q and mixing coefficients K_h, K_v (u, v, Q and K_h, K_v comprise the independent parameters of the model) as well as boundary values for temperature and salinity at the open ocean boundaries taken from the data, the steady-state budget equations for mass, heat and salt are formulated for

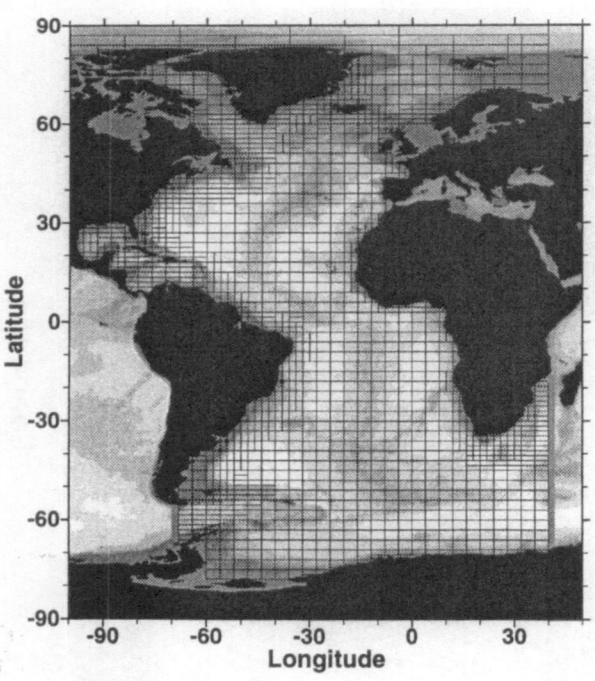

Fig. 2. Map of horizontal model grid. Open ocean boundaries are indicated by gray-shaded bars.

each box and the resulting sets of linear equations are solved for the vertical velocities w and the (steady-state) model potential temperatures θ_m and salinities s_m ("forward step"). Mass, heat and salt conservation is exact. Up to this point the model is similar to traditional box models (Broecker et al. 1960; Oeschger et al. 1974; Sarmiento 1983) even though the present model has a much higher resolution and a much larger number of boxes compared to the older box models.

The new feature of the Schlitzer (1993a; 1993b) concept comes into play when the quality of the current model state $p=[u, v, Q, K_h, K_v]$ is evaluated and model temperatures and salinities are compared with observations θ_d and s_d. For this, a cost function F is defined which, among other contributions, (see Schlitzer (1993a; 1993b) for a full list of terms and see Table 1 for an overview of a priori volume fluxes enforced using cost function terms) accumulates normalized, squared model-data misfits for temperature and salinity. A large value of F indicates large model-data deviations whereas a small value of F is indicative for a close agreement between model simulation and observations. As with traditional box models the overall goal is to minimize F (thereby achieving optimal agreement between simulation and data) by finding new, modified sets of model parameters (here: u^*, v^*, Q^* and K^*_h, K^*_v) that lead to better new simulated model values of temperature θ^*_m and salinity s^*_m. Where-

as this step traditionally had to be performed manually (a formidable task for a complicated model like the present one), here it is performed automatically as part of the model calculations ("adjoint step", see Gill et al., (1981), Le Dimet and Talagrand (1986), Thacker (1988), Schlitzer (1993a, 1993b)). During the adjoint model step the structure of the model-data misfits is taken into account and a new, better model state $p^*=[u, v, Q, K_h, K_v]$ is constructed automatically at a computational cost comparable with the cost of the "forward" step.

The improvement or optimization of the current model state is a gradual, iterative process. Each time a better set of model parameters p^* consisting of horizontal flows, mixing coefficients and surface heat fluxes is found, the corresponding simulated temperatures and salinities θ^*_m and s^*_m are calculated. Then, the new, smaller model-data misfits are analyzed and an even better model state is obtained. When the decrease of F or the magnitude of the parameter modifications during one iteration are smaller than prescribed bounds, it is assumed that the minimum of F has been reached and the computations stop. An overview of this iterative process is given in the itemized flow-chart below.

0. *Initialization.* The horizontal model velocities (u, v) are set to geostrophic velocities calculated from the hydrographic data of Fig. 1. Following suggestions from the literature the reference level is not constant over the model domain, but

Table 1: *A priori* volume transports enforced by cost function terms (soft constraints) for all model experiments. Small tolerances and large weights imply a close reproduction of the *a priori* transport values by the model whereas large tolerances and small weights allow larger deviations.

Description	Transport [Sv]	Tolerance [Sv]	Weight
Bering Straits inflow into Arctic Ocean	1.0	0.1	10^5
Upper layer (0-500 m) inflow into Mediterranean	3.0	0.2	10^4
Net inflow into Mediterranean	0.04	0.01	10^6
Florida Current	30.0	1.0	50.
Drake Passage through-flow	130.0	2.0	10.

rises from about 3500 m at 60°S to about 1000 m at 50°N. Note that Ekman velocities based on winds from Trenberth et al. (1989) are added in the top two model layers. Air-sea heat fluxes Q (Oberhuber 1988) and mixing coefficients K_h and K_v (Olbers et al. (1985) and Olbers and Wenzel (1989)) are taken from the literature.

1. *Forward model run.* The independent parameters (u, v, Q, K_h, K_v) together with exact mass, heat and salt equations are used to calculate the dependent parameters (vertical velocities w, model temperatures θ_m, model salinities s_m).

2. *Adjoint model run.* The value of the cost function F and the gradient of F with respect to the independent parameters are calculated applying the adjoint method ("method of Lagrange multipliers") as an efficient means for obtaining gradients (see Gill et al. 1981; Thacker 1988 and Schlitzer 1993a; 1993b).

3. *Updating the independent parameters.* The gradient of F is passed to a descent algorithm (Gilbert and Lemaréchal 1989) to obtain a new, improved model state. In case a stopping criteria (e.g., sufficiently small decrease in F or sufficiently small modifications of the independent parameters) is satisfied the calculations terminate. Otherwise a new iteration is started at step 1.

It is important to note that in addition to the requirement to correctly reproduce the observed hydrographic fields the model velocities are required to approximately preserve the vertical velocity shear obtained by geostrophic calculations. The model may shift the velocity profile at a given location by a constant offset (the initially unknown reference velocity) without penalty, however, modifications to the shape of the velocity profiles lead to contributions to the cost function F. In this way, minimization of F forces the model velocity profiles to remain close to the geostrophic profiles and information in the hydrographic data on the oceanic pressure field are exploited by the model.

Model Results

In the following, results from three model experiments are presented and discussed. For experiment A only standard terms in the cost function F are employed (see Schlitzer (1993a; 1993b) for a full list of terms), whereas for experiments B and C constraints that force the zonally integrated meridional heat transport at various latitudes in the Atlantic to prescribed, a priori values are added (see Table 2). The prescribed meridional heat fluxes correspond to averages of literature values for experiment B but are close to maximal published values for case C. These experiments are carried out to investigate the sensitivity of the model and to determine the range of flow fields and transport rates for which satisfactory simulations of temperature and salinity can be obtained.

Cost Function

A summary of values of the total cost function F and individual terms for the three model experiments is shown in Table 3. Entries for experiment B are absolute values whereas entries for experiments A and C are normalized to case B (standard case) for easy comparison. It is found that experiments A (no heat transport constraints) and B ("average" meridional heat fluxes) lead to about the same final cost function value but that experiment C ("high" meridional heat fluxes) is significantly higher by a factor of 1.32. Inspection of the individual terms shows that for case A the distributions of air-sea freshwater and heat fluxes (terms 3 and 6) as well as the horizontal fields of vertical velocities w (term 4; calculated level by level) are smoother than for the standard case B. On the other hand, the agreement between model simulated property fields and observations (terms 7 to 14) is better for the standard case B compared with case A. The increased property misfits for experiment A are most pronounced for salinity and are largest in the South Atlantic south of 30°S (factor 1.28 compared to case B). Experiment C exhibits cost function terms significantly larger than those for the standard case B for both the smoothness terms and the terms measuring property misfits. For the entire model domain the systematic temperature deviations (term 9) are a factor of 1.45 larger compared with case B and systematic salinity deviations are a factor of 1.29 higher. Note that for all experiments A, B and C the deviations of the vertical velocity shear in the model from the geostrophic shear are

Table 2: *A priori* values for meridional heat transports [PW] at different latitudes for experiments B and C. The *a priori* values are enforced in the model by additional cost function terms (soft constraints).

Latitude	Experiment B	Experiment C
51°N	0.35	0.40
39°N	0.7	0.80
27°N	1.0	1.20
15°N	0.95	1.20
6°N	0.8	1.10
6°S	0.5	0.90
18°S	0.4	0.75
30°S	0.3	0.60

Table 3: Weight factors and values of total cost function and individual terms normalized to values of standard case B for different numerical experiments. Terms denoted "boxwise comparison" penalize property misfits box by box whereas for terms denoted "neighborhood" the mean misfit for a neighborhood of a box is calculated and this mean misfit is penalized. The latter terms are sensitive to systematic temperature and salinity deviations. See text for description of experiments and further comments on various cost function terms.

	Term	Weight	B	A	C
Σ	Total	–	$3.83 \cdot 10^5$	1.00	1.32
1:	Velocity shear	$1 \cdot 10^{-12}$	$9.12 \cdot 10^{-4}$	0.96	1.02
2:	E-P dat	2	$1.81 \cdot 10^4$	0.61	1.21
3:	E-P smoothness	$1 \cdot 10^{-1}$	$2.85 \cdot 10^3$	0.77	1.09
4:	w smoothness	1	$3.17 \cdot 10^4$	0.88	1.20
5:	Q data	5	$1.27 \cdot 10^4$	0.56	1.10
6:	Q smoothness	0.2	$2.10 \cdot 10^4$	0.75	1.15
7:	θ boxwise comparison	0.1	$6.96 \cdot 10^4$	1.10	1.31
8:	s boxwise comparison	0.1	$5.53 \cdot 10^4$	1.12	1.25
9:	θ neighborhood	0.01	$8.34 \cdot 10^4$	1.02	1.45
10:	s neighborhood	0.01	$6.73 \cdot 10^4$	1.23	1.29
11:	θ boxwise comparison < 30°S	–	$1.96 \cdot 10^4$	1.14	1.27
12:	s boxwise comparison < 30°S	–	$1.68 \cdot 10^4$	1.15	1.21
13:	θ neighborhood < 30°S	–	$2.60 \cdot 10^4$	1.18	1.35
14:	s neighborhood < 30°S	–	$2.36 \cdot 10^4$	1.28	1.25

small (see below) and of about the same size for all cases.

The terms in Table 3 representing systematic deviations between model temperatures or salinities and data for the whole model domain and for the region south of 30°S (items 9, 10, 13 and 14) are displayed in Fig. 3 versus the meridional heat flux at 30°S in the South Atlantic for each of the experiments. Also shown is a parabola fit through these points. Note that experiment A (no heat transport constraints) produces a poleward heat transport of -0.56 PW whereas experiments B and C closely match the prescribed values of +0.3 PW (B) and +0.6 PW (C), respectively. The parabola fit suggests that optimal agreement between model simulated and observed temperatures and salinity is achieved for a meridional heat flux in the South Atlantic at 30°S close to zero. The shape of the parabola indicates that near-optimal temperature and salinity simulations can be obtained for a wide range of model solutions. Based on a close inspection and analysis of model-data temperature and salinity differences for the three experiments a range of acceptable meridional heat transports at 30°S between about -0.5 PW and +0.4 PW is found. Experiment C is rejected mainly because of relatively large systematic temperature deviations. In the upper 400 m of the South Atlantic between 38°S and the equator experiment C is about 1°C too warm whereas in case B mean model temperatures agree with observations within 0.15°C.

Property Fields

As examples of model property fields, the salinity distributions of the standard case B in 200 m depth and along a meridional section at 30°W are displayed in Fig. 4 together with the corresponding distributions for the data. Figs. 4a and 4b show that the model correctly simulates the large-scale fea-

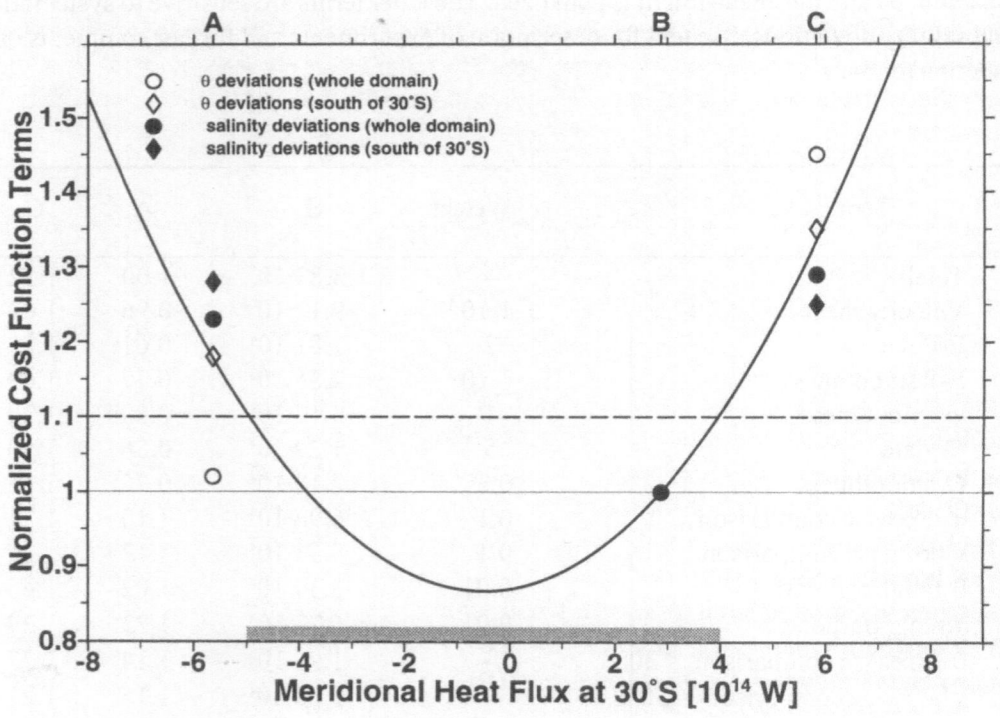

Fig. 3. Values of cost-function terms measuring temperature and salinity misfits of the model for different experiments (normalized to the standard case B) versus meridional heat transport across 30°S. The range of acceptable heat transports is indicated by a gray-shaded bar.

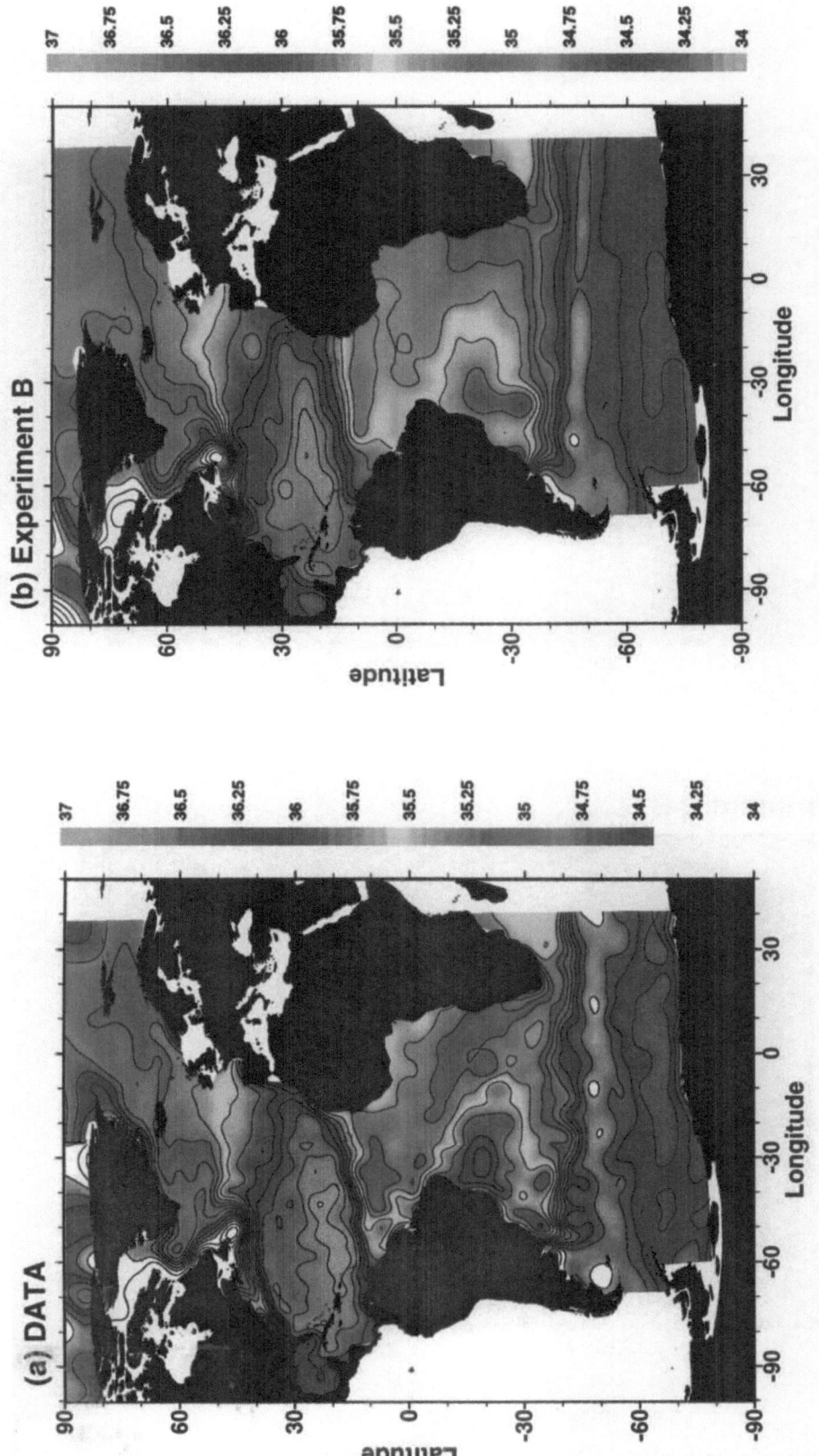

Fig. 4. Distribution of salinity at 200 m depth (a) data and (b) model (standard run B).

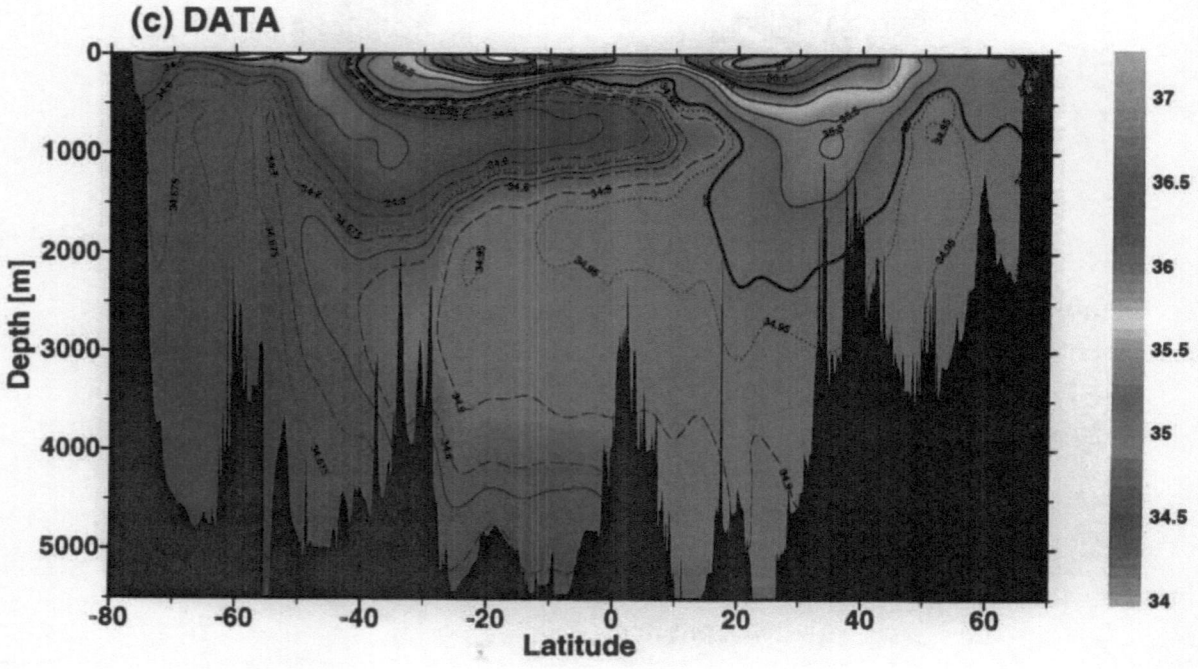

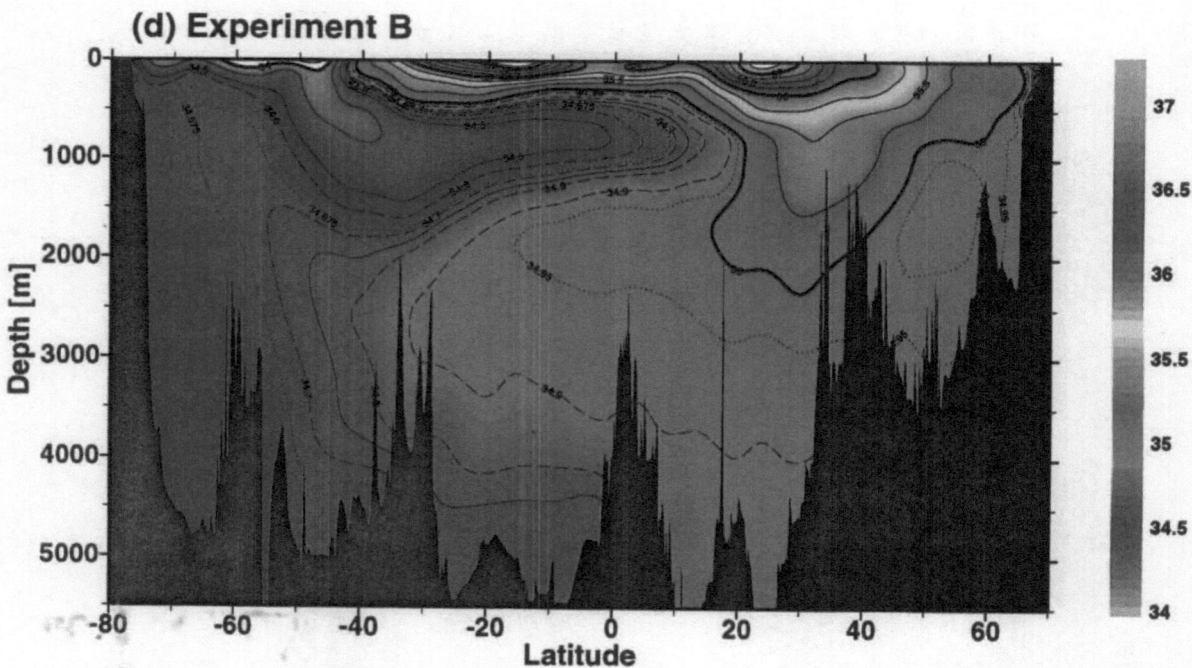

Fig. 4. Meridional section of salinity along at 30°W (c) data and (d) model.

tures in the salinity field in 200 m depth and that it reproduces the observed data values ranging from about 37 ppm in the center of the subtropical salinity maximum at 15°N to about 34 ppm in the salinity minimum along a zonal band at 50°S very closely. Even details like the salinity variations in the North Atlantic are modelled realistically. The meridional sections in Figs. 4c and 4d show that the model also produces the correct vertical layering and meridional extent of the major water masses in the Atlantic. Notably, features like the subtropical salinity maxima (the North Atlantic salinity maximum reaching deeper than South Atlantic maximum), the effect of the Mediterranean outflow in about 1200 m depth at 30°N, the subduction and spreading of low-saline Antarctic Intermediate Water (AAIW) in about 800 m depth in the South Atlantic, the southward reaching tongue of higher saline North Atlantic Deep Water (NADW) between about 1500 and 3500 m depth and the layer of lower salinity Antarctic Bottom Water (AABW) spreading northward are all correctly reproduced. Comparison of position and shape of the salinity isolines reveals a close quantitative agreement between model values and data.

Horizontal Flows

Fig. 5 shows geostrophic velocity profiles at seven locations in the South Atlantic together with the final velocities of the standard case B (note that the geostrophic flows are used to initialize the model velocities; see above). It is found that the vertical velocity shear of the final model flows remains close to geostrophic shear, and the modifications applied by the model are generally smaller than the temporal change of geostrophic velocity shear obtained from repeated observations. Most profiles show only small velocity offsets compared with geostrophic flows indicating that the selection of upward sloping reference depths with latitude between about 3000 and 2000 m in the South Atlantic (for exact reference depths see the zero-crossings of geostrophic profiles in Fig. 5) is appropriate. Exceptions are profile (c) in the eastern Cape Basin where velocities are shifted by about -0.4 cm s^{-1} in the final solution and now provide southward flow of NADW and profile (d) in the ACC east of Drake

Passage where velocities are increased by about +1 cm s^{-1} and now are eastward at all depths. In areas where currents are narrower than the grid size of the model (e.g., profile (g) in the westward flowing, coastal branch of the Agulhas) the averaging of original hydrographic data, in general, leads to an underestimation of actual flow velocities (Olbers et al. 1985; Schlitzer 1993a; 1993b), and velocity modifications of the model in these regions are relatively large mainly because of mass conservation requirements.

Final, optimal model flows for the standard case B in 200 and 800 m depth in the South Atlantic are shown in Fig. 6. Dominant features in the shallow velocity field (Fig. 6a) are the eastward flowing Antarctic Circumpolar Current (ACC) and the Agulhas Current. In Drake Passage the ACC reaches speeds of up to 32 cm s^{-1} (average: about 20 cm s^{-1}) whereas in the South Atlantic (after turning northward and broadening) average velocities are decreasing to about 10 cm s^{-1}. Along the Argentine coast the Falkland/Malvinas Current flows northward and encounters subtropical waters that are transported southward with the Brazil Current in the Confluence region at about 40°S. The Brazil Current feeds the South Atlantic Current which flows eastward along about 40°S. Part of the South Atlantic Current folds into the Benguela Current and is transported north-westward with the South Equatorial Current. North of the triangular-shaped (anti-cyclonic) subtropical gyre consisting of Brazil Current, South Atlantic Current, Benguela Current and the southern branch of the South Equatorial Current, a cyclonic cell is found formed by the northern branch of the South Equatorial Current, the eastward flowing South Equatorial Counter Current and southward flows in the eastern Angola Basin. All the model flows found at 200 m depth (including the cyclonic Weddell Gyre) are in agreement with the general flow patterns of Reid (1989) and Peterson and Stramma (1991).

Model velocities in 800 m depth (largely representative of the circulation of AAIW in the South Atlantic) (Fig. 6b) are smaller than the surface flows (about 50%) but exhibit a similar spatial structure. As in Fig. 6a, highest velocities at this deeper level are observed in the ACC and Agulhas Current region. To the north of the ACC, centered

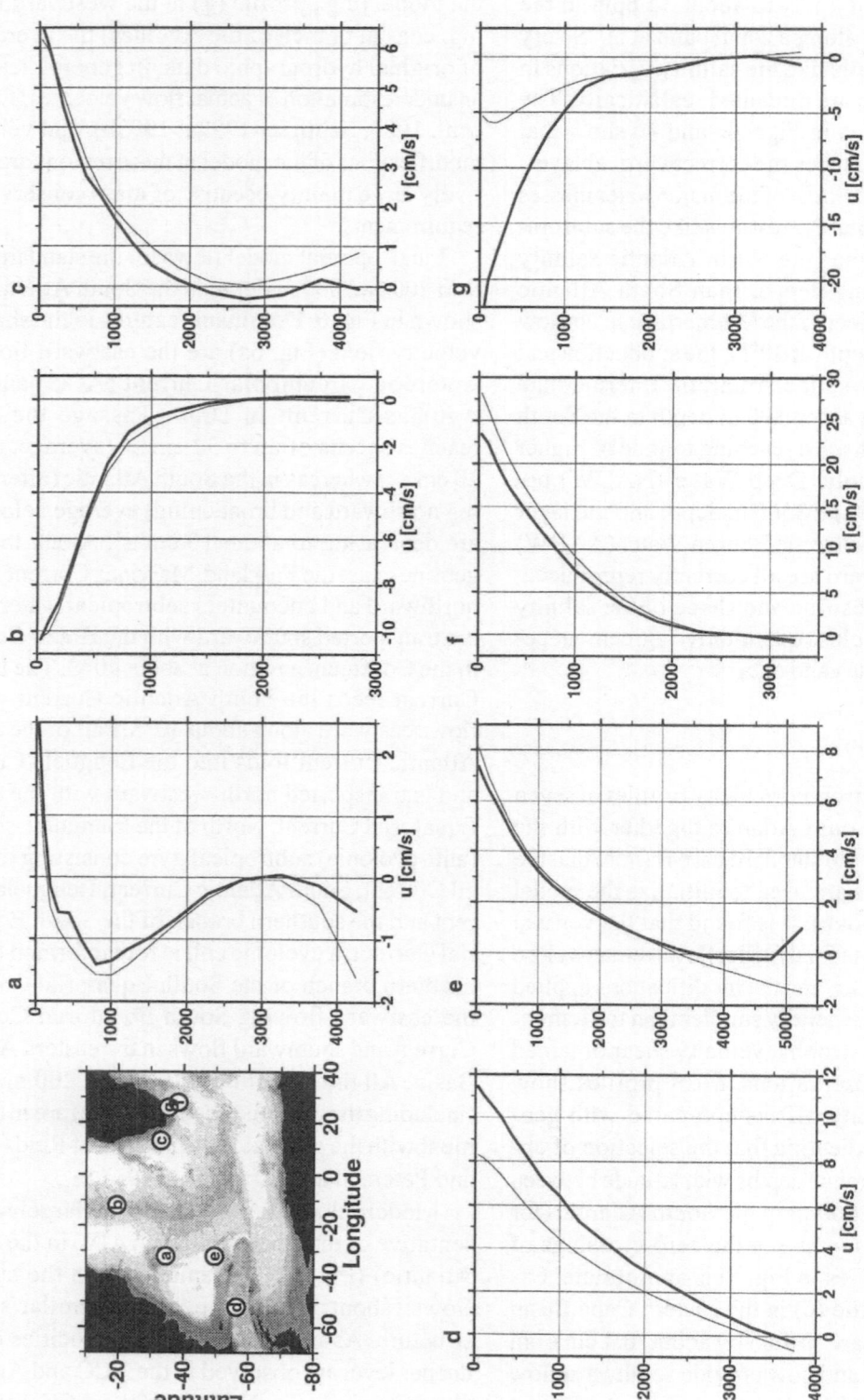

Fig. 5. Profiles of zonal (*u*) and meridional (*v*) velocities from geostrophic calculations (thin lines) and from standard-run B model flows (thick lines). Locations of individual profiles are indicated in the map. Note that the reference level of the initial geostrophic calculations is increasing with latitude (ca. 3500 m at 60°S and ca. 1000 m at 50°N).

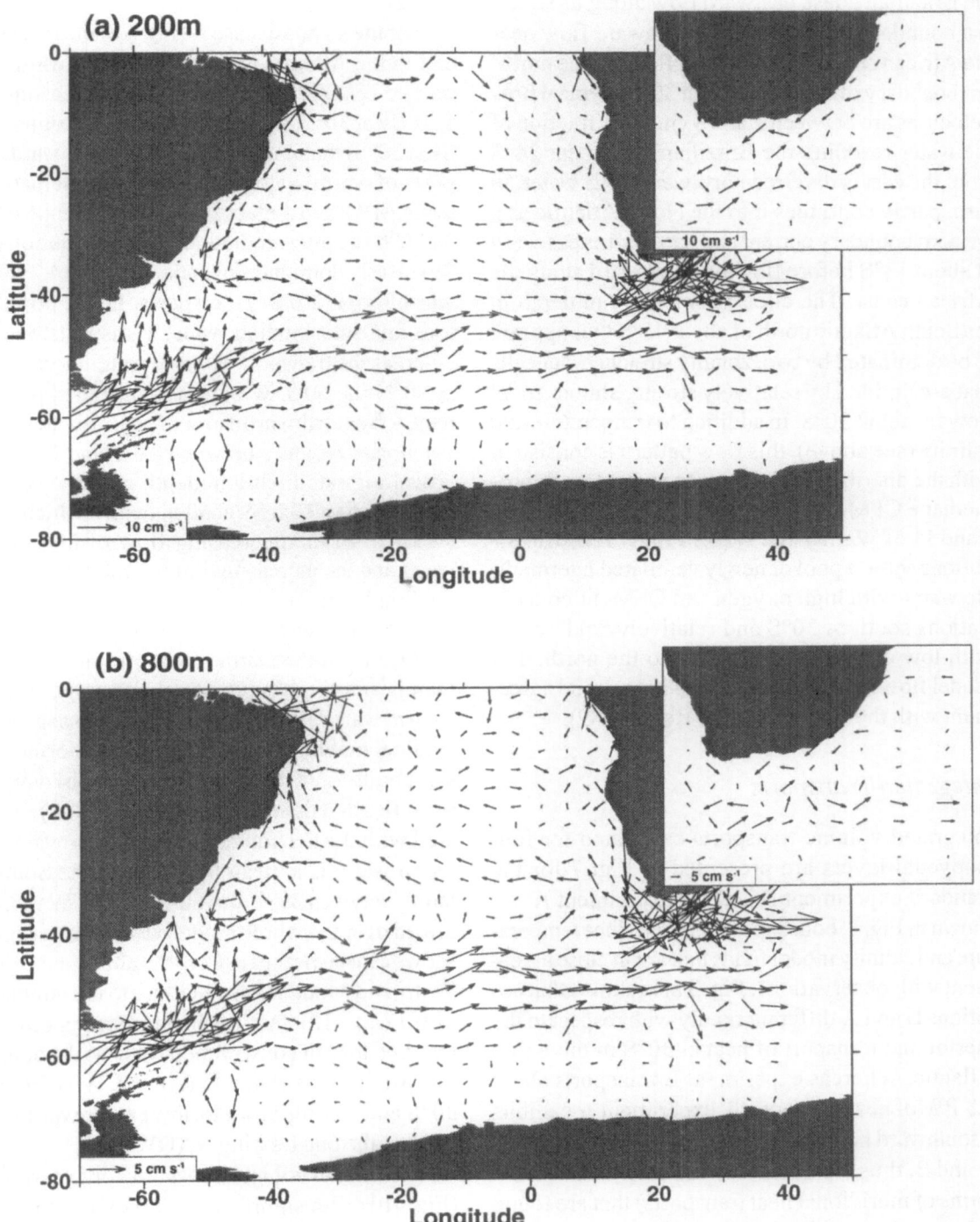

Fig. 6. Horizontal model flows of the standard-case B (a) in 200 m depth and (b) in 800 m depth. Arrows are vector averages of nearby zonal and meridional flow velocities.

at about 37°S, there is an almost rectangular, anti-cyclonic cell stretching across the entire South Atlantic. This cell consists of southward flow along the Brazilian coast, eastward flow along its southern boundary at about 44°S, northward flow near the African coast and westward flows at the northern boundary of the cell at about 30°S. Typical flow velocities are between 2 and 4 cm s⁻¹. A fraction of the water reaching the Brazilian coast near 24°S from the east is diverted northward. This water, in turn, partly continues into the North Atlantic as a western boundary current and partly turns eastward at about 13°S before flowing southward along the African coast. The circulation at 800 m depth in the South Atlantic north of about 45°S thus appears to be dominated by two, zonally stretched-out cells that are divided by relatively strong, almost zonal flows at about 30°S. In addition to temperature and salinity (see above), this flow pattern is consistent with the distributions of oxygen and chlorofluoro-methane CFM-11 in the South Atlantic (see Figs. 4 and 11 of Warner and Weiss 1992). These distributions show a pool of newly ventilated intermediate water with high oxygen and CFM-11 concentrations south of 30°S and relatively "old" water with low oxygen and CFM-11 to the north. The model flows at 800 m depth also are in good agreement with the flow pattern of Reid (1989).

Integrated Transports

Integrated volume transports calculated for four isopycnal layers are presented in Fig. 7 for the standard experiment B and for experiment A. As shown in Fig. 3, both experiments produce temperature and salinity model fields that are in close agreement with observations. Fig. 3 also shows that solutions B and A differ markedly with respect to the meridional transport of heat at 30°S in the South Atlantic. Whereas experiment B transports about 0.3 PW of heat to the north, experiment A exhibits a southward heat transport of about -0.5 PW. Cases A and B, thus represent two extreme scenarios (in terms of meridional heat transports) that are found to be consistent with historic hydrographic data and with the principle of geostrophy. The four isopycnal layers for which transport rates are given represent major water masses found in the South Atlantic (1:

surface water warmer than about 8-10°C, 2: AAIW, 3: NADW, 4: AABW) and closely correspond to subdivisions of Rintoul (1991) and Gordon et al. (1992).

For the standard case B (Fig. 7a) the meridional cell in the Atlantic consists of a southward transport of NADW of about 18 Sv (increasing from 17.1 Sv at 30°N to 17.8 Sv at the equator and 18.7 Sv at 30°S) that is compensated by northward transports of warm water, Antarctic Intermediate Water (AAIW) and Antarctic Bottom Water (AABW). At 30°S the intermediate water contribution (11.9 Sv) clearly dominates over the AABW (4.2 Sv) and warm water (2.0 Sv) transports. To the north, bottom and intermediate water transports decrease whereas the transport of warm water increases and at 30°N the NADW export is balanced predominantly by northward flow of warm water. In the equatorial Atlantic between 21°S and 15°N a net upwelling rate in 360 m depth of 6.3 Sv is found (compared to 10.6 Sv at 60 m depth) which is consistent with the decreasing flow of intermediate water and the increasing flow of warm water. From the South Atlantic south of 30°S there is a net export of warm water into the Indian Ocean of 21.9 Sv. Overall, more warm water is leaving this region than is entering through Drake Passage (the term "warm water" for waters in Drake Passage is misleading; maximal temperatures in its northern part are about 7.5°C and only because of very low salinities densities are less than $\sigma_0 = 26.8$). Associated with the net conversion of colder waters into warm water is a mean heat gain of the South Atlantic between 30°S and 60°S of 17.5 W m⁻². Integrated over the whole model domain experiment B shows a mean heat gain from the atmosphere of 0.45 W m⁻² and a total freshwater loss to the atmosphere of 0.17 Sv. This net freshwater loss is obviously required for salt conservation because the dominant inflows into the Atlantic (AAIW, AABW across 30°S and Bering Strait inflow) are lower in salinity than the out-flowing NADW.

Meridional volume transports for experiment A (Fig. 7b) differ significantly from the values of the standard case B. Overall, experiment A shows a weaker meridional overturning cell with only about 13.5 Sv of NADW flowing southward. The northward transport of AAIW (18 Sv) is considerably

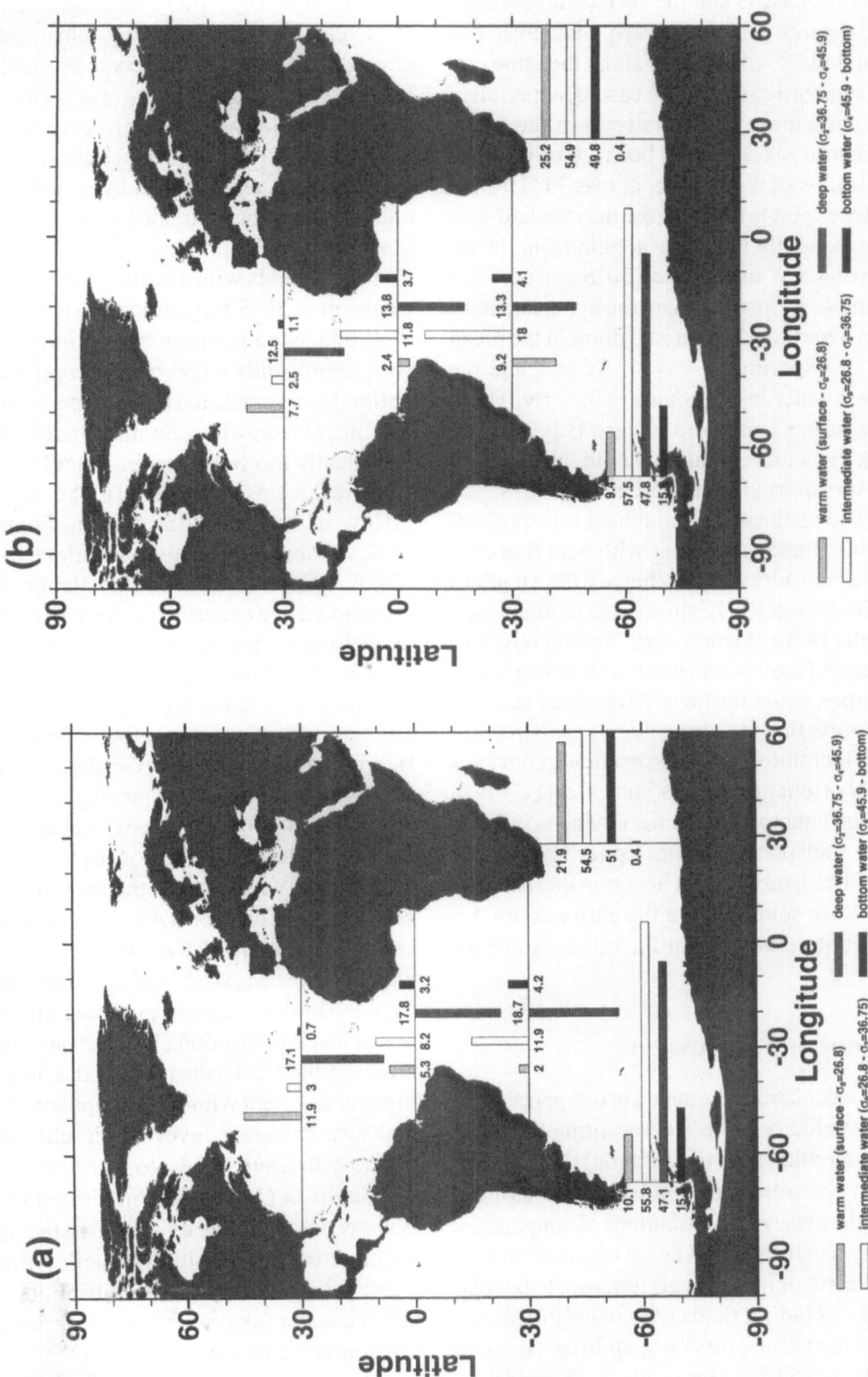

Fig. 7. Integrated volume transports [Sv] for four isopycnal layers representing major water masses in the Atlantic for (a) the standard case B and (b) for the case A exhibiting southward heat transport across 30°S. Note that any residual for the zonal sections in the Atlantic arises from inflow through Bering Strait and air-sea freshwater fluxes.

larger than for case B and the surface layer transport is now large and southward (-9.2 Sv). The northward flow of AABW is of about the same size as for the standard case B. Like case B, experiment A also exports more warm water from the South Atlantic than it receives, but because of the considerable inflow of warm water across 30°S the net warm water export is less than for the standard case B and consequently the net heat gain from the atmosphere of 4.8 W m^{-2} between 30°S and 60°S for experiment A is smaller than for experiment B. Integrated over the whole model domain the mean heat gain of experiment A is 5.7 W m-2 and the overall freshwater loss amounts to 0.27 Sv. These values are larger compared to case B because in solution A even more relatively cold, low salinity AAIW is transported northward across 30°S.

Fig. 8 shows the meridional heat transports of experiments B and A together with heat flux estimates from the literature. Whereas the standard case B with its relatively strong meridional overturning cell (18 Sv of southward flowing NADW) and moderate (2 Sv) contribution of warm water to the compensating northward transport exhibits heat transports that are about average when compared with literature values, experiment A deviates considerably from the other estimates, especially in the south and equatorial Atlantic. Owing to its large southward transport of warm, upper layer water (9.2 Sv at 30°S), meridional heat transports of experiment A are southward in the entire South Atlantic, in contradiction with the other values included in Fig. 8.

Conclusions and Discussion

The model calculations presented in this paper show that steady velocity fields (representing the long-term mean circulation in the Atlantic) that are consistent with geostrophic dynamics and successfully reproduce the observed distributions of temperature and salinity can be found. The set of solutions that satisfy these major model goals turns out to be relatively large including fields with widely differing meridional heat transports ranging from +0.4 PW to the north to -0.5 PW to the south. Associated with changes in meridional heat transports are changes in zonally integrated layer transports that can eas-

ily be obtained by small modifications of absolute flow velocities still remaining consistent with geostrophic calculations. In order to produce the seemingly large difference of warm water transports across 30°S between standard case B (+2 Sv) and experiment A (-9.2 Sv), typically, velocities in the ocean interior are changed by less than 0.1 cm s^{-1} and boundary current speeds are modified by at most 1.5 cm s^{-1}.

Model fields with a northward meridional heat transport at 30°S larger than 0.6 PW (experiment C) produce temperature and salinity simulations with significantly larger misfits compared with the optimal solutions and show temperature values in the upper layers of the South Atlantic that are systematically too warm by about 1°C. This systematic effect seems to be related to the relatively large inflow of warm Indian Ocean Central Water (IOCW) observed in these solutions. Acceptable IOCW inflows into the South Atlantic determined with model dye experiments (described elsewhere) range from 3.7 Sv (experiment A) to 6.5 Sv (standard case B). These values are much smaller than the estimate of Gordon (1986) (13.5 Sv) but are in good agreement with numbers based on chlorofluoromethane data (4 Sv, Gordon et al. (1992)) or the observations of Agulhas rings (5 Sv, Byrne et al. (1994); 6.2 Sv, Ballegooyen et al. (1994)).

The model results demonstrate that in order to produce a large northward transport of upper-layer, warm water in the North Atlantic compensating the southward export of NADW, an inflow of warm water of about equal size entering the South Atlantic from the Indian Ocean as proposed in the conveyor belt scheme of Gordon (1986) (warm water path) is not required and, when enforced, actually is found to be inconsistent with hydrographic data. Contrary to the warm water conveyor belt scheme and more in line with results of Rintoul (1991) and Boddem and Schlitzer (1995), both model solutions A and B show a dominance of AAIW to the upper layer northward transport in the South Atlantic (cold water path) and only small contributions of warm, surface water (as an extreme, experiment A even transports warm water to the south).

The apparent contrast between the warm and relatively salty upper layer waters in the North Atlantic and the low temperatures and salinities of

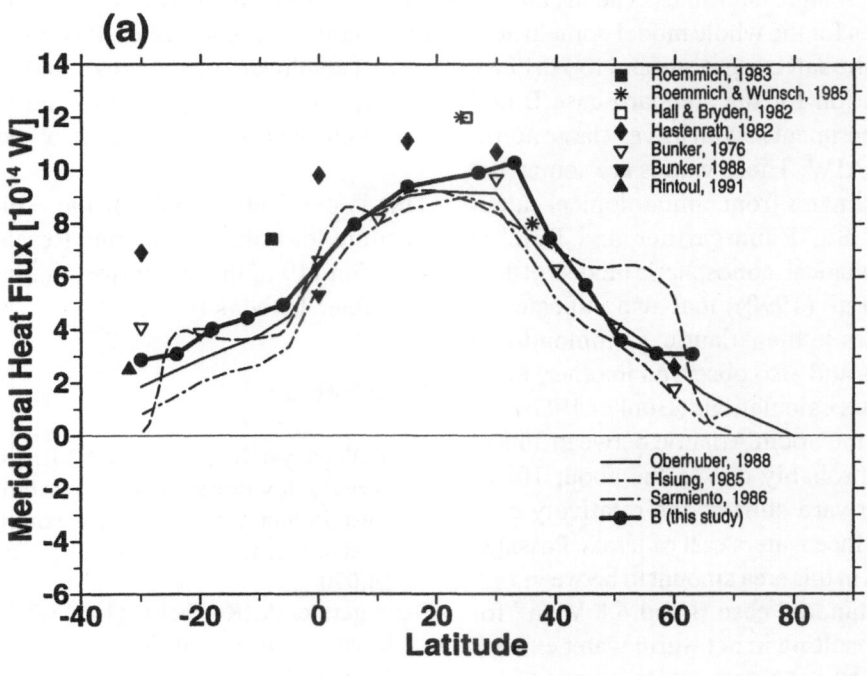

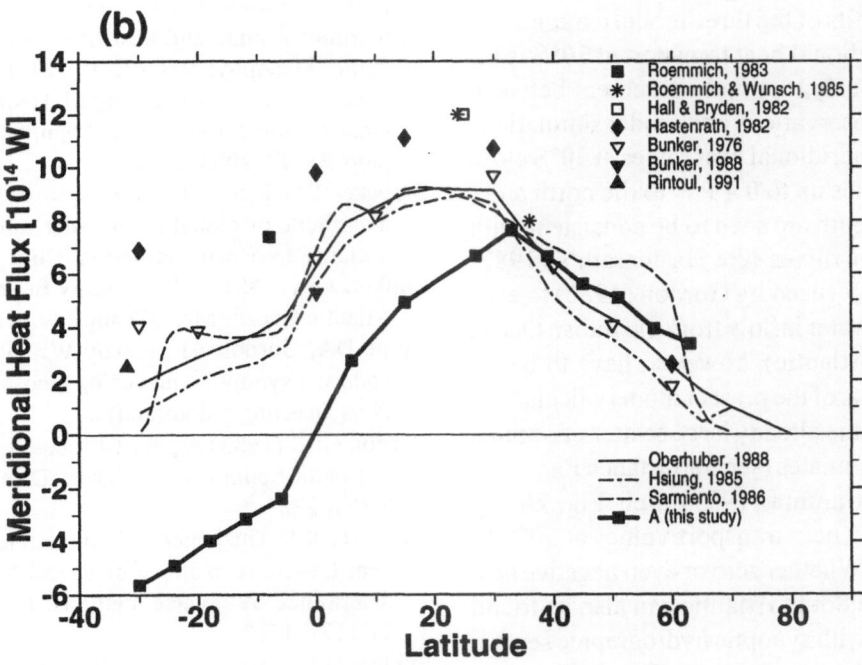

Fig. 8. Meridional heat transports versus latitude (a) for the standard case B and (b) for experiment A. Estimates from the literature are included for comparison.

its major source (according to this study) in the South Atlantic is resolved when taking into account air-sea freshwater and heat fluxes. The net air-sea freshwater fluxes for the whole model domain necessary to close the salt budget amount to +0.17 Sv excess evaporation for the standard case B and +0.27 Sv for experiment A with its very large northward flow of AAIW. These values are somewhat smaller than estimates from climatological calculations (+0.45 Sv; Baumgartner and Reichel (1975)) or hydrological, atmospheric models (+0.26 Sv; Broecker et al. (1990)) that also indicate net freshwater losses in the Atlantic. Common to all model solutions and also observed in other, independent heat-flux calculations (Bunker 1988) is a net heat gain of the South Atlantic between 30 and 60°S which is probably due to the about 10° of latitude equatorward shift of the relatively cold circumpolar surface waters east of Drake Passage. Mean heat gains in this area amount to between 17.5 W m^{-2} for the standard case B and 4.8 W m^{-2} for experiment A resulting in net warm water exports (in contrast to the net warm water inflow of the conveyor belt scheme) in both cases.

Fig. 3 showing the magnitude of temperature and salinity misfits of the three model experiments versus the meridional heat transport at 30°S is intriguing. Obviously, optimal agreement between hydrographic observations and model simulations is obtained for meridional heat fluxes at 30°S close to zero, but values up to 0.4 PW to the north and -0.5 PW to the south are seen to be consistent with data. Large heat fluxes like Hastenrath's (1982) value of 0.69 PW (used by Gordon (1986) to estimate the warm water inflow from the Indian Ocean into the South Atlantic), however, have to be rejected. The results of the present model calculations further increase the already large scatter of oceanic heat transport estimates (see for instance Fig. 8 and Peterson and Stramma (1991) their Fig. 28) by adding poleward heat transport values at 30°S to the existing list. Whether zero or even negative heat transports in the South Atlantic can also be found to be consistent with synoptic hydrographic section data remains to be investigated. The studies of Fu (1981) and Rintoul (1991) who use the same set of hydrographic data at 32°S and similar numerical techniques but arrive at considerably different heat

transports at 32°S (0.77 and 0.23 PW, respectively) reveal the large error margins of oceanic heat flux values based on hydrographic data and suggest that reasonable reference velocities corresponding to zero or southward heat fluxes could be found. Sensitivity studies that explore upper and lower bounds of estimated heat flux values are obviously needed.

This is publication 1007 of the Alfred Wegener Institute for Polar and Marine Research and contribution 119 of the Sonderforschungsbereich 261 at Bremen University.

References

van Ballegooyen RC, Gründlingh ML, Lutjeharms JRE (1994) Eddy fluxes of heat and salt from the southwest Indian Ocean into the southeast Atlantic Ocean: a case study. J Geophys Res 99:14,053–14,070

Baumgartner A, Reichel E (1975)The World Water Balance. Oldenbourg–Verlag, Munich, Federal Republic of Germany

Bennett AF(1978) Poleward heat fluxes in southern hemisphere oceans. J Phys Oceanogr 8:785–798

Boddem J, Schlitzer R (1995) Interocean exchange and meridional mass and heat fluxes in the South Atlantic. J Geophys Res 100:15,821-15,834

Broecker WS, Gerard R, Ewing M, Heezen BC (1960) Natural radiocarbon in the Atlantic Ocean. J Geophys Res 65:2903–2931

Broecker, WS, Peng T-H, Jouzel J, Russel G (1990) The magnitude of global fresh-water transports of importance to ocean circulation. Clim Dyn 4:73–79

Bunker, AF (1988) Surface energy fluxes of the south Atlantic Ocean. Mon Weath Rev 116:809–823

Byrne DA, Gordon AL, Haxby WF (1994) Agulhas eddies: a synoptic view using Geosat ERM data. J Phys Oceanog (submitted)

Deacon GER (1933) A general account of the hydrology of the South Atlantic Ocean. Discovery Reports 7:171–238

Fu L-L (1981) The general circulation and meridional heat transport of the subtropical South Atlantic determined by inverse methods. J Phys Oceanogr 11:1171–1193

Fukumori I, Martel F, Wunsch C (1991) The hydrography of the North Atlantic in the early 1980s. An atlas. Prog Oceanog 27:1–110

Gilbert J Ch, Lemaréchal C (1989) Some numerical

experiments with variable-storage quasi-Newton algorithms. Mathematical Programming 45:407-435

Gill PE, Murray W, Wright MH (1981) Practical Optimization. Academic Press, London

Gordon AL, Weiss RF, Smethie WM, Warner MJ (1992) Thermocline and Intermediate Water Communication between the South Atlantic and Indian Oceans. J Geophys Res 97:7223–7240

Gordon AL (1986) Interocean exchange of thermocline water. J Geophys Res 91:5037–5046

Gordon AL, Molinelli EJ, Baker TN (1986) Southern Ocean Atlas. Amerind Publishing Co., New Dehli

Hastenrath S (1982) On meridional heat transports in the world ocean. J Phys Oceanog 12:922–927

Le Dimet F, Talagrand O (1986) Variational algorithms for analysis and assimilation of meteorological observations : theoretical aspects. Tellus 38:97–110

Oberhuber JM (1988) An atlas based on the COADS data set: the budgets of heat, buoyancy and turbulent kinetic energy at the surface of the global ocean. Technical Report 15, Max-Planck-Institut für Meteorologie, Hamburg

Oeschger H, Siegenthaler U, Schotterer U, Gugelmann A (1974) A box diffusion model to study the carbon dioxide exchange in nature. Tellus 27:168–192

Olbers D, Wenzel M, Willebrand J (1985) The inference of north Atlantic circulation patterns from climatological hydrographic data. Rev Geophys 23:313–356

Olbers D, Wenzel M (1989) Determining diffusivities from hydrographic data by inverse methods with applications to the Circumpolar Current. In: Oceanic Circulation Models: Combining Data and Dynamics. Kluwer Academic Publishers, Dordrecht, 95–139

Peterson RG, Stramma L (1991) Upper-level circulation in the South Atlantic ocean. Prog Oceanogr 26:1–73

Reid JL (1989) On the total geostrophic circulation of the South Atlantic Ocean: flow patterns, tracers, and transports. Prog Oceanogr 23:149–244

Rintoul SR (1991) South Atlantic interbasin exchange. J Geophys Res 96:2675–2692

Sarmiento JL (1983) A tritium box model of the north Atlantic thermocline. J Phys Oceanogr 13:1269–1274

Schlitzer R (1993a) Determining the mean, large-scale circulation of the Atlantic with the adjoint method. J Phys Oceanogr 23:1935–1952

Schlitzer R (1993b) An adjoint model for the determination of the mean oceanic circulation, air-sea fluxes and mixing coefficients. Habilitation Thesis, University Bremen

Stramma L, Peterson RJ (1990) The South Atlantic Current. J Phys Oceanog 20(6):846–859

Thacker WC (1988) Three lectures on fitting numerical models to observations. Technical Report GKSS 87/E/65, GKSS Forschungszentrum, Geesthacht

Trenberth KE, Olsen JG, Large WG (1989) A global ocean wind stress climatology based on ECMWF analyses. Technical Report NCAR/TN-338+STR, National Center for Atmospheric Research, Boulder

Warner MJ, Weiss RF (1992) Chlorofluoromethanes in south Atlantic Antarctic Intermediate Water. Deep-Sea Res 39:2053–2075

WOCE (1988) World Ocean Circulation Experiment Implementation Plan. Vol I, World Meteorological Organization, Wormley, Report 242

Wunsch C (1984) An eclectic Atlantic Ocean circulation model. Part I: The Meridional Flux of Heat. J Phys Oceanog 14:1712–1733

Wüst G (1935) Schichtung und Zirkulation des Atlantischen Ozeans: Die Stratoshäre. Wiss. Ergeb. Dtsch. Atlantischen Exped. Forschungs- und Vermess. Meteor 1925-1927, 6(1st part, 2), p. 180 (English translation by the Al-Ahram Center for Scientific Translations, WJ Emery, 112, Amerind, New Delhi 1978)

Long-term Observation of Particle Fluxes in the Eastern Atlantic: Seasonality, Changes of Flux with Depth and Comparison with the Sediment Record

G. Fischer and G. Wefer

Fachbereich Geowissenschaften, Universität Bremen,
Klagenfurterstraße, 28359 Bremen, GERMANY

Abstract: Long-term particle fluxes have been measured with time-series sediment traps off Cape Blanc (CB), in the southern Guinea Basin (GBS), in the northern one (GBN) and off Namibia (Walvis Ridge, WR). Seasonality was most strongly expressed at the GBS and WR sites. Production half-time (length of time to generate one half of the annual productivity; Berger and Wefer 1990) ranged from 4.6 to 1.8 months suggesting almost constant (e.g at Cape Blanc) to highly-peaked (e.g. at GBS) production systems. Off Cape Blanc, mean annual total flux was 45.3 g m^{-2} to a water depth of 3204 m, exhibiting substantial variation of 62% (deviation from mean value). Holocene sediment accumulation rates amounted to 17 g m^{-2} yr^{-1}. At GBN, mean total flux was 36.1 g m^{-2} at 3939 m of water and the interannual variation reached only 11%. South of the equator at GBS, a total flux of 36.4 g m^{-2} to a water depth of 3382 m was determined. Bulk sediment accumulation rates in the Gulf of Guinea ranged between 17 and 23 g m^{-2}, but revealed higher values at GBS. At the Walvis Ridge, we obtained a mean total deep-water flux of 31.1 g m^{-2} with 42% interannual variation; sediment accumulation rates amounted to 15 g m^{-2}. Average annual organic carbon flux to the seafloor was 1.6 (Cape Blanc), 2-2.1 (Guinea Basin) and 3.2 m^{-2} (Walvis Ridge); these values are typical for open-ocean-(coastal) upwelling transition systems. They revealed no relationship to literature-derived annual primary production values. Organic carbon preservation was generally poor and estimated as 7.5% (Cape Blanc), 2.9%-3.0% (Guinea Basin) and 1.9% (Walvis Ridge) of the carbon fluxes into the traps, respectively. Carbon accumulation was not related to the deep-water fluxes nor to primary production estimates taken from literature. Mean annual total, lithogenic and biogenic opal fluxes generally increased with depth in the water column (lithogenic fluxes about two-fold), obviously due to the contribution of a substantial fraction of fine-grained, resuspensed material originating at topographic elevations. In contrast, organic carbon fluxes decreased with water depth at all sites following an exponential decline. Surprisingly, the accumulation of refractory lithogenic material in the underlying sediments was only 22-56% of the deep-water fluxes and resembled the subsurface fluxes more closely. We assume that the additional resuspended fraction originated at topographic elevations and was not incorporated yet into the sedimentary record but is being transported in suspension and dispersed in the ocean.

Introduction

During the last decade, many sediment trap experiments have been carried out worldwide resulting in a large number of seasonal flux data sets. However, much less data are available from longer-term flux monitorings spanning more than three years. As interannual flux variability is assumed to be signifi-cant at least for some ocean areas (e.g. Deuser et al. 1988), longer-term averages are needed to give a reliable estimate of average fluxes to the seafloor which may be related to the sediment accumulation rates. Moreover, such long-term observations of particle fluxes which are coupled to near-surface

From WEFER G, BERGER WH, SIEDLER G, WEBB DJ (eds), 1996, *The South Atlantic: Present and Past Circulation.* Springer-Verlag Berlin Heidelberg, pp 325-344

water properties can eventually provide information about larger-scale climatic variability in the oceans and atmosphere.

To study the seasonality and interannual variability of carbon flux and associated biogenic and non-biogenic components in the water column as well as the stable isotope records of important paleoceanographic proxies (e.g., for SST reconstructions), we have undertaken an intensive trapping program focussing on the eastern Atlantic high production areas: As a result, we were able to offer a host of new data on the coastal upwelling-open-ocean transition systems off northwest Africa and Namibia and the eastern equatorial upwelling area. From there, longer-term records spanning three to four years are available.

In this paper, we present new data and also summarize results which have been already published. Flux data and the stable oxygen isotope records of planktonic foraminifera and pteropods sampled off Cape Blanc have been shown in Kalberer et al. (1994) and Fischer et al. (1996). Data on seasonal and interannual flux variations from the eastern and western equatorial Atlantic have been presented by Fischer and Wefer (1995; 1996); seasonal diatom fluxes have been described in detail by Lange et al. (1994) and Treppke et al. (1996a). Boltovskoy et al. (1996) studied the radiolarian flux pattern in this area and have compared it to the sediment record (Boltovskoy et al. 1993a). From the Walvis Ridge site, biogenic fluxes as well as the diatom and silicoflagellate fluxes have been discussed in detail by Treppke et al. (1996b). Investigations on the seasonal distribution of cocco-lithophorids were carried out by Cepek and Wefer (1996).

Here, we will focus mainly on the longer-term fluctuations of fluxes rather than on seasonal variations. Seasonality over the three-to four-year sampling period will be quantified applying a seasonality index introduced by Berger and Wefer (1990) which was defined in terms of the number of months which yield one half of the total flux. Year-to-year flux variations will be discussed as well as C_{org}-flux alterations in the water column and the deep-water lithogenic fluxes. Finally, we will compare average annual values calculated for the four-year sampling period with sediment accumulation rates.

Methods

For most of the flux studies presented here we used the classical cone-shaped traps with 0.5 m² opening and 13/20 sampling cups (MARK VI and Kiel SMT 230/234). Only for the CB2 deployment, a MARK V (1.17m²) was used (Table 1). To retard microbial activity in the sampling cups, we used $HgCl_2$ as poison. Pure NaCl was added to the cups to increase the salinity (up to 40‰). After recovery, zooplankton "swimmers" were removed with foreceps and by sieving through a 1 mm screen; subsequently, the material was split. Standard analysis was done on the < 1 mm size fraction which clearly dominated particle flux. We performed organic carbon, nitrogen and carbonate analysis on freeze-dried material using a Heraeus-CHN-analyzer. Biogenic opal was determined with a sequential leaching technique developed by Müller and Schneider (1993). Stable carbon isotope analysis was done on a Finnigan Delta S mass spectrometer attached to a CHN-analyzer (see Fischer and Wefer (1991) for more details on sampling processes and analysis).

Description of the study area

The long-term study sites were located within or adjacent to coastal and open-ocean upwelling sites in the eastern Atlantic where total primary production is assumed to be within 90 and 125 gC m⁻² yr⁻¹ according to maps presented by Berger (1989) (Fig. 1, for deployment data see Table 1).

The Cape Blanc site was located in the cold Canary Current, in the "secondary upwelling" zone (Mittelstaedt 1991). The "primary upwelling" coastal zone is formed by fairly persistent trade winds and upwelling of nutrient-rich subsurface waters throughout the year, with periods of stronger upwelling occurring in spring and fall (Van Camp et al. 1991). Especially off Cape Blanc, "giant filaments" with high phytoplankton biomass are continuously present and reach several hundered kilometers offshore approaching the trap location.

At the eastern tropical Atlantic (Guinea Basin, sites GBN and GBS), high phytoplankton biomass (> 1 mg Chla m⁻³) and production and low SST's

AREA Deployment	Trap type (opening)	Position	Water depth (m)	Trap depths (m)	Sampling duration	Samples x days
LONGER-TERM DEPLOYMENTS						
CAPE BLANC (CB)						
CB1	Mark VI (0.5 m²)	20°45.N 19°45.W	3646	2195	22.03.88- 08.03.89	13 x 27
CB2	Mark V (1.17 m²)	21°09.N 20°41.W	4092	3502	15.03.89- 24.03.90	22 x 17
CB3	Kiel SMT 230 (0.5 m²)	21°08.N 20°40.W	4094	730 3557	08.04.90- 30.04.91 29.04.90- 08.04.91	18 x 21.5 16 x 21.5
CB4	Kiel SMT 230 (0.5 m²)	21°09.N 20°41.W	4108	733 3562	05.03.91- 19.11.91	20 x 10 20 x 10
NORTHERN EQUATORIAL UPWELLING AREA (GBN)						
GBN3	Kiel SMT 230 (0.5 m²)	01°48.N 11°08.W	4481	853 3921	01.03.89- 16.03.90 01.03.89- 25.02.90	20 x 19 19 x 19
GBN6	Kiel SMT 230 (0.5 m²)	01°47.N 11°08.W	4522	859 3965	04.04.90- 07.04.91 04.04.90- 30.03.91	4 x 18, 1 x 297 20 x 18
EA2	Kiel SMT 230 (0.5 m²)	01°47.N 11°15.W	4399	953	13.04.91- 29.11.91	20 x 11.5
SOUTHERN EQUATORIAL UPWELLING AREA (GBS)						
GBZ4	Kiel SMT 230 (0.5 m²)	02°11.S 09°54.W	3912	696	01.03.89- 16.03.90	20 x 19
GBZ5	Kiel SMT 230 (0.5 m²)	02°12.S 09°56.W	3920	597 3382	01.04.90- 30.03.91 04.04.90- 30.03.91	2 x 4.75, 12 x 29.6 20 x 18
EA4	Kiel SMT 230 (0.5 m²)	02°11.S 10°06.W	3906	1068	13.04.91- 29.11.91	20 x 11.5
WALVIS RIDGE (WR)						
WR1	Kiel SMT 230 (0.5 m²)	20°04.S 09°10.E	2217	1640	04.33.88- 16.03.89	4 x 18, 1 x 304
WR2	Kiel SMT 230 (0.5 m²)	20°03.S 09°09.E	2196	599 1648	18.03.89- 13.03.90	20 x 18 8 x 18, 1 x 216
WR3	Kiel SMT 230 (0.5 m²)	20°02.S 09°10.E	2208	1648	25.03.90- 09.04.91	20 x 19
WR4	Kiel SMT 230 (0.5 m²)	20°08.S 08°58.E	2263	1717	21.04.91- 17.12.91	20 x 12
EASTERN EQUATORIAL TRANSECT						
EA1	Kiel SMT 230 (0.5 m²)	03°10.N 11°15.W	4524	984	13.04.91 - 29.11.91	20 x 11.5
EA2 see above	Kiel SMT 230 (0.5 m²)	01°47.N 11°15.W	4371	953	13.04.91 - 29.11.91	20 x 11.5
EA3	Kiel SMT 234 (0.5 m²)	00°05.S 10°46.W	4141	1097	13.04.91 - 29.11.91	20 x 11.5
EA4 see above	Kiel SMT 230 (0.5 m²)	02°11.S 10°06.W	3906	1068	13.04.91 - 29.11.91	20 x 11.5
EA5	Kiel SMT 230 (0.5 m²)	04°20.S 10°16.W	3490	947	13.04.91 - 29.11.91	20 x 11.5
ONE-YEAR DEPLOYMENTS						
CV1	Kiel SMT 234 (0.5 m²)	11°29.N 21°01.W	4968	1003 4523	05.10.92- 04.04.93	all: 18 x 9.5, 1 x 5.5, 1 x 4.5
WA1	Kiel SMT 234 (0.5 m²)	04°00.S 25°34.W	5530	652 1232 4991	17.10.92- 21.03.93	20 x 7.75 20 x 7.75 20 x 7.75

Table 1. Deployment data for the long-term mooring sites (Cape Blanc, Guinea Basin, Walvis Ridge), the north-south transect in the eastern tropical Atlantic (EA1-5, 1991) and CV1 (Cape Verde Islands) and WA1 (western tropical Atlantic).

(sea-surface temperatures) occur across the equator in boreal summer from about 3°N to 7°S (Longhurst 1993). Intensified zonal wind stress in the
western tropical Atlantic causes an uplift of the
thermocline in the eastern part of the ocean (Longhurst 1993) supplying nutrients to the surface layer
and resulting in phytoplankton blooms. During the
rest of the year, a typical tropical situation (TTS,
sensu Herbland 1983) prevails when nitrate concentrations are limiting in the mixed layer and primary production is controlled by the depth of the
nutricline. During boreal winter and spring when
the trade wind separating calms (ITCZ=Inner-tropi-
cal Convergence Zone) approach the equator at
about 10°W, the northernmost sites (Fig. 1) are
influenced by the NE trade winds.

The Benguela upwelling, an eastern boundary
system, is characterized by trade wind-driven upwelling of cold, nutrient-rich subsurface waters
concentrated in several cells along the coast. Our

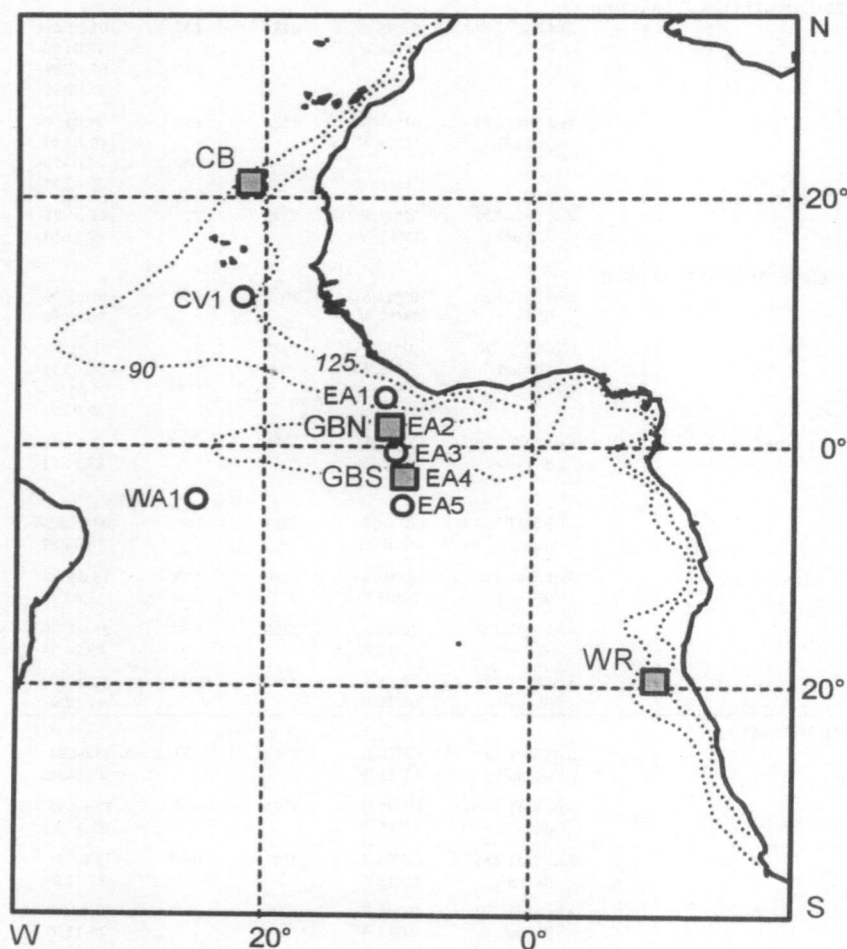

Fig. 1. Study areas in the eastern Atlantic. Sites with longer-term records (large
squares) were Cape Blanc (CB), northern and southern Guinea Basin (GBN,
GBS) and Walvis Ridge (WR). Data from the 1991 north-south transect in the
eastern equatorial Atlantic (EA1-EA5) and two one-year records are shown
with open circles (CV1=Cape Verde Islands; WA1=western equatorial Atlantic). Two contour lines of primary production (90 and 125 gC m^{-2} yr^{-1}) taken
from the "Dahlem map" which was established by Berger (1989) are also indicated. For deployment data see Table 1.

study site (WR, Fig. 1) was located in the coastal-open-ocean mixing zone with upwelling filaments streaming offshore. According to Servain et al. (1985), the wind stress causing the upward movement of nutrient-rich water masses and high primary production in this area has maxima occurring in austral spring and in fall.

More detailed descriptions of the investigation sites, oceanographic and atmospheric settings as well as trap-near currents are presented in Wefer and Fischer (1993), Fischer et al., (1996), Fischer and Wefer (1995; 1996), Lange et al. (1994) and Treppke et al. (1996b).

Results and Discussion

Seasonal patterns

The seasonal pattern off Cape Blanc shows dominant winter-spring and summer flux maxima with highly variable amplitudes (Fig. 2a). The background flux at both depths was relatively high reflecting almost continuous upwelling and production in this area (Schemainda et al. 1975). Nonbiogenic material, mostly aeolian quartz constituted almost one third of the total and covaried with the total fluxes (Fischer et al., 1996). Total flux was dominated by calcium carbonate comprising about one half. The timing of peak total and carbonate fluxes followed the decrease in trade wind speeds during the warm season in late spring and early summer (Fischer et al. 1996). Seasonality expressed as the seasonality index (SI=number of months which yield one half of the total flux; Berger and Wefer 1990) showed some variation in the deeper trap records (1.3-2.8; Fig. 3a; Table 4). The shape of the different curves range from a system with almost constant productivity and flux (SI=1.3) to a pure sinusoidal production-flux curve (SI around 3) which is in accordance to productivity measurements performed by Schemainda et al., (1975). Seasonality was slightly higher in the upper trap depths.

The three-year record from the GBN sites (Fig. 2b) shows total flux peaks dominated by carbonate-bearing primary and secondary producers mainly in summer which represents the time of the shallowing of the thermocline and upwelling along

the equator (Longhurst 1993). Equatorial upwelling was further documented by the dominant occurrence of diatom species characteristic of the upwelling such as members of the *Nitzschia bicapitata* group (Lange et al., 1994). However, summer sedimentation which appears to increase slightly from 1989 to 1991 was always dominated by organisms with carbonate tests such as planktonic foraminifera and coccolithophorids. Spring maxima with terrestrial organic and inorganic components supplied by the north-eastern trades during the southernmost location of the Intertropical Convergence Zone (ITCZ) were also observed. The background flux level was relatively high and comparable to Cape Blanc. The seasonality index derived from the deeper trap records was low and varied only slightly around 2; the upper trap SI values ranged from 2.2 to 2.8 (Fig. 3b; Table 4).

At GBS, background fluxes were relatively low and peak fluxes occurred mainly in boreal spring showing a slight increase from 1989 to 1991 (Fig. 2c). These maxima corresponded to a deep chlorophyll maximum and primary production located at a thermal ridge south of the equator as described by Voituriez and Herbland (1981). Significant summer sedimentation was only observed in 1990; in 1989, no summer sedimentation was found. Seasonality of fluxes in the deeper water was low and comparable to the GBN site (SI=2; Fig. 3c; Table 4). However, in the upper water column, the seasonality index ranged from 3 in 1990 to even 4.3 in 1989. Such a high SI corresponding to a production half-time of two month suggests a production system intermediate between sinusoidal-type to highly peaked (Berger and Wefer 1990).

At the Walvis Ridge site, we observed very low background flux values and a distinct bimodal flux pattern with peaks in austral fall and spring (Fig. 2d), dominated both by coccolithophorids (Cepek and Wefer 1996) and foraminifera. In austral spring when the trade wind-driven coastal upwelling was most intense and SST's were lowest (Treppke et al., 1996b), highest flux values were determined. The diatom flux pattern which paralleled the total flux variations was directly related to wind stress fluctuations and inversely related to alkeonone-derived SST's. Cold SST excursions reaching 14.6-14.9°C combined with high wind stress had the longest duration during austral spring and the wind maxima in fall and winter/spring appear to have preceeded total and opal flux maxima and SST minima by approximately 1-2 months (Treppke et al. 1996b) The seasonality index derived from the lower trap samples varied between 2 and 3.7; the upper trap (one deployment) revealed a value of 3.5 (Fig. 3d; Table 4). Applying the scheme of Berger and Wefer (1990), this would represent a production system with sinusoidal fluctuations.

Because of the expected large regional variability in fluxes in the eastern tropical Atlantic, we deployed five mooring arrays (north-south transect at 10°W, EA1-EA5) with traps in approximately 1000 m water depth (Table 1). The results depicted in Fig. 4 show that spring sedimentation was present at all sites with a higher contribution of lithogenic components north of the equator. Surprisingly, a distinct summer sedimentation event representing high primary production was only observed north of the equator (Fig. 5; Fischer and Wefer, 1995) but not further south as one might expect from satellite-derived chlorophyll distributions (Longhurst 1993) and primary productivity measurements (Voituriez and Herbland 1981). Fischer and Wefer (1995) offer several explanations for this discrepancy and suggest that this phenomenon was a result of restricted particle sinking close to the equatorial divergence and preferential downward movement of particulates further north of the equator in the vicinity of the convergent frontal zone.

Flux changes in the deeper water column

In this chapter, we discuss primarily flux alterations of organic carbon and lithogenic components in the deeper water column (below 500 m to about 500 m above the seafloor). For this purpose, we determined flux values for each collection period of the various deployments (Table 1) rather than annual data. With our data, we were unable to determine any mid-water carbon flux maxima (Walsh et al. 1988). We observed a C_{org}-decrease at all sites below the oxygen minimum layer. There, most of the particulates originated from the surface are probably already transformed to dissolved components via biological consumption and dissolution (Pace et al. 1987). Generally, therefore, degradation is

assumed to be lower in the deep ocean than in the subsurface layer as the proportion of the highly degradable fraction decreases with depth. Alternatively, some unknown fraction could have been supplied by lateral advection and resuspension (see Biscaye et al. 1994, for a summary). But since all traps were located 500 m above the seafloor (see chapter below), we assume that the proportion of the laterally transported organic component was minimal. Instead, the general increase of fluxes with depths appears to be due to rather old rebound particles (see Walsh et al. 1988), mostly composed of fine-grained and refractory lithogenic components resuspended from topographic elevations and/or dispersed along pycnoclinal surfaces as suggested by Walsh et al. (1988).

We assumed an exponential decrease of the organic carbon flux with depth (e.g. Lorenzen et al. 1983) for the oligotrophic site WA1 (Fig. 6, equation 1) which is also applied for the other sites. However, one might also use a power function for the description of the depth dependence (Banse 1990).

$$WA1: f(C_{org}) = 3.37 - 0.78 \log (depth); r^2 = 0.983 \quad (1)$$

$f(C_{org})$ is the organic carbon flux in g m^{-2} and depth is given in meters. This equation is equivalent to:

$$WA1: f(C_{org}) = 3.37 - 0.34 \ln (depth) \quad (2)$$

The WA1 site was characterized by fluxes representing typical oligotrophic open-ocean conditions. All other sites (except WR2) exhibited an annual carbon export typical for open-ocean-upwelling transition systems; the values at WR2 were somewhat higher but were still not typical for a highly productive coastal upwelling system (Berger et al. 1989).

The rather broad range of depth exponents seems to be partly a consequence of differences between zooplankton feeding and organic carbon production as discussed in Bishop et al. (1987). The coefficient of 0.34 in equation 2 with depth appears to be rather low compared to literature data (see Bishop 1989, for a summary). However, for those study sites characterized by higher primary production and stronger seasonality, e.g. for CB4, we

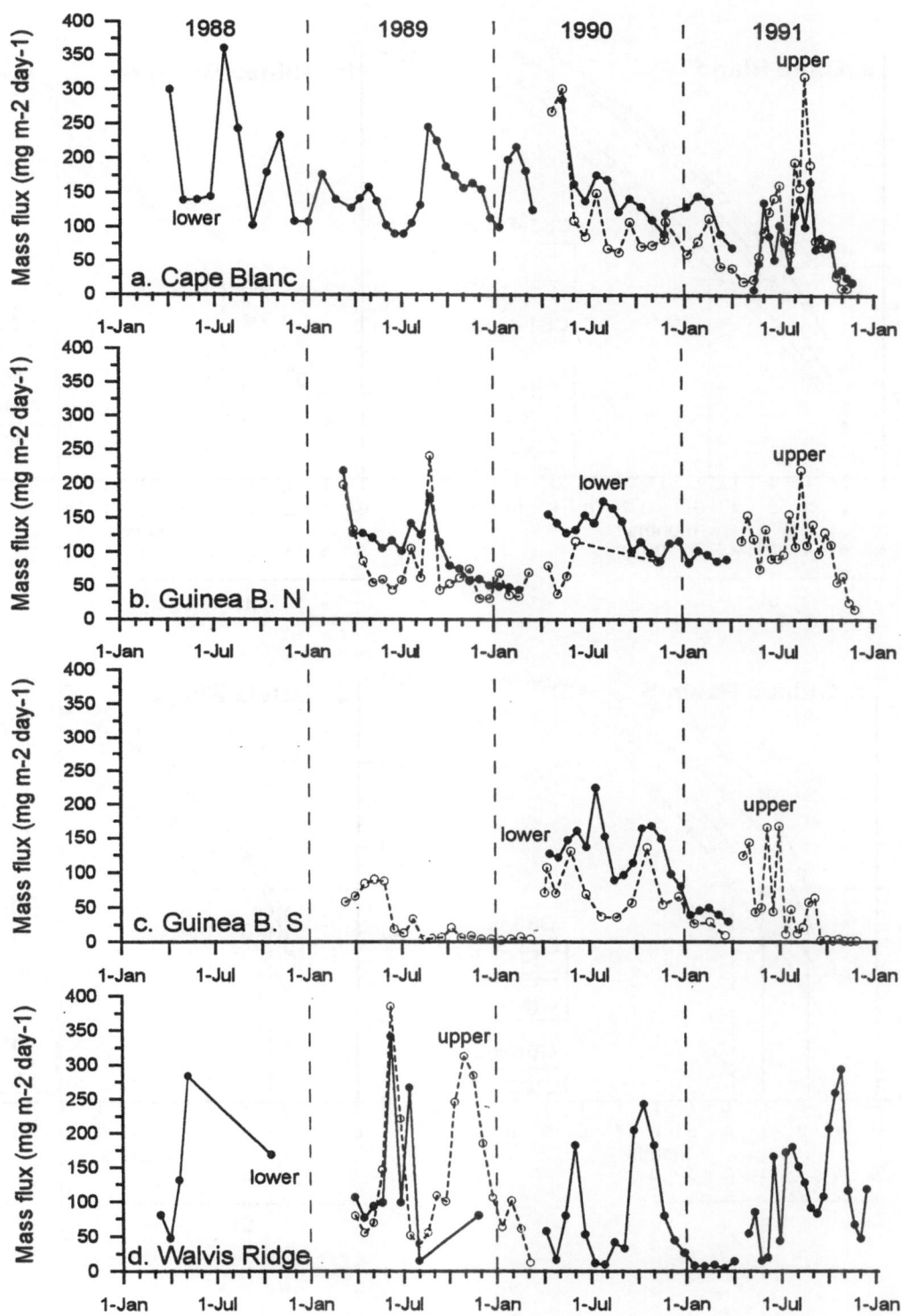

Fig. 2. Total mass flux patterns derived from Cape Blanc (a), the Guinea Basin (north (a) and south (b)) and the Walvis Ridge (d). For deployment data see Table 1. Solid lines show the upper trap, dashed lines the lower trap flux data.

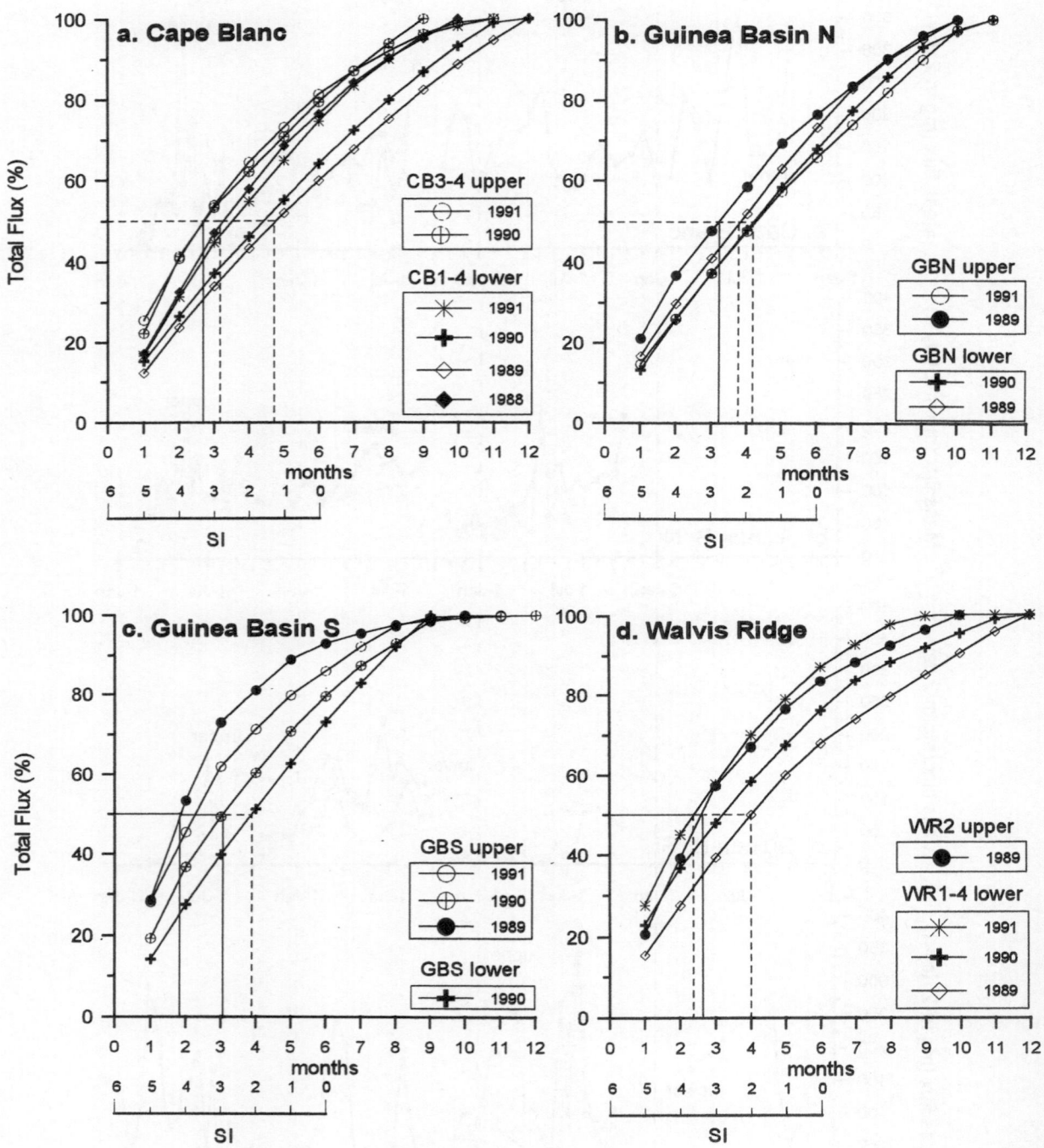

Fig. 3. Particle flux seasonality in the eastern Atlantic showing the seasonality index (SI) and production half-time values (according to Berger and Wefer 1990) for the Cape Blanc (a), the two Guinea Basin sites (b and c) and the Walvis Ridge site (d).

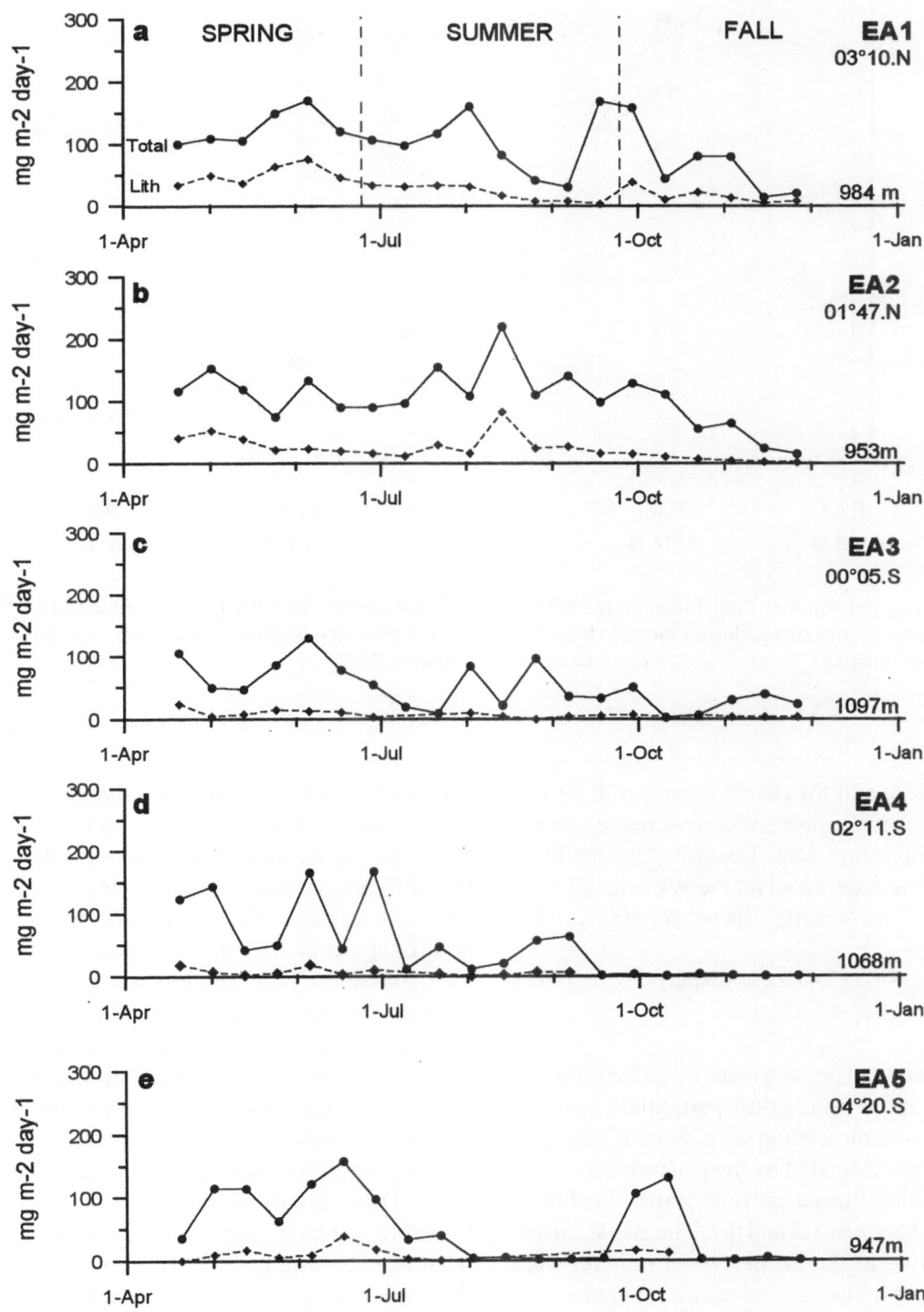

Fig. 4. North-south transect of total (upper curves) and lithogenic fluxes (lower curves) in the eastern equatorial Atlantic (EA1-5, 1991) at 10°W (Fischer and Wefer, 1995). For deployment data see Table 1.

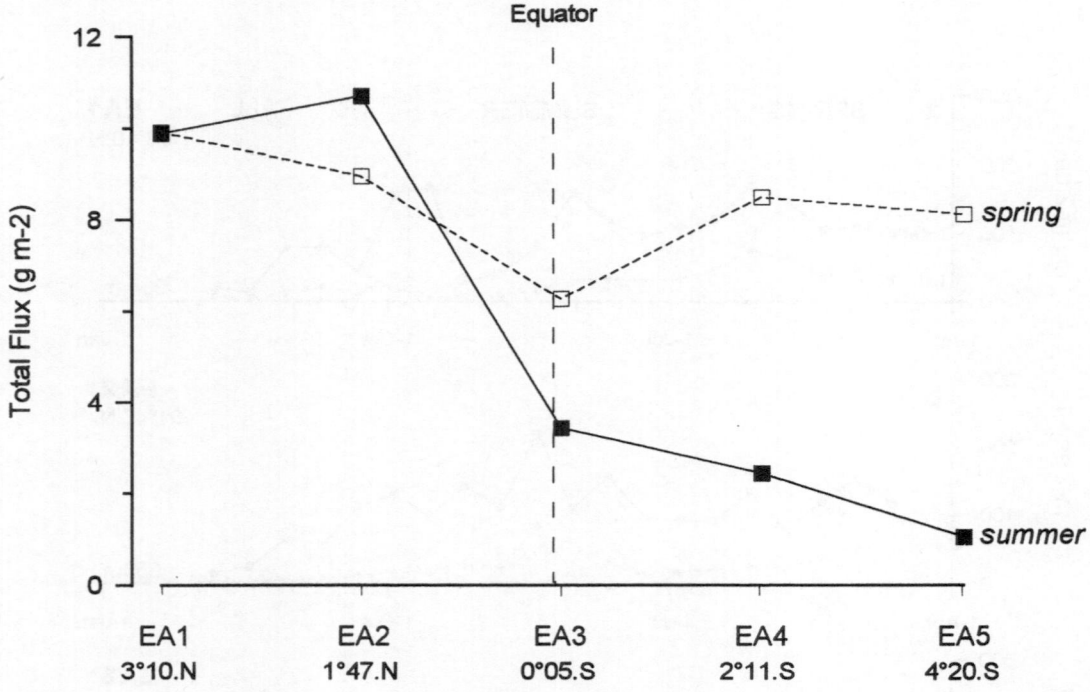

Fig. 5. Spring and summer total fluxes in the eastern equatorial Atlantic in 1991 (EA1-5; Fischer and Wefer, 1995). Note the higher fluxes during the cold boreal summer season which occurred north of the equator. Spring fluxes were almost similar at all sites except close to the equator (EA3).

obtained 0.76, and for GBN3 a value of 0.59 was calculated; these values are in reasonable accordance with literature data. The strongest depth dependence was determined for the WR2 deployment where a high seasonality was observed (Fig. 3d):

$$\text{WR2: } f(C_{org}) = 12.77 - 1.22 \ln (\text{depth}) \qquad (3)$$

Degradation of organic material is characterized by a more rapid removal of particulate nitrogen relative to organic carbon with depth (Pace et al. 1987). This is indicated by the general increase in the C/N-ratios during particle transit (Table 2) which was between 1.2 and 0.7. The stable carbon isotope ratios in the deeper water column were obviously not subject to a substantial alteration. We found a tendency to higher average annual values of about 0.8-0.1‰ with depth (Table 2). The annual $\delta^{13}C_{org}$-values versus the C/N-ratios revealed no significant correlation indicating that terrestrial

contribution did not account for the observed isotope variations (18.5-22.8‰; Table 2). Surprisingly, the values north of the equator (GBN) were slightly higher compared to GBS, although this area should be influenced by the input of terrestrial carbon (Lange et al. 1994) with a lower carbon isotope signature. This trend probably reflects increased CO_2 availability south of the equator. This assumption is supported by stable isotope ratios of plankton values determined on late summer samples which were about -19‰ north of the equator and -22‰ south of the equator.

Fluxes of lithogenic components are shown in Fig. 7. They increased with depth, except for WR2, probably reflecting lateral transport of material from near-by topographic elevations and/or the shelf areas. A very small increase was found at the oligotrophic pelagic site in the northern Brasil Basin (WA1) suggesting that the lateral flux component, particularly with respect to the organic carbon fluxes (see above), appears to be negligible.

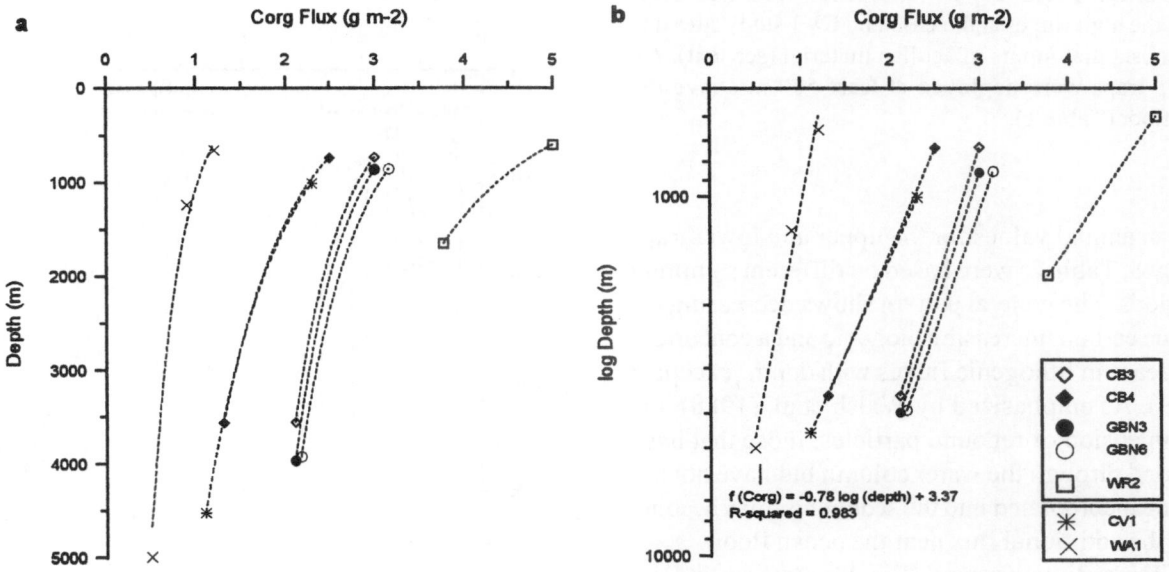

Fig. 6a. Organic carbon fluxes versus depth and the assumed exponential decrease in the water column. An exponential function was applied to the data of WA1; the same organic carbon decline was applied to the other data shown. Fluxes were calculated for the entire sampling periods of CB3,4, GBN3, GBN6, WR2, CV1 and WA1. All deep traps were located approximately 500 m above the seafloor (deployment data and sampling intervals see Table 1). Note the low carbon fluxes at the oligotrophic WA1 site, the intermediate fluxes off Cape Blanc and in the northern Guinea Basin and the higher values at the Walvis Ridge.

Fig. 6b. Organic carbon fluxes versus depth shown as a log-plot. An exponential function which is biologically meaningful (Banse 1990) was applied to describe the depth dependence at WA1 (oligotrophic northern Brasil Basin, Table 1, Fig. 1).

Therefore we have chosen WA1 to compute a relationship of C_{org}-flux with depth.

Variable supply of lithogenic material is observed off Cape Blanc (CB3, 4). For the CB4 deployment period in 1991, we measured low non-biogenic influx with little aeolian contribution (hemipelagic site, Fig. 7) and, even a slight decrease of lithogenic fluxes with depth (Fischer et al. 1996). In contrast, a substantial increase with depth was determined for the Cape Blanc 3 deployment (CB3: 1990) and, in addition, for the Cape Verde site (CV1). Both sites were characterized by a substantial contribution of rather coarse-grained material supplied by the NE trade winds. Using the lines of CB3 and CV1 drawn through the x-axis in com-

parison to the CB4 line (low non-biogenic contribution, Fig. 7), we estimate that the wind-derived component may have been about one-half of the total lithogenic component during the CB3 and CV1 deployments.

Interannual variations, deep-water fluxes and the sediment record

Flux variations from 1988 to 1991, the composition of particles and the C/N, C_{org}/opal and $^{13}C_{org}$-ratios are listed in Table 2. The average annual total, biogenic opal, organic carbon, carbonate and lithogenic flux values and the interannual variability of the total fluxes are shown in Fig. 8. Note that the

Fig. 7. Lithogenic fluxes versus depth. Note the general increase with depths (except for WR2 and CB4) and the high fluxes at the CB3 and CV1 study sites due to substantial inputs of aeolian material (see text). All deep traps were deployed at least 500 m above the seafloor (Table 1).

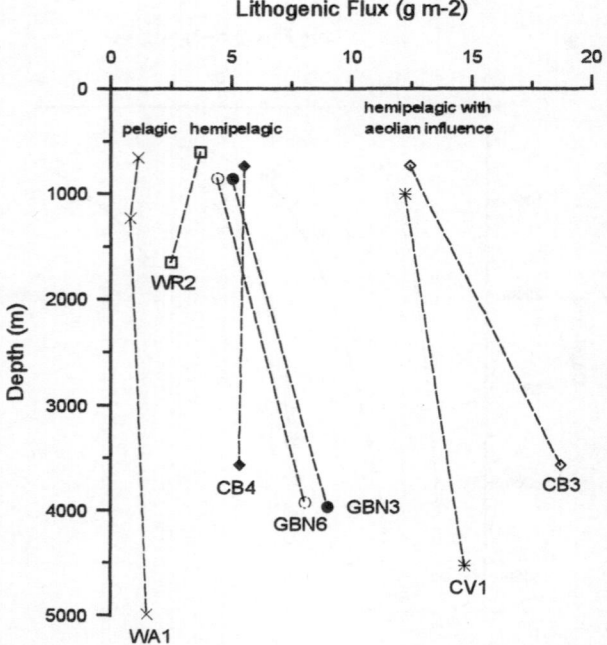

mean annual values for the upper and lower traps (Fig. 8, Table 2) were based on different sampling periods. The general pattern shows decreasing organic carbon, increasing biogenic and a concurrent increase in lithogenic fluxes with depth (except at WR). As emphasized by Walsh et al. (1988), the resupension of rebound particles, those that have settled through the water column but have not become incorporated into the sediments, may account for the additional flux near the ocean floor.

The sediment record (Fig. 8), provides indication of a reduction of flux components at the sediment-water interface, particularly the organic carbon and the biogenic opal fluxes (unpubl. data from C. Rühlemann, Bremen) which constituted only 1.9-10% of the deeper-water flux values (Table 3). It appears that the accumulation rates except for the degradable fractions (e.g. opal, C_{org}) correspond much better to the upper trap values than to the deep-water traps (except at WR). The accumulation of refractory lithogenic components which are not subject to strong dissolution at the sediment-water interface, do not equal the deep-water lithogenic fluxes; these amounted only to 22-56% of the deep-water fluxes. This discrepancy may be due to a substantial increase in deep-water fluxes due to the additional contribution of a primarily fine-grained resuspensed fraction (Walsh et al. 1988a, b, Gardner and Richardson 1992) which was probably about equal to the true downward flux (Table 2). On the other hand, we found that carbonate accumulation, although assumed to be influenced by dissolution in the deeper water column (especially below 4500 m) was only slightly lower than the deep-water fluxes (70-89% at GBN, GBS and WR) with the exception of the Cape Blanc site (33%). We assume that carbonate, mostly foraminifera and coccolithophorids do obviously not contribute larger amounts to the resuspensed material dispersed in the deep water column.

At the Cape Blanc site, we obtained the highest average deep-water fluxes (45.3 g m^{-2} yr^{-1}) and the highest year-to-year variations (62%). The distinct interannual flux variations were accompanied by substantial year-to-year SST variations (20.8 to 24.4°C) deduced from the δ^{18}O-isotope ratios of planktonic foraminifera and pteropods (Fischer et al. 1996). Holocene sediment accumulation rates calculated from a near-by sediment core (12329-6; Koopmann, 1979) were significantly lower (17 g m^{-2} yr^{-1}). We calculated the carbonate accumulation rates using a carbonate content of 60% for the sediments in this region (see Fütterer, 1983). Carbonate (around 7 g m^{-2} yr^{-1}) and lithogenic matter accumulation rates (9.9 g m^{-2} yr^{-1}) were lower compared to the deeper trap values (17.6 and 21.3 g m^{-2} yr^{-1}; Fig. 7; Table 2). However, if we consider the strong interannual variability, these values were in the lower range of the calculated deep-water fluxes.

Organic carbon accumulated in the surface sediments was estimated to 0.12 g m^{-2} yr^{-1} and, thus, was 7.5% of the average fluxes to a depth of 3985 m (1.6 gC m^{-2} yr^{-1}) and about 0.16% of primary

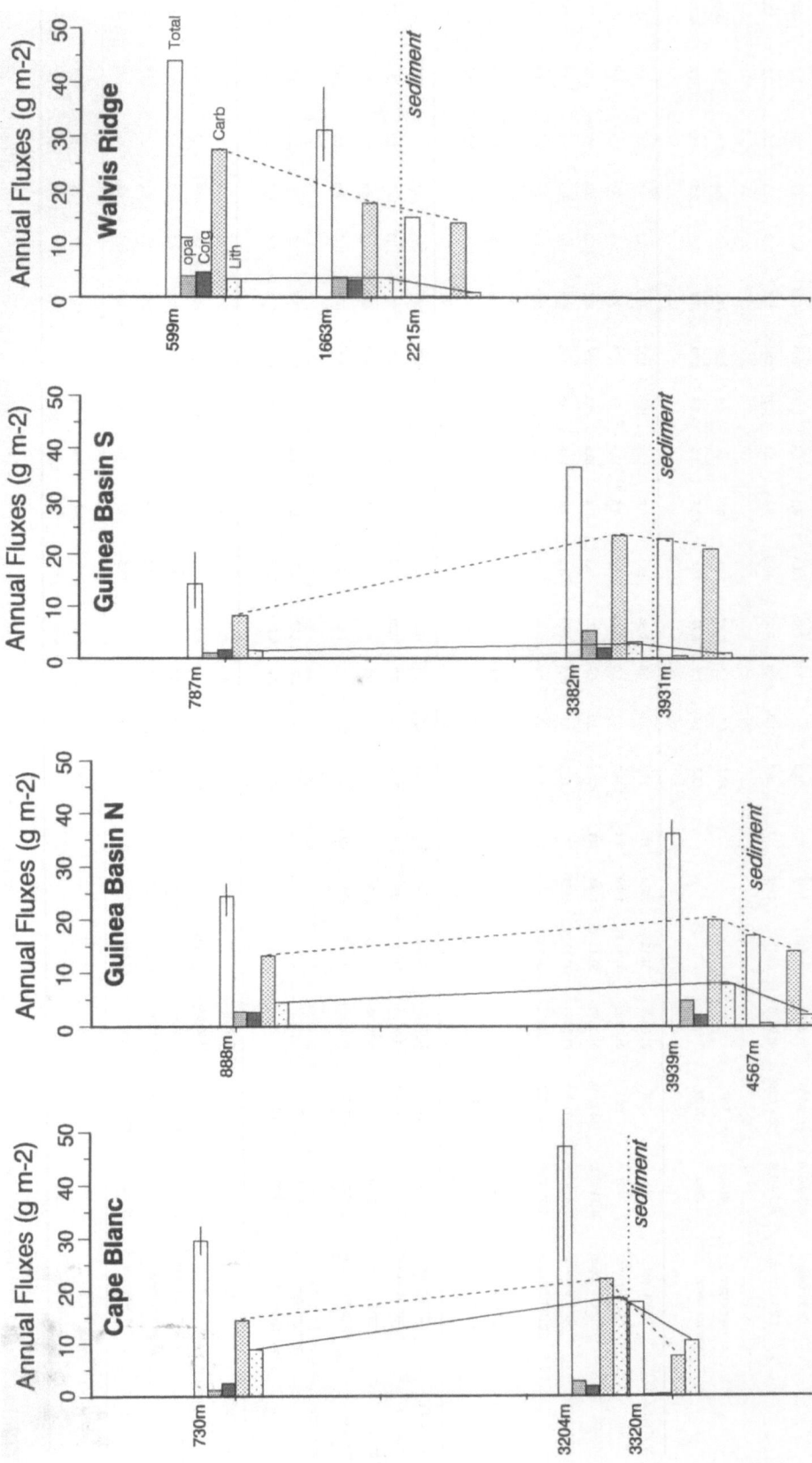

Fig. 8. Calculated average annual fluxes versus depth at the Cape Blanc, Guinea Basin and Walvis Ridge study sites (Table 2). Interannual variations of total fluxes are indicated by a line. For comparison, surface sediment accumulation rates are also shown; sediment data were taken from Koopmann (1979; CB: 12329-6, Bickert (1992; GBN: CeoB 1101-4), Meinecke (1992; GBS; GeoB 1105-3), Schmidt (1992; WR: GeoB 1028-4). For the determination of opal accumulation rates, unpublished values from C. Rühlemann were used. The legend is shown in the upper part of the Walvis Ridge data.

Location Position	Trap	Water depth m	Trap depth m	Sampling interval from	to	YEAR	Days	FLUXES Ttl	Opal	Corg	Nges	CaCO3	Lithog	Percent Opal	Corg	Nges	CaCO3	Lithog	C/N atom	Corg/Opal w/w	CCR	13C/12C %o PDB
Cape Blanc																						
21°08.N																						
20°40.W																						
	CB1 lower	3646	2195	22.03.88 -	31.12.88	1988	285	53,6	4,4	1,2	0,26	22,9	21,9	8,2	4,1	0,49	42,7	40,9	9,8	0,5	0,8	-20,1
	CB1/2 lower	3646/4092	2195/3302	01.01.89 -	31.12.89	1989	358	52,1	2,4	1,6	0,19	26,7	19,7	4,7	3,1	0,38	51,2	37,8	9,6	0,7	0,5	-20,9
	CB3 upper	4094	730	08.04.90 -	31.12.90	1990	268	32,5	1,2	1,8	0,24	17,8	9,8	3,7	5,6	0,75	54,9	30,2	8,6	1,5	0,8	-21,3
	CB2/3 lower	4092/4094	3502/3557	01.01.90 -	31.12.90	1990	329	49,8	2,1	1,7	0,19	25,0	19,4	4,1	3,4	0,40	50,2	38,9	9,9	0,8	0,6	-21,1
	CB3/4 upper	4094/4108	730/733	01.01.91 -	19.11.91	1991	320	26,8	1,4	3,4	0,49	11,3	8,1	5,4	12,8	1,84	42,3	30,1	8,1	2,4	2,5	-20,7
	CB3/4 lower	4094/4108	3557/3562	01.01.91 -	19.11.91	1991	298	25,5	1,3	2,0	0,26	10,6	9,4	5,1	7,9	1,04	42,1	36,9	8,9	1,5	1,5	-18,5
	average upper	*4101*	*730*					*29,7*	*1,3*	*2,6*	*0,37*	*14,6*	*9,0*	*4,4*	*8,8*	*1,23*	*49,1*	*30,2*	*8,4*	*2,0*	*1,7*	*-21,0*
	average lower	*3985*	*3204*					*45,3*	*2,6*	*1,6*	*0,23*	*21,3*	*17,6*	*5,6*	*3,6*	*0,50*	*47,1*	*38,9*	*9,6*	*0,9*	*0,9*	*-20,2*
N° Guinea B.																						
01°48.N																						
11°07.W																						
	GBN3 upper	4481	853	01.03.89 -	31.12.89	1989	353	25,3	3,1	2,8	0,34	13,1	3,8	12,1	10,9	1,34	51,6	15,2	9,5	0,9	1,8	-21,9
GBN	GBN3 lower	4481	3921	01.03.89 -	31.12.89	1989	304	34,2	4,3	2,0	0,25	18,5	7,5	12,5	5,8	0,73	54,1	21,8	9,4	0,5	0,9	-21,3
	GBN3/6 upper	4481/4522	853/859	01.01.90 -	31.12.90	1990	348	20,9	2,2	1,9	0,27	13,1	3,4	10,5	9,1	1,30	62,7	16,3	8,4	0,9	1,4	-21,6
	GBN3/6 lower	4481/4522	3921/3956	01.01.90 -	31.12.90	1990	327	38,0	5,1	1,9	0,24	21,2	8,0	13,5	5,0	0,63	55,6	21,0	9,3	0,4	0,7	-21,7
	GBN6/EA2 upper	4522/4399	859/953	01.01.91 -	29.11.91	1991	327	27,0	2,6	2,8	0,39	13,0	5,9	9,6	10,4	1,44	48,2	21,9	8,2	1,1	1,8	-21,3
	average upper	*4467*	*888*					*24,4*	*2,6*	*2,5*	*0,33*	*13,1*	*4,4*	*10,8*	*10,2*	*1,37*	*53,6*	*17,9*	*8,7*	*1,0*	*1,7*	*-21,6*
	average lower	*4467*	*3939*					*36,1*	*4,7*	*2,0*	*0,25*	*19,9*	*7,8*	*13,0*	*5,4*	*0,68*	*55,0*	*21,5*	*9,4*	*0,5*	*0,8*	*-21,5*
S° Guinea B.																						
02°11.S																						
09°55.W																						
	GBZ4 upper	3912	696	01.03.89 -	31.12.89	1989	304	9,5	0,7	1,1	0,13	5,3	1,2	7,7	11,6	1,37	56,0	13,1	10,1	1,5	1,7	-22,1
GBS	GBZ4/5 upper	3912/3920	696/597	01.01.90 -	31.12.90	1990	351	20,4	1,4	2,7	0,30	11,2	2,3	6,9	13,3	1,47	55,0	11,3	10,7	1,9	2,0	-22,8
	GBZ5 lower	3920	3382	04.04.90 -	31.12.90	1990	270	36,4	5,4	2,1	0,28	23,6	3,2	14,9	5,8	0,77	64,8	8,6	8,8	0,4	0,8	-22,0
	GBZ5/EA4 upper	3920/3906	597/1068	01.01.91 -	29.11.91	1991	321	12,9	0,8	1,2	0,17	8,1	1,3	6,2	9,3	1,32	62,8	10,1	8,6	1,6	1,3	-21,8
	average upper	*3913*	*787*					*14,3*	*1,0*	*1,7*	*0,20*	*8,2*	*1,6*	*6,8*	*11,7*	*1,40*	*57,5*	*11,2*	*9,8*	*1,7*	*1,7*	*-22,5*
	average lower	*3920*	*3382*					*36,4*	*5,4*	*2,1*	*0,28*	*23,6*	*3,2*	*14,8*	*5,8*	*0,77*	*64,8*	*8,8*	*8,8*	*0,4*	*0,8*	*-22,0*
Walvis Ridge																						
20°04.S																						
09°10.E																						
	WR1 lower	2217	1640	04.03.88 -	31.12.88	1988	301	38,7	5,2	4,2	0,45	21,2	3,9	13,4	10,9	1,15	54,7	10,1	11,1	0,8	1,7	-21,8
	WR2 upper	2196	599	18.03.89 -	31.12.89	1989	289	43,9	3,9	4,6	0,68	27,5	3,4	8,8	10,5	1,55	62,5	7,8	7,8	1,2	1,4	-21,3
	WR1/2 lower	2217/2196	1640/1648	01.01.89 -	31.12.89	1989	363	30,5	2,9	3,3	0,47	19,1	1,9	9,4	10,9	1,60	62,6	6,2	8,2	1,2	1,4	-21,2
	WR2/3 lower	2196/2208	1648/1648	01.01.90 -	31.12.90	1990	353	25,6	2,9	2,9	0,42	14,0	3,3	11,5	11,5	1,60	54,8	12,8	8,2	1,0	1,7	-20,6
	WR3/4 lower	2208/2263	1648/1717	01.01.91 -	17.12.91	1991	339	29,5	3,9	2,3	0,41	16,5	5,4	13,3	7,6	1,40	55,9	18,4	6,4	0,6	1,1	-20,5
	upper	*2196*	*599*					*43,9*	*3,9*	*4,6*	*0,68*	*27,5*	*3,4*	*8,9*	*10,5*	*1,55*	*62,6*	*7,7*	*7,8*	*1,2*	*1,4*	*-21,3*
	average lower	*2221*	*1663*					*31,1*	*3,7*	*2,1*	*0,44*	*17,7*	*3,6*	*12,0*	*10,2*	*1,41*	*57,0*	*11,7*	*8,5*	*0,9*	*1,5*	*-21,0*

Table 2. Annual fluxes, C/N-, C_{org}/opal ratios and CCR (carbon rain ratios) and the stable carbon isotopes of organic carbon and mean values for the years 1988 to 1991.

Study site	Traps / sediment	Average depth (m)	Literature	Annual primary production (g C m-2)	Total	Opal	Corg	Carbonate	Lithogenics
Cape Blanc			*Schemainda et al. (1975)*	75					
			Berger (1989)	90–125			75		
	upper	730			29,7	1,3	2,6	14,6	9
	lower	3204			45,3	2,6	1,6	21,3	17,6
	% lower trap/PP						**1,3–2,1**		
	sediment	3320	Koopmann (1979, 12329-6)		17	0,04	0,12	7	9,9
	% sed./lower trap				**38**	**2**	**7,5**	**33**	**56**
Guinea Basin (north)			*Schemainda et al. (1976)*	70–90					
			Berger (1989)	90			80		
	upper	888			24,4	2,6	2,5	13,1	4,4
	lower	3939			36,1	4,7	2	19,9	7,8
	% lower trap/PP						**2,5**		
	sediment	4567	Bickert (1992, GeoB 1101-4)		17	0,48	0,06	14	2
	% sed./lower trap				**47**	**10**	**3,0**	**70**	**26**
Guinea Basin (south)			*Schemainda et al. (1976)*	70–90					
			Berger (1989)	90			80		
	upper	787			14,3	1	1,7	8,2	1,6
	lower	3382			36,4	5,4	2,1	23,6	3,2
	% lower trap/PP						**2,6**		
	sediment	3931	Meinecke (1992, GeoB 1105-3)		23	0,51	0,06	21	1
	% sed./lower trap				**63**	**9**	**2,9**	**89**	**31**
Walvis Ridge			*Berger (1989)*	90–125					
	upper	599			43,9	3,9	4,6	27,5	3,4
	lower	1663			31,1	3,7	3,2	17,7	3,6
	% lower trap/PP						**2,6–3,6**		
	sediment	2215	Schmidt (1992, GeoB 1028-4)		15	0,12	0,06	14	0,8
	% sed./lower trap				**48**	**3**	**1,9**	**79**	**22**

Table 3. Compilation of annual primary production data derived from literature, annual shallow and deep-water fluxes (this study) and sediment accumulation rates. Accumulation rates of lithogenic components were estimated as the difference between bulk and the sum of all other components.

production (75 gC m^{-2} yr^{-1}; Schemainda et al. 1975; Table 3). A compilation of productivity data presented by Berger (1989) reveals values somewhat higher for this site (90-125 gC m^{-2} yr^{-1}) which represent almost coastal ocean productivities. However, we must consider a high variability of production due to the occurrence of large filaments off Cape Blanc (Van Camp et al. 1990). The productivity and the carbon accumulation data translate into a carbon preservation of 0.13-0.1% which is intermediate between a coastal and an open-ocean system (Berger et al. 1989). However, we should consider that some unknown proportion of the settling organic matter was composed of more refractory terrestrial carbon (mostly pollen) as suggested by higher C/N- and lower $\delta^{13}C_{org}$-ratios (Fischer et al. 1996). Thus, the truly marine organic carbon deposition of 0.13-1% may be slightly overstimated. In the northern Guinea Basin (GBN), average total fluxes amounted to 24.4 g m^{-2} yr^{-1} (upper traps) and 36.1 g m^{-2} yr^{-1} (lower traps) with much less interannual variation (10.5 and 25%) compared to the Cape Blanc site (Table 2). Again, deep-water fluxes were significantly higher except for organic carbon (Fig. 8). Bulk sediment accumulation rates in 4567 m water depth amounted to 17g m^{-2} yr^{-1} (Bickert 1992; GeoB 1101-4) and were lower by a factor of two than the deep-water fluxes (Table 3). This was mainly the effect of a substantial reduction in biogenic opal accumulation (10% of the deep water fluxes), carbonate (70%) and lithogenic components (26%). Accumulation of organic carbon was only 0.06 g m^{-2} yr^{-1}, a value assumed to be typical for an open-ocean system (Berger et al. 1989); this translates into 3% of the deep-water flux only (Table 3). Schemainda et al. (1976) determined an annual primary production of 70-90 gC m^{-2}; a similar value was obtained by Berger (1989). Applying these values, organic carbon deposition versus production amounts to only 0.09 to 0.07% at this site (Table 3).

In the southern Guinea Basin (GBS), we obtained an annual deep-water flux calculated on the basis of only one collection period (GBZ5, Table 1) almost similar to the GBN site (Table 2, Fig. 8), except for the lithogenic components being lower

by approximately a factor of two. This was because the GBN site is influenced by the NE trades which supply aeolian material to the northernmost equatorial sites (Fig. 4). An interannual total flux variation to the upper traps was estimated to 76% (Fig. 8). Bulk sediment accumulation rates in 3931 m depth (Meinecke 1992; GeoB 1105-3) were higher than further north at GBN and amounted to 23 g m^{-2} yr^{-1}(Fig. 8, Table 3). This does not correspond to the upper trap fluxes at GBS which were somewhat lower compared to GBN. However, the deeper trap values, although based only on a one-year deployment (GBZ5), were slightly higher at GBS and, are thus in better accordance with the sediment record (Fig. 8). We obtained a typical open-ocean organic carbon accumulation value of 0.06 g m^{-2} yr^{-1} for the GBS site which is similar to GBN (Table 3). Due to only slightly higher deep-water fluxes of 2.1 gC m^{-2} yr^{-1} , we estimated a similar preservation of 2.9%. The proportion of C_{org}-accumulation versus primary production was 0.09 to 0.07%.

At the Walvis Ridge site, the average deep-water flux value was 31.1 g m^{-2} yr^{-1} with 42% interannual variation (Table 2, 3). In 1989, in the upper water column, fluxes of all components were higher but these values were derived from a one-year sampling period only. Bulk sediment accumulation rates (Schmidt, 1992; GeoB 1028-4, 2214m) were lower by a factor of two compared to the lower trap total fluxes (Fig. 8, Table 3). Opal preservation was around 3% and was thus similar to Cape Blanc. Organic carbon accumulation was 0.06 g m^{-2} yr^{-1} (see Guinea Basin) and was 1.9% of the average deeper flux value (Table 3). This value corresponds to an open-ocean production system (Berger et al. 1989) rather than to a coastal upwelling area. Primary production taken from Berger (1989) amounts to 90-125 gC m^{-2} yr^{-1} resulting in a carbon deposition of only 0.05 to 0.07% which is in the lowest range of all studied sites. However, we have to consider that production in this coastal-open-ocean transition zone is significantly influenced by large filaments streaming offshore (Lutjeharms and Stockton 1987) which result in a substantial spatial and temporal variabilty of biomass, production and fluxes.

Site	SI upper traps	SI lower traps	Production system	Estimated Prim. Prod. after Berger (1989) g m-2 yr-1	Carbon flux upper traps g m-2 yr-1	Carbon flux lower traps g m-2 yr-1	Carbon accumulation g m-2 yr-1	Characterization (production system) Berger et al. 1989	Characterization (carbon deposition) Berger et al. 1989
	Berger and Wefer (1990)		Berger and Wefer (1990)					Berger et al. 1989	Berger et al. 1989
Cape Blanc	3.3	1.3-2.8	constant to sinusoidal	90-125	2.6	1.6	0.12	open-ocean - coastal	open-ocean - coastal
Guinea Basin (north)	2.2-2.8	2.0	sinusoidal	70-90	2.5	2.0	0.06	open-ocean	open-ocean
Guinea Basin (south)	3.0-4.3	2.0	sinusoidal to highly peaked	70-90	1.7	2.1	0.06	open-ocean	open-ocean
Walvis Ridge	3.5	2.0-3.7	sinusoidal	90-125	4.6	3.2	0.06	open-ocean - coastal	open-ocean

Table 4. Summary of seasonality indices (Berger and Wefer 1990), primary production (Berger 1989) and carbon flux data. A characterization of the investigation areas is indicated according to literature-derived total biological production and the carbon deposition (carbon accumulation versus deep-water fluxes and primary production, see Table 3).

Summary and conclusions

In Table 4 the seasonal variations (SI index, Berger and Wefer 1990), primary production, mean annual carbon fluxes and the C_{org}-accumulation in the underlying sediments are summarized. In addition, the sites were characterized according to the production system and the carbon deposition in the sediments (Berger et al. 1989). We observed that seasonality was highly variable both regionally and interannually (SI =1.3-4.3, that means constant to highly peaked production, Berger and Wefer 1990) and obtained the highest seasonality index values deduced from the lower traps for the two coastal mesotrophic sites off Cape Blanc and at the Walvis Ridge. However, in the southern Guinea Basin, one year (1989) was characterized by a very high SI of 4.3 (upper trap) corresponding to a highly peaked production system (Berger and Wefer 1990).

We obtained an exponential decrease of the organic carbon fluxes with depth which was strongest at the mesotrophic Walvis Ridge 2 site and lowest at the oligotrophic Western Equatorial Atlantic 1 site. The average annual deep-water carbon flux was not related to annual primary production taken from literature (Berger et al. 1989). However, we have to consider that these values are estimates and that our study sites were located in highly variable production regimes, in particular, those sites located adjacent to the coastal upwelling zones which are characterized by several independent production cells and highly variable upwelling filaments streaming far offshore the coast. Carbon accumulation in the sediment was not correlated either to biological production nor to the deep-water fluxes. This may be due to lateral transport processes in the deeper water column (especially adjacent to the coastal upwelling sites) and at the seafloor.

Acknowledgements

We thank the crews of RV Meteor and RV Polarstern for their help during the deployments and recoveries of the moorings. We are further indepted to M. Scholz, V. Diekamp, C. Slickers for laboratory work and S. Neuer for helpful comments and suggestions to the manuscript. We would also like to thank V. Ittekkot and W.H. Berger for their helpful suggestions to improve the manuscript. This research was funded by the Deutsche Forschungsgemeinschaft (Sonderforschungsbereich 261 at Bremen University, Contribution 99).

References

Banse K (1990) New views on the degradation and disposition of organic particles as collected by sediment traps in the open sea. Deep-Sea Research 37:1177-1195

Berger WH (1989) Global maps of ocean productivity. In: Berger WH, Smetacek VS, Wefer G (eds) Productivity of the oceans: present and past. Wiley, Chichester, pp 429-455

Berger WH, Smetacek VS, Wefer G (1989). Ocean productivity and paleoproductivity-an overview. In: Berger WH, Smetacek VS, Wefer G (eds) Productivity of the oceans: present and past. Wiley, Chichester, pp 1-34.

Berger WH, Wefer G (1990) Export production: seasonality and intermittency, and paleoceanographic implications. Palaeoceanogr, Palaeoclimatol, Palaeoecol 89:245-254

Bickert T (1992). Rekonstruktion der spätquartären Bodenwasserzirkulation im östlichen Südatlantik. Berichte, Fachbereich Geowissenschaften 27:1-205

Biscaye PE, Csanady GT, Falkowski PG, Walsh JJ (eds) (1994). Shelf edge exchange processes in the southern middle Atlantic Bight. Deep-Sea Res 43 (2/3): 703 p

Bishop JKB, Stephien JC, Wiebe PH (1987) Particulate matter distributions, chemistry and flux in the Panama Basin: response to environmental forcing. Progr Oceanogr 17:1-59

Bishop JKB (1989) Regional extremes in particulate matter compostion and flux: effects on the chemistry of the ocean interior. In: Berger WH, Smetacek VS, Wefer G (eds) Productivity of the oceans: present and past. Wiley, Chichester, pp 117-137

Boltovskoy D, Oberhänsli H, Wefer G (1996) Radiolarian assemblages in the eastern tropical Atlantic: patterns in the plankton and in sediment trap samples. J Mar Systems (8, 31-51)

Cepek M., Wefer G (1996) Seasonal distribution of recent coccolithophorids in a sediment trap at the Walvis Ridge. Mar Micropal (in press)

Deuser WG, Muller-Karger FE, Hemleben C (1988) Temporal variations of particle fluxes in the deep Subtropical North Atlantic: Eulerian versus Langrangian effects. J Geophys Res 93:6857-6862

Fischer G, Wefer G (1991) Sampling, preparation and analysis of marine particulate matter. In: Hurd DC, Spencer DW (eds) The analysis and characterization of marine particles, Geophysical Monograph, Vol 63, pp 391-397

Fischer G, Wefer G (1995) Downward particulate matter fluxes in the eastern and western equatorial Atlantic: a comparison. In: Tsunogai, Iseki K., Koike I, Oba T. (eds) Proceedings of the IGBP Symposium on "Global Fluxes of Carbon and its related Substances in the Coastal Sea-Ocean-Atmosphere System. Hokkaido University, Sapporo, Japan, November 1994, pp 317-331

Fischer G, Wefer G (1996) Seasonal and interannual particle fluxes in the eastern equatorial Atlantic from 1989 to 1991: ITCZ migrations and upwelling. In: Ittekkot V, Schäfer P, Honjo S, Depetris PJ (eds) Particle Flux in the Ocean, SCOPE workshop, J. Wiley and sons, pp 199-214

Fischer G, Donner B, Ratmeyer V, Davenport R, Wefer G (1996) Distinct year-to-year flux variations off Cape Blanc during 1988-1991: relationship to δ^{18}O-deduced sea-surface temperatures and trade winds. J Mar Res 54:1-27

Fütterer D (1983) The modern upwelling record off northwest Africa. In: Thiede J, Suess E (eds) Coastal Upwelling (Part B: Sedimentary records of ancient coastal upwelling, Plenum Press, New York, pp 105-121

Gardner WD, Richardson MJ (1992) Particle export and resuspension fluxes in the western North Atlantic. In: Rowe GT, Pariente V (eds) Deep-sea food chains and the global carbon cycle. Kluwer Academic Publishers, Amsterdam, pp 339-364

Herbland A (1983) Le maximum de chlorophylle dans l'Atlantique tropical oriental: description, écologie, interpretation. Océanographie tropicale 18:249-293

Kalberer M, Fischer G, Pätzold J, Donner B, Segl M, Wefer G (1994) Seasonal sedimentation and stable isotope records of pteropods off Cape Blanc. Marine Geology 113:305-320

Koopmann B (1979) Saharastaub in den Sedimenten des subtropisch-tropischen Nordatlantik während der letzten 20,000 Jahre. Ph. D. Thesis, Universität Kiel, 107 p

Lange CB, Treppke UF, Fischer G (1994) Seasonal diatom patterns in the Guinea Basin and their relationships to trade winds, hydrography and upwelling events. Deep-Sea Research 41:859-878

Longhurst A (1993) Seasonal cooling and blooming in tropical oceans. Deep-Sea Res 40:2145-2165

Lorenzen CJ, Welschmeyer NA, Copping A (1983) Particulate organic carbon flux in the subarctic Pacific. Deep-Sea Res 30:639-643

Lutjeharms JRE, Stockton PL (1987) The extent and variability of southeast Atlantic upwelling. South African Journal of Marine Res 5:35-49

Meinecke G (1992) Spätquartäre Oberflächenwassertemperaturen im östlichen Äquatorialen Atlantik. Berichte, Fachbereich Geowissenschaften 29:1-181

Mittelstaedt E (1991) The ocean boundary along the northwest African coast. Progr Oceanogr 26:307-355

Müller PJ, Schneider R (1993) An automated leaching method for the determination of opal in sediments and particulate matter. Deep Sea Res 40:425-444

Pace ML, Knauer G, Karl DM, Martin JH (1987) Primary production, new production and vertical flux in the eastern Pacific ocean. Nature 325:803-804

Schemainda R, Nehring D, Schulz S (1975) Ozeanologische Untersuchungen zum Produktionspotential der nordwestafrikanischen Wasserauftriebsregion 1970-1973. Geodätische und geophysikalische Veröffentlichungen 4:1-88

Schemainda R, Nehring D, Schulz S (1976) Ozeanologische Untersuchungen im tropischen Nordatlantik auf 30°W zwischen 2°N-15°N. Geodätische und Geophysikalische Veröffentlichungen 17:1-33

Schmidt H (1992) Der Benguela-Strom im Bereich des Walfisch-Rückens im Spätquartär. Berichte, Fachbereich Geowissenschaften 28:1-172

Servain J, Picaut J, Busalacchi AJ (1985) Interannual and seasonal variability of the tropical Atlantic ocean depicted by sixteen years of sea-surface temperature and wind stress. In: Nihoul JCJ (ed) Coupled ocean-atmosphere model. Elsevier Oceanography Service 40, Oxford, New York, Toronto, pp 211-237

Treppke UF, Lange CB, Wefer G (1996) Vertical fluxes of diatoms and silicoflagellates in the eastern equatorial Atlantic, and their contribution to the sedimentary record. Marine Micropaleont (28, 73-96)

Treppke UF, Lange CB, Donner B, Fischer G, Ruhland G, Wefer G (1996b) Diatom and silicoflagellate fluxes a the Walvis Ridge: an environment controlled by coastal upwelling in the Benguela system. J Mar Res (in press)

Van Camp L, Nykjaer L, Mittelstadt E, Schlittenhardt P (1991) Upwelling and boundary circulation off Northwest Africa as depicted by infrared and visible satellite observations. Progr Oceanogr 26:357-402

Voituriez B, Herbland A (1981) Primary production in the tropical Atlantic ocean mapped from the oxygen values of EQUALANT 1 and 2 (1963). Bull

Mar Sci 31:853-863

Walsh I, Dymond J, Collier R (1988a) Rates of recycling of biogenic components of settling particles in the ocean derived from sediment trap experiments. Deep-Sea Res 35:43-58

Walsh I, Fischer K, Murray D, Dymond J (1988b)

Evidence for resuspension of rebound particles from near-bottom sediment traps. Deep-Sea Res 35:59-70

Wefer G, Fischer G (1993) Seasonal patterns of vertical particle flux in equatorial and coastal upwelling areas of the eastern Atlantic. Deep-Sea Res 40:1613-1645

Silica Signals in the South Atlantic

A.J. van Bennekom

Netherlands Institute for Sea Research
PoBox 59, 1790AB Den Burg, Texel, HOLLAND

Abstract: Distribution patterns of potential temperature (θ) and of H_4SiO_4 concentrations show that the bottom water in the major part of the Angola Basin originates from the Romanche Fracture Zone. At present the bottom water of the Angola Basin is predominantly lower NADW with a 10-20 % admixture of a southern component. South of about 29°S a gradually increasing percentage of the bottom water originates from the Cape Basin, entering through gaps in the Walvis Ridge. Near the deep entrances into the Guinea Basin and the southern tip of the Angola Basin, vertical gradients of temperature and salinity are large. Small changes in the relative production rates of AABW and NADW, assuming that their properties remain the same, could cause a large change in bottom water composition of the Guinea and Angola Basins.

The concentrations of bio-SiO_2 in sediments and of H_4SiO_4 in interstitial waters increase in the NE Angola Basin and parts of the Guinea Basin. This reflects enhanced productivity by diatoms, caused by various kinds of upwelling. Upward diffusion of H_4SiO_4 from interstitial waters causes a H_4SiO_4 "excess" of up to 10 μM in deep waters above the Zaire deep sea fan.

Bio-SiO_2 in the sediments of the Zaire deep-sea fan contains up to 6% of aluminium, which contracts the amorphous structure and lowers its solubility and dissolution rate, thereby greatly increasing its preservation. The incorporation of Al into bio-SiO_2 mainly occurs in the surface sediment. Gibbsite and possibly kaolinite are considered likely sources of the Al.

Introduction

The present paper discusses distribution patterns of dissolved silica in deep and bottom waters of the Guinea, Angola and Cape Basins, both in relation with water masses and with biogenic silica in sediments. For the water column it is an "update" of an earlier study by van Bennekom and Berger (1984), taking into account water column data collected in 1983 during the Ajax (AJAX 1985) and the RV Oceanus-133 expeditions, discussed by Warren and Speer (1991), and some data from SAVE expeditions (SAVE, 1988), and part of the results (both water column and sediments) from a Dutch expedition with RV Tyro in 1989 (Jansen et al. 1990).

The South Atlantic is a transit area for 3 water masses: Antarctic Intermediate Water (AAIW, minimum salinity, core at 750-1000 m depth), the North Atlantic Deep Water complex, divided (Wüst 1933) in Upper-NADW (maximum salinity, core at 1500-2000 m depth) and Lower-NADW (maximum oxy-

gen concentration, core at about 3500 m) and Antarctic Bottom Water (AABW). In the western basins properties are more extreme because substantial parts of the transports are concentrated in Deep Western Boundary Currents along South America. Deep and bottom waters are mixtures of L-NADW and AABW. Below the TDD (Two Degree Discontinuity) layer, shoaling from 3800m depth at the equator to 3200m at 20°S (Broecker et al. 1976), the percentage of AABW increases steeply, but the TDD layer also contains AABW (11%), recirculated through the NW Atlantic.

The deepest connection between the western and eastern Atlantic is the Romanche Fracture Zone, recently shown by Mercier et al. (1995) to have a controlling sill depth of 4350 m, well below the TDD. Above this sill the water has a salinity of 34.778 (PSU) and 0.92°C potential temperature (θ). Downstream, mixing with waters from above in-

From WEFER G, BERGER WH, SIEDLER G, WEBB DJ (eds), 1996, *The South Atlantic: Present and Past Circulation.* Springer-Verlag Berlin Heidelberg, pp 345-354

creases these values, untill 34.845 and 1.5°C at the entrance of the Guinea Basin, 400 km more to the east. In the Guinea Basin some stratification in deep waters, with decreasing θ and salinity and increasing H_4SiO_4 to the bottom remains, while the deep waters of the Angola Basin are nearly homogeneous (van Bennekom and Berger 1984; Warren and Speer 1991). With salinities of 34.67 for AABW and 34.90 for TDD water (Broecker et al. 1976), the average salinities of 34.86 and 34.88 in bottom waters of Guinea and Angola Basins respectively, would mean 18 and 9% of AABW. Taking into account that TDD water contains 11% of recirculated AABW, the total amounts of AABW in bottom waters of the Guinea and Angola Basin are 28 and 19%. The rest is formed in the North Atlantic Ocean.

"Dissolved silica" (in fact silicic acid, H_4SiO_4) is not a completely conservative tracer for bottom waters, but it will be shown that in the central parts of the Angola Basin, where the sediments contain little bio-SiO_2, H_4SiO_4 can be used as an additional water mass tracer. On the other hand, the sediments in the NE Angola Basin contain more bio-SiO_2, which by dissolution increase H_4SiO_4 concentrations in pore waters, causing a flux into bottom waters. This "excess" H_4SiO_4 in deep waters is read from graphs of H_4SiO_4 against a conservative property, for which θ is chosen.

Results

Water column

From the Romanche sill to about 29°S, θ in bottom waters gradually increases (Fig. 1), while H_4SiO_4 decreases in the Guinea Basin and remains essentially constant in the bottom waters of the central Angola Basin (Fig. 2). This means that the bottom waters in the major part of the Angola Basin originate from the Romanche Fracture Zone, via the Guinea Basin. This is not so clear at shallower depths. Warren and Speer (1991) gave evidence that the NADW layer from 2-4 km depth is transported northward in the Angola Basin, along the Mid Atlantic Ridge.

According to the revised edition of the GEBCO map 5.12 (1994) the deepest saddle (> 4500 m) of

the Guinea Rise, separating the Guinea and Angola Basins is found at about 4°S, 1°E. This is supported by the fact that the lowest θ in bottom waters (1.89°C) and the lowest salinity is found in the extreme NE corner of the Angola Basin. The AJAX section does not pass over this saddle. The behaviour of H_4SiO_4 is influenced by a measurable silica excess (discussed below).

South of 29°S, θ decreases and H_4SiO_4 increases, which has been attributed to transport of AABW from the Cape Basin into the southern tip of the Angola Basin (Connary and Ewing 1974; van Bennekom and Berger 1984). The revised edition of GEBCO chart 5.12 shows a ridge <4000 m at about 33.5°S, which apparently prevents bottom waters of θ <1.5 and H_4SiO_4 > 70 µM to penetrate further north. Shannon and Chapman (1991) found Cape Basin water with lower θ and higher H_4SiO_4 at 32-33°S in the southern tip of the Angola Basin, transported via the "Walvis Kom" through a sinuous gap in the Walvis Ridge. All waters below 3000 m depth in the southern tip of the Angola Basin fall on the same straight line in a H_4SiO_4-θ graph (Fig. 3), only slightly different from the H_4SiO_4-θ relationship in the Cape Basin. The difference could be caused by vertical mixing with waters from higher levels leaking through gaps in the Mid Atlantic Ridge.

The AABW influence clearly extends to 30°S, St 217 of the SAVE expedition (SAVE, 1988) but not further north, At 27°S, H_4SiO_4-θ graphs are, from 3000 m to the bottom, identical with stations in the central Angola Basin (Fig. 3), therefore I conclude that no deep water from the southern tip reaches into the Angola Basin north of 27°S. It seems unlikely to me that there could be a hitherto unknown sill at 28-29°S, separating the Angola Basin from its southern tip, as suggested by Warren and Speer (1991). It seems more likely, as also mentioned by these authors that the southwesterly circulation of deep waters along the Walvis ridge blocks northward penetration of AABW influence. Deep waters from the Angola Basin penetrate into the Cape Basin at 2-3000 m depth through the Namib Col (Speer et al. 1995) and may also recir-culate via the Walvis Kom through an about 3500 m deep saddle at 30°S. The intersection of H_4SiO_4-θ graphs in Fig. 3 for the southern

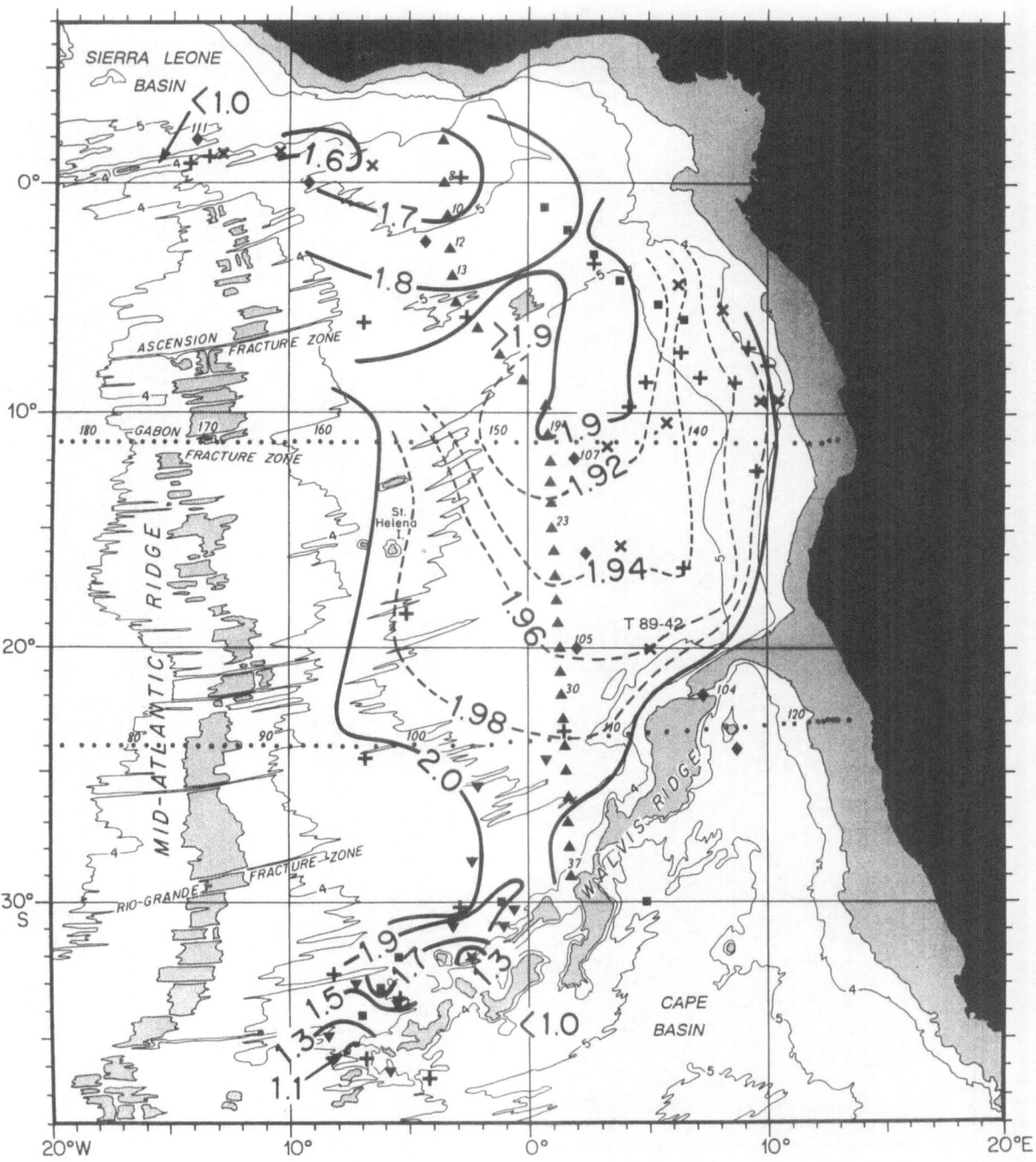

Fig. 1. Potential temperature in bottom waters of the Guinea and Angola Basin. Isobaths from Warren and Speer (1991) Station positions: Geosecs (◆); Ajax (▲); Oceanus-133 (●); Save (■); Connary and Ewing (▼) Shannon and Chapman (⋆); Tyro 1978-1980 (✚); Tyro 1989 (✖).

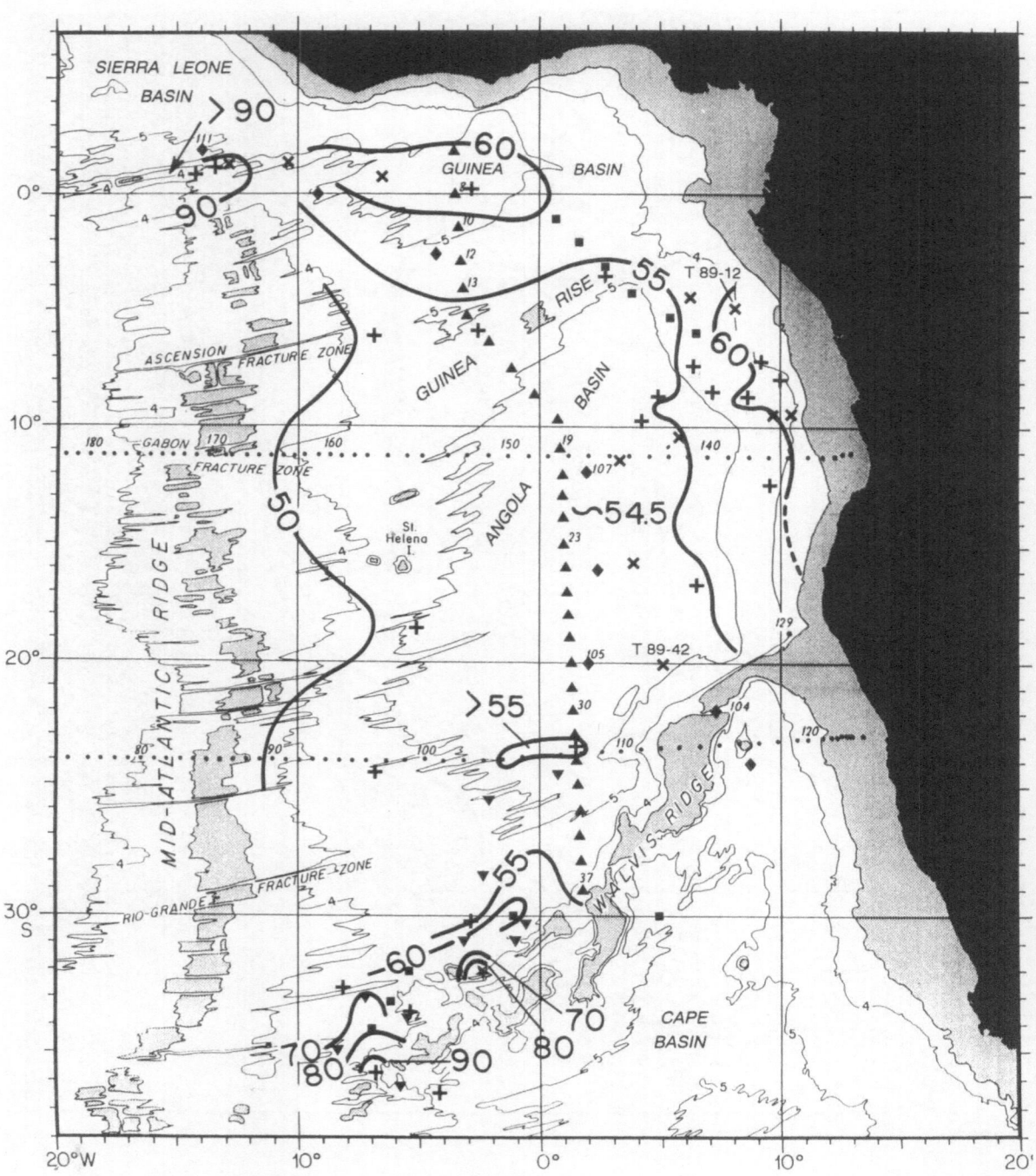

Fig. 2. Concentrations of H_4SiO_4 (μM) in bottom waters of the Guinea and Angola Basin. Stations and symbols as in Fig. 1.

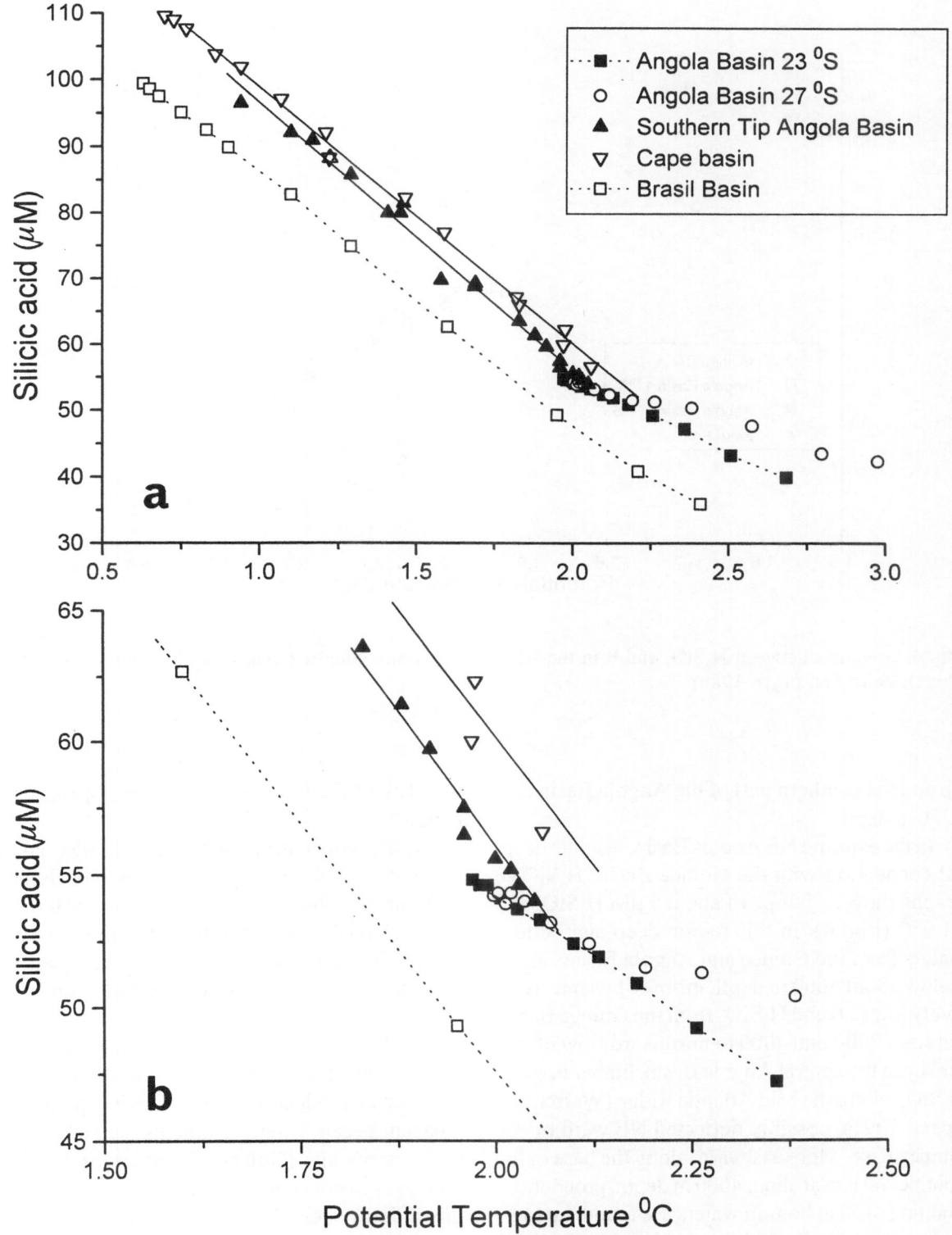

Fig. 3. Relation between H_4SiO_4 and θ in the southern Angola Basin and adjacent areas. (b is an enlargement of a).

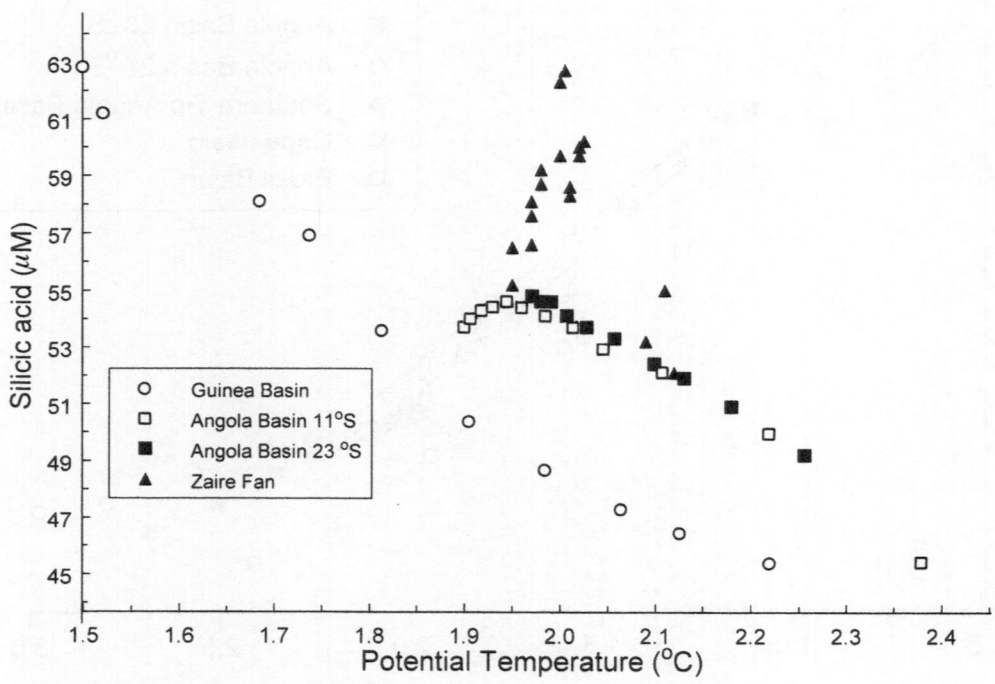

Fig. 4. Relation between H_4SiO_4 and θ in the NE Angola Basin. Guinea Basin; Angola Basin 11 and 23°S (Ajax); Zaire Fan (Tyro 1989)

tip and the southern part of the Angola Basin is at 3250 m depth.

In the extreme NE Angola Basin, near the deepest connection with the Guinea Basin, H_4SiO_4-θ graphs show a "jump" of about 5 μM H_4SiO_4 at θ >1.9°C (Fig. 4). In this region deep and bottom waters from the Guinea and Angola Basins meet: Below about 4000 m depth inflow of waters, relatively low in θ and H_4SiO_4 from the Guinea Basin, between 2000 and 4000 m northward flow of waters from the central Angola Basin, higher in θ and H_4SiO_4, along the Mid Atlantic Ridge (Warren and Speer, 1991), possibly deflected NE ward by the Guinea Rise. More eastward, along the base of the continental rise at about 4000 m depth, pronounced maxima of Si in bottom waters are found: up to 64 μM at 5°S (van Bennekom and Berger 1984), 60 μM at 9.4°S (Tyro 89 results), 61 μM at 11°S (Warren and Speer 1991) and 59 μM at 18°S (sta-

tion 129 of the Oceanus 133 cruise, Speer, pers. comm.).

H_4SiO_4 concentrations > about 55 μM have a "silica excess" due to in situ processes, more H_4SiO_4 at the same θ than in either the Guinea or the Angola Basin (Fig. 4). Quantitatively, this excess is uncertain; reference graphs vary between the H_4SiO_4 - θ graphs for the Angola and Guinea Basins.

The H_4SiO_4 excess in bottom waters was attributed by van Bennekom and Berger (1984) to the high diatom production in the Zaire river plume and adjacent ocean, where sediments are rich in biosiliceous remains (Goll and Bjørklund 1974). This increases Si concentrations in pore waters and by upward diffusion also in bottom waters, compatible with the data at 5-8°S. High diatom production is not known at 9.4 and 11°S, well outside the Zaire river plume. Maybe the high H_4SiO_4 at these latitudes is advected from the North; a deep poleward undercurrent is also suggested by Gordon and Bosley (1991) from the distribution of salinity maxima. Another possibility is that H_4SiO_4 maxima in bottom waters at the base of the continental rise is a more general phenomenon. Recirculation of deep wand bottom waters and vertical mixing distribute the excess of H_4SiO_4 to the central Angola Basin.

Puzzles in paleo-interpretations

The layering of water masses at the northern and southern entrances to the Guinea and Angola Basins suggests that small changes in the strength of the northern and southern Atlantic deep water sources could have a marked influence on the composition of bottom waters in these basins (van Bennekom and Berger 1991). The present-day layering involves minute differences in specific gravity, caused by subtle details in climatic conditions in polar regions and various recirculations. One may safely assume that the layering will be different for different climatic conditions, such as prevailed during the Last Glacial Maximum.

Sediments

Interstitial waters

Box cores were subsampled, sliced and squeezed at in situ temperatures. In 1989 resolution was 1 cm, results in van Bennekom and Berger (1984) were based on a resolution of about 3 cm, which is too coarse, especially for siliceous sediments. Pore waters in the upper cm of the Zaire fan sediments contain 210-250 µM H_4SiO_4; 70-90 µM is found in the sediments of the southern and western Angola Basin. The gradients from 1 to 5 cm depth are approximately linear and surprisingly constant: 25-35 µM/cm. This indicates little dissolution below 1 cm depth, which is also concluded from profiles of bio-SiO_2 and the low dissolution rates in sediments of the Zaire fan (van Bennekom et al. 1989).

The results of station T89-12 in Fig. 5 show the difficulties in calculating the steep gradient of H_4SiO_4 (and hence the flux) over the water-sedi-

ment interface accurately with these techniques. Most of the bio-SiO_2 dissolution apparently occurs at the interface. For the calcareous sediments with little accumulation of bio-SiO_2 as for St T89-42, the jump at the interface is much less pronounced (Fig. 5). This gives less ambiguity for the interface concentration and more accurate flux calculations.

It proved impossible to calculate fluxes over the sediment/water interface with a one-dimensional dissolution-diffusion model using a constant dissolution rate. The jump from bottom water to the uppermost cm of the interstitial water concentration is not well represented. It must be concluded that the dissolution rate decreases when biosiliceous particles are incorporated in the sediments. This was also concluded for the equatorial Pacific by McManus et al. (1995). An approximate estimate of the dissolution flux at the interface, 50 µmol. cm^{-2} .a^{-1}, is obtained from the gradient over the upper cm for T89-12. This is about 10 times more than the dissolution flux for 1-5 cm depth, and comparable with fluxes from diatomaceous sediments in e.g. the Southern Ocean (Rutgers van der Loeff and van Bennekom 1989).

Biogenic silica

It has been and is difficult to determine bio-SiO_2 in sediments, because different methods do not give the same results. It may be impossible to find a chemical leaching agent that completely separates bio-SiO_2 of all species and ages from all clay minerals, as indicated by the results of Mortlock and Froelich (1989) and Müller and Schneider (1993). This especially holds for the Zaire deep-sea fan, where bio-SiO_2 is very insoluble (van Bennekom et al. 1989). Our experience has been that not all bio-SiO_2 dissolves after 7 hours leaching at 85°C with 0.5 M sodium carbonate, but that a "strong" leach with 1M NaOH at 85°C dissolves some clay minerals nearly completely after 2 hours, as appears from XRD spectra (van der Gaast, personal communication).

Surface sediments of the Zaire Fan contain 30 to 40% amorphous silica, determined from the bulge in XRD spectra (van der Gaast and Jansen 1984; van der Gaast, pers. comm. 1995). Müller and Schneider (1993) found only 8% in the surface

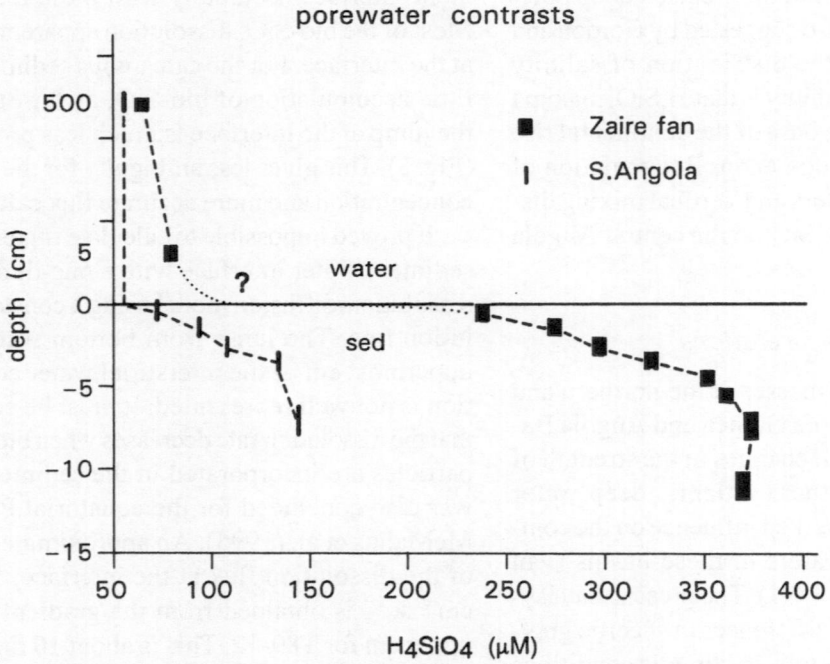

Fig. 5. Concentrations of H_4SiO_4 near the sediment water interface at stations T89-12, 4100 m depth Zaire fan 5°12'S; 7°58.4'E and T89-42, 5300 m depth; 20°S; 4°45' E.

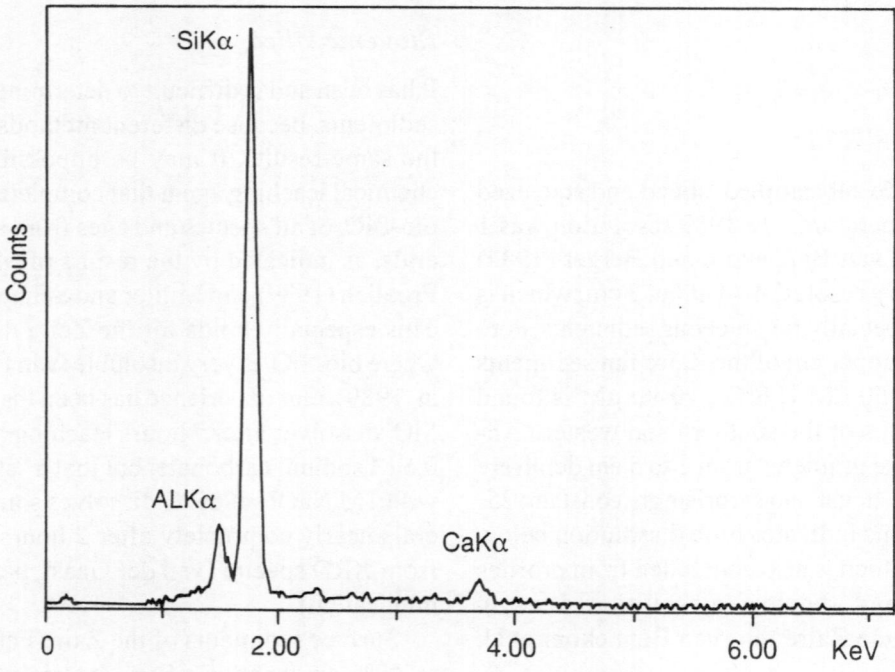

Fig. 6. Typical chemical composition of bio-SiO_2 from the Zaire fan, obtained by microprobe and energy dispersive recording.

Table 1. Atomic Si/Al ratios in siliceous skeletons from surface (0-1 cm) sediments of the Zaire Fan, water depths 4000 - 4500 m

Diatoms	*Thallassionema nitzschoides*	0.19	$\pm$ 0.03	(N=10)
	Cyclotella sp	0.14	$\pm$ 0.01	(N=8)
	Centricates	0.16	$\pm$ 0.03	(N=8)
	Silicoflagellates/Radiolarians	0.12	$\pm$ 0.02	(N=6)

sediments at 3000 m depth by leaching with hot 1M NaOH; we found about 10% with NaOH for St T89-28 (5400 m depth), but only 3% at St T89-16 (825 m). Further work, comparing different methods for the same samples from the Zaire fan is in progress; preliminary results show that large differences between the various methods remain. The silica budget given by van Bennekom and Berger (1984) is inaccurate.

Part of these problems might be caused by the unusual chemistry, structure and properties of bio-SiO_2 in the Zaire deep sea fan sediments. Van Bennekom et al. (1989) found up to 6% of Al and some Ca or K in bio-SiO_2 of all diatom and radiolarian species (e.g. Fig. 6). Part of this Al is structurally incorporated, hence the Ca and K, which compensate for the negative charge when Al substitutes Si. This contracts the amorphous structure which has a low solubility and a very low dissolution rate.

To detect variations of Al-concentrations in different siliceous organisms, they were investigated with a microprobe, coupled to a crystal spectrometer (Table 1). Only Si and Al were measured; an Al/Si ratio of 0.16 is equivalent to 6% of Al in bio-SiO_2 (van Bennekom et al. 1989). Variations are small: the thick frustules of *Cyclotella* sp and the thick skeletons of silicoflagellates and radiolarians have only slightly lower Al/Si than the thinner frustules of *Thallassionema nitzschoides* and centricate diatoms. This indicates that incorporation of Al is a quick process, since the bio-SiO_2 in the upper cm of sediment contains nearly the same amount of Al as in deeper levels investigated by van Bennekom et al. (1989).

The aluminium needed has to be dissolved from other minerals and subsequently adsorbed and incorporated into bio-SiO_2. Gibbsite is the most likely candidate; it is advected by the Zaire river and in some areas of the Zaire fan it completely disappears from the sediments (van der Gaast, personal communication).

Relations between fluxes and silica excess

In the southern and western Angola Basin fluxes of H_4SiO_4 from pore waters range from 0.1 to 0.5 $\mu mol.cm^{-2}.a^{-1}$. Over a depth of 1000 m, where H_4SiO_4 is about constant, this gives an increase of 0.001 to 0.005 μM per year and it takes at least 100 years before this excess is significant. On the other hand, minimum fluxes in the Zaire Fan range from 20 to 50 $\mu mol.cm^{-2}.a^{-1}$. The excess in the water column is found over a depth range of 400 m, the maximum excess of 10-15 μM can result after 10-30 years.

Conclusions

- In most of the Angola Basin H_4SiO_4 can be used as a water mass tracer along with θ and salinity; around the Zaire deep-sea fan upward diffusion from siliceous oozes increases the bottom water concentration significantly.
- Between 37 and 29°S Cape Basin waters enter the southern Angola Basin, but they do not penetrate further north.
- In the Zaire fan sediments, bio-SiO_2 quickly incorporates Al and Ca. This causes a low solubility and dissolution rate.

- In modelling the flux of H_4SiO_4 across the sediment/water interface the decrease of the dissolution rate of bio-SiO_2, going from the surface to the first cm of sediment has to be taken into account. Dissolution at the very surface is an order of magnitude larger than the "supported" flux from within the sediment.

Acknowledgements

I thank Dr. F. Jansen, chief scientist during the Tyro 1989 expedition for cooperation on board, and the Netherlands Council for Oceanic Research for financial support. R. Kloosterhuis is acknowledged for the silicic acid concentrations during the Tyro 1989 expedition and A.C.M. Clerkx for the Al/Si ratios by crystal spectrometry. B. Verschuur made the drawings. The index map of Fig. 1 was obtained by courtesy of Dr. B.A. Warren. The paper benefitted from discussions with Drs. J. Reid and W.H. Berger.

This is NIOZ-publication no. 3008.

References

AJAX (1985) Physical, chemical and in situ CTD data from the Ajax expedition in the South Atlantic Ocean. SIO/TAMU 275 pp. Unpublished data report

Broecker WS, Takahashi T, Li YH (1976) Hydrography of the central Atlantic I. The two-degree discontinuity. Deep-Sea Res 23:1083-1104

Connary SD, Ewing M (1974) Penetration of Antarctic Bottom Water from the Cape Basin into the Angola Basin. J Geophys Res 79:463-469

Goll RM, Bjørklund KR (1974) Radiolaria in surface sediments of the South Atlantic. Micropal 20:38-75

Gordon AL, Bosley KT (1991) Cyclonic gyre in the tropical South Atlantic. Deep-Sea Res 38 Suppl 1:S323-S343

Jansen JHF, De Lange GJ, van Bennekom AJ (1990) (Pale)oceanography and Geochemistry of the Angola Basin. NIOZ unpubl. rep. 1990-4:1-63

McManus J, Hammond DE, Berelson WM, Kilgore TE, DeMaster DJ, Ragueneau OG, Collier RW (1995) Early diagenesis of biogenic opal: dissolution rates, kinetics, and paleoceanographic implications. Deep-Sea Res II 42:871-903

Mercier H, Speer KG, Honnorez J (1994) Flow pathways of bottom water through the Romanche and Chain Fracture Zones. Deep-Sea Res 41:1457-1478

Mortlock RA and Froelich PN (1989) A simple method for the rapid determination of biogenic opal in pelagic marine sediments. Deep-Sea Res 36:1415-1426

Müller PJ and Schneider R (1993) An automated leaching method for the determination of opal in sediments and particulate matter. Deep-Sea Res 40:425-444

Rutgers van der Loeff MM, van Bennekom AJ (1989) Weddell Sea contributes little to silicate enrichment in Antarctic Bottom Water. Deep-Sea Res 36:1341-1357

SAVE (1992) Chemical, Physical, and CTD Data Reports, of the South Atlantic Ventilation Experiment. Legs 1, 2, 3 and Legs 4, 5. ODF Publ 231 and 232, SIO ref's 92-9, 729 pp and 92-10, 625 pp, La Jolla, USA

Shannon LV, Chapman P (1991) Evidence of Antarctic Bottom Water in the Angola Basin at 32°S. Deep-Sea Res 38:1299-1304

Speer KG, Siedler G, Talley L (1995) The Namib Col Current. Deep-Sea Res I 42:1933-1950

van Bennekom AJ, Berger GW (1984) Hydrography and silica budget of the Angola Basin. Neth J Sea Res 17:149-200

van Bennekom AJ, Jansen JHF, van der Gaast SJ, Van Iperen JM, Pieters J (1989) Aluminium-rich opal: an intermediate in the preservation of biogenic silica in the Zaire (Congo) deep-sea fan. Deep-Sea Res 36:173-190

van der Gaast SJ, Jansen JHF (1984) Mineralogy, opal, and manganese of middle and late Quaternary sediments of the Zaire (Congo) deep-sea fan: Origin and climatic variation. Neth J Sea Res 17:313-341.

Warren BA, Speer KG (1991) Deep circulation in the eastern South Atlantic Ocean. Deep-Sea Res 38 Suppl 1:S281-S322

Wüst G (1933) Schichtung und Zirkulation des atlantischen Ozeans I. Das Bodenwasser und die Gliederung der atlantischen Tiefsee. Wiss. Ergebn. dt. atlant. Exped. "Meteor" 6 (I,2): 1-288

On the Bathymetry of the Hunter Channel

J. Pätzold[1], K. Heidland[2], W. Zenk[3] and G. Siedler[3]

[1]Fachbereich Geowissenschaften, Universität Bremen, 28359 Bremen, Germany
[2]Alfred-Wegener-Institut für Polar- und Meeresforschung, P.O. Box 120161,
27515 Bremerhaven, Germany
[3]Institut für Meereskunde an der Universität Kiel, 24105 Kiel, Germany

Abstract: The Rio Grande Rise delineates a large-scale topographic barrier in the South Atlantic separating the Brazil Basin to the north from the Argentine Basin in the south. In addition to the Vema Channel the Hunter Channel represents an important conduit for the equatorward flow of Antarctic Bottom Water. Motivated by the lack of reliable topographic charts from the Hunter Channel Region (34°S, 28°W) we have compiled multibeam soundings from a number of different bathymetric surveys taken on board RV *Meteor*, resulting in a set of three-dimensional images from the region. They are presented in context with remarks on the name "Hunter Channel", together with selected hydrographic observations and long-term near-bottom current meter records.

Introduction

The Rio Grande Rise is the most important barrier to the equatorward flow of Antarctic Bottom Water (AABW) in the western South Atlantic. It originates from hot spot volcanic activity during sea floor spreading of the South Atlantic (Morgan 1972). The primary pathway across this barrier has long been recognized as the Rio Grande Gap (Wüst 1933) or better known today as Vema Channel (Hogg et al. 1996). Here, long-term observations with moored current meters indicate a strong flow of bottom waters with potential temperatures below 2°C (Tarbell et al. 1994). Hogg et al. (1982) calculated a mean northward mass transport of 4 0.4 x 10⁶m³s⁻¹ between the Argentine Basin and the Brazil Basin. Hogg's results have been more recently confirmed by the geostrophic analysis of repeated hydrographic sections between the Brazilian shelf and the Rio Grande Rise (Speer and Zenk 1993; Zenk and Hogg 1996). Stations were sampled by RV *Meteor* within the frame of the Deep Basin Experiment (DBE), a subprogram of the internationally coordinated World Ocean Circulation Experiment (WOCE).

The Hunter Channel is another even more sparsely studied area for the equatorward export of Antarctic Bottom Water between the eastern Rio Grande Rise and the Mid-Atlantic Ridge. While the Vema Channel reveals a channel morphology with a clearly defined sill depth (Zenk et al. 1993) the bathymetry of the Hunter Channel is more complex and is far less known. The Hunter Channel is a poorly defined feature on published maps of the western South Atlantic (IHO/IOC/CHS 1978; Cherkis et al. 1989; NGCD 1993) due to the scarcity of available bathymetric profiles in the area. A bathymetric survey to map the Hunter Channel with the multibeam sonar system HYDROSWEEP was carried out during a number of recent *Meteor* cruises.

Historical Aspects

The term 'Hunter Channel' dates back to 1971 when Burckle and Biscaye first suggested this name in an abstract published by the Geological Society of America (Burckle and Biscaye 1971). However, the bathymetric chart which we believe to be the most up-to-date one of the South Atlantic (Cherkis et al. 1989) still shows the name 'Hunter Gap' instead of a 'Hunter Channel' (Fig. 1).

From WEFER G, BERGER WH, SIEDLER G, WEBB DJ (eds), 1996, *The South Atlantic: Present and Past Circulation.* Springer-Verlag Berlin Heidelberg, pp 355-361

P. Biscaye (1991) communicated the story of finding a name for the Hunter Channel to us. According to this source Biscaye and Dasch (1971) studied the distribution of the strontium isotope ratios on recent sediments in the Argentine and Brazil Basin. They attempted to see "if one could use $^{87}Sr/^{86}Sr$ ratios of bulk sediments as tracers" for sediment transport. In their samples they found indications for sediments of Argentine Basin origin on the northern side of the Rio Grande Rise east of the Vema Channel. Since they expected the Coriolis force to deflect advected sediments towards the western banks of the Vema Channel, there was reason to expect the existence of another outlet for Antarctic Bottom Water farther to the east. This conjecture was confirmed in spring of 1971 during a cruise with RV *Atlantis II* when Burckle and Biscaye found diatoms of Antarctic origin - *Fragilariopsis kerguelensis*, since renamed *Nitschia kerguelensis* - transported in bottom waters of what is now known as the Hunter Channel. Today's name is derived from Hunter College in New York City where geology students looked for the predicted gap in numerous rolls of precision depth recorders as a laboratory exercise. Burckle and Biscaye agreed to name the newly identified channel "for Hunter College whose students had participated in the search for it". The college itself is named after its first president Dr. Thomas Hunter of the Normal College of the City of New York (1870 to 1905, renamed Hunter College in 1914)(Limmer 1987). Despite the obscure first notice of the name Hunter Channel, it got picked up quickly and was published on subsequent bathymetric charts, the first of which was the "Bathymetric Atlas of the Atlantic, Caribbean and Gulf of Mexico" by Uchupi (1971).

Methods

Bathymetric surveys of the Hunter Channel were carried out during *Meteor* expeditions M 15/2 (Pätzold et al. 1993; Siedler and Zenk 1992), M 22 (Siedler et al. 1993), and M 28/2 (Zenk and Müller 1995) using the multibeam sweeping survey echosounder HYDROSWEEP constructed by STN Atlas Elektronik, Bremen. This device is a wide-angle, fanshaped sonar system allowing the two-

dimensional mapping of the ocean topography with high resolution (Gutberlet and Schenke 1989). A large-swath coverage of up to twice the water depth can be accomplished by using a quasi-real-time cross-fan calibration of the measured slant ranges. The system relies on two different measurement modes. In the survey mode the fan-shaped sonar beam is directed perpendicular to the ship's long axis, and in the calibration mode the sonar fan beam parallels the ship. The basic principle of the system is based on a method to calculate mean water sound velocities by comparison of the two measured depth profiles in conjunction with an accurate measurement of the sound velocity under ship's keel. In memoriam to its inventor this patented method is called the Ziese principle.

Precise positioning of the ship is a main prerequisite for the conduction of a bathymetric survey with the HYDROSWEEP system. The navigation of RV *Meteor* was mainly based on the Global Positioning System (GPS) and also the Integrated Navigation System (INS) with Doppler Sonar and TRANSIT satellite fixes. GPS is a military navigation system under the primary resonsibility of the U.S. Department of Defence. The service available to the civil community is called Standard Positioning System (SPS). The accuracy available to SPS users for a single receiver is ± 100 m. With a sophisticated correction procedure, the accuracy with the integrated navigation system is on the order of ± 200 m. Both types of navigation data were integrated and used for final processing. For detailed postprocessing, the measured depths and positional data were stored on magnetic tape. Postprocessing was carried out at the Alfred Wegener Institute for Polar and Marine Research at Bremerhaven. The bathymetric survey of the Hunter Channel area was compiled on the basis of about 110 km of continuous HYDROSWEEP profiling. The survey covers an area of about 800 km^2 between 34°45′S and 34°10′S and 29°W and 26°15′W.

Results

The Hunter Channel is a bathymetric depression located between a narrow ridge extending from the Mid-Atlantic Ridge to the east and the southeast-

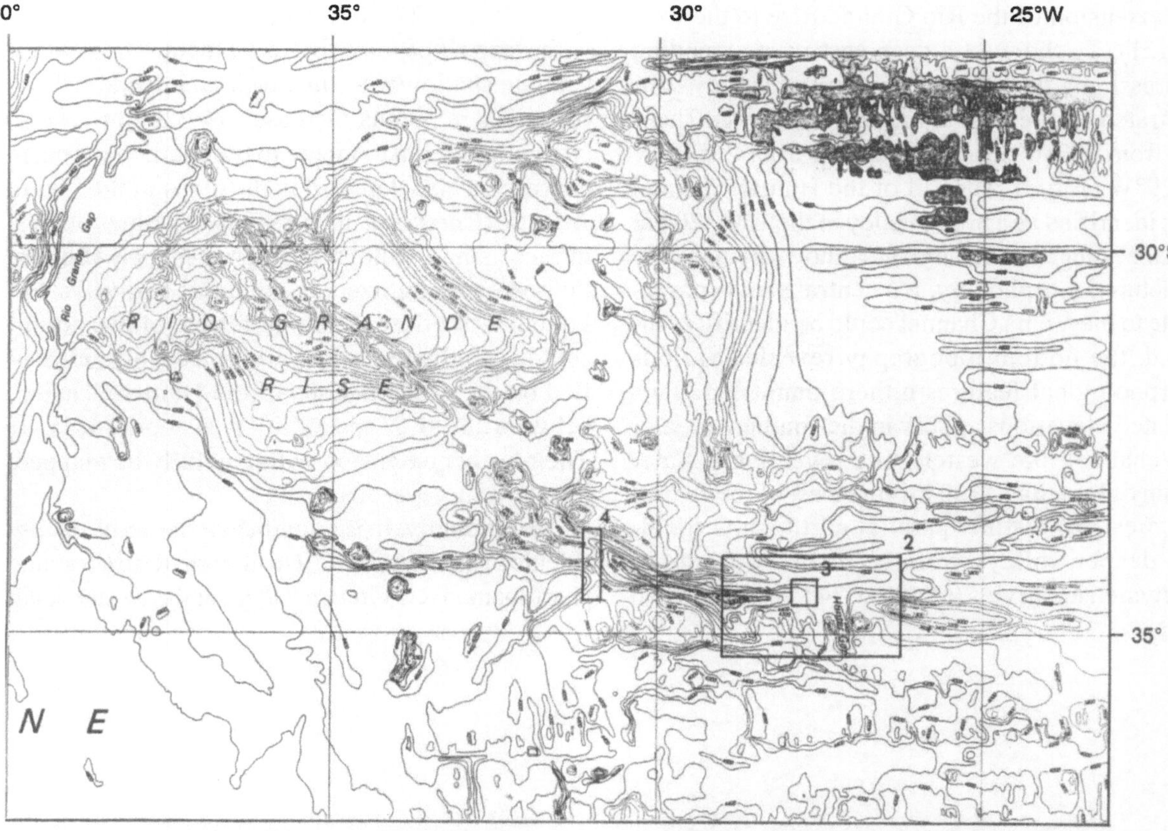

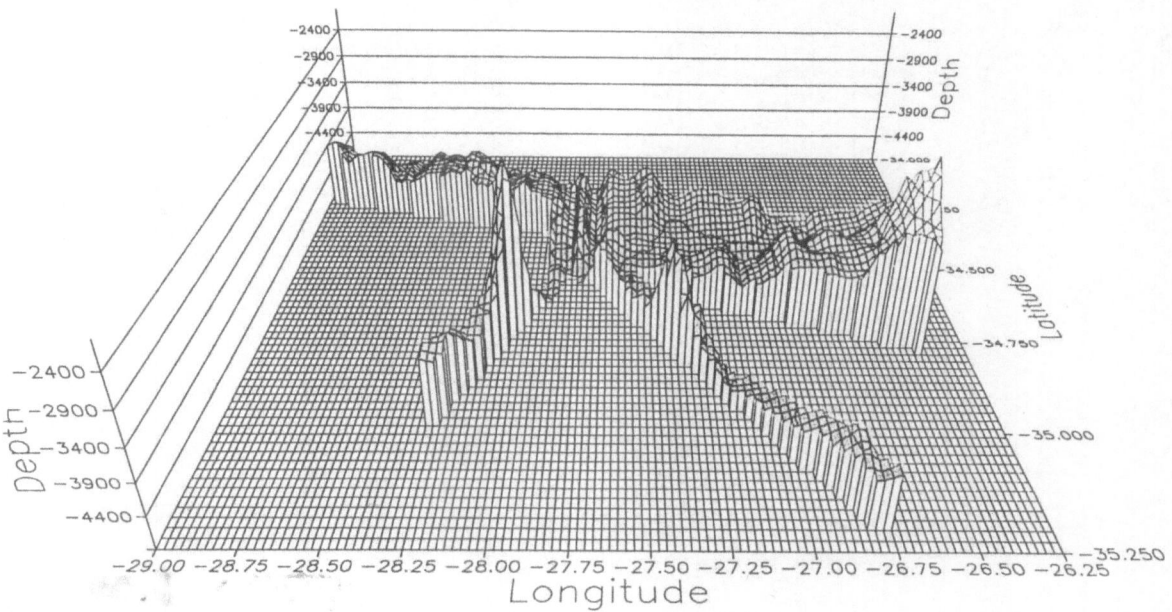

Fig. 1. Bathymetry of the Hunter Channel (here named 'Hunter Gap') as depicted on the bathymetric map of the South Atlantic by Cherkis et al. (1989). Numbers refer to figures 2, 3 and 4.

Fig. 2. Perspective view from the south into the Hunter Channel. Composite picture of bathymetric surveys of various RV *Meteor* cruises.

ern extension of the Rio Grande Rise to the west (Fig. 1). To the west a deep east-west extending depression adjoins the channel. The survey concentrated on the central area of the Hunter Channel from 34°10′S to 35°15′S and from 26°30′W to 29°W (Fig 2). The sill of the Hunter Channel was identified as a broad valley in the 150-220 km distance range. The central area shows an unevenly structured morphology. No central canal comparable to the Vema Channel could be identified. Instead, the bottom topography revealed various morphological features hithero unnoticed. Two isolated elevations appear in the southern area of the channel, the western one revealing a water depth of less than 2300 m. Just west of the central sill area two distinct parallel north-south trending deeper valleys were identified during the bathymetric survey (at about 34°30′S, between 27°35′W and 27°50′W)(Fig. 3). A narrow ridge with steep slopes rises between them. The valleys become shallower to the north and deeper to the south where depths increase to >4700 m. These two valleys seem to merge to the south into a deep depression also depicted in the maps of the South Atlantic (Cherkis et al. 1989; Fig. 1). Since depths in these small valleys were greater than 4000 m they were recognized as potential pathways for bottom water flow. A similar feature of two north-south extending smaller valleys were also identified on the western slope of the Hunter Channel between 28°30′W and 28°40′W at about 34°45′S. Their farther extension will hopefully be mapped in future surveys.

Additionally, in the southeastern meridionally oriented structure of the Rio Grande Rise (in some charts named Rio Grande Ridge) at the western end

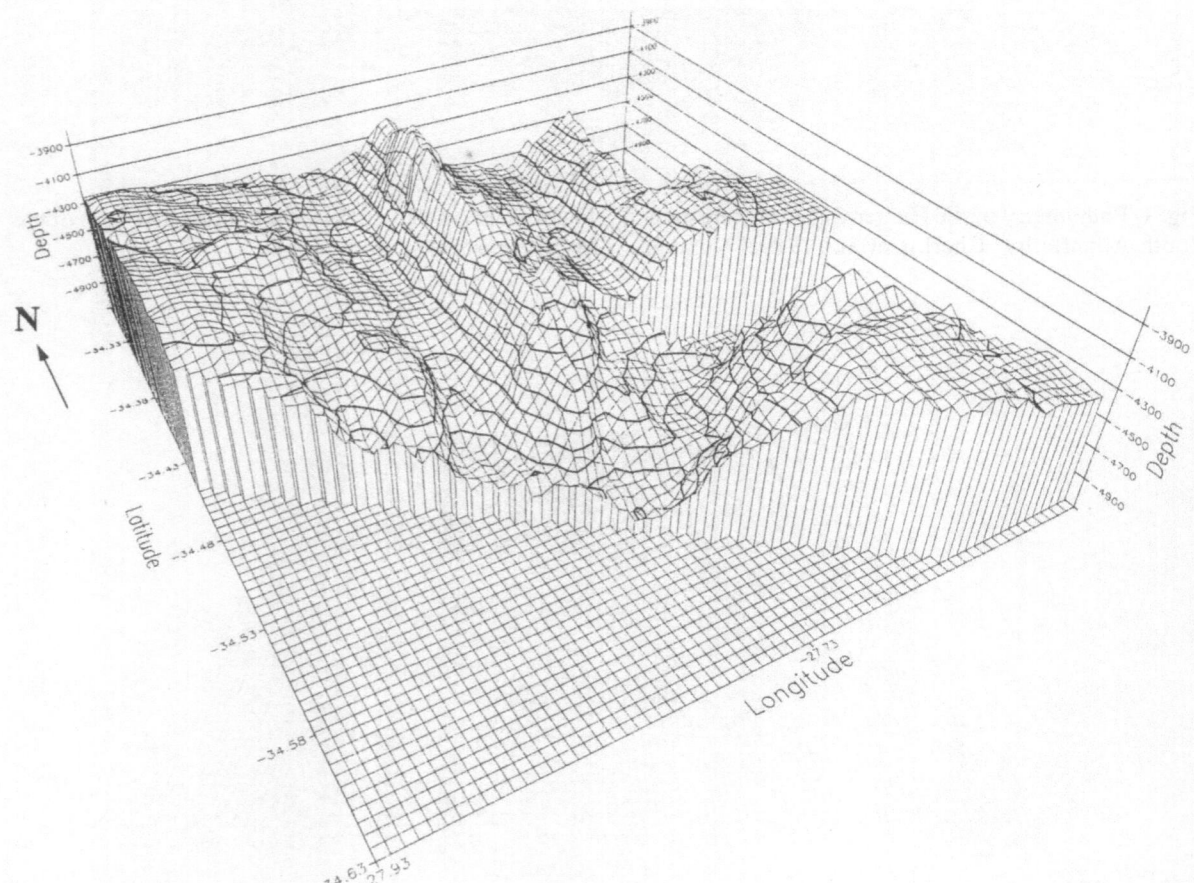

Fig. 3. Three-dimensional view of a detailed mapping of two north-south extending valleys near the central area of the Hunter Channel. View from southwest.

of the deep depression a detailed bathymetric survey of a sill at 34°40′S and 30°95′W has been carried out in a suspected area of bottom water flow (Fig. 4). The sill is located just southwest of the deep east-west trending depression with depths greater than 5000 m. This sill lies at a depth of 3600 m, which seems too shallow to allow passage of Antarctic Bottom Water flow. However, a 5100 m deep canyon with steep slopes was found north of this sill area. The depression appears to be an isolated fragment of a large-scale fracture zone emanating from the Mid-Atlantic Ridge. Its extension to the west is unknown. However, temperature recordings at that site make it an unlikely area for bottom-water flow.

Discussion and Summary

New sets of fine-resolution hydrographic measurements in the Hunter Channel have been obtained during a number of cruises, which allowed its geostrophic bottom flow to be estimated for the first time (Fig. 5; Speer et al. 1992). The northward flow of water colder than 2°C through the Hunter Channel is estimated to be $0.7 \times 10^6 \mathrm{m^3 s^{-1}}$, which is roughly one-fifth of that entering the Brazil Basin through the Vema Channel (Speer and Zenk 1993). Hence, the Hunter Channel flow represents a significant component for the bottom-water exchange between the Argentine Basin and the Brazil Basin. More recently, direct current observa-

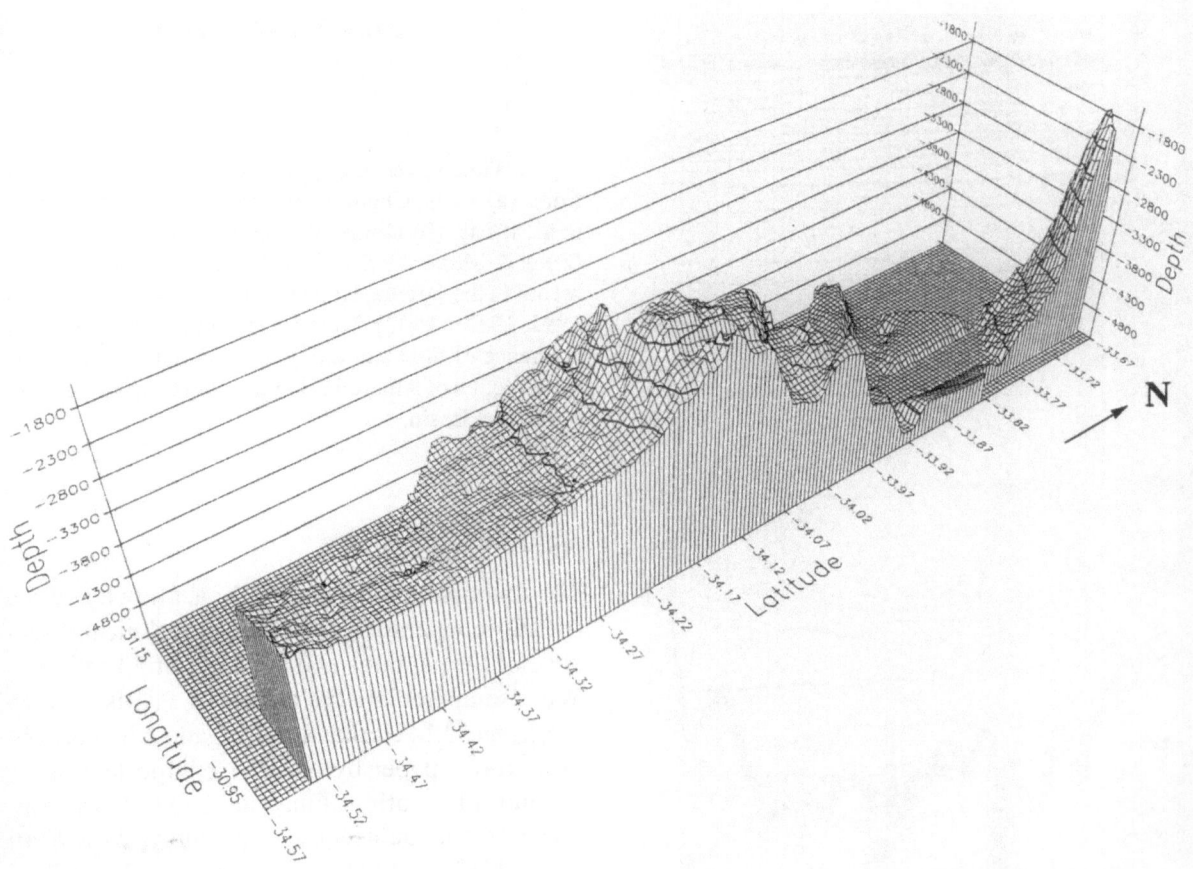

Fig. 4. Detailed mapping in the southeastern area of the Rio Grande Rise, west of the Hunter Channel. View from southeast.

tions (Zenk and Müller 1995) with recording instruments moored for nearly two years at various spots of the Hunter Sill have indicated an almost directional flow (Fig. 6) resembling the situation above the sea floor of the Vema Channel (Tarbell et al. 1994). The 200-km width of the Hunter Channel is roughly ten times that of the Vema Channel to the west. The depth of the Hunter Channel is to 4300 m, which is only about 300 m shallower than the Vema Channel. Narrower south-north extending valleys between 27°35′W and 27°50′W at 34°30′S are deeper (>4700 m), but do not seem to be additional open pathways. The sill of the Hunter Channel is just northeast of an east-west extending deeper fracture zone which has been identified by geophysical studies and which originates from early rifting

stages in the South Atlantic (Cande et al. 1988, 1989). Further detailed mapping is needed to determine where these valleys begin and end, and their role (if any) in guiding the flow of Antarctic Bottom Water.

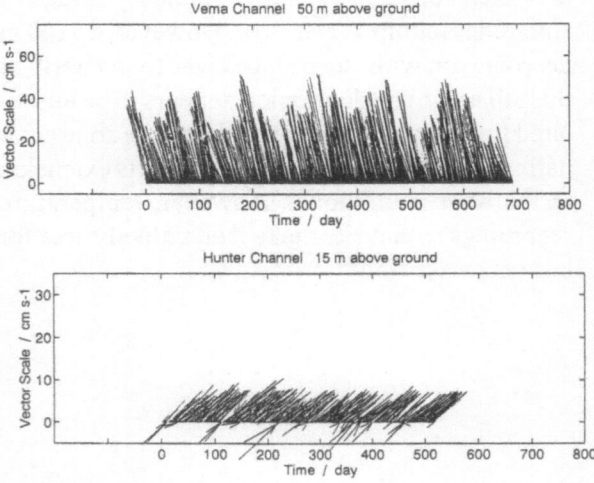

Fig. 6. Time series of current vectors closest to the sea floor. (a) Vema Channel, 50 m above ground (Tarbell et al. 1994), (b) Hunter Channel, 15 m above ground (Zenk & Müller 1995). North direction is upward. The record in the {Vema, Hunter} Channel starts on {12 Jan 1991, 15 Dec 1992}. Note the steadiness and the strong influence of the local topography in both channels on the direction of Antarctic Bottom Water export from the Argentine Basin.

Acknowledgements

The authors wish to gratefully acknowledge the friendly cooperation and efficient technical assistance of the captains and the crew of RV *Meteor*. We also appreciate the efforts of P. Biscaye, N. Hogg, and J.L. Hernandez-Delgado who contributed to this paper by providing important background information. Financial support was provided by the Deutsche Forschungsgemeinschaft (Si 111/38, Si 111/39, Ze 145/7) and the Bundesministerium für Bildung, Wissenschaft, Forschung und Technologie (FKZ 03F0050). This is contribution no. 104 of Sonderforschungsbereich 261 at Bremen University.

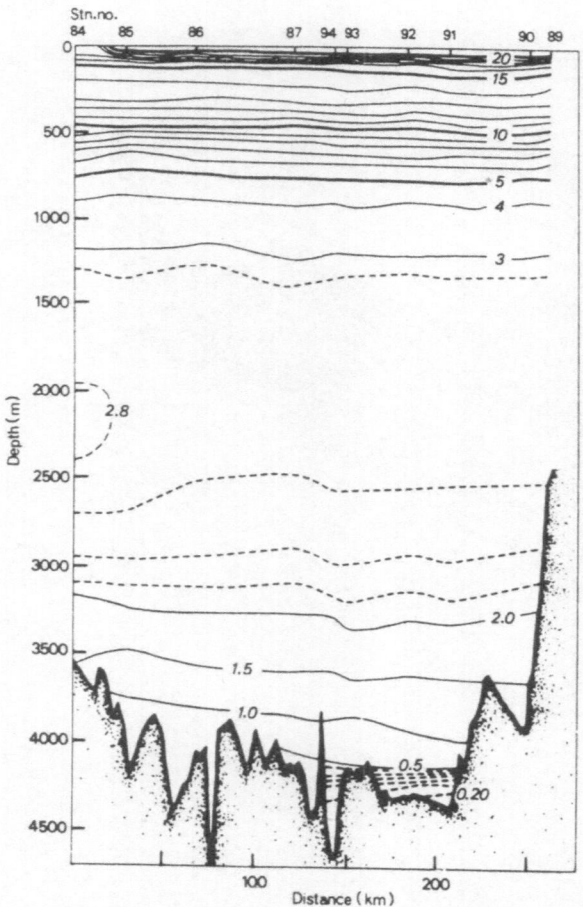

Fig. 5. Potential temperature (°C) section across the Hunter Channel obtained during RV *Meteor* cruise 15 in February 1991 (from Speer et al. 1992).

References

Biscaye PE, Dasch EJ (1971) The rubidium, strontium, strontium-isotope system in deep-sea sediments: Argentine Basin. J Geophys Res 76:5087-5096

Burckle L, Biscaye PE (1971) Sediment transport by Antarctic Bottom Water through the eastern Rio Grande Rise. Geol Soc Am Abstr Progr 3:518-519

Cande SC, LaBrecque JL, Haxby WF (1988) Plate kinematics of the South Atlantic: cron C34 to present. J Geophys Res 93 B11:13,479-13,492

Cande SC, Le Brecque JL, Larson RL, Pitman III WC, Golovchenko X, Haxby WF (1989) Magnetic Lineations of the World's Ocean. Am Ass Petrol Geol, Tulsa, Olka, map

Cherkis NZ, Fleming HS, Brozena JM (1989) Bathymetry of the South Atlantic Ocean, 3°S to 40°S. Geol Soc Am Map Chart Ser MCH069

Gutberlet M, Schenke HW (1989) HYDROSWEEP: New era in high precision bathymetric surveying in deep and shallow water. Mar Geodesy 13:1-23

Hogg N (1996) The South Atlantic deep circulation with emphasis on the Brazil Basin.In: Wefer G, Berger WH, Siedler G, Webb D (eds) The South Atlantic: Present and Past Circulation, Springer, Berlin Heidelberg, this volume

Hogg N, Biscaye P, Gardner W, Schmitz WJ Jr (1982) On the transport and modification of Antarctic Bottom Water in the Vema Channel. J Mar Res 40(Suppl):231-263

IHO/IOC/CHS (1978) GEBCO - general bathymetric chart of the oceans. Sheet 5.12: South Atlantic Ocean. Can Hydrogr Off, Ottawa, 5th ed

Limmer R (1987) The man who became president. The Hunter Magazine 6:12-16

Morgan WJ (1972) Deep mantle convection: plumes and plate motions. Am Assoc Petrol Geol Bull 56:203-213

NGCD (1993) Global relief data, CD-ROM, National Geophysical Data Center

Pätzold J, Bickert T, Brück L, Gaedicke C, Heidland K, Meinecke G, Mulitza S (1993) Bericht und erste Ergebnisse über die Meteor-Fahrt M 15/2, Rio de Janeiro-Vitória, 18.1.-7.2.1991. Berichte, Fachbereich Geowissenschaften, Universität Bremen, 17, 46 pp

Siedler G, Zenk W (1992) WOCE Südatlantik 1991, Reise Nr. 15, 30. Dezember-23. März 1991. METEOR-Berichte, Universität Hamburg, 92-1, 126 pp

Siedler G, Balzer W, Müller TJ, Onken R, Rhein M, Zenk W (1993) WOCE South Atlantic 1992, Cruise No. 22, 22 September 1992-31 January 1993. METEOR-Berichte, Universität Hamburg, 93-5, 131 pp

Speer KG, Zenk W (1993) The flow of Antarctic Bottom Water into the Brazil Basin. J Phys Oceanogr 12:2667-2682

Speer K, Zenk W, Siedler G, Pätzold J, Heidland K (1992). First resolution of flow through the Hunter Channel in the South Atlantic. Earth Planet Sci Lett 113:287-292

Tarbell S, Meyer R, Hogg N, Zenk W (1994) A moored array along the southern boundary of the Brazil Basin for the Deep Basin Experiment - Report on a joint experiment 1991-1992. Ber Inst Meereskunde Kiel 243, 97 pp

Uchupi (1971) Bathymetric atlas of the Atlantic, Caribbean, and Gulf of Mexico. Woods Hole Oceanogr Inst 71-72, 10 sheets

Wüst G (1933) Schichtung und Zirkulation des Atlantischen Ozeans. Das Bodenwasser und die Gliederung der Atlantischen Tiefsee, In: Wissenschaftliche Ergebnisse der Deutschen Atlantischen Expedition auf dem Forschungs- und Vermessungsschiff Meteor 1925-1927, 6: 1, 106 pp

Zenk W, Hogg N (1996) Warming trends of Antarctic Bottom water flowing into the Brazil Basin. Deep-Sea Res, in press

Zenk W, Müller TJ (1995) WOCE Studies in the South Atlantic, Cruise No. 28, 29 March-14 June 1994. METEOR-Berichte, Univ.Hamburg, 95-1, 193pp

Zenk W, Speer KG, Hogg NG (1993). Bathymetry at the Vema Sill. Deep-Sea Res 40:1925-1933

Expeditions into the Past: Paleoceanographic Studies in the South Atlantic

W.H. Berger [1] and G. Wefer [2]

[1] Scripps Institution of Oceanography, La Jolla, Ca. 92093, USA
[2] Universität Bremen, Geowissenschaften, 28334 Bremen, Germany

Abstract: The South Atlantic is tightly coupled to the North Atlantic climate amplifying system. At present, enormous amounts of heat are delivered across the equator to the north, with surface and subsurface waters. The return flow occurs at depth, within the coldwater sphere. During the last glacial the Atlantic Heat Conveyor was much less efficient, that is, the North Atlantic heat piracy is a positive feedback on climate change. This positive feedback is an important ingredient in the orbitally driven climate cycles. The current (that is, late Quaternary) conditions in the South Atlantic are the result of a long evolution of climate and geographic boundary conditions, which started with the opening of the basin at the end of the Jurassic and in the early Cretaceous, by continental breakup and seafloor spreading. Todays margins contain the ancient deposits of a narrow trough with restricted access, including evaporites. Warm-ocean sediments accumulated during the Cretaceous, including organic-rich deposits indicative of poorly oxygenated deep waters. Sinking of the sea floor from cooling of the lithosphere, and ridges produced as hot spot tracks (from Tristan da Cunha on the Mid-Atlantic Ridge) determined the main features of the bathymetry. The leitmotifs of Cenozoic evolution are general cooling (from mountain building and associated regression, and from reduction of atmospheric CO_2), the closure of the world-encircling tropical Tethys Ocean and opening of passages in the south, linking ocean basins through a circumpolar Cold Ring. Overall regression and associated polar deepwater production forced new patterns of biogenous deposition which resulted in a large-scale global drop of the Carbonate Compensation Depth (CCD) about 40 million years ago. At the same time, the isotopic ratio in the element strontium in seawater (as captured by calcareous fossils) started a long trend toward more radiogenic values, indicating increased supply of continental material. The major reorganization in deepsea sediments (the Auversian Facies Shift) in the late Eocene is also expressed as the onset of deposition of rather pure pelagic carbonates, with opaline and organic-rich sediments being increasingly restricted to ocean margins. Continued cooling eventually led to large-scale deepwater formation in high latitudes, which is expressed in the first great cooling step in the deep sea, at the end of the Eocene. The second great cooling step saw the buildup of ice on Antarctica, roughly 15 million years ago, presumably after considerable reduction of atmospheric CO_2. The third great cooling step consists of ice buildup around the North Atlantic, a step that moved the system into modern climate dynamics. Concerning the third step, it is commonly surmised that the closing of the Panama Straits was responsible for its timing (about 3 million years ago). We propose (Panama Hypothesis) exactly the reverse: in fact, the emergence of the Isthmus greatly favored North Atlantic heat piracy, so that northern glaciations were delayed by several million years. After initial onset of northern glaciations (7 to 6 million years ago) it took another 3 million years of mountain building and CO_2 reduction to attain sustained glaciations (3 million years ago). This period of delay is the well-know warm period of the early Pliocene. The story of the onset of northern glaciations is further complicated by the fact that cooling first enhances NADW production, before the onset of northern glaciations, and then obstructs it, presumably by reduction of evaporation and by sea ice formation. The identification of the switch point of NADW production, from negative to positive feedback, is vital for the understanding of the ocean's role in climatic change in the late Neogene.

From WEFER G, BERGER WH, SIEDLER G, WEBB DJ (eds) 1996, *The South Atlantic: Present and Past Circulation.* Springer-Verlag Berlin Heidelberg, pp 363-410

Introduction: The Atlantic Heat Conveyor

From the point of view of paleoceanography, the modern ocean begins with the formation of a strong asymmetry between North Atlantic and North Pacific, with the deep North Atlantic collecting carbonate and the North Pacific diatomaceous silica. This asymmetry is a result of the onset of vigorous production of North Atlantic Deep Water (NADW), which fills the North Atlantic with nutrient-poor surface water, and leaves the North Pacific with nutrient-rich (old) deep water. As seen in deep-sea sediments, the process started roughly 10 million years ago, and has since become more intense, in response to northern hemisphere cooling, and perhaps also as a result of the closing of the Tethys Ocean and of the concomitant opening of the circum-polar Antarctic Ocean.

In consequence of these developments, the South Atlantic has become, during the late Neogene, a thoroughfare for a meridional conveyor which moves warm water into the North Atlantic and cold water out (Fig. 1). This Atlantic Heat Conveyor, which is tied into the Circumpolar Ring Current ("Cold Ring"), increased in strength as a result of overall cooling, as mentioned. However, as cooling proceeded beyond a certain point, NADW production *decreased* again. This means that there is an optimum condition for the production of NADW, presumably due to a favorable combination of heat delivery across the equator and of wind fields responsible for evaporation in the North Atlantic (including, perhaps, the Mediterranean). When these conditions are less than optimal, either due to warming or to cooling, NADW production is diminished. For many problems in paleoceanography involving the grand asymmetries in the global ocean (Pacific vs. Atlantic, North vs. South), it is necessary to understand which side of the NADW-optimum the system is working in. If in the warm side, NADW production delivers negative feedback and stabilizes conditions. If in the cold side, NADW production constitutes positive feedback, as is the case in the late Quaternary.

The fact that NADW production is weakened during the glacial periods of the late Quaternary is readily recognized in fluctuations of carbonate preservation, of benthic foram distributions, and of $\delta^{13}C$ in benthic foraminifers in the South Atlantic. At depth, the position of the boundary between Antarctic Bottom Water (AABW) and NADW is a useful proxy for reconstructing the evolution of the relative importance of the two water masses. This boundary is seen on the sea floor as a foram lysocline (a zone of rapid change of preservation of shells), and as a discontinuity in physical properties. Its depth level fluctuates, with a tendency for deepening when NADW production is strong. Also, these various proxies display changes in east-west asymmetry across the Mid-Atlantic Ridge, the asymmetry being stronger when NADW flows more vigorously, within the western trough of the South Atlantic basin.

The various elements of the conveyor system are recorded in the paleontology and the chemistry of ocean sediments in the South Atlantic, from which reconstruction can proceed. The record includes proxies of surface water temperature, of productivity, and of deepwater circulation. We shall make reference to each of these topics, in the contexts of glacial-postglacial contrast, of ice age cyclicity, and of long-term trends. The great majority of the contributions in this volume treat present-day processes or relatively recent history, which provides the proper interface between physical oceanography and paleoceanography. For balance we emphasize, in addition, a long-range view (that is, millions of years). This view provides a glimpse of the importance of geographic boundary conditions and of the considerable dynamic range of the phenomena under discussion.

Legacies of the Meteor Expedition

Asymmetries: global, north-south, east-west

Exploration of the South Atlantic has contributed importantly to the development of both oceanography and paleoceanography. Many crucial insights of modern physical oceanography were developed on the basis of the quasi-synoptic gridded data of the classic METEOR Expedition

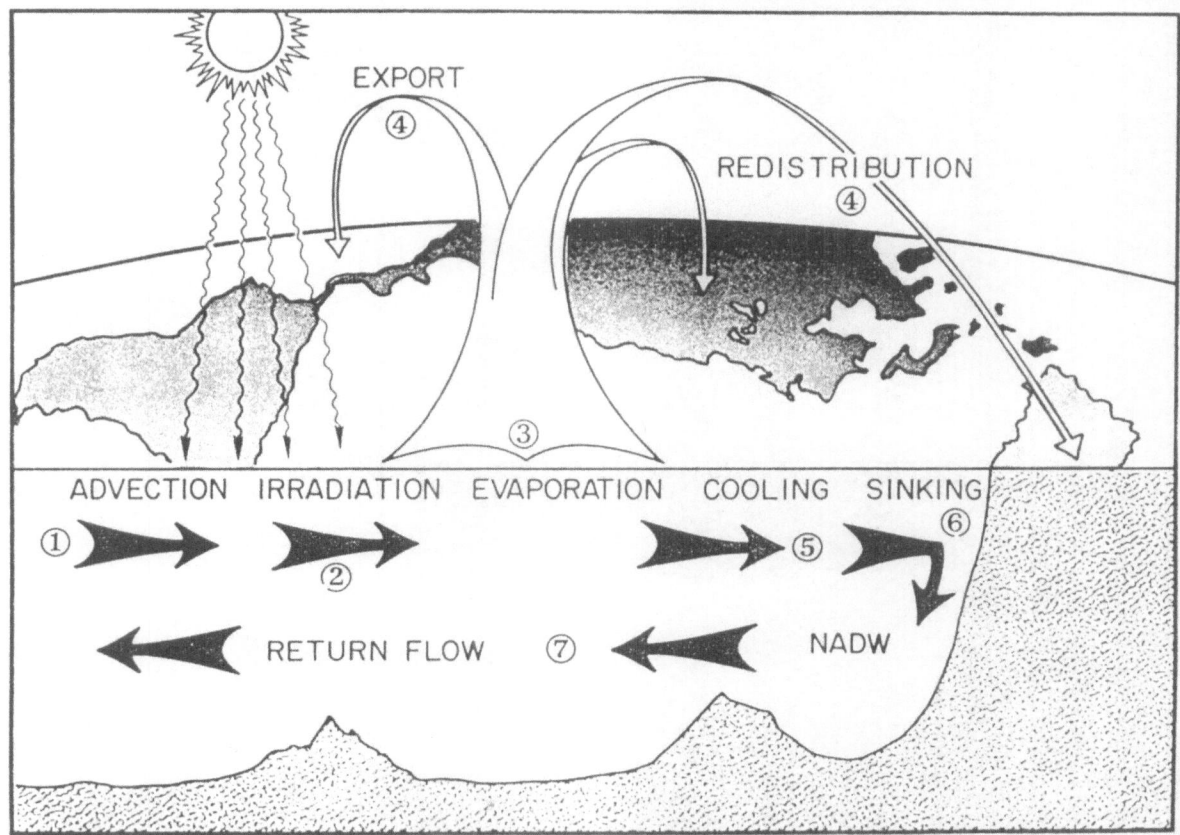

Figure 1. The Atlantic Heat Conveyor. Conceptual model of import and export of heat in the North Atlantic associated with NADW formation. (1) Advection of heat from the Indian Ocean (ca. 50×10^{13} watt). (2) Cross-equatorial flow (ca. 15 sv, 144×10^{13} W). (3) Evaporation of 60,000 km³ (corresponding to 2 sv, 460×10^{13} W). (4) Export of 9000 km³ to the Pacific (0.3 sv, 70×10^{13} W), and redistribution of latent heat to the atmosphere over adjacent continents (70×10^{13} W), rest to atmosphere over ocean (320×10^{13} W). (5) Cooling in high latitudes, and (6) sinking of surface waters to make North Atlantic Deep Water. (7) Return of cold water to the Southern Ocean, via the South Atlantic. From Berger et al. (1987).

into the central and southern Atlantic (1925-1927). Prime examples are the core-layer concept of Georg Wüst (1890-1977), who mapped the stratification of the South Atlantic, and the determination of geostrophic velocities of bottom water flow by Albert Defant (1884-1974)(Wüst 1935; Wüst and Defant 1936). Ever since, the N-S sections of western and eastern basins in the South Atlantic have provided textbook examples for deepwater stratification and bottom water flow, and for the importance of bottom topography in blocking or allowing access to deep water masses (Fig. 2).

In particular, we owe to the *Meteor* expedition the insight that the salty NADW which fills much

of the deep South Atlantic flows vigorously southward, in the western basin. Wüst recognized the sharp lower boundary of NADW with the frigid underlying Antarctic Bottom Water (AABW) as the "2°C-discontinuity surface", a kind of abyssal thermocline, where water moves in opposite directions. In essence, the northward penetrating AABW is eroded at this surface, and this results in a gradual thinning of AABW toward the north, and finally its disappearance north of the equator. The eroded frigid water is mixed upward into the (warmer) lower NADW and returns south. Access to the eastern basin is restricted, with AABW only penetrating east at the equator, through the

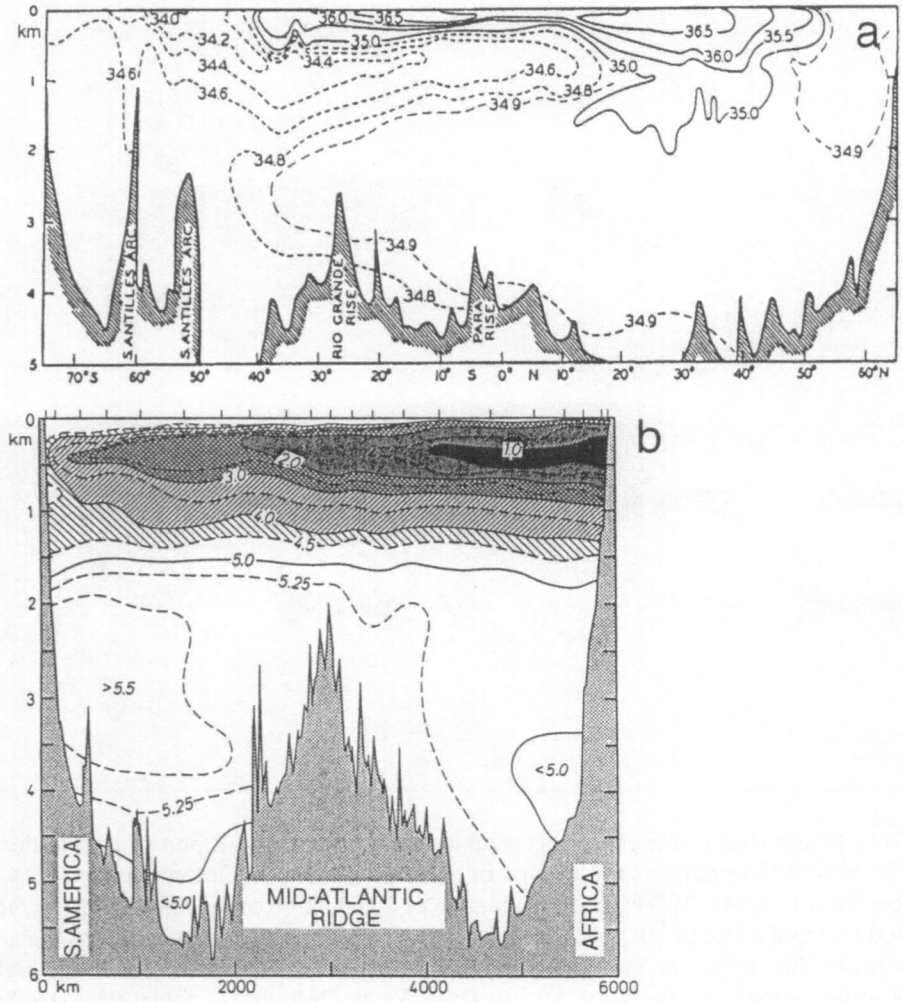

Figure 2. Water masses in vertical sections through the South Atlantic. a) N-S salinity section through the western trough of the Atlantic basin, showing southward flow of high salinity waters at depth, and northward flow of low salinity waters in the South Atlantic, within the thermocline (after Wüst, in Sverdrup et al. 1942, modified). b) E-W oxygen section, South Atlantic near 9°S, based on measurements by RV *Meteor*, Aug.-Sept. 1926 (after Wattenberg, in Dietrich et al. 1975).

Romanche Fracture Zone. Thus, Wüst's deep discontinuity surface is not well developed in the eastern basin. The water there is relatively old, and depleted in oxygen (Fig. 2b).

The vigorous southward flow of NADW (15 to 20 sv) is balanced by northward flow at shallow depths, much of it above 1000 m (roughly 50/50 for surface water and interior water; see Keir 1988; 1990). The upper water masses, naturally, are much warmer than the NADW, so that enormous

amounts of heat are being delivered from south to north, across the equator. This supply stabilizes the dynamics of the North-South asymmetry: much of the additional heat is used in evaporation, increasing the salinity of surface waters in preparation for deep convection (Fig. 1).

The North Atlantic pays for its heat piracy with nutrient deficiency: Heat arrives in nutrient-depleted waters. Thus, the large-scale production of NADW sets up a global asymmetry between Atlantic and the rest of the global ocean, such that the deep Atlantic is depleted in nutrients, while the Circumpolar Ocean (Cold Ring) and the Pacific become enriched.

The global-ocean asymmetry in deepwater properties is reflected in the patterns of biogenous deep-sea sedimentation: the ocean operates as a giant fractionation machine (Berger 1970). The deep Atlantic, because its waters have surface-water properties, is favorable for the deposition of carbonate, while the Pacific and the Cold Ring are unfavorable toward carbonate sedimentation (because of high carbon dioxide values in the aged deep water). In this fashion the intensity of NADW production expresses itself in the asymmetry of carbonate deposition, between Atlantic and Pacific. A breakdown of the asymmetry in carbonate deposition (as in the last glacial) implies a breakdown of NADW production. The existence of the asymmetry, and the fact that there is an interchange between Pacific and Atlantic, was first recognized in the 1960s (e.g., Olausson 1965, working with the cores from the Swedish Deep Sea Expedition 1947-49). An analogous (but inverse) case can be made for the asymmetry in opal deposition, with North Pacific and Cold Ring collecting opal while NADW production keeps silicate concentrations low in the North Atlantic, thus preventing accumulation there.

Messages from Globorotalia menardii

In addition to oceanographers and marine biologists a young geologist, Wolfgang Schott (1905-1989), participated in the Meteor Expedition. In working up the short gravity cores (1 m long, typically), he concentrated on the coarse fraction of calcareous ooze, which consists almost entirely of planktonic foraminifers. Painstakingly, he counted about 300 specimens in each sample, sequentially taken along each core, identifying each specimen by species and characterizing each sample by the percentages of species contained (Schott 1935). In so doing, he pioneered the new field of quantitative paleontology and laid the groundwork for modern paleotemperature estimation from microfossil assemblages. He also observed that the foraminiferal assemblages below Wüst's 2°C-dis-continuity were poorly preserved, that is, they were greatly enriched in fragments and ruins of the more resistant species. With this observation he prepared the ground for subsequent definition of a preservational boundary (lysocline), and its vertical motion in response to changes in deep circulation.

The observation which struck Schott as most significant, however, concerned *Globorotalia menardii*, a disk-shaped tropical foraminifer with keeled chambers of increasing size arranged spirally, in a plane (Fig. 3, insert). Schott found that this species, while abundant in warm waters and in surficial sediments on the sea floor, is absent a short distance below the sediment surface. He surmised (correctly, as we now know) that the disappearance of *G. menardii* downcore marks the boundary between our interglacial (the Holocene) and the last glacial. Thus, he was able to provide the first reasonable estimate of sedimentation rate for calcareous ooze (using periglacial varve counts by de Geer for guidance regarding time since melting of the Scandinavian ice).

Further downcore, however, Schott found that *G. menardii* re-appeared in older sediments: He had discovered the evidence for warm-cold cycles in the tropical Atlantic, based on deep-sea sediments.

Schott's discoveries regarding *G. menardii* raise several fundamental questions about the paleoceanography of the central and southern Atlantic. First of all, it is not entirely clear why this species should become extinct throughout the Atlantic, if in fact the temperature changed as little as indicated by the abundance of the other warm-water species. CLIMAP (1976; 1981) put the change somewhere near -2°C or less, in the tropical Atlantic (Fig. 4a). These reconstructed temperatures are well within the range of the modern habitat of G. menardii (Fig. 3), thus, we must ask why this species disappears in the glacial ocean. Is the reconstruction wrong? Or are there other factors besides temperature that are important in controlling species distribution? If so, what are they? How do such factors impact our paleotemperature estimates?

The simplest possible answer is that the CLIMAP values are incorrect, and that tropical

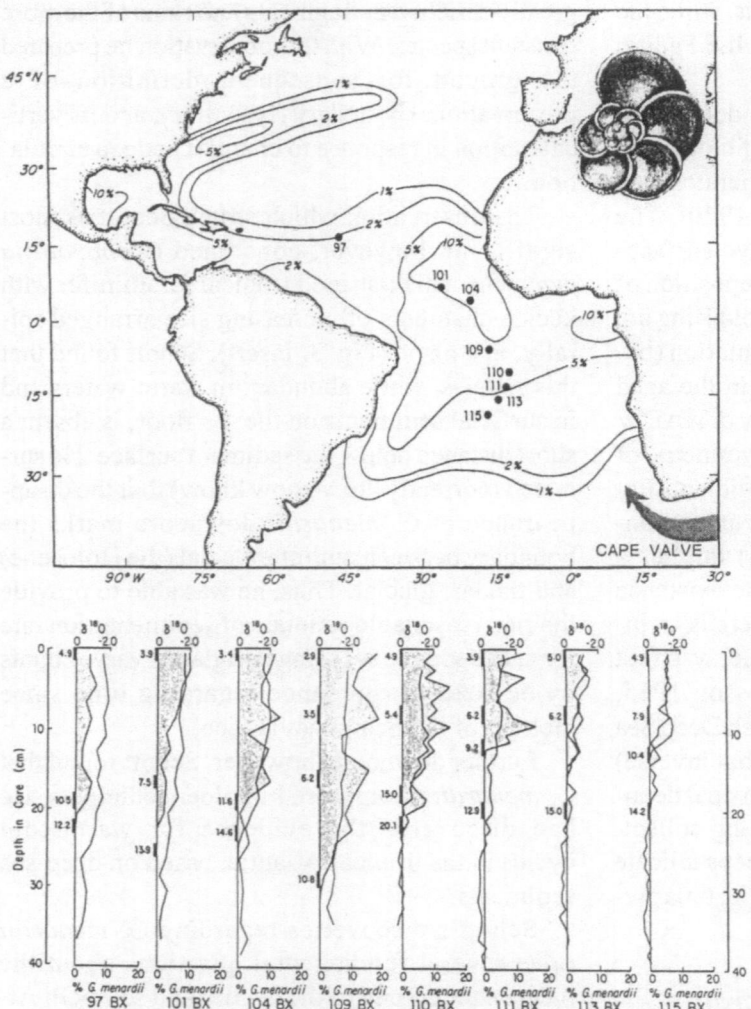

Figure 3. Biogeography of *G. menardii* in the tropical Atlantic. a) Distribution of *G. menardii* (percent abundance). b) Level of disappearance downcore, within box cores from the Mid-Atlantic Ridge (profiles). Map and data from Berger et al. (1985). Insert: *G. menardii* specimen from the Atlantic Ocean (Boltovskoy 1967). Reseeding of the species takes place in the Agulhas Retroflection, around the Cape of Good Hope ("Cape Valve").

temperatures dropped by 4 or 5°C during the last glacial maximum (cf. Rind and Peteet 1985; Broecker and Denton 1989; Guilderson et al. 1994; Stute et al. 1995). However, alkenone data seem to support the CLIMAP values (Schneider et al. 1995), as does re-evaluation of foraminiferal statistics (Prell 1985) and of foraminiferal $\delta^{18}O$ values (Stott and Tang 1996).

Our own data, as well, suggest that the temperature drop in the tropics was modest, at least in the summer (Fig. 5a). If temperatures changed considerably, we should find that the $\delta^{18}O$ range in surface-water species should be much larger than the 1.2 permil expected from the effects of changing

polar ice mass. The difference in range is just large enough to accommodate roughly 2°C, judging from a stacked $\delta^{18}O$ curve for 4 tropical box cores (locations in Fig. 3). The uncertainty to either side of the 1.2 permil may be taken as 0.1 permil. (A change of one degree Celsius produces a change of 0.2 permil in the isotope record.) It may be noted, in Fig. 5a, that *G. menardii* first appears at about 11,000 radiocarbon years ago, in South Atlantic warm waters, well before there is any evidence that the equatorial regions are warming.

The species here used as signal carrier, *Globigerinoides ruber,* pink variety, lives in the

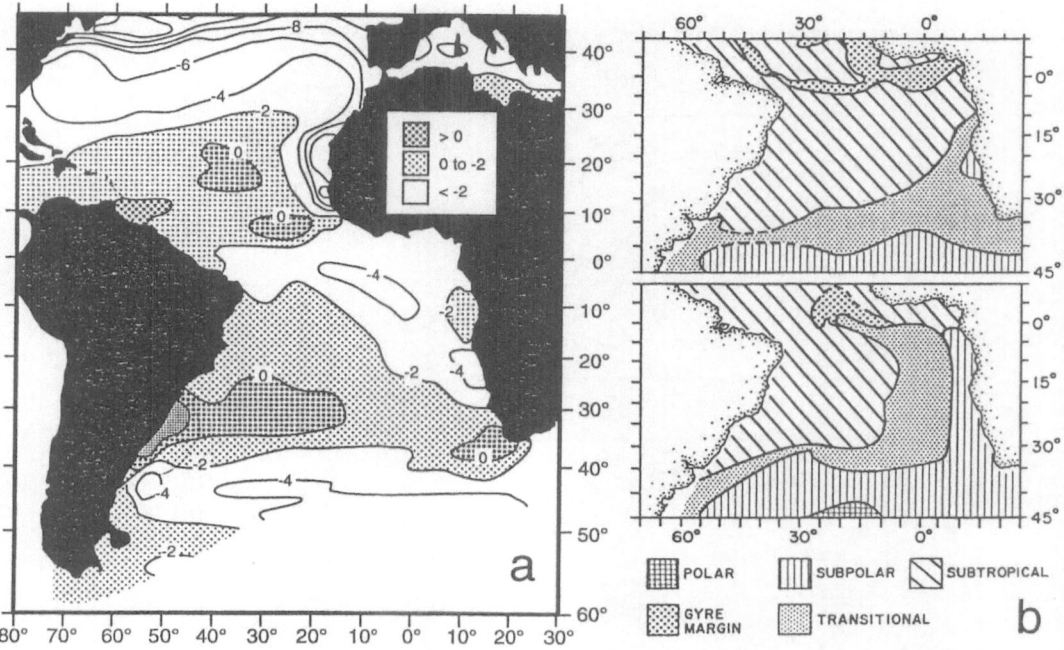

Figure 4. Glacial-Holocene contrasts in surface waters of the central and southern Atlantic. a) Temperature difference between modern ocean (month of August) and reconstructed summer temperatures for the last glacial maximum. Ruled areas: difference is less than 2°C. b) Changes in biogeography of planktonic foraminifers. Sources: CLIMAP Project Members (1981).

uppermost surface waters and is abundant only during the warm season, in well-stratified waters. Thus, it is a biased reporter (as are all other species, in their own way). Nothing is said, in these data, about the difference in winter temperatures, between modern and glacial ocean. Clearly, for a better understanding of these dynamics, the seasonal ranges must be studied. This has become an important topic, especially in the last decade (Mix et al. 1986; Hemleben et al. 1989; Wefer and Berger 1991)

The fact is, we do not know why *G. menardii* disappeared. What we do know is that we need a mechanism for re-seeding *G. menardii* during interglacials. The only way to do this is around the Cape of Good Hope, where hopeful individuals originating in the Indian Ocean can immigrate into the Atlantic, to attempt to start new populations, after glacial periods (Fig. 3, "Cape Valve"). Here warm waters enter the South Atlantic from the Indian Ocean, in the region of the Agulhas

Retroflection (see Lutjeharms, this volume). The presence of this warm water source solves one problem, but raises others: Is *G. menardii* being re-seeded all the time, but only flourishes during times of warmth? If so, why is it absent in some of the interglacials (see Imbrie et al. 1973)? Or is the Cape Valve closed at times, denying warm water flow to the South Atlantic? If so, is this the reason why less heat is transported from the South Atlantic to the North Atlantic, during glacials? Or are there other, more fundamental reasons?

The relationship between the re-appearance of *G. menardii* and the $\delta^{18}O$ record of *G. ruber* (pink) in the central South Atlantic is actually quite complicated. In our example (near-by box cores show similar patterns), there is a strong apparent *cooling* in surface waters during the transition from glacial to postglacial time (Fig. 5b). This is seen in the strong lag that the $\delta^{18}O$ record of *G. ruber* displays, with respect to the expected general change in isotopic values (plotted upside down, to

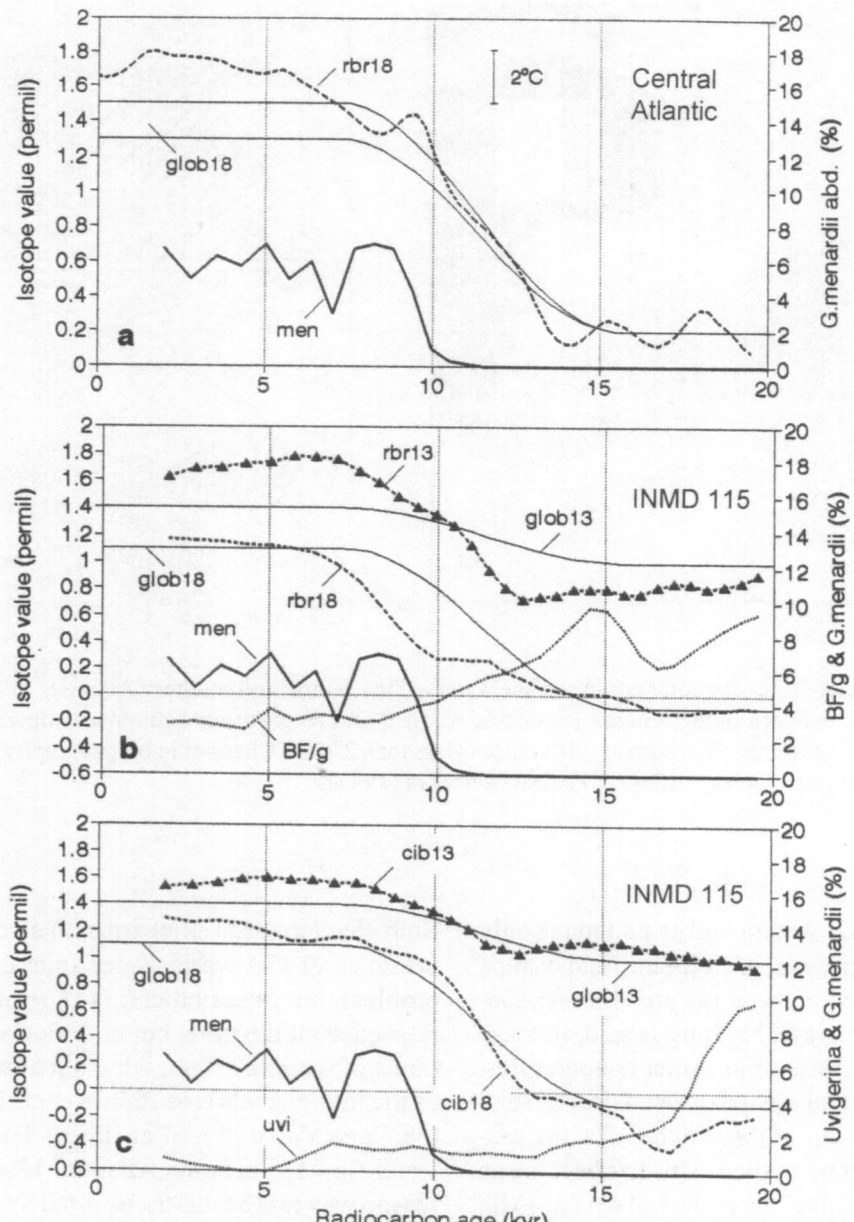

Figure 5. Relationships between the reappearance of *G. menardii* and various paleoceanographic proxies. a) δ¹⁸O of *G. ruber* (pink) in the central Atlantic (rbr18) closely follows the expected global change from the ice effect (glob18). The temperature effect is small and has no relationship to the *G. menardii* abundance in the central South Atlantic (men, abd. in box core INMD115). b) δ¹⁸O of *G. ruber* (pink) from INMD box core 115 lags the expected change (glob18), showing cooling during deglaciation in the central gyre region. The number of benthic foraminifers per gram (BF/g) drops throughout the transition, while δ¹³C of *G. ruber* (pink) increases greatly over background. The changes seen should be unfavorable for *G. menardii,* yet, the species quickly reaches maximum abundance (men). c) Deepwater changes (cib18, cib13, δ¹⁸O and δ¹³C of *Cibicidoides wuellerstorfi*) set in at about the same time as *G. menardii* reappears, as expected if NADW production is linked to an open Cape Valve. The first indication of a change in the deep water is the abundance of *Uvigerina peregrina dirupta* (uvi), a benthic species sensitive to oxygen content and to productivity. Data from Berger et al. (1985; 1987) and unpublished. Data are smoothed (except for *G. menardii* abundance).

keep curves parallel to temperature). Perhaps this apparent cooling reflects the tax taken by the North Atlantic, which is in need of much heat to help melt the polar ice; alternatively, it reflects greatly increased evaporation and concomitant increase of surface water salinity (and enrichment in ^{18}O). At the same time, there is a marked reduction in productivity at the site of INMD115, suggesting expansion of the central gyre. The same expansion is indicated in the drastic increase in the $\delta^{13}C$ values of this surface-dwelling foraminifer, its habitat being increasingly isolated from the waters below (which carry a more negative signal). The range is truly extraordinary: about 3 times the global one (curve labelled rbr13 vs. curve glob13; Fig. 5b).

Expansion of the central gyre during the transition period is further indicated by the changing abundance patterns of planktonic foraminifers: *G. dutertrei* and *G. inflata* (gyre margin and transition species, respectively) markedly decrease in abundance in our box core, while the pink *G. ruber* increases (F.L. Parker, unpublished data). These findings are entirely as expected from previous work (Fig. 4b). It is unlikely that *G. menardii* prefers a cooler, less productive habitat to a warmer, more productive one. Its reappearance and strong showing at this site, therefore, reflects dynamics other than just those of the local conditions. The same may be true, of course, for abundance patterns of many other species.

If the delivery of warm water around the Cape is tied to NADW production (Gordon 1985) and if this is also the source of new populations of *G. menardii* after regional extinction, we should see evidence for a common startup of *G. menardii* and NADW. In fact, we do (Fig. 5c). While this proves nothing regarding the importance of the Atlantic Heat Conveyor, it is reassuring that there is no discrepancy with Gordon's model on this account.

The discovery of the *G. menardii* conundrum by Wolfgang Schott, some 60 years ago, marks the beginning of paleoceanography. In the meantime, we have learned to ask more questions. But many questions surrounding the strange biogeographic patterns of *G. menardii* remain unsolved.

Major current patterns and ice age cycles

What makes the South Atlantic special

The upper water circulation of the South Atlantic, at first glance, looks much like the pattern in other ocean basins. Driven by trade winds and the westerlies, the major currents move anticlockwise around the central gyre, the center of which occurs near 30°S, off Argentina. To the south it is bounded by the eastward flowing South Atlantic Current, to the east by the Benguela Current, which moves north and west, to join the South Equatorial Current, and to the west the gyre margin is formed by the Brazil Current, flowing southward (Fig. 6). In the equatorial region, the eastward moving Equatorial Counter Current runs within the doldrums, between the westward flowing South Equatorial and North Equatorial Current.

Yet, for all this similarity to the standard textbook pattern, the South Atlantic is highly unusual, with regard to heat transport by upper waters. We have seen that the South Atlantic is the link between the global ocean and the great climate amplification center of the world ocean, that is, the North Atlantic basin. The South Atlantic delivers immense amounts of tropical heat to this center, by cross-equatorial transport of warm water (Stommel 1980; Hastenrath 1977). The South Atlantic is unique in this respect: Generally the net movement of heat in warm water is away from the equator, in the tropics (Fig. 7, World Ocean).

Worldwide, much of the heat transport in the tropics is in fact toward the equator, by converging tradewind-driven surface currents. The poleward transport, which results in net outflow of heat, occurs in narrow, powerful jets (e.g., Kuroshio, Gulf Stream). In the South Atlantic, this jet is relatively weak. Both topography and a strong northward displacement of the Intertropical Convergence Zone (ITCZ) play a role. As a result of the bow-shaped east coast of South America, the South Equatorial Current is split into two branches, the north-west flowing Guyana Current and the south-east flowing Brazil Current (Fig. 6). The Guyana Current moves across the equator and adds to the warm-water pile-up in the western

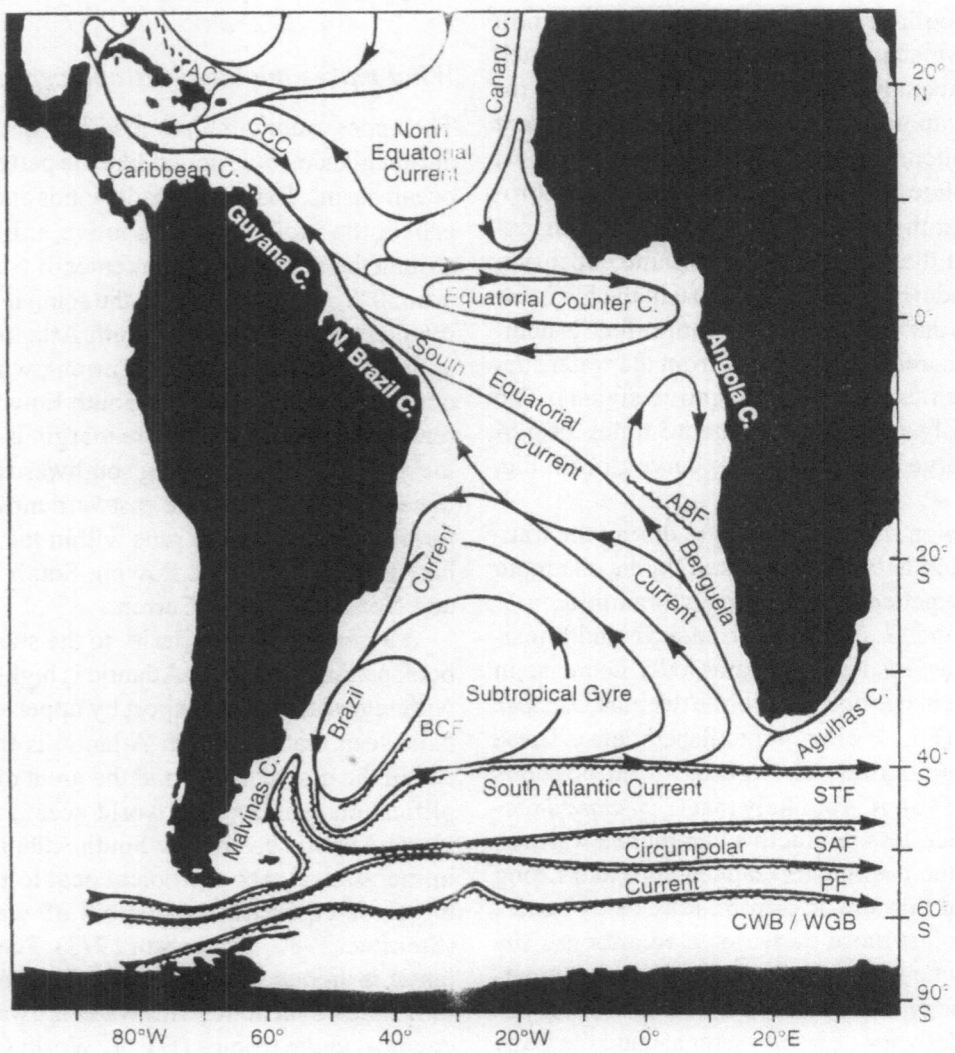

Figure 6. Surface currents of the central and South Atlantic. Sources :
Peterson and Stramma (1991), Tomczak and Godfrey (1994).

tropical Atlantic, including the Caribbean (helping to generate hurricanes there). The northerly
position of the ITCZ (especially during summer)
enhances the cross-equatorial transport of moist
energy-laden air, and also tends to move the North
Equatorial Current off the equator, so that there is
no opposition to the movement of warm waters
from the south. In addition, as mentioned, much
transport occurs in subsurface waters, helping to
balance the southward flow of NADW.

When the North-South asymmetry is reduced,
during the glacial maximum, the heat transport
pattern in the South Atlantic becomes more similar to the general pattern (Fig. 7, glacial Atlantic).
As a consequence, the asymmetry of the global
transport (which reflects the Atlantic anomaly) is
diminished. Some compensation occurs in the
North Pacific (according to Miller and Russell
1989), presumably because the North Pacific receives less fresh water so that its estuarine-type

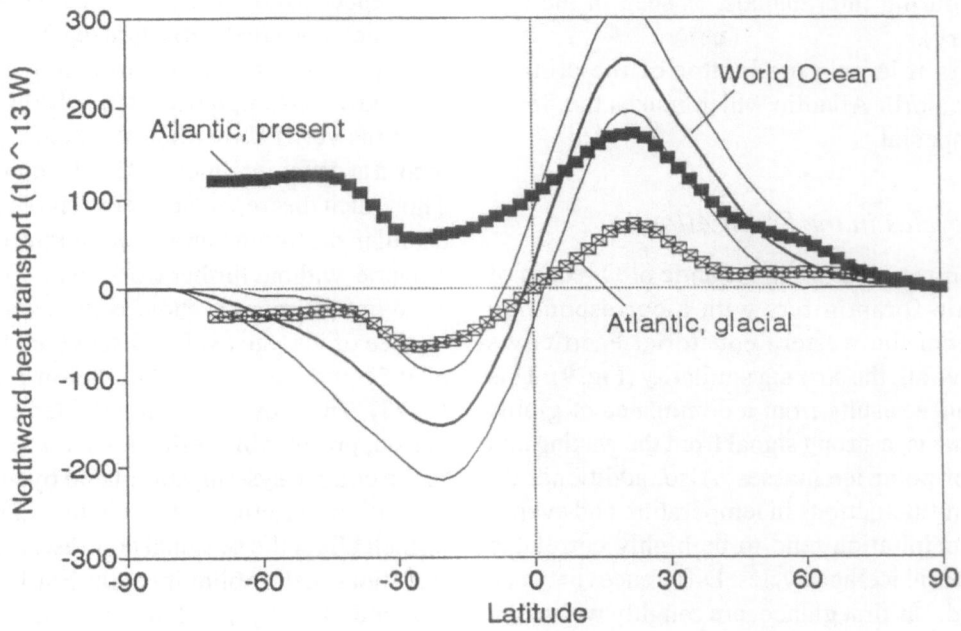

Figure 7. Meridional heat transport in the oceans, present and past. World ocean transport (with envelope showing uncertainty), generalized from various sources (as given in Miller and Russell 1989). Present Atlantic and glacial Atlantic: as calculated by Miller and Russell (1989). Note anomalous northward transport in the present South Atlantic, and conformance to world pattern in glacial.

circulation gives way to greater deepwater production. The entire planet, of course, becomes more symmetrical whenever the northern hemisphere enters the ice age state: The southern hemisphere tends to remain in this state, since the Antarctic ice persists during interglacials. It is not known how accurate the values of the calculated glacial heat transport are. Given the substantial uncertainties for the present situation, the results must have rather large error bars.

How can the South Atlantic be so generous with its export of heat, and yet retain a large and warm subtropical water mass? One reason, apparently, is heat import from the Indian Ocean, as we have seen when discussing the biogeographic dynamics of *G. menardii*. Subtropical and thermocline waters from the Indian Ocean find their way around the Cape of Good Hope (Fig. 8). Portions of this water are captured by the South Atlantic

Current and move right back into the Indian Ocean. But a substantial fraction is imported into the South Atlantic, some as heat-bearing eddies which persist for years (Byrne et al. 1995). Such warm eddies, of course, can act as life-rafts for tropical plankton on its way north to a more clement habitat.

Conditions are not ideal for moving water and heat from Indian Ocean to the South Atlantic: The portal is narrow. We may surmise that it narrows even more whenever the subtropical front moves northward, pinching off the access. It seems reasonable to suspect that this happens during reglaciation, when the planet enters into a glacial period. If this is so, we have here a nonlinear positive feedback mechanism, modulating the ice age cycles. It may be this valve point that is responsible for moving the system from "normal" operation (NADW increase with cooling, as seen in the late Neogene) to "reversed" operation (NADW increase during interglacials, as seen in the late Quaternary).

It is its role as a modulator of the climate amplifier North Atlantic which makes the South Atlantic special.

Ice age cycles in the South Atlantic

When comparing a South Atlantic $\delta^{18}O$ record of planktonic foraminifers with a corresponding record from the western equatorial Pacific, we note, above all, the striking similarity (Fig. 9). This concordance results from a dominance of global effects, that is, a strong signal from the waxing and waning of polar ice masses. Also, additional effects from fluctuations in temperature and evaporation-precipitation tend to be highly correlated with the basic ice-age cycles. Differences between the records, at first glance, are roughly within the

noise level of such fluctuations (that is, similar differences would be expected between two cores from the same area). Offsets at equivalent horizons are within the error limits of dating by tuning to Milankovitch templates or to the SPECMAP template (as verified by multiple dating of the same record at different times and by different workers). Thus, such discrepancies between records are not useful in discussing processes unique to the South Atlantic, without further work on multiple records.

What the records show is the well-known sequence of ice-age cycles, discovered by Emiliani (1955) and dated by Shackleton and Opdyke (1973). These cycles are dominated by a 100-kyr period, presumably reflecting a resonance period in the climate system, stimulated by high-latitude insolation variations. The changes in insolation (which affect the seasonal contrast) are a result of variations in the obliquity of the Earth's axis (with a period of 41 kyr) and of precession of the equi-

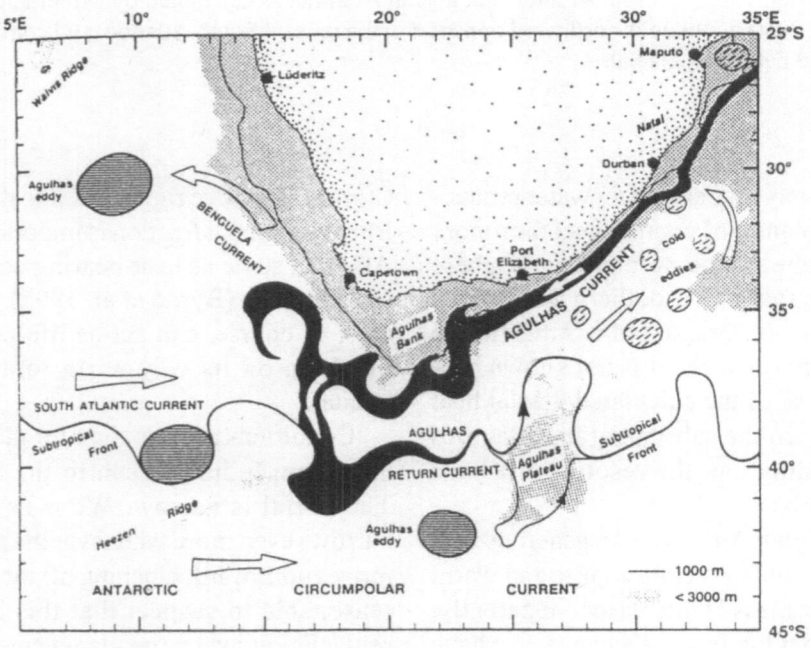

Figure 8. The Cape Valve, showing the Agulhas Current, the return current, and Agulhas eddies. Open arrows: direction of surface geostrophic currents. From Peterson and Stramma (1991), as adapted from Lutjeharms and van Ballegooyen (1988). Courtesy R. Peterson, S.I.O.

noxes (with periods near 23 kyr and 19 kyr). Obliquity determines how high the sun rises over the horizon, in summer. Precession governs how large the sun appears, in the different seasons. The precessional effect has a maximum when perihel (closest approach between Earth and Sun) occurs during summer solstice (highest angle of Sun over the horizon). During such times, according to Milankovitch (1930), polar ice tends to melt, while during times with cool summers it tends to grow.

As shown by Hays et al. (1976) both the obliquity-related period and the precession-related periods are present in these types of records, in addition to the dominant 100-kyr cycle. The record in the western equatorial Pacific has very little precessional power, however. To us, this suggests that precessional power refers mainly to temperature and evaporation-precipitation patterns, rather than to ice mass. Precessional effects tend to dominate in the tropics (e.g., Crowley and North 1991), so that they should be useful in studying the dynamics of processes related to trade winds and monsoon (McIntyre et al. 1989; Imbrie et al. 1992; 1993; Schneider et al. 1995).

An interesting question to ask is to what degree various elements of the South Atlantic circulation system are influenced by processes related to pre-cession (Molfino and Mc Intyre 1990). Let us define a "precessional-response index" (PRx), as the percentage of the sum of precessional-period amplitudes to the sum of both precession- and obliquity-related amplitudes. For "amplitude" of a given period we take the square root of the sum of squares of the appropriate sine and cosine coefficients in the Fourier expansion of a given climate record. "Appropriate" coefficients are defined as those associated with a given period, including its neighbors that are within a factor of 1.1 of this period. We analyze 4 sites: (1) GeoB1523 in the western equatorial region, (2) GeoB1105 in the eastern equatorial Atlantic, (3) GeoB1413 in the central gyre of the South Atlantic, and (4) GeoB1028 in the Benguela system, on Walvis Ridge (for data see Wefer et al., this volume). The Fourier expansions were calculated for $\delta^{18}O$ and $\delta^{13}C$ of G. ruber, and for $\delta^{18}O$ of G. crassaformis for the first 3 sites named, and for $d^{18}O$ and $d^{13}C$ of G. ruber and $d^{18}O$ of G. inflata for GeoB1028. For this latter site, an analysis of the C_{org} series also was done. Results of the Fourier analysis are shown in Fig. 10.

Typically, the precession-response index (PRx) is greater than 60% for all sites and proxies, that is, there is more area under the curve in the (broad)

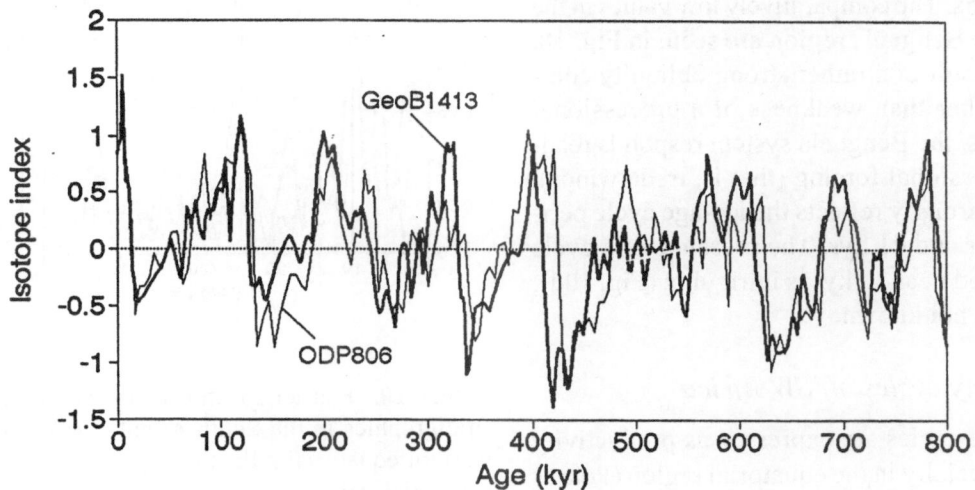

Figure 9. Oxygen isotope record in the central South Atlantic gyre (GeoB1413, *G. ruber* pink) compared with a similar record from the western equatorial Pacific (ODP806B, *G. sacculifer*).Sources: Kemle- von Muecke 1994, and Berger et al. 1995).

precessional band than in the (narrow) obliquity band, when the spectrum is plotted in the manner shown. Visual examination (Fig. 10) confirms the importance of the band centered on 23 kyr, at all sites. The lowest value (56%, Core GeoB1523, Rbr13) is in the western equatorial region; the highest (86%, Core GeoB1413, Rbr13) is in the central gyre. The site with the highest overall PRx values is GeoB1105, in the eastern equatorial Atlantic; the one with lowest is GeoB1028, in the Benguela Current. The pattern tends to confirm expectations; that is, the most affected region, the eastern equatorial Atlantic, is both in the center of the tropics and close to Africa, thus being strongly affected by fluctuations in monsoonal activity. Since the Sun is high in the tropics no matter what the obliquity, it is the apparent size of the solar disk during summer (varying with precession) which determines the strength of the monsoonal signal. The low precessional power in the western equatorial Atlantic, in the $\delta^{13}C$ of *G. ruber,* agrees with similar findings by Curry and Crowley (1987).

The strength of the precessional effect in the $\delta^{13}C$ of *G. ruber* in the central gyre is surprising. It suggests that while temperature varies but little (low PRx values for $\delta^{18}O$), the waxing and waning of the central gyre lens greatly affects productivity at the site of GeoB1413, and that these fluctuations are tied to precession-dominated trade-wind patterns. The comparatively low values in the PRx for the Benguela region are seen, in Fig. 10, to be the result of a rather strong obliquity component, rather than weakness of a precessional effect. Thus, the Benguela system responds readily to precessional forcing (that is, trade winds), but it also strongly reflects the ice-age cycle periods 100 kyr and 41 kyr. The presence of a cycle with a period near 30 kyr is intriguing (Fig. 10d); its origin is not known.

Productivity cycles off SW Africa

The ice-age cycles are expressed as productivity cycles, especially in the equatorial region (Pokras 1987; McIntyre et al. 1989; Mix 1989; Berger and Herguera 1992; Westerhausen et al. 1993; Abrantes et al. 1994). One might expect similar cycles off Southwest Africa, in the Benguela system,

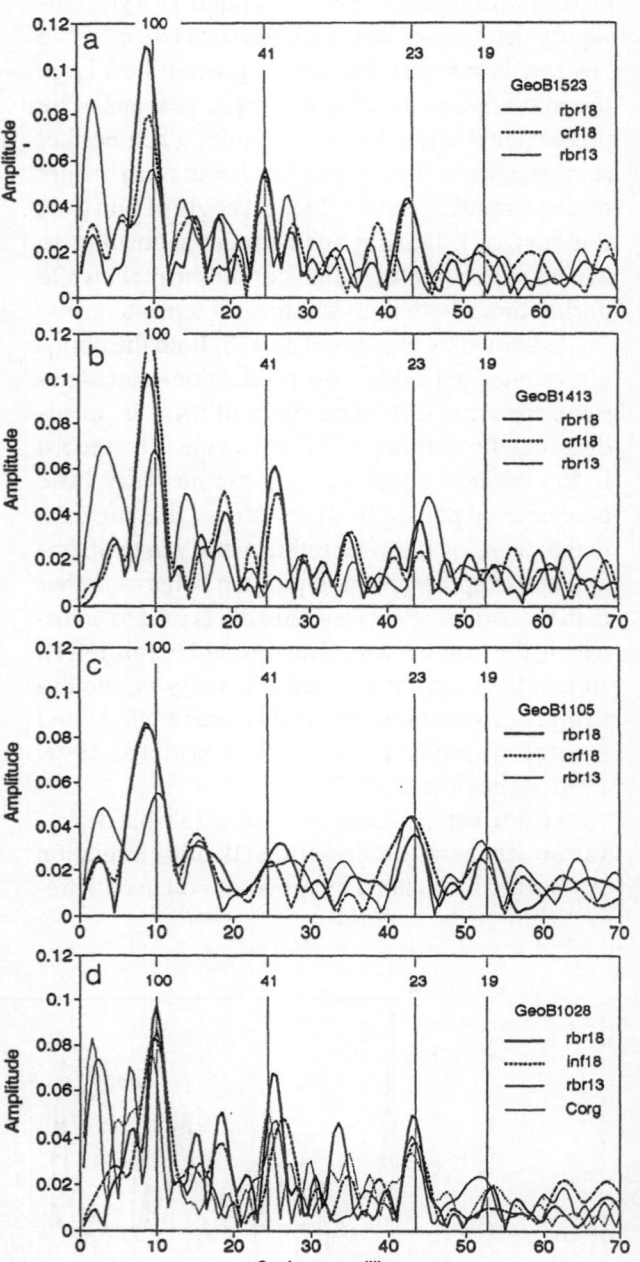

Figure 10. Fourier spectra for four climate-related stratigraphies in the South Atlantic. (a) GeoB1523, western equatorial Atlantic; (b) GeoB1105, eastern equatorial Atlantic; (c) GeoB1413, central gyre; (d) GeoB1028, Benguela system. Data in Wefer et al. (this volume) rbr18, $\delta^{18}O$ of *G. ruber*; crf18, $\delta^{18}O$ of *G. crassaformis*; rbr13, $\delta^{13}C$ of *G. ruber*; inf18, $\delta^{18}O$ of *G. inflata.*

which is one of the five or six great upwelling regions in the world, where production values of more than 200 gCm⁻²kyr⁻¹ occur over considerable areas (Fig. 11).Over much of the region, sediments are characterized by organic-rich sediments, some of which deposited under anaerobic conditions. The significance of this association between coastal upwelling and organic matter deposition, and the possible implications for the origin of petroleum, were recognized half a century ago by Brongersma-Sanders (1947). Many studies have since defined the parameters surrounding deposition below upwelling waters in this region (e.g., Calvert and Price 1971; 1983; Summerhayes et al. 1973; Bremner 1983). Off NW Africa, increased glacial productivity is well established (Mueller and Suess 1979; Mueller et al. 1983). Off Namibia, also, indications are that productivity was substantially increased during glacial time (Summerhayes et al. 1995), presumably because of increased trade wind strength. In addition, productivity apparently was greatly enhanced at the southern border of the South Atlantic, between 40 and 50°S (Kumar et al. 1995).

A puzzling situation exists on the northeastern part of Walvis Ridge, however, where glacial deposits contain less opal than interglacial ones (DSDP Site 532: Hay, Sibuet et al. 1984; Diester-Haass 1985). At present, eddies of cold, upwelled water containing radiolarians and diatoms are transported from the upwelling area near the coast to the area where Site 532 was drilled. Apparently, such transport did not take place in glacial times. During the last glacial maximum, it seems, eddy formation took place farther north, and the Benguela Current flowed parallel to the coast and over the Walvis Ridge to reach the Angola Basin, finally bearing to the west at about 17°S (it does so between 23° and 20°S today; see Fig. 6). Thus, while upwelling may have continued to occur on and off the African shelf, the Benguela Current did not transport that upwelling signal to the Walvis Ridge.

Another possible explanation is that upwelling persisted, or was stronger during glacials, but the upwelled water was poor in nutrients, due to changes in deepsea circulation (Hay and Brock 1992). In a variant of this scenario, the Walvis

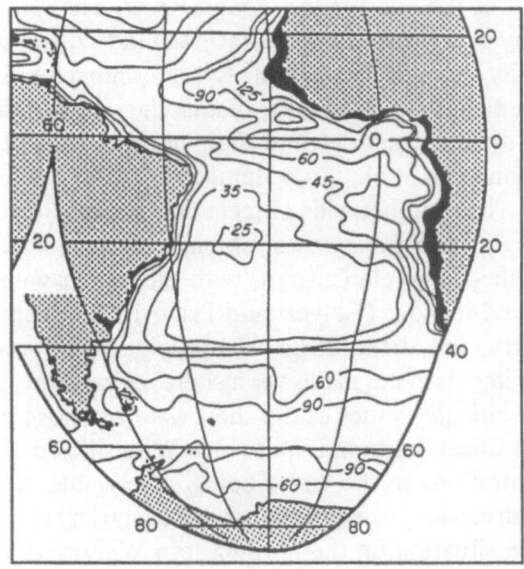

Figure 11. Productivity map for the South Atlantic, based on *in situ* measurements and satellite information. Numbers are primary production in gCm⁻²kyr⁻¹; black areas: >180 gCm⁻²kyr⁻¹. From Berger (1989).

Ridge site did experience intensified productivity from westward expansion of coastal upwelling cells at glacial periods, but this was not reflected in opal deposition. This is the hypothesis favored by the results of Oberhänsli (1991), who proposed increased upwelling during glacials, at DSDP Site 532, for the last 500,000 years, based on the character of foraminiferal assemblages. The situation may be more complicated, of course, than a discussion in terms of glacial-interglacial contrast would suggest (see, e.g., Pokras 1987).

We tend to agree with the assessment of Oberhänsli (1991). When using opal as a proxy for productivity, caution is indicated. For example, off the Congo River opal deposition and organic matter deposition are highly correlated (Fig. 12a). However, off Angola such correlation (if any) is much less in evidence because of the low amount of opal present within the sediment (Fig. 12b). Clearly, the difference in (organic) productivity between the two sites is quite modest, but the same is not true for the difference in opal deposition. Also, for Stage 4 the site off Angola shows the higher rates of organic carbon deposition, but the opal accumulation remains distinctly lower than that of the site off the Congo River. The reason why opal fluxes do not reflect the relative productivity changes between the two sites must involve the chemistry of silicate, that is, the silicification of diatom frustules and their preservation (van Bennekom et al., this volume).

The ratio between silicate and phosphate varies by more than a factor of ten, within the waters of the Benguela Current, with the higher values found inshore (Calvert and Price 1971). Apparently, the dissolution of opal in diatom-rich sediments in the shallower inshore regions elevates the silicate-values of the shelf waters, whose initial silicate content is quite low. With silicate concentrations in the water being so variable, little faith can be placed in opal as a productivity proxy. The situation on the northeastern Walvis Ridge, with opal decreased during glacials, is reminiscent of a similar pattern in the western equatorial Pa-

cific (Lange and Berger 1993). Here, also, diatoms tend to be poorly preserved within glacial sediments, despite the fact that these sediments apparently were laid down under more productive conditions, judging from the abundance of benthic foraminifers.

Idling the North Atlantic Heat Pump

The dynamics of heat piracy by the North Atlantic realm greatly depend on the state of the system with respect to ice buildup. When the northern hemisphere experiences maximum glaciation, asymmetry across the equator is diminished, and so is northward heat flux and hence NADW production. This decrease in production is clearly seen in the change of carbon isotope composition of benthic foraminifers (Fig. 13, compare a and b). The NADW, having but recently left the surface of the ocean, has high $\delta^{13}C$ values. In contrast, the

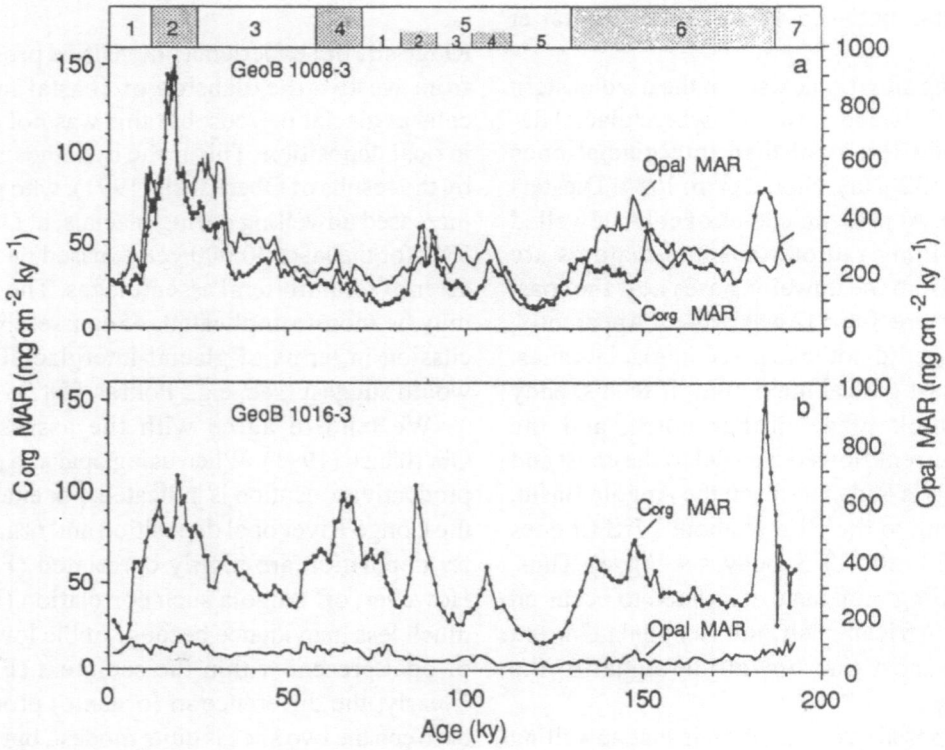

Figure 12. Comparison of C_{org} and opal accumulation rates. Core GeoB1008-3 on Congo fan (ca. 200 km offshore; 6°35S, 10°19E, 3124 m); Core GeoB1016-3 off Angola (ca. 250 km offshore; 11°46S, 11°41E, 3411 m). After Schneider (1991), from Berger et al. (1994).

AABW, having a strong admixture of deep waters (and having had less of a chance to equilibrate with the atmosphere when last near the surface) has low $\delta^{13}C$ values. The contrast between Recent and glacial-age compositions indicates that NADW production was significantly reduced, and the deepest Atlantic was bathed by waters originating around Antarctica. The extent of this replacement can best be studied in the South Atlantic (Fig. 13b).

The observation that NADW production is diminished during glacials raises a paradox. We know that, on the whole, NADW production increased as climate entered the northern ice ages, in the Pliocene (Johnson 1983; Raymo et al.1990; 1992; Whitman and Berger 1992; 1993). In fact,

the turning up of NADW is thought to be the main reason why the North Atlantic becomes a carbonate trap, starting in the Pliocene (Berger 1972). Why then is it that planetary cooling, over millions of years, is favorable to NADW production, but moving the system from an interglacial to a glacial is not? Clearly, the glacial-interglacial contrast is *not* an analog for general planetary cooling in the late Neogene. We shall return to this question as we take up the topics surrounding the origin and evolution of heat taxation of the South by the North, and the changing role of glacial-interglacial fluctuations on this asymmetry. First, however, we have to pay some attention to the changing boundary conditions over long time spans.

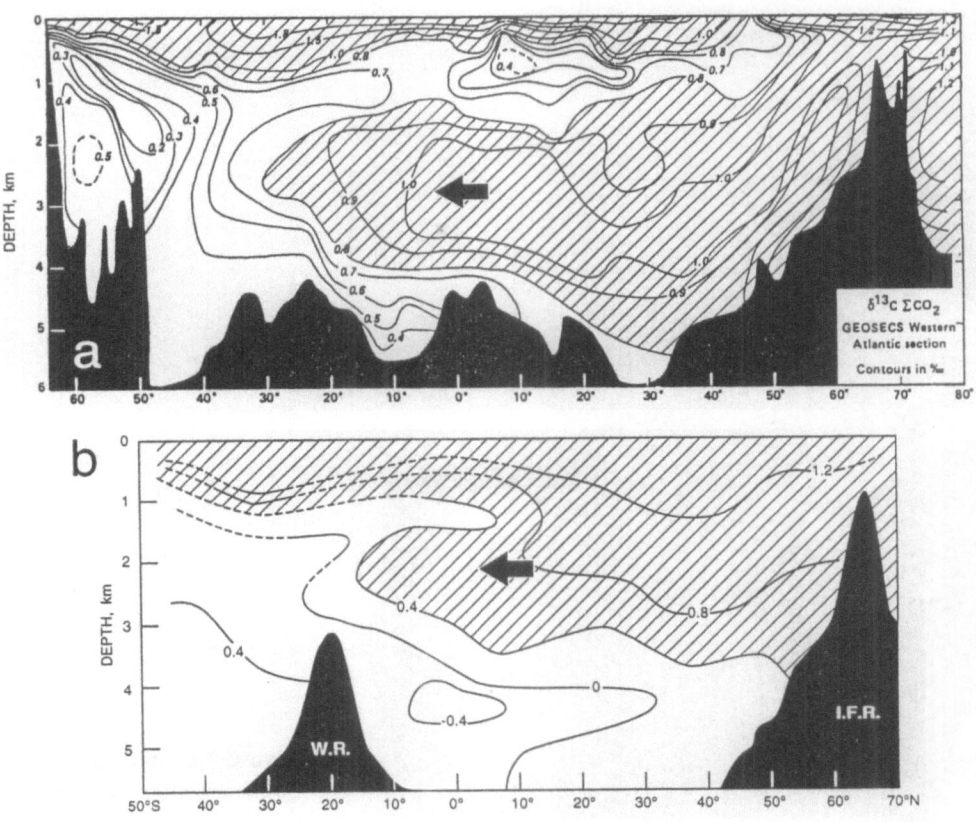

Figure 13. Comparison of $\delta^{13}C$ patterns of the present and the glacial-age Atlantic. a) Distribution of $\delta^{13}C$ values in the present Atlantic, western basin, based on GEOSECS results (Kroopnick 1980). b) Distribution of $\delta^{13}C$ values in the Atlantic 20,000 years ago, during the glacial maximum, based on the composition of benthic foraminifers (*P. wuellerstorfi*) in deep-sea sediments. Data for sediments in Duplessy et al. (1988) and Sarnthein et al. (1994). W.R., Walvis Ridge; I.F.R., Iceland-Faroe Ridge.

The many facets of Plate Tectonics

Seafloor spreading, bottom topography and facies evolution

Perhaps the single most important discovery in the South Atlantic, from any ship, was the confirmation of the hypothesis of seafloor spreading through the results of Leg 3 of the Deep Sea Drilling Project (DSDP) (Maxwell, von Herzen, et al. 1970) (Fig. 14). The biostratigraphers on board the GLOMAR CHALLENGER found that the ages of the basaltic sea floor predicted by the new theory coincided, within a few million years, with the ages of the sediments overlying the basalt. In addition, back to about 70 Ma, the ratios between distance from Ridge center and age were roughly constant, indicating a constant rate of spreading . For most geologists (although by no means for all) this finding constituted proof for the viability and usefulness of the concepts put forward by Dietz (1961) and Hess (1962).

A number of corollaries regarding circulation in the South Atlantic emerge from the Leg 3 discoveries. Because of the spreading, lines of equal seafloor age roughly follow depth, at least where

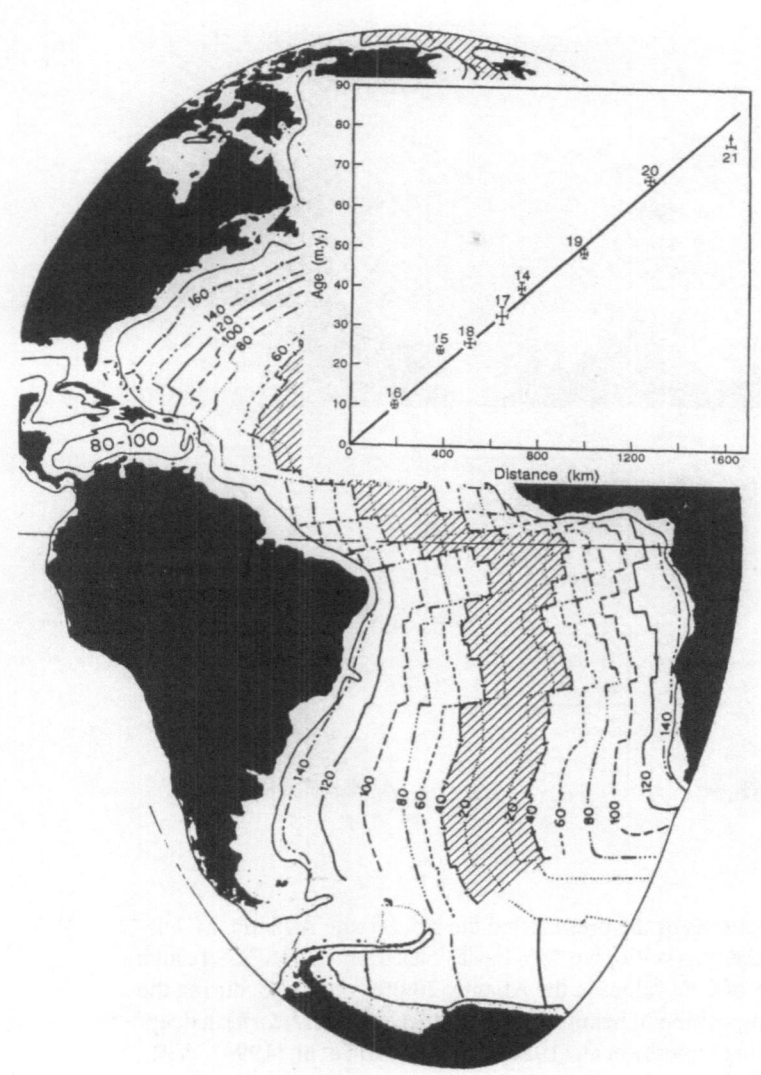

Figure 14. Seafloor spreading in the South Atlantic. XY-graph: Hypothesis of seafloor spreading confirmed by biostratigraphy of DSDP Leg 3 holes (age of fossils in oldest sediment). (Source: Maxwell, von Herzen, et al. 1970.) Map: Opening of South Atlantic reflected by age of sea floor, as seen in magnetic anomalies. (Source: Berger and Winterer 1974).

sediments are thin (Fig. 15). This implies that the twin troughs of the South Atlantic deepen with time, facilitating penetration of bottom water in the north-south direction. When considering, at any one site, the evolution of deepwater characteristics through time, the descent of the site toward greater depths has to be taken into account. Even if there were no change in the temperature profile whatever, sediments deposited millions of years after the origin of the basalt that bears them would show a cooling of bottom waters, relative to the earliest sediments.

The confirmation of seafloor spreading also, implicitly, confirmed a related hypothesis, that of the "nemataphs" of J.Tuzo Wilson (1908-1993). Wilson (1963) had proposed that a stationary hot spot leaves a trail of volcanic edificies on top of a moving lithosphere; the Rio Grande Rise and the Walvis Ridge were his favorite example of such a hotspot centered on the Mid-Ocean Ridge (see Fig. 15 for location). The V-shaped aspect of this pair of ridges (with the ancient contact at the continental margins well north of the central source) he explained as resulting from an overall northward motion of the lithosphere, including both bordering continents. The bottom topography suggests that the delivery of basalt from the central hotspot was not exactly symmetrical in east and west (it is not at present, in any case). Both ridges represent obstacles to bottom water flow.

At present the two ridges, together with the centrally positioned Mid-Atlantic Ridge, constitute obstacles for the northward flow of Antarctic Bottom Water (AABW). Despite being the less massive of the two, the eastern ridge provides the better barrier to the penetration of southern bottom water: the western ridge has a deep channel providing a passage, the Vema Channel, and a somewhat shallower one, the Hunter Channel. In the past, portions of the ridges emerged above sea level. At times in the Cretaceous, the ridges may have entirely isolated the incipient Angola and Brazil basins, with the result that intermittent flooding and evaporation led to the deposition of salt, which is now located deep within the continental margins.

Thus, the evolution of the facies regimes in the South Atlantic (that is, the types of sediments being deposited) was intimately tied to the opening of the basin, through spreading, and to the history

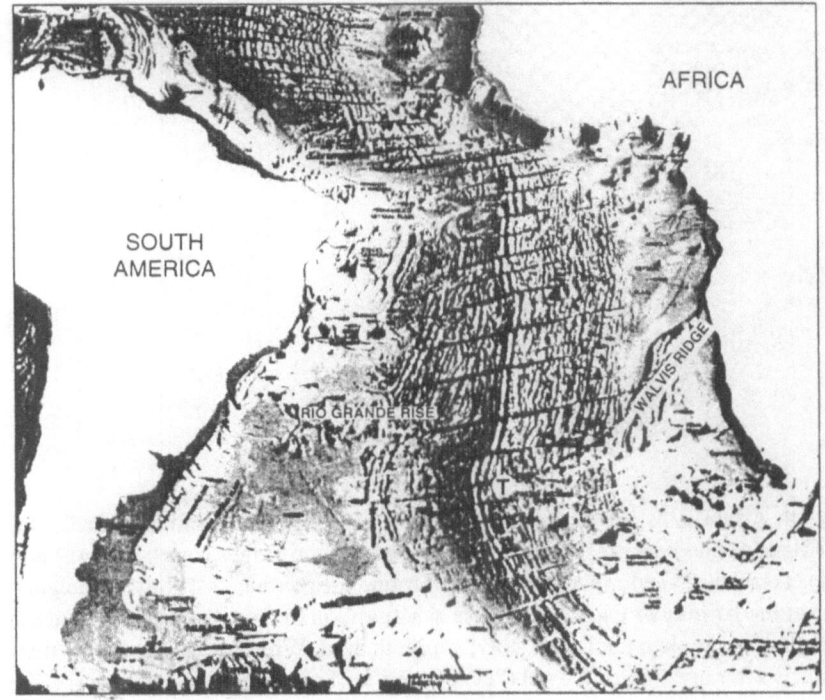

Figure 15. Bottom topography of the South Atlantic. Rio Grande Rise and Walvis Ridge both originate from the same hot spot (marked "T" for Tristan da Cunha Island), through seafloor spreading and by the northward movement of the sea floor with respect to the hot spot. Source: Seibold and Berger (1995), from a painting by H.C. Berann (National Geographic Society), based on work by B.C. Heezen and M. Tharp.

of the twin barriers (Fig. 16). Commonly, carbonate would have accumulated on the Mid-Atlantic Ridge (as today), and clay in the deepening basins (Fig. 16a). The path of the drill sites includes components for spreading, sinking, and northward drift, all of which have to be considered when reconstructing facies evolution (and hence history of productivity and deepwater circulation) (Fig. 16b). Several features are noteworthy in the facies pattern diagram of van Andel et al. (1977): The initial deposition of salt, the difference in sediments of the eastern and western margins (with opal in the east and mud in the west), the widening regions for the deposition of clay, and the drastic increase in the width of the central carbonate area, about 40 million years ago, which is accompanied by increased hiatus formation (white fields in the graph, Fig. 16b).

The warm era ends: The Auversian Facies Shift

What happened 40 million years ago? The overall northward motion of the continents, away from the more or less stationary Antarctica, greatly modified the geographic boundary conditions for global circulation (Fig. 17). As a general trend, important seaways in the tropics, allowing world-encircling flow of tradewind-driven water, tended to close up gradually after the Eocene (black bars in Fig. 17), while gateways in high latitudes opened (open rectangles). As a consequence, the thermohaline system moved slowly but inexorably from a Warm-Ring-dominated world toward the present configuration of a central Cold Ring, with more or less separated basins attached as cul-de-sacs. Naturally, both surface and sub-

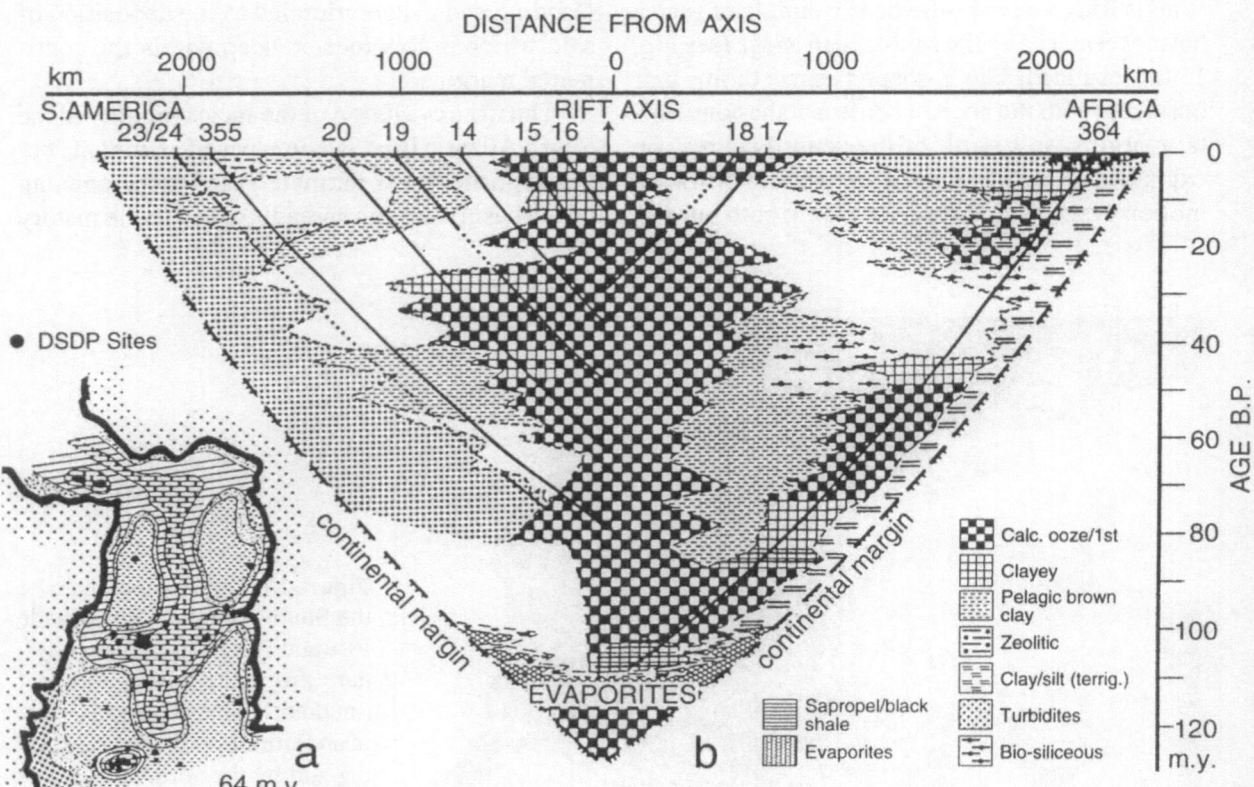

Fig. 16. Facies evolution in the South Atlantic, as a function of spreading, sinking, and northward drift. (a) Sedimentation in the middle-aged Atlantic (one half present age), with carbonate on ridges (brick pattern) and clay in deep basins (broken hachured). Terrigenous sediment accumulates around the continents. (Source: Melguen et al. 1978.) (b) Facies distributions at the latitude of Rio Grande Rise as a function of time since opening of the South Atlantic. Checkered pattern: carbonate; broken hachured: clay; stippled: mud. White areas: erosion or non-deposition. Note the salt deposits shortly after opening of the rift ("evaporites"). (Source: van Andel et al. 1977.)

surface circulation were profoundly affected (e.g., Berggren and Hollister 1974). General cooling is superposed on this purely geographic trend. The interactions between the cooling (especially in high latitudes) and the geographic modification determined the evolution of late Cenozoic circulation and deposition of deepsea sediments.The more or less gradual transition from a warm-ocean to a cold-ocean mode of sedimentation (here called the "Auversian Facies Shift") is roughly centered on a period about 40 million years ago. Important turnovers within the benthic fauna are seen around this time in the South Atlantic and elsewhere (Benson et al. 1985; Oberhänsli et al. 1991) as well as associated changes in benthic $\delta^{18}O$ values in the late Middle Eocene and at the Eocene/Oligocene boundary (Oberhänsli and Toumarkine 1985).

The most striking changes occur in the character of the deepsea carbonate deposition: Before 40 million years(Fig. 16b), carbonate accumulation tended to be restricted to the shallower regions of the deep sea floor, and it commonly alternated with opaline deposits (altered to chert). After the Eocene, we find rather pure carbonates widespread over the sea floor, and opaline deposits tend to be restricted to regions around the continents and to the equatorial divergence. Eventually, in the late Neogene, as much as one half of all the opal was captured within the Cold Ring. Thus, as the ocean moved from a warm mode to a cold mode of operation, it became an efficient fractionation machine. Before that time it was not. This is fundamentally the reason why we find chert deposits throughout the deep sea in the older sediments. (Earlier explanations attempted to explain why chert layers exist in these older sediments, invoking special productivity events and volcanism. Our approach is to change the problem: What needs explanation is the *absence* of siliceous deposits over much of the sea floor, after the mode switch.)

The great change in carbonate deposition which marks the Auversian Facies Shift is strikingly reflected in a world-wide drop of the "carbonate compensation depth" (CCD), that is, the water depth below which but little carbonate accumulates on the sea floor, in any one region. This drop was first identified in the South Atlantic. In addition to the all-important confirmation of seafloor spreading (Fig. 14), the third leg of DSDP (which moved the ship from west to east along 30°S, roughly) provided a rather complete carbonate preservation stratigraphy for the Tertiary period, thanks to the efforts of sedimentologist K.J. Hsü and colleagues (Hsü and Andrews 1970; also see Hsü and Weissert 1985). Hsü and Andrews (1970) found large changes in the intensity of carbonate

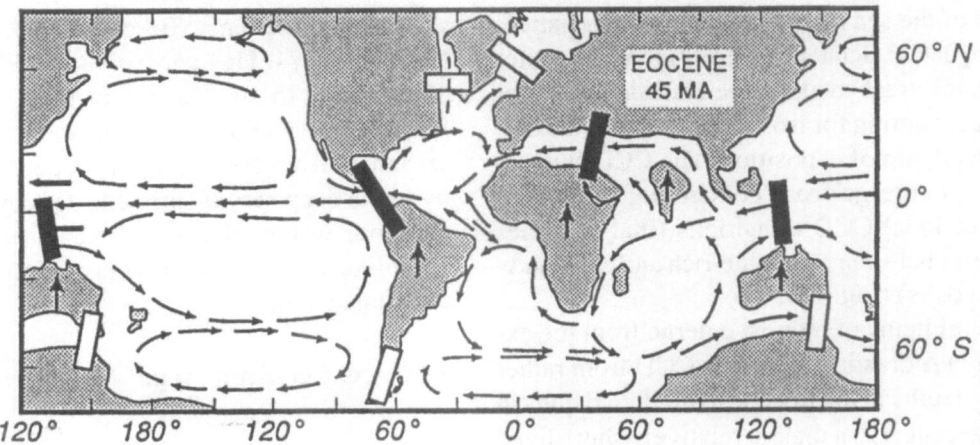

Figure 17. Geography of the middle Eocene (ca. 45 Ma) and major valve points for ocean circulation. Tropical valves are closing (black bars), high latitude valves are opening up (open rectangles) throughout the Tertiary. (Source: Seibold and Berger 1995, modified; base map from Haq 1981.)

dissolution for the Tertiary sequences recovered. Assuming that carbonate preservation is strictly tied to depth, they postulated large changes in the elevation of the Mid-Ocean Ridge, in the South Atlantic. Such a scenario implies that the sea floor is relatively deep when carbonate deposition is reduced, and this, in turn, implies large-scale retreat of the sea from the shelf during such periods. Exactly the opposite is observed: The shelf seas tend to be widespread whenever carbonate accumulation in the deep sea is reduced. The reason, it is thought, is the competition between shelf and deep sea floor for carbonate, which is easily won by the shallow and warm site, where carbonate is more stable (Berger and Winterer 1974; Milliman 1974).

In the modern ocean the elevation of the Mid-Ocean Ridge varies but little throughout the ocean, whereas CCD levels and associated preservation levels vary considerably (Revelle 1944; Bramlette 1961), especially in the South Atlantic, where preservation levels are tied to the boundary between AABW and NADW (Ruddiman and Heezen 1967; Berger 1968). Because the CCD levels are tied to deep water boundaries, it made sense to re-interpret the Leg 3 fluctuations in carbonate preservation in terms of deepwater properties (Berger 1972; Berger and Winterer 1974; van Andel et al. 1977; Melguen et al. 1978; Hsü and Wright 1985). Using the principle that subsidence of the sea floor follows a regular pattern (Menard 1969; Sclater et al. 1971), drilling sites can be back-tracked along their subsidence paths (while accounting for isostasy), to obtain the correct paleodepth of deposition. The CCD fluctuations then emerge from connecting the paleodepths of local CCD transitions (that is, facies boundaries between carbonate-rich and carbonate-poor deposits) (Fig. 18).

Several items of interest emerge from the exercise: (1) A drastic drop in the CCD from rather shallow depths in the Eocene to modern depths in the Oligocene; (2) a drastic relatively short-lived upward CCD excursion in the late Middle Miocene, to a depth near 3 km, leaving little room for any carbonate deposition at all; (3) a steady drop of the CCD beginning in the late Miocene. Hsü and Wright (1985) added considerable detail

to this general picture, employing data from Leg 73. Their reconstruction emphasizes abrupt vertical displacement of the CCD within the last 20 million years (see Fig. 18a, stippled line).

A similar experiment using data from Leg 74 reveals additional detail, regarding the pattern of carbonate accumulation in the Cenozoic (Moore et al. 1985; Fig. 18b). These results, from the Walvis Ridge, illustrate the fact that absence of carbonate does not necessarily mean that the regional CCD is above the site considered. At Site 525, for example, there is no accumulation of carbonate within the late Eocene and the entire Oligocene, presumably an effect of mechanical action of bottom currents, at this relatively shallow depth (or dissolution resulting from supply of organic matter, or both).

Before the South Atlantic pattern can be interpreted properly, it needs to be compared with equivalent patterns from other ocean basins, to ascertain whether it is global or regional. The Eocene-Oligocene CCD drop is a global signal, while the strong excursions within the Miocene have both global and regional elements. The overall shape of the CCD curve more or less parallels large-scale sealevel change, with a major regression setting in during the late Eocene, interrupted by modest transgression in the Miocene. Thus, this general aspect of the CCD is explained by the way shelf seas and the deep ocean share carbonate: The shelf takes precedence (Berger and Winterer 1974; Milliman 1974; Hay and Southam 1977). Denial of carbonate to the deep sea can take two forms: Increased dissolution at depth, as the shelf strips carbonate from the ocean, or reduced productivity in the deep sea, as nitrate is destroyed in oxygen-poor waters and in sediments on the shelf. Most likely the competition is carried out in several dimensions.

The great cooling steps

The first great cooling step: Acquiring cold deep water

We live in an Ice Age. The Antarctic continent is covered by a thick layer of ice, which if melted

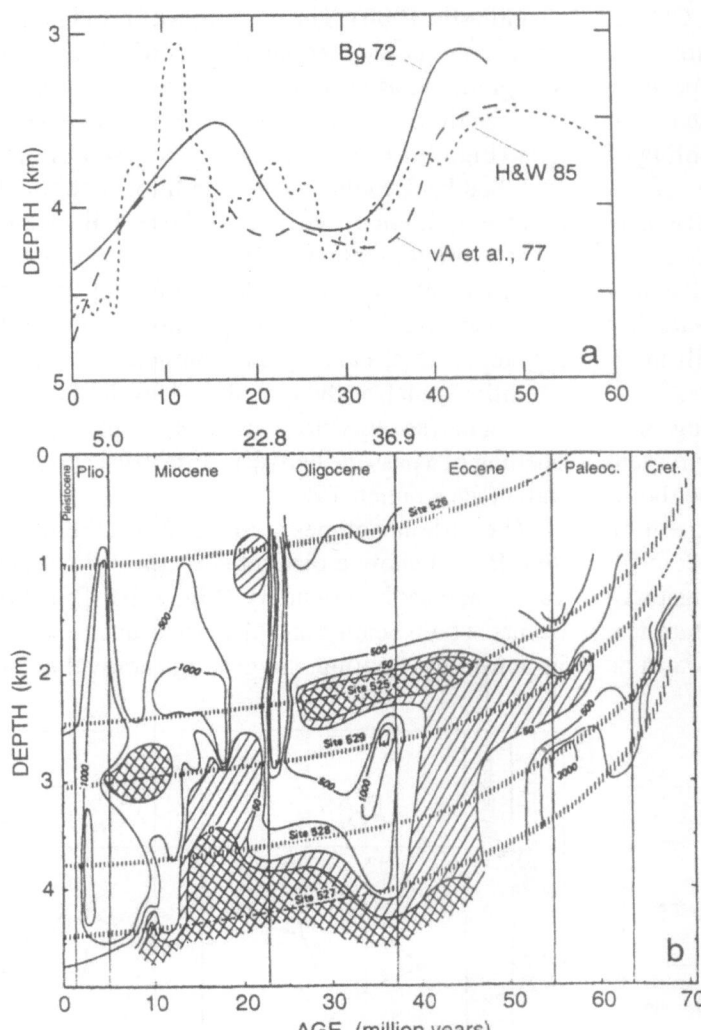

Figure 18. Reconstruction of the CCD in the South Atlantic, using backtracking of drilling sites. (a) Comparison of reconstructions of Berger (1972), van Andel et al. (1977), and Hsü and Wright (1985). (Bg72 reconstruction is the 2.5 m/m.y. contour based on Leg 3 data; H&W85 reconstruction includes Leg 73 data.) (b) Carbonate accumulation rates on Walvis Ridge as a function of reconstructed water depth and age of sediment, based on Leg 74 data (Moore et al.1985).

would raise sea level by some 70 m. Put another way, there is about 6 times more water piled on Antarctica than there is in the entire Mediterranean. The ice on Greenland, on the other hand, is roughly equivalent in mass to the water in the Mediterranean Sea.

How did we get into this situation? What were the effects on circulation in the ocean, and especially in the South Atlantic, and how is this history reflected in the sediments?

On the whole, the Cenozoic is characterized by colder climates than the Mesozoic. However, the position of the Mesozoic-Cenozoic boundary has nothing to do with cooling. It is the result of an historic accident: A large rock careening through the solar system, one of thousands such aimless wanderers, came too close and collided with our planet (Alvarez et al. 1980). Had this collision not occurred, the present ocean with its frigid deep waters, its thin warm layer, and its major upwelling regions, would still be entirely different from the oceans of the warm Cretaceous period, with widespread shelf seas and poorly ventilated deep water. Thus, we would still separate modern times from ancient (whoever "we" might be).

Purely on the appearance of deep-sea sediments, we might choose the final stages of the Eocene as the period showing the greatest and most lasting change in deep-sea facies, that is, the Auversian Facies Shift. At this time, the CCD descends in the deep ocean (Fig. 18), indicating the large-scale re-arrangement of carbonate deposition that comes with major regression. Associated with this depositional mode shift is a major cooling of deep waters (Fig. 19), that is, the origin of the "cold-water sphere" or "psychrosphere" (Benson 1975; Kennett and Shackleton 1976; also see Douglas and Savin 1975; Shackleton and Kennett 1975; Barron 1985). Ever since, the deep water has been getting colder, on the whole, eventually moving the system into the present frigid state.

This end-of-Eocene deepwater cooling, perhaps 6°C or so, was almost certainly contingent upon moving a high-latitude rainbelt off the shores of Antarctica, to make room for the refrigeration of more saline water which could then sink to fill the ocean basins. Thus, it signals an expansion of continental and frigid conditions on Antarctica, and an overall switch from low- to high-latitude deepwater sources (as first envisioned by Chamberlin 1906). The cooling was rapid, and it resulted in a strengthening of the vertical temperature gradient, as seen in the increased separation between the $\delta^{18}O$ values of planktonic and benthic taxa (Fig. 19a). Interestingly, the system never switched back to the earlier conditions, although there is evidence that the southern cold water source varied in intensity (Site 511, eastern Falkland Plateau; Wise et al. 1985). This suggests that an irreversible change was responsible for the cooling step, that is, a change in geographic boundary conditions, such as the growth and uplift of mountain chains (see Raymo et al. 1988), or perhaps the opening of a passage between Antarctica and Australia (see Kennett 1982).

The carbon isotopes at Site 522 (southern Angola Basin) show a distinct shift toward heavier values across the boundary (Fig. 19b). The shift occurs in both benthic and planktonic taxa, and sets in before the cooling, suggesting removal of or-

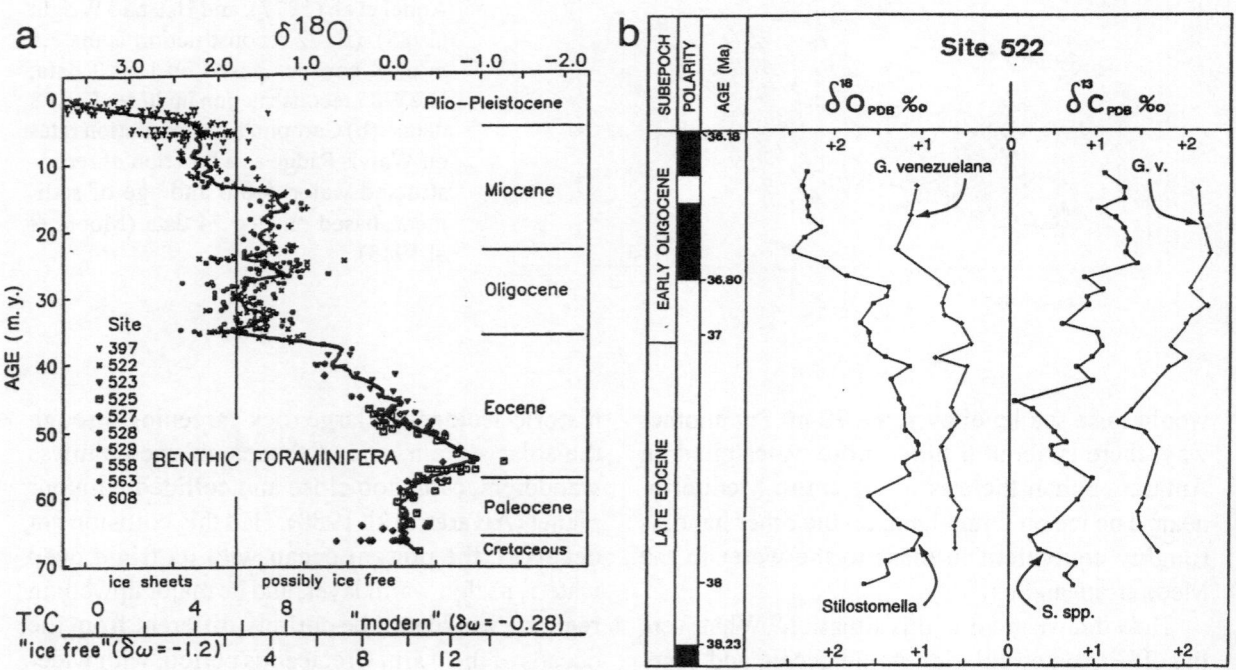

Figure 19. Stable isotopes of foraminifers as indicators of stepwise cooling in the Cenozoic. (a) Composite record for the Atlantic (mostly South Atlantic sites), with paleotemperature scales for two different assumptions about the presence of polar ice caps. Source: Miller et al. (1987). (b) Detailed isotopic record of a planktonic foraminifer (*Globigerina venezuelana*) and a benthic genus (*Stilostomella*) across the Eocene-Oligocene boundary, in the southern Angola Basin. Source: Oberhänsli and Toumarkine (1985).

ganic carbon (which is enriched in ^{12}C) before the deep water became cold. The fact that the $\delta^{13}C$ stays high is partly due to a temperature effect, which is activated through gas exchange between ocean and atmosphere (cf. Charles and Fairbanks 1990). Also, it may reflect continued enhanced extraction of organic carbon, presumably in the ocean margins, or input of terrigenous organic carbon, from weathering (see Schneider and Müller 1995, for discussion). It is noteworthy that the Auversian Facies Shift in carbonate deposition correlates more closely with the shift in carbon isotopes, and a general cooling in the late Eocene, than with the end-of-Eocene cooling step proper.

One of the vexing properties of the $\delta^{18}O$ proxy is that it responds to both ice buildup and temperature change. Much discussion has centered on the question to what degree the cooling step at the end of the Eocene reflects growth of ice on Antarctica. If we assume that the water temperature in the source region for deep water is within a few degrees of 5°C (Fig. 19a), we should expect some ice growth in the interior of the continent, especially at higher elevations. Such ice cover, in fact, would stabilize and extend the high pressure conditions that are necessary to move the Eocene Antarctic rain belt away from the continent, so that excess precipitation did not hinder deepwater formation. We doubt, however, that such ice would make much difference to the $\delta^{18}O$ signals seen at the Eocene-Oligocene transition: The volume is presumably modest (or ice-rafted deposits would be more prominent) and the fractionation is modest also (that is, the $\delta^{18}O$ values would be no more than half of today's because of the lesser temperature gradient). Thus, say, 20% or less of the end-of-Eocene change should be attributable to an ice effect.

Some corollaries of cooling

The Auversian Facies Shift culminates in the end-of- Eocene cooling step: From this time on the variable stratigraphy of the early Cenozoic, with its intriguing chert layers and multicolored clay layers and limestones, is entirely replaced by a superficially rather boring sequence of chalks and calcareous oozes, nicely separated from equally uniform pelagic clays. As K.J. Hsü writes of the work on board the GLOMAR CHALLENGER during Leg 3 of DSDP (Hsü and Weissert 1985, p.1): "Sedimentologists then had the thankless task of describing hundreds and hundreds of metres of pelagic oozes, which all looked alike - a routine that could be done by any high school graduate."

Naturally, the question arises why these sediments should be so much more uniform than the ones in the early Cenozoic. We cannot assume that the sources of dissolved silicate suddenly vanished. On the contrary, inasmuch as rivers deliver about one half of the dissolved silicate (the other half coming from volcanic and hydrothermal processes) we should expect increased delivery with emergence of land areas. Thus, we must postulate increased extraction of silicate, at the rim of the ocean, in the newly forming upwelling regions.

The post-Eocene focussing of silicate extraction (which is also seen in equatorial regions; Berger and Winterer 1974; van Andel et al. 1975) is facilitated by changes both in physical and chemical oceanography. The cooling of high latitudes (the low latitudes are comparatively little affected) strengthens temperature gradients and air pressure differences, and this leads to stronger winds and upwelling along margins. The cooling also allows for better ventilation and oxygen supply to the deep ocean, slowing denitrification. Thus, silicate can be extracted more quickly, upon introduction to the photic zone: It precipitates rapidly at the margins and off rivers.

The new opportunities for rapid growth in upwelling areas favored the development of robust diatoms and resulted in the evolution of deep-sea radiolarians that manage with less opal in their skeletons (Moore 1969). Calcareous plankton underwent a remarkable drop in diversity (Thierstein et al., in Munsch 1988). In the planktonic foraminifera, a flashy fauna rich in morphologically distinct species is replaced by a rather uninspiring assemblage of globular forms that show little change through time (Vincent and Berger 1981; Kennett and Srinivasan 1983).

It appears, then, that the cooling moved much of the action to the ocean margins, both as concerns sedimentation and evolution. In the ancient oceans competition for carbonate was foremost, between shelf and deep sea floor. Now, as we move into a long-term cold phase without shelf seas, the deep sea gets the carbonate but must compete with the narrow margins (including the upper slope) for the available silicate. This competition gets ever more fierce, as the Cold Ring enters the contest. Before we turn to this topic, which greatly impacts the South Atlantic, we should ask: What started the cooling in the first place?

In principle, the factors leading to global cooling are well understood: Mountain building provides for a deepening of the ocean, and a retreat of the shelf seas. Land surfaces are exposed. Ground albedo, that is, the ability to reflect sunlight, immediately increases, diminishing the amount of solar radiation that is converted to heat. On land, seasonality increases. This increases short-term infrared return of energy to space, during summer (without generating latent heat or warm water). In winter, in high latitudes, increased seasonality may lead to snowfall, greatly enhancing ground albedo. The radiatively active trace gases are diminished. Water vapor is decreased because the available surface for evaporation is diminished, and because cool air holds less water. Carbon dioxide is decreased because of increased opportunity for reaction with freshly exposed igneous rocks, and because of increased opportunities for burial of organic carbon in upwelling regions. (This phrasing is dictated by the necessity for geochemical balance - while the flux of C into and out of the atmosphere may not change much, the level at which this flux occurs is lowered by widening the access to sinks.)

Reduction of carbon dioxide in the atmosphere has long been suspected as an important factor in global cooling (see Schneider and Müller 1995, for review). The Urey equation describes the basic weathering process (Urey 1952):

$$CO_2 + CaSiO_3 \rightarrow CaCO_3 + SiO_2 \qquad (1)$$

whereby fresh silicate minerals are destroyed to make carbonate and opaline sediments (coccoliths and diatoms, in a cool ocean). The burial of carbon is described by the photosynthesis equation:

$$CO_2 + H_2O \rightarrow CH_2O + O_2 \qquad (2)$$

To prevent buildup of O_2, iron-rich minerals are oxidized:

$$4 FeSiO_3 + O_2 \rightarrow 2 Fe_2O_3 + 4 SiO_2 \qquad (3)$$

In sum, mountain building and increased upwelling on the cooling planet makes for less CO_2 in the air and for more carbonate and silica in the sediments. In addition, the process produces rust, which, when delivered by air, stimulates productivity in the ocean (further favoring focussing in upwelling regions). Much of the iron ends up in the margin sediments (reduced after burial), some in the reddish-brown pelagic clay, and some in the ferromanganese nodules, a hallmark of cold oceans.

The history of regression, or rather of weathering on land, can be read from the strontium isotopes in calcareous fossils. What is being measured is the ratio between [87]Sr and [86]Sr, the former being a daughter of [87]Rb which is enriched in continental crust. The changes in this ratio suggest increasing weathering of continental crust throughout the second half of the Cenozoic (Fig. 20). Clearly, the first distinct sustained increase in the ratio is at the end of the Eocene, indicating that igneous continental crust becomes increasingly available for weathering at this time. Regression probably set in earlier, but would not have been seen in this proxy, until sufficient amounts of fresh silicate rock were exposed. The trend, and the event at the Auversian Facies Shift (AFS) is enhanced if the estimated input ratio of fluxes is plotted rather than the ratio within the ocean reservoir. This ratio (shown as connected triangles) is calculated by assuming proportional mixing between the values for volcanic input (0.7036; Palmer and Elderfield 1985) and river input (0.7119; Palmer and Edmond 1989), using linear interpolation on the difference between the ratios. (The main assumption is that the river-borne Sr-value does not change, which is probably not quite correct. If it increases, less flux is needed.)

Again we ask, did we enter the Ice Age with this first major cooling step? Probably not. There is some indication of ice-rafted debris around Antarctica (Kennett 1982; Ciesilski and Weaver 1983); however, the deposits stay close to the

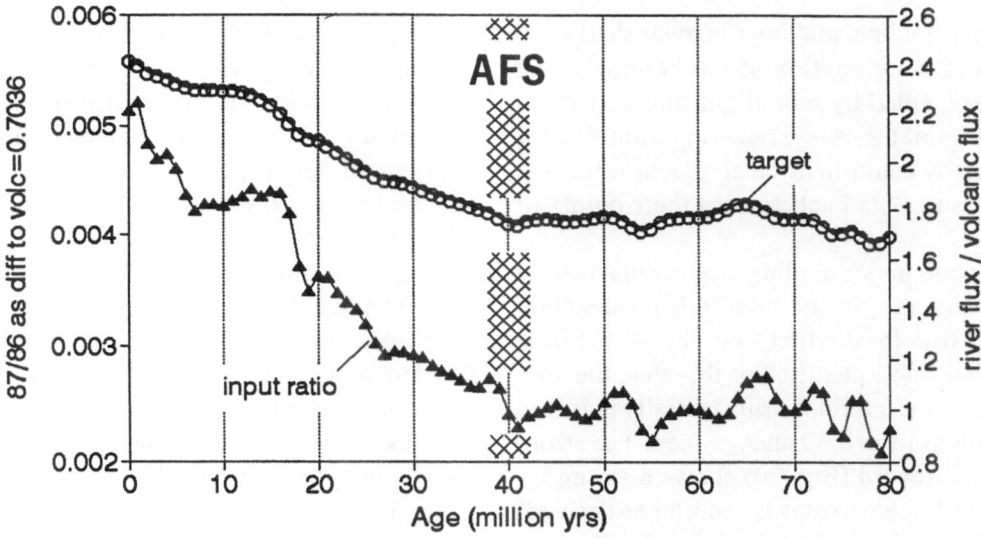

Figure 20. Strontium isotope stratigraphy of the Cenozoic. Target: 87/86 ratio as difference to the volcanic (hydrothermal) value of 0.7036. Input ratio: flux ratio of river input to volcanogenic input, assuming river composition is 0.7116, by linear mixing of endpoint values in a reservoir with 4 million year residence time. Data from Elderfield (1986). AFS, Auversian Facies Shift (segregation of deepsea facies, beginning of sustained deposition of carbonate ooze).

coast, suggesting small bergs released from modest mountain glaciers. If large ice sheets did build during the Oligocene, they were most likely ephemeral (see Zachos et al. 1993, for discussion of this point).

In summary, the first great cooling step occurred in response to an overall drop in sealevel that resulted from tectonics, that is, mountain building. The carbonate supply to the deep ocean increased greatly. A cold-water ocean was established, although all through the Oligocene there was probably still vigorous competition between temperate, salty sources and high-latitude cold-water sources as suppliers of deep water (cf. Kennett and Stott 1991). (Such a competition, and the associated instability, may be responsible for the strange *Braarudosphaera* layers common in the Oligocene carbonates of the South Atlantic. These stress forms indicate monospecific blooms following unusual events; perhaps mid-ocean overturn events.) Increased coastal upwelling removed dissolved silicate from the ocean which profoundly changed the facies distribution of deepsea sediments. Positive feedback on cooling set in,

from organic matter burial, and perhaps from albedo changes in Antarctica. However, the Ice Age had not begun yet, that is, sealevel change owed but little to the buildup of ice.

The second great cooling step: Toward a modern ocean

After the end-of-Eocene cooling step there was no further comparable cooling for a very long time (Fig. 19). In fact, an argument can be made that it was tropical warming in the Miocene which next changed the ocean system, by increasing temperature gradients between high and low latitudes and between shallow and deep water. With the arrival of a strong thermocline, some 20 million years ago, the life habitat (and the life habits) of plankton changed greatly. In planktonic foraminifera, there is a burst of new morphologies and, eventually, a new propensity for encrustation of shells. Presumably, in a water column with less density contrast any pronounced shell thickening would quickly lead to sinking and elimination of any individual weighted down by a thick crust. In an ocean with a strong thermocline, making a heavier shell as an adult allows congregation at the bottom of the mixed layer, aided by moonlight timing, for exchange of gametes. Also, new opportunities for travel arise, by adults in tropical regions riding the deep undercurrents back toward their points of origin.

The second major cooling step occurs near 14 million years ago, in the middle Miocene (Fig. 19a). As is true for the first step, the $\delta^{18}O$ shift is not reversed subsequently. For this step, the consensus is that ice buildup is substantially involved in providing for the $\delta^{18}O$ change seen. The strontium isotope record (Fig. 20) shows a strong increase in radiogenic (that is, continent-derived) strontium preceding the temperature drop in deep waters. Thus, vigorous uplift, and associated chemical weathering of continental igneous rocks, apparently helped trigger the cooling and Antarctic ice buildup. In some reconstructions of sealevel, the middle Miocene has maximum height (Vail and Hardenbol 1979), in others, it is a time of retreat of the ocean from the land (Haq et al. 1987). The strontium record suggests that the second alternative is the more viable.

The mid-Miocene cooling step is preceded by yet another striking anomaly in the stable isotope record of deepsea carbonate: a marked excursion in oceanic $\delta^{13}C$, toward heavier values. This is the so-called Monterey Event. A comprehensive study of benthic isotopes in the South Atlantic, near 30°S, has allowed Woodruff and Savin (1989) to chart the evolution of the Miocene $\delta^{13}C$ record as a function of water depth (Fig. 21a). Centered near 16 Ma, well before the great temperature drop, they find a dramatic increase of $\delta^{13}C$ values, at all depths. This indicates that we are dealing with a global signal. A change in the importance of oceanic carbon sinks, with an increase for organic carbon under upwelling regions, would help explain the phenomenon (Vincent and Berger 1985; Schneider and Müller 1995). In addition, as mentioned, increased opportunity for burial of carbon below upwelling areas would lower the atmospheric pCO_2, and keep it low. The concomitant lowering of the global temperature would provide the negative feedback necessary to stabilize the

system, by reducing the rates of weathering (which takes up CO_2 and provides nutrients to the ocean).

As in the case of the first great cooling step at the end of the Eocene, there is the possibility that changes in ocean passages contributed to the change in climate seen. In this case, the opening of the Drake Passage can be invoked, as well as (perhaps) the closing of a major Indonesian gateway. The history recorded in the strontium isotopes and in the carbon isotopes suggests that more fundamental forces are doing most of the work, that is, mountain building and focussed upwelling. This does not preclude, of course, an important role for changing heat distribution patterns resulting from opening and closing ocean gateways.

That these patterns change drastically as a consequence of the new Ice Age world is nicely demonstrated in the results of Woodruff and Savin (1989): The $\delta^{13}C$ patterns show the vertical segregation between two deep water masses, beginning about 10 million years ago (Fig. 21a), heralding the modern deepwater stratification in the South Atlantic. When comparing $\delta^{13}C$ values between South Atlantic and South Pacific, the age difference between the abyssal waters becomes evident, which results from turning on a North Atlantic deepwater source (Fig. 21b). Again, the process is seen to start 10 million years ago, and the source is seen to increase in importance, involving successively deeper regions within the water column. This confirms that cooling is favorable for NADW production, as long as we are in a pre-Quaternary situation.

The production of NADW sets up a major asymmetry between Atlantic and Pacific, which is reflected in the distribution of deepwater properties and of opaline sediments. The "age" of deep waters, that is, the time since they last saw the atmosphere at the ocean surface, can be traced by the cadmium content in foraminiferal shells. Old water has the higher cadmium, which is a biologically active element, and Cd/Ca patterns can be used to trace deepwater flow, therefore (Boyle and Keigwin 1982). Delaney (1990) compared the Miocene Cd/Ca ratios from benthic foraminifers in a South Atlantic site (DSDP 525, Leg 74) with corresponding ratios in a site in the western equa-

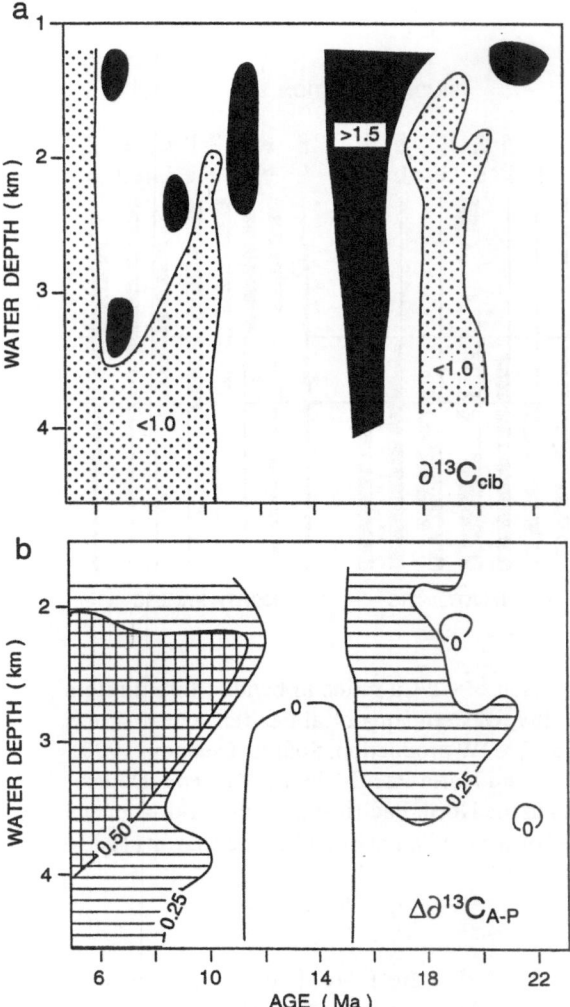

Fig. 21. Carbon isotope evolution in the Miocene deep waters of the South Atlantic (30°S). (a) $\delta^{13}C$ of *Cibicidoides* (benthic foraminifer) as a function of age and water depth. Densely shaded: values greater than 1.5 permil (note the positive excursion centered near 16 Ma). Stippled: less than 1 permil (note deepwater regime after 10 Ma, with vertical separation of water masses of different composition). (b) Differences in $\delta^{13}C$ values between South Atlantic and abyssal South Pacific. Positive values above 0.25 permil shaded. Note the high values beginning about 10 Ma ago at intermediate depths and expanding downward. Source: Woodruff and Savin (1989).

ago, and distinct and growing since 10 Ma ago, in complete agreement with the $d^{13}C$ data of Woodruff and Savin (1989; Fig 21). Opal deposition also supports this scenario (Fig. 22b). Basically, at the end of the Mid-Miocene cooling step, the North Atlantic no longer accumulates much opal, as it fills up with silicate-depleted surface waters through deepwater production. Instead, it turns into a carbonate trap. The North Pacific, in contrast, turns into a silicate trap. This "silica switch" (Keller and Barron 1983; Barron and Baldauf 1989) marks the beginning of the modern ocean.

In the Antarctic realm, the onset of modern times about 10 million years ago is marked by a substantial expansion and intensification of opal deposition (Fig. 23). At the various sites in the Southern Ocean, increased accumulation of diatomaceous sediments is followed, within a few million years by an increase in ice-rafted debris. Opaline deposits and ice-rafted debris are first seen around the Antarctic continent, within the Oligocene. By the mid-Miocene, such sediments occur up to 65°S and somewhat beyond. By the late Miocene, they extend north of the polar front, into the South Atlantic. Larger clasts from bergs start within the Pliocene, for most sites, indicating the buildup of ice sheets on Antarctica which were large enough to carry material considerable distances before melting (Ciesilski and Weaver 1983; Wise et al. 1985). A number of sporadic cooling events were associated with ice-rafting in the Pliocene, as seen in polar front fluctuations at the southern boundary of the South Atlantic (Sites 513 and 514 on the Mid-Atlantic Ridge, 46°S and 47.5°S). These fluctuations are reconstructed from radiolarians and other microfossils (Ciesilski and Weaver 1983; Wise et al. 1985). Near 3 million years ago, the Polar Front Zone (PFZ) moved rapidly northward (north of present latitudes, in fact), signalling the effects from the third major cooling step in the Cenozoic, the step that brought the northern hemisphere into the Ice Age in earnest.

The third great cooling step: The northern hemisphere enters the Ice Age

About 3 million years ago, general cooling had progressed to the stage where land surfaces that

torial Pacific (DSDP 289, Leg 30), to pin down the evolution of the grand asymmetry between Atlantic and Pacific (Fig. 22). Her work shows that the asymmetry becomes noticeable just before 12 Ma

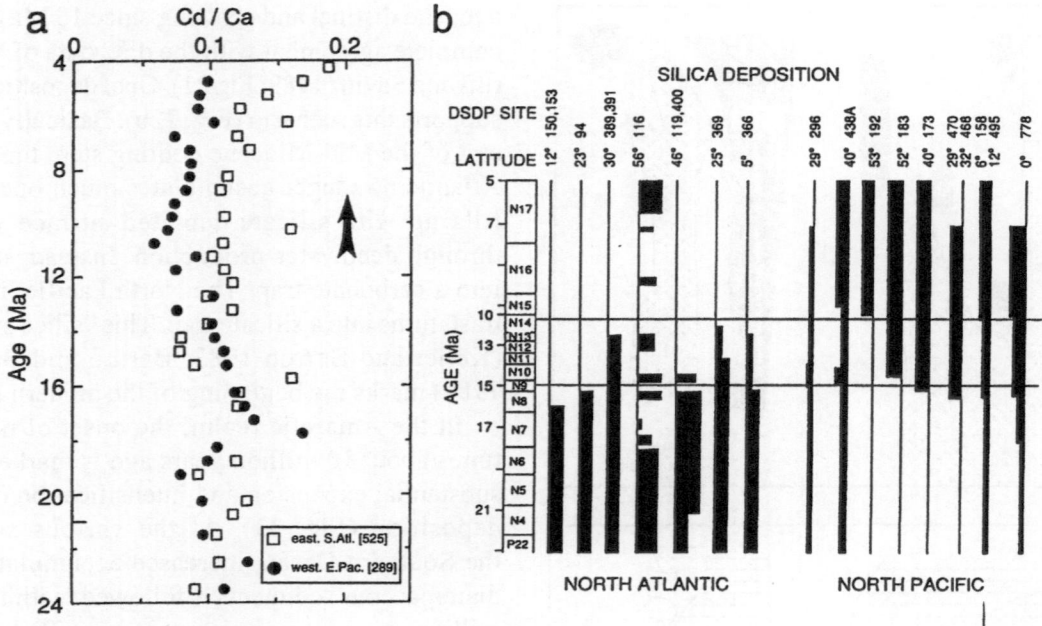

Fig. 22. The onset of NADW production. (a) Comparison of Cd/Ca ratios in benthic foraminifers from Site 525, eastern South Atlantic, and from Site 289, western equatorial Pacific. Arrow shows the beginning of the separation of values, attributable to NADW production. Source: Delaney (1990), redrawn. (b) Opal deposition patterns in North Atlantic and North Pacific, in the Miocene. Deposition ceases in the North Atlantic and greatly increases in the North Pacific during the Mid-Miocene cooling step. Source: Woodruff and Savin (1989), based on a compilation by Keller and Barron (1983).

are well away from the pole (Greenland, northern North America, Scandinavia) were able to acquire and hold large ice sheets. However, these ice sheets were (and are) precariously close to being melted during periods when the sun is unusually high over the horizon in summer. This tendency for instability is increased after isostasy has depressed the continents carrying the ice (Emiliani and Geiss 1958; and many since). The result is cyclic waxing and waning of ice masses on the northernmost land areas, in tune with summer insolation (Milankovitch 1930; Hays et al. 1976). The beginning of this regime is clearly reflected in ice-derived debris in deep-sea sediments of the North Atlantic (Fig. 24), as first seen in Sites 111 and 116 of DSDP Leg 12 (Berggren 1972). Results since then have confirmed this picture, except for the discovery of sporadic earlier attempts at ice buildup in Greenland (Jansen and Sjøholm 1991; Larsen et al. 1994).

Why did the Boreal Ice Age start just then? Why do we see oscillations? What was (and is) the role of the ocean in modifying the ice age cycles (if any)? And how can information from the South Atlantic help in solving these fundamental problems? These are some of the fundamental questions we need to address.

Concerning the general cooling trend, there is little reason to postulate any new and different factors beyond those that were responsible for the climatic trend of the last 40 million years: Deepening of the ocean (from mountain building and from general aging of the oceanic crust) and associated regression and exposure of reflective land surfaces, movement of continental surfaces into higher latitudes in the northern hemisphere, plateau uplift and mountain building and associated changes in atmospheric circulation, downdraw of atmospheric CO_2 from increased availability of carbon sinks (weathering of feldspar, amphiboles and other silicates, and deposition of organic carbon in focussed upwelling regions), continued isolation of ocean basins and evolution of the Cold Ring connection. These various phenomena and postulated associated mechanisms of climatic change have been widely discussed (e.g., Turekian 1971; Donn and Shaw 1977; Berger and Crowell 1982; Sundquist and Broecker 1985; Budyko 1977; Ruddiman and Kutzbach 1989; Crowley and North 1991; Raymo 1991; Hay 1992).

The occurrence of a cooling step and the onset of ice age cycles is readily understood by invoking feedback from albedo changes. A general continued cooling trend eventually reaches the threshold of freezing, setting in motion the powerful positive albedo feedback from snow and ice. This feedback introduces instability into the system: on one side of the labile equilibrium ice keeps on building up until the available area is covered, on the other the ice disappears to expose the dark substrate. Thus, we should expect oscillations

when the treshold of freezing is just exceeded, as observed. Simple climate models (e.g., North and Crowley 1985) bear out this expectation. Ocean circulation can introduce both negative and positive feedback into this simple picture, through heat transport and through modification of atmospheric CO_2.

The effects of the third great cooling step on deep circulation in the South Atlantic were complex (e.g., Hodell et al. 1985; Williams et al. 1985; Turnau and Ledbetter 1989). Responses greatly depend on water depth and location of the site studied. On the whole, developments were characterized by (pulsed) increases in the activity of AABW, AAIW, and finally NADW, with a con-

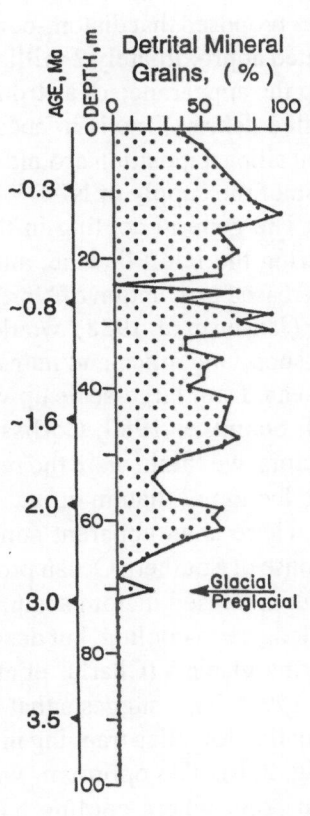

Figure 24. Onset of the Boreal Ice Age. Sudden increase in ice-derived debris from the North American continent, seen in DSDP Site 116, NW Atlantic. From Berggren (1972).

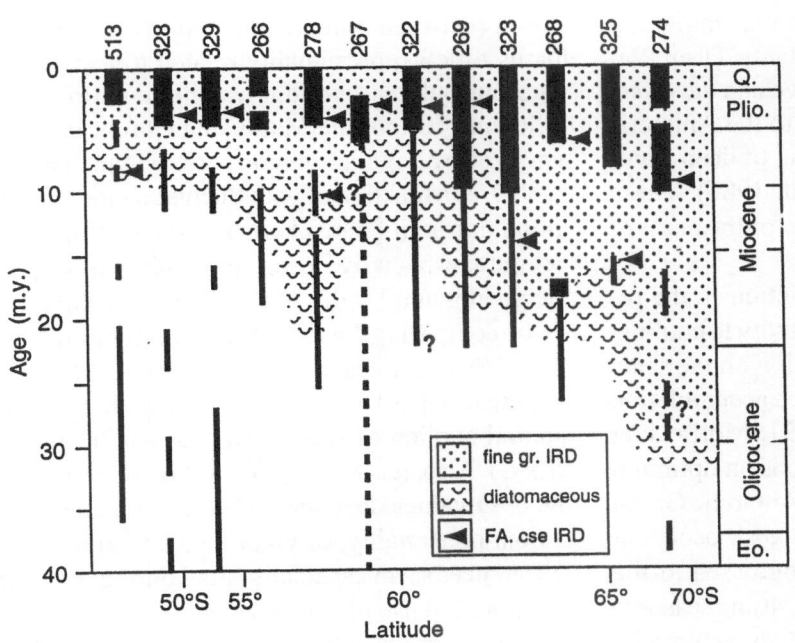

Figure 23. Distribution of ice-rafted debris in DSDP sites from the Southern Ocean, Atlantic sector. Heavy line: IRD limit for coarse material. Fine line: IRD limit for fine material. Dotted line: first appearance of abundant diatoms. From Ciesilski and Weaver (1983) (see in Wise et al 1985).

tinuing trend toward establishing modern deepwater stratification.

Already thirty years ago, Opdyke et al. (1966) suggested that "southern-hemisphere glaciation was initiated about 2.5 million years ago", that is, they noticed a strong increase in ice-rafted debris in their piston cores, shortly after 3 Ma (for details, see Barker, Kennett, et al. 1990; Bleil and Thiede 1990; Ciesielski, Kristoffersen, et al. 1991). Near 2.5 Ma, the $\delta^{18}O$ values of *N. pachyderma* increased by 0.6 to 0.7°/$_{oo}$ (Hodell and Ciesielski 1991), and the accumulation rates of biogenic silica increased substantially (Froelich et al. 1991). The date is remarkably similar to the age proposed for the onset of large-scale northern glaciations (Backman 1979; Shackleton et al. 1984; Sarnthein and Thiedemann 1989). Opdyke et al. (1966) further proposed that diatom-ooze deposition was initiated approximately 2 million years ago, following the appearance of a strong component of ice-rafted debris. This is in accord with the concept that silica deposition around Antarctica is a function of the supply of NADW (Fig. 25).

The general cooling in the northern Atlantic during the late Pliocene, and the concomitantly increased production of North Atlantic Deep Water (Jansen et al. 1988), would have had the effect of supplying silica and nutrients to the Antarctic Ocean, from large-scale upwelling of deep water (cf. Schnitker 1980; Corliss et al. 1986). These factors, we think, were the reason for the onset of production of diatom ooze.

There is an apparent contradiction in the response of Southern Ocean productivity to cooling, with increased diatom accumulation on the scale of long-term cooling, but decreased accumulation during glacials (Charles et al. 1991; Mortlock et al. 1991). This suggests that there is an optimum situation for silica trapping in the Antarctic Ocean (Fig. 25b). The optimum, we suggest, occurs at that point where cooling has progressed to the present interglacial situation, with strong seasonal variations in sea ice cover. This leads to sporadic stratification during sea ice melting in summer, resulting in high export fluxes (Wefer 1989; Hebbeln and Wefer 1991). During glacial periods the influence of NADW wanes, and sea ice expands to near the present polar front. Thus, diatom supply south

of the front decreases markedly (Berger and Wefer 1991).

According to Hodell and Venz (1992), conditions within the Cold Ring water masses were rather stable prior to about 3.2 million years ago, with substantial fluctuations setting in within the late Pliocene, during the late part of the Gauss Chron (also see Williams et al. 1985). They consider that this time interval saw the greatest changes in Neogene climate in the northern Antarctic and Subantarctic regions (Fig. 26). Surface waters cooled considerably, and the Polar Front Zone (PFZ) migrated northward, associated with expansion of perennial sea ice cover.

The record of DSDP Site 704 (47°S, 7°W) SW of the Cape of Good Hope (Fig. 26), is complicated by the fact that it straddles a fluctuating frontal zone, the Subantarctic convergence. Nevertheless, the main features are similar to features elsewhere (e.g., Rio Grande Rise; Hodell et al. 1985; Williams et al. 1985) and are of general significance. A striking drop in temperature, combined with ice buildup in the northern hemisphere, produces the $\delta^{18}O$ step seen within the late Gauss Chron (between 3 and 2.5 million years ago). A hiatus follows, presumably related to a series of vigorous pulses of bottom water circulation. The following period, the Matuyama Chron, is characterized by large-amplitude periodic fluctuations in $\delta^{18}O$, as a function of both fluctuating ice masses in the northern hemisphere and fluctuating temperature, especially in surface waters (as seen in the greater amplitudes of the planktonic signal).

In comparing benthic and planktonic record (Fig. 26), one notes that the magnitude of the late Gaussian step is larger in the plankton, that is, substantial cooling of surface waters is indicated (2 to 3°C). Also, it is seen that the difference between the $\delta^{18}O$ values decreases after the step, an indication, presumably, of vigorous vertical mixing. Differences seem especially small during many of the more intense glacial periods.

The extensive data set of Hodell and Ciesielski (1991), on Site 704, allows some interesting statistical experiments. When comparing $\delta^{18}O$ and $\delta^{13}C$ records of planktonic and benthic taxa by correlation analysis within successive time windows, we find that the most "predictable" time span is

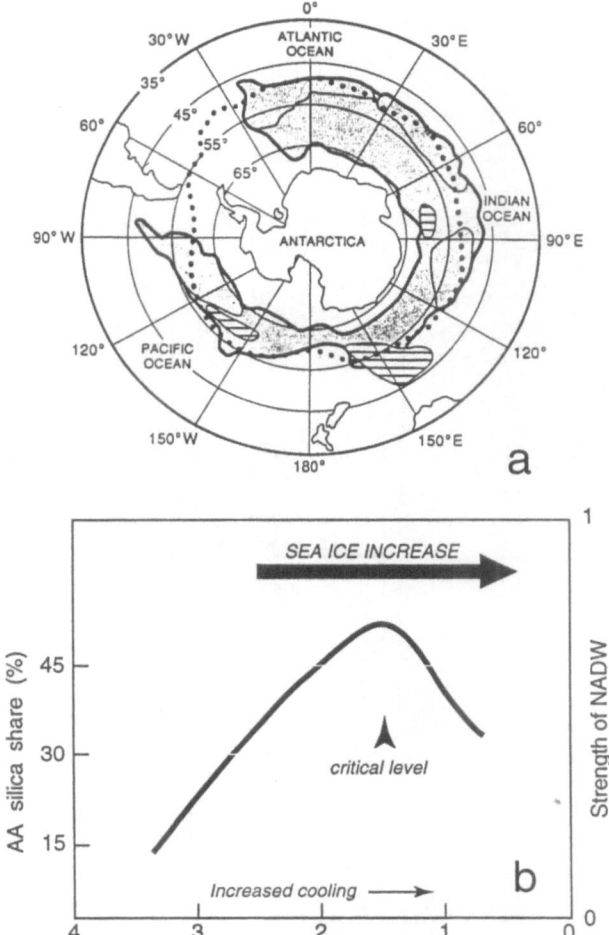

Figure 25. Opaline silica ring around Antarctica, and the postulated tie to NADW production. (a) Region with siliceous ooze. Ruled: current-swept areas with little or no accumulation. Line of heavy dots: Antarctic Polar Front; (b) model involving NADW production (see text). From Berger and Wefer (1991); data from DeMaster (1981) based on data compiled by D.W. Cooke and J.D. Hays.

the period between about 0.5 and 1 million years ago, and the least "predictable" the one between 2.5 and 3 Ma, that is, the late Gauss Chron. ("Predictable" means relatively high correlation coefficients between differenct types of records, in this context.) We suggest that during the Gauss Chron, the system enters a labile state, where negative feedback on cooling (from NADW production) competes with positive feedback (from albedo change, as ice builds up in the boreal realm). The

resulting pulses of deepwater events are widely recognized as acoustic reflectors in Pacific pelagic sediments ("Panama Series"; Berger, Kroenke, Mayer et al. 1993), and as hiatuses elsewhere, including the southern South Atlantic and the Southern Ocean.

On Rio Grande Rise, benthic oxygen isotopes show a substantial increase in $\delta^{18}O$ values within the Gauss Chron (Hodell et al. 1985; Williams et al. 1985), just as in the record of Site 704 (Fig. 26) and in numerous other records spanning the Pliocene. The change is greatest at depth, on the slope leading into the Vema Channel (0.6 permil at 4000 m), suggesting cooling of bottom water. Hodell et al. (1985) postulate a pulse-like increase in the formation of AABW at that time. They suggest a connection to global (northern) cooling and to the closure of the Central American Seaway, or both. They also propose that increased production of NADW may have enhanced AABW formation, with NADW providing for admixture of saline waters into the Cold Ring.

From these studies, and others, it appears that events in the Southern Ocean and the southern South Atlantic are well integrated into the climatic changes of the third great cooling step, changes that are so prominent in the North Atlantic realm. Three types of north-to-south links are commonly considered: (1) Ice buildup and -decay in the northern hemisphere control global sea level, providing for global albedo feedback, and increasing shelf areas suitable for bearing glacial ice in the Antarctic; (2) changes in atmospheric CO_2 affect the radiation balance on a global scale, no matter what their origin; and (3) the production of NADW and its delivery into the southern Cold Ring changes the heat budget in the Southern Ocean (delivery of heat), and in the South Atlantic (extraction of heat). Other links may exist as well, involving ocean circulation, wind fields, ocean productivity and thermocline dynamics.

We have suggested that opal deposition in the Southern Ocean is linked to NADW production, and that both opal deposition and NADW production may go through an optimum condition, within the late Pliocene or early Quaternary (Berger and Wefer 1991). In this context, the findings of Hodell and Venz (1992) regarding the record of ODP Site

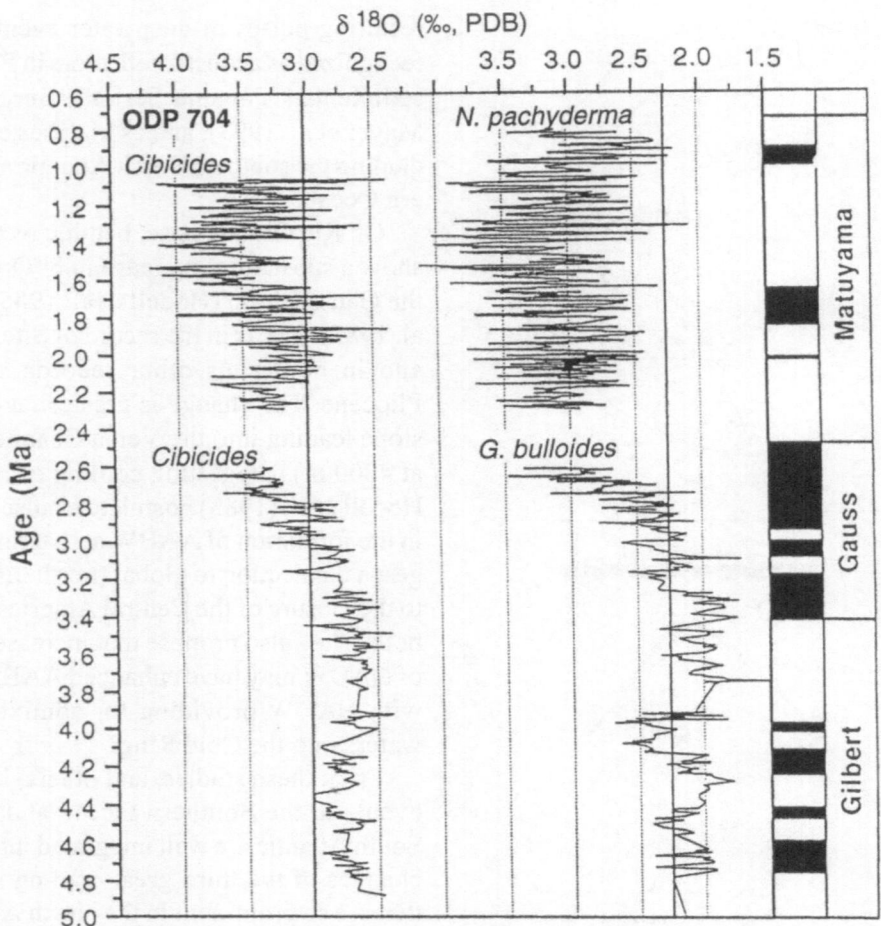

Figure 26. The great change in stability, at the boundary between South Atlantic and Southern Ocean, associated with the third great cooling step 3 million years ago, as seen in fluctuations of d¹⁸O values of planktonic foraminifers (*Neogloboquadrina pachyderma*, *Globigerina bulloides*) and benthic foraminifers (*Cibicides* spp.). From Hodell and Venz (1992).

704 (Meteor Rise) are of special interest. These authors report a strong correlation of all sedimentary parameters in Hole 704B, beginning about 1.5 million years ago, as well as a strong cyclicity centered on 41 kyr (i.e., the period of obliquity variation). They propose that orbital forcing at this period was enhanced by positive feedback from NADW production, that is, by NADW suppression during glacials, once large northern ice sheets existed. It is this positive feedback, they suggest, which "may explain the tightly coupled response

that developed between the Southern Ocean and the North Atlantic" beginning in the early Quaternary (stage 52) (Hodell and Venz 1992, p. 265). They present evidence for glacial-time NADW suppression from a comparison of deepwater δ¹³C values of Site 704, with contemporaneous records in North Atlantic and Pacific.

It appears, then, that we must take into account a fundamental change in the operation of the Atlantic Conveyor, once a certain threshold is exceeded in the overall cooling process. Before this threshold, which is reached sometime near the Pliocene-Pleistocene boundary, NADW production increases during cooling, while after it (when considerable ice is present in the northern hemisphere) NADW production decreases upon cooling. Evidence for such a switch in NADW response to cooling is found in carbon isotope gradients between North Atlantic, Southern Ocean, and Pacific (Hodell and Ciesielski 1990; 1991; Raymo et al. 1990; 1992; Whitman and Berger 1993). Since NADW production is intimately tied to the entire heat budget of the South Atlantic, we should also find widespread evidence for a change in amplitudes and phase relationships of paleoproxies in going from the Pliocene to the Quaternary.

Some of the best late Neogene records in the South Atlantic were recovered from the Benguela Current system, on Walvis Ridge. The Ridge provides access to pelagic carbonates in a coastal-ocean setting, as well as an opportunity, on its flanks, to sample water mass properties at various depths. Thus, it has been the target of several expeditions (DSDP Leg 40; Bolli, Ryan, et al. 1978; DSDP Leg 74, Moore et al. 1985; Leg 75, Hay, Sibuet, et al. 1984). An important result was the discovery (during Leg 40) of high rates of deposition of opaline fossils and of organic matter, beginning about 10 million years ago and increasing since (Siesser 1980), a finding supported by subsequent studies (Meyers et al. 1983; Diester-Haass et al. 1990). These trends suggest that sustained upwelling in the Benguela upwelling system started in the early late Miocene and increased into the Quaternary. If so, it may be surmised that the strength of SE trades increased likewise, over this time interval, presumably as a result of increased temperature gradients from pole to equator.

Detailed carbonate and organic-carbon stratigraphies for the time interval of the great boreal cooling step are available for DSDP Site 532, sampled by hydraulic piston coring, at a water depth of 1131 m (Leg 75; Gardner et al. 1984; Dean et al. 1984). The sediments are characterized by pronounced fluctuations in carbonate content which, on the whole, parallel distinct light-dark color cycles. There is a striking change in the character of the cycles about 3 million years ago, from low-amplitude rapid alternation to high-amplitude cycles with a longer period (the change is from ca.30 kyr to 55-58 kyr, according to Gardner et al. 1984). This is as expected if the Benguela system is linked to the polar climatic fluctuations, via trade wind strength, and perhaps via opening and closing of the Cape Valve, by north-south migration of the frontal zone, which controls admission of warm water from the Indian Ocean (Agulhas retroflection, Fig.8). A marked increase in the (negative) correlation between carbonate and organic carbon occurs at the Pliocene-Pleistocene boundary, about 1.8 Ma ago (Gardner et al. 1984, p.918). We take this as a signal that strong positive feedback begins to pervade the system, that is, that the Atlantic Conveyor now tends to shut down during cold periods.

The record of opal deposition at this site (Dean and Parduhn 1984) indicates a substantial increase in the delivery of diatoms about 3 million years ago, reaching a maximum in the early Quaternary, followed by a decrease in mid-Quaternary time, between 1 and 1.5 million years ago (Fig. 27). The pattern is reminiscent of that found in the Antarctic realm, as discussed above. Organic matter deposition, on long time scales, resembles that of opal deposition, till about 1 Ma ago. After that time, organic matter content shows periods of strong sustained recovery, without similar recovery in opal content. It appears, then, that upwelled water, while supplying the phosphate which generates the organic matter export, is no longer supplying sufficient silicate for opal deposition after 1 Ma ago: thermocline waters are depleted in silicate. It is not clear to us why this should be so; we suspect that there is a connection to the emergence of high-amplitude 100-kyr cycles which depend on strong positive feedback from the Atlantic Conveyor.

As mentioned earlier when discussing glacial-interglacial cycles on Walvis Ridge, opal deposition is reduced during glacials, even though there is evidence that upwelling is increased. The denial of silicate to this region (for whatever cause) during the intense glacial periods of the last 1 million

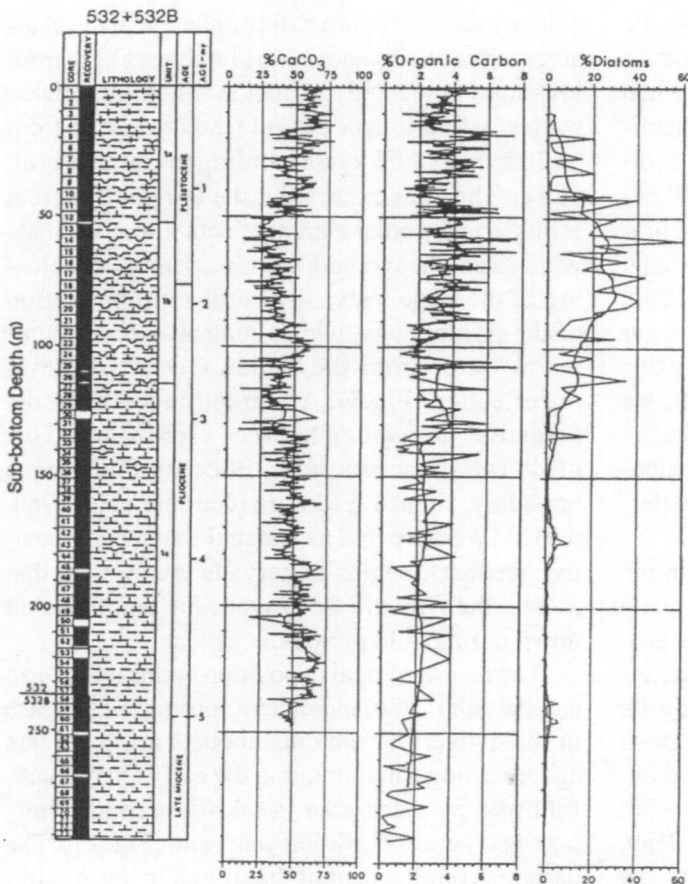

Figure 27. Depositional cycles in biogenous sediments on Walvis Ridge, in the Benguela System. Note overall trend in diatom abundance, with maximum in early Quaternary (Dean and Gardner 1985; Hay and Brock 1994).

years is the reason, presumably, for the trend of late Quaternary decrease of opal deposition on Walvis Ridge. It is significant that before the Pliocene (>5 Ma) opal deposition was high during cold periods, that is, the phase between temperature and opal abundance was reversed sometime within the Pliocene (Diester-Haass et al. 1992; Hay and Brock 1992). We cannot exclude the possibility of regional rearrangements of the path of the Benguela Current (as suggested by Diester-Haass 1985, and other workers) and the possibility that much of the organic matter is redeposited from the shelf, during glacials, and is therefore out of phase with opal (Diester-Haass et al. 1990). However, we suggest that more fundamental processes, having to do with the sign of feedback between circulation and climate, are responsible for these puzzling patterns.

Panama Paradox and NADW Feedback Switch: Conceptual models

The Panama Hypothesis

We have argued, in several contexts, that NADW production first increases with cooling, in the Pliocene, and that it then decreases with cooling, after ice buildup in the northern hemisphere. Thus, we postulate a change in the mode of climate feedback from the Atlantic Heat Conveyor. In the final section of this expedition into the past we formalize this concept and explore some of its ramifications. First, however, we have to deal with the fact that the Central American Seaway closes just about at the time as the northern hemisphere is ready to enter the ice ages. This historic coincidence complicates matters, and provides opportunities for misconceptions.

The "Panama Hypothesis", which we introduce herewith, is the assertion that heat supply from the South Atlantic to the North Atlantic was greatly enhanced by the closing of the Panama Seaway, and that this event delayed the onset of the northern ice ages, contrary to present consensus.

The last step in the overall closure of the Tethys Ocean - the leitmotif of Cenozoic paleoceanography - is the closure of the Central American Seaway by the emergence of the Panama Isthmus. This event separated tropical Atlantic from tropical Pacific, with profound implications for climate and for evolution (Haq 1984; Hay 1988). The exact sequence of events is under discussion (Keller et al. 1989; Coates et al. 1992). Apparently, closure was completed within the late Pliocene, between about 3.5 to 2.5 Ma (Keigwin 1978; Marshall 1988; Coates et al. 1992). The final closure was preceded by a long-term shallowing of the Panama Seaway, with a sill near 1000 m present within the middle Miocene (Keller and Barron 1983; Duque-Caro 1990). Perhaps the most striking evidence for effective closure, as far as shallow-water exchange, is the divergence of planktonic foraminifer species on both sides of the Isthmus, about 3.5 million years ago (Fig. 28). The tropical species *Globorotalia multicamerata* evolved in the early Pliocene, about 5 to 4.5 million years ago. It is found both in the Pacific and in the Atlantic. The related species *Globorotalia miocenica* evolved in the Atlantic about 3.5 million years ago; it is absent in the Pacific (Kennett and Srinivasan 1983).

The fact that the northern ice ages started about 3 million years ago combined with the fact that the formation of the Central American Isthmus was completed by that time invites speculations about possible causal connections. There is no question that differences between Atlantic and Pacific water masses are greatly enhanced by the presence of the Panama Isthmus barrier. This is shown, unsurprisingly, when providing for a connection through Central America, within general ocean circulation models (Maier-Reimer et al. 1990; Mikolajewicz et al. 1993). One can then reasonably argue that the unusually high salinity of North

Atlantic surface waters (a precondition for making NADW) can only arise because of the Central American barrier, and that this barrier is ultimately responsible, therefore, for a greatly enhanced production of NADW. In turn, this production sets in motion the North Atlantic heat piracy, through exchange of cold deep water for warm surface water, with the South Atlantic. The process feeds warm waters to the Nordic realm, where increased evaporation makes possible the buildup of ice sheets. By this chain of reasoning (which has become part of paleoceanographic consensus), "the closing of the Central American isthmus appears to be the best candidate for the event that triggered the onset of northern hemisphere glaciation" (Hay 1992, p. 309).

It will be noted that this chain of logic contains a paradox - which we call the **Panama Paradox** - in that increased heat supply to the North Atlantic results in ice buildup around the North Atlantic. The mechanism invoked is increased precipitation. The reason the precipitation falls as snow rather than rain, presumably, is the fact that the general cooling trend has reached the appropriate threshold for freezing. In this scenario, then, the effect

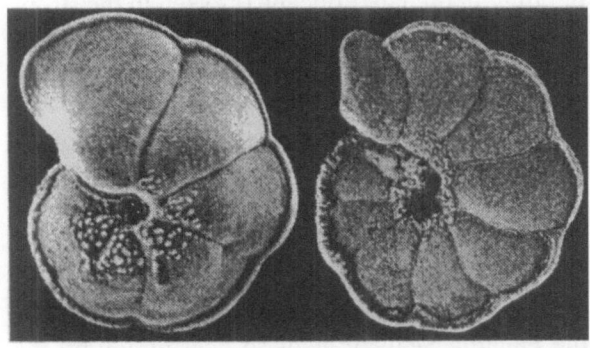

Figure 28. Closure of the Central American gateway as seen in the evolution of two tropical planktonic foraminifers, *Globorotalia miocenica* Palmer and *Globorotalia multicamerata* Cushman and Jarvis. *G. multicamerata* evolved in the early Pliocene and is found on both sides of the Isthmus. *G. miocenica* appears 3.5 million years ago; it is restricted to the Atlantic. From A. Boersma, in Haq and Boersma (1978).

of increased NADW production (from closing the Panama Straits) is to accelerate the process of ice buildup in the boreal realm.

We believe that the exact opposite is correct.

The conventional Panama scenario uses newly gained heat in the North Atlantic to make snow, emphasizing a need for rapid ice buildup. Such a need is unsubstantiated. A more elementary treatment of the problem would emphasize the requirement for low temperatures: No matter at what rate the snow is deposited, the important thing is that it stays on the ground. If we take this approach, the evolution of the Panama land bridge **delays** rather than furthers the onset of the boreal ice ages. Our alternative Panama scenario postulates that boreal cooling reaches the necessary threshold near the end of the Miocene, but that ice buildup is subsequently prevented by the warming of the North Atlantic, through heat piracy made possible by the emergence of the Central American Isthmus. This scenario calls for end-of-Miocene cooling and initial northern glaciation, early Pliocene warming, and resumption of cooling in the late Pliocene, as the heat piracy effect is overwhelmed by the general trend (Fig. 29). The general trend is accelerated by additional uplift and mountain building in the late Pliocene (as seen in the strontium isotope record; Capo and DePaolo 1990), which increases albedo and decreases atmospheric pCO_2.

To reiterate: We consider that the closing of the Panama Straits does nothing for initiating the Boreal Ice Age. This ice age starts because of the overall planetary cooling which has proceeded for the last 40 million years, mainly as a result of mountain building and regression. The closure of the Panama Seaway leads to heat piracy by the North Atlantic, at the expense of the South Atlantic. This process is designed to *prevent* the Boreal Ice Age, rather than generate it.

Our "Panama Hypothesis" (delay of northern glaciation by interruption of the general cooling trend) explains many of the features which dominate the late Neogene record: "premature" late Miocene glaciation on Greenland (Jansen and Sjøholm 1991; Larsen et al. 1994), the reversal of the late Miocene drop in sea level which triggered the drying up of the Mediterranean (Ryan et al. 1973), and the early Pliocene warming which is well represented in the oxygen isotope record (e.g., Jansen et al. 1993). It also addresses the rather abrupt onset of the boreal ice age cycles, with ever-increasing amplitudes (Shackleton et al. 1984; Sarnthein and Thiedemann 1989; Jansen et al. 1993; Berger and Jansen 1994). The evolution of these amplitudes, we think, greatly depend on the Panama Effect (North Atlantic heat piracy at the expense of the South Atlantic) which can be turned off by extensive glaciation, and hence becomes a positive feedback on glacial-interglacial contrast.

Concept of NADW optimum: the NADW feedback switch

In order to discuss the role of the ocean in modifying the northern ice age cycles, we must re-examine some elementary concepts regarding fluctuations in the production of NADW.

The production of NADW depends on evaporation, for the increase of salinity of surface waters, and on cooling, for the decrease of temperature. Both, increase of salinity and decrease of temperature are necessary to achieve the density at the surface that is necessary for deep convection and deepwater formation. Evaporation is enhanced by the availability of heat, cooling by the lack thereof. Thus, there must be an optimum of heat supply (correlated with geographic and seasonal contrast in heat supply) which maximizes deep convection. On the warm side of the optimum, cooling is perceived as a negative feedback on change; on the cold side, as a positive feedback. During the early Pliocene, when the Panama effect operates to delay glaciation (Fig. 29), we expect to be on the warm side of the optimum, in a negative feedback mode ("Panama mode"). During the late Quaternary, when there is widespread formation of sea ice in the boreal realm, we expect to be in the positive feedback mode ("Nordic mode").

The Nordic mode of operation of NADW production is the one generally postulated in Quaternary research. It has been assumed for some time (e.g., Boyle and Keigwin 1982; Curry and

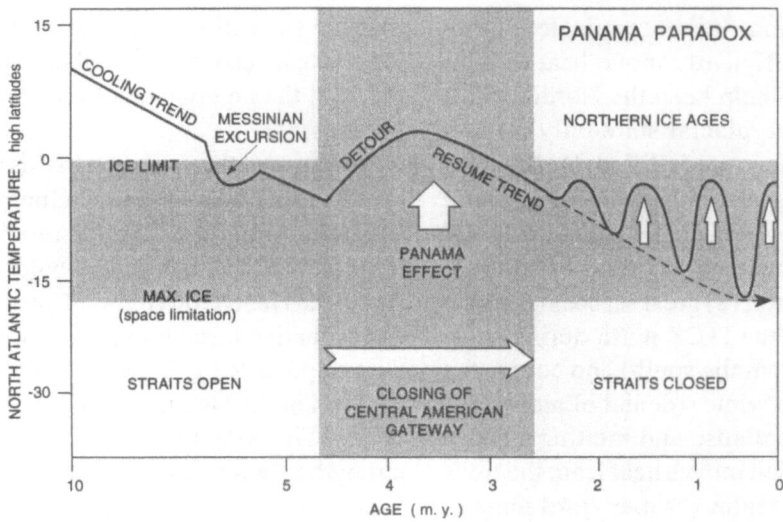

Figure 29. Conceptual scheme for onset of northern hemisphere glaciations and ensuing ice age fluctuations, and relationship to closing the Panama Straits. The "Panama Paradox" arises when the closing of the Straits is invoked as a trigger for the onset, rather than as a delahying factor. The system trajectory starts in the upper left corner, with a state too warm to make boreal ice sheets. It proceeds to the right and downward (toward lower temperatures) as a result of general planetary cooling (albedo, CO_2). Ice first builds up when the ice limit is reached, with positive feedback from the drying up of the Mediterranean ("Messinian excursion"). Subsequently, the closing of the Panama Seaway reverses the cooling trend ("Panama Effect") by enabling North Atlantic heat piracy. The overall cooling trend takes over again in the middle Pliocene, taking the system beyond the ice limit. Attempts to remove the ice, by the southern heat subsidy, only succeed (partially) during times of unusually warm northern summers. Amplitudes of the ice age fluctuations increase as the general cooling moves the system toward maximum ice buildup (set by availability of suitable land and shelf surface).

Lohmann 1983; Duplessy et al. 1988) that NADW production is high during interglacials and low during glacial periods and that this enhances and blocks heat piracy by the North Atlantic. Thus, the varying intensity of the Atlantic Conveyor (warm upper waters into the North Atlantic, cold deep waters out) acts as a positive feedback on climate change on the time scale of late Quaternary ice age fluctuations (e.g., Broecker and Denton 1989; Imbrie et al. 1993).

In our Panama Hypothesis the reverse feedback is assumed. The closure of the isthmus allows retention of high-salinity surface waters, but the continued overall cooling trend is responsible for lowering the temperature of these waters sufficiently for deep convection, increasing the intensity of the Atlantic Conveyor and hence the efficiency with which the North Atlantic extracts heat from the South Atlantic. In this fashion, the

Panama effect (isolation of the North Atlantic, increased salinity and NADW production) attempts to keep the boreal realm in an interglacial or nonglacial state. However, eventually it is too weak to do so, given the general cooling, and only when aided by high summer insolation in northern latitudes (Milankovitch 1930) is the heat supply just large enough to allow the system to reach an interglacial state. Sufficiently great cooling circumvents the Panama Effect, by pushing the ITCZ toward the equator, and by shutting down NADW production through covering the Nordic Seas with sea ice. This greatly reduces the heat import from the South Atlantic, allowing attainment of the state that is "proper" for the present planetary conditions, i.e., the glaciated state. NADW production then operates in the Nordic mode.

Once the Atlantic Conveyor is in the Nordic mode, it is exceedingly difficult to leave the glacial state: the glaciation must be terminated before the Conveyor can efficiently move heat to high northern latitudes to help keep the Nordic realm warm. Thus, the interglacial state only returns when several conditions are fulfilled: (1) Much of the northern ice is unstable because of isostatic sinking, in part below sealevel, because of great thickness, and because of warming by internal refreezing of meltwater; (2) great seasonal contrast in insolation moves the ITCZ north during summer (heat subsidy from the south) and covers the ocean with sea ice in winter (denial of moisture); (3) large-scale ice collapse and melting starts a meltwater pump which moves heat from the tropics into the Nordic realm ("super-fjord mechanism"; Berger and Jansen 1995). Stabilization of the interglacial situation then depends on the NADW heat pump, which attempts to make the southern heat subsidy permanent.

Summary and Conclusion

The history of the South Atlantic is a function of plate tectonics - opening in the late Cretaceous and gradual widening, separation into parallel troughs by the spreading center, partitioning by hot-spot traces - and of the evolution of the global climate. This evolution is characterized by an overall change of a shelf-sea dominated warm world to a land-mass dominated world with ice caps. The first striking change occurred about 40 million years ago, with the "Auversian Facies Shift", a profound change in the facies distribution of deepsea sediments, which proved irreversible. The shift is marked by a large drop in the carbonate compensation depth (first recognized in the South Atlantic) and is shortly succeeded by a drop in deepwater temperatures.

A central leitmotif of Cenozoic paleoceanography is the closure of the circumtropical Tethys Ocean, which accompanies a general cooling trend. As a result of closure of the Tethys, and of first restricting and then closing the Panama Passage, a strong asymmetry developed between North Atlantic and North Pacific, with the deep North Atlantic collecting carbonate and the North Pacific diatomaceous silica. This asymmetry marks the beginning of the modern ocean. It starts

roughly 10 million years ago, and has been growing since, in response to northern hemisphere cooling, and the emergence of the Central American Isthmus.

Because of the evolution of this asymmetry, the South Atlantic has become a thoroughfare for an asymmetry-driven conveyor, moving warm water into the North Atlantic, and cold water out. This Atlantic Heat Conveyor is part of the global thermohaline circulation system centered on the Circumpolar Ring Current ("Cold Ring"), which evolved in the Neogene (the last 25 million years or so). The Atlantic Heat Conveyor increased in strength as a result of overall cooling in the Late Cenozoic, but decreased again beyond a certain level of cooling, presumably due to a decrease in evaporation and (perhaps) development of sea-ice. Thus, it is weakened during glacial periods of the Quaternary, as recognized in fluctuations of carbonate preservation, benthic foram distribution, and $\delta^{13}C$ of benthic foraminifers in the South Atlantic. At depth, the boundary between Antarctic Bottom Water (AABW) and North Atlantic Deep Water (NADW) is the proxy most commonly used to reconstruct developments. This boundary is seen as a foram lysocline, and a discontinuity in physical properties. Its depth level fluctuates, with a tendency for deepening in the late Neogene, but for shallowing during Quaternary glacials. Also, it displays a varying east-west asymmetry across the Mid-Atlantic Ridge, the asymmetry being stronger when NADW flows more vigorously.

The crucial importance of the South Atlantic in the climate evolution of the Late Cenozoic derives from the following: (1) it acts as a major supplier of warm air and surface waters to the North Atlantic, into the Intertropical Convergence Zone (which is north of the equator but with varying distance from the equator); (2) it harbors a major upwelling system off southeast Africa, where organic carbon is being sequestered on a large scale; and (3) its southern boundary is part of the circumpolar system, with varying production of thermocline waters and varying sea-ice cover functioning as feedback mechanisms to climatic change involving NADW production (which is sensed as warm influx into the Cold Ring). The various elements of the system are recorded in the paleontology and the chemistry of ocean sediments, from which reconstruction of the three major features of South Atlantic circulation proceeds.

A parsimonious way to think about the Quaternary ice age fluctuations is to assume that the Southern Hemisphere stays in the Ice Age it entered millions of years ago, while the climate of the Northern Hemisphere oscillates, stimulated by an amplifying system which includes the Atlantic Heat Conveyor. Two opposing factors cooperate to provide for the NADW production which drives the conveyor. One is the fundamental asymmetry between South and North which is set up by the second great cooling step, the Antarctic step, which loads one pole with an ice cap, but not the other. This results in moving the heat equator northward and thus starts the heat piracy of the North Atlantic realm, which ultimately results in salt buildup in North Atlantic surface waters, through increased evaporation. The other is the progressive cooling in the northern hemisphere, which tends to oppose the asymmetry, but which provides for the refrigeration necessary to make deep and bottom water. Thus, we have a system with two factors (evaporation and cooling) which act to neutralize each other but whose product yields the effect sought. Such a system goes through an optimum as it proceeds along the axis of either dimension.

The changing character of the Atlantic Heat Conveyor, as a negative feedback on cooling before the boreal ice buildup, and as a positive feedback within the Quaternary, is crucial for the understanding for many of the major features of climate-related sedimentation in the South Atlantic in the late Neogene, that is, the last 8 million years or so.

This is contribution no. 136 of Sonderforschungsbereich 261 at Bremen University.

References

Abrantes FF, Winn K, Sarntheim M (1994) Late Quarternary paleoproductivity variations in the NE and equatorial Atlantic: Diatom and C org evidence. In: Zahn R et al. (eds) Carbon Cycling in the Glacial Ocean: Constraints on the Ocean's Role in Global Change. Springer Verlag, Berlin Heidelberg, pp 425-441

Alvarez LW, Alvarez W, Asaro F, Michel HV (1980) Extraterrestrial cause for the Cretaceous-Tertiary extinction. Science 208:1095-1108

Backman J (1979) Pliocene biostratigraphy of DSDP Sites 111 and 116 from the North Atlantic Ocean and the age of Northern Hemisphere glaciation. Stockholm Contrib Geol 32:115-137

Barker PF, Kennett JP (1990) Proc ODP Sci. Results 113, College Station,´, TX (Ocean Drilling Program)

Barron EJ (1985) Explanation of the Tertiary global cooling trend. Palaeogeography, Palaeoclimatology, Palaeoecology 50:45-61

Barron JA, Baldauf JG (1989) Tertiary Cooling Steps and Paleoproductivity as Reflected by Diatoms and Biosiliceous Sediments. In: Berger WH, Smetacek VS, Wefer G (eds) Productivity of the Ocean: Present and Past. Dahlem Konferenzen. John Wiley & Sons, Chichester, pp 341-354

Benson RH (1975) The origin of the psychrosphere as recorded in changes of deep-sea ostracode assemblages. Lethaia 8:69-83

Benson RH, Chapman RE, Deck LT (1985) Evidence from the ostracoda of major events in the South Atlantic and world-wide over the past 80 million years. In: Hsü KJ, Weissert HJ (eds) South Atlantic Paleoceanography. Cambridge University Press, Cambridge (UK), pp 325-350

Berger WH (1968) Planktonic foraminifera: selective solution and paleoclimatic interpretation. Deep-Sea Res 15:31-43

Berger WH (1970) Biogenous deep-sea sediments: fractionation by deep-sea circulation. Bull Geol Soc Am 81:1385-1402

Berger WH (1972) Deep-sea carbonates: dissolution facies and age depth constancy. Nature 236:392-395

Berger WH (1989) Global maps of ocean productivity. In: Berger WH, Smetacek VS, Wefer G (eds) Productivity of the Ocean: Present and Past. Dahlem Konferenzen. John Wiley, Chichester, pp 429-455

Berger WH, Crowell JC (eds) (1982) Climate in Earth History. Studies in Geophysics. National Academy Press, Washington, D. C. 198 pp

Berger WH, Herguera JC (1992) Reading the sedimentary record of the ocean's productivity. In: P.G. Falkowski and A.D. Woodhead (eds.) Primary Productivity and Biogeochemical Cycles in the Sea. Plenum Press, New York, pp 455-486

Berger WH, Jansen E (1994) Mid-Pleistocene climate shift: The Nansen connection. In: Johanessen OM,

Muench RD, Overland JE (eds) The Polar Oceans and Their Role in Shaping the Global Environment: The Nansen Centennial Volume. AGU Geophysical Monograph 84:295-311

Berger WH, Jansen E (1995) Younger Dryas episode: Ice collapse and superfjord heat pump. In: Troelstra SR, van Hinte JE, Ganssen GM (eds) The Younger Dryas, North-Holland, Amsterdam, pp 61-105

Berger WH, Wefer G (1991) Productivity of the glacial ocean: Discussion of the iron hypothesis. Limnol Oceanogr 36:1899-1918

Berger WH, Winterer EL (1974) Plate stratigraphy and the fluctuating carbonate line. In: Hsü KJ, Jenkyns H (eds) Pelagic Sediments on Land and Under the Sea. Spec Publ Intl Assoc Sediment 1:11-48

Berger WH, Killingley JS, Metzler CV, Vincent E (1985) Two-step deglaciation: ^{14}C-dated high-resolution d^{18}O records from the tropical Atlantic Ocean. Quaternary Research 23:258-271

Berger WH, Burke S, Vincent E (1987) Glacial-Holocene transition: climate pulsations and sporadic shutdown of NADW production. In: Berger WH Labeyrie LD (eds) Abrupt Climatic Change. D. Reidel, Dordrecht, pp 279-297

Berger WH, Herguera JC, Lange CB, Schneider R (1994) Paleoproductivity: flux proxies versus nutrient proxies and other problems concerning the Quaternary productivity record. In: Zahn R, Kaminski M, Labeyrie LD, Pederson TF (eds) Carbon Cycling in the Glacial Ocean: Constraints on the Ocean's Role in Global Change. Springer, Berlin Heidelberg, pp 385-412

Berger WH, Kroenke LW, Mayer LA (1993) Proc ODP Sci Results, 130, College Station, TX (Ocean Drilling Program), 867 pp

Berger WH, Yasuda M, Bickert T, Wefer G (1995) Brunhes-Matuyama boundary: 790 k.y. date consistent with ODP Leg 130 oxygen isotope records based on fit to Milankovitch template. Geophysical Research Letters 22:1525-1528

Berggren WA (1972) Late Pliocene - Pleistocene glaciation. In: Laughton AS, Berggren et al. Initial Reports Deep Sea Drilling Project 12. Washington DC (US Government Printing Office) pp 953-963

Berggren WA, Hollister CD (1974) Paleogeography, paleobiogeography and the history of circulation in the Atlantic Ocean. In: Hay WW (ed) Studies in Paleo-Oceanography. Soc Econ Paleont Mineral, Spec Publ 20 pp126-186

Bleil U, Thiede J (eds) (1990) Geological History of the Polar Oceans: Arctic versus Antarctic. NAT ASI Series C, Vol. 8. Kluwer Academic Publishers,

Dordrecht Boston London, pp 823

Bolli HM, Ryan WBF et al. (1978) Initial Reports of the Deep Sea Drilling Project 40, Washington, DC. (U.S. Government Printing Office)

Boltovskoy E (1967) Living planktonic foraminifera of the eastern part of the tropical Atlantic. Revue de Micropaléontologie 11:85-98

Boyle EA, Keigwin LD (1982) Deep circulation of the North Atlantic over the last 200,000 years: Geochemical evidence. Science 218:784-787

Bramlette MN (1961) Pelagic Sediments. In: Oceanography. Amer Assoc Advancement of Sci Publ No 67, pp 345-366

Bremner JM (1983) Biogenic sediments on the south west African (Namibian) continental margin. In: Thiede J, Suess E (eds) Coastal Upwelling: Its Sediment Record, Part B. (NATO conference series IV) Plenum Press, New York, pp73-103

Broecker WS, Denton GH (1989) The role of ocean-atmosphere reorganizations in glacial cycles. Geochimica et Cosmochimica Acta 53(10):2465-2501

Brongersma-Sanders M (1947) On the desirability of a research into certain phenomena in the region of upwelling water along the coast of South West Africa. Koninklijke Nederlandsche Adademie van Wetenschappen, Proceedings 50(6):1-8

Budyko MI (1977) Climate Change. American Geophysical Union, Washington D.C.

Byrne DA, Gordon AL, Haxby WF (1995) Agulhas eddies: A synoptic view using Geosat ERM data. J Phys Oceanography 25:902-917

Calvert SE, Price NB (1971) Upwelling and nutrient regeneration in the Benguela Current, October 1968. Deep-Sea Res 18:505-523

Calvert SE, Price NB (1983) Geochemistry of Namibian shelf sediments. In: Suess E, Thiede J (eds) Coastal Upwelling: Its Sediment Record, Part A (NATO conference series IV) Plenum Press, New York, 337-375

Capo RC, DePaolo DJ (1990) Seawater strontium isotopic variations from 2.5 million years ago to the present. Science 249:51-55

Chamberlin TC (1906) On a possible reversal of deep-sea circulation and its influence on geologic climates. Jour Geology 14:363-373

Charles CD, Fairbanks RG (1990) Glacial to interglacial changes in the isotopic gradients of Southern Ocean surface water. In: Bleil U, Thiede J (eds.) Geological History of the Polar Oceans: Arctic versus Antarctic. Kluwer Academic, Dordrecht pp 519-538

Charles CD, Froelich PN, Zibello MA, Mortlock RA, Morley JJ (1991) Biogenic opal in Southern Ocean sediments over the last 450,000 years: Implications for surface water chemistry and circulation. Paleoceanography 6:697-728

Ciesielski PF, Weaver FM (1983) Neogene and Quaternary paleoenvironmental history of Deep Sea Drilling Project Leg 71 sediments, Southwest Atlantic Ocean. In: Ludwig WJ, Krasheninnikov et al., Init Repts DSDP, 71, part 1, Washington (US Govt Printing Office), pp 461-477

Ciesielski PF, Kristoffersen Y (1991) Proc ODP, Sci Results, 114, College Station TX (Ocean Drilling Program)

CLIMAP Project Members 1976. The surface of the ice-age Earth. Science 191:1131-1137

CLIMAP Project Members 1981. Seasonal reconstructions of the Earth's surface at the last glacial maximum. Map and Chart Series MC-36, Geol. Soc. Amer., Boulder Colorado

Coates AG, Jackson JC, Collins LS, Cronin TM, Dowsett HJ, Bybell LM, Jung P, Obando JA (1992) Closure of the Isthmus of Panama: The near-shore marine record of Costa Rica and western Panama. Geol Soc Am Bull 104:814-828

Corliss WB, Martinson DG, Keffer T, (1986) Late Quaternary deep-ocean circulation. Geol Soc Am Bull 97:1106-1121

Crowley TJ, North GR (1991) Paleoclimatology. Oxford University Press, New York, pp 339

Curry WB, Crowley TJ(1987). The $\delta^{13}C$ of equatorial Atlantic surface waters: Implications for ice age pCO_2 levels. Paleoceanography 2:489-517

Curry WB, Lohmann GP (1983) Reduced advection into Atlantic Ocean deep eastern basins during last glaciation maximum. Nature 306:577-580

Dean W, Gardner J (1985) Cyclic variations in calcium carbonate and organic carbon in Miocene to Holocene sediments, Walvis Ridge, South Atlantic Ocean. In: Hsü KJ, Weissert HJ (eds) South Atlantic Paleoceanography, Cambridge University Press, Cambridge (U.K.), 61-78

Dean WE, Hay WW, Sibuet JC (1984) Geologic evolution, sedimentation, and paleoenvironments of the Angola Basin and adjacent Walvis Ridge: Synthesis of results of Deep Sea Drilling Project Leg 75. In: Hay WW, Sibuet JC et al. Init Repts DSDP 75:509-542. Washington (U.S. Govt Printing Office)

Dean WE, Parduhn NL (1984) Inorganic geochemistry of sediments and rocks recovered from the southern Angola Basin and adjacent Walvis Ridge, Sites 530 and 532, Deep Sea Drilling Project 75. In: Hay WW, Sibuet JC et al. Init Repts DSDP 75: 923-958. Washington (US Govt Printing Office)

Delaney ML (1990). Miocene benthic foraminiferal Cd/Ca records: South Atlantic and western equatorial Pacific. Paleoceanography 5:743-760

DeMaster DJ (1981). The supply and accumulation of silica in the marine environment. Geochim Cosmochim Acta 45:1715-1732

Diester-Haass L (1985) Late Quaternary upwelling history off southwest Africa (DSDP Leg 75, HPC 532). In: Hsü KJ, Weissert HJ South Atlantic Paleoceanography. Cambridge University Press, Cambridge (U.K.), pp 47-55

Diester-Haass L, Meyers PA, Rothe P (1990) Miocene history of the Benguela Current and Antarctic ice volumes: Evidence from rhythmic sedimentation and current growth across the Walvis Ridge (Deep Sea Drilling Project Sites 362 and 532). Paleoceanography 5:685-707

Diester-Haass L, Meyers PA, Rothe P (1992) The Benguela Current and associated upwelling on the southwest African Margin: a synthesis of the Neogene - Quaternary sedimentary record at DSDP sites 362 and 352. In: Summerhayes CP, Prell WL, Emeis KC (eds) Upwelling Systems: Evolution Since the Early Miocene. Geological Society Special Publication No. 64, pp 331 - 342

Dietrich G, Kalle K, Krauss W, Siedler G (1975) Allgemeine Meereskunde, 3. Auflage. Borntraeger Berlin Stuttgart, 593 pp

Dietz RS (1961) Continent and ocean basin evolution by spreading of the sea floor. Nature 190:854-857.

Donn WL, Shaw DM (1977) Model of climate evolution based on continental drift and polar wandering. Geol Soc Am Bull 88:390-396

Douglas RG, Savin SM (1975) Oxygen and carbon isotope analyses of Teriary and Cretaceous microfossils from Shatsky Rise and other sites in the North Pacific Ocean. In: Init Rep DSDP 32:509-520

Duplessy JC, Shackleton NJ, Fairbanks RG, Labeyrie L, Oppo D, Kallel N (1988) Deepwater source variations during the last climatic cycle and their impact on the global deepwater circulation. Paleoceanography 3:343-360.

Duque-Caro H (1990) Neogene stratigraphy, paleoceanography and paleobiogeography in northwest South America and the evolution of the Panama Seaway. Palaeogeography, Palaeoclimatology, Palaeoecology 77:203-234

Elderfield H (1986) Strontium isotope stratigraphy. Palaeogeography, Palaeoclimatology,

Palaeoecology 57:71-90

Emiliani C (1955) Pleistocene temperatures, J Geol 63:538-578

Emiliani C, Geiss J (1958) On glaciation and their causes. Geologische Rundschau 46:576-601

Froelich PN et al. (1991) Biogenic opal and carbonate accumulation rates in the subantarctic South Atlantic. Proc Ocean Drill Program Sci Results 114:515-550

Gardner JV, Dean WE, Wilson CR (1984) Carbonate and organic-carbon cycles and the history of upwelling at Deep Sea Drilling Project Site 532, Walvis Ridge, South Atlantic Ocean. Initial Reports Deep Sea Drilling Project 75:905-921

Gordon A (1985) Indian-Atlantic transfer of thermocline water at the Agulhas retroflection. Science 227:1030-1033

Guilderson TP, Fairbanks RG, Rubenstone JL (1994) Tropical temperature variations since 20,000 years ago: Modulating interhemispheric climatic change. Science 263:663-665

Haq BU (1981) Paleogene paleoceanography: Early Cenozoic oceans revisited. Oceanologica Acta 4 Supplement pp 71-82

Haq BU (1984) Paleoceanography: A synoptic overview of 200 million years of ocean history. In: Haq BU, Milliman JD. Marine geology and oceanography of Arabian Sea and coastal Pakistan. Van Nostrand Reinhold Co, New York pp 201 - 231

Haq BU, Boersma A (eds) (1978) Introduction to marine micropaleontology. Elsevier North-Holland, New York, pp 376

Haq BU, Hardenbol J, Vail PR (1987) Chronology of fluctuating sea levels since the Triassic. Science 235:1156-1167

Hastenrath S (1977) Relative role of atmosphere and ocean in the global heat budget: tropical Atlantic and eastern Pacific. Quart J Roy Meteorol Soc 103:519-526

Hay WW (1988) Paleoceanography: A review for the GSA Centennial Geol Soc Am Bull 100:1934-1956

Hay WW (1992) The cause of the late Cenozoic northern hemisphere glaciations: a climate change enigma. Terra Nova 4:305-311

Hay WW, Brock JC (1992) Temporal variation in intensity of upwelling off southwest Africa. In: Summerhayes CP, Prell WL, Emeis KC (eds) Upwelling Systems: Evolution Since the Early Miocene. Geological Society Special Publication No 63, pp 463-497

Hay WW, Southam JR (1977) Modulation of marine sedimentation by the continental shelves. In: N.R. Andersen and A. Malahoff (eds.) The Fate of Fossil Fuel CO2 in the Oceans, Plenum Press, New York, pp. 569-604

Hay WW, Sibuet JC et al. (1984) Initial Reports of the Deeep Sea Drilling Project, 75, Washington (U.S. Govt Printing Office)

Hays JD, Imbrie J, Shackleton NJ (1976) Variations in the Earth's orbit: Pacemaker of the ice ages. Science 194:1121-1132

Hebbeln D, Wefer G (1991) Sedimentation in the Fram Strait: effects of ice coverage and ice-rafted material. Nature 350:409-411

Hemleben C, Spindler M, Anderson OR (1989) Modern Planktonic Foraminifera. Springer-Verlag, New York Berlin Heidelberg, New York, pp 363

Hess HH (1962) History of ocean basins. In: Engel AEJ, James HL, Leonard BF (eds) Petrologic Studies: A Volume to Honor A.F. Buddington. Geol Soc America, pp 599-620

Hodell DA, Ciesielski (1990) Southern Ocean response to the intensification of northern hemisphere glaciation at 2.4 Ma. In: Bleil U, Thiede J (eds) Geological History of the Polar Oceans: Arctic Versus Antarctic. Kluwer, Amsterdam, pp 707-728

Hodell DA, Ciesielski PF (1991) Stable isotopic and carbonate stratigraphy of the late Pliocene and Pleistocene of Hole 704A: Eastern subantarctic South Atlantic. Proc Ocean Drill Program Sci Results 114:409-435

Hodell DA, Venz K (1992) Toward a high-resolution stable isotopic record of the Southern Ocean during the Pliocene-Pleistocene (4.8 to 0.8 MA). In: Kennett JP, Warnke DA (eds) The Antarctic Paleoenvironment: A Perspective on Global Change Part One Vol 56 (Antarctic Research Series) American Geophysical Union, Washington D.C., pp 265-310

Hodell DA, Williams DF, Kennett JP (1985) Late Pliocene reorganization of deep vertical water-mass structure in the western South Atlantic: Faunal and isotopic evidence. Geol Soc America Bull 96:495-503

Hsü KJ, Andrews JE (1970) Lithology. In: Maxwell AE et al. Initial Repts of DSDP 3:445-453

Hsü KJ, Weissert HJ (eds) (1985) South Atlantic paleoceanography. Cambridge University Press, Cambridge, pp 350

Hsü KJ, Wright R (1985) History of calcite dissolution of the South Atlantic Ocean. In: Hsü KJ and Weissert HJ (eds.) South Atlantic Paleoceanography, Cambridge University Press, Cambridge (U.K.), pp 149-187

Imbrie, J et al. (1973) Paleoclimatic investigation of a late Pleistocene Caribbean deep-sea core: Comparison of isotopic and faunal methods. Quaternary Research 3:10-38

Imbrie J et al. (1992) On the structure and origin of major glaciation cycles , 1. Linear responses to Milankovitch forcing. Paleoceanography 7:701-738

Imbrie J et al. (1993) On the structure and origin of major glaciation cycles, 2. The 100,000-year cycle. Paleoceanography 8:699-735

Jansen E, Sjöholm J (1991) Reconstruction of glaciation over the past 6 Myr from ice-borne deposits in the Norwegian Sea. Nature 349:600-603

Jansen E, Bleil U, Henrich R, Kringstad L, Slettemark B (1988) Paleoenvironmental changes in the Norwegian Sea and the northeast Atlantic during the last 2.8 m.y.: Deep Sea Drilling Project/Ocean Drilling Program Sites 610, 642, 643 and 644. Paleoceanography 3:563-581

Jansen E, Mayer LA,Backman J, Leckie RM, Takayama T (1993) Evolution of Pliocene climate cyclicity at Hole 806B (5-2 Ma): Oxygen isotope record. Proceedings Ocean Drill. Program, Scient. Res. 130:349-362

Johnson DA (1983) Paleocirculation of the South Atlantic. In: Barker PF, Johnson DA, et al. Initial reports of the Deep Sea Drilling Project, Volume 72 pp 977-994. Washington D.C. (U.S. Government Printing Office)

Keigwin LD (1978) Pliocene closing of the Isthmus of Panama, based on biostratigraphic evidence from nearby Pacific Ocean and Caribbean Sea cores. Geology 6:630-634

Keir RS (1988) On the late Pleistocene ocean geochemistry and circulation. Paleoceanography 3:413-445

Keir RS (1990) Reconstructing the ocean carbon system variation during the last 150,000 years according to the Antarctic nutrient hypothesis. Paleoceanography 5:253-276

Keller G, Barron JA (1983) Paleoceanographic implications of Miocene deep-sea hiatuses. Geol. Soc. America Bulletin 94:590-613

Keller G, Zenker CE, Stone SM (1989) Late Neogene history of the Pacific-Caribbean gateway. J. South American Earth Sciences 21:73-108

Kemle-von Mücke S (1994) Oberflächenwasserstruktur und -zirkulation des Südostatlantiks im Spätquartär. Berichte, Fachber Geowiss Univ Bremen 55:1-151

Kennett JP (1982) Marine Geology. Prentice Hall, Englewood Cliffs, 813p

Kennett JP, Shackleton NJ (1976) Oxygen isotopic evidence for the development of the psychrosphere 38 Myr. ago. Nature 260:513-515

Kennett JP, Srinivasan MS (1983) Neogene Planktonic Foraminifera. A Phylogenetic Atlas. Hutchinson Ross Publ. Comp., Stroudsburg, Pennsylvania

Kennett JP, Stott LD (1991) Abrupt deep-sea warming, paleoceanographic changes, and benthic extinctions at the end of the Paleocene. Nature 353: 225-229

Kroopnick P(1980) The distribution of ^{13}C in the Atlantic Ocean. Earth Planet Sci Lett 49:469-484

Kumar N, Anderson RF, Mortlock RA, Froelich PN, Kubik P, Dittrich-Hannen B, Suter M (1995) Increased biological productivity and export production in the glacial Southern Ocean. Nature 378:675-680

Lange CB, Berger WH (1993) Diatom productivity and preservation in the western equatorial Pacific: the Quaternary record. Proceedings of the Ocean Drilling Program, Scientific Results 130:509-523

Larsen HC, Saunders AD, Clift PD, Beget J, Wei W, Spezzaferri S and ODP Leg 152 Scientific Party (1994) Seven Million Years of Glaciation in Greenland. Science 264:952-955

Lutjeharms JRE, van Ballegooyen RC (1988) The Retroflection of the Agulhas Current. J Phys Oceanogr 18:1570-1583

Maier-Reimer E, Mikolajewicz U, Crowley T (1990) Ocean general circulation model sensitivity experiment with an open Central American isthmus. Paleoceanography 5:349-366

Marshall LG (1988) Land mammals and the Great American Interchange. American Scientist 76:380-388

Maxwell AE, von Herzen RP et al. (1970) Initial Reports Deep Sea Drilling Project Vol. 3, US Government Printing Office, Washington DC

McIntyre A, Ruddiman WF, Karlin K, Mix AC (1989) Surface water response of the equatorial Atlantic Ocean to orbital forcing. Paleoceanography 4:19-55

Melguen M, Le Pichon X, Sibuet J-C (1978) Paléoenvironnement de l'Atlantique sud. Bull. Soc. géol. France (7) v. 20 (4) 471-489

Menard HW (1969) Elevation and subsidence of oceanic crust: Earth and Planetary Sci Letters 6:275-284

Meyers PA, Brassell SC, Huc AY (1984) Geochemistry of organic carbon in South Atlantic sediments from Deep Sea Drilling Project Leg 75. In: Hay WW, Sibuet JC et al Initial Reports of the Deep Sea Drilling Project 75. Washington (US Government Printing Office) pp 967-982

Mikolajewicz U, Maier-Reimer E, Crowley TJ, Kim KY (1993) Effect of Drake and Panamanian Gateways on the Circulation of an Ocean Model. Paleoceanography 8:409-426

Milankovitch M (1930) Mathematische Klimalehre und astronomische Theorie der Klimaschwankungen. Handbuch der Klimatologie, Bd 1, Teil A. Bornträger, Berlin, 176pp

Miller JR, Russell GL (1989) Ocean heat transport during the last glacial maximum. Paleoceanography 4:141-155

Miller KG, Fairbanks RG, Mountain GS (1987) Tertiary oxygen isotope synthesis, sea level history, and continental margin erosion. Paleoceanography 2:1-19

Milliman JD (1974) Marine Carbonates. Springer Verlag, Berlin. 375 pp

Mix AC (1989) Influence of productivity variations on long term atmospheric CO_2. Nature 337:541-544

Mix AC, Ruddiman WF, McIntyre A (1986) Late Quaternary paleoceanography of the tropical Atlantic, 2: The seasonal cycle of sea surface temperatures, 0-20,000 years b.p. Paleoceanography 1:339-353

Molfino B, McIntyre A (1990) Precessional forcing of nutricline dynamics in the equatorial Atlantic. Science 249:766-769

Moore TC (1969) Radiolaria: change in skeletal weight and resistance to solution. Geol Soc Am Bull 80:2103-2108

Moore TC, Rabinowitz PD, Borella PE, Shackleton NJ, Boersma A (1985) History of the Walvis Ridge. A précis of the results of DSDP Leg 74. In: K.J. Hsü and H.J. Weissert (eds.) South Atlantic Paleoceanography, Cambridge University Press, Cambridge (U.K.), pp 57-60

Mortlock RA, Charles CD, Froelich PN, Zibello MA, Saltzmann J, Hays JD, Burckle LH (1991) Evidence for lower productivity in the Antarctic Ocean during the last glaciation. Nature 361:220-223

Müller PJ, Suess E (1979) Productivity, sedimentation rate, and sedimentary organic matter in the oceans - I. Organic carbon preservation. Deep-Sea Res 26A:1347-1362

Müller PJ, Erlenkeuser H, von Grafenstein R, (1983) Glacial-Interglacial cycles in oceanic productivity inferred from organic carbon contents in eastern North Atlantic sediment cores. In: Coastal Upwelling: Its Sediment Record, B. edited by J. Thiede and E. Suess (Plenum Press, New York), pp 365-398

Müller PJ, Schneider R, Ruhland G (1994) Late Quaternary PCO2 variations in the Angola Current: Evidence from organic carbon $\delta^{13}C$ and alkenone temperatures. In: Zahn R, Pedersen TF, Kaminski MA, Labeyrie L (eds) Carbon Cycling in the Glacial Ocean: Constraints on the Ocean's Role in Global Change. Springer-Verlag, Berlin Heidelberg, pp 343-366

Munsch GB (ed) (1988) Report of the Second Conference on Scientific Ocean Drilling "Cosod II", European Science Foundation, Strasbourg

North GR, Crowley TJ (1985) Application of a seasonal climate model to Cenozoic glaciation. J Geol Soc (London) 142:475-482

Oberhänsli H (1991) Upwelling signals at the northeastern Walvis Ridge during the past 500,000 years. Paleoceanography 6:53-71

Oberhänsli H, Toumarkine M (1985) The Paleogene oxygen and carbon isotope history of Sites 522, 523, and 524 from the central South Atlantic. In: Hsü KJ, Weissert HJ (eds) South Atlantic Paleoceanography, Cambridge University Press, Cambridge (U.K.), pp 125-147

Oberhänsli H, Müller-Merz E, Oberhänsli R (1991) Eocene paleoceanographic evolution at 20-30°S in the Atlantic Ocean. Palaeogeography, Palaeoclimatology, Palaeoecology 83:173-215

Olausson E (1965) Evidence of climatic changes in North Atlantic deep-sea cores, with remarks on isotopic paleotemperature analysis. Progr Oceanogr 3:221-252

Opdyke ND, Glass B, Hays JD, Foster J (1966) Paleomagnetic study of Antarctic deep-sea cores. Science 154:349-357

Palmer MR, Edmond JM (1989) The strontium isotope budget of the modern ocean. Earth Planet. Science Lett. 92:11-26

Palmer MR, Elderfield H (1985) Sr isotope composition of sea water over the past 75 Myr. Nature 314: 526-528.

Peterson RG, Stramma L (1991) Upper-level circulation in the South Atlantic Ocean. Prog Oceanogr 26:1-73

Pokras EM (1987) Diatom record of Late Quaternary climatic change in the eastern equatorial Atlantic and tropical Africa. Paleoceanography 2:273-286

Prell WL (1985) The stability of low-latitude sea-surface temperatures: An evaluation of the CLIMAP reconstruction with emphasis on the positive SST anomalies. DOE Rep. TRO25, U.S. Department of Energy, Washington, D.C., 60 pp

Raymo ME, Ruddiman WF, Froelich PN (1988) Influence of late Cenozoic mountain building on ocean geochemical cycles. Geology 16:649-653

Raymo ME (1991) Geochemical evidence supporting T.C.Chamberlin's theory of glaciation. Geology 19:344-347

Raymo ME, Hodell D, Jansen E (1992) Response of deep ocean circulation to initiation of northern hemisphere glaciation (3-2 Ma). Paleoceanography 7:645-672

Raymo ME, Ruddiman WF, Shackleton NJ, Oppo DW (1990) Evolution of Atlantic-Pacific δ13C gradients over the last 2.5 m.y. Earth and Planet Sci Lett 97:353-368

Revelle RR (1944) Marine bottom samples collected in the Pacific Ocean by the CARNEGIE on its seventh cruise. Carnegie Inst Wash Publ 556:1-196

Rind D, Peteet D (1985) Terrestrial conditions at the last glacial maximum and CLIMAP sea-surface temperature estimates: are they consistent? Quat Res 24:1-22

Ruddiman WF, Heezen BC (1967) Differential solution of planktonic foraminifera. Deep-Sea Res 14:801-808

Ruddiman WF, Kutzbach JE (1989) Forcing of late Cenozoic northern hemisphere climate by plateau uplift in southern Asia and the American West. J geophys Res 94:18409-18427

Ryan WBF, Hsü KJ, Cita MB, Dumitrica P, Lort JM, Mayne W, Nesteroff WD, Pautot G, Strander H, Wezel FC (1973) Initial Reports of the Deep Sea Drilling Project, Vol. 13, Washington DC (US Government Printing Office) p 1447

Sarnthein M, Thiedemann R (1989) Toward a high-resolution stable isotope stratigraphy of the last 3.4 million years. Proceedings of the Ocean Drilling Program, Scientific Results, 108:167-185

Sarnthein M, Winn K, Jung SJA, Duplessy JC, Labeyrie L, Erlenkeuser H, Ganssen G (1994) Changes in east Atlantic deepwater circulation over the last 30,000 years: Eight time slice reconstructions. Paleoceanography 9:209-267

Schneider RR (1991) Spätquartäre Produktivitäts-änderungen im östlichen Angola-Becken: Reaktion auf Variationen im Passat-Monsun-Windsystem und in der Advektion des Benguela-Küstenstroms. Berichte, Fachber Geowiss Univ Bremen 21: 1-198

Schneider RR, Müller PJ (1995) What role has upwelling played in the global carbon and climate cycles on a million-year time scale? In: Summerhayes CP et al. (eds) Upwelling in the Ocean: Modern Processes and Ancient Records. John Wiley, Chichester, pp 361-380

Schneider RR, Müller PJ, Ruhland G (1995) Late Quaternary surface circulation in the east equatorial South Atlantic: Evidence from alkenone sea surface temperatures. Paleoceanography 10:197-219

Schnitker D (1980) Quaternary deep-sea benthic foraminifers and bottom water masses. Ann Rev Earth Planet Sci 8:343-370

Schott W (1935) Die Foraminiferen in dem äquatorialen Teil des Atlantischen Ozeans. Deutl Atl Exped Meteor 1925-1927. 3:43-134

Sclater JG, Anderson RN, Bell ML (1971) Elevation of ridges and evolution of the central eastern Pacific. J. geophys. Res. 76 (32): 7888-7915

Seibold E, Berger WH (1995) The Sea Floor, An Introduction to Marine Geology, 3rd Edition. Springer Verlag, Heidelberg, 356pp

Shackleton NJ, Kennett JP (1975) Paleotemperature history of the Cenozoic and the initiation of Antarctic glaciation: oxygen and carbon isotope analyses in DSDP Sites 277, 279 and 281. Deep Sea Drilling Project Initial Reports 29:743-755

Shackleton NJ, Opdyke ND (1973) Oxygen isotope and palaeomagnetic stratigraphy of equatorial Pacific core V28-238: Oxygen isotope temperatures and ice volumes on a 10⁵ year and 10⁶ year scale. Quaternary Res 3: 39-55

Shackleton NJ, Backman J, Zimmerman H, Kent DV, Hall MA, Roberts DG, Schnitker D, Baldauf JG, Desprairies A, Homrighausen R, Huddlestun P, Keene JB, Kaltenback AJ, Krumsiek KAO, Morton AC, Murray JW, Westberg-Smith J (1984) Oxygen isotope calibration of the onset of ice-rafting and history of glaciation in the North Atlantic region. Nature 307:602-623

Siesser WG (1980) Late Miocene origin of the Benguela upswelling system off northern Namibia. Science 208:283-285

Stommel H (1980) Asymmetry of interoceanic fresh-water and heat fluxes. Proc. Nat. Acad. Sci. (U.S.A.) 77:2377-2381

Stott LD, Tang CM (1996) Reassessment of foraminiferal-based tropical sea surface δ¹⁸O paleotemperatures. Paleoceanography 11:37-56

Stute M, Forster M, Frischkorn H, Serejo A, Clark JF, Schlosser P, Broecker WS, Bonani G (1995) Cooling of tropical Brazil (5°C) during the last glacial maximum. Science 269:379-383

Summerhayes CP, Birch GF, Rogers J, Dingle RV (1973) Phosphate in sediments off southwestern Africa. Natur 243:509-511

Summerhayes CP, Prell WL, Emeis KC (eds) (1992) Upwelling Systems: Evolution Since the Early

Miocene. Geol Soc Spec Publ, 64, London, pp 519

Summerhayes CP, Emeis KC, Angel MV, Smith RL, Zeitzschel B (1995) Upwelling in the Ocean: Modern Processes and Ancient Records. Dahlem Workshop Reports. John Wiley & Sons, Chichester, pp 422

Sundquist ET, Broecker WS (eds) (1985) The Carbon Cycle and Atmospheric CO_2: Natural Variations Archean to Present. American Geophysical Union, Washington D.C. pp 627

Sverdrup HU, Johnson MW, Fleming RH (1942) The Oceans, Their Physics, Chemistry, and General Biology. Prentice-Hall, Englewood Cliffs, New Jersey, 1087 pp

Tomczak M, Godfrey JS (1994) Regional Oceanography: An Introduction. Pergamon, New York

Turekian KK (ed) (1971) The late Cenozoic Glacial Ages. Yale Univ. Press, New Haven, Conn.

Turnau R, Ledbetter MT (1989) Deep circulation changes in the South Atlantic Ocean: responses to initiation of northern hemisphere glaciation. Paleoceanography 4:565-583

Urey HC (1952) The Planets, Their Origin and Development. Yale Univ. Press, New Haven, 245pp

Vail PR, Hardenbol J (1979) Sea-level changes during the Tertiary. Oceanus 22:71-79

Van Andel TH, Heath GR, Moore TC (1975) Cenozoic history and paleoceanography of the central equatorial Pacific Ocean. Geol Soc Am, Mem 143:1-134

Van Andel TH, Thiede J, Sclater JG, Hay WW (1977) Depositional history of the South Atlantic ocean during the last 125 million years. J Geol 85:651-698

Vincent E, Berger WH (1981) Planktonic foraminifera and their use in paleoceanography. In: The Sea, Vol. 7 (C. Emiliani, Ed.). Wiley-Interscience, New York, pp 1025-1119

Vincent E, Berger WH (1985) Carbon dioxide and polar cooling in the Miocene: the Monterey hypothesis. In: The Carbon Cycle and Atmospheric CO2: Natural Variations Archean to Present (E. T. Sundquist and W. S. Broecker, Eds.). Amer Geophys Union. Geophys Monogr vol 32, pp 455-468

Wefer G (1989) Particle flux in the ocean: Effects of episodic production. In: Berger WH, Smetacek VS, Wefer G (eds) Productivity of the Ocean: Present and Past. Dahlem Workshop Reports. J. Wiley and

Sons, Chichester New York Brisbane Toronto Singapore, pp 139-154

Wefer G, Berger WH (1991) Isotope paleontology: growth and composition of extant calcareous species. Marine Geology 100:207-248

Westerhausen L, Poynter J, Eglinton G, Erlenkeuser H, Sarnthein M (1993) Marine and terrestrial origin of organic matter in modern sediments of the equatorial east Atlantic: The $\delta^{13}C$ and molecular record. Deep-Sea Res. 40:1081-1121

Whitman JM, Berger WH (1992) Pliocene-Pleistocene oxygen isotope record of Site 586, Ontong Java Plateau. Marine Micropal. 18:171-198

Whitman JM, Berger WH (1993) Pliocene-Pleistocene carbon isotope record of Site 586, Ontong Java Plateau. Proceedings of the Ocean Drilling Program, Scientific Results 130:333-348

Williams DF, Thunell RC, Hodell DA, Vergnaud-Grazzini C (1985) Synthesis of late Cretaceous, Tertiary, and Quaternary stable isotope records of the South Atlantic based on Leg 72 DSDP core material. In: Hsü KJ, Weissert HJ (eds) South Atlantic Paleoceanography. Cambridge University Press, Cambridge UK, pp 205-241

Wilson JT (1963) Evidence from islands on the spreading of the ocean floor. Nature 197:536-538

Wise SW, Gombos AM, Muza JP (1985) Cenozoic evolution of polar water masses, southwest Atlantic Ocean. In: Hsü KJ, Weissert HJ (eds) South Atlantic Paleoceanography. Cambridge University Press, Cambridge UK, pp 283-324

Wüst G (1935) Schichtung und Zirkulation des Atlantischen Ozeans. Das Bodenwasser und die Stratosphäre. Deutsche Atlantische Expedition *Meteor* 1925-1927, Wiss Erg, Bd 6, 1. Teil, 2. Liefrg, 288 pp

Wüst G, Defant A (1936) Atlas zur Schichtung und Zirkulation des Atlantischen Ozeans. Deutsche Atlantische Exped. *Meteor* 1925-1927, Wiss Erg, Bd 6, Atlas, 103 pp

Woodruff F, Savin SM (1989) Miocene deepwater oceanography. Paleoceanography 4:87-140

Zachos JC, Lohmann KC, Walker JCG, Wise SW (1993) Abrupt climate change and transient climates during the Paleogene: A marine perspective. J Geol 101:191-213

Inverse Modelling of the Glacial Atlantic Circulation System: Investigation of Data Requirements

B. Grieger[1] and R. Schlitzer[2]

[1]*Universität Bremen, Fachbereich Geowissenschaften,*
Postfach 33 04 40, 28334 Bremen, GERMANY
[2]*Alfred-Wegener-Institut für Polar- und Meeresforschung,*
Postfach 12 01 61, 27515 Bremerhaven, GERMANY

Abstract: A series of model experiments to investigate the data requirements of an inverse model of the Atlantic is described. Results of a first attempt to reconstruct the circulation system of the Atlantic for the last glacial maximum (about 20,000 B. P.) are shown. The amount of data used is quite small compared with an application to the present day Atlantic, for which the model was originally developed. Although the results reflect the difficult data situation, some features of the flow field of the glacial Atlantic are recovered. With some extensions the inverse modelling approach seems to be very promising.

Introduction

Various attempts have been made to simulate the glacial ocean using 3-dimensional dynamical models, see e. g. Maier-Reimer and Mikolajewicz (1989), Lautenschlager et al. (1992). To run these models one has to provide boundary conditions at the sea surface. For the paleo ocean these quantities are not very well known. One possible set of boundary conditions, which is often used to simulate the modern ocean, is defined by the wind stress (momentum flux) and the temperature and salinity at the sea surface. However, the wind stresses cannot be reconstructed from paleoceanographic data and there exists only a coarse data set on sea surface temperatures (CLIMAP 1981) and even fewer data on sea surface salinities.

Another set of boundary conditions is given by the wind stress and fluxes of heat and freshwater (precipitation minus evaporation) at the sea surface which may all be extracted from an atmospheric model provided the boundary conditions for an atmospheric model are known (sea surface temperatures, land albedo, ice sheet topography and insolation). But the ocean models are extremely sensi-

tive to flux boundary conditions (Maier-Reimer et al. 1993). Small errors in the modelled surface fluxes can completely change the resulting modelled ocean state.

A possibility to avoid the problems of dynamical forward models described above is the so-called inverse modelling approach. The main advantage of the inverse approach is that boundary conditions need not necessarily be provided and that any available data for a quantity which is modelled can be incorporated to constrain the model output. As more data are used, the model becomes more independent of the boundary conditions and may even subsist on an incomplete formulation of the physics.

The inverse model used here is an adjoint box model of the Atlantic and was developed by Schlitzer (1993a, b). In section 2 the basic model design is outlined while in section 3 the different data used for three model runs are described. The results of the model runs are presented and discussed in section 4. In section 5 a series of sensitivity experiments is described and section 6 summarizes the conclusions.

From WEFER G, BERGER WH, SIEDLER G, WEBB DJ (eds), 1996, *The South Atlantic: Present and Past Circulation.* Springer-Verlag Berlin Heidelberg, pp 411-422

Basic model design

The model domain consists of a grid of boxes, where the zonal and meridional extension does not need to be equal for all boxes. They are smaller where small scale structure in the flow pattern is expected and larger where the flow pattern is believed to be smooth, compare Fig. 3. The boxes are stacked in 20 layers with layer thickness varying from 60 m at the surface to 500 m at the bottom. One single model box is illustrated in Fig. 1.

If u, v and w are the components of the flow velocity vector in zonal, meridional and vertical direction, respectively, the steady-state heat balance equation including advective and diffusive transports and air-sea heat fluxes can be expressed through

$$
\begin{aligned}
&\Sigma_i A_i \left(u_i\, \theta_i - K_h \Delta\theta/L_i \right) &+ \\
&\Sigma_j A_j \left(v_j\, \theta_j - K_h \Delta\theta/L_j \right) &+ \qquad (1)\\
&\Sigma_k A_k \left(w_k\, \theta_k - K_v \Delta\theta_k/L_k \right) &- \quad Q = 0
\end{aligned}
$$

where the three sums represent zonal, meridional and vertical components of heat transport. Summation is over all interfaces of a box. A is the area of an interface, L is the distance between the box centers on both sides of an interface, K_h and K_v are iso- and diapycnal mixing coefficients, θ is the temperature transported by a flow, $\Delta\theta$ is the temperature difference between the two boxes and Q is the heat flux across the air-sea interface which only appears in the heat budgets of surface boxes. Given the temperatures at the open boundaries (Drake passage and meridional boundary between the African Cape and Antarctica), the air-sea heat flux and the three-dimensional flow vector field over the complete model region — i. e. the Atlantic from surface to bottom — the temperatures can be calculated at every grid point. The calculation involves the solution of a large but sparse linear equation system which is constructed from Eq. (1). If the salinities at the open boundaries and the fresh water fluxes are given the salinities can analogously be calculated for every model box. This is the so-called forward or direct part of the model, compare Fig. 2.

In fact, the flow vector field which has been assumed to be given is not known. On the contrary, there are usually only measurements of temperature and salinity, just those quantities which are calculated by the forward model. The inverse modelling strategy is the following: from an initial guess of the flow field and the surface fluxes, temperatures and salinities are calculated by the forward part of the model. The result is compared with observations and the squares of the deviations enter into a cost function. The flow vectors, fresh water fluxes and surface heat fluxes are then adjusted to enhance the reproduction of observed temperatures and salinities (reducing the cost function). This second step is performed with the so-called adjoint part of the model, which is constructed by derivatives of the cost function with respect to the unknown quantities.

Additionally, the differences from observed values (if there are any) of the reconstructed quantities — flow velocities and surface fluxes — can be taken into account by incorporating them in the cost function, see Fig 2. Where top-to-bottom measurements of temperatures and salinities are available, geostrophic velocity shears can be calculated. These shears can be compared with the reconstructed shears, with deviations also entering into the cost function.

The inverse problem described above is ill-posed and has to be stabilised by smoothness conditions, such as horizontal smoothness of vertical velocities. The result of a proper inverse modelling exercise gives the smoothest flow vector field that is compatible with the observations within their measurement errors. A rigorous mathematical treatment of inverse problems was given by Turchin et al.(1971).

Data used

The data requirements of an inverse model are quite different from those of a direct (forward) dynamical model. To run a 'forward' ocean model, the following data *must* be provided:
   - sea surface temperatures (or surface heat fluxes) over the *complete* model domain,
   - sea surface salinities (or fresh water fluxes) over the *complete* model domain,

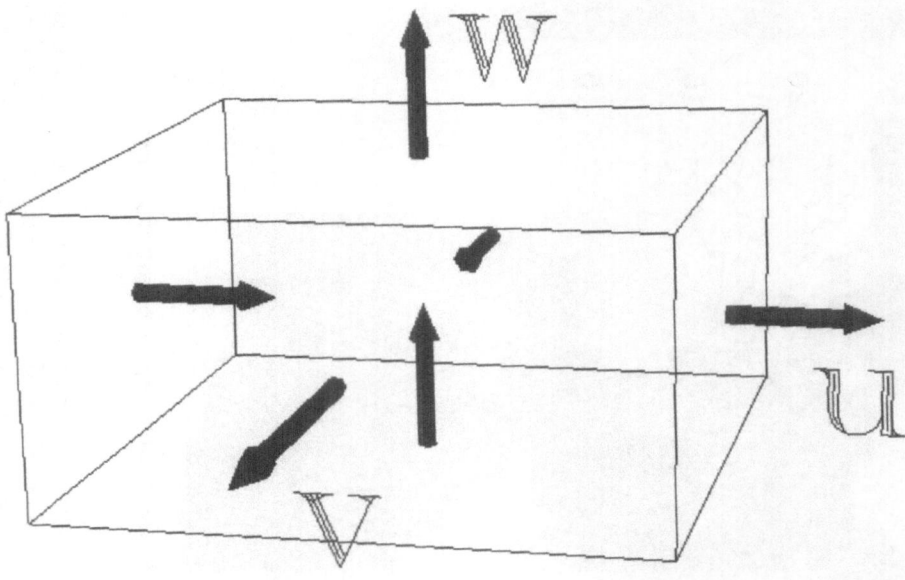

Fig. 1. One box of the model grid. Temperatures and salinities are defined in the center of the box, while zonal, meridional and vertical flow velocities (u, v and w, respectively) are defined at the centers of the box surfaces.

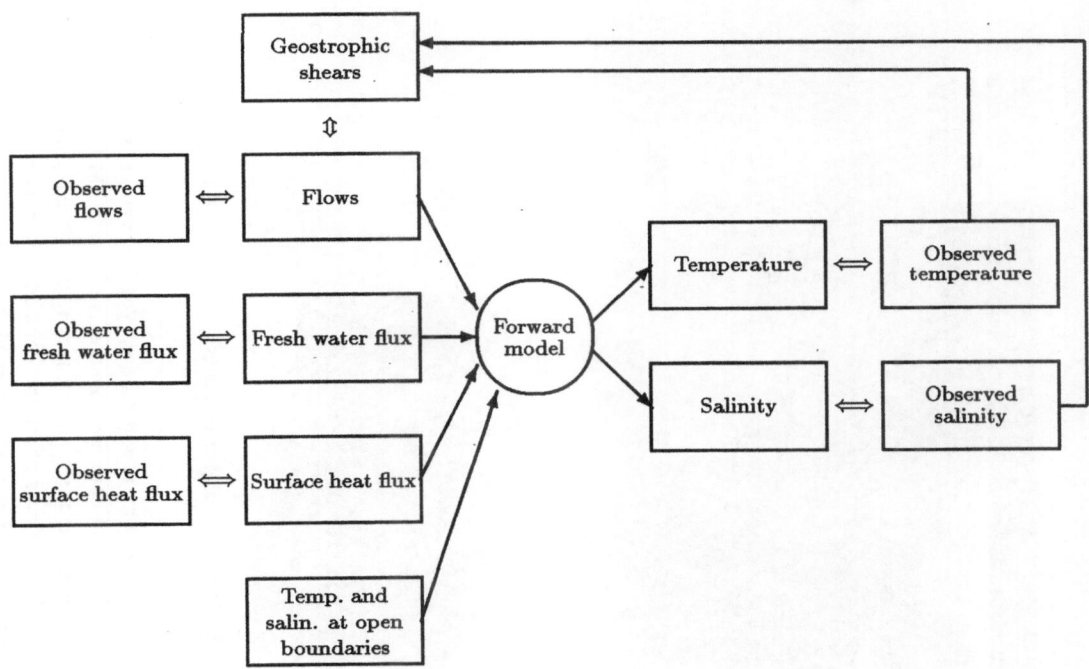

Fig. 2. The inverse modelling approach. Single line arrows denote 'forward' calculations, while double arrows denote comparisons between model and data. The resulting deviation is incorporated into the cost function. See text for further explanation.

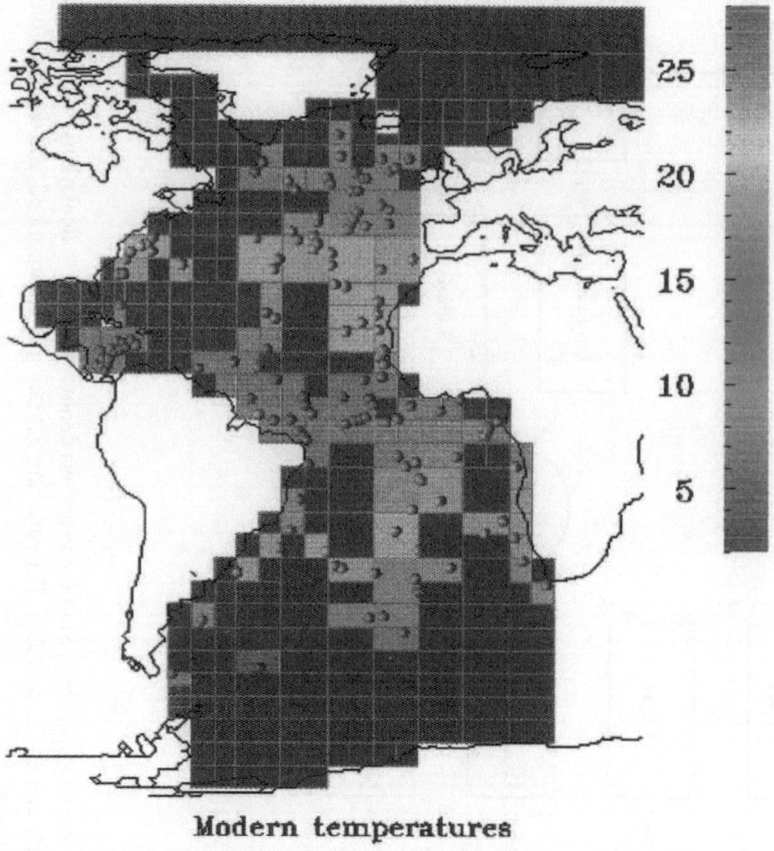

B. Grieger and R. Schlitzer

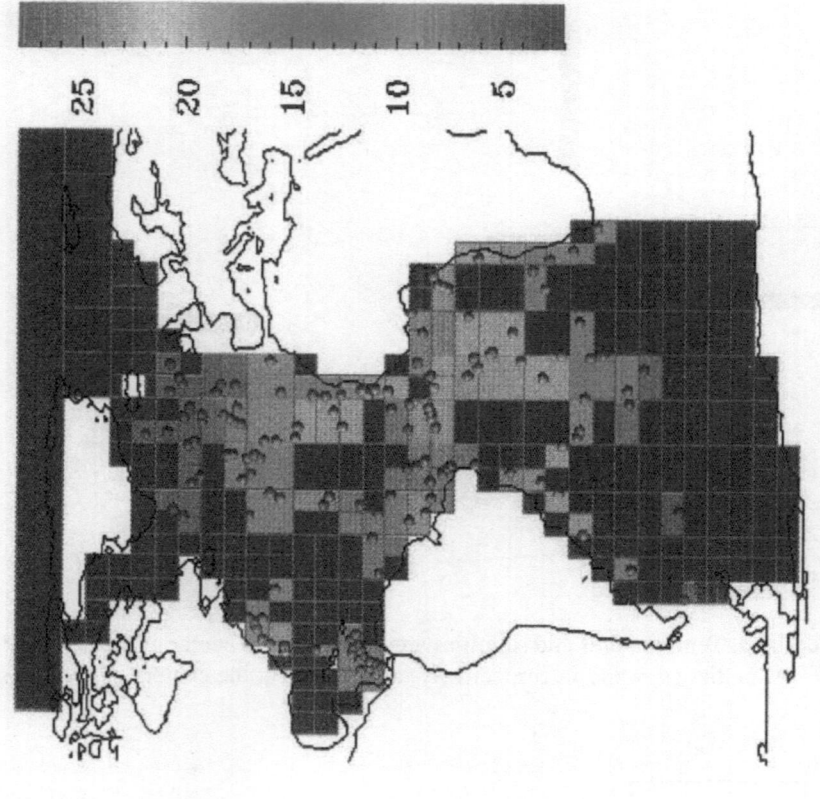

Fig. 3. Model boxes for the Atlantic and modern yearly mean sea surface temperatures estimated from top-core measurements of the CLIMAP sedimentary cores with the MAT (modern analog technique). Spheres indicate the core locations and their colours show the derived temperatures. Each model box with at least one core inside is assigned the mean temperature of all inside cores. Colour bar anotation is in degree Celsius.

Fig. 4. Yearly mean sea surface temperatures for the last glacial maximum, estimated from the CLIMAP sedimentary cores with the MAT (modern analog technique). Spheres indicate the core locations and their colours show the derived temperatures. Each model box with at least one core inside is assigned the mean temperature of all inside cores. Colour bar anotation is in degree Celsius.

- surface wind stresses over the *complete* model domain.

For an inverse ocean model, the following data *can* be used:

- temperatures at any model point,
- salinities at any model point,
- flow velocities at any model point,
- surface heat fluxes,
- fresh water fluxes.

The more data, the better the result, i. e. the more structure can be recovered in the reconstructed flow velocity field. But none of the data are necessary by itself to run the model.

The data which were used by Schlitzer (1993a, b) to model the present day Atlantic are summarised in Tab. 1. The bulk of the data consists of temperatures and salinities for almost every model box (about 9000 data values).

For the last glacial maximum the only data used in this work are sea surface temperatures derived from foraminifera abundance ratios in marine sedimentary cores. The amount of data is very small compared with the present day situation. There are no data at high latitudes and the coverage in low and mid latitudes is quite coarse. Moreover only surface temperatures are used, while for the present day Atlantic top-to-bottom measurements exist.

To estimate the reliability of a result based on such a small amount of data, three model runs are compared:

- The STANDARD run models the present ocean using all available data.
- The CONTROL run models the present ocean using the same amount of data as used for the PALEO run.
- The PALEO run models the glacial ocean.

The deviation between CONTROL run and STANDARD run is a measure for the error of the model caused by the reduction of the amount of data used. It is assumed that the error of the PALEO run has about the same magnitude as the deviation between the CONTROL run and the STANDARD run.

The CONTROL run and the PALEO run are performed using only sea surface temperatures obtained from the set of drilling cores which was used in the CLIMAP reconstruction (CLIMAP 1976, 1981) by a recalculation with the MAT (modern analogue technique) (Prell 1985). For the CONTROL run, the sea surface temperatures are derived from the top-core measurements of the CLIMAP drilling cores (see Fig. 3), while for the PALEO run the measurements for the last glacial maximum are used (see Fig. 4). All runs use modern temperatures at the open boundaries.

The initial state for the STANDARD run is a 'best guess' circulation field, see Schlitzer (1993a, b), while the CONTROL and the PALEO run are initialised with the result of the STANDARD run.

Model results

From all the quantities which are calculated by the model only sea surface temperatures, surface flow velocities and zonally integrated transports are shown here. Additionally to the three dimensional flow field, surface heat fluxes and surface water fluxes are optimized by the model. For the CONTROL and the PALEO run, no data for surface fluxes are used, and therefore the results are dominated by the smoothness conditions. The surface heat and water fluxes exhibit almost no structure.

Over the region which is covered by the CLIMAP data, the sea surface temperatures resulting from the CONTROL run (see Fig. 6) are very similar to those resulting from the STANDARD run (see Fig. 5). On the other hand, in high latitudes, where no data are present in case of the CONTROL run, compare Fig. 3, the model keeps the temperatures as smooth as possible, and therefore they are much to high.

The model sea surface temperatures for the last glacial maximum, which result from the PALEO run, are shown in Fig. 7. The deviation from modern temperatures (compare Fig. 5) coincides with the changes expected for glacial conditions, but the values in high latitudes should not be taken too seriously, as the poor results of the CONTROL run in these areas indicate.

The amount of data used for the PALEO run is very small, and the resultant circulation system, see Fig. 10, does not differ significantly from the CONTROL run, see Fig. 9. Considering the resultant flow vectors, the mean difference between the

Temperature, Salinity	Several large data atlasses, top-to-bottom measurements, data for almost every model box
Geostrophic shear	Calculated from temperature and salinity data
Fresh-water flux	Evaporation and precipitaion estimates, Hellermann (1973)
Air-sea heat flux	Budgets of heat, buoyancy and turbulent kinetic energy based on the COADS data set, Oberhuber (1988), only lower latitudes
A *priori* flows	Upper-layer (0-520 m) meridional transport across equator close to 20 sv Drake-passage throughflow close to 130 Sv Larsen ice shelf flow close to 0.5 Sv Gulf-stream at cape cod close to 120 Sv Florida current close to 30 Sv Bering straits inflow into arctic close to 1.0 Sv Brazil current close to 20 Sv Iceland-Scotland shallow inflow into Norwegian basin close to 15 Sv Upper layer (0-520 m) inflow into Meditterranean close to 3 Sv Lower layer (520-1500 m) outflow from Meditterranean close to -3 Sv Top-bottom flow into Mediterranean close to 0 Sv

Table 1: The data used to model the present day Atlantic (STANDARD run).

PALEO run and the CONTROL run is smaller than the difference between the CONTROL run and the STANDARD run, compare Fig. 8. Therefore it must be concluded that with the present model version and with the amount of data used the overall Atlantic circulation system can not be reconstructed satisfactorily.

Nevertheless, if the three runs (PALEO, CONTROL and STANDARD) are compared locally, it can be seen that significant changes of the currents in some regions are necessary to reproduce the data. Local significant changes are defined here as those deviations between the PALEO run and the CON-TROL run which are larger than the deviations between the CONTROL run and the STANDARD run at the same locations, compare Figs. 9 and 10. In this sense a significant change occurs off the Iberian peninsula, which might be a result of a bend off of the Gulf stream. The only other significant anomaly features of the PALEO surface flows can be found in the eastern South Atlantic and south of Greenland.

The interpretation of inverse modelling results differs from the interpretation of conventional modelling results. Any structures recovered in the proper solution of an inverse problem are reliable, but a

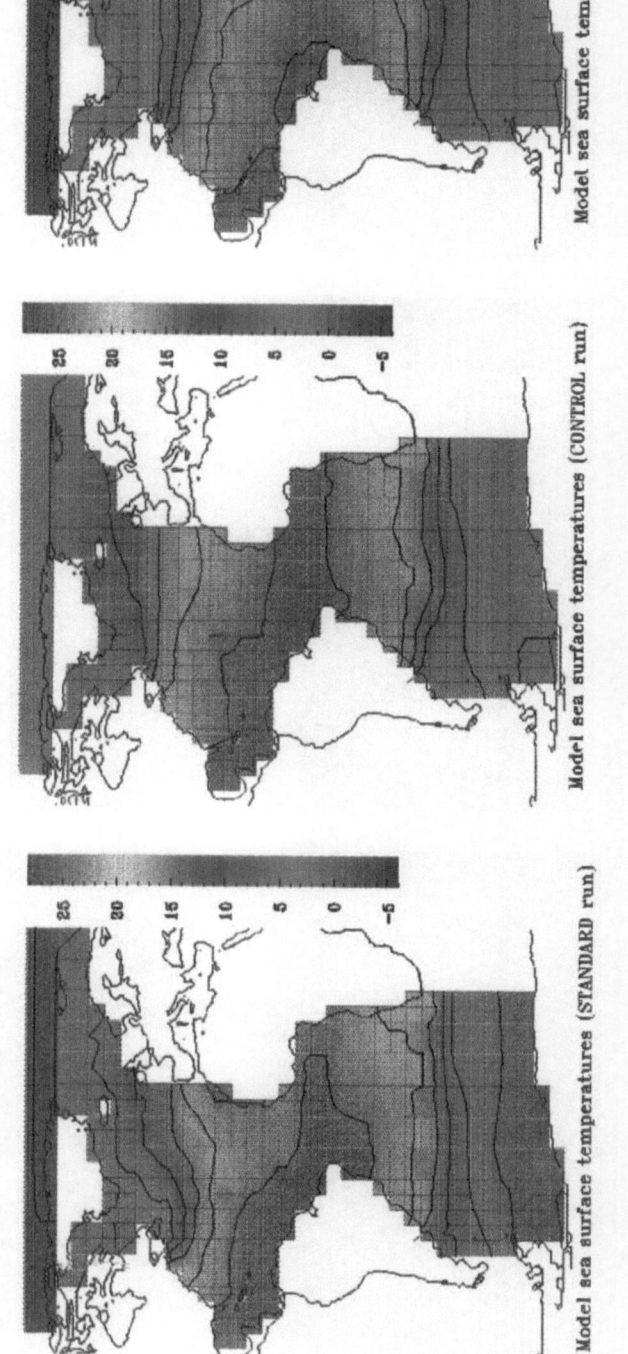

Fig. 5. Model sea surface temperatures resulting from the STANDARD run using all available data for the present day Atlantic. Colour bar anotation is in degree Celsius, isolines are drawn at the anotation values.

Fig. 6. Model sea surface temperatures resulting from the CONTROL run using only sea surface temperatures from the CLIMAP sedimentary cores for the present day Atlantic. Colour bar anotation is in degree Celsius, isolines are drawn at the anotation values.

Fig. 7. Model sea surface temperatures resulting from the PALEO run using only sea surface temperatures from the CLIMAP sedimentary cores for the last glacial maximum. Colour bar anotation is in degree Celsius, isolines are drawn at the anotation values.

Model sea surface temperatures (STANDARD run)

Model sea surface temperatures (CONTROL run)

Model sea surface temperatures (PALEO run)

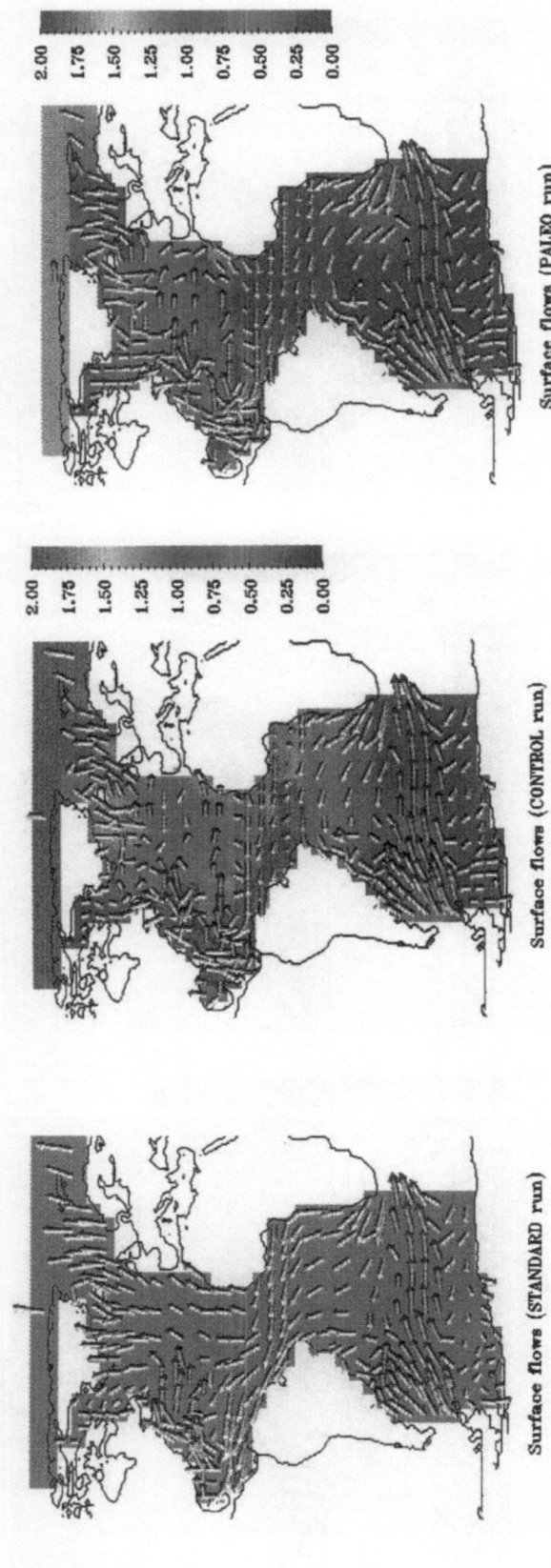

Fig. 8. Surface flow velocities in relative units resulting from the STANDARD run. There is one vector in the center of each box. The field has been smoothed by averaging over adjacent boxes. For plotting the velocities have been raised to the power 1/3 to enhance the visibility of weak flows.

Fig. 9. Surface flow velocities (in the units of Fig. 8) resulting from the CONTROL run. The colour codes the difference from the STANDARD run in units of its standard deviation. Blue colour indicates regions where the resultant flows of the CONTROL run are close to the STANDARD run, while red indicates regions where the CONTROL run differs more than two standard deviations from the STANDARD run, i. e. where it is in error.

Fig. 10. Surface flow velocities (in the units of Fig. 8) resulting from the PALEO run. The colour codes the difference from the CONTROL run in units of its standard deviation.

smooth solution only indicates that the information content of the data is not sufficient to recover any structure. Therefor the similarity between PALEO run and CONTROL run in a large area — almost the complete South Atlantic — does not mean that differences between the paleo and the modern South Atlantic can be excluded. With the amount of data used, the inverse modelling approach is just not able to recover changes in this area.

The zonally integrated transports resulting from the three model runs are shown in Figs. 11, 12 and 13. While the STANDARD run is able to reconstruct quite well the main features of the Atlantic deep water circulation, the results of the CONTROL run and the PALEO run are very poor in this respect, both exhibiting no North Atlantic deep water production. A closer look reveals that the PALEO run even yields a weak upwelling in the North Atlantic. If we trust the anomalies to the CONTROL run - although the absolute value is completely wrong - this may also be interpreted as a weakening of the deep water production during glacial times.

Sensitivity investigations

To investigate the dependence of the results on the amount of data used and on the initial state, a series of experiments was performed, see Table 2. In the main sequence of experiments — which is 1a, 2, 3, 4, 5, 7, 8a — the amount of data used is reduced step by step. The results of experiments 1a, 2, 3 and 4 are very similar. Experiment 5 exhibits some local changes of the deepwater circulation, while the overall pattern persists. The result of experiment 7 shows a significant weakening of the deep water circulation, which breaks together completely in experiment 8a. Therefore we conclude that the usage of a three dimensional data field (top to bottom measurements) is necessary to reconstruct the deep water circulation. On the other hand, the changes of the surface flow field are comparatively small through this sequence of experiments.

Experiment 6 differs from experiment 7 by using the a priory flows (compare Table 1) but omitting the three dimensional temperature field and only incorporating the modern CLIMAP temperatures. The resultant circulation exhibits almost no

North Atlantic deepwater production, which again demonstrates the importance of the temperature data.

The experiments 1a and 1b use the same data but have different initial states. The resultant circulation patterns are very similar. Only a few flow vectors differ significantly. It is worth noting that — using the full data set — the overall circulation pattern of the Atlantic can be recovered starting with an ocean at rest.

Similarly the experiments 8a, 8b and 8c only differ in their initial states. While the results of 8a and 8b are almost identical, the result of experiment 8c is completely different. The latter shows almost no horizontal flows and only vertical transports implying unrealistic fresh water fluxes. Therefore it can be concluded that the result depends on the initial state if the amount of data used is small while the initial state is less important if a large amount of data is used.

Conclusions

The inverse model used here was originally developed to reconstruct the flow velocity field of the modern Atlantic. Since a large amount of data existed, the model physics could be kept very simple. Applied to the glacial Atlantic, the results are not satisfactory. Improvements can either be achieved by incorporating more data (proxy reconstructions or results from other models) or by including more physics in the model.

The poor reconstruction of the deep water circulation might be improved by demanding zonally integrated transports, which can be obtained from the results of 2-dimensional models for the zonal mean circulation The possibility of such an extension is already implemented in the present model version, since one component of the cost function consists of the deviation between model transports and a priori transports through an arbitrary set of box surfaces.

Another possibility to improve the results is to use more sea surface temperatures, possibly some temperatures in deeper ocean layers and some sea surface salinities. There are a lot more sea surface temperatures available from the literature than were

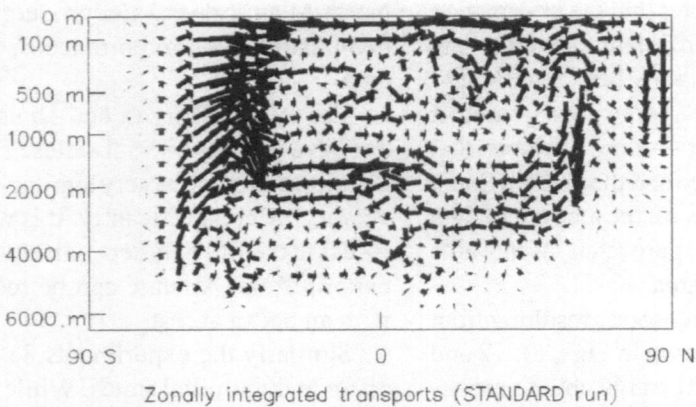

Fig. 11. Zonally integrated transports in relative units resulting from the STANDARD run. There is one row of vectors for each model level.

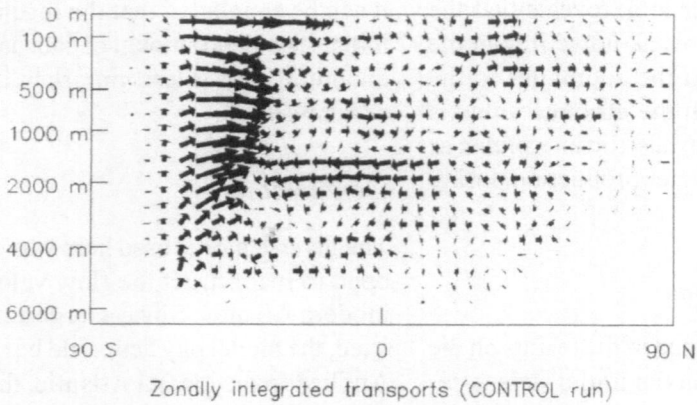

Fig. 12. Zonally integrated transports in relative units resulting from the CONTROL run. There is one row of vectors for each model level.

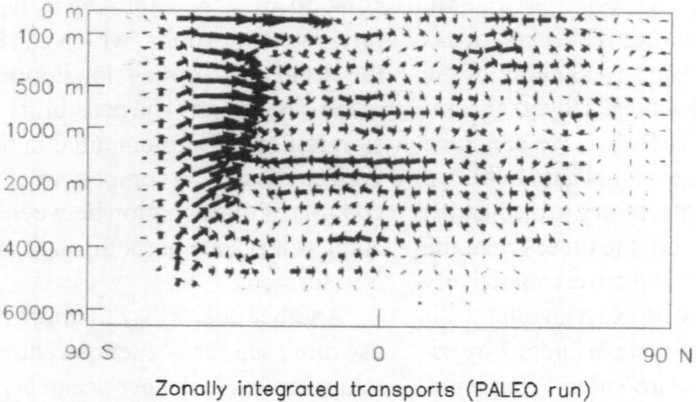

Fig. 13. Zonally integrated transports in relative units resulting from the PALEO run. There is one row of vectors for each model level.

Exp. No.	Geo-strophic shear	Surface heat flux	Fresh water flux	Salinity	Tempera-ture	A priori flows	Initial state
1a	yes	yes	yes	yes	All data	yes	Best guess ocean
1b	yes	yes	yes	yes	All data	yes	Ocean at rest
2	no	yes	yes	yes	All data	yes	Result of exp. 1a
3	no	no	yes	yes	All data	yes	Result of exp. 2
4	no	no	no	yes	All data	yes	Result of exp. 3
5	no	no	no	no	All data	yes	Result of exp. 4
6	no	no	no	no	CLIMAP	yes	Result of exp. 5
7	no	no	no	no	All data	no	Result of exp. 6
8a	no	no	no	no	CLIMAP	no	Result of exp. 7
8b	no	no	no	no	CLIMAP	no	Ocean at rest
8c	no	no	no	no	CLIMAP	no	Result of exp. 1a

Table 2: The different amounts of data used in the series of sensitivity experiments. Experiment 1a corresponds to the STANDARD run and experiment 8c to the CONTROL run.

used in this first attempt. As additional constraints on the modelled quantities, data for surface heat fluxes and fresh water fluxes may be taken from the results of atmosphere general circulation models.

Even if all possibly available glacial data are taken into account, ist amount will still be much smaller than the amount of data used for the STANDARD run. Therefore more physics has to be included in the model to increase the constraints on the solution before a satisfactory reconstruction of the overall Atlantic circulation can be expected. In the present model version, the reconstructed cur-

rents have to satisfy only the mass balance. There are no other physical conditions on the dynamics.

It is planned to incorporate the principle of geostrophy into the model. Since no top-to-bottom measurements of temperature and salinity are available for the glacial Atlantic, geostrophic shears cannot be calculated form the data. But the modelled temperatures and salinities can be checked for geostrophic consistency with the modelled flows, with deviations entering into an extended cost function. This would yield a kind of 'hybrid model', combining the advantages of direct and inverse ocean modelling.

Acknowledgments

We would like to thank G. Meinecke for the preparation of the CLIMAP data and K. Herterich for valuable comments on the manuscript. This is contribution no. 135 of Sonderforschungsbereich 261 at Bremen University.

References

CLIMAP Project Members (1976) The surface of the ice age Earth. Science 191:1131-1137

CLIMAP Project Members (1981) Seasonal reconstructions of the Earth surface at the last glacial maximum. GSA Map and Chart Ser MC-36, Geol Soc Am, Boulder, Colorado

Lautenschlager M, Mikolajewicz U, Maier-Reimer E (1992) Applications of ocean models for the interpretation of atmospheric general circulation model experiments on the climate of the last glacial maximum. Paleoceanogr 7:769-782

Maier-Reimer E, Mikolajewicz U (1989) Experiments with an OGCM on the cause of the younger Dryas. Report 9, Max-Planck-Institut fuer Meteorologie, Hamburg

Maier-Reimer E, Mikolajewicz U, Hasselmann K (1993) Mean Circulation of the Hamburg LSG OGCM and Its Sensitivity to the Thermohaline Surface Forcing. J Phys Oceanogr 23:731-757

Prell WL (1985) The stability of low-latitude sea surface temperatures: An evaluation of the CLIMAP reconstruction with emphasis on the positive SST anomalies. Spec Pub TRO25, US Dep of Energy, Washington DC

Turchin VF, Kozlov VP, Malkevich MS (1971) The use of mathematical-statistics methods in the solution of incorrectly posed problems. Soviet Physics Uspekhi 13:681

Schlitzer R (1993a) An Adjoint Model for the Determination of the Mean Oceanic Circulation, Air-Sea fluxes and Mixing Coefficients. Habilitation thesis, University of Bremen

Schlitzer R (1993b) Determining the mean, large-scale circulation of the Atlantic with the adjoint method. J Phys Oceanogr 23:1935-1952

Chemical Hydrography of the South Atlantic During the Last Glacial Maximum: Cd vs. δ^{13}C

E. Boyle & Y. Rosenthal

*Department of Earth, Atmospheric and Planetary Sciences,
Massachusetts Institute of Technology,
Cambridge, MA 02139, U.S.A.*

Abstract: There has been a major contradiction between benthic foraminiferal Cd/Ca and δ^{13}C data concerning the labile nutrient chemistry of the Southern Ocean during the Last Glacial Maximum (LGM). Cd data indicates that LGM South Atlantic nutrient concentrations were as low as they are today, indicative of a persistent influx of nutrient-depleted North Atlantic Deep Water (NADW). δ^{13}C data indicates that LGM South Atlantic nutrient concentrations were much higher than at present (even higher than anywhere else in the ocean at that time), and these data have been interpreted as signifying the complete shutdown of the export of NADW into the global ocean. This paper examines both true geochemical differences and various confounding foraminiferal artifacts for both tracers. While many different processes and artifacts affect both tracers in the margin, we conclude the discrepancy is mainly due to the „Mackensen Effect" of low foraminiferal δ^{13}C as a result of high carbon flux to the sediments, and that LGM Atlantic Sector Southern Ocean nutrient concentrations remained similar to the levels encountered today.

Introduction

Geochemical paleoceanographers have been attempting to reconstruct the nutrient hydrography from the geochemical composition of benthic foraminifera during the past decade. The goal of this effort is to outline the basic spreading patterns of deep ocean water masses and to help understand the role of ocean chemical redistributions in glacial/interglacial atmospheric CO_2 cycles. There have been some notable successes during this effort, particularly with regard to Atlantic chemical hydrography which has been shown to differ significantly during the Last Glacial Maximum (LGM) from that of the modern ocean. However, there has been a major contradiction between the two benthic foraminiferal tracers (δ^{13}C and Cd/Ca) in the Southern Ocean. This discrepancy has formed a substantial barrier to progress in reconstructing the overall deep water circulation scheme, and significant effort has been expended in attempting to understanding these differences. The purpose of this paper is to examine if the apparent differences between these tracers can be traced either to real differences in the LGM oceanic behavior of δ^{13}C and Cd compared to the modern ocean, or alternatively can be understood as a result of artifacts plaguing one or both tracers. While it turns out that differences in tracer behavior <u>and</u> sedimentary foraminiferal artifacts can be found for both tracers, we will argue that the Cd data has been giving a more faithful account of LGM Southern Ocean chemistry (in support of persistent operation of the North Atlantic „conveyor belt"). As an epilogue, we will summarize some new radionuclide data that supports this conclusion.

The Southern Ocean δ^{13}C-Cd Conundrum

The South Atlantic Ocean is more sensitive than any other oceanic region to variations in the struggle between North Atlantic Deep Water (NADW) and Antarctic Bottom Water (AABW) for global hydrographic dominance. At present, roughly 15 Sverdrups of both are formed. Although some AABW

From WEFER G, BERGER WH, SIEDLER G, WEBB DJ (eds), 1996, *The South Atlantic: Present and Past Circulation.* Springer-Verlag Berlin Heidelberg, pp 423-443

escapes almost directly into the North Atlantic, considerable mixing of these water masses occurs within the Antarctic Circumpolar Current, and today the Circumpolar Deep Water (CPDW) contains a significant admixture of both components (Mantyla and Reid 1983). Given the strong contrast in the initial chemical properties between these sources, it is reasonable to suppose that fluctuations in the relative intensities of production would be reflected in changes in Southern Ocean chemical makeup (Oppo et al. 1990; Charles and Fairbanks 1992). South Atlantic hydrography is also sensitive to the relative strength of these deep waters because of the penetration of almost-unmodified NADW as far south as 30°S. The steep property gradients occurring southward and above and below the water mass core level (Fig. 1) are not likely to remain stationary after a major change in source production rates. It should be expected that fluctuations in the intensities of the Antarctic and NADW sources may be accompanied by relatively large North-South movements of these frontal zones.

Fig. 2 and 3 present these concepts as cartoons. Fig. 2 illustrates the Charles and Fairbanks (1992) notion that the chemical composition of Circumpolar Deep Water is a flux-weighted mixture of young nutrient-depleted North Atlantic Deep Water and recirculated old Pacific Deep Water. As suggested by Charles and Fairbanks, if the NADW flux was lower in the past, then the chemical composition of Southern Ocean waters may reflect variations of North Atlantic Deep Water flux. Alternatively, Fig. 3 suggests an analogy between the steep chemical gradients seen at the distal end of pure NADW and the intensity of a firehose spray. Near the mouth of the firehose, the intense flow overwhelms other water masses and the chemical composition of NADW is almost independent of the total flow; near the end of the spray zone, where NADW mixes with northward-flowing southern source waters, chemical gradients are steep and very sensitive to relatively small fluctuations of the total flow. Almost immediately, one has to confess that these cartoons are absurdly oversimplified representations of the factors that control intra- and inter-ocean chemical gradients. Nonetheless, each probably contains a grain of truth for fluctuations about present-day conditions in flow intensity of the

NADW. If chemical boundary conditions and flows were vastly different at times in the past, such inferences may be more questionable.

The cartoon character of these representations is less of a flaw at present than is the substantial disagreement that exists between the two proxy tracers for the paleochemical properties characteristic of southern and northern source waters:

(1) There are considerable published data on the carbon isotopic composition of the shells of the bottom-dwelling foraminifera *Cibicidoides wuellerstorfi* that lived during the Last Glacial Maximum (LGM). These data indicate that Southern Ocean deep waters had a carbon isotopic composition that was more depleted in ^{13}C relative to ^{12}C (i.e. more negative in $\delta^{13}C$) than deep waters of the eastern tropical Pacific (Curry et al. 1988; Oppo et al. 1990; Charles and Fairbanks 1992). In other words, based on this $\delta^{13}C$ evidence, these waters should be expected to have had lower oxygen concentrations and higher phosphorus and nitrate concentrations than the Pacific deep waters. Curry et al. (1988) pointed out that Antarctic deep water cannot be a mixture of Pacific and North Atlantic Deep Water, and would require some additional source of nutrient-depleted bottom water in the Indian or Pacific Ocean. Charles and Fairbanks (1992) overlooked this point in their subsequent paper.

(2) In sharp contrast, published data on the cadmium composition of the shells of several species of benthic foraminifera from Southern Ocean cores indicate that the cadmium concentration of these waters was similar to what it is today. Because the cadmium concentration of seawater samples is strongly correlated to those of phosphorus (Boyle 1988; Boyle 1994b; and references therein); this correlation is due to mutual involvement in the biological pump), this result appears to flatly contradict the carbon isotope data. Either one (or both) proxy tracers are misleading us, or we simply do not understand their oceanic behavior well enough to understand what they are telling us.

The goal of the paper is to summarize what has been discovered in recent years during attempts to unravel this data contradiction conundrum. Along the way, we have obtained a deeper understanding of the oceanic behavior of the tracers, and a better

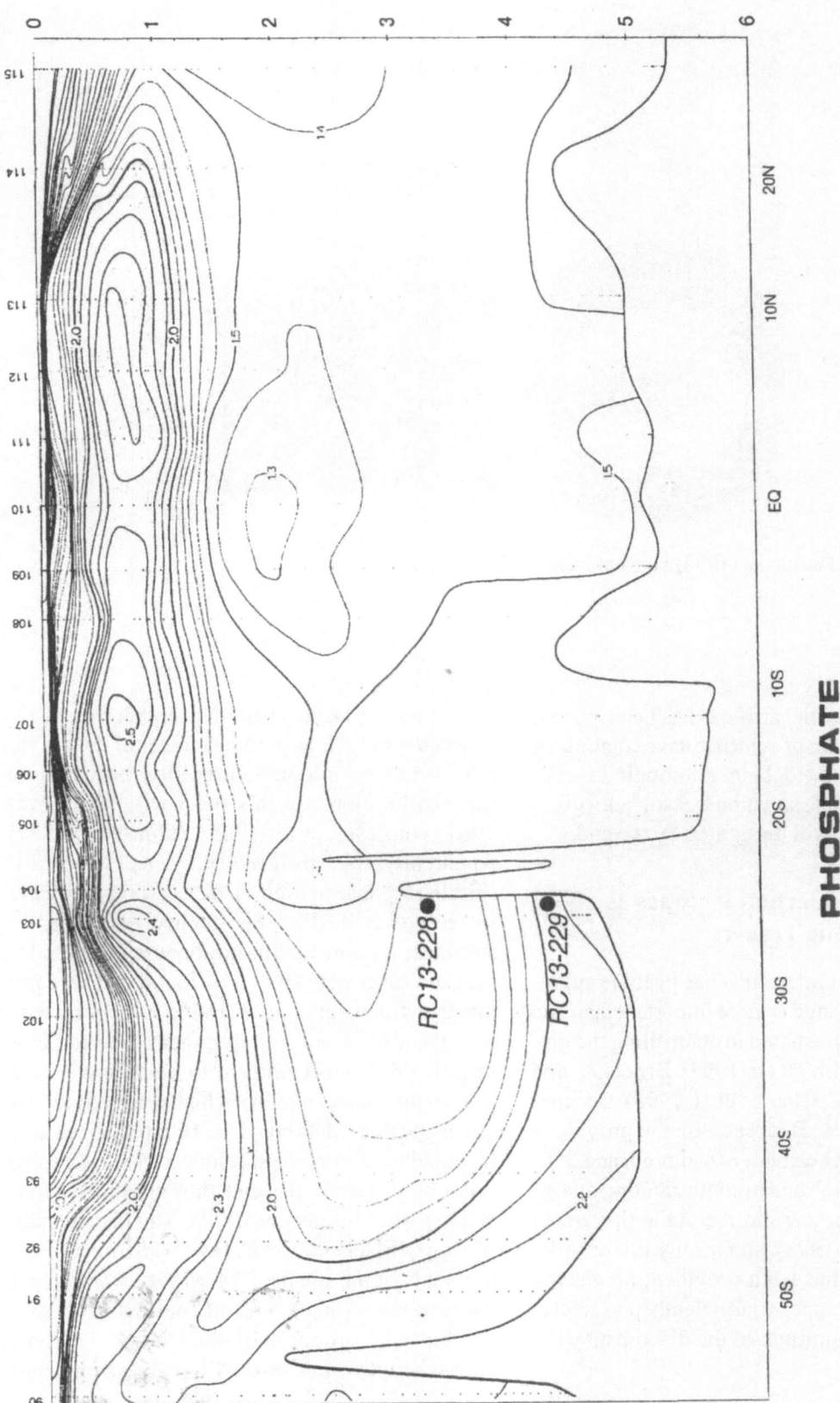

Fig. 1. GEOSECS phosphorus section for the eastern South Atlantic (Bainbridge 1980). Hydrographically-equivalent positions of two cores featured in this paper indicated.

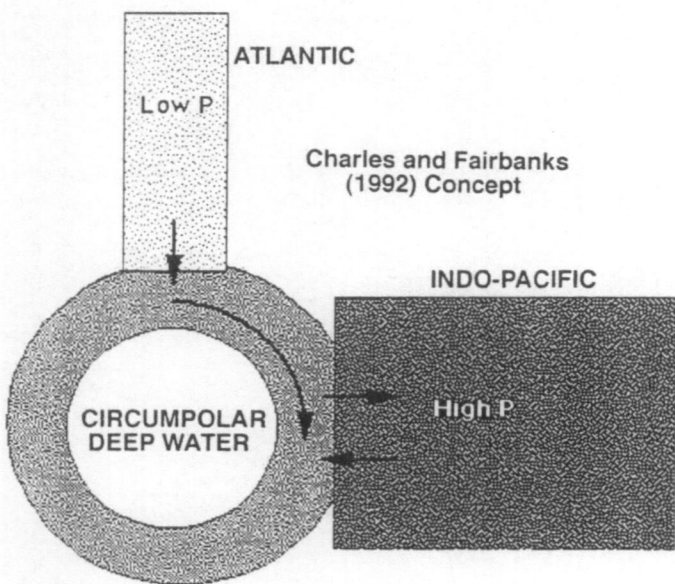

Fig. 2. Charles and Fairbanks (1992) concept cartoon.

appreciation of potential artifacts has been gained. Furthermore, new lines of evidence have come along that may influence the debate. Although it is premature to claim that the situation is resolved, a glimmer of a resolution will be sighted by the end.

Reassessment of Carbon Isotopes as Paleoceanographic Tracers

During the past several years it has become appreciated that gas exchange is more important than had been previously appreciated in controlling the global $\delta^{13}C$ distribution (Keir 1991; Broecker and Maier-Reimer 1992; Charles et al. 1993). Because this process does not affect cadmium, it provides a way to decouple the behavior of cadmium and $\delta^{13}C$. Despite that potential, and notwithstanding one attempt to use this process to reconcile the tracer conflict (Broecker 1993), so far only minor relief to the conundrum has been provided; no gas exchange mechanism appears sufficiently powerful to account for the magnitude of the discrepancy between the tracers.

It is now at least debatable whether the species C. *wuellerstorfi* is as perfect a tracer of deep water $\delta^{13}C$ as has been assumed recently based on the early calibrations for this species (Graham et al. 1981; Duplessy et al. 1984; Zahn et al. 1987). Apparently, this species displays a $\delta^{13}C$ negative shift in response to high surface carbon productivity (Berger et al. 1981; Sarnthein et al. 1988). The strongest support for this notion was presented by Mackensen et al. (1993), who found that both protoplasm-filled specimens (living or very recently living) and fossil core top specimens showed a large negative $\delta^{13}C$ shift relative to measured bottom water values under regions of high productivity. The „living benthic" data could not be rationalized away (as could fossil core top specimens which can sometimes be suspected because they could have lived at times when the deepwater $\delta^{13}C$ was different than it is now). Mackensen et al. (1993) also made measurements of the species *Nutallides umbonifera* in the same region and found that species did not show any hint of a productivity-induced $\delta^{13}C$ offset at sites where C. *wuellerstorfi* showed strong nega-

The North Atlantic Deep Water
Fire Hose

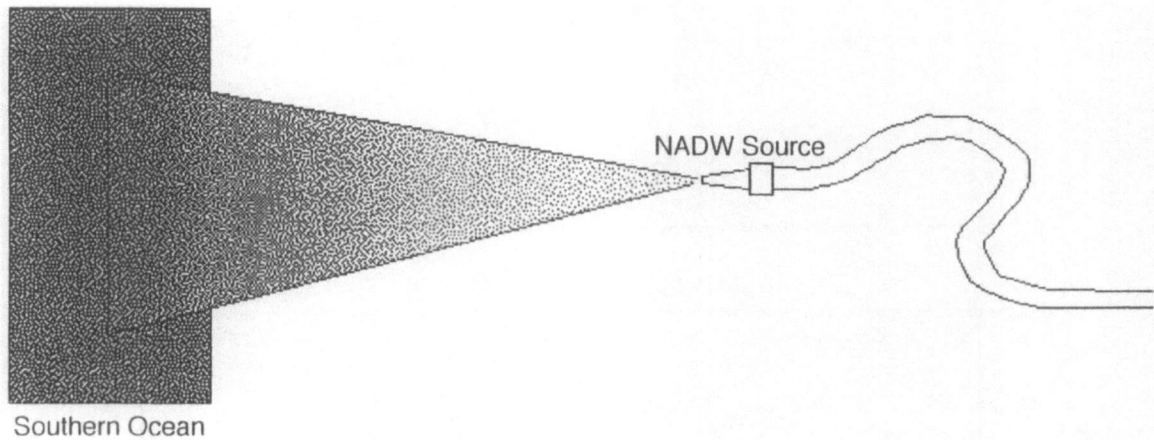

Fig. 3. NADW Firehose concept cartoon. Compare to Fig. 1.

tive shifts (Fig. 4). This species may be less prone to productivity-induced $\delta^{13}C$ offsets. Perhaps this species will become the new gold standard by which carbon isotope data are judged. Although regrettably the geographical distribution of this species is not so widespread as that of *C. wuellerstorfi*, fortunately from the point of view of Southern Ocean paleoceanography, it is often abundant in this region. Clearly much more downcore $\delta^{13}C$ data from this species should be obtained.

In fairness, it should be noted that *C. wuellerstorfi* $\delta^{13}C$ has had some good news as well as bad news. For example, there seemed to be a significant apparent problem when Herguera et al. (1992) observed that core top *C. wuellerstorfi* from the Ontong-Java Plateau were significantly enriched in ^{13}C (~0.4‰) compared to the bottom water (based on Kroopnick's (1985) water column data). McCorkle and Keigwin (1994) made new measurements of water column $\delta^{13}C$ on the Ontong-Java Plateau and found $\delta^{13}C$ values consistently about 0.4‰ more positive than Kroopnick's data. They suggested that some of Kroopnick's (1985) data

was in error. If so, at that site, and perhaps in others, some apparent disagreement between *C. wuellerstorfi* $\delta^{13}C$ vs. estimated bottom water $d^{13}C$ might be due to error in water measurements. In view of this finding, the fairest summary is probably to say that *C. wuellerstorfi* is a reliable indicator of bottom water $\delta^{13}C$ under conditions of low to moderate productivity (such as on the Ontong-Java Plateau), but that it may show a negative shift under regions of moderate to high productivity.

If we accept that *C. wuellerstorfi* can sometimes misrepresent bottom water composition due to productivity-induced artifacts, some loose ends in paleoceanographic $\delta^{13}C$ can be understood in hindsight. For example, Oppo et al. (1990) shifted the *C. wuellerstorfi* $\delta^{13}C$ data from equatorial Pacific core KNR73-3PC (Boyle and Keigwin 1985/6) by -0.3‰ because it was heavier than *C. wuellerstorfi* data from eastern Tropical Pacific cores V19-30 and TR163-31b (Curry et al. 1988). Although Oppo et al. did not assert that this was an analytical error, it is clear that they regarded the KNR73-3PC data set as problematical compared to the oth-

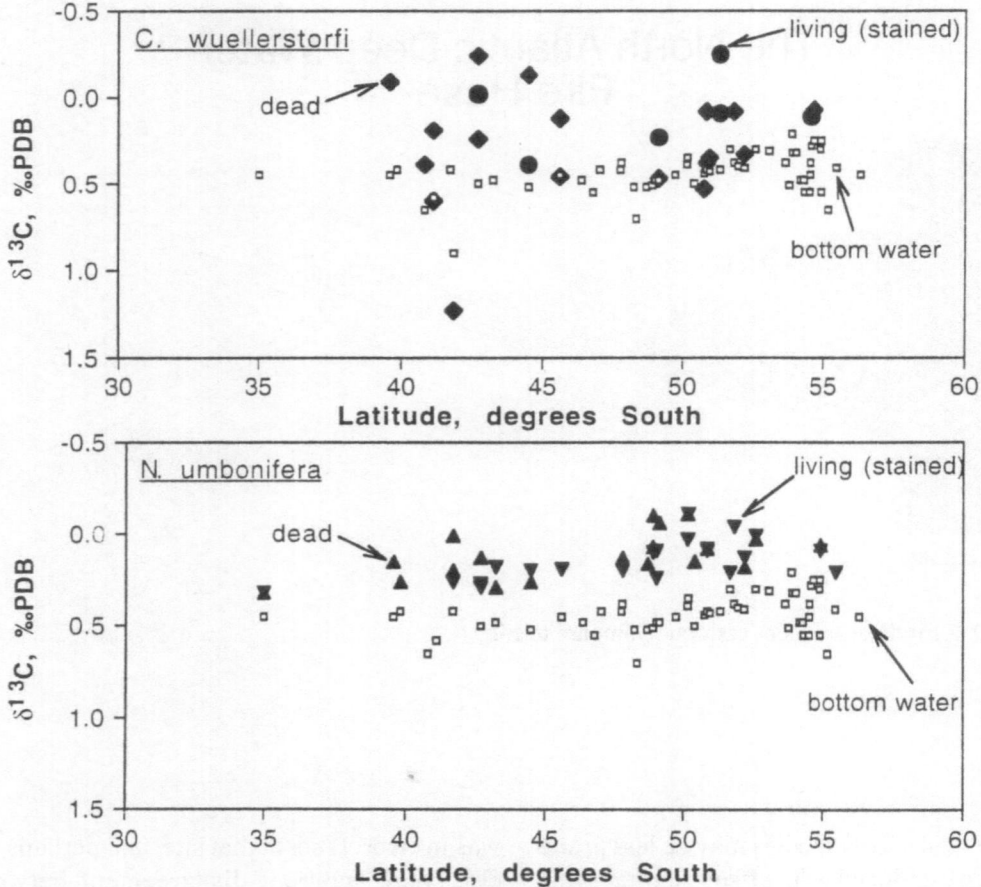

Fig. 4. Mackensen (1993) *C. wuellerstorfi* and *N. umbonifera* δ¹³C data compared to bottom water.

ers. Later, Mix et al. (1990) published *C. wuellers-torfi* δ¹³C data from another equatorial Pacific core (RC13-110) near KNR73-3PC that matched KNR73-3PC almost exactly (Fig. 5). The good agreement between the two equatorial Pacific data sets suggests that there is no basis for supposing that the KNR73-3PC carbon isotope record is flawed. Keeping in mind the *C. wuellerstorfi* „Mackensen Effect" (complementing the *Uvigerina* „Zahn Effect"), we note that cores V19-30 and TR163-31b are located under a higher productivity region than KNR73-3PC and RC13-110 (especially during the last glacial maximum, Pedersen 1983). We suggest that the lighter δ¹³C observed in the eastern Tropical Pacific sites is due to a pro-

ductivity-induced „Mackensen Effect" at those sites and that the lower-productivity Equatorial Pacific sites reflect the bottom waters more faithfully.

In the same vein, Keigwin et al. (1991) showed that in core EN120-GGC2 (33°40'N, 57°37'W, 4479m), *C. wuellerstorfi* showed large abrupt changes in δ¹³C that were not seen in coexisting *N. umbonifera*. In view of the results of Mackensen et al. showing that *N. umbonifera* does not show productivity-induced offsets, the differences between these two species in this core may be due to as a productivity-induced *C. wuellerstorfi* artifact rather than a true δ¹³C bottom water signal.

As we will discuss later, this effect may also contribute to the apparent contradiction between

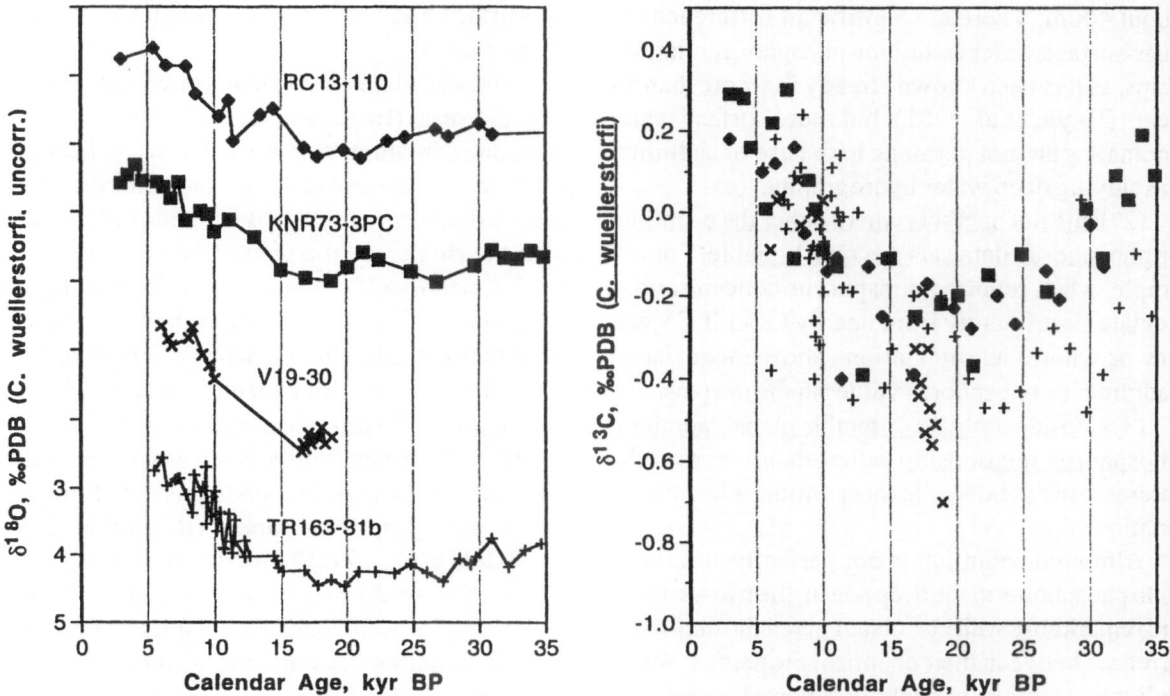

Fig. 5. Comparison of δ¹³C data from equatorial Pacific cores (KNR73-3PC, RC13-110) with eastern tropical Pacific cores (V19-30, TR163-31B). Symbols on left and right are the same for each core.

light *C. wuellerstorfi* δ¹³C values and cadmium evidence in the Southern Ocean; Wefer and Bickert (this volume) have evidence that South Atlantic cores from low-productivity sites do not show the extremely light δ¹³C values seen at sites underlying higher-productivity waters. This matter will be discussed after an exploration of possible artifacts in foraminiferal Cd.

Cadmium: A Search for Artifacts and Independent Verification

In examining the issues related to paleo-cadmium measurements, there are four areas that need to be addressed: (1) Is the deep water oceanographic behavior of cadmium sufficiently different from that of phosphorus to make it a misleading indicator of past nutrient chemistry? (2) Are there artifacts in the cleaning procedure that might lead to mis-esti-

mations of foraminiferal Cd? (3) Could mean oceanic Cd have differed sufficiently to make for a significant mis-inference of past phosphorus levels based on the modern cadmium-phosphorus correlation? (4) Are there artifacts in the uptake and preservation of Cd in foraminiferal shells that might lead to mis-estimations of bottom water Cd?

On the Deep Water Cadmium-Phosphorus Relationship

We have reviewed this issue recently and feel that it needs little additional comment (Boyle 1994b). Although there have been assertions that cadmium and phosphorus show large differences in the ocean (e.g. Saager 1994), most of these differences vanish when the following three criteria are applied as a filter:

(1) <u>Consider only deep water data</u>, say below about 400m. There are significant differences in near-surface water cadmium-phosphorus relationships, as has been known already for more than 14 years (Boyle et al. 1981), but these surface water anomalies are not germane to the use of cadmium in studying deep water hydrography.

(2) <u>Rule out data sets where either the cadmium or phosphorus data sets are questionable</u>. For example, when reported phosphorus concentrations deviate significantly from nearby GEOSECS values or where nearby stations show inconsistent cadmium or phosphorus values in the deep waters.

(3) <u>Consider only the ensemble global cadmium-phosphorus relationship</u> rather than focus on the microscopic details of local cadmium-phosphorus relationships.

Although cadmium is not perfectly correlated with phosphorus in the deep ocean, the two elements are remarkably well correlated given the major differences between their chemical properties. Water column cadmium-phosphorus anomalies are not a credible explanation for the LGM carbon isotope/cadmium conundrum.

A New Cleaning Artifact for Foraminiferal Cd Analysis and it's Significance in the Southern Ocean

One of the major difficulties in using the calcium carbonate crystal-lattice bound cadmium as a paleoceanographic tool is the very low concentrations of cadmium in the shell compared to the amount in the surrounding mud. If foraminifera are not cleaned carefully, one might mistake cadmium in a contaminant phase for cadmium in the crystal lattice of the shell (Boyle 1981). In order to minimize contaminant phases, foraminiferal shells are cleaned by the following series of treatments:

(i) ultrasonic treatments (to shake off any loosely-adhering contamination),

(ii) hot basic oxidizing solution (to oxidize and solubilize exterior organic matter),

(iii) hot basic reducing solution (to reduce and solubilize iron and manganese oxides precipitated on the surface of the foraminifera)

(iv) multiple weak acid leaches (to further strip the surface and any adsorbed contaminants from the surfaces).

This procedure has been modified only slightly over the years (Boyle and Keigwin 1985/6). However, one contaminant phase is known to be resistant to this cleaning treatment. Diagenetic manganese carbonate overgrowths cement contaminants onto the shells under a phase that is not removed by this cleaning treatment. To avoid this artifact, manganese is analyzed on each shell sample, and the foraminiferal cadmium data are considered suspect when manganese exceeds a threshold range [about 100-150 (μmol Mn)/(μmol Ca)].

Recently, we have discovered discovered a second source of contamination that is not eliminated by the old cleaning treatment (Rosenthal, 1994; Rosenthal et al., 1995b). In a 10 cm interval of Indian Ocean sector Southern Ocean core (MD88-769, 46°4.16'S, 90°6.67'E, 3420m), we encountered foraminiferal cadmium concentrations in both planktonic and benthic foraminifera that were much higher than would be seen anywhere in specimens from the modern ocean (Fig. 6). Suspecting a new source of contamination, we analyzed the bulk sediment for cadmium and found that its cadmium concentration sharply peaked more than an order of magnitude above the background levels (Fig. 6). After further study it became evident that this high-Cd layer is created by sedimentary diagenesis. Cadmium sulfide precipitates when dissolved cadmium from the pore waters (originating both from bottom water and from cadmium released from decomposing organic matter in the sediments) encounters traces of dissolved sulfide produced by advanced anoxic degradation of organic matter:

$$2CH_2O + SO_4^= \longrightarrow S^= + 2CO_2 + 2H_2O$$

$$Cd^{++} + S^= \longrightarrow CdS \text{ (solid)}$$

Because cadmium sulfide is amongst the most insoluble of sulfide minerals, it can precipitate even when sulfide levels are below that of iron sulfide solubility and „sniffing" detection (Rosenthal et al. 1995b). Because the nanomolar Cd concentrations

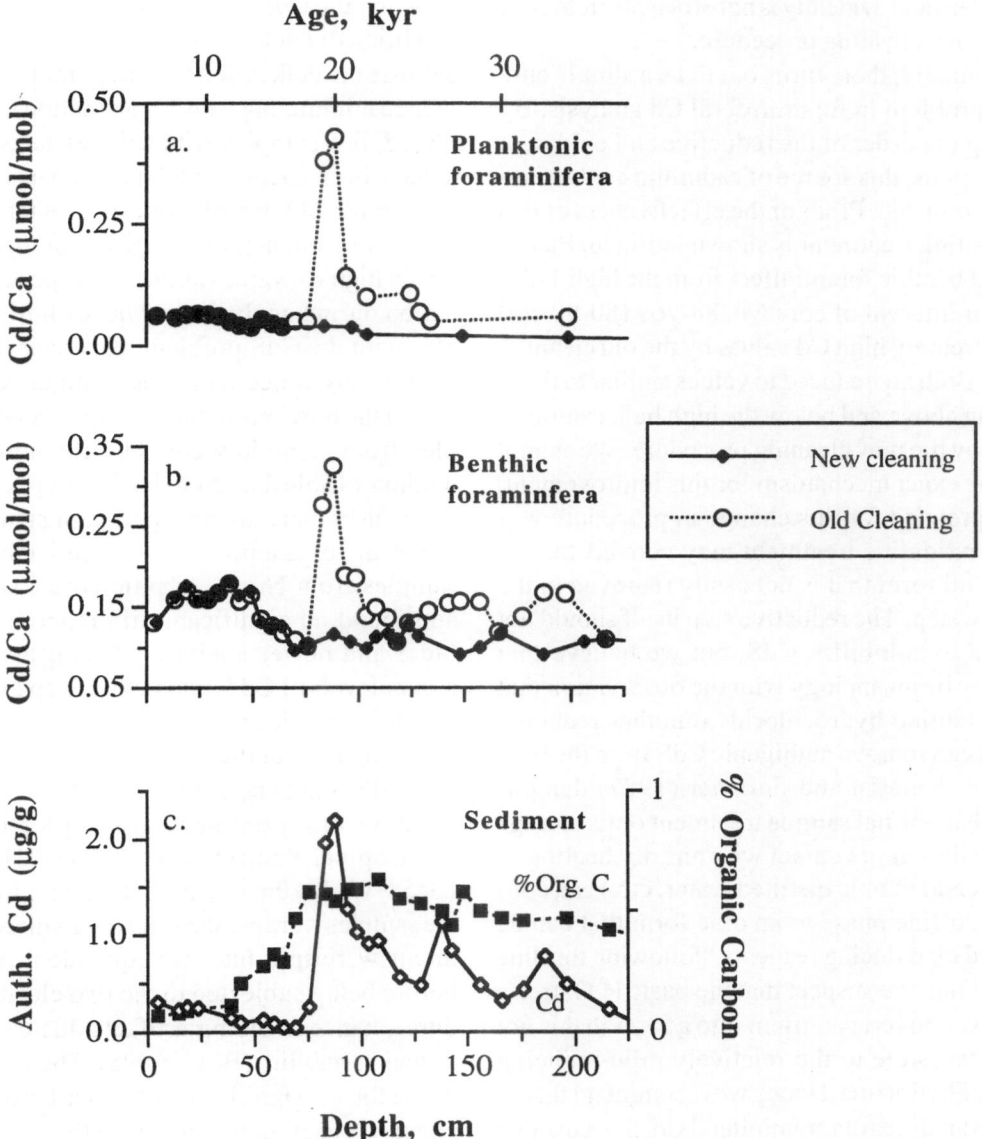

Fig. 6. Foraminiferal and bulk sedimentary data from MD88-769 illustrating Cd concentration process and effectiveness of new cleaning procedure.

in pore waters are small compared to the total free sulfide, this process does not affect the pore water sulfide distribution ([S⁻] is controlled more by reactions with available iron and in some cases oxidation to native sulfur by contact with oxygen). The cadmium sulfide precipitation zone can be quite narrow. When this narrow precipitation zone is coupled to a stalled redox front, cadmium can build up to levels up to two orders of magnitude above that of normal pelagic sediments in a narrow depth zone (Rosenthal et al. 1995b). This cadmium sulfide precipitates on the surfaces or in the pores of foraminifera and evidently is not efficiently removed by the above cleaning procedure.

Fortunately, there turns out to be a simple cure for this problem in foraminiferal Cd analysis. By reversing the order of the reductive and oxidative cleaning steps, this source of cadmium contamination is eliminated. Proof of the effectiveness of this new cleaning treatment is shown in Fig. 6. Planktonic and benthic foraminifera from the high-bulk-cadmium interval of core MD88-769 (80-90 cm) show extremely high Cd values by the old cleaning method. Both are reduced to values similar to those occurring above and below the high bulk cadmium interval by the new cleaning procedure. We cannot prove the exact mechanism for this improvement, but the premise for this change in procedure was that the oxidative treatment may convert the Cd into a solid form that is not easily removed in the reductive step. The reductive step itself should not expected to solubilize CdS, but we believe that there may be an analogy with the observation that hydroxylamine hydrochloride (another reducing reagent) can remove authigenic CdS from the bulk sediment. Kersten and Forstner (1987) demonstrated that normal sample treatment (e.g., storage of the sediment in contact with air; dry heating in air; dispersal in oxic distilled water, etc.) can convert the sulfide phase to an oxic form that can be dissolved by reducing reagents. Following this line of reasoning, we suspect that the basic H_2O_2 treatment may convert cadmium into a form that is not easily accessible to the relatively mild reducing powers of hydrazine. Hence we recommend that in all future analyses of foraminiferal Cd, the reductive cleaning step should precede the oxidative step.

To what extent does this artifact affect published records for foraminiferal cadmium? Obviously it makes a major difference to the samples located in the high Cd zone of MD88-769, but just as obviously the high foraminiferal levels in this zone were a tip-off that something was wrong. Fortunately for the integrity of published data, there is a coincidence in high carbonate cores: the sharp contact between oxic and suboxic conditions that lead to CdS precipitation also favor precipitation of manganese carbonate deeper in the core. In avoiding high-$MnCO_3$ sites, we no doubt avoided many situations

in which the CdS problem occurs. This favorable condition did not occur at the Southern Ocean site because the bulk sediment was dominated by silica (hence minimizing $MnCO_3$ formation). In all likelihood, few samples with CdS artifacts are likely to have been encountered in carbonate cores without the $MnCO_3$ tip-off. In a few other cases, the subjective (but acknowledged) editing of anomalously high Cd values as due to supposed contamination during analysis may have eliminated samples with the CdS problem. However, in order to address any concern that there might be major errors in the previous literature, we re-analyzed samples from some key cores by the new cleaning method (Table 1). On balance, it appears that in most cases there are no significant artifacts in the old data. For example, we have re-analyzed LGM samples from North Atlantic core CHN82-4PC, and found no significant difference between the older and newer analyses. This proves that the higher levels of Cd found in that deep LGM Atlantic sites is not due to CdS contamination. Furthermore, analysis of the bulk sediments for Cd shows that CdS is not significant in that core. Similarly, we have analyzed several samples from eastern tropical Pacific core TR163-31P (3°37'S, 83°58'W, 3210m) by both methods. In this case, the samples were crushed and randomized fine fragments were split into two equivalent subsamples before being subjected to the two cleaning procedures, hence avoiding artifacts due to inter-individual variability (Boyle 1995). There is little evidence for any significant difference between the old and new cleaning treatments in this core.

However, in a few cases where CdS precipitation is important (e.g., RC13-228 and RC13-229: Boyle 1992; Oppo and Rosenthal 1994), it appears that a slight downward revision of the originally reported results will be necessary (see Tables 2 and 3). Cd data from RC13-228 were originally published by Boyle (1992), but most samples have now been re-analyzed by the new cleaning protocol here to eliminate the CdS artifact. The benthic foraminiferal Cd data from the old and new cleaning treatments are compared in Fig. 7. There is a small but significant lowering of foraminiferal Cd resulting from the new cleaning treatment. If one consid-

Table 1. Comparison of samples cleaned by old and new methods for various cores.

Core	Location, Water Depth	Sample Depth cm	Species	Cd/Ca (old method) μmol/mol	Cd/Ca (new method) μmol/mol
Crushed -and-split samples:					
TR163-31p	3°37'S,83°58'W,3210m	12	Uvi	0.259, 0.238	0.226
		38	Uvi	0.184	0.176
		95	Uvi	0.207	0.162
		635	Uvi	0.174	0.174
		655	Uvi	0.160	0.188
		668	Uvi	0.160	0.161
		708	Uvi	0.224	0.213
		738	Uvi	0.150	0.151
RC13-228	22°20'S,11°12'E,3204m	142	Uvi	0.132	0.104
		167	Uvi	0.135	0.106
		502	Uvi	0.094	0.073
		652	Uvi	0.104	0.097
		172	Uvi	0.121	0.106
		175	Uvi	0.128	0.119
EN66-16GGC	5°28'N,21°8'W,3152m	47	wue	0.191	0.170
Separate picks:					
CHN82-4PC	41°43'N,32°51'W,3427m	61	Uvi	0.114,0.119,0.075	0.109

ers only the lowest Cd values from the replicate analyses arrived at by the old cleaning method, they are almost the same as analyses using the new cleaning procedure. Apparently, the CdS artifact is unevenly distributed (affecting some specimens more strongly than others), so that replicate picks of foraminifera sometimes avoid the problem entirely. It also appears that replicates agree better when using the new cleaning procedure, suggesting that the new cleaning treatment eliminates a source of sample-to-sample variability.

However, even though previously published data from RC13-228 and RC13-229 have suffered slightly from a CdS cleaning problem, the bottom line in the case of the South Atlantic cores is that eliminating this effect does not shift the results by very much. First, in RC13-228, foraminiferal Cd/Ca is lowered by about 0.02 μmol/mol; this shift is small compared to the modern-day 0.15 μmol/mol gradient between the North Atlantic and Pacific. Second, the magnitude of the CdS artifact in the old data is relatively uniform with depth (because there is high authigenic Cd throughout the core). Therefore, it is still the case that Cd at these sites does not show large glacial/interglacial cycles. Third, elimination of this artifact increases the magnitude of the discrepancy between the δ^{13}C data and the Cd data; LGM Cd is now even lower than it was previously. Hence the discovery of this problem, beneficial as it may be for the integrity of future foraminiferal Cd data, intensifies rather than eliminates the Southern Ocean Cd-δ^{13}C conundrum.

Does the CdS Precipitation Process Allow for a Significant Decoupling of Oceanic δ^{13}C and Cd?

Is this cadmium sulfide removal process a significant sink for oceanic Cd? Could variations in the intensity of this process lead to variations in the oceanic Cd content and decouple the Cd-P relationship temporally? As shown by Rosenthal et al. (1995a), this CdS precipitation process is probably the major sink of cadmium from the ocean. Removal of Cd as CdS into subantarctic sediments is more intense during glacial periods compared to interglacial periods. Many sediments thoughout the world were more organic carbon-rich and reducing during glacial times than interglacial times (Pedersen 1983; Sarnthein et al. 1988). So it is plausible that enhanced removal of Cd from the ocean occurs during glacial periods, and this removal may not be coupled with enhanced phosphorus removal (Ingall and Jahnke 1994). Even if Cd and P are well-correlated in the ocean at any instant, the correlation could have a lower slope at times when there is less Cd in the oceans as a whole. This scenario has been explored in more detail by Rosenthal et al. (1995a). Briefly summarizing that work, it appears that because Cd has a long residence time in seawater and because glacial stage enhancement of sedimentary Cd uptake is limited, only small fluctuations in the global mean oceanic cadmium content are possible. In particular, on the time scale of

Table 2. Bulk sedimentary Cd data for core RC13-228 (parts per billion, ppb).

Depth, cm	Authigenic Cd (nanograms Cd/gram sediment)	Depth, cm	Authigenic Cd (nanograms Cd/gram sediment)
0-5	453	125-130	493
5-10	171	130-135	877
10-15	123	135-140	859
15-20	147	140-145	701
20-25	639	145-150	653
25-30	587	150-155	620
30-35	346	155-160	663
35-40	459	160-165	957
40-45	711	165-170	866
45-50	483	170-175	754
50-55	480	175-180	378
55-60	102	180-185	518
55-60	377	185-190	459
65-70	405	190-195	409
70-75	333	195-200	421
75-80	354	200-205	359
80-85	520	205-210	692
85-90	366	210-215	537
90-95	148	215-220	507
95-100	643	220-225	371
100-105	390	225-230	938
105-110	868	230-235	839
110-115	301	235-240	768
115-120	246	240-245	720
120-125	739		

(a) Note that "normal" pelagic sediments should have no more than about 100 ppb of Cd, so this core is consistently enriched in Cd throughout most of the upper 250 cm analyzed.

a glacial-to-interglacial transition, the shift is likely to be less than 5%. So this factor is <u>not</u> significant for the discrepancy between Southern Ocean carbon isotope and cadmium data during the last glacial maximum, although it might play a role over longer geological periods.

Dissolution or other artifacts in the Cd record?

McCorkle et al. (1994) measured Cd, Ba, and Sr in fossil core top *C. wuellerstorfi* shells at eleven sites on the Ontong-Java Plateau (western equatorial Pacific); they found fossil Cd, Ba and Sr was lower in deeper cores than in shallower cores. They suggested that the low foraminiferal trace element

values were caused by post-depositional response to dissolution due to the increasing undersaturation of bottom water with increasing depth, with foraminiferal Cd somehow selectively removed through dissolution. They further speculated that this artifact might account for some of the discrepancies between Cd and δ^{13}C. For example, they suggest that the low levels of foraminiferal cadmium in the Southern Ocean (compared to what would be inferred on the basis of δ^{13}C) might be due to enhanced post-depositional dissolution during the LGM.

This matter requires more discussion than is possible here and will have to be dealt with in greater detail elsewhere. However, for the sake of completeness for this Southern Ocean presentation,

Table 3. RC13-228 benthic foraminiferal Cd/Ca (μmol/mol), old vs. new cleaning methods.

Depth (cm)	Old Cleaning Method — Uvigerina spp.	C. kullenbergi	C. wuellerstorfi	lowest	New Cleaning Method — Uvigerina	C wuellerstorfi	Average	Std. Dev.	n	r
7	0.119	0.073	0.104	0.073						
12	?0.150	0.111		0.111						
17	0.109	0.085	0.109	0.085						
22	0.115, 0.097	0.057, 0.087	0.103	0.057		0.080	0.080		1	
27	0.106	0.093, ?0.267	0.125	0.093	0.089	0.101	0.095	0.008	2	
32	0.122	0.087, ?0.151	0.128	0.087		0.080	0.080		1	
37	0.098	0.090	0.110	0.090						
42	0.130	0.105	.	0.105		?0.156, 0.079	0.079		1	1
47	?0.157	0.092		0.092	0.110		0.110		1	
52	0.121	0.114	0.121	0.114		0.092	0.092		1	
57	0.121	0.114	0.111	0.111						
62	0.140	0.106	0.101	0.101		0.114	0.114		1	
67	0.128	0.120	0.118	0.118	0.133		0.133		1	
72	0.136, 0.130	0.112, 0.122	0.109	0.109		0.115	0.115		1	
77	0.130	0.083	0.103	0.083	0.096	0.090	0.093	0.004	2	
82	0.124	0.126		0.124						
87	0.126, 0.143	0.103	0.098	0.098	?0.047	0.091	0.091		1	1
92	0.129, 0.136	0.093	0.102	0.093		0.106	0.106		1	
97	0.127		0.113	0.113	0.104	0.082	0.093	0.016	2	
102	0.152, ?0.561, 0.132	0.130		0.130		0.111	0.111		1	
107	0.113, 0.112, 0.143	0.087	0.093, 0.121	0.087	0.106	0.072, 0.085	0.088	0.017	3	
112	0.110, 0.092, ?0.339	0.099		0.092						
117	0.112, 0.144, 0.092	0.128		0.092	0.104	0.076	0.090	0.020	2	
122	0.112, ?0.422, ?0.388, 0.100	0.066	0.072	0.066						
127	0.148		0.119	0.119	0.095	0.078	0.087	0.012	2	
132	0.123, 0.148		0.087	0.087						
137	0.130, 0.137	0.091	0.087	0.087	0.081	0.077	0.079	0.003	2	
142	0.147		0.133	0.133	0.104		0.104		1	
147	0.156		0.122	0.122						
152	0.112, 0.136		0.137	0.112	0.097	0.097	0.097	0.000	2	
157	0.122		0.119	0.119	0.104		0.104		1	
162	0.146	?0.055	?0.176	0.146	0.108		0.108		1	
167	0.136		0.132	0.132	0.106		0.106		1	
172	0.177		0.125	0.125	0.106, 0.111	0.097	0.105	0.007	3	
175	?0.417	?0.065	0.102	0.102	0.119		0.119		1	

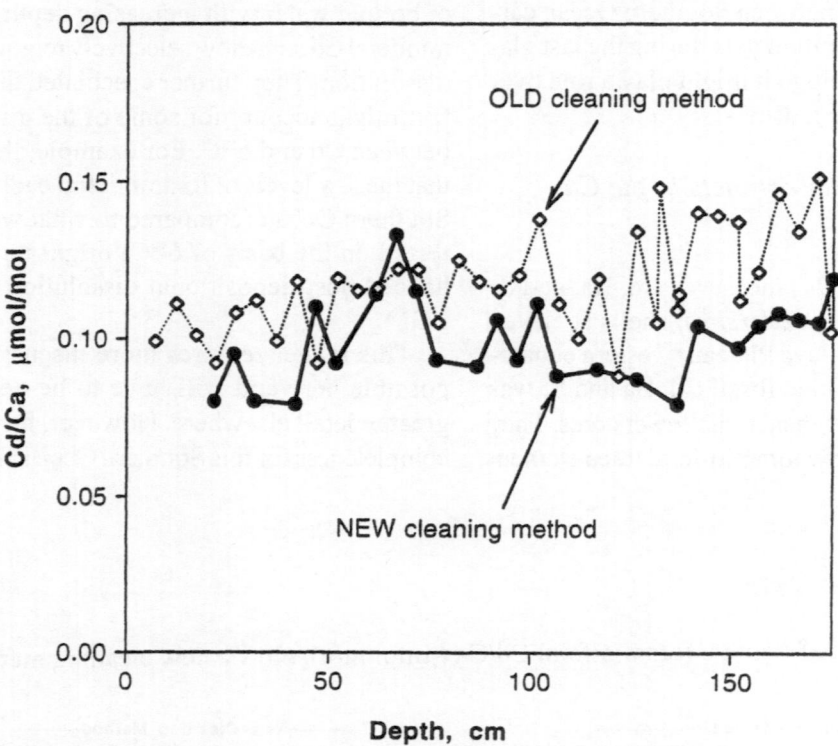

Fig. 7. New cleaning treatment vs. old cleaning treatment for RC13-228.

the argument will be summarized: (1) We also have data showing that in deep Pacific waters, Cd in *C. wuellerstorfi* shows lower Cd than it does in less undersaturated waters (Boyle 1994a). However, contrary to McCorkle et al. (1994), we find that the onset of lower Cd in *C. wuellerstorfi* is coincident with relatively high degrees of undersaturation (>15 micromoles/kg undersaturation in [$CO_3^=$]). So although we confirm the <u>observation</u> of McCorkle et al. of a Cd decrease in core top *C. wuellerstorfi* in the deep Pacific, we find that onset of the effect requires intense undersaturation. (2) It is not clear why *C. wuellerstorfi* Cd is lower in the deep cores. McCorkle et al. focus on post-depositional dissolution and argue against other possibilities. We believe that the other possibilities are equally likely: (a) a direct response of Cd coprecipitation by living foraminifera in response to bottom water un-

dersaturation, and (b) temporal changes in bottom water composition (i.e., the fossil foraminifera may have been deposited at times in the past when bottom water Cd was lower). (3) Contrary to the arguments of McCorkle et al., we consider it implausible that Cd and Sr can be selectively extracted from solid solution sites within calcium carbonate. We are particularly skeptical of any dissolution effect for Sr because Sr/Ca in *C. wuellerstorfi* varies almost linearly with depth from 300 m to 5000m throughout the ocean (Rosenthal 1994; Lloyd-Kinstrand et al. 1994) (4) McCorkle et al. (1994) only report on eleven samples of a single species. They can only speculate whether this effect might occur in other species. On the contrary, we have data showing that this effect does not occur in all species; in particular, it does not occur in *N. umbonifera*. Hence their speculations about all published Cd data (that includes data from several species) wander considerably beyond the limits of the data they presented.

Post-depositional dissolution effects have been established for planktonic foraminifera for several elements (e.g. Mg, Lorens et al. 1977; F, Rosenthal and Boyle 1993; U, Russell et al. 1994), but it would be a mistake to draw a close analogy between dissolution effects in planktonic and benthic foraminifera. Dissolution artifacts in planktonic foraminifera can arise in several ways which are not relevant for benthic foraminifera. We believe that selective extraction of individual elements from within the crystalline matrix of the shell (as implied by McCorkle et al.) is the least likely origin of dissolution artifacts in planktonic foraminifera. The primary flux of planktonic foraminifera to the seafloor consists of a mix of individuals that have grown at different times under conditions that may vary greatly: the temperature conditions, food supply and growth rate may differ greatly from individual to the next. These environmental differences are linked to varying structural characteristics of the shell, which in turn are linked to different chemical compositions. Although the mix of individuals falling to the seafloor should be regionally homogeneous, differences in dissolution intensity as a function of the depth of the seafloor will selectively extract the more dissolution-susceptible individuals from the preserved planktonic assemblage. When dissolution-resistant and dissolution-susceptible individuals differ systematically in their chemical composition, the selective removal of dissolution-susceptible individuals will alter the chemical composition of the assemblage. Progressive dissolution of an assemblage results in predictable chemical trends. Furthermore, different structural types of calcite are deposited by individuals during their lifetime. In particular, some species are known to deposit a thick less porous „crust" during gametogenesis, (which often occurs in deeper waters of different physical properties). Lohmann and Schweitzer (1990) and Lohmann and Lohmann (personal communication) have demonstrated that the isotopic and chemical composition of this crust differs significantly from that of the pre-existing shell. In view of the structural differences between the pre-gametogenic shell and the gametogenic

crust, it should be expected that crusted specimens dissolve more slowly than non-crusted individuals. With chemical differences between primary and crust calcite, it is to be expected that post-depositional dissolution will alter the chemical characteristics of foraminiferal assemblages, as has been appreciated for some time (e.g. Vincent and Berger 1981).

None of these mechanisms apply to benthic foraminifera. During the lifetime of a single benthic individual (perhaps a few weeks to a few months), there will be no significant changes in the physical characteristics of the environment (temperature, etc. are constant). Because the deep ocean can only slowly change its chemical and physical characteristics (and the range of the changes that do occur is relatively small), a population of benthic individuals encounters far less variability than does a comparable number of planktonic individuals. Many of the mechanisms leading to intra- and inter-individual differences in planktonic foraminifera simply do not exist for benthic foraminifera. We believe that the occurrence of dissolution-related artifacts in planktonic foraminifera is of questionable relevance as to whether these effects may occur in benthic foraminifera.

One cannot entirely rule out the possibility of post-depositional artifacts in benthic foraminifera, however. (1) It is possible that differences in shell size or thickness (for whatever reason) might be correlated with changes in chemical composition. However, no clear size-correlated Cd differences have been observed for benthic foraminifera (Boyle 1995). (2) Chemical tools for the analysis of foraminifera chave rather coarse spatial resolution (a few micrometers for minor elements such as Sr, Mg, and Na; and many tens of micrometers or greater for trace elements), so we cannot be certain that benthic foraminiferal calcite is not chemically heterogeneous or zoned at a very small scale. If shells are zoned, it is possible that dissolution of outer surfaces might lead to a residue with different chemical composition than the initial sample. And of course, the low core-top Cd in the deep Pacific may be instead a primary response of living benthic foraminifera. So further work both on carbonate-undersaturation effects on benthic trace metal incorporation is worthwhile.

Support for the basic validity of published Cd records is provided by Cd analyses of the aragonitic benthic foraminifera *Hoeglundina elegans*. As shown by Boyle et al. (1995), this species has a different uptake pattern for Cd than coexisting calcitic species. The *H. elegans* partition coefficient is lower (D_p(Hoe)~1.0 compared to D_p(Calc)~2.9 for deep cores) and shows minimal depth-dependence for Cd uptake (calcitic species show a strong depth-dependence between about 1100-2900m). So *H. elegans* casts an independent „vote" on what bottom water Cd concentrations were at times in the past. In addition *H. elegans* is not encountered in highly corrosive undersaturated waters (note that its aragonitic shell is more soluble than calcitic shells). The presence of this species suggests that

dissolution cannot have been severe. In Fig. 8, we compared *H. elegans* and calcitic bottom water estimates from mid-depth cores from the North Atlantic (CHN82-4PC, 41°43'N, 32°51'W, 3427m) and eastern tropical Pacific (TR163-31P, 3°37'S, 83°58'W, 3210m). The aragonitic *H. elegans* and calcitic estimates agree well in both cases. A conspiratorial outlook is required to believe that all of species are telling us the same lie. Simply put, there is no good reason to question the published results on the basic Atlantic-Pacific fractionation first reported for cadmium (Boyle and Keigwin 1982; 1985/6).

Of particular relevance to the paleochemistry of the South Atlantic, note that cadmium in LGM benthic foraminifera from deep core RC13-229

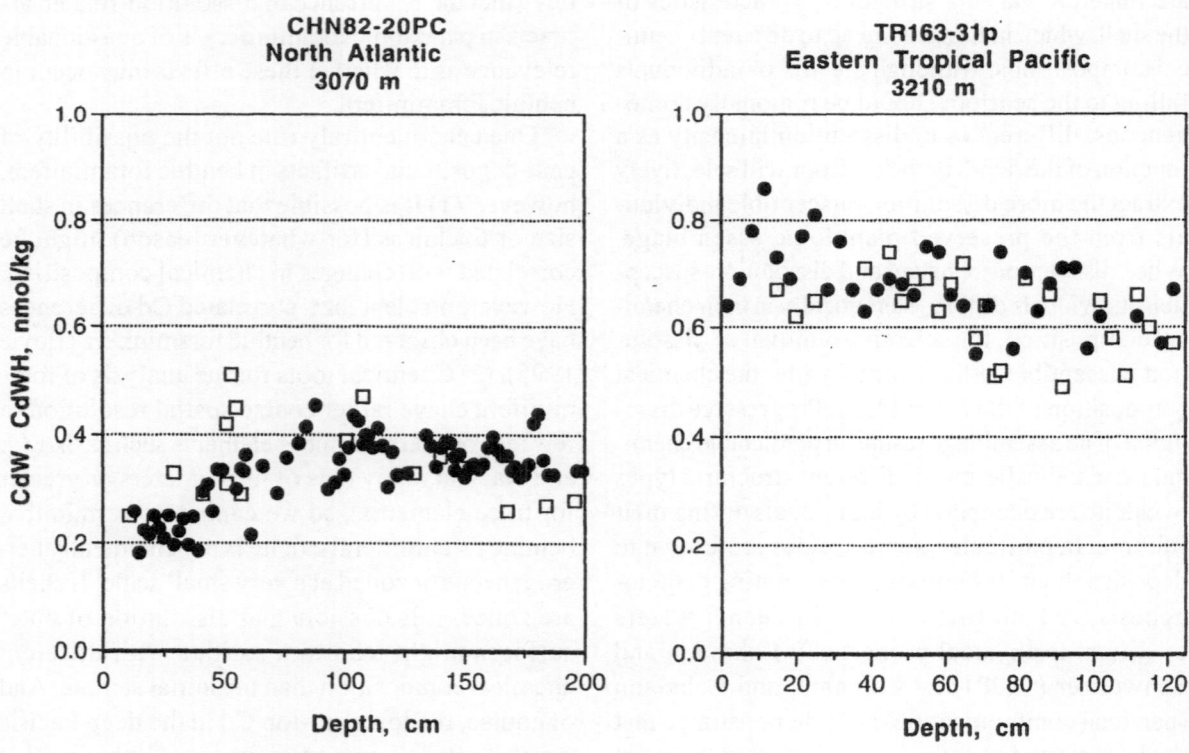

Fig. 8. CHN82-4PC and TR163-31p; comparison of *H. elegans* vs. calcitic Cd estimates. It is assumed that D_p=2.9 for calcitic species and D_p=1.0 for *H. elegans*.

(4191m) is significantly higher than in samples from shallower core RC13-228 (3204m), and neither core shows any discernable systematic glacial/interglacial difference. If there is a significant dissolution effect, it should be the strongest in the deepest cores where undersaturation will be stronger. So if dissolution has reduced foraminiferal cadmium in the deepest core, then the original cadmium concentration of the bottom water would have to be <u>much</u> higher in LGM waters overlying the deep core compared to the shallower cores. Such a drastic increase in nutrients at 4191m compared to 3204m is not supported by the carbon isotope difference between the shallower and deeper cores (less than 0.1‰; Curry et al. 1988; Oppo et al. 1990). Oppo and Rosenthal (1994) also noted that bulk carbonate percentage fluctuations in RC13-229 are not correlated with the Cd record, further evidence against a dissolution artifact in that record. Dissolution artifacts cannot account for the discrepancy between Cd and δ^{13}C at these sites.

South Atlantic Paleochemical Evidence: Status of the Southern Ocean Cd-δ^{13}C Conundrum

As noted earlier, there is a fundamental Cd and carbon isotope data conflict. South Atlantic cores RC13-228 and RC13-229 illustrate this problem. These cores were chosen because: (1) RC13-228 is located near the deeper southernmost extent of the low phosphorus NADW tongue (see Fig. 1) and hence should be expected to be sensitive to relatively small changes in NADW flux (a la the „firehose" model). (2) Nearby core RC13-229 underlies nearly pure Circumpolar Deep Water (see Fig. 1) and should be expected to reflect changes in its composition (a la the „Charles and Fairbanks" model).

Cd data from RC13-228 were originally published by Boyle (1992), but here it has been completely re-analyzed by the new cleaning protocol here to eliminate the CdS artifact. Cd data from RC13-229 are from Oppo and Rosenthal (1994), who used a combination of strategic new-protocol re-analyses and editing of high values from old-protocol analyses to minimize the CdS problem. The carbon isotope data from these cores are from

Curry et al. (Curry et al. 1988) and Oppo et al. (Oppo et al. 1990).

As can be seen in Fig. 9, Cd shows the appropriate relative core top values (low Cd in NADW-influenced RC13-228; high Cd in CPDW-influenced RC13-229). Although the core-top carbon isotope values do not reflect the expected vertical gradient, both values probably should be considered to be within the expected practical error of δ^{13}C calibrations. Cd data from both the 3204m and 4191m sites indicate minimal changes in nutrient chemistry during the past ~30,000 years and indicates persistence of the modern-day vertical gradient between these sites. In sharp contrast, the δ^{13}C data from both sites suggest a dramatic LGM nutrient elevation, to values as high or higher than observed in contemporaneous eastern equatorial Pacific sites. These data are representative of other published Southern Ocean paleochemical sites (Curry et al. 1988; Labracherie et al. 1989; Oppo et al. 1990; Charles and Fairbanks 1992).

Could this difference arise from real differences in the oceanic δ^{13}C-Cd relationship? A search for mechanisms separating Cd and δ^{13}C has not gone empty-handed. Gas exchange is now known to be more important in the global δ^{13}C distribution than previously realized, and small changes in oceanic Cd resulting from enhanced CdS uptake have been proposed. But none of these known processes can create the magnitude of the discrepancy between the tracers.

The search for artifacts in foraminiferal proxy tracers has also turned up candidates: (1) Cd in *C. wuellerstorfi* might be compromised by strong carbonate undersaturation, (2) some published Cd data have to be revised downwards slightly because of insufficient CdS cleaning, and (3) the „Mackensen Effect" of high-productivity negative *C. wuellerstorfi* δ^{13}C shifts has been established. Nonetheless, we have argued that carbonate undersaturation effects cannot account for the discrepancy between δ^{13}C and Cd in the Southern Ocean, and awkwardly enough, solving the CdS cleaning problem has only made for a larger discrepancy between the tracers.

In the end, we suspect that the „Mackensen Effect" is the cause of the discrepancy. It has been argued that the close correlation between the δ^{13}C variations in benthic *C. wuellerstorfi* and planktonic

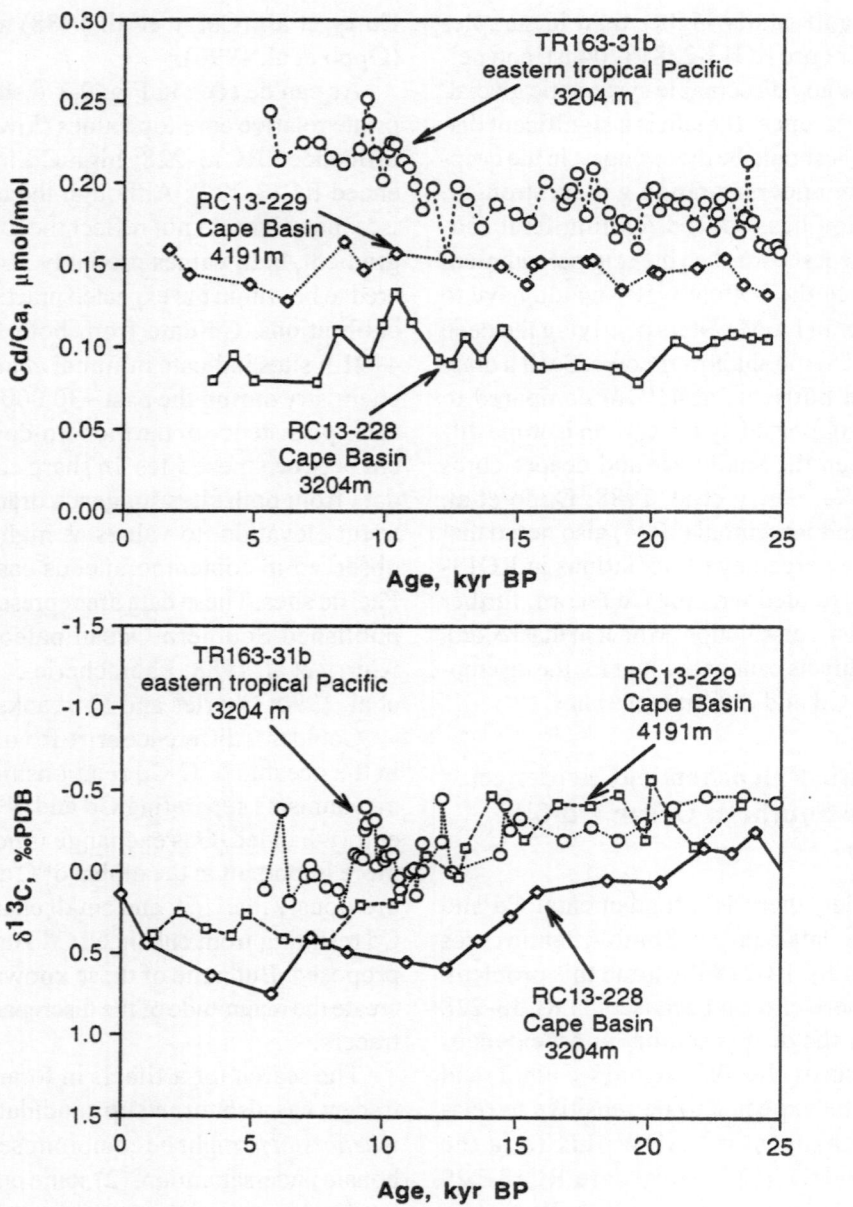

Fig. 9. Comparison of data from South Atlantic cores RC13-228 and RC13-229 with eastern tropical Pacific core TR163-31b.

Neoglobquadrina pachyderma (sinistral) argues against the δ¹³C signal being an artifact (this is the same „it isn't a conspiracy" argument used above for the comparison of *H. elegans* and calcitic Cd). Nonetheless, in a separate work in this volume, Bickert and Wefer may have found the „smoking gun" for the role of the „Mackensen Effect" on

LGM δ¹³C in the Southern Ocean. Noting that the site of RC13-229 underlies waters of high LGM productivity, they analyzed a core underlying the same water mass (Circumpolar Deep Water) but in a region of low productivity (further from the continental margin). At that site, δ¹³C of *C. wuellerstorfi* does not show the large negative δ¹³C shift characteristic of RC13-229 and other published Southern Ocean cores. In fact, the δ^{13}C shift is similar to that of Pacific deep cores (that is entirely attributed to shifts in the global ocean average δ^{13}C due to continental organic carbon oxidation). The data of Bickert and Wefer suggest that the nutrient chemistry of these waters have remained unchanged; this result is completely in accord with the cadmium results presented above.

In general there probably is a selection bias towards high-productivity sites because benthic foraminiferal abundance is correlated with productivity (Herguera and Berger 1991). High-benthic-abundance sites are sought out to minimize sample consumption and picking time. Additionally, productivity is correlated with accumulation rate, hence offering higher temporal resolution. It is possible that these selection biases have led the community to choose core sites where the „Mackensen Effect" is most likely to occur.

Although it would be premature to assert that the problem is resolved to everyone's satisfaction, Bickert and Wefer may be showing us the light at the end of the tunnel.

Epilogue: ^{231}Pa/^{230}Th Evidence for the Persistence of NADW

One recent line of evidence has just become available that could settle the matter of whether NADW reaches the Southern Ocean or not. Yu (Yu, 1994) and Yu, Bacon, and Francois (in press) have inferred from sedimentary ^{231}Pa/^{230}Th measurements that ^{231}Pa is missing from Atlantic sediments, and that a corresponding excess is found in Antarctic sediments. This situation prevails during modern and glacial times. ^{231}Pa and ^{230}Th are generated at precisely known rates from the decay of ^{235}U and ^{234}U in seawater. These radioactive isotopes attach to sinking particles within a matter of decades to several hundred years and spend most of their time within the oceanic sediment column until they are lost to radioactive decay. ^{231}Pa/^{230}Th in the sedimentary flux therefore must equal the production ratio because the reservoir in the ocean water column is small. However, in the sediments of both the modern and LGM Atlantic Ocean, ^{231}Pa/^{230}Th is consistently lower than the production ratio. Yu et al. observe that Southern Ocean sediments have ^{231}Pa/^{230}Th ratios significantly above the production ratio and they note that quantitatively, the amount missing from the Atlantic is equal within (somewhat large) errors to the excess in the Antarctic. They propose that in the modern ocean, Atlantic ^{231}Pa is preferentially transported to the Antarctic by the NADW conveyor belt where it is scavenged by the veil of particles falling under the high productivity zones paralleling the Circumpolar Current. Because LGM ^{231}Pa/^{230}Th data are similar to that of the modern ocean, they propose that ^{231}Pa export by NADW into the Circumpolar Current must have continued during the LGM.

It is difficult as of yet to specify what flux of NADW is necessary to allow most of the ^{231}Pa to be exported into the Southern Ocean. A good estimate will require an ocean model incorporating ^{231}Pa and ^{230}Th behavior. It might be thought that some of the excess ^{231}Pa in the Antarctic might come from the Pacific. However, so far it does not seem that there is missing ^{231}Pa in the Pacific (a statement that needs to be balanced against a relative paucity of data in the Pacific, particularly the South Pacific). So even though there is a return flow of Pacific Deep Water towards the Southern Ocean, the flow is simply too slow to transport the ^{231}Pa out of the Pacific before it attaches to a particle and falls to the bottom. The residence time of deep water in the Pacific today is about 500 years, and for the Atlantic, about 100 years (Stuiver et al. 1983). So the Atlantic gets ^{231}Pa out just in time to avoid losing it to Atlantic sediments, whereas the Pacific moves too slowly for the ^{231}Pa to escape. Vigorous NADW is the key to Atlantic loss of ^{231}Pa. Furthermore, it has been argued that productivity in the glacial ocean was higher during the LGM (e.g., see Sarnthein et al. 1988), in which case it becomes even harder to export ^{231}Pa from the Atlantic. In any event, even if some of the high Antarctic ^{231}Pa comes from the Pacific, one still has to account for the missing ^{231}Pa in the Atlantic. Unless some „missing sink" of Atlantic-generated ^{231}Pa can be found, then „drop dead" scenarios of glacial NADW are eliminated by this new evidence. The ^{231}Pa/^{230}Th evidence indicates that a significant flux of NADW into the Southern Ocean has been a feature of both the modern and LGM ocean.

Acknowledgements

E. Boyle thanks the conference organizers for their generous support for his participation at this conference. We also thank Michael Bacon and Ein-Fen Yu for allowing us to discuss their unpublished [231]Pa/[230]Th work, and Wolf Berger for his numerous suggestions improving the manuscript. Core curation was supported by NSF grants to the Lamont-Doherty Geological Observatory and Woods Hole Oceanographic Institution and CNRS (France) support to the Centre des Faibles Radioactivites. This research was sponsored by NSF grant OCE 9402198-OCE.

References

Bainbridge AE (1980) GEOSECS Atlantic Ocean Expedition, Vol. 2, Sections and Profiles. U.S. Government Printing Office, Washington D.C.

Berger WH, Vincent E, Killingley S (1981) Stable isotope composition of benthic foraminifera from the equatorial Pacific. Nature 289:639-643

Boyle E (1994a) Isotopic and elemental tracers in calcium carbonate fossils. Mineralog Magazine 58A:111-112

Boyle EA (1981) Cadmium, zinc, copper, and barium in foraminifera tests. Earth Planet Sci Lett 53:11-35

Boyle EA (1988) Cadmium: chemical tracer of deep-water paleoceanography. Paleoceanogr 3:471-489

Boyle EA (1992) Cd and C13 paleochemical ocean distributions during the stage 2 glacial maximum. In: Ann Rev Earth Planet Sci 20:245-287

Boyle EA (1994b) A comparison of carbon isotopes and cadmium in the modern and glacial maximum ocean: can we account for the discrepancies? In: Zahn R, Pedersen TF, Kaminski MA, Labeyrie L (eds) Carbon Cycling in the Glacial Ocean: Constraints on the Ocean's Role in Global Change. Springer, vol 17, pp 167-194

Boyle EA (1995) Limits on Benthic Foraminiferal Chemical Analyses as Precise Measures of Environmental Properties. J Foram Res 25:4-13

Boyle EA, Huested SS, Jones SP (1981) On the distribution of Cu, Ni, and Cd in the surface waters of the North Atlantic and North Pacific Ocean. J Geophys Res 86:8048-8066

Boyle EA, Keigwin LD (1982) Deep circulation of the North Atlantic over the last 200,000 years: geochemical evidence. Science 218:784-787

Boyle EA, Keigwin LD (1985/6) Comparison of Atlantic and Pacific paleochemical records for the last 250,000 years: changes in deep ocean circulation and chemical inventories. Earth Planet Sci Lett 76:135-150

Boyle EA, Labeyrie L, Duplessy J-C (1995) Cadmium in Aragonitic Hoeglundina: Confirmation of the Reliability of Cd in Depth-Dependent Calcitic Benthic Foraminifera and Chemical Hydrography in the Last Glacial Maximum Northern Indian Ocean. Paleoceanogr 10:881-900

Broecker WS (1993) An oceanographic explanation for the apparent carbon isotope-cadmium discordance in the glacial Antarctic? Paleoceanogr 8:137-140

Broecker WS, Maier-Reimer E (1992) The influence of air and sea exchange on the carbon isotope distribution in the sea. Glob Biogeochem Cycles 6:315-320

Charles CD, Fairbanks RG (1992) Evidence from Southern Ocean Sediments for the effect of North Atlantic deep-water flux on climate. Nature 355:416-419

Charles CD, Wright JD, Fairbanks RG (1993) Thermodynamic influences on the marine carbon isotope record. Paleoceanogr 8:691-698

Curry WB, Duplessy JC, Labeyrie LD, Shackleton NJ (1988) Changes in the distribution of C13 of deep water CO2 between the last glaciation and the Holocene. Paleoceanogr 3:317-342

Duplessy JC, Matthews RK, Prell W, Ruddiman WF, Caralp M, Hendy CH (1984) C-13 record of benthic forminifera in the last interglacial ocean: implications for the carbon cycle and global deep water circulation. Quat Res 21:225-243

Graham DW, Corliss BH, Bender ML, Keigwin LD (1981) Carbon and oxygen isotopic equilibria of recent deep-sea benthic foraminifera. Mar Micropal 6:483-497

Herguera JC, Berger WH (1991) Paleoproductivity from benthic foraminifera abundance: glacial to postglacial change in the west-equatorial Pacific. Geology 19:201-206

Herguera JC, Jansen E, Berger WH (1992) Evidence for a bathyal front at 2000-m depth in the glacial Pacific, based on a depth transect on Ontong Java Plateau. Paleoceanogr 7:273-288

Ingall E, Jahnke R (1994) Evidence for enhanced phosphorus regeneration from marine sediments overlain by oxygen depleted waters. Geochim Cosmochim Acta 58:2571-2576

Keigwin LD, Jones GA, Lehman SJ, Boyle EA (1991) Deglacial meltwater discharge, North Atlantic

Deep Circulation, and Abrupt Climate Change. J Geophys Res 96:16811-16826

Keir RS (1991) The effect of vertical nutrient redistribution on surface ocean d¹³C. Glob Biogeochem Cycles 5:351-358.

Kersten M, Forsttner U (1987) Effect of sample pretreatment on the reliability of solid speciation data of heavy metal - implications for the study of early diagensis. Mar Chem 22:299-312

Kroopnick PM (1985) The distribution of C-13 in the world oceans. Deep-Sea Res 32:57-84

Labracherie M, Labeyrie LD, Duprat J, Bard E, Arnold M, Pichon J-J, Duplessy J-C (1989) The last deglaciation in the Southern Ocean. Paleoceanogr 4:629-638

Lloyd-Kinstrand L, Rosenthal Y, Boyle E (1994) Depth-dependent incorporation of metals by benthic foraminifera: toward sorting out the pressure, temperature, and dissolution effects. Trans Am Geophys Union 75:331

Lohmann GP, Schweitzer PN (1990) Globorotalia truncatulinoides' growth and chemistry as probes of the past thermocline: 1. shell size. Paleoceanography 5:55-75

Lorens RB, Williams DF, Bender ML (1977) The early nonstructural chemical diagenesis of foraminiferal calcite. J Sediment Petrol 47:1602-1609

Mackensen A, Hubberten HW, Bickert T, Fischer G (1993) d¹³C in benthic foraminiferal tests of Fontbotia wuellerstorfi (Schwager) relative to the d¹³C of dissolved inorganic carbon in Southern Ocean deep water: implications for glacial ocean circulation models. Paleoceanogr 8:587-610

Mantyla AW, Reid JL (1983) Abyssal characteristics of the world oceans. Deep-Sea Res 30:805-833

McCorkle DC, Keigwin LD (1994) Depth profiles of δ¹³C in bottom water and core top C. wuellerstorfi on the Ontong-Java Plateau and Emperor Seamounts. Paleoceanogr 9:197-208

McCorkle DC, Martin PA, Lea DW, Klinkhammer GP (1994) Evidence of a dissolution effect on benthic foraminiferal Cd/Ca and Ba/Ca. Trans Am Geophys Union 75:54

Mix AC, Pisias NG, Zahn R, Rugh W, Lopez C (1990) Carbon-13 in Pacific deep and intermediate waters, 0-370 kyr bp.: implications for ocean circulation and Pleistocene CO2. Paleoceanogr 6:205-226

Oppo DW, Fairbanks RG, Gordon AL, Shackleton NJ (1990) Late Pleistocene Southern Ocean C13 variability. Paleoceanogr 5:43-54

Oppo DW, Rosenthal Y (1994) Cd/Ca changes in a deep Cape Basin core over the past 730,000 years:

Response of circumpolar deepwater variability to northern hemisphere ice sheet melting? Paleoceanogr 9:661-676

Pedersen TF (1983) Increased productivity in the eastern equatorial Pacific during the last glacial maximum (19,000 to 14,000 yr B.P.). Geology 11:16-19

Rosenthal Y (1994) Late quaternary paleochemistry of the southern ocean: evidence from cadmium variability in sediments and foraminifera. PhD dissertation, Mass. Inst. Tech./Woods Hole Ocean. Inst.

Rosenthal Y, Boyle EA (1993) Factors controlling the fluoride content of planktonic foraminifera: an evaluation of its paleoceanographic applicability. Geochim Cosmochim Acta 57:335-346

Rosenthal Y, Boyle EA, Labeyrie L, Oppo D (1995a) Glacial enrichments of authigenic Cd and U in sub-Antarctic sediments: a climatic control on the elements' oceanic budget? Paleoceaongr 10:395-413

Rosenthal Y, Lam P, Boyle EA, Thomson J (1995b) Authigenic cadmium enrichments in reducing sediments: precipitation and post-depositional mobilization. Earth Planet Sci Lett 132:99-111

Russell AD, Emerson S, Nelson BK, Erez J, Lea DW (1994) Uranium in foraminiferal calcite as a recorder of seawater uranium concentration. Geochim Cosmochim Acta 58:671-681

Saager P (1994) On the relationships between dissolved trace metals and nutrients in seawater. PhD dissertation, University of Amsterdam

Sarnthein M, Winn K, Duplessy J-C Fontugne MR (1988) Global variations of surface ocean productivity in low- and mid-latitudes : influence on CO2 reservoir of the deep ocean and atmosphere during the last 21,000 years. Paleoceanogr 3:361-399

Stuiver M, Quay PD, Ostlund HG (1983) Abyssal water carbon-14 distribution and the age of the world oceans. Science 219:849-852

Vincent E, Berger WB (1981) Planktonic foraminifera and their use in paleoceanography. In: Emilani C (ed) The Sea, vol.7. Wiley-Interscience, New York, pp 1025-1119

Yu E-F (1994) Variations in the particulate flux of Th-230 and Pa-231 and paleoceanograhic applications of the Pa-231/Th-230 ratio. PhD dissertation, Woods Hole Oceanographic Institution/Massachusetts Institute of Technology

Yu E-F, Francois R, Bacon MP (in press) Radiochemical constraints on ocean circulation during the last glacial maximum. Nature

Zahn R, Sarnthein M., Erlenkeuser H (1987) Benthic isotope evidence for changes of the Mediterranean outflow during the late Quaternary. Paleoceanogr 2:543-560

High Latitude Deep Water Sources During the Last Glacial Maximum and the Intensity of the Global Oceanic Circulation

J.C. Duplessy[1], L. Labeyrie[1], M. Paterne[1], S. Hovine[2],
T. Fichefet[2], J. Duprat[3] and M. Labracherie[3]

[1] Centre des Faibles Radioactivités, Laboratoire mixte CNRS-CEA,
91198 Gif sur Yvette cedex, FRANCE
[2] Institut d'Astronomie et de Géophysique Georges Lemaitre,
Chemin du Cyclotron 2, B-1348 Louvain La Neuve, BELGIQUE
[3] Département de Géologie et Océanographie, Université de Bordeaux-1,
Av. des Facultés, 33405 Talence, FRANCE

Abstract: Micropaleontological and oxygen isotope analyses of planktonic foraminifera from North Atlantic Ocean and Southern Ocean sediment cores have been used to reconstruct temperature and salinity of surface waters during the last glacial maximum. Whereas the Norwegian-Greenland Sea and the high latitudes of the North Atlantic experienced a large negative anomaly, the Southern Ocean maintained salinity values similar or slightly higher than the modern ones around Antarctica, thus favouring winter convection and deep water formation in the Southern Hemisphere. These data have been used to force the zonally averaged, three-basin ocean model of Louvain-La-Neuve. The model reproduces the main trends of the geochemically constrained glacial Atlantic circulation and suggests that the glacial production of Antarctic Bottom Water was slightly higher than the modern one, whereas that of North Atlantic Deep Water was reduced by about 40%.

Introduction

The glacial-interglacial oscillations of the Pleistocene resulted in large variations of surface water temperature and circulation (CLIMAP 1981). Global deep water circulation patterns also experienced marked changes (Duplessy et al. 1980; Boyle and Keigwin 1982; Shackleton et al. 1983; Duplessy and Shackleton 1985; Oppo and Fairbanks 1987; Duplessy et al. 1988; Kallel et al. 1988; Sarnthein et al. 1994; also see Boyle, this volume, and Curry, this volume). Variations in deep water circulation are controlled by changes in the conditions prevailing at the surface (sea surface temperature and salinity, wind stress) because deep and bottom waters are formed by sinking of surface water in very limited areas of the high-latitude oceans, where surface water becomes sufficiently dense during winter to sink to abyssal depths.

The purpose of this paper is to explore the link between deep water circulation and high latitude sea surface changes during the last glacial maximum. The reconstruction of the Atlantic deep water circulation patterns (Duplessy et al. 1988; Sarnthein et al. 1994) will be compared to the CLIMAP (1981) reconstruction of sea surface temperatures (SST) and to the reconstruction of sea surface salinity (SSS) in both the North Atlantic Ocean (Duplessy et al. 1991) and the Southern Ocean, for which a first reconstruction is presented in this paper. We shall then use the 2-D Ocean General Circulation Model (OGCM) of Louvain La Neuve (Hovine and Fichefet 1994) to understand the impact of salinity changes in both the North Atlantic Ocean and the Southern Ocean for driving the global deep oceanic circulation.

From WEFER G, BERGER WH, SIEDLER G, WEBB DJ (eds), 1996, *The South Atlantic: Present and Past Circulation*. Springer-Verlag Berlin Heidelberg, pp 445-460

The Glacial Deep Atlantic Circulation

Flow lines in the deep ocean may be reconstructed from changes in the $\delta^{13}C$ values of the total CO_2 dissolved in the deep water, because $\delta^{13}C$ decreases as a water mass moves away from its most recent area of ventilation, as a result of the progressive in situ oxidation of organic matter settling from the surface. The present geographic distribution of $\delta^{13}C$ in the deep ocean is closely related to the oxygen and nutrient content of the various water masses and is strongly dependent on circulation patterns (Kroopnick 1985). The geographic distribution of the carbon isotopic composition of the total CO_2 dissolved in the deep waters of the Atlantic Ocean during the last glacial maximum has been reconstructed from the $\delta^{13}C$ values of *Cibicides* that lived during the peak of oxygen isotope stage 2 in numerous cores collected at various locations and depths in the whole Atlantic Ocean (Fig. 1).

Compared to the modern pattern, the salient feature in the section reconstructed for the Last Glacial Maximum (LGM) is the presence of low $\delta^{13}C$ values below 2500 m water depth between 20° and 45°N, indicating that most of the deep eastern Atlantic was filled with bottom water originating from the Southern Ocean (Fig. 1). The relative contribution of water from the southern hemisphere as compared to that from the northern hemisphere in the deep Atlantic was therefore much greater than today. Higher $\delta^{13}C$ values are found above 2500 m due to the presence of a Glacial North Atlantic Deep Water mass, which found its density equilibrium at a depth significantly shallower than today. This water mass formed in the North Atlantic (see below) and received a significant contribution of Glacial Mediterranean Outflow Water in the eastern basin (Zahn, 1986).

The North Atlantic Deep Water Source

Deep waters form in area of intense winter cooling, because here surface waters become then as dense as the deeper waters. This process results in large-scale open ocean convection of which the net result is the sinking of a significant amount of cold dense surface water to the abyss. Under modern conditions, the sources of NADW are mainly located in two basins: The overflow of the Norwegian-Greenland Sea Deep Water results in the formation of the Lower NADW, while Upper NADW forms in the Labrador Sea, because the density of their surface waters is high enough to permit convection during winter cooling (Fig. 2). A detailed discussion on the conditions of deep water formation is given in Labeyrie et al. (1992).

In order to determine the potential areas for deep water formation during the last glacial maximum, Duplessy et al. (1991) developed a method to reconstruct surface water salinity of the North Atlantic Ocean, based on a comparison between transfer function estimates of SST and the oxygen isotope ratio of two common planktonic foraminiferal species, *N. pachyderma* (left coiling) and *G. bulloides*. These authors first used core-top analyses to demonstrate that, under modern conditions, the paleo-temperatures determined from the isotopic composition of foraminiferal shells are linearly linked to the summer SST only within a specific temperature range, the "optimum temperature range" of the taxon. They then used this information to derive an estimate of the isotopic composition and salinity of the North Atlantic surface water during the last glacial maximum. The statistical error involved in such a reconstruction is large, of the order of 0.3‰ for sea water $\delta^{18}O$ and 0.6 for salinity when these estimates rest only on the analysis of a single sediment core level (Duplessy et al. 1991).

The resulting reconstruction for LGM shows a significant difference with the modern pattern and the presence of a sharp salinity gradient associated with the polar front (Fig. 3). However, high-salinity water was present near 30-40°W, north of the polar front. North of this tongue of high saline water, surface salinity was lower than 35 and exhibited very low values in ice melting areas (coast of Greenland and Scandinavia) and in the Bay of Biscay. The salinity anomaly at any one location in the glacial North Atlantic Ocean may be defined as the difference between the LGM values minus the modern value and corrected this difference from the mean oceanic salinity change (Δ) due to the growth of continental ice sheets :

salinity (anomaly)

=

salinity (LGM) - salinity (modern) - Δ (ice volume).

In order to compute the anomaly map (Fig. 4), we adopted a Δ value of 1, as a consequence of a 120 m sea level drop during the maximum extension of continental ice sheets (Fairbanks 1989). Negative anomalies were generally found north of the polar front. This is particularly marked in the Bay of Biscay which received directly the outflow of European rivers, carrying meltwater from snow and from the southern margin of the European ice sheets. Negative anomalies are also seen in the Norwegian-Greenland Sea as a consequence of the disappearance of the North Atlantic Drift during glacial conditions (Kellogg 1976; Ruddiman and McIntyre 1973; Alvinerie et al. 1978). Both the negative salinity anomalies off west Ireland and Faeroe and that in the Bay of Biscay imply a south-

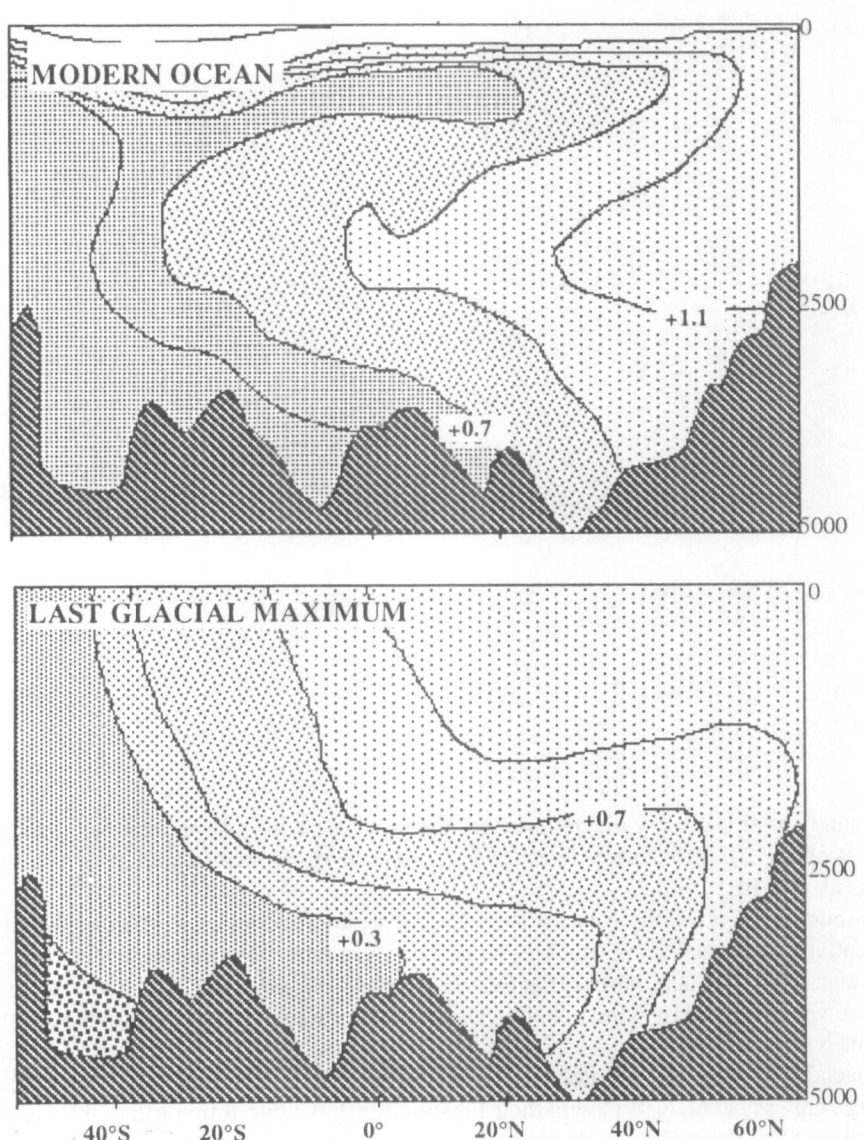

Fig. 1. Outline of δ¹³C distributions across the Atlantic Ocean for the modern period and the last glacial maximum (from Labeyrie et al. 1992). Reconstructions for the LGM are highly schematic; there are no data in the North Atlantic above 800 m depth and in the South Atlantic above 1800 m depth.

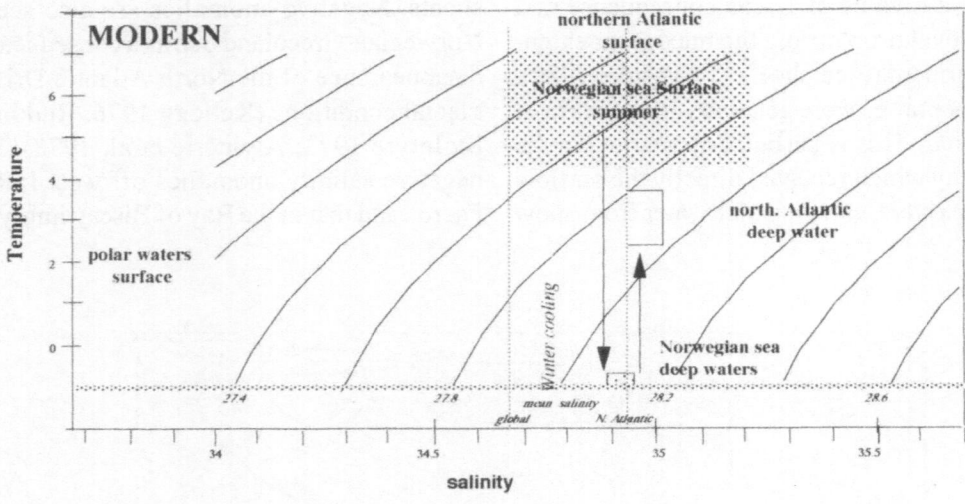

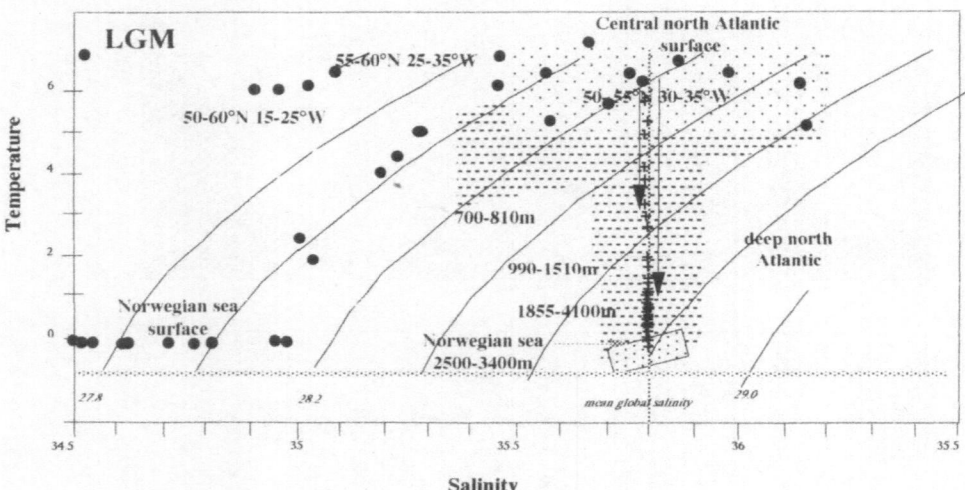

Fig. 2. Temperature-salinity-density diagrammes for the modern period (A) and the last glacial maximum (B) (from Labeyrie et al. 1992). The density lines are calculated as σ_0 (density anomaly at sea surface pressure) following Cox et al. (1970).

Fig. 2a. Mean modern global salinity is taken as 34.68 and mean modern North Atlantic salinity as 34.93. The approximate locations of the northern Atlantic, Norwegian Sea and polar surface waters (in summer), North Atlantic deep waters (NADW) and Norwegian Sea deep waters are reported. Arrows schematize the changes occurring for deep water formed by winter cooling and sea ice formation in the western Norwegian Sea, and the evolution towards NADW.

Fig. 2b. Last Glacial Maximum reconstruction. The data which serve to define the LGM surface waters (●) are plotted following Duplessy et al. (1991) with their location. Each benthic foraminifera $\delta^{18}O$ is plotted as a +, at the deep water temperature derived from Shackleton (1973) equilibrium fractionation equation, mean deep North Atlantic $\delta^{18}O$ +1.37 ‰ vs SMOW and salinity 35.8 (Labeyrie et al. 1992). The corresponding core depth range is indicated. The coarse-dot pattern schematizes the uncertainty in the deep water characteristics which derives from the lack of independent constraints on the deep water salinity range. The arrow symbolizes the possible winter decrease of temperature for surface waters potentially at the source of the deep waters.

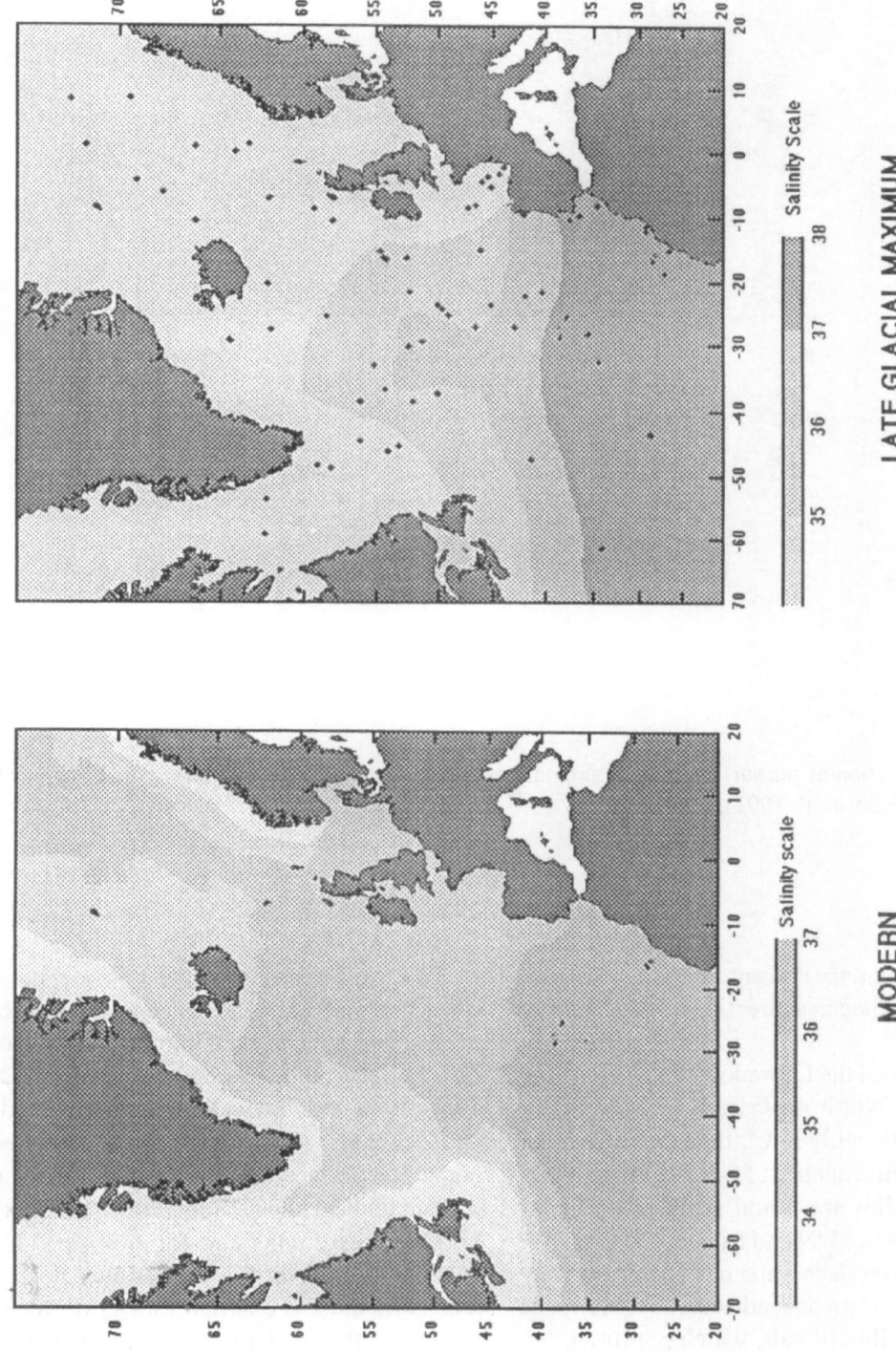

Fig. 3. Sea surface salinity in the North Atlantic Ocean for the modern period and the last glacial maximum (from Duplessy et al. 1991).

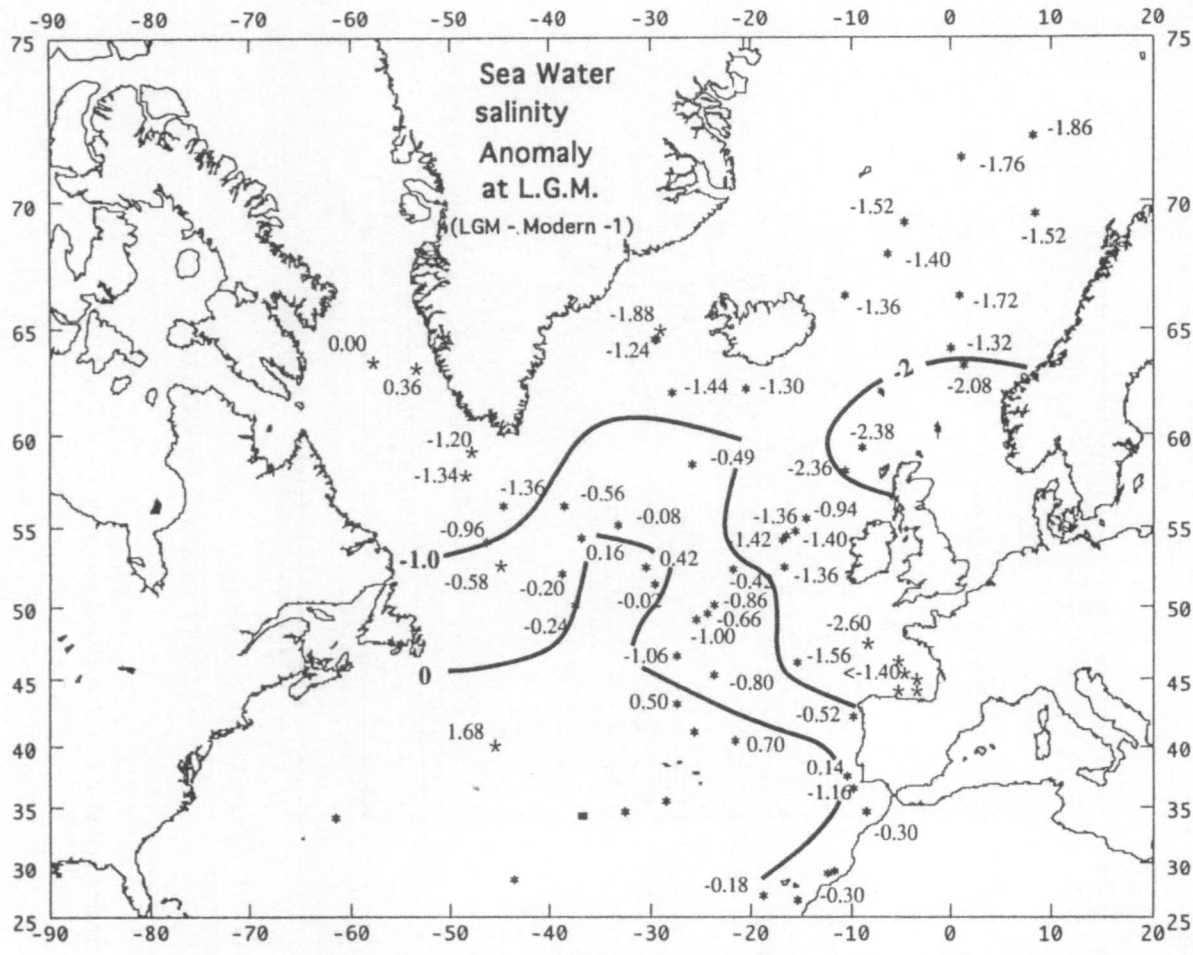

Fig. 4. Reconstruction of sea surface salinity anomaly in the North Atlantic Ocean during the last glacial maximum (from Duplessy et al. 1991).

ward eastern boundary current, carrying freshwater from high latitude in a direction opposite to that of today.

The presence of the Labrador current along the eastern coast of North America is marked by salinity values noticeably lower than those found in the central Atlantic at about 50-55°N. In contrast, positive anomalies are found north of the polar front, from 45°N to 55°N and from 30°W to 40°W. The presence of a surface water mass comparatively saltier than the modern one indicates the occurrence of a significant flux of salt, which permitted the formation of deep water.

To compare surface and deep water characteristics, and discuss possible deep water sources and mixing, we have to take into account the density distribution of the different water masses. Density depends on both the water temperature and salinity (Cox et al. 1970). For the paleo-ocean, we estimated Sea Surface Temperature (SST) from micropaleontological transfer functions and salinity from both SST estimates and foraminiferal isotopic analysis as described in Duplessy et al. (1991). A reconstructed T/S diagram for LGM is compared to that for the modern ocean in Fig. 2. Since the mean ocean salinity was about 1 per mil larger than today and the Atlantic more saline than the Pacific Ocean (Broecker 1989), surface waters with salinity values close to 35.8 were required to produce the densities and temperatures of both intermediate and deep waters by winter cooling. The corresponding cores are located within a relatively small area of subpolar waters, between 52 and 54°N and 20 to 40°W, about midway between North America and Europe. At the opposite, the cores north, east, and west of that area, correspond all to significantly lower surface water salinities (less than 35.2). Even cooling of these waters to freezing temperatures and sea ice formation could not increase their density above the density of the 700 m depth water (σ of 28.2 to 28.4).

The Southern Ocean Bottom Water Source

In order to determine the potential areas of bottom water formation in the Southern Ocean, we shall apply the same strategy as that used in the North Atlantic, and generate SSS estimates for LGM in addition to the SST map made by CLIMAP (1981), which rests on both foraminiferal and radiolarian faunal variations. Sea surface salinity may be derived from planktonic foraminifera δ18O once corrected for the effect of temperature on isotopic fractionation (using the transfer function estimates of Summer SST) and for the global changes in sea water δ18O (Labeyrie et al. 1987; Duplessy et al. 1991).

Calibration of the planktonic foraminiferal oxygen isotopic composition

We calibrated N. pachyderma and G. bulloides by analysing 66 core tops distributed throughout the Southern Ocean and raised from locations where February (warmest summer month) SST ranges from -1.3°C to 13°C (Levitus 1982). Modern sea water δ18O values were estimated from the GEOSECS measurements and the isotopic temperatures calculated using the Shackleton (1974) paleotemperature scale. Levitus (1982) summer SST are compared with the isotopic paleotemperature estimates derived from the δ18O value of N. pachyderma and G. bulloides in Fig. 5. This compari-

son shows that within the temperature range 4-10°C, the isotopic temperature given by both N. pachyderma and G. bulloides is close to T* = February SST -1°C. In the Southern Ocean, N. pachyderma and G. bulloides have therefore a temperature of calcification which is similar to that of G. bulloides in the North Atlantic, but is warmer than that of N. pachyderma. This difference is most probably due to the fact that N. pachyderma has a deep habitat (around 100 m) and that the North Atlantic is much more stratified than the Southern Ocean, in which surface waters are intensively mixed by strong westerlies blowing in the 40°S - 50°S latitudinal belt.

Below 4°C, N. pachyderma exhibits isotopic temperatures somewhat warmer than the mean atlas values. The best fit line of the measurements plotted in Fig. 5 is:

T = 0.6 x February SST + 0.7 (For February SST < 4°C)*

These results suggest that, in very cold water, N. pachyderma develops only when it can find a water mass of which the temperature is close to 0-1°C.

Reconstruction of summer sea surface salinity

We used the δ18O values of N. pachyderma (left coiling) in 31 cores from the Atlantic and Indian sectors of the Southern Ocean to reconstruct the sea water δ18O value and salinity during the peak of oxygen isotope stage 2 (last glacial maximum). Results are reported in Figs. 6 and 7. In a band of about 10° latitude around Antarctica, which was permanently covered with sea ice (CLIMAP 1981), data are available only close to Antarctica (Melles 1991) or close to the northern edge of the sea ice cover. Sea surface δ18O values were close to 1‰ vs SMOW, corresponding to salinity values around 35, which prevailed probably below the sea ice cover or within polynias that maintained during LGM (Melles 1991). North of this area, in a narrow band, δ18O and salinity experienced both a significant decrease reflecting strong summer melting of icebergs and sea ice in the polar front (Note the local occurrence of extremely low δ18O values, which cannot be produced by sea ice melting, but result from the melting of iceberg carrying ice

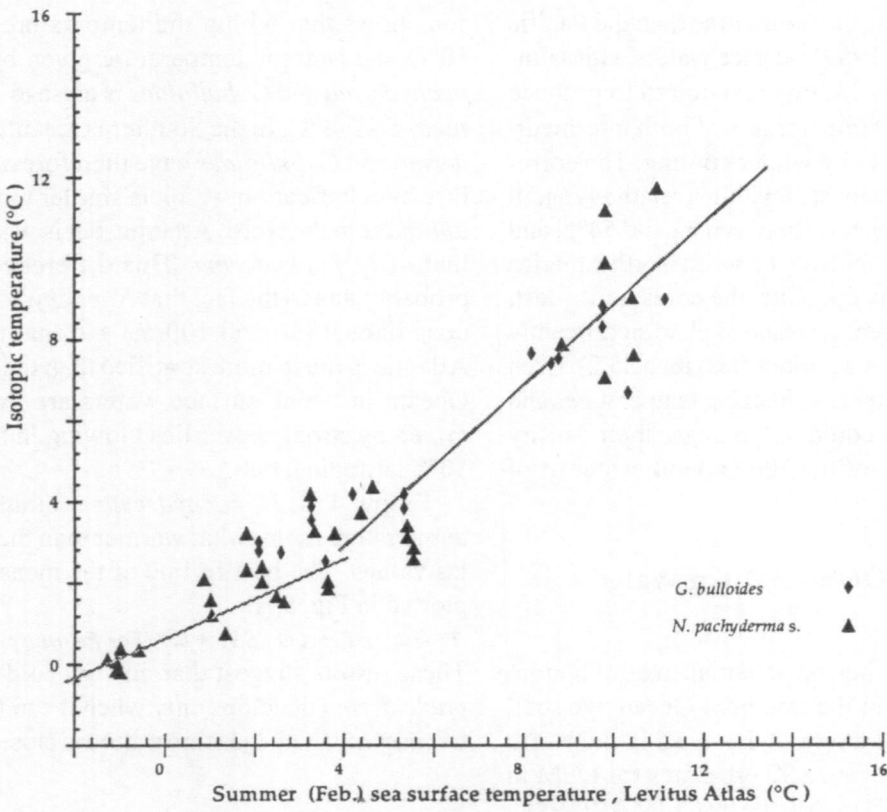

Fig. 5. Comparison between February sea surface temperatures (Levitus 1982) and isotopic temperatures of *N. pachyderma* (left coiling) and *G. bulloides* in core tops from the Southern Ocean.

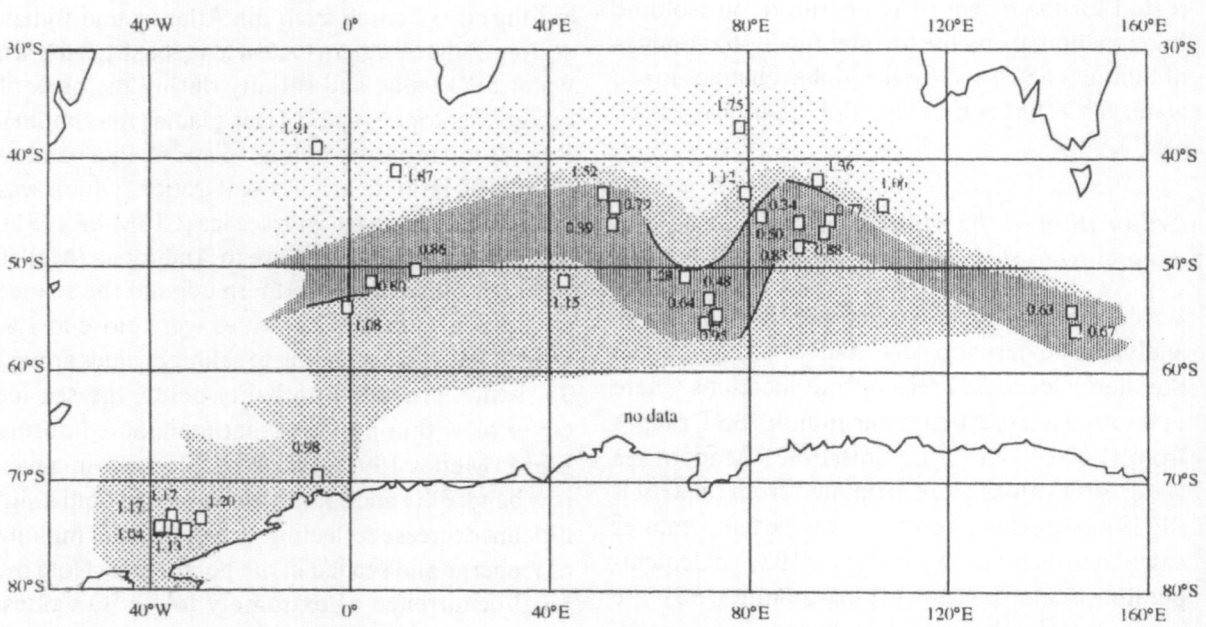

Fig. 6. Reconstruction of sea surface water δ¹⁸O in the Southern Ocean during the last glacial maximum.

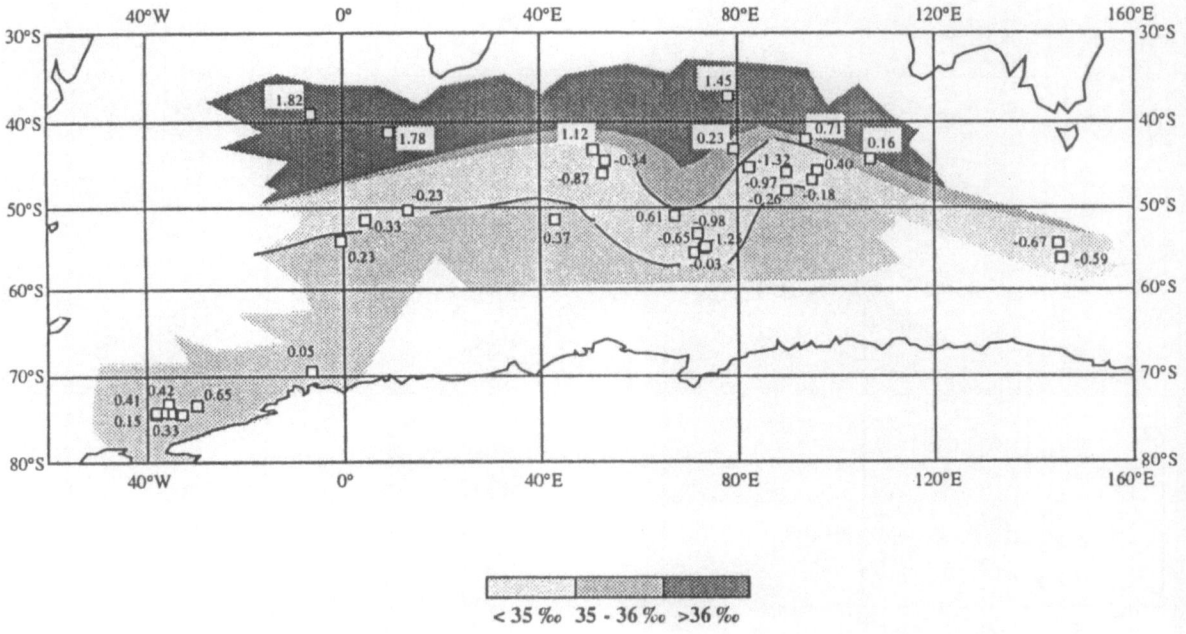

Fig. 7. Reconstruction of sea surface water salinity in the Southern Ocean for the Last Glacial Maximum. Data are expressed as S-35‰.

formed at very low temperature, in Antarctica). North of this front, salinity increased because the Evaporation-Precipitation budget was positive. The salinity anomaly (computed in Fig. 8 using the same formula as for the North Atlantic Ocean (Duplessy et al. 1991)) was slightly positive all around Antarctica, indicating that the glacial sea surface salinity was comparatively higher than today. This anomaly was strongly negative in the polar front, as a consequence of active ice melting during summer. It then increased northward to reach values larger than 1 in the subtropical band. This reconstruction is very robust and is not significantly affected by inaccuracies in the calibration of the $\delta^{18}O$ value of the planktonic foraminifera *N. pachyderma*. The pattern is basically unchanged if we make the alternative assumption that under sea ice, the calcification temperature is 0°C. In that case, the salinity anomaly is close to zero.

Comparison with the North Atlantic and consequences for deep water formation

Figs. 3 and 7 show that at latitudes lower than the polar front, the salinity increased in both hemispheres toward the tropics, resulting in large positive anomalies in the tropical and subtropical gyres. However, the reconstruction of sea surface salinity for the Southern Ocean differs noticeably from that of the North Atlantic, because the Norwegian-Greenland Sea and the high latitudes of the North Atlantic experienced a large negative anomaly during glacial conditions (Fig. 4), whereas the Southern Ocean maintained salinity values similar or slightly higher than the modern ones (after correction for the global ice volume effect). The mean salinity gradient between the tropics and the polar areas was therefore larger than today under glacial conditions. It was also larger in the Northern Hemisphere than in the Southern Hemisphere.

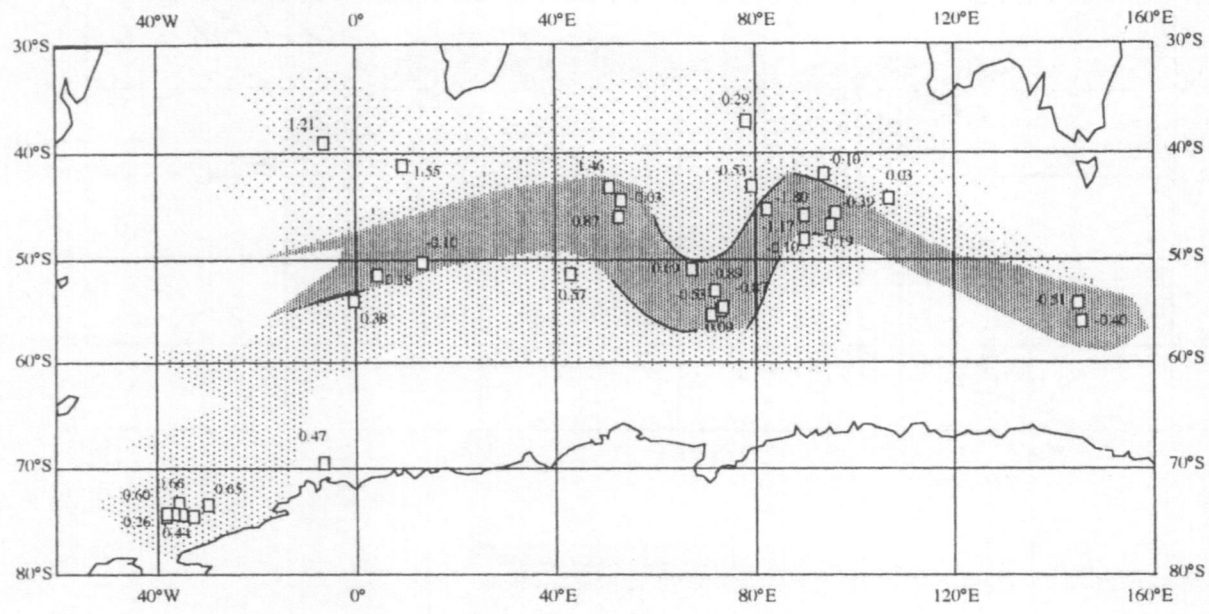

Fig. 8. Reconstruction of sea surface salinity anomaly in the Southern Ocean during the last glacial maximum.

The low salinity band associated with the southern polar front is an original feature of the Southern Ocean. In the North Atlantic, north of the polar front, salinity decreased continuously northward. In the Southern Ocean, surface salinity south of the polar front was higher than in the ice-melting zone associated with the front. As southern polynias were also active during the glaciation (Melles 1991), formation of bottom water along the coast of Antarctica by density increase resulting from the rejection of brines during winter sea ice growth may have been easier than today. Therefore glacial sea surface conditions favoured bottom water formation in the Southern Ocean, whereas they were more difficult for the North Atlantic, which was able to experience deep convection only in a small area from 45°N to 55°N and from 30°W to 40°W.

Simulations with a 2-D ocean general circulation model

The model used here is a global ocean circulation model consisting of three coupled, zonally averaged basins that represent the Atlantic, Pacific, and Indian oceans. The Atlantic and Pacific basins extend from 90°N to 80°S, while the Indian basin runs from 25°N to 80°S (Fig. 9). The latitudinal resolution is 5°, and vertically there are 19 layers covering a total depth of 5,000 m and expanding in thickness from 50 m near the surface to 500 m near the bottom. Each basin has its own bathymetry and a latitude-dependent angular width. Lateral exchanges of heat and salt between the basins are permitted between 85°N and 90°N and between 40°S and 65°S. A summary of the model equations can be found in Hovine and Fichefet (1994).

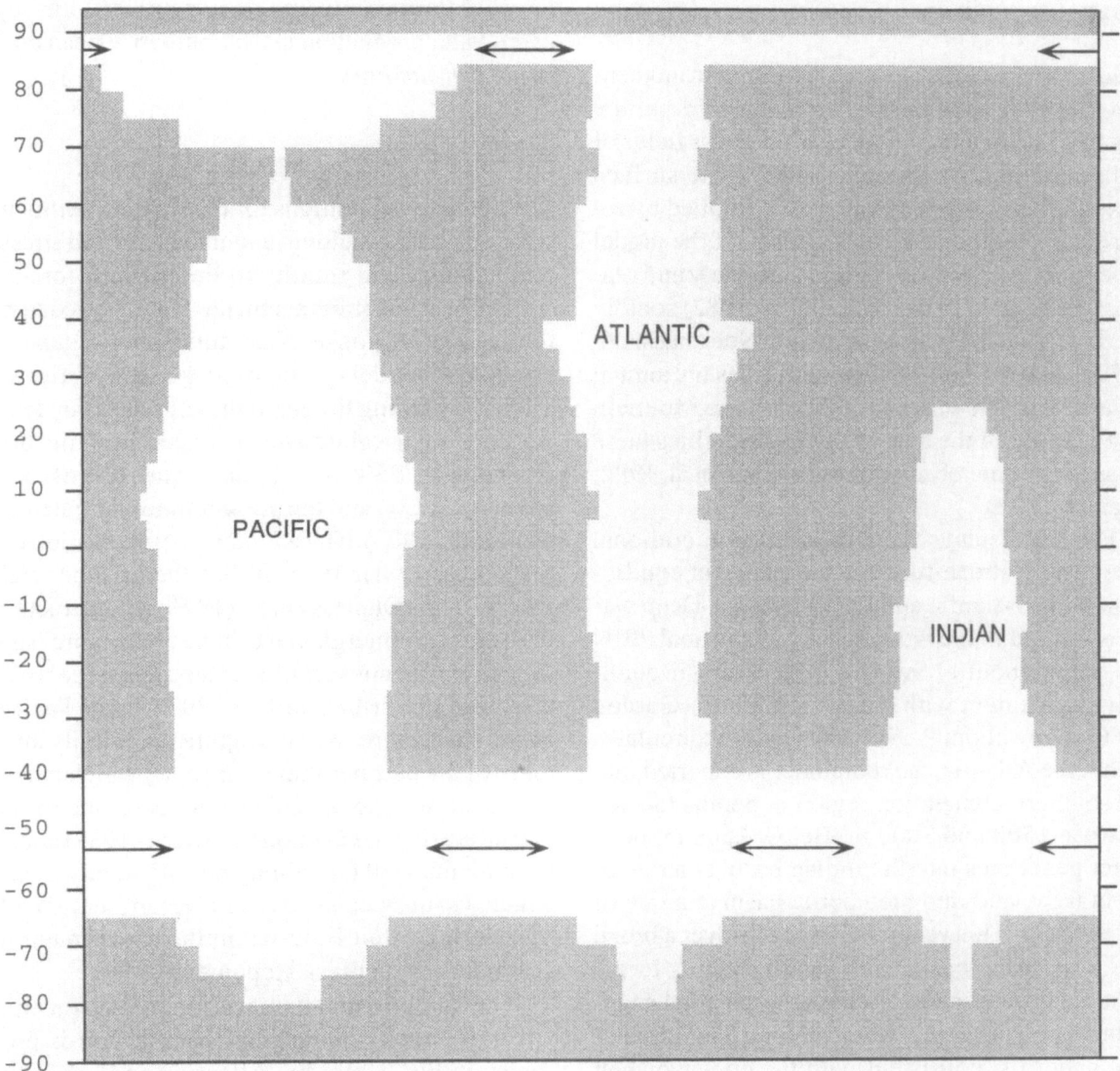

Fig. 9. Geometry of the three zonally averaged ocean basins. Each basin has latitude-dependent angular width and bottom topography. The latitudinal resolution is 5°. Lateral exchanges of heat and salt between the basins are allowed between 85 and 90°N and between 40 and 65°S.

To understand the impact of salinity changes in both the North Atlantic Ocean and the Southern Ocean for driving the global deep oceanic circulation, two experiments with the ocean model described above are discussed here: A control run with present-day surface boundary conditions, and a run with conditions that prevailed during the LGM. In both cases, the model has been integrated for 10,000 years under restoring boundary conditions from a state of rest with uniform temperature and salinity. The acceleration techniques of Bryan (1984) were employed during the first 2,000 years to speed up the convergence towards equilibrium.

Control experiment

In the control run, the atmosphere-to-ocean momentum flux over each basin were deduced from the monthly mean climatological wind-stress fields of Hellerman and Rosenstein (1983). The surface fluxes of heat and freshwater were implied by restoring the temperature and salinity of the model uppermost level to the respective monthly and seasonal mean surface data of Levitus (1982) zonally averaged over the basin considered. Note that south of 70°S, we used as restoring salinities the annual mean data at 100 m depth of Gordon and Molinelli (1986) to avoid the bias of Levitus salinities there towards summer observations (Stocker et al. 1992; England 1992).

Fig. 10 displays the annual mean meridional overturning streamfunction simulated at equilibrium in the Atlantic and Pacific basins. Deep water is formed in the Atlantic between 60 and 70°N at a rate of about 17 Sv (1 Sv = 10^6 m^3 s^{-1}), in qualitative agreement with current estimates (Gordon 1986). Only about 9 Sv of this water recirculates within the Atlantic, the remainder is exported into the southern connection region at depths located between 1500 and 3000 m. Below 1500 m, deep water penetrates into the Indian basin at a rate of about 6 Sv, and into the Pacific basin at a rate of about 12 Sv. This water then upwells over a broad region in these basins and finally returns to the Atlantic basin via the circumpolar channel as intermediate- and upper-water masses. This circulation pattern is consistent with the conveyor-belt picture suggested by Gordon (1986). The annual meridional overturning streamfunction integrated over the three oceanic basins (Fig. 10) reveals that the global thermohaline circulation consists of basically two cells and that about 30 Sv of Antarctic Bottom Water (AABW) are created off the Antarctic continent.

Hovine and Fichefet (1994) have showed that the seasonal variations of the modelled circulation are consistent with those obtained by comprehensive tri-dimensional ocean general circulation models and that perturbations in the surface fresh-water flux may produce transitions to two other equilibria characterised respectively by deep-water formation in the North Pacific with upwelling in the North Atlantic (*inverse conveyor belt*) and by deep-water production in the Southern Ocean only (*southern sinking*).

LGM experiment

The boundary conditions for the LGM experiment were obtained by adding anomalies of wind stress, temperature, and salinity to the modern forcing fields. The wind-stress anomalies were derived from the ice-age response of an atmospheric general circulation model (Lautenschlager and Herterich 1990). Regarding the anomalies of relaxation temperature, their values were deduced from the differences of SST in February and August between the LGM and the present time estimated by CLIMAP (1981). The anomalies of relaxation salinity prescribed north of 40°S in the Atlantic originate from the Duplessy et al. (1991) reconstruction of the glacial-interglacial changes of summer surface salinity in this sector. The same anomalies have also been prescribed north of 70°N in the Pacific basin. Elsewhere, we have applied a salinity anomaly of +1 part per thousand (ppt), which corresponds to the global salinity increase due to the glacial sea-level reduction (Fairbanks 1989) and to the magnitude of the salinity modifications reconstructed in the glacial Southern Ocean (see above). The Bering Strait is closed in this experiment in accordance with glacial topography.

The annual mean meridional overturning streamfunction simulated at equilibrium is presented in Fig. 10 for the Atlantic and Pacific basins, and in the three basins taken as a whole. As reported by Fichefet et al. (1994), the most prominent effects produced by the inclusion of glacial boundary conditions in the model is the cessation of deep-water formation in the highest latitude of the North Atlantic, which follows from the stratification due to the lower relaxation salinities there. The area of deep water formation had moved southward. Around 50°N, the surface-water salinity is sufficiently high to permit deep convection to a maximum depth of 2,500 m as a source of glacial North Atlantic Deep Water (NADW). This gives rise to a thermohaline overturning of moderate strength (the maximum value of the stream function amounts to about 11 Sv) which is essentially

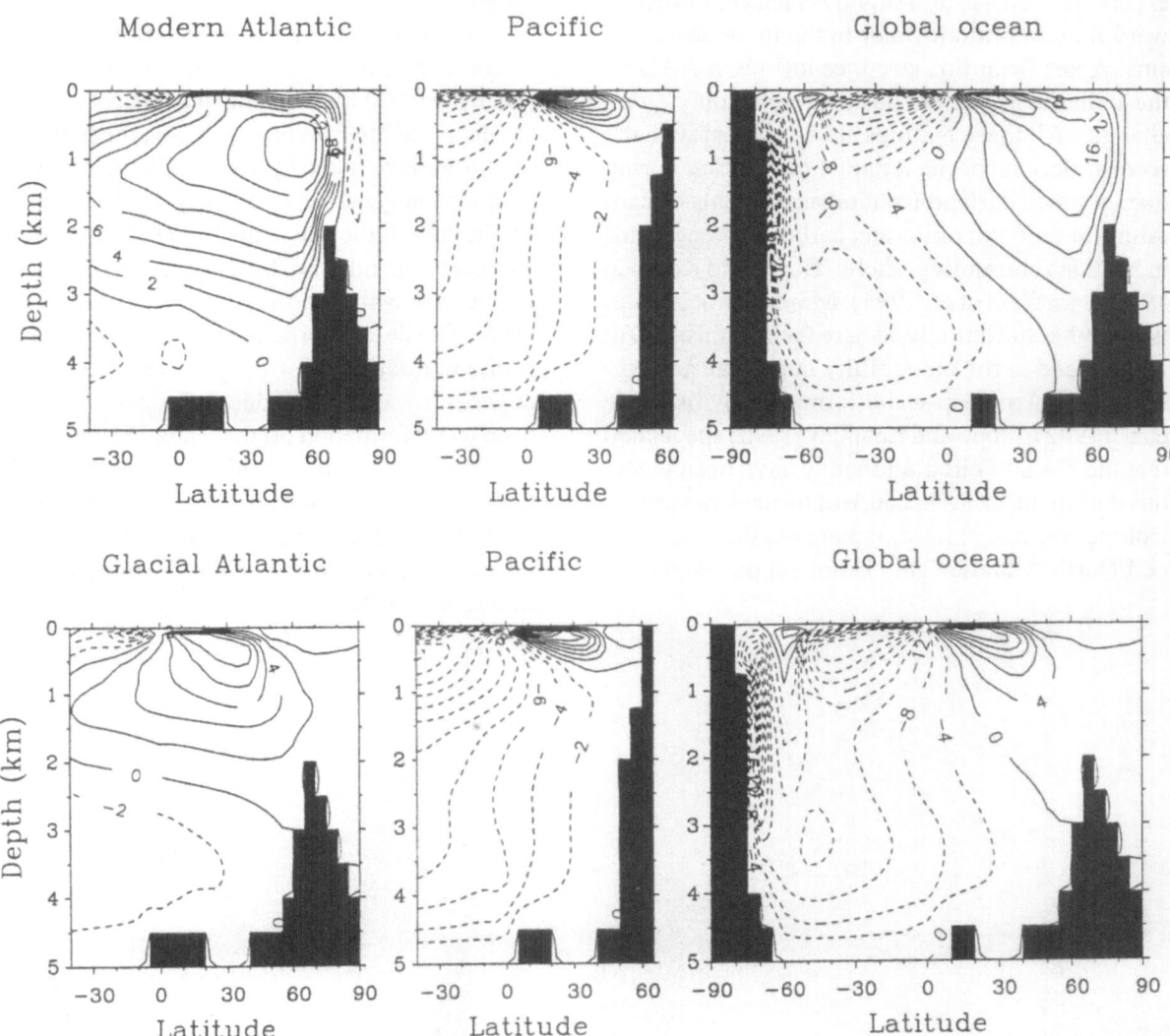

Fig. 10.
Upper: Contours of the annual mean meridional overturning stream function in the Atlantic, Pacific, and in the three basins taken as a whole for the modern simulation. Flow is clockwise around solid contours. Contour interval is 2 Sv for the Atlantic and Pacific sections and 4 Sv for the zonally averaged three-basin section (1 Sv = 10^6 m^3 s^{-1}).
Lower: Contours of the annual mean meridional overturning stream function in the Atlantic, Pacific and in the three basins taken as a whole for the LGM simulation. Flow is clockwise around solid contours. Contour interval is 2 Sv for the Atlantic and Pacific sections and 4 Sv for the zonally averaged three-basin section (1 Sv = 10^6 m^3 s^{-1}).

confined north of 40°S (only 1 Sv of glacial NADW escapes into the southern connection zone). Below 2,500 m, most of the Atlantic basin is filled with AABW, so that the relative contribution of water from the Southern Hemisphere as compared to that from the Northern Hemisphere is much greater than in the control case. The global overturning stream-function shows that in our model, the rate of AABW formation increases by about 10% in the LGM experiment. This in turn slightly enhances the northward flow of bottom water in the three ocean basins. Apart from this stronger inflow of AABW, the Indian and Pacific basins exhibit only small changes. All these features are consistent with the reconstructions of the Atlantic circulation during glacial times derived from measurements of cadmium-to-calcium ratios and carbon-isotope ratios in benthic foraminifera shells (Boyle and Keigwin 1982; Sarnthein et al. 1994). Moreover, it is noteworthy that the latitudes where the glacial NADW is produced in the model fully match the density-based model of deep-water formation by Labeyrie et al. (1992). Oppo and Lehman (1993) speculated that the NADW circulation may have been vigorous during the LGM because of the lack of carbon-isotope contrast at intermediate depths in the glacial North Atlantic. This is not supported by our simulation which shows a marked weakening of the conveyor.

Fig. 11 displays the differences of annual mean potential temperature in the Atlantic and Pacific between the LGM experiment and the control case. In the deep and intermediate Atlantic, the potential temperature is reduced by 1.5 - 2.5°C in the LGM experiment, in general agreement with the geological estimates (Labeyrie et al. 1987). This cooling is primarily produced by the replacement of the relatively warm and modern NADW by cold AABW. The deep Pacific presents a general temperature reduction of 1 - 1.5°C caused by the stronger inflow of AABW due to the increasing rate of deep water formation off the Antarctic continent.

In conclusion, our data indicate that the main features of the global deep oceanic circulation during the LGM can be simply explained by the difference of temperature and salinity fields in the high latitudes of both hemispheres.

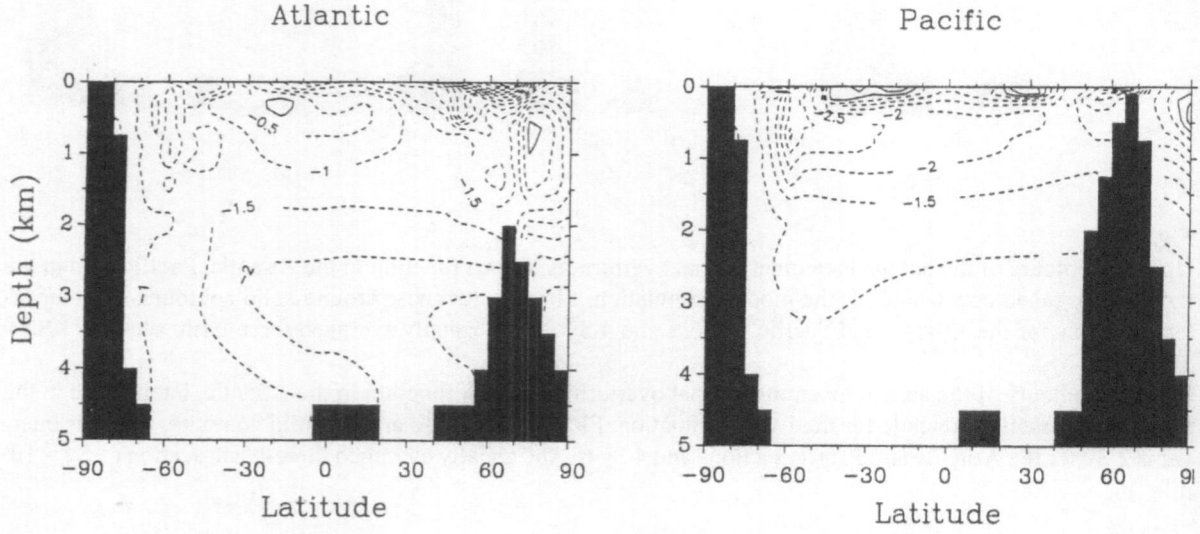

Fig. 11. Latitude-depth distribution of the difference in annual mean potential temperature in the Atlantic and Pacific between the LGM simulation and the modern simulation. Broken line denotes cooling; units are in °C.

Acknowledgements

Thanks are due to W. Berger for a careful review. This work was supported by CNRS, CEA and INSU (PNEDC).

This is CFR contribution N°1733

References

Alvinerie J, Caralp M, Latouche C, Moyes J., Vigneaux M (1978) Apport à la connaissance de la paléo-hydrologie de l'Atlantique nord-oriental pendant le Quaternaire terminal. Oceanologica Acta 1:87-98

Broecker WS (1989) The salinity contrast between the Atlantic and Pacific Oceans during glacial times. Paleoceanogr 4:207-212

Boyle EA, Keigwin L (1982) Deep circulation of the North Atlantic over the last 200,000 years: Geochemical evidence. Science 218:784-787

Bryan K (1984) Accelerating the convergence to equilibrium of ocean-climate models. J Phys Oceanogr 14:666-673

CLIMAP Project Members (1981) Seasonal reconstruction of the earth's surface at the last glacial maximum. Geol Soc Am Map Chart Ser, MC-36

Cox RA, McCartney MJ, Culkin F (970) The specific gravity /salinity/ temperature relationship in natural sea water. Deep-Sea Res 17:679-689

Duplessy JC, Moyes J, Pujol C (1980) Deep water formation in the North Atlantic Ocean during the last ice age. Nature 286:479-482

Duplessy JC, Shackleton NJ (1985) Response of global deep-water circulation to the Earth's climatic change 135,000-107,000 years ago. Nature 316:500-507

Duplessy JC, Shackleton NJ, Fairbanks RG, Labeyrie LD, Oppo D, Kallel N (1988) Deepwater source variations during the last climatic cycle and their impact on the global deepwater circulation. Paleoceanogr 3:343-360

Duplessy JC, Labeyrie LD, Juillet-Leclerc A, Maitre F, Duprat J, Sarnthein M (1991) Surface salinity reconstruction of the NorthAtlantic Ocean during the last glacial maximum. Oceanologica Acta 14:311-324

England MH (1992) On the formation of Antarctic Intermediate and Bottom Water in ocean general circulation models. J Phys Oceanogr 22:918-926

Fairbanks RG (1989) A 17,000 year glacio-eustatic sea level record : influence of glacial melting rates on the Younger Dryas event and deep ocean circulation. Nature 342:637-642

Fichefet T, Hovine S, Duplessy JC (1994) Thermohaline circulation of the Atlantic Ocean during the last glacial maximum. Nature (in press)

Friedrich H, Levitus S (1972) An approximation equation of state for numerical models of ocean circulation. J Phys Oceanogr 2:514-517

Gordon AL (1986) Interocean exchange of thermocline water. J Geophys Res 91:5037-5046

Gordon AL, Molinelli EJ (1986) Thermohaline and chemical distributions and the atlas data set. In: Southern Ocean atlas. Amerind Pub Co Pvt Ltd, New Delhi

Hellerman S, Rosenstein M (1983) Normal monthly windstress over the World Ocean with error estimates. J Phys Oceanogr 13:1093-1104

Hovine S, Fichefet T (1994) A zonally averaged,three-basin ocean circulation model for climate studies. Climate Dynamics 10:313-331

Kallel N, Labeyrie LD, Juillet-Leclerc A, Duplessy JC (1988) A deep hydrological front between intermediate and deep-water masses in the glacial Indian Ocean. Nature 333:651-655

Kellogg TB (1976) Late Quaternary climatic changes: Evidence from deep sea cores from Norwegian and greenland Seas. In: Cline RM, Hays JD (eds) Investigation of Late Quaternary paleoceanography and paleoclimatology. Geol Soc Am Mem 145:77-110

Kroopnick P (1985) Distribution of ^{13}C and SCO_2 in the world oceans. Deep-Sea Res 32:57-77

Labeyrie LD, Duplessy JC, Blanc PL (1987) Variations in the mode of formation and temperature of oceanic deep waters over the past 125,000 years. Nature 327:477-482

Labeyrie LD., Duplessy JC, Duprat J, Juillet-Leclerc A, Moyes J, Michel E, Kallel N, Shackleton NJ (1992) Changes in the vertical structure of the North Atlantic Ocean between glacial and modern times. Quat Sci Rev 11:401-413

Lautenschlager M, Herterich K (1990) Atmospheric response to ice age conditions: Climatology near the Earth's surface. J Geophys Res 95:22,547-22,55?

Levitus S (1982) Climatological Atlas of the World Ocean, NOAA Prof. Paper 13, U.S. Dept of Commerce, Washington, DC, 173 pp

Melles M (1991) Late Quaternary paleoglaciology and paleoceanography at the continental margin of the southern Weddell Sea. Antarctica Berichte zur Polarforschung, 81, Alfred Wegener Institut für

Polar -und Meeresforschung, Bremerhaven

Oppo DW, Fairbanks RG (1987) Variability in the deep and intermediate water circulation of the Atlantic Ocean during the past 25,000 years: Northern Hemisphere modulation of the Southern Ocean. Earth and Planet Sci Lett 86:1-15

Oppo DW, Lehman SJ (1993) Mid-depth circulation of the subpolar North Atlantic during the last glacial maximum. Science 259:1148-1152

Ruddiman WF, McIntyre A (1973) Time-transgressive deglacial retreat of polar waters from the North Atlantic. Quat Res 3:117-130

Sarnthein M, Winn K, Jung SJA, Duplessy JC, Erlenkeuser H, Ganssen G (1994) Changes in the East Atlantic deep-water circulation over the last 30,000 years: Eight time-slice reconstructions. Paleoceanogr 9:209-267

Shackleton NJ (1974) Attainment of isotopic equilibrium between ocean water and the benthonic foraminifera genus *Uvigerina* : isotopic changes in the ocean during the last glacial. Gif/Yvette, Colloque CNRS 219, Paris

Shackleton NJ, Imbrie J, Hall M (1983) Oxygen and carbon isotope record of east Pacific core V 19-30:

Late Quaternary Surface Circulation of the South Atlantic: The Stable Isotope Record and Implications for Heat Transport and Productivity

G. Wefer[1], W.H. Berger[2], T. Bickert[1], B. Donner[1], G. Fischer[1], S. Kemle-von Mücke[1], G. Meinecke [1], P.J. Müller[1], S. Mulitza[1], H.-S. Niebler[3], J. Pätzold[1], H. Schmidt[1], R.R. Schneider[1], and M. Segl[1]

[1]*Universität Bremen, Fachbereich Geowissenschaften, 28334 Bremen, Germany*
[2]*Scripps Institution of Oceanography, UCSD, La Jolla, California 92093*
[3]*Alfred-Wegener-Institut für Polar- und Meeresforschung, 27515 Bremerhaven, Germany*

Abstract: The central problem of late Quaternary circulation in the South Atlantic is its role in transfer of heat to the North Atlantic, as this modifies amplitude, and perhaps phase, of glacial-interglacial fluctuations. Here we attempt to define the problem and establish ways to attack it. We identify several crucial elements in the dynamics of heat export: (1) warm-water pile-up (and lack thereof) in the western equatorial Atlantic, (2) general spin-up (or spin-down) of central gyre, tied to SE trades, (3) opening and closing of Cape Valve (Agulhas retroflection), (4) deepwater E-W asymmetry. Means for reconstruction are biogeography, stable isotopes, and productivity proxies. Main results concern overall glacial-interglacial contrast (less pile-up, more spin-up, Cape Valve closed, less NADW during glacial time), dominance of precessional signal in tropics, phase shifts in precessional response. To generate working hypotheses about the dynamics of surface water circulation in the South Atlantic we employ Croll's paradigm that glacial - interglacial fluctuations are analogous to seasonal fluctuations. Our general picture for the last 300 kyrs is that, as concerns the South Atlantic, intensity of surface water (heat) transport depends on the strength of the SE trades. From various lines of evidence it appears that stronger SE trades appeared during glacials and cold substages during interglacials, analogous to conditions in southern winter (August).

Introduction

The South Atlantic is unique among major ocean basins in showing a meridional heat transport in the upper ocean from pole to equator; all other basins show poleward transport (Hastenrath 1982). The surface-near currents in the South Atlantic are part of a world-wide circulation system, wherein warm water masses are transported to the northern North Atlantic, across the equator. A portion of the heat thus gained is used for evaporation, resulting in the delivery of highly saline waters to the northern North Atlantic. Upon cooling, these saline water masses sink and mix with deep waters in the Nordic seas. In this fashion, dense North Atlantic Deep Water (NADW) is formed, which flows southward in the deep ocean (Warren 1981),

crossing the equator in the deep western basin of the Atlantic. Eventually, the NADW mixes with Antarctic waters and the mixture is transported by the Circumpolar Current to the other oceans. The South Atlantic, therefore, is a key region for the distribution of deep waters to the other oceans, that is, for the global thermohaline circulation system (Gordon 1986, 1988; Rintoul 1991).

The heat transport from the South Atlantic to the North Atlantic is a crucial factor in today's climate, especially in northern Europe. Major changes in the global circulation system occur during transitions from glacial to interglacial conditions (Boyle and Keigwin 1982; Shackleton et al. 1983), with important implications for heat transport across the

From WEFER G, BERGER WH, SIEDLER G, WEBB DJ (eds), 1996, *The South Atlantic: Present and Past Circulation*. Springer-Verlag Berlin Heidelberg, pp 461-502

equator (Ruddiman and McIntyre 1984). Changes in such heat transport have been invoked as a possible cause for abrupt climatic changes during deglaciation (Broecker et al. 1985). It is assumed that the Atlantic circulation cell is reduced or interrupted during glacial times (Charles and Fairbanks 1990), that is, there is a decrease in the formation of NADW. Increased sea-ice coverage in the NADW source areas may be largely responsible for the decrease. Such sea-ice coverage is greatly enhanced by a reduction of the heat transport from the South Atlantic to the North Atlantic.

We are interested in studying the modification of glacial-interglacial fluctuations (amplitude, phase) in the late Quaternary through changing delivery of tropical heat to the North Atlantic. To attack this problem, we need to reconstruct crucial elements of surface water circulation, which is tied to heat transport.

The first crucial element of the upper circulation system of the South Atlantic is the Walvis Ridge and Benguela Current area. This region monitors properties of the gyre margin and hence spin-up of the central gyre. Also, temperatures in the outer portion of the Benguela Current reflect the Agulhas contribution and hence opening or closing of the Cape Valve.

The second crucial element is the eastern equatorial Atlantic, where warm water is removed, either toward western pile-up, or toward the central gyre if warm water flow turns southward. Thus, we think that the strength of SE trades is recorded in water properties of the eastern equatorial Atlantic, in warm layer thickness and in productivity.

The third crucial element of South Atlantic surface water circulation is the western equatorial Atlantic, where warm water pile-up occurs. The area is centered north of the equator, and therefore available as source of heat for the North Atlantic. In the modern ocean the optimum situation for pile-up is reached late in summer and early in fall, when we have stronger SE trades, and a more northerly Intertropical Convergence Zone (ITCZ) than during the rest of the year (Voituriez 1981).

The crucial element of deep circulation is the east-west asymmetry in deep water properties, which monitors strength or thickness of NADW.

There are other key elements such as gyre center, Brazil Current, position of Subarctic Front etc. but these cannot be considered in detail here.

The aim of this paper is to describe the variability in the surface currents in the South Atlantic during the last ca. 300,000 years and to discuss possible connections to the global ocean circulation and implications for trans-equatorial heat transport. Tradewinds dominate the dynamics of South Atlantic surface circulation. Thus, we can learn from the seasonal variation of the strength of tradewinds and the effects on upwelling in the Benguela system, on gyre spin-up and gyre pile-up, and on the equatorial east-west see-saw of the thermocline.

To get information on wind strength and subsequent thermocline/nutricline shoaling we use measurements related to primary production. For temperature reconstructions we use changes in species abundance of planktonic foraminifera. Thickness and position of warm water layers were deciphered with oxygen isotope gradients found in foraminiferal shells. We present isotope data and faunal data from cores along the south-to-north-flowing trans-South-Atlantic currents from the Benguela Current (Walvis Ridge) via the Equatorial Current systems to the North Brazil Current (Ceara Rise). The results we report were obtained within the framework of doctoral theses at the Geosciences Department of Bremen University (GeoB) (Schneider 1991; Bickert 1992; Meinecke 1992; Schmidt 1992; Mulitza 1994; Kemle-von Mücke 1994) using cores raised by R/V Meteor (Fig. 1). Before we turn to the reconstruction of these histories, we present an elementary overview of the major patterns whose changes we wish to trace through geologic time.

Oceanographic Background

General

In the South Atlantic, as in other major basins, the surface water current system is dominated by a subtropical anticyclonic gyre. The gyre is a result of the general atmospheric circulation in low to temperate latitudes, which is dominated by

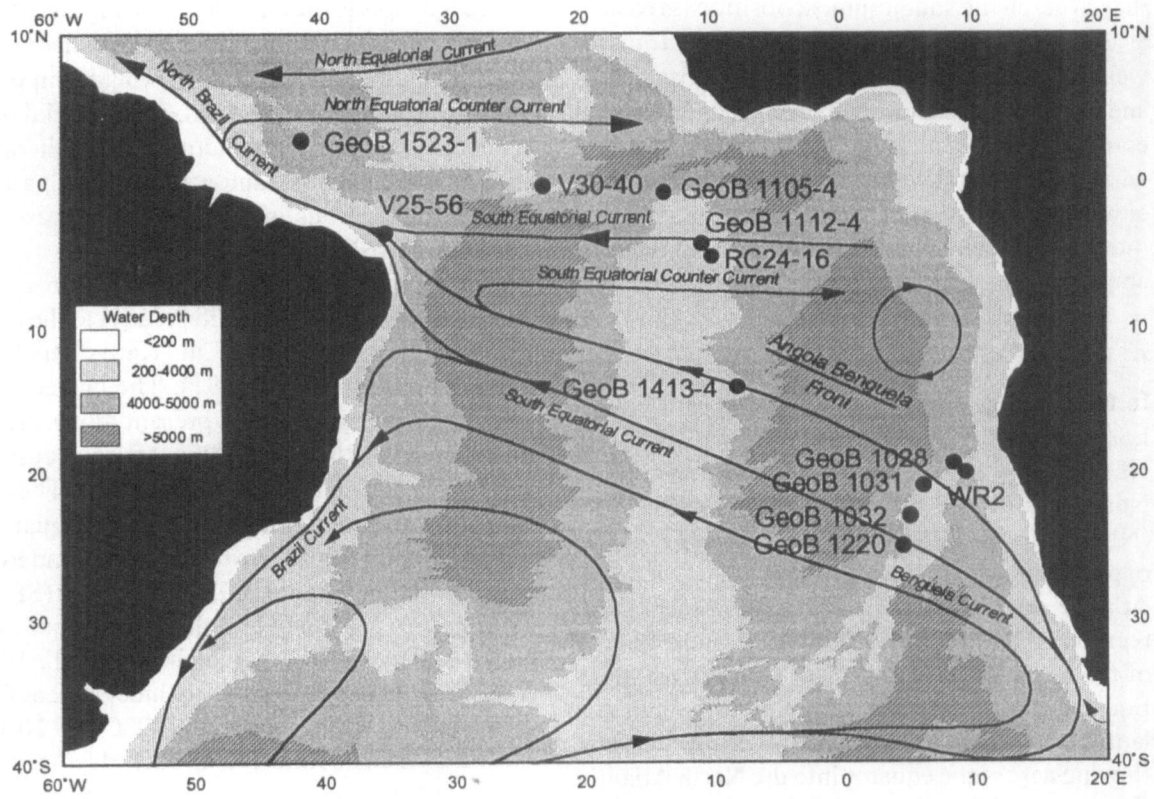

Fig. 1 Major upper-level geostrophic currents and fronts in the South Atlantic (after Peterson and Stramma 1991, modified). Locations of cores and sediment trap WR 2 discussed in this paper shown also.

tradewinds and westwinds. Owing to the Coriolis force, air moving from the subtropical high pressure area to the equatorial low pressure zone is diverted to the west; thus, the SE-trades emerge. Poleward of the gyre strong westwinds promote eastward flow. These two flows, and the shape of the basin, set up the gyre. The center is displaced westward, due to effects from meridional vorticity transfer. During southern winter, the subtropical high pressure area is more strongly developed and situated farther to the northwest. The result is an intensification of SE-trades (which tend to shove heat toward the equator). In opposite seasons, and in opposite hemispheres, trade-winds alternate in their strengths, so that in general strong SE-trades co-occur with weak NE-trades and vice versa. Strong NE-trades, on the whole, tend to „push

back"; obstructing the flow of heat from south to north of the equator.

The two trade wind systems are separated by the intertropical convergence zone (ITCZ), a zone of rising warm and moist air, north of the equator. The importation of warm, moist air masses from south of the equator is a major element in the cross-equatorial heat flow. The area of the ITCZ shows the highest surface water temperatures in the Atlantic (Höflich 1974). Rising moist air masses result in cloud formation and precipitation. The position of the ITCZ changes seasonally with the intensities of the trade-winds. When SE-trades are strong and NE-trades are weak, during boreal summers, the ITCZ has its northernmost position at ca. 15°N, and heat import from the South Atlantic into the North Atlantic is at a maximum. During February/March the southernmost position is reached at 3°N (Servain & Legler 1986); heat transfer weakens. Thus, the seasonal see-saw between NE- and SE trades enhances the background seasonal contrast in the South and North Atlantic. In some-what analogous fashion, in the late Quaternary, enhanced cross-equatorial heatflow during interglacials, and reduced flow during glacials, amplifies the glacial-to-interglacial contrast.

Equatorial currents

In the equatorial area, trade-wind-driven South-East and North-East Equatorial Currents (SEC and NEC) move surface waters westward, while North- and South Equatorial Counter Currents (NECC and SECC) move surface water in the opposite direction (Fig. 1). Relatively warm South Atlantic surface water from the subtropical gyre, reenforced by the waters from the oceanic branch of the Benguela Current (BOC), moves as South Equatorial Current (SEC) through the South Atlantic from south-east to north-west, much of it flowing across the equator into the North Atlantic (Peterson and Stramma 1991; Gordon and Bosley 1991) (Fig 1). The SEC consists of two branches, a wide mainstream flowing near and somewhat south of 10°S, and a smaller, faster flowing branch at 2° to 4°S (Peterson and Stramma 1991). The two branches are separated by the SECC (Reid 1964; Richardson and Walsh 1986).

At about 10°S off the coast of Brasil, the SEC splits into two branches, forming the southward flowing Brazil Current (BC) and the northward moving North Brazil Current (NBC) (or North Brazil Coastal Current, NBCC). The NBC is the stronger one of the two branches (Stramma and Peterson 1990) which (between 3° to 10°N) con-tributes water to the eastward flowing North Equa-torial Counter Current (NECC) (Richardson and Reverdin 1987). The seasonally appearing, east-ward flowing NECC interacts with the northern branch of the SEC, in places. In the mixing area of the two currents, a strong convergence of wa-ter masses results in downwelling of surface wa-ters (Philander and Pacanowski 1986a) with a con-comitant depression of the thermocline. Within the thermocline, water is transported equatorwards,

where it supports the eastward flowing Equatorial Under-Current (EUC) (Fig. 2). For paleoceano-graphic studies, these dynamics provide important indicators of the strength of cross-equatorial flow: A deep depression of the thermocline in the west-ern convergence area, and a strong EUC, are in-dicative of vigorous influx of warm surface wa-ters from the SE, through the SEC.

The EUC is a fast-flowing undercurrent, ex-tending between 5°N and 5°S parallel to the equa-tor in a depth of 50 to 125 m (Katz et al. 1981; Peterson and Stramma 1991). The current runs throughout the year and is present along the en-tire equatorial Atlantic. Off the African coast, the current ceases and feeds surface currents. The east-ward flow of thermocline water in the equatorial region is enhanced by two additional undercur-rents, the South Equatorial Under-Current (SEUC) (Molinari et al. 1981) and the North Equatorial Under-Current (NEUC) (Cochrane et al., 1979) (Fig. 2). Both undercurrents are independent from the overlying surface currents SECC and NECC. The SEUC can be followed in water depths of 150 to 200 m from the coast off Brazil to the northern Angola Basin as a 100 km wide oxygen-rich zone (Molinari et al. 1981).The contact between EUC and SEC at 0° and 2°S forms the equatorial diver-gence zone, where upwelling of colder water masses from the thermocline and from below the thermocline takes place. As a result, surface tem-peratures are reduced in the equatorial upwelling area, in July (Fig. 3). A similar reduction, inter-estingly, is not seen in February (Fig. 3). This sea-sonal contrast flags a crucial region for paleoceanographic study, where sensitivity to cli-matic change is expected to be substantial. The phase of the contrast indicates that *southern* hemispere seasons dominate the eastern equato-rial temperature pattern.

The depth of the thermocline along the equa-tor varies considerably, geographically and sea-sonally. On average, the mixed layer is thick in the west, and thins toward the east, as a result of west-ward flow of warm surface waters (Fig. 4A). This pattern (east-west contrast) covaries with annual changes in wind and current intensities. During boreal spring/austral fall (March/April), wind intensities and zonal components are low at the

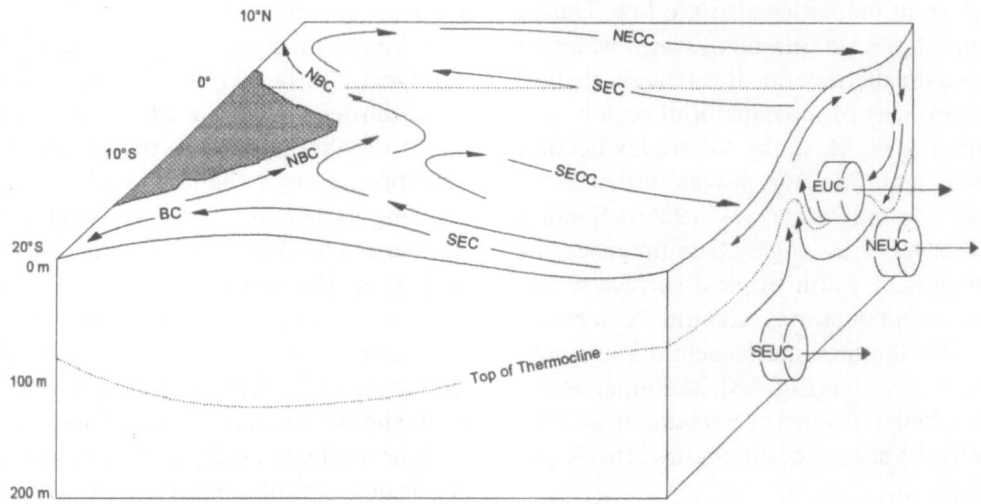

Fig. 2 Schematic of the surface layer in the equatorial Atlantic between 10°N and 20°S. Shown are the surface circulation and the meridional circulation cell of the equatorial divergence zone as well as the position of the top of the thermocline (bottom of mixed layer). NECC, North Equatorial Counter Current; SEC, South Equatorial Current; SECC, South Equatorial Counter Current; BC, Brazil Current; NBC, North Brazil Coastal Current; EUC, Equatorial Under-Current; SEUC, South Equatorial Undercurrent. (Compiled from various sources).

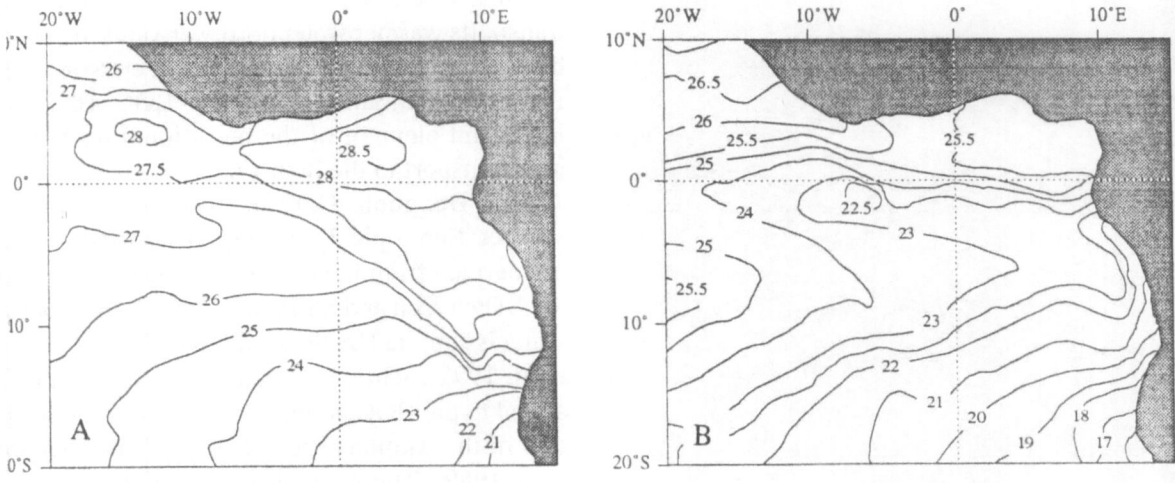

Fig. 3 Surface water temperatures in the equatorial Atlantic for February (A) and July (B) (after Mazeika 1968). Note that the eastern equatorial region is relatively warm in February, that is, temperatures follow seasons of the southern hemisphere.

equator. Current velocities also are low. During this period, there is no pile-up of warm waters in the west, and the thermocline is relatively shallow in the western part of the equatorial region (Fig. 4B). Beginning in May, the SE-trades become more intense and the ITCZ moves further to the north (Philander and Pacanowski 1986b). South of 3° N, the westward flowing SEC commences vigorous transport of warm tropical surface waters into the western equatorial Atlantic. As a result, the depth of the thermocline deepens during boreal summer/austral winter (Fig. 4B). Maximum asymmetry is reached in August or September, and this then must be the season of maximum cross-equatorial transport.

There is a tendency for the thermocline to shallow in the eastern equatorial Atlantic whenever it deepens in the west (Fig. 4B). Equatorial upwelling is most intense when the SE-trades are strongest (Philander and Pacanowski 1986b). Thus, during southern winter (July, August), the top of the thermocline in the eastern equatorial Atlantic rises to a shallow depth of 20 to 30 m, allowing cool thermocline water to mix with warm surface waters (Fig. 3). Generally, the position of the thermocline reflects the strength of SE-trades, which are strongest in August and September, that is, when the sun is moving south across the equator and the ITCZ starts migrating south. This situation should be analogous to onset of a glacial, with the northern hemisphere getting cooler. Early in this onset (as at present) increased cross-equatorial heatflow would be expected, als long as the ITCZ is still well north of the equator. This would slow the plunge into the next ice age (stabilizing the system in its warm phase).

Benguela Current

The eastern limb of the South Atlantic subtropical gyre is formed by the Benguela Current which comprises two different oceanographic regimes: the Benguela coastal current (BCC) and the Benguela oceanic current (BOC) (Figs. 1 and 5). The BCC transports cold water from the wind-dominated coastal area northward, while the BOC transports warmer water northward and westward by geostrophic flow (Stramma and Peterson 1989; Peterson and Stramma 1991). The BOC is the most important element of the meridional northward heat transport in the South Atlantic.

The Benguela Current system has several sources: Subtropical and temperate surface waters are derived from the South Atlantic Current near 40°S (Peterson and Stramma 1991), Indian Ocean water is supplied by the Agulhas Current (Gordon et al. 1992), while subantarctic surface water is added to the BOC at the Subtropical Front northeast of the Agulhas Ridge (38°S, 15°E, Shannon et al. 1989). The warm water inflow from the Indian Ocean is thought to be an important component of the global thermohaline circulation system (Gordon 1986; 1988), now commonly referred to as the „Global Ocean Conveyor" (Broecker and Denton 1989) (although the image of a conveyor

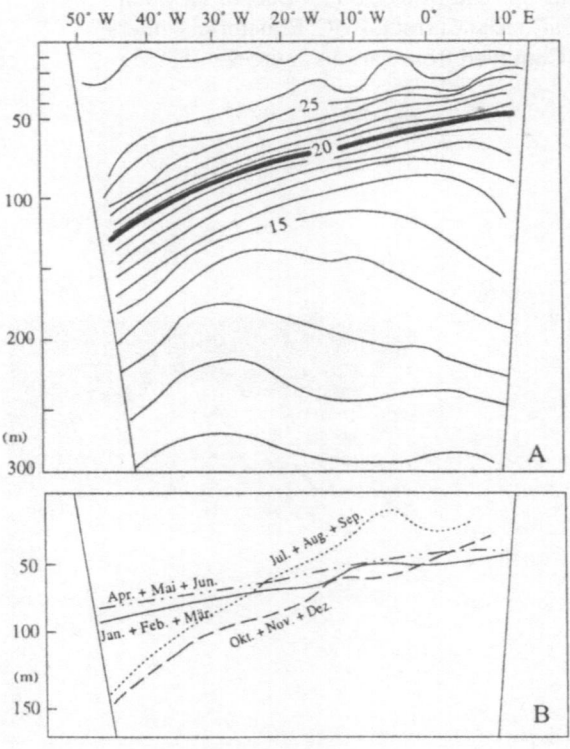

Fig. 4 Isotherms and position of the thermocline (20°C) (A) in the equatorial Atlantic as well as seasonal variation of the thermocline (B). (Modified after Voituriez 1981).

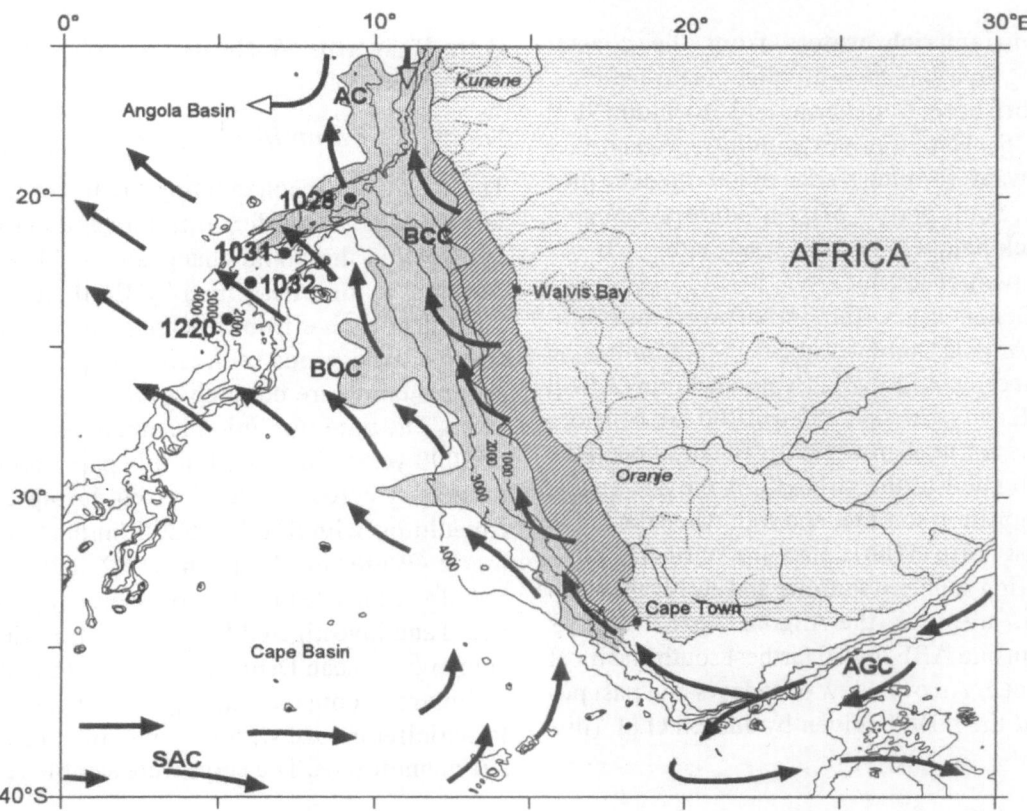

Fig. 5. The major upwelling regions off South Africa are shown, with the area of coastal upwelling (dark pattern) and the extension of the upwelling filaments in the mixing zone between Benguela coastal (BCC) and oceanic currents (BOC) (light pattern), during southern hemisphere winter (August; after Lutjeharms and Stockton 1987, modified). AC, Angola Current; SAC, South Atlantic Current; AGC, Agulhas Current.

belt tends to over-simplify the complex branching of the global thermohaline circulation). It has been suggested that most of the North Atlantic Deep Water (NADW) leaving the Atlantic Ocean is balanced by a warm upper layer return flow from the Indian Ocean (and ultimately the Pacific) around the southern tip of Africa (Gordon 1986; 1988) and that this warm water source turns on and off in tune with the glacial/interglacial fluctuations (Berger et al. 1987; McIntyre et al. 1989). A competing model to the "warm water route" postulates an inflow directly from the Pacific Ocean through the Drake Passage, the so-called "cold water route" (Rintoul 1991).

From the point of view of paleoceanography, the warm-water route is attractive as a valve which is easily opened and closed, by moving the subantarctic regime northward and southward. Thus, a threshold situation can be created for both cold-to-warm and warm-to-cold transitions. Some influence of this route is clearly reflected in the re-seeding of *Globorotalia menardii* in the Atlantic, after regional extinction during glacials (Berger et al. 1985). Thus, the dichotomy of cold-water route versus warm-water route is hardly an either-or proposition.

The upwelling area of the BCC is fed from the thermocline by South Atlantic Central Water (SACW), which originates at the Subtropical-Subantarctic Front by mixing and sinking of subtropical and subantarctic surface water (Lutjeharms and Valentine 1987). Filaments of cold, nutrient-rich waters from the coastal upwelling area extends well offshore (as far as ca. 600 km offshore; Lutjeharms and Stockton 1987) (see Fig. 5). Here it mixes with low-productivity oceanic water forming a zone of intermediate productivity. Some portion of the mixture presumably sinks back below the mixed layer, along a BCC-BOC convergence zone.

The Benguela Coastal Current meets the south-flowing Angola Current at about 15°S, forming the east-west oriented Angola Benguela Front (ABF). The ABF can be traced 250 to 1200 km offshore (Meeuwis & Lutjeharms 1990). The position of the ABF is related to the strength of the SE-trades. During northern summer and fall, when the ITCZ is farthest north ABF is also in a northerly position . During southern summer and early fall, when SE-trades are weak, the Angola Current is at its maximum and ABF is also farthest south and most pronounced. An overview of today's and past positions of the ABF is given by Jansen et al. (this volume).

Data Base and Methods

Samples, foraminiferal species

The means for reconstructing the history of surface circulation are found in biogeographic patterns and in the grand changes in stable isotope patterns, as first described by Emiliani (1955). Such data derive from the analysis of deep-sea cores of calcareous sediments. Typical rates of sedimentation are between 2 and 3 cm per 1000 years. The time resolution is mostly on the order of 5000 years. Data used in this paper are based on gravity cores (Table 1) taken during various expeditions with R/V METEOR in the South Atlantic (Wefer et al., 1988, 1989, 1990, 1992; Schulz et al. 1991). The cores were opened on board and investigated following a modified version of the Ocean Drilling Program protocol. The sediments consist mainly of fine-grained foraminiferal ooze with varying amounts of clay and nannofossils. The cores were sampled at 5 cm

area	core	position	water depth (m)	average sedimentation rate (cm/ky)
Benguela System	GeoB 1028	20°06,2′S, 09°11,1′E	2215	3,5
	GeoB 1031	21°52,8′S, 07°06,1′E	3105	1,5
	GeoB 1032	22°54,9′S, 06°02,2É	2505	1,5
	GeoB 1220	24°02,0 S, 05°18,4 E	2265	1,5
Gyre Margin	GeoB 1413	15°40,8 S, 09°27,3 W	3789	1,07
Equatorial Divergence Zone	GeoB 1105	01°39,9 S, 12°25,7 W	3225	4,8
South Equatorial Current	GeoB 1112	05°46,7 S, 10°45,0 W	3125	3,0
Western Boundary Current	GeoB 1523	03°49,9 N, 41°37,3 W	3292	1,9

Table 1: Cores with isotopic data from the South Atlantic used in this study. Positions of the cores are depicted in Fig. 1. Data available from gwefer@zfn.uni-bremen.de and from NOAA, Boulder, Colorado.

intervals and the samples were wet-sieved at 150 μm and dried at 60°C.

For this paper, cores were selected according to the following criteria: sedimentation should be undisturbed, water depths should be above the present lysocline (Berger 1981). Also, major current regimes should be represented.

The backbone of paleoceanographic reconstructions are the species abundance patterns and stable isotope composition of planktonic foraminifera. Abundances of species depend mainly on temperature and fertility of the surface waters which they inhabit. Foraminifera inhabit the mixed layer and the thermocline; concentration is highest at depths shallower than 100 m and decreases rapidly with increasing depth. Stable isotope composition of the shells contains information on temperature and isotopic composition of the surrounding seawater (which has both global and regional information). When in thermodynamic equilibrium, calcite and seawater differ in their $^{18}O/^{16}O$ ratios and this difference decreases with increasing temperature. In applying this method, difficulties arise from the fact that the isotopic composition of seawater wherein foraminifers grew is unknown for the past. For the glacial-interglacial fluctuations changes in the average oxygen isotope composition of ocean water are near 1.2 ‰ (Berger 1981; Hemleben et al. 1989; Wefer and Berger 1991).

Given that the general pattern of warm surface water distribution is crucial in assessing cross-equatorial heat transfer - how is this pattern reflected on the seafloor? One important signal is simply the distribution of warm-water and cold-water planktonic foraminifera, distributions which were first mapped by Schott (1935), and by many since (Phleger et al. 1953; Berger 1968; Imbrie and Kipp 1971; CLIMAP 1976). Another important set of clues are derived from oxygen isotopes (Emiliani 1954; 1955). It is now well-understood that $\delta^{18}O$ differences between shallow- and deep-living foraminifera record temperature gradients in the surface water, independent of polar ice volume (e.g. Vincent et al. 1985; Whitman and Berger 1992). A basin-wide overview of upper-water temperature gradients, as recorded in foraminifer

shells recovered from the sea floor, demonstrates the overall distributional pattern of the warm-water layer (Fig. 6). Oxygen isotope compositions of different foraminiferal species collected from 194 high quality box- and multicorer surface-sediment samples are contained in this overview. Samples were taken between the tropical Atlantic (10°N) and the Atlantic sector of the Antarctic Circumpolar Current (60°S). The oxygen isotope composition of the shallow-living species *Globigerinoides ruber* and *Globigerinoides sacculifer*, *Globigerina bulloides* and *Neo-globoquadrina pachyderma* clearly reflects the overall decrease in surface water temperature from the tropics to the Polar Front, and locates the maximum gradient near 45°S (a boundary also seen in surface water nutrient distributions). The deeper-dwelling species *Globorotalia truncatulinoides* and *Globorotalia crassaformis* show lowest $\delta^{18}O$-values in the subtropics (15°-35°S), in accordance with higher temperatures of the subsurface waters. Mulitza (1994) calculated an average calcification depth of 250 m for *G. truncatulinoides* and of 400 m for *G. crassaformis*.

It is clear from the nature of the record of temperature gradients that the range of $\delta^{18}O$ values in planktonic foraminifer species, at any one location, contains important information about the thickness of the (warm) mixed layer and therefore about the depth of the thermocline (Fig. 6). Well-separated modes in $\delta^{18}O$ values of shallow- and deep-living species (N of 15°S in Fig. 6) denote a deep thermocline; merging modes denote thermocline shallowing. This principle, reconstructing changes about the depth of the thermo-cline through time, might only work in the middle of the tropics and along the equator, where there is no chance of lateral invasion of a front. For meridional boundary currents with a well-defined front (such as the Benguela Current) a small movement of the front can result in a large change in $\Delta\delta^{18}O$, at any one site.

In choosing planktonic foraminifers for analysis, we considered both abundance and water depth where the shell is precipitated. From earlier work (summaries in Vincent and Berger 1981; Hemleben et al. 1989) and from recent results from

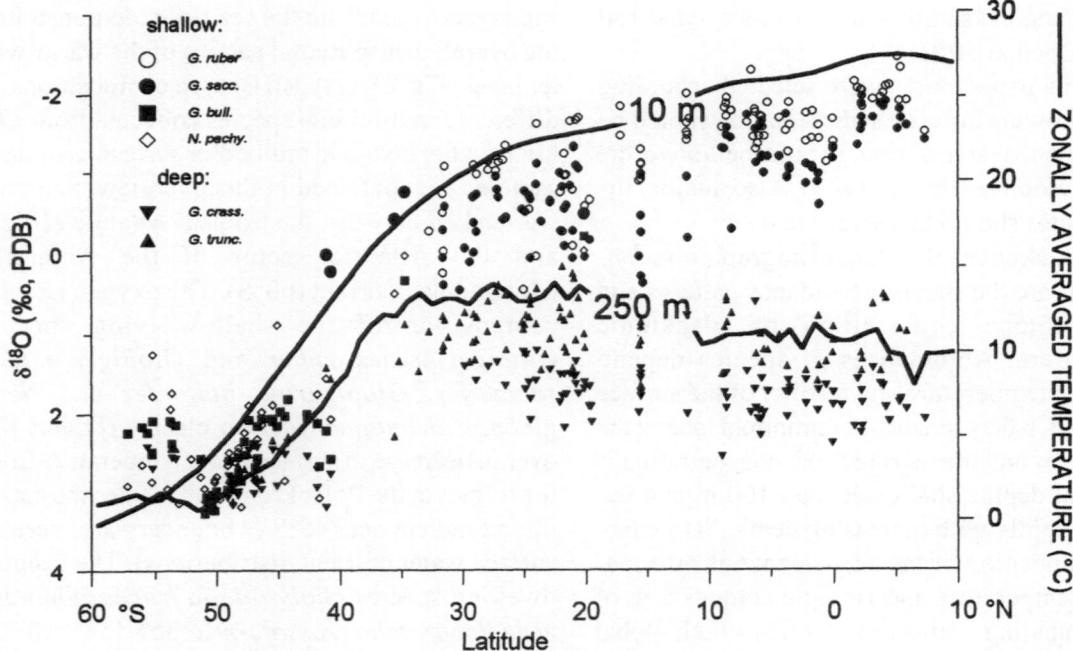

Fig. 6. δ¹⁸O values of planktonic foraminifera with different depth habitats, in surface samples from the South Atlantic. Heavy lines: zonally averaged temperatures (after Levitus 1982) for the surface and 250 m water depth. After Mulitza et al. (manuscript subm.).

plankton tows and long-term sediment-trap experiments during Bremen expeditions (Oberhänsli et al. 1992; Wefer and Fischer 1993; Kemle-von Mücke 1994) we have abundant data on the life habitat of foraminifera.

The following provides a brief overview over the species used in this study and the assumptions concerning their habitat:

Globigerinoides ruber (white)

G. ruber (wh.) spends most of its lifetime within the uppermost 25 - 35 m of the water column (see Hemleben et al. 1989, for a review). Kemle-von Mücke (1994) found the maximum of plasma-containing tests within the uppermost 50 m, in her study of plankton net samples. Ganssen (1983) found a good correlation between δ¹⁸O-values in the tests of this species and summer surface-water temperatures, in the upwelling area off northwest Africa. Therefore, this species can be taken

as representative of surface-water temperatures during the warm season.

Globorotalia inflata

G. inflata is typical for transitional waters. Its depth range is not well known (Hemleben et al., 1989); however, oxygen isotope data indicate that this species produces most of its calcite in the surface mixed layer (Fairbanks et al. 1982). According to Ganssen (1983) and Deuser and Ross (1989), there is good agreement between δ¹⁸O - derived temperatures and those measured during the winter.

Globorotalia truncatulinoides

G. truncatulinoides is a deep-living species whose range extends through the upper 1000 m of the water column. It is found in abundance where there is seasonal deep mixing (Lohmann 1992).

Lohmann and Schweizer (1990) suggest that reproduction takes place near 600 m depth. A crust is formed at depth, in this species. Hemleben et al. (1985), working near Bermuda, estimated that crust formation must occur in waters colder than 10°C.

Globorotalia crassaformis

In the equatorial Atlantic, *G. crassaformis* prefers water depths between 100 and 300 m (Kemle-von Mücke 1994); that is, the level of the oxygen minimum zone. ^{18}O-values of foraminifera collected in the study area with a multiple opening-closing net showed that *G. crassaformis* precipitates most of its carbonate in a water depth of about 300 m. Jones (1967) earlier reported negative correlation between *G. crassaformis* abundance and oxygen content. Generally, in the tropical ocean, oxygen minimum zones occur where the thermocline is shallow. Thus, this deeper living species, where abundant, points to a shallow thermocline with a strong oxygen minimum.

According to our own investigations and data of others (especially Fairbanks et al. 1982; Lohmann 1992; Ravelo and Fairbanks 1992), many reconstructions of surface water conditions in the late Quaternary that are available in the literature are, in fact, based on species of planktonic foraminifera with different depth habitats. Such reconstruction implies invariant relationships between surface and subsurface conditions. To avoid this assumption, we use *G. ruber* (white and pink) as recorders of the uppermost layer. The species *G. inflata* is used as proxy carrier for conditions in the upper thermocline. For conditions in the lower thermocline, we use *G. crassaformis* and *G. truncatulinoides* (left).

Isotope determination and ice effect

Our methods at the Geoscience Department of the University of Bremen (GeoB) for determination of isotope composition differ little from those of other laboratories. Depending on size, 4 to 30 tests of foraminifera are picked from any one sample. Only tests that are undamaged are selected. The sample size for isotopic measurements is 30-80 µg carbonate. The carbonate is reacted with orthosphoric acid at 75°C. Isotope ratios are measured using a Finnigan MAT 251 Micromass Spetrometer with a Kiel Automated Carbonate Device. Precision is regularly checked with an internal carbonate standard (Solnhofen Limestone); over a one-year period (in 1990) the standard error (1) was <0.07‰ for δ^{18}O, and <0.05‰ for δ^{13}C. Conversion to the international PDB scale is performed using NBS standards NBS 18, 19 and 20.

Paleotemperatures based on δ^{18}O -values are calculated using the original paleotemperature equation of Epstein et al. (1953):

$$T_w \, (°C) = 16.5 - 4.3 \, (\delta O_c - \delta O_w) + 0.14 \, (\delta O_c - \delta O_w)^2$$

where T_w(°C) is water temperature at the time of calcite precipitation, δO_c (‰PDB) is the δ^{18}O of the calcite measured and δO_w (‰ PDB) is the δ^{18}O of the ocean water. δO_w (‰) is related to Standard Mean Ocean Water (SMOW) and can be calculated from δO_w-salinity-relationships; 0.22‰ were subtracted for conversion to PDB (Craig 1961).

At several locations in the tropical Atlantic (0° to 30°S), the δ^{18}O of the water was measured. The results are in accordance with the δ^{18}O-salinity-relationship of Craig and Gordon (1965) and Duplessy et al. (1991).

In order to obtain the δ^{18}O-record for temperature, the ice-volume effect on δO_w was estimated using the stacked record of Prell et al. (1986), which was then subtracted from a given isotopic record, as needed. The ice-volume effect was calculated by expressing the Prell record in terms of changes in sea level (in m), with the last glacial/interglacial sea level rise taken to be 120 m (Fairbanks, 1989), and assuming that the relationship between changes in the oxygen isotopic composition of sea water and sea level change is 0.1‰ per 10 m (Chappell and Shackleton 1986). The corresponding glacial/interglacial change in δ^{18}O is 1.2 ‰, in good agreement with earlier estimates (1.2‰, Berger and Gardner 1975; 1.1‰, Labeyrie et al. 1987; 1.05 ‰, Duplessy et al. 1988; 1.3 ‰, Zahn and Mix 1991). The residuals - after adjustment for the ice effect - contain information about variations of temperature as well as of evaporation and precipitation patterns. We assume that the paleotemperature effect is entirely dominant, so that the other effects can be neglected. Since these paleotemperature calculations are based on several assumptions, they must be interpreted with caution. Results for temperature *gradients* will be more reliable than results for absolute values, because they do not depend on the ice-effect correction. (This correction may be in error by as much as 0.2‰, in any 10,000 yr interval, which corresponds to 1°C.)

Strategy of presentation

We make use of multiple species information from upper water masses in an attempt to reconstruct mixed layer and thermocline properties. The $\Delta\delta^{18}O$-between foraminifera living in different water depths reflects differences in upper water column, i.e., stratification (see Fig. 6). This approach allows us to reconstruct thickness of mixed layer (and thereby productivity) and difference in mixing intensity and thermocline depth. Our working hypothesis is that glacial-interglacial changes are more or less analogous to seasonal changes.

Regarding oxygen isotopes, foraminiferal shells tend to be precipitated close to equilibrium with the surrounding seawater (Wefer and Berger 1991). In contrast, the carbon isotope signals in planktonic foraminifera are difficult to interpret, because they reflect a mix of (1) exchange of the ocean's carbon reservoir with atmosphere, biosphere, soil, and sediments, (2) changes in surface water productivity, (3) shifts in water-mass structure and circulation, and (4) organism-specific fractionation effects due to changes in microhabitat and/or ontogenic fractionation ("vital effects") (Berger and Vincent 1986).

For each site, for a time span of 300 kyr or less, we present $\delta^{18}O$ and $\delta^{13}C$ time series of shallow and deep living foraminifera.$\Delta\delta^{18}O$- and $\Delta\delta^{13}C$-values provide information on strength of thermocline, oxygen minimum and thickness of mixed layer. We describe long-term trends, glacial-postglacial changes and relationships to precessional forcing. We start with the cores from the Benguela area and

subsequently compare Benguela dynamics with records from eastern and western equatorial areas and from the gyre region.

Benguela System

As we have emphasized in the introduction, a central theme of South Atlantic paleoceanography is the equatorward transport of heat and the loss of much of this heat to the North Atlantic. To attempt to understand this complex phenomenon on a glacial-interglacial time scale, we need to reconstruct equatorward flow of surface waters. We start with cores from the Benguela system, which is influenced by tradewind-driven upwelling and by warm water influx from the Agulhas retroflection. Our expectation is that upwelling is strongest during glacial times, especially during onset of a glacial period. Strong trade winds move the Angola-Benguela Front northward (Jansen et al., this volume).

The four GeoB cores used to reconstruct Late Quaternary conditions in the Benguela system lie on an east-west transect on Walvis Ridge (Fig. 5, Table 1). Site GeoB 1028 is located near the continental margin (about 200 km off the coast) in the mixing area between the Benguela Oceanic and Coastal Currents (BOC/BCC). Sites GeoB 1031, GeoB 1032 and GeoB 1220 lie under the Benguela oceanic regime with decreasing influence of upwelling, respectively. Temperature, salinity and oxygen for the uppermost 1000 m of the water column of stations GeoB 1028 and GeoB 1032 are shown in Fig. 7. Temperature and salinity show a similar trend between surface and 1000 m of water depth. Temperature decreases from 23°C at GeoB 1032 (Fig. 7A) or 19°C at GeoB 1028 (Fig. 7B) to about 5°C at 600 m water depth. Both stations show the core of Antarctic Intermediate water (AAIW) in about 800 m water depth, as a salinity minimum. Oxygen profiles differ markedly. At GeoB 1032 oxygen values are high throughout the uppermost 1000 m while at the site nearer to the coast (GeoB 1028) a distinct oxygen minimum layer is present, centered at a depth of 300 m.

We first turn to the detailed multi-species and multi-size record of GeoB 1032, an open-ocean site located on the Walvis Ridge, near 23°S

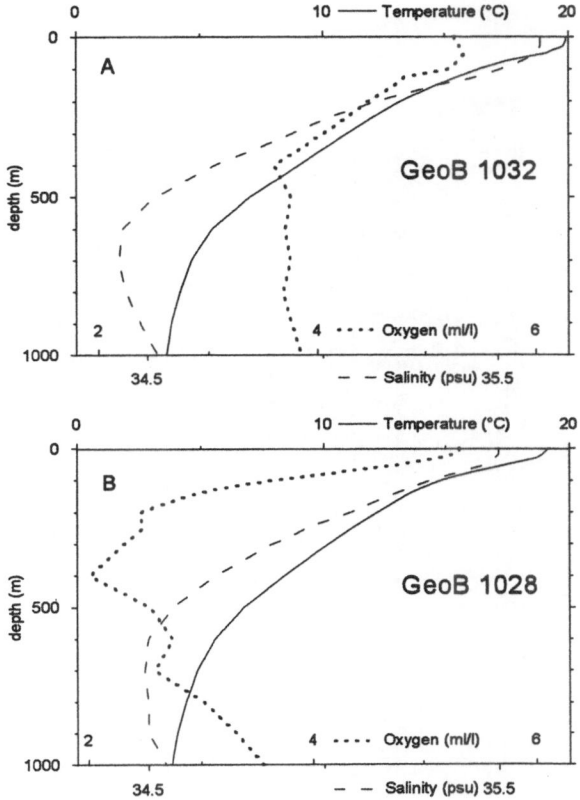

Fig. 7 Temperature, salinity and oxygen content from Stations GeoB 1032 (A), and GeoB 1028 (B) and Walvis Ridge. The records of stations GeoB 1031 and GeoB 1220 are similar to that of Site GeoB 1032. Interpolated from Levitus (1982).

(Fig. 8). To capture warm-water surface-water conditions we use *G. ruber* (white); for winter surface conditions *G. inflata,* and for subsurface conditions *G. truncatulinoides* (Fig. 8A).

The quality of the record appears to be excellent. The pattern closely resembles that of the stacked record of Prell et al. (1986), suggesting that the core contains a complete section free of turbidites or hiatuses. The record spans the last 280 kyr. The record of *G. ruber* (wh.) shows the familiar isotopic signature of the Late Quaternary (Emiliani 1955), with stages 1 through 8. On the whole, the records of warm-water (*G. ruber*), cold-water (*G. inflata*) and thermocline-dwelling (*G.*

truncatulinoides) forms are parallel, but there are interesting discrepancies, especially in the the vicinity of the major climate transitions. Three major terminations (8 to 7, 6 to 5, 2 to 1) are clearly recognized. It is at these terminations where complications arise.

In general, $\delta^{18}O$ differences between *G. inflata* and *G. ruber* (expressed as "seasonal contrast" in Fig. 8A) vary between 0.2 and 1.7‰; they are larger during interglacials, especially during warm events, e.g. substages 1.1, 5.5, 7.3 and 7.5. Likewise, $\Delta\delta^{18}O$ differences between *G. truncatulinoides* and *G. ruber* (expressed as "temperature-gradient" in Fig. 8A) are larger during warm intervals, e.g. during stages 1, 3, 5 and 7. The large range of variations in the difference between *G. truncatulinoides* and *G. ruber* owes much to extreme excursions during times of deglaciation, when the difference tends to go to zero. No such effect is seen in the difference between *G. inflata* and *G. ruber.*

Estimates of the history of seasonal and vertical temperature distribution can be obtained by comparing warm- and cold-water taxa and by comparing warm-water and thermocline forms as well as by reconstructing the isotopic record within different size classes of the same foraminiferal species, if a deep-living taxon such as *G. truncatulinoides* (left) is used. Lohmann and Schweitzer (1990) showed, in their North Atlantic study, that this species starts its growth in the surface waters and continues growing during descent to deeper (colder) water masses. In agreement with these observations, the $\delta^{18}O$ curves of GeoB 1032-3 showed heavier values with increasing test size (Fig. 8B).

Small tests of *G. truncatulinoides* (150-212 μm) have lower $\delta^{18}O$-values than large ones (425-500 μm), by more than 1 ‰ in places. Generally, values of the three size classes studied are similar to the data of the three species precipitating calcite in different water depths (Figs. 8A,B). Values of size class 150 - 212 μm are comparable to *G. ruber* data showing the shallowest calcite-precipitation depth. However, a difference persists, showing that the small forms of *G. truncatulinoides* precipitate in waters colder than those for *G. ruber.*

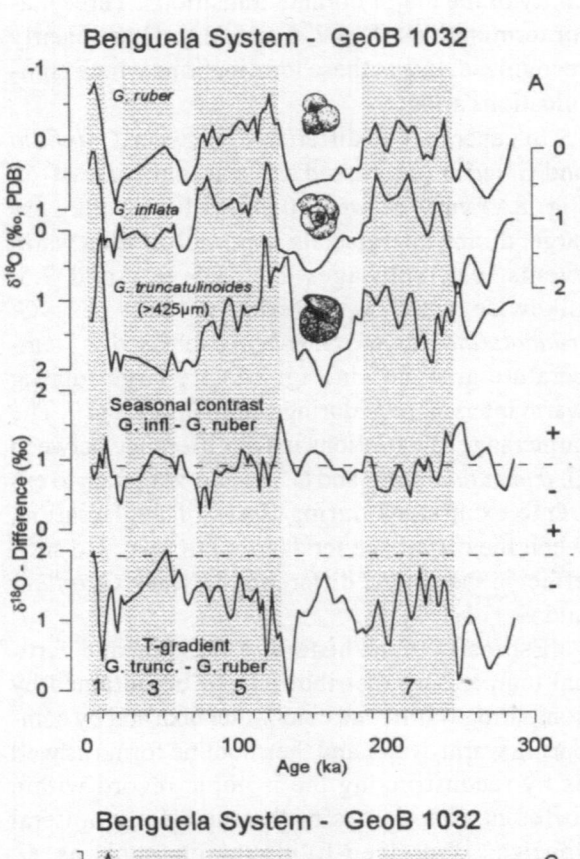

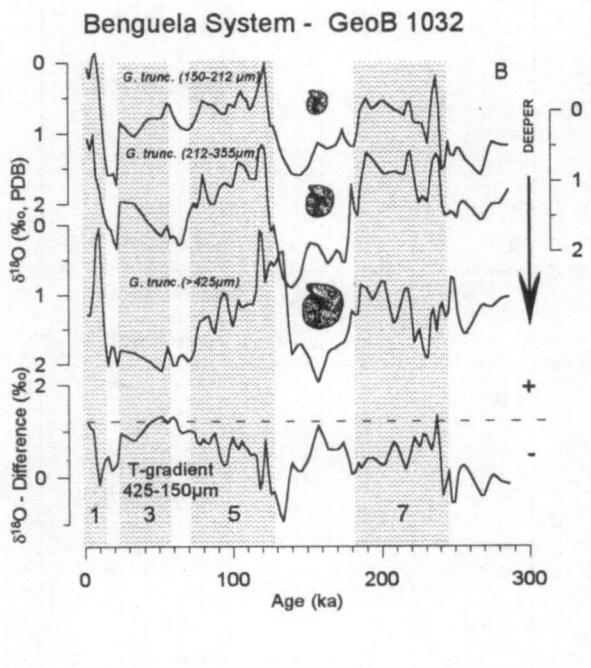

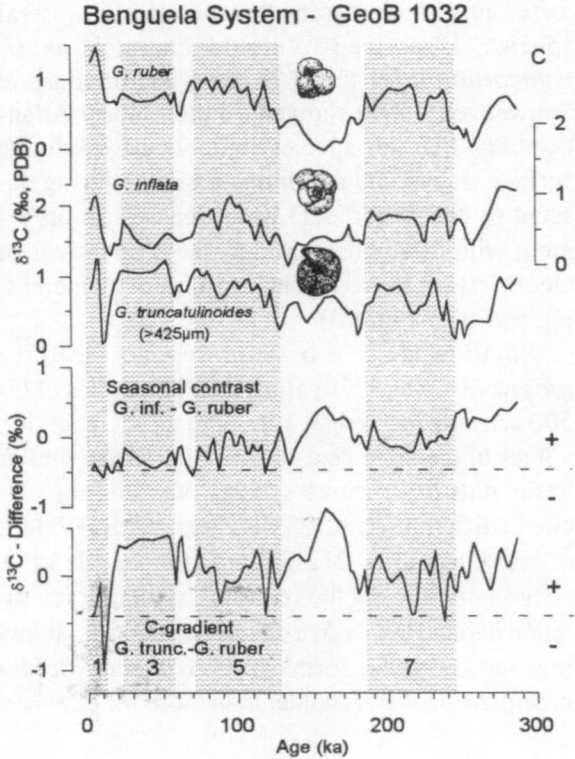

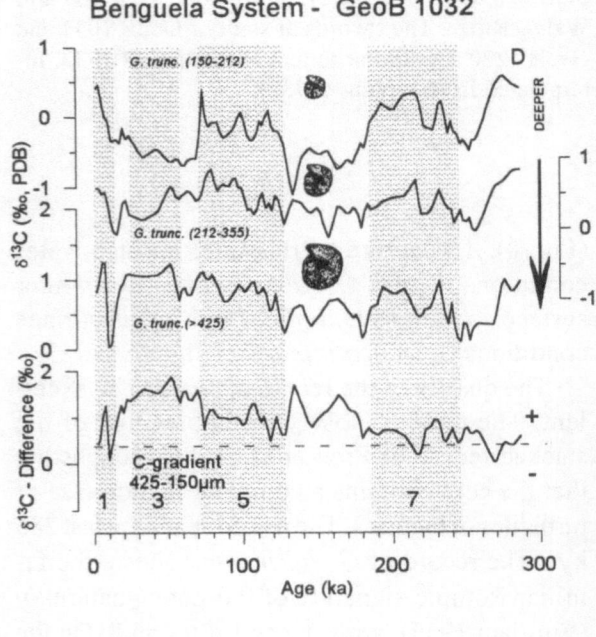

The values of the size class 212 - 355 μm are similar to those of *G. inflata*, but also follow those for the large *G. truncatulinoides*.

Differences between the largest (425 - 500 μm) and smallest (150 - 212 μm) size class of *G. truncatulinoides* and between *G. truncatulinoides* (425 - 500 μm) and *G. ruber* are parallel (Figs. 8A and B, bottom); they may be taken as a recorder of the temperature structure of the surface waters. A decrease in the vertical gradient, as reflected in these difference curves, presumably records a thickening of the warm surface layer. This would result in *G. truncatulinoides* (425 - 500 μm) picking up more negative $\delta^{18}O$ values, assuming its depth habitat stays the same.

Although the pattern of differences between *G. inflata* and *G. ruber* is similar to that seen in the multi-size-class record of *G. truncatulinoides*, the proxies are by no means identical (Figs. 8A and B). In particular, when *G. truncatulinoides* are used to define warm-to-cold differences, the range

Fig. 8. Stable isotope stratigraphy of planktonic foraminifers in core GeoB 1032, oceanic Benguela Current, Walvis Ridge.

A, $\delta^{18}O$ values of *G. ruber* (wh.), *G. inflata* and *G. truncatulinoides* (left) (425-500 μm); $\delta^{18}O$ differences between *G. inflata* and *G. ruber*, and between *G. truncatulinoides* (425-500 μm) and *G. ruber*.

B, $\delta^{18}O$ curves of different size classes (150-212 μm, 212-355 μm, 425-500 μm) of the deep-living planktonic species *G. truncatulinoides* (left); $\delta^{18}O$ differences between two size classes of *G. truncatulinoides* (left) (425 - 500 μm and 150 - 212 μm).

C, $\delta^{13}C$ values of *G. ruber*, *G. inflata* and *G. truncatulinoides* (left) (425-500 μm); $\delta^{13}C$ differences between *G. inflata* and *G. ruber*, and between *G. truncatulinoides* (425-500 μm) and *G. ruber*.

D, $\delta^{13}C$ curves of different size classes (150-212 μm, 212-355 μm, 425-500 μm) of the deep-living planktonic species *G. truncatulinoides* (left); $\delta^{13}C$ differences between two size classes of *G. truncatulinoides* (left) (425 - 500 μm and 150 - 212 μm).

is much larger than when *G. inflata* is employed for the same purpose. Two possibilities arise: (1) The vertical gradient (which presumably reflects upwelling activity) is largely decoupled from the history of seasonality (as reflected by the *G. inflata* - *G. ruber* proxy); (2) *G. inflata* adjusts its season or depth of growth to reduce the range of differences. In either case, this raises problems for a simple interpretation of the difference between $\delta^{18}O$-values of the shallow-living *G. ruber* (wh.) and *G. inflata* in terms of summer-winter differences or temperature differences between surface waters and thermocline waters.

Taken at face value, the $\delta^{18}O$ values for *G. truncatulinoides* (425-500 μm) minus shallow-water species or size classes (Figs. 8A and B) suggest that the mixed layer thickened greatly during early deglaciation, that is, the thermocline descended. The descent was brief and soon reversed. A marked slowdown of the windfield, or possibly a direct effect from a sharp increase in meltwater component within the mixed layer, could conceivably produce such an effect. The implication would be, in either case, that the Benguela current was weak during deglaciation. This would be unfavorable to any hypothesis deriving heat for the North Atlantic (for warming and melting of ice) from Benguela-related transport of surface waters toward the equator.

In Fig. 8C, $\delta^{13}C$ values of *G. ruber*, *G. inflata* and *G. truncatulinoides* (425-500 μm) are shown together with $\Delta \delta^{13}C$ values between *G. inflata* and *G. ruber*, and *G. truncatulinoides* and *G. ruber*. On the whole, the records of the three species are parallel. In general, all three curves show higher values during interglacials compared to glacial values. Differences between glacial and interglacial periods are larger than the global shift of the $\delta^{13}C$-value in ΣCO_2 of seawater between the last glacial maximum and today, which is about 0.4 ‰ (Bickert and Wefer, this volume). There is an interesting trend toward higher values from past to present in the *G. ruber* and *G. truncatulinoides* curves, but not in the *G. inflata* curve. A trend is also seen in one of the $\Delta \delta^{13}C$ curves: The difference between *G. inflata* and *G. ruber* shows a trend to more negative values with time, from about 0.4 to almost -0.5 ‰. The $\Delta \delta^{13}C$ curve of *G.*

truncatulinoides-G. ruber shows no long-term trend, but strong variability which is related to the glacial/interglacial cycles in the 100-kyr band. The sense of the variations suggests stronger oxygen minima during interglacial conditions (stages 1, 5, 7) than during glacial periods (2, 3, 4, 6). It is intriguing that shallow warm-water and deep-living species show similar trends, both changing δ¹³C to more positive values, while the shallow cold water species *G. inflata* does not display this trend.

The trend to higher values seen in the δ¹³C curve of *G. ruber* (Fig. 8C) is not observed in the curve of the shallow-living *G. truncatulinoides* (Fig. 8D). In general, values of *G. truncatulinoides* (150-212 μm) are lighter compared to *G. ruber*, showing no trend to higher or lower values with time. With increasing test size δ¹³C values of *G. truncatulinoides* become greater, perhaps due to decreasing vital effects with age (size) (Berger et al. 1978). The reason for the difference between the *G. ruber* and *G. truncatulinoides* curves might be that the small forms of *G. truncatulinoides* precipitate in waters colder than those for *G. ruber* (see oxygen isotopes, Figs.8A and 8B).

What can we learn from these records concerning the intensity of upwelling in a glacial-interglacial cycle? Not much, from these data alone. The problem is that both the cooling of surface waters and the warming of deep waters would be reflected in a decrease of Δδ¹⁸O, for example. Thus this Δ proxy is not clearly tied to a unique physical process, and needs to be supplemented by other observations. A similar argument can be made for the Δδ¹³C proxy, in this complex setting, involving the interaction of coastal and pelagic systems, and strong and varying oxygen minima.

Turning to the other records from the Walvis Ridge (GeoB 1028, 1031, 1220), we note somewhat similar patterns as for GeoB 1032 (Fig. 9). Only data from *G. ruber* and *G. inflata* are available for these cores. Perhaps the most striking similarity (surprisingly) is between GeoB 1032 and GeoB 1028, the site nearest to the coast. However, in GeoB 1028, Δδ¹⁸O-values (*G. inflata* - *G. ruber* (wh.)) show a larger variability (Fig. 9A), especially at higher frequencies compared to the records in the oceanic cores GeoB 1032 (Fig. 8A), GeoB 1031, and GeoB 1220 (Figs. 9C and 9E). In

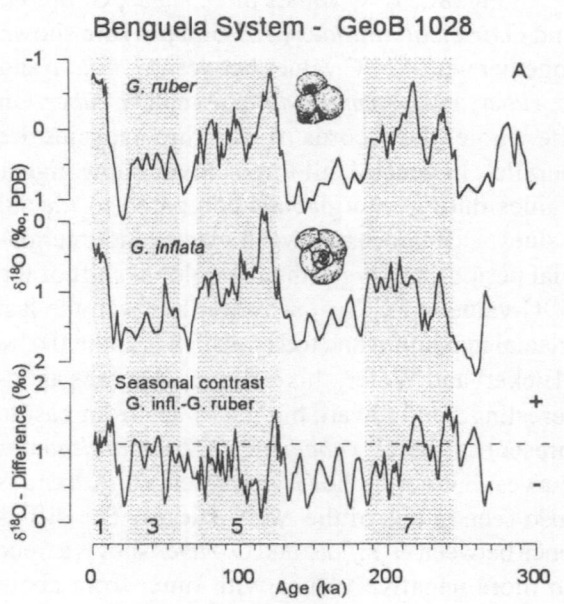

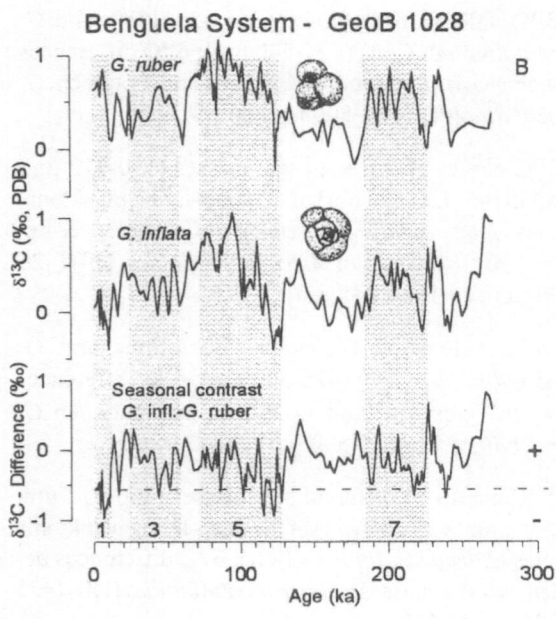

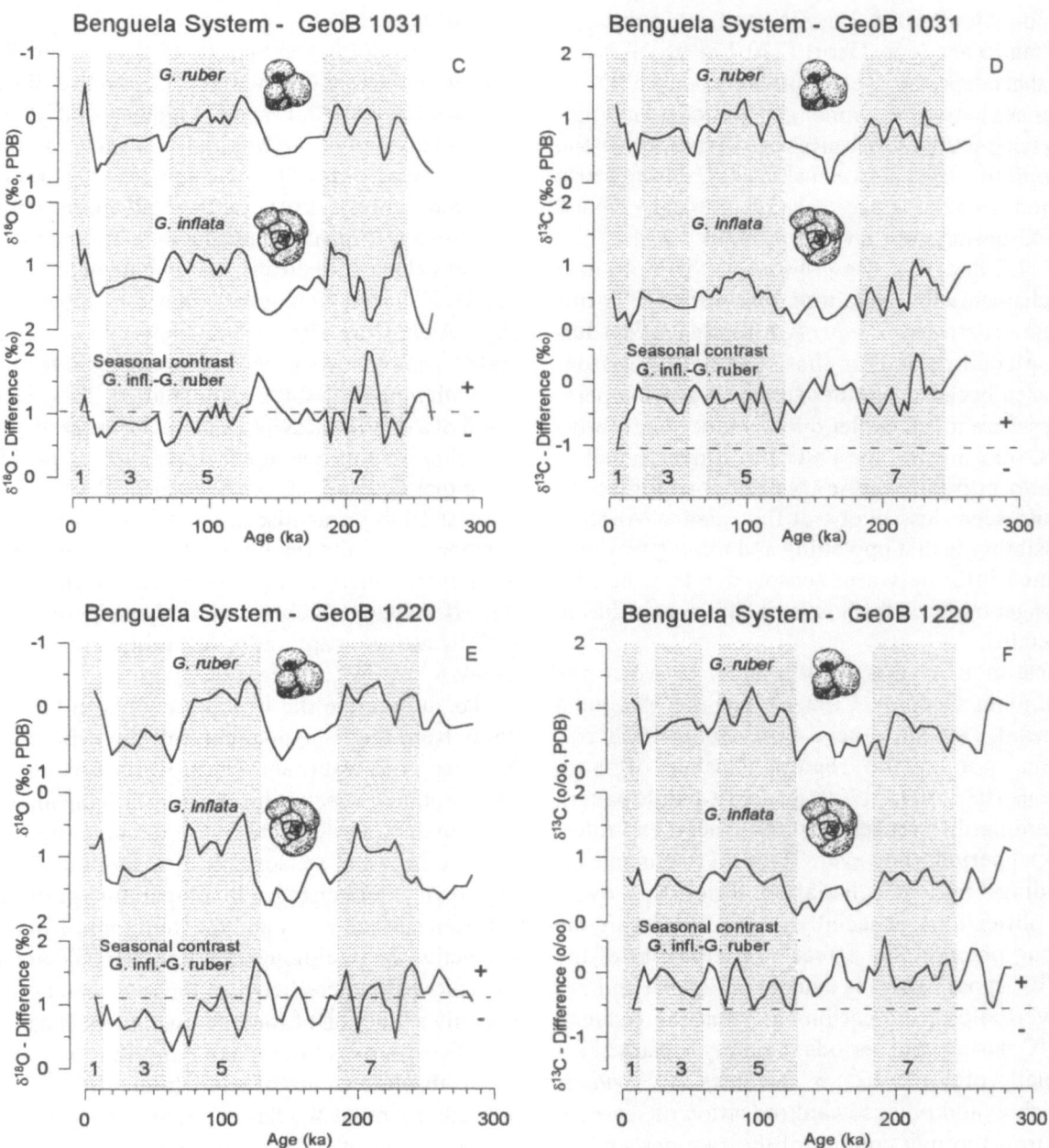

Fig. 9. Oxygen isotope records of *G. ruber* (wh.) and *G. inflata* in GeoB Cores 1028 (Fig. 9A), 1031 (Fig. 9C) and 1220 (Fig. 9E); carbon isotope records of *G. ruber* (wh.) and *G. inflata* in GeoB cores 1028 (Fig. 9B), 1031 (Fig. 9D), and 1220 (Fig. 9F) (from the Walvis Ridge. The δ18O and δ13C differences between the two species analyzed are also shown.

large part, this must be attributed to the higher sedimentation rate (and hence better resolution) in GeoB 1028.

There is some indication that δ^{18}O-differences tend to be smaller during glacial periods, e.g. Stage 2 (Core GeoB 1028, Fig. 9A) and cold substages of Stage 5 or 6 (core GeoB 1220, Fig. 9E). Assuming that evaporation-precipitation related δ^{18}O differences between summer and winter are negligible, temperature differences between G. ruber and G. inflata varied between about 6°C during warm periods (e.g. substages 1.1, 5.5, 7.3 and 7.5) and 1-2°C during cool ones (substages 2.2, 4.2, 5.4, 6.4, 7.2 and 7.4; Figs. 8 and 9). The obvious mechanisms for explaining diminished δ^{18}O-differences between G. ruber (wh.) and G. inflata during glacial periods is that seasonality was decreased because summer surface water temperatures were much cooler due to reduced advective BOC (closing of Cape Valve). The inverse, that G. inflata grew in warmer waters during glacials would seem less likely, at first glance. Another possibility is that upwelling and mixing was prolonged into the warm season due to generally stronger trades in the Southern hemisphere during glacials.

Neither in the mixing area between the Benguela Oceanic Current and the Benguela Coastal Current, where Core GeoB 1028 was taken, nor in the regions further offshore (GeoB 1032, 1031, 1220) is there a simple pattern regarding differences between glacial and interglacial periods, however. Thus, all statements regarding "typical" glacial conditions are oversimplifications, especially when considering the strong precessional effect on periodicity of the $\Delta\delta^{18}$O signal, which is evident in Fig. 9A and 9E. Nevertheless, indications are that the reduced $\Delta\delta^{18}$O during cold periods is mainly a result of especially high ("cold") δ^{18}O-values of G. ruber, which would point toward reduction of seasonal contrast through cooling of the summer surface layer, and hence (presumably) increased upwelling.

Two records from the Benguela oceanic regime (GeoB 1031 and GeoB 1220, Figs. 9C, E) have low average δ^{18}O-differences (G. inflata - G. ruber) during the last 100 kyr. The magnitude of the off-set is near 0.2‰, that is during the last 100 kyr there has been a decrease in the temperature difference by about 1°C between the habitats of the two species (warm water vs. cold surface water and thermocline water), assuming constant salinity difference.

Carbon isotope values of G. ruber and G. inflata from Cores GeoB 1028, 1031 and 1220 are depicted in Figs. 9B, D and F, together with $\Delta\delta^{13}$C values between G. inflata and G. ruber. On the whole, the records of the two species are parallel in all three cores. As in Core 1032, all curves show higher values during interglacials compared to glacial values. The differences in glacial to interglacial δ^{13}C values are greater than the global δ^{13}C shift. As in Core 1032, $\Delta\delta^{13}$C between G. inflata and G. ruber shows a trend to more negative values with time in all three cores studied. This is the result of a trend to heavier values of G. ruber, while G. inflata retains average isotope ratios. Considering that G. ruber shows heavier δ^{18}O values in the last 100 kyr, as discussed, this trend toward heavier δ^{13}C is not readily explained. It may hint at an overall increase in nutrients (pre-formed) in upwelled water which would imply an increase of δ^{13}C in surface wates, after full utilization of nutrients.

To summarize the isotope records in GeoB-cores from the Benguela current, the difference between cool-water and warm-water species G. inflata and G. ruber (Figs. 8 and 9) is considered to be mainly an estimate of the temperature difference between seasons. It may contain to an unknown extent changes in temperature gradient between surface water and the thermocline in the respective areas, although it is understood that the isotopic differences between the two species are not only a function of water temperatures, but also of δ^{18}O (salinity) of the water. Amplitudes as well as the absolute values of temperature differences are rather similar for the two regimes, Benguela Oceanic Current (BOC, Cores GeoB 1220, 1031 and 1032) and the mixing zone of Benguela Oceanic and Coastal Currents (BOC/BCC Core GeoB 1028). There is a long-term trend in the records of Cores GeoB 1220 and 1031 showing a decrease in δ^{18}O differences during the last 100 kyr. The trends in this difference can be accounted for by assuming that the environment of *G. ruber* cooled more than the habitat of *G. inflata*. Parallel to this δ^{18}O trend G. ruber changes to heavier δ^{13}C values, while *G. inflata* remains more or less unchanged. This implies long-term changes in the contribution of cold surface water to the Benguela Current system, presumably largely from filaments derived from coastal upwelling.

The meaning of the decrease in the seasonality contrast (δ^{18}O difference between *G. inflata* and *G. ruber*) during the last 100 kyr (Figs. 9C, E) becomes clearer when we compare it with a long-term change in the coiling ratios of *Neogloboquadrina pachyderma* that is, in the relative abundance of left-coiling *N. pachyderma*

(sinistral). In the cores for which data are available, there is a general increase in the abundance of the left-coiling (cold-water) variety, beginning about 170 kyr ago (Fig. 10). In the South Atlantic, the *N. pachyderma* population north of the Antarctic Polar Front is dominated by the right-coiled variety. South of this front left-coiled *N. pachyderma* contributes more than 50 % to the total *N. pachyderma* population, and in higher southern latitudes even up to 100 % (Bé 1977). High proportions of left-coiled *N. pachyderma* populations are also found in the upwelling area off Namibia (Giraudeau 1993; Ufkes and Zachariasse 1993). The left-coiling variety of *N. pachyderma* presumably traces surface water in-

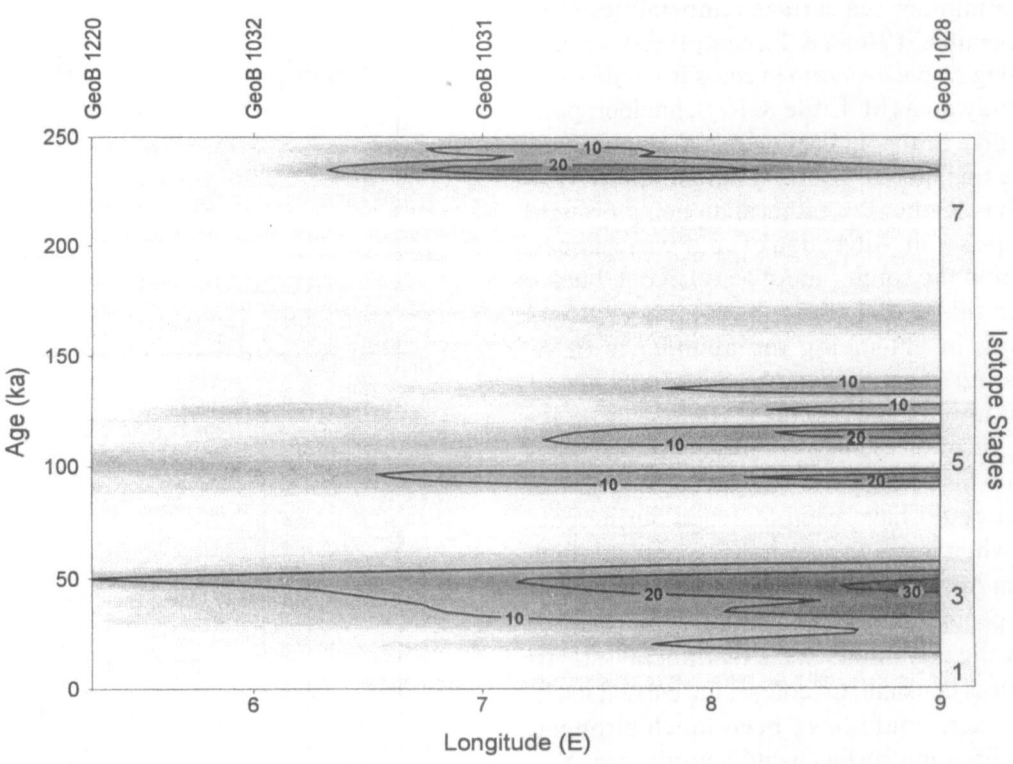

Fig. 10. Percentage of left-coiled *N. pachyderma* (sinistral) of total *foraminifera* in Cores GeoB 1028, 1031, 1032 and 1220.

put from high southern latitudes into the BOC/ BCC mixing zone or, alternatively, it indicates nearby coastal upwelling from deep within the thermocline. In both cases, colder water is transported into the study area.

Supportive evidence for this scenario of a larger contribution of coastal upwelling water comes from the faunal record. The isotope trends coincide with the appearances of sinistral *N. pachyderma* indicating increased availability of cold water. We think that left-coiling *N. pachyderma* growing in the coastal upwelling area (Ufkes and Zachariasse 1993) are transported by filaments to the core sites. Offshore transport of upwelling filaments is a common feature in this area (Lutjeharms and Stockton 1987) (Fig. 5). The seasonal flux pattern of *N. pachyderma* (sin.) at Walvis Ridge shows a maximum in spring, between September and November (Fig. 11), corresponding to the maximum total flux peak and following minimum sea surface temperatures (August/September) (Wefer & Fischer, 1993). Lack of left-coiling *N. pachyderma* in cores from the south of the study area (M. Little & R. Schneider, pers. comm.) give further indication that *N. pachyderma* (sin.) are transported within filaments offshore to the Walvis Ridge area rather than being brought in from the south. Advection of cold-water assemblages from the south cannot be ruled out, but we consider this as less important than regional upwelling in enhancing the abundance of *N. pachyderma* (sin.), at this subtropical latitude.

What implications do these findings have for the strength of the Benguela Current and for cross-equatorial heat transport? Benguela Current heat transport could have been lessened during cold periods when *N. pachyderma* (left) content is high and when $\Delta\delta^{18}O$ (between warm-habitat and cool-habitat species) is small, indicating cooler surface water, while conditions in the thermocline stayed more or less constant. Of course, the current itself, while colder, could have been much stronger, moving warming surface waters northward. Yet, without a strong warm contribution from the (pinched-off) Agulhas retroflection, the northward transport of heat deficit (upwelled and southern water) might have outweighed the transport of heat picked up on the way, from solar input. Inasmuch

as a reduction of surface-to-thermocline differences reflects cooling of surface waters, upwelling probably was indeed increased, and the Agulhas contribution was decreased.

The record of $\Delta\delta^{18}O$ on Walvis Ridge, by itself, is ambiguous. For periods when a decrease in the value of this proxy reflects thickening of the mixed layer (that is, warming at depth), we should not be able to defend the argument just made. In a sense, the same change in $\Delta\delta^{18}O$ may be due to opposite causes (cooling of surface waters, warming at depth), so that additional information is needed (on productivity and on surface water temperature). Diester-Haass (1985) and Diester-Haass et al. (1988) found reduced silica supply during glacial time on Walvis Ridge, based on coarse fraction analysis (DSDP Site 532; see Fig. 5 for position).

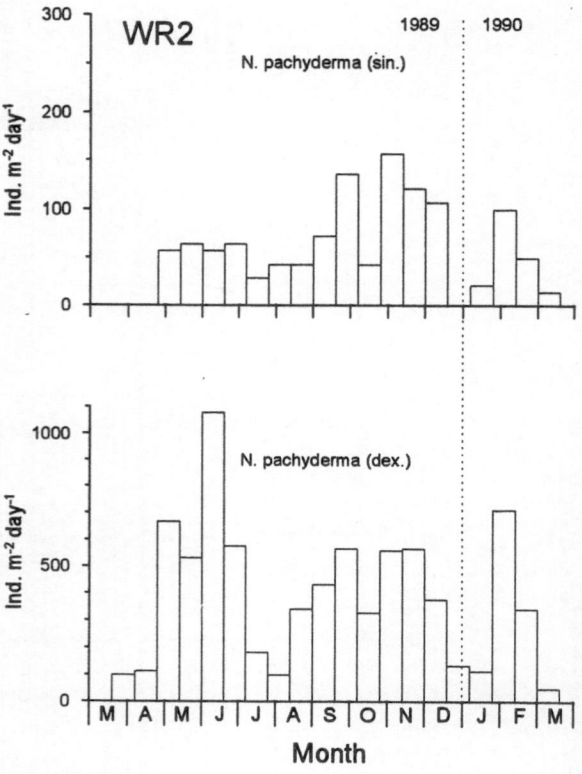

Fig. 11 Seasonal flux rates of *N. pachyderma* left and right coiling varieties at Walvis Ridge (position of Core GeoB 1028) in 1989/1990.

If we follow her conclusion that productivity was reduced even though *N. pachyderma* sin. increased, northward advection of cold water (rather than upwelling) would seem important in causing the isotopic and faunal changes observed in our cores from the Benguela current system. Alternatively, the increased abundance of *N. pachyderma* sin. indicates nearby upwelling (as we surmise) and the reduction of opal on Walvis Ridge, during glacial periods, indicates preservational effects. A general drawing down of dissolved silica in the global ocean, due to increased opal capture by the margins (Lange and Berger 1993) could lead to decreased silicification of diatoms and to increased dissolution. Summerhayes et al. (1995) assume that diatoms are unreliable indicators of productivity in slope sediments (Site 532), presumably because the ecology of the shelf edge system is not favourable to increased diatom production. Maximum diatom production is restricted to upwelling cells of the coastal upwelling system. As Diester-Haass et al. (1986) point out, glacial rates of sedimentation tend to be reduced from increased carbonate dissolution. In turn, this increase in dissolution points to increased supply of organic matter (providing for lowered pH in interstitial waters, through combustion). This is another argument that decreased supply of opal during glacial time is not an effect of lowered productivity.

A two-to four-fold increase in total organic carbon (TOC) content in glacial sediments compared to interglacial deposits in cores from the Walvis Ridge (core GeoB 1028) and off Angola (core GeoB 1016) is shown by Schneider et al. (this volume). Superimposed on these glacial/interglacial fluctuations are variations related to the orbital precessional cycle (periodicities in the TOC content are about 20 to 25 kyrs). From these data increased productivities are inferred for glacial and cold periods within the precessional cycle. In Core GeoB 1028 from the Walvis Ridge a trend toward higher productivity is recognizable, TOC and amplitudes of TOC between glacial and interglacial times increase during the last 150 kyrs. This is another indication that offshore transport of coastal upwelling filaments increased during the last ca. 150 kyrs.

Gyre System and Eastern Equatorial Atlantic

What does the record from the eastern equatorial Atlantic tell us about changing conditions? What can we learn about the history of cross-equatorial heat transport? A transect consisting of three gravity cores (Table 1, Fig. 1) from the eastern equatorial Atlantic is available to estimate sea surface temperatures and temperature gradients during the last several hundred thousand years. Each core location represents special oceanographic conditions in the upper part of the water column. These conditions range from the complex current system of the equatorial area, with convergence and divergence zones, to the more stable central gyre with its characteristically deep thermocline. Differences in nutrient contents between the three sites studied are shown in phosphate concentration at 100 m water depth (Conkright et al. 1994) (Fig. 12A). In general, nutrient content at central water is documented in $\delta^{13}C$ values of deep-living planktonic foraminifera *G. crassaformis* (Fig. 12B).

Core GeoB 1413 is from the edge of the Gyre System, where a deep thermocline and a distinct oxygen minimum with a center at about 400 m is present (Fig. 13). Isotope records for the shallow warm-water species *G. ruber* and the deep-living species *G. crassaformis* are available (Fig. 14), showing the typical glacial- interglacial variability in $\delta^{18}O$. Although the $\Delta\delta^{18}O$ record of core GeoB 1413 (Fig. 14A, bottom) has no readily recognizable relationship to glacial-interglacial variation, it may contain important information about upper water structure. A strong precessional effect seems indicated.

Carbon isotope values of *G. ruber* and *G. crassaformis* are depicted in Fig. 14B; values vary between 0 and 1.5 ‰ for *G. ruber* and 0.4 and 1.5 ‰ for *G. crassaformis*. What is most striking is the small variability in the $\delta^{13}C$ curve of *G. crassaformis*, especially when compared to the $\delta^{13}C$ record of *G. ruber*. As expected, during interglacial periods values are higher than during glacial times. Absolute differences between the two species are small, indicating similar nutrient levels within the uppermost ca. 300 m of water. The fact that $\delta^{13}C$ variation in *G. crassaformis* is dis-

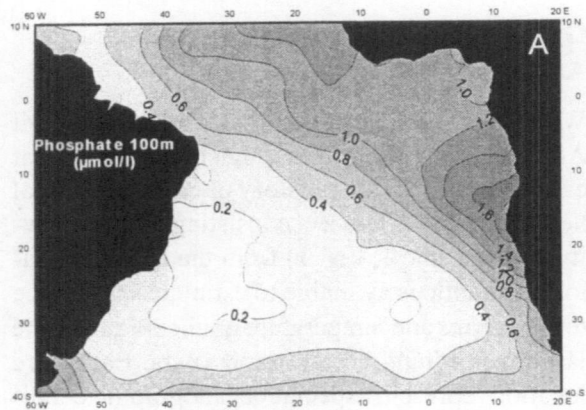

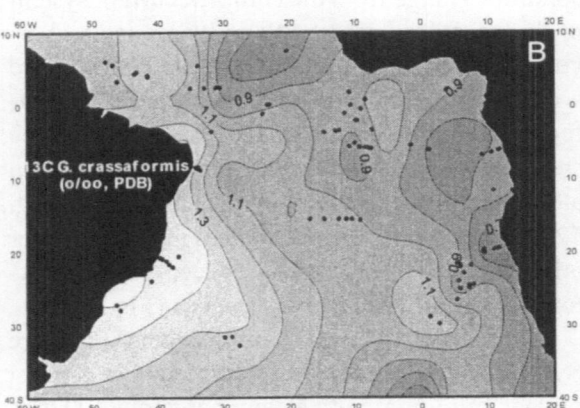

Fig. 12 (A) Phosphate content in mol/l at 100 m water depth (Conkright et al. 1994); (B) $\delta^{13}C$ values of planktonic foraminifera *G. crassaformis* from core-top samples.

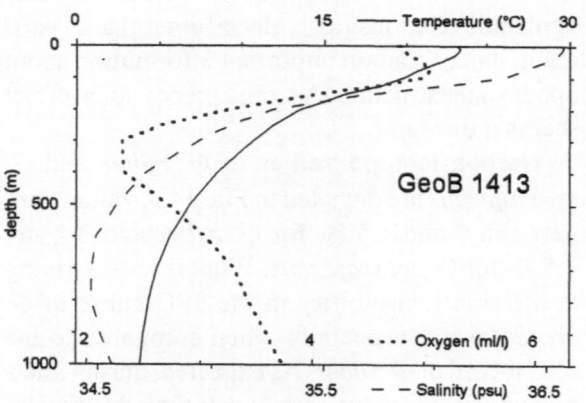

Fig. 13 Temperature, salinity and oxygen content of station GeoB 1413, gyre area. Interpolated from Levitus (1982).

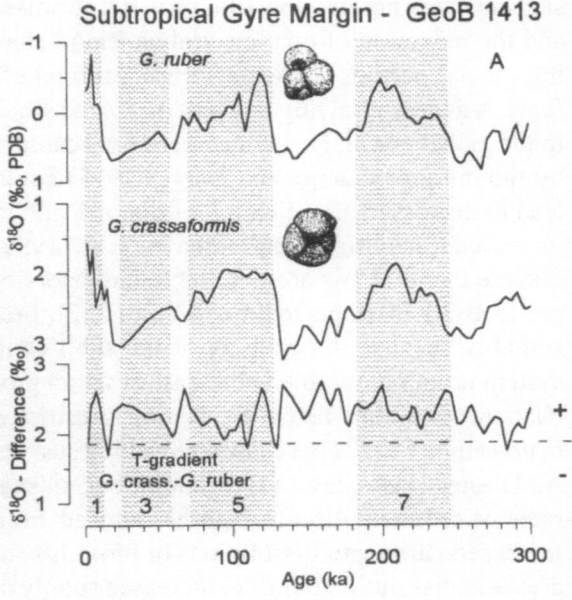

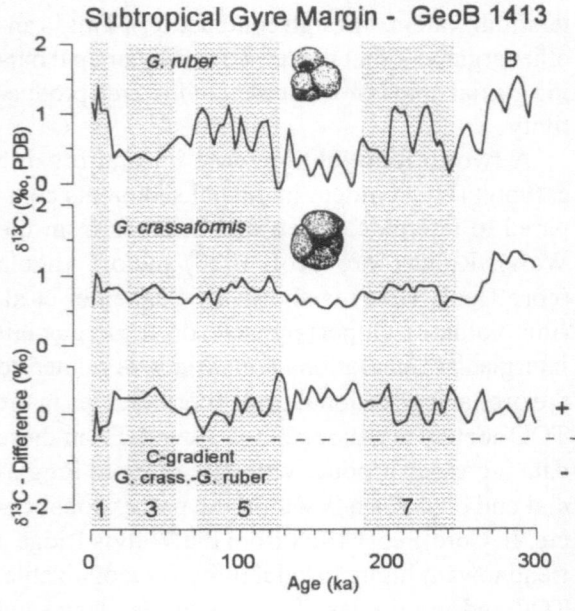

Fig. 14 Stable isotope records of *G. ruber* (white) and *G. crassaformis* and isotope differences between the species analyzed of core GeoB 1413. A, oxygen isotopes; B, carbon isotopes.

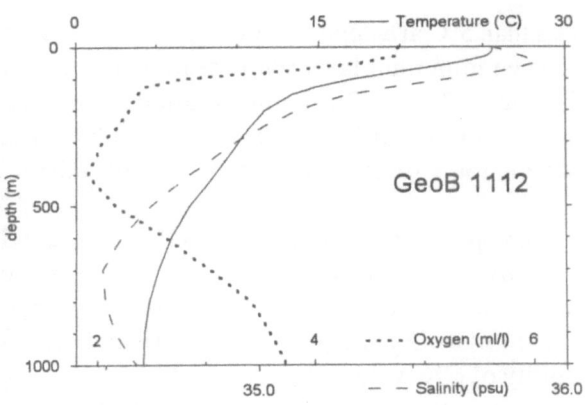

Fig. 15 Temperatur, salinity and oxygen content at station GeoB 1112, uppermost 1000 m, SEC. Interpolated from Levitus (1982).

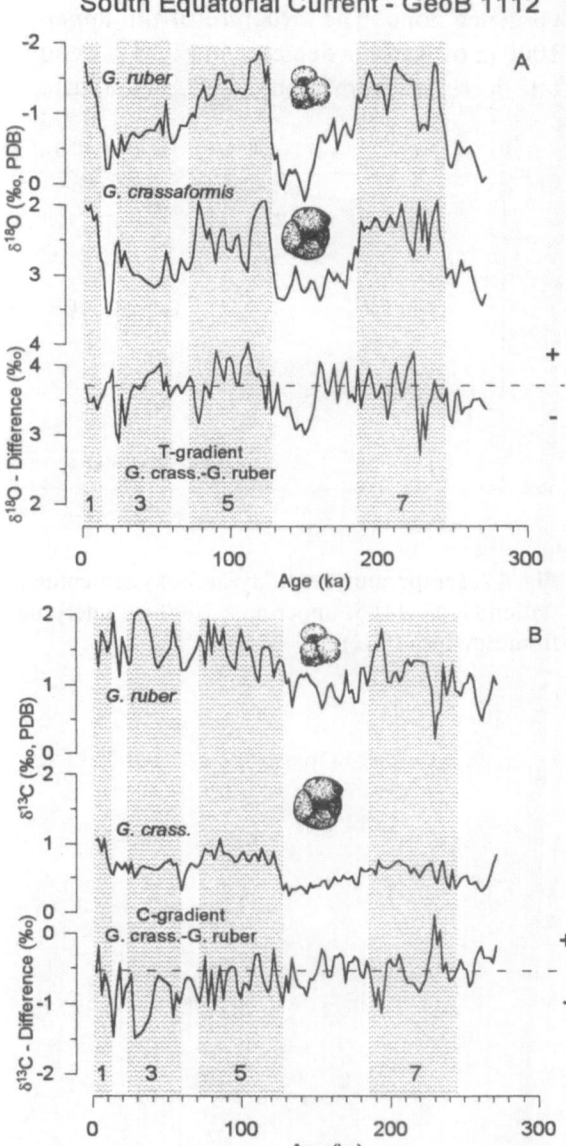

Fig. 16 Oxygen and carbon isotopes of *G. ruber* (pink) and *G. crassaformis* from Core GeoB 1112 and isotope differences of the species studied, South Equatorial Current.

tinctly below amplitude of global signal (0.4 ‰) means that thermocline water is more positive in glacial time and strongly hints at nutrient-depletion of thermocline waters during glacial time. A precessional signal seems obvious in the *G. ruber* record, but is entirely missing in *G. crassaformis*.

Core GeoB 1112, somewhat farther north, is from the region of the SEC, an equatorial area of relatively low productivity. Temperatures, salinities, and O_2 values of the upper water column are depicted in Fig. 15; foraminiferal stable isotopes in Fig. 16. The uppermost 80 m of water are mixed as indicated in temperature and salinity distribution. Between about 200 and 500 m water depth an oxygen minimum layer is present. As in the other cores oxygen isotopes show the typical glacial- interglacial pattern, covering the last ca. 280 kyrs (Fig. 16A). $\Delta\delta^{18}O$ values (*G. crassaformis* - *G. ruber*) vary between 2.8 and 4.3 ‰. Smaller differences are observed for glacial periods, due to higher (colder) *G. ruber* values. The rapid fluctuations in the temperature gradient suggest a doubling of the precessional effect (analogous to spring and fall productivity cycles).

Values of $\delta^{13}C$ values of *G. ruber* and *G. crassaformis* are shown in Fig. 16B. Both curves show a trend for increase with time. Differences in $\delta^{13}C$ between the two curves show a trend to more negative values. As in core GeoB 1413, variability of $\delta^{13}C$ is subdued in the deep-living *G. crassaformis*, although not quite as much as at the

gyre-near site. Values, on the whole, are slightly more negative, indicating loss of oxygen between sites GeoB 1413 and GeoB 1112. Again, the fluctuations in the C-gradient suggest a doubling of precessional effects.

Core GeoB 1105 was taken about 400 kilometers north of GeoB 1112, in the equatorial divergence zone. The structure of the uppermost 1000 m of water is depicted in Fig. 17. Temperature decreases from about 27.5°C at the surface to

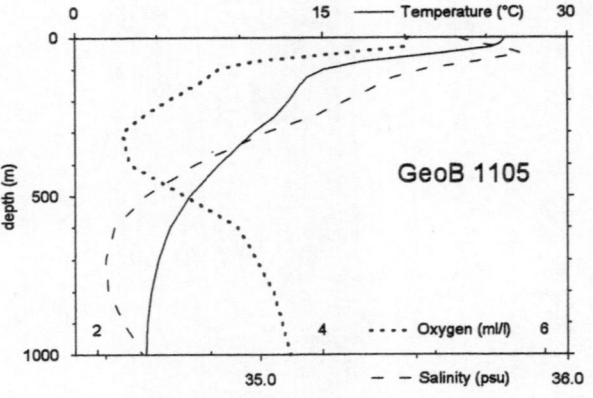

Fig. 17 Temperature, salinity and oxygen content at station GeoB 1105, uppermost 1000 m. Interpolated from Levitus (1982).

less than 5°C at a depth of 1000 m. The core of the oxygen minimum layer with values below 2 ml/l is at about 300 m. Due to high rainfall in the area, salinity in the surface water is lowered by about 0.7 ‰, compared to values at 50 m below the surface.

Isotope values of *G. ruber* (pink), *G. inflata* and *G. crassaformis* covering the last 240 kyrs are shown in Fig. 18. Oxygen values are similar to those of GeoB 1112 (Fig. 16). Surprisingly, variability of δ¹⁸O values is greater in the deep-living species *G. crassaformis* than in the shallow-water and warm-season species *G. ruber* (Fig. 18A). Oxygen isotope differences between *G. crassaformis* and *G. ruber* vary between 2 and 4‰ (Fig. 18A, bottom). Generally, larger differences are observed during interglacials or during warm substages of glacial periods. There is little indica-

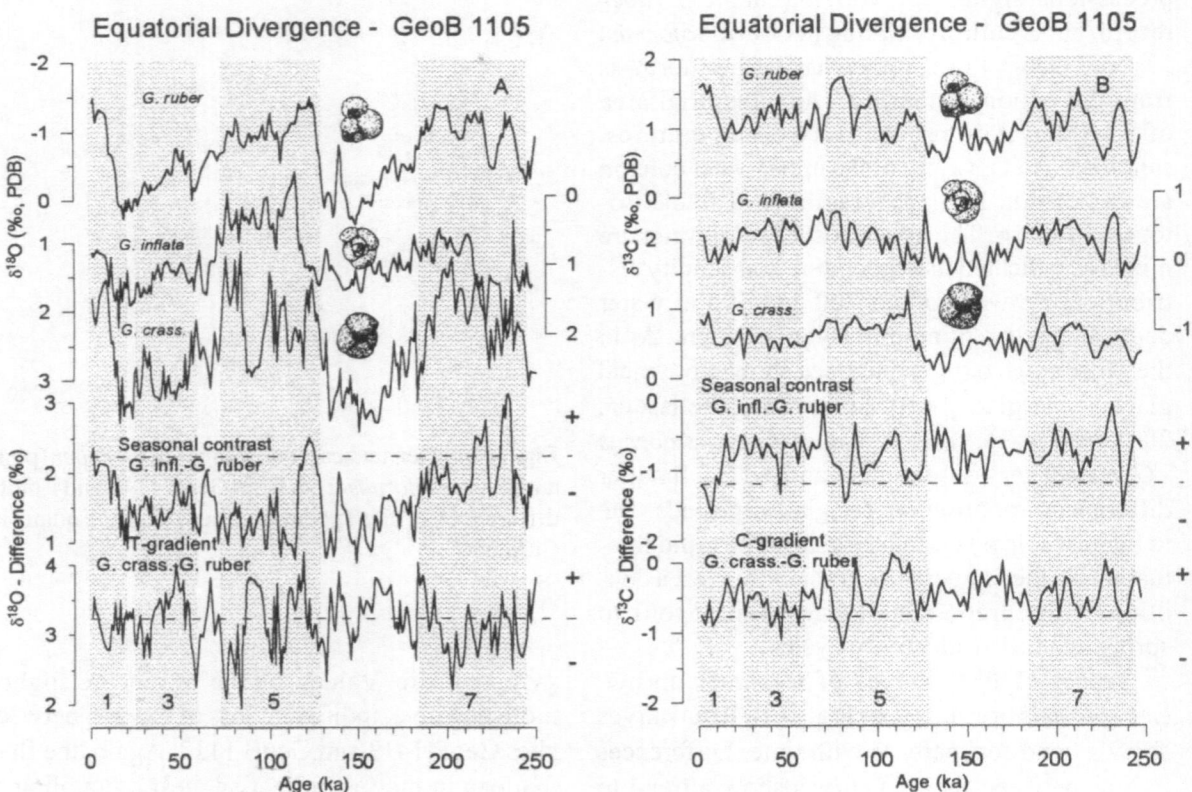

Fig. 18 Stable oxygen (A) and carbon isotopes (B) of *G. ruber* (pink), *G. inflata,* and *G. crassaformis* and isotope differences of the species studied from core GeoB 1105, equatorial divergence zone.

tion that the major ice cycles have a strong effect on temperature-gradient. Instead, there is a strong precessional signal, presumably representing monsoonal and trade-wind dynamics.

Carbon isotope values of Core GeoB 1105 are shown in Fig. 18B; values vary between 0.5 and 1.7 for *G. ruber*, -0.4 and 0.8 for *G. inflata* and 0.1 and 1.3 for *G. crassaformis*. Similar to Cores GeoB 1413 (Fig. 14) and GeoB 1112 (Fig. 16) the fluctuations in the seasonal contrast and C-gradient suggest a doubling of precessional effects.

In addition to isotope records, Cores GeoB 1105 and GeoB 1112 yielded reconstructions of faunal sea surface temperatures (see Meinecke 1992). Both the factor regression method of Imbrie and Kipp (1971) (traditional transfer function, "TTF") and the modern analog technique (MAT, Hutson 1980; Prell 1985) were employed for calculation of summer and winter temperatures, based on counts of planktonic foraminifera (Figs. 19 and

20; for positions see Fig. 1 and Table 1). Transfer function estimates for the last glacial yielded summer temperatures 4 - 5°C colder and winter temperatures 7 - 8°C colder than present values for the eastern equatorial Atlantic. Also, the residual $\delta^{18}O$-values of *G. ruber* (pink) and *G. inflata* (after correction for ice effect) were transformed into $\delta^{18}O$-temperatures using the equation of Epstein et al. (1953) (Fig. 21). *G. ruber* (pink) yielded a sea-surface warm-season temperature and *G. inflata* provided cold-season or subsurface temperatures within the thermocline.

The large cooling during the cold season, as seen in the MAT reconstruction, is responsible for a greatly increased seasonal contrast during glacial periods (Figs. 19 and 20). Cold-water species at the equator are almost invariably upwelling species. Thus, the finding (by statistics) of strong seasonal contrast implies strong seasonal upwelling, and this interpretation is strengthened

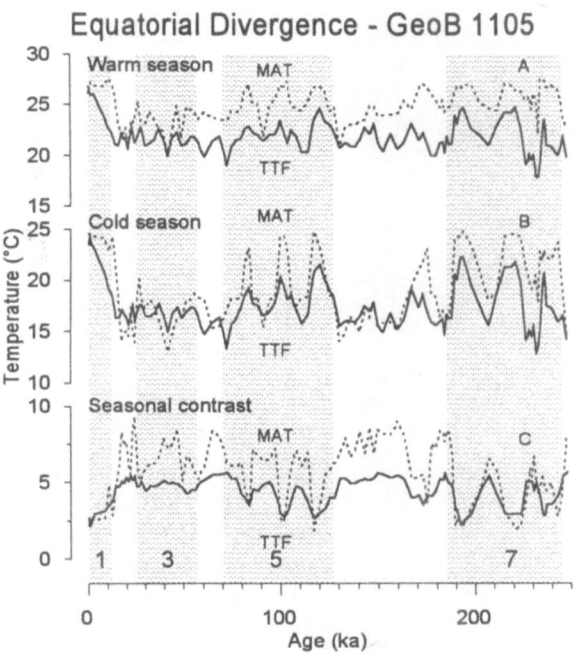

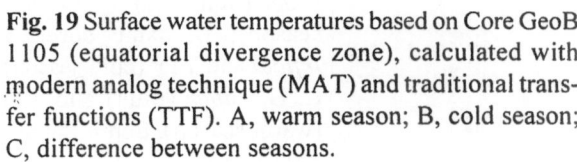

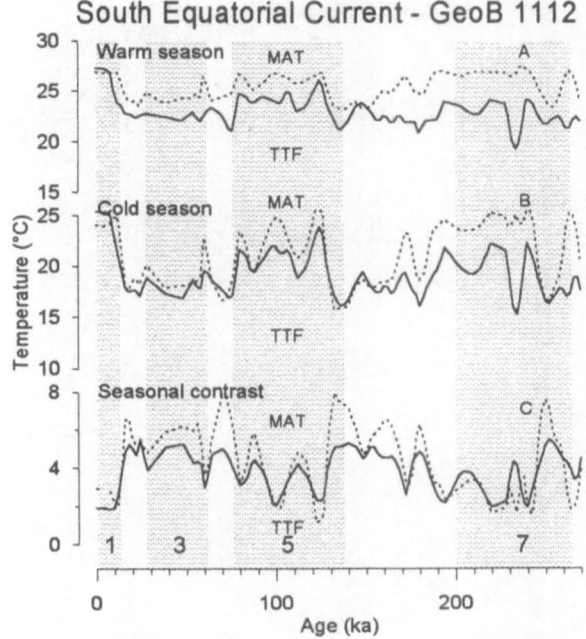

Fig. 19 Surface water temperatures based on Core GeoB 1105 (equatorial divergence zone), calculated with modern analog technique (MAT) and traditional transfer functions (TTF). A, warm season; B, cold season; C, difference between seasons.

Fig. 20 Surface water temperatures based on Core GeoB 1112 (South Equatorial Current), calculated with modern analog technique (MAT) and traditional transfer functions (TTF). A, warm season; B, cold season; C, difference between seasons.

when considering productivity indicators (Schneider et al., this volume; also see e.g. Mix 1989). It is not clear, immediately, what this implies for cross-equatorial heat export. On the whole, the lower surface temperatures, by themselves, would suggest less availability of heat to be transported northward. The main question concerns the position of the ITCZ; this can only be reconstructed from sites in the North Atlantic.

The comparison of the two types of temperature reconstruction, TTF and MAT, is of interest. TTF, on the whole, yields distinctly lower values than MAT, and there is some indication that MAT tends to produce higher amplitudes than TTF. In addition, seasonality values are greater for MAT. In general, the reconstruction by MAT seems less conservative than the one provided by TTF, which tends to be damped. The higher sensitivity of MAT produces phase shifts as well, with MAT-summer and winter temperatures leading TTF-tempera-

tures in many cases. A similar pattern was observed by Sikes and Keigwin (1994).

There is good agreement of the Holocene and glacial warm season temperatures of the equatorial divergence zone with the CLIMAP data. Temperatures here retrieved with TTF are about 1°C colder, but interglacial-glacial summer temperature differences are almost identical (ca. 4°C). Differences are larger for cold season temperatures in the equatorial divergence zone. Holocene winter temperatures of MAT and TTF are both about 2°C lower than CLIMAP temperatures. Glacial winter temperatures of MAT and TTF likewise are colder than the CLIMAP data but differences are greater than for the Holocene. As a result, interglacial-to-glacial ranges of winter temperatures of TTF are near 7.5°C compared to the 6°C change estimated by CLIMAP.

A comparison between faunal paleo-temperatures and isotopic (residual) temperatures

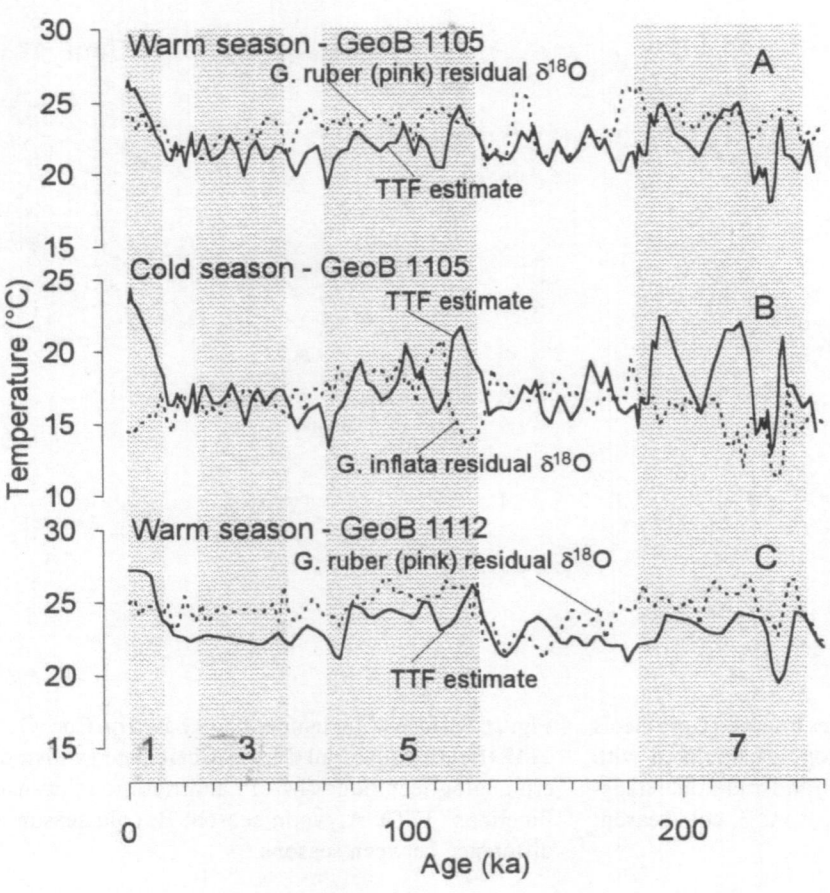

Fig. 21 Temperature variations for the last 250,000 years, based on δ18O-residuals in the tropical Atlantic, compared with temperatures derived from transfer function (TTF). A, Equatorial divergence zone, *G. ruber* (pink) and TTF (warm season); B, same core, *G. inflata* and TTF (cold season); C, *G. ruber* (pink) and TTF (warm season), South Equatorial Current.

is instructive (Fig. 21). In the equatorial divergence zone (Fig. 21A) the residual $\delta^{18}O$ -temperatures of *G. ruber* (pink), which are interpreted as the sea-surface temperature centered on the warmest month of summer, result in a 3°C lower summer temperature for the last glacial (21°C instead of 24°C). This difference of 3°C is somewhat less than that based on TTF. Schneider et al. (this volume) calculated a glacial to interglacial temperature range of 23.0 and 26.5°C, based on alkenones in Core GeoB 1105. Comparison of alkenone temperatures from core top samples with measured sea-surface temperatures (after Levitus 1982) gave the best agreement with spring/summer temperatures at 10 m water depth (I. von Storch, unpublished data). Going back in time, the correlation between TTF and residual $\delta^{18}O$ of *G. ruber* (pink) is poor. The same is true, on the whole, for GeoB 1112 (South Equatorial Current, Fig. 21C). The reason for this poor correlation is not clear; it does, however, raise questions about the usefulness of temperature estimates based on residual $\delta^{18}O$ values. Some scepticism also seems appropriate regarding TTF and MAT estimates, especially in areas where foram abundances might be more related to vertical water column hydrography and nutrient distribution than to sea surface temperature (Ravelo and Fairbanks 1991; Sikes and Keigwin 1994). The temperature based on the $\delta^{18}O$-residual of *G. inflata* shows little or no relationship to the TTF estimates for either warm or cold season (Fig. 21B). In fact, a case could be made for negative correlation: During interglacial periods, residual $\delta^{18}O$ -values for *G. inflata* show colder temperatures than during glacial periods.

It appears that subtracting the global ice effect from the *G. inflata* $\delta^{18}O$ signal results in over-correction, that is, in subsurface waters the global ice effect may be partially compensated by opposing salinity-related effects. Alternatively, *G. inflata* compensates for glacial-interglacial fluctuation, seeking warmer waters during glacials. Did the mixed layer thicken? Or did *G. inflata* change its depth habitat? Did it grow more vigorously during the warm season than at present, perhaps fed by the export from increased upwelling? Additional data are necessary before such questions can be properly addressed.

The observation that different paleo-temperature methods yield different results for glacial-to-interglacial differences has been reported earlier by Sikes and Keigwin (1994). They calculated glacial-to-interglacial temperature changes of 1.8°C for a core from the equatorial Atlantic based on alkenones, 3.8°C with MAT and 5.6°C from $\delta^{18}O$-values of *G. sacculifer*. In contrast to these results, Mix et al. (1986) found a smaller range in the $\delta^{18}O$-derived temperatures compared to TTF-temperatures in the same general region. These results, and our own, emphasize the need for caution in the use of temperature estimates when reconstructing climate dynamics and call for multiple estimation of paleotemperature.

One important way to constrain the changing dynamics of a current system (especially one associated with divergence) is the reconstruction of productivity. In fact, the possibility arises that changes in intensity of physical mixing can be recovered from changes in export production if variation in nutrient concentrations can be constrained (Berger et al. 1994b; Herguera and Berger 1994).

In Core GeoB 1105 from the equatorial divergence zone, a good correlation between MAT-estimates of cold-season sea-surface temperature (SST) and paleoproductivity is found as seen in percent C_{org} values (Fig. 22). Cold sea-surface temperatures and a strong seasonality during glacial times (Fig. 19), indicating strong seasonal upwelling, are expected to result in enhanced paleo-productivity. Increased upwelling during cold periods may be reflected in the $\delta^{13}C$ values of *G. ruber*, whose amplitudes exceed global background (0.4%) despite an opposing temperature effect from air-sea exchange.

We assume that there is a strong coupling of upwelling to the zonal wind component of the SE-trades, so that evidence for increased upwelling is evidence for increased tradewind activity. The zonality of the SE-trades is itself affected by the African SW-monsoon system which in turn is influenced by long-term variations in insolation. Reducing the SW-monsoon strengthens the SE-trades and therefore the upwelling in the equatorial divergence, which results in higher productivity. The assumption that upwelling is coupled to the SE-trades is supported by cross-correlation analysis showing significant 23 ky coherency between the time series of sea-surface temperature and paleoproductivity (Schneider et al., 1995; also see Schneider et al., this volume).

To determine to what extent the dynamics of the western equatorial region is dominated by the glacial-interglacial climatic overprint (ice-age dynamics or boreal dynamics) and to what extent it reflects regional dynamics tied to tropical

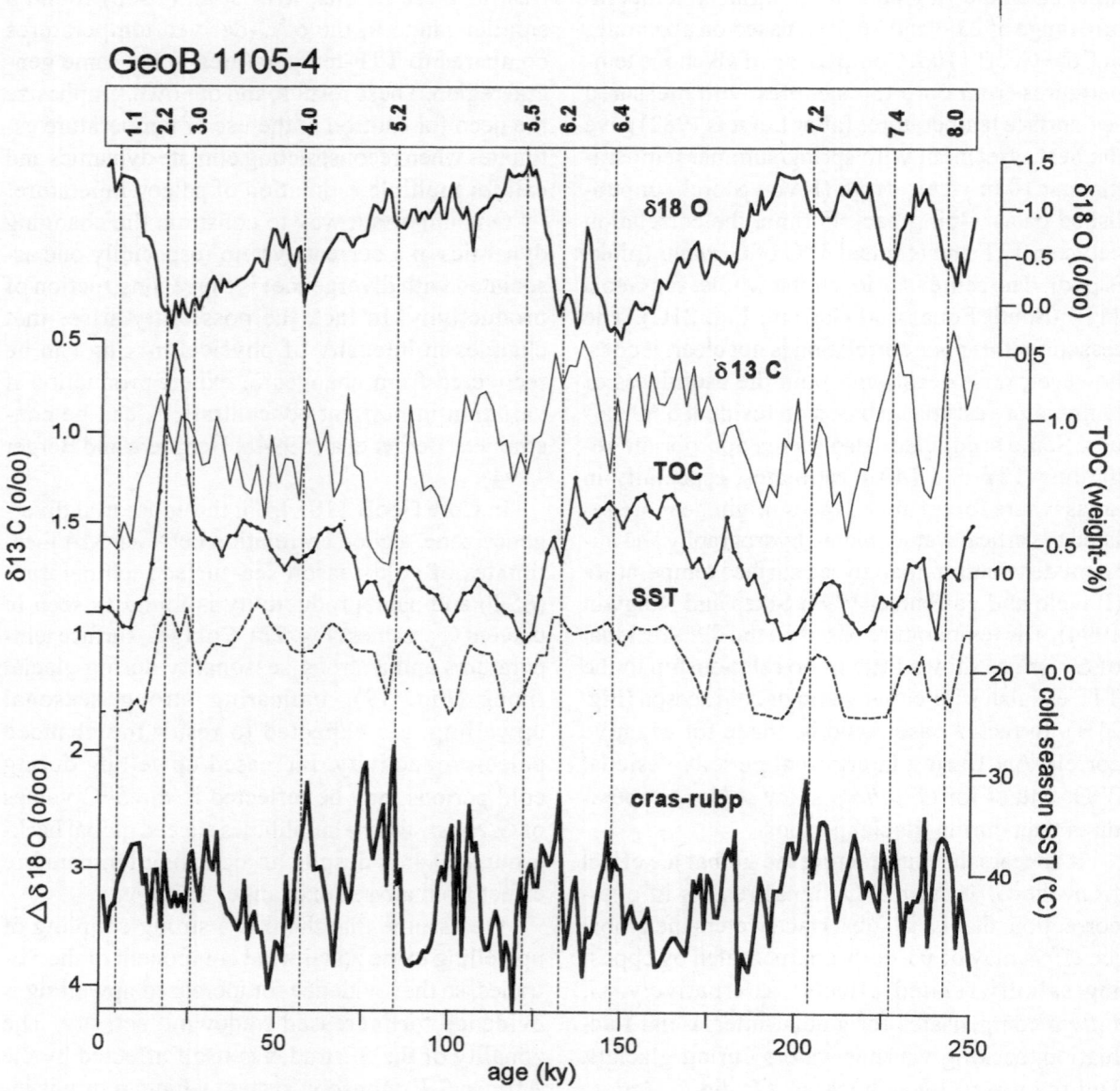

Fig. 22 Temperature and productivity proxies for the equatorial divergence zone, Core GeoB 1105. Upper two: $\delta^{18}O$ and $\delta^{13}C$-values of *G. ruber* (pink) ; next down: C_{org} values; next down: Cold season temperatures, calculated with MAT; at bottom: difference in $\delta^{18}O$ of *G. ruber* (pink) and *G. crassaformis* , a measure of the temperature gradient between surface waters and 300 m depth.

insolation (as e.g. monitored by the July 15°N insolation; Berger and Loutre 1991), we show the Fourier spectrum of Core GeoB 1105 from the eastern equatorial region (Fig. 23A). Ice-age dynamics has relatively little precessional information as shown in records from the western equatorial Pacific (Berger et al. 1994a), so that precessional information large reflects tropical and subtropical wind fields (tradewind and monsoon). In our conceptual model we consider the (ice-related and NADW-related) boreal dynamics as being largely independent of the intertropical dynamics. Therefore, it is interesting to find out to what degree upwelling and productivity dynamics at the equator follow one or the other of the main climate controls.

Fourier analysis of $\delta^{18}O$ of *G. ruber* shows the expected Milankovitch frequencies (Milankovitch 1930)(Fig. 23A, rbr18) with the obliquity-period somewhat subdued, in favor of precession and eccentricity. A hint of a double-precession is present. This spectrum suggests that precessional power (intertropical dynamics) is at least of equal importance to boreal dynamics in producing the *G. ruber* $\delta^{18}O$ record. For the $\delta^{13}C$ record (rbr13) precession is even more important, suggesting that this proxy is tied to wind-driven processes (i.e. upwelling). The spectrum of TOC is much like that of $\delta^{18}O$, as is SST, except that in the latter the 41-period disappears into the noise level at intermediate periods. When $\delta^{18}O$ of *G. ruber* is subtracted from $\delta^{18}O$ of *G. crassaformis*, both 100-kyr cycle and 41-kyr cycle disappear, and what remains is power at 23 kyr, mainly. This suggests that temperature gradients between surface and subsurface waters are entirely dominated by intertropical dynamics, and have little or nothing to do with boreal dynamics.

The last-made point is illustrated by comparing insolation at 15°N, in July, with the temperature-gradient proxy ($\delta^{18}O$ *G. crassaformis - G. ruber*, Fig. 23B top). Whenever Jul15N is high (that is, when the ITCZ is far to the north, in northern summer) the temperature-gradient is large, that is, the mixed layer is thick and warm (see Fig. 4).

When we plot the temperature-gradient, we eliminate the ice signal from the north. What remains is the South Atlantic contribution, which is dominated by precession (trade wind strength). As expected, TOC is inversely correlated with the temperature-gradient (Fig. 23B crf18-rbr18) and also with Jul15N-insolation.

All three cores from the equatorial divergence zone (GeoB 1105) (Fig. 18), the south equatorial current (GeoB 1112) (Fig. 16), and the subtropical gyre (GeoB 1413) (Fig. 14) tend to show cooling of surface waters (and presumably shoaling of the thermocline) during cold events of glacial and interglacial stages, based on several proxies. Likewise, the cores from Walvis Ridge show a cooling of surface waters during cold events (indicated by heavier values of *G. ruber*) while thermocline conditions remained unchanged. Our interpretation is that shoaling of the thermocline indicates, on the whole, a reduction of the warm water layer over the tropical eastern South Atlantic, during glacial periods or cold periods, tied to precessional forcing.

Changes in thermocline depth are tied to variations in the strength of equatorial upwelling, SEC circulation and subtropical gyre circulation (McIntyre et al. 1989; Molfino and McIntyre 1990; Meinecke 1992) caused by variations of southeastern wind zonality (Philander and Pacanowski 1984). At the margin of the subtropical gyre (GeoB 1413) (Fig. 14) there is a general decrease in the $\delta^{18}O$ differences between *G. crassaformis* and *G. ruber* for the last 120 kyr (Fig. 14A) which suggests a gradual deepening of the thermocline documenting an increased advection of warm surface water in the gyre through time. At the southwest Walvis Ridge (GeoB 1031 and 1220) (Figs. 9B and 9C) the $\delta^{18}O$ difference between *G. inflata* and *G. ruber* also shows a gradual decrease of the seasonal contrast over the same period. Thus, the processes responsible for the buildup of the gyre's warm water lens act over the entire basin. We interpret this observation as a spin-up in the intensity of gyre-margin circulation with a SW-migration of the gyre margin, and hence an overall increase in meridional heat transport.

Western equatorial Atlantic

We next leave the eastern equatorial Atlantic and turn to the west. The western equatorial Atlantic is the main exit for the transfer of warm water from

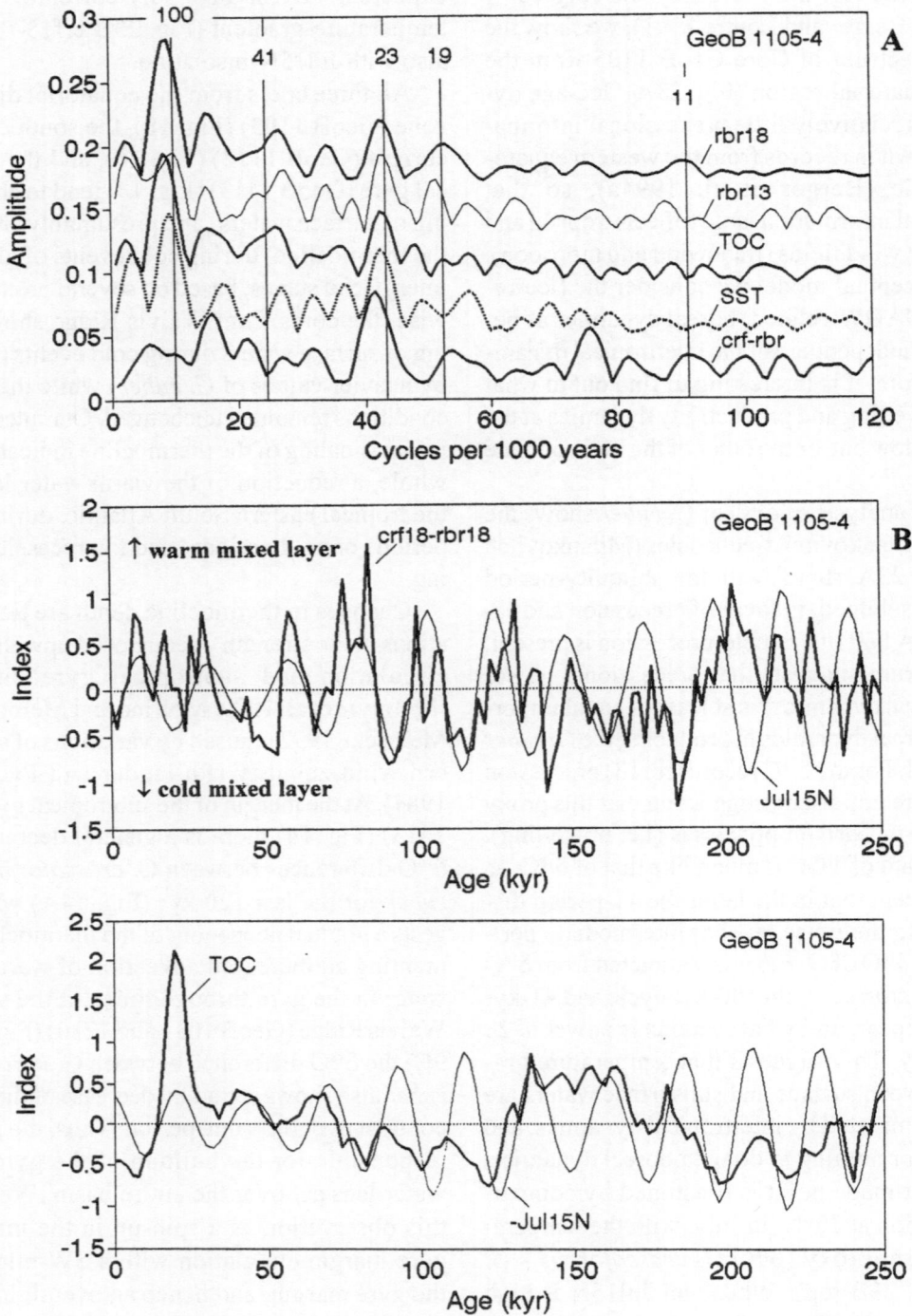

Fig. 23 Importance of precessional forcing in eastern equatorial dynamics. (A) Spectral analysis (straight Fourier expansion, amplitude is square root of sums of squares of sine and cosine coefficients) of $\delta^{18}O$ of *G. ruber* (rbr18), $\delta^{13}C$ of *G. ruber* (rbr13), total organic carbon (TOC), sea surface temperature (SST) and $\delta^{18}O$ *G. crassaformis* - *G. ruber* from Core GeoB 1105-4 from eastern equatorial Atlantic). (B) Comparison between insolation at 15°N, in July, with $\delta^{18}O$ difference between *G. crassaformis* and *G. ruber* (top) and total organic carbon (bottom).

the South Atlantic to the North Atlantic. This is a key area, therefore, for assessing the heat export. We assume, in principle, that export is potentially large when water temperature is high and the warm water layer is thick in this region. Inversely, when these conditions do not obtain, heat export is likely reduced. In reconstructing conditions, therefore, we are especially interested in surface water temperatures and in the temperature gradient below the mixed layer. These parameters will indicate, we think, whether or not large-scale heat transfer was likely. Fig. 24 shows profiles of temperatures, salinity and oxygen content at the site studied. A well mixed 100 m thick surface layer is present. The center of the oxygen minimum layer is at about 250 m of water depth.

Isotopic compositions were determined for a number of species of planktonic foraminifera living at different water depths (*G. ruber* (pink), *G. truncatulinoides* and *G. crassaformis*), in samples from a gravity core taken in the western-most part

of the equatorial circulation system. The oxygen isotope records (reproduced in Fig. 25A) show typical glacial-interglacial changes reflecting both the waxing and waning of the northern hemisphere

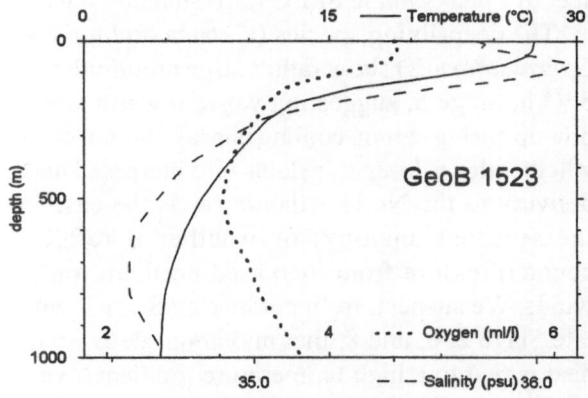

Fig. 24 Temperature, salinity and oxygen content at Site GeoB 1523, western equatorial Atlantic. From Levitus (1982).

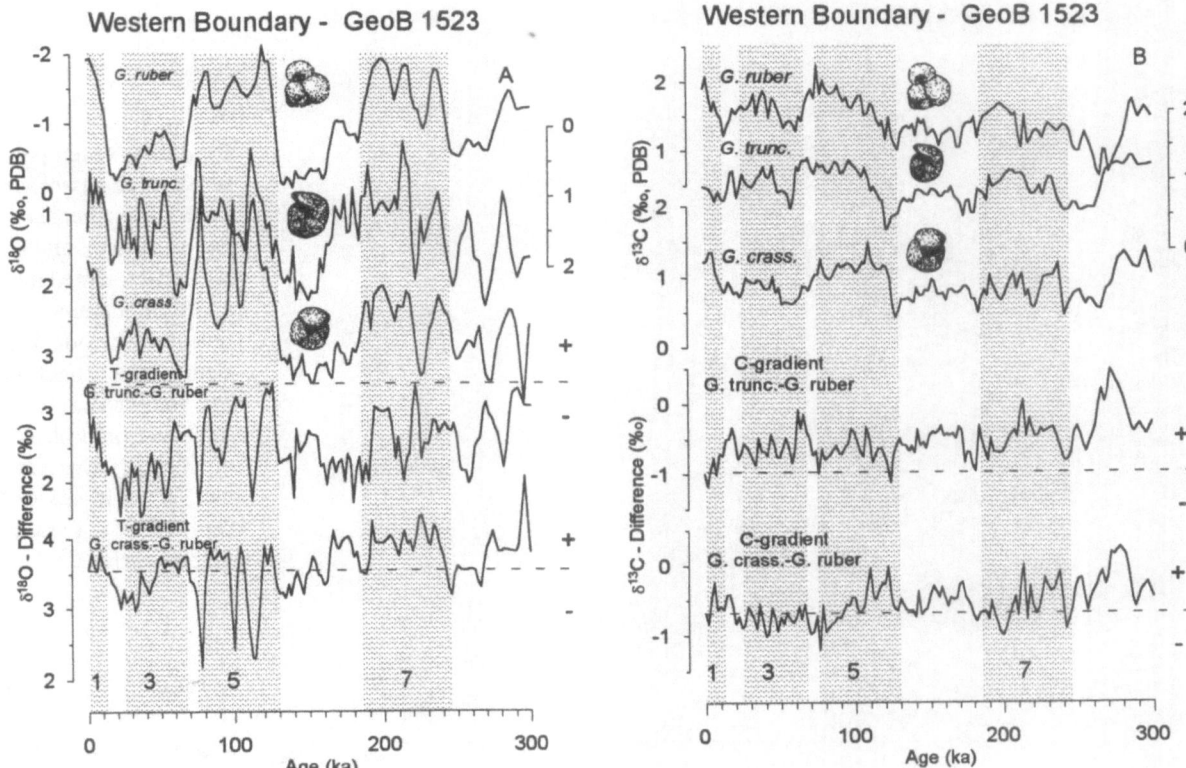

Fig. 25 δ¹⁸O-and δ¹³C-values of planktic foraminifera living at different water depths and differences between the species studied. Core GeoB 1523, western equatorial Atlantic.

ice sheets and associated temperature changes. Isotope stages and substages can be easily identified down to Stage 9 (ca. 730 cm depth in the core) (Fig. 25A). Distinct maxima and minima of the $\delta^{18}O$ curves of *G. ruber* are correlated with the respective peaks in the SPECMAP standard curve.

The deep-living species (*G. truncatulinoides, G. crassaformis*) show rather large amplitudes in $\delta^{18}O$ in Stage 5, suggesting western warm-water pile-up during certain cooling phases. It is not clear whether this pile-up translates into increased heat delivery to the North Atlantic (as is the case in the seasonal analogy) or whether it reflects counterpressure from increased northern trade-winds. We suspect, from comparing Stage 7 with late Stage 2, 6, and 8, that maximum delivery of heat is tied to a high temperature gradient, overall. The amplitude of *G. ruber* (pink) oxygen isotope values suggests a range of >2 permil, that is, a temperature change of about 4 to 5°C in the surface waters, between glacial and interglacial maxima.

The $\delta^{13}C$ records show the typical variability between glacial and interglacial values, with more negative values during glacial times and vice versa (Fig. 25B). Noteworthy is the trend to larger values of *G. ruber*; average values change from 0.8 to 1.6‰ between 270 kyrs and the Holocene. The average values of the two deep-living species remain more or less unchanged through time. The subdued variability in deep-living forms with regard to $\delta^{18}O$ amplitudes, means that some compensation may be at work, that is, ^{13}C is added during cold intervals and vice versa.

A detailed temperature record, based on $\delta^{18}O$-residuals, obtained by subtracting the SPECMAP stack from the GeoB 1523 record, is shown in Fig. 26. On the basis of $\delta^{18}O$-values of *G. sacculifer* and *G. ruber* surface waters appear to have been 4-5°C colder during glacial times (Stages 2, 4, 6, 8, 10) than in interglacial periods (Fig. 26). The estimates are of the right magnitude to account for the lowering of specific tree lines and of snowlines in tropical Latin-America (e.g. Rind and Peteet 1985;

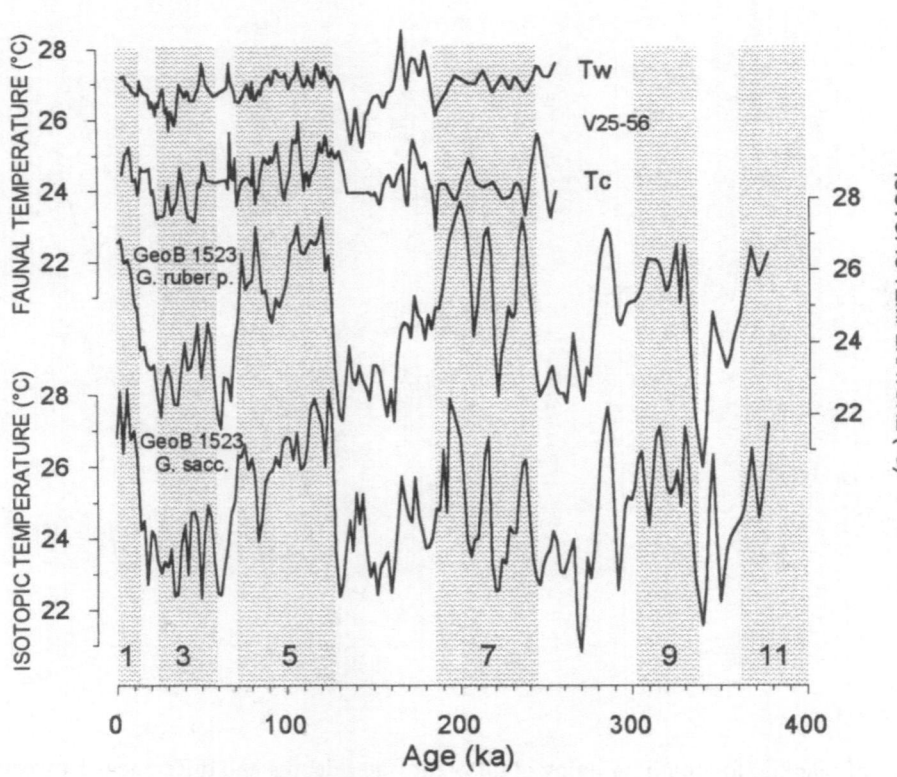

Fig. 26 $\delta^{18}O$ temperatures of the two surface-water species *G. ruber* (pink) and *G. sacculifer,* Core GeoB 1523. For comparison, temperatures derived with transfer functions are also shown (Core V25-56 from Imbrie et al. 1989; T$_w$, warm season; T$_c$, cold season). Gray fields indicate glacial periods.

Stute et al. 1995). Sea surface temperatures for cold and warm seasons, based on transfer functions (Imbrie et al. 1989), are given for comparison. These estimates show only 1-2°C lower temperatures during glacial times. Ravelo (1991) and Sikes and Keigwin (1994) discuss this discrepancy and suggest that foraminiferal assemblages are not reliable as recorders of surface water temperatures in the tropical Atlantic. We agree. In fact, error ranges for transfer temperatures greatly increase toward both the cold and the warm end of the temperature range; they are smallest for intermediate temperatures. The problem has both statistical and ecologic aspects, and is quite well understood, in principle.

The temperature differences between surface-living and deep-living foraminifera change during the late Quaternary (Fig. 27). Temperature varia-

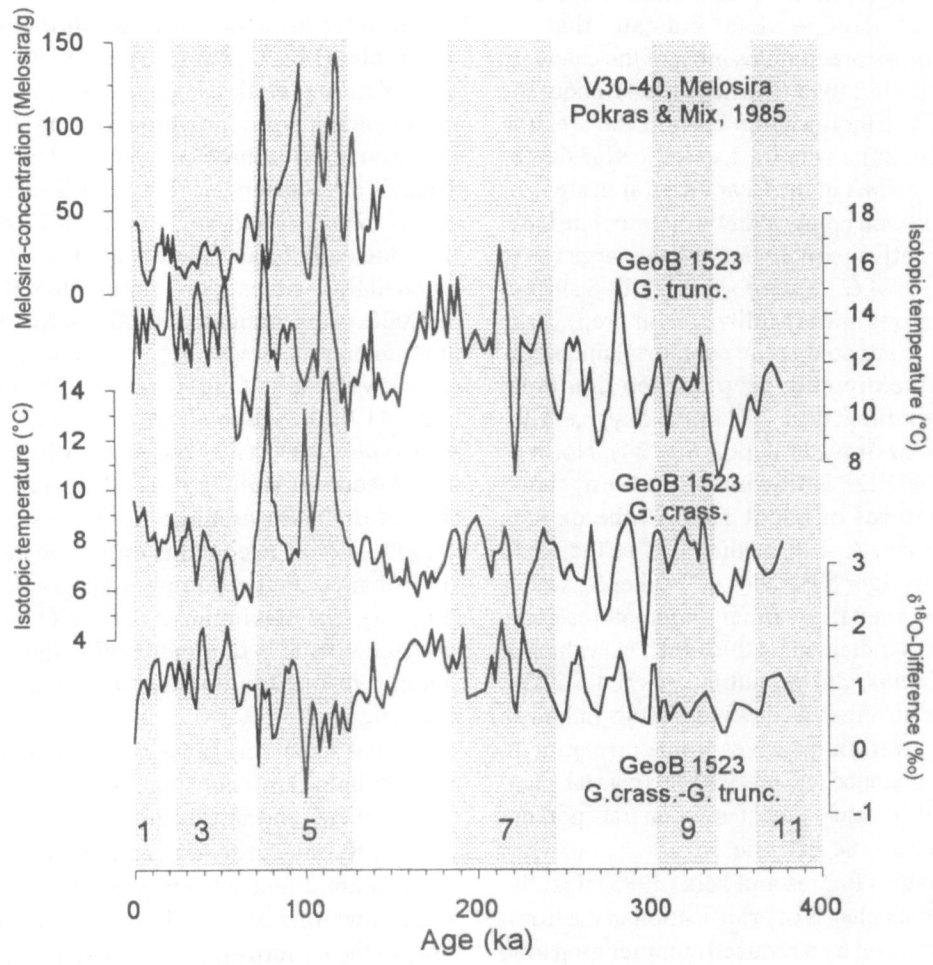

Fig. 27 Temperature records (δ^{18}O-residuals) of the deep-living foraminifera *G. truncatulinoides* and *G. crassaformis* and differences between the species studied from core GeoB 1523, Ceara Rise. Concentrations of the fresh water diatom *Melosira* in Core V30-40 (equatorial Atlantic) (Pokras and Mix 1985) are shown for comparison. The presence of this form indicates wind transport. Vertical grey stripes: boreal summer insolation minima in the precession cycle ($\pm$ 2500 years).

tions of the deep-living species are much higher (ca. 10°C) than those of the shallow-dwelling (compare Figs. 26 and 27). *G. crasssaformis* shows extreme warm temperatures only during Stage 5 at about 78,000, 100,000 and 114,000 years BP, close to the boreal summer insolation minimum. Presumably the thermocline was so deep during these times that temperatures were up to 5°C warmer than today at the calcification depth of *G. crassaformis* (ca. 400 m) . We assume that the overall differences in the temperatures recorded by the two deep-living foraminifera reflect overall differences in depth of calcification. Sediment samples and isotope data indicate that *G. truncatulinoides* precipitates most of the calcite in a water depth of 250 m (Mulitza 1994), about the position of the thermocline base in the western Atlantic (Fig. 24). In contrast, calcification depths of *G. crassaformis* in the Ceara Rise area are near 400 m; this is well below today's thermocline base.

An alternative explanation for the large range in $\delta^{18}O$ values of *G. crassaformis* is a N-S shift of central water regimes (Mulitza ms in prep). Core GeoB 1523 is situated in the equatorial undercurrent with an extremely deep oxygen minimum zone (temperatures near 7°C in the oxygen minimum at 400 m of water depth, Fig. 24). North of the EUC, the NEC is flowing to the west, showing temperatures of about 15°C in the oxygen minimum zone. A shift of the NEC to the south caused by stronger NE trades or weaker SE trades could also explain the warmer temperatures in the deeper waters indicated by the light $\delta^{18}O$ values of *G. crassaformis,* during sub-stages 5.4, 5.2, 5.0. Support for stronger tradewinds being linked to thermocline warming derives from correlation to *Melosira* abundance in Core V30-40 (central equatorial Atlantic) which indicates eolian transport out of Africa across the Atlantic.

According to Pokras and Mix (1985) this diatom documents phases of aridification in the tropical Africa, caused by a reduced summer monsoon (Prell and Kutzbach 1987). Within the error of the isotope stratigraphy (±2500 years, Imbrie et al. 1984) *Melosira* maxima correlate with high temperatures at depth, as derived from oxygen isotopes of *G. truncatulinoides* and *G. crassaformis* (Fig. 27). The correlation between the abundance

of wind-blown *Melosira* and temperatures near and below the thermocline base supports the model of McIntyre et al. (1989) which postulates that surface water circulation during the late Quaternary is in phase with the precession cycle. More specifically, the correlation supports the concept that warm-water pile-up in the west-equatorial Atlantic is a function of easterly trades (which also bring *Melosira*). The question that remains in this context is whether NE- or SE-trades are ultimately responsible for the *Melosira* signal, and whether this makes a difference to the argument concerning cross-equatorial heat export.

The $\delta^{18}O$ differences between the two deep-living species (Fig. 27, bottom) reinforce the impression of strong relationships between precessional forcing and thermocline properties in this region. Differences fluctuate between 2.1 and -0.7‰. Today's $\delta^{18}O$-differences between *G. crassaformis* and *G. truncatulinoides,* observed in sediments from the equatorial Atlantic, are near 0.4‰. The much higher values found for much of the last ca. 380,000 years indicate situations for which we have no present-day analog. Cross-spectral analysis shows coherent significant cycles for 23,000 and 41,000 years between the $\delta^{18}O$ *G. truncatulinoides - G. crassaformis* differences and the ETP curve (Mulitza 1994). The phase relationship of different climatic indicators within the 23,000-year cycle contains important clues to the significance of the thermocline gradient proxy (Fig. 28). The maximum of the $\Delta\delta^{18}O$ (crf-trc) parameter virtually coincides with the minimum summer insolation, that is, maximum rate of ice buildup.

At that same time (near minimum boreal summer insolation), we note maximum wind transport (of *Melosira*), and shortly after we have maximum seasonality seen in species composition of planktonic foraminifera, which goes parallel to maximum upwelling and minimum surface temperatures in the eastern equatorial realm (Core RC24-16; McIntyre et al. 1989). It appears, then, that the main factor in increasing $\Delta\delta^{18}O$(crf-trc) is a thickening of the warm water pool in the west-equatorial Atlantic (warming of the habitat of *G. truncatulinoides*), as one approaches the maximum zonality of the trades (which is correlated with

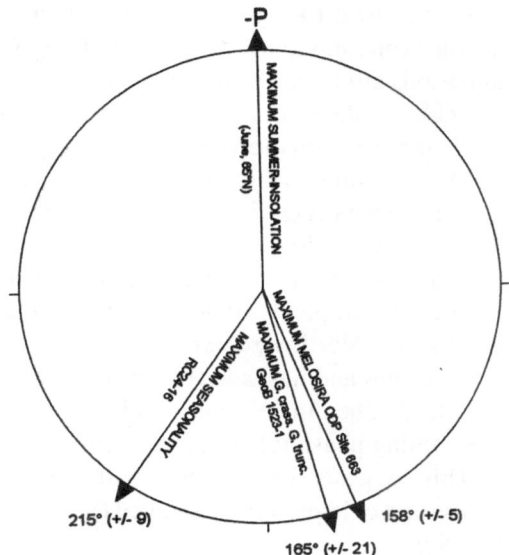

Fig. 28 Phase relationships in the 23,000 years cycle; between Milankovitch forcing δ¹⁸O-differences of deep-living planktonic foraminifera *G. crassaformis* and *G. truncatulinoides* in Core GeoB 1523, concentrations of fresh water diatom *Melosira* in ODP Site 663 (DeMenocal et al. 1993), maximum seasonality in the eastern equatorial Atlantic from transfer functions of planktonic foraminifera (Core RC24-16 from McIntyre et al. 1989).

upwelling in the east). This pile-up of warm water apparently occurs during the boreal cooling phase. If so, this would support the concept that increased heat export to the North Atlantic slows cooling at the end of an interglacial (Berger et al. 1987).

One of the longest *Melosira* time series as a record of phases of aridification in tropical Africa is from ODP site 663 from the equatorial Atlantic (1°S;11°W) (DeMenocal et al. 1993). The phase of the *Melosira* cycle in ODP 663 is identical, within the present time resolution, to that of the maximum δ¹⁸O-difference between the two deep-living foraminifera species in core GeoB 1523 from the Ceara Rise (Fig. 27, bottom). The near-coincidence in phase between *Melosira* concentrations in sediments of the eastern equatorial Atlantic and δ¹⁸O-values of deep-living foraminifera in the western equatorial Atlantic show that atmos-

pheric and oceanic circulation in the equatorial Atlantic is tied to intensities of the trades. The phase difference shown is within the error of isotope stratigraphy when correlating between cores. Thus, leads and lags of less than ca. 2000 yrs (30°) are significant only if they occur within the same record.

Discussion - the Search for a Paradigm

A fundamental aspect of surface circulation is the overall distribution of warm surface water. This distribution delineates, roughly, the region contained between the poleward limits of the subtropical gyres. Also, the region where warm surface waters are thick is congruent, on the whole, with the region from which heat is exported, while areas with a thin (seasonal) warm layer and those without a warm layer year-round import heat from elsewhere to achieve balance with infrared radiation to space. The boundary between the two major environments (heat excess vs. heat deficiency) lies south of 35°S (Fig. 6) and is marked by a sharp temperature gradient which may be interpreted as a surfacing of the thermocline. On a glacial-interglacial scale, the position of the maximum gradient (subtropical-subpolar convergence) can change its latitude. It is commonly assumed that it migrates equatorward during glacial conditions.

For the purposes of this discussion, we are interested how the distribution of warm surface water reflects cross-equatorial heat transport. We shall make the assumption that a pile-up of warm surface water in the western part of the equatorial Atlantic signals the efficient working of such heat transport. Thus, the main clue to the heat transport dynamics, in our view, will be found in the region of Ceara Rise. Furthermore, whenever warm water piles up in the west, along the equator, the eastern part of the equatorial region is robbed of its share. Thus, we expect that, during times of efficient heat transfer, the eastern equatorial region tends to be anomalously cool, while when this transfer is decreased, it tends to be anomalously warm.

As concerns the southern end of the upper limb of the heat conveyor, the water that moves across the equator, into the North Atlantic, is largely

replaced by invasion of thermocline-and surface
water in the southern part of the Benguela Current
(Fig. 1). Right now, and presumably during interglacial periods in general, warm water penetrates
from the Indian Ocean with the Agulhas
retroflection, as described earlier. During glacial
periods, we assume, this influx is pinched off by
the northward motion of the subantarctic front. We
expect an increased participation, therefore, of cold
subantarctic water in the Benguela Current, during glacials, as well as increased contributions
from coastal upwelling, which occurs in all other
eastern boundary currents studied. With warm
water supply pinched off, and cold water supply
enhanced, the Benguela Current would transport
heat deficit northward, as well as the heat collected
from insolation. The question is, whether, on the
whole, the net heat transport is decreased whenever the Benguela Current carries cold water from
the South and from upwelling. Presumably, clues
to this problem are obtained when comparing the
Benguela System with the Gyre System and the
eastern and western equatorial Atlantic, regarding
long-term and glacial - interglacial changes, as
well as precessional cycles and deep water patterns.

To generate working hypotheses about the dynamics of surface water circulation in the South
Atlantic during the Late Quaternary we employ
Croll's paradigm that glacial-interglacial fluctuations are analogous to seasonal fluctuations. Croll
(1875) thought that glacial periods alternate in
North and South just as seasons do. While we now
know that this is wrong, there is much to be learned
by applying the analogy to precession-related variations because the seasonal cycle is in a way like
a precessional cycle, the two hemispheres having
opposite phase.

The central idea of Croll's concept is that the
South Atlantic is in an ice age, while the North
Atlantic realm is not. This concept is useful, in a
more general form: The South Atlantic stays in the
ice age, while the great fluctuations are largely a
boreal phenomenon. It means that tropical and
subtropical processes are relatively much more important in the dynamics of the South Atlantic than
in the North Atlantic, where boreal forcing dominate.

As far as heat transport across the equator is
concerned (maximum in spring and fall, Fig. 29),
cold periods are a lot like northern winter, warm
periods like northern summer. Heat transport in the
Atlantic is northward most of the year except during boreal winter when there is significant
southwart transport (Hsiung et al. 1989). Fig. 28
shows a computation of latitudinal and annual
variation of net meridional heat transport within
the North and tropical Atlantic Ocean (Lamb &
Bunker 1982). Northward transport at the equator
(positiv values and unshaded) is strongest during
August to October. In our analogy this corresponds
to the cooling phase following the peak interglacial. This suggests that heatflow will eventually
decrease as a cold period is achieved (boreal winter analog).

In the equatorial region it is difficult, evidently,
to choose the time of the year (northern or south-

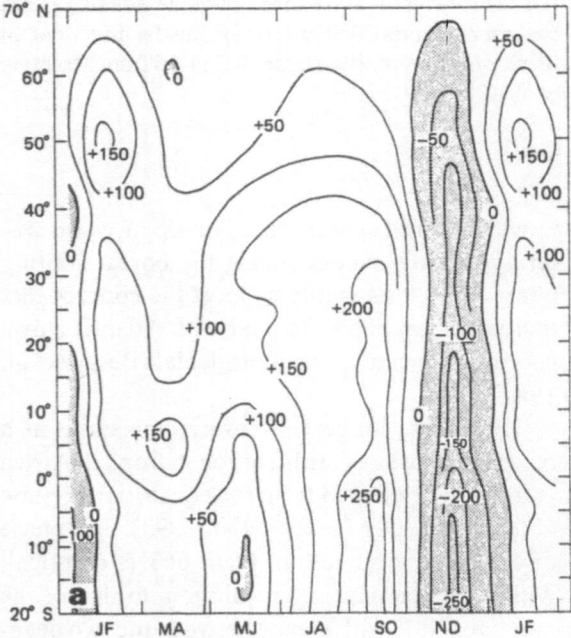

Fig. 29 Latitudinal and annual variation of net
meridional heat transport within the North and tropical
Atlantic Ocean (including Gulf of Mexico and Caribbean Sea) in 10^{13}W. Northward transport are positiv and
unshaded. (From Lamb and Bunker 1982).

ern winter) that is to deliver analogies for glacial conditions. In some respects the glacial equatorial system will be like northern winter, with the ITCZ moving south. In others, it will be like southern winter, with greatly increased SE trades.

Late in southern winter, a large difference exists in the thermocline between east and west (steep thermocline) (Fig. 4), indicating pile-up of warm water in the west (depressing the thermocline) and removing it from the east (seen in the rising of the thermocline). From this analogy, we would expect increased productivity during cold periods, as indeed seen in the eastern equatorial Atlantic (Fig. 23). In going to the west-equatorial Atlantic, however, we must note that this region is strongly influenced by seasonal change in the North Atlantic, as well, and our simple analogy must break down. Maximum western pile-up of warm water is in northern late summer and fall, that is, southern late winter and spring. This situation corresponds to the cooling phase in the boreal precessional cycles. Thus, both SE and NE trades cooperate to produce the pile-up, heatflow is strong, and the ITCZ is kept north after the sun has already reduced its northern input. This process, related exclusively to precession, would seem to slow reglaciation.

Our general picture for the last 300 kyrs is that circulation in the South Atlantic is forced by the SE trades moving water from the South through the Benguela Current to the west (SEC) into the North Atlantic. Intensity of water transport (and heat) depends on the strength of the SE tradewind. However, the actual movement of warm water across the equator, in the western region, also depends on the position of the ITCZ. While this zone is well north when SE trades are strong (southern winter), we cannot assume that this is so during glacial conditions, because of the NE trades pushing back (Flohn, 1985). From the sedimentary record strongest SE trades appeared during cold phases. Migration of the gyre margin to the west might reflect a spin-up effect due to increased flow of BOC and SEC during cold periods. During cold periods, upwelling in the equatorial area (documented in accumulation rates of organic carbon) and coastal zone off Namibia & Angola is increased (seen in accumulation rates of organic

carbon and *N. pachyderma* (sin) content, indicating offshore movement of upwelling filaments). This interpretation assumes greater transport from the South Atlantic toward the equator during cold periods. Although colder water is transported during periods with strong SE trades (August situation), heat transport may be just as great as now. However, we must assume that much of this heat does not leave the South Atlantic, because the ITCZ (and the heat equator) are close to the equator, decreasing cross-equatorial transport. Our data give no indication that contribution through the Agulhas retroflection changed significantly during the last 300 kyrs; it is not clear at this point how such a contribution could be identified, given the complexities of the Benguela System.

Acknowledgments

We would like to thank K. Herterich and F. Schott for discussions. We gratefully acknowledge B. Meyer-Schack and I. Pésza for help with stable isotope analyses. Thanks also go to A. Grimm-Geils, S. Middendorf and V. Diekamp for technical assistance. This is contribution no. 117 of Sonderforschungsbereich 261 at Bremen University.

References

Bé AWH (1977) An ecological, zoographic and taxonomic review of recent planctonic foraminifera. In: Ramsey ATS (ed) Oceanic micropaleontology pp 1-101 (Academic Press)

Berger A, Loutre MF (1991) Insolation values for the climate of the last 10 million years. Quat Sci Rev 10:297-317

Berger WH (1968) Planktonic foraminifera: selective solution and paleoclimatic interpretation. Deep-Sea Res 15:31-43

Berger WH (1981) Paleoceanography: the deep-sea record. In: Emiliani C (ed) The Sea, Vol. 7 Wiley-Interscience, New York, pp 1437-1519

Berger WH, Gardner JV (1975) On the determination of Pleistocene temperatures from planktonic foraminifera. J Foram Res 5 (2):102-113

Berger WH, Vincent E (1986) Deep-Sea carbonates: reading the carbon-isotope signal. Geol Rundschau 75:249-269

Berger WH, et al. (1978) Stables isotopes in deep-sea

carbonates: Box core ERDC-92, West Equatorial Pacific. Oceanologica Acta 1(2):203-206

Berger WH, Burke S, Vincent E (1987) Glacial-Holocene transition: climate pulsations and sporadic shutdown of NADW production. In: Berger WH Labeyrie LD (eds) Abrupt Climatic Change. D. Reidel, Dordrecht, pp 279-297

Berger WH, Killingley JS, Metzler CV, Vincent E (1985) Two-step deglaciation: ^{14}C-dated high-resolution d^{18}O records from the tropical Atlantic Ocean. Quaternary Res 23:258-271

Berger WH, Yasuda MK, Bickert T, Wefer G, Takayama T (1994a) Quaternary time scale for Ontong Java Plateau: Milankovitch template for Ocean Drilling Program Site 806. Geology 22:463-467

Berger WH, Herguera JC, Lange CB, Schneider R (1994b) Paleoproductivity: flux proxies versus nutrient proxies and other problems concerning the Quaternary productivity record. In: Zahn R, Kaminski M, Labeyrie LD, Pederson TF (eds) Carbon Cycling in the Glacial Ocean: Constraints on the Ocean's Role in Global Change. Springer, Berlin Heidelberg, pp 385-412

Bickert T (1992) Rekonstruktion der spätquartären Bodenwasserzirkulation im östlichen Südatlantik über stabile Isotope benthischer Foraminiferen. Berichte, Fachbereich Geowissenschaften, Univ. Bremen, 27, 205 pp

Boyle EA, Keigwin LD (1982) Deep circulation of the North Atlantic over the last 200,000 years: Geochemical evidence. Science 218:784-787

Broecker WS, Denton GH (1989) The role of ocean-atmosphere reorganizations in glacial cycles. Geochim Cosmochim Acta 53:2465-2501

Broecker WS, Peteet DM, Rind D (1985) Does the ocean-atmosphere system have more than one stable mode of operation? Nature 315:21-26

Chappell J, Shackleton NJ (1986) Oxygen isotopes and sea level. Nature 324:137-140

Charles CD, Fairbanks RG (1990) Glacial to interglacial changes in the isotopic gradients of the Southern Ocean surface water. - In: Bleil U, Thiede J (eds) Geological History of the Polar Oceans: Arctic versus Antarctic. Kluwer Academic, Dordrecht, pp 519-538

CLIMAP-Project Members (1976) The surface or the ice-age earth. Science 191: 1131-1137

Cochrane JD, Kelly FJ Jr, Olling CR (1979) Subthermocline countercurrents in the western equatorial Atlantic Ocean. J Phys Oceanogr 9:724-738

Conkright ME, Boyer TP, Levitus S (1994) Quality control and processing of historical oceanographic nutrient data. NOAA Technical Rep NESDIS 79, Washington, DC, pp 1-75

Craig H (1961) Standard for reporting concentrations of deuterium oxygen-18 in natural waters. Science 133:1833-1834

Craig H, Gordon LI (1965) Deuterium and oxygen-18 variations in ocean and marine atmosphere. In: Tongori E (ed) Stable Isotopes in Oceanographic Studies and Paleotemperatures, Spoleto, Consiglio Naz. delle Ricerche, Labor die Geol Nuc, Pisa , pp 9-130

Croll J (1875) Climate and Time in their Geological Relations. A Theory of Secular Changes of the Earth's Climate. Daldy, Isbister, & Co., London, 577 pp

DeMenocal PB, Ruddiman WF, Pokras EM (1993) Influences of high- and low-latitude Processes on African terrestrial climate: Pleistocene eolian records from equatorial Atlantic Ocean Drilling Program Site 663. Paleoceanography 8(2):209-242

Deuser WG, Ross EH (1989) Seasonally abundant planktonic foraminifera of the Sargasso Sea: Sucession, deep-water fluxes, isotopic compositions and palaeoceanographic implications. J Foram Res 19(4):268-293

Diester-Haass L (1985) Late Quaternary sedimentation on the eastern Walvis Ridge, SE Atlantic (HPC 532 and four piston cores). Mar Geol 65:145-189

Diester-Haass L, Meyers PA, Rothe P (1986) Light-dark cycles in opal-rich sediments near the Plio-Pleistocene boundary, DSDP Site 532, Walvis Ridge continental terrace. Marine Geology 73:1-23

Diester-Haass L, Heine K, Rothe P, Schrader H (1988) Late Quaternary history of continental climate and the Benguela current off southwest Africa. Palaeogeogr Palaeoclimatol Palaeoecol 65:81-91

Duplessy JC, Labeyrie L, Juillet-Leclerc A, Maitre F, Duprat J, Sarnthein M (1991) Surface salinity reconstruction of the North Atlantic Ocean during the last glacial maximum. Oceanol Acta 14:311-324

Duplessy JC, Shackleton NJ, Fairbanks RG, Labeyrie L, Oppo D, Kallel N (1988) Deepwater Source Variations During the Last Climatic Cycle and Their Impact on the Global Deepwater Circulation. Paleoceanography 3:343-360

Emiliani C (1954) Depth habitat of some species of pelagic foraminifera as indicated by oxygen isotope ratio. Amer J Sci 252:149-158

Emiliani C (1955) Pleistocene temperatures. J Geol 63:538-578

Epstein S, Buchsbaum R, Lowenstam HA, Urey HC (1953) Revised carbonate-water isotopic temperature scale. Bull Geol Soc Am 64:1315-1325

Fairbanks RG (1989) A 17,000-year glacio-eustatic sea level record: Influence of glacial melting rates on the Younger Dryas event and deep-ocean circulation. Nature 342:637-642

Fairbanks RG, Sverdlove M, Free R, Wiebe PH, Bé AWH (1982) Vertical distribution and isotopic fractionation of living planktonic foraminifera from the Panama Basin. Nature 298:841-844

Flohn H (1985) Das Problem der Klimaänderungen in Vergangenheit und Zukunft. Wissenschaftliche Buchgesellschaft, Darmstadt, 228 pp

Ganssen G (1983) Dokumentation von küstennahen Auftrieb anhand stabiler Isotope in rezenten Foraminiferen vor Nordwest-Afrika. Meteor Forschungs Ergebnisse 37:1-46

Giraudeau J (1993) Planktonic foraminiferal assemblages in surface sediments from the southwest African continental margin. Marine Geology 110:47-62

Gordon AL (1986) Interocean exchange of thermocline water. J Geoph Res 91:5037-5046

Gordon AL (1988) The South Atlantic: an overview of results from 1983-1988 research. Oceanography 1:12-18

Gordon AL, Bosley KT (1991) Cyclonic gyre in the tropical South Atlantic. Deep-Sea Res 38(1):323-343

Gordon AL, Weiss RF, Smethie WM Jr, Warner MJ (1992) Thermocline and intermediae water communication between the South Atlantic and Indian Oceans. J Geophys Res 97(C5):7223-7240

Hastenrath S (1982) On meridional heat transport in the world ocean. J Phys Oceanogr 12:922-927

Hemleben Ch, Spindler M, Breitinger I, Deuser WG (1985) Field and laboratory studies on the ontogeny and ecology of some globorotaliid species from the Sargasso sea off Bermuda. J Foram Res 15:254-272

Hemleben Ch, Spindler M, Anderson OR (1989) Modern planktonic foraminifera. Springer, New York, 363 pp

Herguera JC, Berger WH (1994) Glacial to postglacial drop of productivity in the western equatorial Pacific: Mixing rate versus nutrient concentrations. Geology 22:629-632

Höflich O (1974) The seasonal and secular variation of the meteorological parameters on both sides of the ITCZ in the Atlantic Ocean. GATE Reports, 2, Part VI, 36 pp

Hsiung J, Newell RE, Houghtby T (1989) The annual cycle of oceanic heat storage and oceanic meridional heat transport. J R Meteorol Soc 115:1-28

Hutson WH (1980) The Agulhas Current during the Late Pleistocene: Analysis of modern faunal analogs. Science 207:64-66

Imbrie J, Kipp NG (1971) A new micropaleontological method for quantitative paleoclimatology: Application to a late Pleistocene Caribbean core. In: Turekian KK (ed) The late Cenozoic glacial ages: Yale University Press, New Haven pp 71-181

Imbrie J, Hays JD, Martinson DG, McIntyre A, Mix AC, Morley JJ, Pisias NG, Prell WL, Shackleton NJ (1984) The orbital theory of Pleistocene climate: Support from a revised chronology of the marine ^{18}O record. In: Berger AL, Imbrie J, Hays JD, Kukla J, Saltzman J (eds) Milankovitch and Climate (Part 1) Reidel, Hingham Mass., pp 269-305

Imbrie J, McIntyre A, Mix A (1989) Oceanic response to orbital forcing in the late Quaternary: Observational and experimental strategies. In: Berger A et al. (eds) Climate and Geo-Sciences. Kluwer Academic, Boston Mass., pp 121-164

Jones JI (1967) Significance of distribution of planktonic foraminifera in the Equatorial Atlantic Undercurrent. Micropal 13(4):489-501

Katz EJ, Molinari L, Cartwright DE, Hisard P, Lass HU, DeMesquita A (1981) The seasonal transport of the Equatorial Undercurrent in the western Atlantic (during the Global Weather Experiment). Oceanol Acta 4:445-450

Kemle-von Mücke S (1994) Oberflächenwasserstruktur und -zirkulation des Südostatlantiks im Spätquartär. Berichte, Fachbereich Geowissenschaften, Univ. Bremen, 55, 151 pp.

Labeyrie LD, Duplessy JC, Blanc PL (1987) Variations in mode of formation and temperature of oceanic deep waters over the past 125,000 years. Nature 327:477-482

Lamb PJ, Bunker HF (1982) The annual march of the heat budget of the North and Tropical Atlantic Oceans. J Phys Oceanog 12:1388-1409

Lange CB and Berger WH (1993) Diatom productivity and preservation in the western equatorial Pacific: the Quaternary record. Proceedings of the Ocean Drilling Program, Scientific Results, 130:509-523

Levitus S (1982) Climatological Atlas of the World Ocean. NOAA Prof Pap 13:1-173

Lohmann GP (1992) Increasing seasonal upwelling in the subtropical South Atlantic over the past 700,000 yrs: Evidence from deep-living planktonic foraminifera. Mar Micropal 19:1-12

Lohmann GP, Schweitzer PN (1990) *Globorotalia truncatulinoides* growth and chemistry as probes of the past thermocline: 1. shell size. Paleoceanography 5(1): 55-75

Lutjeharms JRE, Meeuwis JM (1987) The extent and variability of South-east Atlantic upwelling. In: Payne AIL, Gulland JA, Brink KH (eds) The Benguela and comparable ecosystems. South African Journal of Marine Science 5:51-62

Lutjeharms JRE, Stockton PL (1987) Kinematics of the upwelling front off southern Africa. The Benguela and comparable ecosystems. South African Journal of Marine Science 5:35-49

Lutjeharms JRE, Valentine HR (1987) Water types and volumetric considerations of the south-east Atlantic upwelling regime. South African Journal of Marine Science 5:63-71

Mazeika PA (1968) Eastward flow within the equatorial current in the eastern South Atlantic. J Geophys Res 73:5819-5828

McIntyre A, Ruddiman WF, Karlin K, Mix AC (1989) Surface water response of the equatorial Atlantic Ocean to orbital forcing. Paleoceanography 4(1):19-55

Meinecke G (1992) Spätquartäre Oberflächen-wassertemperaturen im östlichen äquatorialen Atlantik. Berichte, Fachbereich Geowissenschaften, Univ. Bremen, 29, 181 p.

Meeuwis JM, Lutjeharms JRE (1990) Surface thermal characteristics of the Angola-Benguela front. South African Journal of marine Science 9:261-279

Milankovitch M (1930) Mathematische Klimalehre und astronomische Theorie der Klimaschwankungen. Handbuch der Klimatologie, Bd 1, Teil A. Bornträger, Berlin. 176pp

Mix AC (1989) Pleistocene Paleoproductivity: Evidence from Organic Carbon and Foraminiferal Species. In: Berger WH, Smetacek VS, Wefer G (eds) Productivity of the Ocean: Present and Past. Dahlem Workshop, Wiley, New York, pp 313-340

Mix AC, Ruddiman WR, McIntyre A (1986) Late Quaternary paleoceanography of the tropical Atlantic, 1: spatial variability of annual mean sea-surface temperatures, 0-20,000 years B.P.. Paleoceanography 1:43-66

Molfino B, McIntyre A (1990) Precessional forcing of nutricline dynamics in the Equatorial Atlantic. Science 249:766-769

Molinari RL, Voituriez B, Duncan P (1981) Observations in the subthermocline undercurrent of the equatorial South Atlantic Ocean: 1978-1980. Oceanologica Acta 4:451-456

Mulitza S (1994) Spätquartäre Variationen der oberflächennahen Hydrographie im westlichen äquatorialen Atlantik. Berichte, Fachbereich Geowissenschaften, Univ. Bremen, 57, 97pp

Mulitza S, Niebler HS, Dürkoop A, Wefer G. Planktonic foraminifera as recorders of past surface water stratification. Geology (manuscript submitted)

Oberhänsli, H. (1991) Upwelling signals at the northeastern Walvis Ridge during the past 500,000 years. Paleoceanography 6:53-71

Oberhänsli H, Bénier C, Meinecke G, Schmidt H, Schneider R, Wefer G (1992) Planktic foraminifers as tracers of ocean currents in the eastern South Atlantic. Paleoceanography 7(5):607-632

Peterson RG, Stramma L (1991) Upper-level circulation in the South Atlantic Ocean. Prog Oceanog 26:1-173

Philander SGH, Pacanowski RC (1984) Simulation of the seasonal cycle in the tropical Atlantic Ocean. Geophys Res Lett 11:802-804

Philander SGH,. Pacanowski RC (1986a) A model of the seasonal cycle in the tropical Atlantic Ocean. J Geophys Res 91:14192-14206

Philander SGH, Pacanowski RC (1986b) The mass and heat budget in a model of the tropical Atlantic Ocean. J Geophys Res 91:14212-14220

Phleger FB, Parker FL, Peirson JF (1953) North Atlantic Foraminifera. Sediment cores from the North Atlantic Ocean. Swed Deep Sea Exped Rpts 7:1-122

Pokras EM, Mix AC (1985) Eolian evidence for spatial variability of late Quaternary climates in tropical Africa. Quat Res 24:137-149

Prell WL (1985) The stability of low-latitude sea surface temperatures: An evaluation of the CLIMAP reconstruction with emphasis on the positive SST anomalies. Spec Publ TR025 U.S. Dept. Energy, Washington DC, 60 pp

Prell WL, Imbrie J, Martinson DG, Morley JJ, N.G. Pisias, N.J. Shackleton and H.F. Streeter (1986) Graphic correlation of oxygen isotope stratigraphy: Application to the Late Quaternary. Paleoceanography 1:137-162

Prell WL, Kutzbach JE (1987) Monsoon variability over the past 150,000 years. J Geophys Res 92:8411-8425

Ravelo AC (1991) Reconstructing the tropical Atlantic seasonal thermocline using planktonic foraminifera. PhD, Columbia University, New York, 168 pp

Ravelo AC, R.G. Fairbanks (1992) Oxygen isotopic composition of multiple species of planktonic foraminifera; records of the modern photic zone temperature gradient. Paleoceanography 7:815-831

Reid JL (1964) Evidence of a South Equatorial Counter Current in the Atlantic Ocean in July 1963. Nature 203:182

Richardson PL, Reverdin G (1987) Seasonal Cycle of Velocity in the Atlantic North Equatorial Countercurrent as Measured by Surface Drifters, Current Meters, and Ship Drifts. J Geophys Res 92:3691-3708

Richardson PL, Walsh D (1986) Mapping climatological seasonal variations of surface currents in the tropical Atlantic using ship drifts. J Geophys Res 91:10537-10550

Rind D, Peteet D (1985) Terrestrial conditions at the last glacial maximum and CLIMAP seasurface temperature estimates: Are they consistent? Quat Res 24:1-22

Rintoul SR (1991) South Atlantic interbasin exchange. J Geophys Res 96:2675-2692

Ruddiman and McIntyre (1984) An Evaluation of Ocean-Climate Theories on the North Atlantic. In: Berger A, Imbrie J, Hays J, Kukla G, Saltzman B (eds) Milankovitch and Climate. Part 2, NATO ASI Series C, D Reidel Publishing Comp, Dordrecht Boston Lancaster, pp 671-686

Schmidt H (1992) Der Benguela-Strom im Bereich des Walfisch-Rückens im Spätquartär. Berichte, Fachbereich Geowissenschaften, Univ. Bremen, 28, 172 pp

Schneider R (1991) Spätquartäre Produktivitätsänderungen im östlichen Angola-Becken: Reaktion auf Variationen im Passat-Monsun-Windsystem und in der Advektion des Benguela-Küstenstroms. Berichte, Fachbereich Geowissenschaften, Univ. Bremen, 21, 198 pp

Schneider RR, Müller PJ, Ruhland G (1995) Late Quaternary surface circulation in the east equatorial South Atlantic: Evidence from alkenone sea surface temperatures. Paleoceanography 10:197-219

Schott W (1935) Die Foraminiferen in dem äquatorialen Teil des Atlantischen Ozeans. Deutsche Atl. Exped. Meteor 1925-1927. 3:43-134

Schulz HD and Fahrtteilnehmer (1991) Bericht und erste Ergebnisse der METEOR-Fahrt M 16/2, Recife-Belem, 28.4-21.5.1991. Berichte, Fachbereich Geowissenschaften, Univ. Bremen, 19, 149 pp

Servain J, Legler DM (1986) Empirical orthogonal function analysis of tropical Atlantic sea surface temperature and wind stress. J Geophys Res 91:181-191

Shackleton NJ, Hall MA, Line J, Shuxi C (1983) Carbon isotope data in core V 19-30 confirm reduced carbon dioxide concentration of the ice age atmosphere. Nature 306:319-322

Shannon LV, Lutjeharms JRE, Agenbag JJ (1989) Episodic input of subantarctic water into the Benguela region. Suid-Afrikaanse Tydskrif vir Wetenskap 85:317-322

Sikes EL, Keigwin LD (1994) Equatorial Atlantic sea surface temperature for the last 30 kyr: A comparison of $U^{k'}_{37}$, $\delta^{18}O$ and foraminiferal assemblage temperature estimates. Paleoceanography 9:31-45

Stramma L, Peterson RG (1989) Geostrophic transport in the Benguela Current region. J Phys Oceanogr 19:1440-1448

Stramma L, Peterson RG (1990) The South Atlantic Current. J Phys Oceanogr 20:846-859

Stute M, et al. (1995) Cooling of tropical Brazil (5°C) during the last glacial maximum. Science 269:379-383

Summerhayes CP, Emeis KC, Angel MV, Smith RL, Zeitzschel B (1995) Upwelling in the Ocean: Modern Processes and Ancient Records. Dahlem Workshop Reports. John Wiley & Sons, Chichester New York Bisbane Toronto Singapore pp 422

Ufkes E, Zachariasse W (1993) Origin of coiling differences in living neogloboquadrinids in the Walvis Bay region, off Namibia, southwest Africa. Micropal 39:283-287

Vincent E, Berger WH (1981) Planktonic foraminifera and their use in paleoceanography. In: Emiliani C (ed) The Oceanic Lithosphere, The Sea, Vol. 7 John Wiley & Sons, New York, pp 1025-1119

Vincent E, Killingley JS, Berger WH (1985) Miocene oxygen and carbon isotope stratigraphy of the tropical Indian Ocean. In: Kennett JP (ed) The Miocene Ocean: Paleoceanography and Biogeography. Geol Soc Amer Memoir 163:103-130

Voituriez B (1981) Les sous-courants équatoriaux nord et sud et la formation des dômes thermiques tropicaux. Oceanologica Acta 4:497-506

Warren BA (1981) Deep circulation of the world ocean. In: Warren BA, Wunsch C (eds) Evolution of Physical Oceanography. MIT-Press, Cambridge, Mass, pp 6-41

Wefer G, Berger WH (1991) Isotope paleontology: growth and composition of extant calcareous species. Mar Geol 100:207-248

Wefer G, Fischer G (1993) Seasonal patterns of vertical particle flux in equatorial and coastal upwelling areas of the eastern Atlantic. Deep-Sea Res 40:1613-1645

Wefer G und Fahrtteilnehmer (1988) Bericht über die

502   G. Wefer et al.

METEOR-Fahrt M6/6, Libreville-Las Palmas, 18.2.-23.3.1988. Berichte, Fachbereich Geowissenschaften, Univ. Bremen, 3, 97 pp

Wefer G und Fahrtteilnehmer (1989) Bericht über die METEOR-Fahrt M9/4, Dakar-Santa Cruz, 19.2.1989-16.3.1989. Berichte, Fachbereich Geowissenschaften, Univ. Bremen, 7, 103 pp

Wefer G und Fahrtteilnehmer (1990) Bericht über die METEOR-Fahrt M12/1, Kapstadt-Funchal, 13.3.1990-14.4.1990. Berichte, Fachbereich Geowissenschaften, Univ. Bremen, 11, 66 pp

Wefer G und Fahrtteilnehmer (1992) Bericht und erste Ergebnisse der METEOR-Fahrt M20/1, Bremen-Abidjan, 18.11.1991-22.12.1991, Berichte Fachbereich Geowissenschaften, Univ. Bremen, 24, 74 pp

Whitmann JM, Berger WH (1992) Pliocene-Pleistocene oxygen isotope record Site 586, Ontong Java Plateau. Mar Micropal 18:171-198

Zahn R, Mix AC (1991) Benthic foraminiferal δ^{18}O in the ocean's temperature-salinity-density field: constraints on ice age thermohaline circulation. Paleoceanography 6:1-20

Climate Feedback and Pleistocene Variations in the Atlantic South Equatorial Current

A.C. Mix and A.E. Morey

College of Oceanic and Atmospheric Sciences,
Oregon State University, Corvallis, OR 97331, U.S.A.

Abstract: Ice-age cooling of the central equatorial Atlantic Ocean reflects both equatorial upwelling and advection of cool water off the southern-hemisphere eastern boundary, but the largest contribution appears to be advection. This conclusion is based on planktonic foraminiferal assemblages that vary over the last ~300 ky in the tropical Atlantic and Pacific Oceans. Core-top maps of these assemblages reveal relationships to 1) the tropical-subtropical warm pool, 2) equatorial and coastal upwelling, and 3) eastern boundary current advection. The faunal indices are not sensitive to selective dissolution, as they do not correlate with measures of preservation such as calcite fragmentation or water depth. The sequence of changes going into a glacial interval is 1) initial cooling associated with equatorial upwelling driven by trade winds, followed by 2) amplified cooling by intensified meridional winds along the SE Atlantic margin which advected cool Benguela Current water to the central equatorial Atlantic. The Gulf of Guinea maintained its present cyclonic gyre circulation in glacial time, bounded by a front similar to the present Angola-Benguela Front. We prefer a model in which thermal gradients in the southern hemisphere drive changes in the South Equatorial Currents.

Introduction

What caused large ice-age cooling in the South Equatorial Currents? These currents, relatively cool in their eastern reaches in both the Atlantic and Pacific (Fig. 1), are highly sensitive to change on all time frames. Examples of this sensitivity include large oscillations of temperature over annual cycles and inter-annual El Niño - Southern Oscillation phenomena (Philander 1979, 1986). Over longer time intervals, significant cooling occurred in both oceans associated with the peak of the last ice age (CLIMAP 1976, 1981). The South Equatorial Currents, which include upwelling zones near the equator, are major sources of CO_2 to the atmosphere (Tans et al. 1990) and thus their long-term variations are an important element in global climate feedback associated with greenhouse forcing (Jasper et al. 1994).

Although many features of the South Equatorial Currents may change, including current strength, upwelling, thermocline depth, and biological productivity, many studies describe changes in terms of sea-surface temperature. The exact range of glacial-to interglacial temperature contrast in the equatorial oceans, and the cause of the changes, remains controversial. The first detailed map of ice-age sea-surface temperatures, based on microfossil percentages (CLIMAP 1976, 1981) suggested that the South Equatorial Current was cooler than today in the cool upwelling season (August) by 2-4°C in the central Atlantic (or about a 1-3°C annual average change). Molfino et al. (1982) argued that in most cases temperature estimates are robust because different faunal and floral groups with unique ecological preferences and different preservation biases give similar results. An exception to this was off SW Africa, and in the equatorial zone, where temperature estimates from different fossil groups were somewhat discordant.

A controversy about tropical temperatures arose when Rind and Peteet (1985) suggested that tropical and subtropical sea-surface temperatures must have been much cooler than the CLIMAP estimates

From WEFER G, BERGER WH, SIEDLER G, WEBB DJ (eds), 1996, *The South Atlantic: Present and Past Circulation.* Springer-Verlag Berlin Heidelberg, pp 503-525

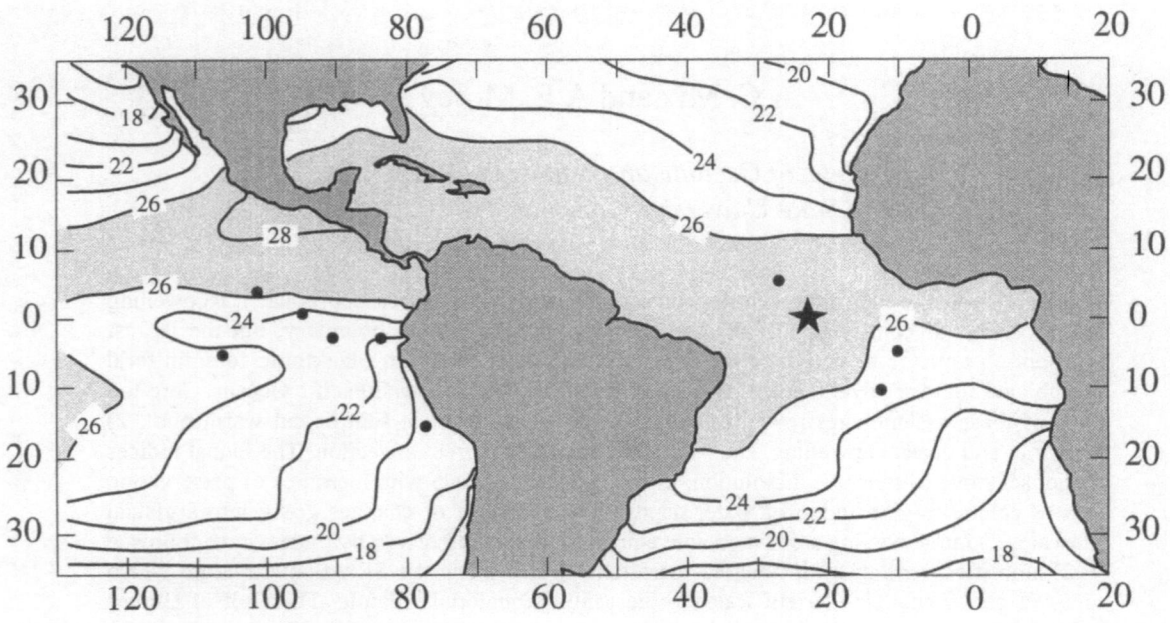

Fig. 1. Modern annual mean sea-surface temperatures (from Levitus 1982) illustrate relatively cool equatorial temperatures associated with equatorial upwelling and advection of cool water off the eastern boundaries. Core locations are shown, with small dots representing sites of down-core foraminiferal species data (Table 1) used to define faunal factors, and the star representing core V30-40, which is analyzed in detail here.

at the last glacial maximum, based on an atmospheric climate model and evidence for tropical mountain glaciation. Guilderson et al. (1994) supported this assertion at subtropical locations, using the Sr/Ca thermometer in corals. Other paleothermometers, however, (U^K_{37}, Sikes and Keigwin 1994; and oxygen isotopes, Broecker 1986; Mix 1992) support the lower contrast of glacial and interglacial temperature patterns in the low latitudes similar to the CLIMAP reconstructions.

Using methods similar to CLIMAP but a more detailed foraminiferal time series data set, Mix et al. (1986a,b) inferred an annual-average difference of about 4°C between glacial and interglacial extremes in the central equatorial Atlantic (up to 6°C in the upwelling season; Mix 1986; McIntyre et al. 1989). This large change was confined to the central equatorial zone; the temperature changes in the western equatorial Atlantic and the subtropics were as small as the original CLIMAP estimate, about

1-2°C. Mix (1989) suggested that some of the faunal and floral changes in the equatorial systems may reflect changing productivity, as well as temperature. The importance of thermocline depth rather than absolute temperature was noted by Ravelo et al. (1990). Oberhänsli et al. (1992) emphasize advection, food, and oxygenation of water as key elements controlling standing stocks of modern foraminifera.

There is as yet no consensus on the true amplitude of tropical sea surface temperature change. In this paper, we focus instead on the oceanographic mechanisms for changing equatorial faunas, using the foraminifera as watermass tracers. If we can identify from the fauna the primary mechanism that causes equatorial cooling, we gain insight into an important climatic feedback, possibly with global consequences related to inter-hemispheric heat transport and the global greenhouse effect.

Modern Oceanographic Summary

Variations in the South Equatorial Currents are driven by, and in turn modify, the low-latitude winds. Sea-surface temperatures are warmest north of the equator, especially in the Western Atlantic (Fig 1). This warm pool is associated with the position of the inter-tropical convergence zone (ITCZ) in the northern hemisphere. The ITCZ approaches the equator in the eastern Atlantic during northern hemisphere winter, when sea-surface temperatures are warmest in the Gulf of Guinea. It occasionally enters the southern hemisphere, in years when the Atlantic SEC is anomalously warm (Philander 1986). Upwelling along the equator responds to the zonal component of the southern trade winds via Ekman divergence. If this zonal component of the winds changes, one would expect to see corresponding changes in equatorial upwelling.

In the eastern equatorial Atlantic the winds are on average from the south (Chelton et al. 1990; Picaut et al. 1985). These southerly winds, which extend as far south as 5-10°S and cover nearly half the width of the Atlantic Basin, are in part a monsoonal effect driven by strong summer heating of the large African land mass north of the equator, but they also reflect the position of the subtropical high. Thus, cool temperatures south of the equator in the Gulf of Guinea reflect a combination of cool eastern-boundary waters advected westward by monsoonal wind forcing (Cane 1979; Philander 1979), and wave-like movements in the thermocline related to seasonal winds in the western Atlantic (Moore et al. 1978; Philander and Pacanowski 1986).

The cool eastern boundary waters of the Benguela Current are presently kept out of the eastern equatorial Atlantic by the Angola-Benguela front (Shannon et al. 1987), the southern boundary of a weak cyclonic gyre in the eastern tropical Atlantic (Gordon and Bosley 1991). Surface currents are dominated by northward and westward flowing Ekman drift (Arnault 1987), but the geostrophic component of currents flows eastward as a subsurface South Equatorial Counter Current (Molinari 1982), and southward in the Angola Current near the African coast (Arnault 1987; Peterson and Stramma 1991). The northward and westward

Ekman transport at the surface is small compared to the subsurface eastward advection of relatively warm, salty, subtropical water near the equator. Following the zone of zero wind-stress curl, the cool Benguela Current core leaves the eastern boundary near 15-20°S, and although it advects westward and northward, it is not strong enough at present to cause extreme cooling in the central equatorial Atlantic, where annual average temperatures are > 26°C at present (Fig. 1).

Hypotheses to Explain Ice-age Variations

A vast range of different opinions exists about the processes driving Pleistocene cooling of the SEC in the Atlantic Ocean, and at present there is no consensus. The options for the dominant process are 1) stronger equatorial upwelling related to zonal (Trade) winds, driven by thermal gradients in the southern hemisphere, 2) stronger upwelling along the tropical eastern boundary, related to southerly winds of the North African Monsoon, or 3) influence of cool Benguela Current water, advected to the equator, linked to meridional winds along Southwest Africa. We assess fossil data from the equatorial Atlantic in light of these hypotheses. Cane (1979), and Philander and Pacanowski (1980, 1981) illustrate with ocean models the response of a hypothetical ocean to the first two hypotheses. In both models, zonal (Trade) winds produce circulation symmetrical around the equator, with a shallow thermocline, implying cool temperatures, in the east and along the equator. Southerly (monsoonal) winds drive an asymmetrical circulation with cool water advected off the eastern boundary in the southern hemisphere and along the equator. Although these responses are quite different in their geographic details, with just a few study sites and only sea-surface temperature as a measure, it might be difficult to tell them apart. A better test is to trace the water masses separately, as we do here.

The Trade Wind Effect

Trade wind velocities respond to hemispheric thermal (and thus pressure) gradients. Krauss (1977) and Nicholson and Flohn (1981) argue that expanded glaciation of the high southern latitudes during the last ice age drove changes in the velocity of the southern trades. This idea was supported by the CLIMAP (1976, 1981) reconstructions, and by time-series studies of sea surface temperatures that associated cool events in the tropics with glacial maxima (Gardner and Hays 1976; Mix et al. 1986a). A problem with this inference was the reconstruction of enhanced seasonal contrast in the glacial tropical Atlantic (Mix et al. 1986b), opposite the reduced seasonal cycle in the southern oceans (Hays 1978). Atmospheric model results do not support a link to continental glaciation. Glaciers do not produce strong changes in the position or intensity of southern-hemisphere wind systems (Manabe and Broccoli 1985). Continental ice sheets alone apparently can not do the job, because the changes in Antarctica were relatively small.

Mix et al. (1986a) argued that a likely causal mechanism for stronger southern-hemisphere winds was greenhouse cooling associated with globally-reduced atmospheric pCO_2 of the last glaciation (Jouzel et al. 1993). Using an atmospheric general circulation model coupled to a mixed-layer ocean model, Broccoli and Manabe (1987) tested this idea. They did indeed find a significant cooling effect around Antarctica, and an increase in southern-hemisphere thermal gradients in the atmosphere due to lower pCO_2. The feedback mechanisms driving this gradient in the model were 1) expanded sea ice in the southern oceans (which amplified high-latitude cooling by reducing thermal inertia of the sea surface), and 2) convection of the tropical atmosphere (which reduced low-latitude cooling by spreading it through the troposphere). The response of sea-surface temperature in the tropics was small in the model, but because the model ocean did not simulate advection or upwelling, the tropical oceans' responses remain untested.

The Monsoon Effect

Another option for modifying the tropical Atlantic is the African monsoon. Solar heating of the large African land mass north of the equator generates relatively low atmospheric pressures. The contrast of warm land with cool waters (and thus relatively high atmospheric pressures) to the south draws winds across the equator. Changes in the temperature of NW Africa would modulate southerly winds in the Gulf of Guinea, and thus advection of cool water off the eastern boundary in the low latitudes of the southern hemisphere.

The African monsoon did vary on glacial-interglacial time frames, based on the occurrence of organic-rich sapropel layers in the Mediterranean Sea (Rossignol-Strick 1983) and wind-blown continental lake deposits in the deep Atlantic (Pokras and Mix 1986; DeMenocal et al. 1993). This process may be partially decoupled from the effect of southern hemisphere thermal gradients on easterly trade winds, if it is driven from the north rather than the south. High early-Holocene lake levels in tropical Africa suggest that northern-hemisphere summer insolation modulates the monsoon (Kutzbach and Street-Perrott 1985; Street-Perrott and Perrott 1993). A low-resolution climate model (Prell and Kutzbach 1987) also supported a significant role for northern-hemisphere insolation, although a higher-resolution model suggested that the insolation effect was small (DeMenocal and Rind 1993). Street-Perrott and Perrott (1990) noted that warmth of the North Atlantic may also have had a strong effect on the monsoon, and this effect was confirmed in a model sensitivity test by DeMenocal and Rind (1993).

Mix et al. (1986b) argued that the African monsoon was weakened during glacial time, and suggested that glacial cooling and an enhanced seasonal cycle in the eastern equatorial Atlantic was due to stronger trade winds and equatorial upwelling linked to reduced monsoonal advection. McIntyre et al. (1989) supported this idea of opposite responses of monsoon winds and trade winds, but argued that three mechanisms (monsoonal winds, equatorial upwelling, and advective heat transport) all contribute to the equatorial Atlantic response. They noted (as did Molfino et al. 1982) increased ice age abundance of a transitional and subpolar fauna in the equatorial Atlantic. More recently, Molfino and McIntyre (1990) analyzed pycnocline depth history using abundance of *Florisphaera profunda*, a species of coccolithophore, and emphasized the role of equatorial upwelling.

Unresolved is the relationship between southerly monsoon winds and trade winds. Maximum upwelling (minimum *F. profunda* of Molfino and

McIntyre 1990) does not correlate precisely with minimum monsoon index in the same samples (maximum eolian flux of freshwater diatom *Melosira sp.*, from Pokras and Mix 1985). With the linkage hypothesized by Mix et al. (1986a,b) and McIntyre et al. (1989), one might expect the fauna or flora associated with equatorial upwelling to vary antithetically with that associated with the eastern boundary currents.

The Benguela Effect

A third mechanism proposed to cool the SEC is to extend the cool Benguela Current into the equatorial zone (Gardner and Hays 1976). The northward advection of eastern boundary water is related to the position and strength of the subtropical high pressure zone. At present, the Benguela Current leaves the African coast near 20°S, near the northern limit of strongest southerly winds along Namibia, and advects westward well south of the SEC. Gardner and Hays (1976) did not specify how Benguela water would have reached the equatorial Atlantic during glacial time, but implied in a schematic drawing a northward extension of the Benguela current to the eastern-basin source of the SEC, between 5° and 15°S. This would imply northward or eastward displacement of the high-pressure zone, and perhaps elimination of the Angola-Benguela Front. It is also possible that Benguela Current water could leave the coast near its present position, and advect to the central equatorial Atlantic, west of a geostrophic cyclonic gyre that presently occupies the eastern tropical South Atlantic.

Methods

To reconstruct past oceanographic conditions, common practice has been to calibrate fossil plankton assemblages in core tops using the modern environment, and to apply these calibrations to faunas from the ancient geologic record. The usual approach is to simplify the core-top fauna into orthogonal Q-mode factors rotated with a varimax criterion (Klovan and Imbrie 1971). Q-mode factor analysis resolves each fossil assemblage into a proportional combination of a smaller number of independent end-members, or factors. Each factor is in turn a linear combination of input species. These assemblages, which generally form oceanographically meaningful map patterns in core tops, are assumed to be ecologically meaningful groupings. The result is a table of factor scores (that define the assemblages as weighted combinations of species), factor loadings (the weighting of each factor for each sample), and communalities (the sum of squares of loadings, representing the fraction of variance in each sample described by the factor model). The next step is often to generate a environmental transfer function which predicts a useful property, such as sea-surface temperature, by regressing modern faunal factor loadings against modern oceanographic properties (Imbrie and Kipp 1971). Finally, the modern factor definitions can be applied to ancient samples to find the relative importance of each factor for samples not included in the calibration. As long as significant evolution or some other process has not modified the ecological groupings, the transfer function can be used to make paleoenvironmental estimates of useful oceanographic indices.

A problem with this approach is that ice-age samples from the eastern equatorial Pacific and Atlantic are not well represented in modern core top assemblages (Moore et al. 1981). This is because the ice-age variation in climates from these areas is larger than the modern range within the tropics. For example, Ravelo et al. (1990) show that the faunal factors defined using only tropical Atlantic core tops explain just 35% of the population variance in similar samples from the last glacial maximum. This is called a "no-analog" condition, which can lead to erroneous estimates of past oceanographic conditions (Hutson 1977).

Here, we improve on this situation by calculating the faunal assemblages using Q-mode factor analysis of down-core samples from the equatorial Atlantic and Pacific Oceans over the past ~300 kyr, rather than from core tops. This guarantees that the assemblages are defined based on species that covary in the past, over a suitably large range of variations. An advantage is that the local down-core faunal assemblages record the regional context of tropical oceanographic processes. Their definitions are not biased by high-latitude core-top faunas that never existed within the tropics. By including Pacific samples with Atlantic samples in these factor definitions, we attempt to decrease the sensitivity of the faunal assemblages to selective dissolution. This works, because glacial cooling in the Atlantic and Pacific Oceans is similar but dissolution is in the opposite sense. We do not attempt to calibrate these assemblages to yield estimates of sea-surface temperature or any other index, but instead use the assemblages directly as tracers of water masses.

The samples used here for down-core faunal analyses (Table 1) come from four cores in the tropical Atlantic (1057 samples) and six cores in the eastern tropical Pacific (715 samples). Locations are illustrated in Fig. 1. Core-top samples come from the database assembled by Prell (1985) with the addition of Atlantic core tops from Mix (1986) and Pacific core-top samples from Coulbourn (1980), Sverdlove (1983), and Mix (unpublished

data). The Pacific results will be discussed in more detail elsewhere (A. Mix, in prep.).

In addition to our definition of factors from down-core samples, two features of our factor analysis differ from common practice. We excluded the dissolution resistant species *G. tumida*, *G. menardii*, and *G. menardii neoflexuosa*. These species were absent from the Atlantic for long periods of the late Pleistocene (Ericson 1968). They repopulated the Atlantic about 11,000 (^{14}C) years ago (Berger et al. 1985) perhaps when they were advected around South Africa during an episode of anomalous warmth in the southern ocean. This migration effect would introduce bias into local climate reconstructions that included these species as part of the index. Second, our analyses are based on ln(species percentage +1) rather than just percentages. This logarithmic transform amplifies the importance of less abundant species, and minimizes the dominance of a few species. One is added to

Table 1:
SOURCE OF 1772 DOWN-CORE PLANKTONIC FORAMINIFERAL SPECIES COUNTS TO DEFINE FACTORS

ATLANTIC:

Core	Latitude	Longitude	Depth	Samples	Source
RC24-16	5°02'S	10°12'W	3543 m	255	McIntyre et al. (1989)
V22-174	10°04'S	12°49'W	2630 m	333	Imbrie et al. (1989)
V30-36	5°21'N	27°19'W	4245 m	217	Mix (1989)
V30-40	0°12'N	23°09'W	3706 m	252	McIntyre etal.(1989)
					1057 TOTAL ATLANTIC

PACIFIC:

Core	Latitude	Longitude	Depth	Samples	Source
ODP-846	3°06'S	90°49'W	3307 m	239	Le et al. (in press)
RC10-62	3°20'N	101°43'W	3120 m	60	Mix (in prep)
RC13-110	0°06'N	95°39'W	3231 m	180	Mix (in prep)
V19-29	3°35'S	83°56'W	3157 m	89	Sverdlove (1983)
Y71-9-101P	6°23'S	106°56'W	3175 m	104	Mix (in prep)
Y71-6-12P	16°26'S	77°34'W	2734 m	43	Mix (in prep)
					715 TOTAL PACIFIC

each percentage value to avoid the log of zero. The log transform is applied to all species percentage data (down-core and core top) prior to calculation of factor loadings or scores. As noted by May (1975) regarding other biological systems, this makes relative abundances of species more Gaussian, which is an underlying assumption in the statistical calculations.

We assume closure (100%) around 26 species and morphotypes, listed in Table 2. Only species without significant taxonomic controversies are included. We included specimens referred to by Kipp (1976) as "P-D" intergrade with *N. dutertrei*, as this is how much of the Pacific core-top data set was counted by several investigators. Also, *G. theyeri* is grouped with *G. scitula*, because the former species was not widely recognized in earlier core-top studies. Morphotypes of *G. sacculifer* with and without a terminal chamber are grouped together, as are pink and white varieties of *G. ruber*.

To gain insight into the meaning of down-core faunal factors, we apply them to core-top samples. No-analog conditions are apparent as low communalities in core top samples. Because they occur in the modern samples rather than in the down-core samples, these poor analogs can be understood more clearly and excluded from the environmental calibrations. In all cases, our down-core samples are well described by our down-core faunal factors. Communalities in the down-core samples analyzed here are always >0.70, and average 0.90. Rather than explaining 35% of the population variance in down-core samples from the equatorial Atlantic and Pacific Oceans, as was the case for the application of core-top factors to down-core samples, they explain 90%. Thus, variations of our down-core factors are much more representative of oceanographic changes in the tropical water masses than were previous estimates based on core-top factors.

One core from the equatorial Atlantic (V30-40) is analyzed in detail. This site is on the equator and more than 1000 km from the eastern boundary. The Atlantic data were reported earlier by Mix et al. (1986a,b) and McIntyre et al. (1989). To express down-core variations as a function of time, age models developed elsewhere based on oxygen isotope stratigraphy (Mix 1992; Imbrie et al. 1984) were applied to the faunal data.

Cross-spectral analysis techniques assess the significance relationships of various time series generated in this study (Jenkins and Watts 1968). The time series were first interpolated at 3-ka intervals with a 9-ka Gaussian smoothing window. The spectra and cross spectra were calculated on a time series from 0-270 ka, with a bandwidth of 0.011 ka-1. The phase ranges and coherence significance were calculated at the 80% confidence level. For the purposes of calculating phase, we follow the SPECMAP conventions of Imbrie et al. (1989), in which the sign of all faunal indices is changed such that higher positive values would correspond to warmer temperatures or full interglacial conditions. For example, because lower values of $\delta^{18}O$ are associated with interglacial episodes, the oxygen isotope index is multiplied by negative one.

Results

Down-Core Faunal Factors and Core-Top Patterns

Q-mode factor analysis of the down-core faunas in the 10 sites used as input here resolves three significant faunal factors, which combine to explain 90% of the pooled population variance (Table 2). Core-top studies of foraminifera over a broad range of latitudes typically resolve five or more factors (Imbrie and Kipp 1971; Kipp 1976). Our results indicate, as did Ravelo et al. (1990), that just three factors are sufficient to describe the long-term variations of planktonic foraminifera in the equatorial current systems. A four-factor solution, which we reject, adds less than 2% additional information, and splits the first factor into two groups that are not ecologically meaningful.

We apply the factor scores from Table 2 to the core-top foraminiferal species percentages to assess the modern distributions of the assemblages. This yields three maps of loadings (strength) of these factors (Fig. 2). When mapping these factor loadings, we exclude any core-top samples with factor communalities less than 0.5, as these are not well described by the fauna present in the equatorial oceans. In nearly all cases, the low-communality core-top samples are from distant regions

Table 2: Q-mode Varimax Factor Scores based on 1772 down-core faunal counts in 10 cores from the tropical Atlantic and Pacific Oceans, calculated from species data expressed as ln(% + 1). Species are grouped by their dominant factor, and scores ≥0.25 are noted in bold type. Larger numbers indicate greater weighting in the factor index.

SPECIES	FACTOR 1 WARM TROPICAL	FACTOR 2 UPWELLING	FACTOR 3 E. BOUNDARY
G. ruber	**0.564**	-0.025	0.086
G. sacculifer	**0.471**	-0.094	0.129
G. glutinata	**0.380**	0.087	-0.170
G. aequilateralis	0.224	-0.055	0.013
P. obliquiloculata	0.226	-0.024	-0.080
G. rubescens	0.203	-0.070	-0.153
G. tenellus	0.182	-0.060	-0.135
G. falconensis	0.143	-0.050	-0.029
G. conglobatus	0.101	-0.047	-0.117
G. calida	0.105	0.051	0.003
G. scitula + G. theyeri	0.091	-0.023	-0.016
G. digitata	0.030	-0.004	0.016
S. dehiscens	0.022	-0.002	-0.030
N. dutertrei	0.170	**0.774**	-0.163
G. bulloides	-0.011	**0.469**	0.098
N. pachyderma (right-coiling)	-0.107	**0.340**	-0.461
O. universa	0.052	0.102	0.097
G. hirsuta	0.000	0.004	0.000
G. inflata	0.060	-0.055	**0.675**
G. truncatulinoides (right-coiling)	0.144	-0.102	**0.297**
G. crassaformis	0.122	-0.024	0.244
N. pachyderma (left coiling)	-0.009	0.020	0.087
G. quinquiloba	-0.004	0.013	0.035
G. truncatulinoides (left-coiling)	0.025	-0.012	0.012
G.conglomerata	0.021	0.040	-0.117
G. hexagona	0.072	0.025	-0.106

[*]G. menardii, G. menardii flexuosa, and G. tumida excluded from the percentage calculation and the factor analysis (see text).

such as the NW Pacific and Antarctic, which have little relevance to interpretation of the tropical data.

Factor 1 contains seven abundant species, in order of importance *G. ruber*, *G. sacculifer*, *G. glutinata*, *G. aequilateralis*, *P. obliquiloculata*, *G. rubescens*, and right-coiling *G. truncatulinoides*. All of these species are common in warm tropical and subtropical environments (Kipp 1976; Parker and Berger 1971). In core-top samples this factor

(Fig. 2a) is clearly associated with warm, salty subtropical and tropical water masses in the western Atlantic, and warm subtropical water masses in the Pacific where the thermocline is deep. Factor 1 also dominates the Panama Basin, where sea surface temperatures are warm, salinities are low, and the thermocline is shallow (Levitus, 1982). Warm temperature is the variable common to all these locations, so we refer to Factor 1 as the "Warm Tropi-

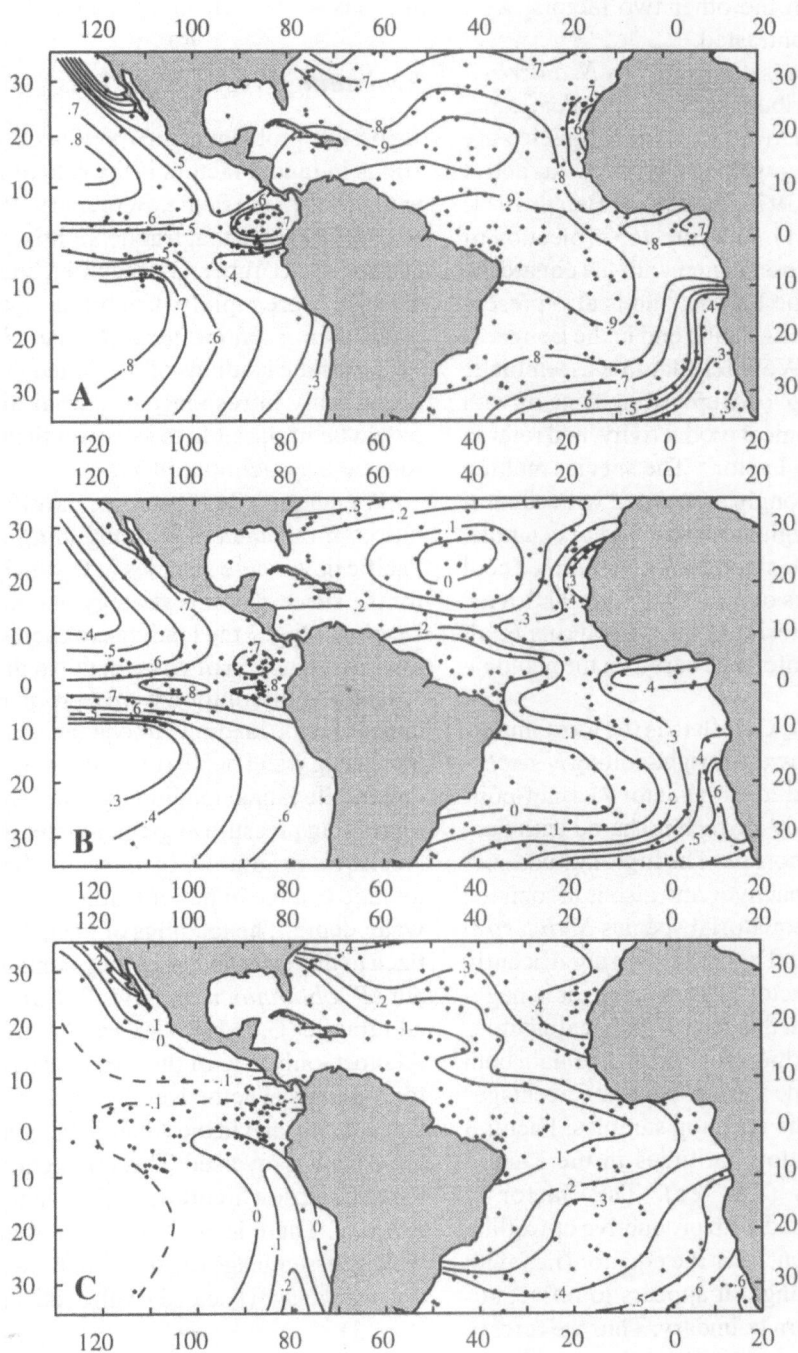

Fig. 2. Faunal factors applied to core-top samples. Values plotted are factor loadings of all samples with communality >0.50. a) Factor 1, Tropical Assemblage, b) Factor 2, Upwelling Assemblage, c) Factor 3, Eastern-Boundary Assemblage.

cal Assemblage". This is the relatively warm end member with which the other two factors, with cooler aspect, are contrasted.

The second factor is dominated by *N. dutertrei*, with secondary contributions from *G. bulloides* and right-coiling *N. pachyderma*. These species are common in tropical upwelling environments and at high latitudes (Kipp 1976; Parker and Berger 1971; Prell and Curry 1981). The core-top projection of Factor 2 (Fig. 2b) is associated with the equatorial upwelling zone in the Pacific, and is also present in the eastern tropical Atlantic and in the Benguela upwelling system. We infer that this assemblage records the strength of cool upwelling systems with relatively high biological productivity, and refer to it as the "Upwelling Factor". The species making up this factor are strongly associated with equatorial upwelling and high productivity. For example, their abundance in plankton tows that sampled cool upwelling conditions of the 1991 "La Niña" event in the central equatorial Pacific (Watkins et al. 1994) supports our inference based on the geologic record.

In the third factor, G. inflata is the most important contributor, followed by right-coiling *N. pachyderma* (which is shared with Factor 2), right-coiling *G. truncatulinoides* (which is shared with Factor 1), and *G. crassaformis*. The high-latitude species left-coiling N. pachyderma also is associated with Factor 3. The equatorial species *N. dutertrei* (which is common in Factor 2) has a significantly negative score in Factor 3. This means it is negatively correlated with this factor. Tropical samples which have high loadings for Factor 3 contain both abundant high-latitude species and low percentages of *N. dutertrei*. In the core-top samples, Factor 3 penetrates into the low latitudes in the eastern boundary currents (Fig. 2c). This factor is not strongly associated with productive upwelling centers. When present near the equator (i.e., with positive factor loadings) it appears to reflect advection off the eastern boundary. Thus we refer to it as the "Eastern Boundary Factor".

Species without significant loadings in any of these three factors are *S. dehiscens, G. digitata, G. hirsuta*, left-coiling *G. truncatulinoides*, and *G. quinquiloba*. These species are not abundant in the tropical water masses, and could be ignored in the analysis without changing the results.

Evaluating Preservation Bias

A possible problem in interpreting down-core variations in faunal factors is the role of preservation in modifying the fauna. Berger (1970) and Berger et al. (1982) note that fragile, lightly calcified species such as G. ruber, G. sacculifer, and G. glutinata dissolve more rapidly than robust species such as *G. tumida, P. obliquiloculata*, and *G. truncatulinoides*. If the loadings of the faunal factors change appreciably in response to partial dissolution, it would be impossible to interpret them in terms of surface-ocean climate change.

We attempted to minimize the effects of dissolution in our factors by combining Atlantic and Pacific down-core samples in the analysis. We test for the success of this strategy in two ways. First, we plot in Fig. 3 the loadings of each factor in core tops, as a function of water depth in the Pacific and Atlantic. If dissolution were a problem, we would expect factor loadings to change as a function of increasing water depth (resistant faunal loadings up and fragile faunal loadings down). This is because increasing pressure at greater water depths makes calcite more soluble. In all cases, there is no systematic change in factor loadings as a function of water depth. The loadings of Factor 1 are desensitized to this effect because they contain both resistant (*P. obliquiloculata* and *G. truncatulinoides*) and fragile (*G. ruber, G. sacculifer*) components. At most, only 5 % of the variance in core top factor loadings (Factor 1 in the Pacific) is related to water depth, and in the Atlantic the sense of change with depth is reversed. This is insignificant relative to the large down-core variations in the cores studied here. There is no correlation of either the upwelling assemblage (Factor 2) or the eastern boundary assemblage (Factor 3) with depth in either ocean (Fig. 3).

A second test of dissolution artifacts is to compare down-core factor loadings with a dissolution index in the same samples (Fig. 4). We use the percentages of foraminiferal fragments as dissolution index, as it is well known that partial dissolution induces fragmentation (Thunell 1976; Le and Shackleton 1992; Le et al. in press). Of the down-core data examined here, we have fragment data from Pacific cores RC10-62, RC13-110, and Y71-9-101P. All these data were generated in a consistent way in our laboratory at Oregon State University. In Fig. 4, we find no significant correlation of any of the faunal factors or communality with fragmentation in the same samples. The strongest relationship is between the warm tropical Factor 1 and the percentage of fragments, with a correlation coefficient $r = -0.19$ (not significantly different from zero). At most, 4% of the variance in Factor 1 loadings (and a smaller percentage of variance in the loadings of the other faunal factors) is linearly related to fragmentation in down-core samples. Again we find no evidence for significant control of the faunal loadings by dissolution. We conclude that our faunal factors are well buffered against dissolution bias, and that we can use them to infer history of upper ocean circulation.

The Last Glacial Maximum

An important test of the causes of large-scale climate change in the South Equatorial Current is to map the loadings of our faunal factors at the last glacial maximum. On this time horizon there is sufficient spatial coverage of samples that we can clearly see water masses. Fig. 5 illustrates the glacial assemblage patterns, which are best interpreted by comparing them to the core-top patterns.

Loadings of Factor 1, the tropical assemblage (Fig. 5a), are slightly reduced at the glacial maximum relative to at present along the equator in the eastern Atlantic. They change little in the western Atlantic, and actually increase at the glacial maximum in the eastern Gulf of Guinea. A strong gradient to less tropical fauna, along the eastern boundary between 15° and 20°S, demonstrates that the front separating cool Benguela waters from the eastern Gulf of Guinea existed in glacial time. This is consistent with the findings of Jansen et al. (this volume) and Schneider et al. (this volume) in cores near the African margin.

The upwelling assemblage of Factor 2 (Fig. 5b) is slightly stronger at the glacial maximum, and

extends farther west along the equator, than at present. Strong concentrations of this factor off NW Africa and SW Africa show that these eastern-boundary upwelling centers were maintained in approximately their present configuration in both glacial and interglacial time.

The changes in the first two faunal factors are significant, but subtle. The largest changes we observe are in Factor 3, the eastern boundary assemblage. At present (Fig. 2c) this fauna is most common at higher latitudes. High loadings are drawn to lower latitudes in the eastern boundary currents, but are presently very low (< 0.1) along the equator. At the glacial maximum (Fig. 5c), this fauna is clearly drawn into the South Equatorial Current, with loadings greater than 0.4 in the Central Equatorial Atlantic. Note that loadings in the Gulf of Guinea remain relatively low at the glacial maximum. We infer that cool waters from the Benguela system leave the eastern boundary between 10° and 20°S, and transit the westward edge of a cyclonic gyre in the Gulf of Guinea (south and west of the convergent Angola-Benguela Front). The strong presence of the eastern boundary fauna along the equator in the central Atlantic implicates this water mass as a source of cooling in the tropics. It is possible that the eastern boundary fauna is advected to the equator in the shallow subsurface, and blooms along the equator where a cool productive upwelling system provides it with a suitable food supply. Even if this late-bloom effect is at work, the only way we can imagine this fauna reaching the equator is through much stronger advection of Benguela water in glacial time.

Down-Core Variations

In Fig. 6 we illustrate variations of the down-core species percentages as a function of age in core V30-40, and in Fig. 7 the down-core faunal factors interpolated and smoothed for the time-series analysis, next to the oxygen isotope variations used for stratigraphic control. Lower values of $\delta^{18}O$ (plotted to the right) indicate "interglacial" episodes of low ice volume at high latitudes, and high values indicate glacial events.

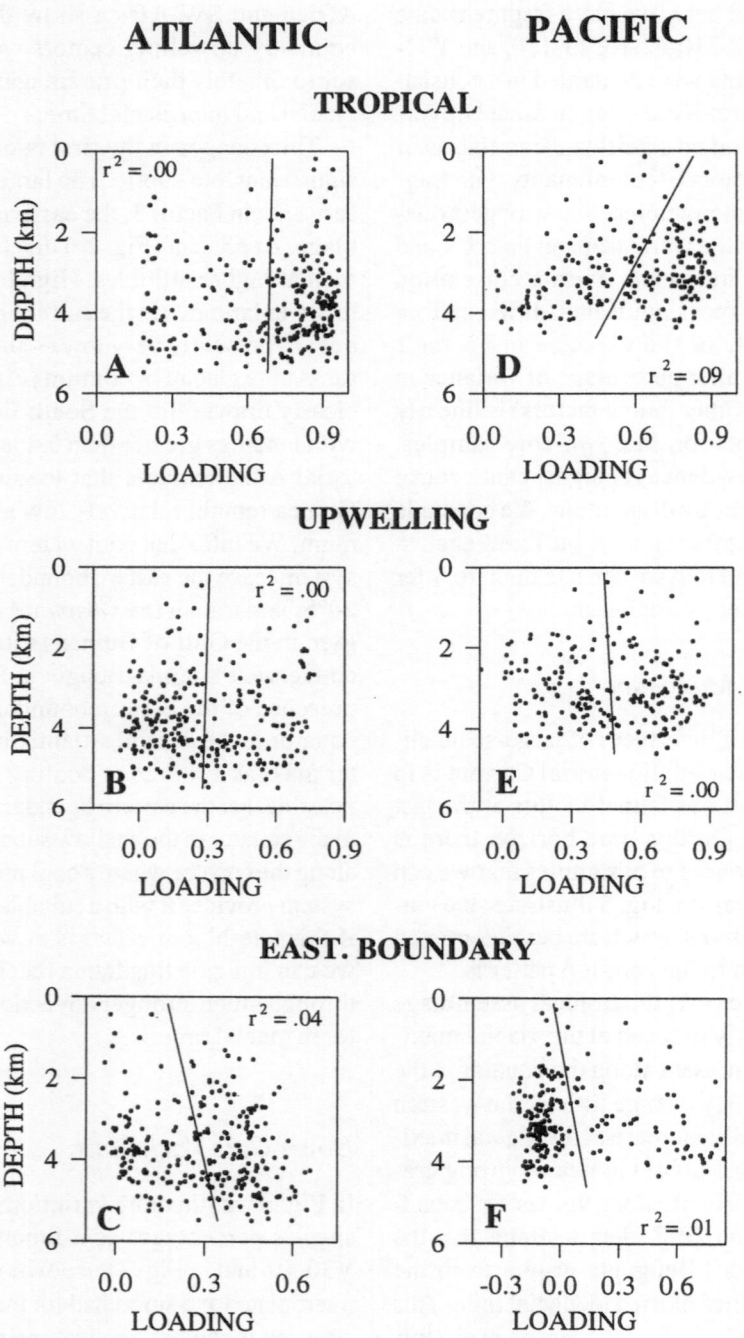

Fig. 3. Factor loadings of the tropical, upwelling, and eastern-boundary assemblages in core top samples are plotted separately in Atlantic and Pacific samples, as a function of water depth of the core sites. The lack of significant correlations with water depth indicate lack of sensitivity of these assemblages to dissolution, which increases with depth.

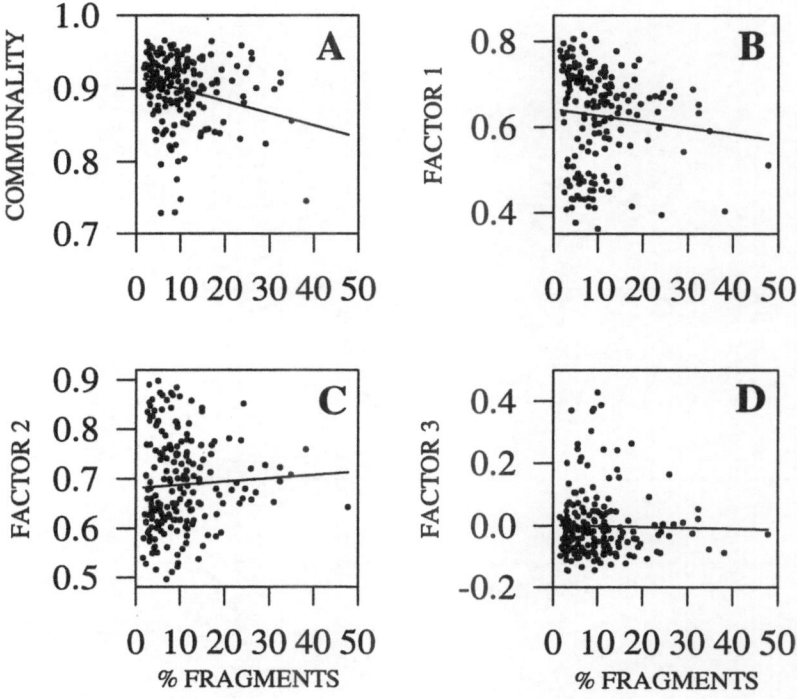

Fig. 4. Factor loadings and communalities as a function of the percentage of foraminferal fragments (relative to total foraminfera) in down-core samples. In all cases, there is no significant correlation, which suggests that these factor loadings are not sensitive to bias caused by partial dissolution.

In core V30-40, the upwelling and eastern boundary factors tend to covary. Both are higher during glacial episodes. This modifies the inference of Mix et al. (1986a) that zonal and meridional components of the equatorial winds vary antithetically due to monsoonal "steering". Loadings of the upwelling factor vary between extremes of about 0.3 and 0.6, while the eastern boundary factor varies from 0 to +0.5. The warm tropical factor changes in the opposite sense, with low loadings (down to 0.6) during glacial events and high values (up to 0.9) during interglacial events. This is consistent with past interpretations of cool glacial conditions in the equatorial Atlantic (CLIMAP 1976; Mix et al. 1986a; McIntyre et al. 1989). The new information here is that the greatest variability is in the eastern boundary assemblage (with total variance about four times that of the upwelling assemblage).

Discussion

Environmental change of South Equatorial Current is more complicated than just a change in equatorial upwelling. The relationships between the different oceanographic processes contributing to cooling in the Atlantic are expressed via spectral and cross-spectral analyses (Fig. 8), which compare the faunal changes in V30-40 associated with the upwelling and eastern boundary faunas to ice volume as reflected in the oxygen isotope record.

Fig. 8 shows that both the upwelling and eastern boundary faunas in the equatorial Atlantic are dominated by variance near the 100-ky period of orbital eccentricity, and the 23-ky period of orbital precession, and at both these periods the faunal changes are coherent with ice volume as recorded by $\delta^{18}O$. We infer from this that both upwelling and

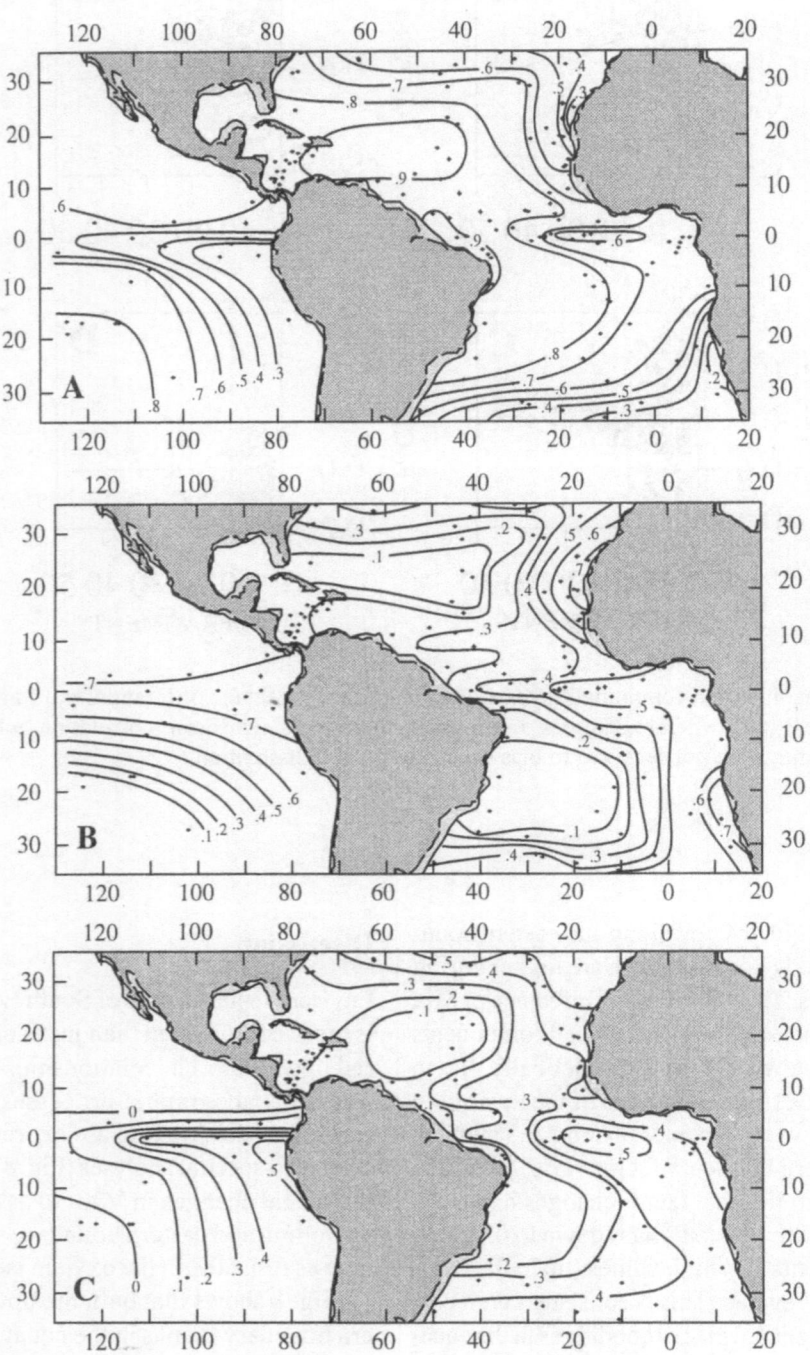

Fig. 5. Faunal factors applied to samples from the last glacial maximum. Values plotted are factor loadings of all samples with communality >0.50. a) Factor 1, Tropical Assemblage, b) Factor 2, Upwelling Assemblage, c) Factor 3, Eastern-Boundary Assemblage. Compare with modern distributions in Fig. 2. Note the large changes in Factor 3, extending to the central equatorial Atlantic, west of an apparent cyclonic gyre in the Gulf of Guinea.

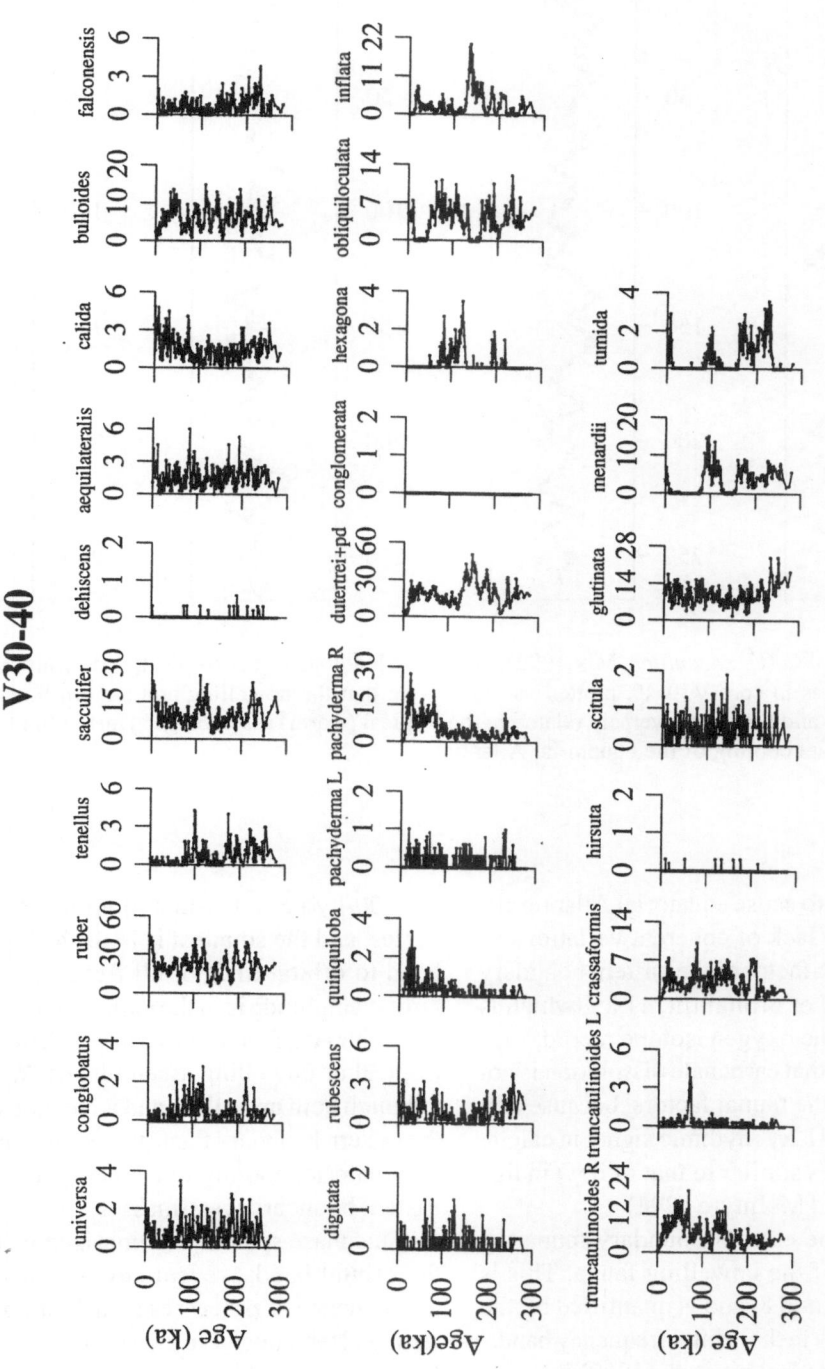

Fig. 6. Down-core species percentages vs. age in equatorial Atlantic core V30-40.

V30-40

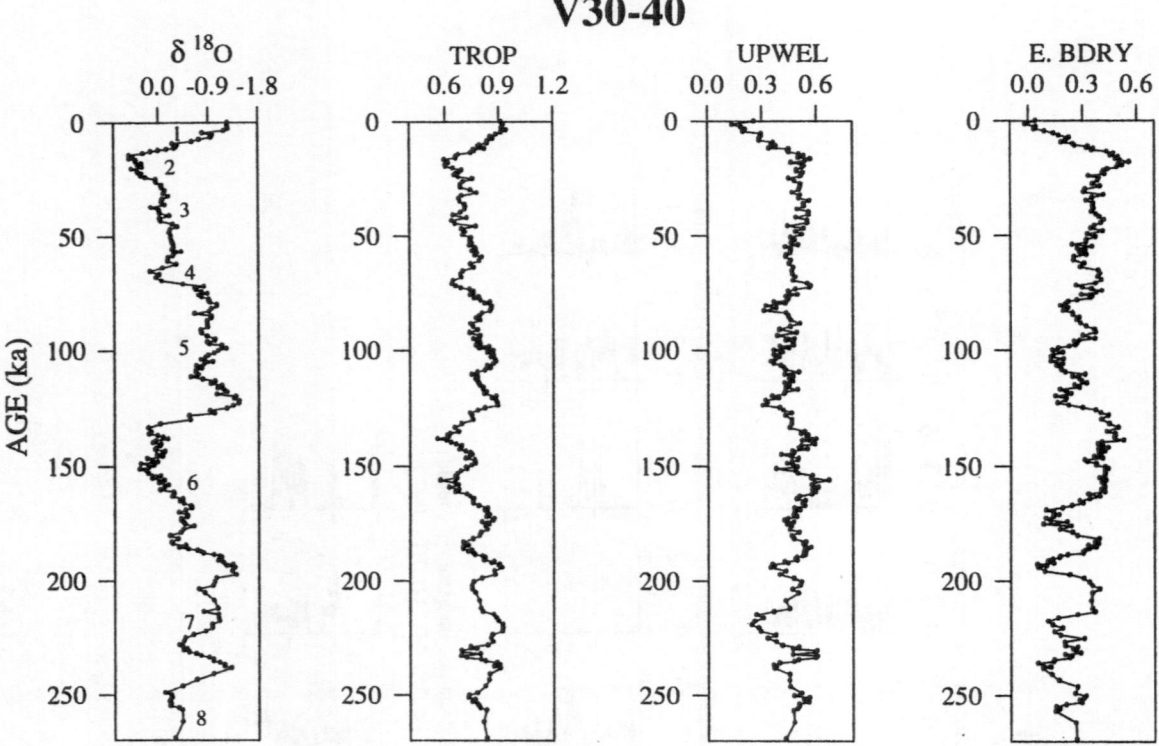

Fig. 7. Down-core $\delta^{18}O$ (*G. sacculifer*, Mix 1992) and factor loadings of the tropical, upwelling, and eastern-boundary assemblages in core V30-40, plotted vs. time. Note that the upwelling and eastern boundary (cool) assemblages covary, and both are inversely related to the tropical (warm) assemblage. Thus both upwelling and advection drive glacial cooling of the equatorial Atlantic.

advection conspire to cause equatorial Atlantic climate changes. The lack of coherent variations of either the upwelling factor or the eastern boundary factor at the period of orbital tilt, 41 ky, which is clearly present in the oxygen isotope record, supports our inference that carbonate dissolution is not strongly affecting the faunal factors, because dissolution imparts a 41-ky rhythmic signal in calcite preservation indices similar to that of $\delta^{18}O$ in this region (Verardo and McIntyre 1994).

Variations in the eastern boundary fauna are larger than those of the upwelling fauna. This is apparent in Figs. 7 and 8, and is quantified by the coherent amplitudes in the orbital frequency bands, calculated following Imbrie et al. (1989) in Table 3. For all assemblages, the largest amplitude is in

the 100-ky period, which dominates the ice age cycles, and the smallest is in the 41-ky period related to orbital tilt. For all frequency bands, the largest amplitude is in the eastern boundary assemblage; about a factor of two larger than those of the tropical or upwelling assemblages. We infer that although both upwelling and advection contribute, the eastern boundary fauna traces the primary cause of equatorial cooling to advection of water from eastern boundary watermasses.

The phase spectra, summarized in Table 4 for the orbital bands, reveal key information about the sequence of processes contributing to climate change. Here, negative phases indicate changes in fauna lead ice-volume changes as recorded by $\delta^{18}O$. Note that the phases are calculated such that zero

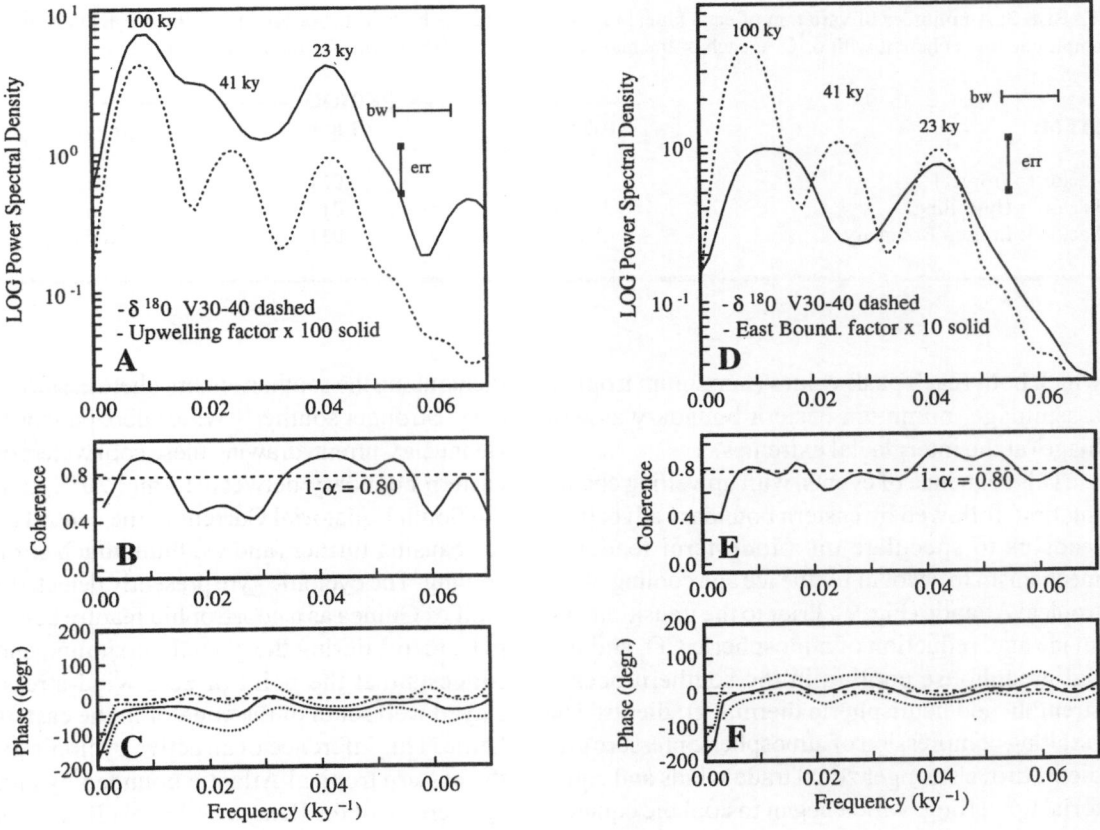

Fig. 8. Cross spectra, upwelling and eastern boundary faunal factors relative to $-\delta^{18}O$. In all cases, time series are 270 ky long, interpolated at 3 kyr intervals. Bandwidth is 0.011 kyr[-1]. a) Log of power spectral density, upwelling factor and $\delta^{18}O$, note that values for the upwelling factor are muliplied by 100 for scaling. b) Coherency of upwelling factor vs. $\delta^{18}O$. c) Phase angle, with negative values indicating a lead of minimum upwelling fauna ahead of full interglacial conditions. d) log of power spectral density, eastern boundary factor and $\delta^{18}O$. Values for the eastern boundary factor are multiplied by 10 for scaling. e) Coherency of eastern boundary factor vs. $\delta^{18}O$. f) Phase angle, with negative values indicating a lead of minimum eastern boundary fauna ahead of full interglacial conditions. Both the upwelling and eastern boundary assemblages have power concentrated near the 100-kyr (orbital eccentricity) and 23-kyr (orbital precession) bands.

phase indicates allignment of interglacial events with contribution of each index to warming (i.e., high tropical assemblage, low upwelling or eastern boundary assemblages). In the 23-kyr band, all factors lead ice volume by 30-40° (i.e., 2-3 kyr), and the phase error bars do not allow us to separate the different processes. Faunal changes are, however, in phase with atmospheric pCO_2 as re-

corded in the Vostok ice core (using the age model of Jouzel et al. 1993). This may suggest a link between greenhouse cooling and wind forcing in this band. In the 100-kyr band, minimum upwelling leads interglacial conditions by about 20° (i.e., ~6 kyr). The upwelling phase is similar to that of pCO_2 (low pCO_2, high upwelling), but leads changes in the tropical and eastern boundary assemblages,

TABLE 3: Amplitudes of variation of each faunal assemblage. Values here, in factor loading units, are the portion of amplitude that coherent with $\delta^{18}O$ in each of the main orbital bands. Bandwidth for the integration is 0.09 ky^{-1}.

ITEM:	PERIOD		
	100 kyr	41 kyr	23 kyr
Factor 1: Tropical	.062	.021	.055
Factor 2: Upwelling	.066	.019	.050
Factor 3: Eastern Boundary	.066	.031	.068

which both reach peak values (maximum tropical assemblage, minimum eastern boundary assemblage) at the interglacial extremes.

This sequence of events, with upwelling changing first, followed by eastern boundary advection, leads us to speculate on a long-term feedback mechanism to account for the ice age cooling of the tropical Atlantic (Fig. 9). Prior to the transition into an ice age, reduction of atmospheric CO_2 and initial greenhouse cooling in the southern oceans strengthened hemispheric thermal gradients. The resulting compression of atmospheric pressure gradients drove stronger zonal trade winds and equatorial upwelling, which began to cool the equator. Oceanic cooling relative to a warm African land mass shifted the thermal equator northward, and intensified the land-sea atmospheric pressure gra-

dients along the southern-hemisphere eastern boundary. Stronger southerly winds here enhanced the Benguela Current, drawing these cool waters off the eastern boundary between 15 and 20°S, to reach the South Equatorial Current in the central Atlantic, causing further (and we think much stronger) cooling. The cyclonic gyre presently detected in the Gulf of Guinea as a geostrophic feature continued to be active during the glacial maximum, and this suggests that the point of zero wind-stress curl stayed well south of the equator in the eastern Atlantic. This inference of an active cyclonic gyre in the eastern tropical Atlantic bounded by and ice-age version of the Benguela-Angola Front was also made by Schneider et al. (1995). Our down-core record in the central equatorial Atlantic is west of the cyclonic gyre, which may have varied independ-

TABLE 4: Phase Angle of orbital parameters (Berger and Loutre, 1991), ice-core pCO_2 (Jouzel et al., 1993), and equatorial Atlantic faunal assemblages relative to $-\delta^{18}O$ in core V30-40. Zero phase means index is coincident with interglacial maximum, negative values indicate index leads interglacial maximum. "nc" indicates no significant coherency. Phase of orbital parameters is equivalent to that of northern-hemisphere insolation strength on June 21 for the 41 kyr and 23 kyr bands (Berger and Loutre, 1991).

ITEM:	PERIOD		
	100 kyr	41 kyr	23 kyr
Orbital parameters	-13 ± 12	-80 ± 12	-87 ± 9
Vostok pCO_2	-26 ± 12	2 ± 13	-33 ± 23
Factor 1: Maximum Tropical	-4 ± 13	-19 ± 29	-36 ± 7
Factor 2: Minimum Upwelling	-23 ± 12	nc	-37 ± 12
Factor 3: Minimum East. Boundary	10 ± 22	nc	-21 ± 11

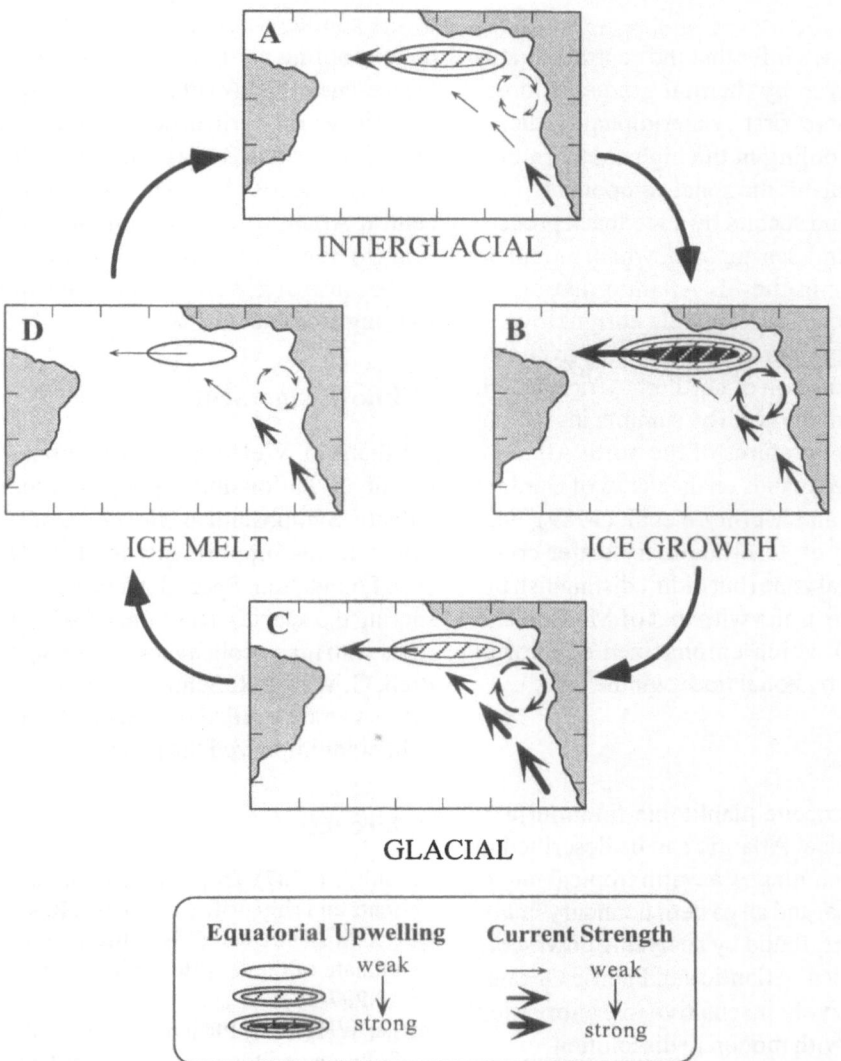

Fig. 9. Schematic diagram illustrating the inferred sequence of events in a "generic" ice age cycle in the South Atlantic near-surface curculation. a) Interglacial event: moderately strong equatorial upwelling. Benguela Current water weak or absent in the central equatorial Atlantic. b) As polar glaciation begins, southern ocean cooling strengthens hemispheric thermal gradients, driving stronger zonal trade winds and equatorial upwelling, which cool the equator. c) Near glacial maximum, thermal equator northward, and land-sea atmospheric pressure gradients enhanced the Benguela Current, drawing cool waters off the eastern boundary between 15 and 20°S, to reach the South Equatorial Current in the central Atlantic, amplifying cooling. The cyclonic gyre presently detected in the Gulf of Guinea as a geostrophic feature continued to be active during the glacial maximum, and this suggests that the point of zero wind-stress curl stayed well south of the equator in the eastern Atlantic. d) Ice melting phase, warmth in the southern ocean reduces thermal gradients, trade winds and equatorial upwelling diminish. Advection of Benguela water lessens, and the central equatorial Atlantic reaches peak warmth prior to full interglacial time.

ently in response to local monsoonal winds over the Gulf of Guinea.

Summarizing, we infer that the central equatorial system is driven by thermal gradients in the southern hemisphere, first by meridional gradients associated with cooling in the high southern latitudes which modulates the zonal component of the southern trades, and second by a feedback process associated with land-sea contrast, which modulate meridional winds in the SE Atlantic that draws water offshore from the Benguela current toward the equator. This system may be separate from that affected by the monsoon of northern Africa, which may be driven from the north by summer insolation, and changing temperatures of the north Atlantic. This supports and expands on the views of Gardner and Hays (1976) and McIntyre et al. (1989), who thought that cold eastern-boundary water could reach the equatorial zone (but didn't distinguish the local cyclonic gyre), and with that of Molfino and McIntyre (1990) which emphasized equatorial upwelling driven by zonal trade winds.

Conclusions

Changes in Pleistocene planktonic foraminiferal faunas in the tropical Atlantic can be described in terms of three assemblages, a warm tropical fauna, an upwelling fauna, and an eastern boundary fauna. These assemblages, found by analyzing down-core faunas in the tropical Atlantic and Pacific Oceans, appear to be relatively insensitive to preservation effects associate with moderate dissolution.

Pleistocene variations of the "upwelling" and "eastern boundary" assemblages in the central equatorial Atlantic are concentrated in 100-kyr and 23-kyr cycles similar to orbital eccentricity and precession. Most of the variance, both in time series and in the difference between core top and glacial maximum maps, is in the eastern boundary fauna, suggesting that advection of cool Benguela Current water is a key process in large-scale climate change at the equator. The sequence of faunal events at the equator, with early changes in the upwelling assemblage preceding glaciation, followed a few thousand years later by changes in the eastern boundary assemblage, suggests a succession of events in a coupled atmosphere and ocean.

We infer that early changes in the zonal component of the southern trades, perhaps driven by greenhouse cooling at higher southern latitudes, began the ice-age cycle. A later effect of a cool ocean next to a still-warm African continent intensified meridional winds that drew cool Benguela water from its source off SW Africa, toward the equator in the central Atlantic. The eastern Gulf of Guinea appears to be excluded from this large-scale variability, because of a local cyclonic gyre that maintains a strong front near 15°S.

Acknowledgements

We thank G. Wefer and the organizing committee for the invitation and support to attend the South Atlantic Symposium at University of Bremen. This research was supported by the U.S. National Science Foundation. Special thanks to A. McIntyre for sharing the species data from core V30-40. Discussions with many colleagues, including N. Pisias, W. Prell, G. Wefer, R. Schneider, and F. Jansen added to this work. Detailed reviews by W. Berger and R. Schneider improved the paper.

References

Arnault S (1987) Tropical Atlantic Geostrophic currents and ship drifts. J Geophys Res 92:5076-5088

Berger A, Loutre MF (1991) Insolation values for the climate of the last 10 million years. Quat Sci Rev 10:297-317

Berger WH (1970) Planktonic Foraminifera: Selective Solution and the Lysocline. Mar Geol 8:111-138

Berger WH, Bonneau MC, Parker FL (1982) Foraminifera on the deep-sea floor: lysocline and dissolution rate. Oceanol Acta 5:249-258

Berger WH, Killingley JS, Metzler CV, Vincent E (1985) Two-step deglaciation: ^{14}C-dated high-resolution d^{18}O records from the tropical Atlantic Ocean. Quat Res 23:258-271

Broccoli AJ, Manabe S (1987) The influence of continental ice, atmospheric CO_2, and land albedo on the climate of the last glacial maximum. Climate Dynamics 1:87-99

Broecker WS (1986) Oxygen isotope constraints on surface ocean temperatures. Quat Res 26:121-134

Cane, ME (1979) The response of an equatorial ocean to simple wind stress patterns: II. Numerical Results. J Mar Res 37:253-299

Chelton DG, Mestas-Nunez AM, Freilich M (1990) Global Wind Stress and Sverdrup circulation from the Seasat scatterometer. J Phys Oceanogr 20:1175-1205

CLIMAP Project Members (1976) The Surface of the Ice-Age Earth. Science 191:1131-1137

CLIMAP Project Members (1981) Seasonal Reconstructions of the Earth's Surface at the last glacial maximum. Geol Soc of Am, Map and Chart Series, MC-36:1-18

Coulbourn WT, Parker FL, Berger WH (1980) Faunal and solution patterns of planktonic foraminifera in surface sediments of the North Pacific. Marine Micropaleo 5:329-399

DeMenocal PB, Ruddiman WF, Pokras EM (1993) Influences of high- and low-latitude processes on African climate. Paleoceanogr 8:209-242

DeMenocal PB, Rind D (1993) Sensitivity of Asian and African climate to variations in seasonal insolation, glacial ice cover, sea-surface temperature, and Asian orography. J Geophys Res 98:7265-7287

Ericson DB (1968) Pleistocene climates and chronology in deep-sea sediments. Science 162:1227-1234

Gardner JV, Hays JD (1976) Responses of sea-surface temperature and circulation to global climate changes during the past 200,000 years in the eastern equatorial Atlantic Ocean. Mem Geol Soc Amer 145:221-246

Gordon AL, Bosley KT (1991) Cyclonic gyre in the tropical South Atlantic. Deep-Sea Res 38, Suppl. 1:S323-S343

Guilderson TP, Fairbanks RG, Rubenstone JL (1994) Tropical temperature variations since 20,000 years ago: Modulating inter-hemispheric climate change. Science 263:663-665

Hays JD (1978) A review of the late Quaternary climatic history of Antarctic Seas. In: van Zinderen Bakker, EM (ed) Antarctic Glacial History and World Palaeoenvironments, Balkeema, Rotterdam, pp 57-71

Hutson WH (1977) Transfer functions under no-analog conditions: Experiments with Indian Ocean Planktonic Foraminifera. Quat Res 8:355-367

Imbrie J, Kipp NG (1971) A new micropaleontological method for quantitative paleoclimatology: Application to a late Pleistocene Caribbean core. In: Turekian KK (ed) Late Cenozoic Glacial Ages. New Haven (Yale University Press), pp 71-182

Imbrie J, Hays JD, Martinson DG, McIntyre A, Mix AC, Morley JJ, Pisias NG, Prell WL, Shackleton NJ (1984) The orbital theory of Pleistocene climate: support from a revised chronology of the marine

$d^{18}O$ record. In: Berger AL, Imbrie J, Hays JD, Kukla G, Saltzman B (eds) Milankovitch and Climate, Part 1: Hingham (D. Reidel), pp 269-305

Imbrie J, McIntyre A, Mix A (1989) Oceanic response to orbital forcing in the late Quaternary: Observational and Experimental Strategies. In: Berger A, et al. (eds) Climate and Geosciences. Kluwer Academic, Dordrecht, pp 121-164

Jasper JP, Hayes JM, Mix AC, Prahl FG (1994) Photosynthetic fractionation of ^{13}C and concentrations of dissolved CO_2 in the central equatorial Pacific during the last 25,000 years. Paleoceanogr 9:781-798

Jenkins GM, Watts DG (1968) Spectral analysis and its applications. Holden day, San Francisco, 525 pp

Jouzel J, Barkov NI, Barnola JM, Bender M, Chapellaz J, Genthon C, Kotlyokov VM, Lipenkov V, Lorius C, Petit JR, Raynaud D, Raisbeck G, Ritz C, Sowers T, Stievenard M, Yiou F, Yiou P (1993) Extending the climatic records over the penultimate glacial period. Nature 364:407-412

Kipp NG (1976) New transfer function for estimating past sea surface conditions from sea-bed distribution of planktonic foraminiferal assemblages in the north Atlantic. In: Cline RM, Hays JD (eds) Investigation of Late Quaternary Paleoceanography and Paleoclimatology. Geol Soc Amer, Memoir 145:3-41

Klovan JE, Imbrie J (1971) An algorithm and FORTRAN IV program for large-scale Q-mode factor analysis. J Intern Math Geol 3:61-67

Krauss EB (1977) Subtropical droughts and cross-equatorial energy transports. Mon Weather Rev 105:1009-1018

Kutzbach JE, Street-Perrott FA (1985) Milankovitch forcing of fluctuations in the level of tropical lakes from 18 to 0 kyr B.P. Nature 317:130-134

Le J, Mix AC, Shackleton NJ (in press) Late Quaternary Planktonic foraminfers in ODP Site 846. In: Pisias NG, Mayer L, Janecek T (eds) Proc ODP, Sci. Results 138, Ocean Drilling Program, College Station, TX

Le J, Shackleton NJ (1992) Carbonate dissolution fluctuations in the western equatorial Pacific during the late Quaternary. Paleoceanography 7:21-42

Levitus S (1982) Climatological atlas of the world ocean. NOAA Prof. Paper 13, U.S. Govt Printing Office, Washington DC, pp 1-173

Manabe S, Broccoli AJ (1985) Influence of continental ice sheets on the climate of an ice age. J Geophys Res 90:2167-2190

May RM (1975) Patterns of species abundance and diversity. In: Cody ML, Diamond JM (eds) Ecology and evolution of communities. Belknap Press, Boston MA, pp 81-120

McIntyre A, Ruddiman WF, Karlin K, Mix AC (1989) Surface water response of the equatorial Atlantic Ocean to orbital forcing. Paleoceanogr 4:19-55

Mix AC (1986) Late Quaternary paleoceanography of the Atlantic Ocean: Foraminiferal faunal and stable isotopic evidence. PhD Dissertation, Columbia University, New York, 738 pp

Mix AC (1989) Pleistocene paleoproduc-tivity: evidence from organic carbon and foraminiferal species. In: Berger WH, Smetacek VS, Wefer G (eds) Productivity of the Ocean: Present and Past. Wiley, New York, pp 313-340

Mix AC (1992) The marine oxygen isotope record: Constraints on the timing and extent of ice-growth events (120-65 ka). In: Clark PU, Lea PD (eds) The last interglacial-glacial transition in North America. Geol Soc Amer Special Paper 270, Boulder, CO, pp 19-30

Mix AC, Ruddiman WF, McIntyre A (1986a) Late Quaternary paleoceanography of the tropical Atlantic, 1: Spatial variability of annual mean sea-surface temperatures, 0-20,000 years B.P. Paleoceanogr 1:43-66

Mix AC, Ruddiman WF, McIntyre A (1986b) Late Quaternary paleoceanography of the tropical Atlantic, 2, The seasonal cycle of sea-surface temperatures, 0-20,000 years B.P. Paleoceanogr 1:339-353

Molfino B, Kipp NG, Morley JJ (1982) Comparison of foraminiferal, coccolithophorid, and radiolarian paleotemperature equations: Assemblage coherency and estimate concordancy. Quat Res 17:279-313

Molfino B, McIntyre A (1990) Precessional forcing of nutricline dynamics in the equatorial Atlantic. Science 249:766-769

Molinari RL (1982) Observations of eastward currents in the tropical South Atlantic Ocean: 1978-1980. J Geophys Res 87:9707-9714

Moore DW, Hisard P, McCreary J, Merle J, O'Brien J, Picaut J, Verstraete J, Wunsch C (1978) Equatorial Adjustment in the eastern Atlantic. Geophys Res Lett 5:638-640

Moore TC Jr, Hutson WH, Kipp N, Hays JD, Prell WL, Thompson P, Boden G (1981) The biological record of the ice age ocean. Palaeogeogr, Palaeoclimatol, Palaeoecol 35:357-370

Nicholson SE, Flohn H (1981) African climatic changes in late Pleistocene and Holocene and the general atmospheric circulation. IAHS Publ 131, pp 295-301

Oberhänsli H, Bénier C, Meinecke G, Schmidt H,

Schneider R, Wefer G (1992) Planktonic foraminifers as tracers of ocean currents in the eastern South Atlantic. Paleoceanogr 7:607-632

Parker F, Berger WH (1971) Faunal and solution patterns of planktonic foraminifera in surface sediments of the South Pacific. Deep-Sea Res 18:73-107

Peterson RG, Stramma L (1991) Upper-level circulation in the South Atlantic Ocean. Progr Oceanogr 26:1-73

Philander SGH (1979) Variability of the tropical oceans. Dynamics of Atmospheres and Oceans 3:191-208

Philander SGH (1986) Unusual conditions in the tropical Atlantic Ocean in 1984. Nature 322:236-238

Philander SGH, Pacanowski RC (1980) The generation of equatorial currents. J Geophys Res 85:1123-1136

Philander SGH, Pacanowski RC (1981) The oceanic response to cross-equatorial winds (with application to coastal upwelling in low latitudes). Tellus 33:201-210

Philander SGH, Pacanowski RC (1986) A model of the seasonal cycle in the tropical Atlantic Ocean. J Geophys Res 91:14,192-14,206

Picaut J, Servain J, LeCompte P, Seva M, Lukas S, Rougier G (1985) Climatic Atlas of the tropical Atlantic wind stress and sea surface temperature 1964-1979. Publ, Laboratoire d'Oceanogr Phys, Univ Bretagne Occidental and Joint Inst. Marine Atmosph Res, University of Hawaii

Pokras EM, Mix AC (1986) Earth's precession cycle and Quaternary climate change in tropical Africa. Nature 326:486-487

Prell WL (1985) The Stability of Low-Latitude Sea-Surface Temperatures: An Evaluation of the CLIMAP Reconstruction with Emphasis on the Positive SST Anomalies. Technical report. TRO25, 60pp. Department of Energy, Washington D.C.

Prell WL, Curry WB (1981) Faunal and isotopic indexes of monsoonal upwelling: Western Arabian Sea. Oceanol Acta 4:91-98

Prell WL, Kutzbach JE (1987) Monsoon variability over the past 150,000 years. J Geophys Res 92:8,411-8,425

Ravelo AC, Fairbanks RG, Philander SGH (1990) Reconstructing tropical Atlantic hydrography using planktonic foraminifera and an ocean model. Paleoceanogr 5:409-431

Rind D, Peteet D (1985) Terrestrial conditions at the last glacial maximum and CLIMAP sea-surface temperature estimates: Are they consistent? Quat Res 24:1-22

Rossignol-Strick M (1983) African monsoons, an immediate response to orbital insolation. Nature 303:46-49

Schneider RR, Muller PJ, Ruhland G (1995) Late Quaternary surface circulation in the east equatorial South Atlantic: Evidence from alkenone sea surface temperatures. Paleoceanogr 10:197-220

Shannon LV, Agenbag JJ, Buys MEL (1987) Large- and mesoscale features of the Angola-Benguela Front. South African J Mar Sci 5:11-34

Sikes EL, Keigwin LD (1994) Equatorial Atlantic sea-surface temperature for the last 30 kyr: A comparison of U^K_{37}, $d^{18}O$, and foraminiferal assemblage temperature estimates. Paleoceanogr 9:31-46

Street-Perrott FA, Perrott RA (1990) Abrupt climate fluctuations in the tropics: the influence of Atlantic Ocean circulation. Nature 343:607-612

Street-Perrott FA, Perrott RA (1993) Holocene vegetation, lake levels, and climate of Africa. In: Wright HE Jr et al. (eds) Global Climates since the last glacial maximum. Univ. Minnesota Press, Minneapolis, pp 318-356

Sverdlove MS (1983) Planktonic foraminiferal ecology of the eastern equatorial Pacific Ocean: Including a paleoceanographic reconstruction of the Panama Basin for the last 320,000 years. PhD dissertation, Univ. Cincinnati, Cincinnati, OH, USA, 317 pp

Tans PP, Fung IY, Takahashi T (1990) Observational constraints on the global atmospheric CO_2 budget. Science 247:1431-1438

Thunell (1976) Optimum indices of calcium carbonate dissolution in deep-sea sediments. Geology 4:525-528

Verardo DJ, McIntyre A (1994) Production and destruction: Control of biogenous sedimentation in the tropical Atlantic 0-300,000 years B.P. Paleoceanography 9:63-86

Watkins J, Mix AC, Wilson J (1994) Physical and biological control of living planktic foraminifers of the JGOFS Equatorial Pacific transect. EOS 75:374 (abstr)

Late Quaternary Surface Temperatures and Productivity in the East-Equatorial South Atlantic: Response to Changes in Trade/ Monsoon Wind Forcing and Surface Water Advection

R.R. Schneider, P.J. Müller, G. Ruhland, G. Meinecke,
H. Schmidt and G. Wefer

Fachbereich Geowissenschaften, Universität Bremen, 28334 Bremen, Germany

Abstract: In order to reconstruct Late Quaternary variations of surface oceanography in the east-equatorial South Atlantic, time series of sea-surface temperatures (SST) and paleoproductivity were established from cores recovered in the Guinea and Angola Basins, and at the Walvis Ridge. These records, based on sedimentary alkenone and organic carbon concentrations, reveal that during the last 350,000 years surface circulation and productivity changes in the east-equatorial South Atlantic were highly sensitive to climate forcing at 23- and 100-kyr periodicities. Covarying SST and paleoproductivity changes at the equator and at the Walvis Ridge appear to be driven by variations in zonal trade-wind intensity, which forces intensification or reduction of coastal and equatorial upwelling, as well as enhanced Benguela cold water advection from the South. Phase relationships of precessional variations in the paleoproductivity and SST records from the distinct sites were evaluated with respect to boreal summer insolation over Africa, movements of southern ocean thermal fronts, and changes in global ice volume. The 23-kyr phasing implies a sensitivity of eastern South Atlantic surface water advection and upwelling to West African monsoon intensity and to changes in the position of the subtropical high pressure cell over the South Atlantic, both phenomena which modulate zonal strength of southeasterly trades. SST and productivity changes north of 20°S lack significant variance at the 41-kyr periodicity; and at the Walvis Ridge and the equator lead changes in ice volume. This may indicate that obliquity-driven climate change, characteristic for northern high latitudes, e.g fluctuations in continental ice masses, did not substantially influence subtropical and tropical surface circulation in the South Atlantic. At the 23-kyr cycle SST and productivity changes in the eastern Angola Basin lag those in the equatorial Atlantic and at the Walvis Ridge by about 3500 years. This lag is explained by variations in cross-equatorial surface water transport and west-east countercurrent return flow modifying precessional variations of SST and productivity in the eastern Angola Basin relative to those in the mid South Atlantic area under the central field of zonal trade winds. Sea level-related shifts of upwelling cells in phase with global climate change may be also recorded in SST and productivity variability along the continental margin off Southwest Africa. They may account for the delay of the paleoceanogreaphic signal from continental margin sites with respect to that from the pelagic sites at the equator and the Walvis Ridge.

Introduction

Aside from the common interglacial-glacial cyclicity, the response of the mixed layer in the equatorial Atlantic to climate change during the Late Quaternary is predominantly forced by the precessional component of insolation. In particular, the east-equatorial region shows marked temporal variations in estimated sea surface temperatures (SST) and foraminiferal assemblages with a significant periodicity of 23 kyrs (McIntyre et al. 1989, Mix & Morey, this volume). Similar dominant precessional variance is revealed in the proxies records for past changes in thermocline depth

From WEFER G, BERGER WH, SIEDLER G, WEBB DJ (eds), 1996, *The South Atlantic: Present and Past Circulation*. Springer-Verlag Berlin Heidelberg, pp 527-551

(Molfino & McIntyre 1992, Wefer et al., this volume) and in biological productivity (Lyle 1988, Gingele & Dahmke 1994, Verardo & McIntyre 1994). The 23-kyr signal in all equatorial records responds more or less in phase with changes in southern hemisphere sea -surface temperatures and boreal summer insolation at low latitudes, whilst leading changes in northern hemisphere sea-surface temperature and continental ice volume. It was therefore assumed that equatorial Atlantic surface-water variability is the product of both orbitally driven variations of trade wind zonality, modulated by African monsoon intensity, and changes of cold water advection by the eastern boundary current from the Cape Basin (McIntyre et al. 1989). To test this hypothesis sediment cores from the eastern Angola Basin and from the Walvis Ridge, located beneath the modern trade-wind driven upwelling and the Benguela Current pathway, were investigated in order to reconstruct changes in SST and paleoproductivity for the last 200,000 years (Schmidt 1992, Müller et al. 1994, Schneider et al. 1994, 1995). These records constrain the strong influence of precessional forcing via boreal summer insolation, trade wind zonality and eastern boundary current advection for the entire eastern South Atlantic as far south as 20°S. In this contribution we summarize different approaches of using alkenone SST estimates and sedimentary organic carbon as indicators for Late Quaternary changes in cold water advection and upwelling off the Southwest African margin. For this purpose two SST records from the subtropical eastern Atlantic were extended back to 300-400 kyrs B.P. and are compared with a new alkenone SST and paleoproductivity record from beneath the east-equatorial upwelling zone.

Modern Surface Circulation and Productivity

The hydrography of surface and subsurface waters in the equatorial South Atlantic was reviewed in detail by Peterson & Stramma (1991) and Van Bennekom & Berger (1984). The modern surface water circulation between 0° and 20°S and east of 15°W is characterized by a cyclonic gyre circulation including the northwest-directed Benguela Oceanic Current (BOC), the eastward South Equatorial Counter Current (SECC), and the Angola Current (AC) which flows southward along the Angola Margin (Fig. 1). The opposing directions of the Angola Current and the Benguela Coastal Current (BCC) result in a convergence of equatorial warm and colder subtropical surface waters between 14° and 17° S off Angola. North of this convergence called the Angola-Benguela-Front (ABF, Shannon et al. 1987), the BCC flows as a subsurface current feeding the oceanic upwelling with nutrient-rich waters from the Benguela system. High biological productivity in the eastern South Atlantic is induced by strong trade winds in austral winter at the equatorial upwelling zone and at the centers of coastal upwelling off Nambia (17° to 30°S). Also seasonal shoaling of nutrient-rich subsurface waters in the Angola cyclonic gyre area leads to enhanced oceanic productivity (Fig. 1) . The strong effect of wind stress on eastern South Atlantic surface waters can be depicted from the close relationship between mean annual SST distributions and wind vectors (Fig. 2).

Material and methods

Cores and Data

The sediment cores of this study were recovered from the eastern flank of the mid ocean ridge at the equator (GeoB 1105-4, 01°39.9'S / 12°25.7'W, water depth 3225 m), from the Congo Fan (GeoB 1008-3, 6°34.9'S / 10°19.1'E, water depth 3124 m), from the continental margin off Angola (GeoB 1016-3, 11°46.2'S / 11°40.9'E, water depth 3411 m), and from the Walvis Ridge (GeoB 1028-5, 20°06.2'S / 09°11.1'E, water depth 2209 m). The cores are from the different surface current and upwelling regimes in the east-equatorial South Atlantic (Fig. 1) and were previously investigated for their stable isotope geochemistry, foraminiferal assemblages and fluxes of sedimentary biogenic components (Schneider 1991, Meinecke 1992, Schmidt 1992, Müller & Schneider 1993, Müller

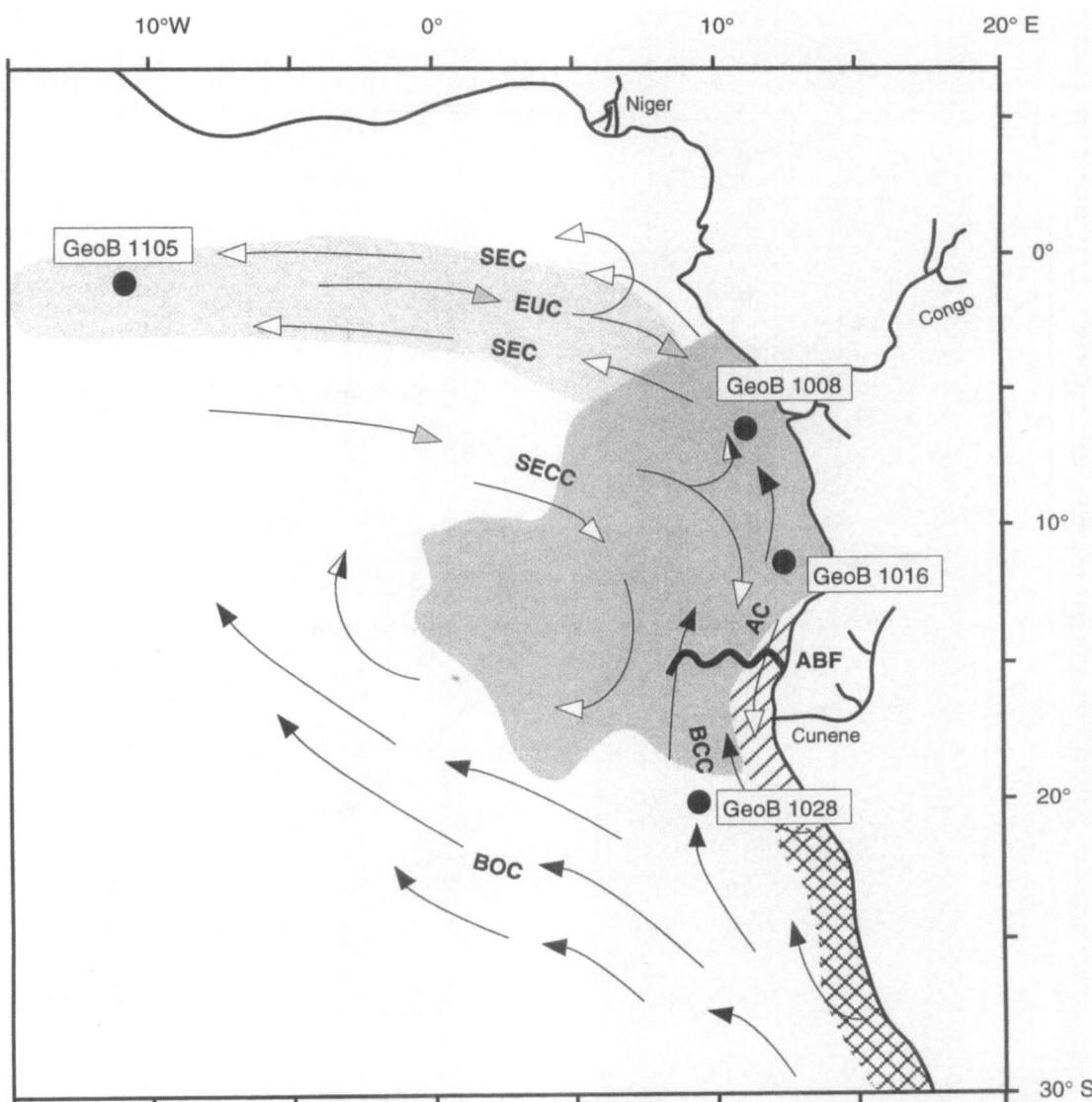

Fig. 1. Core locations (black dots), surface currents (black arrows - cold; white arrows - warm water; grey arrows - subsurface currents), and areas with high primary productivity (dotted area - equatorial divergence; grey schaded area - oceanic upwelling with thermocline shoaling and frontal mixing; hatched area - seasonal coastal upwelling; and cross hatched area - permanent upwelling) in the eastern South Atlantic.

et al. 1994, Schneider et al. 1994). Here we compare the records of alkenone SST estimates and sedimentary organic carbon fluxes in the time and frequency domain for the last 300 to 400 kyrs. All data and stratigraphic information are available on disk from the first author or from the SEPAN Paleoclimate Data Center (AWI, Bremerhaven).

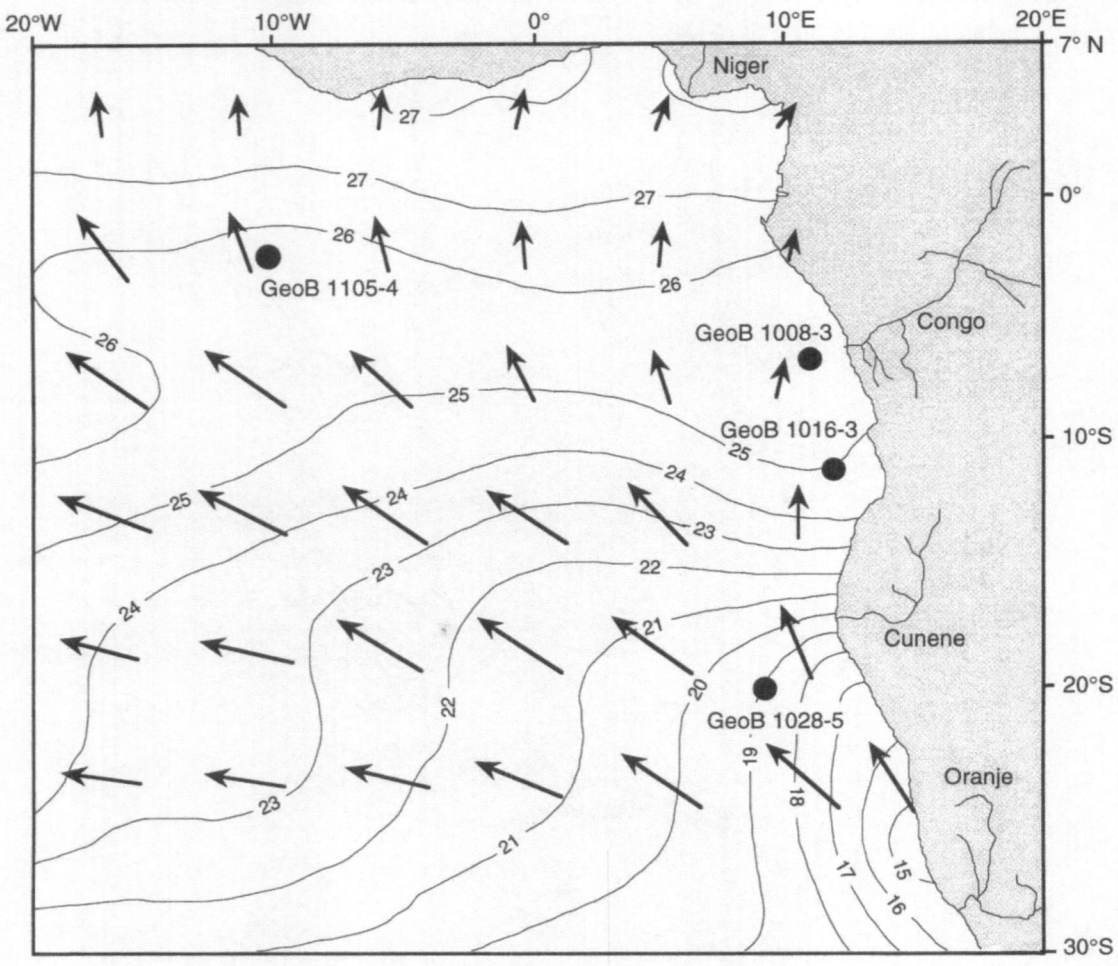

Fig. 2. Annual mean wind speed squared (redrawn from Picaut et al. 1985) and annual mean SST (Levitus 1982) in the eastern South Atlantic. Vector lengths on an arbitrary scale indicate relative wind speed changes. Black dots denote core locations.

Oxygen Isotopes, Stratigraphy and Time Series Analyses

Oxygen isotope analyses of *Globigerinoides ruber* were carried out on 10 to 12 specimens per sample from the size fraction of 200 to 350 microns, following standard techniques on a FINNIGAN MAT 251 mass spectrometer. Further details are given in Meinecke (1992) and Schneider (1991).

Isotope samples were taken at depth intervals of 5 cm in all cores. The age models for the sediments retrieved were derived from graphic correlation of the $\delta^{18}O$ records with the SPECMAP standard record (Imbrie et al. 1984). In general, the $\delta^{18}O$ records of *G. ruber* exhibit the typical Late Quaternary pattern resulting from changes in seawater $\delta^{18}O$ due to the buildup and retreat of polar ice caps

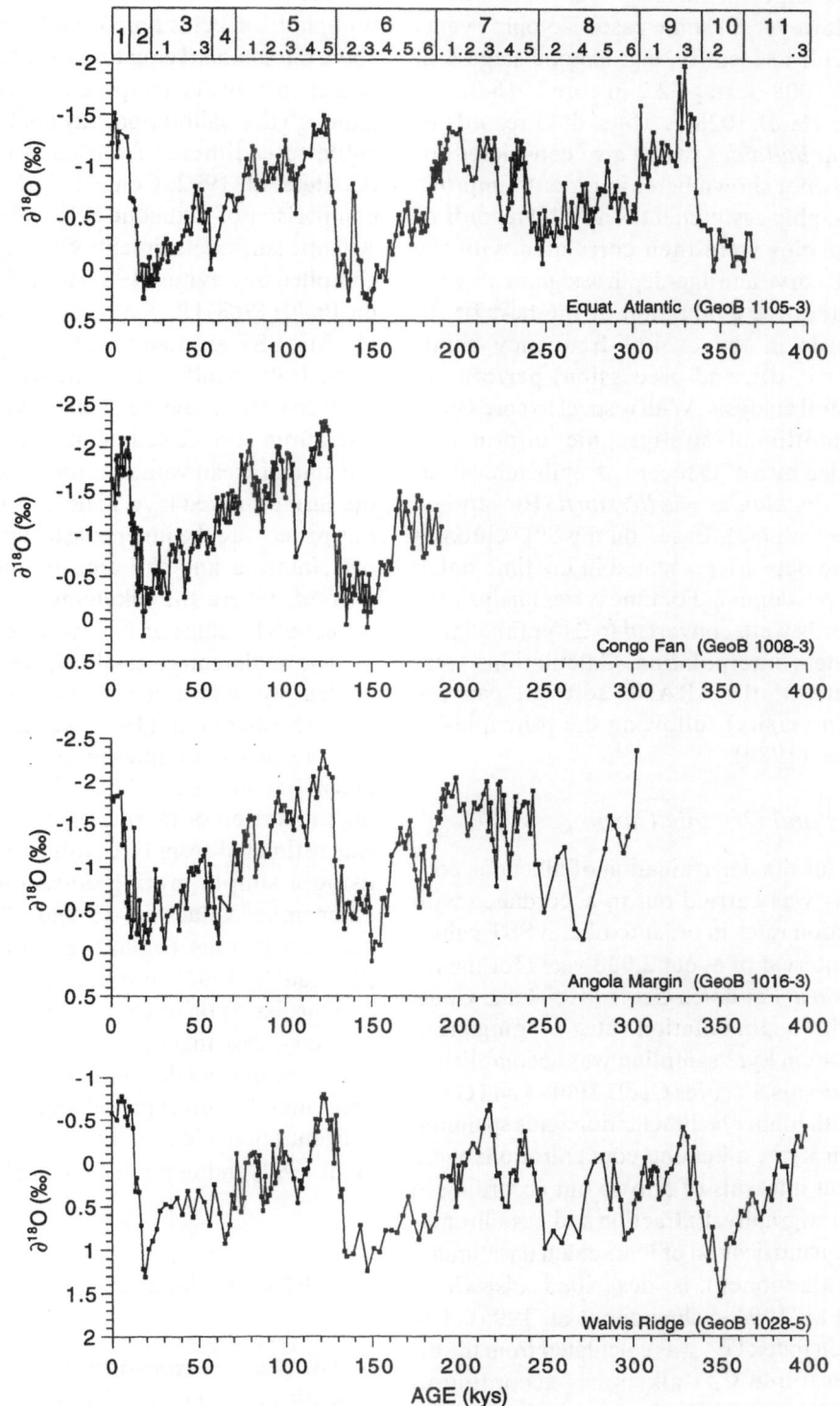

Fig. 3. Time series of stable oxygen isotopes of *Globigerinoides ruber* for cores used in this study. See Figure 1 for locations. Head bar denotes SPECMAP oxygen isotope stages and substages (Imbrie et al. 1984).

and from changes in surface water temperature (Fig. 3). However, in some cases isotopic events were difficult to identify, e.g. isotope stage 4 in core GeoB 1008-3, stage 2.2 in core 1016-3, and 7.1 in core GeoB 1028-5. Thus, $\delta^{18}O$ records of *Globigerina bulloides* where also considered for these cores (not shown here) in order to improve the stratigraphic assignment. The $\delta^{18}O$ records of the two species were then correlated with the SPECMAP curve and age-depth assignments were fixed by iterative evaluation of the best fit for both records in the orbital frequency bands (excentricity, tilt, and precession) performing cross-spectral analysis. With respect to core GeoB 1105-4, additional stratigraphic information was provided by a $\delta^{18}O$ record of epibenthic foraminifera *Cibicidoides wuellerstorfi* (Bickert and Wefer, this volume). Based on the $\delta^{18}O$ chronology all core data are presented in the time rather than the depth domain. For time series analysis the proxy records were converted to 2 kyr time intervals by linear interpolation. Calculations were carried out with the ARAND software package (Brown University) following the principles of Imbrie et al. (1989).

Alkenones and Organic Carbon

Sampling for the determination of alkenone concentrations was carried out in accordance with sedimentation rates in order to obtain SST values at a time interval of about 2,000 years for the respective cores. For cores GeoB 1105-4 and GeoB 1028-5 with sedimentation rates ranging from about 2 to 4 cm kyr^{-1} sampling was accomplished at 5 cm intervals. In cores GeoB 1008-3 and GeoB 1016-3, with higher sedimentation rates spanning 5 to 12 cm kyr^{-1}, alkenone concentrations were measured at intervals of 5 to 15 cm according to the $\delta^{18}O$ stratigraphy. Extraction and gas-chromatographic quantification of long-chain unsaturated ketones (alkenones) is described elsewhere (Müller et al. 1994, Schneider et al. 1995). The unsaturation index U^k_{37} was calculated from the di- and tri-unsatured C37-alkenones according to Brassel et al. (1986) and translated into SST using the calibration derived by Prahl et al. (1988):

$$SST (C°) = (U^k_{37}-0.039) / 0.034.$$ This relationship

is based on laboratory cultures of the coccolithophorid species *Emiliania huxleyi* and has become the standard equation for conversion of U^k_{37} values into water temperatures (Brassel 1993), although this calibration may not be strictly applicable to sediments from each oceanic region (Conte et al. 1992, Conte & Eglington 1993). A comparison of alkenone SST values from South Atlantic surface sediments with modern SST data compiled by Levitus (1982) reveals, however, that the Prahl (1988) U^k_{37} temperature calibration yields realistic SST estimates for this region (Schneider et al. 1995, Müller and Ehrhardt, unpublished). Between 10°N and 55°S the alkenone temperatures from surface sediments are similar to modern annual mean values of the uppermost 10 m of the surface mixed layer. The only exception from this pattern are the inner coastal upwelling centres off Namibia and the area of the circumpolar current, where the alkenone estimates tend to reflect SST values of the warm season.

The sedimentary total organic carbon (TOC) content was determined at 5 cm intervals according to Müller et al. (1994) at the same depths in the cores where samples for oxygen isotopes were taken. In core GeoB 1008-3 from the Congo Fan, the proportion of marine organic carbon (MOC) was estimated from $\delta^{13}C$ values of TOC ($\delta^{13}C_{org}$) using a simple mixing equation between two endmember values of -19 and -27‰ for marine and terrigenous organic carbon, respectively (Schneider, 1991). For the other three cores we assume, based on $\delta^{13}Corg$ values ranging between -18 and -22‰, that the TOC is overwhelmingly of marine origin and that the $\delta^{13}C_{org}$ variations in these cores mainly reflect past changes in surface water concentration of dissolved carbon dioxide (Müller et al. 1994, and unpublished data).

Results and Discussion

Sea Surface Temperatures: Changes in Upwelling and Cold Water Advection

The alkenone temperature records (Fig. 4) show that SST in the east-equatorial South Atlantic had

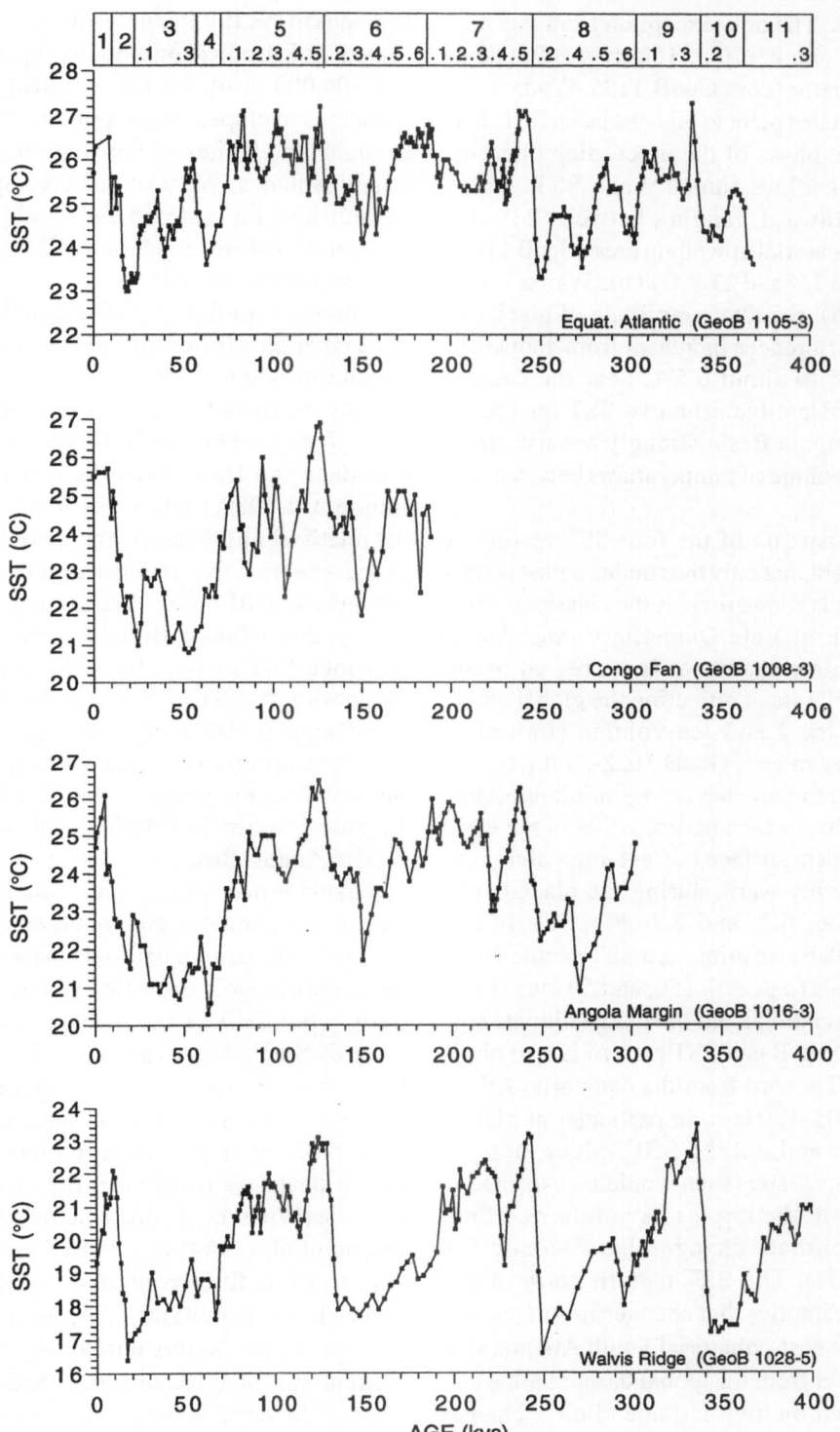

Fig. 4. Time series of alkenone-derived sea-surface temperatures in the east-equatorial South Atlantic

generally been lower in glacial compared to inter-glacial periods. The only exception from this pattern occurs at about 170 to 180 kyrs B.P. (stage 6.6) at the equator (core GeoB 1105-4) when the alkenone estimates point to higher glacial SST than during the late phase of the preceeding interglacial. While absolute annual mean SST values decrease southward, ranging between 23° and 27°C in the equatorial upwelling area (GeoB 1105-4) and between 16° and 23.5°C at the Walvis Ridge (GeoB 1028-5), the total amplitude of glacial to interglacial differences increases from about 4°C at the equator to about 6.5°C near the coastal upwelling off Namibia. The two SST records in the eastern Angola Basin strongly covary, spanning the same range of temperatures between 21° and 26.5°C.

From comparison of the four SST records it becomes evident that only the southernmost record from the Walvis Ridge reveals the classical „sawtooth" pattern of Late Quaternary interglacial to glacial climate change as is expressed in the SPECMAP $\delta^{18}O$ stack reflecting the global variations in sea level and ice-volume (Imbrie et al. 1984). Only in core GeoB 1028-5 all late glacial periods are characterized by minimum temperatures in the surface ocean, while in the eastern Angola Basin surface ocean temperatures remained relatively warm during the glacial sub-stages 8.2, 6.6, 6.2, and 2.2 (Fig. 5a). In the eastern Angola Basin minimum SST occured during mid glacials (e.g. 270, 150, and 50 kyrs B.P.). Part of this deviation from the global climate signal in the Angola Basin SST pattern is also obvious in the SST record from the equatorial Atlantic (GeoB 1105-4). Here, in particular at glacial substages 6.6 and 6.2 the SST values indicate warmer surface waters than would be expected if SST changes in the tropics have followed Late Quaternary climate change characteristic for higher latitudes. The SST pattern north of the Walvis Ridge implies that changes in surface circulation in the east-equatorial South Atlantic significantly differ from the global thermohaline circulation driven by high-latitude climate change (Broecker and Denton 1989, Imbrie et al. 1992) and having minimum SST in the Atlantic during peak glacial times (CLIMAP 1981). The long-term

trend in Angola Basin SST changes better corresponds to the 100-kyr undulation of precessional variations in tropical insolation (Fig. 5b). Over the last 300,000 years the low frequent part of SST variance envelopes the amplitude change in precessional oscillations of boreal summer insolation at low latitudes. We take this correspondence as an indicator for a strong linkage between east-equatorial surface circulation and tropical to sub-tropical climate change.

Previous studies provided considerable evidence that the SST changes observed in the east-equatorial South Atlantic predominantly reflect varying intensities of wind-driven upwelling and advection of cold water from the Southern Ocean (Gardner and Hays 1976, Mix et al. 1986 a,b, Imbrie et al. 1989, McIntyre et al. 1989). However, the relative importance of the two oceanographic processes for the reconstructed temperature changes is still being discussed (see Mix and Morey, this volume). Based on comparison of the alkenone SST records from the eastern Angola Basin with the SST record from the Walvis Ridge it was suggested that a modern-type cyclonic gyre circulation (Fig. 2) has continuously existed over the last 200,000 years (Schneider et al. 1995). Relatively warm SST during peak glacials in the eastern Angola Basin, as indicated in Fig. 5a, were explained as on average warmer annual mean SST due to a seasonally enhanced counter-current transport of warm equatorial surface waters into the eastern Angola Basin, in response to enhanced trade wind drift in the central South Atlantic towards South America (Schneider et al. 1995). Moreover, stronger trades during maximum glacial conditions in the northern hemisphere could have blocked cross-equatorial heat flow, upon which heat was trapped in the east-equatorial Atlantic (Wefer et al., this volume). According to Jansen et al. (1984) and Jansen and van Iperen (1989) it was further inferred that during maximum glacial conditions, advection of Benguela Current surface waters into the South Equatorial Current was directed northwestward to the central South Atlantic as in the modern ocean, instead of being focussed to a more northerly direction along the Angola continental margin as proposed in some reconstructions of glacial surface circulation in the

South Atlantic (Gardner & Hays 1976, Morley and Hays 1979, Diester-Haass et al. 1985, Pokras 1987). Maximum cold water advection into the eastern Angola Basin occurred at about 40 to 50 ka B.P. (Schneider et al. 1995, Jansen et al., this volume), a period when the amplitude of precessional insolation changes was at its minimum during the last 300 kyrs.

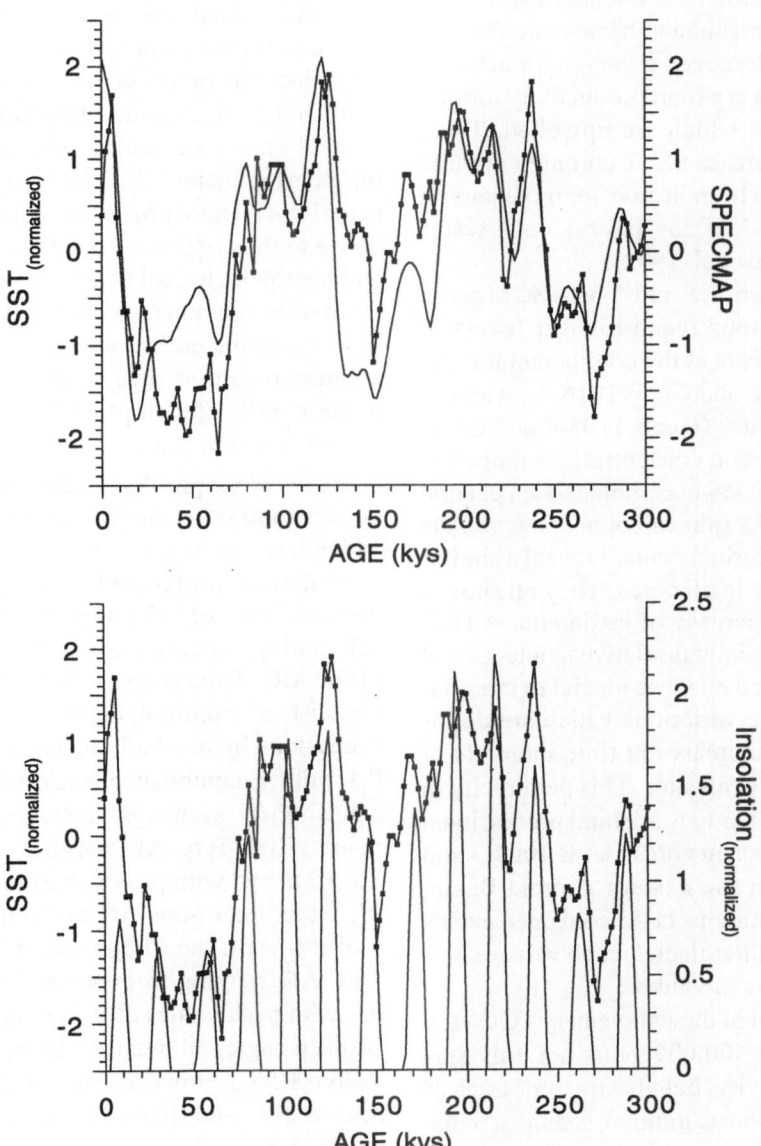

Fig. 5. Comparison of SST changes in the eastern Angola Basin (GeoB 1016-3) with (a) the SPECMAP $\delta^{18}O$ stack reflecting changes in global ice volume and sea level, and (b) boreal summer insolation at 10°N for August (after Berger 1978). For comparison all records are normalized to their mean and scaled in standard deviation units.

Organic Carbon: The Productivity Record

In order to unravel how variations in equatorial and coastal upwelling may have contributed to the Late Quaternary SST fluctuations in the east-equatorial South Atlantic, changes in sedimentary organic carbon fluxes were determined in the same cores used for SST reconstructions. Enhanced upwelling injects both cold water and nutrients into the surface mixed layer. Moreover, organic carbon burial is related to changes in export productivity formed on excess nutrients which are upwelled. Thus, periods of both decreased SST and high organic carbon burial should be indicative for more intense upwelling rather than for lateral cold water advection (e.g. Lyle et al. 1992).

Varying between 0.5 and 5 wt.-%, organic carbon concentrations reach highest levels in hemipelagic sediments at the continental margin (cores GeoB 1008-3 and GeoB 1016-3), whereas at the more distal sites (GeoB 1105-4 and GeoB 1028-5) organic carbon concentrations range between 0.2 and 1.4 wt.-% in carbonate-rich pelagic sediments (Fig. 6). Despite minor mismatches, the records of organic carbon content reveal a similar temporal variability in all cores. They all show a two- to four-fold increase of sedimentary TOC content in glacial sediments relative to interglacial levels. Superimposed on these glacial to interglacial fluctuations are variations which are dominated by periodic increases at time intervals of about 20 to 25 kyrs at all sites. This periodicity in the TOC signal similar to the orbital precessional cycle is most obvious in cores GeoB 1008-3 and GeoB 1016-3 from the eastern Angola Basin, where TOC concentrations in the cold interstadials of the penultimate interglacial reach values similar to maximum glacial values.

For interpretation of the sedimentary TOC variations over the last 400,000 years not only surface water productivity, but also the processes of terrigenous and carbonate dilution, sediment redistribution, as well as organic carbon preservation and remineralization must be evaluated as to their relative importance in contributing to this signal. Several studies have addressed this problem for east-equatorial Atlantic sediments over the last decade (Lyle 1988, Sarnthein et al. 1988,

Schneider 1991, Bickert 1992, Meinecke 1992, Sarnthein et al. 1992, Schmidt 1992, Westerhausen et al. 1993, Müller et al. 1994, Struck et al. 1994) and have presented strong evidence which favours changes in productivity over preservation/remineralisation processes to be primarily responsible for late Quaternary sedimentary TOC variations. Additional support for this hypothesis came from a suite of other proxy records for past changes in productivity or upwelling: biogenic barium content (Gingele & Dahmke 1994, Rutsch et al. 1995) or $\delta^{13}C$ values of benthic and planktonic foraminifera (Schneider 1991, Bickert 1992, Schneider et al. 1994). These proxies reveal close correspondence to the organic carbon variations in pelagic and hemipelagic sediments, and also exclude the input of terrigenous organic matter or redistribution of organic carbon as important for the sedimentary record at sites located on the middle to lower continental slope and at pelagic sites far distant from the coast.

To avoid the problem of dilution by terrigenous detritus and other major biogenic components, which have to be taken into account when TOC concentrations are considered, we have quantified the variability of TOC flux into the sediment by calculating organic carbon accumulation rates (TOC AR). Time series of TOC AR (Fig. 7) were established by multiplying the organic carbon concentrations by dry bulk density (DBD in g cm^{-3}, P.J. Müller unpublished data) and by linear sedimentation rates (cm kyr^{-1}) between stratigraphic tie points. High TOC AR, varying between 0.2 and 1.2 gC m^{-2} y^{-1} with peak values of 1.5 to 2 gC m^{-2} y^{-1}, were determined in the continental margin sedi-ments off the Congo and off Angola. Lower TOC AR, ranging from about 0.05 to 0.2 gC m^{-2} y^{-1}, with peak values of 0.4 gC m^{-2} y^{-1}, characterise the pelagic sediments at the equator and on the Walvis Ridge. However, these relatively low values are still a significant indicator of high primary productivity under oceanic upwelling conditions since TOC AR in pelagic sediments beneath low-productive regions rarely exceed values of about 0.05 gC m^{-2} y^{-1} (Berger et al. 1989).

The on average lower TOC concentrations and lower TOC AR in carbonate-rich sediments at sites

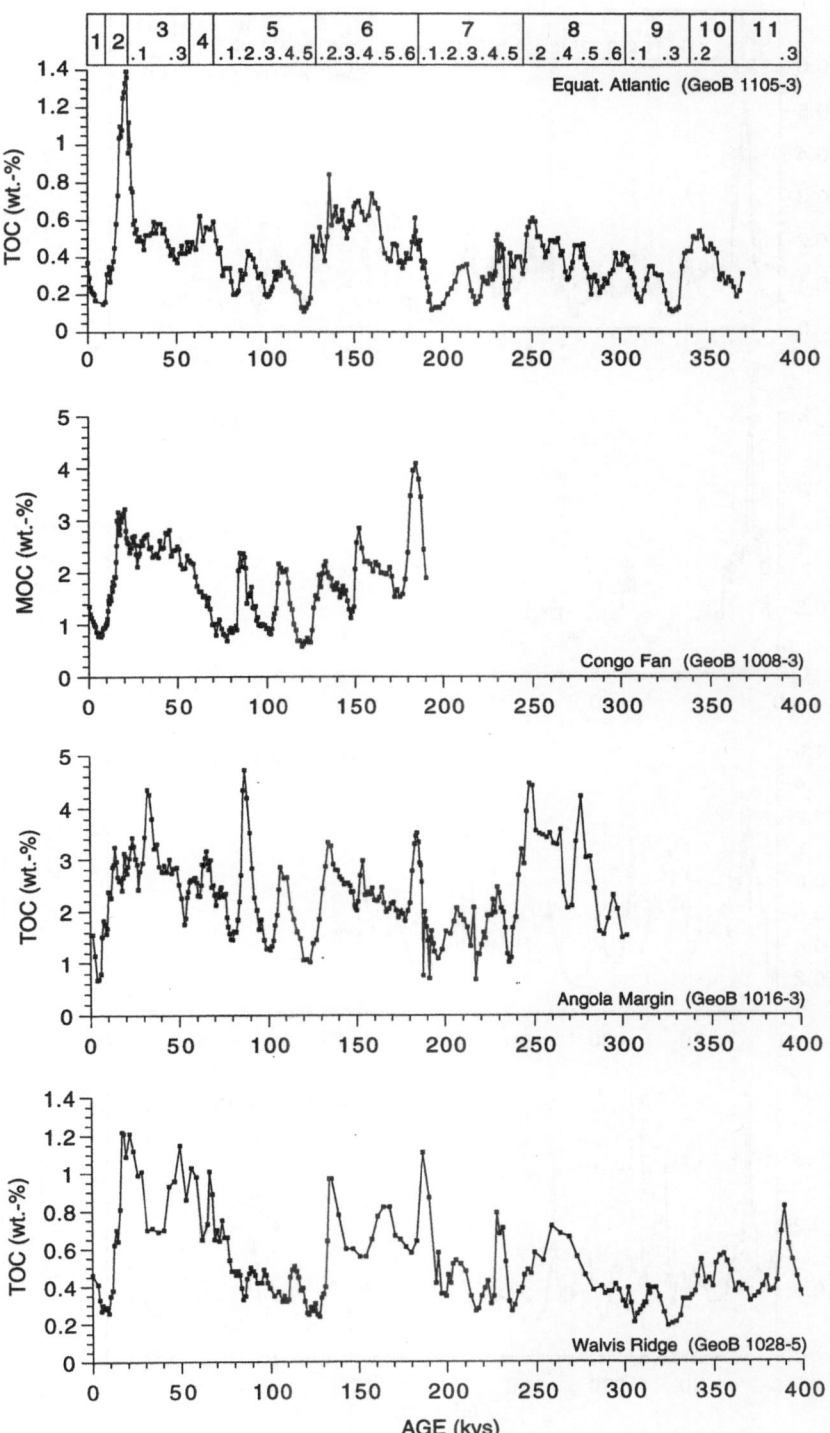

Fig. 6. Time series of total organic carbon (TOC) contents in late Quaternary sediments in the east-equatorial South Atlantic. MOC signifies the portion of marine organic carbon quantified from $\delta^{13}C_{org}$ for the Congo Fan core GeoB 1008-3 (see text).

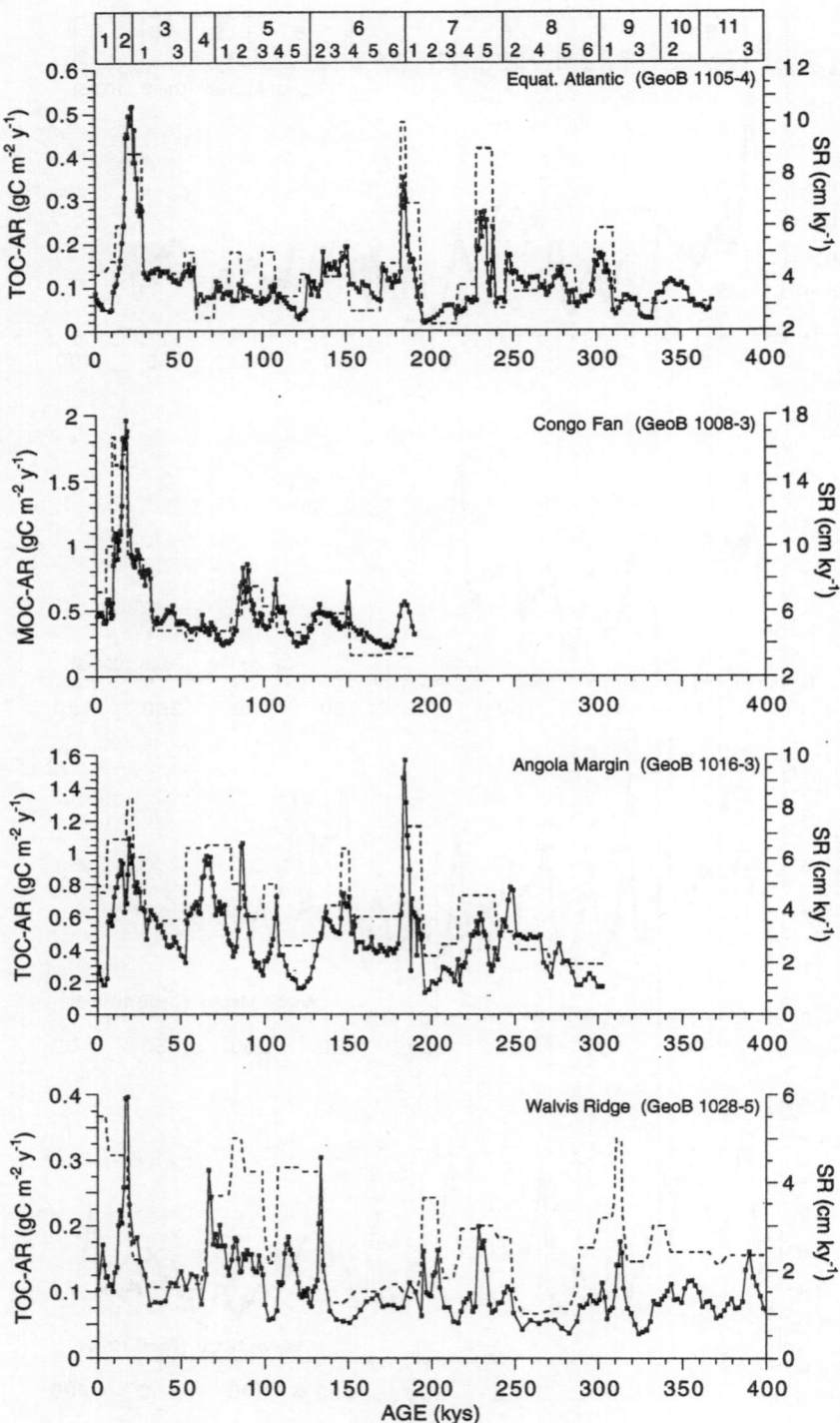

Fig. 7. Time series of organic carbon accumulation rates (TOC AR, MOC AR for GeoB 1008-3) in Late Quaternary sediments in the east-equatorial South Atlantic. Stippled lines denote sedimentation rates linearly interpolated between age control points.

GeoB 1105-4 and GeoB 1028-5 with respect to continental-margin sediments at sites GeoB 1008-3 and GeoB 1016-3 (Figs. 6 and 7) are attributed (i) to generally lower export production rates in the equatorial upwelling and in the Benguela filamentous zone compared to the eastern Angola Basin (Fig. 1), and (ii) to probably better preservation of organic matter in hemipelagic sediments relative to pelagic sites. The latter is due to higher sedimentation rates of terrigenous material at the continental margins. Under constant primary production rates a ten-fold increase in sedimentation rate could enhance the portion of organic carbon buried in the sediment by a factor of two (Müller & Suess 1979), because a faster deposition reduces the residence time of organic matter at the sediment-water interface, where most of the organic carbon reaching the sea floor is remineralized. Thus, when sedimentary TOC AR are used as an estimate for absolute values of past production rates and compared between different upwelling systems, TOC AR should first be corrected for preservational effects due to intra-core changes and inter-core differences of bulk sedimentation rates .

However, if the major interest is to infer the temporal variations of paleoproductivity from organic matter, the pattern of TOC concentrations seem to provide more reliable results than variations in accumulation rates (see Berger et al. 1989). A comparison of the TOC percentage records (Fig. 6) with those of TOC AR, demonstrates that calculation of accumulation rates discriminates the precession-related variations of TOC content in favour of a few peaks representing short intervals of very high TOC accumulation. The most pronounced peak in TOC AR occurs at about 20 kyrs B.P. in all cores (Fig. 7). Only in core GeoB 1016-3, where linear sedimentation rates remain fairly constant over the whole length of the core (Schneider 1991), does the temporal pattern of TOC AR closely match the variations of TOC content. For the Congo Fan sediments in core GeoB 1008-3 Schneider et al. (1994) assessed the degree of correlation between $\delta^{13}C$ changes in planktonic foraminifera, considered as a proxy for past nutrient levels in surface waters, and changes in paleoproductivity as indicated by organic carbon concentrations or accumulation rates. It was found that a much better correlation exists between $\delta^{13}C$ changes of planktonic foraminifera and variations of TOC percentages than between the nutrient proxy signal and variations in TOC AR. This discrepancy could result from the fact that accumulation records are very sensitive to errors in age-depth alignment by correlation of individual $\delta^{18}O$ records to the SPECMAP standard stack and/or in the absolute reliability of the standard stack itself (see discussion for South Atlantic sediments in Charles et al. 1988, Lyle 1988, Bickert 1992). Depending on the standard time scale used for correlation (Imbrie et al. 1984, Martinson et al. 1987), bulk sedimentation rates can differ by about 60% if isotopic substages are differentiated (Lyle 1988). Moreover, calculation of TOC AR most often yields maximum values at intervals where bulk sedimentation rates are highest. Thus, at continental margins where terrigenous input is the major sediment constituent, calculation of TOC AR may result in misleadingly high values, because the sealing effect enhancing TOC preservation at phases of more rapid sedimentation (see discussion above) is not accounted for. On the other hand, in pelagic sediments where bulk sedimentation rates are primarily controlled by the accumulation of calcite, as is the case for cores GeoB 1105-4 and 1028-5, the processes of carbonate productivity and dissolution may also govern the accumulation pattern of other minor biogenic components. Detailed investigations on equatorial Atlantic sediment cores from different water depths have shown that with increasing calcite dissolution the precessional paleoproductivity signal derived from sedimentary TOC is muted, while obliquity and eccentricity periods are enhanced with depth via dissolution effects originating from deep-water masses formed at high-latitudes (Verardo and McIntyre, 1994). As a consequence, for the exploration of correlation between past SST and productivity changes within the time and frequency domain, records of TOC concentration are regarded to reflect the Late Quaternary pattern of productivity changes in the east-equatorial South Atlantic more properly than TOC AR.

Periodic Variations of SST and Productivity over the last 350,000 Years

Since our interest is to investigate whether SST and productivity changes in the east-equatorial South Atlantic have covaried at periodicities associated with orbital cycling and how they may be related to other climate response mechanisms of low and high latitudes, we compare them with (1) the summed variance of Earth's eccentricity, tilt and precession (ETP: Imbrie et al. 1984), (2) July insolation changes at 10°N, and (3) the stacked $\delta^{18}O$ record (SPECMAP). In this comparison (1) represents the primary orbital forcing of global climate change (after Imbrie et al. 1984), which has strong impact on atmospheric and oceanic circulation, (2) is regarded as the major forcing for low-latitude monsoonal wind strength, and (3) is taken as an estimate for relative changes in global ice volume and sea level, which is the main high-latitude response to orbital forcing and has its own feedback mechanisms to climate change. This kind of approach has been conducted and described in detail before (e.g. Clemens & Prell 1991, Imbrie et al. 1989, 1992, 1993, DeMenocal et al. 1994) and is thus not further explained here.

For the evaluation of temporal congruence between the SST and TOC time series, as well as for their correlation to respective time series of the presumed forcing mechanisms, it is more convenient to perform cross-spectral instead of linear regression analysis. Cross-spectral analysis evaluates the degree of correlation in the frequency domain. This method generates values of coherency between two time series if they contain significant variance at similar frequencies. Calculated as a function of frequency after setting phase to zero, coherency values can be regarded as linear correlation coefficients for the dominant periodic signals co-occuring in the variance spectra of different time series (Jenkins & Watts 1968, Imbrie et al. 1989). Cross-spectral analysis of time series not only includes determination of coherency, but also estimates the lead or lag between similar periodic variations in two different time series. These lead or lag values (reported as phase angles in Tables 1 and 2) provide means to evaluate whether the paleoceanographic changes investigated here

preceeded, paralled or succeeded similar periodic oscillations in insolation or ice volume which exert strong impact on atmospheric and surface ocean circulation. Phase relationships thus can give insights to the linkage between SST and productivity changes in the east-equatorial South Atlantic and the climate system.

Power spectra of the individual SST, TOC % and TOC AR records, indicating the distribution of variance in certain frequency bands, are shown in Figure 8. According to the shortest time series of core GeoB 1008-3 (0-190 kyrs), low resolution (bandwith: 0.013 cycles/kyr, number of lags 50, no pre-whitening) was chosen for identification of the spectral peaks in order to keep the method similar for all records. If spectral peaks were not clearly separated from each other, a higher number of lags was used for calculation of autocovariances from the records longer than 300 kyrs. This procedure enhanced the resolution of spectral estimates by decreasing the bandwidth and improved the sharpness of the spectral peaks. Well-defined peaks for frequency bands where a significant amount of variance is centered are marked with the respective periodicity value in kiloyears (Fig. 8). Coherency values between spectral peaks and orbital cycles, represented by the ETP record, are listed in Table 1. For spectral peaks which could not be clearly distinguished from the rest of the spectrum and for those where variance density is not coherent with orbital forcing, periodicity values are given in brackets (see Table 1) and are not considered in the discussion of possible climate forcing in the east-equatorial South Atlantic.

Alkenone SST changes in the equatorial upwelling zone are dominated by periodicities of 100 and 23 kyrs, while significant variance at the 41-kyr cycle is absent (Fig. 8). Similar to SST changes, spectral analysis yields well-defined peaks at periodicities of 100 and 23 kyrs for the equatorial TOC percentage record. SST and TOC variations are coherent and in phase at these periodicities (Table 2). Significant variance is also obvious at the 41-kyr period for TOC concentrations at the equator, which is not observed in the SST variance density spectrum. When TOC AR are considered, the 41-kyr dominates over the

23-kyr cycle in core GeoB 1105-3. Moreover, the 100-kyr cycle is less well-defined in the TOC AR record, similar to the TOC AR spectra in the other cores. For example, in core GeoB 1016-3 calculation of TOC AR obviously diminishes the distinct 100-kyr peak observed in the TOC concentration spectrum. This effect is attributed to spurious linear sedimentation rates which result in high TOC AR occurring at terminations or across substage boundaries (Fig. 6, see discussion above). The variance density increase at the 41-kyr period in TOC AR relative to TOC concentrations in core GeoB 1105-3 presumably results from dissolution variations in sedimentary $CaCO_3$ which overprint the productivity signal of other minor biogenic constituents when mass accumulation rates are calculated for the pelagic carbonate-rich sediments (Verardo & McIntyre 1994). Following the suggestion of Verardo & McIntyre (1994) the 41-kyr

periodicity in the TOC records must be regarded as a remote deep-water signal originating from high   latitudes. It is imported to the equatorial sediments through changes in the contribution of water masses corrosive to carbonate, via variations in deep-water formation at high latitudes. The assumption that the 41-kyr cycle in the equatorial TOC records is not a productivity signal is supported by our data because this periodicity is not characteristic for the SST changes in core GeoB 1105-3. We assume that upwelling intensity at the equator varied especially in response to changes in eccentricity and precession, as was suggested earlier (Lyle, 1988, McIntyre et al. 1989, Molfino & McIntyre 1990).

The strong response of equatorial dynamics to changes in orbital eccentricity and precession is also reflected in the variance spectra of SST and

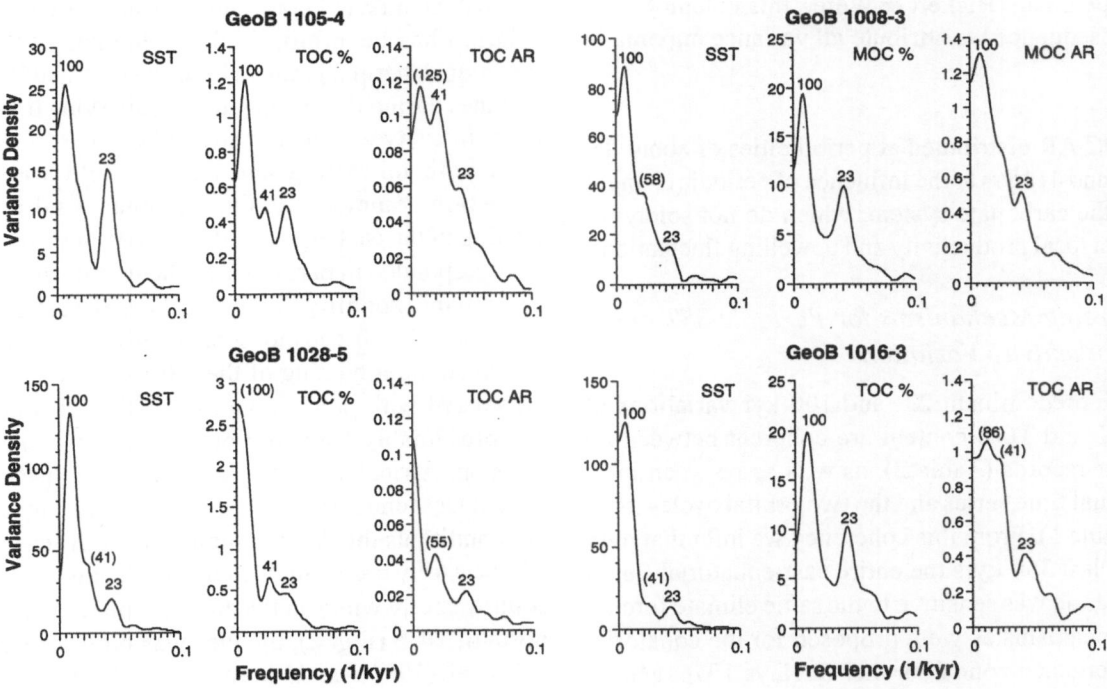

Fig. 8. Variance density plots (percent variance versus frequency) of SST, TOC, and TOC AR records. The spectra were estimated using an autocovariance function at the 80% confidence level with variable numbers of data points (Table 1), interpolated to fixed time intervals of 2 kyrs and linearly detrended.

TOC records from the other sites in the east-equatorial South Atlantic (Fig. 8). Variance peaks at 100- and 23-kyr periodicities can be clearly identified in all records, while periodic 41-kyr changes are absent in TOC percentage changes in cores GeoB 1008-3 and 1016-3 and are only poorly defined in the variance spectra of SST and TOC AR records in core GeoB 1028-5. Again it should be noted, that, with increasing carbonate content in the three cores along the South African margin, calculation of TOC AR shifts variance centered at the 23-kyr cycle towards the 41-kyr periodicity. In core GeoB 1008-3 from the Congo Fan where CaCO$_3$ contents range between 0.5 and 16 wt-% (Schneider et al. 1991) this effect is smaller than in core GeoB 1016-3 with carbonate concentrations varying between 0.2 and 30 wt-%. On the Walvis Ridge, where the carbonate content in Late Quaternary sediments ranges between 72 and 95 wt.-% (Müller, unpublished data), variance in TOC AR is centered at the 55-kyr period, which is characteristic of changes in carbonate dissolution in the Cape Basin (Bickert & Wefer, this volume). As a consequence we attribute all variance maxima in

TOC AR distributed at periodicities of about 85, 55 and 41 kyrs to the influence of periodic changes in the carbonate system, which do not solely reflect local productivity and upwelling fluctuations.

Forcing Mechanisms for Periodic SST and Productivity Variations

The predominant 23- and 100-kyr variations of SST and TOC content are coherent between all four records (Table 2), as well as between individual time series and the two orbital cycles itself (Table 1). From this coherency we infer that over the last 350 kyrs the entire east-equatorial South Atlantic was sensitive to the same climate forcing mechanisms as were proposed for the equatorial divergence zone (Gardner & Hays 1976, Mix et al. 1986 a,b, McIntyre et al. 1989, Verardo & McIntyre 1994, Schneider et al. 1995, Mix & Morey, this volume). The processes suggested to control SST and productivity changes are (1) vari-

ations in equatorial and coastal upwelling, and (2) changes in the advection of cold and nutrient-enriched Benguela Current waters. Both are probably related to (i) changes in southeasterly trade wind intensity, driven by the meridional thermal gradient in the southern hemisphere, and (ii) variations in more zonal versus more meridional trade winds, modulated by the position of the subtropical high over the South Atlantic in relation to the African heat low. The latter by its intensity has driven African southwesterly monsoon strength and has probably changed also the latitudinal positions of zonal thermal fronts in the eastern South Atlantic (e.g. Angola Benguela front, Subtropical and Subantarctic fronts; see Jansen et al., this volume).

We must assume that these processes did not vary independently from each other. Wind stress controls the upwelling of subsurface cold waters at the equator and thermocline shoaling in the eastern South Atlantic, as well as the advection of cold surface waters with the BOC, BCC and SEC to the tropical Atlantic (Fig. 2). Thus, if Hadley cell circulation has been modified by changes in the meridional thermal gradient between high and low latitudes during the late Quaternary, varying trade wind intensity would have strengthened or weakened both upwelling and surface cold water advection coincidently. On the other hand upwelling in the east-equatorial South Atlantic may have responded in particular to changes in austral trade wind zonality linked to boreal summer insolation via African monsoon intensity. Variable boreal summer heating of the African land mass associated with precessional insolation changes has presumably triggered periodic 23-kyr variations in African monsoon intensity (see discussion in DeMenocal et al. 1993). Since under modern conditions the African monsoon modifies the southeasterly trade winds to become southerly or southwesterly winds in the area east of 15°W and north of 20°S (Fig. 2), the idea was promoted by Mix et al. (1986b) and McIntyre et al. (1989), that precessional variations in African monsoon intensity affecting the zonality of trade winds, produced the pronounced 23-kyr cycle observed in Late Quaternary winter SST estimates for the equato-

TABLE 1. Summary of Cross-Spectral Analyses of SST, TOC, and TOC AR time series versus orbital forcing (ETP)

| | Orbital Frequency Bands | | | | | | | | | Series | CL |
| | 1/100 kys | | | 1/41 kys | | | 1/23 kys | | | Length | 80% |
Time Series	k	Ø	Error	k	Ø	Error	k	Ø	Error	kyr	
Maximum SST											
GeoB 1105-4	.87	+11	±14		n.c.		.78	+52	±26	368	.58
GeoB 1008-3		r.l.			n.c.		.95	+88	±10	190	.80
GeoB 1016-3	.92	-2	±11	(.76	+50	±20)	.93	+95	±10	302	.64
GeoB 1028-5	.81	-1	±16	(.72	+63	±22)	.94	+51	±8	402	.55
Minimum TOC %											
GeoB 1105-4	.79	+17	±17	.67	+72	±25	.78	+46	±18	368	.58
GeoB 1008-3		r.l.			n.c.		.93	+87	±12	190	.80
GeoB 1016-3	.90	+11	±13		n.c.		.87	+73	±14	302	.64
GeoB 1028-5	(.70	-8	±22)	.80	+64	±16	.80	+26	±15	418	.54
Minimum TOC AR											
GeoB 1105-4		n.c.		.75	+56	±21	.70	+15	±22	368	.58
GeoB 1008-3		r.l.			n.c.		.85	+89	±19	190	.80
GeoB 1016-3		n.c.		(.81	+45	±18)	.66	+70	±27	302	.64
GeoB 1028-5		n.c.			n.c.		.74	+71	±19	418	.54
Minimum Ice Volume (SPECMAP)											
δ^{18}O stack	.81	+12	±15	.84	+78	±13	.91	+87	±9	750	.52
Southern Indian Ocean Fronts											
SST (44°-46°S)	.81	-6	±18	.86	+51	±15	.72	+56	±24	448, 501	.65

Note coherencies k and phases of proxy records are shown with respect to ETP. ETP represents summed variance of orbital eccentricity, tilt and precession following the SPECMAP convention described in Imbrie et al. (1989), where zero phases indicate orbital constellation forcing interglacial conditions (Maxima in eccentricity and tilt, and minimum in precession index (21 June perihelion)). Error refers to phase estimates in degrees, where positive values indicate lag with respect to orbital forcing and vice versa. Phase angles can be converted to time by dividing values in degrees by 360° and then multiplying by the orbital period in kys. All analyses were computed using a bandwith of 0.013 cycles/kyr and time series interpolated to 2-kyr-intervals. CL is test statistic for non-coherency at the 80% significance level; n.c. indicates no significant coherency; r.l. hints to record length prohibiting statistical evaluation of the 100-kyr cyle. Numbers in parentheses indicate coherency with orbital period but the lack of a distinct peak in the variance density spectrum at the respective frequency band (compare with Fig. 8). Numbers for the SPECMAP δ^{18}O stack from Imbrie et al. (1989), for the Indian Ocean SST (Howard and Prell 1992: cores E45-29 and E49-18 where coherent with orbital cycles).

TABLE 2. Summary of Cross-Spectral Analyses of SST versus TOC time series.

| Time Series | Orbital Frequency Bands | | | | | | Series Length | CL |
| | 1/100 kys | | | 1/23 kys | | | | 80% |
	k	∅	Error	k	∅	Error	kyr	
Maximum SST versus minimum TOC								
GeoB 1105-4	.76	+3	±19	.83	+8	±16	248	.58
GeoB 1008-3	r.l.			.95	-2	±17	190	.80
GeoB 1016-3	.92	-16	±17	.83	+20	±17	302	.64
GeoB 1028-5	(.81	+6	±15)	.82	+25	±16	402	.55

For explanation of abbreviations and symbols see Table 1.

rial divergence. But the zonal vector and the latitudinal position of trade winds in the past was probably also influenced by the northward position and intensity of the South Atlantic subtropical high, which in turn may have been controlled by the thermal gradient between high and low latitudes, as was suggested by results from atmospheric general circulation model (AGCM) experiments (Rind 1986, DeMenocal & Rind 1993).

The interference of processes regarded to be important for surface-water changes in the east-equatorial Atlantic makes it very difficult to separate one of them as the dominant forcing mechanism for either SST or productivity changes, or for both in conjunction. One method to evaluate possible dominance of one of the processes over the other is to consider their correspondence to variations of paleotemperatures and paleoproductivity by phase relationships at those periodicities which are coherent between assumed forcing and surface water response. It is then further assumed that SST and productivity have directly responded to those forcing mechanisms which reveal zero phasing with the time series of paleoceanographic proxies (e.g. Imbrie et al. 1989, 1992, Clemens & Prell 1989). Since our records reveal coherency consistently only at the precessional cycle, we emphasise this periodicity in our discussion about phasing of SST and productivity with respect to probable forcing mechanism.

Commonly, orbital constellation which is regarded to exert forcing toward interglacial conditions is set to zero phase . With respect to the 100- and 23-kyr cycles there exists no firm evidence which orbital geometry represents the maximum value of interglacial forcing (Imbrie et al. 1989). Thus by convention maximum eccentricity and minimum in precession index, which is equivalent to 21 June perihelion or maximum boreal summer insolation during June, are locked to zero phase (Imbrie et al. 1989). According to the SPECMAP oxygen isotope time scale (Imbrie et al. 1984) global ice volume changes lag eccentricity and precession by about 3,000 (+12° phase angle) and 5,500 years (+87° phase angle), respectively (Tab. 1).

Monsoonal influence

Changes in monsoonal circulation over the east-equatorial Atlantic have probably been tightly coupled to variations in boreal summer insolation in low latitudes over Africa and thus have been dominated by precessional variations in the Earth's orbit (Prell & Kutzbach 1987, Pisias & Mix 1985, DeMenocal et al. 1993). The phasing of changes in monsoonal wind intensity over West Africa with respect to eccentricity and precession is not well constrained from terrestrial and marine records, due to the lack of proxies which would enable the reconstruction of monsoonal wind strength. Most of the assumptions made about past changes in West African monsoonal circulation were derived from proxies which reflect precipitation changes on land. However, higher precipitation rates are not necessarily indicative of stronger monsoonal winds. Therefore, the question remains unsolved, whether higher lake levels in central Africa (Kutzbach & Street-Perrott 1985), higher abundances of freshwater diatoms in marine sediments (Pisias & Mix, 1985, DeMenocal et al. 1993), or mediterranean sapropels off the Nile river (Rossignol-Strick 1983) can be taken as evidence for increased monsoon wind strength causing more southerly to southwesterly winds over the east-equatorial Atlantic. Results from different AGCM experiments suggest both cases, either that increased African irradiation forces stronger monsoonal winds (Prell & Kutzbach 1987) or that changes in low-latitude insolation do not significantly affect the wind strength (DeMenocal & Rind 1993). Here we follow the results from Kutzbach & Prell (1987) assuming that, if paleo-monsoon wind strength has been influenced somehow by insolation changes, it would have been the direct response to precessional changes in low-latitude insolation over central Africa. To reach maximum monsoon wind strength, African irradiation should then be highest during summer, which requires minimum earth-sun distance (minimum precession index) in July (Prell & Kutzbach 1987). Because the movement of the perihel from one month to the next takes about 2,000 years (23,000 kyrs / 12 month), the vector for maximum July in-solation, the presumed forcing geometry for maximum monsoon, is lagging zero phase (21 June perihel by convention) by a phase angle of 30°.

Southern hemisphere influence

To assess the phasing of changes in the southern hemisphere thermal gradient with respect to orbital cycles, we rely on SST changes in the southern Indian Ocean, assuming that they predominantly reflect past movements of the Subtropical Convergence and Antarctic Polar Front. In the Southern Indian Ocean maximum SST indicate southernmost positions of the Southern Ocean thermal fronts and vice versa (CLIMAP 1981, Morley 1989, Howard & Prell 1992). The latitudinal position of these thermal fronts in the modern surface ocean is determined by seasonal sea-ice extent and southward extension of the subtropical gyre, as well as by the position and strength of westerly winds (summary in Howard & Prell 1992). Therefore, past movements of these fronts may also be regarded as indicators for latitudinal compression or expansion of atmospheric circulation cells, which control zonal trade wind strength, as well as the locations of the subtropical high pressure system and the trade wind belt. The study of SST changes in the southern Indian Ocean (Howard & Prell 1992) implies that the meridional thermal gradient in the southern hemisphere has varied in phase with eccentricity and that this gradient followed the precession index by about 3,500 years (phase angle of +56°, Table 1). For these orbital periodicities, the 23 and the 100 kyr cycles, Southern Indian Ocean SST lead changes in ice volume as represented by the SPECMAP $\delta^{18}O$ record (Tab. 1). This phase pattern from the Southern Indian Ocean is supported by a paleotemperature record from the eastern South Atlantic located between the present Subtropical and Subantarctic Front (U. Brathauer, in press). For comparison, the phase relationships of proposed forcing mechanisms and SST and productivity changes (TOC wt-%) in the east-equatorial South Atlantic are shown relative to each other (Fig. 9), as vectors indicating the phase angles (from Tab. 1) with respect to the precessional cycle.

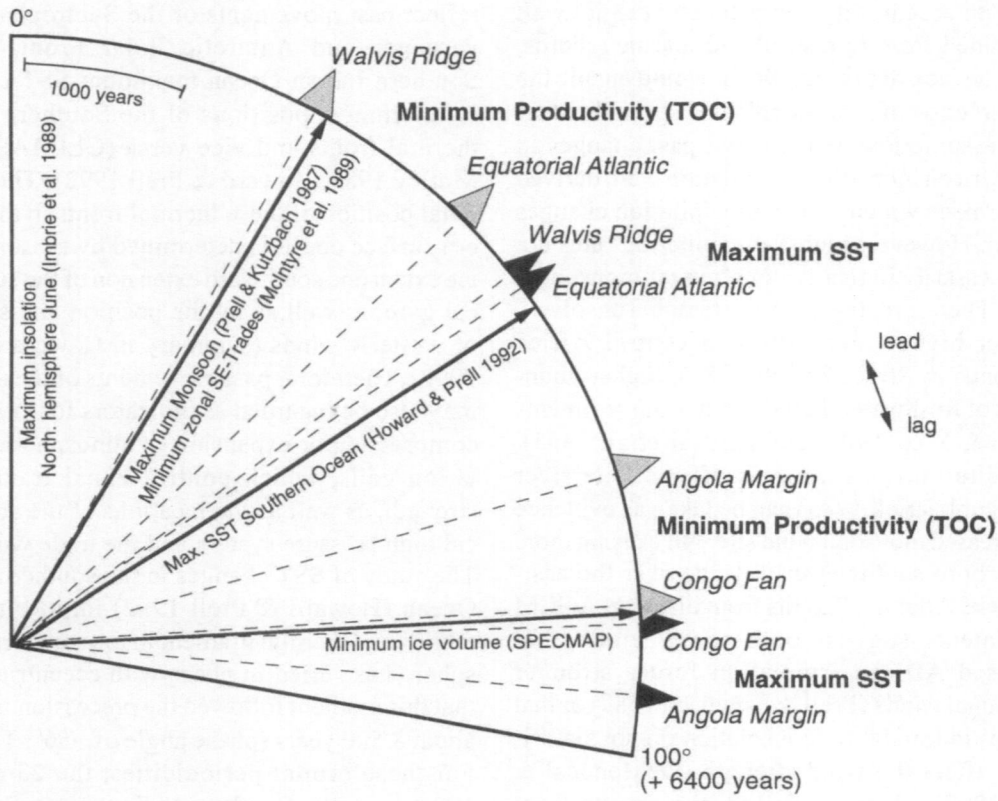

Fig. 9. Diagram showing phasing for the precessional cycle (in degrees) between signals of posited forcing mechanisms and proxies delinating surface ocean response. Minimum in precession index (21 June perihel) is set to zero by convention (Imbrie et al. 1989). Solid lines show phases of forcing relative to precession, while oceanic response (SST: black arrows; productivity: gray arrows) is given by stippled lines. Forcing mechanisms: Maximum boreal summer insolation for July at 10°N (Berger et al. 1978) inducing maximum monsoonal flow over West Africa (Prell & Kutzbach 1987); maximum Indian Ocean SST's at 44° to 46° S, indicating southernmost position of southern ocean thermal fronts (Howard & Prell, 1992), and minimum ice volume (sea level highstand) as expressed by minimum values in the SPECMAP δ¹⁸O stack (Imbrie et al. 1984). Clockwise displacement indicates lag in degrees, which can be converted to time by multiplying the phase value by the periodicity considered and divided by 360°. Overlap of SST and productivity changes (in the range of phase estimate errors: Table 1) with one of the posited forcing mechanisms is taken as evidence that the surface ocean has responded to the respective mechanism.

Oceanic response in the east-equatorial South Atlantic

Minimum glacial SST values were about 3.5° and 6°C lower than maximum Holocene values in the equatorial upwelling zone and along the Southwest African margin, respectively (Fig. 4). This is two to three times the glacial to interglacial difference reported for the west-equatorial Atlantic based on the same method (Sikes & Keigwin 1994). Since it is assumed that cold water advection from the South has influenced in particular the central and western equatorial Atlantic over the last 350 kyrs (Mix & Morey, this volume, Schneider et al. 1995), this greater glacial to interglacial SST difference in our easterly records implies that a substantial portion of very low glacial SST is due to increased upwelling in the east-equatorial Atlantic. Moreover, alkenone SST are likely to reflect annual mean temperature values and it seems plausible that, because strong upwelling is presumably restricted to austral winter, our alkenone SST values underestimate the temperature decreases related to variations in upwelling. This is indicated in the SST estimates derived from planktonic foraminifera assemblages in a sediment core neighbouring GeoB 1105-4 (McIntyre et al. 1989: core RC 24-16), where late Quaternary winter SST estimates differ by about 8°C, while summer estimates only reveal changes of about 2°C. From the eastward increase in the glacial to interglacial alkenone SST contrast in the equatorial Atlantic and from the close correlation between SST and TOC variations (Table 2) we conclude that variation in wind-driven upwelling instead of cold-water advection becomes the more important forcing mechanism for glacial-interglacial cycles of SST and productivity at the equator east of 15° W. This is corroborated by a general west-east increase of TOC AR in sediment cores on a transect across the mid-ocean ridge indicating higher overall export productivity and stronger upwelling in the east-equatorial Atlantic than in the central and western part over the last 350,000 years (Bickert 1992, Meinecke 1992).

Within the 100-kyr cycle SST and productivity variations in the equatorial Atlantic are more or less in phase with ice volume changes (SPECMAP $\delta^{18}O$) slightly lagging eccentricity (Tab. 1). We infer from this phasing that, with respect to glacial-interglacial changes and despite possible low-latitude monsoonal forcing and southern hemisphere influence related to the precessional cycle, global climate boundary conditions have dominated surface hydrography in the equatorial Atlantic. Only SST variations in the eastern Angola Basin (cores GeoB 1008-3 and 1016-3) reveal a tendency to precede ice volume changes and to be more in phase with eccentricity. An explanation for this tendency could be (i) a strong subsurface return flow associated with the South Equatorial Countercurrent, transporting relatively warm surface waters towards the continental margin of Southwest Africa , and/or (ii) a heat storing effect in the east-equatorial South Atlantic as a consequence of a weaker cross-equatorial heat flow during peak glacial times (see discussion above). In particular, minimum surface temperatures in the Angola Basin were reached earlier than maximum ice age conditions (see also Fig. 5) which statistically results in a negative phase angle for SST changes relative to changes in ice volume at the 100-kyr cycle. On the other hand, warming in the Angola Current earlier than in the Benguela Oceanic Current and in global climate may be also explained by a decoupling of Angola Basin gyre circulation from the variations in the conveyor surface circulation and a more direct response to monsoonal wind strength of the eastern Angola Basin circulation which is characterised predominately by precessional variations (Schneider et al. 1995).

For the 23-kyr cyle the SST and TOC records from the equator (GeoB 1105-4) and from the Walvis Ridge (GeoB 1028-5) reveal a phase pattern which is different from that obtained for the continental margin sites (Fig. 9). While precession-related SST and productivity fluctuations in the eastern Angola Basin (GeoB 1008-3, and 1016-3) are nearly in phase with the precessional vector of global ice volume changes, the phases from the sites located beneath the modern field of strong southeasterly trade winds (GeoB 1105-3 and 1028-5, Fig. 2) suggest that SST and productivity changes in the east-equatorial South Atlantic lead ice volume changes. Periods of maximum temperatures and minimum productivity at the equator and the Walvis Ridge seem to be aligned with the time intervals when monsoonal circulation was near its maximum during intensified boreal summer insolation, and when the subtropical high over the South Atlantic and the southern ocean frontal system were at their southernmost positions (Fig. 9). We infer from this 23-kyr phasing, that the combination of a very intense heat low over central Africa and a subtropical high located far south over the South Atlantic weakened the zonal intensity of trade winds and induced stronger southerly to southwesterly (monsoonal) winds every 23,000 years. As a consequence, equatorial upwelling intensity and propagation of coastal upwelling filaments to the outer Walvis Ridge, as well as cold water advection towards the central Atlantic with the BOC, were reduced. For the region along the Angola Margin south of 5° S, it was suggested that precessional changes in upwelling and advection of cold, nutrient-rich, subsurface waters with the BCC were delayed compared to the open ocean sites due to cyclonic gyre circulation in the eastern Angola Basin (see above and Schneider et al. 1995).

A totally different explanation for the fact that SST and productivity changes seen in continental margin cores seem to be closely tied to precessional changes in global ice volume could be the influence of sea-level fluctuations. Raising and descending sea levels may have shifted the centers of coastal upwelling perpendicular to the coast line. Then fluctuations in temperatures and sedimentary TOC content from a distinct marginal site would reflect relocations of the upwelling cells relative to the coast instead of indicating variations in upwelling intensity. A similar in-phase 23-kyr variation of a paleo-upwelling proxy (abundance of *G. bulloides*) with the global ice volume, or sea-level, was reported from continental margin cores beneath the coastal upwelling off Oman. In contrast, records of the same proxy retrieved at greater distance from the coast in the Arabian Sea reveal a clear lag of upwelling changes with respect to global ice volume (Anderson & Prell 1993), which implies that coastal upwelling records may be strongly affected by lateral sea level-triggered shifts of coastal upwelling cells. For the moment

we have no answer to this problem and have to await more comparisons of coastal and pelagic records affected by the same wind system.

Conclusions

The phase angle between vectors for maximum interhemispheric monsoonal circulation and minimum austral thermal gradient in Figure 9 may indicate a temporal separation of the two proposed forcing mechanism (e.g. Mix and Morey, this volume). However, it seems not appropriate to try to assess the relative importance of one of these forcing mechanisms for one or both of the oceanographic processes (upwelling or advection) by phase relationships. It must be kept in mind that stratigraphic errors within the individual time series, used to describe timing of forcing and to relate oceanic response to it, can account for ±2,500 years (Imbrie et al. 1984) which is in the range of phase differences considered here. Additionally, the statistical phase errors (Tab. 1) also provide a level of uncertainty in the exact phasing which is higher than what would be needed to clearly distinguish whether advection and upwelling have responded either to monsoonal low-latitude or southern hemisphere thermal gradient forcing.

Nevertheless, the following assumptions can be made from cross-spectral analysis of the temporal variations of SST and sedimentary TOC with orbital cycles. The obliquity period, though to be characteristic for high latitude climate change (Imbrie et al. 1989, 1992, Howard & Prell, 1992), is not present in the SST and productivity signals from the east-equatorial South Atlantic. The east-equatorial Atlantic records contain only significant variance at 100- and 23-kyr periodicities for SST and productivity changes, similar to previous results retrieved from equatorial Atlantic records (Mix et al. 1986, McIntyre et al. 1989, Verardo & McIntyre et al. 1994). At open ocean sites precession-related hydrographic changes precede changes in ice volume and vary almost in phase with changes in monsoonal strength and/or frontal movements in the southern ocean. We infer from this that low-latitude insolation in the northern hemisphere and precession-related changes in the thermal gradient in the southern hemisphere can be regarded as the primary forcing mechanism for SST and productivity changes in the entire eastern South Atlantic over the last 350,000 years. Influence of northern high-latitude forcing, e.g. continental ice-mass fluctuations, is transferred to the tropical and substropical South Atlantic predominantly by variations in surface water advection within the Benguela Oceanic Current (BOC) which is part of the global conveyor circulation. This is consistent with results from AGCM experiments conducted under ice age boundary conditions (Manabe & Broccoli 1985, Rind et al. 1986, 1987, DeMenocal & Rind 1993). All records indicate a tendency for TOC changes slightly leading SST at the 23-kyr periodicity. This may indicate that wind-driven upwelling responds faster to orbital forcing than advection of cold water from the Southern Ocean, a suggestion also made by Mix & Morey (this volume) based on a 270,000 year record of foraminiferal abundance fluctuations.

Acknowledgements

We thank the crew and scientists aboard R.V. METEOR for their help with coring and sampling operations during several cruises to the South Atlantic, and M. Segl, Birgit Meyer-Schack, Hella Buschhof, and Dietmar Grotheer for technical assistance. We are also grateful to W.H. Berger and J.H.F. Jansen who made thoughful suggestions to the first manuscript. M.G. Little helped to improve the english language. This research was funded by the Deutsche Forschungsgemeinschaft (Sonderforschungsbereich 261 at Bremen University, Contribution No. 96) and the Bundesminister für Bildung und Forschung (BMBF), Bonn.

References

Anderson DM, Prell WL (1993) A 300 kyr record of upwelling off Oman during the Late Quaternary: Evidence of the Asian southwest monsoon. Paleoceanography 8:193-208

Berger, WH, Smetacek VS, Wefer G. (1989) Ocean productivity and paleoproductivity: An overview. In: Berger WH, Smetacek VS, Wefer G (eds) Productivity in the ocean: Present and Past. John Wiley and Sons, Chichester 1989, pp 1-34

Bickert, T (1992) Rekonstruktion der spätquartären Bodenwasser-Zirkulation im östlichen Südatlantik über stabile Isotope benthischer Foraminiferen. Berichte, Fachbereich Geowissenschaften 27, Universität Bremen, 205 pp

Brassell SC (1993) Applications of biomarkers for delineating marine paleoclimatic fluctuations during the Pleistocene. In: Engel MH, Macko SA (eds) Organic Geochemistry. Plenum Press, New York, pp. 699-738

Brassell SC, Eglinton G, Marlowe IT, Pflaumann U, Sarnthein M (1986) Molecular stratigraphy: a new tool for climatic assessment. Nature 320:129-133

Brathauer U (1996) Radiolarian-based reconstruction of Late Quaternary climate change in the Atlantic Sector of the Southern Polar Ocean. PhD. Thesis, Reports on Polar Research, Alfred-Wegener Institut for Polar and Marine Research, in prep

Broecker WS, Denton GH (1989) The role of ocean-atmosphere reorganizations in glacial cycles. Geochim Cosmochim Acta, 53: 2465-2501

Charles CD, Froelich PN, Zibello MA, Mortlock RA, Morley JJ (1991) Biogenic opal in Southern Ocean sediments over the last 450,000 years: Implications for surface water chemistry and circulation. Paleoceanography 6:697-728

Clemens SC, Prell WL (1990) Late Pleistocene variability of Arabian Sea summer monsoon winds and continental aridity: Eolian records from the lithogenic component of deep-sea sediments. Paleoceanography 5:109-145

CLIMAP (1981) Seasonal reconstructions of earth's surface at the last glacial maximum. GSA Map and Chart Ser, MC-36, Geol Soc Am, Boulder

Conte MH, Eglinton G (1993) Alkenone and alkenoate distributions within the euphotic zone of the eastern North Atlantic: correlation with production temperature. Deep-Sea Res I 40:1935-1961

Conte MH, Eglinton G, Madureira LAS (1992) Long-chain alkenones and alkyl alkenoates as palaeotemperature indicators: their production, flux and early sedimentary diagenesis in the Eastern North Atlantic. Org Geochem 19:287-298

DeMenocal PB, Ruddiman WF, Pokras EM (1993) Influences of high- and low-latitude processes on African terrestrial climate: Pleistocene eolian records from equatorial Atlantic Ocean Drilling Program site 663. Paleoceanography 8:209-242

DeMenocal PB, Rind D (1993) Sensitivity of Asian and African climate to variations in seasonal insolation, glacial ice cover, sea-surface temperature, and Asian orography. J Geophys Res 98:7265-7287

Diester-Haass L (1985) Late quaternary sedimentation on the eastern Walvis Ridge, SE Atlantic (HPC 532 and four piston cores). Mar Geol 65:145-186

Gardner JV, Hays JD (1976) Response of sea-surface temperature and circulation to global climatic change during the past 200,000 years in the eastern equatorial Atlantic Ocean. Mem Geol Soc Am 145: 221-246

Gingele FX, Dahmke A (1994) Discrete barite particles and barium as tracers of paleoproductivity in South Atlantic sediments. Paleoceanography 9:151-168

Howard WR, Prell WL (1992) Late Quaternary surface circulation of the Southern Indian Ocean and its relationship to orbital variations. Paleoceanography 7:79-118

Imbrie J, Hays JD, Martinson DG, McIntyre A, Mix AC, Morley JJ, Pisias NG, Prell WL, Shackleton NJ (1984) The orbital theory of Pleistocene climate: Support from a revised chronology of the marine $\delta^{18}O$ record. In: Berger AL, Imbrie J, Hays JD, Kukla J, Saltzman J (eds) Milankovitch and Climate, Part I. Reidel, Hingham, Mass, pp 269-305

Imbrie J, McIntyre A, Mix AC (1989) Oceanic response to orbital forcing in the Late Quaternary: Observational and experimental strategies. In: Berger AL, Schneider S, Duplessy JC (eds) Climate and Geosciences. Kluwer, Dordrecht, pp 121-164

Imbrie J, Boyle EA, Clemens SC, Duffy A, Howard WR, Kukla G, Kutzbach J, Martinson DG, McIntyre A, Mix AC, Molfino B, Morley JJ, Peterson LC, Pisisas NG, Prell WL, Raymo ME, Shackleton NJ, Toggweiler JR (1992) On the structure and origin of major glacial cycles, 1, Linear response to Milankovitch forcing. Paleoceanography 7:701-738

Jenkins GM, Watts DG (eds) (1968) Spectral analysis and its application. Holden Day, San Francisco, 525 pp

Kutzbach JE, Street-Perrott FA (1985) Milankovitch forcing of fluctuations in the level of tropical lakes from 18 to 0 kys BP. Nature, 317:130-134

Levitus S (1982) Climatological atlas of the world ocean. NOAA Prof Pap 13, 173 pp. US Govt Print Off, Washington, DC

Lyle M (1988) Climatically forced organic carbon burial in equatorial Atlantic and Pacific Oceans. Nature 335:529-532

Lyle MW, Prahl FG, Sparrow MA (1992) Upwelling and productivity changes inferred from a temperature record in the central equatorial Pacific. Nature 355:812-815

Martinson DG, Pisias NG, Hays JD, Imbrie J, Moore TC, Shackleton NJ (1987) Age dating and the orbital theory of the ice ages: development of a high-resolution 0 to 300,000-year chronostratigraphy. Quat Res 27:1-29

McIntyre A, Ruddiman WF, Karlin K, Mix AC (1989) Surface water response of the equatorial Atlantic Ocean to orbital forcing, Paleoceanography 4:19-55

Meinecke G (1992) Spätquartäre Oberflächenwassertemperaturen im östlichen äquatorialen Atlantik. Berichte, Fachbereich Geowissenschaften 29, Universität Bremen, 181pp

Manabe S, Broccoli AJ (1985) The influence of continental ice sheets on the climate of an ice age. J Geophys Res 90(D1):2167-2190

Mix AC, Ruddiman WF, McIntyre A (1986a) Late Quaternary paleoceanography of the tropical Atlantic, 1: Spatial variability of annual mean sea surface temperatures, 0-20,000 years B.P. Paleoceanography 1:43-66

Mix AC, Ruddiman WF, McIntyre A (1986b) Late Quaternary paleoceanography of the tropical Atlantic, 2: The seasonal cycle of sea surface temperatures, 0-20,000 years B.P. Paleoceanography 1:339-353

Molfino B, McIntyre A (1990) Precessional forcing of nutricline dynamics in the Equatorial Atlantic. Science 249:766-769

Morley JJ (1989) Variations in high-latitude oceanographic fronts in the southern Indian Ocean: An estimation based on faunal changes. Paleoceanography 4:547-554

Morley JJ, Hays JD (1979) Comparison of glacial and interglacial oceanographic conditions in the South Atlantic from variations in calcium carbonate and radiolarian distributions. Quat Res 12:396-408

Müller PJ, Suess E (1979) Productivity, sedimentation rate, and sedimentary organic matter in the oceans. I. Organic carbon preservation. Deep-Sea Res 26A:1347-1362

Müller PJ, Schneider R (1993) An automated leaching method for the determination of opal in sediments and particulate matter. Deep-Sea Res 40:425-444

Müller PJ, Schneider R, Ruhland G (1994) Late Quaternary PCO_2 variations in the Angola Current: Evidence from organic carbon $\delta^{13}C$ and alkenone temperatures. In: Zahn R, Kaminski MA, Labeyrie L, Pedersen TF (eds),Carbon Cycling in the Glacial Ocean: Constraints on the Ocean's Role in Global Change. Springer, Heidelberg, pp. 343-366

Peterson RG, Stramma L (1991) Upper-level circulation in the South Atlantic Ocean. Prog Oceanography 26:1-73

Picaut J, Servain J, LeComte P, Seva M, Lukas S, Rougier G (1985) FOCAL Climatic Atlas of the Tropical Atlantic: Wind Stress and Sea Surface Temperature 1964-1979. Université de Bretagne Occidentale, Laboratoire d'Oceanograhie Physique, Brest, France

Prahl FG, Muehlhausen LA, Zahnle DL (1988) Further evaluation of long-chain alkenones as indicators of paleoceanographic conditions. Geochim Cosmochim Acta 52:2303-2310

Prell WL, Kutzbach JE (1987) Monsoon variability over the past 150,000 years. J Geophys Res 92:8411-8425

Pokras EM, Mix AC (1985) Eolian evidence for spatial variability of Late Quaternary climates in tropical Africa. Quat Res 24:137-149

Rind D (1986) The dynamics of warm and cold climates. J Atmos Sci 43:3-24

Rind D (1987) Components of the Ice Age circulation. J Geophys Res 92(D4):4241-4281

Rossignol-Strick M (1985) Mediterranean Quaternary sapropels, an immediate response of the African monsoon to variation of insolation. Paleogeogr Paleoclimatol Paleoecol 49:237-263

Rutsch HJ, Mangini A, Bonani G, Dittrich-Hannen B, Kubik PW, Suter M, Segl M (1995) ^{10}Be and Ba concentrations in the West African sediments trace productivity in the past. Earth Planet Sci Lett 133: 129-143

Sarnthein M, Winn K, Duplessy JC, Fontugne M (1988) Global variations of surface ocean productivity in low and mid latitudes: influence on CO_2 reservoirs of the deep ocean and atmosphere during the last 21,000 years. Paleoceanography 3:361-399

Sarnthein M, Pflaumann U, Ross R, Tiedemann R, Winn K (1992) Transfer functions to reconstruct ocean paleoproductivity, a comparison. In: Summerhayes CP, Prell WL, Emeis KC (eds) Upwelling Systems: Evolution since the Early Miocene. Geological Society Special Publication 64:411-427

Schmidt H (1992) Der Benguela-Strom im Bereich des Walfisch-Rückens im Spätquartär. Berichte, Fachbereich Geowissenschaften 28, Universität Bremen, 172 pp

Schneider R (1991) Spätquartäre Produktivitätsänderungen im östlichen Angola Becken: Reaktion auf Variationen im Passat-Monsun Windsystem und in der Advektion des Benguela-Küstenstroms. Berichte, Fachbereich Geowissenschaften 21, Universität Bremen, 198 pp

Schneider RR, Müller PJ, Wefer G (1994) Late Quaternary paleoproductivity changes off the Congo deduced from stable carbon isotopes of planktonic foraminifera. Palaeogeogr Palaeoclimatol Palaeoecol 110: 255-274

Schneider RR, Müller PJ, Ruhland G (1995) Late Quaternary surface circulation in the east equatorial Atlantic: Evidence from alkenone sea surface temperatures. Paleoceanography 10:197-219

Schneider RR, Müller PJ, Kroon D, Price B, Alexander I (1996) Monsoon related Zaire (Congo) discharge fluctuations and influence of fluvial silicate supply on marine productivity in the east equatorial Atlantic over the last 200,000 years. (submitted)

Shannon LV, Agenbag JJ, Buys MEL (1987) Large- and mesoscale features of the Angola-Benguela Front. In: Payne AIL, Gulland JA, Brink KH (eds) The Benguela and Comparable Ecosystems. S Afr J Mar Sci 5: 11-34

Sikes EL, Keigwin LD (1994) Equatorial Atlantic sea surface temperature for the last 30 kyr: A comparison of U^k_{37}', $\delta^{18}O$ and foraminiferal assemblage temperature estimates. Paleoceanography 9:31-45

Struck U, Sarnthein M, Westerhausen L, Barnola JM, Raynaud D (1993) Ocean-atmosphere carbon exchange: impact of the „biological pump" in the Atlantic equatorial upwelling belt over the last 330,000 years. Palaeogeogr. Palaeoclimatol. Palaeoecol. 103: 41-56

Van Bennekom AJ, Berger GW (1984) Hydrography and silica budget of the Angola Basin. Neth J Sea Res 17:149-200

Verardo DJ, McIntyre A (1994) Production and destruction: Control of biogenous sedimentation in the tropical Atlantic 0-300,000 years B.P. Paleoceanography 4: 63-86

Westerhausen L, Poynter J, Eglinton G, Erlenkeuser H, Sarnthein M (1993) Marine and terrigenous origin of organic matter in modern sediments of the equatorial East Atlantic: the $\delta^{13}C$ and molecular record. Deep-Sea Res I 40:1087-1121

Late Quaternary Movements of the Angola-Benguela Front, SE Atlantic, and Implications for Advection in the Equatorial Ocean

J.H.F. Jansen[1], E. Ufkes[1] and R.R. Schneider[2]

[1]*Netherlands Institute for Sea Research, P.O. Box 59,
1790 AB Texel, THE NETHERLANDS*
[2]*Fachbereich Geowissenschaften, Universität Bremen,
Postfach 330440, D-28334 Bremen, GERMANY*

Abstract: Planktic foraminifera data from three cores of the Angola-Zaïre margin are used to reconstruct palaeopositions of the Angola-Benguela Front (ABF) between the warm Angola Current and cold Benguela Current for the last 180,000 years. Strong northward shifts occurred in stages 4 and 3.3-3.1, but not in stages 6 and 2. The Benguela Current did not penetrate into the Gulf of Guinea. The southernmost positions, not far from the present one, were occupied in stages 5.5 and 1, but also in stage 6.3. The record of the shifts contains significant variance in the 23 ky^{-1} orbital frequency band and there are indications for a strong 100-ky^{-1} frequency component. There is also much variance at 15 ky^{-1}, the sum frequency of the 23-ky^{-1} and the absent 41-ky^{-1} cycles. This cyclicity is a real feature in the records.

The Benguela Current system performs a combination of two types of precessional (23 ky^{-1}) movements: shifts of the ABF to the south and north precede swings between more zonal and more meridional directions respectively of the Benguela Oceanic Current by 6.6 ky. The extreme positions of the ABF may be described by the 100-ky^{-1} eccentricity component which is also documented in cores from Walvis Ridge and Cape Basin. Both the 23-ky and 100-ky shifts of the ABF are in phase with the meridional movements of the Subtropical Convergence zone (STC) in the southern Indian Ocean, and with advection variations in the equatorial Atlantic. The movements of the STC are probably driven by meridional displacements of the belt of westerly winds over the southern hemisphere. These caused displacements of the circumpolar fronts, the SE trades and the Benguela Current system, and made the cool advection at the equator fluctuate.

The northernmost positions of the ABF coincide with strong sea-surface temperature minima in the Arabian Sea and the equatorial Pacific Ocean. These minima are probably the result of increased advection of relatively cool surface water from the south which is also caused by the northward eccentricity-driven movement of the westerlies and the associated oceanic polar fronts.

Introduction

The surface circulation in the eastern South Atlantic Ocean is dominated by the South Equatorial Counter Current with its coastal branch the Angola Current, and the Benguela Current which is the eastern limb of the subtropical anti-cyclonic gyre (Van Bennekom and Berger 1984; Peterson and Stramma 1991) (Fig. 1). Between the south-flowing Angola Current and the north-flowing cold water of the Benguela Current there is a convergence region called the Angola Benguela frontal zone (Meeuwis and Lutjeharms 1990). Here the Benguela Current is deflected to the northwest. The Benguela Current is an important component of the inter-hemispheric heat-transport of southern hemisphere surface water across the equator (Gordon 1986).

The Earth's climate was greatly influenced by this global thermohaline transport system which is believed to have fluctuated in response to Milankovitch forcing (Broecker et al. 1985; Imbrie et al. 1989, 1992, 1993). As a consequence, changes in

From WEFER G, BERGER WH, SIEDLER G, WEBB DJ (eds), 1996, *The South Atlantic: Present and Past Circulation*. Springer-Verlag Berlin Heidelberg, pp 553-575

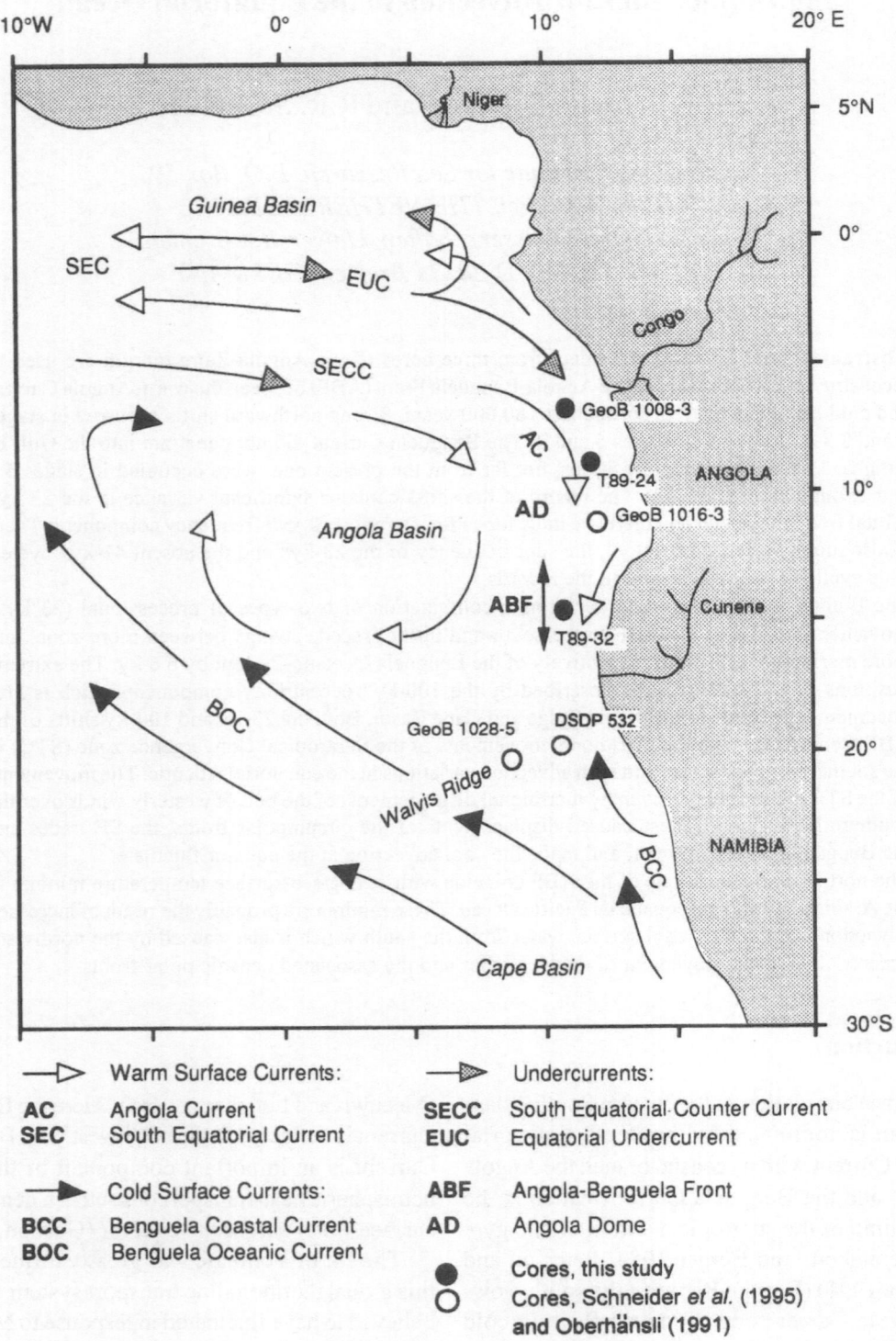

Fig. 1. Generalized surface and shallow subsurface circulation in the southeastern Atlantic Ocean and locations of the cores. The hydrography is summarized from Jansen et al. (1984), Shannon et al. (1987), Stramma and Peterson (1989), and Gordon and Bosley (1991).

the Benguela Current system are changes in the global transport system. Because the Benguela Current is wind driven, its history will also reflect variations in intensity and position of the southeast trade winds.

Palaeoceanographic studies of the surface circulation in the South Atlantic Ocean assumed that the offshore deflection of the Benguela Current moved to the north during glacial periods in connection with the global increase of oceanic circulation (McIntyre and Kipp et al. 1976). For the last glacial maximum, several investigators reconstructed an inflow of Benguela Current water into the Gulf of Guinea which even reached the equator (Gardner and Hays 1976; Morley and Hays 1979; Pokras 1987; McIntyre et al. 1989). Others indicated only a limited intensification of the Benguela Current with a minor shift of the frontal zone to the north by a few degrees (Jansen et al. 1984; Bjørklund & Jansen 1984; Zachariasse et al. 1984; Jansen 1985, 1990; Diester-Haass 1985; Van Leeuwen 1989; Jansen and Van Iperen 1991), or no significant shift at all (Schneider et al. 1995). Oberhänsli (1991) traced incursions of Angola Current water towards the northeastern Walvis Ridge during the interglacial stages 5.5, 7 and 11, but also, sporadically, in the glacial stages 6 and 8 and not in stage 9.

The aim of this paper is to answer the question whether the direction and position of the Benguela Current has varied during the late Quaternary and if so, what mechanism is responsible and how does it affect other oceanic regions. We will therefore present the planktic foraminifera records of three cores from the Angola continental margin, of which one is located in the centre of the present Angola-Benguela frontal zone. The cores are studied in comparison with previous results of surface water and sediment samples from the Angola and Guinea Basins. The present research is part of a series of studies on the palaeoceanography of the SE Atlantic and the palaeoclimate of equatorial Africa, based on piston cores collected on board the RV „Tyro" during three expeditions of the Netherlands Institute for Sea Research in 1978, 1980 and 1989 to the Angola and Guinea Basins. It also contributes to the research program of the Department of Geosciences, Bremen University, regarding the Quaternary oceanography of the South Atlantic (SFB 261).

Material and methods

Cores and stratigraphy

Two cores of this study, the piston cores T89-24 and T89-32, were collected on the continental slope off Angola with the RV "Tyro" during fall 1989 (Jansen et al. 1990) (Fig. 1, Table 1). Core T89-32 comes from the present Angola-Benguela Front. The core descriptions will be provided elsewhere (Ufkes et al., in preparation). The third core, the gravity core GeoB 1008-3, was recovered during the RV "Meteor" cruise in February/March 1988 from the Congo (Zaïre) deep-sea fan (Wefer et al. 1988) (Fig. 1, Table 1). The stratigraphy of the cores T89-24

Table 1. Positions, water depth and lengths of the cores

Core	Type	Longitude	Latitude	Water depth	Core length
T89-24	PC	8°54.7' S	12°03.1' E	2136 m	18.89 m
T89-32	PC	14°58.2' S	10°41.6' E	3330 m	17.66 m
GeoB 1008-3	GC	6°34.9' S	10°19.1' E	3124 m	12.04 m
GeoB 1016-3	GC	11°46.2' S	11°40.9' E	3411 m	12.41 m
GeoB 1028-5	GC	20°06.2' S	9°11.1' E	2209 m	10.79 m
DSDP 532	HPC	19°44.6' S	10°31.1' E	1331 m	17.09 m

PC = pistoncore, GC = gravity core, HPC = hydraulic piston core

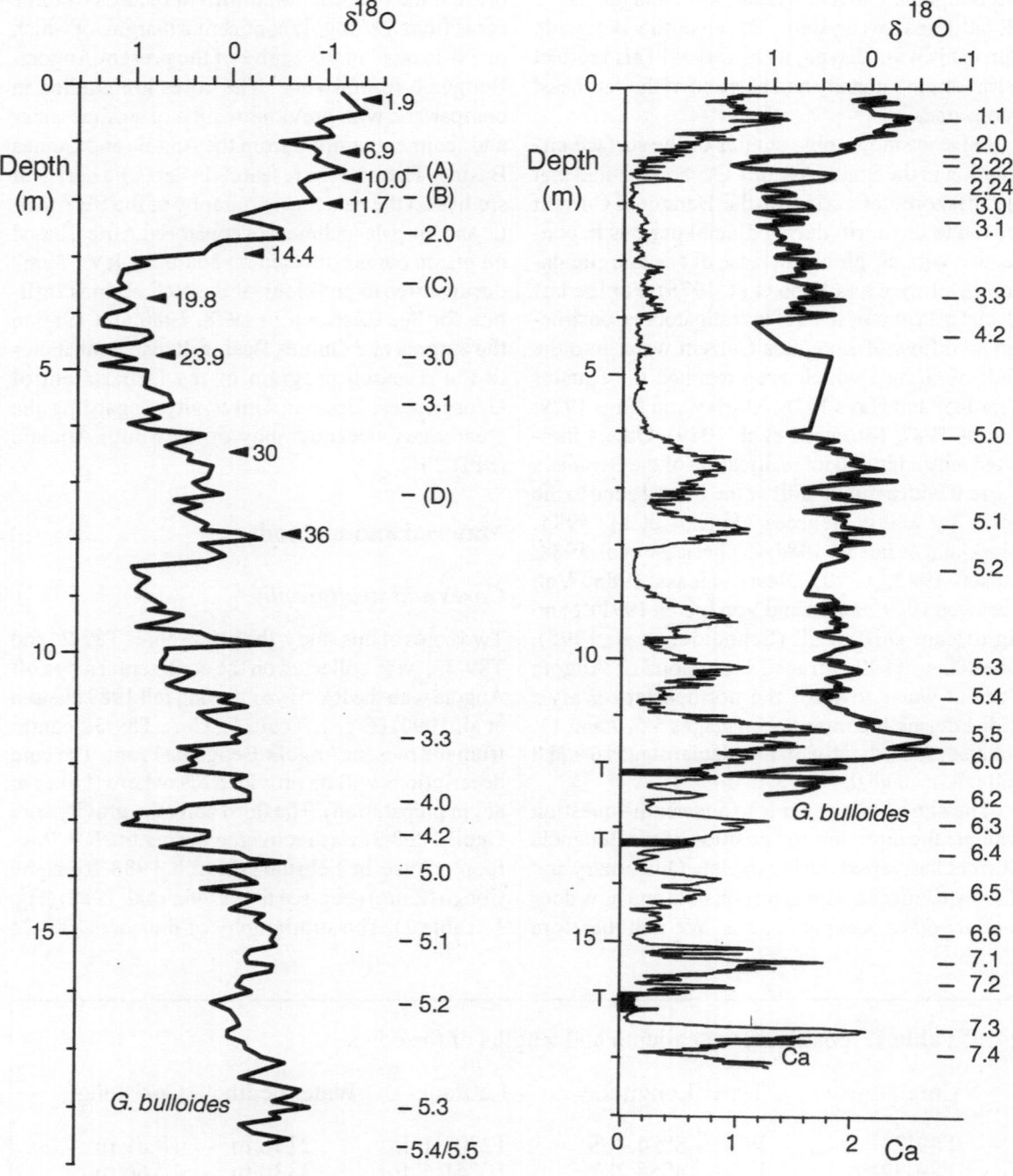

Fig. 2a. Core T89-24: δ¹⁸C of *G. bulloides*, ¹⁴C AMS datings (ky BP), and stratigraphic interpretation. The ¹⁴C ages (Table 2) are corrected with -400 y for the apparent age of ocean surface water. A,B,C and D are levels correlated with the cores GeoB 1008-3 and GeoB 1016-3 (Schneider 1991; Fig. 22).

Fig. 2b. Core T89-32: δ¹⁸O of *G. bulloides*, Ca concentrations (semi quantitative XRF measurements, values in counts per second) and stratigraphic interpretation. T: turbidite

Table 2. ^{14}C AMS datings for core T89-24

Depth (cm)	Age	$\delta^{13}C$	Date nr
29	2 320 ± 60	-0.21	UtC-2488
119	7 320 ± 70	1.13	UtC-2489
161	10 430 ± 90	0.21	UtC-2490
206	12 120 ± 120	-0.02	UtC-2491
279	14 850 ± 130	-0.22	UtC-2492
379	20 280 ± 160	-0.39	UtC-2494
476.5	24 300 ± 400	-0.43	UtC-2814
640	30 000 + 700/-600	-0.58	UtC-2813
818	36 300 +1300/-1100	-0.35	UtC-2812

Measurements on bulk carbonate. Ages in years BP, calculated from the ^{14}C activity normalized to $\delta^{13}C$ = -25‰; $\delta^{13}C$ in ‰ with respect to PDB; The ^{14}C ages are not corrected for the reservoir age of the sea in which the shells have grown. The measurements were carried out in the Robert J. van de Graaff Laboratory of the Utrecht University.

and T89-32 is based on $\delta^{18}O$ measurements of hand-picked planktic foraminifera of the species *Globigerina bulloides* and AMS ^{14}C datings (T89-24) (Fig. 2, Table 2). The stratigraphy of core GeoB 1008-3 is based on $\delta^{18}O$ measurements of *G. bulloides* and *Globigerinoides ruber*, pink variety (Schneider et al. 1994, 1995). The $\delta^{18}O$ measurements were performed in the isotope laboratory of the Department of Earth Sciences of the University Bremen, the ^{14}C datings in the Robert J. van de Graaff Laboratory of the University of Utrecht. We adopted the SPECMAP time scale by Imbrie et al. (1984) for the age model. For the period before 138 ky BP in core T89-32, no oxygen-isotope data are available yet. There is, however, a very good correlation between $\delta^{18}O$ and calcium carbonate over the last 138 ky, which allows us to apply the carbonate data for this time-interval (Fig. 2b). Cross-spectral analyses of $CaCO_3$ with $-\delta^{18}O$ shows coherent variance in all three orbital cycles. The two signals are nearly in phase for the 41-ky^{-1} frequency band ($-\delta^{18}O$ precedes $CaCO_3$ by 1600 ± 700 y) and well in phase for the 23-ky^{-1} band (260 ± 760 y). Consequently, we used the $CaCO_3$ data in the 23-ky^{-1} band for the statigraphy over the period prior to 138 ky.

Planktic foraminifera

The planktic foraminifera data were counted in the size range 150-600 µm, and for each sample at least 200 specimens were counted. The numbers of counted samples are 189 for core T89-24 and 92 for T89-32 (upper 12.41 m) (Ufkes et al., unpublished results) and 120 for GeoB 1008-3 (Schneider, unpublished results). The original data are to be published elsewhere (Ufkes et al., in preparation).

Time-series analyses

The records of core T89-32 were linearly interpolated to 2-ky intervals, and we conducted cross-spectral analyses with respect to the $-\delta^{18}O$ SPECMAP stack according to the SPECMAP ARAND method developed at Brown University (Imbrie et al. 1989). We also used foraminifera data of DSDP site 532, kindly provided by Hedi Oberhänsli. These data were analysed after linear interpolation at 5-ky intervals.

Modern surface circulation and foraminiferal assemblages

Surface circulation

Overviews of the surface circulation are given by Van Bennekom and Berger (1984), Meeuwis and Lutjeharms (1990), and Peterson and Stramma (1991) (Fig. 1). The large-scale circulation in the South Atlantic is controlled by a semi-permanent atmospheric high-pressure system in the central South Atlantic (the South Atlantic Anticyclone, around 30°S 10°W) and a low-pressure system over the continent moving from southern equatorial Africa (around 15°S) during the austral summer towards equatorial regions and the Sahara during the boreal summer. This pattern induces more meridionally directed trades nearby the continent and more zonal trades further offshore, and onshore monsoonal winds north of ca 15°S.

The prevailing southerly and southeasterly winds drive an offshore surface drift and intense coastal upwelling of cold, nutrient-rich water. The drift, the Benguela Current (BC), has a main limb, the Benguela Oceanic Current (BOC) that leaves the African coast at 20°S from where it moves in north-western directions. A coastal branch of the BC, the Benguela Coastal Current (BCC) moves up to 16-14°S. Here also terminates the intense coastal upwelling (Meeuwis and Lutjeharms 1990). South of the equator, an offshoot of the eastward flowing warm South Equatorial Counter Current (SECC), the Angola Current (AC), flows south along the African coast, where its meets the cold Benguela waters at 16-14°S, the Angola-Benguela Front (ABF). North of the front, the BCC can be detected as a narrow subsurface tongue up to 5°S (Van Bennekom and Berger 1984). The combined BC, SECC, and AC form a large cyclonic gyre system centred at about 10°W (Peterson and Stramma 1991).

The most common orientation of the front is west to east, it penetrates on average 250 km offshore, with traces up to 1000 km (Meeuwis and Lutjeharms 1990). It is well-developed in the upper 50 m, and can be detected in the salinity field to depth of at least 200 m (Peterson and Stramma 1991). There is a link between the position of the

ABF and the atmospheric South Atlantic Anticyclone (SAA) (Shannon et al. 1987). During the austral summer and autumn, when the SAA moves southward, the flow of warm Angola Current water is at its maximum and the ABF is also farthest south, widest and most pronounced. During winter and early spring, when the SAA moves farther north, more cold BC water flows in, the ABF moves north, and becomes narrower and weaker (Meeuwis and Lutjeharms, 1990). The front may follow the movements of the SAA very closely. During two weeks, in July 1973 and January 1974, Shannon et al. (1987) observed movements of the SAA that were followed by SST changes of 1-3°C within a few days. The SST changes possibly point to changes in the position of the ABF.

At the equator, a narrow subsurface jet flows eastward: the Equatorial Undercurrent (EUC). The southern limb of the EUC induces a permanent thermal dome below the thermocline at roughly 200 m depth off the Angola coast centred near 10°S, 9°E. This „Angola Dome" is most active during the austral summer. Equatorial doming occurs at the equator between 10°W and 5°E, mainly during the boreal summer (Peterson and Stramma 1991).

Surface distribution of planktic foraminifera

A previous investigation of planktic foraminifera in the surface waters of the eastern South Atlantic revealed that there is a good agreement between the distribution patterns of the foraminifera and the surface water masses (Ufkes 1996). For this study, we use data about the species that are characteristic of the water masses in the region of the Angola-Benguela Front (ABF). Six assemblages of foraminifera are relevant and four of these are mono-specific (Table 3). One consists of *Globorotalia inflata*, a species usually found in unstratified water masses with intermediate food levels. This species is characteristic of the water of the ABF itself. The other three mono-specific assemblages are found in the colder waters south of the ABF. *Globigerina bulloides* is related to poorly stratified eutrophic waters and is most abundant within the ABF and in the BC. The third contributor, the right coiled *Neogloboquadrina pachyderma*, points to high food levels but also to relatively low temperatures.

Table 3. Foraminifera assemblages

Species	Occurrence
1 *Globorotalia menardii* *G. hexagona* *Neogloboquadrina dutertrei*	Warm, thermocline; AC
2 *Globigerinoides ruber* (white) *G. ruber* (pink) *G. sacculifer* *Globigerinella siphonifera*	Warm, mixed layer; AC
3 *Globorotalia inflata*	Front; ABF
4 *Globigerina bulloides*	Poorly stratified; ABF + BC
5 *Neogloboquadrina pachyderma* (dextral)	Well stratified, cool; ABF + BC
6 *N. pachyderma* (sinistral)	Well stratified, cold; BCC

AC = Angola Current, ABF = Angola-BenguelaFront, BC = Benguela Current, BCC = Benguela Coastal Current.

It is found in the ABF and the BC south of it. The left coiled *N. pachyderma* requires temperatures below 9°C (Reynolds and Thunnell 1986). In the surface waters of this region, however, it can stand 15°C, which is attributed to the high food levels caused by the coastal upwelling (Ufkes and Zachariasse 1993). Left coiled *N. pachyderma* is present only in BCC water (Ufkes and Zachariasse 1993; Giraudeau 1993).

The final two assemblages are warm-water assemblages. One of these, the thermocline assemblage, consists of *Globorotalia menardii*, *G. hexagona* and *Neogloboquadrina dutertrei*. All three species live in a warm but steep thermocline. The other assemblage, the mixed-layer assemblage, consists of the classic oligotrophic surface-water species *Globogerinoides ruber* white and pink, *G. sacculifer* and *Globigerinella siphonifera*. The latter species flourishes also in a pre-upwelling situation with a poorly developed thermocline. The two assemblages were only encountered north of the

ABF, apart from a few exceptional findings further south.

Movements of the Angola-Benguela Front

Core T89-32

The downcore distributions of the six foraminiferal assemblages mirror north-south movements of the ABF. The foraminifera in from core T89-32, located in the actual front, show several levels with very high abundances, over 30%, of *N. pachyderma* (left) which nowadays only occur far south of the front , in the Benguela upwelling region (Fig. 3). At these levels, 97 ky (stage 5.3), 88 ky (5.2), and 50-41 and 39-34 ky BP (3.3 to 3.1), the two warm-water assemblages have nearly completely disappeared. These features cannot be attributed to dissolution of warm species because the total associations at these levels are not dissolution associations. There are two possible explanations for the cold sig-

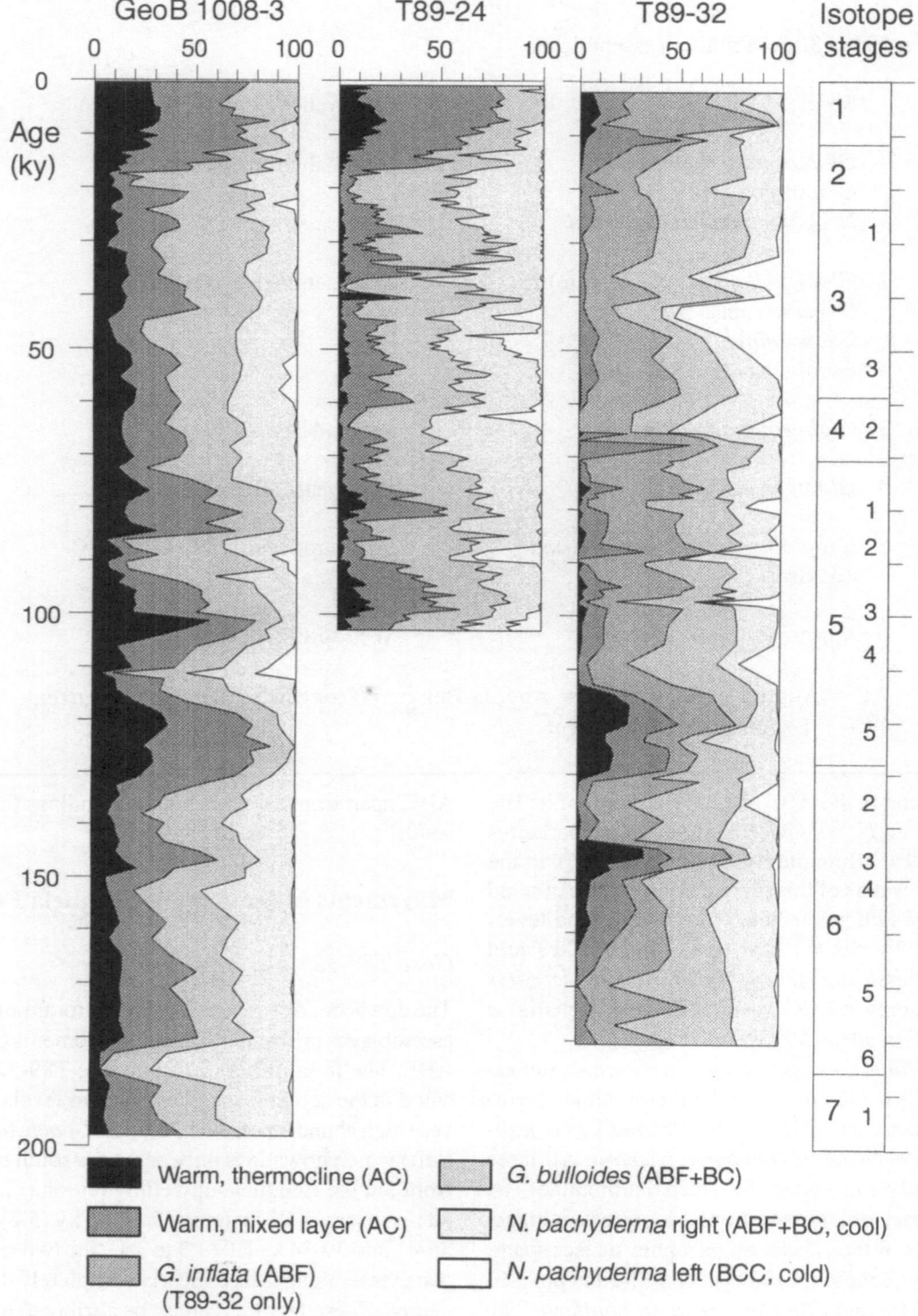

Fig. 3. Downcore distributions of the planktic foraminifera assemblages in the cores GeoB 1008-3, T89-32 and T89-24. For definition of the assemblages see Table 3. AC=Angola Current, ABF = Angola-Benguela Front, BC = Benguela Current, BCC = Benguela Coastal Current. For locations of the cores see Fig. 1 and Table 1.

nature at these levels: admixture with cold water by upwelling or advection from the south, and replacement of the warm AC water by BC water. Since the marker species of the AC have nearly completely disappeared at the levels in question, we conclude that the BC had occupied the site of the core and that the ABF was located further north. The strongest shifts must have occurred in the stages 5.3 and 3.3-3.1, because the ABF species (*G. inflata*) was also strongly decreased (<15%) at that time. Lower peaks of *N. pachyderma* (left), slightly over 15%, are observed at 143-131 ky (stage 6.2), 108 ky (5.4), 79 (5.1) and 65 ky BP (4.2). These point to minor northward shifts of the front, probably less far than the other three.

The ABF was situated far south three times during the last 180 ky, as is demonstrated by the high percentages (> 40%) of the two warm-water assemblages together. This happened 150-143 ky (stage 6.3), 131-113 ky (5.5), and 12-0 ky BP (1), and probably also about 68 ky ago (stage 5-4 transition). The latter shift is less certain because the observation depends on only one sample. During stage 6.3 and 5.5, the percentages were even higher than during stage 1, which suggests that the ABF lay further south than today. Two less conspicuous warm peaks (ca 30%) represent foraminiferal associations comparable to the present one, and consequently indicate positions nearby the actual front occurring 165-155 ky (stage 6.5-6.4) and 86-81 ky ago (stage 5.2, just after the cold peak). The rest are minor warm-water peaks (ca 20%), at 101 ky, 95 ky, 76 ky, 55 ky and 21 ky BP, that were formed when the was ABF slightly north of the actual front.

Core T89-24

The plot of core T89-24 contains five of the six assemblages of core T89-32 (Fig. 3). We omitted *G. inflata* because it does not inform us about the position of the ABF at this location where, at the present time, *G. inflata* mirrors primarily vertical mixing during the austral winter. The record of the core shows five strong peaks of *N. pachyderma* (left) suggesting an ABF north of 9°S, the latitude of T89-24 (Fig. 3). This happened ca 100 ky, 89 ky, 63 ky, and 47 and 44 ky BP, coincident with the northward shifts of the stages 5.3, 5.2, 4.2, and 3.3

respectively in core T89-32 . The shift of stage 3.3, 50-41 ky BP, apparently has fallen apart in two peaks in T89-24. In the last four peaks, the warm-water assemblages were forced back to below 15%, while their actual contribution is about 85% at this location. We conclude that these peaks mirror shifts of the ABF to a position north of T89-24. The first *N. pachyderma* (left) peak, prior to 100 ky BP, is large, but the warm species have not disappeared or are back again. This indicates that the front was not far from T89-24, but south of it.

Over the last 103 ky, the contribution of the warm-water species has never been as large as in the last 12 ky. This is in accordance with the southern position of the front as inferred from the record of T89-32. Enlarged contributions of warm-water species at 100-91 ky and around 80 ky can be correlated with the less conspicuous and minor peaks in T89-32.

Core GeoB 1008-3

A number of small peaks of *N. pachyderma* (left) occur in the record of this core (Fig. 3). They are, however, much smaller than in the cores T89-32 and T89-24. Apart from the first peak, 112 ky BP, they do not exceed 10% of the association, and they never suppress the warm water species to the low levels that are found in T89-32 and T89-24. Consequently, the peaks do not represent BC water but, probably, cold subsurface water below the Congo low-salinity plume (Ufkes et al., in preparation), and the BC did never reach 6½°S, the latitude of GeoB 1008-3, during the last 200 ky.

Comparison with previous studies

From the positions of the ABF relative to the locations of the above three cores, as discussed in the preceeding paragraphs, the history of the movements of the ABF over the last 170 ky can be inferred. We present our reconstruction in Fig. 4.

Additional information for the reconstruction comes from alkenone temperatures measured on core GeoB 1008-3 (at 6½°S), another core from the continental slope off Angola (GeoB 1016-3 at 12°S), and one from Walvis Ridge (GeoB 1028-5 at 20°S) by Schneider et al. (1995). The recorded

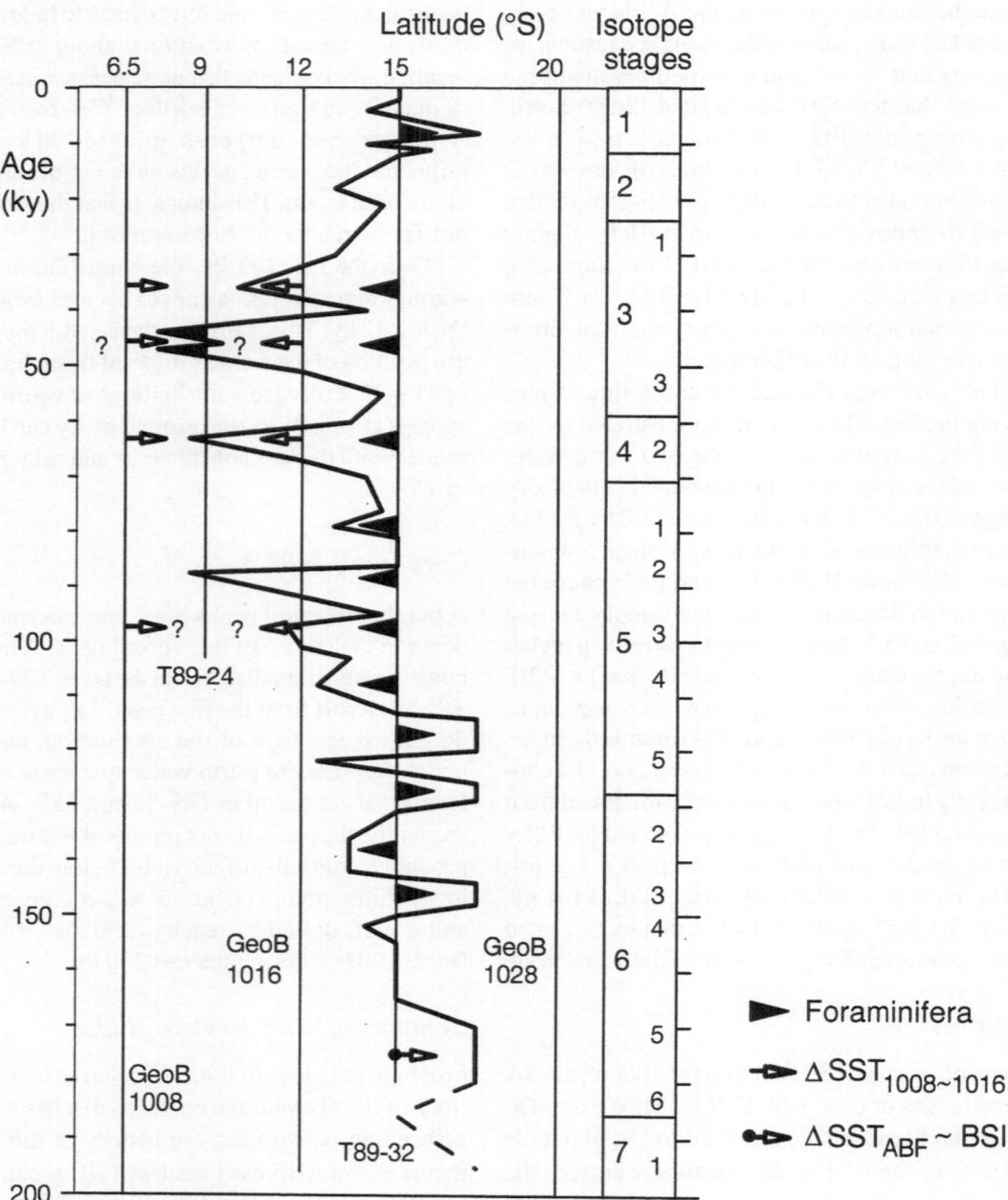

Fig. 4. Reconstruction of the average positions of the Angola-Benguela Front over the last 180 ky. The arrows indicate the positions of the ABF with respect to the locations of the cores, based on the foraminifera contents (this study), the differences between the alkenone SST in the cores GeoB 1008-3 and GeoB 1016-3 (in Schneider et al. 1995), and the comparison of ΔSST_{ABF} with BSI (Schneider et al. 1995). For locations of the cores see Fig. 1 and Table 1.

temperatures of the two Angola Basin cores display a close similarity one to the other, and are consistently 2-6°C higher than on Walvis Ridge. The reconstructed shifts, however, can be recognised as relatively small differences in SST between the two Angola Basin records (see Fig. 5 in Schneider et al. 1995). These show several intervals with higher temperatures in GeoB 1008-3 than in GeoB 1016-3. At present, the difference in annual mean SST between the two Angola Basin sites is about 0.5°C (Levitus 1982; Meeuwis and Lutjeharms 1990). This suggests that larger differences reflect an ABF situated between 6½ and 12°S. Superimposed on the general northward increase in SST, the ABF itself causes an extra step in annual mean temperature of about 2°C (Levitus 1982; Fig. 4 in Schneider et al. 1995). During the shifts around 65 and 36 ky BP, GeoB 1008-3 was about 2°C warmer than GeoB 1016-3, which corroborates the ABF position between 6½ and 12°S at these times. The measured alkenone temperature difference was about 1°C during the shifts around 100 ky and 50-41 ky BP. This suggests that the average ABF was at a short distance from one of the cores at 12°S (100 ky) or 6½°S (50-41 ky). Considering that the error of the alkenone temperature measurements is generally equivalent to about 0.5°C (Prahl and Wakeham 1987; Schneider et al. 1995), we conclude that the two alkenone temperature records corroborate our reconstruction of the ABF between 12 and 6½°S at 65 ky and 36 ky BP, and does not contradict it at 100 ky and 50-41 ky BP.

There are two more periods displayed with higher alkenone temperatures at 6½°S than at 12°S, suggesting a front location north of 12°S during 80-70 ky and ca 10 ky BP (Schneider et al. 1995). These positions, however, are not represented as a climatic cooling in the foraminiferal records of Fig. 3. In contrast to the shifts discussed earlier, these periods show strong parallel temperature changes at the two locations, which makes the temperature differences very sensitive to uncertainties in the chronology. The differences at 80-70 and 10 ky BP, therefore, do not necessarily point to a front between 6½ and 12°S.

Schneider et al. (1995) found that the temperature differences between the records from Walvis Ridge (GeoB 1028-5) and the Angola Basin (GeoB 1016-3), regarded as the differences over the Angola-Benguela Front (ΔSST_{ABF}), show a strong cyclicity in the 23 ky^{-1} frequency band which is in phase with minimum boreal summer insolation (BSI) (Fig. 5). They explain this cyclicity with a model of monsoon-modulated trade-wind zonality (McIntyre et al. 1989; Imbrie et al. 1989; Meinecke 1992). During periods of minimum BSI, the monsoon activity over Africa weakened (Prell and Kutzbach 1987), and more zonally directed trade winds enhanced the southeastward transport of warm tropical water of the South Equatorial Counter Current and Angola Current. The result was enlarged temperature contrasts over the ABF. Comparison of the ΔSST_{ABF} record with its 23-ky filtered signal and BSI revealed a clear correspondence between the three parameters. Consequently, the monsoon-modulated trade-wind zonality model is supposed to explain the ΔSST_{ABF} record. The explanation requires that the ΔSST_{ABF} has always represented the SST difference between the BOC and the AC or, in other words, that the ABF lay continuously between GeoB 1028-5 (20°S) and GeoB 1016-3 (12°S) (Fig. 1). The correspondence, however, of the ΔSST_{ABF} signal with the BSI signal failed around 120 and 95 ky and between 70 and 40 ky BP when ΔSST_{ABF} was much smaller than would be predicted by the BSI signals (Fig. 5). Exactly then, the ABF had moved toward its northernmost positions according to the faunal record (Fig. 4). Close inspection reveals that the maximum deviations of ΔSST_{ABF} from BSI occurred 120 and 95 ky, 70-60 ky, and 50-40 ky BP, all coinciding with northward movements of the front as indicated by the foraminifera. Apparently, these movements have suppressed the SST contrast over the two sites.

There were five periods when ΔSST_{ABF} was larger than predicted by the BSI signal: around 200 ky (stage 7.1), 182-170 ky (stage 6.6-6.5), 150-140 ky (6.3), around 127 ky BP (5.5) and 6 ky BP (1). The latter three deviations mirror positions of the ABF further south than at present, which corroborates our reconstruction of the shifts. The deviations predict previous movements far to the south around 200 ky (stage 7.1) and 175 ky BP (6.6-6.5). We conclude that the ΔSST_{ABF} record is a combination of a monsoon-modulated trade-wind zonality signal

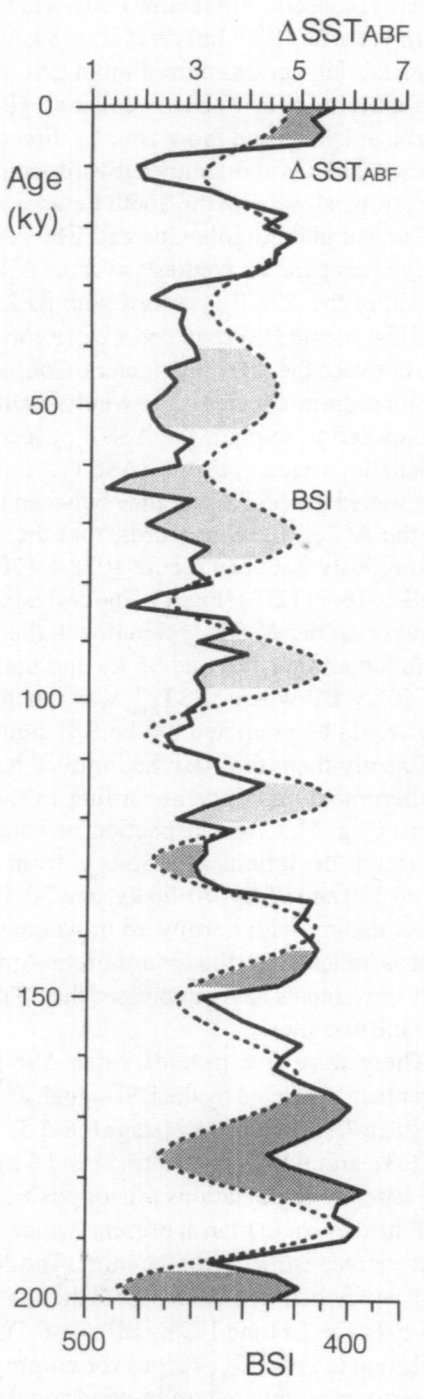

and a signal of the north-south movements of the Angola-Benguela Front.

In a study of marine diatom records in four cores from the Congo (Zaïre) deep-sea fan, Jansen and Van Iperen (1991) demonstrated that the SECC water has never disappeared from the region off the Congo River mouth during the last 220 ky. The BC, therefore, did not reach 8°S (core T80-6) in this period. The diatoms point to increased oceanic upwelling during stage 6, with maxima around 180 ky and 150-145 ky BP, the stage 6-5 transition, stage 5.3, and during the late stage 2 plus stage 1. These maxima can be correlated with the major southward shifts of the ABF. The oceanic upwelling signal, however, was attributed to the Angola Dome, representing increased influence of the Benguela Current and northward shifts of the ABF, which is in contradiction to our present reconstructions. We believe now that the relevant diatom association should be re-interpreted: it mirrors probably the equatorial doming (Peterson and Stramma 1991) rather than the Angola Dome; the Congo-fan cores are located just in between the two thermal domes. If this re-interpretation is true, the oceanic upwelling signal indicates southwards movements of the front, in accordance with our reconstruction. The maximum upwelling 180 ky BP then confirms the movement far to the south during stage 6.6-6.5 predicted by the alkenone temperature differences between GeoB 1028-5 and GeoB 1016-3 discussed above.

On the continental margin of Namibia, a period of extremely strong upwelling occurred during stage 3 with maxima dated at 61, 46, 42 and 35 ky BP (core PC12, 22°S) (Summerhayes et al. 1995). At

Fig. 5. Alkenone SST differences between the cores GeoB 1016-3 and GeoB 1028-5 (ΔSST_{ABF}, in °C, solid line) and boreal summer insolation at 10°N (BSI, in W.m^{-2}, dotted line), according to Schneider et al. (1995). Light shaded intervals indicate ΔSST_{ABF} values smaller than predicted by BSI, dark shaded intervals indicate values larger than predicted. These intervals point to positions of the ABF nearby or north of the location of GeoB 1016-3 and nearby GeoB 1028-5 respectively. For locations of the cores see Fig. 1 and Table 1.

these times, maximum accumulation of organic carbon was accompanied with low sea surface temperatures as recorded by U^k_{37} temperatures and nannofossil associations. Since the core PC12 is located in the northern part of the actual upwelling region which has its centre off Lüderitz at 27°S, the maxima may be attributed to northward movements of the zone of maximum upwelling. They are then correlated with the northward ABF shifts 65, 47, 44 and 36 ky BP respectively. These low temperatures are also visible in the SST record of core RC13-228 from the basis of the continental slope west of PC12 (Morley and Shackleton 1984). This record, produced with transfer functions based on radiolarian assemblages, shows also remarkably high SST values around the stage 6-5 transition when the ABF had moved far north. We infer that both SST records reflect latitudinal movements of the Benguela Current system.

The southernmost shifts of the ABF are also documented by the occurrence of the Angola Current species *Neogloboquadrina dutertrei* at the northeastern Walvis Ridge at 19°45' S (DSDP site 532) (Oberhänsli 1991). Here, *N. dutertrei* reached peak abundances during the stages 11, 8, 7, around the stage 6-5 transition, and in stage 1. The peaks represent incursions of Angola Current water to the south, and are culminations of longer warm-water periods. The latter two periods stress the southern positions of the ABF in stage 7.1, 6.6-6.5, 6.3, 5.5 and 1. According to Oberhänsli (1991), the temperatures at site 532 were probably lowest of all during the upper part of stage 5 and the stages 4-2. This observation affirms the northward positions in stage 5.2, 4.2 and 3.3-3.1.

Time-series analyses

The movements of the ABF do not closely follow the global $\delta^{18}O$ record of glacial-interglacial variations (Imbrie et al. 1984). Northward shifts took place during the minima in the "glacial" stages 6.2 and 4.2, but not in stage 6.4 and 2. There were also shifts to the north in the "interstadials" 5.4 and 5.2 and the stadial 5.3, but the strongest ones occurred in stage 3. The southernmost extensions were recorded in the interglacial maxima of stages 5.5 and 1, but also in the interstadial stage 6.3.

We developed two parameters for the north-south movements of the Angola-Benguela Front; the Angola-Benguela Ratio (ABR) and the Benguela Index (BI). The Angola-Benguela Ratio weights the species living exclusively or preferably north of the ABF against the "cold" *N. pachyderma* left + right coiled. The "warm" species are the species of the two warm-water assemblages, the thermocline and the mixed-layer assemblage as defined in the section: Surface distribution of planktic foraminifera. The formula is:

$$ABR = warm/(warm+cold)\ species.$$

It appears that the ABR record (Fig. 6) is not very sensitive to the reconstructed northward shifts in Fig. 4. For this reason, we calculated also the BI. BI represents the cold Benguela Current species *N. pachyderma*, but gives a double weight to the left coiled form. Its formula is:

$$BI = (2/3\ x\ left+1/3\ x\ right)\ coiled\ N.\ pachyderma.$$

Comparison of the indices ABR and BI (Fig. 6) with the reconstruction of the ABF movements (Fig. 4) demonstrates that BI is indeed most sensitive for northward shifts of the ABF and ABR is a better signal for the southward movements.

Cross-spectral analyses of BI and ABR in core T89-32 reveal signals in the 23-ky^{-1} precession cycle which are coherent with the SPECMAP stack (Imbrie et al. 1984) (Fig. 7; Table 4). The two signals, ABR and -BI (the "warm" counterpart of BI), are approximately in phase with minimum global ice volume ($\theta = 87°$), although a slight lead is not excluded. Besides, they also contain variance in the 100-ky^{-1} eccentricity cycle. There is no power in the 41-ky^{-1} obliquity band, but the records contain much variance at 15-ky^{-1} which is the sum frequency of the 23-ky^{-1} and 41-ky^{-1} cycles. Spectral analyses with cleaning methods that correct for interference of overtones and comparison of the 15-ky^{-1} filtered records with the original curves shows that the 15-ky^{-1} cyclicity is not a mathematical artifact, but a real feature in the records. It is true that the cores are too short to determine a long periodicity in the eccentricity band, but visual inspection of the records show that the long-term variations may well be described by a 100-ky^{-1} cyclicity which is about in phase with the 100-ky^{-1} SPEC-MAP signal (Fig. 8b) (Imbrie et al. 1989). This cyclicity is corroborated by the *N. dutertrei* record from Walvis Ridge (DSDP site 532) (Oberhänsli 1991). Cross-spectral analyses of this record with the SPECMAP stack shows significant variance in the 400 and 100 ky^{-1} eccentricity frequencies, the 100 ky^{-1} signal being in phase with global ice volume ($\theta=12°$)(Fig. 7, Table 4). The same phasing is observed in the SST eccentricity component in core RC13-228 from the continental slope off Namibia in the Cape Basin (Imbrie et al. 1989).

The BI signal deviates from the general pattern of combined precession and eccentricity signals before about 160 ky, when northward excursions are predicted by the ice-volume signals (Fig.8c). The deviation can be explained by interference of the 23-ky^{-1} signal with a 19-ky^{-1} signal (Fig. 8d). In contrast to ABR, BI appears to contain a double, 23 and 19-ky^{-1} precessional band as is suggested by spectral analyses with an increased number of lags. The minima and maxima of the 23-ky filtered records of ABR and -BI closely follow the 23-ky filtered SPECMAP record, except during the period before 160 ky BP which is influenced by the 19-ky^{-1} cycle discussed above (Fig. 9).

Summarising, besides a strong 15-ky^{-1} signal, the ABR and -BI records are composed of a 23-ky^{-1} and probably also a 100-ky^{-1} component. The 100-ky^{-1} component is also documented in the *N. dutertrei* record from Walvis Ridge and the SST record from Cape Basin. In both periodicities, the parameters for the ABF shifts are approximately in phase with minimum global ice volume, although for the 23-ky^{-1} component a slight lead is not excluded.

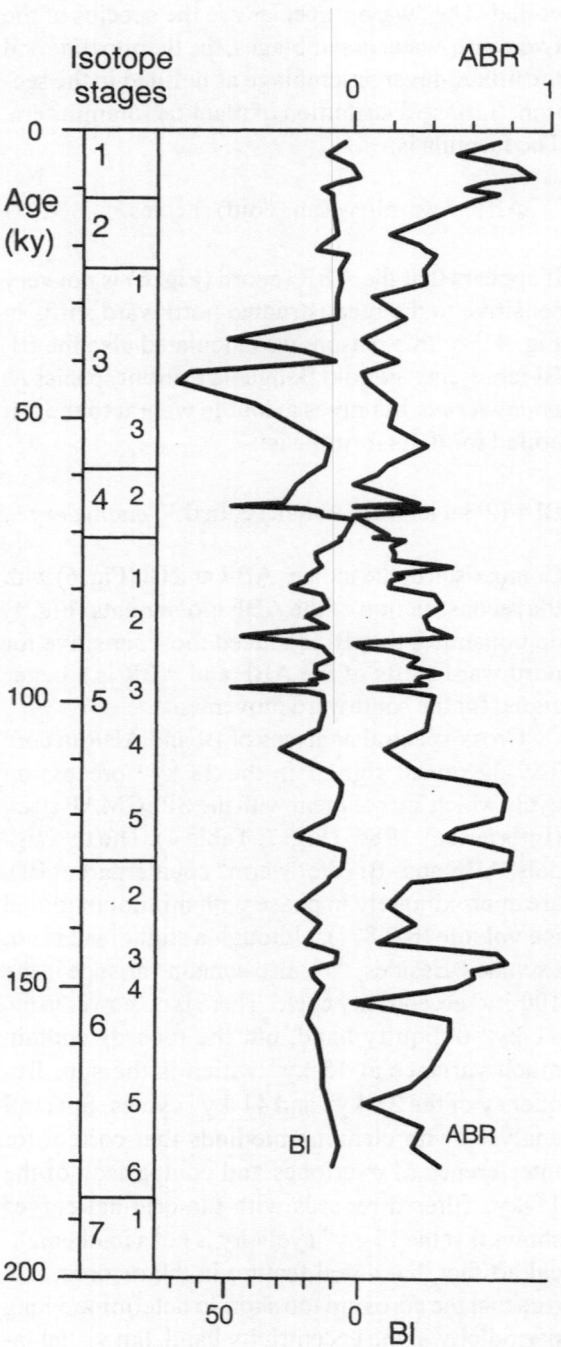

Fig. 6. Downcore distributions of the Angola-Benguela Ratio (ABR) and Benguela Index (BI) for core T89-32. ABR = warm/(warm+cold)species, BI = (2/3 x left + 1/3 x right) coiled *N. pachyderma*.

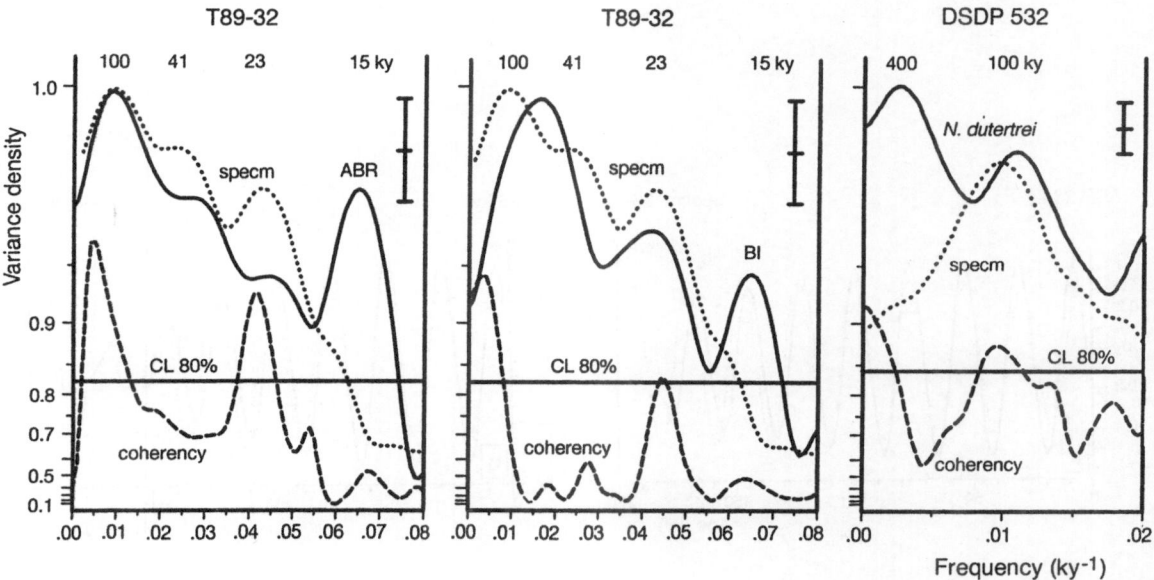

Fig. 7. Results of cross-spectral analyses of the Angola-Benguela Front parameters of core T89-32 (left and middle) and DSDP 532 (right) with the SPECMAP stack (Imbrie et al. 1984, 1989). ABR = Angola-Benguela Ratio, BI = Benguela Index, specm (dotted line) = δ^{18}O SPECMAP stack. Vertical bars give 80 % confidence interval. Coherency (dashed line) is given at an arbitrary scale, with the 80 % confidence level indicated by a horizontal line (CL 80%). *Neogloboquadrina dutertrei* data of DSDP 532 are from Oberhänsli (1991).

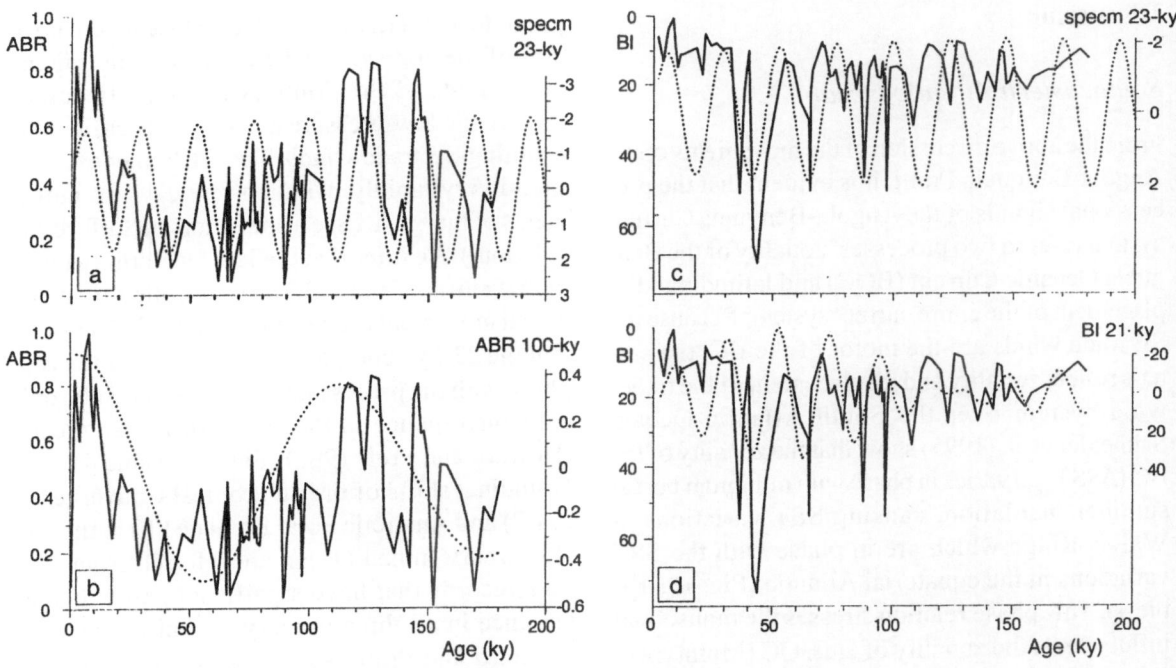

Fig. 8. Comparison of the records of the Angola-Benguela Ratio (ABR) and Benguela Index (BI) (solid lines) for core T89-32 with several filtered components (dotted lines). a: ABR with the 23-ky filtered SPECMAP stack.

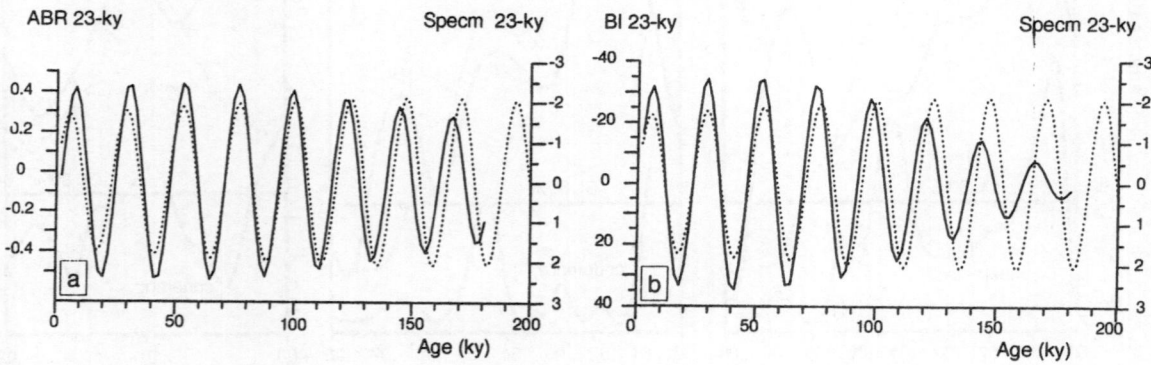

Fig. 9. Comparison of 23-ky filtered ABR (a) and BI (b) signals (solid lines) with the 23-ky filtered SPECMAP stack (dotted lines). Window 23-ky: 25-21 ky^{-1}.

Discussion

Southeastern Atlantic Ocean

From the above discussion of the movements of the Angola-Benguela Front, it is evident that the precessional signals of the Angola-Benguela Current system refer to two processes: zonality of the Benguela Oceanic Current (BOC) and latitudinal displacement of the entire current system. Because the SE trade winds are the motor of the BC, the signals reflect zonality and displacement of the trade-wind system over the South Atlantic Ocean. Schneider et al. (1995) show that the zonality of the BC (ΔSST_{ABF}) varies in phase with minimum boreal summer insolation, causing SST variations at Walvis Ridge which are in phase with the SST variations in the equatorial Atlantic (Fig. 10; Table 4). This phase relation stresses the monsoonal influence on the zonality of the BOC (McIntyre et al. 1989; Imbrie et al. 1989). Apparently, the site of the Walvis Ridge core, at the joint of the swinging limb of the Benguela Current (BC), is very sensitive to changes in the direction of the BOC.

In the modern South Atlantic Ocean, the position of the ABF is coupled with the meridional movements of the South Atlantic Anticyclone (SAA), the low-pressure cell over Africa, and the resulting SE trade winds. The ABF may even respond very rapidly, within a few days, to movements of the SAA. (See also section on Surface circulation.) We infer that the late Quaternary meridional shifts of the ABF are controlled by meridional movements of the trade-wind belt. It appears that the 23-ky^{-1} component of the ABF shifts is in phase with the precessional shifts of the Subtropical Convergence in the southern Indian Ocean (Howard and Prell 1992) (Fig. 10; Table 4). The latitudinal shifts of the Subtropical Convergence (STC) and Antarctic Polar Front (APF) in the late Quaternary Indian Ocean are inferred from four SST records that have significant and coherent variance in all three primary orbital frequencies (Howard and Prell 1992). The records are from cores that were collected in between the modern STC (40°S) and APF (50°S). According to the reconstruction by Howard and Prell (1992), the site of the northernmost core (E49-21 at 42°S) was

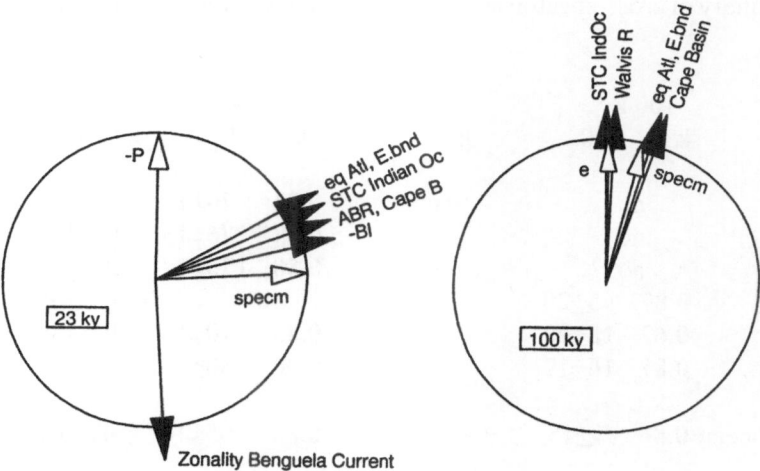

Fig. 10. 23-ky and 100-ky phase clocks of properties of the Benguela Current system and related features. Zero positions of the phase clocks are defined at minimum precession index (-P) indicating maximum boreal summer insolation (at 21 June) and maximum eccentricity (e), according to the SPECMAP convention (Imbrie et al. 1989). ABR = Angola-Benguela Ratio; BI = Benguela Index (Conformably to the SPECMAP conventions the "warm" signal -BI is plotted); Cape B = SST Cape Basin (core RC13-228, Imbrie et al. 1989); specm = SPECMAP δ^{18}O stack; Zonality Benguela Current = ΔSST_{ABF} as described by Schneider et al. (1995); STC Indian Oc = southward shifts of the Subtropical Convergence in the southern Indian Ocean from SST variations of core E49-21 (Howard and Prell 1992); eq Atl, E. bnd = minimum Eastern Boundary Current assemblage in the equatorial Atlantic (core V30-40, Mix & Morey, this volume).

passed back and forth only by the STC during the last 480 ky. The next two cores (45 and 46°S) were influenced and probably passed several times by both fronts, the STC and APF. The southernmost core (48°S) was not influenced anymore by surface water from north of the STC, but the statistical error of the phase of its 23-ky⁻¹ signal is large. Moreover, the movements of the APF precede the STC movements (Morley 1989), so the purest signal of the frontal shifts is found in the SST record of the northernmost core E49-21, representing the alternation of only two water masses. Like the ABF, the STC and APF are wind-forced phenomena. Their positions are controlled by the positions of the zone of zero wind-stress-curl. Since the maximum westerly winds are associated with this zone, the shifts of the two oceanic fronts are attributed to variations in the wind strength in subtropical to subpolar latitudes or north-south migrations of the entire belt of maximum westerlies. The wind stress is particu-

larly enhanced by zonal asymmetries in sea ice fields in the Southern Ocean (Howard and Prell 1992).

The precessional shifts of the ABF do not only coincide the with the shifts of the STC in the southern Indian Ocean. They coincide also with the latitudinal movements of the Benguela Current system in the Cape Basin as inferred from the SST variations of core RC13-228 from the upwelling region off Namibia (Morley and Shackleton 1984; Imbrie et al. 1989) (Fig. 10; Table 4). This points to a single collective origin: variations in strength or position of the westerly wind belt. The variations caused shifts of the circumpolar oceanic fronts, and meridional movements of the SAA and trade wind belt, at least over the South Atlantic. There is one important difference between the two regions. The shifts of the fronts in the Indian Ocean show a strong 41-ky⁻¹ obliquity variability which is weak in the ABF record and the SST record of the Cape Ba-

Table 4: Summary of cross-spectral analyses of the time series of this study and literature data.

Time series	100 ky		41 ky		23 ky		Series length	CL 80%
	k	θ	k	θ	k	θ		
ABR	nc		nc		0.93	70±13	180 ky	0.95
-BI	nc		nc		0.83	76±13	180 ky	0.82
ΔSST_{ABF}	nc		nc		0.88	176±17	200 ky	0.79
Walvis Ridge	0.87	3±20	nc		nc		478 ky	0.84
Cape Basin	0.62	18±28	nc		0.85	70±15	305 ky	0.60
eq. Atlantic E. bound.	0.87	16±19	nc		0.96	60± 9	257 ky	0.75
STC Indian Ocean	0.86	-1±15	0.94	74±10	0.85	65±16	471 ky	0.65

Coherence (k) is given with respect to $\delta^{18}O$ SPECMAP stack, phases (θ) with respect to maximum eccentricity, maximum obliquity, and minimum precession index (Imbrie et al. 1989); CL 80%: test statistics for non-zero coherency at the 80% confidence level; nc: no significant coherency. ABR = Angola-Benguela Ratio; BI = Benguela Index; ΔSST_{ABF} describes the zonality of the Benguela Current (Schneider et al. 1995); Walvis Ridge: *N. dutertrei* taken from core DSDP 532 (Oberhänsli 1991); Cape Basin: SST taken from core RC13-228 (Imbrie et al. 1989); eq Atlantic, E. bound. Minimum Eastern Boundary Current assemblage in the equatorial Atlantic taken from core V30-40 (Mix & Morey, this volume). STC Indian Ocean: SST South Tropical Convergence in the Indian Ocean taken from core E49-21 (Howard and Prell 1992).

sin. Considering that fluctuations in the 41-ky[-1] obliquity cycle have the strongest effects in high latitudes and little effect on tropical insolation and that the 23-ky[-1] precession fluctuations are more important in low latitudes, we propose that the strength of the westerlies is primarily controlled by the obliquity cycle, while the position of the westerlies reacts much stronger in the 23-ky[-1] precession band. Precessional movements of the westerlies wind belt would then also have caused the precessional displacements of the SAA, the belt of trade winds and, consequently, the shifts of the Angola Benguela Front. The 23-ky[-1] shifts of the ABF are approximately in phase with the global ice volume (Fig. 10; Table 4), so they are probably induced by precessional growth and decay of the ice fields in the Southern Ocean.

The 100-ky[-1] eccentricity fluctuations of the ABF to the south during stage 5.5 and the LGM and to the north around 160 and 50 ky BP, may also be explained by meridional shifts of the subtropical and polar fronts. The fluctuations are about in phase with the frontal movements in the Southern Ocean as recorded in the southern Indian Ocean (Fig. 10; Table 4). There, the SST record of core E49-21, representing the meridional shifts of the STC, shows a slight but insignificant lead with respect to $-\delta^{18}O$ (Howard and Prell 1992). During periods of minimum orbital eccentricity and maximum ice volume, increased sea-ice fields in the Southern Ocean pushed the westerly-wind and trade-wind belts northwards (Imbrie et al. 1993), which moved also the Angola Benguela Front to the equator. As a consequence, the eccentricity-driven waxing and waning of the southern hemisphere ice fields probably also controls the latitudinal movements of the Angola Benguela front in the 100-ky[-1] frequency band.

Implications for equatorial advection

Equatorial Atlantic Ocean

The northward shifts of the ABF cause cooling in the equatorial Atlantic Ocean. In a study of planktic foraminifera, Mix and Morey (this volume) describe an eastern boundary current assemblage representing the advection of cool Benguela Current water towards the equator. This assemblage shows significant variance in the 23 and 100 ky^{-1} frequency bands, but not in the 41 ky^{-1} band. In the first two bands, the assemblage fluctuates in phase with the ABF shifts (Fig. 10; Table 4). Apparently, the inferred displacement of the subtropical fronts vary the intensity of the surface circulation in the South Atlantic and the advective supply of cool Benguela Current water towards the equatorial Atlantic Ocean.

Arabian Sea

The above strong movements of the ABF are also recognized in records from the Arabian Sea. Here extremely low SST values occurred during stage 3.3-3.1 (Kroon et al. 1990; Ten Haven and Kroon 1991; Zahn and Pedersen 1991; Rostek et al. 1993; Emeis et al. 1995). This feature was attributed to strong monsoonal upwelling (Prell and Van Campo 1986; Clemens et al. 1991; Steens et al. 1992). The SST low, however, does not correlate well with productivity and upwelling indicators like δ^{13}C of planktic foraminifera, $\Delta\delta^{18}$O (surface minus thermocline dwellers) and organic carbon content (Zahn and Pedersen 1991; Steens et al. 1992; Emeis et al. 1995) or pollen originating from Arabia (Prell and Van Campo 1986). Moreover, the SST decreased also in the southeastern Arabian Sea, a region uninfluenced by upwelling (Rostek et al. 1993). We propose that the SST low was caused by advection of cold water instead of upwelling. An alternative explanation attributes the low SSTs to prolonged NE monsoons during winter (Ten Haven and Kroon 1991; Emeis et al. 1995). The strongest NE monsoon and lowest snowline at the Tibetan Plateau, however, occurred during the Last Glacial Maximum (Van Campo et al. 1982; Emeis et al. 1995).

This timing does not explain the low SSTs in stage 3, so we prefer the advection hypothesis to explain these low SSTs in the Arabian Sea.

Visual inspection of the records in Zahn and Pedersen (1991), Steens et al. (1992) and Emeis et al. (1995) show long-term fluctuations with minima in both SST and upwelling coincident with minimum eccentricity and maximum global ice volume, like it happened 44 ky ago (Berger and Loutre 1991). We infer that, like in the Atlantic sector, the Antarctic and subtropical fronts had moved to the north, and the density gradients in the southern Indian Ocean had steepened during these minima. The eccentricity-driven steeper gradients enhanced the advection of cold subsurface water towards the equator and, eventually, also to the Arabian Sea.

The transequatorial advection happens through the Somali Current which today maintains a flow of cold subsurface water across the equator of 6 Sverdrup (Schott et al. 1990). Schott et al. suggest that the southward transequatorial return of this flow occurs in the diffusive surface mixed layer of about 28°C. If this is true, all the cold subsurface Somali Current water is heated up by 15°C by insolation and, consequently, an increase of the subsurface Somali Current flow from 6 to 7 Sv would reduce SST from 28 to 26°C. The total SST changes in the Arabian Sea outside the upwelling region are about 3°C (Rostek et al. 1993), so the above heat budget calculation is consistent with our hypothesis that increased advection through the cold subsurface Somali Current caused low SSTs in the Arabian Sea during stage 3.3-3.1.

Equatorial Pacific

The equatorial Pacific Ocean went through comparable phenomena during the late Quaternary. In a comparison of alkenone temperatures and organic carbon burial fluxes in a core from 1°N, 139°W, Lyle et al. (1992) made a distinction between SST variations caused by equatorial upwelling and advection. They argued that periods of enhanced upwelling should produce a sedimentary interval with both low SSTs and high organic carbon burial ($C_{org}AR$), whereas periods of enhanced advection of cold subpolar waters to the equator should manifest as low SST and lower $C_{org}AR$. In a part of the record, $C_{org}AR$ roughly mirrors SST and in another part it roughly parallels SST, thus representing upwelling and advection respectively according to Lyle et al.. The decision, however, whether the two parameters mirror or parallel depends greatly on how much detail is considered while observing the curves. Focusing on the extremes of the two records, maximum SSTs were recorded about 130 and 10 ky and minimum SSTs about 160 and 40 ky BP, both in periods with average $C_{org}ARs$. We infer that the minima represent strong advection and the maxima weak advection of cold surface water, and that these eccentricity-driven advection fluctuations caused the extremes of the SST record at the equatorial Pacific, roughly in phase with variations in the extension of sea-ice the Southern Ocean.

Conclusions

1 Planktic foraminifera data from the Angola-Zaïre continental margin indicate north-south shifts of the Angola-Benguela Front (ABF) during the last 180 ky.

2 The ABF shifts do not simply follow glacial to interglacial variations. The ABF has been north of its present position (15°S) for most of the past 200 ky: the northernmost positions, north of 9°S, occurred in the stages 4 (63 ky) and 3 (50-41 ky BP), and less extreme shifts in stage 5.3 (ca 100 ky) and around 36 ky BP. No strong northward shifts, however, were recorded in stage 2 and the glacials in stage 6. Shifts to the south occurred during the stage 6.6-6.5 transition and in stages 5.5 and 1, but also in stage 6.3.

3 The Benguela Current did not penetrate into the Gulf of Guinea during the last 220 ky.

4 The ABF shifts are also displayed by parts of the differences in U^k_{37} temperature north and south of the actual front (ΔSST_{ABF}), measured by Schneider et al. (1995).

5 Spectral analyses of two parameters for the shifts of the front, ABR and -BI, reveals a 23-ky^{-1} precessional component and there are indications for a strong 100-ky^{-1} eccentricity component. There is no power in the 41-ky^{-1} obliquity band.

6 There is also much variance in the ABR and BI records at 15 ky^{-1} which is the sum frequency of the 23-ky^{-1} and the absent 41-ky^{-1} cycles. This 15-ky^{-1} cyclicity is a real feature in the records.

7 In the 23-ky^{-1} band, ABR and -BI lead ΔSST_{ABF} variations by 6.6 ky. Consequently, the Benguela Current system performs a combination of two types of movements: lateral shifts of the ABF to the south precede swings to more zonal directions of the Benguela Oceanic Current, shifts to the north precede swings to more meridional directions.

8 The 23-ky shifts are in phase with SST variations in the Cape Basin and with meridional shifts of the Subtropical Convergence (STC) in the southern Indian Ocean. They are also in phase with global ice-volume change or slightly lead it. This indicates that north-south movements of the position of the westerlies result in displacements of the subpolar and subtropical fronts and the SE trade winds, thus causing the shifts of the ABF.

9 The extremes in the positions of the ABF are described by a 100-ky^{-1} eccentricity component. This component is also documented in cores from Walvis Ridge and Cape Basin. These 100-ky shifts are in phase with the eccentricity driven movements of the STC in the Indian Ocean, and with global ice volume. We infer that the waxing and waning of the southern hemisphere sea-ice fields controls the shifts of the subtropical convergence, the circumpolar fronts and the ABF for both the 23 and 100-ky^{-1} frequency bands.

10 The ABF moves also in phase with advection variations in the equatorial Atlantic Ocean (Mix and Morey, this volume). This happens in the 23 and 100 ky-1 frequency bands, but not in the 41 ky-1 band. Apparently, the inferred displacement of the subtropical fronts vary the intensity of the surface circulation in the South Atlantic, and the advective supply of cool Benguela Current water towards the equatorial Atlantic Ocean.

11 The northernmost displacements of the ABF are correlated with minimum sea-surface temperatures in the Arabian Sea and the equatorial Pacific Ocean. These temperature minima are

probably also caused by the northward movements of the polar fronts, resulting in increased advection of relatively cool surface water from the south.

Acknowledgements

We thank captain De Jong and the crew of the RV „Tyro" for their help during the coring operations. We also thank Jack Schilling who designed, built and run the excellent NIOZ piston corer, and Ellen Okkels for her patience and loyal help on board and in the laboratory. We thank Monka Segl (Bremen) for the stable isotope measurements and Klaas van der Borg (Robert J. van de Graaff Laboratory, Utrecht University) for the AMS ^{14}C datings. We thank Hedi Oberhänsli who kindly provided the foraminifera data of core DSDP 532. We thank Leo Maas, Torsten Bickert and Alan Mix for fruitful discussions, and Wolf Berger for critically reading the manuscript. The „Tyro" cruise was financed by the Netherlands Marine Research Foundation (SOZ), The Hague.
This is contribution no. 134 of Sonderforschungsbereich 261 at Bremen University.

References

Berger A, Loutre MF (1991) Insolation values for the climate of the last 10million years. Quat Sci Rev 10(4): 297-317

Bjørklund KR, JHF Jansen (1984) Radiolaria distribution in Middle and Late Quaternary sediments and palaeoceanography in the eastern Angola Basin. Neth J Sea Res 17(2-4):299-312

Broecker WS, Peteet DM, Rind D (1985). Does the ocean-atmospheric system have more than one stable mode of operation? Nature 315:21-26

Diester-Haass L (1985) Late Quaternary sedimentation on the eastern Walvis Ridge, SE Atlantic (HPC 532 and four piston cores). Mar Geol 65(1/2):145-198

Emeis K, Anderson DM, Doose H, Kroon D, Schulz-Bull D (1995) Sea-surface temperatures and the history of monsoon upwelling in the northwest Arabian Sea during the last 500,000 years. Quat Res 43(3):355-361

Gardner JV, Hays JD (1976) Responses to sea-surface temperature and circulation to global climatic change during the past 200 000 years in the eastern equatorial Atlantic Ocean. Mem Geol Soc Am 145:221-246

Giraudeau J (1993) Planktonic foraminiferal assemblages in surface sediments from the southwest African continental margin. Mar Geol 110(1/2):47-62

Gordon AL (1986) Interocean exchange of thermocline water. J Geophys Res 91:5037-5046

Gordon AL, Bosley KT (1991) Cyclonic gyre in the tropical South Atlantic. Deep-Sea Res 38:S323-S343

Howard WR, Prell WL (1992) Late Quaternary surface circulation of the southern Indian Ocean and its relationship to orbital variations. Paleoceanogr 7(1):79-117

Imbrie J, Hays JD, Martinson DG, McIntyre A, Mix AC, Morley JJ, Pisias NG, Prell WL, Shackleton NJ (1984) The orbital theory of Pleistocene climate: support from a revised chronology of the marine $\delta^{18}O$ record. In: Berger A, Imbrie J, Hays J, Kukla G, Saltzman B (eds) Milankovitch and climate, Part 1. Reidel, Dordrecht, pp 269-305

Imbrie J, McIntyre A, Mix A (1989) Oceanic response to orbital forcing in the late Quaternary: observational and experimental strategies. In: Berger A, Schneider S, Duplessy J (eds) Climate and Geosciences. Kluwer Academic Publishers, Dordrecht, pp 121-164

Imbrie J, Boyle EA, Clemens SC, Duffy A, Howard WR, Kukla G, Kutzbach J, Martinson DG, McIntyre A, Mix AC, Molfino B, Morley JJ, Peterson LC, Pisias NG, Prell WL, Raymo ME, Shackleton NJ, Toggweiler JR (1992) On the structure and origin of major glaciation cycles. 1. Linear responses to Milankovitch forcing. Pale-oceanogr 7(6):701-738

Imbrie J, Berger A, Boyle EA, Clemens SC, Duffy A, Howard WR, Kukla G, Kutzbach J, Martinson DG, McIntyre A, Mix AC, Molfino B, Morley JJ, Peterson LC, Pisias NG, Prell WL, Raymo ME, Shackleton NJ, Toggweiler JR (1993) On the structure and origin of major glaciation cycles 2. the 100,000-year cycle. Paleoceanogr 8(6):699-735

Jansen JHF, Van Weering TCE, Gieles R, Van Iperen J (1984) Middle and Late Quaternary oceanography and climatology of the Zaire-Congo fan and adjacent eastern Angola Basin. Neth J Sea Res 17(2-4):201-249

Jansen JHF (1985) Middle and Late Quaternary carbonate production and dissolution, and paleoceanography of the eastern Angola Basin, South Atlantic Ocean. In: Hsü KJ, Weissert HJ (eds) South

Atlantic Paleoceanography. Cambridge University Press, Cambridge, pp 25-46

Jansen JHF (1990) Glacial-interglacial oceanography of the southeastern Atlantic Ocean and the paleoclimate of west central Africa. In: Lanfranchi R, Schwartz D (eds) Paysages quaternaires de l'Afrique centrale atlantique. ORSTOM, Paris, pp 110-123

Jansen JHF, De Lange GJ, Van Bennekom AJ, et al. (1990) (Pale)oceanography and geochemistry of the Angola Basin (South Atlantic Ocean); Cruise Report R.V.Tyro, 30 September - 19 November 1989. Neth Inst Sea Res Rep 1990-4:1-65

Jansen JHF, Van Iperen JM (1991) A 220,000-year climatic record for the east equatorial Atlantic Ocean and equatorial Africa: evidence from diatoms and opal phytoliths in the Zaire (Congo) deep-sea fan. Paleoceanogr 6(5):573-591

Kroon D, Beets K, Mowbray S, Shimmield G, Steens T (1990) Changes in northern Indian Ocean monsoonal wind activity during the last 500ka. Mem Soc Geol It 44:189-207

Levitus S (1982) Climatological atlas of the world ocean. US Govt Print Off, Washington DC

Lyle MW, Prahl FG, Sparrow MA (1992) Upwelling and productivity changes inferred from a temperature record in the central equatorial Pacific. Nature 355(6363):812-815

McIntyre A, Kipp NG, with Bé AWH, Crowley T, Kellogg T, Gardner JV, Prell W, Ruddiman WF (1976) Glacial North Atlantic 18 000 years ago: a CLIMAP reconstruction. Mem Geol Soc Am 145:43-76

McIntyre A, Ruddiman WF, Karlin K, Mix AC (1989) Surface water response of the equatorial Atlantic Ocean to orbital forcing. Paleoceanogr 4(1):19-55

Meeuwis JM, Lutjeharms JRE (1990) Surface thermal characteristics of the Angola-Benguela front. S Afr J Mar Sci 9:261-279

Meinecke G (1992) Spätquartäre Oberflächenwassertemperaturen im östlichen äquatorialen Atlantik. Berichte, Fachbereich Geowissenschaften, Universität Bremen 29:1-181

Mix AC, Morey AE (1996) Climate Feedback and Pleistocene Variations in the Atlantic South Equatorial Current. In: Wefer G, Berger WH, Siedler G, Webb D (eds) The South Atlantic: Present and Past Circulation. Springer-Verlag, Berlin Heidelberg

Morley J, Hays JD (1979) Comparison of glacial and interglacial oceanographic conditions in the South Atlantic from variations in calcium carbonate and radiolarian distributions. Quat Res 12(3):396-408

Morley JJ, Shackleton NJ (1984) The effect of accumulation rate on the spectrum of geologic time series: evidence from two South Atlantic sediment cores. In: Berger A, Imbrie J, Hays J, Kukla G, Saltzman B (eds) Milankovitch and climate, Part 1. Reidel, Dordrecht, pp 467-480

Morley JJ (1989) Variations in high-latitude oceanographic fronts in the southern Indian Ocean: an estimation based on faunal changes. Paleoceanogr 4(5):547-554

Oberhänsli H (1991) Upwelling signals at the northeastern Walvis Ridge during the past 500,000 years. Paleoceanogr 6(1):53-71

Peterson RG, Stramma L (1991) Upper-level circulation in the South Atlantic Ocean. Prog Oceanog 26(1):1-73

Pokras EM (1987) Diatom record of late Quaternary climatic change in the eastern equatorial Atlantic and tropical Africa. Paleoceanogr 2(3):273-286

Prahl FG, Wakeham AS (1987) Calibration of unsaturation patterns in long-chain ketone compositions for palaeotemperature assessment. Nature 330:367-369

Prell WL, Van Campo E (1986) Coherent response of Arabian Sea upwelling and pollen transport to late Quaternary monsoonal winds. Nature 323(6088): 526-528

Prell WL, Kutzbach JE (1987) Monsoon variability over the past 150,000 years. J Geophys Res 92(D7):8411-8425

Reynolds L, Thunell RC (1986) Seasonal production and morphological variation of *Neogloboquadrina pachyderma* (Ehrenberg) in the northeast Pacific. Micropal 32:1-18

Rostek F, Ruhland G, Bassinot FC, Müller PJ, Labeyrie LD, Lancelot Y, Bard E (1993) Reconstructing sea surface temperature and salinity using $\delta^{18}O$ and alkenone records. Nature 364(6435):319-321

Schneider R (1991) Spätquartäre Produktivitätsänderungen im östlichen Angola-Becken: Reaktion auf Variationen im Passat-Monsun-Windsystem und in der Advektion des Benguela-Küstenstroms. Berichte, Fachbereich Geowissenschaften, Universität Bremen 21:1-198

Schneider RR, Müller PJ, Wefer G (1994) Late Quaternary paleoproductivity changes off the Congo deduced from stable carbon isotopes of planktonic foraminifera. Palaeogeogr, Palaeoclimatol, Palaeoecol, 110(3-4):255-274

Schneider R, Müller PJ, Ruhland G (1995) Late Quaternary surface circulation in the east equatorial South Atlantic: Evidence from alkenone sea surface temperatures. Paleoceanogr 10(2):197-219

Schott F, Swallow JC, Fieux M (1990) The Somali Current at the equator: annual cycle of currents and transports in the upper 1000 m and connection to neighbouring latitudes. Deep-Sea Res 37(12):1825-1848

Shannon LV, Agenbag JJ, Buys MEL (1987) Large- and mesoscale features of the Angola-Benguela front. S Afr J Mar Sci 65:11-34

Steens TNF., Ganssen G, Kroon D (1992) Oxygen and carbon isotopes in planktonic foraminifera as indicators of upwelling intensity and upwelling-induced high productivity in sediments from the northwestern Arabian Sea. In: Summerhayes CP, Prell WL, Emeis K (eds) Upwelling systems: evolution since the Early Miocene. Geol Soc Spec Publ 64:107-119

Stramma L, Peterson RG (1989) Geostrophic transport in the Benguela Current region. J Phys Oceanogr 19:1440-1448

Summerhayes CP, Kroon D, Rosell-Melé A, Jordan RW, Schrader H, Hearn R, Villanueva J, Grimalt JO, Eglinton G (1995). Variability in the Benguela Current upwelling system over the past 70,000 years. Progr Oceanog 35:207-251

Ten Haven HL, and Kroon D (1991) Late Pleistocene sea surface water temperature variations off Oman as revealed by the distribution of long-chain alkenones. Proc Ocean Drilling Program, Scientific Results 117:445-452

Ufkes E, Zachariasse W (1993) Origin of coiling differences in living neogloboquadrinids in the Walvis Bay region, off Namibia, southwest Africa. Micropal 39(3):283-287

Ufkes E (1996) Planktonic foraminifera in the surface waters of the eastern South Atlantic during springtime. Journal of Foraminiferal Research (submitted)

Ufkes E, Jansen JHF, Wefer G (in prep) A late Quaternary foraminifera record from the Angola-Benguela Front, southeastern Atlantic.

Ufkes E, Jansen JHF, Schneider R (in prep) Late Quaternary paleoceanography of the Angola Basin, SE Atlantic: evidence from planktic foraminifera.

Van Bennekom AJ, Berger GW (1984) Hydrography and silica budget of the Angola Basin. Neth J Sea Res 17(2-4):149-200

Van Campo E, Duplessy JC, Rossignol-Strick M (1982) Climatic conditions deduced from a 150-kyr oxygen isotope-pollen record from the Arabian Sea. Nature 296(5852:56-59

Van Leeuwen RJW (1989) Sea-floor distribution and Late Quaternary faunal patterns of planktonic and benthic foraminifers in the Angola Basin. Utrecht Micropal Bull 38:1-288

Wefer G et al. (1988) Bericht über die Meteor-Fahrt M6-6, Libreville-Las Palms, 18.2.1988-23.3.1988. Berichte, Fachbereich Geowissenschaften, Universität Bremen 3:1-97

Zachariasse WJ, RR Schmidt, Van Leeuwen RJW (1984). Distribution of Foraminifera and calcareous nannoplankton in Quaternary sediments of the eastern Angola Basin in response to climatic and oceanic fluctuations. Neth J Sea Res 17(2-4):250-275

Zahn R, Pedersen TF (1991) Late Pleistocene evolution of surface and mid-depth hydrography at the Oman Margin: planktonic and benthic isotope records at Site 724. Proc Ocean Drilling Program, Scientific Results 117:291-308

Late Quaternary Deep Circulation in the Western Equatorial Atlantic

W.B. Curry

Department of Geology and Geophysics,
Woods Hole Oceanographic Institution,
Woods Hole, MA 02543, U.S.A.

Abstract: Variations in benthic foraminiferal $\delta^{13}C$ from a suite of cores located in the equatorial western Atlantic document that the production rate of North Atlantic Deep Water has varied considerably during the last 400,000 years. The cores are located on the eastern slope of Ceara Rise and monitor the chemical variation in deep waters over the depth range of 3200 to 4300 m. Peak production apparently occurred during interglacial stages with isotopic stage 7.3 (216ka) exhibiting the greatest bathymetric extent of NADW. Minima in NADW bathymetric range occurred during glacial maxima, when significant northward penetration of deep water from the Southern Ocean occupied large portions of the western Atlantic. Minimum production of NADW and maximum penetration of southern-source deep water occurred during glacial stages 4, 8 and 10. During these periods, there is no evidence for NADW at depths below 3200 m.
Climate change and deep water production are linked but the records are complicated because maxima in NADW production do not necessarily occur during the most extreme interglacials. Also there is a water depth-related difference in the phase relationships between ice volume and deep water production. In the deepest sites, changes in deep water production (as denoted by benthic foraminiferal $\Delta\delta^{13}C$) appear to lead changes in ice volume (as denoted by benthic foraminiferal $\delta^{18}O$) for both the eccentricity (100 kyr) and obliquity (41 kyr) periods of orbital geometry, providing partial confirmation of the Imbrie et al. (1992) model of a deep water-climate link. But the shallower cores do not lead ice volume at the 100 kyr period. Furthermore, NADW production lags ice volume in the precessional period at all water depths, unlike the records used in the Imbrie et al. (1992) model. Higher frequency, sub-orbital oscillations in benthic foraminiferal $\delta^{13}C$ imply that NADW production is occurring at frequencies like those seen in ice core records. These oscillations in NADW production appear to be restricted to intervals when ice volume was greater than today's.

Introduction

Variations in deep water production have been implicated as a driving force in the growth and decay of major ice sheets because deep water also plays a major role in the transport of heat from the northern hemisphere to the southern hemisphere (Broecker and Denton, 1989). On the basis of observed phase relationships between ice volume and several deep water circulation proxies, the SPECMAP group proposed a four stage model of the glacial cycle that relies on early responses in deep water production at high latitudes of the North Atlantic to produce early and large changes in climate synchronously in both hemispheres (Imbrie et al. 1992;

1993). The SPECMAP model suggests that during a preglacial phase, decreased production of the lower branch of North Atlantic Deep Water (the "Nordic" heat pump) results in a decrease in the heat flux to the southern ocean which causes an early cooling at high latitudes of the southern hemisphere, early sea ice growth, the northward migration of oceanic fronts and partial drawdown of atmospheric CO_2. All of these changes in surface water hydrography, which were forced by the early decrease in deep water production, contribute to ice growth. During the glacial phase, production of the upper branch of NADW (the "Boreal" heat pump) in-

From WEFER G, BERGER WH, SIEDLER G, WEBB DJ (eds), 1996, *The South Atlantic: Present and Past Circulation.* Springer-Verlag
Berlin Heidelberg, pp 577-598

creases, but production in the lower branch remains suppressed. Early warming at high latitudes of the North Atlantic provides the means to produce lower NADW again during the deglacial phase of the cycle. With both the upper and lower heat pumps actively transporting heat to the south, a nearly simultaneous warming occurs in the southern hemisphere, sea ice melts, fronts migrate to the south and atmospheric CO_2 increases. Rapid deglaciation follows.

This elegant model of the connection between deep water production and climate changes relies heavily on a few records of deep water circulation, not all of which are ideally located to monitor the critical deep water components (Fig. 1). The primary evidence for the early response for NADW export to the southern ocean comes from two South Atlantic locations: a benthic foraminiferal Cd/Ca record from the Rio Grande Rise (Boyle 1984) and a benthic foraminiferal $\delta^{13}C$ record from the Cape Basin (Oppo and Fairbanks 1987; Oppo et al.

1990). The deep water proxy in each of these records suggests that changes in NADW production lead ice volume: NADW production decreases just prior to glacial inception and NADW production increases just prior to deglaciation. Specifically for the Rio Grande Rise cores (AII107-131, AII107-65), minima in Cd/Ca occur during deglaciations; for the Cape Basin core (RC13-229), maxima in $\delta^{13}C$ enrichment ($\Delta\delta^{13}C$) with respect to the Pacific occur during deglaciations. In contrast to these South Atlantic records, the primary record of NADW production in the North Atlantic (DSDP Site 607: Raymo et al. 1990) has the opposite phasing: maxima in NADW production lag minimum ice volume.

There are ambiguities associated with each of these proxy records of deep water production. For instance, the Cd/Ca record from the Rio Grande Rise is a patched record based on two cores with very low sedimentation rates (AII107-65GGC < 2 cm/kyr; AII107-131GGC < 1 cm/kyr). Conse-

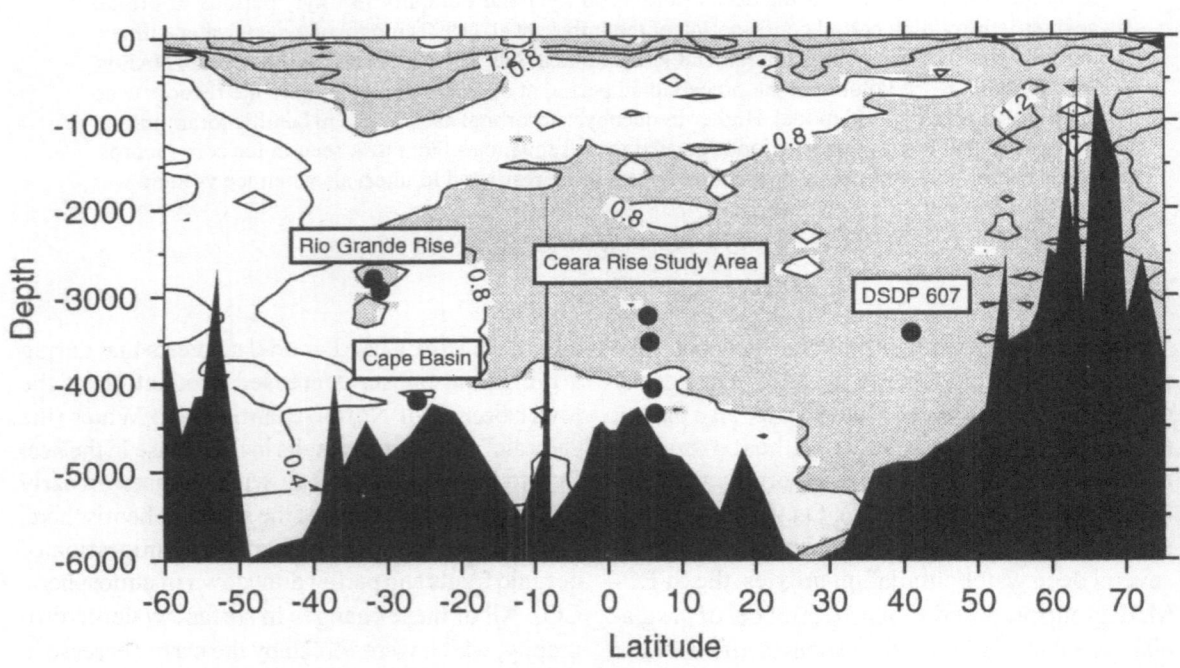

Fig. 1. Western Atlantic transect of $\delta^{13}C$ of $\Sigma CO2$ from GEOSECS (Kroopnick 1985). The shaded regions mark the regions with $\delta^{13}C$ values greater than 0.8‰ (PDB), which illustrates the geographic extent of NADW. Core locations from the Ceara Rise study area are located near 5°N. The principal cores comprising the Imbrie et al. (1992) deep water data set are also identified.

quently, the effects of bioturbation are likely to be large and the phasing between Cd/Ca and ice volume may be affected. For the Cape Basin record, the ambiguity lies in the $\Delta\delta^{13}C$ calculation: throughout much of its record, measured $\delta^{13}C$ in the Cape Basin core is *lower* than $\delta^{13}C$ in the Pacific (Oppo et al. 1990). Simple geometries of water mass flow cannot account for southern ocean $\delta^{13}C$ which is lower than Pacific $\delta^{13}C$ (Curry et al. 1988). Thus the $\Delta\delta^{13}C$ record for the Cape Basin cannot solely reflect increases and decreases in the production of NADW; perhaps it is also affected by local $\delta^{13}C$ overprints caused by high productivity over the site of RC13-229. Furthermore, a new Cd/Ca record from the Cape Basin (RC13-229: Oppo and Rosenthal 1994) suggests that NADW production lags ice volume for the 100 kyr and 41 kyr periods, opposite the phasing of Cd/Ca and ice volume observed at the Rio Grande Rise. This new Cd/Ca record directly contradicts a key aspect of the SPECMAP model, early NADW response in the southern ocean, and implies that the $\Delta\delta^{13}C$ record in RC13-229 cannot be a good proxy for NADW production. Finally, because it is so close to the source, the location of DSDP Site 607 makes it less sensitive to changes in the production of NADW; very large reductions in NADW can occur before they will be measured at this site.

Our understanding of deep water circulation and its link to climate change would be enhanced greatly by new records from regions which are sensitive to changes in deep water production, contain high sedimentation rates and provide useful deep water proxy measurements with unambiguous interpretations. Ceara Rise has all of these qualities and is an ideal location to significantly expand the database on which the SPECMAP model can be tested and refined. In this paper a history of deep water circulation for the last 400,000 years will be reconstructed using measurements of benthic foraminiferal $\delta^{13}C$ made in a series of piston cores recovered from the slopes of Ceara Rise. In 1992, these piston cores were collected from the northeast slopes of Ceara Rise as part of the site survey for Ocean Drilling Program Leg 154 (Fig. 2). The cores contain abundant, well-preserved calcium carbonate microfossils and have sedimentation rates in excess of 4 cm/kyr. Consequently they are nearly ideal cores for recon-

structing past changes in benthic foraminiferal chemistry, particularly for changes in ocean circulation occurring on Milankovitch time scales (10^4-10^5 years).

Study Area

Circulation in the deep western Atlantic, the main site for meridional deep water transport, is dominated by the interactions between North Atlantic Deep Water flowing toward the south and Antarctic Bottom Water (AABW) and Circumpolar Deep Water (CPDW) flowing toward the north. Today the density characteristics of these water masses are such that NADW divides the southern water masses into an upper and lower branch. Consequently NADW occupies the depth interval between about 2000 and 4000 meters; below 4000 meters, lower CPDW and AABW are encountered. The mixing zone between NADW and CPDW-AABW is marked by a benthic thermocline and halocline, as well as sharp gradients in nutrient content, in $\delta^{13}C$ of ΣCO_2, and in the corrosiveness of the water to calcium carbonate. In the equatorial regions of the western Atlantic, Ceara Rise intersects these water masses over the depth range of 2800 to 4500 meters, providing an ideal location for reconstructing the history of changes in water mass chemistry and water mass boundary. Lower NADW (the proxy for "Nordic" heat pump) is characterized by high salinity water that is colder than about 2.4°C but warmer than 2.0°C (potential temperature); this places it below 3200 m and above 4000 m in the vicinity of Ceara Rise (Fig. 3).

Research Strategy

The $\delta^{13}C$ tracer

Benthic foraminifera record important geochemical tracers (Cd, $\delta^{13}C$) of ocean circulation in their calcium carbonate tests (see for instance, Belanger et al. 1981; Graham et al. 1981; Boyle and Keigwin 1982; Duplessy et al. 1984; Curry et al. 1988). The usefulness of these tracers is based on their relationships to nutrient cycles, and to the carbon chemistry of the oceans and deep water circulation (for a summary, see Curry et al. 1988). The carbon iso-

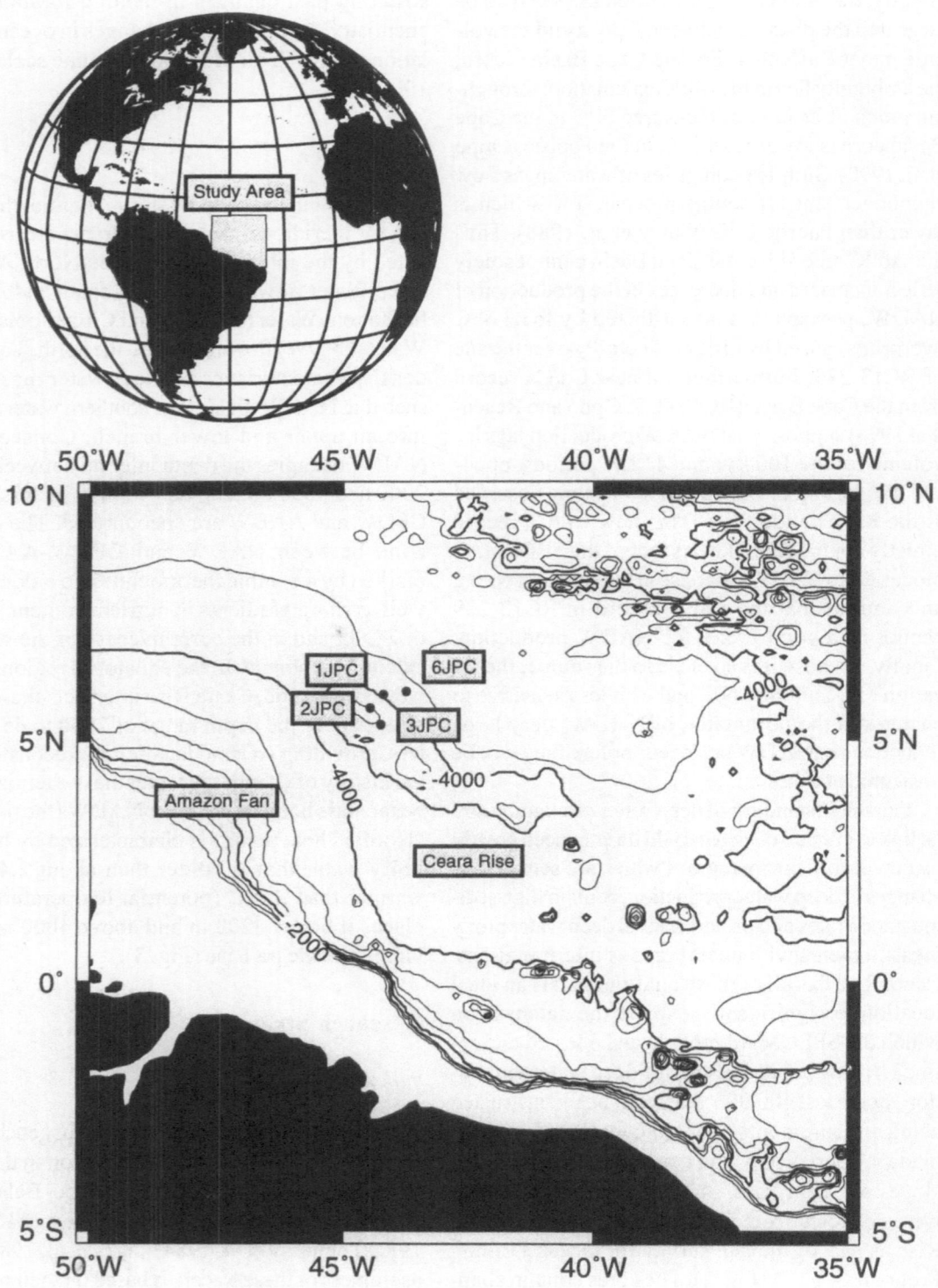

Fig. 2. Location map showing the positions of the four cores used in this study.

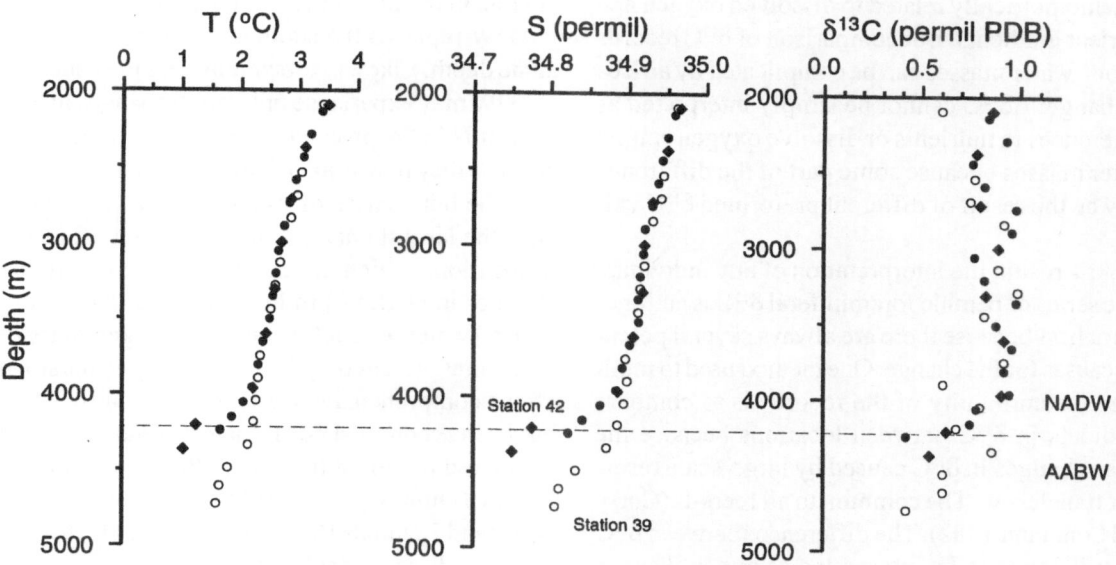

Fig. 3. Hydrographic properties from GEOSECS stations 39 (o), 40 (•), and 42 (♦) which straddle the position of Ceara Rise. The benthic thermocline, halocline and the gradient in $\delta^{13}C$ of ΣCO_2 which mark the water mass transition are apparent below 4100 m. Lower NADW, the Nordic heat pump of Imbrie et al. (1992), is located between 3200 and 4000 m.

topic composition of ΣCO_2 is linked to deep water circulation patterns because its distribution in sea water is affected by photosynthesis and respiration. Deep water that is formed with a significant surface water component, as is today's NADW, is enriched in the heavier carbon isotope because photosynthesis has preferentially removed ^{12}C from the ΣCO_2 of the surface water. Since nearly all of the carbon that is fixed by photosynthesis is subsequently returned to the water column by oxidation, deep water masses acquire successively lower in $\delta^{13}C$ values along their flow path. Along with this decrease in $\delta^{13}C$, increases in nutrient concentration and decreases in dissolved oxygen concentration occur which closely correspond to the Redfield stoichiometry of marine organic matter. Thus gradients in $\Delta\delta^{13}C$ provide key information about the direction of flow of a deep water mass and the chemical gradients within it.

The interpretation of benthic foraminiferal $\delta^{13}C$ records is complicated by several factors. First, there are changes in the ocean's mean value of $\delta^{13}C$ that are caused by transfer of carbon between the ocean and any of several transient reservoirs for carbon such as the terrestrial biosphere. Thus, the change in plant mass on land affects the carbon isotopic composition of sea water because the carbon in the plant material is intimately linked to the ocean-atmosphere reservoir. Second, the physics of air-sea exchange of CO_2 and carbon isotopes is temperature-dependent. Thus, there are ways in which the preformed $\delta^{13}C$ value can be changed without changing the nutrient concentration of the sea water (Broecker and Maier-Reimer 1992; Charles et al. 1993). Because the temperature range is small in the deep ocean, this effect is likely to be small (Broecker and Maier-Reimer 1992). In any case, air-sea exchange cannot affect deep water $\delta^{13}C$ values after the water mass has left the surface, so gradients in $\delta^{13}C$ within water masses are still stoichiometrically related to dissolved oxygen and nutrient gradients. But comparison of $\delta^{13}C$ records among water masses can be complicated by air-sea exchange; $\Delta\delta^{13}C$ cannot be simply interpreted as differences in nutrients or dissolve oxygen among water masses because some part of the difference may be the result of different preformed $\delta^{13}C$ values.

As a result, the interpretation of any individual time series of benthic foraminiferal $\delta^{13}C$ is ambiguous at best because there are always several potential causes for the change. One method used to minimize the ambiguity of the records is to compare gradients in $\delta^{13}C$ between locations because the mean changes in $\delta^{13}C$ caused by large-scale reservoir transfers will be common to all records (Curry and Lohmann 1982). The differences between $\delta^{13}C$ records can then be interpreted as either changes in the initial chemistry of the deep water or changes in the nutrient content of the deep water. Synoptic carbon isotopic reconstructions can then be used as snapshots of past water mass geometry and water mass boundaries.

The depth transect approach

The 0.8‰ $\delta^{13}C$ value for ΣCO_2 in the western Atlantic Ocean approximates the boundary between low-nutrient, high-$\delta^{13}C$ NADW and high-nutrient, low-$\delta^{13}C$ southern-source deep waters (see Figs. 1 and 3). Because these water masses have different modes of formation, the boundaries between them contain strong carbon isotopic and nutrient gradients. The depth and slope of the transition zone is controlled by the intensity of production and the different physical properties of the two opposing water masses. As the properties of water masses change through time or as their production rates respond to changes in earth's climate, the depth range of the water masses will vary. As a consequence, the $\delta^{13}C$ changes recorded by benthic foraminifera in cores from the western Atlantic are strongly depth dependent and no single core can be used to monitor the history of deep water. For example, cores which are presently located within

southern-source deep water will record a change in benthic foraminiferal chemistry resulting from a change in the production of NADW if and only if NADW replaces the southern-source water mass at the depth. Likewise, locations near the source of NADW may experience only small changes in $\delta^{13}C$ even if NADW production is drastically reduced because they may remain bathed by NADW. Cores near the boundaries of water masses will experience the largest changes in $\delta^{13}C$ because they are in locations which are most sensitive to smaller changes in NADW production. Thus, the history of deep water production and the past geometry of water masses can only be completely understood after a comprehensive water depth reconstruction has been accomplished. To date this has only been attempted for about the last 30,000 years of earth history (Duplessy et al. 1988; Curry et al. 1988; Oppo and Lehman 1993; Sarnthein et al. 1994).

Ceara Rise is ideally located to trace changes in the depth distribution of deep water properties because it intersects the mixing zone between the principal deep water masses flowing in the western Atlantic. The mixing zone between the NADW and southern-source deep waters deepens from about 4000 meters at 20°S in the South Atlantic to near 5000 m at 30°N in the North Atlantic (Fig. 1) as AABW and CPDW penetrate into the western basins of the North Atlantic, flowing beneath the NADW. In the vicinity of Ceara Rise, the water mass transition occurs at about 4300 m today (Fig. 3). By selecting cores which straddle this present water mass transition, one can monitor past changes in the position of NADW and the southern-source deep waters by monitoring the vertical gradient of benthic foraminiferal $\delta^{13}C$. Because we compare only synchronous measurements of $\delta^{13}C$, the vertical gradients in $\delta^{13}C$ will be unaffected by global changes in the ocean's mean carbon isotopic composition and will be related to changes in the properties of the water masses or in their water column positions. If the initial chemistry of the water masses can be constrained, then we can interpret the variations through time as changes in deep water circulation and make meaningful reconstructions of water mass nutrient, dissolved oxygen and ΣCO_2.

Methods

Six taxa of benthic foraminifera (*Cibicides wuellerstorfi* (1637 analyses), *Nuttalides umbonifera* (149 analyses), undifferentiated *Cibicides* spp. (109 analyses), *C. cicatricosus* (6 analyses), *C. kullenbergi* (5 analyses), and *Planulina robertsoniensis* (3 analyses)) were analyzed for stable oxygen and stable carbon isotopic composition using standard procedures. The foraminiferal tests were picked from raw samples that had been washed in a calgon-hydrogen peroxide solution, wet sieved over a 63 μm sieve and dried in an oven at 50°C. Benthic foraminiferal tests for isotopic analysis were picked from the >250 μm size fraction. These samples were analyzed using a Finnigan MAT252 mass spectrometer with the Kiel automated carbonate preparation device. The samples were not roasted prior to analysis. Individual tests were used to produce the isotopic measurements and at least two measurements were made at each stratigraphic level in the cores. We converted the isotopic data to PDB by using the intermediate standard NBS19, assuming 1.95‰ for its $\delta^{13}C$ and -2.20‰ for its $\delta^{18}O$ value.

The external precision of the isotopic analyses is ±0.03‰ for $\delta^{13}C$ and ±0.07‰ for $\delta^{18}O$ based on more than 1200 measurements of NBS19. The precision of each measurement is size dependent, varying from ±0.08‰ for $\delta^{13}C$ and 0.15‰ for $\delta^{18}O$ for samples 10 to 20 μg in size to 0.02‰ for $\delta^{13}C$ and 0.06‰ for $\delta^{18}O$ for samples greater than about 70 μg. Individual tests of *Cibicides* have a range of masses but are nearly always >30 μg and usually more than 80 μg. We rejected analyses that produced less than 0.8V of signal on the mass 44 beam on the MAT252, which in our present system configuration corresponds to an equivalent mass of calcite of about 10 μg. Samples which did not react completely were rejected by this criterion. Normal analytical procedures include the analysis of three pairs of different standards (NBS19, B1 and calcitic deep sea coral (Atlantis) we use as a laboratory working standard) within each carousel of 46 samples. The standards have isotopic compositions which differ by more than 5‰ in $\delta^{18}O$ and bound the range of isotopic compositions typically found in Holocene planktonic and benthic foraminifera (-2 to 3‰ PDB). The precision of B1 and Atlantis are worse than for NBS19 because they are both biologically precipitated calcite and thus suffer from greater heterogeneity; both have precisions which are better than ±0.09‰ for $\delta^{13}C$ and $\delta^{18}O$.

Four cores were analyzed to produce the depth transect (locations in Fig. 2, Tab. 1). The cores were sampled at approximately 10 cm spacing throughout their entire length. In this region sedimentation rates average about 4 to 5 cm/kyr, so the sample spacing corresponds, on average, to a resolution of about 2000 to 2500 years. However, sedimentation rates are lower in the deeper portions of each core, so the sample spacing degrades to about 4000 to 5000 years. In addition, intervals in several of the cores have been sampled at 2 to 3 cm spacing, which corresponds to about 500 to 750 year resolution. All of the *Cibicides* data are plotted in the figures without corrections. The *Nuttalides umbonifera* isotopic data have been corrected for isotopic differences from *Cibicides* based on paired analyses in core KNR31-GPC9 (Keigwin et al. 1994), a core with a sedimentation rate high enough to believe the assumption that *Cibicides* and *Nuttalides* from the same sample are coeval. For each *N. umbonifera* data point in the figures (where they appear commingled with the *Cibicides* data), the carbon isotopic composition has been increased by 0.2‰ and the oxygen isotopic composition has been decreased by 0.2‰ to conform with the *Cibicides* isotopic data. The data listed in the tables are the measured values without corrections. The isotopic data are available by anonymous FTP from the author (IP address 128.128.16.21, directory pub/ceara_rise) and are also deposited at World Data Center A for Paleoclimatology at the National Geophysical Data Center in Boulder, Colorado.

Results

The isotopic data for the four cores in this study are presented in Figs. 4-7 in order of increasing water column depth. The $\delta^{18}O$ records exhibit typical glacial-interglacial changes documenting the waxing and waning of the northern hemisphere ice sheets. Isotopic stages can be recognized generally

Table 1 Core locations and water depths for piston cores used in this study

Core	Latitude	Longitude	Water depth (m)
EW9209-3JPC	5° 18.8′	44° 15.6′	3288
EW9209-2JPC	5° 38.1′	44° 28.2′	3528
EW9209-1JPC	5° 54.4′	44° 11.7′	4056
EW9209-6JPC	5° 58.6′	43° 44.5′	4340

down to stage 12, but only down to stage 10 in EW9209-6JPC, the shortest and deepest core. The $\delta^{13}C$ records exhibit glacial-interglacial changes which are larger than found at most locations because of their sensitive location to NADW-AABW variability. Carbon isotopic values are always higher during interglacial isotopic stages and lower during glacial stages. The amplitude of the $\delta^{13}C$ signal is more than 1.5‰; highest values are found during the Holocene, parts of stage 5 and during stage 7.3, while lowest values are found during stages 4, 8 and 10.

The isotopic measurements of the individual tests document that there is greater within-sample variability for $\delta^{13}C$ than for $\delta^{18}O$. Paired measurements of $\delta^{18}O$ throughout the record usually agree to within 0.2‰, except during the rapid transition at isotopic stage boundaries. But throughout the record the $\delta^{13}C$ values exhibit significantly greater variability both within samples and between stratigraphic levels. It is not unusual to observe paired values within a sample to differ by more than 0.5‰ and sometimes by more than 1.0‰. In cores with very high sedimentation rates, tests of individual shells of *Cibicides* species exhibit a range of 0.3 to 0.5‰ for both $\delta^{13}C$ and $\delta^{18}O$ (Curry et al. 1993; in prep.) which is probably the minimum variability one might expect due to various biological and environmental factors that may affect the calcification process. The variability observed in these samples exceeds this level for $\delta^{13}C$ but not for $\delta^{18}O$,

which suggests that the wide range of values of $\delta^{13}C$ may be recording real differences in the chemistry of the benthic environment. That the variability occurs within samples suggests that it results in part from bioturbation. In slower sedimentation rate cores this is clearly true because glacial and interglacial specimens can be identified (Lohmann and Lohmann 1991; this vol.). One implication of the within-sample variability in high sedimentation rate cores is that there may be important changes in benthic foraminiferal $\delta^{13}C$ and deep water chemistry) that are occurring on periods that are less than orbital scale and perhaps closer to our sample spacing (~2kyr).

Abundance distributions of benthic foraminiferal isotopic compositions in these cores exhibit modal distributions in both $\delta^{13}C$ and $\delta^{18}O$ (Fig. 8), rather than continuous variability. For $\delta^{18}O$ the modes are dominated by values at about 3.7‰ and 3.0‰, with a smaller mode at about 2.5‰. This distribution suggests that, within these cores, glacial ^{18}O values are more common than interglacial values. This distribution pattern is probably caused by a combination of the rarity of extreme interglacial episodes like the Holocene and isotope stage 5e, and by the higher than average glacial sedimentation rate. When sampled at an equal depth spacing, more glacial samples are encountered. For $\delta^{13}C$ there are generally three modes, 1.0, 0.3, and -0.2‰, which are about equal in abundance. The mode at about 1.0‰ corresponds to values that

EW9209-3JPC

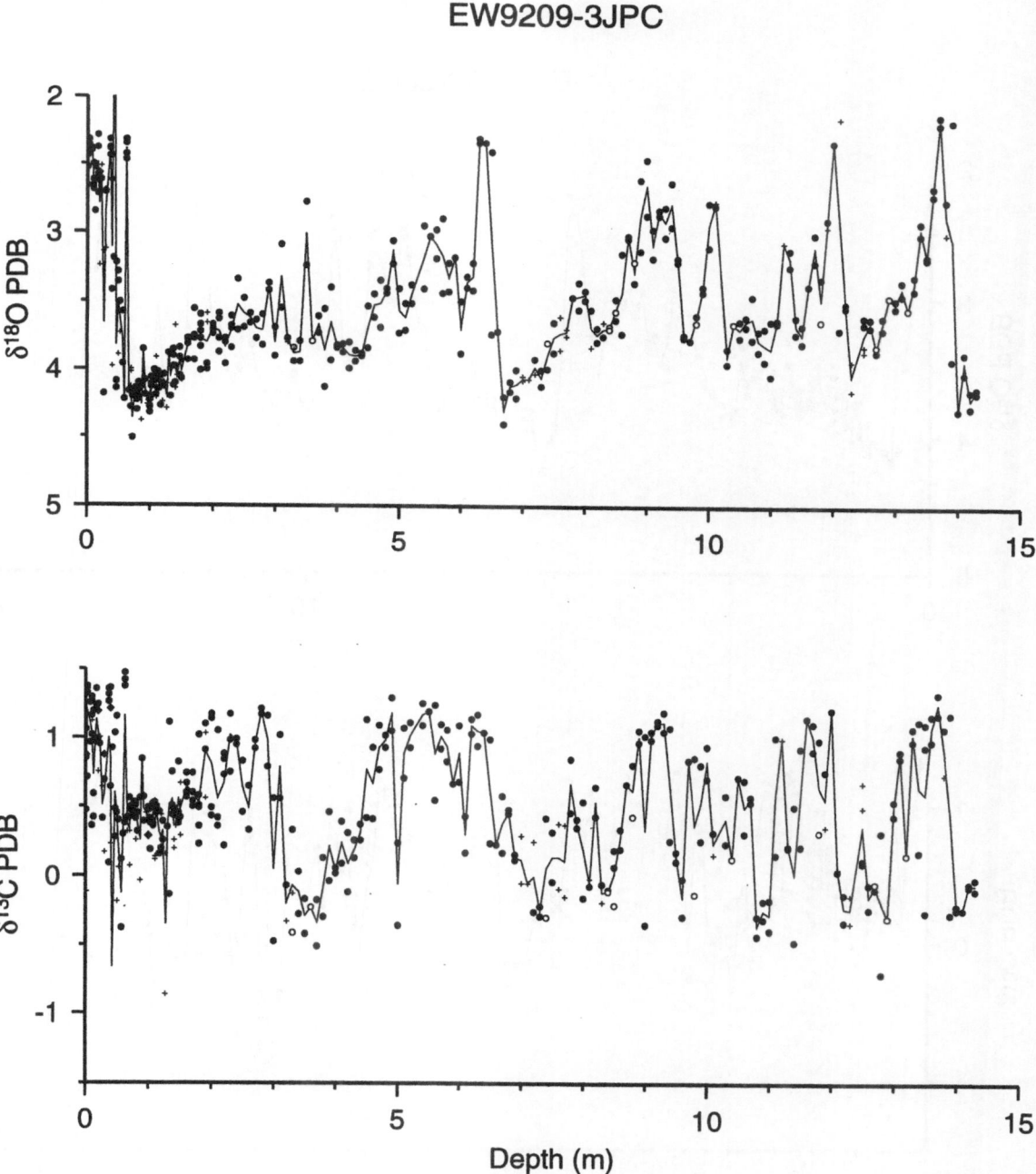

Fig. 4. δ¹⁸O and δ¹³C measurements for benthic foraminifera in core EW9209-3JPC. *Cibicides wuellerstorfi* (•), *C. kullenbergi* (+), *C. cicatricosus* (+), *Cibicides* spp. (+) and *Nuttalides umbonifera* (o) measurements were used to compose the figure. *N. umbonifera* were offset by +0.2 for δ¹³C and -0.2 for δ¹⁸O to make their data equivalent to those of *Cibicides*. The line traces the average value at each stratigraphic depth.

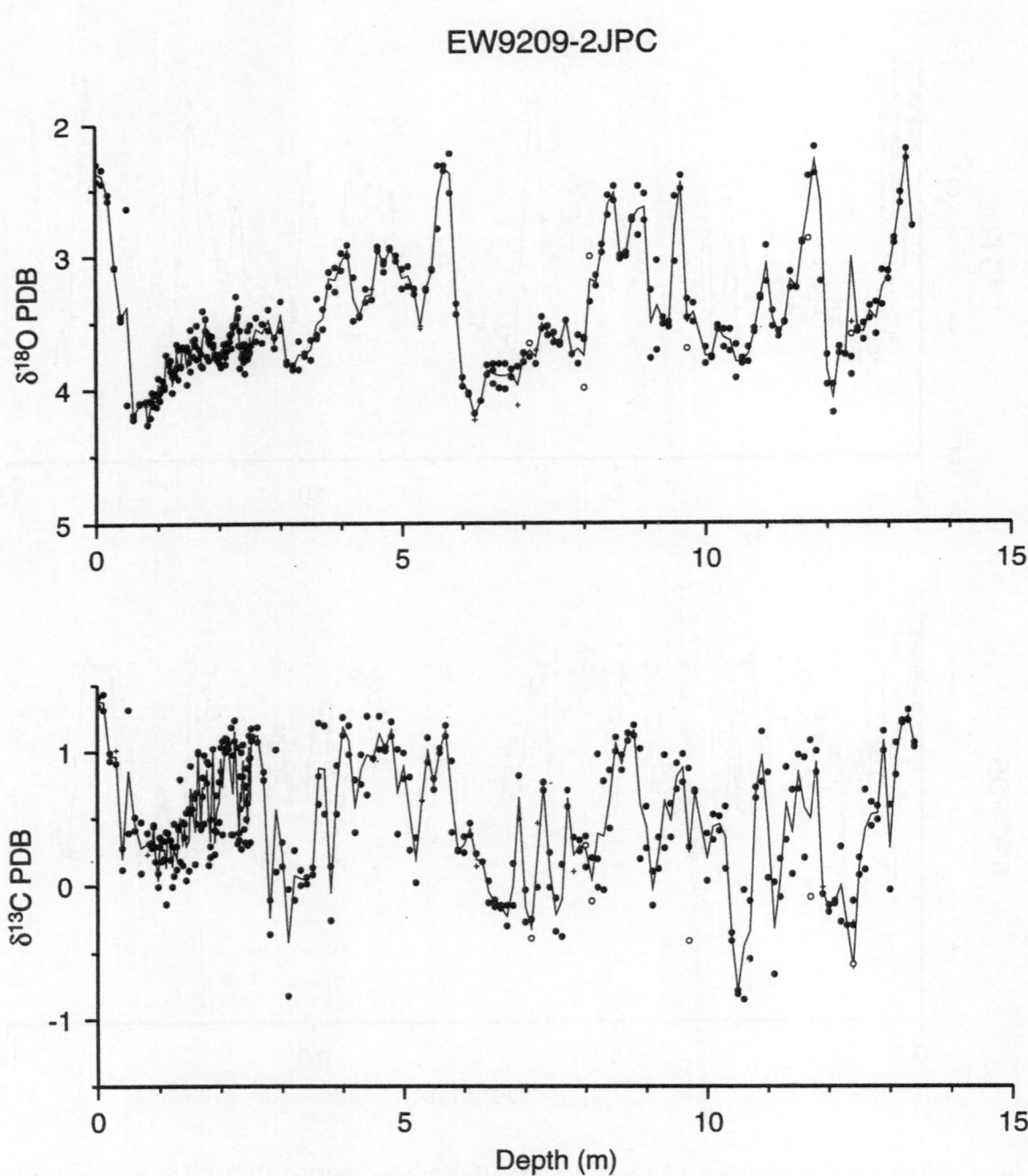

Fig. 5. $\delta^{18}O$ and $\delta^{13}C$ measurements for benthic foraminifera in core EW9209-2JPC. The symbols are the same as in Fig. 4.

EW9209-1JPC

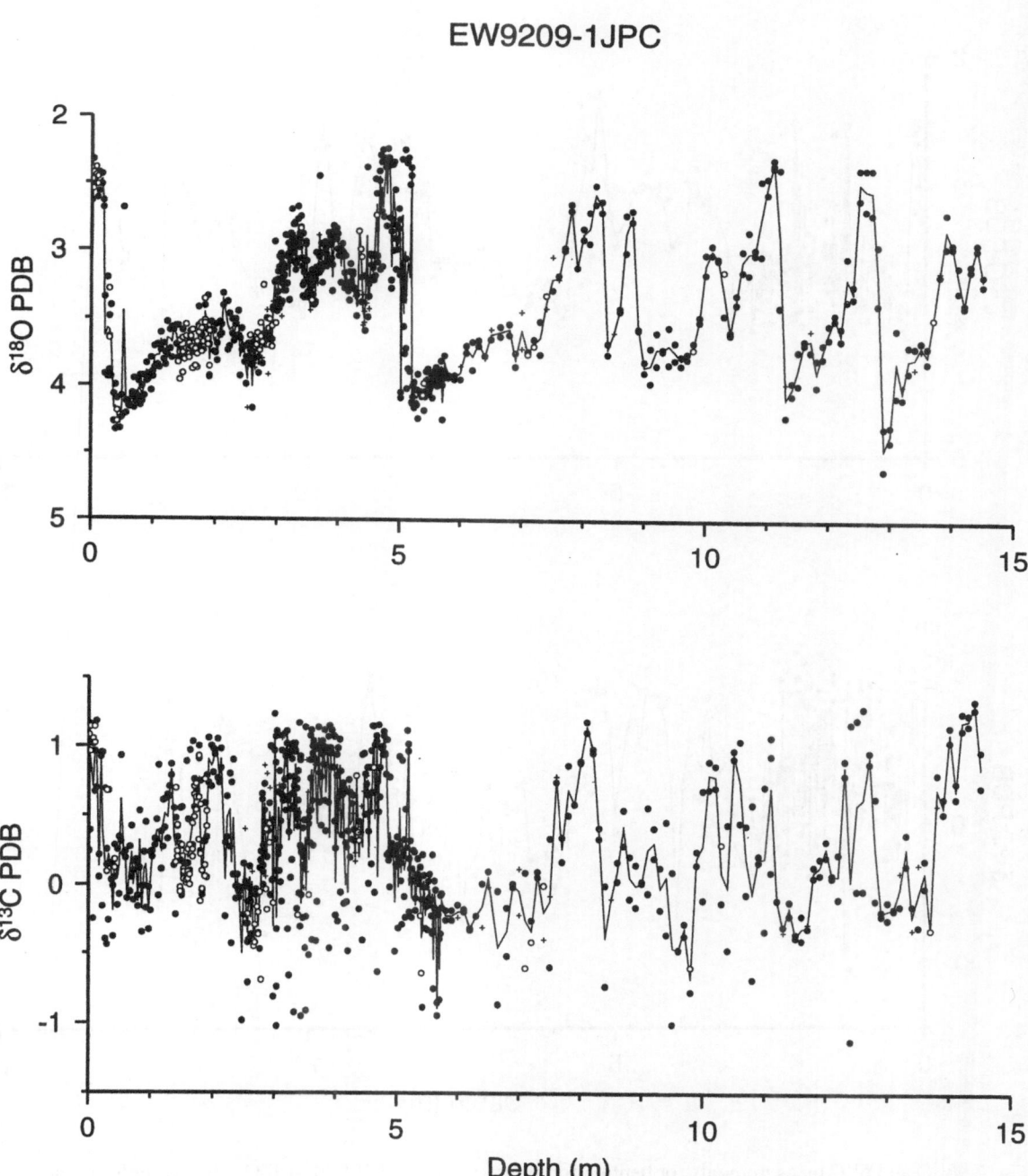

Fig. 6. $\delta^{18}O$ and $\delta^{13}C$ measurements for benthic foraminifera in core EW9209-1JPC. The symbols are the same as in Fig. 4.

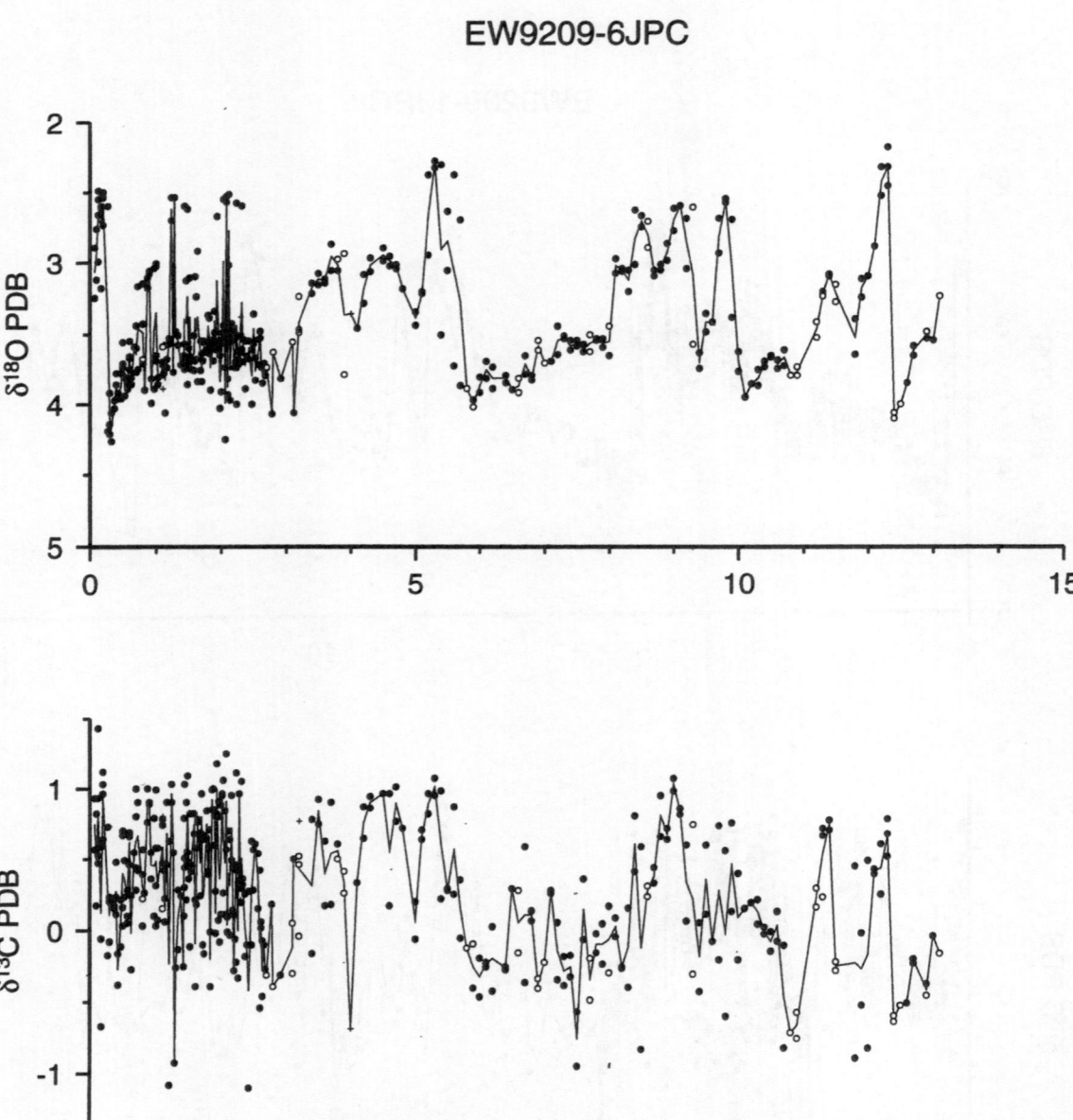

Fig. 7. $\delta^{18}O$ and $\delta^{13}C$ measurements for benthic foraminifera in core EW9209-6JPC. The symbols are the same as in Fig. 4.

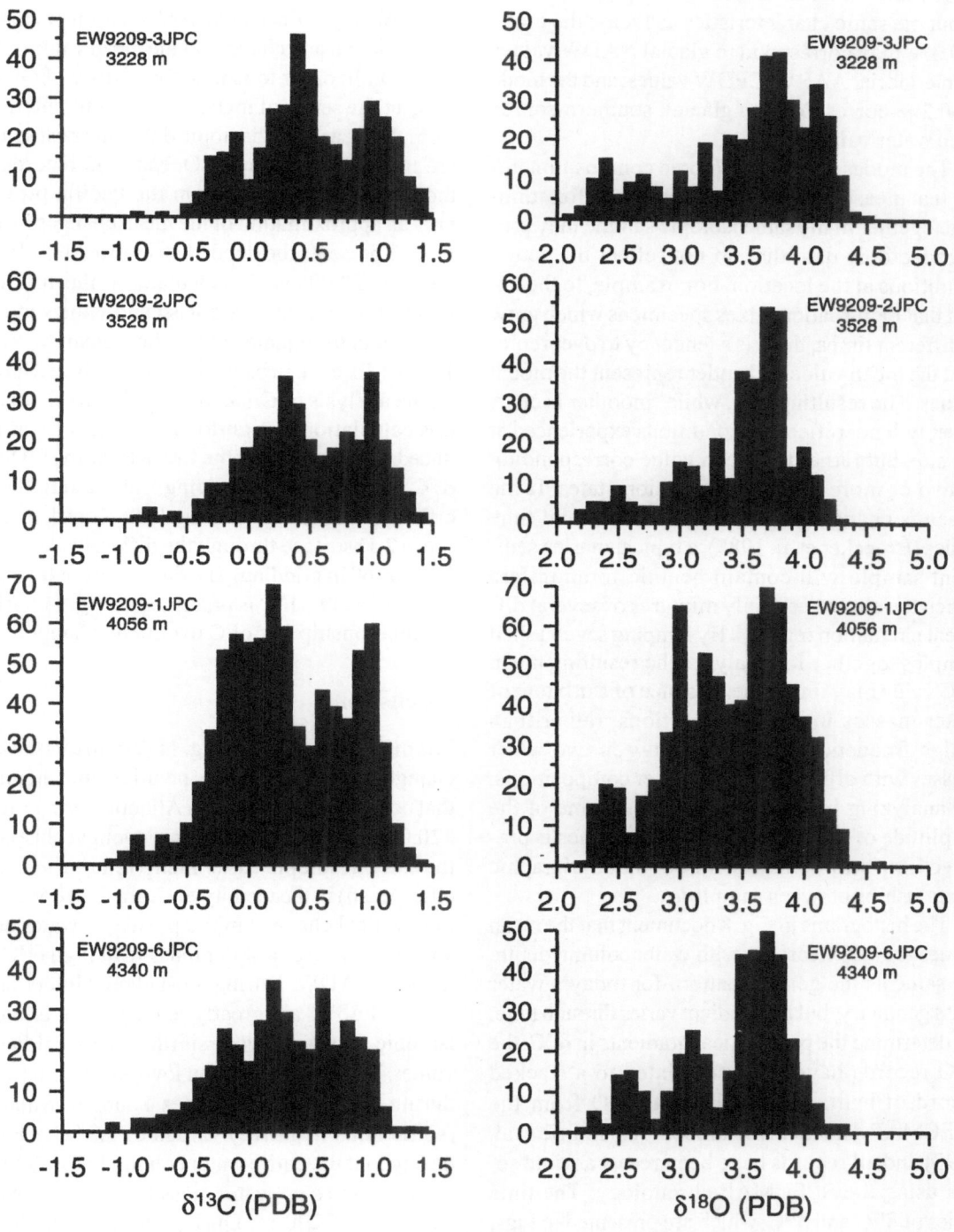

Fig. 8. Histograms of isotopic measurements for $\delta^{13}C$ and $\delta^{18}O$ for individual shells of benthic foraminifera. The histograms document that the data do not have a smooth distribution, but rather display prominent modes. For $\delta^{13}C$, the modes fall at about 1.0, 0.3, and -0.2‰. Overall there is a gradual decrease in mean $\delta^{13}C$ with increasing water depth which results from the transition from high-$\delta^{13}C$ NADW to lower-$\delta^{13}C$ southern-source water masses.

would occur if NADW were being produced with about the same characteristics as today; the mode at 0.3‰ may correspond to glacial NADW values or interglacial AABW-CPDW values; and the mode at -0.2‰ corresponds to glacial, southern-source deep water values.

The modes of carbon isotopic composition imply that measuring several specimens of foraminifera together in the same isotopic sample may produce isotopic data that do not reflect the actual conditions at the location. For example, to the extent that bioturbation mixes specimens which grew at different times, there is a tendency to over represent the mean value and under represent the modal values. The resulting data, while smoother in character, will not reflect the conditions experienced at the site, but rather the mean value corresponding to two or more extreme circulation states. If the ocean is changing on century to millennial time scales (Broecker et al. 1985), a typical marine sediment sample will contain benthic foraminifera which are almost certainly mixtures of several different circulation regimes. By lumping several shell samples together for analysis, the resulting mean $\delta^{13}C$ value may imply the presence of a mixture of water masses in equal proportions, rather than a high frequency oscillation between two water masses with different end-member compositions. By analyzing individual specimens, some of the amplitude of the variability of the regimes is preserved, but with some sacrifice of the stratigraphic relationships between samples.

The histograms in Fig. 8 document that the mean values for $\delta^{13}C$ decrease with water column depth, the same as the general pattern for today's water mass geometry, but the gradient varies through time. To determine the past vertical gradients in $\delta^{13}C$, the $\delta^{18}O$ records have been correlated to a stacked record of benthic foraminiferal $\delta^{18}O$ from the SPECMAP data archive number four (Imbrie et al. 1993) and all records have been recast as time series using the SPECMAP chronology. The time series of $\delta^{18}O$ and $\delta^{13}C$, which are presented in Figs. 9 and 10 respectively, document that most of the orbital-scale cyclicity in isotopic composition is common to all cores. Likewise, the prominent minima and maxima in $\delta^{13}C$ are found in all cores.

The magnitude of the changes in $\delta^{13}C$ are large and probably reflect some combination of changes in circulation and changes in the mean value of $\delta^{13}C$ of ΣCO_2. In order to isolate the history of circulation, a time series of mean $\delta^{13}C$ must be subtracted from the Ceara Rise isotopic data. The best sources are high-quality, Pacific Ocean $\delta^{13}C$ records, although a single record from the Pacific provides only an approximation of the mean changes in $\delta^{13}C$. Fig. 11 presents the depth distribution of $\Delta\delta^{13}C$ for the past 420,000 years, calculated as the difference between values at Ceara Rise and values at ODP Site 849 in the equatorial Pacific Ocean (Mix et al. 1995). Mix et al. produced the record based largely on the analysis of *Cibicides wuellerstorfi*, making this calculation straightforward. Fig. 11 was produced by interpolating the Ceara Rise and ODP849 $\delta^{13}C$ records at 1 kyr spacing, subtracting the Pacific record from each Ceara Rise record, then fitting a 2-D surface through the difference data with 3 kyr by 60 m gridding. The time series of $\delta^{18}O$ from core EW9209-3JPC is presented in Fig. 11 to show the relationship of $\Delta\delta^{13}C$ to climate change.

Discussion

The difference record in Fig. 11 documents the large changes in the water mass position and chemistry that occurred in the western Atlantic during the last 420,000 years. There are large, cyclical changes in the carbon isotopic enrichment (relative to the Pacific Ocean) in the vicinity of Ceara Rise documenting cyclical changes in the presence, water depth and proportion of a water mass with high $\delta^{13}C$ like modern NADW. During most glacial intervals the $\delta^{13}C$ and $\Delta\delta^{13}C$ are greatly reduced, that is carbon isotopic values are very similar to coeval Pacific values. The most prominent lows in $\Delta\delta^{13}C$ occurred during oxygen isotope stages 4 and 10, when low (sometimes negative) values of $\Delta\delta^{13}C$ are seen throughout the entire water column below 3200 m. At these times no water mass boundary is detectable below 3200 m. During stages 6 and 8, low $\Delta\delta^{13}C$ values also occur throughout much of the water column, but higher values can be observed in the shallowest core, suggesting some bathymetric gradient in $\delta^{13}C$ and the presence of a water mass

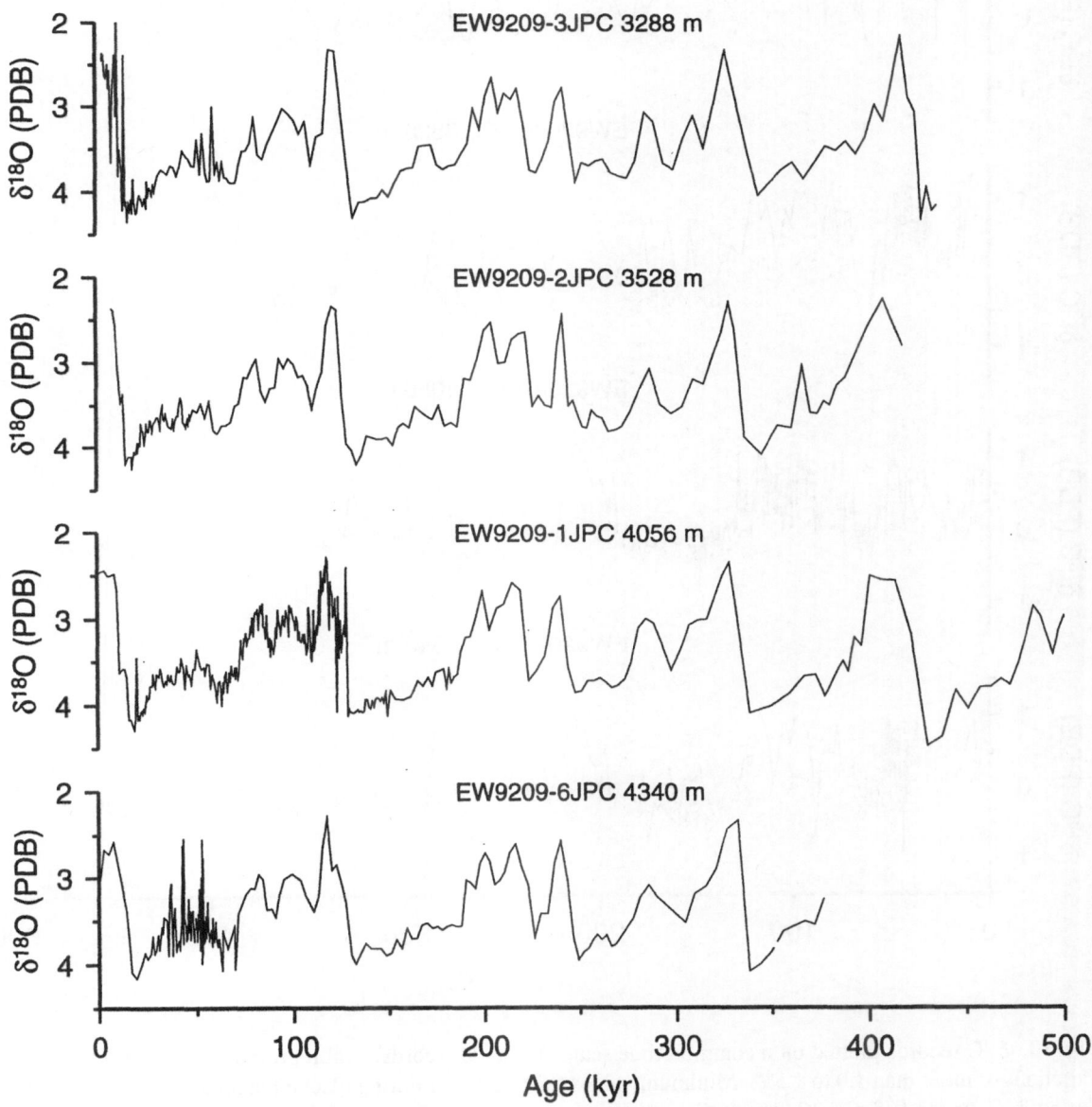

Fig. 9. $\delta^{18}O$ records plotted on a common time scale. The cores were correlated to the SPECMAP benthic stack record of $\delta^{18}O$, which is part of SPECMAP archive number 4.

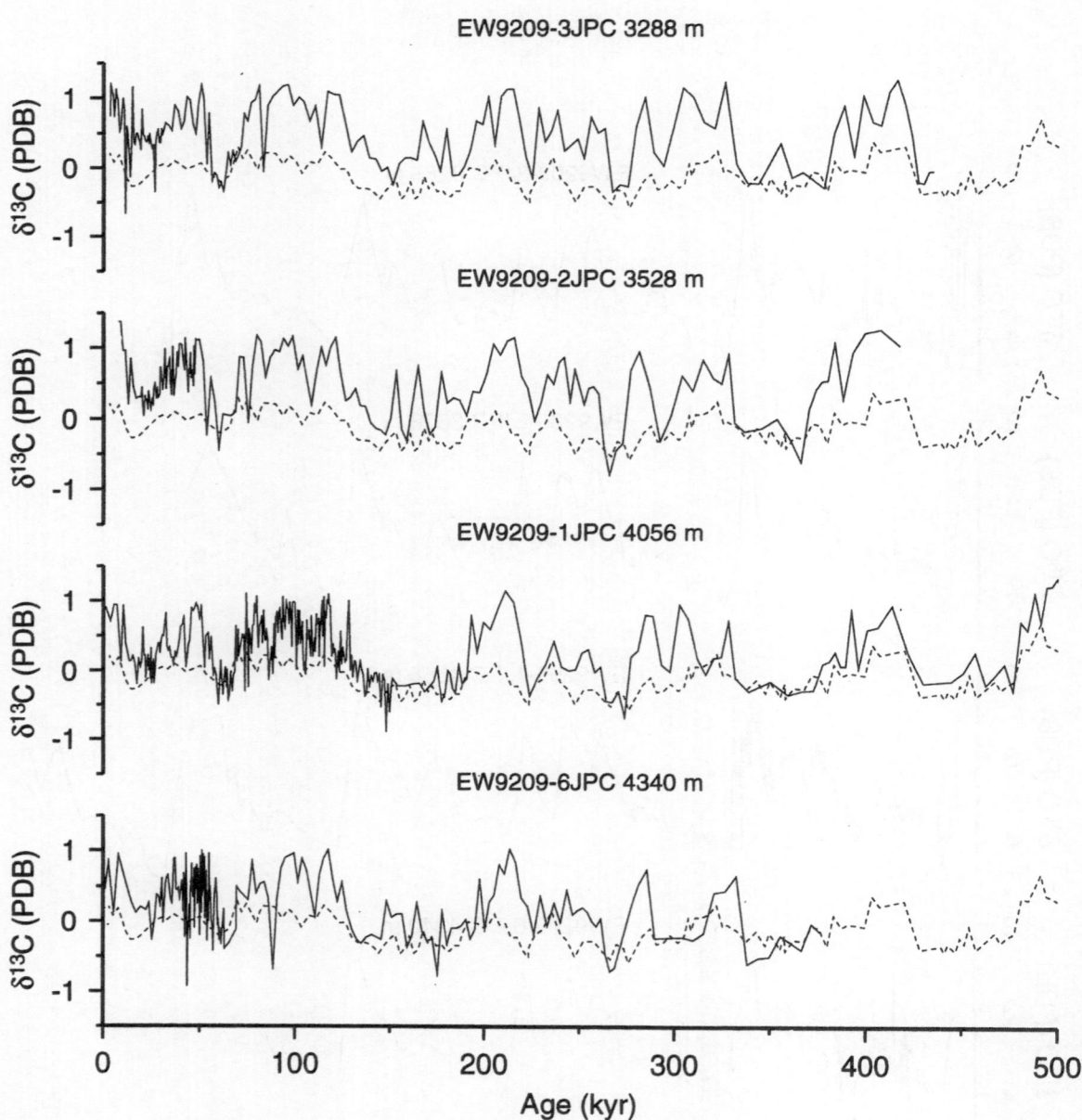

Fig. 10. $\delta^{13}C$ records plotted on a common time scale. The $\delta^{13}C$ records exhibit prominent glacial-interglacial variations of more than 1.0 to 1.5‰. Minimum values for $\delta^{13}C$ occur during glacial maxima, with lowest values during $\delta^{18}O$ stages 4, 8 and 10. The $\delta^{13}C$ record from Pacific ODP Site 849 (Mix et al. 1995) is plotted as the dashed line on each record. The enrichment in $\delta^{13}C$ of the Atlantic records with respect to the Pacific is the proxy used to denote the presence of NADW. During several glacials, the observed $\delta^{13}C$ values in the deep Atlantic are lower than coeval Pacific values.

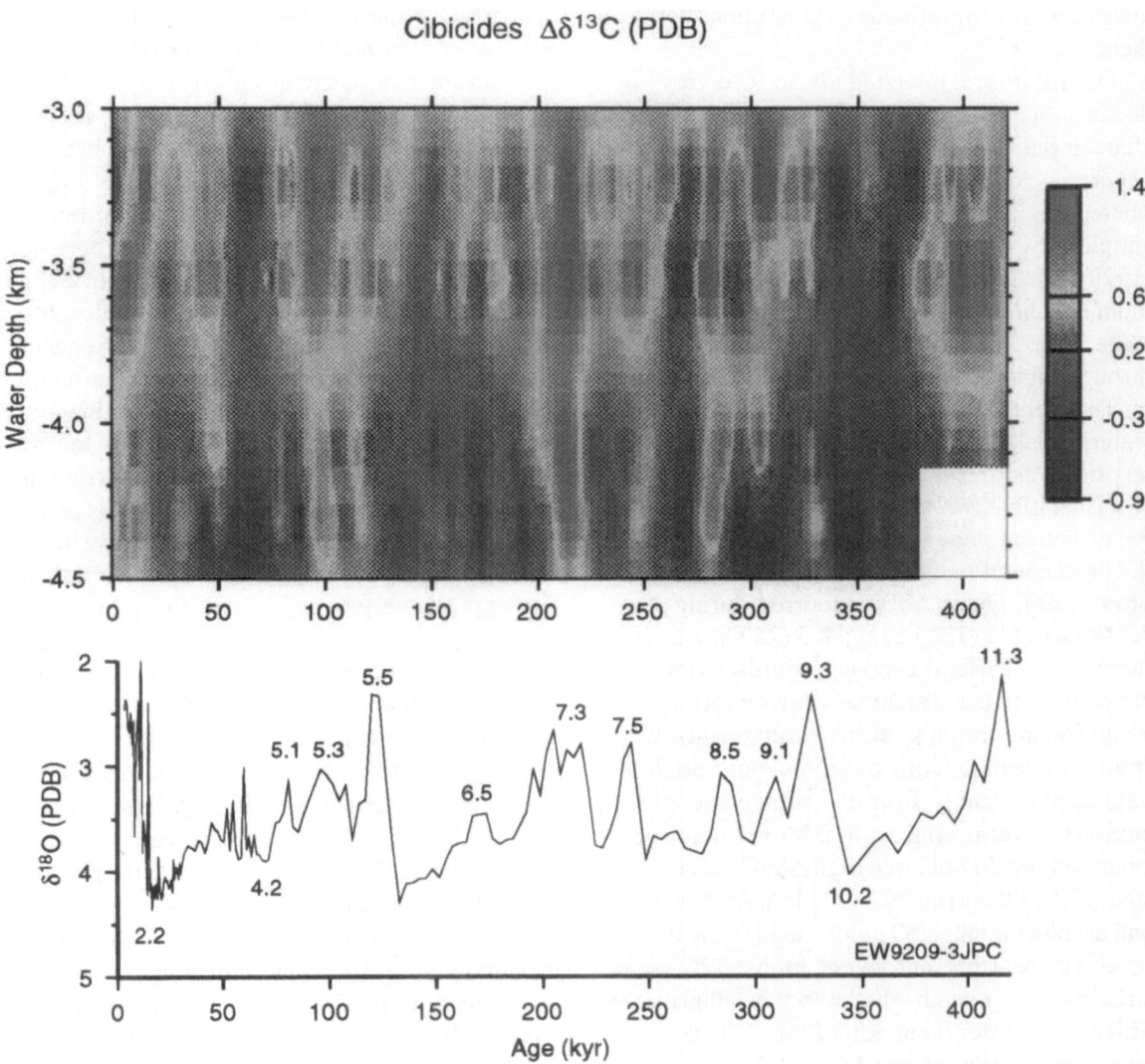

Fig. 11. Depth distribution of $\Delta\delta^{13}C$ versus time, interpolated from the four cores used in this study. The isotopic data were corrected for global changes in $\delta^{13}C$ by subtracting the Mix et al. (1995) $\delta^{13}C$ from ODP Site 849 in the Pacific Ocean. This normalization is meant to remove secular changes in $\delta^{13}C$ that result from the short-term transfer of carbon between the ocean and terrestrial reservoirs. The resulting difference values monitor the enrichment in ^{13}C values that reflects NADW production. The record documents large changes in the occurrence and water depth of NADW. The greatest volumetric extent of NADW occurred during $\delta^{18}O$ stage 7.3 (216 ka). Maximum northward penetration of southern-source deep water occurred during stages 4, 8 and 10.

boundary near 3200 m. By comparing the depth distribution of $\Delta\delta^{13}C$ during stages 2 and 4 the complexity of deep water production becomes apparent: despite having less continental ice, stage 4 appears to have greater suppression of NADW production than stage 2. The bathymetric gradient during stage 2 implies the presence of two water masses with a significant $\delta^{13}C$ gradient between them.

During several interstadials $\Delta\delta^{13}C$ is very high, documenting the presence of a water mass with $\delta^{13}C$ characteristics much like today's NADW. But the occurrence of this ^{13}C-enriched water mass is not related to glacial-interglacial climate change in a simple way, as was first suggested by Raymo et al. (1990). The greatest bathymetric extent and maximum enrichment occur during oxygen isotope substage 7.3 (~216 ka), when $\Delta\delta^{13}C$ values are high throughout the entire water column down to 4350 m. This depth distribution of $\Delta\delta^{13}C$ implies that a water mass with NADW carbon isotopic characteristics was present at all water depths at Ceara Rise, that is NADW was volumetrically larger than today. During stage 7.3 there are no indications for the presence of a water mass low in $\Delta\delta^{13}C$. Other prominent highs in $\Delta\delta^{13}C$ occurred during stages 5.3 (99 ka), 5.5 (122 ka) and 8.5 (287 ka). Each of these depth distributions is very similar to the modern pattern, with a water mass having formed with a significant component of warm surface water from the North Atlantic overlying one with lower values, presumably from the southern ocean. In contrast, several minima in $\delta^{18}O$ are noteworthy because they do not have high $\Delta\delta^{13}C$ values: substages 7.5 (238 ka) and 9.3 (331 ka). At these times and despite having $\delta^{18}O$ values suggesting that sea level was near present levels, high $\Delta\delta^{13}C$ are restricted to water depths shallower than 4000 m. Significantly greater incursion of low-$\delta^{13}C$ deep water occurred during stages 7.5 and 9.3. If deep water circulation intensity and glacial-interglacial climate change were related in a simple way, then minima in $\delta^{18}O$ should be correlated with maxima in $\Delta\delta^{13}C$. The most obvious counter-example to this pattern can be seen by comparing $\Delta\delta^{13}C$ during stage 3 with $\Delta\delta^{13}C$ during stages 9.3 or 7.5. Higher values of $\Delta\delta^{13}C$ occur over a wider depth interval during stage 3 than either stage 7.5 or 9.3 even though the

earth was much more glaciated during stage 3. Like the results of Raymo et al. (1990), the observations point to a decoupling of climate and the intensity of NADW production.

The observations here suggest that the production of NADW in the Atlantic is not restricted to interglacial intervals; rather there is significant evidence for the presence of a water mass with NADW carbon isotopic characteristics at Ceara Rise during intervals with significantly greater ice volume than today. Conversely, during extreme interglacial conditions, there is very little evidence for anything but NADW at Ceara Rise. Fig. 12 presents a 2-D frequency distribution of benthic foraminiferal carbon and oxygen isotopic measurements, sorted into 0.1‰ by 0.1‰ bins, from all four cores used in this study. The two patterns described above can be seen in this figure. First, the frequency distribution of $\delta^{13}C$ when $\delta^{18}O$ suggests minimum ice volume (<3‰) clearly shows the dominance of higher, NADW-like $\delta^{13}C$ values: very few low values of $\delta^{13}C$ occur when $\delta^{18}O$ values are at extreme interglacial levels. The implication of this pattern is that deep water from the south is mostly (but not entirely) excluded from the western tropical Atlantic during peak interglacials. At the other extreme, the frequency of occurrence of low $\delta^{13}C$ significantly increases during glacial periods, indicating the greater dominance of southern sources of deep water when the earth is glaciated. But unlike the extreme interglacials, when only single mode of high ^{13}C values is observed, there is a broad range (~1‰) in $\delta^{13}C$ when $\delta^{18}O$ is enriched. The very wide range of $\delta^{13}C$ during glacial periods implies the presence of a water mass like modern NADW, either at shallower depths because of a change in the relative production of the two water masses or varying in its intensity of production on shorter time scales than can be easily recovered from typical marine sediments.

High frequency oscillations in NADW production were observed in very high sedimentation rate cores from the Blake Bahama Outer Ridge (Keigwin et al. 1994) and Bermuda Rise (Keigwin and Jones 1994). These cores come from sediment drifts and have sedimentation rates 2 to 4 times greater than at Ceara Rise. Keigwin et al. (1994) observed quasi-periodic variations in $\delta^{13}C$ during

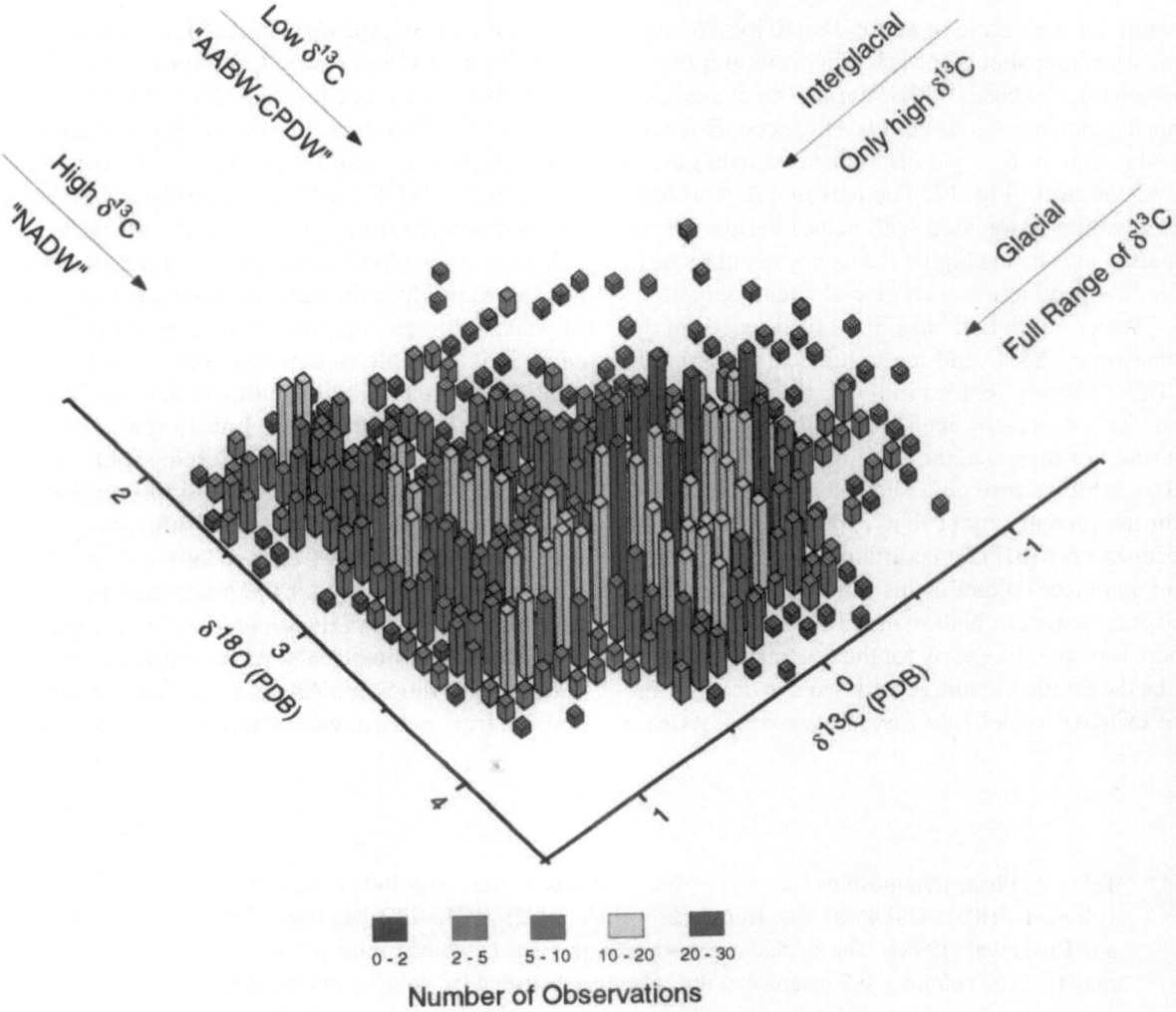

Fig. 12. Two dimensional histogram of $\delta^{18}O$ and $\delta^{13}C$ documenting the high amount of variability in $\delta^{13}C$ that occurs during glacial climatic conditions: a 1‰ range in $\delta^{13}C$ values occurs when the earth is glaciated. In contrast, extreme interglacials are dominated, nearly exclusively, by high $\delta^{13}C$ values. This pattern implies that NADW production may oscillate more during glacial periods than during interglacial periods, suggesting fundamental differences in the stability of climate with and without ice sheets.

stage 5 which had three maxima and minima during each precession cycle suggesting that the higher frequency changes in $\delta^{13}C$ have a characteristic time scale of about 7000 years. The variations in $\delta^{13}C$ appeared to correlate with large changes in air temperature over the GISP and GRIP ice cores in Greenland, apparently linking rapid climate changes in the North Atlantic region to changes in

NADW production. Variability in $\delta^{13}C$ on these time scales can also be observed at Ceara Rise within isotope stage 3 and stage 6. The best example in this data set comes from the upper section of EW9209-1JPC, where we have sampled the record are greater resolution than in the other cores (see Fig. 10). During stage 3, four prominent maxima and minima in $\delta^{13}C$ are apparent and have a characteristic time scale of about 7 to 10 kyr. What is most striking about the variability is its amplitude, which often exceeds 0.7‰. Variability of this scale during glacial periods could easily account for the wide range of $\delta^{13}C$ values associated with glacial $\delta^{18}O$ values in Fig. 12. The narrow range of high $\delta^{13}C$ values associated with peak interglacial climate suggests that higher frequency oscillations in NADW production are a glacial phenomenon.

Water depth and latitudinal differences in the phasing of $\Delta\delta^{13}C$ and ice volume in the Atlantic (Tab. 2) imply that no individual record can adequately represent deep water circulation in this basin. For instance, the shallower cores on Ceara Rise exhibit a zero phase between $-\delta^{18}O$ and $\Delta\delta^{13}C$ for the eccentricity (100 kyr) period, while in the deepest core (6JPC) maximum $\Delta\delta^{13}C$ clearly leads minimum ice volume at this period. Thus, the record of deep water circulation in the deepest Ceara Rise core is responding early for the eccentricity period, like the South Atlantic records used to develop the SPECMAP model. Likewise, the two deepest Ceara

Rise cores (1JPC, 6JPC) also lead ice volume for the obliquity (41 kyr) period, again more like the South Atlantic cores used by SPECMAP. In contrast, $\Delta\delta^{13}C$ in all Ceara Rise cores lags ice volume in the precession band, like the North Atlantic record from DSDP Site 607 but unlike the South Atlantic records used by SPECMAP. ($\Delta\delta^{13}C$ calculations and %NADW calculations for the Ceara Rise cores produce the same depth-related pattern of phase with ice volume. $\Delta\delta^{13}C$ is used here because it is possible to calculate this record with much higher temporal resolution than %NADW). Thus there is support for the Imbrie et al. (1992) model in two periods of orbital forcing (obliquity and eccentricity), but not in the third (precession).

Curiously the new phase relationships presented here agree with the new Cd/Ca record of Oppo and Rosenthal (1994) only for the precession period. This implies that the Cd/Ca and $\Delta\delta^{13}C$, two widely used proxies for deep water reconstructions, have variability in the South Atlantic which may be uncoupled from nutrient variations (Oppo and Rosen-

Table 2. Phase relationships (in degrees) between deep water proxy indicators and ice volume (-δ18O). DSDP607 data from Imbrie et al. (1992); RC13-229 data from Oppo and Rosenthal (1994). The shaded regions highlight cores in which the deep water proxy may lead ice volume. NS means that the coherence between ice volume and the deep water proxy are not significant at the 80% confidence level. The phases identified with (*) were from the adjacent periods (44.4 and 21.1 kyr), where coherence was significant at the 80% confidence level.

Core	100kyr	41kyr	23kyr
%NADW DSDP6067	16±15	5±16	41±17
Δδ13C 3JPC	-3±17	NS	39±19
Δδ13C 2JPC	-5±14	-27±25*	46±25*
Δδ13C 1JPC	11±18	-62±27	32±17
Δδ13C 6JPC	-52±19	-17±20	14±21
Δδ13C RC13-229	-34±18	-36±16	-56±10

thal 1994); they may be complicated by gas exchange processes at the surface (Broecker and Maier-Reimer 1992; Charles et al. 1993), by variations in the stoichiometry of organic matter production (Rau et al. 1989; Goericke and Fry 1994), or by variations in the oceanic budget for Cd (Rosenthal 1994).

Summary

A signal corresponding to variations in NADW is found at Ceara Rise over a wide range of climate conditions for the last 400,000 years. The production of NADW has varied on orbital time scales, but the relationship between NADW production and climate is not simple. During peak interglacial intervals, NADW dominates to such an extent that southern-source deep waters are nearly excluded from the Ceara Rise region. At times during glacial periods, the influence of southern-source deep waters dominates below 3200 m. However, during other glacial periods, significant bathymetric gradients in $\delta^{13}C$ are observed which suggest that two water masses are present over this depth interval. In addition, the growing body of evidence for high-frequency, sub-orbital oscillations in deep water production is supported here by the high-frequency variations in $\delta^{13}C$ that occur most prominently during glacial stages 3 and 6. As in the observations of Keigwin et al. (1994), the temporal scale of this variability is about 7 to 10 kyr. In typical marine sediments (with sedimentation rates of 2 cm/kyr, for example) it would be difficult to measure changes on these time scales because bioturbation would effectively blur the record. But by measuring multiple samples of individual tests of benthic forami-nifera it is possible to characterize this higher-frequency variability in deep water chemistry. That this variability is modal in character suggests that many of the variations in carbon isotopic composition at Ceara Rise result from the presence or absence of NADW from this region rather than mixtures of water masses with intermediate compositions (Broecker et al. 1985). There are also depth-related differences in the phasing of ice volume and deep water production, which support the SPECMAP observations for the eccentricity and obliquity periods of climate change; the 23 kyr precessional period may be more complicated than originally suggested. That the early response of deep water production is at the depth levels of lower NADW supports the Imbrie et al. (1992) hypothesis that this component of deep water plays a significant role in the transfer of heat between the northern and southern hemispheres at least for the eccentricity and obliquity periods of orbital forcing. But the absence of a lead for the precession period may make it necessary to reevaluate the model presented by Imbrie et al. (1992) or to reconsider whether other factors besides NADW production can be affecting the deep water proxies indicators.

Acknowledgements

Support for this research was provided by the National Science Foundation award OCE-9116303. The cores are archived at the Woods Hole Oceanographic Institution; we gratefully acknowledge support from NSF, the Office of Naval Research and the United States Geological Survey for support of the WHOI core repository. This research benefitted from many discussions with my colleagues at WHOI, D. Oppo and G. Lohmann. Thoughtful reviews were provided by W. Berger, T. Bickert and M. Raymo. This is WHOI Contribution No. 9020.

References

Belanger PE, Curry WB, Matthews RK (1981) Core-top evaluation of benthic foraminiferal isotopic ratios for paleoceanographic interpretation. Palaeogeogr, Palaeoclimatol, Palaeoecol 33:205-220

Boyle EA (1984) Cadmium in benthic foraminifera and abyssal hydrography: Evidence for a 41 kyr obliquity cycle. In: Hansen JE and Takahashi T (eds) Climate Processes and Climate Sensitivity. AGU, Washington DC, pp 360-368

Boyle EA, Keigwin LD (1982) Deep circulation of the North Atlantic over the last 200,000 years: Geochemical evidence. Science 2128:784-787

Broecker WS, Peteet DM, Rind D (1985) Does the ocean-atmosphere system have more than one stable mode of operation? Nature 315:21-26

Broecker WS, Denton GH (1989) The role of ocean-atmosphere reorganizations in glacial cylces. Geochimica et Cosmochimica Acta 53:2465-2501

Broecker WS, Maier-Reimer E (1992) The influence of air and sea exchange on the carbon isotopic distribution in the sea. Global Biogeochemical Cycles 6:315-320

Charles CD, Wright JD, Fairbanks RG (1993) Thermodynamic influences on the marine carbon isotope record. Paleoceanography 8:691-697

Curry WB, Lohmann GP (1982) Carbon isotopic changes in benthic foraminifera from the western South Atlantic: Reconstruction of glacial abyssal circulation patterns. Quat Res 18:218-235

Curry WB, Duplessy J-C, Labeyrie LD, Shackleton NJ (1988) Changes in the distribution of $\delta^{13}C$ of deep water CO_2 between the last glaciation and the Holocene. Paleoceanography 3:317-342

Curry WB, Slowey NC, Lohmann GP (1993) Oxygen and carbon isotopic fractionation of aragonitic and calcitic benthic foraminifera on Little Bahama Bank, Bahamas. EOS Transactions of the American Geophysical Union 74:368

Duplessy J-C, Shackleton NJ, Matthews RK, Prell W, Ruddiman WF, Caralap M, Hendy C (1984) ^{13}C record of benthic foraminifera in the last interglacial ocean: Implications for the carbon cycle and the global deep water circulation. Quat Res 21:225-243

Goericke R, Fry B (1994) Variations of marine plankton $\delta^{13}C$ with latitude, temperature and dissolved CO_2 in the world ocean. Global Biogeochemical Cycles 8:85-90

Graham DW, Corliss BH, Bender ML, Keigwin LD (1981) Carbon and oxygen isotopic disequilibria of Recent deep-sea foraminifera. Mar Micropaleontol 6:483-497

Imbrie J, Boyle EA, Clemens SC, Duffy A, Howard WR, Kukla G, Kutzbach J, Martinson DG, McIntyre A, Mix AC, Molfino B, Morley JJ, Peterson LC, Pisias NG, Prell WL, Raymo ME, Shackleton NJ, Toggweiler JR (1992) On the structure and origin of major glaciation cycles 1. Linear responses to Milankovitch forcing. Paleoceanogr 7:701-738

Imbrie J, Berger A, Boyle EA, Clemens SC, Duffy A, Howard WR, Kukla G, Kutzbach J, Martinson DG, McIntyre A, Mix AC, Molfino B, Morley JJ, Peterson LC, Pisias NG, Prell WL, Raymo ME, Shackleton NJ, Toggweiler JR (1993) On the structure and origin of major glaciation cycles 2. The 100,000-year cycle. Paleoceanography 8:699-735

Keigwin LD, Curry WB, Lehman SJ, Johnsen S (1994) The role of the deep ocean in North Atlantic climate change between 70 and 130 kyr ago. Nature 371:323-326

Keigwin LD, Jones GA (1994) Western North Atlantic evidence for millennial-scale changes in ocean circulation and climate. J Geophys Res 99:12397-12410

Kroopnick P (1985) The distribution of ^{13}C of ΣCO_2 in the world oceans. Deep-Sea Res 32:57-84

Lohmann GP, Lohmann KC (1991) Reconstructions of glacial-interglacial $\delta^{18}O$-$\delta^{13}C$ profiles through deep and intermediate waters of the western South Atlantic from analysis of individual shells. EOS Transactions of the American Geophysical Union 72:160

Mix AC, Pisias NG, Rugh W, Wilson J, Morey A, Hagelberg TK (1995) Benthic foraminiferal stable isotope record from Site 849, 0-5 Ma: Local and global climate changes. Proceedings of the Ocean Drilling Program, Scientific Results 138:371-412

Oppo DW, Fairbanks RG (1987) Variability in the deep and intermediate water circulation of the Atlantic Ocean: Northern Hemisphere modulation of the Southern Ocean. Earth and Planet Sci Letts 86:1-15

Oppo DW, Fairbanks RG, Gordon AL, Shackleton NJ (1990) Late Pleistocene Southern Ocean $\delta^{13}C$ variability. Paleoceanography 5:43-54

Oppo DW, Lehman SJ (1993) Mid-depth circulation of the subpolar North Atlantic during the Last Glacial Maximum. Science 259:1148-1152

Oppo DW, Rosenthal Y (1994) Cd/Ca changes in a deep Cape Basin core over the past 730,000 years: Response of circumpolar deepwater variability to northern hemisphere ice sheet melting? Paleoceanogr 9:661-675

Rau GH, Takahashi T, Des Marais DJ (1989) Latitudinal variations in plankton $\delta^{13}C$: Implications for CO_2 and productivity in past oceans. Nature 341:516-518

Raymo ME, Ruddimann WF, Shackleton NJ, Oppo DW (1990) Evolution of Atlantic-Pacific $\delta^{13}C$ gradients over the last 2.5 m.y. Earth and Planet Sci Letts 97:353-368

Rosenthal Y (1994) Late Quaternary paleochemistry of the Southern Ocean: Evidence from cadmium variability in sediments and foraminifera. MIT/WHOI Joint Program in Oceanography, Ph.D. Thesis, 184 pp

Sarnthein M, Winn K, Jung SJA, Duplessy J-C, Labeyrie L, Erlenkeuser H, Ganssen G (1994) Changes in east Atlantic deepwater circulation over the last 30,000 years: Eight time slice reconstructions. Paleoceanography 9:209-267

Late Quaternary Deep Water Circulation in the South Atlantic: Reconstruction from Carbonate Dissolution and Benthic Stable Isotopes

T. Bickert and G. Wefer

Fachbereich Geowissenschaften, Universität Bremen,
28334 Bremen, Germany

Abstract: Carbonate dissolution data (sand contents) and $\delta^{13}C$ records of the epibenthic foraminifer *Cibicides wuellerstorfi* from 12 gravity cores are used to reconstruct the history of deep water circulation in the South Atlantic for the last 360,000 years. The cores were selected from depth-sections in four basins (Brasil-, Guinea-, Angola- and Cape Basins) in water depths between 2900 m and 4600 m. The depth-transect approach allows removal of mean global shifts as well as local productivity effects from the paleo-property records and extraction of variations which are due to changes in deep water chemistry and/or circulation in the South Atlantic. As a result of the reduction of NADW during the last glacial maximum the Southern Component Water was higher in the water column and extended farther north than it does today. This glacial water mass can be divided into an upper part (USCW) with $\delta^{13}C$ values between 0.2‰ and 0.7‰ and a lower part (LSCW) characterized by values of -0.2‰ to 0.2‰. The boundary, marked also by the calcite lysocline, was at 3800 m water depth near the equator and rose slightly toward the Southern Ocean. The asymmetry observed in bottom water circulation today (LCDW in western basins and in the Cape Basin, NADW in eastern basins below 4000 m) was not present. From comparison to a deep western Pacific core (ODP 806B; Bickert et al. 1993) there is evidence that the nutrient-enriched but oxygen-depleted LSCW resembles the glacial Pacific Deep Water. This is also true for the older glacial stages 4, 6, 8 and 10.

Introduction

Deep sea carbonate sediments cover about one half of the total oceanic floor (Berger et al. 1976; Biscaye et al. 1976; Kolla et al. 1976) and act as a large and reactive reservoir for carbon dioxide (Broecker and Peng 1982, 1987; Sundquist and Broecker 1985). Understanding temporal and spatial changes in carbonate preservation is of key importance for testing the numerous models which seek to explain past changes in atmospheric pCO_2 through changes in the oceanic carbon cycle (Keir and Berger 1983; Boyle 1988; Broecker and Peng 1989; Keir 1990; Archer and Maier-Reimer 1994).

Since the early work of Arrhenius (1952) it has been clearly established that carbonate contents of pelagic sediments show cyclic variations associated with Pleistocene glacial-interglacial changes (Berger 1973; Volat et al. 1980; Moore et al. 1982; Crowley 1985; Farrell and Prell 1989; Le and Shackleton 1992; Howard and Prell 1994). But while in the Atlantic and Southern Oceans generally dissolution is intensified during glacial periods (Gardner 1975; Crowley 1983; Balsam and McCoy 1987; Howard and Prell 1994), the Pacific carbonate fluctuations are related to the rate of change in climate, with enhanced dissolution during the transition from interglacial to glacial times and increased preservation during deglaciation periods (Berger 1973; Farrell and Prell 1989; Hebbeln et al. 1990; Wu et al. 1990; Le and Shackleton 1992; Yasuda et. al. 1993). This asymmetry between the Atlantic and the Pacific dissolution pattern has been attributed to changes in basin-to-basin fractionation resulting from variations in North Atlantic Deep Water (NADW) formation (Berger 1970; Volat 1980; Crowley 1985).

From WEFER G, BERGER WH, SIEDLER G, WEBB DJ (eds), 1996, *The South Atlantic: Present and Past Circulation.* Springer-Verlag Berlin Heidelberg, pp 599-620

The deep South Atlantic is an ideal place to observe changes in the depth distribution of deep water properties because it reveals the mixing zone between NADW and southern-source deep water masses (Reid 1989). But deep water mass properties are not the only controlling factor for carbonate dissolution. Studies of Emerson and Bender (1981), and more recently, benthic flux chamber experiments carried out by Jahnke et al. (1994), showed that even in supralysoclinal waters, which are supersaturated with respect to calcite, sediments undergo a significant carbonate dissolution. These authors propose that this dissolution results from the decay of organic matter, reducing carbonate ion concentration in the pore water. Furthermore, long records of carbonate fluctuations exhibit long-term trends in dissolution (e.g. the Mid-Brunhes dissolution cycle) which are thought to be associated with global changes in the carbon reservoir of the oceans (Vincent 1981; Farrell and Prell 1991; Bassinot et al. 1994; Bickert et al. 1996). Thus, for reconstructing the deep water circulation in the past, both the productivity effect and the global shift have to be removed from dissolution records to extract the true water mass signal. Since the carbon isotopic composition of seawater, which is monitored in the calcitic shells of benthic foraminifers, is linked to the carbon system, the factors controlling the $\delta^{13}C$ distribution in the past should be much the same as for carbonate dissolution, although their relative importance should differ.

The purpose of this study is to present sand fraction and benthic carbon isotope records from the deep South Atlantic Ocean to decipher the effects controlling the deep water properties in the past and to reconstruct the deep water circulation for the last 360,000 years.

Study area and material

We selected 12 gravity corers (5 to 15 m in length) and the associated giant box corers retrieved during the METEOR cruises M6-6, M9-4, and M12-1 (Wefer et al. 1988, 1989, 1990) in the eastern South Atlantic (Fig.1 and Table 1). These pelagic carbonate sections are distributed in four depth transects in the main basins of the South Atlantic (Brasil-, Guinea-, Angola- and Cape Basins) in water depths ranging from 2900 to 4600 m. The strategy was to study cores of a small area in each basin over the depth range of change in principal water mass boundaries. The sediment input is believed to be the same at each site of a transect, while the only changing factor is the water depth. Thus, we assume that vertical gradients in the measured parameters are independent of local effects and are related to changes in the properties of the water masses or in their water column positions. By taking only the within-transect differences into account, mean global changes can be eliminated from each record, and variations through time thus can be interpreted as changes in deep water chemistry and/or circulation.

Today the circulation in the deep western South Atlantic, the main flow path of deep water, is dominated by interactions between North Atlantic Deep Water (NADW) flowing toward the south and Circumpolar Deep Water (CDW) flowing to the north. The density characteristics of these water masses are such that NADW divides the southern water mass into an upper and lower branch (Reid 1989). Consequently the relatively warm and saline NADW occupies the depth interval between 2000 and 4000 m, while below 4000 m lower CDW (LCDW) is encountered. The mixing zone between NADW and LCDW is marked by gradients in temperature, salinity, nutrient concentrations, and in the corrosiveness of the water with respect to calcium carbonate (Fig. 2). For the eastern basins of the South Atlantic, i.e. the Guinea and the Angola Basins, the inflow of LCDW is restricted by the Mid-Atlantic Ridge in the west and the Walvis Ridge in the south to only small quantities passing the sills through the Romanche Fracture Zone (Van Bennekom and Berger 1984; Warren and Speer 1991) and the Walvis Passage (Shannon and Chapman 1992). Therefore the deepest parts of these basins are filled almost exclusively by NADW. The Cape Basin, although located eastwards of the Mid-Atlantic Ridge, is dominated by LCDW below 4000 m due to a bottom water passage, which allows LCDW to enter the basin from the south.

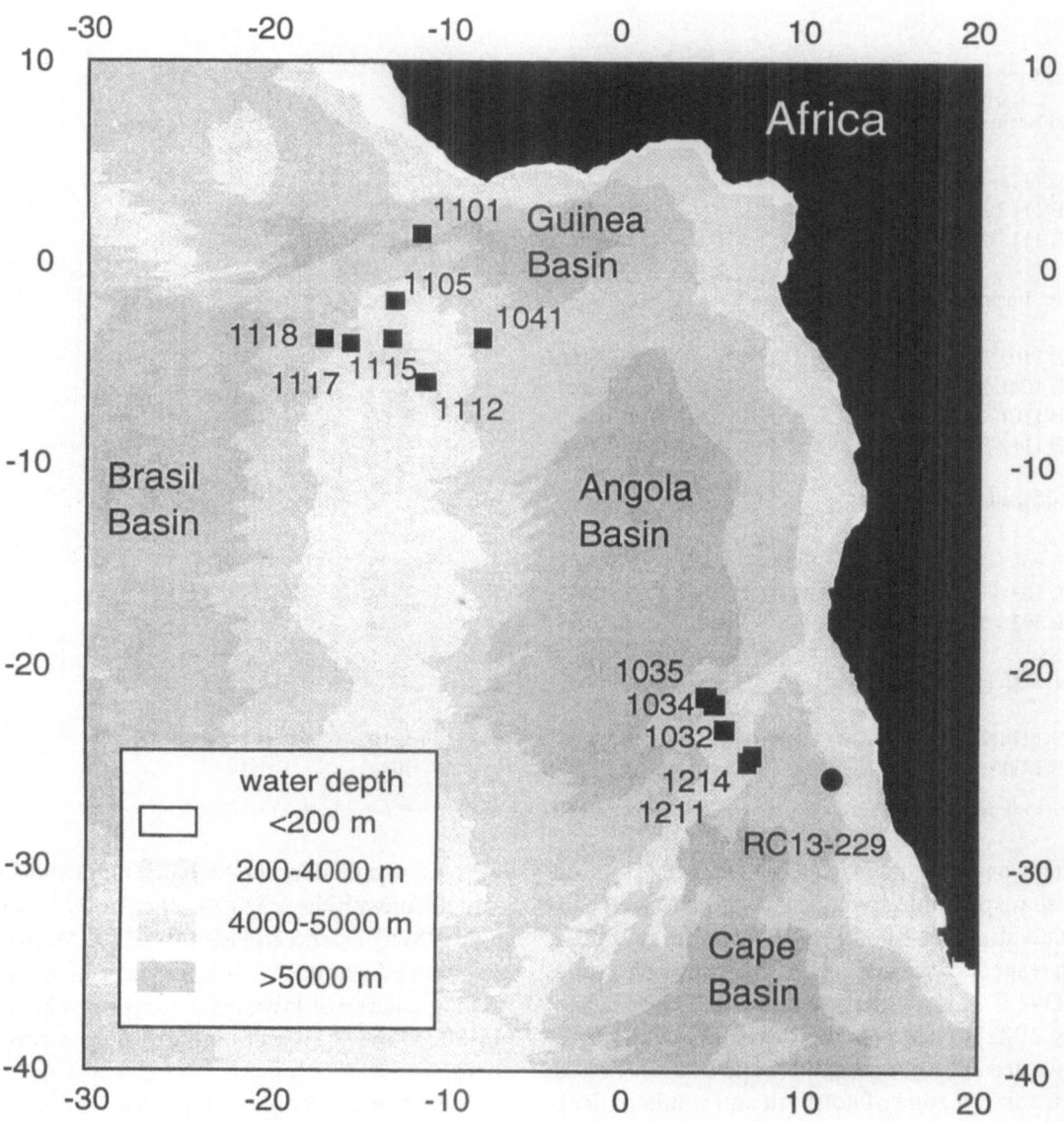

Fig. 1. Location map showing the positions of the 12 cores used in this study, and the position of core RC 13-229 (Oppo and Fairbanks 1990) for comparison.

Table 1. Locations, water depths and average sedimentation rates of the cores presented in this study.

Box/Gravity core	Latitude	Longitude	Water depth (m)	average Sed.-rate (cm / ky)
Brasil Basin (Equatorial South Atlantic)				
GeoB 1115-4/3	3°33.7'S	12°33.6'W	2945	2.95
GeoB 1117-3/2	3°48.9'S	14°53.8'W	3984	3.01
GeoB 1118-2/3	3°33.6'S	16°25.7'W	4671	2.21
Guinea Basin (Equatorial South Atlantic)				
GeoB 1105-3/4	1°39.9'S	12°25.7'W	3225	4.16
GeoB 1041-1/3	3°28.5'S	07°36.0'W	4033	2.14
GeoB 1101-4/5	1°39.5'N	10°58.8'W	4588	1.46
GeoB 1112-3/4	5°46.2'N	10°44.7'W	3122	2.54
Angola Basin (Walvis Ridge)				
GeoB 1032-2/3	22°54.9'S	06°02.2'E	2505	1.46
GeoB 1034-1/3	21°44.1'S	05°25.3'E	3772	0.95
GeoB 1035-3/4	21°35.2'S	05°01.7'E	4453	1.30
Cape Basin (Walvis Ridge)				
GeoB 1214-2/1	24°41.4'S	07°14.5'E	3210	1.14
GeoB 1211-1/3	24°28.5'S	07°32.0'E	4084	1.17

This asymmetry in bottom water distribution is also responsible for the modern pattern of carbonate dissolution. LCDW is undersaturated with respect to the carbonate concentration, while NADW is slightly supersaturated (Broecker and Peng 1982). Therefore the lysocline of calcite in the western basins and in the Cape Basin is linked to the mixing zone of northern and southern deep water masses at about 4000 m, while in the eastern basins, due to the lack of LCDW, the lysocline is located much deeper in a water depth between 4700 to 4900 m (Biscaye 1976; Thunell 1982; see Fig. 2). The carbonate content of the surface sediments at the selected sites exceeds 90 %, except for the deepest site in the Brasil Basin (GeoB 1118, 70%; Bickert 1992), which is bathed today the LCDW. A few percent opal in the equatorial

cores are supplied by the high primary productivity in the upwelling zone (75-90 gC m^{-2}y^{-1}, Berger et al. 1987; Verardo and McIntyre 1994). In contrast, cores from the Walvis Ridge, which are located in an area of low surface layer productivity (45-50 gC m^{-2}y^{-1}), exhibit almost pure carbonate oozes.

The carbon isotopic composition of ΣCO_2 is linked to deep water circulation patterns because its distribution in sea water is affected by photosynthesis and respiration. Deep water that is formed with a significant surface water component, like NADW today, is enriched in ^{13}C, because photosynthesis has preferentially removed ^{12}C from the ΣCO_2 of the surface water. LCDW as an old water mass has lower δ^{13}C values, since nearly all of the organic matter that is produced by pho-

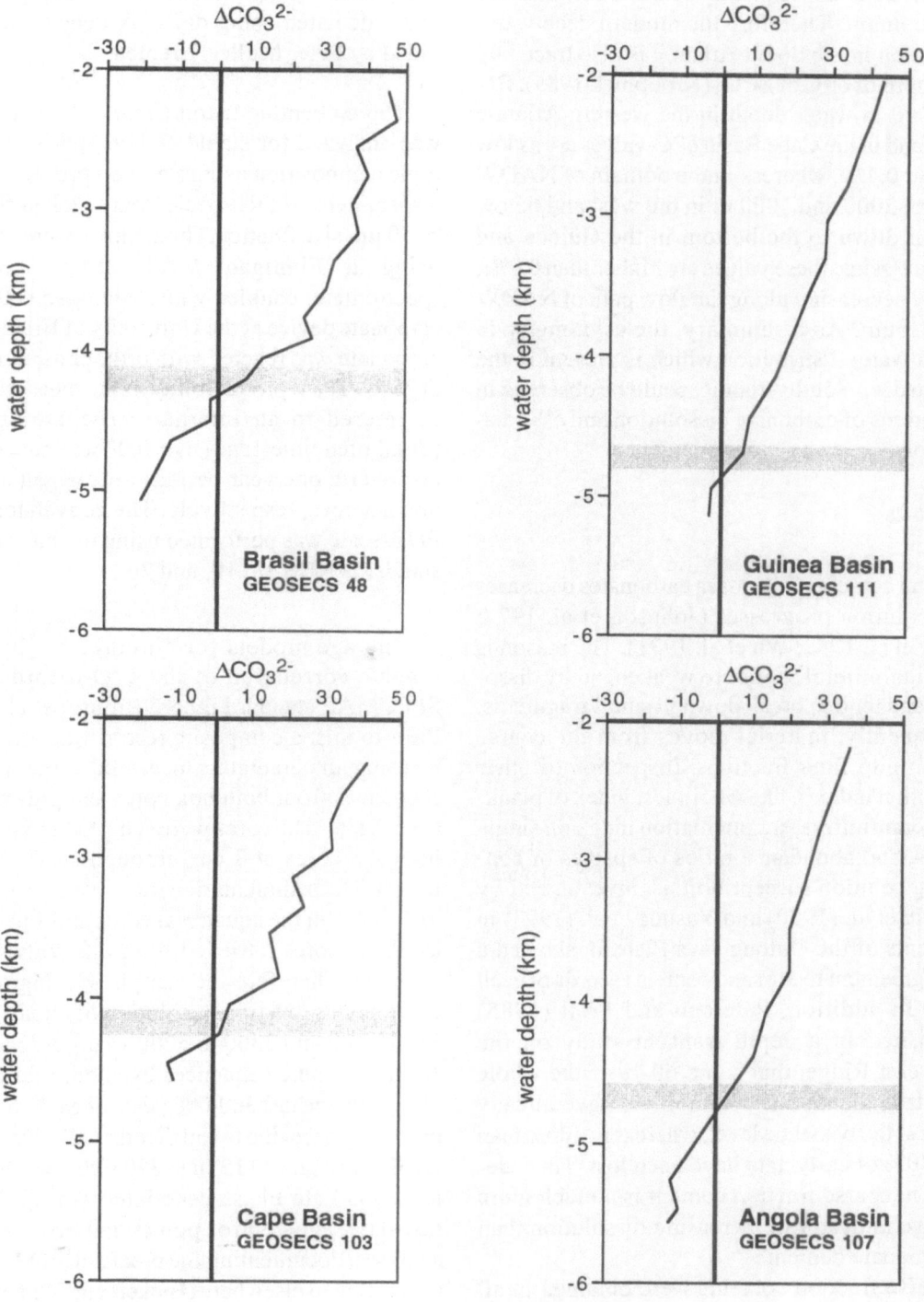

Fig. 2. ΔCO_3^{2-} depth profiles of GEOSECS stations 48, 103, 107, and 111, which are located in the 4 basins of the South Atlantic mentioned in this study. The shaded areas mark the modern depth of lysocline with respect to calcite.

tosynthesis is subsequently remineralized in the water column. Therefore the modern deep water circulation in the South Atlantic is also traced by the pattern of $\delta^{13}C$ of ΣCO_2 (Kroopnick 1985). Below 4000 m water depth in the western Atlantic basins and in the Cape Basin $\delta^{13}C$-values are as low as about 0.4‰, whereas in the domain of NADW between 2000 and 4000 m in the west and below 2000 m down to the bottom in the Guinea and Angola Basins, these values are higher than 0.9‰, slightly decreasing along the flow path of NADW to the south. As a summary, the asymmetry in bottom water distribution which is present in the modern deep South Atlantic, could be observed in the patterns of carbonate dissolution and $\delta^{13}C$ distribution.

Methods

The sand content of deep-sea carbonates decreases as dissolution progresses (Johnson et al. 1977; Berger et al. 1982; Wu et al. 1991). The reason is that foraminiferal shells are weakenend by dissolution and tend to break down in small fragments. Subsequently, material moves from the coarse fraction into finer fractions. Inspection of other dissolution indices, like whole-test index of planktonic foraminifera, fragmentation index on single species, and abundance ratios of species of contrasting solution susceptibilities, investigated by Hebbeln et al. (1990) and Yasuda et al. (1993) in sediments of the Ontong Java Plateau, showed a good agreement to the sand content records of each study. In addition, Peterson and Prell (1985) established in a depth transect study on the Ninetyeast Ridge, that some 60 % of the whole sand-sized planktonic foraminifers have already broken at the lysocline level, whereas no more than 20 to 30% of carbonate have been lost. Thus, decrease in coarse fraction content is a much more sensitive indicator of increasing dissolution than the carbonate content.

Coarse fraction contents were obtained in all cores through wet-sieving of 10 cm3 bulk sediment over a 63-μm mesh sieve. Samples were then dried and wheighed to determine the wheight per-

cent of sand-size particles. The percentage of sand was calculated using dry bulk density data provided by Peter Müller, Bremen.

The epibenthic taxon *Cibicides wuellerstorfi* was analyzed for stable carbon and oxygen isotopic composition using standard procedures. 2 to 4 specimens of this species were picked from the >250 μm size fraction.The samples were analyzed using a Finnigan MAT 251 micromass-spectrometer coupled with a Finnigan automated carbonate device at the University of Bremen. The carbonate was reacted with orthophosphoric acid at 75°C. The reproducibility of the measurements, as refered to an internal carbonate standard (Solnhofen limestone), is ± 0.07 ‰ and ± 0.05 ‰ (1σ over a one year period) for oxygen and carbon isotopes, respectively. The conversion to the PDB-scale was performed using the international standards NBS 18, 19, and 20.

The age models for all cores are based on graphic correlation of the $\delta^{18}O$ records to the SPECMAP standard record (Imbrie et al. 1984). Prior to this, a composite record was established by using all parameters measured in the sediment sequences from both box corer and gravity corer at each site. The cores were sampled at 5 cm spacing (box cores at 3 cm) throughout their entire length. The sedimentation rates average about 1.5 to 4 cm/ky in the equatorial cores and 1 to 1.5 cm/ky in the cores retrieved from the Walvis Ridge (Tab. 1). Therefore the sample spacing of 5 cm corresponds to a time resolution of about 1250 to 3300 years and 3300 to 5000 years, respectively. Although some sequences extending far back in time, only the last 360,000 years of each record are presented here due to the fact that all cores except GeoB 1112 and 1115 (last 270,000 years) span at least this Late Pleistocene interval (Fig. 3). The raw data, age control points and cross spectral analyses, documenting the precision of the dating, are presented elsewhere (Bickert and Wefer 1996). All data sets here discussed are available via internet from the first author upon request (bickert@allgeo.uni-bremen.de).

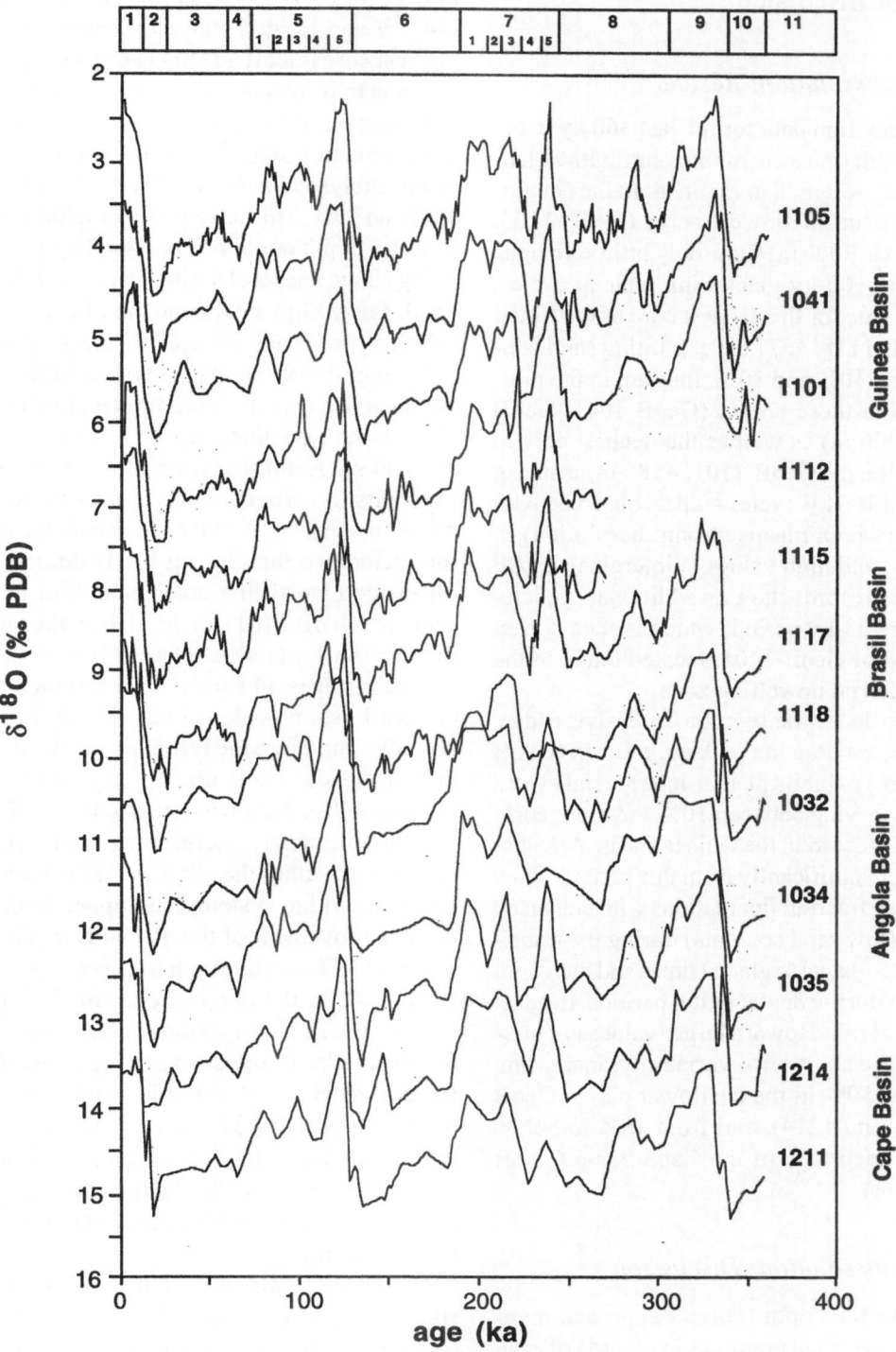

Fig. 3. Plot of oxygen isotope records, measured on the benthic foraminifer taxon *C. wuellerstorfi* , as a function of time in order of the 4 depth transects (each record with an offset of +1‰). The shaded areas mark glacial time intervals.

Results and Discussion

Carbonate dissolution: Results

The coarse fraction data for the last 360 ky in order of the depth transects of the four South Atlantic basins are presented in Figure 4. In the equatorial transects the shallowest cores (GeoB 1105, GeoB 1115, ca. 3000 m) show only little variations related to the 100-ky-glacial-interglacial cycles. The same is true for the deepest core of the Brasil Basin (GeoB 1118, 4671 m) exhibiting sand contents between 10% and 30%. Instead, in the mid-depth cores in these basins (GeoB 1041, GeoB 1117, ca. 4000 m) as well as the deepest core in the Guinea Basin (GeoB 1101, 4588 m) a strong evidence of a 100 ky-cycle could be observed with low values (which means strong dissolution) in glacial times and high values in interglacials. All six equatorial records show an additional cyclicity related to the 23-ky period, which is seen at best in the record of GeoB 1105, located today in the central equatorial upwelling zone.

In the two depth-transects on the Walvis Ridge, only the deepest core in the Angola Basin (GeoB 1035, 4453 m) exhibits glacial-interglacial cycles with values varying between 10% and 50%, comparable to the record of the Guinea Basin. All other records differ significantly from this pattern. They exhibit coarse fraction fluctuations with enhanced dissolution (low sand contents) during the transition from interglacial to glacial times and increased preservation during deglaciation periods. In addition, a general trend toward larger values is superimposed to the short-term variability, increasing from 30% to 50% in the shallower cores (GeoB 1032, 1034, and 1214), and from 10% to below 30% in the deep core of the Cape Basin (GeoB 1211, 4084 m).

Carbonate dissolution: Discussion

According to the depth-transect approach mentioned above, vertical gradients in records of each transect are independent of global shifts or local effects and must be related to changes in the properties of the water masses or in their water column positions. In the Brasil Basin, where the shallower cores (GeoB 1115, 1117) are bathed today in NADW and located above the lysocline and the deepest core (GeoB 1118) is affected by dissolution due to its position in the corrosive LCDW, it is evident that during glacial periods the mid-depth sand contents (GeoB 1117) are close to the values of the sublysocline record (GeoB 1118) between 10% and 20%. This increase in dissolution at 4000 m water depth could be explained by a lysocline rising above this level during glacials. On the other hand, fairly high sand contents of some warmer substages in glacial times ($\delta^{18}O$-stages 3.3, 6.3, 6.5, 8.5) suggest that the glacial lysocline level could not have risen much above 4000 m. In older interglacials the lysocline dropped to its modern level at 4200 m. A similar variability of the vertical gradients in carbonate dissolution could be observed in the records of the Guinea Basin, but with the difference that during the Holocene and in older interglacials the sand contents of the deepest core (GeoB 1101) are as high as the values of the two shallower sites (GeoB 1105, 1041). During these stages, like today, it is assumed that the Guinea Basin was almost exclusively filled with NADW. Therefore the lysocline was located very deep in the water column. This asymmetry in bottom water distribution is equalized in glacial times by the volumetric increase of southern water masses exceeding the sill levels of the surrounding ocean ridge systems. The upper limit of the vertical movement of the glacial lysocline could be obtained from the depth transect in the Angola Basin. While the deepest core of this transect (GeoB 1035, 4453 m) exhibits glacial-interglacial cycles similar to the deepest core of the Guinea Basin (GeoB 1101), the mid-depth core of this profile (GeoB 1034, 3772 m) is unaffected by these major climatic cycles. It is therefore assumed, that during the entire late Pleistocene interval, considered in this study, the lysocline did not rise above about 3800 m.

These observations are consistent with earlier studies on carbonate dissolution in the Atlantic Ocean (Crowley 1983; Curry and Lohmann 1990; Balsam and McCoy 1987; Francois 1990; Verardo and McIntyre 1994), but not with recent findings of Howard and Prell (1994). They conclude from calcite carbonate records of the deep Cape Basin

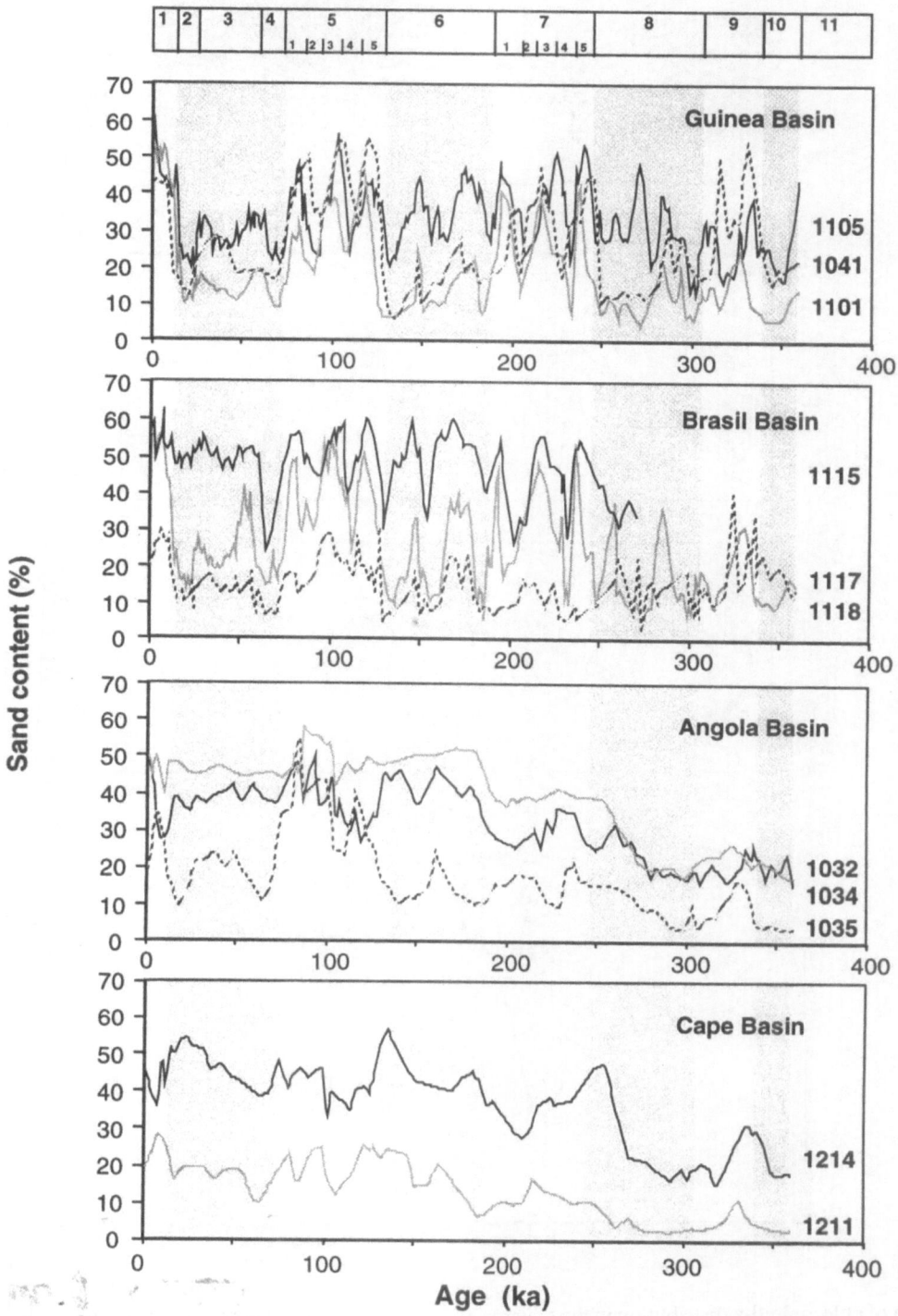

Fig. 4. Plots of sand content records (wheight-%) as a function of time in order of the 4 depth transects. The shaded areas mark glacial time intervals.

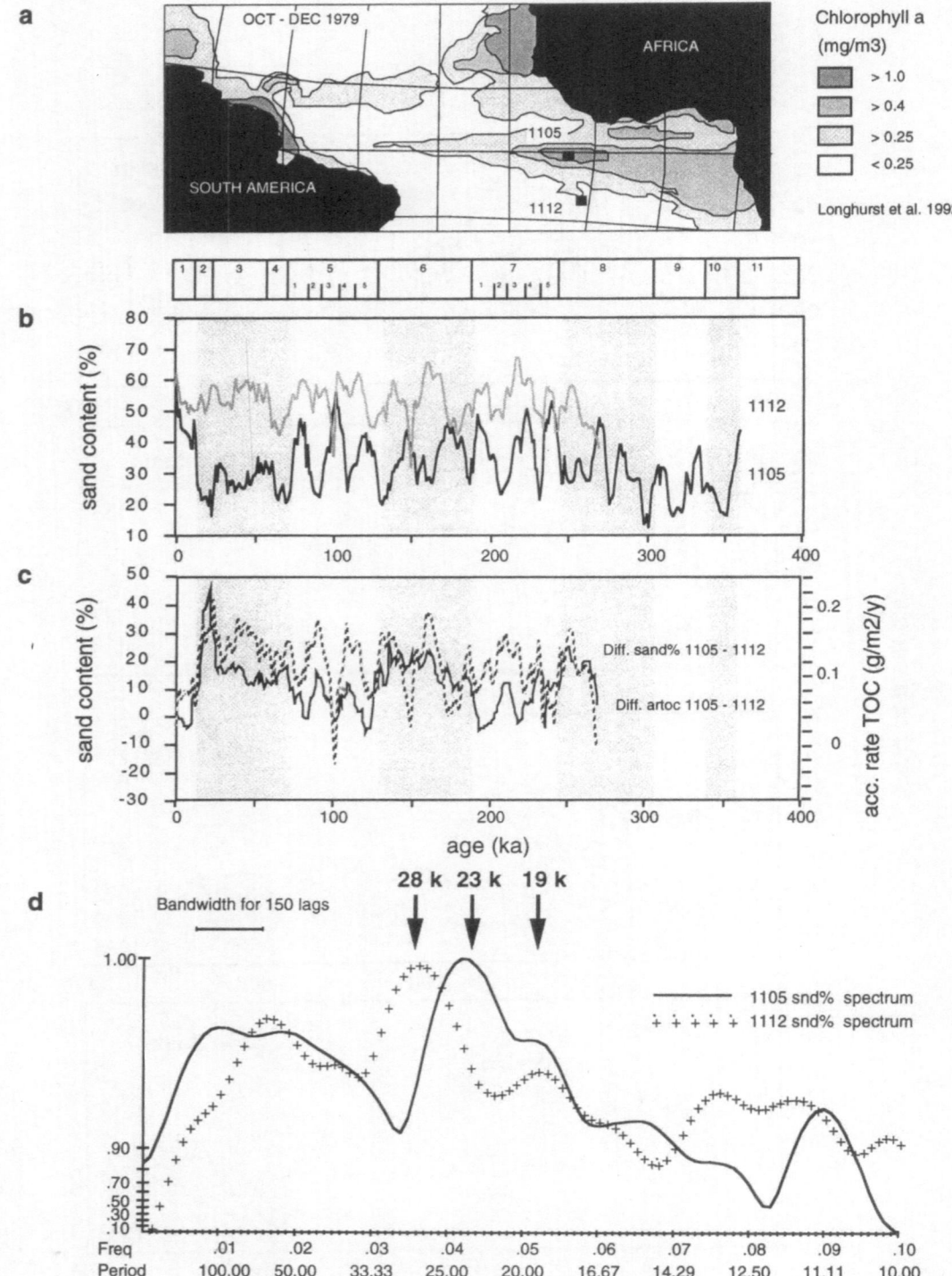

Fig. 5. a) Map of chlorophyll-a distribution in the equatorial Atlantic (Longhurst et al., 1992) and the positions of cores GeoB 1105 and 1112. b) Sand content records of both cores versus time. c) Comparison of the difference of the sand contents to the difference of the accumulation rates of TOC. d) Power spectra of the sand content records of the two cores. Arrows mark the main periods (in ky) observed in these spectra.

(RC 13-228, -229) that the lysocline stood at least about 600 m shallower than today during glacial stages 2 and 4 and about 900 m shallower during early stage 6 and stage 8. But these two cores are located at the slope of the southwest African continental margin and close to the high productive Namibia coastal upwelling area. As indicated by other parameters such as carbon isotopes (discussed below), it is evident that changes in the influx of terrigenous material and changes in the intensity of the upwelling have significant influence on the carbonate accumulation, dilution, and dissolution, making interpretations of changes in deep water chemistry difficult or impossible.

The striking high-frequency cyclicity in the coarse fraction records of the equatorial cores are of special interest. They are best developed in core GeoB 1105, exhibiting amplitudes up to 30%. The sand contents and the TOC accumulation rates of core GeoB 1105, which is located in the central area of the equatorial upwelling zone (both now and in glacial times; see Mix 1989), may be compared to the records of GeoB 1112, which is positioned south of this high productive area (Fig. 5). The record of GeoB 1105 shows a strong 23-ky period in both parameters, while there is only little power in this frequency band in the other core. Instead, the sand record of GeoB 1112 exhibits strong power at the 28-ky period, which is not related to any primary Milankovitch cycle, but which is known from records from the equatorial Atlantic with regard to opal or TOC contents (Verardo and McIntyre 1994). The coherency of the 1105 records with the Earth's precession cycles suggests that the 23-ky cycles of the sand record are productivity-driven dissolution cycles in agreement with the model of Emerson and Bender (1981). The sand contents are related to the productivity cycles such that low sand contents correspond with high accumulation rates in C_{org} and vice versa. A review of the phase angles between the sand records of all equatorial cores to their oxygen isotope records in the precession frequency band gives a lead of 52° (about 3.3 ky) for the dissolution maxima of core 1105 (located in the central equatorial upwelling zone) relative to the maxima

of the ice volume (Tab. 2). This value is consistent with the angles for the maxima in Corg accumulation (65°; Bickert 1992; Schneider et al. this volume) and other paleoproductivity indicators like the maximal seasonality in equatorial surface temperatures (52°; McIntyre et al. 1989) and the maximal zonality of the trade winds (57°; Prell and Kutzbach 1987). The other equatorial records, especially the deeper ones, show phase angles which are closer to the maximum in ice volume. This might be due to the stronger influence of glacial/interglacial changes in water mass corrosiveness with increasing water depth, consistent with recent observations of Verardo and McIntyre (1994).

Due to the very low productivity in the area of the Walvis Ridge transects it is unlikely that such a productivity-related effect would have any influence on the carbonate dissolution in this area. To explain the somewhat different pattern in the coarse fraction records of the cores GeoB 1032, 1034, 1214, and 1211, a sand record of ODP Site

Table 2. Cross-spectral analysis between downcore sand contents and oxygen isotopes in the precession frequency band (150 lags). Positive phase angles indicate a lead of sand contents relative to oxygen isotopes. In brackets are values, which are not significant at the 80% confidence level.

Box/Gravity	Time interval (ky)	Coherency	Phase angle (°)	Phase Error (± °)
Brasil Basin (Equatorial South Atlantic)				
GeoB 1115-4/3	0-270	0.92	7.2	9.4
GeoB 1117-3/2	0-360	0.92	5.4	8.3
GeoB 1118-2/3	0-360	0.68	6.3	19.1
Guinea Basin (Equatorial South Atlantic)				
GeoB 1105-3/4	0-360	0.91	51.8	8.9
GeoB 1041-1/3	0-360	0.93	14.3	7.4
GeoB 1101-4/5	0-360	0.80	37.1	14.0
GeoB 1112-3/4	0-270	(0.51)	(-71.6)	(16.0)

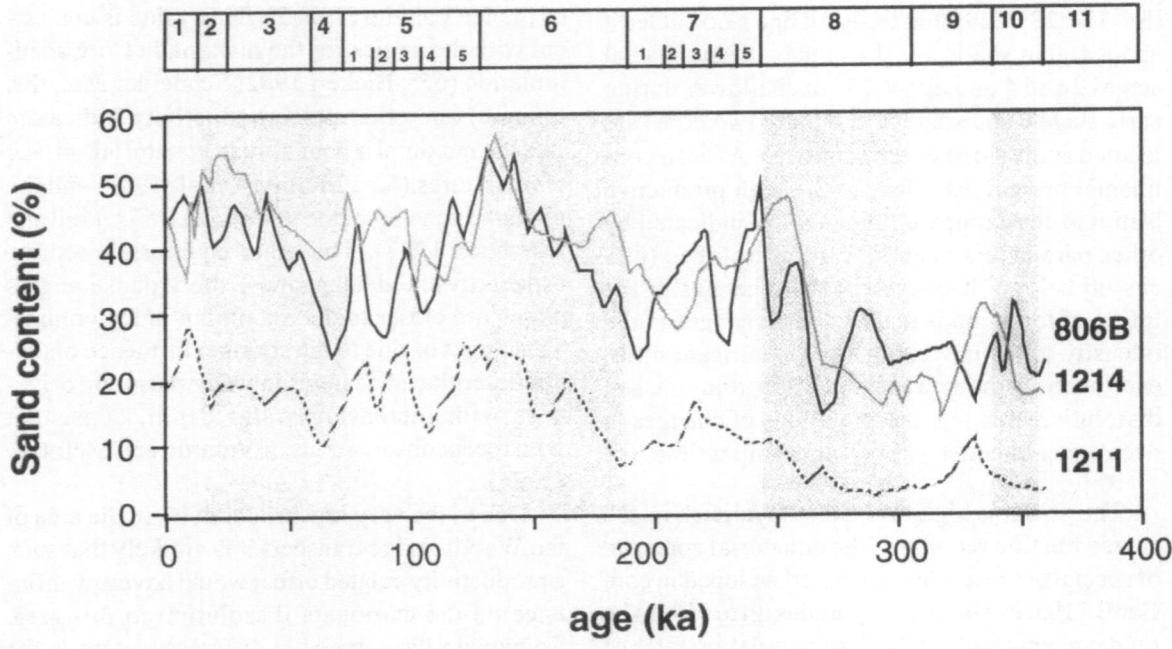

Fig. 6. Sand content records of GeoB 1214 (dotted) and GeoB 1211 (slashed) of the deep Cape Basin compared the sand content record of ODP Site 806B, Ontong Java Plateau, West Pacific (Yasuda et al. 1993).

806B (0°N, 159°E, 2529 m, Yasuda et al. 1993) from the western equatorial Pacific is presented in Fig. 6 together with the one of GeoB 1214, Cape Basin, at about 3220 m water depth. Both records exhibit coarse fraction fluctuations with enhanced dissolution (low sand contents) during the transition from interglacial to glacial times and increased preservation during deglaciation periods, which is found to be typical for the dissolution variability in the Pacific Ocean (Berger 1973; Farrell and Prell 1989; Hebbeln et al. 1990; Wu et al. 1990; Le and Shackleton 1992; Yasuda et. al. 1993). In addition, a general trend toward larger values is superimposed to the short-term variability, increasing from 30% to 50% in the both cores. The general congruency of the sand records of the Cape Basin (GeoB 1214, GeoB 1211), and also of the shallower cores of the Angola Basin (GeoB 1032, 1034) to the Pacific record would suggest that this variability is due to global changes in the oceanic carbon reservoir. Shackleton (1977) proposes that the rebuilding of the continental biomass retrieves

carbon from the ocean and raises therefore the carbonate preservation in the ocean. Broecker (1982) favours the deposition of organic material on the shelves during the rising sea level which leads to a better preservation. Also, a redistribution of nutrients between the intermediate and the deep ocean, as modeled by Boyle (1988), could have lead to a preservation optimum in deglaciation times. Le and Shackleton (1992) state that the debate about the origin of the preservation spike is still going on.

On the other hand, Wu and Berger (1991) showed on a depth transect on the Ontong Java Plateau that carbonate dissolution affects sediments significantly only below 3000 m water depth and that winnowing of fine grained particles is important on the top of the plateau down to 2500 m, which is about the depth of Site 806. Winnowing due to intense bottom currents also explains the finding of Berger and Stax (1994) from the Neogene carbonate stratigraphy of ODP Leg 130 Sites that the loss of carbonate at depth (as derived from differences in accumulation rates) is much greater than suggested by the change in carbonate percentages (calculated under the assumption that carbonate dissolution is the cause of loss), a phenomenon which they refer to as the loss paradox. Interpreting therefore the sand percent records of the Walvis Ridge to be influenced rather by the varying intensity of bottom currents than by carbonate dissolution, it is evident that not the climate variability itselfs, but its rate of change drives the intensity of winnowing (Bickert et al., 1996). Highest sand contents, indicating stronger bottom water activity, occur during deglaciations, and lowest sand contents, indicating weak bottom current intensity, occur during inglaciations. This pattern would go with the four-stage model of deep-water circulation proposed by Imbrie et al. (1992), which postulates a maximum overturn of the global thermohaline circulation at the transition from a glacial to an interglacial state and a minimum overturn at the transition from an interglacial to a glacial state.

The question arises, whether such global variations in the sand content should be recognized also in the equatorial cores of the Atlantic. And indeed, if only the interglacial peaks of the sand records are considered, the same long-term trend toward larger values, referred to as the Mid-Brunhes Cycle (Vincent 1981; Farrell and Prell 1991; Bassinot et al. 1994; Bickert et al. 1996), could be seen, as in the Pacific record. But the global short-term patterns of sand percentage variability, associated with deglaciation and inglaciation, are overprinted by strong dissolution due to the changes in water masses and due to the additional productivity influence at these sites.

The Pacific pattern of sand contents in the Walvis Ridge profiles is in agreement with the data sets of the deep Indian Ocean, given by Bassinot et al. (1994), but it contradicts the interpretation of carbonate preservation data of the Southern Ocean, given by Howard and Prell (1994). Their carbonate records follow a tight glacial-interglacial pattern and are coherent with changes in ice volume and a $\delta^{13}C$ proxy for the percentage of NADW in the deep Atlantic at the Milankovitch periods. But all cores, from which these records are obtained, are located in highly productive ar-

eas close to frontal systems in surface waters, suggesting that carbonate chemistry in this deep circumpolar region is influenced by other factors in addition to changes in the relative contribution of NADW to the region. Consideration of these other factors may be needed to reconcile the data sets.

$\delta^{13}C$: Results

The carbon isotope data for the last 360 ky, measured in the benthic foraminifer species C. wuellerstorfi, are presented in Figure 7 in order of the depth transects of the four South Atlantic basins. All records exhibit a glacial-interglacial-cyclicity with low $\delta^{13}C$ values during ice ages and high values in warm ages. But the magnitude of the late Quaternary variations is different at each core position. The lowest glacial-interglacial amplitudes of 0.6 to 0.8‰ occur in the two cores of the Cape Basin, where the $\delta^{13}C$ values of the deeper core GeoB 1211 are generally about 0.4‰ lower than the values of the shallower core. More than twice are the amplitudes of the equatorial records. GeoB 1105 and 1115, the shallower cores, show glacial to interglacial ranges of 1.2‰, the mid-depth cores GeoB 1041 and 1117 and the deepest core from the Guinea Basin show amplitudes up to 1.6‰. On the other hand, GeoB 1118 as the deepest core in the Brasil Basin, exhibits smaller variations between 0.8 and 1.0‰, but the $\delta^{13}C$ values are generally the lowest ratios of all records.

The comparison of the carbon isotope records of the Brasil Basin cores, where the two shallower sites (GeoB 1115, 1117) are bathed today in NADW, while the deepest core GeoB 1118 is influenced by the LCDW, shows that in glacial times the mid-depth record gets closer to the "LCDW" record and exhibits therefore an about 0.4‰ greater amplitude compared to the shallower record. A similar pattern could be obtained from the $\delta^{13}C$ records from the Guinea Basin, except for the fact that in interglacials the deepest record (GeoB 1101) exhibits as high values as the two shallower sites of this transect, corresponding to the modern hydrographic situation that the Guinea Basin as well as the Angola Basin is filled almost exclusively with NADW.

T.Bickert and G.Wefer

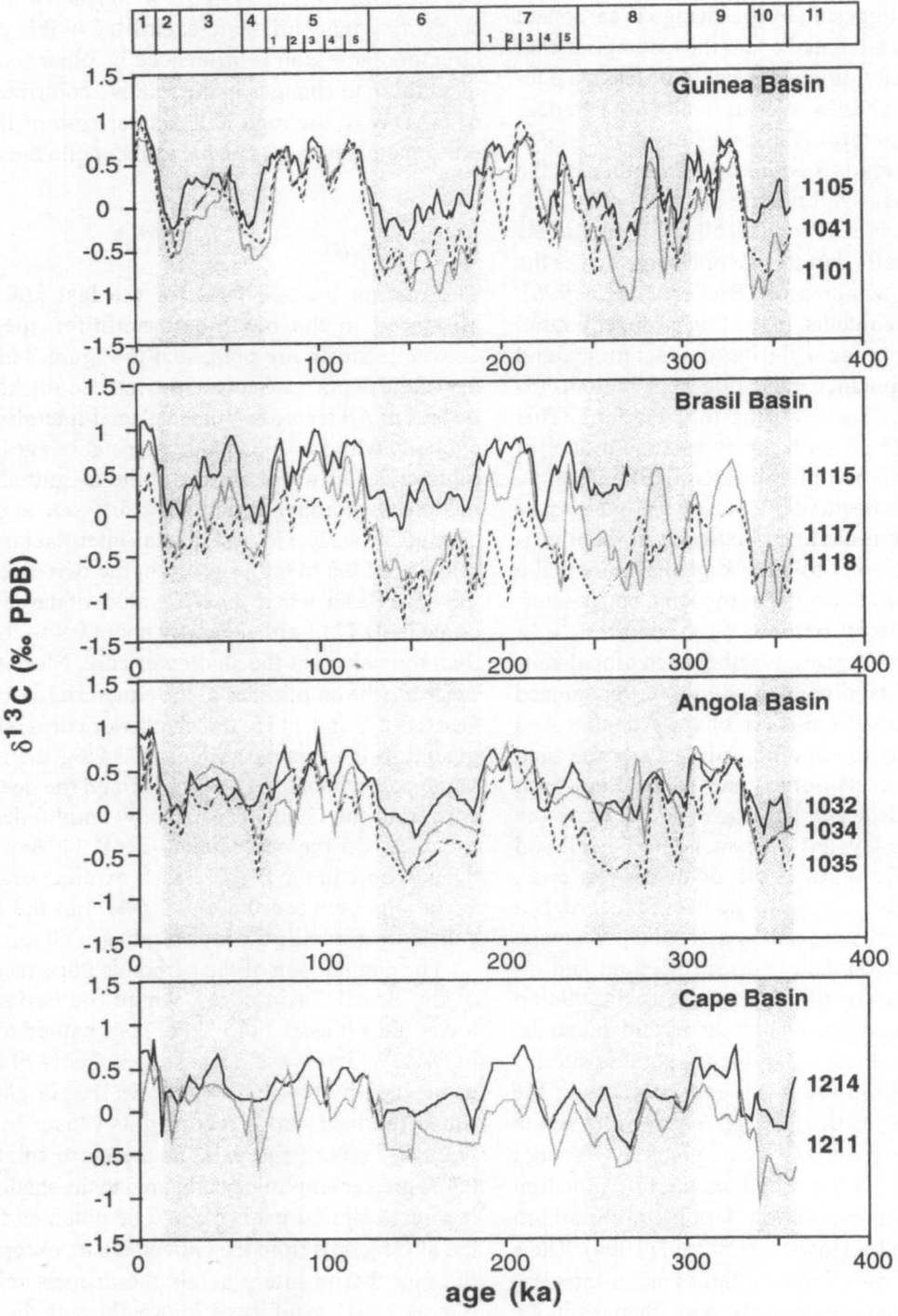

Fig. 7. Plot of carbon isotope records, measured on the benthic foraminifer taxon *C. wuellerstorfi* , as a function of time in order of the 4 depth transects.

$\delta^{13}C$: Discussion

Three factors must be considered for interpreting late Quaternary variations in benthic carbon isotope ratios:

- Global changes in the $\delta^{13}C$ of SCO_2 of seawater (Curry et al. 1988; Duplessy et al. 1988).
- productivity-related porewater effects (Mackensen et al. 1993).
- Changes in the depth distribution or in the properties of deep water masses at a given core position. This includes a possible thermodynamic imprint due to air-sea fractionation in the sourcearea of a deep water mass (Broecker and Maier-Reimer 1992; Charles et al. 1993).

For the global shift of the $^{13}C/^{12}C$ ratio in ΣCO_2 of seawater between the last glacial maximum and today Duplessy et al. (1988) give a value of 0.32‰, Curry et al. (1988) a value of 0.46‰. While the latter value is an average value of cores from the Atlantic, Indic, and Pacific Ocean deeper than 2000 m and wheighed by the volume of each ocean, Duplessy et al. (1988) try to consider an additional variability of $\delta^{13}C$ in intermediate water masses. Therefore the value given by Curry et al. (1988) seems to be a maximal estimation for this global shift.

A comparison of the carbon isotope records of GeoB 1211, located in the deep Cape Basin (4084 m), to the record of ODP 806B from the western equatorial Pacific, which is assumed to monitor exclusively the global variability in $\delta^{13}C$ (Bickert et al., 1993), shows that the interglacial values of the Cape Basin are approximately 0.2‰ higher compared to the Pacific values, while in glacial times the values of both records are nearly the same (Fig. 8). The interglacial $\delta^{13}C$ gradient of LCDW to the Pacific Deep Water is due to the contribution of NADW mixed into the LCDW at the water depth of core GeoB 1211 (Reid 1989; Broecker et al. 1991). The absence of this gradient in glacial times is consistent with the results from carbonate dissolution mentioned above, which reveals that the chemical properties of the glacial bottom water mass in the Cape Basin resembles the deep and bottom water in the Pacific Ocean.

For the estimation of the productivity-related pore water effect two examples are presented in Fig. 9. The first one is again a comparison of the records of GeoB 1105, which is located in the central upwelling area south of the equator, and GeoB 1112, located south of the highly productive area. Although both cores are located at about the same water depth, the glacial $\delta^{13}C$ values of the upwelling core are up to 0.5 ‰ lower than the values of core 1112. The pattern of the difference of carbon isotopes between the two sites is similar to the difference in accumulation rate of TOC, and is characterized by a strong power in the 23-ky period. As in the comparison of the dissolution records of these cores, there is a clear relationship between the deviation of the carbon isotopes to the proxy of productivity. This effect on $\delta^{13}C$ is explained to be caused by the decay of organic matter, reducing $^{13}C/^{12}C$ ratio in the pore water, which influences to some degree the carbon isotopic composition of the C. wuellerstorfi shells (Mackensen et al. 1993). Today this effect can be observed in the high productivity belts related to the Southern Ocean frontal systems, which are characterized by strong seasonality and rapid phytodetritus accumulation.

The second example is obtained from the comparison of the deep Cape Basin core GeoB 1211 (4084 m), located under a very low productive surface layer (Berger 1989), to core RC 13-229 (Oppo and Fairbanks 1990), positioned today close to the edge of the Namibia coastal upwelling area. This core is only in a distance of 500 nautic miles of the position of GeoB 1211 and located at about the same water depth (25°30′S, 11°20′E, 4191 m). The stable carbon isotopes have been determined also on C. wuellerstorfi shells. In Figure 8 it is evident that the $\delta^{13}C$ values of RC 13-229 are generally lower than those of GeoB 1211, only slightly in interglacials, but up to 0.4‰ in glacial times. Because of both the short distance between the cores and the equal water depth, this difference cannot be related to water mass variability. Instead, we attribute this difference again to the influence of the decay of organic matter in the high productive area of Namibia upwelling.

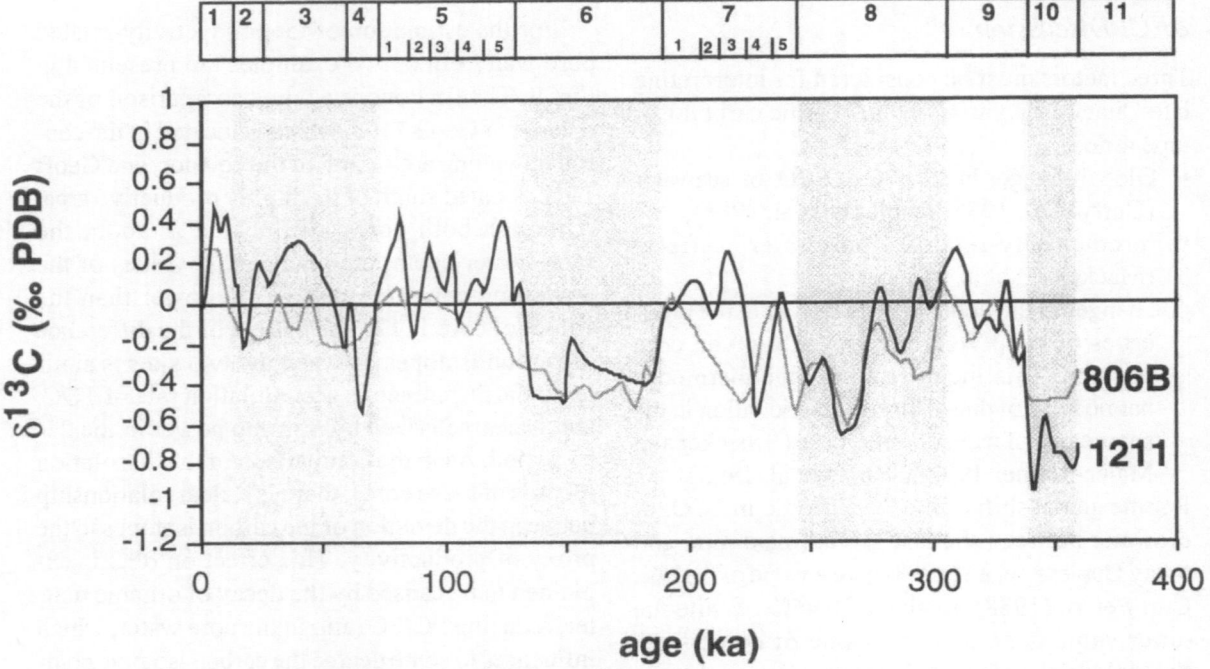

Fig. 8. Carbon isotope record of core GeoB 1211 of the deep Cape Basin compared to the record of ODP Site 806B, Ontong Java Plateau (Bickert e al. 1993). Both records are measured on the benthic foraminifer taxon *C. wuellerstorfi* .

As a consequence, taking this productivity-related effect on carbon isotopes into account would diminish the discrepancy between the results of $\delta^{13}C$ and Cd/Ca from RC 13-229, recently discussed by Oppo and Rosenthal (1994). They pointed out that, while there is a strong glacial to interglacial variability in the carbon isotope record, almost no variability could be observed in the Cd/Ca record. Instead, taking the low productivity record of GeoB 1211 and subtracting the global variability of $\delta^{13}C$ from this record, a similar small variability remains, which would be consistent with the Cd/Ca results of the Cape Basin core. Furthermore, instead of an inversion of the Atlantic-to-Pacific $\delta^{13}C$ gradient in glacial times (i.e. lower values in the Atlantic) as observed by Oppo and Rosenthal, there is then no interoceanic gradient, as mentioned above. Nevertheless, the still existing gradient in glacial nutrients (i.e. lower Cd/Ca ratios in the Atlantic compared to the Pacific) remains unexplained (see, however, Boyle this volume).

Both effects on carbon isotopes - the global variability and the productivity-related deviation - are together not sufficient to explain the high glacial to interglacial amplitudes of up to 1.6‰, which is observed especially in the carbon isotope records of the equatorial cores. An additional effect is called for, namely a change in water mass distribution and/or chemistry.

Starting with the records of the Brasil Basin (Fig. 7), where the shallower cores (GeoB 1115, 1117) are bathed today in NADW and the deepest core (GeoB 1118) is completely covered by LCDW, it is evident that during glacial periods the mid-depth carbon isotope record (GeoB 1117) is close to the values of the deepest record (GeoB 1118) and shows therefore the highest glacial to interglacial amplitudes of 1.2‰ (stages 2/1 transition) and 1.6‰ (stages 10/9 and 8/7 transitions). Estimating the sum of global shift and porewater effect for the last termination to be about 0.8‰, an additional amplitude of 0.4‰ remains, which equals about the today's difference in $\delta^{13}C$ between

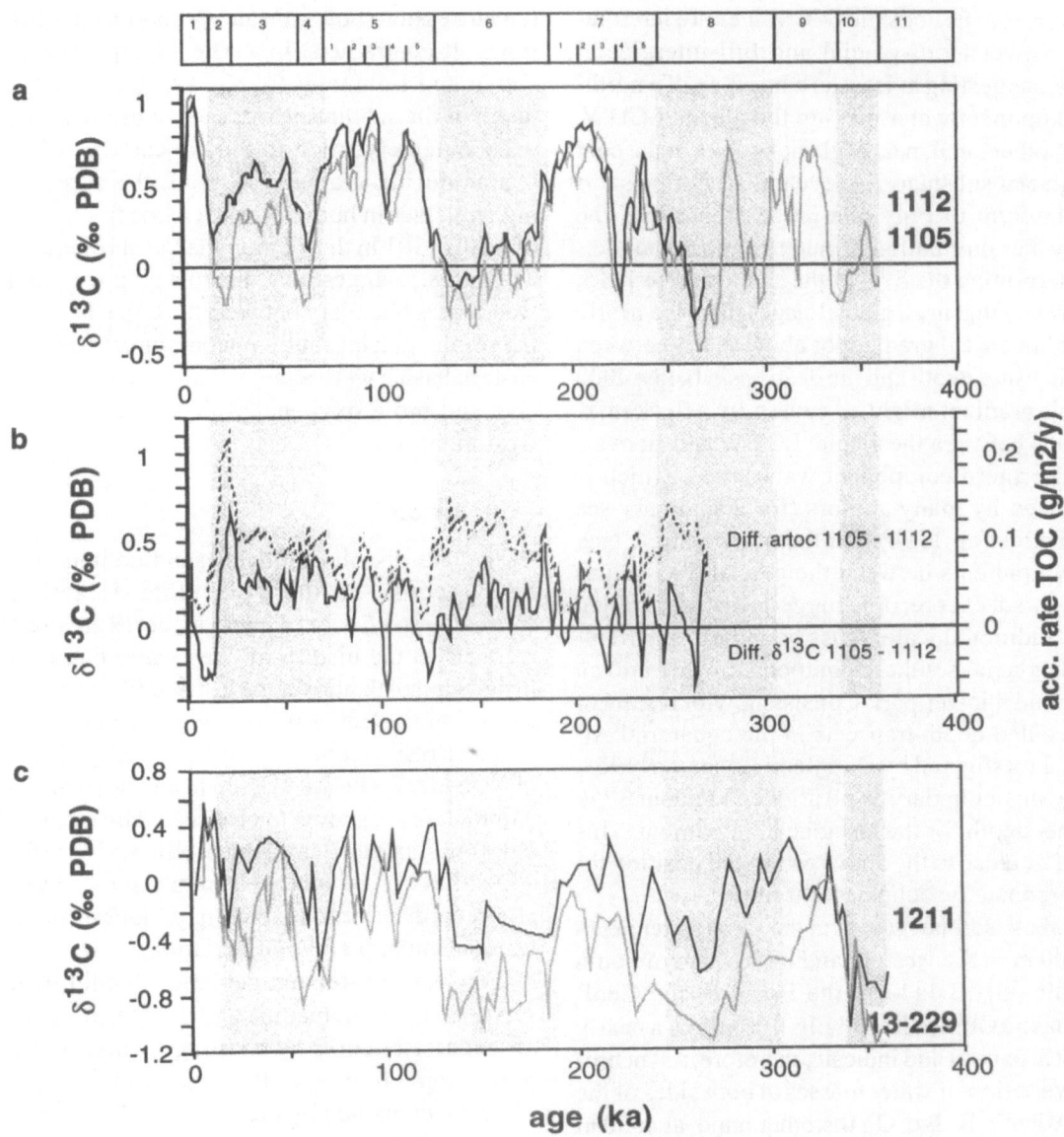

Fig. 9. Upper graph shows the carbon isotope records measured in *C. wuellerstorfi* of cores GeoB 1105 and GeoB 1112. Below a comparison of the difference of the carbon isotopes to the difference of the accumulation rates of TOC. The bottom graph compares the carbon isotope record of GeoB 1211 to the record of RC 13-229 (Oppo and Fairbanks 1990).

NADW (0.9‰) and LCDW (0.5‰) in the western South Atlantic (Kroopnick 1985). It is therefore commonly explained by a glacial reduction of NADW, substituted by a southern component water mass, as reported by many authors (Boyle and Keigwin 1987; Labeyrie et al. 1987; Oppo and Fairbanks 1987, 1990; Curry et al. 1988; Duplessy et al. 1988; Raymo et al. 1990, Sarnthein et al. 1994). This change in deep water circulation is also obvious for older glacials such as stages 4, 6, 8, and 10, except for a few warmer periods in glacial times, such as substages 3.3, 8.3, and 8.5. In

these warmer periods the $\delta^{13}C$ values are intermediate between full-glacial and full-interglacial
values, suggesting at least an admixture of a northern component water mass to the glacial LCDW.
On the other hand, nearly glacial values in the cold
interglacial substage 7.4 reveal that, for the short
time interval of only one precession cycle, the
deep-water circulation turned to a glacial mode.

Interpreting the $\delta^{13}C$ of the shallow core 1115,
it is evident that in all glacial stages there is a nearly
constant vertical gradient of about 0.4‰ between
3000 m water depth and the deep basin below 4000
m. This gradient might be caused by a thick mixing zone between the glacial LCDW and an overlying northern component water mass, which is
postulated by many authors (for a summary see
Sarnthein et al. 1994). On the other hand, a lack
of any gradients between the glacial $\delta^{13}C$ values
of the two deeper records suggests that there might
be an additional water mass boundary, subdivid-
ing the glacial southern component water into an
upper and a lower part. Consistent with results of
the detailed depth transects in the equatorial Atlantic, investigated by Curry and Lohmann (1983,
1990), this boundary is positioned at about 3750
m water depth for the last glacial maximum. This
would be close to the above estimated position for
the carbonate lysocline at that time.

Carbon isotope records at the same water depth
of 4000 m in the western and in the eastern South
Atlantic (GeoB 1117 in the Brasil Basin, GeoB
1041 in the Guinea Basin; Fig. 10a) show a nearly
identical pattern and indicate, therefore, a synchronous variation of water masses at both sides of the
Mid Atlantic Ridge. On the other hand, at 4600 m
water depth (GeoB 1118 in the Brasil Basin, GeoB
1101 in the Guinea Basin; Fig. 10b) the interglacial values are approximately 0.6‰ higher in the
eastern basin compared to the western basin in
agreement with the modern hydrographic asymmetry in bottom water distribution. The nearly
identical values in glacial times suggest an equali-
zation of bottom water distribution as the thickening of southern water masses allows spilling
over the sill levels of the surrounding ocean ridge
systems. This result is consistent with the pattern
observed in carbonate dissolution records (Fig.
10c), which reveals the same nivellement of west-

ern and eastern bottom water properties in glacial
times. It contradicts the earlier interpretation of
Curry and Lohmann (1983, 1990) that the sill
depth of the submarine ridge systems reduces the
deep water advection to the the eastern Atlantic
basins during glacials. Instead, the somewhat
higher values in both $\delta^{13}C$ and coarse fraction data
of GeoB 1101 in the warmer glacial substages 3.3,
8.3, and 8.5 suggest that due to a slightly sinking
water mass boundary between the upper and lower
part of the glacial southern component water, the
eastern basins were again filled with a less corrosive and more oxygenated deep water than the
western basins.

Conclusions

Carbonate dissolution data (sand contents) and
$\delta^{13}C$ records of the epibenthic foraminifer
C. wuellerstorfi from 12 gravity cores are used to
reconstruct the history of deep water circulation
in the South Atlantic for the last 360,000 years. The
cores were selected from depth-sections in four
basins (Brasil-, Guinea-, Angola- and Cape Basins)
in water depths between 2900 m and 4600 m. Both
parameters are shown to be affected by mean global shifts as well as local productivity effects
which have to be removed from the paleo-property records prior to extracting variations in deep
water chemistry and/or circulation.

Productivity-related deviations could be observed in the sand records and the $\delta^{13}C$ records of
all equatorial cores as a strong response to the
precessional forcing of the equatorial upwelling.
They are explained by the effects of organic matter degradation on pore water chemistry. Global
patterns in both parameters are at best observed in
the core transect of the Walvis Ridge into the Cape
Basin. From a comparison to a deep western Pacific core (ODP 806B) it is evident that the nutrient-enriched but oxygen-depleted deep water in
the Cape Basin resembles the glacial Pacific Deep
Water. This is also true for the older glacial stages
4, 6, 8 and 10.

As a result of the reduction of NADW during
the last glacial maximum, the Southern Component Water was higher in the water column and
extended farther north than it does today. In recon-

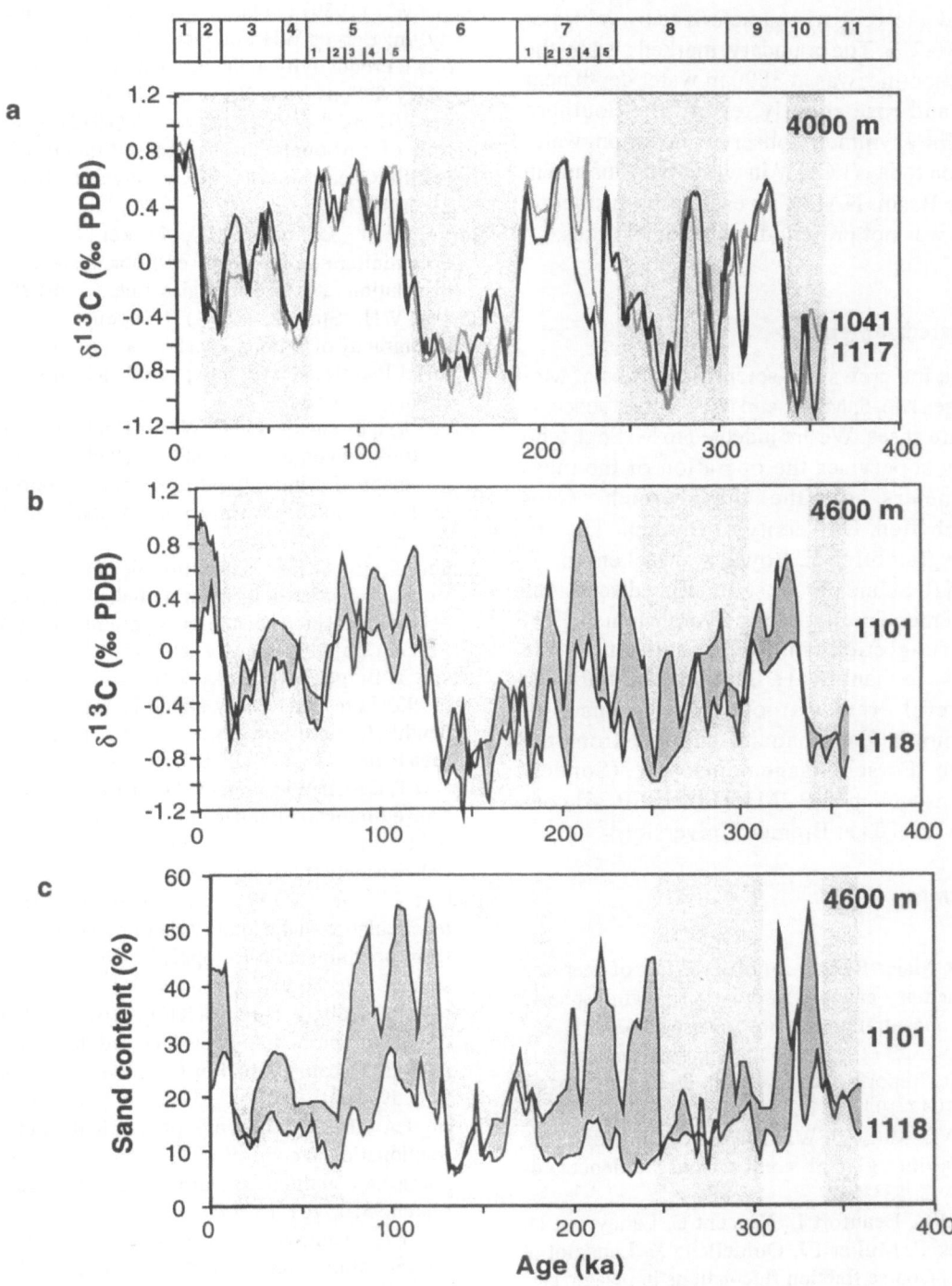

Fig. 10. Comparison of carbon isotope records of cores in the Brasil Basin and the Guinea Basin in 4000 m (upper graph) and 4600 m (middle) water depth. The lower graph compares the sand content records at 4600 m water depth.

structing this water mass it can be divided into an upper part with $\delta^{13}C$ values between 0.2‰ and 0.7‰ and a lower part characterized by values of -0.2‰ to 0.2‰. The boundary, marked also by the calcite lysocline, was at 3800 m water depth near equator and rose slightly toward the Southern Ocean. The asymmetry observed in bottom water circulation today (LCDW in western basins and in the Cape Basin, NADW in eastern basins below 4000 m) was not present during glacial times.

Acknowledgements

We thank the crews and scientific parties of Meteor cruises M6-6, M9-4, and M12-1 for a successful venture at sea. We are indebted to M. Segl, who carefully supervises the operation of the mass spectrometers of the Fachbereich Geowissenschaften, University of Bremen. The authors are grateful to E. Boyle, A. Mackensen, A. Mix, and R. Schneider for stimulating discussions at the Bremen South Atlantic Symposium in 1994 and for suggestions improving this study. We would like to thank W. H. Berger and L. Labeyrie for careful reviews of this manuscript. We acknowledge financial support from the Deutsche Forschungsgemeinschaft (Sonderforschungsbereich SFB 261). This is SFB 261 contribution no. 103 at Bremen University.

References

Archer D, Maier-Reimer E (1994) Effect of deep-sea sedimentary calcite preservation on atmospheric CO_2 concentration. Nature 367: 260-263

Arrhenius GOS (1952) Sediment cores from the east Pacific. Reports of the Swedish Deep Sea Expedition, 1947-1948 5: 1-202

Balsam WL, McCoy FWJ (1987) Atlantic sediments: Glacial/interglacial comparisons. Paleoceanography 2: 531-542

Bassinot FC, Beaufort L, Vincent E, Labeyrie LD, Rostek F, Müller PJ, Quidelleur X, Lancelot Y (1994) Coarse fraction fluctuations in pelagic carbonate sediments from the tropical Indian Ocean: A 1500-kyr record of carbonate dissolution. Paleoceanography 9: 579-600

Berger WH (1973) Deep-sea carbonates: Pleistocene

dissolution cycles. Journal of Foraminiferal Research 3: 187-195

Berger WH (1989) Global maps of ocean productivity. In: Berger WH Smetacek VS, and Wefer G. (eds), Productivity in the oceans: Present and Past. Wiley & Sons, New York, pp 429-455

Berger WH, Adelseck CG, Mayer LA (1976) Distribution of carbonate in surface sediments of the Pacific Ocean. Journal of Geophysical Research 81: 2617-2627

Berger WH, Bonneau MC, Parker FL (1982) Foraminifera on the deep-sea floor: lysocline and dissolution rate. Oceanologica Acta 5: 249-258

Berger WH, Stax R (1994) Neogene carbonate stratigraphy of Ontong Java Plateau (western equatorial Pacific): three unexpected findings. Terra Nova 6: 520-534

Berger WH, Fischer K, Lai C, Wu G (1987) Ocean productivity and organic carbon flux. (Part I: Overview and maps of primary productivity and export production). San Diego: University of California, SIO Reference 87-30: 67

Bickert T (1992) Rekonstruktion der spätquartären Bodenwasserzirkulation im östlichen Südatlantik über stabile Isotope benthischer Foraminiferen. Ber. FB Geo Univ. Bremen 27: 205 pp

Bickert T, Berger WH, Burke S, Schmidt H, Wefer G (1993) Late Quaternary stable isotope record of benthic foraminifera at sites 805 and 806, Ontong Java Plateau. Proc. ODP, Sci. Results 130: 411-420

Bickert T, Berger WH, Wefer G (1996) The deep western equatorial Pacific in Quaternary times: Results from Leg 130 (Ontong Java Plateau). Paleoceanography, subm.

Bickert T, Wefer G (1996) Late Quaternary deep water circulation in the South Atlantic: Reconstruction from benthic stable isotopes. Paleoceanography, subm.

Biscaye PE, Kolla V, Turekian KK (1976) Distribution of calcium carbonate in surface sediments of the Atlantic Ocean. Journal of Geophysical Research 81: 2595-2603

Boyle EA (1988) The role of vertical chemical fractionation in controlling late Quartenary atmospheric carbon dioxide. Journal of Geophysical Research 93: 15701-15714

Boyle EA, Keigwin L (1987) North Atlantic thermohaline circulation during the past 20,000 years linked to high-latitude surface temperature. Nature 330: 35-40

Broecker W, Blanton S, Smethie WM, Ostlund G (1991) Radiocarbon decay and oxygen utilization in the deep Atlantic Ocean. Global Biogeochemical Cycles 5: 87-117

Broecker WS, Maier-Reimer E (1992) The influence of air and sea exchange on the carbon isotope distribution in the sea. Global Biogeochemical Cycles 6: 315-320

Broecker WS, Peng TH (1982) Tracers in the Sea. Lamont Doherty Geol. Obs. Publication, Columbia University, New York, 689 pp

Broecker WS, Peng TH (1987) The role of CaCO₃ compensation in the glacial to interglacial atmospheric CO₂ change. Global Biogeochemical Cycles 1: 15-29

Broecker WS, Peng TH (1989) The cause of the glacial to interglacial atmospheric CO₂ change: A polar alkalinity hypothesis. Global Biogeochemical Cycles 3: 215-239

Charles CD, Wright JD,Fairbanks RG (1993) Thermodynamic influences on the marine carbon isotope record. Paleoceanography 8: 691-697

Connary SD, Ewing M (1974) Penetration of Antarctic Bottom Water from the Cape Basin into the Angola Basin. Journal of Geophysical Research 79: 463-469

Crowley TJ (1983) Depth-dependent carbonate dissolution changes in the eastern North Atlantic during the last 170,000 years. Marine Geology 54: 25-31

Crowley TJ (1985) Late Quaternary carbonate dissolution changes in the North Atlantic and Atlantic/Pacific comparisons. In: Sundquist E, Broecker W (eds), The carbon cycle and atmospheric CO₂: Natural variations Archean to Present. AGU, Washington D. C., pp

Curry WB, Duplessy JC, Labeyrie LD, Shackleton NJ (1988) Changes in the distribution of δ¹³C of deep water TCO₂ between the last glaciation and the Holocene. Paleocanography 3: 317-341

Curry WB, Lohmann GP (1983) Reduced advection into Atlantic Ocean deep eastern basins during last glaciation maximum. Nature 306: 577-580

Curry WB, Lohmann GP (1990) Reconstructing past particle fluxes in the tropical Atlantic Ocean. Paleoceanography 5: 487-506

Duplessy JC, Shackleton NJ, Fairbanks RG, Labeyrie L, Oppo D, Kallel N (1988) Deepwater source variations during the last climatic cycle and their impact on the global deepwater circulation. Paleoceanography 3: 343-360

Emerson S, Bender M (1981) Carbon fluxes at the sediment- water interface of the deep-sea: calcium carbonate preservation. Journal of Marine Research 39: 139-162

Farrell JW, Prell WL (1989) Climatic change and CaCO₃ preservation: an 800,000 year bathymetric reconstruction from the central equatorial Pacific Ocean. Paleoceanography 4: 447-466

Hebbeln D, Wefer G, Berger WH (1990) Pleistocene dissolution fluctuations from apparent depth of deposition in core ERDC-127P, West-Equatorial Pacific. Marine Geology 92: 165-176

Howard WR, Prell WL (1994) Late Quaternary CaCO₃ production and preservation in the Southern Ocean: Implications for oceanic and atmospheric carbon cycling. Paleoceanography 9: 453- 482

Imbrie J, Boyle EA, Clemens SC, Duffy A, Howard WR, Kukla G, Kutzbach J, Martinson DG, McIntyre A, Mix AC, Molfino B, Morley JJ, Peterson LC, Pisias NG, Prell WL, Raymo ME, Shackleton NJ, Toggweiler JR (1992) On the structure and origin of major glaciation cycles, 1, Linear responses to Milankovitch forcing. Paleoceanography 7: 701-738

Imbrie J, Boyle EA, Clemens SC, Duffy A, Howard WR, Kukla G, Kutzbach J, Martinson DG, McIntyre A, Mix AC, Molfino B, Morley JJ, Peterson LC, Pisias NG, Prell WL, Raymo ME, Shackleton NJ, Toggweiler JR (1993) On the structure and origin of major glaciation cycles. 2. The 100,000-year cycle. Paleoceanography 8: 699-735

Imbrie J, Hays JD, Martinson DG, McIntyre A, Mix AC, Morley JJ, Pisias NG, Prell WL,Shackleton NJ (1984) The orbital theory of Pleistocene climate: support from a revised chronology of the marine d¹⁸O record. In: Berger A, Imbrie J, Hays J, Kukla G,Saltzman B (eds), Milankovitch and climate, Part I. D. Reidel, Dordrecht, pp 269-305

Jahnke RA, Craven DB, Gaillard JF (1994) The influence of organic matter diagenesis on CaCO₃ dissolution at the deep-sea floor. Geochimica et Cosmochimica Acta 58: 2799-2809

Johnson TC, Hamilton EL,Berger WH (1977) Physical properties of calcareous ooze: Control by dissolution at depth. Marine Geology 24: 259-277

Keir RS, Berger WH (1983) Atmospheric CO₂ content in the last 120,000 years: the phosphate-extraction model. J Geophys Res 88: 6027-6038

Keir RS (1990) Reconstructing the ocean carbon system variation during the last 150,000 years according the Antarctic nutrient hypothesis. Paleoceanography 5: 253-276

Kolla V, Bé AWH,Biscaye PE (1976) Calcium carbonate distribution in the surface sediments of the Indian Ocean. J Geophys Res 81: 2605-2616

Kroopnick P (1985) The distribution of ¹³C of ΣCO₂ in

the world oceans. Deep-Sea Research 32: 57-84

Labeyrie LD, Duplessy JC, Blanc PL (1987) Variations in mode of formation and temperature of oceanic deep waters over the past 125,000 years. Nature 327: 477-482

Le J, Shackleton NJ (1992) Carbonate dissolution fluctuations in the western equatorial Pacific during the Late Quaternary. Paleoceanography 7: 21-42

Mackensen A, Hubberten HW, Bickert T, Fischer G, Fütterer DK (1993) $\delta^{13}C$ in benthic foraminiferal tests of *Fontbotia wuellerstorfi* (SCHWAGER) relative to $\delta^{13}C$ of dissolved inorganic carbon in Southern Ocean deep water: implications for glacial ocean circulation models. Paleoceanography 8: 587-610

Mix AC (1989) Pleistocene paleoproductivity: evidence from organic carbon and foraminiferal species. In: Berger WH, Smetacek VS, Wefer G (eds), Productivity of the ocean: present and past. J. Wiley & Sons, Chichester, pp 313-340

Oppo DW, Fairbanks RG (1987) Variability in the deep and intermediate water circulation of the Atlantic Ocean during the past 25,00 years: Northern Hemisphere modulation of the Southern Ocean. Earth and Planetary Science Letters 86: 1-15

Oppo DW, Fairbanks RG (1990) Atlantic Ocean thermohaline circulation of the last 150.000 years: relationship to climate and atmospheric CO_2. Paleoceanography 5: 277-288

Oppo DW, Rosenthal Y (1994) Cd/Ca changes in a deep Cape Basin core over the past 730,000 years: Response of circumpolar deepwater variability to northern hemisphere ice sheet melting? Paleoceanography 9: 661-676

Östlund HG, Craig C, Broecker WS, Spencer D (1987) GEOSECS Atlantic, Pacific, and Indian Ocean Expedition, Shorebased data and graphics (GEOSECS Atlas Ser. vol. 7. U.S. Government Printing Office, Washington, 200 pp

Raymo ME, Ruddiman WF, Shackleton NJ, Oppo DW (1990) Evolution of Atlantic-Pacific $\delta^{13}C$ gradients over the last 2.5 m.y. Earth and Planetary Science Letters 97: 353-368

Reid JL (1989) On the total geostrophic circulation of the South Atlantic Ocean: Flow patterns, tracers, and transports. Progress in Oceanography 23: 149-244.

Sarnthein M, Winn K, Jung SJA, Duplessy JC, Labeyrie L, Erlenkeuser H, Ganssen G (1994) Changes in east Atlantic deepwater circulation over the last 30,000 years: Eight time slice reconstructions. Paleoceanography 9: 209-268

Shackleton NJ (1977) Tropical rainforest history and the equatorial Pacific carbonate dissolution cycles.

In: Anderson NR, Malahoff A (eds), Fate in fossil fuel CO_2 in the oceans. Plenum, New York, pp 401-427

Shannon LV, Chapman P (1991) Evidence of Antarctic Bottom Water in the Angola Basin at 32°S. Deep-Sea Research 38: 1299-1304

Sundquist ET, Broecker WS (1985) The carbon cycle and atmospheric CO_2: Natural variations Archean to Present. AGU, Washington D. C., 627 pp

Thunell RC (1982) Carbonate dissolution and abyssal hydrography in the Atlantic Ocean. Marine Geology 47: 165-180

Van Bennekom AJ, Berger GW (1984) Hydrography and silica budget of the Angola Basin. Netherlands Journal of Sea Research 17: 149-200

Verardo DJ, McIntyre A (1994) Production and destruction: Control of biogenous sedimentation in the tropical Atlantic 0-300,000 years B.P. Paleoceanography 9: 63-86

Vincent E (1981) Carbonate stratigraphy of Hess Rise, Central North Pacific and paleoceanographic implications. DSDP, Initial Reports 62: 571-606

Volat JL, Pastouret L, Vergnaud-Grazzini C (1980) Dissolution and carbonate fluctuations in Pleistocene deep-sea cores: a review. Marine Geology 34: 1-28

Warren BA, Speer KG (1991) Deep circulation in the eastern South Atlantic Ocean. Deep-Sea Research 38, Suppl 1:281- 322

Wefer G, Bleil U, Müller PJ, Schulz HD, Fahrtteilnehmer (1988) Bericht über die Meteor-Fahrt M6/6, Libreville - Las Palmas, 18.2. - 23.3.1988. Ber. FB Geow. Univ. Bremen 3: 97 pp

Wefer G, Bleil U, Schulz HD, Fahrtteilnehmer (1989) Bericht über die Meteor-Fahrt M9/4, Dakar - Santa Cruz, 19.2. - 16.3.1989. Ber. FB Geow. Univ. Bremen 7: 103 pp

Wefer G, Fahrtteilnehmer (1990) Bericht über die Meteor- Fahrt M12/1, Kapstadt - Funchal, 13.3. - 14.4.1990. Ber FB Geow Univ Bremen 11: 66 pp

Wu G, Berger WH (1991) Pleistocene $\delta^{18}O$ record from Ontong Java Plateau: effects of winnowing and dissolution. Marine Geology 96: 193-209

Wu G, Herguera JC, Berger WH (1990) Differential dissolution: Modification of late Pleistocene oxygen isotope records in the western Equatorial Pacific. Paleoceanography 5: 581-594

Yasuda M, Berger WH, Wu G, Burke S, Schmidt H (1993) Foraminiferal preservation record for the last million years: Site 805, Ontong Java Plateau. Proc ODP, Sci Results 130: 491-508

Clay Mineral Fluctuations in Late Quaternary Sediments of the Southeastern South Atlantic: Implications for Past Changes of Deep Water Advection

B. Diekmann[1], R. Petschick[1,2], F.X. Gingele[1],
D.K. Fütterer[1], A. Abelmann[1], U. Brathauer[1],
R. Gersonde[1], A. Mackensen[1]

1) *Alfred Wegener Institute for Polar and Marine Research,
P.O. Box 12 01 61, 27515 Bremerhaven, GERMANY*
2) *Present address: Geologisch-Palaeontologisches Institut,
Universität Frankfurt, Senckenberganlage 32-34,
60054 Frankfurt/Main, GERMANY*

Abstract: Downcore clay mineral fluctuations in Late Quaternary sediment cores from the southeastern South Atlantic and adjoining Southern Ocean are of low amplitude. North of the Antarctic Circumpolar Current/Weddell Gyre boundary, small-scale variations, particularly of clay mineral ratios, essentially monitor cyclic changes of deep water advection in response to climatic oscillations.

Kaolinite and chlorite are of most reliable palaeoceanographic significance. Chlorite originates from southern high-latitudes (Antarctic Peninsula, Patagonia) and is transported and distributed by northward advection of southern-source deep water. Northern-source deep water carries pedogenic kaolinite from the tropical regions to the Southern Ocean. The opposite advection of kaolinite and chlorite by both deep water masses is displayed by a latitudinal zonation of kaolinite/chlorite ratios with higher values to the north. Downcore kaolinite/chlorite variations exhibit raised ratios during warm climatic stages compared to glacial periods. Cross spectral analyses reveal that kaolinite/chlorite ratios are coherent and in phase with global ice volume in the 41-kyr and 100-kyr periods of obliquity and eccentricity.

By relating kaolinite/chlorite ratios in modern deep-sea sediments to the present-day deep water distribution, a model of past deep water advection may be inferred from downcore variations of kaolinite/chlorite ratios. Thus, northern-source water reaches not farther south than 40°S during glacial maxima. During interglacial optima northern-source deep water is injected into the Southern Ocean to latitudes around 55°S.

Introduction

The South Atlantic plays a profound role in global thermohaline deep water circulation as it forms the pathway for the major deep water masses, the North Atlantic Deep Water (NADW), the Circumpolar Deep Water (CPDW), and the Antarctic Bottom Water (AABW). These water masses are fed into the Southern Ocean and distributed world-wide via the Antarctic Circumpolar Current (ACC), which passes through the South Atlantic sector of the Southern Ocean.

Changes of past deep water circulation in the South Atlantic are commonly inferred from the geochemical record of benthic foraminifera from deep-sea sediment cores, mainly from downcore variations of $\delta^{13}C$ (Curry et al. 1988, Duplessy et al. 1988, Oppo and Fairbanks 1990, Charles and

From WEFER G, BERGER WH, SIEDLER G, WEBB DJ (eds), 1996, *The South Atlantic: Present and Past Circulation*. Springer-Verlag Berlin Heidelberg, pp 621-644

Fairbanks 1992, Hodell 1993, Mackensen et al. 1994, Bickert and Wefer this volume) and Cd/Ca ratios (Boyle 1988, 1994, Oppo and Rosenthal 1994, Lea 1995). Both proxies are correlated with the nutrient inventory of distinct water masses. However, the reliability of $\delta^{13}C$ and Cd/Ca records as tracers of past deep water circulation has to be questioned in areas underlying productive surface layers, especially in the Southern Ocean (Broecker 1993, Mackensen et al. 1993, Boyle 1994, Oppo and Rosenthal 1994). Some authors consider benthic foraminiferal assemblages (Mackensen et al. 1994, Schmiedl 1995) and carbonate preservation indices/carbonate concentrations as proxies of palaeo deep water circulation (Curry and Lohmann 1990, Howard and Prell 1994, Bickert and Wefer this volume).

There is a general consensus that the weakening of NADW production and flow during glacial stages resulted in a shallowing and less southward extension of NADW compared to the interglacial and modern situation. Reduced NADW injection into the Southern Ocean was accompanied by suppressed generation of southern-source bottom water (Kellogg 1987, Mackensen et al. 1994, Schmiedl 1995). In the Weddell Sea (Grünig 1991, Pudsey 1992, Brehme 1992) and the Vema Channel (Ledbetter 1984, Ledbetter 1986a) grain size data confirm a more vigorous outflow of AABW during interglacials. In the southern Brazil Basin (Massé et al. 1994) and the Argentine Basin (Ledbetter, 1986b) strongest AABW current velocities appear at interglacial-glacial boundaries.

In this study we use downcore clay mineral fluctuations as tracers of past deep water advection in the southeastern South Atlantic and adjoining Southern Ocean. We attempt to reconstruct the variable palaeo-latitudinal extensions of northern and southern-source deep water. Since established proxies of deep water circulation partly contradict each other and mainly depend on the availability of calcareous sediment components, a great advantage of clay mineral investigations is their feasibility even in regions with poor carbonate preservation, particularly in the highly productive Southern Ocean with its opal-rich sediments.

Background

Clay mineral assemblages in the South Atlantic are mainly of terrigenous origin and reflect the modern climatically controlled pattern of weathering on the adjacent continents (Biscaye 1965, Petschick et al. in press). Primary clay mineral input is driven by the large river systems of Africa (Pastouret et al. 1978, Eisma et al. 1978, van der Gaast and Jansen 1984, Bremner and Willis 1993) and South America (Gibbs 1967, Barretto et al. 1975, Tintelnot et al. 1994), NE- and SE-trade winds from African deserts (Chester et al. 1972, Behairy et al. 1975, Windom 1975, Gingele 1992, in press, Bremner and Willis 1993), as well as glaciogenic supply from the continental margin of Antarctica (Ehrmann et al. 1992, Grobe and Mackensen 1992, Petschick et al. in press).

In the South Atlantic clay mineral transport and distribution to the deep sea is most effectively controlled by deep water advection (Petschick et al. in press). The determinant control of ocean currents on the clay mineral composition of South Atlantic deep-sea sediments has already been pointed out by Robert (1980) and Robert and Maillot (1983), which they attributed to enhanced deep water circulation since the Early Pliocene. In older sediments the clay mineral signal is mainly linked to the structural evolution of the Atlantic Ocean and bordering continents (Robert 1980, 1987).

So far, only a few attempts have been undertaken to use clay mineral assemblages as a proxy of Late Quaternary changes in deep water advection. Investigations were focused on the Vema Channel and Rio Grande Rise in the western South Atlantic (Chamley 1975, Melguen et al. 1978, Jones 1984). Chamley (1975) concluded that the strong latitudinal and reciprocal distribution patterns of kaolinite and chlorite, first described by Biscaye (1965), should characterize advection of NADW to the south and AABW to the north, respectively. Jones (1984) could demonstrate that chlorite associated with displaced Antarctic diatoms (Jones and Johnson 1984) is enriched during interglacial periods in sediments below 4,000 m due to a more vigorous flow of AABW. However, he denied the use of kaolinite as tracer of NADW.

The purpose of this contribution is to assess the palaeoceanographic significance of downcore clay mineral fluctuations in Late Quaternary sediments of the southeastern South Atlantic Ocean and to demonstrate the utilisation of kaolinite/chlorite ratios as an alternative proxy of past deep water advection.

Study area

Most core sites are located between 35°-65°S and 0°-10°E and stretch along a N-S transect through the Agulhas Basin, the western tip of the SW Indian Ridge, and the eastern part of the Weddell Basin (Fig. 1a, c). This transect crosses the frontal system of the Antarctic Circumpolar Current (ACC) and ends up in the Weddell Gyre (Fig. 1b). Gravity cores and one kasten core were recovered from water depths between 2,500 and 5,300 m during several cruises of RV 'Polarstern' (Table 1). Within the ACC region two additional gravity cores were taken from the Mid-Atlantic Ridge at 5°W, between 41° and 44°S, in 3,100-3,800 m water depths. The northernmost gravity cores were recovered during two cruises of RV 'Meteor' across the top of the Walvis Ridge in water depths between 3,200 and 4,450 m (Table 1).

AABW, CPDW, and NADW drive the modern abyssal circulation in the study area. The stratification of these water masses is illustrated in Fig. 2. AABW forms the deepest water mass and is characterized by very low potential temperatures (0 to -0.7°C) and moderate salinity (Reid 1989). Its source is considered to be in the marginal seas of Antarctica, mainly the Weddell Sea, and the adjacent Southern Ocean (Carmack and Foster 1975, Foldvik and Gammelsrød 1988). In the study area AABW only occurs in the deepest parts of the eastern Weddell Basin below 4,000 m (Whitworth and Nowlin 1987). CPDW is the main deep water mass within the ACC where it occurs in water depths below 500 m. It represents an admixture of Weddell Sea, North Atlantic, Pacific and Indian Ocean waters with potential temperatures ranging from 0.1 to 2.0°C (Reid, 1989). North of about 45°S the CPDW splits off into Lower and Upper CPDW due to injection of NADW at intermediate depths (2,000-3,500 m). Lower CPDW passes the SW

Indian Ridge to the north and fills the deepest parts of the Agulhas and Cape Basins below 4,000 m (Whitworth and Nowlin 1987). Relatively warm, saline, oxygen-rich, and nutrient-poor NADW overlies Lower CPDW between 1,200 to 4,000 m in the Agulhas and Cape Basin. North of the Walvis Ridge NADW also fills the deepest parts of the Angola Basin.

Apart from core positions at the Walvis Ridge and core position PS2495-3 at the Mid-Atlantic Ridge, which are bathed by NADW, all other core locations are presently subject to deep water mass advection of southern origin.

Stratigraphy

Stratigraphic age interpretations and determinations of isotope stage boundaries are mainly based on the correlations of benthic and planktic foraminiferal $\delta^{18}O$ records and siliceous microfossil fluctuations. The latter were mainly applied on sediment cores from the deeper sites of the southeastern part of the study area, which are barren of biogenic carbonate. Diatom and radiolarian biofluctuation stratigraphies and biostratigaphies used are according to Hays et al. (1976), Burckle (1982), Burckle et al. (1978), and Gersonde et al. (1990). Isotope stages refer to the $\delta^{18}O$ record of the SPECMAP stack (Imbrie et al. 1984). Additional chronostratigraphic information was deduced from magnetostratigraphic, ^{14}C AMS, and $^{230}Th_{ex}$ measurements. Most of the applied age models have been published elsewhere (see references, Table 1).

Methods

Sediment samples were taken in 10-cm intervals from a total of 160 core metres. Sample treatment and preparation of preferentially oriented clay mounts (<2μm grain size fraction) followed the methods applied and described in detail by Petschick et al. (in press). Bulk samples were treated with 3-% hydrogenic peroxide solution and 10-% acetic acid for disaggregation and removal of organic carbon and carbonate. Grain size separation was achieved by sieving and Stokes-law settling procedure in Atterberg tubes. The separated clay fraction was saturated with Mg cations (50 % $MgCl_2$

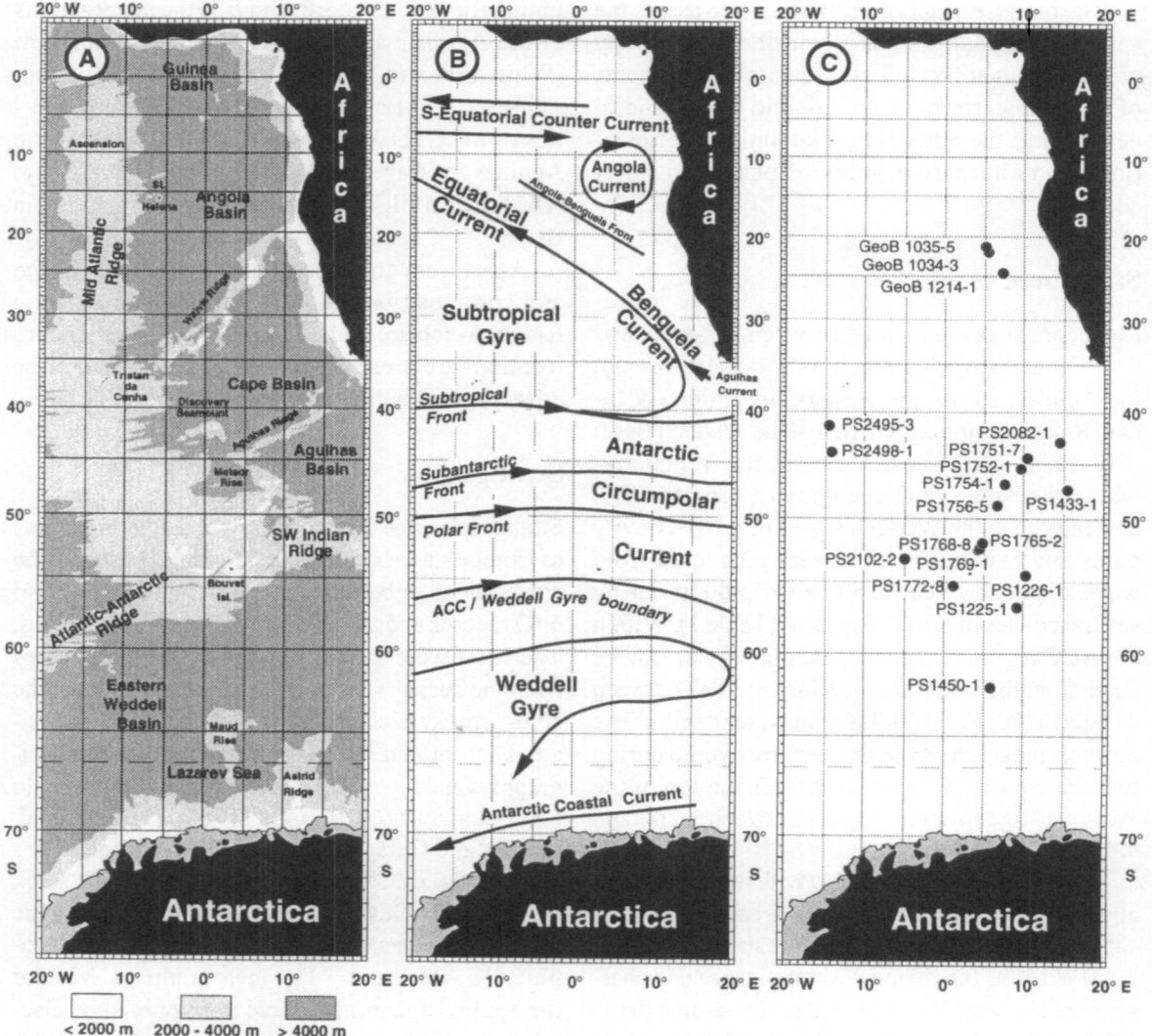

Fig. 1. Study area in the southeastern South Atlantic: (A) major topographic features, (B) modern ocean circulation pattern with frontal system of the Antarctic Circumpolar Current (after Peterson and Stramma 1991), (C) locations of investigated sediment cores.

solution) in order to provide a homogenous charging and d-spacing of the basal lattice layers within the expandable clay minerals.

XRD measurements were run on a Philips PW 1820 device, using CoKa radiation (40 kV, 40 mA). Each sample was analysed between 2° and 40° 2Θ, with a step size of 0.02° in the air-dry state and after ethylene glycol solvation. Additionally, a slow scan between 28° and 30.5° 2Θ with steps of 0.005° 2Θ

was performed on the glycolated mounts to obtain a better resolution of the 3.54/3.58 Å kaolinite/chlorite double peak.

The evaluation of the diffractograms was performed on a Apple Personal Computer using the 'MacDiff' software (Petschick unpublished). Abundances of individual clay mineral groups were estimated semi-quantitatively by calculating the peak areas of their main basal reflections in the

Core (Device)	Latitude Longitude	Water depth	Cruise (Report)	References for age model
GeoB 1034-3 (SL)	21° 44.1' S 05° 26.0' E	3772 m	M 6/6 (Wefer et al. 1988)	Bickert (1992)
GeoB 1035-5 (SL)	21° 34.7' S 05° 01.4' E	4453 m	"	Bickert (1992)
GeoB 1214-1 (SL)	24° 41.4' S 07° 14.4' E	3210 m	M 12/1 (Wefer et al. 1990)	Bickert (1992)
PS1225-1 (SL)	57° 02.6' S 09° 14.0' E	5408 m	ANT-II/4 (Köhnen 1984)	Abelmann and Gersonde (1988), Abelmann et al. (1990)
PS1226-1 (SL)	54° 31.8' S 10° 17.1' E	4033 m	"	Abelmann and Gersonde (1988), Abelmann et al. (1990)
PS1433-1 (SL)	47° 32.6' S 15° 21.9' E	4801 m	ANT-IV/4 (Fütterer 1987)	Abelmann and Gersonde (1988), Abelmann et al. (1990)
PS1450-1 (SL)	62° 20.7' S 05° 48.7' E	5310 m	"	Westall (unpublished)
PS1751-7 (SL)	44° 29.6' S 10° 28.1' E	4760 m	ANT-VIII/3 (Gersonde and Hempel 1990)	Abelmann and Gersonde (unpublished)
PS1752-1 (SL)	45° 37.3' S 09° 35.8' E	4519 m	"	Bárcena et al. (1992)
PS1754-1 (SL)	46° 46.2' S 07° 36.7' E	2471 m	"	Niebler (1995), Mackensen (unpubl.) Frank et al. (submitted)
PS1756-5 (SL)	48° 53.9' S 06° 42.8' E	3787 m	"	Abelmann et al. (unpubl. rpt. 1992), Frank et al. (submitted)
PS1765-2 (KAL)	51° 49.6' S 04° 50.9' E	3749 m	"	Abelmann et al. in Gersonde and Hempel (1990), pp. 97-102
PS1768-8 (SL)	52° 35.6' S 04° 28.5' E	3270 m	"	Niebler (1995), Frank et al. (submitted)
PS1769-1 (SL)	52° 36.7' S 04° 27.5' E	3230 m	"	Abelmann and Gersonde (unpublished)
PS1772-8 (SL)	55° 27.5' S 01° 09.8' E	4135 m	"	Abelmann et al. (unpubl. rpt. 1992), Frank et al. (submitted)
PS2082-1 (SL)	43° 13.2' S 14° 33.7' E	4610 m	ANT-IX/4 (Bathmann et al. 1992)	Mackensen et al. (1994)
PS2102-2 (SL)	53° 04.4' S 04° 59.1' W	2390 m	"	Niebler (unpublished)
PS2495-3 (SL)	41° 17.2' S 14° 30.0' W	3131 m	ANT-XI/2 (Gersonde 1995)	Mackensen (unpublished)
PS2498-1 (SL)	44° 09.2' S 14° 13.7' W	3783 m	"	Gersonde (unpublished), Mackensen (unpublished)

Table 1. Core locations with references to cruise reports and age models.

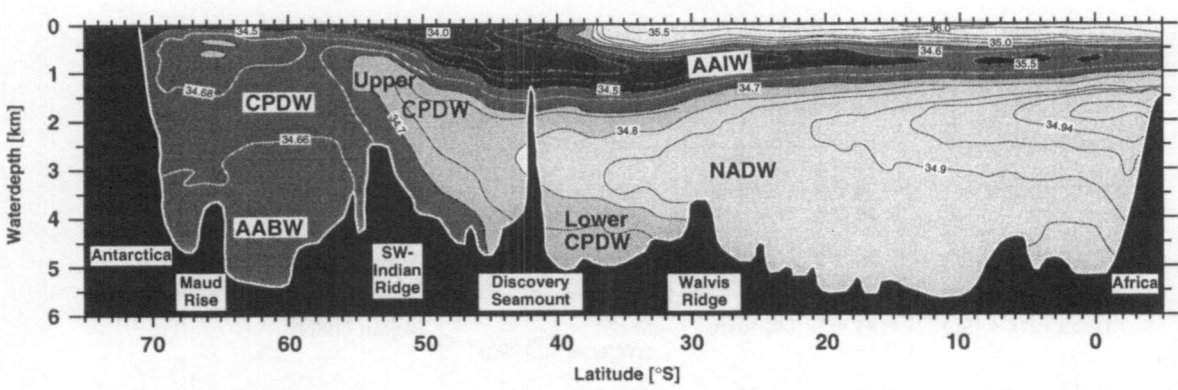

Fig. 2. Present-day stratification of main ocean water masses in the South Atlantic as displayed by salinity contour lines (‰) at a north-south transect along the Greenwich Meridian (modified from Reid 1989). (AABW) Antarctic Bottom Water, (CPDW) Circumpolar Deep Water, (AAIW) Antarctic Intermediate Water, (NADW) North Atlantic Deep Water.

glycolated state: smectite (17 Å), illite (10 Å), kaolinite/chlorite (7 Å). Relative proportions of kaolinite and chlorite were deduced from the kaolinite/chlorite ratios yielded by the 3.54/3.58 Å peak doublet of the slow scans (Petschick et al. in press). Using the weighting factors of Biscaye (1965), percentages of each clay mineral group were computed: smectite (1x), kaolinite (2x), chlorite (2x), illite (4x).

Illite chemistry was inferred from the 5 Å/10Å peak intensity (Esquevin 1969). Ratios above 0.45 represent Al-rich illites (muscovite, sericite), those below 0.25 Fe-, Mg-rich illites (<0.1 biotite)

The degree of lattice ordering and crystallite size of smectite and illite, usually referred to as 'crystallinity index', is expressed as the Integral Breadth (IB) of the glycolated 17 Å (smectite) and 10 Å (illite) peaks, respectively. The IB represents the breadth ($\Delta° 2\Theta$) of an rectangle of the same area and height as the peak. We define the following crystallinity categories (ranges of IB values) for smectite: (<1.5) well crystalline, (1.5-2.0) moderately crystalline, (>2.0) poorly crystalline; and illite: (<0.4) very well crystalline, (0.4-0.6) well crystalline, (0.6-0.8) moderately crystalline, (>0.8) poorly crystalline.

Downcore clay mineral fluctuations

Downcore clay mineral fluctuations are generally of low amplitude. Variations of clay mineral ratios, particularly kaolinite/chlorite ratios, show clearer downcore tendencies and are frequently correlated with glacial-interglacial cycles displayed by planktic and benthic foraminiferal $\delta^{18}O$ records (Figs. 3-6). Regional differences of clay mineral distributions (clay mineral provinces), as seen in modern surface sediments (Petschick et al. in press), prevailed throughout the Late Quaternary. This is well documented by the comparison of clay mineral distributions in sediment cores from the ACC region and adjacent regions to the north and south (Figs. 3-7).

ACC region (Mid-Atlantic Ridge, Agulhas Basin, SW Indian Ridge)

Results: Moderately crystalline smectite dominates the clay mineral assemblage of the ACC. Maximal interglacial values are observed in the Agulhas Basin and at the SW Indian Ridge (60-75 %), which decrease to around 50 % during glacials (Figs. 4, 5). At the Mid-Atlantic Ridge smectite proportions

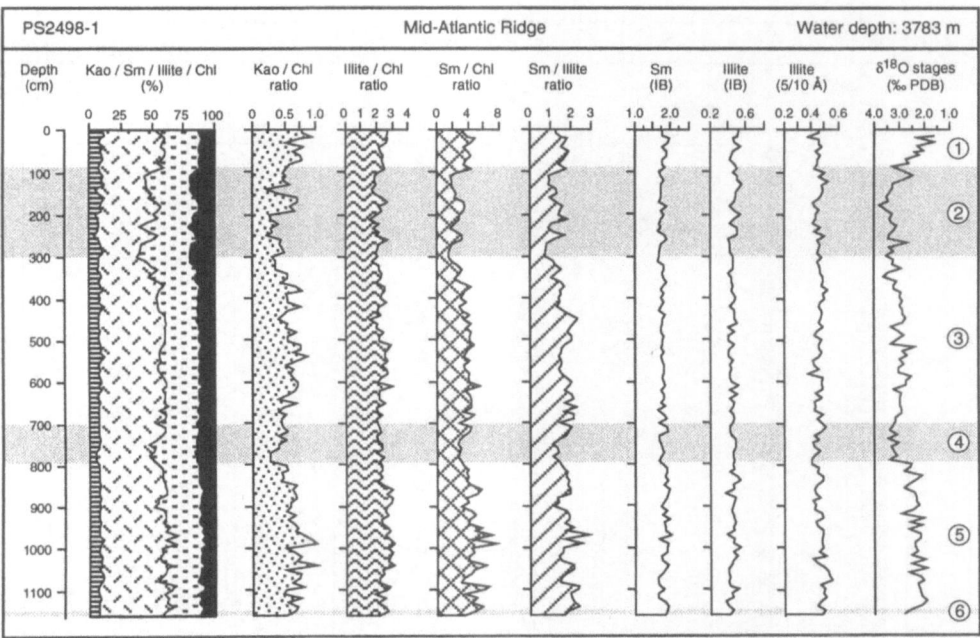

Fig. 3. Downcore clay mineral fluctuations in sediments of gravity core PS2498-1 from the Mid-Atlantic Ridge. Isotope stages are deduced from the planktic foraminiferal δ¹⁸O record of *Globigerina bulloides* (Gersonde unpublished).

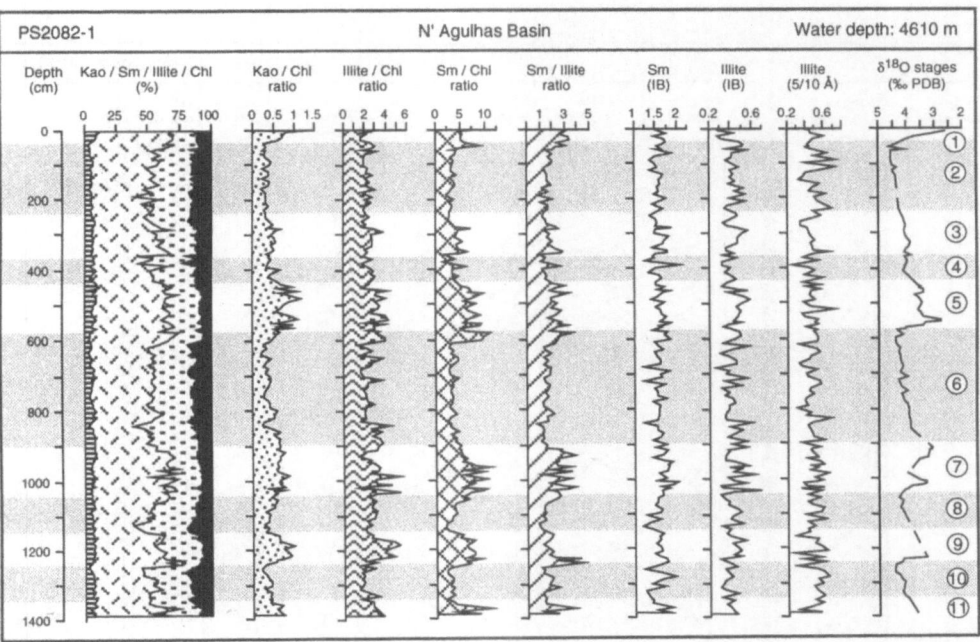

Fig. 4. Downcore clay mineral fluctuations in sediments of gravity core PS2082-1 from the northern Agulhas Basin. Isotope stages are deduced from the benthic foraminiferal δ¹⁸O record of *Cibicidoides* spp. (Mackensen et al. 1994) and are corroborated by downcore fluctuations of *Cycladophora davisiana* (Brathauer and Abelmann 1994).

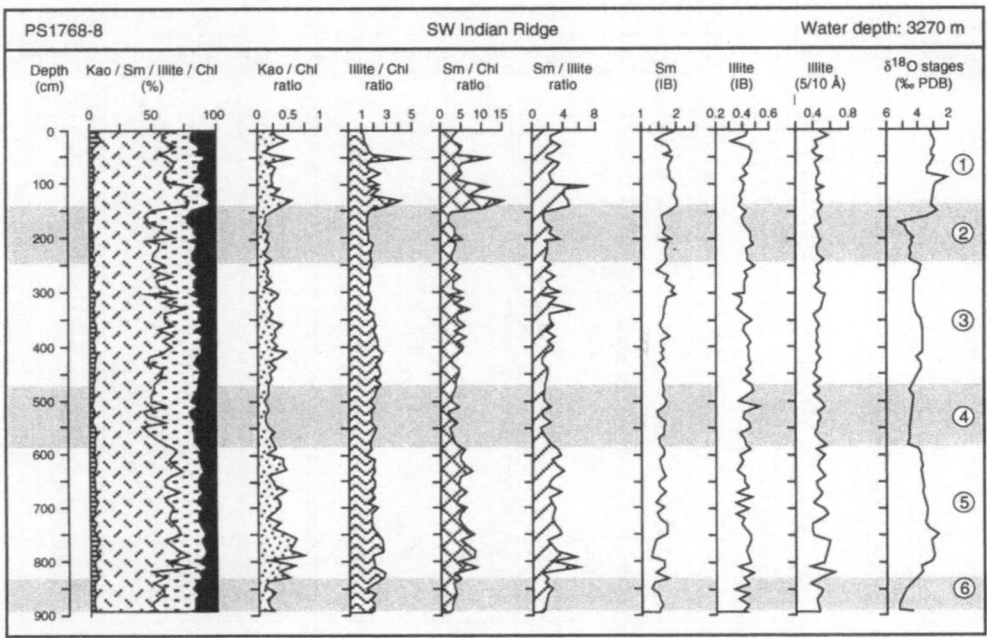

Fig. 5. Downcore clay mineral fluctuations in sediments of gravity core PS1768-8 from the SW Indian Ridge. Isotope stages are deduced from the planktic foraminiferal $\delta^{18}O$ record of *Globigerina bulloides* (Niebler 1995), ^{14}C AMS dating and $^{230}Th_{ex}$ flux modelling (Frank et al. submitted).

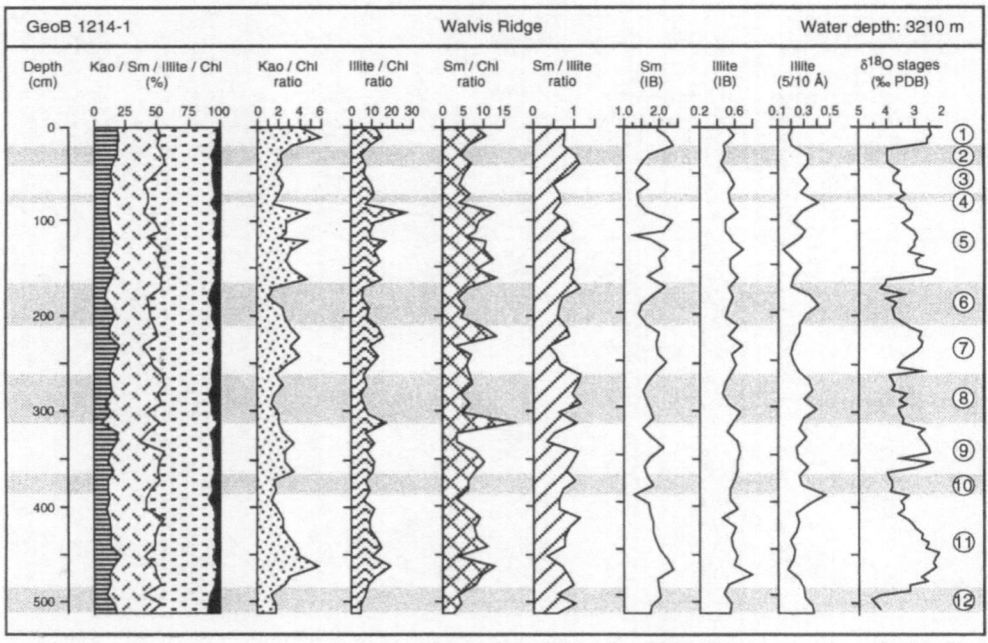

Fig. 6. Downcore clay mineral fluctuations in sediments of gravity core GeoB 1214-1 from the Walvis Ridge. Isotope stages are deduced from the benthic forminiferal $\delta^{18}O$ record of *Cibicidoides wuellerstorfi* (Bickert 1992).

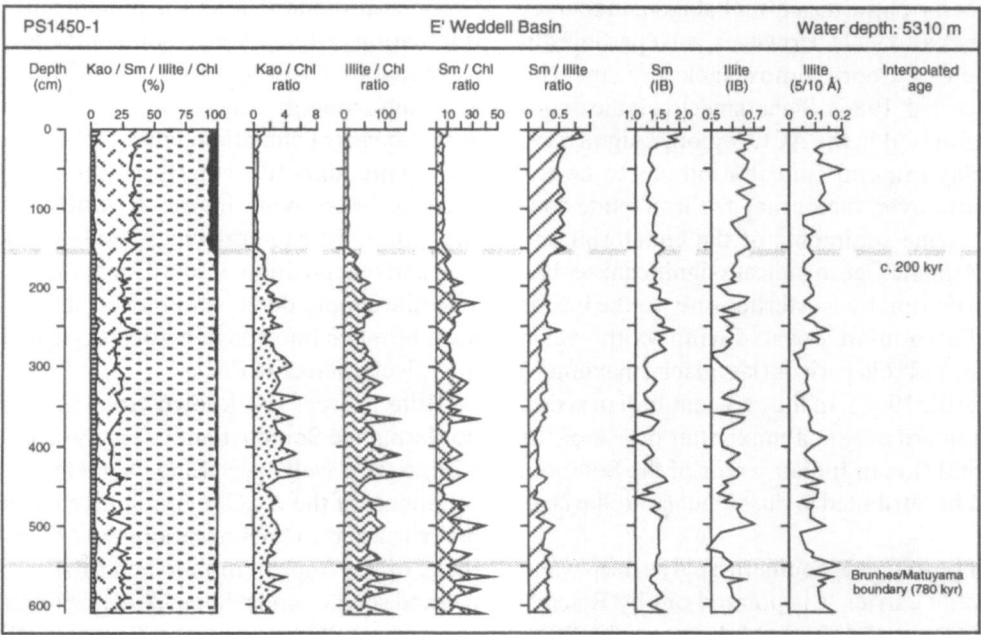

Fig. 7. Downcore clay mineral fluctuations in sediments of gravity core PS1450-1. Palaeomagnetic dating by Westall (unpublished). Age of the Brunhes/Matuyama chron boundary refers to the Geomagnetic Polarity Time Scale of Cande and Kent (1992).

vary between 45 and 55 % during interglacials and 30-45 % during colder periods (Fig. 3).

Temporal fluctuations of illite concentrations generally show an opposite trend to smectite (glacial: 20-30%, interglacial: 35-40%). This tendency is also displayed by smectite/illite ratios which range between 1 (glacial) and 3-4 (interglacial) in the Agulhas Basin and at the SW Indian Ridge, and at a lower amplitude between 1 (glacial) and 2 (interglacial) at the Mid-Atlantic Ridge. Illites within the ACC region are generally Al-rich and very well to well crystalline.

Relatively high percentages of chlorite occur. Glacial proportions (5-10%) are frequently up to one magnitude higher than the interglacial proportions (10-20%). This trend is also mirrored by downcore variations of kaolinite/chlorite, illite/chlorite, and smectite/chlorite ratios, which show raised values during warm intervals.

Kaolinite represents a minor component of the clay mineral spectrum and shows an opposite distribution pattern to chlorite (glacial: 2.5-5.0%,

interglacial 5.0-7.5%). Downcore fluctuations of kaolinite/chlorite ratios with increased values during interglacials are an expression of the inverse correlation between kaolinite and chlorite concentrations.

Discussion: Clay minerals in modern sediments of the ACC region are almost exclusively supplied by southern water masses of the ACC and the NADW which is injected into the ACC at intermediate depths (Petschick et al. in press). Long-distance aeolian clay input by winds of the southern westerlies is of minor importance. Modern air-borne dust fluxes to the Southern Ocean are very low (Duce et al. 1991) and were also low during glacials, even if greater fluxes by a factor of 15 are assumed during the last glacial maximum, as documented in the Vostok ice core from the East Antarctic ice shield (Petit et al. 1990). Nevertheless, isotopic signatures (Nd and Sr) in detrital particles of the last glacial maximum in the Dome C ice core (East Antarctic ice shield) indicate provenance and long-distance transport of air-borne dust from Patagonian sources (Grousset et al. 1992). However, dust particles in the Dome C ice core almost lack any smectite (Gaudichet et al. 1986). Since, smectite is the dominant clay mineral in the ACC region, a significant aeolian clay mineral contribution has to be excluded. Moreover, factor maps of iron oxide contents in marine sediments of the equatorial and southern Atlantic Ocean indicate significant aeolian dust contribution by westerlies only to the basins off the Patagonian coast during both warm (Holocene) and cold periods (last glacial maximum) (Balsam et al. 1995). In the adjacent Indian sector of the Southern ocean, a maximum of 5% of the total detrital flux to Indian sector of the Southern ocean can be attributed to dust input (Bareille et al. 1994).

The importance of southern-source deep water as a smectite carrier was pointed out by Biscaye (1965), Kolla et al. (1976), Melguen et al (1978), and Petschick et al. (in press). Extensive exposures of mafic volcanic rocks within the Andean mobile belt (South America, Antarctic Peninsula and adjoining islands of the Scotia Arc) are regarded as important sources of detrital smectite. Some smectite may originate from the erosion of Mesozoic and Cenozoic sediments of the Falkland Plateau (Ciesielski et al. 1982, Robert and Maillot 1983). In the vicinity of the young volcanoes of Deception and Bouvet islands significant quantities of non-terrigenous smectite (nontronite) are probably provided by subaquatic alteration (halmyrolysis) of volcanic rocks and tephra (Petschick et al. in press). Actually, higher smectite proportions in Late Quaternary sediment from the SW Indian Ridge and the Agulhas Basin in respect to sediments from the Mid-Atlantic Ridge are in accordance with a more frequent occurrence of volcanic ashes at these sites.

The southern Pacific Ocean is an additional potential smectite source, where low sedimentation rates favour widespread halmyrolysis of submarine basalts (Honnorez 1981, Cole and Shaw 1983). Moreover, Late Cretaceous and younger sedimentary rocks, located at the Pacific continental margin of Antarctica contain abundant smectite (Zemmels and Cook 1976). Smectite from the southern Pacific may be introduced through the Drake Passage into the South Atlantic by the ACC.

Several explanations for glacial-interglacial fluctuations of smectite abundances are likely. Chemical weathering of mafic rocks in the Andean belt, and smectite formation, probably increase during stages of climatic amelioration and can raise pedogenic smectite supply to the ACC water masses. Moreover, increased bottom current strengths during interglacial periods enhance erosion and redeposition of fossil smectite. Increased smectite supply due to stronger NADW injection must be taken into consideration to explain raised interglacial smectite fluxes.

Illite sources are known from all continents bordering the South Atlantic. The typical Al-rich, well to very well crystalline character of illite in sediments of the ACC region indicates dominant provenance from low-metamorphic sericite-bearing rocks of the Andean mobile belt of West Antarctica and southernmost South America (Petschick et al. in press). We assume that downcore illite fluctuations are essentially linked to dilution effects caused by variable glacial-interglacial smectite fluxes.

Due to a lack of chemical weathering, terrigenous debris from southern high-latitude source areas does not yield any pedogenic kaolinite. Some fossil kaolinite from East Antarctica is encountered in sediments of the eastern Weddell Basin and restricted to that region, as discussed later on. Therefore it is suggested that northern-source deep water is essentially responsible for kaolinite transfer from the equatorial regions to the ACC region.

Chlorite in modern surface sediments of the South Atlantic can be exclusively traced back to low-metamorphic rocks of the mobil Andean belt of southernmost South America (Siegel et al. 1981) and the Antarctic Peninsula (Yoon et al. 1992, Petschick et al. in press), from where it is distributed by southern water masses, mainly AABW and Lower CPDW (Biscaye 1965, Jones 1984, Petschick et al. in press). Thus chlorite is enriched in sediments of the ACC region and may be regarded as a single-source clay mineral tracing southern-source deep water flow. Minima of chlorite proportions with concomitant maxima of illite/chlorite, smectite/chlorite, and kaolinite/chlorite ratios during warm periods reflect greater fluxes of all other clay mineral species in relation to chlorite. Reasons for enhanced interglacial smectite fluxes were pointed out previously. Raised kaolinite/chlorite ratios apparently reflect increased kaolinite input to the ACC region by northern-source deep water. Since illite and chlorite percentages show common interglacial minima and both clay minerals originate from the same southern source area higher illite/chlorite ratios can only be explained by additional illite, probably delivered by northern-source deep water.

Walvis Ridge

Results: Illite shows the highest proportions within the clay mineral spectrum followed by smectite. Downcore smectite abundances in core GeoB 1214-1 (Fig. 6) fluctuate between 40 and 55%, those of illite between 30 and 40%, without any climatic cyclicity. However, smectite and illite display better crystallinity indices in glacial compared to interglacial intervals. Moreover, illites in glacial segments are characterized by a Fe-, Mg-rich composition in contrast to Al-rich illites in interglacial segments.

Chlorite is the minor clay mineral component at the Walvis Ridge. Its downcore fluctuation is clearly attributed to glacial-interglacial cycles with higher proportions (6-8%) during glacial stages and lower proportions (2-4%) during warm periods. These chlorite variations are also mirrored by lower glacial illite/chlorite ratios and smectite/chlorite ratios in relation to interglacial ratios.

Kaolinite abundances range within 10 to 15% with occasional peaks of up to 20% during warmer periods. Thus kaolinite shows a reciprocal trend to chlorite. Pronounced downcore variations of kaolinite/chlorite ratios with ratios >2 during warm periods and ratios <2 during glacials are an expression of the opposite kaolinite and chlorite variations, as also seen in the ACC region.

Discussion: Offshore southeastern trade winds ('berg winds') and ocean circulation control modern clay mineral deposition at the top of the Walvis Ridge (Gingele 1992, Petschick et al. in press). Illite mainly originates from the Namib and Kalahari deserts and from illite-rich soils of Southern Africa, from where it is transported by the 'bergwinds' and

to a lesser degree by the Orange River to the ocean (van der Gaast and Jansen 1984, Gingele 1992, Bremner and Willis 1993, Gingele in press). This Illite and some smectite reaches the Walvis Ridge by long-distance aeolian transport or within the water masses of the Benguela Current.

Northeastern trade winds and low-latitude African rivers contribute clay minerals to the East Atlantic which are incorporated into the NADW and transported to the south. The northeastern trade winds carry high amounts of kaolinite eroded from fossil kaolinite-bearing soils of the Sahel zone (Aston et al. 1973). The Congo/Zaire River (Eisma 1978, van der Gaast and Jansen 1984) and Niger River (Pastouret et al. 1978) discharge considerable amounts of pedogenic clay minerals, mainly kaolinite besides moderate amounts of low-crystalline smectite and some degraded Al-rich illite. High smectite runoff is observed at the Kunene River mouth due to a vast catchment area in the semiarid to semihumid regions of central to southern Africa (Bremner and Willis 1993).

Although smectite and illite concentrations show no distinct glacial-interglacial variations poorer crystallinity indices of both clay minerals and the Al-rich composition of illite indicate a higher input from tropical sources via northern-source deep water during warm periods. A more pronounced influence of northern-source deep water on clay mineral deposition during interglacials is also evident by raised kaolinite concentrations. In contrast, well crystalline Fe-, Mg-rich illite associated with well crystalline smectite, which are usually both altered by chemical weathering, indicate a higher clay mineral supply from the arid regions of southern Africa during glacials. This may be explained by an intensification of the southeastern trade wind system and/or attenuation of northern-source deep water flow to the Walvis Ridge.

An African source area which provides significant amounts of chlorite is unknown. Since the crest of the Walvis Ridge is presently bathed by NADW, low chlorite concentrations in modern (and interglacial) sediments are an expression of little chlorite input from relatively shallow southern-source water masses, namely Upper CPDW and Antarctic Intermediate Water. During glacial times a dramatic increase of chlorite concentrations coupled with coinciding decreases of kaolinite/chlorite, illite/chlorite, smectite/chlorite ratios demonstrate raised relative chlorite fluxes. Bathing of the Walvis Ridge by deep chlorite-bearing southern-source water is the most obvious possibility to facilitate substantial chlorite fluxes to the top of the Walvis Ridge. A rise of the lower northern-source deep water boundary and consequent overspill of southern-source deep water into the Angola Basin due to an attenuation of northern-source deep water flow has actually been deduced from carbonate dissolution data and the benthic foraminiferal $\delta^{13}C$ record (Bickert 1992).

Eastern Weddell Basin

Results: In contrast to the other core sites of the study area Late Quaternary sedimentation rates were low in the eastern Weddell Basin. Condensed siliciclastic mud sequences are almost barren of biogenous components (Abelmann and Gersonde in Fütterer 1987, pp. 182-184). Although the chronostratigraphic resolution is poor (palaeomagnetic dating), it is obvious that clay mineral fluctuations in sediment core PS1450-1 exhibit some long-term tendencies (Fig. 7).

The upper sediment core section, representing approximately the last 200 kyr (interpolated age), is characterized by nearly constant proportions of individual clay minerals: well to moderately crystalline illite (60%), well to moderately crystalline smectite (30%), chlorite (6-8%), kaolinite (1-2%).

Below the c. 200 kyr boundary an abrupt decrease of chlorite down to 1-2% is coupled with a slight increase of kaolinite to 2-4% and displayed by raised kaolinite/chlorite ratios. Illite concentrations reach highest values in the lower section (75-85%). Smectite concentrations show an opposite trend with minimum values down to 10%.

Very low 5 Å/10 Å XRD peak height ratios (<0.1) attest the dominance of very Fe, Mg-rich ('biotitic') illites in the lower section. Illites in the upper section tend to slightly increased values (0.1-02) indicating somewhat higher Al-contents.

Discussion: At present, core position PS1450-1 is located at the confluence of the eastern limb of the cyclonic Weddell Gyre with the westward flowing Antarctic Coastal Current. Water masses of the Weddell Gyre contribute clay minerals from West and East Antarctic sources, those of the Antarctic Coastal Current merely from East Antarctica (Petschick et al. in press). Fe, Mg-rich illites indicate provenance from biotite-bearing metamorphic rocks of the East Antarctic craton, whereas chlorite is mainly derived from low-metamorphics of West Antarctica. Smectite is provided by altered Jurassic volcanic rocks covering the East Antarctic Craton and by Meso- and Cenozoic volcanic rocks of West Antarctica. Kaolinite in the eastern Weddell Sea is supplied from external sources farther east (Petschick et al. in press). Possible kaolinite-bearing sources are the Permian Amery Formation in the hinterland of the Prydz Bay as well as Tertiary sediments from submarine ridges (Maud Rise, Astrid Ridge, Gunnerus Ridge), which are encountered in shallow depths below Quaternary deposits and mostly consist of smectite (Ehrmann et al. 1991). However, submarine exposures of these Tertiary sediments hitherto have not been reported.

Downcore clay mineral distributions in sediment core PS1450-1 suggest that water masses of the Weddell Gyre and the Antarctic Coastal Current controlled clay mineral input throughout the past c. 200 kyr. In older sediments, clay mineral supply was dominated by the Antarctic Coastal Current as indicated by a dramatic decrease of chlorite and increase of fossil kaolinite and biotitic illite. This may be explained by a westward shift of the eastern Weddell Gyre limb or an attenuation of Weddell Gyre circulation.

In contrast to the ACC region and the Walvis Ridge, long-term constancy of clay mineral proportions reflect stable oceanographic conditions which were not affected by glacial-interglacial changes. Moreover, the striking difference in illite chemistry (Al-rich versus Fe, Mg-rich) between sediments of the ACC region and the eastern Weddell Basin gives evidence that the ACC deflects water masses and detrital particles eastwards, and inhibits advection of Fe, Mg-illites (and other clay minerals) to the north. Thus the Weddell Gyre has to be regarded as a closed system in terms of clay mineral deposition.

Palaeoceanographic significance of kaolinite/chlorite ratios

In the previous sections we have pointed out that, apart from the eastern Weddell Basin, downcore clay mineral fluctuations in the southeastern South Atlantic and adjoining Southern Ocean essentially monitor periodic changes of deep water advection in response to glacial-interglacial cycles. The palaeoceanographic significance of individual clay minerals is variable. Since smectite and illite may originate from all continents bordering the South Atlantic, it is difficult to interpret their temporal and areal distribution patterns.

North of the Weddell Gyre region kaolinite and chlorite represent appropriate tracers of deep water advection. Chlorite as a high-latitude single-source component indicates advection of southern-source deep water, whereas kaolinite acts as the low-latitude counterpart which is supplied almost exclusively by northern-source deep water. This is indicated by increasing chlorite and decreasing kaolinite concentrations from the equatorial regions to the Southern Ocean.

A much clearer picture of this factual situation arises from the variability of kaolinite/chlorite ratios in the investigated sediment cores (Fig. 8). The comparison of two components by their ratio offers the advantage of eliminating dilution effects by other components. Apart from the eastern Weddell Basin site, interglacial intervals are generally characterized by raised kaolinite/chlorite ratios. Moreover a northward tendency to increasing glacial and interglacial kaolinite/chlorite ratios is evident with lowest values in the ACC region (glacial: 0.1-0.25, interglacial: 0.35-0.5) and highest values at the Walvis Ridge (glacial: 1.5-2.0, interglacial: 4.0-8.0). A depth-related variation of kaolinite/chlorite ratios in sediment cores from similar latitudes and various water depths is not evident.

In order to determine the relationship between temporal fluctuations of kaolinite/chlorite ratios and global climatic cyclicity by statistical means, we have conducted a cross-spectral analysis between kaolinite/chlorite ratios in sediment core PS2082-1 and the $\delta^{18}O$ record of the reversed SPECMAP Stack (Imbrie et al. 1984), using ARAND programs

(OSU Computer Center staff 1973, Howell 1989). Techniques are described in detail by Imbrie et al. (1989). The SPECMAP stack is regarded as a proxy of global ice volume variations. The SPECMAP $\delta^{18}O$ sign was reversed by a factor of -1 before calculating a cross spectrum in order to relate minima of ice volume to orbital configurations causing maximal insolation.

Sediment core PS2082-1 from the Agulhas Basin offers a suitable age frame to deduce time frequencies in the frequency range of orbital parameters. Average sedimentation rates of 3.5 cm/kyr and a total age of 400 kyr provide an appropriate temporal resolution. The applied age model depends on isotope stratigraphy (Mackensen et al. 1994) which is corroborated by *Cycladophora davisiana* biofluctuation stratigraphy (Brathauer and Abelmann 1994). Moreover, the benthic foraminiferal $\delta^{18}O$ record of sediment core PS2082-1 reveals high coherencies and little phase differences to frequencies of the $\delta^{18}O$ SPECMAP stack (Table 2).

Cross spectral analysis between kaolinite/chlorite ratios and the reversed SPECMAP $\delta^{18}O$ record shows that the variance of kaolinite/chlorite ratios in the 100-kyr^{-1} and 41-kyr^{-1} frequencies of eccentricity and obliquity are coherent with ice volume (Fig. 9, Table 2). However, there is no coherent variance of kaolinite/chlorite ratios with ice volume in the the 23-kyr^{-1} and 19-kyr^{-1} bands of precession. The phase spectrum reveals that kaolinite/chlorite ratios are nearly in phase with global ice volume in the 100-kyr^{-1} and 41-kyr^{-1} bands (Fig. 10, Table 2).

Reconstruction of glacial-interglacial deep water advection

Petschick et al. (in press) have shown that the latitudinal zonation of kaolinite/chorite ratios in surface sediments of the South Atlantic reflect the extensions of deep water masses in the South Atlantic. Thus, kaolinite/chlorite ratios decrease continuously to the south, and values around 0.5 delineate the southernmost limit of direct NADW injection into the Southern Ocean.

Fig. 11 summarizes the temporal and latitudinal variability of kaolinite/chlorite ratios on a north-south transect through the study area demonstra-

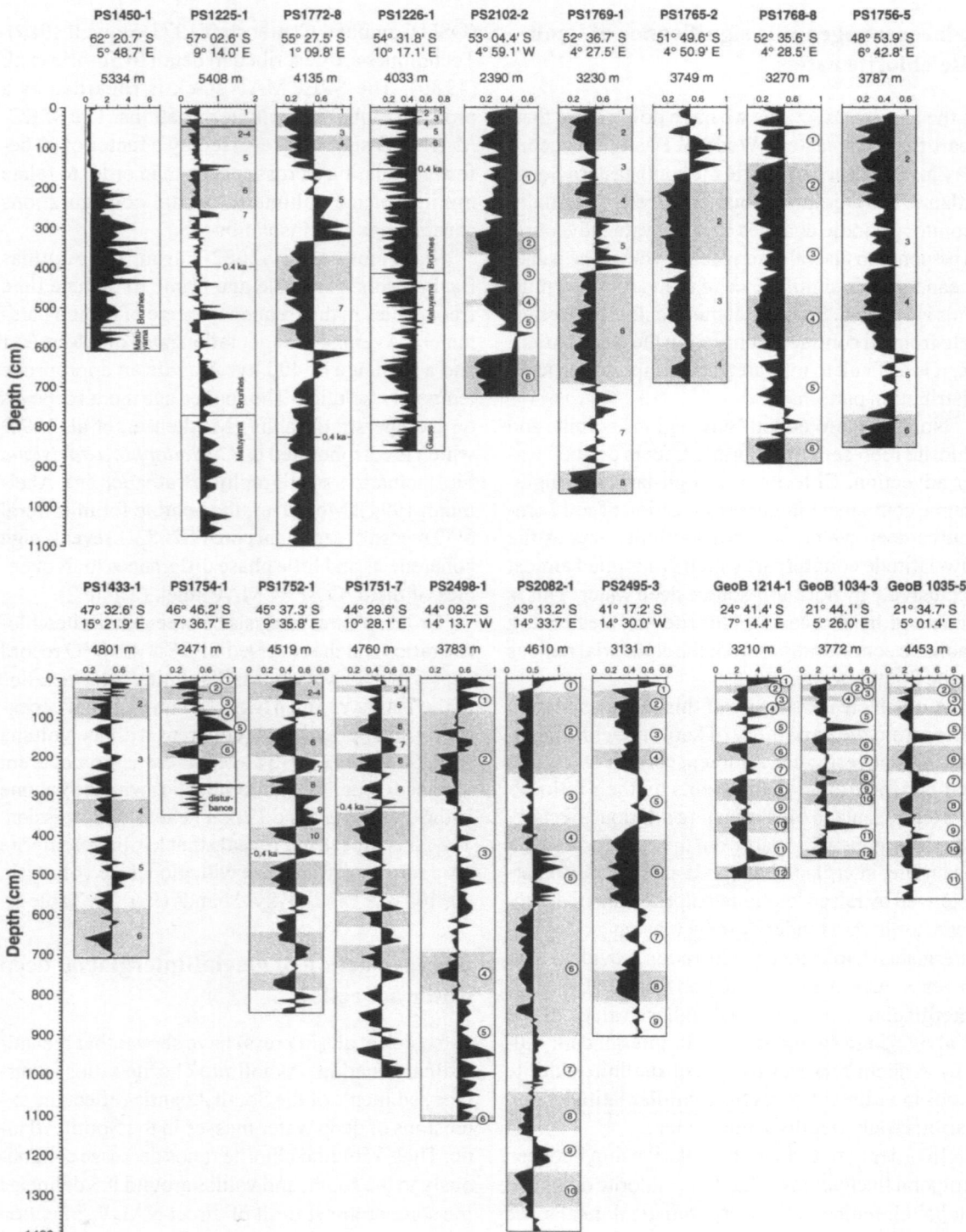

Fig. 8. Kaolinite/chlorite ratio variations (3.58/3.54 Å XRD peak ratios) within 19 sediment cores from various sites of the southeastern South Atlantic. The encircled numbers next to the curves refer to glacial-interglacial isotope stages inferred from the planktic or benthic foraminferal δ¹⁸O record. Isotope stages which were essentially deduced from siliceous microfossil measurements are indicated by normal numbers.

Reversed SPECMAP δ¹⁸O stack vs. variable (*)	100-kyr⁻¹ band k	°Phi	41-kyr⁻¹ band k	°Phi	23-kyr⁻¹ band k	°Phi	k₀	T (kyr)	Δt (kyr)
1. Kaolinite/chlorite ratio (PS2082-1)	0.96	-6±7(11)	0.87	+11±14(22)	-	-	0.64 (0.80)	388	2.0
2. Benthic δ¹⁸O record (PS2082-1)	0.99	-3±(6)	0.99	+7±(6)	0.96	+1±(12)	(0.88)	388	2.0
3. '%NADW index' (Raymo et al. 1990)	0.87	+16±15	0.86	+5±16	0.83	+41±17	0.66	400	3.0
4. Sea surface temperatures 44° S (Hays et al. 1976)	0.92	-47±11	0.93	-14±10	0.86	-35±15	0.66	400	2.0
5. Cd/Ca North Atlantic (*) (Boyle and Keigwin 1985/86)			0.98	+112±25				215	

Bandwidth for calculations of variables 1-4 is 0.01 kyr⁻¹. Items tabled are: (k) coherency, (°Phi) phase angle with 80% confidence interval, 95% confidence intervals are given in parentheses, (k₀) 80% test statistics for nonzero coherency, for 95% values are given in parentheses, (T) maximal age used in the calculation, (Δt) sample interval. (*) Variable 5 was calculated versus ETP with 90% confidence interval (Boyle and Keigwin 1985/86). Positive phase differences indicate that the variable lags the SPECMAP -δ¹⁸O variable (variable 5 lags ETP). The δ¹⁸O SPECMAP stack was taken from Imbrie et al. (1984). Variables 3 and 4 were calculated by Imbrie et al. (1992, 1993). The '%NADW index' (variable 3) is a δ¹³C proxy for NADW in the North Atlantic relative to Pacific deep water.

Table 2. Data of cross-spectral analyses.

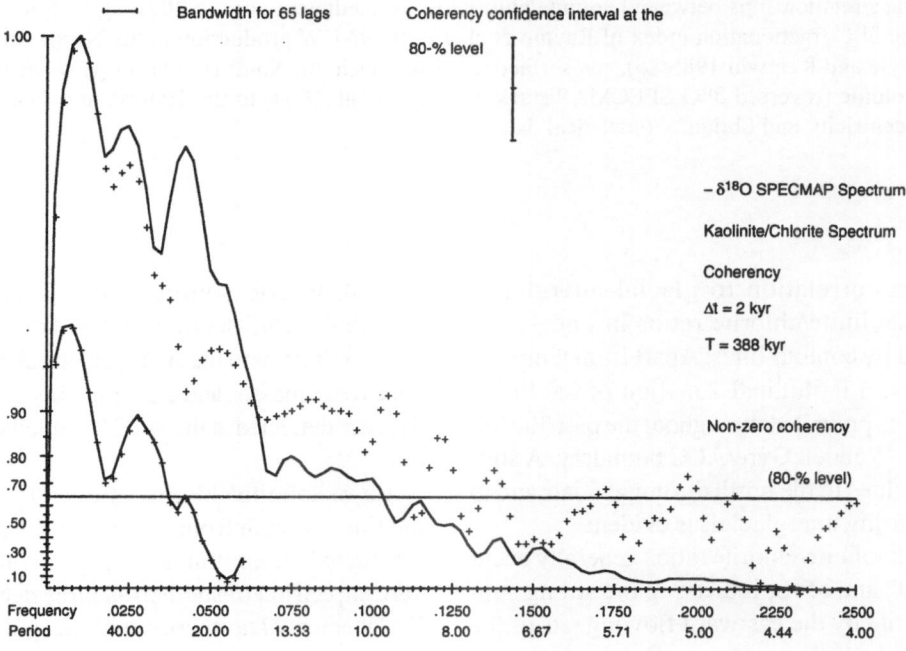

Fig. 9. Cross-spectral analysis between the δ¹⁸O SPECMAP stack (Imbrie et al. 1984) and kaolinite/chlorite ratios in sediment core PS2082-1.

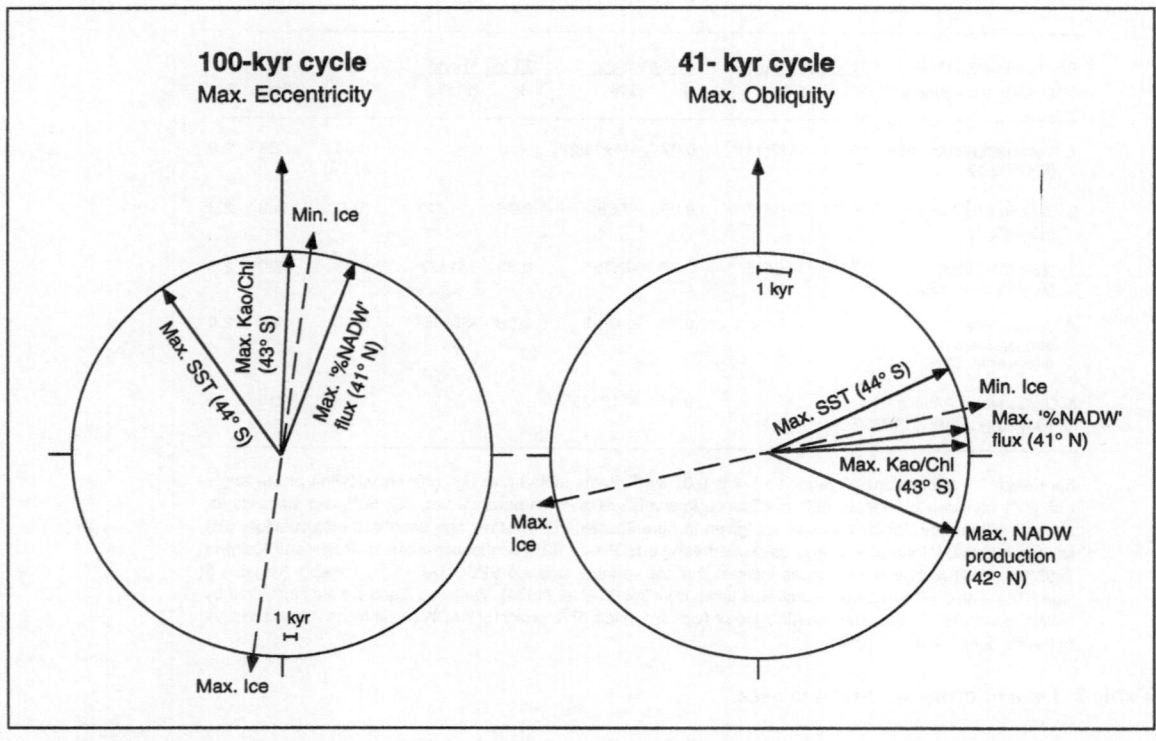

Fig. 10. Phase relationships between kaolinite/chlorite ratios (sediment core PS2081-1), '%NADW' flux (Atlantic-Pacific δ¹³C fractionation index of Raymo et al. 1990), NADW production in the North Atlantic (Cd/Ca proxy of Boyle and Keigwin 1985/86), sea-surface temperatures in the Southern Ocean (Hays et al. 1976), and global ice volume (reversed δ¹⁸O SPECMAP stack of Imbrie et al. 1984) to the 100-kyr and 41-kyr periods of maximal eccentricity and obliquity (statistical data in Table 2).

ting strong correlation to glacial-interglacial changes. Kaolinite/chlorite ratios in Fig. 11 are represented by contour lines. Apart from temporal fluctuations, a latitudinal zonation of kaolinite/chlorite ratios prevailed throughout the past 300 kyr north of the Weddell Gyre/ACC boundary. A shift of higher values to the north during glacials and to the south during interglacials is evident.

Lowest kaolinite/chlorite ratios generally occur between 50° and 55°S, indicating the strong supply of chlorite by the eastward flowing jets of the ACC. Glacial values are <0.3, whereas interglacial values vary around 0.5. South of 57°S these values increase abruptly without any glacial-interglacial contrasts, delineating the ACC/Weddell Gyre boundary, south of which fossil kaolinite is supplied

from Antarctic sources. As discussed in the 'Weddell Basin' section, an input of southern-derived kaolinite into the ACC region can be excluded, since water masses and clay minerals of the Weddell Gyre are deflected at the ACC/Weddell Gyre boundary.

Low kaolinite/chlorite ratios (<0.5) during glacial times stretch from 57° to 45°-40°S during the last glacial maximum, showing the intense northward expansion of southern-source deep water and displacement of northern-source deep water. In turn, the profound shift of values >1 to high latitudes between 45° and 40°S documents the southward injection of kaolinite-rich northern-source deep water into the ACC during warmer periods.

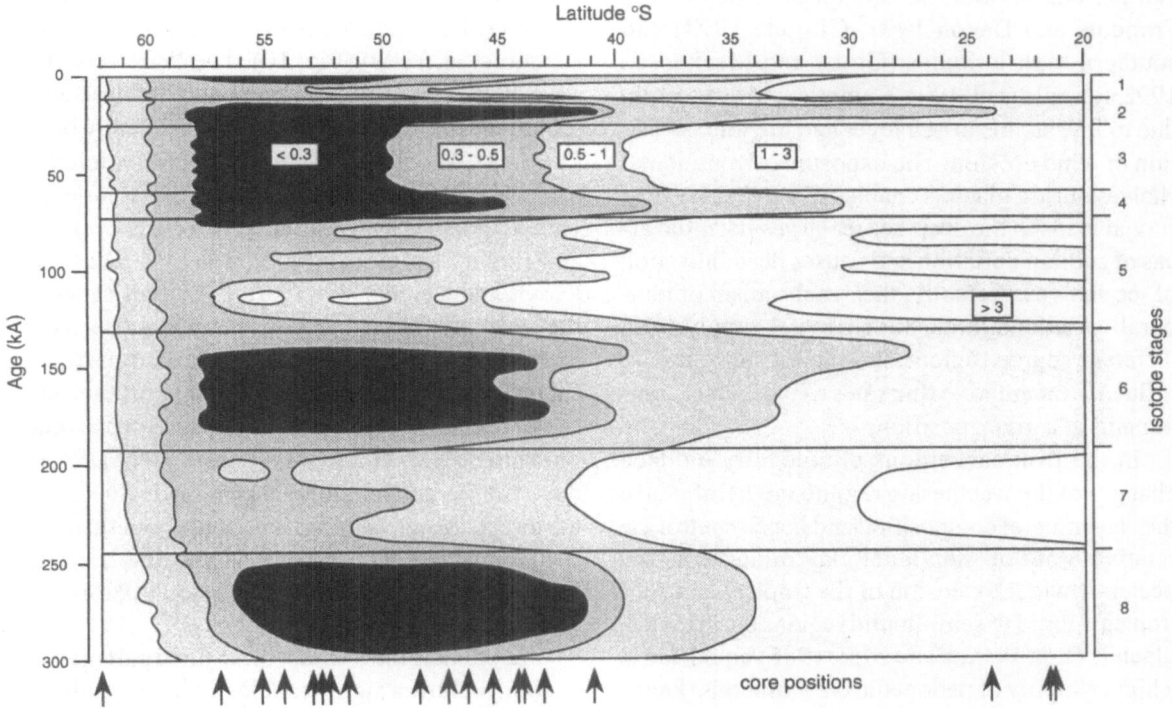

Fig. 11. Temporal and latitudinal variations of kaolinite/chlorite ratios along a north-south transect through the southeastern South Atlantic. Contour lines are interpolated from values presented in Fig. 9 by projection of the core sites (arrows) to a meridional reference.

Assuming that a kaolinite/chlorite ratio of about 0.5 represents the southernmost present-day spreading of NADW to 45°S latitude, the meridional shifts of this value through the glacial-interglacial cycles should be regarded as a proxy of past deep water mass extension. Thus, the interpretation of the kaolinite/chlorite ratios suggests a southward advance of northern-source deep water not farther south than 40°S during glacial maxima. This 40°S rank was also proposed by Duplessy et al. (1988) for the last glacial maximum. In turn, during interglacial optima northern-source deep water was fed into the ACC far south to latitudes around 55°S, near the ACC/Weddell Gyre boundary.

We are not able to decipher vertical variations of past deep water stratification from kaolinite/chlorite ratios. However, at the top of the Walvis Ridge, which at present is bathed by NADW, only a rise of the lower boundary of the northern-source deep water boundary above the crest of the Walvis Ridge can explain the significant drop of kaolinite/chlorite ratios from maximal interglacial values of 8.0 to glacial values of 1.5. Such a shallowing of northern-source deep water, also claimed by Bickert (1992), achieves bathing of the Walvis Ridge by chlorite-bearing southern-component deep water.

Other constraints than deep water mass advection could account for variable clay mineral fluxes in the South Atlantic and influence kaolinite/chlorite ratios. One important aspect is the variability of total terrigenous input to the ocean. Terrigenous fluxes in the South Atlantic were higher during glacial periods in both the equatorial regions (e. g. Francois and Bacon 1991, Gingele 1992) and southern high-latitudes (Grobe and Mackensen 1992, Bareille et al. 1994, Camerlenghi et al. 1995) due to low stands of sea level and the intensification of wind erosion. The exposure of continental shelves during glacials enables direct discharge of fluvial loads to the deep-sea, enlarges potential areas of aeolian deflation and causes destabilisation of ice shelves. Actually, the synchronism of temporal variations in absolute detrital supply from different source regions should not substantially influence the relative flux rates of individual components at a given position.

In the tropical regions climatically induced changes of the weathering regime might influence the clay mineral composition and hence control the relative input of individual clay minerals to the ocean. A wider extension of the tropical and subtropical humid to semi-humid regions during interglacials (e. g. Pokras and Mix, 1985) could cause a higher supply of pedogenic clay minerals (kaolinite and smectite). In turn, glacial aridification reduces the degree of hydrolysis in soils and thus the formation of pedogenic clay minerals (Chamley 1989). Reduced pedogenic clay mineral formation, nevertheless, can be compensated by wind erosion of fossil pedogenic clay minerals, as presently observed in the Sahel zone (Aston et al. 1973). Climatic changes in aridity are usually linked to the 23-kyr cycles of precessional forcing (e. g. Pokras and Mix 1985, Prell and Kutzbach 1987, Gingele in press). Indeed, fluctuations of kaolinite/feldspar ratios in sediments off the Congo/Zaire mouth indicate variations in fluvial kaolinite discharge in tune with precessional rhythms during the last 200 kyr (Schneider and Mueller 1995). If this precessional signal of variable kaolinite input in the tropical region virtually affects advective kaolinite fluxes in deep water masses of the southeastern South Atlantic, this should be reflected by downcore fluctuations of kaolinite/chlorite ratios. Time series analyses, however, do not yield any indication of kaolinite/chlorite variations in the 23-kyr^{-1} frequency band (Fig. 9, Table 2).

Instead, temporal variations of kaolinite/chlorite ratios are driven by the 41-kyr and 100-kyr cycles of obliquity and excentricity and are nearly in phase with global ice minima and the maximal '%NADW index' (Atlantic-Pacific δ^{13}C fractionation index of Raymo et al. 1990) (Fig. 10). The '%NADW index', however, is also coherent and in phase with global ice minima in the 23-kyr^{-1} frequency band (Table 2). According to the Cd/Ca proxy, relative NADW formation rates in the North Atlantic are only modulated with a significant 41-kyr cyclicity and slightly lag global ice minima by 3.6±2.8 kyr (Boyle and Keigwin 1985/86). The significance of the 100-kyr^{-1} frequency has probably not been detected by Boyle and Keigwin since their used Cd/Ca record was too short in age to support the 100-kyr periodicity statistically. The close phase relationships between the parameters '%NADW index', Cd/Ca, and kaolinite/chlorite ratio document the direct linkage between production and fluxes of northern-source deep water (Cd/Ca proxy, 'NADW index') and its latitudinal extensions (kaolinite/chlorite ratio), respectively.

Important is the fact that these three parameters, generally show a significant lag to both South Atlantic sea-surface temperatures (Imbrie et al. 1992, 1993) and maximum insolation during maxima of obliquity (Fig. 10). This contradicts the assumption that northern-source deep water flux rapidly transmits the insolation signals of the northern hemisphere directly to the southern hemisphere (Imbrie et al. 1992) and thus immediately controls temperatures in the southern oceans (e. g. Hodell 1993).

Conclusions

Downcore clay mineral fluctuations in Late Quaternary sediments from the south-eastern South Atlantic and adjoining Southern Ocean are of low amplitude. Our results suggest that small-scale clay mineral fluctuations in deep-sea sediments north of the ACC/Weddell Gyre boundary primarily monitor periodic changes of deep water advection in response to glacial-interglacial oscillations (Fig. 12). At the Walvis Ridge an additional aeolian influence on clay mineral input and fluxes by the south-eastern trade wind system is evident.

In contrast, long-term constancy of assemblages and relative proportions of clay minerals indicate stable Late Quaternary oceanographic conditions in the eastern Weddell Basin. High amounts of very

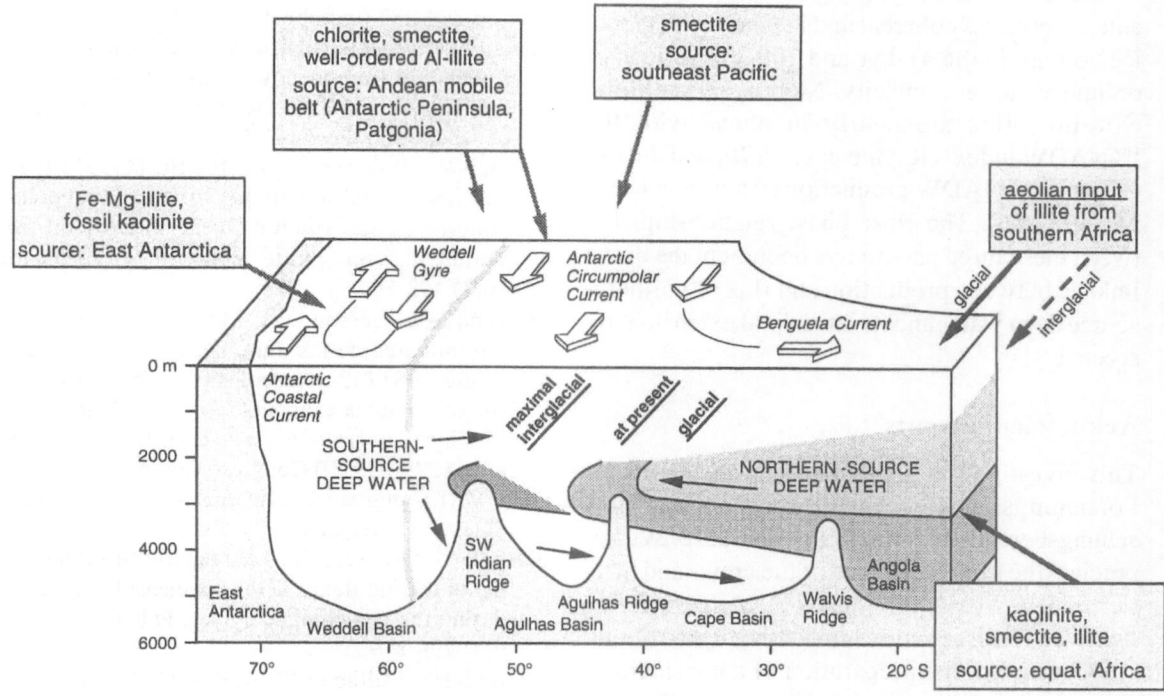

Fig. 12. Simplified model of deep water circulation in the South Atlantic with clay mineral sources and transport pathways. The extension of northern-source deep water at present and during glacial and interglacial periods is inferred from downcore variations of kaolinite/chlorite ratios.

Fe-, Mg-rich 'biotitic' illite and some fossil kaolinite point to a dominant clay mineral supply from East Antarctic sources by the Weddell Gyre and the Antarctic Coastal Current. The prevalence of Al-rich illites in sediments beyond the ACC/Weddell Gyre boundary shows that no clay mineral exchange took place between both current systems.

North of the ACC/Weddell Gyre boundary, kaolinite and chlorite are appropriate tracers of deep water advection. Kaolinite from the equatorial region traces northern-source deep water flow to the Southern Ocean, whereas chlorite from southern high-latitudes (Antarctic Peninsula, Patagonia) outlines southern-source deep water advection to the north. This is clearly displayed by a latitudinal zonation of kaolinite/chlorite ratios. The present-day kaolinite/chlorite gradient with higher values towards the north prevailed throughout the last 300

kyr. A shift of higher values to the north is evident during glacial periods.

Although modified weathering conditions in response to climatic changes might had influence on the formation and absolute input of clay minerals (particularly kaolinite) to the South Atlantic, our results confirm the idea that deep water advection is the dominant factor controling relative fluxes of kaolinite and chlorite. Kaolinite/chlorite ratios are interpreted as a proxy of the past latitudinal extension of northern-source deep water to the south. Thus, northern-source water reaches not farther south than 40°S during glacial maxima. During interglacial optima northern-source deep water is injected into the ACC south to latitudes around 55°S near the ACC/Weddell Gyre boundary. The latter remained at an almost constant position throughout the last 300 kyr.

Cross spectral analyses reveal that the kaolinite/ chlorite proxy is coherent and in phase with global ice volume in the 41-kyr and 100-kyr periods of obliquity and eccentricity. Moreover, kaolinite/ chlorite ratios are nearly in phase with the '%NADW index' (Raymo et al. 1990) and the Cd/ Ca proxy of NADW production (Boyle and Keigwin 1985/86). The close phase relationships between these three parameters document the direct linkage between production and flux of northern-source deep water and its latitudinal extension, respectively.

Acknowledgements

This investigation was funded by the Deutsche Forschungsgemeinschaft through Sonderforschungsbereich 261 and grant Ku 683/2. We appreciate the kind assistance of the crews and masters of RV 'Polarstern' and RV 'Meteor' during numerous cruises to the South Atlantic and Southern Ocean. For the preparation of a considerable amount of samples we are indebted to U. Bock, R. Cordelair, R. Fröhlking, I. Klappstein, and H. Rhodes. Finally we acknowledge the constructive comments and reviews of W.H. Berger and H. Chamley.

This is contribution No. 118 of the Sonderforschungsbereich 261 and publication No. 1008 of the Alfred Wegener Institute for Polar and Marine Research.

References

Abelmann A, Gersonde R (1988) *Cycladophora davisiana* stratigraphy in Plio-Pleistocene cores from the Antarctic Ocean (Atlantic sector). Micropaleontol 34:268-276

Abelmann A, Gersonde R, Spieß V (1990a) Pliocene-Pleistocene paleoceanography in the Weddell Sea - siliceous microfossil evidence. In: Bleil U, Thiede J (eds) Geological History of the Polar Oceans: Arctic versus Antarctic. Kluwer Acad Publ, Dordrecht, pp 729-759 (NATO ASI Series C, vol 308)

Abelmann A, Gersonde R, Hubberten H-W, Niebler S, Ott G, Mackensen A (1992) Der Südatlantik im Spätquartär: Rekonstruktion von Stoffhaushalt und Stromsystemen, Arbeits- und Ergebnisbericht.

Sonderforschungsbereich 261, Bremen, pp 194-195 and 249-250 (unpublished report)

Aston SR, Chester R, Johnson LR, Padgham RC (1973) Eolian dust from the lower atmosphere of the eastern Atlantic and Indian Ocean, China Sea and Sea of Japan. Mar Geol 14:15-28

Balsam WL, Bliesner BL, Deaton BC (1995) Modern and last glacial maximum eolian sedimentation patterns in the Atlantic Ocean interpreted from sediment iron oxide content. Paleoceanogr 10(3):493-507

Bárcena MA, Gersonde R, Flores JA (1992) Datos preliminares de las diatomeas subantarticas del sondeo PS1752-1 (sector Atlantico del Oceano Antartico) de la campana ANT-VIII/3. In: López-Martinez J (ed) Geologia de la Antartida Occidental, pp 229-239 (III Congreso Geológico de Espana y VIII Congreso Latinoamericano de Geologia, Salamanca, Espana)

Bareille G, Grousset FE, Labracherie M (1994) Origin of detrital fluxes in the southeast Indian Ocean during the last climatic cycles. Paleoceanography 9(6):799-819

Barretto HT, Milliman JD, Amaral CAB, Francosconi O (1975) Northern Brazil. In: Millimann JD, Summerhayes CP (eds) Upper Continental Margin Sedimentation off Brazil. Schweizerbart, Stuttgart, pp 11-43 (Contributions to Sedimentology, vol 4)

Bathmann U, Schulz-Baldes M, Fahrbach E, Smetacek V, Hubberten HW (1992) Die Expeditionen ANTARKTIS IX/1-4 des Forschungsschiffes "Polarstern" 1990/91. Bremerhaven (Berichte zur Polarforschung, vol 100)

Behairy AK, Chester R, Griffiths AJ, Johnson LR, Stoner JH (1975) The clay mineralogy of particulate material from some surface seawaters of the eastern Atlantic Ocean. Mar Geol 18:M45-M56

Bickert T (1992) Rekonstruktion der spätquartären Bodenwasserzirkulation im östlichen Südatlantik über stabile Isotope benthischer Foraminiferen. Bremen (Berichte, Fachbereich Geowissenschaften, Universität Bremen, vol 27)

Bickert T, Wefer G (1996) Late Quaternary deep water circulation in the South Atlantic: Reconstruction from carbonate dissolution and benthic stable isotopes. In: Wefer G, Berger WH, Siedler G, Webb D (eds) The South Atlantic: Present and Past Circulation. Springer-Verlag, Berlin Heidelberg

Biscaye PE (1965) Mineralogy and sedimentation of recent deep-sea clay in the Atlantic Ocean and adjacent seas and oceans. Geol Soc Am Bull 76:803-832

Boyle EA (1988) Cadmium: chemical tracer of deepwater paleoceanography. Paleoceanogr 3:471-489

Boyle EA (1994) A comparison of carbon isotopes and cadmium in the modern and glacial maximum ocean: Can we account for the discrepancies? In: Zahn et al (eds) Carbon cycling in the glacial ocean: constraints on the ocean's role in global change. Springer, Berlin Heidelberg New York, pp 167-193 (NATO ASI Series C, vol 117)

Boyle EA, Keigwin LD (1985/86) Comparison of Atlantic and Pacific paleochemical records for the last 215,000 years: changes in deep ocean circulation and chemical inventories. Earth and Planet Sci Lett 76:135-150

Brathauer U, Abelmann A (1994) Late Pleistocene sea surface temperature variations in the Southern Ocean (Atlantic Sector). In: The South Atlantic: Present and Past Circulation, Bremen/Germany, 15-19.08.94, Abstracts. Bremen, p 28 (Berichte, Fachbereich Geowissenschaften, Universität Bremen, vol 52)

Brehme I (1992) Sedimentfazies und Bodenwasserstrom am Kontinentalhang des nordwestlichen Weddellmeeres. Bremerhaven (Berichte zur Polarforschung, vol 110)

Bremner JM, Willis JP (1993) Mineralogy and geochemistry of the clay fraction of sediments from the Namibian continental margin and the adjacent hinterland. Mar Geol 115:85-116

Broecker WS (1993) An oceanographic explanation for the apparent carbon isotope-cadmium discordancy in the glacial Antarctic. Palaeoceanogr 8:137-140

Burckle, LH (1982) First appearance datum of *Hemidiscus karstenii* in the late Pleistocene of the subantarctic region. Antarctic Journal 17(5):142-143

Burckle LH, Clarke DB, Shackleton NJ (1978) Isochronous last-abundant-appearance datum (LAAD) of the diatom *Hemidiscus karstenii* in the sub-Antarctic. Geology 6:243-246

Camerlenghi A, Rebesco M, Pudsey CJ (1995) High resolution terrigenous sedimentary record of the continental rise of the Antarctic Peninsula Pacific margin (initial results of the 'Sedano' program). Abstracts 5th International Conference on Palaeoceanography, Halifax, Nova Scotia, Canada, p 155

Cande SC, Kent DV (1992) A new geomagnetic polarity time scale for the Late Cretaceous and Cenozoic. J Geophys Res B97(10):13,917-13,951

Carmack EC, Foster TD (1975) On the flow of water out of the Weddell Sea. Deep-Sea Res 22:711-724

Chamley H (1975) Influence des courants profonds au large du Brésil sur la sédimentation argileuse récente. In: 9ème Cong Int Sédimentol, Nice, vol 8, pp 13-17

Chamley H (1989) Clay Sedimentology. Springer, Berlin Heidelberg New York

Charles CD, Fairbanks RG (1992) Evidence from Southern Ocean sediments for the effect of North Atlantic deep-water flux on climate. Nature 355:416-419

Chester R, Elderfield H, Griffin J J, Johnson LR, Padgham RC (1972) Eolian dust along the eastern margins of the Atlantic Ocean. Mar Geol 13:91-105

Cisielski PF, Ledbetter MT, Ellwood BB (1982) The development of Antarctic glaciation and the Neogene paleoenvironment of the Maurice Ewing Bank. Mar Geol 46:1-51

Cole TG, Shaw HF (1983) The nature and origin of authigenic smectites in some Recent marine sediments. Clay Minerals 18:239-252

Curry WB, Lohmann GP (1990) Reconstructing past particle fluxes in the tropical Atlantic Ocean. Paleoceanogr 5:487-506

Curry WB, Duplessy JC, Labeyre LD, Shackleton NJ (1988) Changes in the distribution of $\delta^{13}C$ of deep water ΣCO_2 between the last glaciation and the Holocene. Paleoceanogr 3(3):317-341

Duce, RA et al (1991) The atmospheric input of trace species to the world ocean. Global Biogeochem Cycles 5:193-260

Duplessy JC, Shackleton NJ, Fairbanks RG, Labeyrie LD, Oppo D, Kallel N (1988) Deep water source variations during the last climatic cycle and their impact on the global deep water circulation. Paleoceanogr 3(3):343-360

Ehrmann WU, Grobe H, Fütterer DK (1991) Late Miocene to Holocene glacial history of East Antarctica revealed by sediments from Sites 745 and 746. Proc Ocean Drill Proj Sci Results 119: 239-260

Ehrmann WU, Melles M, Kuhn G, Grobe H (1992) Significance of clay mineral assemblages in the Antarctic Ocean. Mar Geol 107:249-273

Eisma D, Kalf J, van der Gaast SJ (1978) Suspended matter in the Zaire estuary and the adjacent Atlantic ocean. Netherl J Sea Res 12:382-406

Esquevin J (1969) Influence de la composition chimique des illites sur le cristallinité. Bull Centre Rech Pau SNPA 3:147-154

Foldvik A, Gammelsrød T (1988) Notes on Southern Ocean hydrography, sea-ice and bottom water formation. Palaeogeogr Palaeoclimatol Palaeoecol 67:3-17

Francois R, Bacon MP (1991) Variations in terrigenous input into the deep equatorial Atlantic during the past 24,000 years. Science 251:1473-1476

Frank M, Gersonde R, Rutgers van der Loeff M, Kuhn

G, Mangini A (submitted) Late Quaternary sediment dating and quantification of lateral sediment redistribution applying $^{230}Th_{ex:}$ A study from the eastern Atlantic sector of the Southern Ocean. (submitted to Geol Rdsch)

Fütterer DK (1987) Die Expedition ANTARKTIS-IV mit FS "Polarstern" 1985/86, Bericht von Fahrtabschnitten ANT-IV/3-4. Bremerhaven (Berichte zur Polarforschung, vol 33)

Gaudichet A, Petit J-R, Lefevre R, Lorius C (1986) An investigation by analytical transmission electron microscopy of individual insoluble microparticles from Antarctic (Dome C) ice core samples. Tellus 38B:250-261

Gersonde R (1995) Die Expedition ANTARKTIS-XI/II mit FS "Polarstern" 1993/94. Bremerhaven (Berichte zur Polarforschung, vol 163)

Gersonde R, Hempel G (1990) Die Expeditionen ANTARKTIS-VIII/3 und VIII/4 mit FS "Polarstern" 1989. Bremerhaven (Berichte zur Polarforschung, vol 74)

Gersonde R, Abelmann A, Burckle LH, Hamilton N, Lazarus D, McCartney K, O'Brien B, Spieß V, Wise SW (1990) Biostratigraphic synthesis of Neogene siliceous microfossils from the Antarctic Ocean, ODP Leg 113 (Weddell Sea). Proc Ocean Drill Proj Sci Results, Pt B, 113:915-936

Gibbs RJ (1967) The geochemistry of the Amazon river system, Part I: The factors that control the salinity and the composition and concentration of the suspended solids. Geol Soc Am Bull 78:1203-1232

Gingele FX (1992) Zur klimaabhängigen Bildung biogener und terrigener Sedimente und ihrer Veränderung durch die Frühdiagenese im zentralen und östlichen Südatlantik. Bremen (Berichte, Fachbereich Geowissenschaften, Universität Bremen, vol 26)

Gingele FX (in press) Holocene climatic optimum in Southwest Africa - evidence from the marine clay mineral record. Palaeogeogr Palaeoclimatol Palaeoecol

Grobe H, Mackensen A (1992) Late Quaternary climate cycles as recorded in sediments from the Antarctic continental margin. In: Kennett JP, Warnke DA (eds) The Antarctic Paleoenvironment: A Perspective on Global Change, Part One. American Geophysical Union, Washington DC, pp 349-376 (Antarctic Research Series, vol 56)

Grousset FE, Biscaye PE, Revel M, Petit J-R, Pye K, Joussaume S, Jouzel J (1992) Antarctic (Dome C) ice-core dust at 18 k.y. B.P.: Isotopic constraints on origins. Earth Planet Sci Lett 111: 175-182

Grünig S (1991) Quartäre Sedimentationsprozesse am

Kontinentalhang des Süd-Orkney-Plateaus im nordwestlichen Weddellmeer (Antarktis). Bremerhaven (Berichte zur Polarforschung, vol 75)

Hays JD, Imbrie J, Shackleton NJ (1976) Variations in the earth's orbit: Pacemaker of the ice ages. Science 194:1121-1132

Hodell DA (1993) Late Pleistocene palaeoceanography of the South Atlantic sector of the Southern Ocean: Ocean Drilling Program Hole 704A. Paleoceanogr 8(1):47-67

Honnorez J (1981) The aging of the oceanic crust at low temperature. In: Emiliani C (ed) The oceanic lithosphere. John Wiley and Sons, New York, pp 525-587 (The Sea, vol 7)

Howard WR, Prell WL (1994) Late Quaternary $CaCO_3$ production and preservation in the Southern Ocean: Implications for oceanic and atmospheric carbon cycling. Paleoceanogr 9:453-482

Howell P (1989) ARAND programs for Macintosh, Brown University

Imbrie J, Hays JD, Martinson DG, McIntyre A, Mix AC, Morley JJ, Pisias NG, Prell WL, Shackleton NJ (1984) The orbital theory of Pleistocene climate: support from a revised chronology of the $\delta^{18}O$ record. In: Berger A, Imbrie J, Hays J, Kukla G, Saltzman B (eds) Milankovitch and Climate. Reidel, Dordrecht, pp 269-305 (NATO ASI Series C, vol 126)

Imbrie J, McIntyre A, Mix A (1989) Oceanic response to orbital forcing in the late Quaternary: observational and experimental strategies. In: Berger A, Schneider S, Duplessy JC (eds) Climate and Geo-Sciences, A Challenge for Science and Society in the 21st Century. Kluewer, Dordrecht, pp 121-164

Imbrie J, Boyle E, Clemens S, Duffy A, Howard W, Kukla G, Kutzbach J, Martinson DG, McIntyre A, Mix A, Molfino B, Morley JJ, Peterson L, Pisias NG, Prell WL, Raymo M, Shackleton NJ, Toggweiler J (1992) On the structure and origin of major glaciation cycles, 1. Linear responses to Milankovitch forcing. Paleoceanography 7:701-738

Imbrie J, Berger A, Boyle E, Clemens S, Duffy A, Howard W, Kukla G, Kutzbach J, Martinson DG, McIntyre A, Mix A, Molfino B, Morley JJ, Peterson L, Pisias NG, Prell WL, Raymo M, Shackleton NJ, Toggweiler J (1993) On the structure and origin of major glaciation cycles, 2. The 100,000-year cycle. Paleoceanogr 8:699-735

Jones GA (1984) Advective transport of clay minerals in the region of the Rio Grande Rise. Mar Geol 58:187-212

Jones GA, Johnson DA (1984) Displaced Antarctic diatoms in Vema Channel sediments: Late Pleistocene/Holocene fluctuations in AABW flow. Mar Geol 58:165-186

Kellogg TB (1987) Glacial-interglacial changes in global deep water circulation. Paleoceanogr 2(3):259-271

Köhnen H (1984) Die Expedition ANTARKTIS II mit FS "Polarstern" 1983/84, Bericht vom Fahrtabschnitt 4, Punta Arenas - Kapstadt (ANT-II/4). Bremerhaven (Berichte zur Polarforschung, vol 19)

Kolla V, Henderson L, Biscaye PE (1976) Clay mineralogy in the western Indian ocean. Deep-sea Res 23:949-961

Ledbetter MT (1984) Bottom-current speed in the Vema Channel recorded by particle size of sediment fine-fraction. Mar Geol 58:137-149

Ledbetter MT (1986a) A Late Pleistocene time-series of bottom-current speed in the Vema Channel. Palaeogeogr Palaeoclimatol Palaeoecol 53:97-105

Ledbetter MT (1986b) Bottom-current pathway in the Argentine Basin revealed by mean silt particle size. Nature 321:423-425

Mackensen A, Hubberten H-W, Bickert T, Fischer G, Fütterer DK (1993) The $\delta^{13}C$ in benthic foraminiferal tests of *Fontbotia Wuellerstorfi* (Schwager) relative to the $\delta^{13}C$ of dissolved inorganic carbon in Southern Ocean deep water: Implications for glacial ocean circulation models. Paleoceanogr 8:587-610

Mackensen A, Grobe H, Hubberten H-W, Kuhn G (1994) Benthic foraminiferal assemblages and the $\delta^{13}C$-signal in the Atlantic sector of the southern ocean: glacial-to-interglacial contrasts. In: Zahn et al (eds) Carbon cycling in the glacial ocean: constraints on the ocean's role in global change. Springer, Berlin Heidelberg New York, pp 105-144 (NATO ASI Series C, vol 117)

Massé L, Faugères J -C, Bernat M, Pujos A, Mézerais M -L (1994) A 600,000-year record of Antarctic Bottom Water activity inferred from sediment textures and structures in a sediment core from the southern Brazil Basin. Paleoceanogr 9(6):1017-1026

Melguen M, Debrabant P, Chamley H, Maillot H, Hoffert M, Courtois C (1978) Influence des courants profonds sur les faciès sédimentaires du Vema Channel (Atlantique sud) à la fin du Cénozoique. Bull Soc géol Fr 20(7):121-136

Niebler HS (1995) Rekonstruktion von Paläo-Umweltparametern anhand von stabilen Isotopen und Faunen-Vergesellschaftungen planktischer Foraminiferen im Südatlantik. Bremerhaven (Berichte zur Polarforschung, vol 167)

Oppo DW, Fairbanks RG (1990) Atlantic Ocean thermohaline circulation of the last 150,000 years: relationship to climate and atmospheric CO_2. Palaeoceanogr 5(3):277-288

Oppo DW, Rosenthal Y (1994) Cd/Ca changes in a deep Cape Basin core over the past 730,000 years: Response of circumpolar deepwater variability to northern hemisphere ice sheet melting? Paleoceanogr 9(5):661-675

OSU Computer Center staff (1973) ARAND routines and programs, Oregon State University

Pastouret L, Chamley H, Delibrias G, Duplessy J-C, Thiede J (1978) Late Quaternary climatic changes in Western tropical Africa deduced from deep-sea sedimentation off the Niger delta. Oceanol Acta 1:217-232

Peterson RG, Stramma L (1991) Upper-level circulation in the South Atlantic Ocean. Progr Oceanogr 26:1-73

Petit JR, Mounier L, Jouzel J, Korotkevich YS, Kotlyakov VI, Lorius C (1990) Palaeoclimatological implications of the Vostok core dust record. Nature 343:56-58

Petschick R, Kuhn G, Gingele FX (in press) Clay mineral distribution in surface sediments of the South Atlantic: sources, transport, and relation to oceanography. Mar Geol (1996)

Pokras EM, Mix AC (1985) Eolian evidence for spatial variability of Late Quaternary climates in tropical Africa. Quat Res 24:137-149

Prell WL, Kutzbach JE (1987) Monsoon variability over the past 150,000 years. J Geophys Res 92:8411-8425

Pudsey CJ (1992) Late Quaternary changes in Antarctic Bottom Water velocity inferred from sediment grain size in the northern Weddell Sea. Mar Geol 107:9-33

Raymo ME, Ruddiman WF, Shackleton NJ, Oppo DW (1990) Evolution of Atlantic-Pacific $\delta^{13}C$ gradients over the last 2.5 m.y. Earth Planet Sci Lett 97:353-368

Reid JL (1989) On the total geostrophic circulation of the South Atlantic Ocean: Flow patterns, tracers, and transports. Progr Oceanogr 23:149-244

Robert C (1980) Climats et courants cénoziques dans l'Atlantique Sud d'après l'étude des minéraux argileux (legs 3, 39 et 40 DSDP). Oceanol Acta 3(3):369-376

Robert C (1987) Clay mineral associations and structural evolution of the South Atlantic: Jurassic to Eocene. Palaeogeogr Palaeoclimatol Palaeoecol 58:87-108

Robert C, Maillot H (1983) Palaeoenvironmental significance of clay mineralogical and geochemical data, southwest Atlantic, Deep Sea Drilling Project Legs 36 and 71. Init Rpts Deep Sea Drill Proj 71:317-343

Schmiedl G (1995) Rekonstruktion der spätquartären Tiefenwasserzirkulation und Produktivität im östlichen Südatlantik anhand von benthischen Foraminiferenvergesellschaftungen. Bremerhaven (Berichte zur Polarforschung, vol 160)

Schneider RR, Mueller PJ (1995) Late Quaternary Congo discharge fluctuations and influence of fluvial nutrient supply on palaeoproductivity in the eastern Angola Basin. Abstracts 5th International Conference on Palaeoceanography, Halifax, Nova Scotia, Canada, p 137

Siegel FR, Pierce JW, Kostick DS, Ronca, LB (1981) Suspensates and bottom sediments in the Chilean Archipelago. Modern Geology 7:217-229

Tintelnot M, Jennerjahn T, Hübner M, Irion G, Ittekkot V, Morais JO, Brichta A (1994) Sedimentological and geochemical indicators of river supply along the Brazilian continental margin. In: The South Atlantic: Present and Past Circulation, Bremen/ Germany, 15-19.08.94, Abstracts. Bremen, pp 154-156 (Berichte, Fachbereich Geowissenschaften, Universität Bremen, vol 52)

Van der Gaast SJ, Jansen JHF (1984) Mineralogy, opal, and manganese of middle and late Quaternary sediments of the Zaire (Congo) deep-sea fan: origin and climatic variation. Neth J Sea Res 17:313-341

Wefer G, Bleil U, Müller PJ, Schulz HD, Fahrtteilnehmer (1988) Bericht über die Meteor-Fahrt M 6/ 6, Libreville - Las Palmas, 18.2.-23.3.1988. Bremen (Berichte, Fachbereich Geowissenschaften, Universität Bremen, vol 3)

Wefer G, Fahrtteilnehmer (1990) Bericht über die Meteor-Fahrt M 12/1, Kapstadt - Funchal, 13.3.-14.4.1990. Bremen (Berichte, Fachbereich Geowissenschaften, Universität Bremen, vol 11)

Whitworth T III, Nowlin WD Jr (1987) Water masses and currents of the Southern Ocean at the Greenwich meridian. J Geophys Res C92(6):6462-6476

Windom HL (1975) Eolian contributions to marine sediments. J Sedimentol Petrol 45:520-529

Yoon HI, Hab MW, Park BK, Han SJ, Oh JK (1992) Distribution, provenance, and dispersal pattern of clay minerals in surface sediments, Bransfield Strait, Antarctica. Geo-Mar Lett 12:223-227

Zemmels I, Cook HE (1976) X-ray mineralogy data from the Southeast Pacific basin, DSDP Leg 35. Init Rpts Deep Sea Drill Proj 35:747-754

Springer
and the
environment

At Springer we firmly believe that an international science publisher has a special obligation to the environment, and our corporate policies consistently reflect this conviction.
We also expect our business partners – paper mills, printers, packaging manufacturers, etc. – to commit themselves to using materials and production processes that do not harm the environment. The paper in this book is made from low- or no-chlorine pulp and is acid free, in conformance with international standards for paper permanency.

 Springer